BIOLOGY
Concepts and Applications

SECOND EDITION

General Advisors/Contributors

JOHN ALCOCK
Arizona State University

AARON BAUER
Villanova University

ROBERT COLWELL
University of Connecticut

GEORGE COX
San Diego State University

PAUL HERTZ
Barnard College

DANIEL FAIRBANKS
Brigham Young University

JOHN JACKSON
North Hennepin Community College

EUGENE KOZLOFF
University of Washington

ROBERT LAPEN
Central Washington University

WILLIAM PARSON
University of Washington

CLEON ROSS
Colorado State University

SAMUEL SWEET
University of California, Santa Barbara

STEPHEN WOLFE
University of California, Davis

Developmental Editors

ELMARIE HUTCHINSON
AND BEVERLY MCMILLAN

Primary Biological Illustrator

RAYCHEL CIEMMA

BIOLOGY
Concepts and Applications
SECOND EDITION

CECIE STARR
Belmont, California

Wadsworth Publishing Company
Belmont, California
A Division of Wadsworth, Inc.

BIOLOGY EDITOR: Jack C. Carey

EDITORIAL ASSISTANT: Kristin Milotich

ART DIRECTOR AND DESIGNER: Gary Head and Michele Mangelli, Gary Head Design

PRODUCTION EDITOR: Mary Douglas, Rogue Valley Publications

PRODUCTION SERVICES MANAGER: Sandra Craig

PRINT BUYER: Randy Hurst

EDITORIAL PRODUCTION: Myrna Engler, Marilyn Evenson, Michele Mangelli, Mary Roybal, and Ed Serdziak

PERMISSIONS AND PHOTO RESEARCH: Marion Hansen

ARTISTS: Lewis Calver, Raychel Ciemma, Robert Demarest, Hans & Cassady, Inc. (Hans Neuhart), Darwin Hennings, Vally Hennings, Leonard Morgan, Palay/Beaubois, Precision Graphics (Jan Flessner), Victor Royer, Nadine Sokol, Kevin Somerville, Lloyd Townsend, Jennifer Wardrip

PRODUCTION ARTISTS: Jessie Bunn, Natalie Hill, Carole Lawson, Jill Turney

COVER DESIGN: Gary Head, Gary Head Design

COVER PHOTOGRAPH: © Jim Brandenburg—Minden Pictures

COMPOSITOR: American Composition & Graphics, Inc. (Jim Jeschke, Jody Ward)

COLOR PROCESSING: H & S Graphics, Inc. (Tom Anderson, Nancy Dean)

PRINTING: R. R. Donnelley & Sons Company

This book is printed on acid-free recycled paper.

I(T)P™

International Thomson Publishing
The trademark ITP is used under license.

Printed in the United States of America
2 3 4 5 6 7 8 9 10—97 96 95 94

Library of Congress Cataloging-in-Publication Data
Starr, Cecie.
    Biology: concepts and applications/Cecie Starr.—2nd ed.
    p. cm.
    Includes bibliographical references and index.
    ISBN 0-534-17616-X
    1. Biology. I. Title.
QH307.2.S73 1993   93-42032
574—dc20

PREFACE

Twenty years! Imagine devoting twenty years of your life to scribbling and sketching. Imagine interacting that long with more than 2,000 biology instructors and researchers who take the time again and again to prune, praise, and pummel your writing and illustrations, and so help mold them into books.

I thought about this while putting the final touches on the new edition of *Biology*. I thought about the students who will be reading this book, joining the ranks of a couple of million other students who may never know how many first-rate minds were tapped to give them this part of their education. Then I thought about all the biologists of the past century who studied so many aspects of the world, who tested their ideas and wrote up their findings. And it came to me that twenty years is a mere blip of time, compared to this great continuum of gathering and passing on the gift of knowledge.

And this is extraordinary knowledge! With it, students everywhere can cut their own intellectual paths through environmental, medical, and social thickets. With it, they can come to understand the past and to predict possible futures for ourselves and all other organisms.

OBJECTIVES FOR THIS EDITION

Those of us who author biology textbooks start out with the same basic objectives. The writing itself must be engaging, logical, clear, and never patronizing. The book as a whole should be framed in light of the most inclusive principles of biochemistry, inheritance, and evolution. The power of evolutionary theory to explain life's unity and diversity must be conveyed through examples of problem solving and experiments. For each major field of inquiry, key concepts must be identified, and the topics selected must give students a good picture of current understandings and research trends. The structure and function of organisms must be explained in enough detail to help students build a working vocabulary about the parts and processes of life.

Those of us who put together this second edition of *Biology* met four additional objectives. First, our attempts to create self-contained topic spreads in earlier books were so well received, we structured the entire second edition this way. Second, we refined and updated the examples of applications. Third, at the start of each chapter, we wrote a lively or sobering short story that leads students to key concepts in a comfortable, nonthreatening way. Finally, as time-consuming and intellectually draining as the task may be, we added many more original visual summaries. With near-universal enthusiasm, students let us know that these are remarkably effective at explaining difficult concepts.

REVISION HIGHLIGHTS

Self-Contained Topic Spreads

When students must flip back and forth among pages to find illustrations that accompany the text, they often become confused and frustrated. In the second edition of *Biology*, they won't have to resort to this kind of searching. *Each text section—including its referenced illustrations—is laid out on the same page or two-page spread.* All the material is right there, in a cohesive, self-contained unit. With this format, students can easily focus on and digest one topic at a time.

Consider aerobic respiration, a topic that makes most students blanch with fear. Then look at Chapter 6, with its self-contained page layouts on glycolysis, the Krebs cycle, and electron transport phosphorylation. The organization of the text description and illustrations facilitates understanding. The text proper concentrates on the main points that carry the chapter story to its logical conclusion; it is not cluttered with details of the reactions. Once students grasp the main points, they can focus on visually inviting illustrations with simple, step-by-step descriptions that walk them through the details. Icons help keep the overall story in focus by reminding students where the reactions take place in cells.

We did not make the mistake of letting format dictate content; that would be like the tail wagging the dog. We *wrote* the book first. For every chapter, we developed and refined the text and illustrations as part of a continuous, integrated story. We carefully developed headings and subheadings, so that students can track the relative importance of sections to one another. Any good story has this hierarchy of information, with major and minor characters, background development, and high points where everything comes together.

Without this, information has all the excitement, flow, and drama of an encyclopedia.

After we were satisfied with the content, we assembled the topic spreads. This entailed minor paragraph tinkerings and major creative adjustments in illustration sizes and photographs. We introduced subtle variations in page layouts to avoid the monotonous, encyclopedic look that can quickly bore the reader. Students get the best of both worlds: content that is appropriate and interesting—*and* that is laid out so the focus is on one topic at a time.

Illustrations

Often we are asked how we come up with such beautiful, pedagogically sound illustrations for our biology books. All it takes, really, is a twenty-year obsession with writing and creating illustrations at the same time, as inseparable units. Researching, developing, and integrating illustrations with the text takes almost as much time as writing the text itself.

Visual Summaries Illustrations having step-by-step written descriptions that are positioned within the art itself are one of the hallmarks of our books. When we break down information into a series of illustrated steps, students find this far less threatening than a complex, "wordless" diagram. *In this edition, all major concepts now have their own visual summaries.*

A few examples are the visual summaries for mitosis (page 104), meiosis (page 113), protein synthesis (page 163), antibody-mediated immunity (page 443), and neural function (page 476).

New to this edition are visual summaries for anatomical drawings—they integrate structure and function. Students don't have to jump back and forth from text, to tables, to an illustration in order to get an overview of how an organ system is put together and what its assorted parts do. Even the individual descriptions are arranged to reflect the system's structural and functional organization.

Zoom-Sequence Illustrations Many illustrations progress from macroscopic to microscopic views. Figure 5.2 is a good example; it shows where the reactions of photosynthesis proceed. The zoom sequence tying together Figures 24.11 through 24.13 on skeletal muscle contraction is another.

Color Coding Careful use of color helps students track information on hard-to-visualize molecular topics. Throughout the book, for instance, proteins are color coded green, carbohydrates pink, lipids yellow and gold, DNA blue, and RNA orange. Compare the coding

in Chapter 2 with that in Chapters 11 and 12 to see how we follow through on this. The *Learning Guide for Students* on page x lists all the color codes.

Icons Icons, or small pictorial representations, relate the topic of an illustration to the big picture. For example, small but detailed representations of a cell remind students that a particular event proceeds in the nucleus or the cytoplasm (as in Chapters 7 and 12). Some icons in the text itself even remind students of the evolutionary relationship of one animal phylum to others (as in Chapter 19).

Improvements in the Writing

When you look through this edition, you will see stunning new illustrations, vignettes, and *Focus* essays. Don't let them distract you from the *line-by-line* judgment calls made with respect to the text proper. Where the first edition was a bit too terse, we added enough explanation to clarify the material. We tossed out the verbosity and cleaned up the logic; it is as simple as that.

Actually, it is not as simple as that. We worked equally hard at making the entire book *interesting* to read. Years ago, we had an idea that students could be enticed into the core science of biology with lively writing, memorable analogies, and engaging bits of natural history. The prevailing view was that such an idea was inappropriate for something as serious as science. And so, for fifteen years, we focused primarily on making the writing clear and the science accurate.

Today, students often pick up biology textbooks with apprehension. If the words do not engage them, they sometimes end up hating the book *and* the subject. Instructors still ask for a scientifically accurate book—but now they also ask for one that puts the life back in.

We could not be more pleased. Because we devoted so many years to writing with precision and objectivity, we have a strong sense of when the core material should be left alone and when it can be loosened up. Interrupting a description of, say, the mechanisms of mitosis with a distracting anecdote does the struggling student no good. Plunking humorous asides into a chapter on the correlation between geologic and organismic evolution trivializes a magnificent story. By contrast, it certainly is appropriate to liven up paragraphs on, say, the splendid variation in human traits (page 122) and the functions of skin (page 384).

Balancing Concepts and Applications

At strategic points in *Biology*, examples of applications parallel the flow of concepts—not so many as to be distracting, but enough to keep minds perking along with the core material. Striking the right balance between concepts and applications is less chancy when you have twenty years of feedback from instructors of two million students.

The applications are indexed separately, on the back end papers, for easy reference. Many are *Focus* essays on scientific methods, human health, the environment, and bioethics. These provide in-depth information for the interested student but do not interrupt the text flow. For example, the *Focus on Bioethics* essay on page 187 asks students to look beyond the clinical trials for human gene therapy, as described in the chapter, to the question of whether we as a species are ready to engage in such gene tinkerings. Besides this, many brief applications of concepts are incorporated into the text when they can illuminate rather than interrupt the conceptual flow.

Chapters 16 (human evolution) and 37 (human impact on the biosphere) are entirely applications-oriented.

Vignettes

Taking up valuable reading time with interesting stories is pointless—unless they lead students to the big concepts. Take a look at how the vignettes on growing old (Chapter 4) and on "killer bees" (Chapter 6) reinforce the concept of life's biochemical unity. Consider how another describes the effects of crack cocaine on the nervous system (Chapter 30) and puts this in evolutionary perspective. Or consider how the vignette on the trashing of Antarctica (Chapter 35) drives home the concept of life's interconnectedness. Classroom responses indicate that the accompanying photographs are arresting enough to make students sit up and take notice.

Science in Action

In our earlier books, many examples of biologists at work helped students develop an understanding of critical thinking—as have the descriptions of experimental evidence for the concepts being discussed, such as those in the sections on plant growth and development (Chapter 22). Many other examples of experiments are listed in the index (under the entry *Experiments*).

We even selected and wrote certain chapters to give students a sense of science in action. Examples are Chapters 9 (Mendelian genetics), 11 (DNA structure and function), and 38 (an evolutionary approach to animal behavior). This edition continues to integrate material on science in action. For example, John Alcock contributed two *Focus on Science* essays. The first is an entertaining account of scientific methods that con-

cludes with a real-life example (pages 12 and 13). The other describes how DNA fingerprinting was used to help explain self-sacrificing behavior—not among insects, but among a fascinating group of mammals (page 642).

As another example, the *Focus on Science* essay on page 256 helps students understand that biology is not a closed book, so to speak. Even when research suddenly brings a sweeping story into sharper focus—in this case, the very origins of the great prokaryotic and eukaryotic kingdoms of life—questions abound about the details. Students who read today about the theories of Lynn Margulis and others may well be among the new scientific detectives who eventually find answers to such fascinating questions.

STUDY AIDS

In this edition, a *list of key concepts* follows the introductory vignette for each chapter. Boldfaced *summary statements* of concepts within the text itself help keep readers on track. Several end-of-chapter study aids reinforce the key concepts. Each chapter has a *summary* in list form, conceptual *review questions, self-quiz questions* (objective questions), *selected key terms*, and *recommended readings*. Page numbers tie review questions and key terms to relevant text pages.

Genetics problems help students grasp the principles of inheritance. The *glossary* includes all boldfaced terms in the text, as well as pronunciation guides and origins of words when such information will make formidable words less so. The *index* is quite comprehensive; students find a door to the text more quickly through finer divisions of topics.

The first appendix is a detailed *classification scheme* that students can use for reference purposes. The second has *metric-English conversion charts*. The third has *detailed answers* to the genetics problems, and the fourth, *answers* to self-quizzes. The final appendix has structural formulas for *major metabolic pathways* for interested students and instructors who prefer the added detail. The appendixes and glossary are printed on paper of different tints to preclude frustrating searches for where one ends and the next starts.

A COMMUNITY EFFORT

One, two, or a smattering of authors can write accurately and often well about their field of interest, but it takes more than this to deal with the full breadth of the biological sciences. For us, it takes an educational network that includes more than 2,000 teachers, researchers, and photographers in the United States, Canada, England, Germany, France, Australia, Sweden, and elsewhere. There simply is no way to describe the thoughtful effort that each gave to our books; I can only salute their commitment to quality in education.

John Alcock, Aaron Bauer, Rob Colwell, George Cox, Daniel Fairbanks, Eugene Kozloff, Bill Parson, Cleon Ross, Sam Sweet, and Steve Wolfe are exceptional advisors and contributors of resource manuscripts. John Jackson, honored in 1993 as the best undergraduate teacher in Minnesota, and Kate Denniston, Tom Garrison, Ron Hoham, and Tyler Miller are ongoing guardians of teachability. Joining us this year is Robert Lapen, whose Chapter 27 manuscript cuts through the fog-shrouded lexicon of immunology. Also joining us is Paul Hertz, whose detailed suggestions improved the ecology unit.

Beverly McMillan and Elmarie Hutchinson are gifted developmental editors; Bev also drafted some great vignettes and essays and helped keep me sane.

Raychel Ciemma, a friend of long standing, has become a partner in the creation of original art.

As publishing lurches into the electronics age, the talented and supportive Mary Douglas always manages to overcome 17 million problems. The exuberantly creative Gary Head and Michele Mangelli were instrumental in designing the book and in page makeup. I owe much to Marion Hansen and Kristin Milotich, both unstinting in their professional and emotional support. Kathie Head, Pat Brewer, and Sandra Craig kept things moving. So did Randy Hurst, Myrna Engler, Mary Roybal, Jennie Redwitz, Bob Demarest, Natalie Hill, Carole Lawson, Jill Turney, Ed Serdziak, and the other individuals listed on the copyright page. Precision Graphics, H & S Graphics, Inc. (Tom Anderson especially) and American Composition & Graphics, Inc. accommodated my vision of excellence, which sometimes sends people up the wall and clinging to the ceiling.

Twenty years! Jack Carey and I have worked together with the same vision for that long. He remains my closest counselor, my abiding friend.

CECIE STARR, November 1993

Twenty-Seven Supplements

Full-color transparencies and *35mm slides* of almost all book illustrations are labeled with large, boldface type. All diagrams are on *CD-ROM*. A *Test Items* booklet has more than 4,000 questions by outstanding test writers. Questions are available in electronic form on *IBM* and *Macintosh*.

An *Instructor's Resource Manual* has, for each chapter, an outline, objectives, list of boldface and italic terms, and a detailed lecture outline. It includes suggestions for lecture presentations, classroom and laboratory demonstrations, suggested discussion questions, research paper topics, and annotations for filmstrips and videos. *Lecture outlines* in the Instructor's Resource Manual are available on a *Mac* or *IBM* disk for those who wish to modify the material.

Collaborating with Carolina Biological Supply, Bill Surver of Clemson University developed a new *text-specific videodisc*. Its material corresponds to each chapter of *Biology: Concepts and Applications*. Almost all of the book's illustrations are either broken down into a step-by-step format or animated in full motion. They are available with fill-in-the-blank labels for testing purposes. New motion sequences were prepared. Interviews with prominent scientists highlight discoveries and new research that may help solve significant medical or societal problems. There are 3,500 photographic stills. A directory and bar code guide comes with the disc. A separate *Videodisc Correlation Directory and Barcode Guide* correlates the text with three generic general biology videodiscs. *HyperCard Stacks for the Mac* and *Toolbook for the IBM* correlate the text with the same videodiscs.

A new, active-oriented *Study Guide and Workbook* asks students to respond in writing to almost all questions. Questions are arranged by chapter section. Each chapter has a set of critical thinking questions. *Chapter objectives* of the Study Guide are available on disk as part of the testing file for those who wish to modify or select portions of the material. An *Electronic Study Guide* consists of multiple choice questions different from those in the Test Items booklet. After responding to each question, an on-screen prompt lets students review their answers and learn why they are right or wrong. A 100-page *Answer Booklet* has answers to the book's end-of-chapter review questions. A special version of *STELLA II*, a software tool for developing critical thinking skills, is available to adopters, together with a workbook. One thousand glossary items are available on *Flash Cards*.

Wadsworth offers eight additional-readings supplements: *Contemporary Readings in Biology • Science and the Human Spirit: Contexts for Writing and Learning • A Beginner's Guide to Scientific Method • The Game of Science • Environment: Problems and Solutions • Green Lives, Green Campuses • Watersheds: Classic Cases in Environmental Ethics • Environmental Ethics.*

Perry and Morton's new *Laboratory Manual* has 38 experiments and exercises, with hundreds of full-color labeled photographs and diagrams. Many experiments are divided into parts that can be assigned individually, depending on time available. All have objectives, discussion (introduction, background, and relevance), a list of materials for each part of an experiment, procedural steps, pre-lab questions, and post-lab questions. An accompanying *Instructor's Manual* covers quantities, procedures for preparing reagents, time requirements for each portion of the exercise, suggestions to make the lab a success, and vendors of materials with item numbers. It offers additional investigative exercises that can be copied for laboratory use.

REVIEWERS

ARMSTRONG, PETER, *University of California, Davis*
ATCHISON, GARY, *Iowa State University*
BAKKEN, AIMÉE, *University of Washington*
BARBOUR, MICHAEL, *University of California, Davis*
BAUER, AARON, *Villanova University*
BINKLEY, DAN, *Colorado State University*
BLEEKMAN, GEORGE, *American River College*
BRENGELMANN, GEORGE, *University of Washington*
BRINSON, MARK, *East Carolina University*
BUCKNER, VIRGINIA, *Johnson County Community College*
CALVIN, CLYDE, *Portland State University*
CASE, CHRISTINE, *Skyline College*
CASE, TED, *University of California, San Diego*
CLAIRBORNE, JAMES, *Georgia Southern University*
CONNELL, MARY, *Appalachian State University*
COQUELIN, ARTHUR, *Texas Tech University*
COTTER, DAVID, *Georgia College*
DANIELS, JUDY, *Washtenau Community College*
DAVIS, JERRY, *University of Wisconsin, La Crosse*
DELCOMYN, FRED, *University of Illinois at Urbana-Champaign*
DEMPSEY, JEROME, *University of Wisconsin*
DENGLER, NANCY, *University of Toronto*
DENNISTON, KATHERINE, *Towson State University*
DESAIX, JEAN, *University of North Carolina at Chapel Hill*
DETHIER, MEGAN, *University of Washington*
DICKSON, KATHRYN, *California State University, Fullerton*
DLUZEN, DEAN, *Northeastern Ohio Universities College of Medicine*
DORRIS, PEGGY, *Henderson State University*
DOYLE, PATRICK, *Middle Tennessee State University*
DYER, BETSEY, *Wheaton College*
EDWARDS, JOAN, *Williams College*
ELSER, JIM, *Arizona State University*
ENDLER, JOHN, *University of California, Santa Barbara*
ENGLISH, DARREL, *Northern Arizona University*
ERICKSON, GINA, *Highline Community College*
ERWIN, CINDY, *City College of San Francisco*
EWALD, PAUL, *Amherst College*
FISHER, DAVID, *University of Hawaii, Manoa*
FISHER, DONALD, *Washington State University*
FONDACARO, JOSEPH, *Marion Merrell Dow*
FORTNEY, SUSAN, *Johnson Space Center*
FRISBIE, MALCOLM, *Eastern Kentucky University*
FRYE, BERNARD, *University of Texas, Arlington*
FUNK, FRED, *Northern Arizona University*
GAGLIARDI, GRACE, *Bucks County Community College*
GHOLZ, HENRY, *University of Florida*
GOSZ, JAMES, *University of New Mexico*
HANRATTY, PAMELA, *University of Southern Mississippi*
HARDIN, JOYCE, *Hendrix College*
HARLEY, JOHN, *Eastern Kentucky University*
HARRIS, JAMES, *Utah Valley Community College*
HARTNEY, KRISTINE BEHRENTS, *California State University, Fullerton*
HASSAN, ASLAM, *University of Illinois College of Veterinary Medicine*
HESS, WILFORD, *Brigham Young University*
HEWITSON, WALTER, *Bridgewater State College*
HINCK, LARRY, *Arkansas State University*
HODGSON, RONALD, *Central Michigan University*
HOHAM, RONALD, *Colgate University*
HOOPER, SCOTT, *University of Mississippi*
HUFFMAN, DAVID, *Southwest Texas State University*
HURD, SYLVIA, *Southwest Texas State University*
JACKSON, JOHN, *North Hennepin Community College*
JENSEN, STEVEN, *Southwest Missouri State University*
JUILLERAT, FLORENCE, *Indiana University–Purdue University*
KALLAND, GENE, *California State University, Dominguez Hills*
KAREIVA, PETER, *University of Washington*
KAUFMAN, JUDY, *Monroe Community College*
KAYE, GORDON, *Albany Medical College*
KEIM, MARY, *Seminole College*
KENDRICK, BRYCE, *University of Waterloo*
KREBS, CHARLES, *University of British Columbia*
KREBS, JULIA, *Francis Marion University*
KROHNE, DAVID, *Wabash College*
KUTCHAI, HOWARD, *University of Virginia Medical School*
LASSITER, WILLIAM, *University of North Carolina at Chapel Hill School of Medicine*
LATTA, VIRGINIA, *Jefferson State Junior College*
LEVY, MATTHEW, *Mt. Sinai Medical Center*

LEWIS, LARRY, *Bradford University*
LINDSEY, JERRI, *Tarrant County Junior College*
LITTLE, ROBERT, *Medical College of Georgia*
LUBCHENCO, JANE, *Oregon State University*
MACKLIN, MONICA, *Northeastern State University*
MARR, ELEANOR, *Dutchess Community College*
MARTIN, JAMES, *Reynolds Community College*
MARTIN, TERRY, *Kishwaukee College*
MATSON, RONALD, *Kennesaw State College*
MATTHAI, WILLIAM, *Tarrant County Junior College*
MATTHEWS, ROBERT, *University of Georgia*
McCLINTIC, J. ROBERT, *California State University, Fresno*
McCLURE, JERRY, *Miami University*
McKEAN, HEATHER, *Eastern Washington State University*
McNABB, F. M. ANNE, *Virginia Polytechnic Institute & State University*
MILLER, GLENDON, *Wichita State University*
MILLER, G. TYLER, *Pittsboro, North Carolina*
MOISES, HYLAN, *The University of Michigan*
MOORE-LANDECKER, ELIZABETH, *Rowan College*
MORRISON, WILLIAM, *Shippensburg University*
MORTON, DAVID, *Frostburg State University*
MUCH, DAVID, *Muhlenberg College*
MURPHY, RICHARD, *University of Virginia Medical School*
MYRES, BRIAN, *Cypress College*
NAGARKATTI, PRAKASH, *Virginia Polytechnic Institute & State University*
PEARCE, FRANK, *West Valley College*
PECK, JAMES, *University of Arkansas, Little Rock*
PEET, ROBERT, *University of North Carolina*
PERRY, JAMES, *University of Wisconsin Center–Fox Valley*
PETERSON, GARY, *South Dakota State University*
PIPERBERG, JOEL, *Millersville University*
PLETT, HAROLD, *Fullerton College*
RAYLE, DAVID, *San Diego State University*
REEVE, MARIAN, EMERITUS, *Merritt Community College*
RICKETT, JOHN, *University of Arkansas, Little Rock*
ROBBINS, ROBERT, *National Science Foundation*
ROSE, RICHARD, *West Valley College*
ROST, THOMAS, *University of California, Davis*
RUIBAL, RODOLFO, *University of California, Riverside*
SALISBURY, FRANK, *Utah State University*
SCHAPIRO, HARRIET, *San Diego State University*
SCHLESINGER, WILLIAM, *Duke University*
SCHNERMANN, JURGEN, *University of Michigan School of Medicine*
SCHOENER, THOMAS, *University of California, Davis*
SELLERS, LARRY, *Louisiana Tech University*
SHEA, JUDITH, *Kutztown University*
SHONTZ, NANCY, *Grand Valley State University*
SHOPPER, MARILYN, *Johnson County Community College*
SLOBODA, ROGER, *Dartmouth College*
SMITH, DAVID, *Williams College*
SMITH, JERRY, *St. Petersburg Junior College, Clearwater Campus*
SMITH, MICHAEL, *Valdosta State University*
SMITH, RALPH, *University of California, Berkeley*
SMITH, ROBERT, *West Virginia University*
STANTON, MAUREEN, *Center for Population Biology, University of California, Davis*
STARR, LISA, *Scripps Clinic and Research Foundation*
STEARNS, DONALD, *Rutgers University*
STEINERT, KATHLEEN, *Bellevue Community College*
SUMMERS, GERALD, *University of Missouri*
SUNDBERG, MARSHALL, *Louisiana State University*
SWANSON, ROBERT, *North Hennepin Community College*
TAYLOR, JANE, *Northern Virginia Community College*
TERHUNE, JERRY, *Jefferson Community College, University of Kentucky*
TILMAN, DAVID, *University of Minnesota*
TIZARD, IAN, *Texas A & M University*
TRAMMELL, JAMES, JR., *Arapahoe Community College*
WARING, RICHARD, *Oregon State University*
WARNER, MARGARET, *Indiana University, Krennert Institute*
WEIGL, ANN, *Winston-Salem State University*
WEISS, MARK, *Wayne State University*
WENDEROTH, MARY PAT, *University of Washington*
WHIPP, BRIAN, *St. George's Hospital Medical School, London, U.K.*
WHITE, EVELYN, *Alabama State University*
WHITENBERG, DAVID, *Southwest Texas State University*
WINICUR, SANDRA, *Indiana University, South Bend*
WOLFE, STEPHEN, EMERITUS, *University of California, Davis*
YONENAKA, SHANNA, *San Francisco State University*

LEARNING GUIDE FOR STUDENTS

Before you start reading, take a moment to look over the next few pages, which provide an overview of some of the book's built-in learning devices. Becoming familiar with these unique features can make your march through the material a lot easier.

1. Consistent Color Coding

Certain large molecules are characteristic of life, and they appear repeatedly in much of the art. We've assigned a different color to each type. This will remind you of the structural and functional uses of these molecules, as when you track them through a metabolic pathway.

Color Codes:

Carbohydrates

Lipid heads

Lipid tails

Lipid bilayer of cell membranes

Proteins

DNA, chromosomes

RNAs

ATP

Coenzymes (NADP$^+$, NAD$^+$, FAD)

Energy flow

Examples:

water

water

lipid bilayer

2. Self-Contained Topic Spreads

No doubt you have found yourself flipping back and forth among the pages of a book, trying to find illustrations that accompany a text passage, with your brain trying to put the artificially fractured contents back together. In this book, all of the text under a heading or subheading stays together on the same page or facing pages. The accompanying illustrations are arranged the same way. All the material is right there for you, in a cohesive, self-contained unit, so you can focus easily on one topic at a time.

For example, this two-page spread provides a self-contained overview of the type of internal structures you would find if you were to go looking inside many plant cells. A three-dimensional painting of a cut-apart cell shows how the structures fit together. A micrograph shows what it actually would look like sliced through its midsection. Accompanying diagrams show the cell's main internal components and briefly state the functions of each one.

This example shows a self-contained spread on skeletal muscle contraction. Notice the zoom-sequence of illustrations, starting with a human arm and ending with interactions among the proteins that actually bring about contraction. The entire sequence, from large to small, is illustrated on these two pages. Notice also the boldfaced display sentences, set off with blue lines. These sentences summarize the key concepts that unfold in the text.

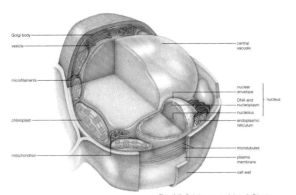

Figure 3.12 Typical components of plant cells. This cutaway diagram corresponds roughly to the micrograph in Figure 3.13.

EUKARYOTIC CELLS

Organelles—The Hallmark of Eukaryotes

The nucleus is the hallmark of eukaryotic cells. (Hence the name *eukaryotic*, meaning "true nucleus.") It is one of many organelles that distinguish these cells from prokaryotes. An **organelle** is an internal, membrane-bounded sac or some other kind of compartment that has a specific metabolic function within a cell.

No human-built apparatus can match the eukaryotic cell for the sheer number of chemical activities that can proceed simultaneously in so small a space. For example, many metabolic reactions are incompatible with others. Think about a plant cell putting together a starch molecule by some reactions and breaking it down by others. The cell would gain nothing if the synthesis and breakdown reactions proceeded at the same time on the same starch molecule! Such reactions proceed smoothly, largely because organelle membranes keep them physically separated.

Organelle membranes also permit compatible, interconnected reactions to proceed at different times. Thus

starch molecules are produced and stored by way of reaction sequences in one organelle—then later released for use in other reaction sequences in the same plant cell.

Organelles physically separate chemical reactions, many of which are incompatible, inside the cell.

Organelles separate different reactions in time, as when molecules are produced in one organelle, then used later in other reaction sequences.

Components of Plant Cells

Figure 3.12 can start you thinking about where organelles are located in a typical plant cell. Keep in mind that calling this cell "typical" is like calling a cactus or crocus a "typical" plant. As is true of animal cells, variations on the basic plan are mind-boggling. With this qualification in mind, also take a look at Figure 3.13. The sketches accompanying the micrograph outline the functions of organelles and other structures you are likely to find in the plant kingdom.

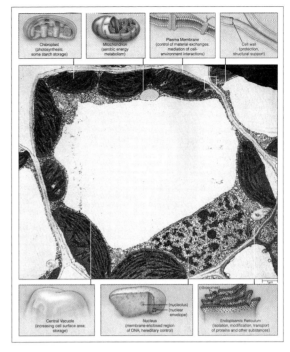

Figure 3.13 Transmission electron micrograph of a plant cell from a blade of Timothy grass, cross-section

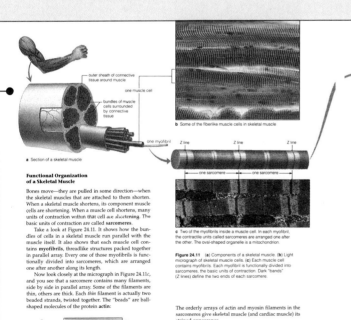

a Section of a skeletal muscle

b Some of the fiberlike muscle cells in skeletal muscle

c Two of the myofibrils inside a muscle cell. In each myofibril, the contractile units called sarcomeres are arranged one after the other. The oval-shaped organelle is a mitochondrion.

Figure 24.11   (a) Components of a skeletal muscle. **(b)** Light micrograph of skeletal muscle cells. **(c)** Each muscle cell contains myofibrils. Each myofibril is functionally divided into sarcomeres, the basic units of contraction. Dark "bands" (Z lines) define the two ends of each sarcomere.

Functional Organization of a Skeletal Muscle

Bones move—they are pulled in some direction—when the skeletal muscles that are attached to them shorten. When a skeletal muscle shortens, its component muscle cells are shortening. When a muscle cell shortens, many units of contraction within that cell are shortening. The basic units of contraction are called **sarcomeres.**

Take a look at Figure 24.11. It shows how the bundles of cells in a skeletal muscle run parallel with the muscle itself. It also shows that each muscle cell contains **myofibrils,** threadlike structures packed together in parallel array. Every one of those myofibrils is functionally divided into sarcomeres, which are arranged one after another along its length.

Now look closely at the micrograph in Figure 24.11c, and you see that a sarcomere contains many filaments, side by side in parallel array. Some of the filaments are thin, others are thick. Each *thin* filament is actually two beaded strands, twisted together. The "beads" are ball-shaped molecules of the protein **actin:**

one actin molecule   } one thin filament

Each *thick* filament consists of molecules of **myosin,** a protein with a head and long tail. The myosin tails are packed together in parallel. The heads, which look like double-headed golf clubs, stick out to the sides:

one myosin molecule   } one thick filament

The orderly arrays of actin and myosin filaments in the sarcomeres give skeletal muscle (and cardiac muscle) its striped appearance.

Think of the parallel orientation of all the myofibrils, muscle cells, and muscle bundles within a skeletal muscle. The orientation of a skeletal muscle's component parts focuses the force of contraction onto the bone in a particular direction.

A skeletal muscle shortens through the combined decreases in length of its sarcomeres, the basic units of contraction.

The parallel orientation of a muscle's component parts directs the force of contraction toward a bone that must be pulled in some direction.

Figure 24.12 Simplified picture of how actin and myosin filaments are arranged in a sarcomere. Interactions between the two kinds of filaments shorten (contract) the sarcomere.

How Skeletal Muscle Contracts

Sliding-Filament Model How do sarcomeres shorten and so bring about contraction of a skeletal muscle? The answer lies with sliding and pulling interactions among the sarcomere's filaments. As Figure 24.12 shows, there are two sets of actin filaments, attached to opposite sides of the sarcomere and extending partway to its center. There is one set of myosin filaments. These don't extend all the way to the sides, but they partially overlap the actin filaments. During contraction, myosin filaments physically slide along and *pull* the two sets of actin filaments toward the center of the sarcomere, which thereby shortens. This is the key premise of the **sliding-filament model** of muscle contraction.

The interactions between myosin and actin filaments occur through **cross-bridge formation.** As indicated by Figure 24.13, each cross-bridge is an attachment between a myosin "head" and a binding site on actin. When a muscle cell is stimulated, myosin heads are energized. They attach to an adjacent actin filament and tilt in a short power stroke toward the sarcomere's center. An input of energy from ATP drives the power stroke. During the stroke, the heads pull the

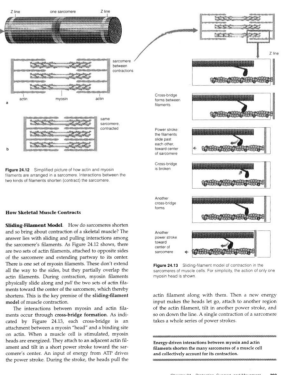

Figure 24.13 Sliding-filament model of contraction in the sarcomeres of muscle cells. For simplicity, the action of only one myosin head is shown.

actin filament along with them. Then a new energy input makes the heads let go, attach to another region of the actin filament, tilt in another power stroke, and so on down the line. A single contraction of a sarcomere takes a whole series of power strokes.

Energy-driven interactions between myosin and actin filaments shorten the many sarcomeres of a muscle cell and collectively account for its contraction.

3. Visual Summaries

Who hasn't stared at the text on a complex topic, stared at illustrations, then stared back at the text, wondering how it all fits together? Visual summaries can give you perspective on the big picture. These illustrations have detailed, often step-by-step descriptions integrated with the diagrams. It takes an incredible amount of time and thought to come up with this type of biological illustration. We've been creating more and more over the years. With this edition, all major concepts now have their own visual summary. When you come across pages with this type of illustration, read through it so your brain cells can start working on the big picture. Read the text, then go back and work your way more slowly, step by step, through the illustration. This will enhance understanding of the material.

This example walks you through one of the immune responses. Those helper T cells initiating the response are targets of the virus that causes AIDS. Understand this diagram and you'll get a handle on why AIDS, discussed later in the chapter, is so devastating.

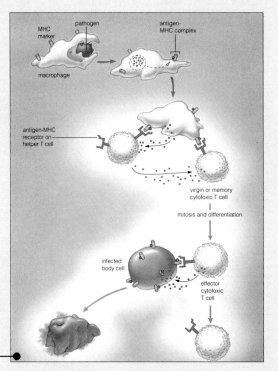

a A macrophage engulfs and digests a pathogen, then cleaves its antigen into fragments that bind to MHC markers. The macrophage becomes an antigen-presenting cell; it displays processed antigen-MHC complexes at its surface.

b Receptors on T cells bind to the complexes. Binding stimulates the macrophage to secrete interleukin-1 (pink dots). This stimulates helper T cells to secrete other interleukins (blue dots). These stimulate virgin or memory cytotoxic T cells to divide and differentiate into large populations of effector and memory cells. Only the effector cytotoxic T cells have cell-killing abilities.

c An effector encounters a target: an infected body cell that has the processed antigens bound with MHC markers at its surface. And it delivers a lethal hit. It releases perforins and toxic substances (green dots) onto its target and so programs it for death.

d The effector disengages from the doomed cell and reconnoiters for new targets. Meanwhile, perforins make holes in the target's plasma membrane. Toxins move into the cell, disrupt organelles, and make the DNA disassemble. The infected cell dies.

Figure 27.7 Example of a cell-mediated immune response, as carried out by activated cytotoxic T cells. An antigen-presenting macrophage provides the signal that starts this immune response.

This is a step-by-step illustration on crossing over, one of the most complex topics in biology. Learn it well; this event is absolutely central to understanding diversity among sexually reproducing organisms, humans included.

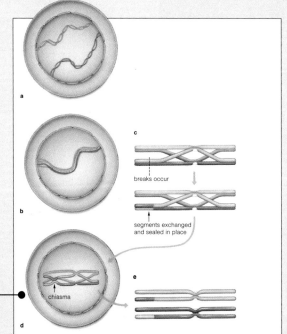

a Chromosomes become duplicated before meiosis begins. Early in prophase I, each duplicated chromosome is in threadlike form, attached at both ends to the nuclear envelope. Its two sister chromatids are so close together they look like a single thread.

b The two chromosomes become zippered together, so that all four chromatids are positioned close together.

c One or more crossovers occur at intervals along the chromosomes. In each crossover, two nonsister chromatids break at identical sites. They swap segments at the breaks, then enzymes seal the broken ends. (For clarity, the two chromosomes are shown in condensed form and pulled apart. Crossing over may seem more plausible when you realize that it occurs while chromosomes are extended like threads and tightly aligned.)

d As prophase I ends, the chromosomes continue to condense, becoming thicker, rodlike forms. They detach from the nuclear envelope and from each other—except at "chiasmata." Each chiasma is indirect evidence that a crossover occurred at some point in the chromosomes.

e Crossing over breaks up old combinations of alleles and puts new ones together in pairs of homologous chromosomes.

Figure 8.4 Key events during prophase I, the first stage of meiosis. For clarity, only a single pair of homologous chromosomes and only one crossover event are shown. Blue signifies the paternal chromosome, and purple signifies its maternal homologue.

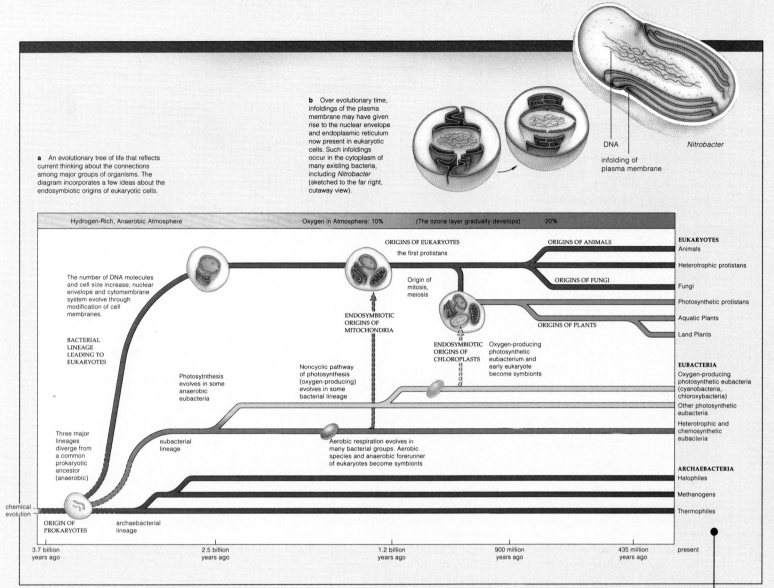

b Over evolutionary time, infoldings of the plasma membrane may have given rise to the nuclear envelope and endoplasmic reticulum now present in eukaryotic cells. Such infoldings occur in the cytoplasm of many existing bacteria, including *Nitrobacter* (sketched to the far right, cutaway view).

DNA

infolding of plasma membrane

Nitrobacter

a An evolutionary tree of life that reflects current thinking about the connections among major groups of organisms. The diagram incorporates a few ideas about the endosymbiotic origins of eukaryotic cells.

Hydrogen-Rich, Anaerobic Atmosphere | Oxygen in Atmosphere: 10% | (The ozone layer gradually develops) | 20%

ORIGINS OF EUKARYOTES
the first protistans

ORIGINS OF ANIMALS

EUKARYOTES
Animals

Heterotrophic protistans

ORIGINS OF FUNGI

Fungi

Photosynthetic protistans

The number of DNA molecules and cell size increase; nuclear envelope and cytomembrane system evolve through modification of cell membranes.

Origin of mitosis, meiosis

Aquatic Plants

Land Plants

ENDOSYMBIOTIC ORIGINS OF MITOCHONDRIA

BACTERIAL LINEAGE LEADING TO EUKARYOTES

ORIGINS OF PLANTS

ENDOSYMBIOTIC ORIGINS OF CHLOROPLASTS

Oxygen-producing photosynthetic eubacterium and early eukaryote become symbionts

EUBACTERIA
Oxygen-producing photosynthetic eubacteria (cyanobacteria, chloroxybacteria)

Photosynthesis evolves in some anaerobic eubacteria

Noncyclic pathway of photosynthesis (oxygen-producing) evolves in some bacterial lineage

Other photosynthetic eubacteria

Heterotrophic and chemosynthetic eubacteria

Three major lineages diverge from a common prokaryotic ancestor (anaerobic)

eubacterial lineage

Aerobic respiration evolves in many bacterial groups. Aerobic species and anaerobic forerunner of eukaryotes become symbionts

ARCHAEBACTERIA
Halophiles

Methanogens

chemical evolution

ORIGIN OF PROKARYOTES

archaebacterial lineage

Thermophiles

3.7 billion years ago | 2.5 billion years ago | 1.2 billion years ago | 900 million years ago | 435 million years ago | present

From a *Focus on Science* essay.

Now this is a really big picture, a unique visual summary of 3.5 billion years of evolution from the first cells to the existing spectrum of life. It is based on current data from comparative biochemistry. Notice the descriptions integrated *inside* the diagram. These help you track the flow of information.

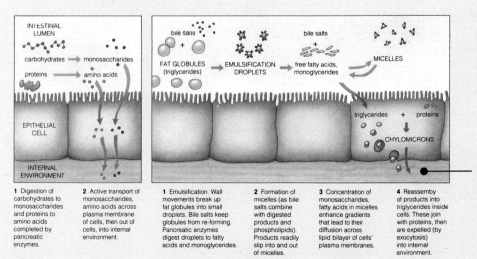

INTESTINAL LUMEN

carbohydrates → monosaccharides

proteins → amino acids

EPITHELIAL CELL

INTERNAL ENVIRONMENT

bile salts

FAT GLOBULES (triglycerides) → EMULSIFICATION DROPLETS → free fatty acids, monoglycerides

bile salts

MICELLES

triglycerides + proteins

CHYLOMICRONS

1 Digestion of carbohydrates to monosaccharides and proteins to amino acids completed by pancreatic enzymes.

2 Active transport of monosaccharides, amino acids across plasma membrane of cells, then out of cells, into internal environment.

1 Emulsification. Wall movements break up fat globules into small droplets. Bile salts keep globules from re-forming. Pancreatic enzymes digest droplets to fatty acids and monoglycerides.

2 Formation of micelles (as bile salts combine with digested products and phospholipids). Products readily slip into and out of micelles.

3 Concentration of monosaccharides, fatty acids in micelles enhance gradients that lead to their diffusion across lipid bilayer of cells' plasma membranes.

4 Reassemby of products into triglycerides inside cells. These join with proteins, then are expelled (by exocytosis) into internal environment.

Figure 25.8 Summary of digestion and absorption processes in the small intestine.

How organ systems function is breathtakingly complex. This is one area where visual summaries will truly help you connect the details.

4. Focus Essays

We've devised two tracks through the book, as its title suggests. First, the text proper deals with theoretical aspects of biology—the concepts and the connections among them. Second, throughout the book, numerous examples of applications show how those concepts can be used to explain something that's going on in the world around you. *Focus* essays are the most prominent of these applications, all of which are indexed on the end papers at the back of the book.

One type of focus essay gives working examples of how biologists think up and then test their ideas. Others provide detail on environmental and bioethical issues. Another provides detail on timely health issues.

Abortion, human gene therapy, pollution versus economic growth—these are a few of the most explosive issues of our time. Where appropriate, we address these bioethical issues in an objective way and invite *you* to think about their implications.

Just about every day of your life, television, newspapers, and magazines bombard you with "facts" and opinions of "experts" on everything from your personal diet and hygiene to planetwide changes in the atmosphere. The *Focus on Science* essays can help you understand the advantages of critical thinking—of evaluating ideas and observations with an objective eye. This study showed that natural selection of self-sacrificing behavior is a concept that can be applied to more than just a few extreme insect societies; here it was successfully applied to naked mole-rats which, like you, are mammals.

Many focus essays highlight behavioral, genetic, and environmental factors that can affect your health. A great deal of misleading or downright inaccurate information exists on these topics. We worked closely with dozens of specialists to provide you with information that you can use with a high degree of confidence.

Similarly, a great deal of conflicting information exists on many environmental issues. We worked with respected environmentalists and ecologists to give you a balanced picture of the major issues.

Focus on Health

A Cascade of Proteins and Cancer

Every second, millions of cells in your skin, gut lining, liver, and other body regions divide and replace their worn-out, dead, and dying predecessors. They do not divide willy-nilly. They cannot divide at all unless they synthesize and stockpile cyclin, a protein. Cyclin binds with cdc2, the first of a series of enzymes that transfer phosphate from ATP to the next enzyme in line or to a final protein target. The enzyme molecules act more than once. Each activates many others, which activate many others, and so on in a growing cascade of reactions that ripple through the cell.

The enzymes switched on first replicate the cell's DNA. Others bring about nuclear division. For instance, some assemble and operate the microtubular spindle that moves chromosomes. Others orchestrate the split into two cells. As the cell divides, enzymes destroy all the cyclin, and this puts the division machinery to rest. Each daughter cell starts stockpiling cyclin. If the cell goes on to divide, cyclin will again be destroyed.

Of all proteins, only cyclin accumulates at a constant rate, disappears abruptly, then accumulates again during cell cycles (Figure *a*). But how does a cell "know" when to start and stop building cyclin? It requires signals from hormones and other regulator molecules. *Such signals lift controls that otherwise suppress the cyclin-driven engine, then put on the brakes by reinstating controls when cell division is completed.*

On rare occasions, controls over cell division are lost. It is not that affected cells divide at a horrendous rate. Rather, as long as conditions for growth stay favorable, the divisions never stop. Any tissue mass of cells that are not responding to normal controls over cell growth and division is a tumor.

When a tumor is benign, it remains in the same place in the body. Surgical removal of the tissue mass removes its threat to the functioning of the surrounding tissues. When a tumor is malignant, its cells can migrate and then grow and divide in other organs. The cells have undergone a mutation that alters or suppresses certain genes. Those genes code for the recognition proteins at the plasma membrane that allow cells to bind together in tissues and

organs. When genes for those proteins are altered or suppressed, the cell can leave its proper place and travel (by way of blood or lymph). The process of invasion is called metastasis.

There are many types of malignant tumors. All are grouped into the general category of **cancer**. Cancer is not just a human affliction. It has been observed in most animal species studied to date. Similar abnormalities

a Chart of the controlled changes in the intracellular levels of cyclin (brown line) in normal cells. Cyclin is the protein that guides cells into mitosis (light blue bands) during the cell cycle.
b Scanning electron micrograph of a cancer cell, surrounded by some of the body's white blood cells that may or may not be able to destroy it.

cytoplasmic extension of cancer cell body

white blood cell

175

Focus on the Environment

From Greenhouse Gases to a Warmer Planet?

Near the earth's surface, atmospheric concentrations of gaseous molecules play a profound role in shaping the average global temperature, which in turn has enormous effect on the global climate. Molecules of carbon dioxide, water, ozone, methane, nitrous oxide, and chlorofluorocarbons are the key players. Collectively, they act somewhat like a pane of glass in a greenhouse (hence their name, "greenhouse gases"). They let wavelengths of visible light reach the earth's surface, but they impede the escape of longer, infrared wavelengths—that is, heat—from the earth into space. They absorb infrared wavelengths, and much of the energy inherent in those wavelengths is reradiated back

toward the earth (Figure *a*). In short, greenhouse gases cause heat to build up in the lower atmosphere, a warming action known as the **greenhouse effect.**

Without greenhouse gases, the earth would be cold and lifeless. But there can be too much of a good thing. Largely as a result of human activities, greenhouse gases are building up to higher levels in the atmosphere (Figure *b*). The increase may be contributing to an alarming trend toward global warming.

What is so alarming about a warmer planet? Suppose the temperature of the lower atmosphere were to rise by only 4°C (7°F). Sea levels could rise by about 2 feet, or 0.6

meter. Why? The warming would increase ocean surface temperatures—and water expands when heated. Global warming also could make glaciers and the Antarctic ice sheet melt faster, so low coastal regions could flood.

Think of a long-term rise in sea level, combined with high tides and storm waves. Waterfronts of Vancouver, Boston, San Diego, Galveston, and other coastal cities would be submerged. So would agricultural lowlands and deltas in India, China, and Bangladesh, where much of the world's rice is grown. Huge tracts of Florida and Louisiana would face saltwater intrusions. Besides this, global warming could disturb regional patterns of precipitation and temperature. Crop yields would decline in currently productive regions, including parts of Canada and the United States, and increase in others.

In the late 1950s, researchers on a mountaintop in the Hawaiian Islands started measuring concentrations of different greenhouse gases, and their monitoring activities are still going on. They chose the remote site because it was free of local contamination and reflected average conditions for the Northern Hemisphere.

Consider what they found out about carbon dioxide alone. Atmospheric levels of carbon dioxide follow the annual cycle of plant growth in the Northern Hemisphere. They are lower in summer, when photosynthesis rates are highest. They are higher in winter, when aerobic respiration continues and photosynthesis slows. In Figure *c* (part 1),

the peaks and troughs around the graph line represent the highs and lows. For the first time, scientists saw the integrated effects of the carbon balances for the land and water ecosystems of an entire hemisphere. The midline of the peaks and troughs in the cycle has steadily increased. Many scientists take this as evidence of a buildup of carbon dioxide that may intensify the greenhouse effect over the next century.

The global burning of fossil fuels is probably contributing most to increasing carbon dioxide levels. Deforestation adds to it; carbon is released when wood burns. Today, vast tracts of tropical forests are being cleared and burned at a rapid rate (refer to Figure 37.10). More importantly, the plant biomass is plummeting—and this affects global absorption of carbon dioxide in photosynthesis.

Many scientists wonder whether atmospheric levels of greenhouse gases will continue to increase until the middle of the twenty-first century—and whether the global temperature will rise by several degrees. If this is indeed a trend already in motion, we will not be able to reverse it now by stopping fossil fuel burning and deforestation. So there is widespread agreement that we should begin preparing for the consequences. For example, we might step up genetic engineering studies to develop drought-resistant and salt-resistant plants. Such plants may prove crucial in regions of saltwater intrusions and climatic change.

a The greenhouse effect

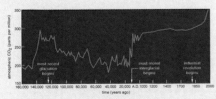

1. Sunlight penetrates the atmosphere and warms the earth's surface.

2. The earth's surface radiates heat (infrared wavelengths) to the atmosphere. Some heat escapes into space. Greenhouse gases and water vapor absorb some infrared wavelengths and reradiate a portion back toward the earth.

3. When greenhouse gases build up in the atmosphere, more heat is trapped near the earth's surface. Ocean surface temperatures rise, more water vapor enters the atmosphere, and the earth's surface temperature rises.

b Shifts in atmospheric concentrations of carbon dioxide, correlated with the most recent glaciation and interglacial period during the past 160,000 years.

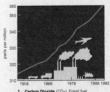

1. **Carbon Dioxide** (CO_2). Fossil fuel burning, factory emissions, car exhaust, and deforestation are contributing to the increased atmospheric concentration.

c Recently documented increases in atmospheric concentrations of four greenhouse gases.

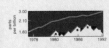

2. **Chlorofluorocarbons** (CFCs). These are used in plastic foams, air conditioners, refrigerators, and industrial solvents (page 615).

3. **Methane** (CH_4). This is produced by anaerobic bacteria in swamps, landfills, and termite activities. It is produced also by bacteria in the digestive tract of cattle and other ruminants.

4. **Nitrous Oxide** (N_2O). This is a natural by-product of denitrifying bacteria. It also is released in great amounts from fertilizers and animal wastes, as in livestock feedlots.

580

581

5. Vignettes

You can view biology as just another course that you have to take in order to graduate. Or you can open your mind to a fantastically rewarding way to gain perspective on the world around you—and on your place in it. No matter what comes to mind—be it stings from bees or jellyfishes, amazingly plump grapes in a grocery store, a friend who suffers from a genetic disease, or firestorms through the canyons of Malibu—you will be armed with knowledge that can help you make sense of it.

Beautifully illustrated short stories—vignettes—start each chapter. They are our attempt to convey the excitement and wonder of biology and its relevance to your own life. Many, including the one below, also convey *how* biologists think critically and conduct their research. Each vignette will lead you, in a nontechnical way, to the chapter's key concepts. Right after the vignette, the concepts are listed as simple summary statements, as indicated below. At each chapter's end, those same concepts are reinforced, in the manner outlined on page xvii.

---

6 ENERGY-RELEASING PATHWAYS

The Killers Are Coming!

In 1990, "killer" bees from South America buzzed across the border between Mexico and the United States. The bees are descended from African queen bees. When provoked, they can be terrifying.

Earlier, some queen bees had been shipped from Africa to Brazil for breeding experiments. Honeybees happen to be big business. Besides producing honey, they are rented out to pollinate commercial orchards. But the honeybees in Brazil seemed sluggish. The idea was to cross-breed them with their African relatives and produce a mild-mannered but zippier pollinator.

Twenty-six African queens escaped. That was bad enough. Then beekeepers got wind of the program. After learning that the first few generations of offspring were jazzed-up but nice honeybees, they imported *hundreds* of African queens and encouraged them to mate with locals. And they set off a genetic time bomb.

Before long, African bees were firmly established in commercial hives—and in wild bee populations. And their traits became dominant. The "Africanized" bees do everything other bees do, but they do more of it faster. Their eggs develop into adults more quickly. Adults fly more rapidly, outcompete other bees for nectar, and even die sooner. When their hives are disturbed, they become far more agitated. Whereas a mild-mannered honeybee might chase an intruding animal fifty yards or so, a squadron of Africanized bees will chase it a quarter of a mile. If they catch up to it, they can sting it to death.

Doing things faster means having a nonstop supply of energy. This bee's stomach can hold 30 milligrams of sugar-rich nectar—enough fuel to fly 60 kilometers (more than 35 miles). Large mitochondria in its flight muscle cells efficiently convert energy stored in sugar to ATP energy. Africanized bees cannot survive where winters are harsh and plants stop blooming for months at a time.

In their ability to release energy stored in sugar and other organic compounds, Africanized bees are like all other organisms. Although energy-releasing pathways differ in some details, all start with certain materials, then end with predictable products. And they *all* yield ATP. *At the biochemical level, there is undeniable unity among all forms of life.* We will return to this idea in the *Commentary* at the chapter's end.

Figure 6.1 A mild-mannered honeybee buzzing in for a landing on a flower, wings beating with energy provided by ATP. If this were one of its Africanized relatives approaching a hive, possibly you would not stay around to watch the landing.

---

38 ANIMAL BEHAVIOR

Deck the Nest With Sprigs of Green Stuff

About a century ago, the European starling hitched a boat ride to North America. Ever since, starlings have been multiplying and evicting great numbers of native birds from scarce nest sites.

Curiously, starlings decorate the usurped nests with many sprigs of wild carrot, freshly plucked. Why do they do this? Does greenery camouflage the nests from predators? Not really. The nests are already concealed, in the dark cavities of tree holes. Does the greenery insulate eggs from the cold? Actually, moist, green plant parts promote heat loss. Well, then, does the greenery combat parasites? After all, mites parasitize birds and infest the nest holes (Figure 38.1). Even a few tiny mites can rapidly produce thousands of descendants. In large numbers, the mites can suck enough blood from a nestling to weaken it and interfere with its growth.

A bit of wild carrot does indeed fumigate a nest—and it measurably increases a starling's chance of producing healthy, surviving offspring. Wild carrot contains a highly aromatic steroid compound. Most likely, the compound functions in the plant's chemical defense against herbivorous animals. By coincidence, it also happens to arrest immature bird mites in their developmental tracks.

As biologists Larry Clark and Russell Mason discovered, fresh wild carrot sprigs prevent mites from becoming sexually mature—and therefore prevent a mite population explosion in the nest. They constructed experimental nests with and without carrot sprigs, and these were occupied by starlings. In time, by Clark's and Mason's count, an average of 750,000 mites occupied undecorated nests. A mere 8,000 or so occupied the green-sprigged ones.

And so starlings lead us into the world of behavioral research. As you will see, some behavioral studies focus on structural and functional mechanisms that enable individuals to behave as they do. Others focus on the adaptive value of some behavior to an individual's reproductive success.

Figure 38.1 (a) A most excellent fumigator in nature—the European starling (*Sturnus vulgaris*), which combats infestations of mites by decorating previously owned nests with fresh sprigs of wild carrot (b).

a

b

KEY CONCEPTS

1. "Behavior" refers to the observable, coordinated responses an animal makes to stimuli. The responses are instinctive, learned, or a combination of both.

2. Forms of behavior have a genetic basis. Certain genes contain instructions that govern the development of the nervous and endocrine systems, by which an animal detects, processes, and issues commands for behavioral responses to stimuli.

3. Like other traits having a genetic basis, forms of behavior have evolved by way of natural selection—the measure of which is reproductive success. Thus sexual selection and other evolutionary processes have favored behavioral mechanisms that enhance the ability of the individual to pass on his or her genes to offspring.

4. Social behavior may be explained in terms of natural selection. It has costs and benefits that can be measured in terms of the reproductive success of the individual.

MECHANISMS UNDERLYING BEHAVIOR

Genetic Effects on Behavior

An animal's nervous system detects and processes information about conditions outside and inside the body, then it commands muscles and glands to make appropriate responses. The observable responses that animals make to stimuli are what we call **animal behavior.**

Genes contribute in an indirect yet major way to behavior. Consider Stevan Arnold's studies of the feeding preferences of different populations of a garter snake species. For snake populations living along the California coast, the food of choice is the banana slug (Figure 38.2). Farther inland, tadpoles and small fishes are preferred. Offer inland snakes a banana slug and they ignore it.

In one set of experiments, Arnold offered captive newborn garter snakes a chunk of slug as the first meal. Offspring of coastal parents usually ate it and even flicked their tongue at cotton swabs drenched in essence of slug. (Snakes "smell" by tongue-flicking, which draws chemical odors into the mouth.) Offspring of inland parents ignored the swabs and only rarely ate the slug meat.

Here was a clear difference between captive baby snakes that had no prior, direct experience with slugs. It suggested that the snakes were programmed to accept

a

b

c

Figure 38.2 (a) Banana slug, food for (b) a grown-up garter snake of coastal California. (c) A newborn garter snake from a coastal population, tongue-flicking at a cotton swab drenched with banana slug fluids.

or reject slugs before they were born; they didn't learn to do so through "taste trials." To test this hypothesis, Arnold crossed coastal male snakes with inland female snakes. He also crossed coastal females with inland males. If the difference between the populations has a genetic basis, then "hybrid" offspring might make an intermediate response to slug chunks and odors.

Arnold's observations matched the predicted results. Compared with typical newborn inland snakes, many baby snakes of mixed parentage tongue-flicked *more* often at slug-scented cotton swabs—but *less* often than typical newborn coastal snakes. Thus, the difference in feeding behavior could stem from differences in genes that affect how odor-detecting mechanisms are put together in the snake embryo during its development.

By influencing the development of the nervous system, genes contribute in an indirect yet major way to behavior.

629

6. Built-In Study Aids

Besides the topic spreads, visual summaries, in-text summary statements, and other features shown in earlier examples, the chapters offer you the following study aids:

End-of-chapter summaries

in list format reinforce the main chapter concepts.

Review questions

are keyed to italicized and boldface sentences to further reinforce main concepts. These conceptual questions are cross-referenced to page numbers in the text. Detailed answers to all review questions are available in a separate booklet.

Key terms

All of the boldfaced terms defined in the chapter are listed. See if you can give definitions for all of them. The italic page numbers will provide fast entry back into the chapter for the ones you aren't sure of.

SUMMARY

1. A gene is a unit of information about a heritable trait. The alleles of a gene are slightly different versions of that information. Through experimental crosses with pea plants, Mendel gathered evidence that diploid organisms have two genes for each trait and that genes retain their identity when transmitted to offspring.

2. Mendel performed monohybrid crosses (between two true-breeding plants showing different versions of a single trait). The crosses provided evidence that a gene can have different molecular forms (alleles), some of which are dominant over other, recessive forms.

3. Homozygous dominant individuals have two dominant alleles (*AA*) for the trait being studied. Homozygous recessives have two recessive alleles (*aa*). Heterozygotes have two nonidentical alleles (*Aa*).

4. In Mendel's monohybrid crosses (*AA* × *aa*), all F_1 offspring were *Aa*. Crosses between F_1 plants resulted in these combinations of alleles in F_2 offspring:

	A	*a*
A	*AA*	*Aa*	*AA* (dominant)
a	*Aa*	*aa*	*Aa* (dominant)

This produced the expected phenotypic ratio of 3:1.

5. Armed with results from his monohybrid crosses, Mendel proposed a theory of segregation. In modern terms, diploid organisms have pairs of genes, on pairs of homologous chromosomes. The two genes of each pair segregate from each other during meiosis, such that each gamete formed ends up with one or the other.

6. Mendel also performed dihybrid crosses (between two true-breeding plants showing different versions of two traits). Results from many experiments were close to a 9:3:3:1 phenotypic ratio:

9 dominant for both traits
3 dominant for *A*, recessive for *b*
3 dominant for *B*, recessive for *a*
1 recessive for both traits

7. Mendel's dihybrid crosses led him to propose a theory of independent assortment. In modern terms, the gene pairs of two homologous chromosomes tend to be sorted into one gamete or another independently of how the gene pairs of other chromosomes are sorted out.

8. Four factors can influence gene expression. First, degrees of dominance exist between some gene pairs. Second, gene pairs can interact to produce some positive or negative effect on a trait. Third, a single gene can influence many seemingly unrelated traits. Fourth, environmental conditions can affect gene expression.

135   Principles of Inheritance

Review Questions

1. State the theory of segregation. Does segregation occur during mitosis or meiosis? 127

2. Define the difference between these terms. 125
  a. gene and allele
  b. dominant allele and recessive allele
  c. homozygote and heterozygote
  d. genotype and phenotype

3. Define true-breeding. What is a hybrid? 124, 125

4. Distinguish between monohybrid and dihybrid crosses. What is a testcross, and why is it useful in genetic analysis? 126, 128

5. State the theory of independent assortment. Does independent assortment occur during mitosis or meiosis? 129

Self-Quiz *(Answers in Appendix IV)*

1. Alleles are _____.
  a. different molecular forms of a gene
  b. different molecular forms of a chromosome
  c. self-fertilizing, true-breeding homozygotes
  d. self-fertilizing, true-breeding heterozygotes

2. A heterozygote has _____ for the trait being studied.
  a. a pair of identical alleles
  b. a pair of nonidentical alleles
  c. a haploid condition, in genetic terms
  d. a and c

3. The observable traits of an organism are its _____.
  a. phenotype   c. genotype
  b. sociobiology   d. pedigree

4. Offspring of a monohybrid cross *AA* × *aa* are _____.
  a. all *AA*   d. 1/2 *AA* and 1/2 *aa*
  b. all *aa*   e. none of the above
  c. all *Aa*

5. Second-generation offspring from a cross are the _____.
  a. F_1 generation   c. hybrid generation
  b. F_2 generation   d. none of the above

6. Assuming complete dominance, offspring of the cross *Aa* × *Aa* will show a phenotypic ratio of _____.
  a. 3:1   c. 9:1
  b. 1:2:1   d. 9:3:3:1

7. Crosses between F_1 individuals resulting from the cross *AABB* × *aabb* lead to F_2 phenotypic ratios close to _____.
  a. 1:2:1   c. 3:1
  b. 1:1:1:1   d. 9:3:3:1

8. Match each genetic term appropriately.
  ___ dihybrid cross   a. *AA* × *aa*
  ___ monohybrid cross   b. *Aa*
  ___ homozygous condition c. *AA BB* × *aa bb*
  ___ heterozygous condition d. *aa*

Genetics Problems *(Answers in Appendix III)*

1. One gene has alleles *A* and *a*. Another has alleles *B* and *b*. For each genotype listed, what type(s) of gametes will be produced? (Assume independent assortment occurs.)
  a. *AA BB*   c. *Aa Bb*
  b. *Aa BB*   d. *Aa Bb*

2. Still referring to Problem 1, what will be the genotypes of offspring from the following matings? With what frequency will each genotype show up?
  a. *AA BB* × *aa BB* c. *Aa Bb* × *aa bb*
  b. *Aa BB* × *AA Bb* d. *Aa Bb* × *Aa Bb*

3. In one experiment, Mendel crossed a pea plant that bred true for green pods with one that bred true for yellow pods. All the F_1 plants had green pods. Which form of the trait (green or yellow pods) is recessive? Explain how you arrived at your conclusion.

4. At one gene location on a human chromosome, a dominant allele controls whether you can curl the sides of your tongue upward (*see photo*). People homozygous for the trait cannot roll their tongue. At a different gene location, a dominant allele controls whether earlobes are attached or detached (see Figure 9.1). These two gene pairs assort independently.
  Suppose a tongue-rolling woman with detached earlobes marries a man who has attached earlobes and can't roll his tongue. Their first child has attached earlobes and can't roll its tongue.
  a. What are the genotypes of the mother, father, and child?
  b. What is the probability that a second child will have detached earlobes and won't be a tongue roller?

5. Go back to Problem 1, and assume you now study a third gene having alleles *C* and *c*. For each genotype listed, what type(s) of gametes will be produced?
  a. *AA BB CC*   c. *Aa BB Cc*
  b. *Aa BB cc*   d. *Aa Bb Cc*

6. Mendel crossed a true-breeding tall, purple-flowered pea plant with a true-breeding dwarf, white-flowered plant. All F_1 plants were tall and purple-flowered. If an F_1 plant self-fertilizes, what is the probability that a randomly selected F_2 plant will be heterozygous for the genes specifying height and flower color?

7. Assume that a new gene has been identified in mice. One of its alleles specifies yellow fur color. A second allele specifies brown fur color. Suppose you are asked to determine whether the relationship between the two alleles is one of simple dominance, incomplete dominance, or codominance. What types of crosses would give you the answer? On what types of observations would you base your conclusions?

8. The ABO blood-typing system has been used to settle cases of disputed paternity. Suppose, as a geneticist, you must testify during a case in which the mother has type A blood, the child has type O blood, and the alleged father has type B blood. How would you respond to the following statements?
  a. *Attorney of the alleged father:* "The mother has type A blood, so the child's type O blood must have come from the father. Because my client has type B blood, he could not have fathered this child."
  b. *Mother's attorney:* "Further tests prove this man is heterozygous, so he must be the father."

9. As in Labrador retrievers (page 132), fur color in mice is governed by genes concerned with producing and distributing melanin. At one gene location, a dominant allele (*B*) specifies dark brown and a recessive allele (*b*) specifies light brown, or tan. At another gene location, a dominant allele (*C*) permits melanin production and a recessive allele (*c*) shuts it down and results in albinism.
  a. A homozygous *bb cc* albino mouse mates with a homozygous *BB CC* brown mouse. State the probable genotypic and phenotypic ratios for the F_1 and F_2 offspring.
  b. If an F_1 mouse from Problem 9a is backcrossed with its albino parent, what phenotypic and genotypic ratios would you expect?

Selected Key Terms

ABO blood typing 131	monohybrid cross 126
allele 125	multiple allele system 131
codominance 130	phenotype 125
continuous variation 134	pleiotropy 130
dihybrid cross 128	probability 126
epistasis 132	Punnett-square method 127
gene 125	testcross 127
genotype 125	theory of independent
heterozygous 125	assortment 129
homozygous dominant 125	theory of segregation 127
homozygous recessive 125	true-breeding 124
incomplete dominance 130	

Readings

Cummings, M. 1991. *Human Heredity.* Second edition. St. Paul: West Publishing Company.

Mendel, G. 1959. "Experiments in Plant Hybridization." Translation in J. Peters (editor), *Classic Papers in Genetics.* Englewood Cliffs, New Jersey: Prentice-Hall.

Orel, V. 1984. *Mendel.* New York: Oxford University Press.

Suzuki, D., et al. 1989. *An Introduction to Genetic Analysis.* Fourth edition. New York: Freeman.

Chapter 9 Observable Patterns of Inheritance   **136**

Self-quizzes

are a quick way to test your recall of important definitions and key concepts with the help of objective questions. Appendix IV lists the answers.

End-of-chapter genetics problems

You will get a better sense of Mendelian patterns of inheritance by working your way through genetics problems, which are available in appropriate chapters. Appendix III gives detailed solutions to assist understanding.

Recommended readings

CONTENTS IN BRIEF

DETAILED TABLE OF CONTENTS

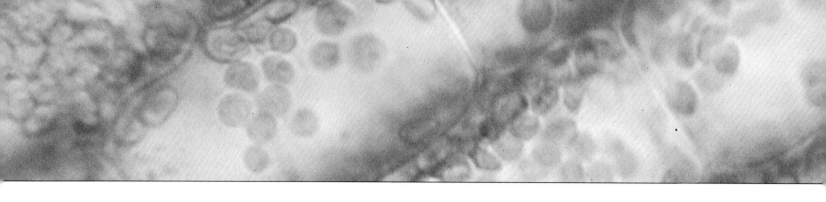

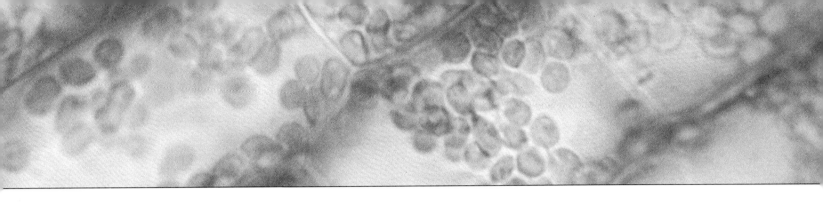

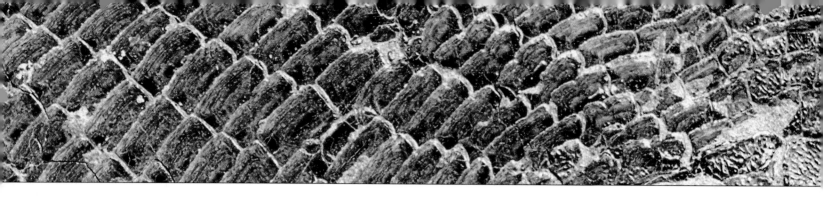

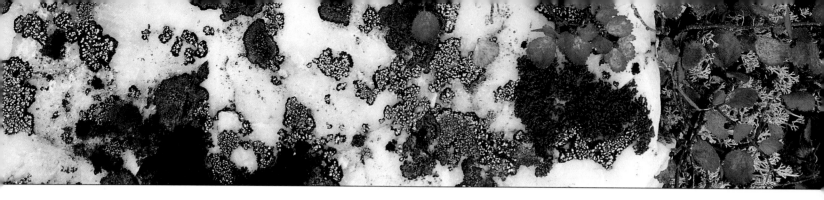

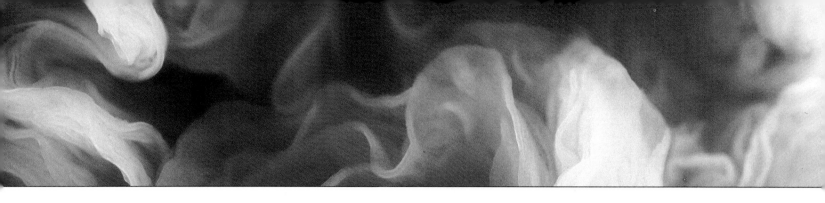

Current configurations of the earth's oceans and land masses—the geologic stage upon which life's drama continues to unfold. Thousands of separate images were pieced together to create this remarkable cloud-free view of our planet.

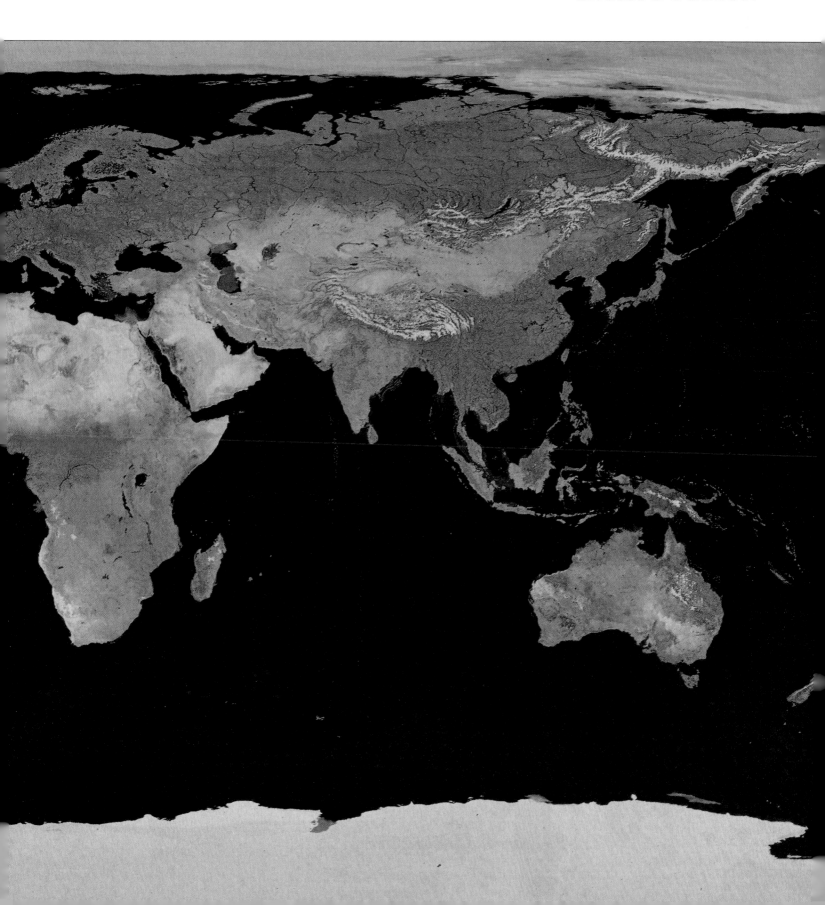

1 METHODS AND CONCEPTS IN BIOLOGY

Biology Revisited

Buried somewhere in that mass of tissue just above and behind your eyes are memories of first encounters with the living world. Still in residence are memories of discovering your hands and feet, your family, the change of seasons, the smell of rain-drenched earth and grass. In that brain are memories of early introductions to a great disorganized parade of insects, flowers, frogs, and furred things—mostly living, sometimes dead. There are memories of questions—"*What is life?*" and, inevitably, "*What is death?*" There are memories of answers, some satisfying, others less so.

By observing, asking questions, and accumulating answers, you have built up a store of knowledge about the world of life. Experience and education have been refining your questions, and no doubt some answers are difficult to come by. Think of a young man whose brain is functionally dead as a result of a motorcycle accident. If breathing and other basic functions proceed only as long as he remains hooked up to mechanical support systems, is he no longer "alive"? Think of an embryo growing inside a pregnant woman, but currently no more than a cluster of a few dozen tiny cells. At what point in its development is it a definably "human" life? If questions like these cross your mind, your thoughts about life obviously run deep.

The point is, this book isn't your introduction to biology—"the study of life"—for you have been studying life ever since information began penetrating your brain. This book simply is biology *revisited*, in ways that may help carry your thoughts to more organized levels of understanding.

Return to the question, *What is life?* Offhandedly, you might reply that you know it when you see it. To biologists, however, the question opens up a story that has been unfolding in countless directions for several billion years! "Life" is an outcome of ancient events by which nonliving materials became assembled into the first living cells. "Life" is a way of capturing and using energy and materials. "Life" is a way of sensing and responding to specific changes in the environment. "Life" is a capacity to reproduce, grow, and develop. And "life" evolves, meaning that details in the body plan and functions of organisms can change through successive generations. This short description only hints at the meaning of life. Deeper insight requires wide-ranging study of life's characteristics.

Throughout this book you will come across examples of how organisms are constructed, how they function, where they live, and what they do. The examples support certain concepts which, taken together, will give you a sense of what "life" is. This chapter provides an overview of the basic concepts. As you continue reading, you may find it useful to return to this overview to reinforce your grasp of details.

Figure 1.1 Think back on all you have known and seen. This is a foundation for your deeper probes into the world of life.

1. There is unity in the living world, for all organisms are alike in some key respects. First, their structural organization and functions depend on certain properties of matter and energy. Second, they obtain and use energy and materials from the environment. Third, they make controlled responses to changing conditions. Fourth, they grow and reproduce, based on instructions contained in DNA.

2. There also is diversity in the living world, for organisms vary immensely in body plans, body functions, and behavior. The diversity has come about mainly through evolutionary processes.

3. Biology, like other branches of science, is based on systematic observations, hypotheses, predictions, and relentless tests. The external world, not internal conviction, is the testing ground for scientific theories.

SHARED CHARACTERISTICS OF LIFE

Energy, DNA, and Life

Picture a frog on a rock, busily croaking. Without even thinking about it, you know the frog is alive and the rock is not. At a much deeper level, however, the difference between them blurs. They and all other things are composed of the same particles (protons, electrons, and neutrons). The particles are organized as atoms, according to the same physical laws. At the heart of those laws is something called **energy**—a capacity to make things happen, to do work. Energetic interactions bind atom to atom in predictable patterns, giving rise to the structured bits of matter we call molecules. Energetic interactions among molecules hold a rock together—and they hold a frog together.

It takes a special molecule called deoxyribonucleic acid, or **DNA**, to set living things apart from the nonliving world. No chunk of granite or quartz has it. DNA molecules contain instructions for assembling each new organism from "lifeless" molecules that contain carbon, hydrogen, and a few other kinds of substances. By analogy, think of what you can do with a little effort and just two kinds of ceramic tiles in a crafts kit. By following the kit's directions, you can glue the tiles together to form organized patterns (Figure 1.2). Similarly, life emerges from lifeless matter with DNA "directions," some raw materials, and inputs of energy.

The structure and organization of nonliving and living things arise from the properties of matter and energy.

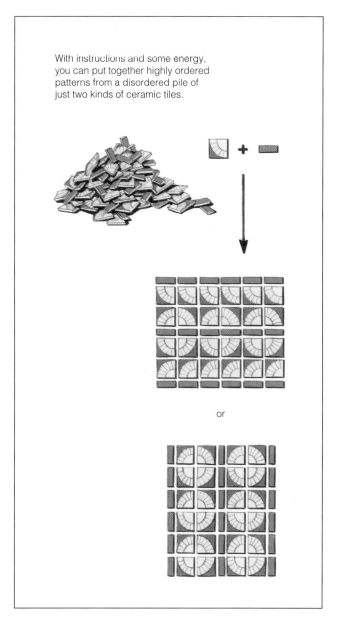

With instructions and some energy, you can put together highly ordered patterns from a disordered pile of just two kinds of ceramic tiles.

or

Figure 1.2 Emergence of organized patterns from disorganized beginnings. Two ceramic tile patterns are shown. You probably can visualize other possible patterns using the same two kinds of tiles. Similarly, the organization of each living thing emerges from pools of simple building blocks, given energy sources and DNA instructions.

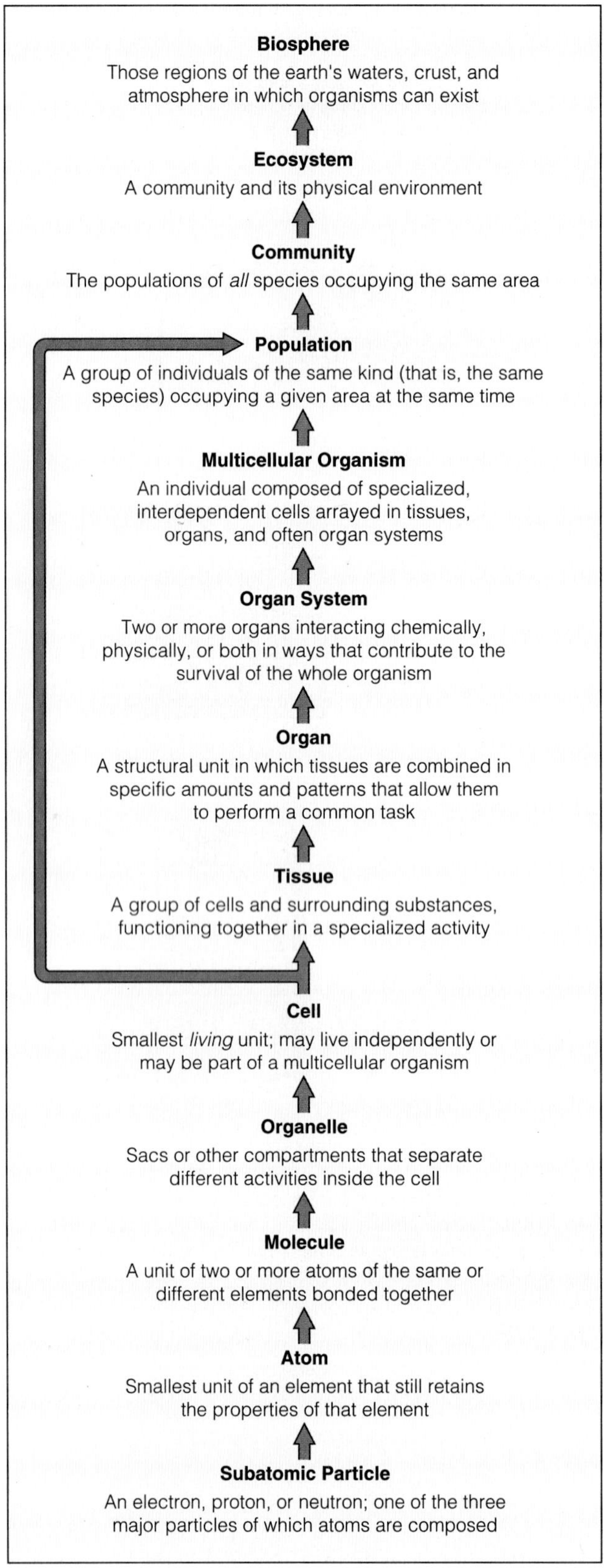

Biosphere
Those regions of the earth's waters, crust, and atmosphere in which organisms can exist

⬆

Ecosystem
A community and its physical environment

⬆

Community
The populations of *all* species occupying the same area

⬆

Population
A group of individuals of the same kind (that is, the same species) occupying a given area at the same time

⬆

Multicellular Organism
An individual composed of specialized, interdependent cells arrayed in tissues, organs, and often organ systems

⬆

Organ System
Two or more organs interacting chemically, physically, or both in ways that contribute to the survival of the whole organism

⬆

Organ
A structural unit in which tissues are combined in specific amounts and patterns that allow them to perform a common task

⬆

Tissue
A group of cells and surrounding substances, functioning together in a specialized activity

⬆

Cell
Smallest *living* unit; may live independently or may be part of a multicellular organism

⬆

Organelle
Sacs or other compartments that separate different activities inside the cell

⬆

Molecule
A unit of two or more atoms of the same or different elements bonded together

⬆

Atom
Smallest unit of an element that still retains the properties of that element

⬆

Subatomic Particle
An electron, proton, or neutron; one of the three major particles of which atoms are composed

Figure 1.3 Levels of organization in nature. Cells represent the first level at which the properties of life emerge.

Levels of Biological Organization

Look carefully at Figure 1.3, which outlines the levels of organization in nature. The properties of life emerge at the level of cells. A **cell** is an organized unit that can survive and reproduce on its own, given DNA instructions and sources of energy and raw materials. In other words, the cell is the basic *living* unit.

That definition obviously fits an organism that is a single, free-living cell, such as an amoeba. Does it fit a **multicelled organism**, which has specialized cells organized into tissues and organs? Yes. You might find this a strange answer. After all, your own cells could never live all by themselves in nature. They must be bathed by fluids inside your body. Yet even human cells can be isolated and kept alive under controlled laboratory conditions. Researchers around the world routinely maintain human cells for use in cancer studies and other experiments.

Referring again to Figure 1.3, we find a more inclusive level of organization, called the **population**. This is a group of single-celled or multicelled organisms of the same kind in a given area. A group of emperor penguins at their breeding grounds in Antarctica is an example. Next is the **community**, which includes all populations of all species (penguins, whales, seals, fishes, and so on) living in the same area. The next level, the **ecosystem**, includes the community and its physical and chemical environment. The most inclusive level of organization is the **biosphere**. The word refers to all regions of the earth's waters, crust, and atmosphere in which organisms live.

The organization characteristics of the world of life starts at the level of cells—which start with instructions contained in DNA molecules.

Metabolism: Life-Sustaining Energy Transfers

You never, ever will find a rock engaged in metabolic activities. Only living cells can do this. **Metabolism** refers to the cell's capacity to (1) extract and convert energy from its surroundings and (2) use energy and so maintain itself, grow, and reproduce. Simply put, metabolism means *energy transfers* within cells.

Think of a rice plant. Many of its cells engage in **photosynthesis**. In the first stage of this process, cells trap sunlight energy, then convert it to another form of energy. In the second stage, cells use the chemical energy to build sugars, starch, and other substances. As part of the process of photosynthesis, molecules of **ATP**, an "energy carrier," are put together. ATP transfers

Figure 1.4 Energy flow (yellow arrows) and the cycling of materials (brown arrows) in the biosphere. Here, grass plants of the African savanna are producer organisms. They directly provide energy for plant-eating zebras (a type of herbivore). They indirectly provide energy for lions and vultures (meat eaters, or carnivores). Decomposers get energy by breaking down the wastes and remains of all these organisms. Plants take up some of the breakdown products.

energy to other molecules that function as metabolic workers (enzymes), building blocks, or energy reserves.

In rice plants, some of the stored energy becomes concentrated in starchy seeds—rice grains. Energy reserves in countless trillions of rice grains provide energy for billions of rice-eating humans around the world. How? In humans, as in most animals and plants, stored energy is released and transferred to ATP by way of **aerobic respiration**, another metabolic process.

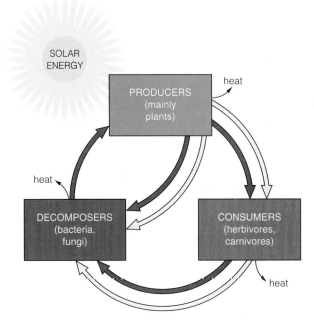

Living things show metabolic activity. Their cells acquire and use energy to stockpile, tear down, build, and eliminate materials in ways that promote survival and reproduction.

Interdependency Among Organisms

With few exceptions, a flow of energy from the sun maintains the great pattern of organization in nature. Plants and some other photosynthetic organisms are the entry point for this flow. They are food **producers** for the world of life. Animals are **consumers**. Directly or indirectly, they feed on energy stored in plant parts. Thus zebras tap directly into the stored energy when they nibble on grass, and lions tap into it indirectly when they chomp on zebras. Bacteria and fungi are **decomposers**. When they feed on tissues or remains of other organisms, they break down sugars and other biological molecules to simple raw materials—which can be cycled back to producers.

And so we have interdependency among organisms, based on a one-way flow of energy *through* them and a cycling of materials *among* them (Figure 1.4).

Such interactions among organisms influence populations, communities, and ecosystems. They even influence the global environment. Understand the extent of the interactions and you will gain insight into amplification of the greenhouse effect, acid rain, and many other modern-day problems.

Webs of organization connect all organisms in nature, in that organisms depend directly or indirectly on one another for energy and raw materials.

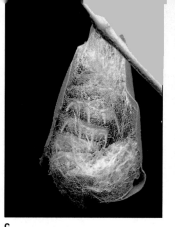

a   b   c   d

Figure 1.5 "The insect"—a continuous series of stages in development. Different adaptive properties emerge at each stage. Shown here, a silkworm moth, from egg (**a**) to larval stage (**b**), to pupal form (**c**), to the splendid adult form (**d**,**e**).

Sensing and Responding to Change

It is often said that only organisms "respond" to the environment. Yet a rock also "responds" to the environment, as when it yields to gravity and tumbles downhill or changes shape slowly under the battering of wind, rain, or tides. The real difference is this: *Organisms have the cellular means to sense changes in the environment and make controlled responses to them.* They do so with the help of **receptors**, which are molecules and structures that can detect specific information about the environment. When cells receive signals from receptors, they adjust their activities in ways that bring about an appropriate response.

Your body, for example, can withstand only so much heat or cold. It must rid itself of harmful substances. Certain foods must be available, in certain amounts. Yet temperatures shift, harmful substances may be encountered, and food is sometimes plentiful or scarce.

Think about what happens after you eat and simple sugar molecules enter your bloodstream. Blood is part of the body's "internal environment" (the other part is the tissue fluid bathing your cells). When the sugar level in blood rises, liver cells step up their secretion of insulin. Most cells in your body have receptors for insulin, a hormone that prods the cells into taking up sugar molecules. With so many cells taking up sugar, the blood sugar level returns to normal.

Suppose you skip breakfast, then lunch, and the blood sugar level falls. Now a different hormone prods liver cells to dig into their stores of energy-rich molecules. Those molecules are broken down to simple sugars, which are released into the bloodstream—and again the blood sugar level returns to normal.

Usually, the internal environment of a multicelled organism is kept fairly constant. When conditions in the internal environment are being maintained within tolerable limits, we call this a state of **homeostasis**.

Organisms have the means to sense and respond to changes in their environment. The responses help maintain favorable operating conditions inside the cell or multicelled body.

Reproduction

We humans tend to think we enter the world rather abruptly and leave it the same way. Yet we and all other organisms are more than this. *We are part of an immense, ongoing journey that began billions of years ago.* Think of the first cell produced when a human sperm penetrates an egg. The cell would not even exist if the sperm and egg had not formed earlier, according to DNA instructions passed down through countless generations. With those time-tested instructions, a new human body develops in ways that will prepare it, ultimately, for helping to produce individuals of the next generation. With **reproduction**—that is, the production of offspring—life's journey continues.

Or think of a moth. Do you simply picture a winged insect? What of the tiny fertilized egg deposited on a branch by a female moth (Figure 1.5)? The egg contains the instructions necessary to become an adult. By those instructions, the egg develops into a caterpillar, a larval stage adapted for rapid feeding and growth. The caterpillar eats and grows until an internal "alarm clock" goes off. Then its body enters a so-called pupal stage of development, which involves wholesale remodeling. Some cells die, others multiply and become organized in different patterns. In time an adult moth emerges. It has organs that contain eggs or sperm. Its wings are brightly colored and flutter at a frequency appropriate for attracting a mate. In short, the adult stage is adapted for reproduction.

None of these stages is "the insect." The insect is a series of organized stages from one fertilized egg to the

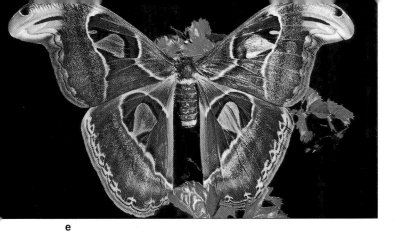

e

next. Each stage is vital for the ultimate production of new moths. The instructions for each stage were written into moth DNA long before each moment of reproduction—and so the ancient moth story continues.

Each organism arises through reproduction.

Each organism is part of a reproductive continuum that extends back through countless generations.

Mutation: Source of Variations in Heritable Traits

Reproduction involves **inheritance**. The word means that parents transmit DNA instructions for duplicating their traits, such as body form, to offspring.

DNA has two striking qualities. Its instructions assure that offspring will resemble parents—and they also permit *variations* in the details of traits. For example, having five fingers on each hand is a human trait. Yet some humans are born with six fingers on each hand instead of five! Variations in traits arise through **mutations**, which are heritable changes in the structure or number of DNA molecules.

Many mutations are harmful. A change in even a bit of DNA may be enough to sabotage the steps necessary to produce a vital trait. In *hemophilia A*, for example, a tiny mutation leads to an impaired ability to clot blood. Bleeding continues for an abnormally long time after even a small cut or bruise.

Yet some mutations are harmless, even beneficial, under prevailing conditions. A classic example is a mutation in light-colored moths that leads to dark-colored offspring. Moths fly by night and rest during the day, when birds that eat them are active. What happens when a light moth rests on a light-colored tree

Figure 1.6 An example of how two different forms of the same trait (coloration of moths) are each adaptive under different environmental conditions.

trunk (Figure 1.6)? Birds simply don't see it. Suppose, as a result of heavy industry, light trunks in a forested region become soot covered—and dark. The dark moths are less conspicuous, so they have a better chance of living long enough to reproduce. Under sooty conditions, the mutated form of the trait is more adaptive.

An **adaptive trait** simply is one that helps an organism survive and reproduce under a given set of environmental conditions.

DNA is the molecule of inheritance in organisms. Its instructions for reproducing traits are passed on from parents to offspring.

Mutations introduce variations in heritable traits.

Although many mutations are harmful, some give rise to variations in form, function, or behavior that turn out to be adaptive under prevailing conditions.

Figure 1.7 Representatives of life's diversity.

Kingdom Monera. (**a**) A bacterium, a microscopically small single cell. Bacteria live nearly everywhere, including in or on other organisms. The ones in your gut and on your skin outnumber the cells of your body.

Kingdom Protista. (**b**) A trichomonad, living as a parasite in a termite's gut. Most protistans are single celled, but they generally are much larger and have much greater internal complexity than bacteria.

Kingdom Fungi. (**c**) A stinkhorn fungus. The kingdom of fungi includes many major decomposers, which break down the remains and wastes of organisms. Without decomposers, communities would gradually become buried in their own garbage.

Kingdom Plantae. (**d**) A grove of California coast redwoods. Like nearly all members of the plant kingdom, they produce their own food through photosynthesis. (**e**) From a plant called a composite, a flower having a pattern that guides bees to nectar. The bees get food, the plants get help in reproducing. Many organisms are locked in mutually helpful interactions.

Kingdom Animalia. (**f**) Male bighorn sheep competing for females. Like all members of the animal kingdom, they cannot produce their own food; they depend on other organisms for it. They generally move about far more than other kinds of organisms.

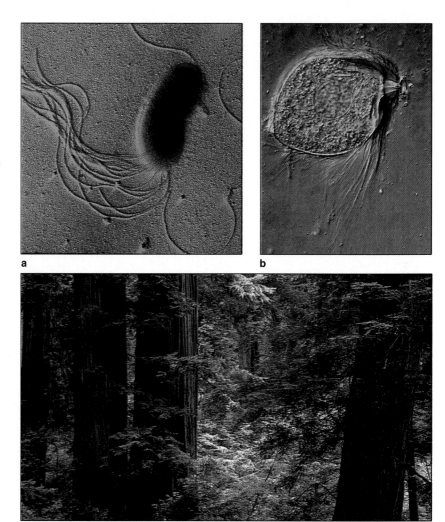

LIFE'S DIVERSITY

So Much Unity, Yet So Many Species

Until now, we have focused on the *unity* of life—on characteristics shared by all organisms. Superimposed on the shared heritage is immense *diversity*. Many millions of different kinds of organisms, or **species**, inhabit the earth. Many millions more lived in the past and became extinct. Attempts to make sense of diversity led to a classification scheme in which each species is assigned a two-part name. The first part designates the **genus** (plural, genera). It encompasses all species related by descent from a common ancestor. The second part designates a particular species within that genus. For instance, *Quercus alba* is the scientific name of the white oak. *Q. rubra* is the name of the red oak. (Once the genus name is spelled out in a document, subsequent uses of it can be abbreviated.)

Life's diversity is further classified by assigning species to groups at more encompassing levels. Genera that share a common ancestor are placed in the same *family*, related families are placed in the same *order*, then related orders in the same *class*. Related classes are placed in a *phylum* (plural, phyla) or *division*, which is assigned to one of five *kingdoms*:

Monerans	Bacteria (singular, bacterium). Single cells, all prokaryotic (their DNA is not enclosed in a membrane-bound compartment called a nucleus). Producers, consumers, decomposers. Kingdom of greatest metabolic diversity.
Protistans	Mostly single cells larger than bacteria. Eukaryotic (DNA is enclosed in a nucleus). Producers, consumers. Diverse life-styles.
Fungi	Mostly multicelled. Eukaryotic. Decomposers, consumers. Their secretions digest food outside the fungal body; their cells absorb digested bits.
Plants	Mostly multicelled. Eukaryotic. Nearly all producers that rely on photosynthesis.
Animals	Multicelled. Eukaryotic. Consumers, typically motile, with diverse life-styles.

Figure 1.7 shows a few members of these kingdoms.

c

e

f

An Evolutionary View of Diversity

Table 1.1 summarizes the characteristics of life described so far. With few exceptions, all living organisms in all five kingdoms display these characteristics. Given that they are alike in so many ways, what could account for their diversity? One key explanation is called evolution by means of natural selection.

Evolution Defined. Suppose a DNA mutation gives rise to a different form of a trait in a few members of a population. We can use the dark moths in a sooty forest as an example. Birds see and eat many of their light-colored relatives, but dark moths escape detection and live to reproduce. So do their dark offspring—and so do *their* offspring. The dark version of the trait is popping up with greater frequency. In time it may become the more common form (and we might end up referring to "the dark moth population"). **Evolution** is taking place—the features that characterize a population are changing through successive generations.

Table 1.1 Characteristics of Living Organisms
1. Complex structural organization based on instructions contained in DNA molecules.
2. Directly or indirectly, dependence on other organisms for energy and material resources.
3. Metabolic activity by the single cell or multiple cells composing the body.
4. Use of homeostatic controls that maintain favorable operating conditions in the body within tolerable limits.
5. Reproductive capacity, by which the instructions for heritable traits are passed from parents to offspring.
6. Diversity in body form, in the functions of various body parts, and in behavior. Such traits are adaptations to conditions in the environment.
7. The capacity to evolve, based ultimately on variations in traits that arise through mutations in DNA.

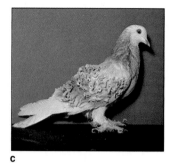

a   b   c   d

Figure 1.8 Representatives of the more than 300 varieties of domesticated pigeons. The varieties are the result of artificial selection practices. Variant forms of traits among captive populations of wild rock doves (**a**) were the starting point for pigeon breeders.

Natural Selection Defined. Long ago, Charles Darwin used pigeons to explain the link between variation in traits and evolution. Domesticated pigeons vary greatly in size, feather color, and other traits (Figure 1.8). As Darwin knew, pigeon breeders "select" certain traits. Suppose the desired traits are black tail feathers with curly edges. Only those pigeons having the most black and the most curl in their tail feathers are allowed to mate. Over time, those forms of the traits become most common as other forms are eliminated from the pigeon population.

Because pigeon breeders do their selecting in an artificial environment rather than in nature, the practice is an example of *artificial* selection. Even so, Darwin recognized it as a way to explain evolution by *natural* selection—that is, selection of adaptive traits in nature. Here are the key points of that explanation, expressed in modern terms:

---

1. Members of a population vary in form, function, and behavior. Much of this variation is heritable.

2. Some forms of heritable traits are more adaptive than others—they improve chances of surviving and reproducing. As a result, those forms become more common in the population.

3. **Natural selection** is simply the result of differences in survival and reproduction that have occurred among individuals that differ in one or more traits.

4. Any population *evolves* when some forms of traits become more or less common, or even disappear, over the generations. In this manner, variations have accumulated in different lines of organisms. Life's diversity is the sum total of those variations.

Later in the book, we will consider the mechanisms by which organisms evolve. In the meantime, keep the points of the preceding list in mind. They are absolutely central to biological inquiry. We will be using them to explain topics throughout the book.

THE NATURE OF BIOLOGICAL INQUIRY

On Scientific Methods

Biology, like science generally, is an ongoing, methodical search for information that helps reveal the secrets of the natural world. Since Darwin's time, the searchers have branched out and established a great number of specialized subdivisions. Biologists now pursue topics that range from the molecular structure of the virus that causes AIDS to the ozone hole in the stratosphere. There is still so much to be learned about each topic that few now claim the whole of nature as their research interest. And no one claims one method alone can be used to study its complexity.

Despite the specialization, scientists everywhere still have practices in common. *They ask questions, make educated guesses about possible answers, then devise ways to test predictions that will hold true if their guesses are good ones.* The following list is a more formal description of how scientists generally proceed with an investigation:

---

1. Ask a question (or identify a problem) about some aspect of the natural world.

2. Develop one or more **hypotheses**, or educated guesses, about what the answer (or solution) might be. This might involve sorting through what has been learned already about related phenomena.

3. Using hypotheses as a guide, make a **prediction**— that is, a statement of what you should be able to observe in nature, if you were to go looking for it. This is often called the "if–then" process. (*If* gravity pulls

e

f

g

h

objects toward the earth, *then* it should be possible to observe apples falling down, not up, from a tree.)

4. Devise ways to test the accuracy of predictions. You might do so by making observations, developing models, and doing experiments.

5. If the tests do not turn out as expected, check to see what might have gone wrong. (Maybe you overlooked something that influenced the results. Or maybe the hypothesis isn't a good one.)

6. Repeat the tests or devise new ones—the more the better. Hypotheses supported by many different tests are more likely to be correct.

7. Objectively report the test results and the conclusions drawn from them.

In broad outline, a scientific approach to studying nature is that simple. You yourself can use this approach to advantage. You can use it to satisfy curiosity about mammoths or moth wings. You can use it to pick your way logically through environmental, medical, and social land mines of the sort described later in the book. And you can use it to understand the past and predict possible futures for life on this planet.

About the Word "Theory"

What is the difference between a hypothesis and a theory? By way of example, let's go back to Darwin's ideas about the evolution of species. When Darwin proposed his ideas more than a century ago, he ushered in one of the most dramatic of all scientific revolutions. The core of his thinking became popularly known as "the theory of evolution."

In science, a **theory** is a related set of hypotheses which, taken together, form a broad-ranging, testable explanation about some fundamental aspect of the nat-ural world. A scientific theory differs from a scientific hypothesis in its *breadth of application*. Darwin's theory fits this description—it is a big, encompassing "Aha!" explanation that, in a few intellectual strokes, makes sense of a huge number of observable phenomena. Think of it. Darwin's theory explains how most of the diversity among many millions of different living things came about!

Yet is any theory an "absolute truth" in science? Ultimately, no. Why? It would be impossible to perform the infinite number of tests required to show that a theory holds true under all possible conditions! Objective scientists say only that they are *relatively* certain that a theory is (or is not) correct.

Even so, "relative certainty" can be impressive. Especially after exhaustive tests by many scientists, a theory may be as close to the truth as we can get with the evidence at hand. After more than a century's worth of thousands of different tests, Darwin's theory still stands, with only minor modification. Most biologists accept the modified theory—although they still keep their eyes open for contradictory evidence.

Scientists must keep asking themselves, "Will some other evidence show my idea to be incorrect?" They are expected to put aside pride or bias by testing ideas, even in ways that might prove them wrong. If an individual doesn't (or won't) do this, *others will*—for science proceeds as a community that is both cooperative and competitive. Ideas are shared, with the understanding that it is just as important to expose errors as it is to applaud insights. Individuals can change their mind when presented with new evidence—and this is a strength of science, not a weakness.

A scientific theory is a testable explanation about the cause or causes of a broad range of related phenomena. Like hypotheses, theories are open to tests, revision, and tentative acceptance or rejection.

Darwin's Theory and Doing Science

A time-tested theory serves as a general frame of reference for studying nature. Consider Charles Darwin's theory of evolution by natural selection. How might you use it to explain the patterned wings of the moth shown in Figure 1.5? According to this theory, the traits of moths and all other organisms exist because they have contributed to reproductive success. So your question might be this: "I wonder how the wing pattern helps the moth leave descendants." Then you hypothesize about the answer.

"Maybe a wing pattern is a mating flag that helps males and females of the same species identify each other." This may be correct, but you don't limit yourself to one hypothesis. Why? Nearly always, there's more than one possible answer to a question about some aspect of nature. So you also come up with an alternative: "Maybe the pattern camouflages moths during the day." In science, *alternative hypotheses are the rule, not the exception.*

Testing Hypotheses. Of any number of alternative hypotheses, how do you identify the most plausible one? The trick is to let each one guide you in making testable predictions. If wing patterns help moths identify mates (the hypothesis), then it follows that moths should mate only when the patterns are visible (the prediction). Moths mate at night. If wing pattern is a mating flag, then on moonless nights, moths shouldn't be able to see it and you won't see moths mating. To test the prediction, you watch moths on a moonlit and then on a moonless night. You notice they mate with or without help from the light of the moon. Here is evidence that the prediction—and, by extension, the hypothesis—might be wrong. Then again, maybe you overlooked something important. For example, maybe moths (like cats) see better than you do in the dark.

The Role of Experiments. Now you decide to test the same hypothesis by experimentation. An **experiment** is a test in which nature is manipulated to reveal one of its secrets. It requires careful design of a set of controls to evaluate possible side effects of the manipulation.

If wing pattern is a mating flag, then moths with altered color patterns might have a tough time attracting a mate. To test this new prediction, you capture new moths, paint an altered pattern on their wings (Figure *a*), then put them in a cage with unaltered moths to see what happens.

You also set up a **control group** to evaluate possible side effects of a test involving an experimental group. Ideally, members of a control group should be identical to those of an experimental group in all respects—except for a key **variable** (the factor being investigated). The number of individuals in both groups also must be large enough so results won't be due to chance alone.

Besides wing pattern, what other variables between the two groups might affect the outcome of your experiment? Maybe paint fumes are as repulsive to a potential mate as a painted-on pattern. Maybe when you paint the moths you somehow rough them up, making them less desirable than those in the control group. Maybe the paint weighs enough to change the flutter frequency of the wings.

So you decide the control group also must be painted with the same kind of paint, using the same brushes, and must be handled the same way. But for this group, you *duplicate* the natural wing color pattern as you paint. Now your experimental and control groups are identical except for the variable under study. If only those moths with altered wing patterns turn out to be unlucky in love, then the greater reproductive success of your control group will help substantiate your hypothesis.

How might you use a time-tested theory to explain something you observe in the natural world? The *Focus* essay provides an example.

The Limits of Science

The call for objective testing strengthens the theories that emerge from scientific studies. Yet it also puts limits on the kinds of studies that can be carried out. Beyond the realm of scientific analysis, some events remain unexplained. Why do we exist, for what purpose? Why does any one of us have to die at a particular moment and not another? Answers to such questions are *subjective*. This means they come from within us, as an outcome of all the experiences and mental connections that shape our consciousness. Because people differ so enormously in this regard, subjective answers do not readily lend themselves to scientific analysis.

This is not to say that subjective answers are without value. No human society can function without a shared commitment to standards for making judgments, even if the judgments are subjective. Moral, aesthetic, economic, and philosophical standards vary

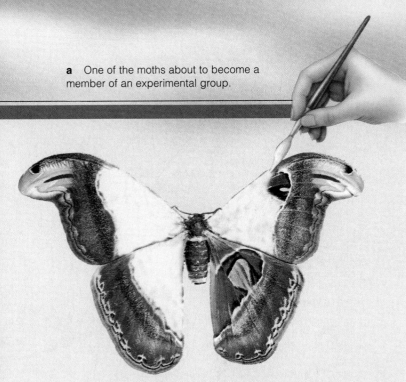

a One of the moths about to become a member of an experimental group.

Generally, experiments are devised to disprove a hypothesis. Why? It would be impossible to prove beyond a shadow of a doubt that a hypothesis is correct. It would take an infinite number of experiments to demonstrate that it holds under all possible conditions.

Have you been thinking that painting moths is a rather fanciful example of a scientific approach? As reported in *Nature* in 1993, Karen Marchetti, a graduate student of the University of California at Davis, wielded a paintbrush on birds in a forest in Kashmir, India. She found evidence for her hypothesis that bright feather color, not patterning, gives male yellow-browed leaf warblers an edge in mating. In early tests, Marchetti put a patternless dab of yellow paint on head feathers. In later tests, she painted larger-than-normal yellow bands on wings of one group and painted out part of the bands of another group. She used transparent paint for a third group. (Can you guess why?) As Marchetti discovered, color-enhanced birds secured larger territories and produced more offspring than toned-down birds.

from one society to the next. But all guide their members in deciding what is important and good, and what is not. All attempt to give meaning to what we do.

Every so often, scientists stir up controversy when they explain part of the world that was considered beyond natural explanation—that is, belonging to the "supernatural." This is sometimes true when moral codes are interwoven with religious narratives. Exploring some longstanding view of the world from a scientific perspective may be misinterpreted as questioning morality, even though the two are not at all the same thing.

For example, centuries ago Nicolaus Copernicus studied the movements of planets and stated that the earth circles the sun. Today the statement seems obvious. Back then, it was heresy. The prevailing belief was that the Creator had made the earth (and, by extension, humanity) the immovable center of the universe! Not long afterward a respected professor, Galileo Galilei, studied the Copernican model of the solar system. He thought it was a good one and said so. He was forced to retract his statement publicly, on his knees, and to put the earth back as the fixed center of things. (Word has it that when he stood up he muttered, "Even so, it does move.")

Today, as then, society has its sets of standards. Today, as then, those standards may be called into question when a new, natural explanation runs counter to supernatural belief. This doesn't mean the scientists who raise the questions are less moral, less lawful, less sensitive, or less caring than anyone else. It simply means one more standard guides their work: *The external world, not internal conviction, must be the testing ground for scientific beliefs.*

Systematic observations, hypotheses, predictions, tests—in all these ways, science differs from systems of belief that are based on faith, force, or simple consensus.

SUMMARY

1. All organisms share the following characteristics:
   a. Their structure, organization, and interactions arise from basic properties of matter and energy.
   b. They survive by relying on metabolic and homeostatic processes.
   c. They have the capacity for growth, development, and reproduction, based on instructions contained in their DNA.

2. There are many millions of different organisms. Each kind of organism is a species. In classification schemes, species are placed in increasingly inclusive groupings, from genus on up through family, order, class, phylum (or division), and kingdom.

3. Diversity among organisms arises through mutation. Mutations introduce changes in the DNA. The changes may lead to variation in heritable traits (traits that parents transmit to offspring, including most details of the body's form and functioning).

4. Different versions of the same heritable trait occur among individuals of a population. These influence the ability to survive and reproduce. Under prevailing conditions, some may be more adaptive than others and will become more common ("selected") in subsequent

generations. Others will become less common and may disappear. Thus the population changes over time; it evolves. These points are central to the theory of evolution by natural selection.

5. There are many specialized scientific methods, corresponding to many different fields of inquiry. The following terms are important to all of them:

a. Theory: A testable explanation of a broad range of related phenomena. An example is the theory of evolution by natural selection.

b. Hypothesis: A possible explanation of a specific phenomenon. Sometimes called an educated guess.

c. Prediction: A claim about what you can expect to see in nature if a theory or hypothesis is correct.

d. Test: An attempt to produce actual observations that match predicted or expected observations.

e. Conclusion: A statement about whether a theory or hypothesis should be accepted, rejected, or modified, based on tests of the predictions derived from it.

6. Scientific theories are based on systematic observations, hypothesizing, predictions, and tests. The external world, not internal conviction, is the testing ground for those theories.

Review Questions

1. Why is it difficult to give a simple definition of life? 2 (For this and subsequent chapters, *italic numbers* following review questions indicate the pages on which the answers may be found.)

2. What characteristics do all organisms have in common? 9

3. What is energy? What is DNA? 3

4. Study Figure 1.3. Then, on your own, arrange and define the levels of biological organization. 4

5. Define metabolic activity. 4–5

6. Make a sketch of the one-way flow of energy and the cycling of materials through the biosphere. 5

7. What is mutation? How is it related to the diversity of life? 7

8. Witnesses in a court of law are asked to "swear to tell the truth, the whole truth, and nothing but the truth." What are some problems inherent in the question? Can you think of a better alternative?

9. Design a test to support or refute the following hypothesis: Body fat appears yellow in certain rabbits—but only when those rabbits also eat leafy plants that contain a yellow pigment molecule called xanthophyll.

Self-Quiz *(Answers in Appendix IV)*

1. The _cell_ is the smallest unit of life.

2. _metabolism_ is the ability of cells to extract and transform energy from the environment and use it to maintain themselves, grow, and reproduce.

homeostasis

3. _____ is a state in which the body's internal environment is being maintained within tolerable limits.

4. If a form of a trait improves chances for surviving and reproducing in a particular environment, it is a(n) _adaptive_ trait.

5. The capacity to evolve is based on variations in heritable traits, which originally arise through _mutation_.

6. You have some number of traits that also were present in your great-great-great-great-grandmothers and -grandfathers. This is an example of _____ .
   a. metabolism   c. a control group
   b. homeostasis   d. inheritance

7. DNA molecules _____ .
   a. contain instructions for traits
   b. undergo mutation
   c. are transmitted from parents to offspring
   d. all of the above

8. For many years in a row, a dairy farmer allowed his best milk-producing cows but not the poor producers to mate. Each generation, milk production increased. This outcome is an example of _____ .
   a. natural selection   c. evolution
   b. artificial selection   d. both b and c

9. A related set of hypotheses that explains some aspect of the natural world is a scientific _____ .
   a. prediction   c. theory
   b. test   d. observation

Selected Key Terms

For this and subsequent chapters, these are the **boldface** terms that occur in the text on the pages indicated by italic numbers. Make a list of these terms, write a definition next to each, then check it against the one in the text. (You will be using these terms later on.)

adaptive trait 7	energy 3	natural selection 10
aerobic respiration 5	evolution 9	photosynthesis 4
animal 8	experiment 12	plant 8
ATP 4	fungus 8	population 4
biosphere 4	genus 8	prediction 10
cell 4	homeostasis 6	producer 5
community 4	hypothesis 10	protistan 8
consumer 5	inheritance 7	receptor 6
control group 12	metabolism 4	reproduction 6
decomposer 5	moneran 8	species 8
DNA 3	multicelled organism 4	theory 11
ecosystem 4	mutation 7	variable 12

Suggested Readings

Committee on the Conduct of Science. 1989. *On Being a Scientist.* Washington, D.C.: National Academy of Sciences. Paperback.

Larkin, T. June 1985. "Evidence vs. Nonsense: A Guide to the Scientific Method." *FDA Consumer* 19:26–29.

FACING PAGE: *Living cells of a plant* (Elodea)*, as seen with the aid of a microscope. Each rectangular cell contains efficient chemical factories called chloroplasts (the green spheres).*

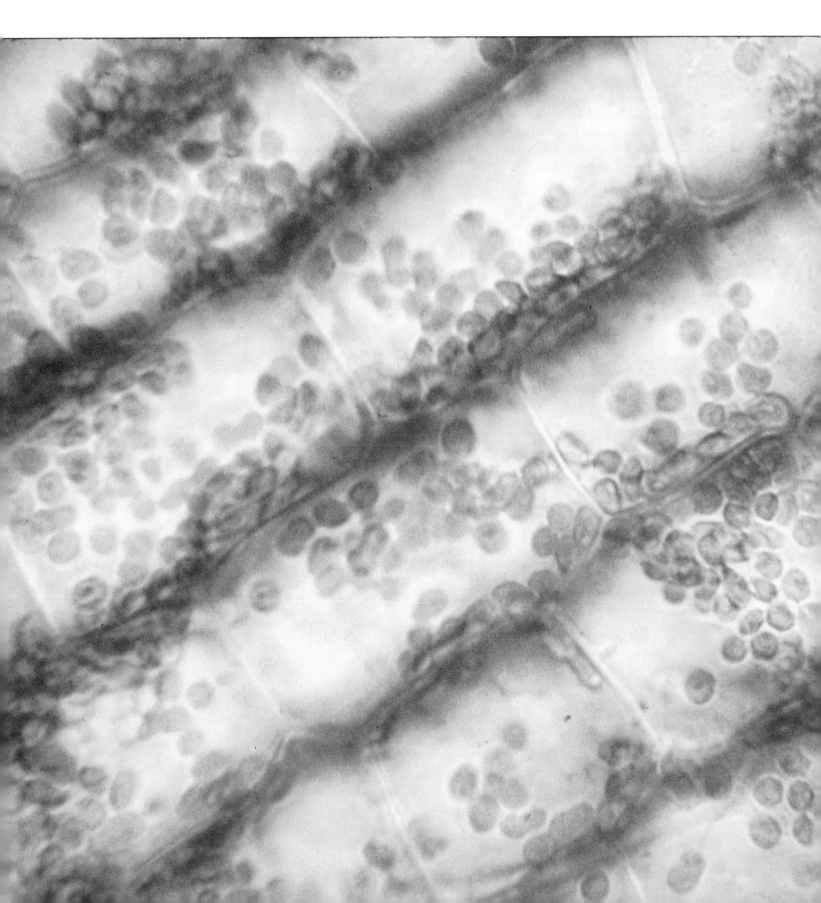

2 CHEMICAL FOUNDATIONS FOR CELLS

The Chemistry In and Around You

Right now you are breathing in oxygen. You would die without it. But where did the oxygen come from? Whether it's noon or midnight, countless plants on the sunlit side of our planet are busily converting energy from the sun to forms of energy they can use—and they release oxygen during the conversions. Aquatic plants have been doing this for more than 900 million years. Some types of bacteria have been doing the same thing for more than 3 *billion* years! It all adds up.

In the past two centuries (a mere blip of evolutionary time), we managed to discover what substances are made of and how they can be converted into different forms. With this amazing knowledge of chemistry, we developed such products as fertilizers, nylons, vaccines, lipsticks, antibiotics, and plastic parts of refrigerators, computers, television sets, jet planes, and cars.

Our chemical "magic" brings benefits *and* problems. Without synthetic fertilizers to help grow crops, more humans than you might imagine would starve to death. But weeds don't know that fertilizers are for crop plants, and animal pests don't know that crops aren't being raised for them. Each year pests ruin or gobble up nearly half of what we grow. In 1945 we began using synthetic pesticides. Among other things, pesticides kill weeds, moths, worms, and rats that threaten our food supplies, health, pets, and ornamental plants. In 1988 alone, Americans spread more than a billion pounds of pesticides through homes, gardens, offices, industries, and farmlands (Figure 2.1).

Most pesticides are neurotoxins that block vital communication signals between nerve cells. Some remain active for days, others for weeks or years. Besides pests, they kill great numbers of pest eaters, including dragonflies and birds. We, too, inhale pesticides, ingest them with food, or absorb them through skin. Many pesticides cause headaches, rashes, and asthma in susceptible people. Some trigger hives, joint pain, even life-threatening allergic reactions in millions of Americans.

Maintaining our crops, industries, and health depends on chemistry. So does our chance of reducing harmful side effects of the application of chemistry. You owe it to yourself and others to gain understanding of chemical substances. By demystifying chemistry's "magic," you will be better equipped to assess its benefits and risks.

Figure 2.1 Cropduster with its rain of pesticides.

1. Like atoms generally, the atoms of carbon and other elements associated with life consist of no more than protons, electrons, and neutrons. Even so, atoms do not necessarily hang on to all their parts. They lose, gain, and share electrons with one another. The structural organization and activities of all living things arise through electron jugglings among atoms.

2. Oxygen, carbon, hydrogen, and nitrogen are the most abundant kinds of atoms in living things. Ionic, hydrogen, and covalent bonds are the main chemical bonds between these and other atoms.

3. Complex carbohydrates, lipids, proteins, and nucleic acids are the large "molecules of life." Cells assemble and use them as structural materials, energy stores, workers (such as enzymes), and books of hereditary information.

4. The large molecules of life are assembled from only four kinds of smaller building blocks—the simple sugars, fatty acids, amino acids, and nucleotides.

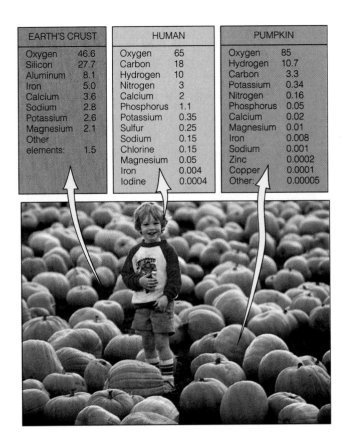

EARTH'S CRUST		HUMAN		PUMPKIN	
Oxygen	46.6	Oxygen	65	Oxygen	85
Silicon	27.7	Carbon	18	Hydrogen	10.7
Aluminum	8.1	Hydrogen	10	Carbon	3.3
Iron	5.0	Nitrogen	3	Potassium	0.34
Calcium	3.6	Calcium	2	Nitrogen	0.16
Sodium	2.8	Phosphorus	1.1	Phosphorus	0.05
Potassium	2.6	Potassium	0.35	Calcium	0.02
Magnesium	2.1	Sulfur	0.25	Magnesium	0.01
Other		Sodium	0.15	Iron	0.008
elements:	1.5	Chlorine	0.15	Sodium	0.001
		Magnesium	0.05	Zinc	0.0002
		Iron	0.004	Copper	0.0001
		Iodine	0.0004	Other:	0.00005

Figure 2.2 Proportions of different elements in the earth's crust, the human body, and a pumpkin as percentages of the total weight of each.

ORGANIZATION OF MATTER

All the solids, liquids, and gases within and around you are forms of matter that consist of one or more elements. An element is any substance that cannot be broken down to a different substance, at least by ordinary means. Ninety-two elements occur naturally on earth. Four kinds—oxygen, carbon, hydrogen, and nitrogen—make up most of your body (Figure 2.2).

Each kind of **atom** is the smallest unit of matter that is unique to a particular element. A **molecule** is two or more joined-together atoms of the same or different elements. Pure water (if there still is such a thing) consists of a stupendous number of water molecules, each composed of one oxygen and two hydrogen atoms.

Water, incidentally, is a compound. A **compound** is a substance in which the relative proportions of two or more elements never vary. Water in rainclouds or a Siberian lake or your bathtub always has twice as many hydrogen atoms as oxygen atoms, period.

The Structure of Atoms

Atomic Building Blocks. The organization and activities of living things begin with charged particles called **protons** and **electrons**. Together with **neutrons**, which have no charge, these are building blocks of atoms. As Figure 2.3 suggests, an atom's core region, or

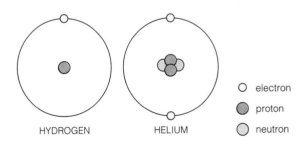

Figure 2.3 Model of atomic structure, using hydrogen and helium as examples. (At this scale, the nucleus actually would be an invisible speck at the atom's center.)

nucleus, consists of some number of protons and (except for hydrogen) neutrons. Protons carry a positive charge (p^+). Electrons move around the nucleus and occupy most of the atom's space. They carry a negative charge (e^-). An atom has just as many electrons as protons, so it has no *net* charge, overall.

17

Table 2.1	Atomic Number and Mass Number of Elements Common in Living Things		
Element	Symbol	Atomic Number	Most Common Mass Number
Hydrogen	H	1	1
Carbon	C	6	12
Nitrogen	N	7	14
Oxygen	O	8	16
Sodium	Na	11	23
Magnesium	Mg	12	24
Phosphorus	P	15	31
Sulfur	S	16	32
Chlorine	Cl	17	35
Potassium	K	19	39
Calcium	Ca	20	40
Iron	Fe	26	56
Iodine	I	53	127

Atomic Number and Mass Number. "Atomic number" refers to the number of protons in the nucleus. It differs for each element. For example, the hydrogen atom has one proton; its atomic number is 1. The carbon atom has six protons; its atomic number is 6. "Mass number" (or atomic weight) is the total number of protons *and* neutrons in the nucleus. The mass number of a carbon atom with six protons and six neutrons is 12 (Table 2.1).

As you will see, atomic and mass numbers give an idea of whether certain atoms can lose, gain, or share electrons. *Such electron activity is the basis for the organization of materials and the flow of energy through the living world.*

Isotopes: Variant Forms of Atoms

The atoms of an element may differ slightly in how many neutrons they contain. Such atoms, which have the same atomic number but a different mass number, are **isotopes**. Thus "a carbon atom" might be carbon 12 (six protons, six neutrons), carbon 13 (six protons, seven neutrons), or carbon 14 (six protons, eight neutrons). These can be written as ^{12}C, ^{13}C, and ^{14}C. All isotopes of an element have the same number of electrons, so they interact with other atoms the same way. Thus cells can use any carbon isotope for a metabolic reaction.

You have probably heard of radioactive isotopes, or "radioisotopes," which are unstable and tend to break apart (decay) into more stable atoms. The *Focus* essay describes some uses of radioisotopes in research, medicine, and charting the history of life.

All atoms of an element have the same number of electrons and protons, but they can vary slightly in the number of neutrons. Variant atoms of the same element are isotopes.

Focus on Science

Using Radioisotopes To Date Fossils, Track Chemicals, and Save Lives

In the winter of 1896, the physicist Henri Becquerel tucked a heavily wrapped rock of uranium into a desk drawer, on top of an unexposed photographic plate. A few days later, he opened the drawer and discovered a faint image of the rock on the plate—apparently caused by energy emitted from the rock. A coworker, Marie Curie, named the phenomenon "radioactivity."

As we now know, radioisotopes are unstable atoms with dissimilar numbers of protons and neutrons. They emit electrons and energy. In this spontaneous process (radioactive decay), an isotope changes to a new, stable one that is not radioactive.

Radioactive Dating. Each type of radioisotope has a certain number of protons and neutrons. And it decays spontaneously at a certain rate into a particular isotope of a different element. "Half-life" is the time it takes for half the nuclei in any given amount of a radioactive element to decay into another element. Half-life can't be modified by temperature, pressure, chemical reactions, or any other environmental factor. That is why radioactive dating is a reliable way to discern the age of different rock layers in the earth—and of fossils contained in the layers. To discover a rock's age, we can compare the amount of one of its radioisotopes with the amount of the decay product for that isotope.

For example, ^{40}potassium has a half-life of 1.3 billion years and decays to ^{40}argon, a stable isotope. The age of anything that contains ^{40}potassium can be determined by measuring the ratio of ^{40}argon to ^{40}potassium. In such ways, researchers have dated ancient fossils (Figure *a*).

a Fossilized frond from a tree fern that lived between 320 million and 250 million years ago.

Similarly, researchers using ^{238}uranium (with a half-life of 4.5 billion years) discovered that the earth formed more than 4.6 billion years ago. The following table lists useful ranges of some radioisotopes employed in dating methods:

Radioisotope (unstable)		Stable Product	Half-life (years)	Useful Range (years)
^{87}rubidium	→	^{87}strontium	49 billion	100 million
^{232}thorium	→	^{208}lead	14 billion	200 million
^{238}uranium	→	^{206}lead	4.5 billion	100 million
^{40}potassium	→	^{40}argon	1.3 billion	100 million
^{235}uranium	→	^{207}lead	704 million	100,000
^{14}carbon	→	^{14}nitrogen	5,730	0–60,000

Tracking Chemicals. Scintillation counters and other devices can detect emissions from radioisotopes. Thus radioisotopes can be used as **tracers**. Tracers reveal a pathway or destination of a substance that has entered a cell, the human body, an ecosystem, or some other "system."

Carbon provides an example. All isotopes of an element have the same number of electrons, so they all interact with other atoms the same way. This means cells can use any isotope of carbon in a carbon-requiring reaction. Such reactions occur in photosynthesis. By putting plant cells in a medium enriched in ^{14}carbon, researchers identified steps by which plants take up carbon and incorporate it into carbohydrates. Tracers also may reveal how plants use naturally occurring nutrients and synthetic fertilizers, and this knowledge may help us improve crop production.

What about medical applications? Consider the human thyroid, the only gland of ours that takes up iodine. If a tiny amount of the radioisotope ^{123}iodine is injected into a patient's blood, the thyroid can be scanned with a photographic imaging device. Figure *b* shows examples of the resulting images.

Saving Lives. Radioisotopes are used in nuclear medicine to diagnose and treat diseases. For example, patients with irregular heartbeats are given artificial pacemakers, powered by energy from ^{238}plutonium. (This dangerous radioisotope is sealed in a case so its emissions won't damage body tissues.) As another example, PET (positron-emission tomography) yields images of metabolically active and inactive tissues. Radioisotopes are attached to glucose or some other biological molecule. Then they are injected into a patient, who is moved into a PET scanner. When cells in certain tissues absorb glucose, radioisotope emissions are used to produce a vivid image of variations or abnormalities in metabolic activity (Figure *c*).

Finally, in some cancer treatments, radioisotopes are used to destroy or impair living cells. In radiation therapy, localized cancers are deliberately bombarded with energy from a ^{226}radium or ^{60}cobalt source.

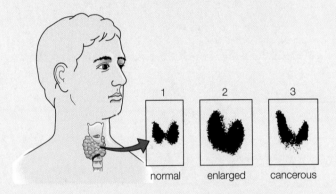

b Scans of human thyroid glands after ^{123}iodine was injected into the bloodstream. The thyroid normally takes up iodine and uses it in hormone production. Uptake by glands that are (1) normal, (2) abnormally enlarged, and (3) cancerous.

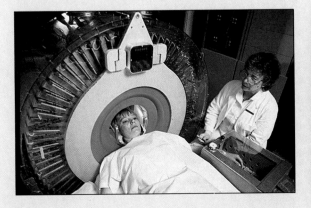

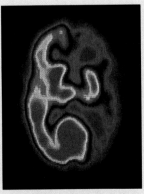

c Patient being moved into a PET scanner. Also shown is a brain scan of a child with a severe neurological disorder. The different colors signify differences in metabolic activity in one half of the brain. The brain's other half shows no activity.

BONDS BETWEEN ATOMS

We turn now to interactions between atoms. Take a moment to review Figure 2.4, which summarizes a few conventions used to describe these events.

The Nature of Chemical Bonds

A chemical bond is a union between the electron structures of atoms. Whether one atom will bond with another depends on the *number* and *arrangement* of its electrons.

Picture three actresses arriving at the Academy Awards ceremony wearing the same bright red designer dress. Each seeks recognition but dreads being caught next to the others. Two might maneuver themselves *near* the center of attention while avoiding each other. But by unspoken agreement, all three never, ever are in the same place at the same time.

Electrons behave roughly the same way. They are attracted to an atom's protons but repelled by other electrons. They spend as much time as possible near the nucleus and far away from each other by moving rapidly in different orbitals. Orbitals simply are regions of space around the nucleus in which electrons are likely to be at any instant. Each orbital has enough room for two electrons, at most.

There is a simple (but not quite accurate) way to think about this. Imagine that all possible orbitals are arranged in different *shells* around the nucleus (Figure 2.5). A shell closest to the nucleus has one orbital; it can

Figure 2.4 Chemical bookkeeping.

We use symbols for elements when writing *formulas*, which identify the composition of compounds. (For example, water has the formula H_2O. The subscript indicates two hydrogen atoms are present for every oxygen atom.) Symbols and formulas are used in *chemical equations*—representations of reactions among atoms and molecules.

In written chemical reactions, an arrow means "yields." Substances entering a reaction (reactants) are to the left of the arrow. Reaction products are to the right. For example, photosynthesis is often summarized this way:

$$6CO_2 \ + \ 6H_2O \ \Longrightarrow \ C_6H_{12}O_6 \ + \ 6O_2$$

6 carbons	12 hydrogens	6 carbons	12 oxygens
12 oxygens	6 oxygens	12 hydrogens	
		6 oxygens	

Notice there are as many atoms of each element to the right of the arrow as there are to the left. Although the atoms are combined in different forms, none is consumed or destroyed. The total mass of all products of any chemical reaction equals the total mass of all its reactants. (That's the general idea of the law of conservation of mass.) Keep in mind, the equations that you use to represent cellular reactions must be balanced this way.

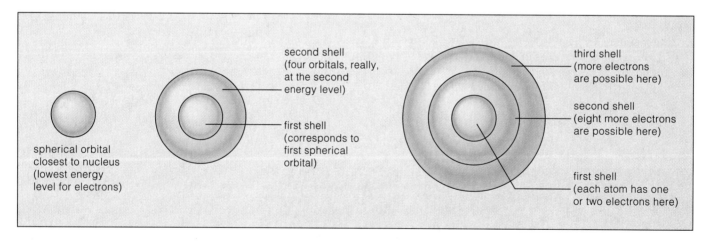

Figure 2.5 Distribution of electrons in atoms. One or at most two electrons occupy a ball-shaped volume of space (an orbital) close to the nucleus. This orbital is at the lowest energy level. At the next (higher) energy level, there can be as many as eight more electrons (two in each of four orbitals). The orbital shapes get tricky, but we can ignore them here. Simply think of the total number of orbitals at a given energy level as being somewhere in a "shell" of the sort shown.

hold no more than two electrons. The next shell has four orbitals; it can hold eight electrons. Successive shells can hold still more electrons (Table 2.2).

Hydrogen, the simplest atom, has one electron in its first and only shell. Sodium, with eleven electrons, has a lone electron in its outermost shell (Figure 2.6). Each of these atoms has a "vacancy" in an orbital in the outermost shell. Such atoms tend to react with other atoms. Hydrogen, oxygen, carbon, and nitrogen—*the most abundant components of organisms*—are like this.

Ionic Bonding

Electrons can be knocked out of atoms, pulled away from them, or added to them. When an atom loses or gains one or more electrons, it becomes positively or negatively charged. In this state, it is an **ion**.

Atoms lose or gain electrons when another atom of the right kind is nearby to accept or donate electrons. Because one loses and one gains, *both* become ionized. Depending on the surroundings, the two ions go their separate ways or stay together through the mutual attraction of opposite charges. An association of two oppositely charged ions is an **ionic bond**. Table salt, or NaCl, has ions of sodium (Na^+) and chloride (Cl^-) linked together this way (Figure 2.7).

When an atom gains or loses one or more electrons, it becomes an ion, with an overall positive or negative charge.

In an ionic bond, a positive and a negative ion are linked by the mutual attraction of opposite charges.

Table 2.2 Electron Distribution for a Few Elements

Element	Chemical Symbol	Atomic Number	First Shell	Second Shell	Third Shell
Hydrogen	H	1	1	—	—
Helium	He	2	2	—	—
Carbon	C	6	2	4	—
Nitrogen	N	7	2	5	—
Oxygen	O	8	2	6	—
Neon	Ne	10	2	8	—
Sodium	Na	11	2	8	1
Magnesium	Mg	12	2	8	2
Phosphorus	P	15	2	8	5
Sulfur	S	16	2	8	6
Chlorine	Cl	17	2	8	7

(Electron Distribution spans the First Shell, Second Shell, and Third Shell columns.)

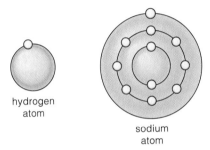

hydrogen atom

sodium atom

Figure 2.6 Distribution of electrons (yellow dots) in hydrogen and sodium atoms. Each atom has a lone electron (and room for more) in its outermost shell. Atoms having such partly filled shells tend to enter into reactions with other atoms.

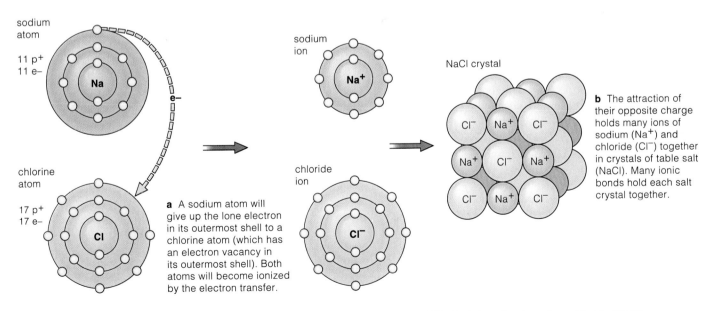

sodium atom

11 p+
11 e–

Na

e–

chlorine atom

17 p+
17 e–

Cl

a A sodium atom will give up the lone electron in its outermost shell to a chlorine atom (which has an electron vacancy in its outermost shell). Both atoms will become ionized by the electron transfer.

sodium ion

Na^+

chloride ion

Cl^-

NaCl crystal

Cl^- Na^+ Cl^-
Na^+ Cl^- Na^+
Cl^- Na^+ Cl^-

b The attraction of their opposite charge holds many ions of sodium (Na^+) and chloride (Cl^-) together in crystals of table salt (NaCl). Many ionic bonds hold each salt crystal together.

Figure 2.7 Ionic bonding in sodium chloride (NaCl).

Covalent Bonding

In a **covalent bond**, one atom cannot pull electrons completely away from another atom, and the two end up *sharing* electrons. A covalent bond can be represented by a single line, as in H—H. (Or dots can represent the shared electrons, as in H· + H· ⟶ H:H.) In a double covalent bond, two atoms share two pairs of electrons. This happens in an O_2 molecule, or O=O. In a triple covalent bond (such as N≡N), two atoms share three pairs of electrons.

Covalent bonds are nonpolar or polar. In a *nonpolar* covalent bond, both atoms exert the same pull on shared electrons. The term nonpolar implies no difference at the two ends of the bond. An example is H—H. The hydrogen atoms, with one proton each, attract the shared electrons equally.

In a *polar* covalent bond, atoms of different elements (which have different numbers of protons) do not exert the same pull on shared electrons. The more attractive atom ends up slightly negative. It is balanced by the other atom, which ends up slightly positive. In other words, the two atoms together have no *net* charge—but the charge is distributed unevenly between the two ends of the bond between them.

A water molecule (H—O—H) has two polar covalent bonds. Its electrons are less attracted to the hydrogens than to the oxygen, which has more protons. The molecule carries no *net* charge, yet it can weakly attract other atoms because of its polarity.

In a covalent bond, atoms share electrons.

If electrons are shared equally, the bond is nonpolar. If they are not shared equally, the bond is polar (slightly positive at one end and slightly negative at the other).

Hydrogen Bonding

In a **hydrogen bond**, an atom of a molecule weakly interacts with a neighboring hydrogen atom that is already taking part in a polar covalent bond. (The hydrogen, with its slight positive charge, is attracted to the other atom's slight negative charge.) Hydrogen bonds can form between two different molecules. They also can form between two different regions of the same molecule where it twists back on itself (Figure 2.8).

Hydrogen bonds are common in organisms. They hold the two strands of DNA together. Individually the bonds are easily broken, but collectively they help stabilize DNA's structure. Hydrogen bonds also give water many of its life-sustaining properties.

Figure 2.8 Examples of hydrogen bonds. Their collective action gives water and other substances some of their properties.

In a hydrogen bond, an atom or molecule interacts weakly with a neighboring hydrogen atom that is already taking part in a polar covalent bond.

Properties of Water

Life originated in water. Many organisms still live in it, and the ones that don't carry water around with them, in cells and tissue spaces. Many metabolic reactions require water as a reactant. The very shape and internal structure of cells depend on water.

Bonding Properties. A water molecule carries no *net* charge, but remember that the charges it does carry are unevenly distributed. A water molecule's oxygen "end" is a bit negative and its other end is a bit positive (Figure 2.9). The polarity allows water molecules to hydrogen-bond with one another and with other polar substances, such as sugars. All polar molecules are attracted to water; they are **hydrophilic** (water-loving).

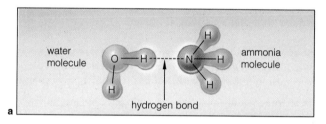

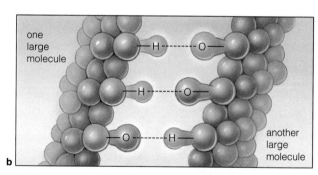

By contrast, water's polarity repels oil and other nonpolar substances, which are **hydrophobic** (water-dreading). Shake a bottle containing water and salad oil, then put it on a counter. In time, hydrogen bonds reunite the water molecules. (They replace bonds that were broken when you shook the bottle.) As they do, they push oil molecules aside and force them to cluster in droplets or in a film at the water's surface. As you will see, such hydrophobic interactions help organize the rather oily, sheetlike layers of cell membranes.

Temperature Stabilization. Life originated in *liquid* water—not water in a frozen or gaseous form. And most of the earth's water tends to stay liquid because of hydrogen bonds between water molecules.

Consider that the molecules of any substance are in constant motion and that energy inputs make them move faster. *Temperature* is a measure of molecular motion. Compared to most other fluids, water requires a greater input of heat energy before its temperature will increase measurably. Hydrogen bonds in water absorb much of the incoming energy, so the motion of individual molecules does not increase as fast. This property helps stabilize temperatures in cells (which are mostly water). It helps cells resist temperature changes that could disrupt enzyme function, among other things.

When water is liquid, its hydrogen bonds are constantly breaking and constantly forming again. With enough heat energy, hydrogen bonds stay broken and molecules at the water's surface escape into the air. We call this "evaporation." When molecules break free and depart in large numbers, they carry away energy and lower the water's surface temperature. You cool off on hot, dry days when sweat (which is 99 percent water) evaporates from your skin.

Below 0°C, hydrogen bonds can't break, and they lock water molecules in the bonding pattern of ice (Figure 2.10). During winter freezes, ice sheets form on ponds, lakes, and streams. They hold in water's heat and help protect aquatic organisms against freezing.

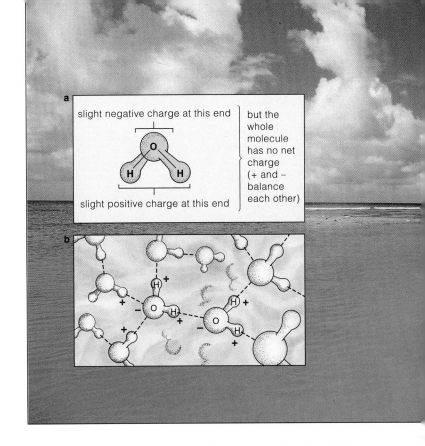

Figure 2.9 A view you never will see on any other planet in the solar system—an ocean of water, a substance absolutely vital for life. (**a**) Polarity of a water molecule. (**b**) Hydrogen bonding between water molecules in liquid water. Dashed lines signify hydrogen bonds.

Figure 2.10 Hydrogen bonding between water molecules in ice. Dashed lines signify the angles of hydrogen bonds.

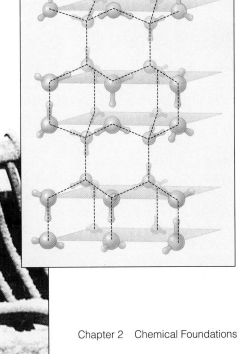

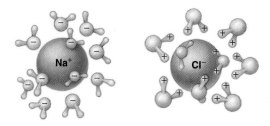

Figure 2.12 Spheres of hydration around charged ions.

Figure 2.11 Water's cohesion, as demonstrated by a water strider "walking" on water molecules held together by hydrogen bonds.

Water's Cohesion. Swimming on a hot summer night can be refreshing, if you don't mind the night-flying bugs that hit the water's surface and float about on it. Because of hydrogen bonds, the water shows cohesion. This means it resists rupturing when placed under tension—that is, stretched—as by weighty bugs (Figure 2.11). At the surface, hydrogen bonds pull the uppermost water molecules inward. A high surface tension results from the ongoing, collective bonding.

In most land plants, cohesion helps move water through pipelines that extend from roots to leaves. When water evaporates from the leaves, hydrogen bonds "pull" more water molecules up into leaf cells.

Water's Solvent Properties. Water is a great solvent, which means ions and polar molecules readily dissolve in it. Dissolved substances are called **solutes**. But what does "dissolved" mean? By way of example, pour some table salt into a glass of water. In time the salt crystals disappear because they separate into Na^+ and Cl^- ions. Each sodium ion attracts the negative end of water molecules. Each chloride ion attracts the positive end of other water molecules. Many water molecules cluster together as "spheres of hydration" around ions and keep them dispersed in fluid (Figure 2.12). *A substance is "dissolved" in water when spheres of hydration form around its individual ions or molecules.* This happens to solutes in cells, body fluids, the sap of maple trees, and so on.

ACIDS, BASES, AND SALTS

Acids and Bases

Some substances release protons when they dissolve in water. Free (unbound) protons are called **hydrogen ions** (H^+). A substance that releases H^+ in water is an **acid**. Think of what happens when you sniff, chew, then swallow fried chicken. Stomach cells are stimulated to secrete hydrochloric acid (HCl), which separates into H^+ and Cl^-. These ions make the fluid in your stomach more acidic—and a good thing, too. Increased acidity switches on enzymes that can digest the chicken proteins. It also helps kill bacteria that may have lurked in or on the chicken. However, if you eat too much fried chicken, you may end up with an "acid stomach." And you might reach for an antacid tablet.

Milk of magnesia is one kind of antacid. When dissolved, it releases magnesium ions and hydroxide ions (OH^-). Hydroxide ions may then *combine with* some of those excess hydrogen ions in your stomach fluid and so help settle things down. Any substance that combines with H^+ is a **base**.

The pH Scale

The **pH scale** is used to measure the concentration of unbound hydrogen ions in blood, water, and other solutions. Figure 2.13 shows the scale, which ranges from 0 (most acidic) to 14 (most basic). A solution with a pH value of 7 is neutral; it has just as many H^+ as OH^- ions. A neutral solution becomes more acidic when H^+ ions are added to it and more basic when H^+ ions are removed from it. Each change by one unit of the pH scale corresponds to a tenfold increase or decrease in H^+ concentration.

To get a sense of the magnitude of the pH scale, put a dab of baking soda (pH 9) on your tongue. Next, sip pure water (pH 7), then vinegar (3), then lemon juice (2.3). The fluid in most of your cells hovers around pH 7. The pH of your blood and most tissue fluids ranges between 7.35 and 7.45. The pH of an aquatic habitat may be higher or lower than it is inside organisms dwelling there. Industrial wastes can be so acidic they affect the pH of rain (Figure 2.14 and Chapter 37).

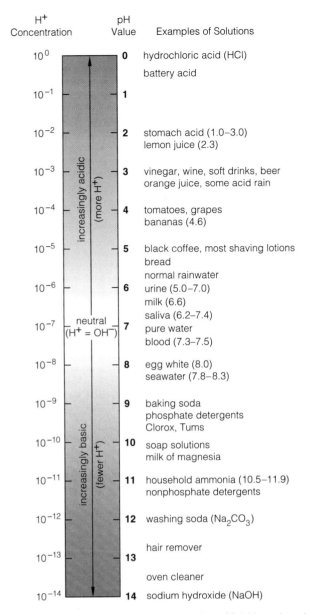

H+
Concentration

pH
Value

Examples of Solutions

10^0	0	hydrochloric acid (HCl)
		battery acid
10^{-1}	1	
10^{-2}	2	stomach acid (1.0–3.0)
		lemon juice (2.3)
10^{-3}	3	vinegar, wine, soft drinks, beer
		orange juice, some acid rain
10^{-4}	4	tomatoes, grapes
		bananas (4.6)
10^{-5}	5	black coffee, most shaving lotions
		bread
		normal rainwater
10^{-6}	6	urine (5.0–7.0)
		milk (6.6)
		saliva (6.2–7.4)
10^{-7}	7	pure water
		blood (7.3–7.5)
10^{-8}	8	egg white (8.0)
		seawater (7.8–8.3)
10^{-9}	9	baking soda
		phosphate detergents
		Clorox, Tums
10^{-10}	10	soap solutions
		milk of magnesia
10^{-11}	11	household ammonia (10.5–11.9)
		nonphosphate detergents
10^{-12}	12	washing soda (Na_2CO_3)
10^{-13}	13	hair remover
		oven cleaner
10^{-14}	14	sodium hydroxide (NaOH)

increasingly acidic (more H+)

neutral ($H^+ = OH^-$)

increasingly basic (fewer H+)

Figure 2.13 The pH scale, in which a liter of fluid is assigned a number according to the number of hydrogen ions in it. The scale ranges from 0 (most acidic) to 14 (most basic). A change of only 1 on the scale means a tenfold change in the H^+ concentration.

Dissolved Salts

Inside organisms, acids commonly combine with bases. The results are ionic compounds called **salts**. Salts often dissolve and form again, depending on pH. Consider how sodium chloride forms, then dissolves:

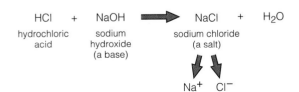

HCl + NaOH ⟶ NaCl + H_2O
hydrochloric   sodium   sodium chloride
acid   hydroxide   (a salt)
                 (a base)

Na^+ Cl^-

Figure 2.14 Sulfur dioxide emissions from a coal-burning power plant. Special camera filters revealed these otherwise invisible emissions. Together with other airborne pollutants, sulfur dioxides dissolve in atmospheric water to form acidic solutions. They are a major component of acid rain.

Many other salts also dissolve into ions. Such ions serve important functions in cells. For example, many help maintain the body's acid-base balance, as described next.

Buffers

Chemical reactions in cells are sensitive to even slight shifts in pH. Yet hydrogen ions are continually being added to the cellular environment and withdrawn from it. Control mechanisms counter the potentially disruptive shifts and help maintain cellular pH.

Controls also maintain the pH of blood and most tissue fluids. Later in the book, you will see how the lungs and kidneys help control the body's overall acid-base balance. For now, simply keep in mind that many of the controls involve buffer molecules. A **buffer** is any molecule that can combine with hydrogen ions, release them, or both, and so help stabilize pH.

Consider how one buffer, bicarbonate (HCO_3^-), helps restore pH when blood becomes too acidic. It combines with excess H^+ to form carbonic acid:

$$HCO_3^- + H^+ \implies H_2CO_3$$
bicarbonate   carbonic acid

Conversely, bicarbonate releases H^+ ions when blood is not acidic enough. Its buffering action is crucial. For instance, some lung diseases interfere with carbon dioxide elimination, so the blood level of carbonic acid (hence of H^+) increases. This abnormal condition, a form of *acidosis*, makes breathing difficult and weakens the body.

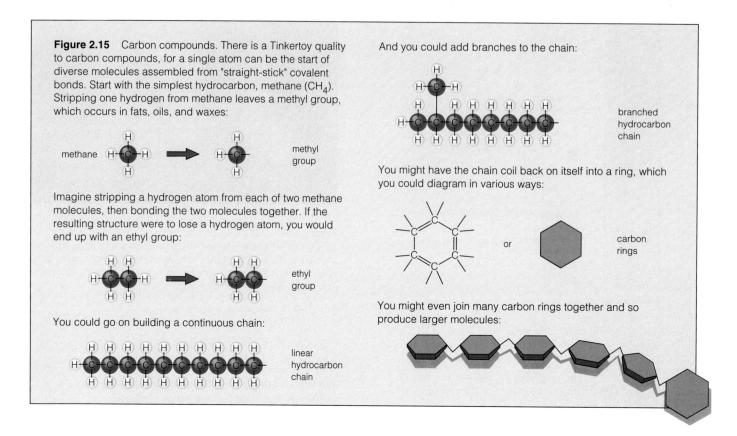

Figure 2.15 Carbon compounds. There is a Tinkertoy quality to carbon compounds, for a single atom can be the start of diverse molecules assembled from "straight-stick" covalent bonds. Start with the simplest hydrocarbon, methane (CH_4). Stripping one hydrogen from methane leaves a methyl group, which occurs in fats, oils, and waxes:

methane

methyl group

Imagine stripping a hydrogen atom from each of two methane molecules, then bonding the two molecules together. If the resulting structure were to lose a hydrogen atom, you would end up with an ethyl group:

ethyl group

You could go on building a continuous chain:

linear hydrocarbon chain

And you could add branches to the chain:

branched hydrocarbon chain

You might have the chain coil back on itself into a ring, which you could diagram in various ways:

or

carbon rings

You might even join many carbon rings together and so produce larger molecules:

CARBON COMPOUNDS

By far, oxygen, hydrogen, and carbon are the most abundant elements in living things. Much of the oxygen and hydrogen is linked together as water molecules. Notable amounts also are linked to carbon—the most important structural element in the body.

Backbones and Functional Groups

A carbon atom can bond covalently to as many as four other atoms (Figure 2.15). Commonly, carbon atoms are joined one after another into a chain or ring, and most have hydrogen atoms attached to them. Because the carbons share electrons equally, the chains and rings are stable backbones for molecules. Any molecule with a carbon backbone is an "organic" compound. Such backbones are not present in water, carbon dioxide, and other simple, "inorganic" compounds.

Other atoms besides hydrogen also bond to carbon backbones. Such atoms are **functional groups**, and they influence the chemical behavior of organic compounds (Figure 2.16). Consider sugars, which belong to the important group of organic compounds called alcohols. Sugar molecules have hydroxyl groups (—OH). Water hydrogen-bonds with —OH groups, so sugars dissolve in water.

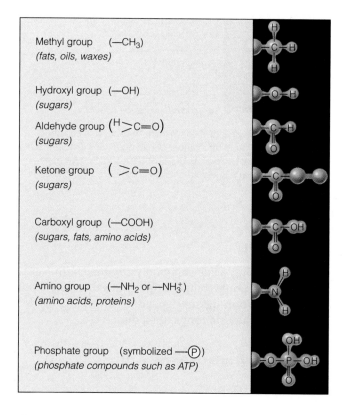

Methyl group   (—CH$_3$)
(fats, oils, waxes)

Hydroxyl group   (—OH)
(sugars)

Aldehyde group (^{H}>C=O)
(sugars)

Ketone group   (>C=O)
(sugars)

Carboxyl group (—COOH)
(sugars, fats, amino acids)

Amino group   (—NH$_2$ or —NH$_3^+$)
(amino acids, proteins)

Phosphate group   (symbolized —Ⓟ)
(phosphate compounds such as ATP)

Figure 2.16 Major functional groups that help give organic compounds their distinctive properties. They are common in the compounds listed in italics.

From Small Organic Compounds to the Molecules of Life

Of all substances, certain organic compounds stand out. Under conditions that now exist on earth, only living cells can construct them. These compounds are known as complex carbohydrates, lipids, proteins, and nucleic acids. *They are the large molecules of life.* Different kinds function as energy sources, structural materials, metabolic workers, or libraries of hereditary information. As you will see, these large molecules are all assembled from only four families of small organic compounds. The compounds are called simple sugars, fatty acids, amino acids, and nucleotides. Each has no more than twenty or so carbon atoms.

Simple sugars, fatty acids, amino acids, and nucleotides serve as energy sources and building blocks for the large molecules of life—complex carbohydrates, lipids, proteins, and nucleic acids.

Condensation and Hydrolysis

Organic compounds are assembled and disassembled with the help of enzymes. **Enzymes** are a class of proteins that speed up reactions between specific substances, usually at functional groups.

A **condensation** reaction results in covalent bonding between small molecules and, often, the formation of water as a by-product. Enzyme action causes one molecule to lose an H atom and another molecule to lose an —OH group; then a bond forms at the exposed sites (Figure 2.17a). The unbound parts, now H^+ and OH^- ions, may combine to form a water molecule.

A **hydrolysis** reaction is like condensation in reverse. Enzyme action splits a molecule into two or more parts by breaking covalent bonds. At the same time, H^+ and OH^- derived from water become attached to the exposed sites (Figure 2.17b). Cells commonly break apart molecules by hydrolysis.

Let's turn now to the key characteristics of the main organic compounds that all cells build as well as break apart.

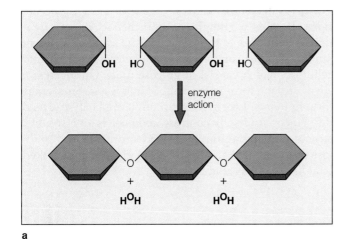

a

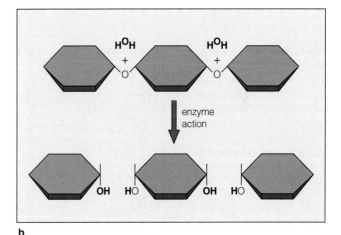

b

Figure 2.17 Condensation and hydrolysis—two kinds of reactions by which cells build and break apart organic compounds.

(**a**) Condensation. In this generalized example, three molecules covalently bond into a larger molecule, and two water molecules are formed.

(**b**) Hydrolysis. Two covalent bonds of a molecule are split, and H^+ and OH^- derived from water molecules become bonded to the molecular fragments.

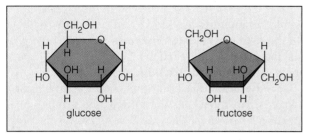

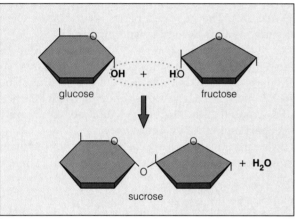

a

b

Figure 2.18 (**a**) Structural formulas for glucose and fructose, two monosaccharides. Notice how an oxygen atom occupies a certain position in each carbon ring. (**b**) Condensation of glucose and fructose into sucrose, a disaccharide.

Figure 2.19 Starch compared with cellulose. Both of these polysaccharides consist only of glucose units, but they have different properties.

(**a**) Part of a chain of amylose, a form of starch. Starch, a sugar-storage form in plants, is readily broken apart by hydrolysis reactions.
(**b**) Linkages between adjacent glucose units cause such chains to twist into a coil. Coiling orients the linkages so that they are accessible to enzymes of hydrolysis.

(**c**) A portion of the glucose chains in cellulose, an insoluble structural material in plant cell walls. (**d**) Different linkages between glucose units cause the chains to stretch out, side by side, and hydrogen-bond to one another at —OH groups. The hydrogen bonds stabilize the chains in tight bundles. These may twist and coil up together as cellulose threads.

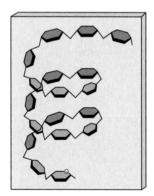

a Starch chain

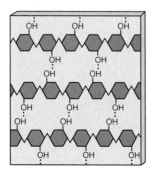

c Cellulose chain

Carbohydrates

A **carbohydrate** is a simple sugar or a molecule that is composed of sugar units. Carbohydrates are the most abundant molecules of life. Cells use them as structural materials, transportable packets of quick energy, and storage forms of energy.

Monosaccharides and Oligosaccharides.   The term "saccharide" comes from a Greek word meaning sugar. A *mono*saccharide, or one sugar unit, is the simplest carbohydrate of all. Such sugars are soluble in water, most are sweet-tasting, and the most common have five or six carbon atoms. Ribose and deoxyribose (in RNA and DNA, respectively) are examples. So is glucose (Figure 2.18*a*). Most cells use glucose as their primary energy source. Glucose also is a precursor (parent molecule) of many organic compounds.

An *oligo*saccharide is a short chain of two or more covalently bonded sugar units. The ones with two sugars—the *di*saccharides—include sucrose, lactose, and maltose. Sucrose, the most plentiful sugar in nature, forms from a glucose and a fructose unit (Figure 2.18*b*). Table sugar is a crystallized form of sucrose extracts from sugarcane and other plants. Milk contains lactose (a glucose and a galactose unit). Seeds contain maltose (two glucose units).

Complex Carbohydrates.   These are *poly*saccharides. Each is a straight or branched chain of sugar units—hundreds or thousands, of the same or different kinds. The most common ones are glycogen, starch, and cellulose, which consist of glucose units. Glycogen is a sugar-storage form in animals. Starch is a sugar-storage form in plants. Like glycogen, starch can be hydrolyzed into glucose units. By contrast, cellulose is an insoluble, structural material in plants. Why are starch and cellulose so different? Each has adjacent glucose units linked in a different way (Figure 2.19).

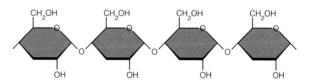

b Linkages between glucose units in starch

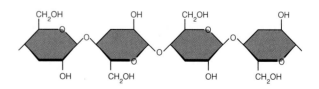

d Linkages between glucose units in cellulose

Lipids

Lipids are greasy or oily compounds that show little tendency to dissolve in water, but they do dissolve in nonpolar solvents such as ether. Like polysaccharides and proteins, lipids can be broken down by hydrolysis reactions. Some lipids function in energy storage. Others are structural materials in membranes, coatings, and other cell structures. Here we will focus on two types: lipids with and without fatty acid components. A **fatty acid** is a long, flexible, hydrocarbon chain that has a —COOH group at one end (Figure 2.20). It is largely insoluble. Three common lipids having fatty acid tails are glycerides, phospholipids, and waxes.

The Glycerides. Glycerides are the body's most abundant lipids and its richest source of stored energy. They have fatty acid tails attached to a glycerol backbone (Figure 2.21). *Mono*glycerides have only one tail, *di*glycerides have two, and *tri*glycerides have three. Many fats and oils are composed of such molecules.

Saturated fats, including butter and lard, tend to be solids at room temperature. The tails of their adjacent molecules snuggle together in parallel. "Saturated" means only single C—C bonds occur in the backbone of the tails, and hydrogen atoms are attached to those carbon atoms at the remaining bonding sites (Figure 2.20*a*).

Unsaturated fats, or oils, tend to be liquid at room temperature. They have one or more double bonds between carbon atoms in their fatty acid tails (Figure 2.20*b,c*). Oils are liquid because the double bonds create kinks that disrupt packing between tails.

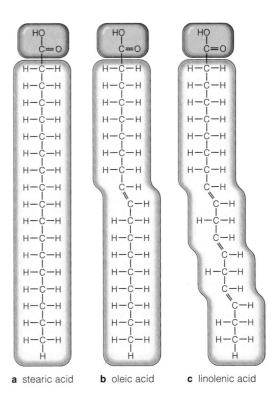

a stearic acid   **b** oleic acid   **c** linolenic acid

Figure 2.20 Structural formulas for three fatty acids. (**a**) Stearic acid's carbon backbone is fully saturated with hydrogen atoms. (**b**) Oleic acid, with its double bond in the carbon backbone, is unsaturated. (**c**) Linolenic acid, with three double bonds, is a "polyunsaturated" fatty acid.

Figure 2.21 Condensation of fatty acids into a triglyceride. Several animals, including penguins, can swim in icy waters for long periods. A very thick insulative layer of triglycerides under the skin helps keep them warm.

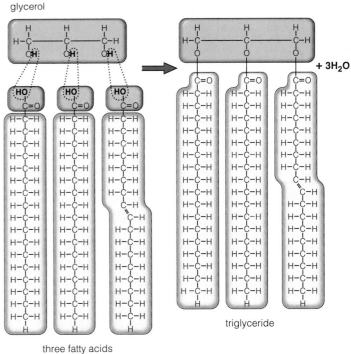

glycerol

$+ 3H_2O$

three fatty acids

triglyceride

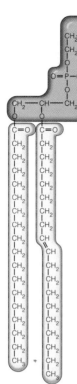

Figure 2.22 Structural formula of a typical phospholipid present in animal cell membranes. The hydrophilic head is shaded orange. Are the hydrophobic tails (gold) saturated or unsaturated?

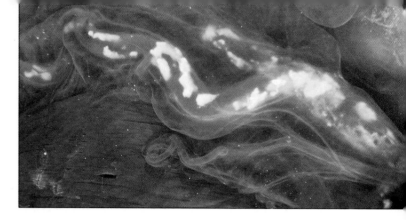

Figure 2.23 Lipid deposits in a heart patient's arteries.

Phospholipids. These lipids are the main structural component of all cell membranes. A phospholipid molecule has a glycerol backbone, two fatty acid tails, and a hydrophilic "head" that incorporates a phosphate group (Figure 2.22).

Waxes. Waxes are another type of lipid with fatty acid components. Wax coatings help keep skin and hair protected, lubricated, and pliable. Waxes help feathers repel water. For many plants, waxes and cutin (another lipid) form a cuticle, a surface covering on aboveground parts that helps restrict water loss.

Steroids. Steroids are among the many lipids with no fatty acid tails that are important in membrane structure and in metabolism. Steroids differ in their functional groups, but they all have the same kind of backbone (four carbon rings):

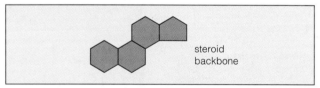

steroid backbone

The steroid cholesterol is a component of animal cell membranes. Vitamin D (essential for bone and tooth development) cannot be synthesized without it. But excess cholesterol in the blood may damage the blood-transporting tubes called arteries. In *atherosclerosis*, cholesterol and other lipids accumulate in the arterial wall (Figure 2.23 and pages 426–427). The deposits initiate formation of fibrous masses, even clots. In time, the obstruction may interfere with blood flow, causing tissue damage or a heart attack.

Steroid hormones influence growth, development, reproduction, and everyday functions. Some athletes and bodybuilders use hormonelike steroids to increase muscle mass, but these have side effects (page 396).

Proteins

By far, the most diverse biological molecules of all are proteins. Among their ranks are structural materials, the stuff of bone and cartilage, hoof and claw. Enzymes, receptors and transporters of signals and substances, movers of cell structures, and other worker molecules in the cell are proteins. So are antibodies, molecular weapons that help defend the body against disease.

By definition, **proteins** are large molecules assembled from twenty or so different kinds of amino acids. An **amino acid** is a small organic compound having an amino group, an acid group, a hydrogen atom, and one or more atoms called its R group. All four parts are covalently bonded to the same carbon atom. Under cellular conditions, the amino group and acid group are ionized, as shown here:

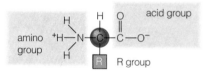

Figure 2.24 shows some amino acids that we will consider later in the book.

Protein Structure. The amino acids of proteins are joined together, one after the other, by peptide bonds. During protein synthesis, these covalent bonds form between the *amino* group of one amino acid and the *acid* (carboxyl) group of another (Figure 2.25). Three or more joined amino acids are a **polypeptide chain**.

Different kinds of amino acids follow one another in a chain. Consider part of the sequence for insulin, a protein hormone that prods cells to take up glucose:

The complete sequence for insulin is unique; no other protein has it. The same is true for each kind of protein. Its unique overall sequence is its *primary* structure. That structure influences a protein's shape, what its function will be, and how it will interact with other substances. It does so in two major ways.

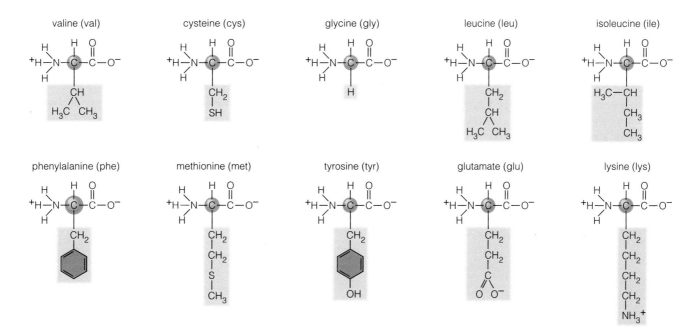

Figure 2.24 Structural formulas for ten of the twenty common amino acids. Green boxes highlight the R groups.

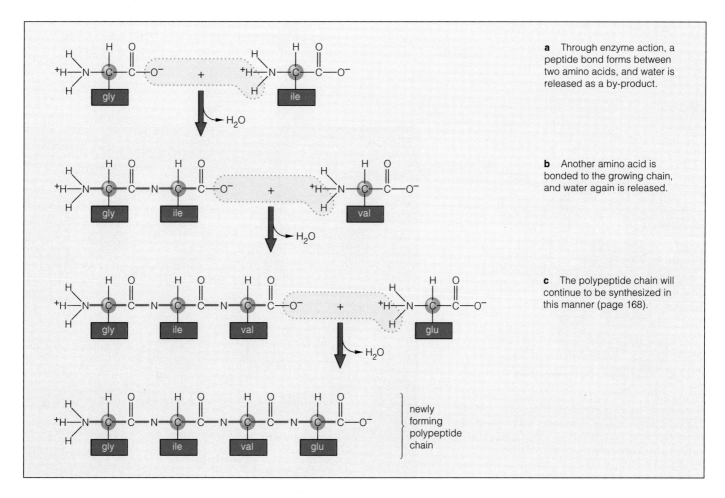

a Through enzyme action, a peptide bond forms between two amino acids, and water is released as a by-product.

b Another amino acid is bonded to the growing chain, and water again is released.

c The polypeptide chain will continue to be synthesized in this manner (page 168).

Figure 2.25 Peptide bond formation during protein synthesis. The amino acids shown are the first four in the sequence for one of two polypeptide chains that make up the protein insulin in cattle.

First, oxygen and other atoms of different amino acids in the sequence take part in hydrogen bonds that keep the chain coiled or extended (Figure 2.26). Thus proteins have *secondary* structure: a coiled or extended pattern, based on regular hydrogen bonding. Second, different R groups in the sequence interact and dictate how the chain bends and twists into its three-dimensional shape. *Tertiary* structure is the protein shape resulting from R-group interactions. Figure 2.27*a* is an example.

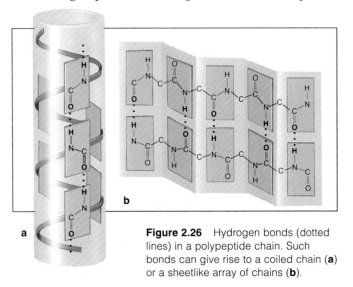

a   **b**

Figure 2.26 Hydrogen bonds (dotted lines) in a polypeptide chain. Such bonds can give rise to a coiled chain (**a**) or a sheetlike array of chains (**b**).

Some proteins have *quaternary* structure, meaning they incorporate two or more polypeptide chains. The resulting protein is globular, fiberlike, or a combination of the two. Hemoglobin is an example (Figure 2.27*b*).

The amino acid sequence in polypeptide chains gives rise to the three-dimensional structure of proteins. That structure governs a protein's chemical behavior.

Protein Denaturation. The loss of any molecule's three-dimensional shape following disruption of weak bonds is called **denaturation**. Being weak, heat sensitive, and pH sensitive, the hydrogen bonds that help hold a protein in its normal, three-dimensional shape can be disrupted. When that happens, its polypeptide chains unwind or change shape, and the protein can no longer function.

The protein albumin is concentrated in the "egg white" of uncooked chicken eggs. When you cook an egg, the heat doesn't affect the strong covalent bonds of albumin's primary structure. But it destroys the weaker bonds holding albumin in its three-dimensional shape. For some proteins, denaturation can be reversed when normal conditions are restored—but albumin isn't one of them. There is no way to uncook a cooked egg.

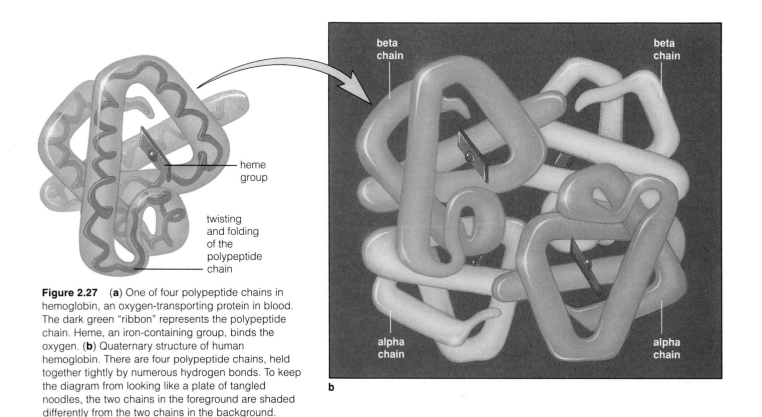

Figure 2.27 (**a**) One of four polypeptide chains in hemoglobin, an oxygen-transporting protein in blood. The dark green "ribbon" represents the polypeptide chain. Heme, an iron-containing group, binds the oxygen. (**b**) Quaternary structure of human hemoglobin. There are four polypeptide chains, held together tightly by numerous hydrogen bonds. To keep the diagram from looking like a plate of tangled noodles, the two chains in the foreground are shaded differently from the two chains in the background.

heme group

twisting and folding of the polypeptide chain

beta chain   beta chain

alpha chain   alpha chain

b

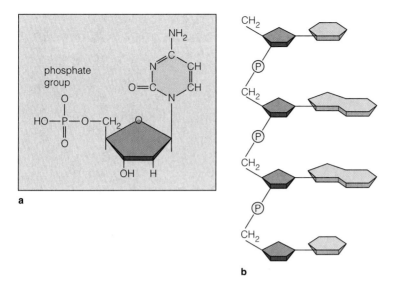

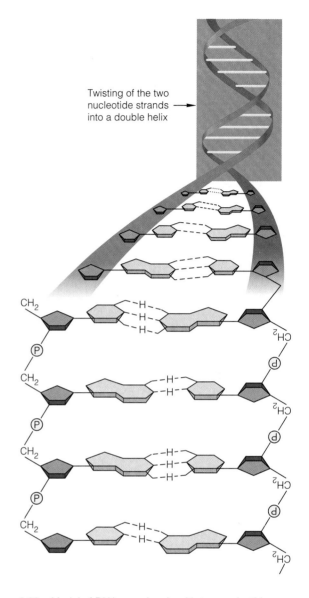

Figure 2.28 (**a**) Example of a nucleotide. Its base is shown in blue, and its sugar, ribose, is shown in red. (**b**) How nucleotides are strung together in a single-stranded nucleic acid molecule.

Nucleotides and Nucleic Acids

Each **nucleotide** has three parts: a five-carbon sugar (ribose or deoxyribose), a phosphate group, and a nitrogen-containing base. The base's carbon backbone is either a single- or double-ring structure (Figure 2.28*a*).

One of the nucleotides is called **ATP** (adenosine triphosphate). It carries energy from one reaction site to another in cells. Other kinds of nucleotides function as **coenzymes**, or "enzyme helpers." They accept hydrogen atoms and electrons from various molecules and transfer them elsewhere. The coenzymes abbreviated NAD⁺ and FAD are examples.

In **nucleic acids**, four different kinds of nucleotides are strung together, forming large single- or double-stranded molecules. Each strand's backbone consists of joined-together phosphate groups of adjacent nucleotides. The nucleotide bases stick out to the side (Figure 2.28*b*). Nucleic acids differ in which nucleotide base follows the one before it in sequence.

You have probably heard of **RNA** (ribonucleic acid) as well as **DNA** (deoxyribonucleic acid). RNA is a single nucleotide strand, most often. DNA is usually a double-stranded molecule that twists helically, like a spiral staircase (Figure 2.29). Hydrogen bonds hold the two strands together. You will read more about these molecules in chapters to come. For now, it is enough to know this: (1) Genetic instructions are encoded in the sequence of bases in DNA, and (2) RNA molecules function in the processes by which genetic instructions are used to build proteins.

Figure 2.29 Model of DNA, a molecule with two nucleotide strands twisted together. This molecule is central to cell survival and reproduction.

Table 2.3 Summary of the Main Carbon Compounds in Living Things

Category	Main Subcategories	Some Examples and Their Functions	
CARBOHYDRATES *contain an aldehyde or a ketone group, and one or more hydroxyl groups*	Monosaccharides (simple sugars)	Glucose	Structural roles, energy source
	Oligosaccharides	Sucrose (a disaccharide)	Form of sugar transported in plants
	Polysaccharides (complex carbohydrates)	Starch Cellulose	Energy storage Structural roles
LIPIDS *are largely hydrocarbon, generally do not dissolve in water but dissolve in nonpolar substances*	Lipids with fatty acids: *Glycerides*: one, two, or three fatty acid tails attached to glycerol backbone	Fats (e.g., butter) Oils (e.g., corn oil)	Energy storage
	Phospholipids: phosphate group, another polar group, and (often) two fatty acids attached to glycerol backbone	Phosphatidylcholine	Key component of cell membranes
	Waxes: long-chain fatty acid tails attached to alcohol	Waxes in cutin	Water retention by plants
	Lipids with no fatty acids: *Steroids*: four carbon rings; the number, position, and type of functional groups vary	Cholesterol	Component of animal cell membranes; can be rearranged into other steroids (e.g., vitamin D and sex hormones)
PROTEINS *are polypeptides (up to several thousand amino acids, covalently linked)*	Fibrous proteins: Individual polypeptide chains, often linked into tough, water-insoluble molecules	Keratin Collagen	Structural element of hair, nails Structural element of bones and cartilage
	Globular proteins: One or more polypeptide chains folded and linked into globular shapes; many roles in cell activities	Enzymes Hemoglobin Insulin Antibodies	Increase in rates of reactions Oxygen transport Control of glucose metabolism Tissue defense
NUCLEIC ACIDS (AND NUCLEOTIDES) *are chains of units (or individual units) that each consist of a five-carbon sugar, phosphate, and a nitrogen-containing base*	Adenosine phosphates	ATP	Energy carrier
	Nucleotide coenzymes	NAD^+, $NADP^+$	Transport of protons (H^+) and electrons from one reaction site to another
	Nucleic acids: Chains of thousands to millions of nucleotides	DNA, RNAs	Storage, transmission, translation of genetic information

SUMMARY

1. Protons, neutrons, and electrons are building blocks of atoms. All atoms of an element have the same number of protons and electrons. Interaction between atoms depends on the number and arrangement of their electrons, which carry a negative charge.

2. Atoms have no net charge. If an atom gains or loses one or more electrons, it becomes an ion, with an overall positive or negative charge.

3. Hydrogen, oxygen, carbon, and nitrogen are the most abundant elements in organisms.

4. In ionic bonds, a positive ion and a negative ion stay together by mutual attraction of their opposite charges. In covalent bonds, atoms share one or more electrons. In hydrogen bonds, an atom or molecule weakly interacts with a neighboring hydrogen atom that is already taking part in a polar covalent bond.

5. Acids release hydrogen ions (H^+) in water. Bases combine with H^+. At pH 7, the H^+ and OH^- concentra-

tions in a solution are equal. Buffers and other mechanisms maintain pH values of blood, tissue fluids, and the fluid inside cells.

6. Water takes part in many reactions and helps give cells shape and internal organization. Because of hydrogen bonds between its molecules, water shows resistance to temperature changes, internal cohesion, and a notable capacity to dissolve other substances.

7. Carbon atoms covalently bonded into chains and rings serve as the backbone of organic compounds. Hydrogen atoms are attached to many of the carbon atoms. So are functional groups, which influence the chemical behavior of organic compounds.

8. Cells assemble simple sugars, fatty acids, amino acids, and nucleotides. These small organic molecules serve as energy sources or building blocks for the large "molecules of life"—the complex carbohydrates, lipids, proteins, and nucleic acids. Table 2.3 summarizes the main characteristics of these molecules.

Review Questions

1. How do ions differ from atoms? *17, 21*

2. Define ionic bond, covalent bond, and hydrogen bond. *21, 22*

3. Name four families of small organic molecules used to assemble carbohydrates, lipids, proteins, and nucleic acids (the large biological molecules). *27*

4. Which of the following is a carbohydrate? a fatty acid? an amino acid? a polypeptide? *28–31*
   a. $^+NH_3$—CHR—COO$^-$ c. (glycine)$_{20}$
   b. $C_6H_{12}O_6$ d. $CH_3(CH_2)_{16}COOH$

5. Describe the four levels of protein structure. *30, 32*

6. Distinguish between the following:
   a. monosaccharide, polysaccharide *28*
   b. triglyceride, steroid *29, 30*
   c. amino acid, protein *30*
   d. nucleotide, nucleic acid *33*

Self-Quiz *(Answers in Appendix IV)*

1. Life's structural organization and activities depend on atoms that _____ electrons.
   a. gain c. share
   b. lose d. all of the above

2. Atoms of an element that vary slightly in their number of neutrons are _____ .
   a. ions c. proton-deficient
   b. isotopes d. nonexistent

3. Whether an atom will take part in a chemical bond depends on the _____ and _____ of its electrons.
   a. size; shape c. number; arrangement
   b. size; number d. number; energy value

4. A hydrophilic substance forms _____ with water and other polar substances.
   a. ionic bonds c. hydrogen bonds
   b. covalent bonds d. isotopes

5. All of the following *except* _____ are small organic molecules that serve as the main building blocks or energy sources in cells.
   a. fatty acids c. DNA
   b. simple sugars d. amino acids

6. In _____ reactions, small molecules become covalently linked into larger ones, and water can also form.
   a. symbiotic c. condensation
   b. hydrolysis d. ionic

7. Proteins called _____ make metabolic reactions proceed much faster than they would on their own.
   a. amino acids c. enzymes
   b. fatty acids d. coenzymes

8. _____ are the building blocks of DNA and RNA.
   a. Simple sugars c. Amino acids
   b. Fatty acids d. Nucleotides

9. Match the substance with the most appropriate description.
   _____ coenzyme   a. carbon backbone with hydrogen
   _____ solute   atoms attached
   _____ steroid   b. combines with H$^+$ in water
   _____ acid   c. simple sugar or polysaccharide
   _____ base   d. releases H$^+$ in water
   _____ hydrocarbon   e. carries energy to reaction sites
   _____ carbohydrate   f. transfers electrons, hydrogen
   _____ ATP   g. any dissolved substance
                         h. no fatty acid tails

Selected Key Terms

acid *24*	electron *17*	molecule *17*
amino acid *30*	enzyme *27*	neutron *17*
atom *17*	fatty acid *29*	nucleic acid *33*
ATP *33*	functional group *26*	nucleotide *33*
base *24*	hydrogen bond *22*	pH scale *24*
buffer *25*	hydrogen ion (H$^+$) *24*	polypeptide
carbohydrate *28*	hydrolysis *27*	chain *30*
coenzyme *33*	hydrophilic *22*	protein *30*
compound *17*	hydrophobic *23*	proton *17*
condensation *27*	ion *21*	RNA *33*
covalent bond *22*	ionic bond *21*	salt *25*
denaturation *32*	isotope *18*	solute *24*
DNA *33*	lipid *29*	tracer *19*

Readings

Goodsell, D. September–October 1992. "A Look Inside the Living Cell." *American Scientist* 80:457–465. Current models of biological molecules.

Science. 26 June 1992. "A New Blueprint for Water's Architecture." (No. 256, p. 1,764)

Scientific American. October 1985. "The Molecules of Life." Entire issue devoted to biological molecules.

3 CELL STRUCTURE AND FUNCTION

It Isn't Easy Being Single

As small as it may be, a cell is a *living* thing engaged in the risky business of survival. Consider how something as ordinary as water can challenge its very existence. Water bathes cells inside and out, donates its molecules to metabolic reactions, and dissolves vital ions. If all goes well, each cell holds on to enough water and dissolved ions—not too little, not too much—to survive. But who is to say that life consistently goes well?

Think of a goose barnacle drifting offshore on a log, at the mercy of ocean currents. The barnacle clings to the wood by a stalk and extends its featherlike, food-gathering appendages into the water (Figure 3.1*a*). The fluid bathing each living cell in the barnacle's body

tissues is salty, rather like the salt composition of sea-water. And tissue fluid is in balance with the salty fluid inside cells. But suppose the log drifts into dilute waters from a melting glacier and stays there (Figure 3.1*b*). For reasons explored in this chapter, salts inevitably move out of the barnacle's body—and out of its cells. The salt–water balance is gradually destroyed, and the cells die. (And so, in time, does the barnacle.)

With this example we begin to see the cell as a tiny, organized bit of life in a world that is, by comparison, unorganized and sometimes harsh. How exquisitely adapted the cell must be to its environment! *The cell must be built in such a way that it can bring in certain substances, release or keep out others, and conduct its internal activities with great precision.*

For this bit of life, precision begins at the plasma membrane—a flimsy surface layer composed of little more than lipids and proteins. Across this membrane, cells exchange substances with their surroundings. Within eukaryotic cells, materials also move across internal membranes that form diverse compartments called organelles. Understand the structure and function of a cell's membranes, and you will gain insight into survival at life's most fundamental level.

Figure 3.1 (**a**) Goose barnacles, which live attached to logs and other floating objects in the seas. The cells of these marine animals are vulnerable to changes in salt concentration. Such changes occur when the barnacles accidentally end up in glacial meltwaters. (**b**) The dark blue seawater in this photograph is quite salty. The lighter blue water flowing out from the glacier in the background is much lower in salts.

a

b

1. Cells are the smallest units that still retain the characteristics of life, including complex organization, metabolic activity, and reproductive behavior.

2. All cells have an outermost, plasma membrane that keeps their interior distinct from the surroundings. They have a region of DNA. They have cytoplasm, a region that is structurally and functionally organized for energy conversions, protein synthesis, cell movements, and other activities necessary for survival.

3. Eukaryotic cells contain organelles, such as a nucleus. These are compartments inside the cell, bounded by internal membranes. Organelles separate different metabolic reactions and allow them to proceed in orderly fashion. Prokaryotic cells (bacteria) do not have comparable organelles.

4. The three main regions of eukaryotic cells are arranged in this general fashion:

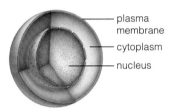

— plasma membrane
— cytoplasm
— nucleus

5. Cell membranes are mostly phospholipids and proteins. The phospholipids form a double layer. This "bilayer" gives membranes their structure and serves as a barrier to water-soluble substances. The proteins carry out most membrane functions, as when they actively or passively transport substances into and out of the cell.

THE NATURE OF CELLS

Basic Cell Features

Cells differ greatly in size, shape, and activities, as you might gather by comparing a bacterium with one of your liver cells. Yet they are alike in three respects. All cells start out life with a plasma membrane, a region of DNA, and a region of cytoplasm.

The **plasma membrane** is an outermost membrane that allows metabolic events to proceed inside the cell in organized, controlled ways, apart from the environment. It does not *isolate* the interior; many substances and signals can move across the membrane. Part of the cell interior is set aside as a region that houses DNA, along with molecules that can copy and read DNA's

hereditary instructions. The **cytoplasm** is everything enclosed by the plasma membrane, *except* for the region of DNA. It consists of a semifluid substance in which particles, filaments, and other internal cell structures are organized.

This chapter introduces two fundamentally different kinds of cells. **Eukaryotic cells** contain organelles (sacs and other compartments formed by internal membranes). The most conspicuous organelle, the **nucleus**, contains the DNA. **Prokaryotic cells** have no nucleus; the DNA simply occupies an irregularly shaped region. Bacteria are the only prokaryotic cells. Outside the realm of bacteria, all other organisms—from amoebas to peach trees and puffball mushrooms to zebras—are composed of eukaryotic cells.

Cell Size and Cell Shape

Can any cell be seen with the unaided human eye? There are a few, including the "yolks" of bird eggs, cells in the red part of a watermelon, and the fish eggs we call caviar. Generally, however, cells are too small to be observed without microscopes, as described in the *Focus* essay on the next page. To give you a sense of the sizes involved, one of your red blood cells is about 8 *millionths* of a meter in diameter. You could fit a string of about 2,000 of them across your thumbnail.

Why are most cells so small? There is a physical constraint on increases in cell size, called the "surface-to-volume ratio." By this relationship, volume increases with the cube of the diameter, but surface area increases only with the square. Simply put, *as a cell expands in diameter, its volume increases more rapidly than its surface area does.*

Suppose we figure out a way to make a round cell grow four times wider. Its volume increases sixty-four times (4^3), and its surface area increases sixteen times (4^2). Unlike fat-storing cells and chicken eggs (which are chockful of fat, food, and so on), our expanded cell is chockful of cytoplasmic machinery. Unfortunately, each unit of plasma membrane must serve four times as much cytoplasm as before! Past a certain point, the inward flow of nutrients and outward flow of wastes will not occur fast enough, and the cell will die.

A very large, round cell also would have trouble moving nutrients and wastes *through* the cytoplasm. By contrast, the random, tiny motions of molecules easily distribute substances through small or skinny cells. If a cell isn't small, it probably is long and thin or has outfoldings and infoldings that increase its surface relative to its volume. *The smaller or more stretched out or frilly-surfaced the cell, the more efficiently materials can cross its surface and become distributed through the interior.*

Microscopy and the Cell Theory

Emergence of the Cell Theory. Early in the seventeenth century, Galileo Galilei arranged two glass lenses in a cylinder. With this instrument, he looked at an insect and came to describe the stunning geometric patterns of its tiny eyes. Thus Galileo was the first to record a biological observation made through a microscope.

At mid-century, Robert Hooke looked at a thin slice of cork through his microscope and saw tiny compartments (Figure *a*). He gave them the name *cellulae* (meaning small rooms)—and this was the origin of the biological term "cell." Hooke actually was looking at walls of dead cells, which is what cork is made of, although he did not think of them as being dead because he did not know cells could be alive.

Given the simplicity of their instruments, it is amazing that the pioneers in microscopy saw as much as they did. Antony van Leeuwenhoek had great skill in lens construction and possibly the keenest vision. He even observed a bacterium—a type of cell so small it would not be seen again for another two centuries!

In the 1800s, improvements in lens design brought cells into sharper focus. Theodor Schwann and Matthias Schleiden concluded that all animal and plant tissues are composed of cells. Each cell, they said, develops as an independent unit even though its life is influenced by the whole plant or animal.

Rudolf Virchow studied cell reproduction and figured out where cells come from. These observations were distilled into the **cell theory**: *All organisms are composed of one or more cells, the cell is the basic living unit of organization for all organisms, and all cells arise from preexisting cells.*

Microscopy Today. As Figure *d* indicates, modern microscopes are astoundingly improved windows on hidden worlds, with some kinds even providing details of molecular structure. The micrographs in Figures *e* through *h* are examples of the kind of details they afford. (A *micrograph* is simply a photograph of an image formed with a microscope.) The specimen is a green alga, and in each case it is magnified 1,300 times.

The *compound light microscope* has glass lenses that bend incoming light rays to form an enlarged image of a specimen. If you wish to observe a *living* cell, it must be small or thin enough for light to pass through. Also, cell parts must differ in color and density from their surroundings—but most are nearly colorless and optically uniform in density. Cells are stained (exposed to dyes that react with some parts but not others), but staining can alter the parts and

a

b

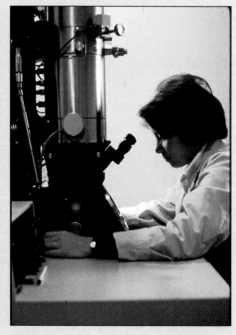

c

Microscopy then and now. (**a**) Robert Hooke's compound microscope and his drawing of dead cork cells. (**b**) Antony van Leeuwenhoek, microscope in hand. (**c**) An electron microscope in a modern laboratory.

kill the cell. Dead cells begin to break down at once, so they are pickled or preserved before staining.

No matter how good a glass lens system is, when the diameter of the object being viewed is magnified by 2,000 times or more, cell parts appear larger but are not clearer. The properties of wavelengths of visible light limit the resolution of smaller details.

Transmission electron microscopes have great magnifying power because they use magnetic lenses to focus electrons (which respond to magnetic force). Electrons are particles, but they also behave like waves—and their wavelengths are much shorter than those of visible light. Details of a cell's internal structure show up best with these microscopes.

With a *scanning electron microscope*, a narrow electron beam is directed back and forth across a specimen's surface, which has been coated with a thin metal layer. Electron energy triggers the emission of electrons in the metal. The emission patterns can be used to form an image. Scanning electron microscopy provides a three-dimensional view of surface features.

1 centimeter (cm) = 1/100 meter, or 0.4 inch	3 cm	chicken egg (the "yolk")
1 millimeter (mm) = 1/1,000 meter	1 mm —	frog egg, fish egg
1 micrometer (μm) = 1/1,000,000 meter	100 μm —	human egg
	10–100 —	typical plant cells
	10–30 —	typical animal cells
	2–10 —	chloroplast
	1–5 —	mitochondrion
	5 —	*Anabaena* (cyanobacterium)
	1 —	*Escherichia coli*
1 nanometer (nm) = 1/1,000,000,000 meter	100 nm —	large virus (HIV, influenza virus
	25 —	ribosome
	7–10 —	cell membrane (thickness)
	2 —	DNA double helix (diameter)
	0.1 —	hydrogen atom

UNAIDED HUMAN EYE — LIGHT MICROSCOPES — ELECTRON MICROSCOPES

$$1 \text{ meter} = 10^2 \text{ cm} = 10^3 \text{ mm} = 10^6 \text{ μm} = 10^9 \text{ nm}$$

d Units of measure used in microscopy. The micrometer is used in describing whole cells or large cell structures. The nanometer is used in describing smaller cell structures and large organic molecules.

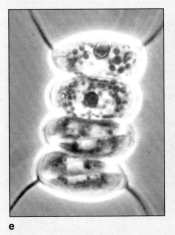  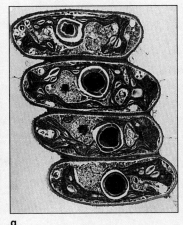 

e f g h 10μm

How different types of microscopes reveal different aspects of the same organism—in this case, a green alga (*Scenedesmus*). All four specimens used in this comparison are at the same magnification. (**e,f**) Phase-contrast and Nomarski microscopy create optical contrasts without staining the cells; both have enhanced the usefulness of light microscopes. (**g**) Details of a cell's internal structure show up best with transmission electron microscopy. (**h**) Scanning electron microscopy provides a three-dimensional view of surface features.

As on other micrographs in this book, the short bar provides a reference for size. Each micrometer (μm) is only 1/1,000,000 of a meter.

CELL MEMBRANES

Membrane Structure and Function

A cell's organization and activities depend on membranes. The main components of cell membranes are phospholipids and proteins. A phospholipid, recall, has a hydrophilic (water-loving) head and two fatty acid tails, which are hydrophobic (water-dreading). We can use the following simplified diagram to represent a phospholipid:

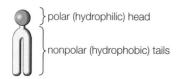

Consider what happens when phospholipids are immersed in water. Hydrophobic interactions may force the molecules to cluster together into two layers, with all the fatty acid tails sandwiched between the hydrophilic heads. This arrangement, a **lipid bilayer**, is the structural basis of all cell membranes:

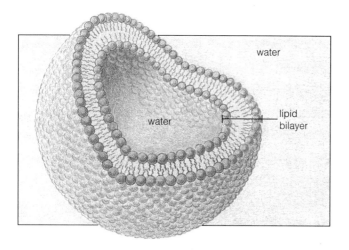

Figure 3.2 is a micrograph of a plasma membrane, which consists largely of a lipid bilayer and proteins. Figure 3.3 shows the **fluid mosaic model** of this type of membrane's structure. The bilayer represents the "fluid" aspect of the model, for its lipid molecules show quite a bit of movement. They move sideways and their tails flex back and forth, so neighboring lipids cannot become packed into a solid layer. Lipids with short tails and unsaturated (kinked) tails also disrupt the packing. The membrane is said to have a "mosaic" quality because it is a composite of lipids and proteins. The proteins are embedded in the membrane or positioned at its surfaces. They carry out most membrane functions.

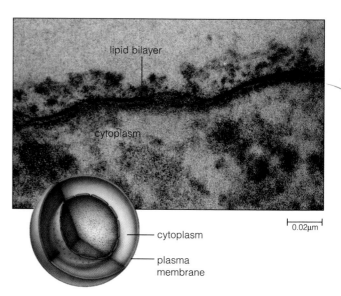

Figure 3.2 Micrograph of the plasma membrane of a eukaryotic cell, thin section.

Some membrane proteins are channels for water-soluble substances. Others are carriers of electrons. Still others are like pumps—they actively transport substances across the lipid bilayer with the help of energy from ATP. In multicelled organisms, certain proteins help cells recognize each other and stick together in a tissue. Receptor proteins latch onto hormones and other substances that trigger alterations in cell activities. For example, enzymes that crank up machinery for cell growth and division are switched on when the hormone somatotropin binds to cell receptors.

The features common to all cell membranes can be summarized this way:

1. Cell membranes are composed mostly of lipids (especially phospholipids) and proteins.

2. The lipid molecules have their hydrophilic heads at the two outer faces of a bilayer and their fatty acid tails sandwiched in between.

3. A lipid bilayer imparts structure to cell membranes and serves as a hydrophobic barrier between two solutions. (A plasma membrane separates the fluids inside and outside the cell. Internal cell membranes separate different fluids in the cytoplasm.)

4. Proteins embedded in the bilayer or positioned at its surfaces carry out most membrane functions.

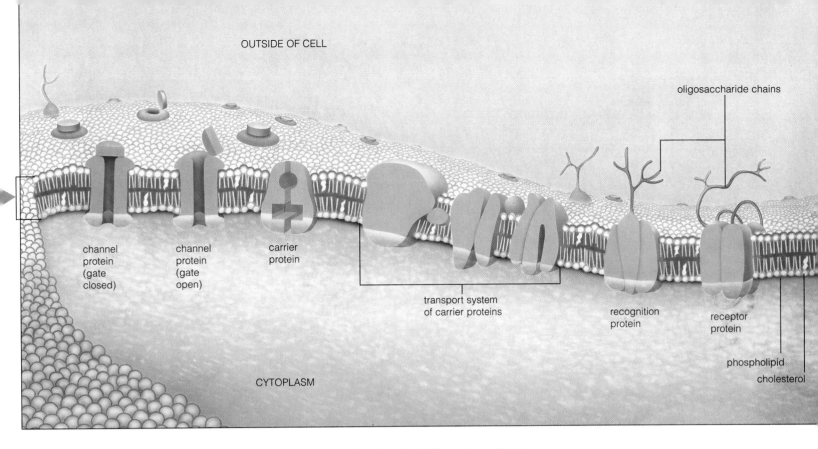

OUTSIDE OF CELL

oligosaccharide chains

channel protein (gate closed)

channel protein (gate open)

carrier protein

transport system of carrier proteins

recognition protein

receptor protein

phospholipid

cholesterol

CYTOPLASM

Figure 3.3 Generalized diagram of a plasma membrane based on the fluid mosaic model of membrane structure.

Diffusion

Cellular life depends on the energy inherent in molecules (or ions), which keeps them in constant motion. It depends also on **concentration gradients**. "Concentration" refers to the number of molecules of a substance in a specified volume of fluid. Add the word "gradient," and this means one region of the fluid contains more molecules than a neighboring region.

In the absence of other forces, molecules move down their concentration gradient. They do so because they constantly collide with one another, millions of times a second. Random collisions send the molecules back and forth, but the *net* movement is away from the place of greater concentration. When a gradient is steep, far more molecules move outward compared to the number moving in. As the gradient decreases, the difference in the number of molecules moving one way or the other becomes less pronounced. When the gradient is gone, individual molecules are still in motion. But the total number moving one way or the other is the same.

The net movement of like molecules down a concentration gradient is called **diffusion**. Diffusion is a key factor in the movement of substances across cell membranes and through fluid parts of the cytoplasm.

Suppose more than one substance is present in the same fluid. This makes no difference. Each substance still diffuses in some direction according to its *own* con-

centration gradient. Put a drop of dye at one side of a bowl of water. Dye molecules diffuse in one direction—to the region where they are less concentrated. And water molecules move in the opposite direction—to the region where *they* are less concentrated (Figure 3.4).

Diffusion is the net movement of like molecules or ions down their concentration gradient. They show a net outward movement from a region where they are most concentrated to a neighboring region where they are less concentrated.

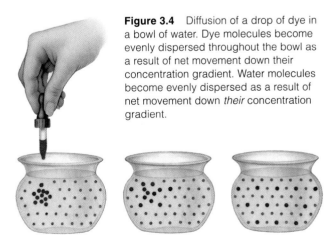

Figure 3.4 Diffusion of a drop of dye in a bowl of water. Dye molecules become evenly dispersed throughout the bowl as a result of net movement down their concentration gradient. Water molecules become evenly dispersed as a result of net movement down *their* concentration gradient.

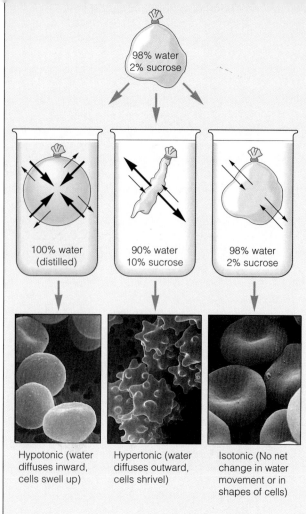

Hypotonic (water diffuses inward, cells swell up)

Hypertonic (water diffuses outward, cells shrivel)

Isotonic (No net change in water movement or in shapes of cells)

Figure 3.5 Osmosis in different environments. In the sketches, membranous bags through which water but not sucrose can move are placed in hypotonic, hypertonic, and isotonic solutions in three containers. In each container, arrow width represents the relative amount of water movement. The sketches explain what happens to red blood cells that are placed in comparable solutions. Red blood cells have no mechanisms to actively take in or expel water.

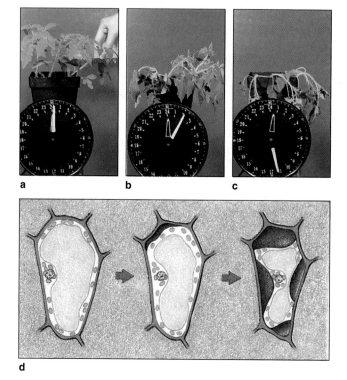

Figure 3.6 Osmosis and plant cells. Soil water typically has fewer solutes than plant cells do and moves into plants by osmosis. When soil is dry, soft plant parts lose water and wilt, and their solutes become more concentrated. In one experiment, 10 grams of table salt dissolved in water was added to soil in a pot containing tomato plants (**a**). Wilting began after about five minutes (**b**) and became severe in less than thirty minutes (**c**).

When soil water is dilute, water moves into plant cells and fluid pressure against the cell walls increases. When pressure is great enough to counter the effects of solute concentrations in the cells, some water molecules are squeezed out even as others are moving in. When the outward and inward movements are equal, internal pressure is constant and soft plant parts stay erect. When soil water is not dilute, the cells lose water, internal fluid pressure drops, and cell walls collapse (**d**).

Water Concentration and Cell Membranes

Some small molecules readily diffuse through the lipid bilayer of a cell membrane or through open channels across it. Glucose and other molecules cannot do this; membrane proteins must assist them across. Because the membrane shows this selective permeability, the concentrations of dissolved substances (that is, solutes) can increase on one side of the membrane and not the other. The resulting solute concentration gradients affect the movement of water into and out of cells.

Water by itself cannot get more or less concentrated. (Hydrogen bonds keep water molecules from crowding closer together or drifting apart.) Water becomes "less concentrated" only when solutes are dissolved in it.

Said another way, *a water concentration gradient is influenced by the number of molecules of all the solutes present on both sides of the membrane.*

When solute concentrations are *equal* on both sides of a cell membrane, there is no net movement of water in either direction. Fluids on both sides are said to be isotonic. When solute concentrations are not equal, one fluid is hypotonic (has *fewer* solutes) and the other is hypertonic (has *more* solutes). Water tends to move from a hypotonic fluid to a hypertonic one. If cells did not have mechanisms for adjusting to such differences, they would shrivel or burst, as Figure 3.5 illustrates.

Fluid pressure can influence the movement of water across membranes. For example, water moving into a plant cell exerts internal pressure on the cell wall. When the pressure increases, it pushes water back out across the membrane. When outward and inward water movements are equal, plant cells stay plump (Figure 3.6).

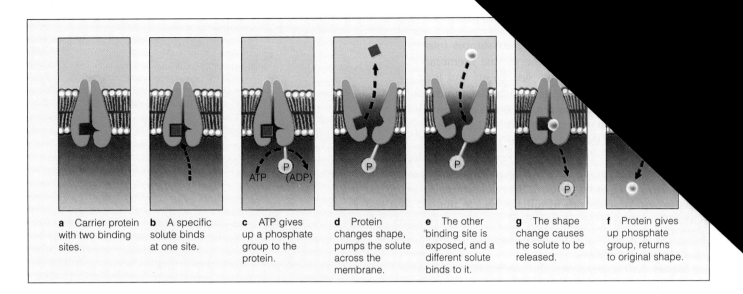

a Carrier protein with two binding sites.

b A specific solute binds at one site.

c ATP gives up a phosphate group to the protein.

d Protein changes shape, pumps the solute across the membrane.

e The other binding site is exposed, and a different solute binds to it.

g The shape change causes the solute to be released.

f Protein gives up phosphate group, returns to original shape.

Figure 3.7 Active transport across animal cell membranes. In this example, transport of one kind of solute is coupled with transport of another kind in the opposite direction. When a carrier protein receives an energy boost from ATP, its shape changes in ways that pump the solutes across the membrane.

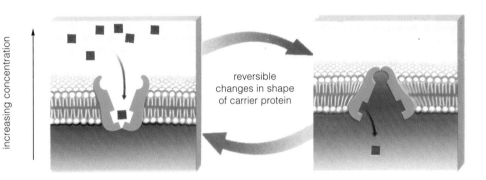

reversible changes in shape of carrier protein

Figure 3.8 Passive transport across a cell membrane. In this example, reversible changes in a membrane protein's shape allow a solute to diffuse through the protein's interior. Changes occur when a solute binds to the protein. In one state, the binding sites face the fluid outside the membrane. In the other state, binding sites face the cytoplasm. The transition from one state to another does not require an energy input. It simply depends on the direction of the solute concentration gradient across the membrane.

Osmosis is the special name for the movement of water across membranes in response to concentration gradients, fluid pressure, or both.

Osmosis is the movement of water across a selectively permeable membrane in response to concentration gradients, fluid pressure, or both.

Available Routes Across Membranes

We turn now to the manner in which solutes move across a cell membrane. Oxygen, carbon dioxide, and other small molecules with no net charge readily diffuse across the membrane's lipid bilayer. But glucose and other large, water-soluble molecules with no net charge almost never diffuse freely across the bilayer. Neither do positive or negative ions. These substances must cross the membrane through the interior of proteins embedded in the bilayer. Certain proteins get energetic about this, and others don't.

In **active transport**, a membrane protein undergoes a series of changes in shape after receiving an energy boost (from ATP, most often). The changes cause it to *pump* specific solutes through its interior and across the membrane (Figure 3.7). In **passive transport**, a protein serves as a channel or carrier for a specific solute without being booted into doing so. The protein simply allows ions or molecules of a substance to diffuse through it, down their concentration gradient. Few, if any, of those substances would cross the membrane without the protein, however passive its role. Figure 3.8 shows an example of passive transport.

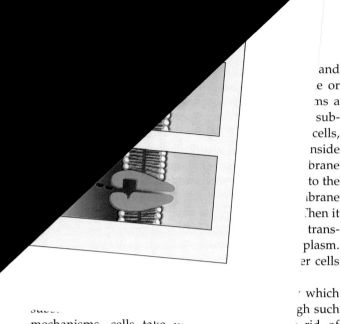

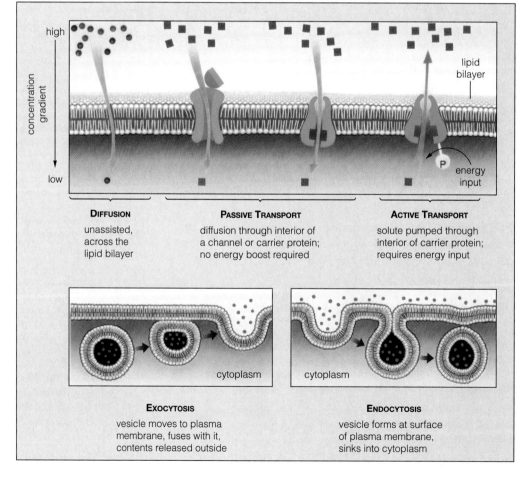

Figure 3.9 Example of endocytosis. Phagocytic cells such as amoebas "eat" other cells by means of endocytosis. (*Phagocyte* means cell eater.) An engulfed cell is digested when lysosomes (sacs of digestive enzymes) fuse with the vesicle that carries it into the cytoplasm.

vesicle

edible cell

amoeba

and
e or
ns a
sub-
cells,
nside
brane
to the
ibrane
Then it
trans-
plasm.
er cells

which
gh such
mechanisms, cells take in mater... rid of wastes, at controlled rates. Transport mechanisms also help maintain cellular volume and pH.

Figure 3.10 Summary of mechanisms by which substances move across cell membranes. Notice that exocytosis and endocytosis occur at the plasma membrane only, not at internal cell membranes.

high

concentration gradient

low

lipid bilayer

P energy input

DIFFUSION
unassisted, across the lipid bilayer

PASSIVE TRANSPORT
diffusion through interior of a channel or carrier protein; no energy boost required

ACTIVE TRANSPORT
solute pumped through interior of carrier protein; requires energy input

cytoplasm

cytoplasm

EXOCYTOSIS
vesicle moves to plasma membrane, fuses with it, contents released outside

ENDOCYTOSIS
vesicle forms at surface of plasma membrane, sinks into cytoplasm

PROKARYOTIC CELLS—
THE BACTERIA

We turn now to specific cell types, starting with bacteria. All bacterial cells are prokaryotic—their DNA is not enclosed in a nucleus. *Prokaryotic* means "before the nucleus." The word implies that bacteria existed on earth before the nucleus evolved in the forerunners of all other cells.

Bacteria are the smallest cells and, in structural terms, the simplest to think about. Most have a rigid or semirigid **cell wall** surrounding the plasma membrane (Figure 3.11). The wall supports the cell and imparts shape to it. The plasma membrane controls the movement of substances into and out of the cytoplasm. It also has built-in machinery for metabolic reactions.

Bacterial cells have a small volume of cytoplasm, with many ribosomes dispersed through it. A **ribosome** consists of two subunits, each composed of RNA and protein molecules. *In all cells, not just bacteria, proteins are synthesized at ribosomes located in the cytoplasm.* At the ribosomal surface, enzymes speed the assembly of polypeptide chains. Each new protein consists of one or more of those chains (page 32).

As a group, bacteria are the most metabolically diverse organisms. They have managed to exploit energy and raw materials in just about every kind of environment. Besides this, ancient members of their kingdom gave rise to all the protistans, plants, fungi, and animals ever to appear on earth. The evolution, structure, and functioning of bacteria are topics that will be addressed later in the book. Here our focus will be on nucleated cells, the eukaryotes.

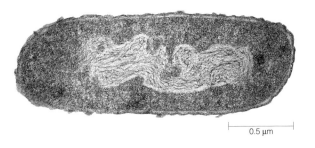

0.5 µm

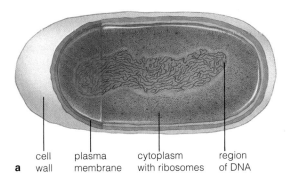

a cell   plasma   cytoplasm   region
   wall   membrane with ribosomes of DNA

Figure 3.11 Prokaryotic body plans. (**a**) Micrograph and sketch of a common bacterium, *Escherichia coli*. (**b**) This *E. coli* cell has been shocked into releasing its circular molecule of DNA.

Your own gut is home to a large population of a normally harmless strain of *E. coli*. In 1993, a dangerous strain contaminated meat that was sold to some fast-food restaurants. The same strain also contaminated hard apple cider sold at a few roadside stands. Cooking the meat thoroughly (or boiling the cider) would have killed the bacterial cells. Where this was not done, people who ate the meat or drank the cider became quite sick, and some died.

Interesting variations exist on the basic body plan. (**c**) Cells of assorted species are shaped like rods, corkscrews, or balls. The ball-shaped cells of *Nostoc*, a photosynthetic bacterium, stick together inside a thick, gelatin-like sheath. (**d**) Like the *Pseudomonas marginalis* cell shown here, many species have surface appendages such as bacterial flagella, which propel the cell through fluid environments. Chapter 17 gives other splendid examples.

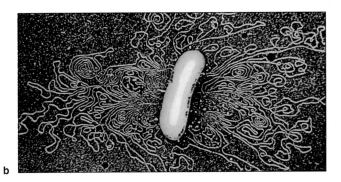

b

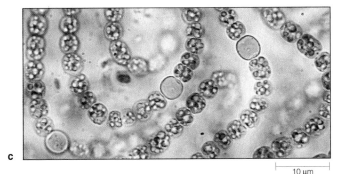

c

10 µm

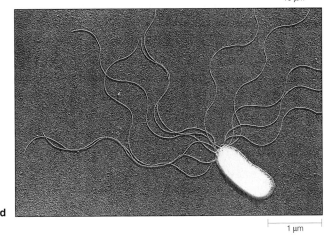

d

1 µm

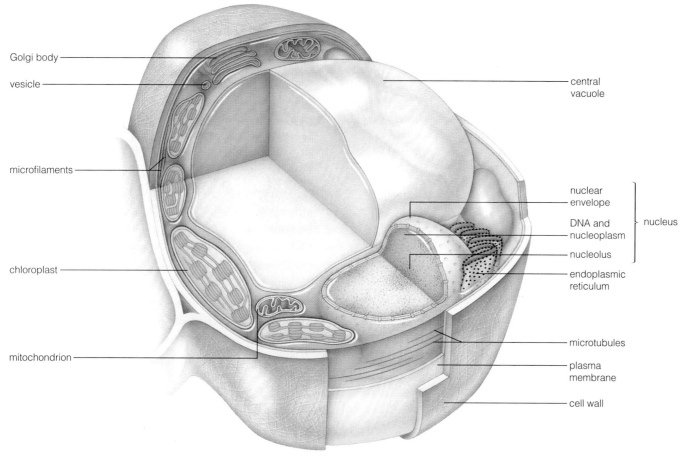

Golgi body

vesicle

microfilaments

chloroplast

mitochondrion

central vacuole

nuclear envelope

DNA and nucleoplasm

nucleolus

endoplasmic reticulum

microtubules

plasma membrane

cell wall

nucleus

Figure 3.12 Typical components of plant cells. This cutaway diagram corresponds roughly to the micrograph in Figure 3.13.

EUKARYOTIC CELLS

Organelles—The Hallmark of Eukaryotes

The nucleus is the hallmark of eukaryotic cells. (Hence the name *eukaryotic*, meaning "true nucleus.") It is one of many organelles that distinguish these cells from prokaryotes. An **organelle** is an internal, membrane-bounded sac or some other kind of compartment that has a specific metabolic function within a cell.

No human-built apparatus can match the eukaryotic cell for the sheer number of chemical activities that can proceed simultaneously in so small a space. For example, many metabolic reactions are incompatible with others. Think about a plant cell putting together a starch molecule by some reactions and breaking it down by others. The cell would gain nothing if the synthesis and breakdown reactions proceeded at the same time on the same starch molecule! Such reactions proceed smoothly, largely because organelle membranes keep them physically separated.

Organelle membranes also permit compatible, interconnected reactions to proceed at different times. Thus starch molecules are produced and stored by way of reaction sequences in one organelle—then later released for use in other reaction sequences in the same plant cell.

Organelles physically separate chemical reactions, many of which are incompatible, inside the cell.

Organelles separate different reactions in time, as when molecules are produced in one organelle, then used later in other reaction sequences.

Components of Plant Cells

Figure 3.12 can start you thinking about where organelles are located in a typical plant cell. Keep in mind that calling this cell "typical" is like calling a cactus or crocus a "typical" plant. As is true of animal cells, variations on the basic plan are mind-boggling. With this qualification in mind, also take a look at Figure 3.13. The sketches accompanying the micrograph outline the functions of organelles and other structures you are likely to find in the plant kingdom.

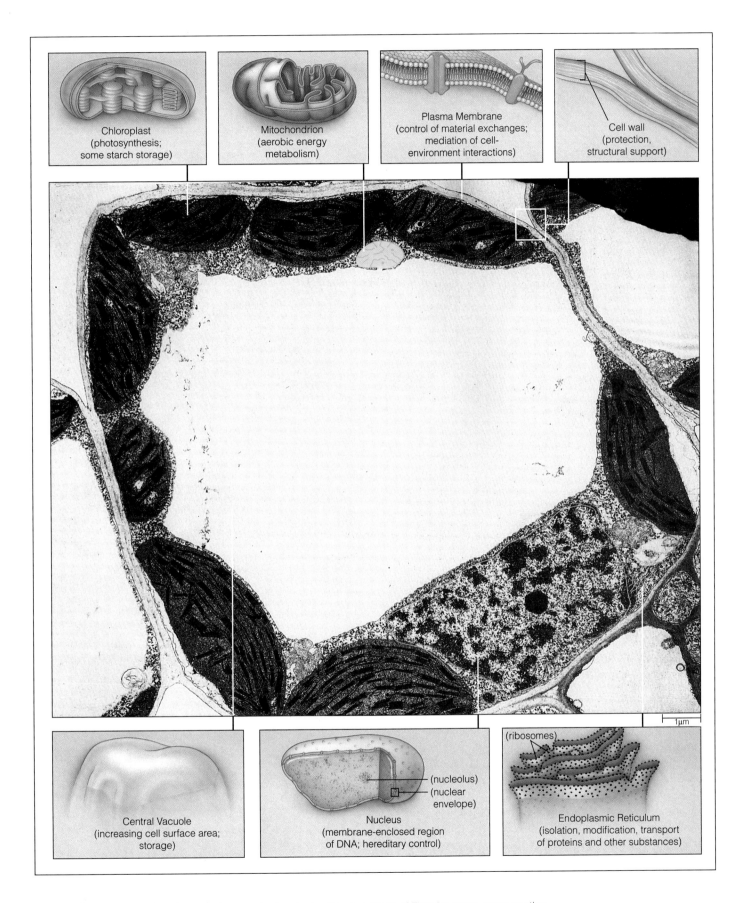

Chloroplast (photosynthesis; some starch storage)

Mitochondrion (aerobic energy metabolism)

Plasma Membrane (control of material exchanges; mediation of cell-environment interactions)

Cell wall (protection, structural support)

1μm

Central Vacuole (increasing cell surface area; storage)

Nucleus (membrane-enclosed region of DNA; hereditary control)

(nucleolus)
(nuclear envelope)

(ribosomes)

Endoplasmic Reticulum (isolation, modification, transport of proteins and other substances)

Figure 3.13 Transmission electron micrograph of a plant cell from a blade of Timothy grass, cross-section.

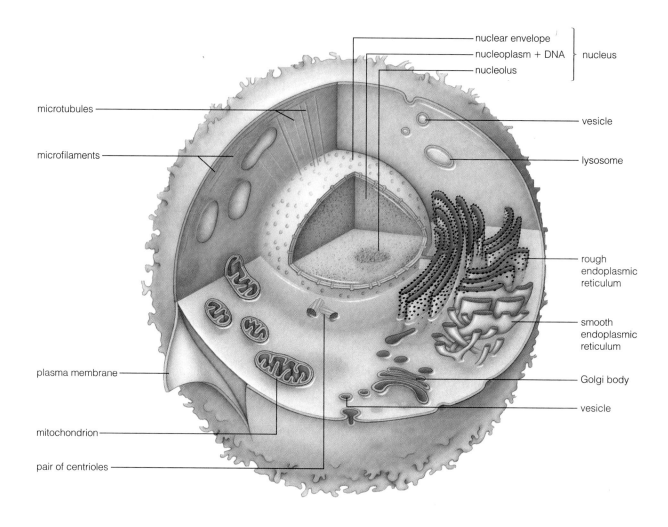

microtubules

microfilaments

nuclear envelope

nucleoplasm + DNA — nucleus

nucleolus

vesicle

lysosome

rough endoplasmic reticulum

smooth endoplasmic reticulum

Golgi body

vesicle

plasma membrane

mitochondrion

pair of centrioles

Figure 3.14 Typical components of animal cells. This cutaway diagram corresponds roughly to the micrograph in Figure 3.15.

Components of Animal Cells

Now take a look at the organelles of a typical animal cell, as shown in Figures 3.14 and 3.15. Right away, you can see that it is similar to the plant cell illustrated earlier, in that it has a nucleus, mitochondria, and the other components listed in Table 3.1. Such similarities point to basic functions that are necessary for survival and reproduction, regardless of the cell type. We will return to this concept throughout the book.

Comparisons of Figures 3.12 through 3.15 also give you an initial idea of the ways in which plant and animal cells are structurally different. For example, you won't ever see an animal cell surrounded by a cell wall. (You might see assorted fungal and protistan cells with one, however.) What other structural differences can you identify?

Table 3.1	Features Typical of Most Eukaryotic Cells
Nucleus	*Physical isolation and organization of DNA*
Ribosomes	*Synthesis of polypeptide chains*
Endoplasmic reticulum	*Initial modification of new polypeptide chains; lipid synthesis*
Golgi bodies	*Further modification of polypeptide chains into mature proteins; sorting, shipping of proteins and lipids for secretion or use in cell*
Diverse vesicles	*Transport or storage of substances; digestion inside cell; other functions*
Mitochondria	*ATP formation*
Cytoskeleton	*Cell movement, shape, and internal organization*

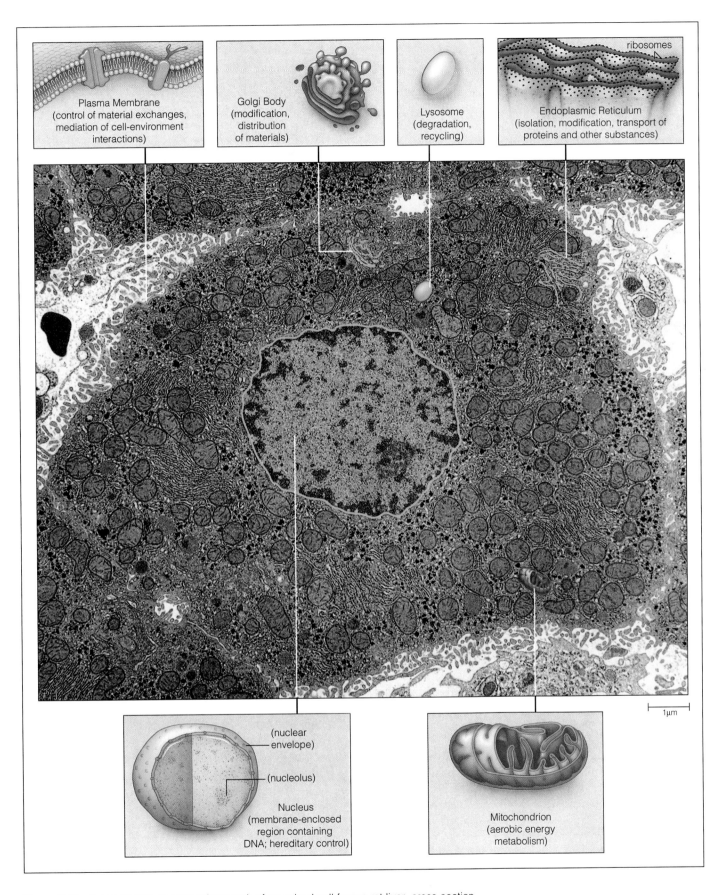

Plasma Membrane (control of material exchanges, mediation of cell-environment interactions)

Golgi Body (modification, distribution of materials)

Lysosome (degradation, recycling)

ribosomes

Endoplasmic Reticulum (isolation, modification, transport of proteins and other substances)

1μm

(nuclear envelope)

(nucleolus)

Nucleus (membrane-enclosed region containing DNA; hereditary control)

Mitochondrion (aerobic energy metabolism)

Figure 3.15   Transmission electron micrograph of an animal cell from a rat liver, cross-section.

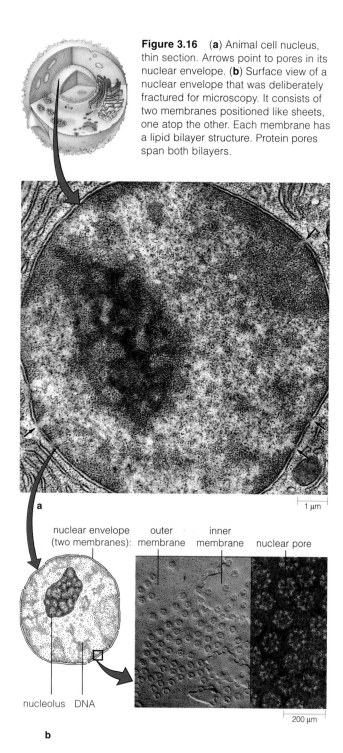

Figure 3.16 (**a**) Animal cell nucleus, thin section. Arrows point to pores in its nuclear envelope. (**b**) Surface view of a nuclear envelope that was deliberately fractured for microscopy. It consists of two membranes positioned like sheets, one atop the other. Each membrane has a lipid bilayer structure. Protein pores span both bilayers.

a

1 μm

nuclear envelope (two membranes):   outer membrane   inner membrane   nuclear pore

nucleolus   DNA

200 μm

b

Table 3.2	Components of the Nucleus
Nuclear envelope	*Double-membraned, pore-riddled boundary between cytoplasm and interior of nucleus*
Nucleolus	*Dense cluster of the RNA and proteins used to assemble ribosomal subunits*
Nucleoplasm	*Fluid portion of the nuclear interior*
Chromosomes	*DNA molecules and numerous proteins attached to them*

The Nucleus

There would be no cells whatsoever without complex carbohydrates, lipids, proteins, and nucleic acids. It takes a special class of proteins (enzymes) to build and use those molecules. Thus *cell structure and function begin with proteins—and instructions for building the proteins themselves are contained in DNA.*

In eukaryotic cells, DNA resides in the nucleus. Figure 3.16 shows the structure of a typical nucleus, and Table 3.2 lists its components. A nucleus serves two

functions. First, it helps control access to hereditary instructions contained in DNA. Second, it keeps DNA separated from all the substances and metabolic machinery in the cytoplasm. This makes it easier to package the DNA when the time comes for a cell to divide. The DNA can be sorted into parcels, one for each new cell that forms.

Nuclear Envelope. Imagine a golf ball sheathed in a double layer of Saran Wrap. The outermost part of the nucleus, the **nuclear envelope**, is something like that; it is two membranes thick. Each membrane is a lipid bilayer studded with various proteins. Pores composed of proteins span both bilayers (Figure 3.16*b*). The pores are passageways through which substances can move into and out of the nucleus in controlled ways.

Nucleolus. As a eukaryotic cell grows, one or more dense masses of irregular size and shape appear in the nucleus. Each mass is a **nucleolus** (plural, nucleoli). Nucleoli are sites where the protein and RNA subunits of ribosomes are assembled. The subunits are shipped from the nucleus into the cytoplasm. There they join together into intact, functional ribosomes. A nucleolus is visible in Figure 3.16*a*.

Chromosomes. Between cell divisions, eukaryotic DNA is threadlike, with many enzymes and other proteins attached to it like beads on a string. Except at extreme magnification, the beaded threads look grainy, as they do in Figure 3.16*a*. Prior to division, however, DNA molecules are duplicated (so each new cell will get a set of hereditary instructions). Before the molecules are sorted into two sets, they fold and twist into condensed structures, proteins and all.

Early microscopists named the grainy material *chromatin* and the condensed structures *chromosomes*. Today we call a DNA molecule and its associated proteins a **chromosome** regardless of whether it is in a threadlike (grainy) or condensed form. In chapters to come, keep in mind that "the chromosome" does not always look the same during the life of a cell.

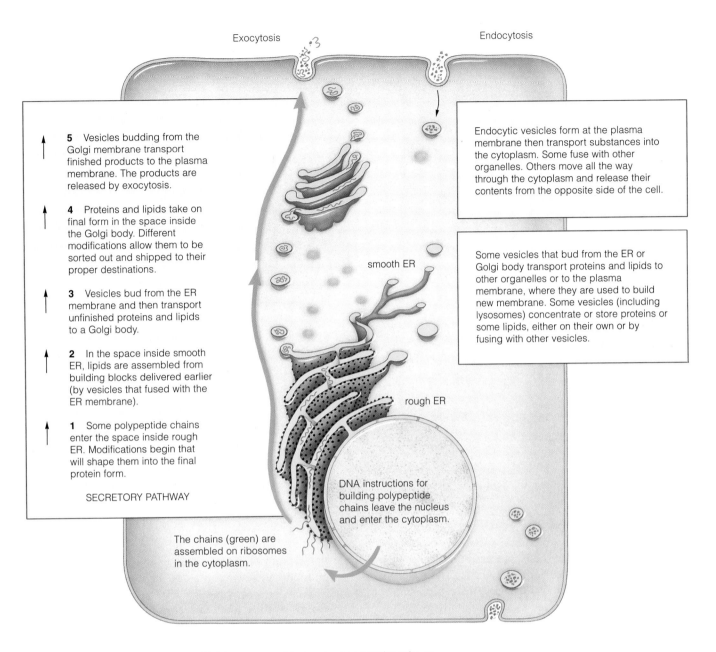

Exocytosis   Endocytosis

5 Vesicles budding from the Golgi membrane transport finished products to the plasma membrane. The products are released by exocytosis.

4 Proteins and lipids take on final form in the space inside the Golgi body. Different modifications allow them to be sorted out and shipped to their proper destinations.

3 Vesicles bud from the ER membrane and then transport unfinished proteins and lipids to a Golgi body.

2 In the space inside smooth ER, lipids are assembled from building blocks delivered earlier (by vesicles that fused with the ER membrane).

1 Some polypeptide chains enter the space inside rough ER. Modifications begin that will shape them into the final protein form.

SECRETORY PATHWAY

Endocytic vesicles form at the plasma membrane then transport substances into the cytoplasm. Some fuse with other organelles. Others move all the way through the cytoplasm and release their contents from the opposite side of the cell.

Some vesicles that bud from the ER or Golgi body transport proteins and lipids to other organelles or to the plasma membrane, where they are used to build new membrane. Some vesicles (including lysosomes) concentrate or store proteins or some lipids, either on their own or by fusing with other vesicles.

smooth ER

rough ER

DNA instructions for building polypeptide chains leave the nucleus and enter the cytoplasm.

The chains (green) are assembled on ribosomes in the cytoplasm.

Figure 3.17 Cytomembrane system. Lipids are assembled and many proteins take on their final form in the system. A cell uses the various lipids and proteins or secretes them.

Cytomembrane System

Polypeptide chains are assembled in the cytoplasm, on many thousands of ribosomes. What happens to the new chains? Many become stockpiled in the cytoplasm. Others pass through a series of organelles called the **cytomembrane system**. These include the endoplasmic reticulum, Golgi bodies, and certain vesicles. In this system, lipids as well as proteins take on final form, then become packaged in vesicles. Some vesicles deliver proteins and lipids to regions where new membrane must be built. Others accumulate and store proteins or lipids for specific uses. Still other vesicles are part of a "secretory pathway." They move to the plasma membrane, fuse with it, and release their contents to the surroundings. Figure 3.17 shows where the secretory pathway fits in the system.

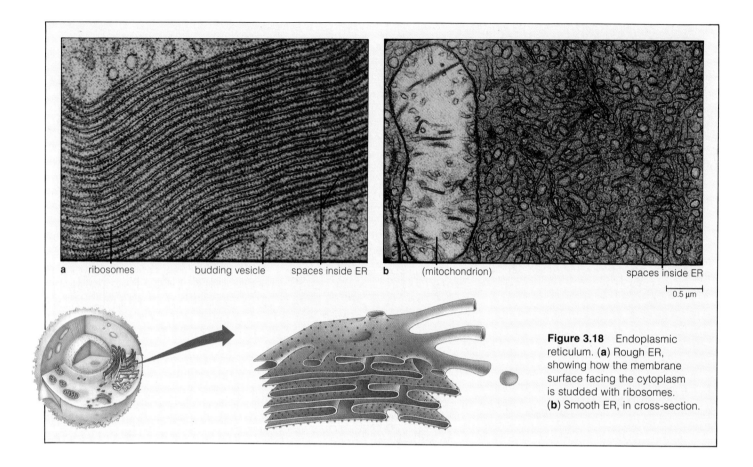

a ribosomes   budding vesicle   spaces inside ER   b (mitochondrion)   spaces inside ER

0.5 µm

Figure 3.18 Endoplasmic reticulum. (**a**) Rough ER, showing how the membrane surface facing the cytoplasm is studded with ribosomes. (**b**) Smooth ER, in cross-section.

Endoplasmic Reticulum.   In animal cells, **endoplasmic reticulum**, or **ER**, is a membrane that begins at the nucleus and curves through the cytoplasm. It has rough and smooth regions, due largely to the presence or absence of ribosomes on the side facing the cytoplasm.

Rough ER often is arranged as stacked, flattened sacs and has many ribosomes attached (Figure 3.18*a*). Polypeptide chains enter spaces inside rough ER. There some acquire side chains (oligosaccharides, mostly). Rough ER is abundant in cells that specialize in secreting digestive enzymes and other proteins.

Smooth ER is free of ribosomes and curves through the cytoplasm like connecting pipes (Figure 3.18*b*). It is the main site of lipid synthesis in many cells. The smooth ER of liver cells inactivates certain drugs and harmful by-products of metabolism. Sarcoplasmic reticulum, a type of smooth ER in skeletal muscle cells, stores and releases calcium ions for muscle contraction.

Peroxisomes are among the vesicles that bud from ER membranes. Enzymes in these sacs break down fatty acids and amino acids. The reactions produce hydrogen peroxide, a potentially harmful substance. Before hydrogen peroxide can do harm, another enzyme con-verts it to water and oxygen or uses it to break down alcohol. If you drink alcoholic beverages, nearly half of the alcohol is degraded in peroxisomes of your liver and kidney cells.

Golgi Bodies and Lysosomes.   Outwardly, a **Golgi body** resembles a stack of pancakes (Figure 3.19). The "pancakes" actually are flattened, interconnected sacs in which lipids and proteins are modified. Specific modifications allow them to be sorted out and packaged in vesicles for transport to specific locations. For example, an enzyme in a Golgi vesicle may attach a phosphate group to a particular protein, and this serves as a mailing tag that routes the protein to its proper destination.

A vesicle forms when a portion of the topmost Golgi membrane bulges out, then breaks away. Among the vesicles are **lysosomes**, the cell's main organelles of digestion. Lysosomes fuse with other vesicles that contain cell substances or foreign particles brought into the cell by endocytosis (Figure 3.17). Different enzymes in lysosomes can break down virtually all proteins, polysaccharides, nucleic acids, and some lipids.

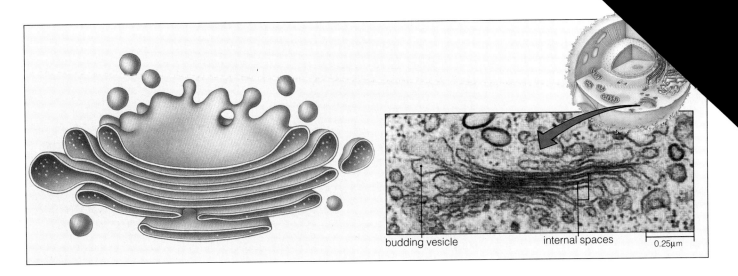

internal spaces

0.25µm

Figure 3.19 Sketch and micrograph of a Golgi body.

Mitochondria

The energy of ATP drives nearly all cell activities. **Mitochondria** (singular, mitochondrion) liberate energy stored in glucose and other substances, then use it to form *many* ATP molecules. Metabolic machinery in these organelles uses oxygen to extract far more energy from glucose than can be done by any other means. (When you breathe in, you are taking in oxygen for mitochondria.) Some cells have a sprinkling of these organelles; others (including energy-demanding muscle cells) may have more than a thousand.

Each mitochondrion has an outer membrane facing the cytoplasm. It also has an inner membrane, usually with many deep, inward folds called cristae (Figure 3.20). The double-membrane system creates two compartments that are used in ATP formation.

In size and biochemistry, mitochondria resemble bacteria. They have their own DNA and some ribosomes. They even divide on their own. Possibly they evolved from ancient bacteria that were engulfed by a predatory, amoebalike cell, yet managed to escape digestion (page 256). Perhaps they were able to reproduce inside the cell and its descendants. If they became permanent, protected residents, they may have lost many structures and functions required for independent life while becoming mitochondria.

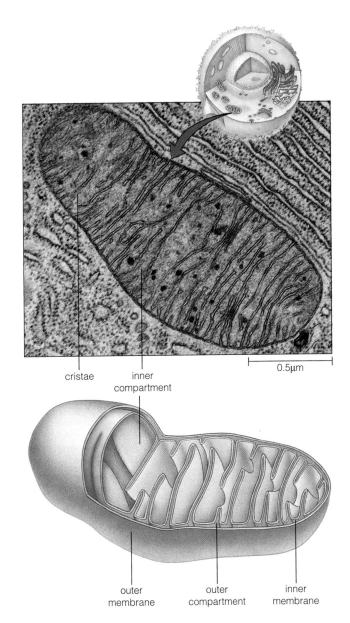

0.5µm

cristae

inner compartment

outer membrane

outer compartment

inner membrane

Figure 3.20 Micrograph and sketch of a mitochondrion.

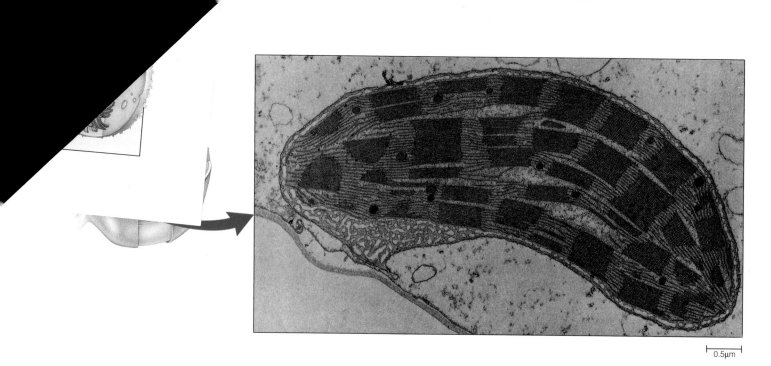

0.5μm

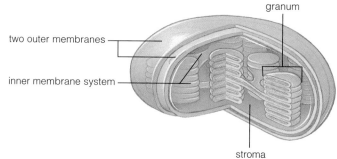

granum

two outer membranes

inner membrane system

stroma

Figure 3.21 Micrograph and generalized structure of a chloroplast. Enzymes and other molecules embedded in the membrane system inside the chloroplast take part in ATP formation. The ATP is used to form sugars and other products in the stroma, a semifluid substance surrounding the inner membrane.

Specialized Plant Organelles

Chloroplasts. The **chloroplast** functions in photosynthesis. This organelle resembles certain photosynthetic bacteria and, like mitochondria, may be derived from ancient bacterial ancestors.

A chloroplast has two membranes, one atop the other, that wrap around a semifluid substance (the stroma). Within the stroma, *another* membrane forms an elaborate system of compartments that connect with one another (Figure 3.21). Commonly, many disk-shaped compartments stack together. The stacks are called grana (singular, granum). Pigments, enzymes, and other molecules of the membrane system trap sunlight energy and have roles in ATP formation.

Enzymes speed the assembly of sugars, starch, and other products of photosynthesis in the stroma. Clusters of new starch molecules (starch grains) may be temporarily stored within the chloroplast.

Chloroplasts often are oval or disk-shaped and may be green, yellow-green, or golden brown. Their color depends on the kinds and amounts of light-absorbing pigment molecules in their membranes. Chlorophyll, a green pigment, is the most abundant type in nature.

Central Vacuoles. Mature, living plant cells often have a fluid-filled **central vacuole** (see Figure 3.12). It usually occupies 50 to 90 percent of the cell's interior, leaving room for only a narrow zone of cytoplasm beneath the plasma membrane. A central vacuole stores amino acids, sugars, ions, and toxic wastes. It also increases cell size and surface area. During growth, fluid pressure builds up in the vacuole and forces the still pliable cell wall to enlarge. The cell itself enlarges under this force, and its increased surface area enhances the rate at which it absorbs minerals.

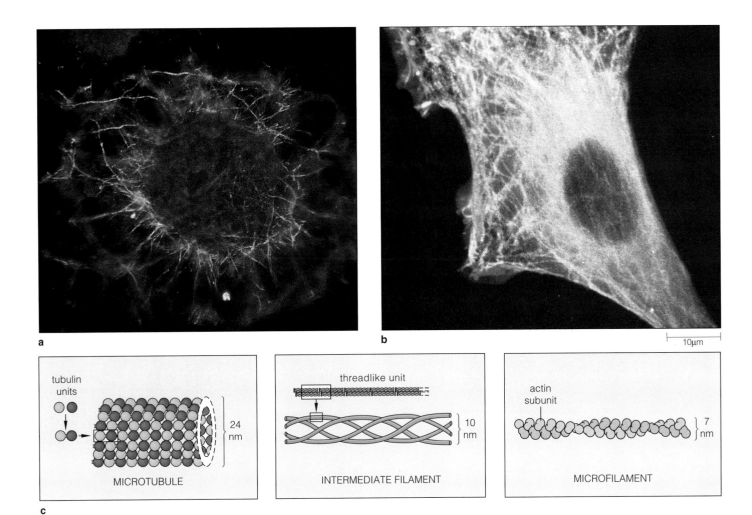

c

The Cytoskeleton

An interconnected system of bundled fibers, slender threads, and lattices extends from the nucleus to the plasma membrane. This is the **cytoskeleton**, which gives eukaryotic cells their shape and internal organization.

The key components of cytoskeletons are assembled from numerous protein subunits. **Microtubules**, the most predominant component, consist of tubulin subunits linked in parallel rows (Figure 3.22). In animal cells, you also will find **microfilaments**. Among these are the actin and myosin filaments that are the basis of muscle contraction. Also present are a variety of **intermediate filaments**, which consist of different proteins in different cell types.

Some parts of the cytoskeleton appear only at certain times in a cell's life. Before a cell divides, for instance, some fibers assemble into a "spindle" structure that moves chromosomes about, then the fibers disassemble when the task is done. Other parts of the cytoskeleton are permanent. Among them are the microtubules of flagella and cilia, described next.

Figure 3.22 Cytoskeleton. The components of the cytoplasm function in the internal organization, shape, and motion of cells. Its microtubules are present in all eukaryotic cells, where they serve in cell division and other vital tasks.

(**a**) Cytoskeleton of a plant cell (African blood lily). The cell took up molecules that had been labeled with fluorescent dyes. The selected molecules bound only to certain proteins and made them glow. The tubulin of microtubules stained green, revealing that these fibers extend from the nucleus (with its purple-stained chromosomes) to the plasma membrane.

(**b**) Cytoskeleton of a fibroblast, a type of cell that gives rise to certain tissues of animals. In this composite of three images, microtubules are green, and microfilaments are blue and red.

(**c**) Molecular structure of microtubules and a representative intermediate filament and microfilament from animal cells.

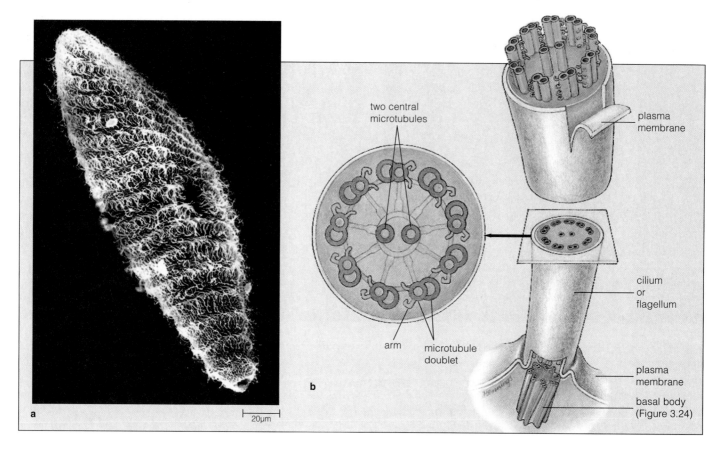

Figure 3.23 (**a**) Scanning electron micrograph of *Paramecium*, a single-celled protistan, showing its profusion of cilia. (**b**) The 9 + 2 array of microtubules in cilia and flagella.

Flagella and Cilia. Many free-living eukaryotic cells, including sperm, have tail-like motile structures called **flagella** (singular, flagellum). Other free-living cells have **cilia** (singular, cilium). Structurally, cilia are similar to flagella, but they are shorter and more abundant at the cell surface. Cells with fixed positions in tissues often have cilia that stir the surroundings. In your own body, cilia on thousands of cells whisk airborne bacteria and other particles out of your respiratory tract.

Flagella and cilia have the same internal organization. Nine pairs of microtubules ring two central microtubules, in a *9 + 2 array* (Figure 3.23).

What Organizes the Cytoskeleton? Each new eukaryotic cell develops a scaffold of microtubules that will dictate its shape. The scaffold's structure and orientation depend on small collections of proteins and other substances in the cytoplasm. Each collection is a microtubule organizing center (MTOC). In most animal cells, an MTOC also includes a pair of centrioles.

Centrioles are cylinders of triplet microtubules (Figure 3.24). Centrioles give rise to the microtubules of fla-

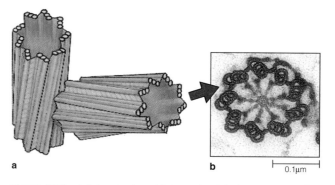

Figure 3.24 (**a**) A pair of centrioles, microtubular structures that give rise to cilia and flagella. (**b**) Electron micrograph of a basal body from a protistan (*Saccinobacculus*), thin section.

gella and cilia. Before either motile structure develops, a centriole becomes positioned near the plasma membrane and serves as the start point for microtubule assembly. After the motile structure is completed, the centriole remains attached to it as a basal body. Figure 3.23 shows this type of arrangement.

In multicelled organisms, the location and orientation of MTOCs influence the plane in which cells divide. During development, cells must divide at a prescribed angle relative to other cells. The successive division planes influence the shape of the developing embryo and the adult form.

Cell Surface Specializations

Single-celled eukaryotes interact directly with the environment. Like bacteria, many are supported or protected by a cell wall around the plasma membrane. Cell walls are porous, so water and solutes can move to and from the plasma membrane. In multicelled species, walls or other surface features allow adjacent cells to interact with one another and with their physical surroundings, as a few examples will illustrate.

Cell Walls in Plants. Cells of leafy plants stick together, wall to wall. In growing plant parts, bundles of cellulose strands form a *primary* cell wall. This wall is pliable, so the young cells enlarge under incoming water pressure.

Later, more layers may be deposited inside the first wall, forming a rigid, *secondary* wall that helps maintain the cell's shape (Figure 3.25). Deposits of pectin cement adjacent walls together (they also thicken jams and jellies). Other deposits, such as waxes, protect and reduce water loss from cells at the plant's surface. Numerous channels cross adjacent walls and connect the cytoplasm of neighboring cells. The number of channels affects how fast substances are transported through the plant.

Intercellular Material in Animals. Cell secretions and other material often intervene between cells of animal tissues. Think of the cartilage at the knobby ends of your leg bones. Cartilage consists of cells scattered in a "ground substance" (of modified polysaccharides) as well as protein fibers (of collagen or elastin). Through this material, nutrients and other substances diffuse from cell to cell. In mature bone and other tissues, intercellular material accounts for much of your body weight.

Cell Junctions in Animals. Three cell-to-cell junctions are common in animals. *Tight* junctions link the cells of epithelial tissues. (This type of tissue lines the body's outer surface, inner cavities, and organs.) The junctions form seals that keep molecules from freely crossing the tissue, as when they keep acids from leaking out of the stomach. *Adhering* junctions are like spot welds. They keep cells together in tissues of the skin, heart, and other organs that are subject to stretching. *Gap* junctions link the cytoplasm of adjacent cells; they are open channels for the rapid flow of signals and substances. We will return to these and other cell-to-cell interactions in later chapters.

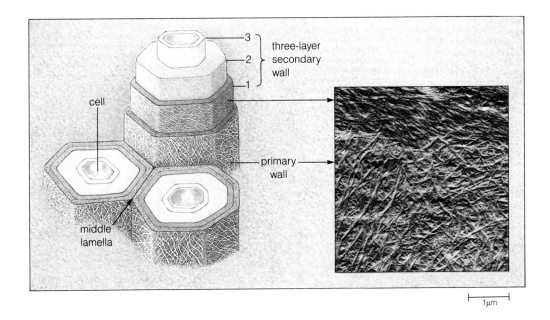

Figure 3.25 Sketch and micrograph of the primary and secondary walls of plant cells.

Table 3.3 Summary of Typical Components of Prokaryotic and Eukaryotic Cells

Cell Component	Function	Prokaryotic	Eukaryotic			
		Moneran	Protistan	Fungus	Plant	Animal
Cell wall	Protection, structural support	✓*	✓*	✓	✓	none
Plasma membrane	Control of substances moving into and out of cell	✓	✓	✓	✓	✓
Nucleus	Physical separation and organization of DNA	none	✓	✓	✓	✓
DNA	Encoding of hereditary information	✓	✓	✓	✓	✓
RNA	Transcription, translation of DNA messages into specific proteins	✓	✓	✓	✓	✓
Nucleolus	Assembly of ribosomal subunits	none	✓	✓	✓	✓
Ribosome	Protein synthesis	✓	✓	✓	✓	✓
Endoplasmic reticulum	Initial modification of many newly forming proteins; lipid synthesis	none	✓	✓	✓	✓
Golgi body	Final modification of proteins, lipids; sorting and packaging them for use inside cell or for export	none	✓	✓	✓	✓
Lysosome	Intracellular digestion	none	✓	✓*	✓*	✓
Mitochondrion	ATP formation	**	✓	✓	✓	✓
Photosynthetic pigment	Light–energy conversion	✓*	✓*	none	✓	none
Chloroplast	Photosynthesis, some starch storage	none	✓*	none	✓	none
Central vacuole	Increasing cell surface area, storage	none	none	✓*	✓	none
Cytoskeleton	Cell shape, internal organization, basis of cell motion	none	✓*	✓*	✓*	✓
Complex flagellum, cilium	Movement	none	✓*	✓*	✓*	✓

*Known to occur in at least some groups.
**Aerobic reactions do occur in many groups, but mitochondria are not involved.

SUMMARY

1. The cell theory has three key points: All living things are made of one or more cells. Each cell is a basic living unit (it lives independently or has the inherent capacity to do so). A new cell arises only from cells that already exist.

2. Each living cell has a plasma membrane surrounding an inner region called the cytoplasm. Eukaryotic cells also have internal membranes that compartmentalize the cytoplasm into different functional zones. The membranous compartments, which include the nucleus, are called organelles. Prokaryotic cells (bacteria) do not have comparable organelles. Table 3.3 summarizes the features of both cell types.

3. Cell membranes consist mostly of phospholipids and proteins. A lipid bilayer gives the membrane its structure and impermeability to water-soluble substances. Proteins in the bilayer or at one of its surfaces carry out most membrane functions.

4. Some membrane proteins transport substances across the membrane. Others are receptors for substances that trigger alterations in cell behavior. Others function in cell-to-cell recognition (they help the cells chemically recognize each other and stick together).

5. Substances move across cell membranes as follows:

   a. Diffusion: natural, unassisted movement of solutes from a region of higher to lower concentration.

   b. Osmosis: movement of water across a membrane in response to concentration gradients, pressure, or both.

   c. Passive transport: diffusion of a solute through the interior of a membrane protein that does not require an energy input to serve as a channel or carrier.

d. Active transport: pumping of solutes, with or against their concentration gradient, through the interior of a membrane protein that requires an energy boost to operate.

6. Oxygen, carbon dioxide, and other small molecules with no net charge diffuse across the lipid bilayer. Ions and large, water-soluble molecules such as glucose are actively or passively transported across through the interior of membrane proteins.

Review Questions

1. State the three key points of the cell theory. *38*

2. Why is it likely that you will never encounter a predatory two-ton living cell on the sidewalk? *37*

3. Suppose you want to observe the three-dimensional surface of an insect's eye. Would you benefit most from a compound light microscope, transmission electron microscope, or scanning electron microscope? *38–39*

4. Describe the three features that all cells have in common. Then, after reviewing Table 3.2, write a paragraph describing the key differences between prokaryotic and eukaryotic cells. *37, 58*

5. Which organelles are part of the cytomembrane system? Sketch their arrangement in an animal cell, from the nuclear envelope to the plasma membrane. *51*

6. Is the following statement true or false: Plant cells contain chloroplasts, but not mitochondria. *46, 47, 58*

7. Distinguish between these terms:
   a. diffusion; osmosis *41–43*
   b. passive transport; active transport *43*
   c. endocytosis; exocytosis *44*

Self-Quiz *(Answers in Appendix IV)*

1. The plasma membrane _____ .
   a. surrounds cytoplasm
   b. separates nucleus from cytoplasm
   c. acts as a nucleus in prokaryotic cells
   d. both a and b

2. Unlike eukaryotic cells, prokaryotic cells _____ .
   a. do not have a plasma membrane
   b. have RNA, not DNA
   c. do not have a nucleus
   d. all of the above

3. The _____ is responsible for cell shape, internal structural organization, and cell movement.
   a. flagellum   c. cytoskeleton
   b. cilium   d. both a and b

4. Cell membranes consist mainly of a _____ .
   a. carbohydrate bilayer and proteins
   b. protein bilayer and phospholipids
   c. phospholipid bilayer and proteins
   d. none of the above

5. Most membrane functions are carried out by _____ .
   a. proteins   c. nucleic acids
   b. phospholipids   d. hormones

6. When a cell is placed in a hypotonic solution, water tends to _____ .
   a. move into the cell
   b. move out of the cell
   c. show no net movement
   d. move by exocytosis

7. _____ can diffuse across a lipid bilayer.
   a. Glucose   c. Carbon dioxide
   b. Oxygen   d. Both b and c

8. Sodium ions diffuse across a membrane, through the interior of a membrane that has received an energy boost. This is an example of _____ .
   a. passive transport c. diffusion
   b. active transport   d. both a and c

9. Match each cell component with its function.
   ____ mitochondrion   a. synthesis of polypeptide chains
   ____ chloroplast   b. movement
   ____ ribosome   c. digestion in cell
   ____ smooth ER   d. initial modification of new
   ____ rough ER   polypeptide chains
   ____ Golgi body   e. modify, sort, ship proteins and
   ____ nucleolus   lipids
   ____ cytoskeleton   f. cell shape, organization,
   ____ lysosome   movement
                         g. photosynthesis
                         h. ATP formation
                         i. ribosome subunit assembly

Selected Key Terms

active transport *43*	diffusion *41*	microfilament *55*
cell theory *38*	endocytosis *44*	microtubule *55*
cell wall *45*	endoplasmic	mitochondrion *53*
central vacuole *54*	reticulum (ER) *52*	nuclear envelope *50*
centriole *56*	eukaryotic cell *37*	nucleolus *50*
chloroplast *54*	exocytosis *44*	nucleus *37*
chromosome *50*	flagellum *56*	organelle *46*
cilium *56*	fluid mosaic model *40*	osmosis *43*
concentration	Golgi body *52*	passive transport *43*
gradient *41*	intermediate	plasma membrane *37*
cytomembrane	filament *55*	prokaryotic cell *37*
system *51*	lipid bilayer *40*	ribosome *45*
cytoplasm *37*	lysosome *52*	vesicle *44*
cytoskeleton *55*		

Readings

Bretscher, M. October 1985. "The Molecules of the Cell Membrane." *Scientific American* 253(4):100–108.

Burgess, J., M. Marten, and R. Taylor. 1987. *Under the Microscope—A Hidden World Revealed.* Cambridge: Cambridge University Press. Paperback.

deDuve, C. 1985. *A Guided Tour of the Living Cell.* New York: Freeman. Two brief, beautifully illustrated volumes.

4 GROUND RULES OF METABOLISM

Growing Old With Molecular Mayhem

Somewhere in those slender strands of DNA in your cells are snippets of instructions for constructing two wonderful proteins. Those proteins go by the names superoxide dismutase and catalase. Both are enzymes—

a

b

they make metabolic reactions proceed much, much faster than they would on their own. And both help keep you from growing old before your time.

The two enzymes help your cells clean house, so to speak. Together, they produce and then hack up hydrogen peroxide (H_2O_2), a normal but toxic outcome of certain oxygen-requiring reactions. Oxygen (O_2) picks up electrons from those reactions, and sometimes it does not get quite as many as it is supposed to. So it takes on a negative charge (O_2^-).

Like other unbound, molecular fragments with the wrong number of electrons, O_2^- is a "free radical." Free radicals are *so* reactive, they even attach to molecules that usually do not take part in random reactions—molecules like DNA.

Enter superoxide dismutase. Under its prodding, two of the rogue oxygen molecules combine with hydrogen ions, forming H_2O_2 and O_2.

Enter catalase. Under *its* prodding, two molecules of hydrogen peroxide react and split into ordinary water and ordinary oxygen: $2H_2O_2 \longrightarrow 2H_2O + O_2$.

As people age, their capacity to produce functional proteins—including enzymes—begins to falter. Among those enzymes are superoxide dismutase and catalase. They are produced in diminishing numbers, crippled form, or both. Now free radicals and hydrogen peroxide can accumulate. Like loose cannons, they careen through cells with tiny blasts at the structural integrity of proteins, DNA, membranes, and other vital parts. Cells suffer or die outright. Those brown "age spots" on an older person's skin are evidence of assaults by free radicals (Figure 4.1). The spots are masses of brown pigment molecules that build up in cells when free radicals take over—all for the want of two enzymes.

With this chapter, we start to examine activities that keep cells alive and functioning smoothly. Sometimes the topics may seem remote from the world of your interests. But they help define who *you* are and who you will become, age spots and all.

Figure 4.1 (**a**) Owner of a good supply of functional enzymes that keep free radicals in check. (**b**) Owner of skin with age spots—visible evidence of free radicals on the loose.

1. Cells trap and use energy for building, stockpiling, breaking apart, and eliminating substances in ways that help them survive and reproduce. These activities are called metabolism.

2. Metabolism proceeds as long as cells acquire and use energy. To stay alive, then, cells must replace the energy they inevitably lose during each metabolic reaction. The sun is the original source of energy replacements through most of the biosphere.

3. Different metabolic pathways, operating in coordinated ways, help maintain, increase, or decrease the relative amounts of substances inside cells.

4. Enzymes increase the rate of specific reactions. They take part in nearly all metabolic pathways. So does ATP, an organic compound that transfers energy from one reaction site to another in cells.

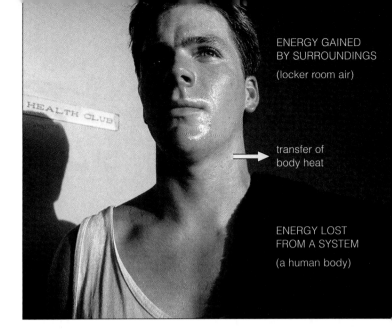

ENERGY GAINED
BY SURROUNDINGS

(locker room air)

transfer of
body heat

ENERGY LOST
FROM A SYSTEM

(a human body)

net energy change = 0

Figure 4.2 Example of how the total energy content of a system *together with its surroundings* remains constant. "System" means all matter within a specific region, such as a human body, a plant, a DNA molecule, or a galaxy. "Surroundings" can be a small region in contact with the system or an area as vast as the entire universe. The system shown here is giving off heat to the surroundings by evaporative water loss from sweat. What one loses, the other gains, so the total energy content of both doesn't change.

ENERGY AND LIFE

Find a light microscope somewhere and use it to peer down on a living cell suspended in a water droplet. The image is eerie—something that small is practically pulsating with movements. Even as you watch it, the cell takes in energy-rich solutes from the water. It busily builds membranes, stores things, checks out DNA, and replenishes pools of enzymes. It is alive; it is growing; it may divide in two. Multiply these proceedings by *65 trillion cells* and you have an idea of what goes on in your own body as you sit quietly, observing that single cell!

Metabolism is the somewhat dreary name for this dynamic activity. It refers to the cell's capacity to acquire energy and use it to build, store, break apart, and eliminate substances in controlled ways. Metabolism is the basis for survival and reproduction—and it all begins with energy.

Energy is a capacity to make things happen, to cause change, to do work. You use energy to wax a car, even to watch a movie. In both cases, cells of your muscles, brain, and other body parts are being put to work. Energy in muscle cells drives the contractions that are holding your body parts in various positions. Energy in brain cells allows them to chatter among themselves and guide what you are doing.

How Much Energy Is Available?

You, like cells, cannot create your own energy from scratch. You must get it from someplace else. That is the message of the **first law of thermodynamics:**

The total amount of energy in the universe remains constant. More energy cannot be created, and existing energy cannot be destroyed. It can only be converted from one form to another.

Consider what this law means. The universe has only so much energy, distributed in a variety of forms. One form can be converted to another, as when corn plants absorb sunlight energy and convert it to the chemical energy of starch. By eating corn, you can extract and convert energy in starch to other forms, such as mechanical energy for your movements. With each conversion, a little energy escapes to the surroundings as heat. (Your body steadily gives off about as much heat as a 100-watt light bulb because of ongoing conversions in your cells.) However, none of the energy vanishes. It just ends up someplace else, as suggested by Figure 4.2.

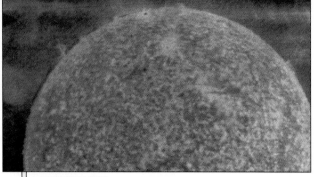

ENERGY LOST
one-way flow of energy away from the sun

ENERGY GAINED
one-way flow of energy from the sun into organisms

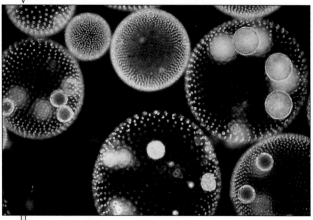

ENERGY LOST
one-way flow of energy from organisms
to the surroundings

Figure 4.3 The one-way flow of energy into the world of life that compensates for the one-way flow out of it. Living cells capture some of the energy being lost from the sun and convert it to useful forms. At each conversion, energy is released and lost to the surroundings, mostly as heat. Green "dots" in the lower photograph are photosynthetic cells (*Volvox*), organized in tiny, spherical colonies. Orange "dots" are cells set aside for reproduction. They use energy to form new colonies inside the parent sphere.

The One-Way Flow of Energy

For cells, energy concentrated in starch and other molecules is high quality, for it lends itself to conversions. A small amount of heat energy spread out in the air is low quality, because cells can't gather it up well and convert it to something they can use.

Bad news for cells of the remote future: The amount of low-quality energy in the universe is increasing. Why? No energy conversion can ever be 100 percent efficient. Some of the energy always goes off as heat. That is the point of the **second law of thermodynamics**:

The spontaneous direction of energy flow is from high-quality to low-quality forms. With each conversion, some energy is randomly dispersed in a form (usually heat) that is not as readily available to do work.

Without energy to maintain it, any organized system tends to become disorganized over time. **Entropy** is a measure of the degree of a system's disorder. Think about the Egyptian pyramids—originally organized, presently crumbling, and many thousands of years from now, dust. The *ultimate* destination of the pyramids and everything else in the universe is a state of maximum entropy. Billions of years from now, *all* of the energy available for conversions will be dissipated, and nothing we know of will pull it together again.

Can it be that life is one glorious pocket of resistance to the depressing flow toward maximum entropy? After all, every time a new organism grows, atoms become linked in precise arrays, and energy becomes more concentrated and organized, not less so! Yet a simple example will show that the second law does indeed apply to life on earth.

The primary energy source for life on earth is the sun, which is steadily losing energy. Plants capture some sunlight energy, convert it in various ways, then lose energy to other organisms that feed, directly or indirectly, on plants. At each energy transfer along the way, some energy is lost, usually as heat that joins the universal pool. Overall, energy still flows in one direction. The world of life maintains a high degree of organization only because it is being resupplied with energy lost from someplace else (Figure 4.3).

A steady flow of sunlight energy into the interconnected web of life compensates for the steady flow of energy leaving it.

THE NATURE OF METABOLISM

Energy Changes in Metabolic Reactions

Photosynthesis and aerobic respiration, the main pathways of energy flow in the world of life, are topics of the next two chapters. The reaction steps of these pathways will make more sense if you keep three concepts in mind.

First, most reactions are reversible. They proceed in the forward direction, from the starting substances (reactants) to products. They also proceed in reverse—from products back to reactants. The reaction in Figure 4.4 is reversible, as the two opposing arrows signify. *Second*, unless other events in the cell keep it from doing so, a reversible reaction approaches equilibrium. At equilibrium, the rates at which the forward and reverse reactions are proceeding are the same (Figure 4.5). The *amounts* of reactant and product molecules at equilibrium may or may not be the same. Depending on the substances, the forward reaction may be producing

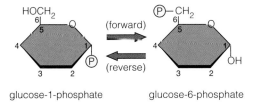

glucose-1-phosphate   glucose-6-phosphate

Figure 4.4 A reversible reaction. Glucose is primed to enter reactions when a phosphate group becomes attached to it. With high concentrations of glucose-1-phosphate, the reaction tends to run in the forward direction. With high concentrations of glucose-6-phosphate, it runs in reverse. (The 1 and 6 of these names simply identify the particular carbon atom of the glucose ring where phosphate is attached.)

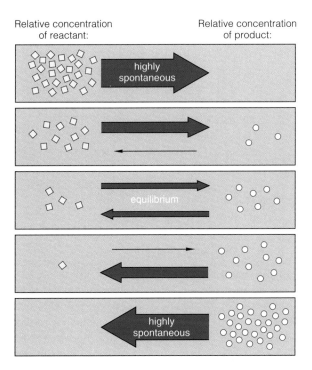

Figure 4.5 Chemical equilibrium. When concentrations of reactant molecules are high, reactions generally proceed most strongly in the forward direction. When concentrations of product molecules are high, reactions proceed most strongly in reverse. At equilibrium, the rates of the forward and reverse reactions are the same.

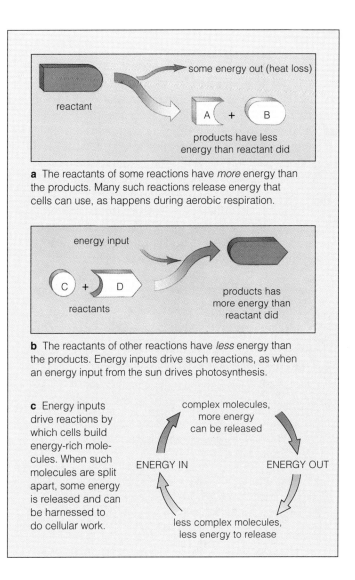

a The reactants of some reactions have *more* energy than the products. Many such reactions release energy that cells can use, as happens during aerobic respiration.

b The reactants of other reactions have *less* energy than the products. Energy inputs drive such reactions, as when an energy input from the sun drives photosynthesis.

c Energy inputs drive reactions by which cells build energy-rich molecules. When such molecules are split apart, some energy is released and can be harnessed to do cellular work.

Figure 4.6 Energy changes in metabolic reactions.

Directions of Metabolic Reactions

Cells never stop building and tearing down substances until they are dead. And they control the directions in which different reactions proceed. Why? Cells use only so many molecules of each substance at a given time, and they have only so much internal space to hold any excess. If they produce more than they use, put into storage, or secrete, the excess might cause problems.

Consider people affected by the genetic disorder *phenylketonuria* (PKU). A defective enzyme keeps their cells from using an amino acid, phenylalanine. In excess amounts, this amino acid enters reactions that produce phenylketones, and these build up to toxic levels. Within months, they damage the brain. Routine screening programs identify affected newborns, who can grow up symptom-free on a phenylalanine-restricted diet.

more or fewer molecules in the same amount of time. *Third*, as Figure 4.6 suggests, the products of a reaction may have more or less energy than the reactants. For instance, when glucose is broken down to carbon dioxide and water, some energy is released. The products have less energy but are more stable (it takes less energy to hold their atoms together).

Metabolic Pathways

All cells normally maintain, increase, and decrease the concentrations of substances by coordinating different metabolic pathways. A **metabolic pathway** is an orderly sequence of reactions, with specific enzymes acting at each step along the way. Pathways are linear or circular (Figure 4.7). Often they connect, with products of one pathway serving as reactants for others.

The main metabolic pathways are biosynthetic or degradative, overall. In *biosynthetic* pathways, small molecules are assembled into proteins, lipids, and other large molecules of higher energy. In *degradative* pathways, large molecules are broken down to products of lower energy. Both types have these participants:

---

1. **Substrates:** Substances able to enter a reaction; also called reactants or precursors.

2. **Intermediates:** Compounds formed between the start and the end of a metabolic pathway.

3. **Enzymes:** Proteins that catalyze (speed up) reactions.

4. **Cofactors:** Small molecules and metal ions that help enzymes or that carry atoms or electrons from one reaction site to another.

5. **Energy Carriers:** Mainly ATP, which readily donates energy to diverse reactions.

6. **End Products:** Substances present at the end of a metabolic pathway.

Let's take a quick look at the roles of these substances.

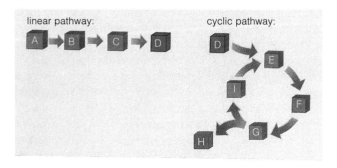

Figure 4.7 Linear and cyclic metabolic pathways.

ENZYMES

Enzymes are catalytic molecules, meaning that they greatly speed up specific reactions. They enhance the rate at which the reactions approach equilibrium.

With few exceptions, enzymes are proteins, and they have four features in common. First, enzymes do not make anything happen that could not happen on its own. They merely make it happen at least a million times faster, usually. Second, enzymes are not permanently altered or used up in the reaction; they can be used over and over again. Third, the same enzyme works for the forward *and* reverse directions of a reaction. Fourth, each type of enzyme is highly selective about its substrates.

Substrates are molecules that a specific enzyme can chemically recognize, bind, and modify in a specific way. Thrombin, an enzyme involved in blood clotting, recognizes and splits only one bond, the peptide bond between two amino acids, arginine and glycine. Its action can be depicted this way:

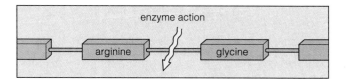

Enzyme Structure and Function

An enzyme has one or more crevices in its surface where substrates interact with it. Such a crevice is an **active site**, the place where a specific reaction is catalyzed (Figure 4.8). Each substrate has a surface region that almost *but not quite* matches chemical groups in an active site. When substrates first settle into the site, the contact strains some of their bonds. Strained bonds are easier to break, and this helps pave the way for new bonds (within the products). Also, interaction with charged or polar groups in the site favors a redistribution of electric charge that primes the substrates for conversion to an activated state.

Substrates reach an activated, transition state when they fit most precisely in the active site (Figure 4.9). The enzyme induces the fit in several ways. For instance, weak but extensive bonding at the active site puts substrates in positions that make them collide and so promotes reaction. If those same molecules were colliding on their own, they would do so from random directions. The reaction rate would not be very impressive, because mutually attractive chemical groups would meet up less frequently.

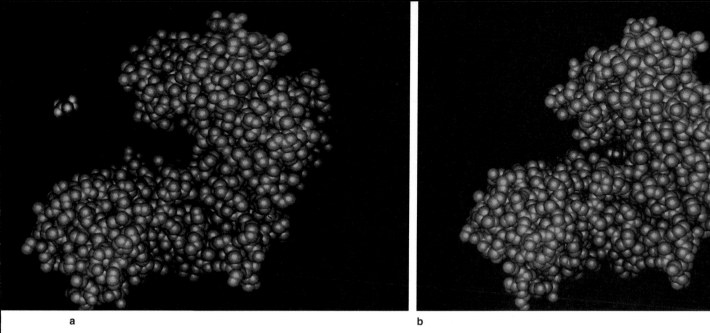

a **b**

Figure 4.8 An enzyme at work. (**a**) Model of an enzyme (hexokinase) and its substrate (a glucose molecule, shown here in red). The cleft into which the glucose is heading is the enzyme's active site. (**b**) When glucose makes contact with this active site, the enzyme temporarily changes shape. The upper and lower parts then close in around the glucose.

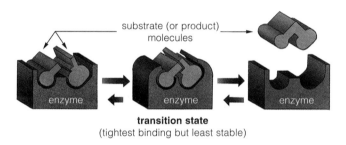

substrate (or product) molecules

enzyme   enzyme   enzyme

transition state
(tightest binding but least stable)

Figure 4.9 Induced-fit model of enzyme-substrate interactions. Only when the substrate is bound in place is the enzyme's active site complementary to it. The most precise fit occurs during a transition state that precedes the reaction. An enzyme-substrate complex is short-lived, partly because it is usually held together by weak bonds.

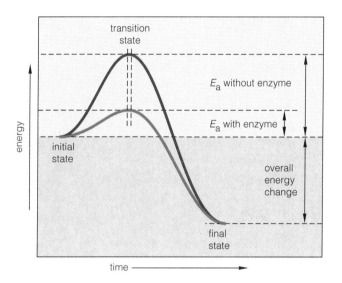

transition state

energy

initial state

E_a without enzyme

E_a with enzyme

overall energy change

final state

time

Figure 4.10 Energy hill diagram showing the effect of enzyme action. An enzyme greatly enhances the rate at which a given number of molecules complete a reaction because it lowers the required activation energy (E_a). Less energy is needed to boost reactants to the crest (transition state) of the lower energy hill.

Once substrates are in the transition state, they react spontaneously, just as a boulder pushed up and over the crest of a hill rolls down on its own. However, they simply won't reach that state unless they collide with some minimum amount of energy. That amount, the **activation energy**, is like an "energy hill" that must be surmounted (Figure 4.10). By putting its substrates on a precise collision course, an enzyme makes the energy hill smaller, so to speak. And this means the reaction will occur more rapidly.

An enzyme enhances the rate at which a reaction occurs. It does this by lowering the amount of activation energy necessary to make its substrates react.

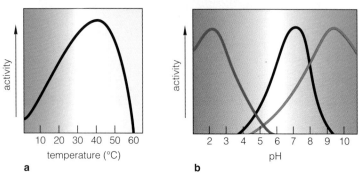

a

b

Figure 4.11 (**a**) Diagram showing how the activity of one kind of enzyme changes with increases in temperature. Siamese cats show observable effects of such changes. Fur on the ears and paws contains more dark-brown pigment (melanin) than the rest of the body. A heat-sensitive enzyme controlling melanin production is less active in warmer body regions, and this results in lighter fur in those regions. (**b**) Diagram showing how the activity of three different enzymes is influenced by pH. One enzyme (brown line) functions best in neutral solutions. Another (red line) functions best in basic solutions, and another (purple line) in acidic solutions.

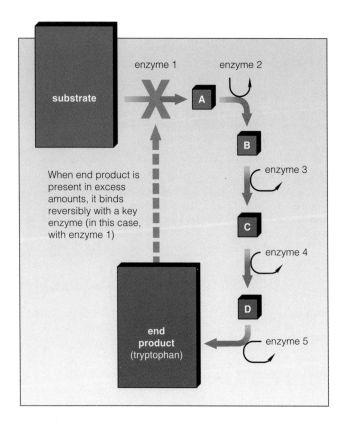

Effects of Temperature and pH on Enzymes

Each type of enzyme functions best within a certain temperature range (Figure 4.11a). When temperatures become too high, reaction rates decrease sharply. The increased thermal energy disrupts weak bonds holding the enzyme in its three-dimensional shape. This alters the active site, and substrates cannot bind to it. Exposure to high temperatures often destroys enzymes—and disrupts metabolism. This happens during dangerously high fevers. Humans usually die when body temperature reaches 44°C (112°F).

Each enzyme also functions best within a certain pH range. Higher or lower pH values generally disrupt its structure and function (Figure 4.11b). Most enzymes function best in neutral solutions (pH 7). Pepsin is one of the exceptions. This protein-digesting enzyme performs its task in gastric fluid, which is extremely acidic.

Control of Enzyme Activity

Earlier you read that cells maintain, increase, and decrease concentrations of substances by coordinating different metabolic pathways. They do this mostly by controlling *enzymes*. Some controls govern enzyme synthesis and so cut down or beef up the number of enzyme molecules that are operating at any time at a key step in a pathway. Other controls stimulate or inhibit enzymes that are already formed.

Internal controls come into play when concentrations of substances change within the cell itself. Think of a bacterium, busily synthesizing tryptophan and other amino acids necessary to construct its proteins.

After a bit, protein synthesis slows and no more tryptophan is needed. But the tryptophan pathway is still in full swing, so the cellular concentration of its end product rises. Now a control mechanism called **feedback inhibition** operates. By this mechanism, when production of a substance triggers a cellular change, the substance *itself* shuts down its further production (Figure 4.12). In this case, unused molecules of tryptophan inhibit a key enzyme in the pathway. (The enzyme happens to be an "allosteric" one. Besides the active site, such enzymes have control sites where specific substances can bind and alter enzyme activity.)

When the pathway is blocked, fewer tryptophan molecules are around to inhibit the key enzyme—so

Figure 4.12 Feedback inhibition of a metabolic pathway. In this example, five enzymes act in sequence to convert a substrate into an end product (tryptophan). When an excessive number of end-product molecules accumulate, some of them bind to the first enzyme in the pathway. This blocks the entire pathway.

production rises. In such ways, feedback inhibition quickly adjusts concentrations of substances in cells.

In humans and other multicelled organisms, control of enzyme activity is just amazing. Cells not only work to keep themselves alive, they work with other cells in coordinated ways that benefit the whole body! **Hormones** are signaling agents in this vast enterprise. Specialized cells release these chemical signals into the bloodstream. Any cell having receptors for a particular hormone will take it up, and its program for constructing a particular protein or some other internal activity will be altered. The hormone trips internal control agents into action, and the activities of specific enzymes change.

Enzyme Helpers

During a metabolic reaction, one or more enzymes speed the transfer of electrons, atoms, or functional groups from one substrate to another. "Cofactors" help catalyze the reaction or serve briefly as transfer agents.

Cofactors include **coenzymes**, which are large organic molecules. **NAD$^+$** (nicotinamide adenine dinucleotide) and **FAD** (flavin adenine dinucleotide) are examples. During glucose breakdown, they pick up unbound protons (H$^+$) and electrons, then transfer them to other reaction sites. **NADP$^+$** (nicotinamide adenine dinucleotide phosphate), another coenzyme, transfers protons and electrons during photosynthesis and other pathways. Coenzymes are derived partly from vitamins. For instance, the vitamin niacin is a precursor for NAD$^+$ and NADP$^+$.

Some metal ions serve as cofactors. Ferrous iron (Fe^{++}) is one. It assists certain membrane proteins (cytochromes) in chloroplasts and mitochondria.

ELECTRON TRANSFERS IN METABOLIC PATHWAYS

If you were to throw some glucose into a wood fire, its atoms would quickly let go of one another and combine with oxygen in the atmosphere, forming CO$_2$ and H$_2$O. The chemical energy of glucose would be lost as heat. Cells do not "burn" glucose all at once. Doing so would waste its stored energy. Instead, atoms are plucked away from glucose in controlled steps, so that *intermediate* molecules form along the route. Each enzyme-mediated step releases only *some* of the energy.

Chemical bonds are interactions between electrons of different atoms. Breaking those bonds puts electrons up for grabs. In chloroplasts and mitochondria, the liberated electrons are sent through **electron transport systems**. The systems are bound in cell membranes. They include enzymes and coenzymes that transfer electrons in a highly organized sequence. One molecule donates electrons, and the next in line accepts them.

(You may hear of "oxidation-reduction reactions," but this awful name simply refers to *electron transfers*. A donor that gives up electrons is said to be oxidized. An acceptor that acquires electrons is reduced.)

Many transfers take place after atoms or molecules absorb enough energy to "excite" electrons—that is, boost them farther from the nucleus. When an excited electron returns to the lowest energy level available, it gives off energy. (The *Focus* essay on the next page describes some interesting evidence of this.) *Electron transport systems intercept excited electrons and make use of the energy they release.* Think of a transport system as a staircase (Figure 4.14). Excited electrons at the top step have the most energy. They drop down, one step at a time. At certain steps, energy being released is harnessed to do work—for instance, to move ions in ways that set up pH and electric gradients across a membrane. Such gradients, you will discover, are central to ATP formation.

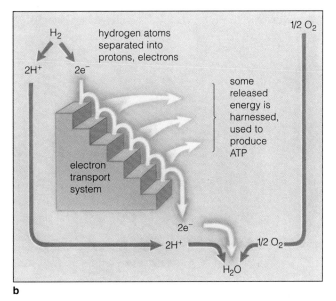

Figure 4.13 (**a**) Release of energy as heat when hydrogen and oxygen are made to react (say, by an electric spark). (**b**) Cells make the same type of reaction occur in many small steps that allow released energy to be harnessed in useful forms. The "steps" are electron transfers, often between molecules that operate together as an electron transport system.

You Light Up My Life—Visible Effects of Excited Electrons

At night, during certain metabolic reactions, fireflies and some other organisms flash with light. These displays of *bioluminescence* occur when enzymes called luciferases excite electrons of highly fluorescent substances (luciferins). Excited electrons quickly return to a lower energy level—and emit light. Four kinds of luciferase genes occur among Jamaican click beetles, also known as kittyboos. Different kittyboos emit green, greenish-yellow, yellow, or orange flashes.

Biochemists can insert copies of the kittyboo genes into different organisms (Figure *a*). Besides being fun to think about, the gene transfers have practical applications. Each year, for example, 3 million people die from a lung disease caused by *Myobacterium tuberculosis*. Different strains of

a Bioluminescent colonies of four strains of a bacterium (*Escherichia coli*). Each cell in a colony is glowing with light of a particular color, "borrowed" from a kittyboo.

this bacterium resist different drugs. Before 1993, a patient could not be treated effectively until the strain causing the infection was identified by tests that took weeks to complete. Now, bacteria from a patient are rapidly cultured and exposed to luciferase genes, which become inserted into the bacterial DNA. Then colonies of the genetically modified bacteria are exposed to different antibiotics. If an antibiotic has no effect, the cells churn out gene products—including luciferase—and the colony glows. If a colony doesn't glow, the antibiotic has stopped cells in their metabolic tracks.

ATP—THE MAIN ENERGY CARRIER

Photosynthetic cells don't run *directly* on sunlight. First they convert sunlight energy to the chemical energy of **ATP** (adenosine triphosphate, a type of nucleotide). Besides this, no cell whatsoever directly uses the energy released during glucose breakdown. First the stored energy is converted to ATP energy.

As Figure 4.14 shows, ATP consists of adenine (a nitrogen-containing compound), ribose (a five-carbon sugar), and a triphosphate (three phosphate groups), all covalently bonded together. The bonding arrangement is not too stable. Many hundreds of different enzymes

easily split off the outermost phosphate group and so release a great deal of useful energy. They use the energy for diverse tasks—as when they construct and break apart molecules, actively transport substances across cell membranes, and trigger muscle contraction.

ATP molecules are like coins of a nation—they are the cell's common currency of energy. Not surprisingly, cells have mechanisms for renewing the molecule after it gives up phosphate. In the **ATP/ADP cycle**, an energy input drives the binding of adenosine diphosphate (ADP) to a phosphate group or to unbound phosphate (P_i), forming ATP. Then the ATP is ready to donate a phospahte group elsewhere and revert back to ADP:

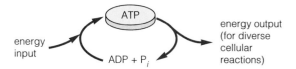

energy input → ATP → energy output (for diverse cellular reactions)
ADP + P_i

Adding phosphate to a molecule is called **phosphorylation**. Keep this molecular event in mind in chapters to follow. Generally speaking, when a molecule becomes phosphorylated by ATP, its store of energy increases, *and it becomes primed to enter a specific reaction*.

Figure 4.14 Structural formula for adenosine triphosphate (ATP).

adenine NH_2

triphosphate

OH OH ribose

ATP delivers energy to or picks up energy from almost all metabolic pathways.

SUMMARY

1. Cells acquire and use energy to build, store, break down, and rid themselves of substances. These activities are called metabolism. They underlie growth, maintenance, and reproduction of all organisms.

2. Energy flows in one direction through the world of life. The sun is the primary energy source. It replaces metabolically generated energy that all cells inevitably lose to their surroundings (as heat, mostly).

3. A metabolic pathway is a stepwise sequence of reactions in cells. In biosynthetic pathways, large molecules are assembled, and energy becomes stored in them. In degradative pathways, large molecules are broken down to smaller ones, and energy is released.

4. These substances take part in metabolic reactions:

   a. Substrates (or reactants): the substances that enter a specific reaction.

   b. Enzymes: proteins that serve as catalysts (they speed up reactions).

   c. Cofactors: coenzymes (including NAD^+) and metal ions that help catalyze reactions or carry electrons, hydrogen, or functional groups stripped from substrates.

   d. Energy carriers: mainly ATP, which readily donates energy to other molecules. Most biosynthetic pathways are driven by ATP energy.

   e. End products: the substances formed at the end of a metabolic pathway.

5. Control mechanisms stimulate or inhibit the activity of enzymes at key steps in metabolic pathways. They help coordinate the flow of substances into, through, and out of cells.

6. Electron transport systems are organized sequences of enzymes and coenzymes. They are built into cell membranes, such as those of chloroplasts and mitochondria. Electrons stripped from substrates are transferred through these systems. During certain transfers, energy is released that can be used to do work—for example, to make ATP.

Review Questions

1. State the first and second laws of thermodynamics. Does life violate the second law? *61, 62*

2. In metabolic reactions, does equilibrium imply equal amounts of reactants and products? Think of a cellular activity that might keep a reaction from approaching equilibrium. *62–63*

3. Describe an enzyme and its role in metabolic reactions. *64, 66*

4. What are the three components of ATP? Why is phosphorylation of a molecule by ATP so important? *68*

Self-Quiz (Answers in Appendix IV)

1. A cell's capacity to acquire and use energy for building and breaking apart molecules is called _____ .

2. Two laws of _____ govern how cells acquire, convert, and transfer energy during metabolic reactions.

3. _____ is the primary source of energy for life on earth.
   a. Food   c. The sun
   b. Water   d. ATP

4. Which is *not* true of chemical equilibrium?
   a. Product and reactant concentrations are always equal.
   b. The rates of the forward and reverse reactions are the same.
   c. There is no further net change in product and reactant concentrations.

5. In a biosynthetic pathway, _____ .
   a. large molecules are broken down to smaller ones
   b. energy is not required for the reactions
   c. large molecules are assembled from simpler ones
   d. both a and b

6. Enzymes _____ .
   a. enhance reaction rates   c. act on specific substrates
   b. are affected by pH   d. all of the above

7. Electron transport systems involve _____ .
   a. enzymes and cofactors   c. cell membranes
   b. electron transfers and   d. a, b, and c
   released energy

8. The main energy carriers in cells are _____ .
   a. NAD^+ molecules   c. ATP molecules
   b. cofactors   d. enzymes

9. Match each substance with its correct description.
   _____ a coenzyme or metal ion   a. substrate
   _____ mainly ATP   b. enzyme
   _____ substance entering a reaction c. cofactor
   _____ protein that catalyzes a reaction   d. energy carrier

Selected Key Terms

activation energy 65	energy 61	intermediate 64
active site 64	energy carrier 64	metabolic pathway 64
ATP 68	entropy 62	metabolism 61
ATP/ADP cycle 68	enzyme 64	NAD^+ 67
coenzyme 67	FAD 67	$NADP^+$ 67
cofactor 64	feedback inhibition 66	phosphorylation 68
electron transport	first law of	second law of
system 67	thermodynamics 61	thermodynamics 62
end product 64	hormone 67	substrate 64

Readings

Fenn, J. 1982. *Engines, Energy, and Entropy.* New York: Freeman. Deceptively simple paperback.

Rusting, R. December 1992. "Why Do We Age?" *Scientific American* 267(6):131–141.

5 ENERGY-ACQUIRING PATHWAYS

Sun, Rain, and Survival

Just before dawn in the Midwest, the air is dry and motionless. The heat that has scorched the land for weeks still rises from the earth and hangs in the air of a new day. There are no clouds in sight. There is no promise of rain. For hundreds of miles, crops stretch out, withered and nearly dead. All the marvels of modern agriculture can't save them. In the absence of one vital substance—water—life in each cell of those many hundreds of thousands of plants has ceased.

In Los Angeles, a student wonders if the Midwest drought will bump up food prices. In Washington, D.C., economists analyze crop failures in terms of tonnage available for domestic consumption and export.

Thousands of kilometers away, in the vast Sahel Desert of Africa, grasses and cattle are dying after a similar unrelenting drought. Children with bloated bellies and spindly legs wait passively for death. Deprived of nourishment for too long, cells of their bodies will never function normally again.

You are about to explore pathways by which cells trap and use energy. At first these pathways may seem to be far removed from your everyday world. *Yet the food that nourishes you and nearly all other organisms cannot be produced or used without them.*

We will return to this point in later chapters, when we address major concerns such as human population

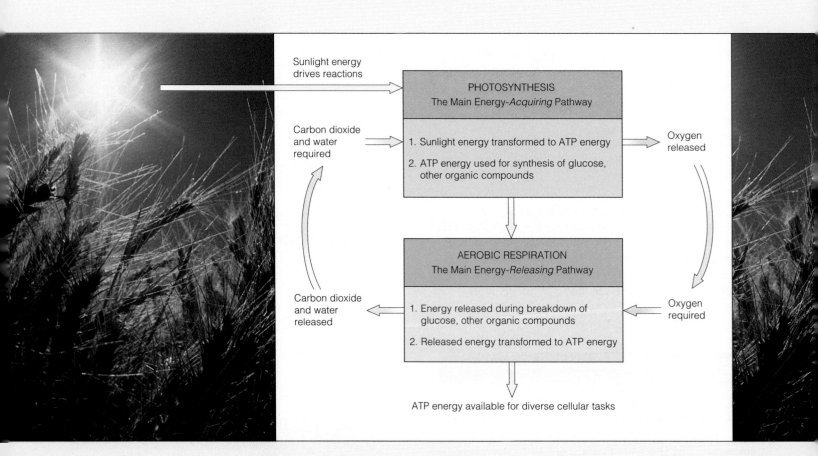

Figure 5.1 Links between photosynthesis and aerobic respiration—the main energy-acquiring and energy-releasing pathways in the world of life.

growth, nutrition, limits on agriculture, genetic engineering of plants, and effects of pollution on crops. Here, our point of departure is the *source* of food—which isn't a farm or a supermarket or a refrigerator. What we call "food" was put together somewhere in the world by living cells from glucose and other organic compounds. Such compounds are built on a framework of carbon atoms, so the questions become these:

1. Where does the carbon come from in the first place?

2. Where does the energy come from to drive the assembly of carbon-based compounds?

Answers to these questions depend on an organism's mode of nutrition.

Organisms classified as **autotrophs** get carbon and energy directly from the environment. They are "self-nourishing," which is what *autotroph* means. Carbon dioxide (CO_2), a gaseous substance all around us in the air and dissolved in water, is the source of carbon. Sunlight is the energy source for the *photosynthetic* autotrophs. These are the world's plants, certain protistans, and certain bacteria. *Chemosynthetic* autotrophs extract energy from inorganic substances such as sulfur. Only a few bacteria can do this.

Animals, fungi, many protistans, and most bacteria are classified as **heterotrophs**. They eat autotrophs, each other, and organic wastes. (*Hetero* means other, as in "nourished by other organisms.") In other words, they get carbon and energy from compounds that were originally put together by autotrophs.

When you think about this grand pattern of who eats whom, one thing becomes clear. *The survival of nearly all organisms ultimately depends on photosynthesis, the main pathway by which carbon and energy enter the world of life.*

After glucose and other simple organic compounds are assembled, cells can use them as building blocks. Or they can break the compounds apart—and so release energy as required for cell activities. The compounds can be dismantled by several pathways. However, *the predominant energy-releasing pathway is called aerobic respiration*. Figure 5.1 is a preview of the links between photosynthesis and aerobic respiration—the focus of this chapter and the next.

1. Carbon-based compounds are the building blocks and energy stores of life. Plants assemble these compounds by photosynthesis. First they trap sunlight energy and convert it to chemical energy (in the form of certain bonds in ATP molecules). Then ATP delivers energy to reactions in which glucose is put together from carbon dioxide and water. Finally, glucose subunits are joined together, forming starch and other molecules.

2. Photosynthesis is the biosynthetic pathway by which most carbon and energy enter the web of life.

3. In plant cells, photosynthesis occurs in organelles called chloroplasts. The pathway starts at a membrane system inside the chloroplast. Its machinery includes light-absorbing pigments, enzymes, and a coenzyme ($NADP^+$), which delivers hydrogen and electrons to the synthesis reactions.

PHOTOSYNTHESIS: AN OVERVIEW

Sources of Energy and Materials for the Reactions

Photosynthesis is an ancient pathway, and it evolved in distinct ways in different organisms. To keep things simple, let's focus on what goes on in lettuce, weeds, and other leafy plants.

The pathway consists of two stages, each with its own set of reactions. In the *light-dependent* reactions, sunlight energy is absorbed and converted to ATP energy. Also, water molecules are split, and an "enzyme helper" (the coenzyme $NADP^+$) picks up the liberated hydrogen and electrons. When carrying its cargo, this coenzyme is abbreviated NADPH.

In the *light-independent* reactions, ATP energy drives the assembly of carbon, hydrogen, and oxygen into molecules of glucose ($C_6H_{12}O_6$). Carbon dioxide (CO_2) provides the carbon and oxygen, and water provides the hydrogen (delivered by NADPH). Often photosynthesis is summarized this way:

$$12H_2O + 6CO_2 \xrightarrow{\text{sunlight}} 6O_2 + C_6H_{12}O_6 + 6H_2O$$

In this summary equation, glucose is shown as an end product to keep the chemical bookkeeping simple. However, the reactions don't really stop with glucose. Glucose and other simple sugars are bonded together at once to form sucrose, starch, and other carbohydrates—the true end products of photosynthesis.

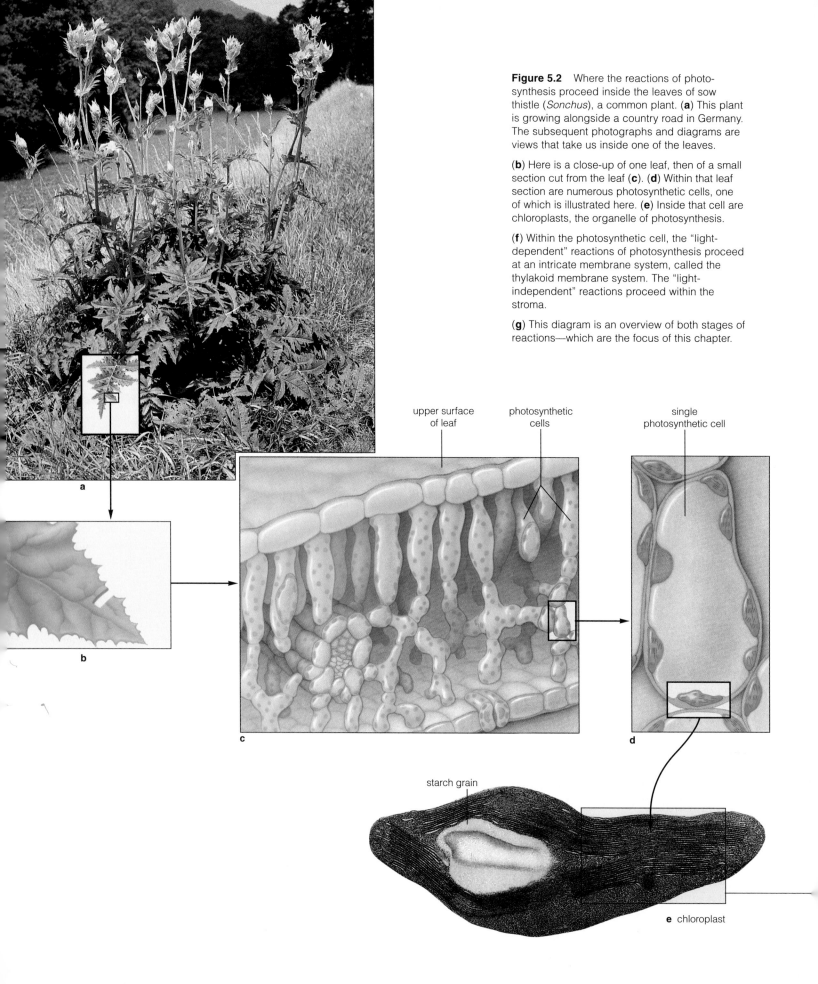

Figure 5.2 Where the reactions of photosynthesis proceed inside the leaves of sow thistle (*Sonchus*), a common plant. (**a**) This plant is growing alongside a country road in Germany. The subsequent photographs and diagrams are views that take us inside one of the leaves.

(**b**) Here is a close-up of one leaf, then of a small section cut from the leaf (**c**). (**d**) Within that leaf section are numerous photosynthetic cells, one of which is illustrated here. (**e**) Inside that cell are chloroplasts, the organelle of photosynthesis.

(**f**) Within the photosynthetic cell, the "light-dependent" reactions of photosynthesis proceed at an intricate membrane system, called the thylakoid membrane system. The "light-independent" reactions proceed within the stroma.

(**g**) This diagram is an overview of both stages of reactions—which are the focus of this chapter.

upper surface of leaf

photosynthetic cells

single photosynthetic cell

a

b

c

d

starch grain

e chloroplast

Where the Reactions Occur

The two stages of photosynthesis proceed at different sites inside the **chloroplast**. This type of organelle, recall, occurs only in eukaryotic photosynthetic cells (page 54).

Each chloroplast has two outer membranes wrapped around a semifluid interior, the **stroma**. An inner membrane weaves through the stroma. Often it takes the form of flattened channels and disks, which are arranged in stacks called grana (singular, granum). The first stage of photosynthesis proceeds at this inner membrane, which is the **thylakoid membrane system**.

The spaces inside the disks and channels connect as a single compartment for hydrogen ions, which are used in ATP production. The second stage of photosynthesis (the reactions by which sugars are assembled) takes place in the stroma.

Figure 5.2 marches down through a chloroplast from sow thistle (*Sonchus*), a common weed. Two thousand of those chloroplasts, lined up single-file, would be no wider than a dime. Imagine all the chloroplasts in just one weed or lettuce leaf, each a tiny factory for producing sugars and starch—and you get an idea of the magnitude of metabolic events required to feed you and all other organisms living together on this planet.

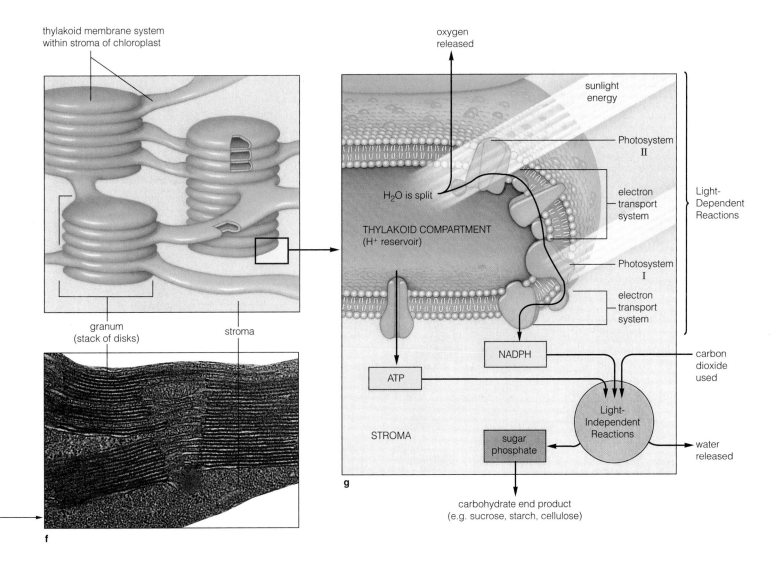

thylakoid membrane system within stroma of chloroplast

granum (stack of disks)

stroma

f

oxygen released

sunlight energy

Photosystem II

H_2O is split

electron transport system

THYLAKOID COMPARTMENT (H^+ reservoir)

Photosystem I

electron transport system

Light-Dependent Reactions

NADPH

ATP

STROMA

sugar phosphate

Light-Independent Reactions

carbon dioxide used

water released

g

carbohydrate end product (e.g. sucrose, starch, cellulose)

Figure 5.3 Wavelengths of visible light. These fall within a much larger range of wavelengths, called the electromagnetic spectrum (**a**). Organisms use wavelengths ranging from about 400 to 750 nanometers for photosynthesis, vision, and other light-requiring processes. Shorter wavelengths (such as those of ultraviolet light and x-rays) are so energetic they break bonds in organic compounds, so they can destroy cells.

(**b**) Wavelengths absorbed by some photosynthetic pigments. The graph colors correspond to the wavelengths each kind of pigment absorbs. Three pigment classes are represented: the chlorophylls (green lines), carotenoids (yellow-orange line), and phycobilins (red and blue lines). The peaks in the ranges of absorption correspond to the measured amount of energy absorbed and used in photosynthesis. The dashed line shows what would happen if we *combined* the amounts of energy that all the different pigments absorb. Taken together, photosynthetic pigments can absorb most of the wavelengths of visible light.

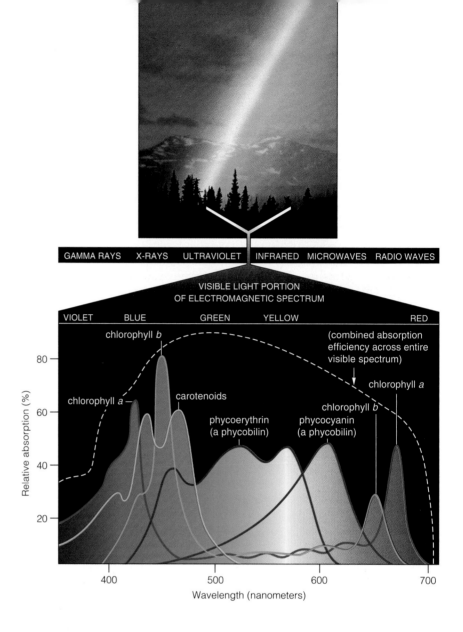

LIGHT-DEPENDENT REACTIONS

Three events unfold during the **light-dependent reactions**, the first stage of photosynthesis. *First*, pigments absorb sunlight energy and give up electrons. *Second*, electron and hydrogen transfers lead to ATP and NADPH formation. *Third*, the pigments that gave up electrons in the first place get electron replacements.

Absorbing Sunlight Energy

Light-Trapping Pigments. Generally speaking, **pigments** are light-absorbing molecules. In thylakoid membranes, clusters of pigments trap photons from the sun. Photons are packets of energy that travel through space in undulating motion, rather like ocean waves. A photon's wavelength (the distance from one wave peak to the next) is related to its energy. The most energetic photons travel as short wavelengths, and the least ener-getic as long wavelengths. We perceive light of different wavelengths as different colors (Figure 5.3).

Just as some ocean waves are more exciting than others to surfers, certain wavelengths are more exciting to a plant's pigments. For example, **chlorophyll** pigments absorb violet-to-blue and also red wavelengths but transmit green ones, so we see them as "green" pigments (Figure 5.4). **Carotenoids** absorb violet and blue wavelengths but transmit red, orange, and yellow.

In all plants, chlorophyll *a* is the main pigment of photosynthesis (Figure 5.3). Chlorophyll *b*, carotenoids, and other pigments absorb energy of different wavelengths and transfer it to the main pigment, enhancing its effectiveness. In green leaves, carotenoids are far less abundant than chlorophylls. Often they become visible in autumn, when many plants stop producing chlorophyll (Figure 5.5). Several other pigments exist, including red and blue **phycobilins** that are abundant in red algae and cyanobacteria.

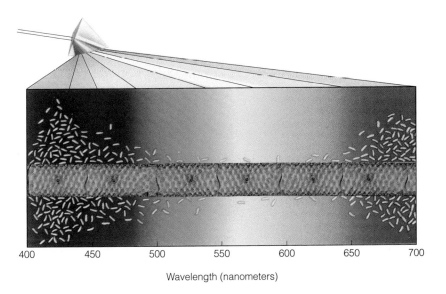

Figure 5.4 T. Englemann's 1882 experiment that correlated photosynthetic activity by a strandlike green alga (*Spirogyra*) with certain wavelengths of light.

Oxygen is a by-product of photosynthesis. Many organisms use oxygen (for aerobic respiration). Among them are certain free-living bacteria that tumble about and so move through their aquatic habitat. Englemann reasoned that oxygen-requiring bacteria living in the same habitats as the alga would congregate where oxygen was being produced. He put a strand of the alga in a drop of water containing such bacteria, then mounted it on a microscope slide. He used a crystal prism to cast a spectrum of colors across the slide. Bacteria congregated mostly where violet and red wavelengths fell across the strand. Those wavelengths were most effective for photosynthesis (and oxygen production).

Wavelength (nanometers)

maple in summer

maple in autumn

Figure 5.5 Changes in leaf color in autumn. Chloroplasts of mature leaves contain chlorophylls, carotenoids (including the yellow carotenes and xanthophylls), and other pigments. The intense green of abundant chlorophylls usually masks the presence of other pigments. In autumn, however, the gradual reduction in daylength and other factors trigger the breakdown of chlorophyll in many species, and more colors show through.

Also in autumn, water-soluble anthocyanins accumulate in the central vacuoles of leaf cells. These pigments appear red if plant fluids are slightly acidic, blue if basic (alkaline), or colors in between at intermediate pH levels. The differences in pH are influenced by soil conditions.

The color of birch, aspen, and other tree species is always the same in autumn. The color of other species, including maple, ash, and sumac, varies around the country. It even varies from one leaf to the next, depending on the pigment combinations.

Photosystems. In thylakoid membranes, different pigments are organized as light-trapping clusters, called **photosystems**. There may be many thousands of these clusters, and each includes 200 to 300 pigment molecules. Most of the pigments simply "harvest" sunlight. When they absorb a photon, one of their electrons gets boosted to a higher energy level:

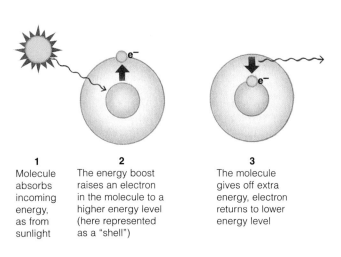

1
Molecule absorbs incoming energy, as from sunlight

2
The energy boost raises an electron in the molecule to a higher energy level (here represented as a "shell")

3
The molecule gives off extra energy, electron returns to lower energy level

When the electron returns to a lower level, it quickly gives up the added energy. The energy bounces among pigments, and a bit is lost at each bounce (as heat). Soon the remainder corresponds to a certain wavelength that a few special chlorophylls can trap. As you will see, only those chlorophylls give up the electrons used in photosynthesis. The chlorophylls transfer them to an electron-accepting molecule poised at the start of a neighboring transport system.

ATP and NADPH: Loading Up Energy, Hydrogen, and Electrons

Recall that **electron transport systems** are organized sequences of enzymes and other proteins bound in a cell membrane. When excited electrons are transferred through these systems, they release their extra energy, some of which is harnessed to drive specific reactions. In thylakoid membranes, two kinds of photosystems give up electrons to different transport systems. They allow plants to make ATP by two different pathways, one cyclic and the other noncyclic.

Cyclic Pathway. One pathway starts with electrons from *photosystem I*. The electron-donating chlorophyll of this photosystem is designated P700. In the **cyclic pathway of ATP formation**, electrons "cycle" from P700, through a transport system, then back to P700. Energy

released during the electron flow is used in the formation of ATP from ADP and unbound phosphate:

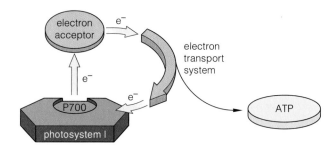

The cyclic pathway is probably the oldest means of ATP production. The first cells to use it were as tiny as existing bacteria, so their body-building programs were scarcely enormous. ATP alone would have provided enough energy to build the organic compounds for such tiny bodies. But it takes far more organic compounds—and vast amounts of hydrogen atoms and electrons—to build larger organisms. Long ago, in the forerunners of multicelled plants, the cyclic pathway's machinery underwent expansion and became the basis of a more efficient ATP-forming pathway. Amazingly, this tiny bit of new machinery changed the course of evolution.

Noncyclic Pathway. Today, the cyclic pathway still operates in trees, weeds, and other leafy members of the plant kingdom. But the **noncyclic pathway of ATP formation** dominates. Electrons aren't cycled through this pathway. They depart (in NADPH), and electrons from *water molecules* replace them.

The pathway goes into operation when the sun's rays bombard *photosystem II*. The incoming photon energy makes this photosystem's special chlorophyll (P680) give up electrons. It also triggers **photolysis**, a reaction sequence that splits apart water molecules into oxygen, hydrogen ions, and electrons. P680 attracts these rather unexcited electrons as replacements for the excited ones that got away (Figure 5.6).

Meanwhile, the excited electrons are transferred through a transport system—then to chlorophyll P700 of *photosystem I*. Electrons arriving at P700 haven't lost all of their extra energy. Because photons are also bombarding P700, the incoming energy boosts the electrons to a higher energy level that allows them to enter a second transport system. One of the enzymes of this system has a helper—$NADP^+$. When the enzyme takes part in the transfers, its helper picks up two electrons and a hydrogen ion, and so becomes NADPH. In this form, hydrogen and electrons are carted away, to sites where organic compounds are built.

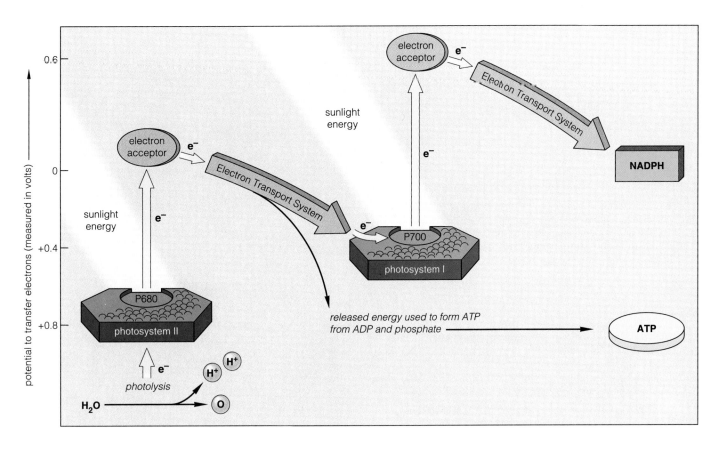

Figure 5.6 Noncyclic pathway of ATP formation, which also yields NADPH. Electrons derived from the splitting of water molecules (photolysis) travel through two photosystems. These work together to boost electrons to an energy level high enough to lead to NADPH formation.

Figure 5.7 Visible evidence of photosynthesis—oxygen emerging from the leaves of *Elodea*, an aquatic plant.

The Legacy—A New Atmosphere. On sunny days, on the surfaces of aquatic plants, you can see bubbles of oxygen (Figure 5.7). The oxygen is a by-product of the noncyclic pathway photosynthesis, which may have evolved more than 2 billion years ago. At first, the oxygen simply dissolved in seas, lakes, wet mud, and other bacterial habitats. By about 1.5 billion years ago, however, large amounts of dissolved oxygen were escaping into what had been an oxygen-free atmosphere. Its accumulation changed the atmosphere forever. And it made possible aerobic respiration, the most efficient pathway for extracting energy from organic compounds. The emergence of the noncyclic pathway ultimately allowed you and all other animals to be around today, breathing the oxygen that helps keep your cells alive.

In the light-dependent reactions, energy from the sun drives the formation of ATP (which carries energy) and NADPH (which carries hydrogen and electrons).

Oxygen, a by-product of photosynthesis, profoundly changed the early atmosphere and made aerobic respiration possible.

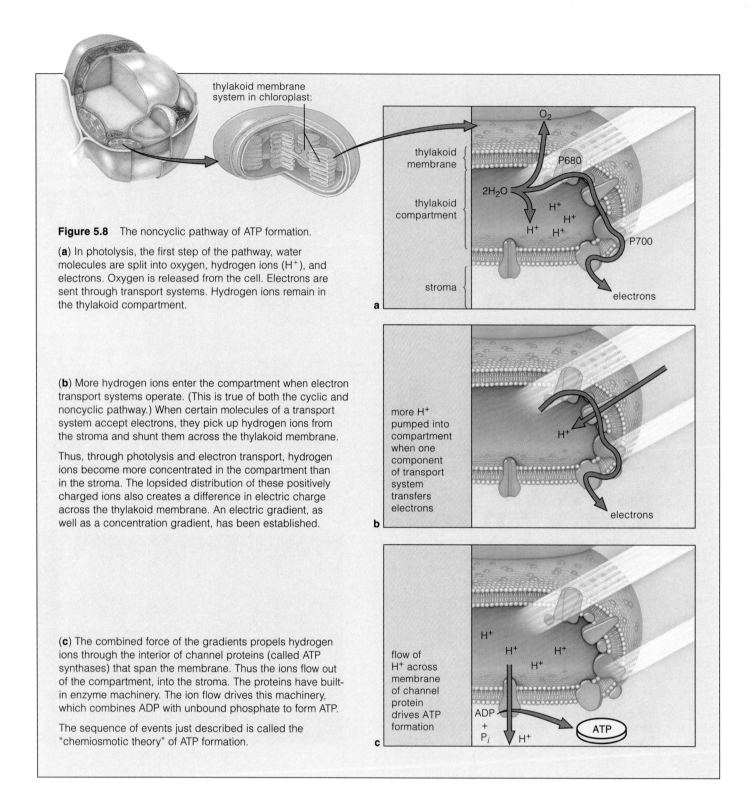

Figure 5.8 The noncyclic pathway of ATP formation.

(**a**) In photolysis, the first step of the pathway, water molecules are split into oxygen, hydrogen ions (H^+), and electrons. Oxygen is released from the cell. Electrons are sent through transport systems. Hydrogen ions remain in the thylakoid compartment.

(**b**) More hydrogen ions enter the compartment when electron transport systems operate. (This is true of both the cyclic and noncyclic pathway.) When certain molecules of a transport system accept electrons, they pick up hydrogen ions from the stroma and shunt them across the thylakoid membrane.

Thus, through photolysis and electron transport, hydrogen ions become more concentrated in the compartment than in the stroma. The lopsided distribution of these positively charged ions also creates a difference in electric charge across the thylakoid membrane. An electric gradient, as well as a concentration gradient, has been established.

(**c**) The combined force of the gradients propels hydrogen ions through the interior of channel proteins (called ATP synthases) that span the membrane. Thus the ions flow out of the compartment, into the stroma. The proteins have built-in enzyme machinery. The ion flow drives this machinery, which combines ADP with unbound phosphate to form ATP.

The sequence of events just described is called the "chemiosmotic theory" of ATP formation.

A Closer Look at ATP Formation. As you may have noticed, we've saved the trickiest question for last. *How*, exactly, does ATP form during the noncyclic (and cyclic) pathways?

As Figure 5.8 shows, when electrons flow through the membrane-bound transport systems, they pick up hydrogen ions (H^+) outside the membrane and dump them into the thylakoid compartment. This sets up H^+ concentration and electric gradients across the membrane. Hydrogen ions split from water increase the gradients. The ions respond by flowing out through the interior of channel proteins that span the membrane. The flow causes ADP and unbound phosphate to combine, the result being one ATP molecule.

LIGHT-INDEPENDENT REACTIONS

The **light-independent reactions** are the "synthesis" part of photosynthesis. ATP molecules deliver the required energy for the reactions. NADPH molecules deliver the required hydrogen and electrons. Carbon dioxide (CO_2) in the air around photosynthetic cells provides the carbon and oxygen.

We say the reactions are light-independent because they do not depend directly on sunlight. They can proceed even in the dark, as long as ATP and NADPH are available.

Capturing Carbon

Let's track a CO_2 molecule that diffuses into the air spaces inside a sow thistle leaf and ends up next to a photosynthetic cell (Figure 5.2*c*). From there, it diffuses across the plasma membrane and on into the stroma of a chloroplast.

The light-independent reactions start when the carbon atom of the CO_2 becomes attached to **RuBP** (ribulose bisphosphate), a molecule with a backbone of five carbon atoms. This is called **carbon dioxide fixation**.

Building the Glucose Subunits

Attaching carbon to RuBP is the first step of a cyclic pathway that produces a sugar phosphate *and* regenerates the RuBP. The pathway is named the **Calvin-Benson cycle**, in honor of its discoverers. A specific enzyme catalyzes each one of its steps. For our purposes, we can simply focus on the carbon atoms of the pathway's substrates, intermediates, and end products. Figure 5.9 shows the carbon atoms as red circles.

The attachment of a carbon atom to RuBP produces an unstable six-carbon intermediate. This splits into two molecules of **PGA** (phosphoglycerate), each with a three-carbon backbone. ATP donates a phosphate group to each PGA. NADPH donates hydrogen and electrons to the resulting intermediate, forming **PGAL** (phosphoglyceraldehyde).

The reaction steps just outlined occur not once but *six* times. In other words, six CO_2 molecules are fixed, and twelve PGAL molecules are produced.

Most of the PGAL becomes rearranged to form new RuBP molecules, which can be used to fix more carbon. But *two* of the PGAL are combined to form a six-carbon sugar phosphate. Phosphate groups attached to such sugars prime them for further reaction (refer to Figure 4.4).

The Calvin-Benson cycle yields enough RuBP molecules to replace the ones used in carbon dioxide fixation. The ADP, NADP+, and phosphate leftovers diffuse back to sites of the light-dependent reactions and are converted once more to NADPH and ATP. The sugar

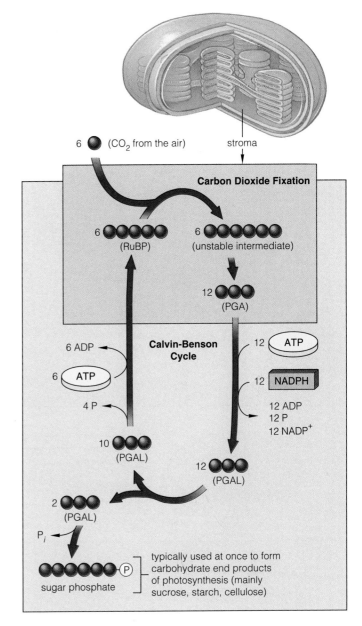

Figure 5.9 Summary of the light-independent reactions of photosynthesis. Carbon atoms of the key molecules are depicted in red. All of the intermediates have one or two phosphate groups. For simplicity, only the phosphate on the resulting sugar phosphate is shown.

phosphate formed in the cycle serves as a building block for sucrose, starch, or cellulose, the plant's main carbohydrates. Synthesis of such compounds by other pathways marks the conclusion of the light-independent reactions.

During the Calvin-Benson cycle, carbon is "captured" from carbon dioxide, a sugar phosphate forms in reactions that require ATP and NADPH, and RuBP (needed to capture the carbon) is regenerated.

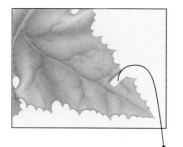

The Reactions, Start to Finish

Be sure to look carefully at Figure 5.10. It summarizes the key reactants, intermediates, and products of both the light-dependent and light-independent reactions of photosynthesis.

When studying this summary, remind yourself of how the reactions "fit" in the world of living things. Think of the mind-boggling numbers of single-celled and multicelled photosynthetic organisms in which the reactions are proceeding at this very moment, on land and in the sunlit waters of the earth. The sheer volume of the reactant and product molecules that the photosynthesizers deal with may surprise you. The *Focus* essay provides a case in point.

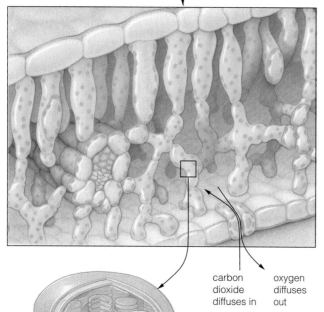

carbon dioxide diffuses in

oxygen diffuses out

one chloroplast from one photosynthetic cell within a leaf

Figure 5.10 Summary of the main reactants, intermediates, and products of photosynthesis, corresponding to the equation:

$$12H_2O + 6CO_2 \xrightarrow{\text{sunlight}} 6O_2 + C_6H_{12}O_6 + 6H_2O$$

Starting with the light-dependent reactions, the splitting of twelve molecules of water yields twenty-four electrons and six molecules of O_2. For every four electrons, three ATP and two NADPH form.

During the light-independent reactions, *each turn* of the Calvin-Benson cycle requires one CO_2, three ATP, and two NADPH. Each sugar phosphate that forms has a backbone of six carbon atoms—so its formation requires six turns of the cycle.

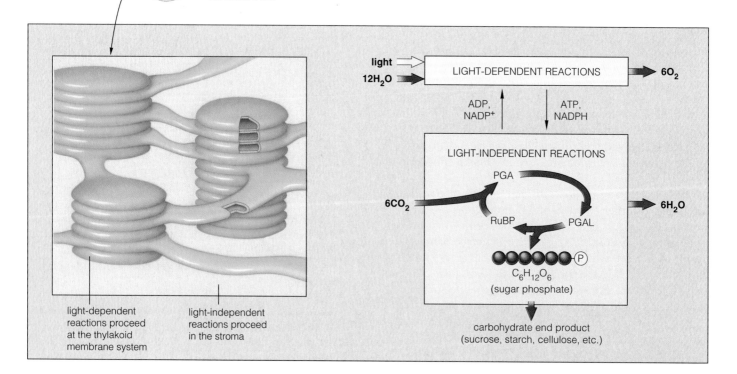

light-dependent reactions proceed at the thylakoid membrane system

light-independent reactions proceed in the stroma

light

$12H_2O$

LIGHT-DEPENDENT REACTIONS

$6O_2$

ADP, NADP$^+$

ATP, NADPH

LIGHT-INDEPENDENT REACTIONS

PGA

$6CO_2$

$6H_2O$

RuBP

PGAL

$C_6H_{12}O_6$ (sugar phosphate)

carbohydrate end product (sucrose, starch, cellulose, etc.)

Pastures of the Seas

Drifting through the surface waters of the world ocean are uncountable numbers of single cells. You can't see them without a microscope—a row of 7 million cells of one species would be less than a quarter-inch long. Yet are they abundant! In some parts of the world, a cup of seawater may hold 24 million cells of one species, and that doesn't even include all the cells of *other* aquatic species.

Many of the drifters are photosynthetic bacteria, protistans, and plants. Together they are the pastures of the seas, the food base for diverse consumers. The pastures "bloom" in spring, when waters become warmer and enriched with nutrients churned up from the deep by winter currents. Then, populations burgeon as cells divide again and again. Biologists had no idea that the number of cells and their distribution were so mind-boggling until satellites provided photographs from space. For example, the satellite image in Figure *b* shows a springtime bloom in the North Atlantic stretching from North Carolina all the way to Spain!

Those single cells have enormous impact on the world's climate. As they collectively photosynthesize, they sponge up nearly half the carbon dioxide we humans release each year (as when we burn fossil fuels and forests). Without them, carbon dioxide in the air would accumulate more rapidly and possibly accelerate global warming, as described in Chapter 35. If our planet warms too much, vast coastal regions may become submerged, and current food-producing nations may be hit hard.

Amazingly, we daily dump industrial wastes, fertilizers, and raw sewage into the ocean and alter the living conditions for those drifting cells. How much of that noxious chemical brew can they tolerate?

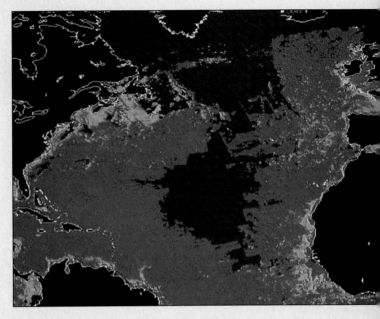

a Photosynthetic activity in winter. (In this color-enhanced image, red-orange shows where chlorophyll concentrations are greatest.)

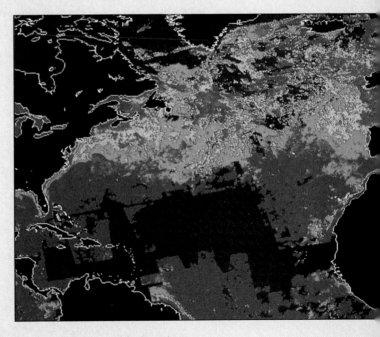

b Photosynthetic activity in spring.

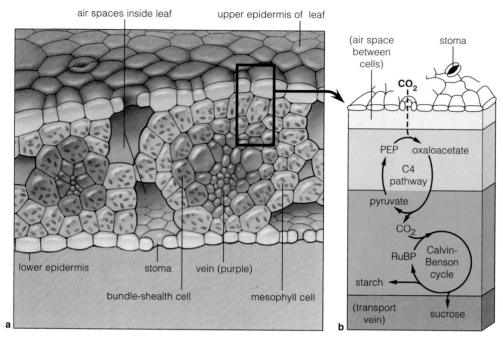

air spaces inside leaf   upper epidermis of leaf

Figure 5.11 C4 pathway. (**a**) Internal structure of a leaf from corn (*Zea mays*), a typical C4 plant. Photosynthetic mesophyll cells (light green) surround bundle-sheath cells (dark green). These in turn surround veins, the transport tubes that carry water into leaves and photosynthetic products away from them. (**b**) A carbon-fixing system precedes the Calvin-Benson cycle in C4 plants.

lower epidermis   stoma   vein (purple)

bundle-shealth cell   mesophyll cell

C4 Plants: Squirreling Away Carbon on Hot, Dry Days

Without CO_2, leafy plants can't grow. Although there is plenty of CO_2 in the air, it is not always abundantly available to the cells *inside the leaves*. Leaves have a waxy cover that retards moisture loss. Water escapes mainly through tiny openings across the leaf's surface. The openings are called stomata (singular, stoma). Most O_2 and CO_2 diffuse into and out of leaves through these openings.

On hot, dry days, stomata are shut. The plant conserves water—but CO_2 can't enter the leaf. Meanwhile, oxygen (a by-product of photosynthesis) builds up in the leaf. The stage is set for a wasteful process called "photorespiration." By this process, oxygen instead of CO_2 becomes attached to the RuBP used in the Calvin-Benson cycle, with different results:

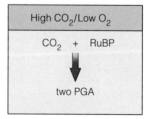

High CO_2/Low O_2

CO_2 + RuBP

↓

two PGA

Calvin-Benson cycle predominates

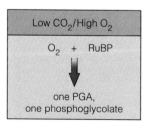

Low CO_2/High O_2

O_2 + RuBP

↓

one PGA, one phosphoglycolate

photorespiration predominates

Formation of sugar phosphates depends on PGA. When photorespiration wins out, less PGA forms and the plant's capacity for growth suffers.

On hot, dry days, crabgrass, sugarcane, corn, and many other plants maintain adequate amounts of CO_2 inside leaves even though their stomata are closed. When they fix carbon, the first compound formed is the four-carbon oxaloacetate. Hence their name, "C4" plants. (Three-carbon PGA is produced in "C3" plants.)

C4 plants fix CO_2 not once but twice, in two different types of photosynthetic cells. *Mesophyll* cells have first crack at the CO_2 and use it to form oxaloacetate. It is a temporary fix. Oxaloacetate is quickly transferred to *bundle-sheath* cells that wrap around every vein in the leaf. In those cells, CO_2 is released, fixed again, and used to build carbohydrates (Figure 5.11).

When temperatures climb, stomata close and oxygen accumulates in the leaves of any type of plant. In hot weather, C3 plants photorespire even more than usual, and their growth slows. But C4 plants do not photorespire as much. Because of the special carbon-fixing system in these plants, the CO_2 concentration still increases (in bundle-sheath cells), so oxygen loses out in the competition for RuBP.

C4 species are abundant where temperatures are highest during the growing season. For example, 80 percent of all native species in Florida are C4 plants—compared to 0 percent in Manitoba, Canada. Kentucky bluegrass and other C3 species have an advantage where temperatures drop below 25°C (they are better adapted to cold). When we mix C3 and C4 species from different regions in our gardens, one or the other kind will do better at least part of the year. That is why a lawn of Kentucky bluegrass thrives during cool spring weather in San Diego, only to be overwhelmed during the hot summer months by a C4 plant—crabgrass.

CHEMOSYNTHESIS

Photosynthesis is so pervasive, it sometimes is easy to overlook other, less common energy-acquiring routes. Bacteria that are classified as chemosynthetic autotrophs obtain energy not from sunlight but rather by pulling away electrons from ammonium ions, iron or sulfur compounds, and other inorganic substances. Many have roles in the cycling of nitrogen and other vital elements. We will return later to their environmental effects. In this unit, we turn next to pathways by which energy is released from glucose and other biological molecules—the chemical legacy of autotrophs.

SUMMARY

1. Plants and other photosynthetic autotrophs use sunlight (as an energy source) and carbon dioxide (as the carbon source) for building organic compounds. Animals and other heterotrophs obtain carbon and energy from organic compounds already built by autotrophs.

2. Photosynthesis is the main biosynthetic pathway by which carbon and energy enter the web of life. It consists of two sets of reactions that proceed from the capture of sunlight energy ("photo") through the assembly reactions ("synthesis").

  a. The light-dependent reactions take place at the thylakoid membrane system of chloroplasts. The reactions produce ATP and NADPH.

  b. The light-independent reactions take place in the stroma around the membrane system. They produce sugar phosphates that are used in building sucrose, starch, and other end products of photosynthesis.

3. These are the key points concerning the light-dependent reactions:

  a. Light energy is absorbed at photosystems (clusters of photosynthetic pigments embedded in the thylakoid membrane). It drives the transfer of electrons from a specific chlorophyll to an acceptor molecule, which donates them to a transport system in the membrane.

  b. In a cyclic pathway of ATP formation, excited electrons leave photosystem I, give up energy in the transport system, and return to that photosystem.

  c. In the noncyclic pathway of ATP formation, electrons from photosystem II pass through a transport system, enter photosystem I, pass through another transport system, then end up in NADPH. Also, water molecules are split into hydrogen ions, oxygen, and electrons. These electrons replace the ones given up.

  d. As electron transport systems operate, hydrogen ions accumulate in the compartment surrounded by the thylakoid membrane. Hydrogen ions split from water accumulate here also. This sets up concentration and electric gradients that drive ATP formation.

4. These are the key points concerning the light-independent reactions:

  a. ATP delivers energy and NADPH delivers hydrogen and electrons to the stroma of chloroplasts, where sugars are synthesized and assembled into starch, cellulose, and other end products.

  b. Sugar phosphates form during the Calvin-Benson cycle. This cyclic pathway begins when carbon dioxide from the air is affixed to RuBP, making an unstable intermediate that splits into two PGA. ATP donates a phosphate group to each PGA. The resulting molecule receives H^+ and electrons from NADPH to form PGAL.

  c. For every six CO_2 molecules that enter the cycle, twelve PGAL are produced. Two of those are used to produce a six-carbon sugar phosphate. The remainder are used to regenerate RuBP for the cycle.

5. In hot climates, some plants have an extra carbon-fixing pathway. The C4 pathway helps circumvent photorespiration, in which oxygen instead of carbon dioxide becomes affixed to RuBP. (Photorespiration undoes much of what photosynthesis accomplishes.)

Review Questions

1. A caterpillar chewing on a weed is speared and eaten by a bird, which in turn is eaten by a cat. Which of these organisms are autotrophs? heterotrophs? *71*

2. Summarize the photosynthesis reactions as an equation. State the key events of the light-dependent reactions, then of the light-independent reactions. *71*

3. Which of the substances listed accumulates in the thylakoid compartment of chloroplasts: glucose, photosynthetic pigments, hydrogen ions, fatty acids? *78*

4. Which of the substances listed are *not* required for the light-independent reactions: ATP, NADPH, RuBP, chlorophyll, carotenoids, free oxygen, carbon dioxide, enzymes? *79*

5. Suppose a plant busily photosynthesizing is exposed to CO_2 molecules that contain radioactively labeled carbon atoms ($^{14}CO_2$). Identify the compound in which the labeled carbon will first appear: NADPH, PGAL, pyruvate, or PGA. *79*

6. How many CO_2 molecules must enter the Calvin-Benson cycle to produce one sugar phosphate? Why? *79, 80*

1. Molecules with a backbone of _____ serve as the main building blocks of all organisms.

2. Photosynthetic autotrophs use _____ from the air as their carbon source and _____ as their energy source.

3. In plants, light-*dependent* reactions occur at the _____ .
   a. cytoplasm
   b. plasma membrane
   c. stroma
   d. thylakoid membrane

4. The light-*independent* reactions occur in the _____ .
   a. cytoplasm
   b. plasma membrane
   c. stroma
   d. grana

5. In the light-dependent reactions, _____ .
   a. carbon dioxide is fixed
   b. ATP and NADPH form
   c. carbon dioxide accepts electrons
   d. sugar phosphates form

6. When a photosystem absorbs light, _____ .
   a. sugar phosphates are produced
   b. electrons are transferred to an acceptor molecule
   c. RuBP accepts electrons
   d. light-dependent reactions begin

7. The Calvin-Benson cycle starts when _____ .
   a. light is available
   b. light is not available
   c. carbon dioxide is attached to RuBP
   d. electrons leave a photosystem

8. In the light-independent reactions, ATP furnishes phosphate groups to _____ .
   a. RuBP
   b. NADP$^+$
   c. PGA
   d. PGAL

9. Match each event in photosynthesis with its correct description.
   _____ RuBP used; PGA formed
   _____ ATP and NADPH used
   _____ NADPH formed
   _____ ATP and NADPH formed
   _____ only ATP formed
   a. cyclic pathway
   b. noncyclic pathway
   c. carbon dioxide fixation
   d. PGAL formation
   e. transfer of H$^+$ and electrons to NADP$^+$

Selected Key Terms

autotroph 71
Calvin-Benson cycle 79
carbon dioxide fixation 79
carotenoid 74
chlorophyll 74
chloroplast 73
cyclic pathway of ATP formation 76
electron transport system 76
heterotroph 71
light-dependent reactions 74
light-independent reactions 79
noncyclic pathway of ATP formation 76
PGA 79
PGAL 79
photolysis 76
photosystem 76
phycobilin 74
pigment 74
RuBP 79
stroma 73
thylakoid membrane system 73

Readings

Daviss, B. February 1992. "Going for the Green." *Discover* 13:20. Artificial systems for photosynthesis.

Hendry, George. May 1990. "Making, Breaking, and Remaking Chlorophyll." *Natural History*, pp. 36–41.

Youvan, D., and B. Marrs. 1987. "Molecular Mechanisms of Photosynthesis." *Scientific American* 256:42–50.

6 ENERGY-RELEASING PATHWAYS

The Killers Are Coming!

In 1990, "killer" bees from South America buzzed across the border between Mexico and the United States. The bees are descended from African queen bees. When provoked, they can be terrifying.

Earlier, some queen bees had been shipped from Africa to Brazil for breeding experiments. Honeybees happen to be big business. Besides producing honey, they are rented out to pollinate commercial orchards. But the honeybees in Brazil seemed sluggish. The idea was to cross-breed them with their African relatives and produce a mild-mannered but zippier pollinator.

Twenty-six African queens escaped. That was bad enough. Then beekeepers got wind of the program. After learning that the first few generations of offspring were jazzed-up but nice honeybees, they imported *hundreds* of African queens and encouraged them to mate with locals. And they set off a genetic time bomb.

Before long, African bees were firmly established in commercial hives—and in wild bee populations. And their traits became dominant. The "Africanized" bees do everything other bees do, but they do more of it faster. Their eggs develop into adults more quickly. Adults fly more rapidly, outcompete other bees for nectar, and even die sooner. When their hives are disturbed, they become far more agitated. Whereas a mild-mannered honeybee might chase an intruding animal fifty yards or so, a squadron of Africanized bees will chase it a quarter of a mile. If they catch up to it, they can sting it to death.

Doing things faster means having a nonstop supply of energy. This bee's stomach can hold 30 milligrams of sugar-rich nectar—enough fuel to fly 60 kilometers (more than 35 miles). Large mitochondria in its flight muscle cells efficiently convert energy stored in sugar to ATP energy. Africanized bees cannot survive where winters are harsh and plants stop blooming for months at a time.

In their ability to release energy stored in sugar and other organic compounds, Africanized bees are like all other organisms. Although energy-releasing pathways differ in some details, all start with certain materials, then end with predictable products. And they *all* yield ATP. *At the biochemical level, there is undeniable unity among all forms of life*. We will return to this idea in the *Commentary* at the chapter's end.

Figure 6.1 A mild-mannered honeybee buzzing in for a landing on a flower, wings beating with energy provided by ATP. If this were one of its Africanized relatives approaching a hive, possibly you would not stay around to watch the landing.

1. All organisms produce ATP by releasing energy stored in glucose and other organic compounds. Energy-releasing pathways begin with the same stage of reactions, called glycolysis. The pathway called aerobic respiration yields the most energy from each glucose molecule.

2. Aerobic respiration has three stages. First, glucose is partly broken down to pyruvate. Second, pyruvate is broken down to carbon dioxide and water. Coenzymes pick up electrons liberated by the breakdown reactions. Third, the coenzymes release their cargo to a transport system. The unbound hydrogen (H$^+$ ions) and electrons are used to form many ATP molecules.

3. Aerobic respiration requires oxygen. The oxygen withdraws electrons from the transport system and joins with H$^+$ to form water.

4. Over evolutionary time, photosynthesis and aerobic respiration have become linked on a global scale. Oxygen released during photosynthesis is required for the aerobic pathway. And the carbon dioxide and water released during aerobic respiration are raw materials used in building organic compounds during photosynthesis:

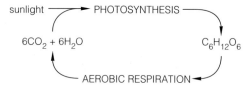

ATP-PRODUCING PATHWAYS

Organisms stay alive by taking in energy. Plants get energy from the sun. Animals get energy secondhand, thirdhand, and so on, by eating plants and one another. Regardless of its source, energy must be put in a form that can drive metabolic reactions. Energy carried by adenosine triphosphate, **ATP**, serves that function.

Plants make ATP during photosynthesis. They and all other organisms also make ATP by releasing stored energy from carbohydrates, lipids, or proteins. Electron transfers of the sort described on page 67 are at the heart of all energy-releasing pathways.

Aerobic respiration is the main energy-releasing pathway. The "aerobic" part of the name means that the pathway cannot be completed without oxygen. With every breath, you provide your busily respiring cells with oxygen. Other energy-releasing pathways are "anaerobic," in that they can be completed without using oxygen. The most common of these are **fermenta-**

tion and **anaerobic electron transport**. Many bacteria and protistans depend exclusively on anaerobic pathways to make ATP. Your own cells depend on aerobic respiration, but they also use a fermentation pathway for short periods when oxygen supplies are low.

All three types of energy-releasing pathways start with the same reactions, in which a glucose molecule is split and rearranged into two pyruvate molecules. This first stage of reactions, called **glycolysis**, proceeds in the cytoplasm, and oxygen has no role in it. Once the first stage is completed, the pathways differ. Most important, the aerobic pathway continues inside a mitochondrion, where oxygen serves as the final acceptor of electrons that are released during the reactions. Anaerobic pathways end in the cytoplasm, and a substance other than oxygen is the final electron acceptor.

As we examine the three types of pathways, keep in mind that the reaction steps do not proceed all by themselves. Enzymes catalyze each reaction step, and the intermediate produced at a given step serves as a substrate for the next enzyme in the pathway.

AEROBIC RESPIRATION

Overview of the Reactions

Of all energy-releasing pathways, aerobic respiration produces the most ATP for each glucose molecule. Whereas fermentation has a net yield of two ATP, the aerobic route may yield *thirty-six* or more. If you were a bacterium, you would not require much ATP. Being large, complex, and highly active, you depend on the high yield of the aerobic route.

When glucose is the starting material, aerobic respiration can be summarized this way:

$$C_6H_{12}O_6 + 6O_2 \longrightarrow 6CO_2 + 6H_2O$$

glucose   carbon
                                                 dioxide

The summary equation tells us only what the substances are at the start and finish of the pathway. In between are *three stages* of reactions.

Figure 6.2 outlines the reactions. The first stage, again, is glycolysis. In the second stage, which includes the **Krebs cycle**, pyruvate is broken down to carbon dioxide and water. Neither stage produces much ATP. But hydrogen and electrons are stripped from intermediates during both stages, and coenzymes deliver these to an electron transport system. Such systems, along with neighboring enzymes, serve as the machinery for **electron transport phosphorylation**. This third and final stage of reactions yields many ATP molecules. As it draws to a close, oxygen accepts the "spent" electrons from the transport system.

Figure 6.2 (a) Overview of the three stages of aerobic respiration, the main energy-releasing pathway. Only this pathway delivers enough ATP for sustained, strenuous activity.

In the first stage (glycolysis), glucose is partially broken down to pyruvate. In the second stage, which includes the Krebs cycle, pyruvate is completely broken down to carbon dioxide. During these two stages, hydrogen and electrons stripped from intermediates are loaded onto coenzymes (NAD^+ and FAD).

In the final stage, electron transport phosphorylation, the loaded coenzymes (NADH and $FADH_2$) give up hydrogen and electrons. As the electrons pass through transport systems, energy is released that indirectly drives ATP formation. Oxygen accepts electrons at the end of the transport system.

From start (glycolysis) to finish, the aerobic pathway typically has a net energy yield of thirty-six ATP.

(b) Like the other main energy-releasing pathways, aerobic respiration starts with glycolysis. But only this pathway is completed in mitochondria.

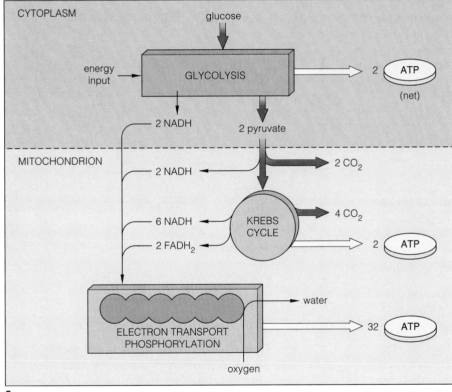

a

Typical Energy Yield: 36 ATP

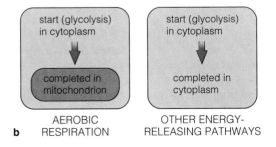

b AEROBIC RESPIRATION   OTHER ENERGY-RELEASING PATHWAYS

Figure 6.3 Glycolysis, the first stage of the main energy-releasing pathways. The reactions take place in the cytoplasm of all prokaryotic and eukaryotic cells.

In this example, glucose is the starting material. The reactions produce two pyruvate, two NADH, and four ATP molecules. Two ATP are invested to start glycolysis, so the *net* energy yield of glycolysis is two ATP.

Depending on the cell type and on conditions in its environment, the pyruvate molecules may be used in the second set of reactions of the aerobic pathway, which includes the Krebs cycle. Or it may be used in different reactions, such as those of fermentation pathways.

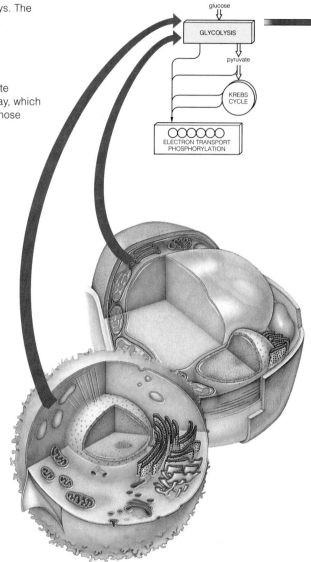

Glycolysis

Recall, from Figure 2.18, that glucose is a simple sugar with a backbone of six carbon atoms. In glycolysis, glucose (or some other carbohydrate) present in the cytoplasm is partially broken down to **pyruvate**, a molecule with a backbone of three carbon atoms.

We can think about the glucose backbone in this simplified way:

As Figure 6.3 shows, the first steps of glycolysis are *energy-requiring*. They proceed only when two ATP molecules each donate energy to the glucose backbone by transferring a phosphate group to it. Such transfers are called phosphorylations. The attachments cause the backbone to split apart into two molecules of **PGAL** (phosphoglyceraldehyde).

PGAL formation marks the start of the *energy-releasing* steps of glycolysis. Each PGAL is converted to an unstable intermediate that gives up a phosphate group to ADP, forming ATP. The next intermediate in the sequence does the same thing (Figure 6.3). Thus a total of four ATP molecules has formed by **substrate-level phosphorylation**—the direct transfer of a phosphate group from a substrate of the reactions to ADP. Remember, though, two ATP were invested to start the reactions. So the net energy yield is only two ATP.

Meanwhile, hydrogen atoms and electrons that were released from each PGAL are transferred to an enzyme helper—the coenzyme **NAD$^+$**. Like other coenzymes, NAD$^+$ is reusable. As you read on page 67, it picks up hydrogen and electrons stripped from a substrate and so becomes NADH. When it gives them up at a different site, it becomes NAD$^+$ again.

In sum, glycolysis converts a bit of the energy stored in glucose to ATP energy. Hydrogen and electrons stripped from glucose have been loaded onto a coenzyme, and these have roles in the next stage of reactions. So do the end products of glycolysis—two molecules of pyruvate.

During glycolysis, a glucose molecule is partially broken down. Two NADH and four ATP form during the reactions, which end with two pyruvate molecules.

After subtracting the two ATP molecules required to start the reactions, the *net* energy yield of glycolysis is only two ATP.

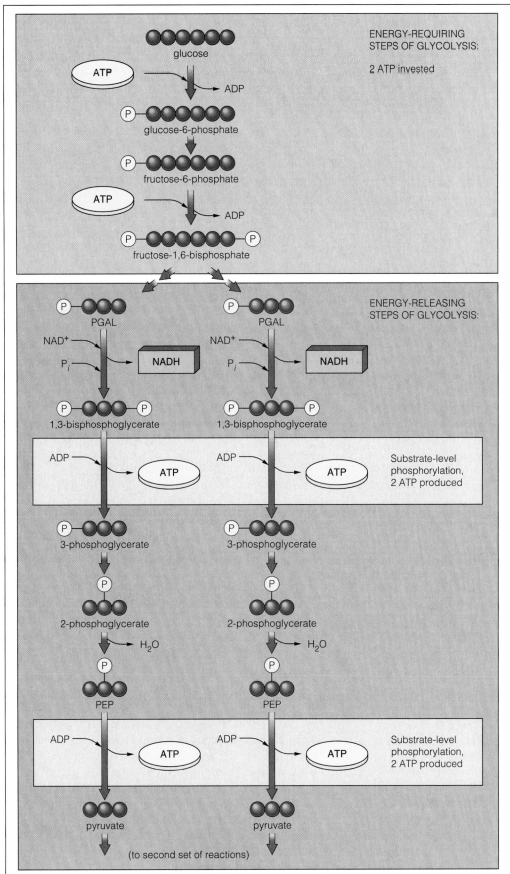

ENERGY-REQUIRING STEPS OF GLYCOLYSIS:

2 ATP invested

glucose

glucose-6-phosphate

fructose-6-phosphate

fructose-1,6-bisphosphate

ENERGY-RELEASING STEPS OF GLYCOLYSIS:

PGAL

NAD⁺

P_i

NADH

1,3-bisphosphoglycerate

ADP

ATP

Substrate-level phosphorylation, 2 ATP produced

3-phosphoglycerate

2-phosphoglycerate

H_2O

PEP

ADP

ATP

Substrate-level phosphorylation, 2 ATP produced

pyruvate

(to second set of reactions)

NET ENERGY YIELD: 2 ATP

a *Glycolysis starts with an energy investment of two ATP.* First, enzyme action promotes the transfer of a phosphate group from ATP to glucose, which has a backbone of six carbon atoms. With this transfer, the glucose molecule becomes slightly rearranged.

b Enzyme action promotes the transfer of a phosphate group from another ATP to the rearranged molecule.

c The resulting fructose-1,6-bisphosphate molecule splits at once into two molecules, each with a three-carbon backbone. We can call these two PGAL.

d During enzyme-mediated reactions, two NADH form after each PGAL gives up two electrons and a hydrogen atom to two NAD⁺. Each PGAL also combines with inorganic phosphate (P_i) present in the cytoplasm, then donates a phosphate group to ADP.

e *Thus two ATP have formed by the direct transfer of phosphate from two intermediate molecules that serve as substrates in the reactions.* With this formation of two ATP, the original energy investment of two ATP is paid off.

f In the next two enzyme-mediated reactions, each of the two intermediate molecules releases a hydrogen atom and an —OH group, which combine to form water.

g The resulting intermediates (two molecules of 3–phosphoenolpyruvate, or PEP), are rather unstable. Each gives up a phosphate group to ADP. *Once again, two ATP have formed by substrate-level phosphorylation.*

h Thus the net energy yield from glycolysis is two ATP for each glucose molecule entering the reactions. The end products of glycolysis are two molecules of pyruvate, each with a three-carbon backbone.

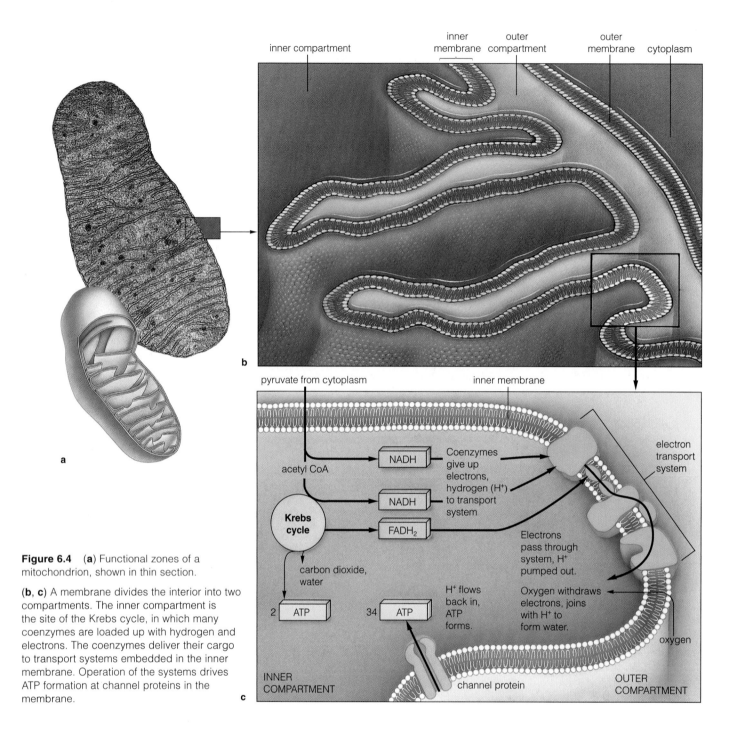

Figure 6.4 (**a**) Functional zones of a mitochondrion, shown in thin section.

(**b, c**) A membrane divides the interior into two compartments. The inner compartment is the site of the Krebs cycle, in which many coenzymes are loaded up with hydrogen and electrons. The coenzymes deliver their cargo to transport systems embedded in the inner membrane. Operation of the systems drives ATP formation at channel proteins in the membrane.

Krebs Cycle

After pyruvate forms in the cytoplasm, it may enter a **mitochondrion**. The second and third stages of aerobic respiration proceed *only* in this organelle. Figures 3.20 and 6.4 show its compartmented structure.

The second-stage reactions occur in the inner compartment. A bit more ATP forms. All of pyruvate's carbon and oxygen atoms end up in carbon dioxide and water. And hydrogen and electrons stripped from it are loaded onto coenzymes.

Take a look at Figure 6.5. During a few preparatory steps, a carbon atom is stripped from each pyruvate molecule, leaving an acetyl group that gets picked up by a coenzyme (forming acetyl-CoA). The acetyl group becomes attached to oxaloacetate, the point of entry into a cyclic pathway called the Krebs cycle. The name honors Hans Krebs, who began working out the pathway's details in the 1930s. Notice that three carbon atoms enter the second stage of reactions (as a pyruvate backbone) and three leave (in three carbon dioxide molecules) during the preparatory reactions and the cycle proper.

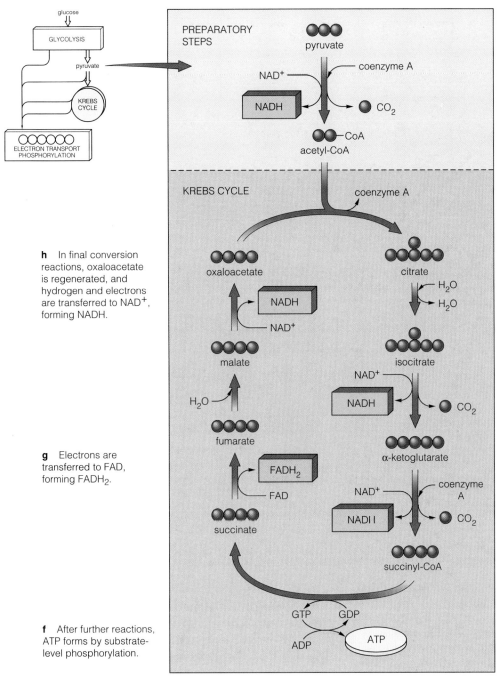

PREPARATORY STEPS

pyruvate

coenzyme A

NAD+

NADH

CO_2

CoA

acetyl-CoA

KREBS CYCLE

coenzyme A

oxaloacetate

citrate

H_2O
H_2O

NADH

NAD+

malate

isocitrate

NAD+

NADH

CO_2

H_2O

fumarate

α-ketoglutarate

FADH$_2$

FAD

NAD+

coenzyme A

NADH

CO_2

succinate

succinyl-CoA

GTP   GDP

ADP   ATP

glucose

GLYCOLYSIS

pyruvate

KREBS CYCLE

ELECTRON TRANSPORT PHOSPHORYLATION

a A pyruvate molecule enters a mitochondrion. It undergoes preparatory conversions before entering cyclic reactions (Krebs cycle).

b First, the pyruvate is stripped of a functional group (COO^-), which departs as CO_2. Next, it gives up hydrogen and electrons to NAD+, forming NADH. A coenzyme joins with the two-carbon fragment, forming acetyl-CoA.

c The acetyl-CoA is transferred to oxaloacetate, a four-carbon compound that is the point of entry into the Krebs cycle. The result is citrate, with a six-carbon backbone.

d Citrate enters conversion reactions in which a COO^- group departs (as CO_2). Also, hydrogen and electrons are transferred to NAD+, forming NADH.

e Another COO^- group departs (as CO_2) and another NADH forms. *At this point, three carbon atoms have been released, balancing out the three that entered the mitochondrion (in pyruvate).*

h In final conversion reactions, oxaloacetate is regenerated, and hydrogen and electrons are transferred to NAD+, forming NADH.

g Electrons are transferred to FAD, forming FADH$_2$.

f After further reactions, ATP forms by substrate-level phosphorylation.

Figure 6.5 The Krebs cycle and the preparatory reactions preceding it. For each three-carbon pyruvate molecule, three CO_2, one ATP, four NADH, and one FADH$_2$ are formed. The steps shown occur *twice* (remember, the glucose molecule was broken down initially to *two* pyruvate molecules).

The Krebs cycle has three functions. First, hydrogen and electrons from substrates are transferred to NAD+ and **FAD** (another coenzyme), which thereby become NADH and FADH$_2$. Second, substrate-level phosphorylations produce two ATP. Third, intermediates are juggled back to the form of oxaloacetate. Cells have only so much oxaloacetate, and it must be regenerated to keep the cyclic reactions going.

The second stage adds only two more ATP to the small yield from glycolysis. But it also loads up many coenzymes with hydrogen and electrons that can be used during the third stage:

Glycolysis:		2 NADH
Pyruvate conversion preceding Krebs cycle:		2 NADH
Krebs cycle:	2 FADH$_2$	6 NADH
Total coenzymes sent to third stage:	2 FADH$_2$ +	10 NADH

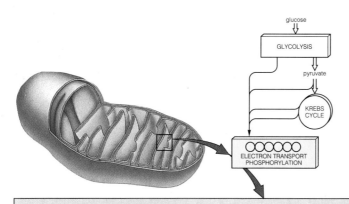

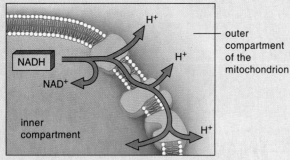

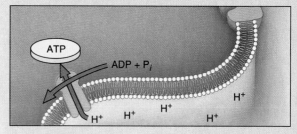

Figure 6.6 Electron transport phosphorylation. These reactions occur at transport systems and channel proteins (called ATP synthases) embedded in the inner mitochondrial membrane. Each transport system consists of enzymes and other proteins (including cytochrome molecules) that operate in sequence. The inner membrane creates two compartments. The reactions begin in the inner compartment, when NADH and FADH$_2$ give up hydrogen (as H$^+$ ions) and electrons to the transport system. Electrons are accepted and passed through the system, but the ions are left behind—in the outer compartment:

Soon there is a higher concentration of H$^+$ in the outer compartment than in the inner one. Concentration and electric gradients now exist across the membrane.

The ions follow the gradients and flow across the membrane, through the interior of the channel proteins. Energy associated with the flow drives the formation of ATP from ADP and unbound phosphate:

Do these events sound familiar? They should. ATP forms in much the same way in chloroplasts (Figure 5.8). The idea that concentration and electric gradients across a membrane drive ATP formation is sometimes called the chemiosmotic theory.

Electron Transport Phosphorylation

ATP production goes into high gear during the third stage of the aerobic pathway. The production machinery runs on electrons and unbound hydrogen (that is, H$^+$ ions) delivered by coenzymes. Electron transport systems and neighboring channel proteins serve as the machinery. These are embedded in a membrane that divides the interior of a mitochondrion into two compartments (Figure 6.4).

Electron transport systems consist of an organized sequence of enzymes and other molecules. As Figure 6.6 shows, electrons pass all the way through these systems, but the H$^+$ gets tossed off into the outer compartment. This sets up H$^+$ concentration and electric gradients across the membrane. H$^+$ follows the gradient and flows back into the inner compartment, through channel proteins. The flow drives ATP formation from ADP and unbound phosphate.

Free oxygen withdraws electrons from the transport system, then combines with H$^+$ to form water molecules. Its action pulls electrons through the system and keeps ATP production going.

The third stage commonly produces thirty-two ATP. This brings the total net yield of the aerobic pathway to thirty-six ATP for each glucose molecule.

In aerobic respiration, glucose is completely broken down to carbon dioxide and water.

NAD$^+$ and FAD accept hydrogen and electrons stripped from substrates of the reactions and deliver them to an electron transport system. Oxygen is the final acceptor of those electrons.

From start (glycolysis in the cytoplasm) to finish (in the mitochondrion), this pathway commonly yields thirty-six ATP for every glucose molecule.

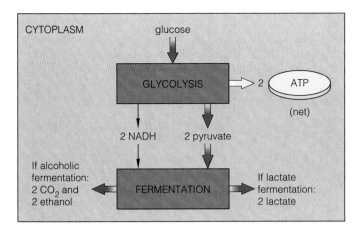

Figure 6.7 Overview of fermentation. In this type of energy-releasing pathway, the net ATP yield is from the initial reactions (glycolysis).

ANAEROBIC ROUTES

Many organisms thrive in marshes, bogs, mud, the animal gut, canned foods, sewage treatment ponds, and other anaerobic settings. Some actually die if exposed to oxygen. Bacteria responsible for many diseases, including botulism and tetanus, are like this. Other organisms, such as the bacterial "employees" of yogurt manufacturers, are indifferent to oxygen. Still others use oxygen but switch to anaerobic pathways when oxygen levels drop.

Three anaerobic pathways are common. They are called alcoholic fermentation, lactate fermentation, and anaerobic electron transport.

Alcoholic Fermentation

Like aerobic respiration, *both fermentation pathways begin with glycolysis*. As Figure 6.7 shows, glucose is partly broken down to two pyruvate molecules, two NADH form, and the net energy yield is two ATP.

In **alcoholic fermentation**, each pyruvate is then rearranged to form acetaldehyde, and carbon dioxide is released. The NADH gives up electrons to acetaldehyde to form the end product, ethanol (Figure 6.8).

Yeasts (single-celled fungi) use this pathway. And commercial bakers use certain yeasts. *Saccharomyces cerevisiae* makes bread dough rise. Bakers mix this yeast with sugar and blend the mixture into dough. As the yeast cells use the sugar, they release carbon dioxide, a gas that expands the dough (makes it "rise"). Oven heat forces the gas out, leaving a porous product.

Beer and wine manufacturers use yeasts on a large scale. Vintners use wild yeasts that live on grapes. They also use cultivated strains of *S. ellipsoideus*, which

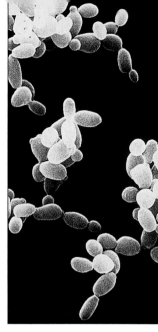

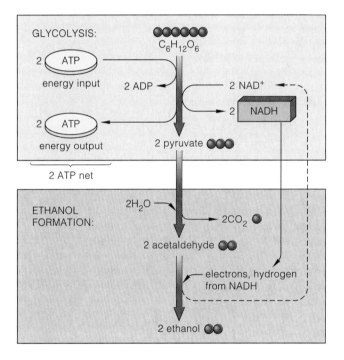

Figure 6.8 Alcoholic fermentation. In this anaerobic pathway, an intermediate of the reactions (acetaldehyde) serves as the final electron acceptor, and ethanol is the end product. The photograph to the left shows the dustlike coating on grapes. The coating contains yeasts, single-celled organisms that use a fermentation pathway. The photograph to the right shows cells of the yeast that makes bread dough rise.

remain active until the alcohol concentration in the vats exceeds 14 percent. Wild yeasts die when it exceeds 4 percent—but 4 percent still packs a punch. Robins get drunk on the naturally fermenting berries of *Pyracantha* shrubs. So do wild turkeys when they gobble up fermenting apples in untended orchards.

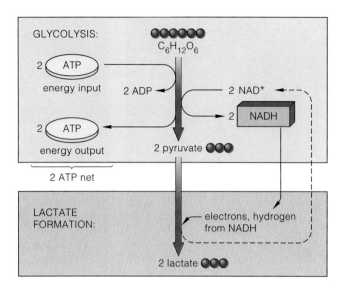

Figure 6.9 Lactate fermentation. In this anaerobic pathway, electrons end up in the reaction product (lactate).

Lactate Fermentation

In **lactate fermentation**, glycolysis produces two pyruvate and two NADH molecules, and two ATP form. But that's about it. Pyruvate itself is the final acceptor of electrons from NADH and so becomes the end product, lactate (Figure 6.9). Sometimes this product is called "lactic acid." In cells, however, the ionized form (lactate) is far more common.

When they produce lactate, certain bacterial cells turn milk or cream sour. When your demands for energy are intense but brief—say, during a short race—your muscle cells switch to lactate fermentation for a quick ATP fix. They can't do this for long—too much of glucose's stored energy would be thrown away for too little ATP. When glucose stores are depleted, muscles fatigue and lose their ability to contract.

This brings us to an important point. Fermentation pathways do not break down glucose completely, so no more ATP is produced beyond the yield from glycolysis. *The final steps of a fermentation pathway serve only to regenerate NAD$^+$.* The energy yield is enough for single-celled anaerobic organisms. It even helps carry some "aerobic" cells through times of stress. But it is not enough to sustain large, multicelled organisms (this being one reason why you never will come across an anaerobic elephant).

The initial stage of reactions of both fermentation pathways (glycolysis) has a net energy yield of two ATP. The remaining reactions simply regenerate the NAD$^+$ required for glycolysis.

Anaerobic Electron Transport

Diverse energy-releasing pathways occur in the natural world, mostly among bacteria. Some influence the cycling of nitrogen, sulfur, and other vital elements through ecosystems (Chapter 35). Anaerobic electron transport is such a pathway. After the electrons required for ATP production pass through transport systems in the bacterial plasma membrane, they often are accepted by an inorganic compound in the environment. While you are reading this, for example, certain anaerobic bacteria living in waterlogged soil are stripping electrons from a variety of compounds. Then they dump the electrons on sulfate (SO_4^-), leaving hydrogen sulfide (H_2S), a very foul-smelling gas.

ALTERNATIVE ENERGY SOURCES IN THE HUMAN BODY

Of the foods we eat, glucose and other carbohydrates are the main ones sent through the ATP-producing pathways. This is true of mammals generally. Our cells use glucose molecules for ATP production for as long as they are available. Our brain cells can use almost nothing else. When the level of glucose in the blood decreases slightly, we break into stores of glycogen, a polysaccharide composed of glucose units. A starving mammal breaks into fat reserves, thus sparing whatever glucose is left for the all-important brain. It breaks down the body's own proteins as a last resort.

Figure 6.10 shows how proteins and fats can be used as alternative energy sources. First, proteins are broken down to amino acid subunits. Then the amino group (NH_3) of each subunit is split away. One way or another, the fragments enter the Krebs cycle, where hydrogen is stripped from the remaining carbon atoms—and transferred to the cycle's coenzymes. The amino groups undergo conversions that produce urea, a nitrogen-containing waste product that is excreted from the body.

A fat molecule, recall, has a glycerol head and one, two, or three fatty acid tails. First the head and tails are separated. The glycerol is converted to PGAL and slipped into the reaction sequence of glycolysis. Each fatty acid's carbon backbone is split into fragments, which are converted to acetyl-CoA and slipped into the Krebs cycle. A fatty acid has many more carbon-bound hydrogen atoms than glucose—and its breakdown produces much more ATP.

This concludes our look at aerobic respiration and other energy-releasing pathways. To gain insight into how these pathways fit into the greater picture of life's evolution and interconnectedness, take a moment to read the *Commentary* that concludes this unit.

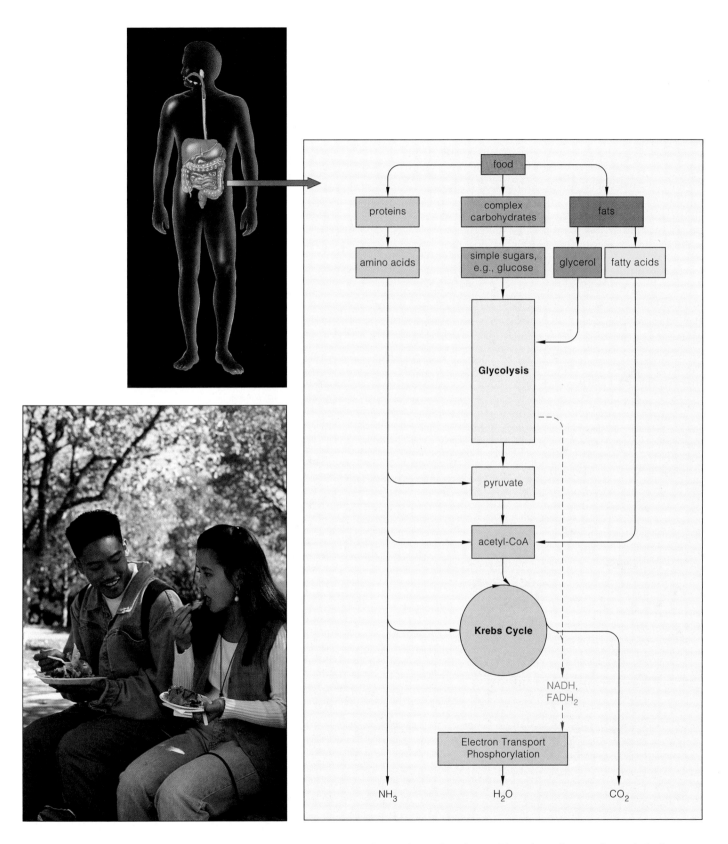

Figure 6.10 Points of entry into the aerobic pathway for complex carbohydrates, fats, and proteins after they have been reduced to their simpler components in the human digestive system.

Perspective on Life

In this unit, you read about photosynthesis and aerobic respiration—the main pathways by which cells trap, store, and release energy. Those pathways became linked on a grand scale over evolutionary time.

Life began about 3.8 billion years ago, when the atmosphere had little free oxygen. Early single-celled organisms probably made ATP by reactions similar to glycolysis. Without oxygen, fermentation pathways must have dominated. About 1.5 billion years later, oxygen-producing photosynthetic cells had emerged, and they turned out to be a profound force in evolution. Oxygen, a by-product of photosynthesis, began to accumulate in the atmosphere. Possibly as a result of mutations in the proteins of electron transport systems, some cells started using oxygen as an electron acceptor. In time, descendants of those fledgling aerobic cells abandoned photosynthesis. Among them were the forerunners of animals and other organisms that engage in aerobic respiration.

With aerobic respiration, life became self-sustaining, for the final products—carbon dioxide and water—are precisely the materials used to build organic compounds in photosynthesis! Thus the flow of carbon, hydrogen, and oxygen through the metabolic pathways of living organisms came full circle:

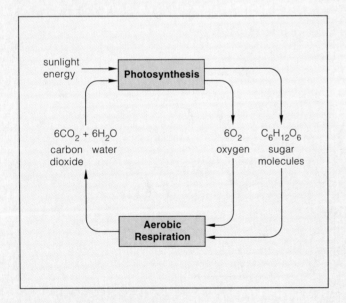

Perhaps you have difficulty perceiving the connection between yourself—a living, intelligent being—and such remote-sounding things as energy and the cycling of carbon, hydrogen, and oxygen. Is this really the stuff of humanity?

Think back, for a moment, to a water molecule. Two hydrogen atoms sharing electrons with an oxygen atom doesn't seem close to our daily lives. But through that sharing, water molecules show polarity and hydrogen-bond with one another. That is a beginning for the organization of lifeless matter that leads to the organization of all living things.

For now you can imagine different molecules in water. The nonpolar ones resist interaction with water; the polar ones dissolve in it. The phospholipids among them spontaneously assemble into a two-layered film. Such lipid bilayers are the basis for all cell membranes, hence all cells. From the beginning, the cell has been the fundamental *living* unit.

The essence of life is not some mysterious force. It is metabolic control. With a cell membrane to contain them, metabolic reactions *can* be controlled. Cells can respond to energy changes and to the kinds of molecules in their environment. Their response mechanisms operate by "telling" protein molecules—enzymes—when and what to build or tear down.

And it is not some mysterious force that creates the proteins themselves. DNA, the slender double strand of heredity, has the chemical structure—*the chemical message*—that allows molecule to reproduce molecule, one generation after the next. Those DNA strands tell trillions of cells in your body how countless molecules must be built and torn apart for their stored energy.

So yes, carbon, hydrogen, oxygen, and other organic molecules represent the stuff of you, and us, and all of life. But it takes more than molecules to complete the picture. Life exists only as long as a constant flow of energy maintains its organization. Molecules are assembled into cells, cells into organisms, organisms into communities, and so on up through the biosphere. It takes energy, primarily from the sun, to maintain all these levels of organization. And energy flows through time in one direction—from organized to less organized forms. Only as long as sunlight flows into the web of life can life continue in all its rich diversity.

In short, life is no more *and no less* than a marvelously complex system of prolonging order. Sustained by energy transfusions, life continues because of a capacity for self-reproduction—a handing down of hereditary instructions in DNA. With DNA, energy and materials can be organized, generation after generation. Even with the death of the individual, life is prolonged. With death, molecules are released and recycled once more, providing raw materials for new generations. In this flow of energy and cycling of material through time, each birth is affirmation of our ongoing capacity for organization, each death a renewal.

SUMMARY

1. Nearly all metabolic reactions run on energy delivered to them by ATP molecules. ATP can be produced by aerobic respiration, fermentation, and other pathways that release chemical energy from glucose and other organic compounds. All three kinds of pathways begin with the same reactions, called glycolysis. Glycolysis occurs only in the cytoplasm.

2. Electrons and hydrogen stripped from glucose are central to the energy-releasing pathways. Coenzymes (NAD^+ and often FAD) deliver electrons and hydrogen to sites of ATP production.

3. In aerobic respiration, oxygen is the final acceptor of electrons stripped from glucose. The pathway proceeds through three stages of reactions: glycolysis, the Krebs cycle (and preparatory conversions), and electron transport phosphorylation. Its net energy yield is commonly thirty-six or more ATP molecules.

   a. During glycolysis, a glucose molecule is partially broken down and two pyruvate, two NADH, and four ATP are produced. The net energy yield is only two ATP (because two ATP had to be invested up front to get the reactions going).

   b. The second stage takes place in mitochondria. Pyruvate is converted to a molecule that can enter a cyclic pathway, the Krebs cycle. By the time the initial conversions and the cyclic reactions are over, glucose has been broken down completely to carbon dioxide and water. The second stage produces ten coenyzmes (eight NADH and two $FADH_2$) and two ATP.

   c. The third stage also takes place in mitochondria. An inner membrane divides a mitochondrion into two compartments. Electron transport systems and channel proteins are embedded in the membrane.

   d. Coenzymes deliver electrons to a transport system. Operation of the system sets up H^+ concentration and electric gradients across the membrane. H^+ flows down the gradients, through channel proteins. Energy associated with the flow drives the formation of ATP from ADP and unbound phosphate. Oxygen withdraws electrons and combines with H^+ to form water.

4. Alcoholic fermentation, lactate fermentation, and anaerobic electron transport use an organic or inorganic compound (not oxygen) as the final electron acceptor.

   a. Like the aerobic pathway, the fermentation pathways start with glycolysis. (Two ATP are invested, glucose is partially broken down to pyruvate, and two NADH and four ATP form.)

   b. In alcoholic fermentation, pyruvate is converted to an intermediate (acetaldehyde), and carbon dioxide is released. That intermediate accepts the "spent" electrons and so becomes ethanol.

c. In lactate fermentation, the pyruvate molecule simply is rearranged, then accepts the "spent" electrons and so becomes lactate.

5. Compared with aerobic respiration, the anaerobic pathways have a small net yield (two ATP, from glycolysis), because glucose is not completely degraded. Following the initial reactions (glycolysis), the remainder of those pathways simply regenerates the NAD^+ required for glycolysis.

Review Questions

1. Which energy-releasing pathway yields the most ATP for each glucose molecule? *86, 92, 93*

2. Think of the various energy-releasing pathways. Which of their reactions occur only in cytoplasm? Which occur only in mitochondria? *86*

3. State the function of coenzymes in the energy-releasing pathways. *88*

4. Briefly describe glycolysis, the first stage of the energy-releasing pathways. *88–89*

5. Describe the two stages of aerobic respiration that follow glycolysis:
   a. the Krebs cycle (and preparatory conversions) *90–91*
   b. electron transport phosphorylation *92*

6. Describe the functional zones of a mitochondrion. Explain where the electron transport systems and carrier proteins required for ATP formation are located. *90, 92*

7. Is the following statement true? Your muscle cells cannot function at all without oxygen. *94*

8. In fermentation, conversions following the net production of two ATP do not yield more energy. What do they accomplish? *94*

Self-Quiz *(Answers in Appendix IV)*

1. Stored energy can be released from glucose and used to produce _____ , an energy carrier.

2. In the first stage of the main energy-releasing pathways, glucose is partly broken down to _____ .

3. ATP is _____ .
   a. a phosphate compound
   b. an energy carrier
   c. produced by all organisms
   d. all of the above

4. Which of the following is *not* produced during glycolysis?
   a. NADH   c. FAD
   b. pyruvate   d. ATP

5. Glycolysis occurs in the _____ .
   a. nucleus   c. plasma membrane
   b. mitochondrion   d. cytoplasm

6. The final acceptor of electrons stripped from glucose during aerobic respiration is _____ .
   a. water   c. oxygen
   b. hydrogen   d. NADH

7. For the aerobic pathway, electron transport systems are located in the _____ .
   a. cytoplasm
   b. inner mitochondrial membrane
   c. outer mitochondrial compartment
   d. stroma

8. The flow of _____ through channel proteins drives the formation of ATP from ADP and phosphate.
   a. electrons   c. NADH
   b. hydrogen ions   d. $FADH_2$

9. Match the events with the metabolic reaction.
   _____ glycolysis   a. ATP, NADH, and CO_2 form
   _____ fermentation   b. glucose to two pyruvate
   _____ Krebs cycle   c. NAD^+ regenerated, two ATP
   _____ electron transport   net
        phosphorylation   d. H^+ flows through channel
                                     proteins, ATP forms

Selected Key Terms

aerobic respiration *86*   Krebs cycle *86*
alcoholic fermentation *93*   lactate fermentation *94*
anaerobic electron transport *86* mitochondrion *90*
ATP *86* NAD^+ *88*
electron transport   PGAL *88*
   phosphorylation *86*   pyruvate *88*
FAD *91*   substrate-level
fermentation *86*   phosphorylation *88*
glycolysis *86*

Readings

Levi, P. October 1984. "Travels with C." *The Sciences*. Journey of a carbon atom through the world of life.

Roberts, L. August 28, 1987. "Discovering Microbes with a Taste for PCBs." *Science* 237:975–977. Bacteria make ATP by breaking down pollutants.

Wolfe, S. 1992. *Molecular and Cellular Biology.* Belmont, California:

FACING PAGE: *Human sperm, one of which will penetrate this mature egg and so set the stage for the development of a new individual in the image of its parents.*

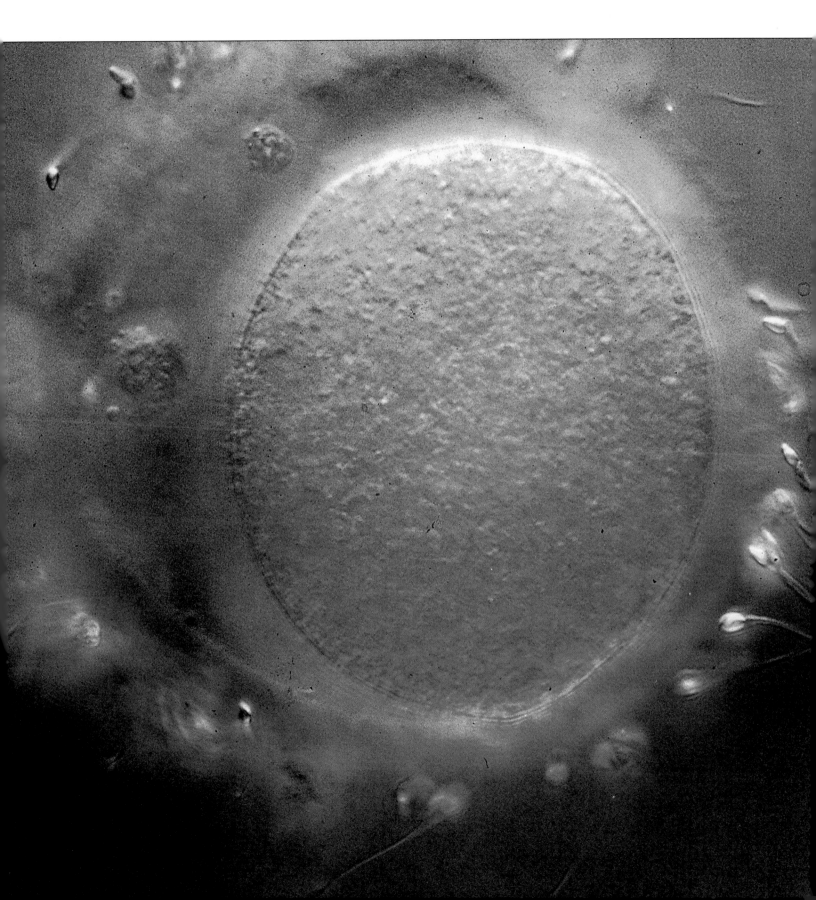

7 CELL DIVISION AND MITOSIS

Silver in the Stream of Time

Five o'clock, and the first rays of the sun dance over the wild Alagnak River of the Alaskan tundra. It is September, and life is both ending and beginning in the clear, frigid waters. By the thousands, mature silver salmon have returned from the open ocean to spawn in their shallow native home. The females are tinged with red, the color of spawners, and they are dying.

On this morning, a female salmon releases translucent pink eggs into a "nest," hollowed out by her fins in the gravel riverbed (Figure 7.1). Within moments, a male salmon sheds a cloud of sperm, and fertilization follows. Trout and other predators eat most of the eggs, but some survive and give rise to a new generation.

Within three years, the pea-size eggs have become streamlined salmon, fashioned from billions of cells. A few of those cells will develop into eggs or sperm. In time, on some September morning, they will take part in an ongoing story of birth, cell divisions and growth, death, and rebirth.

For you, as for salmon and all other multicelled organisms, growth as well as reproduction depends on *cell division*. In your mother's body, a single fertilized egg divided in two, then the two into four, and so on until billions of cells were growing, developing in

specialized ways, and dividing at different times to produce your genetically prescribed body parts. Your body now has roughly 65 trillion cells—and many of the cells are still dividing. Every five days, for instance, cell divisions replace the lining of your small intestine.

Understanding cell division—and, ultimately, how new individuals are put together in the image of their parents—begins with answers to three questions. *First,* what instructions are necessary for inheritance? *Second,* how are those instructions duplicated for distribution into daughter cells? *Third,* by what mechanisms are those instructions divided into daughter cells? We will require more than one chapter to consider cell reproduction and other mechanisms of inheritance. However, the points made in the first part of this chapter can help you keep the overall picture in focus.

Figure 7.1 The last of one generation and the first of the next in the Alagnak River of Alaska.

1. When a cell divides, its two daughter cells must each receive a required number of DNA molecules and some cytoplasm. In eukaryotes, mitosis sorts out the DNA into two new nuclei. A separate mechanism divides the cytoplasm in two.

2. Mitotic cell division is the basis of growth and tissue repair in multicelled eukaryotes. It also is the means by which single-celled species and many multicelled species reproduce asexually.

3. DNA molecules are also called chromosomes. Members of the same species normally have the same total number of chromosomes in their body cells. The chromosomes have different lengths and shapes, and they carry different portions of the hereditary instructions.

4. When cells prepare for division, every chromosome is duplicated. Each now consists of two DNA molecules. Mitosis separates the two for distribution to daughter cells. In this way, each cell gets the same total number and the same types of chromosomes as the parent cell.

5. For many species, body cells have two of each type of chromosome, inherited from two parents. "Chromosome number" is the number of *each type* of chromosome for the species. Mitosis keeps the chromosome number constant from one cell generation to the next.

DIVIDING CELLS: THE BRIDGE BETWEEN GENERATIONS

Overview of Division Mechanisms

In biology, **reproduction** means producing a new generation of cells or multicelled individuals. Reproduction begins with the division of single cells. And the ground rule for cell division is this:

Parent cells must provide their daughter cells with hereditary instructions (encoded in DNA) and enough cytoplasmic machinery to start up their own operation.

DNA, recall, contains instructions for synthesizing proteins. Some proteins are structural materials. Many are enzymes that put together carbohydrates, lipids, and other building blocks of cells. Unless a daughter cell receives the necessary instructions for making proteins, it cannot grow or function properly.

Also, the parent cell's cytoplasm already has operating machinery—enzymes, organelles, and so on. When a daughter cell inherits what looks like a blob of cytoplasm, it really is getting start-up machinery for its operation, until it has time to use its inherited DNA for growing and developing on its own.

The cells of plants, animals, and other eukaryotic organisms divide the DNA by **mitosis** or **meiosis**. Both mechanisms sort out and package DNA molecules into new nuclei for forthcoming daughter cells. In other words, mitosis and meiosis only divide the nuclear material. Different mechanisms divide the cytoplasm and so split a parent cell into daughter cells.

Multicelled organisms grow by mitosis and cytoplasmic division of body cells, which are called **somatic cells**. They also repair tissues that way. Nick yourself peeling a potato, and mitotic cell divisions will replace the cells the knife sliced away. Besides this, many protistans, fungi, plants, and even some animals reproduce asexually by mitotic cell division.

In contrast, meiosis occurs only in **germ cells**, a cell lineage set aside for sexual reproduction. Meiosis must precede the formation of gametes, such as sperm and eggs. As you will see in the next chapter, it has much in common with mitosis, but the end result is different.

What about prokaryotic cells—the bacteria? They reproduce asexually by way of a different mechanism. We will consider the bacteria later, in Chapter 17. For now, take a moment to study Table 7.1, which lists the major division mechanisms.

Table 7.1 Cell Division Mechanisms	
Mechanisms	Used by
Mitosis, cytoplasmic division	*Single-celled eukaryotes (for asexual reproduction)*
	Multicelled eukaryotes (for bodily growth; also for asexual reproduction in many species)
Meiosis, cytoplasmic division	*Eukaryotes (basis of gamete formation and sexual reproduction)*
Prokaryotic fission	*Bacterial cells (for asexual reproduction)*

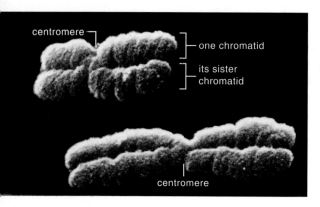

Figure 7.2 Photomicrograph of two chromosomes, each in the duplicated state.

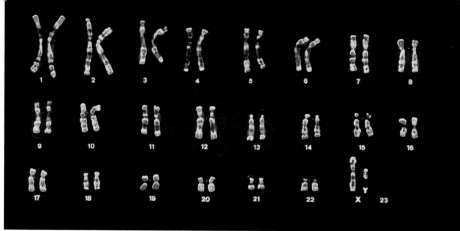

Figure 7.3 Forty-six chromosomes from a human male. Each chromosome is in the duplicated state.

Some Key Points About Chromosomes

Before a cell starts preparing for mitosis, its DNA molecules are stretched out like threads, with many proteins attached to them. Each DNA molecule, along with its attached proteins, is a **chromosome**.

Chromosomes undergo duplication while they are in this threadlike form. Now each consists of *two* DNA molecules, which will stay together until late in mitosis. For as long as they remain attached, the two are called **sister chromatids** of the chromosome:

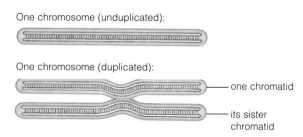

"Sister chromatids" is a name that seems to confuse almost everybody. It may help to stretch their dictionary meaning a bit and think of them as the forthcoming daughters of a parent chromosome.

Notice how the duplicated chromosome narrows down in a small region. This constricted region, the **centromere**, has attachment sites for microtubules that move the chromosome during nuclear division:

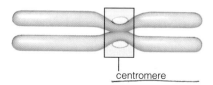

Keep in mind, the preceding sketches are simplifed. As Figure 7.2 suggests, the centromere location is different on different chromosomes. Also, a DNA molecule's two parallel strands don't look like a ladder—they are twisted together repeatedly like a spiral staircase and are much longer than can be shown here.

Mitosis and the Chromosome Number

All normal members of the same species have the same total number of chromosomes in their somatic cells. Humans have forty-six, gorillas have forty-eight, and pea plants have fourteen.

Figure 7.3 shows all forty-six human chromosomes, lined up as twenty-three pairs. Think of them as two sets of books on how to build a house. Your father gave you one set. Your mother had her own ideas about plumbing, storage, and so forth, so she gave you a revised edition. Her set covers the exact same topics but has different things to say about many of them.

Your chromosomes are like the volumes of two sets of books. Each set is numbered 1 to 23. Thus, for example, you have two "volumes" of chromosome 22—that is, a pair of them. Generally, both members of each pair have the same length and shape, and they carry instructions for the same traits. In these respects, they are not like any of the other pairs of chromosomes.

Any cell having two of each type of chromosome is a **diploid cell**. Such cells exist in gorillas, pea plants, and a great many other organisms besides humans.

With mitosis, a diploid parent cell produces two diploid daughter cells. This doesn't mean each merely gets forty-six or forty-eight or fourteen chromosomes. If only the total mattered, one cell might get, say, two pairs of chromosome 22 and no pairs of chromosome 9. However, neither cell would function properly without *one pair of each type of chromosome.*

Chromosome number tells you how many of each type of chromosome are present in a cell. Diploid cells, with two of each type, are $2n$. The n stands for chromosome number.

With mitosis, the chromosome number remains constant, division after division, from one cell generation to the next. Thus, if a parent cell is diploid, its two daughter cells will be diploid also.

MITOSIS AND THE CELL CYCLE

Mitosis is only one phase of the **cell cycle**. Such cycles start at the time new cells are produced, and they end when those cells complete their own division. The cycle starts again for each new daughter cell. As Figure 7.4 indicates, **interphase** is usually the longest part of a cell cycle. During this time, a cell increases its mass, roughly doubles the number of its cytoplasmic components, and finally duplicates its chromosomes.

The cell cycle lasts about the same length of time for cells of a given type. Its length differs among cells of different types. Your brain's nerve cells are arrested at interphase and typically do not divide again. Cells of a new sea urchin may double in number every two hours.

Adverse conditions can disrupt the cell cycle. When deprived of a vital nutrient, for instance, the free-living cells called amoebas will not leave interphase. Even so, when a cell proceeds past a certain point in interphase, the cycle normally continues regardless of outside conditions, owing to built-in controls over its duration.

Good health depends on the proper timing and completion of events in the cell cycle. Mistakes in the duplication or distribution of even one chromosome during interphase may lead to a genetic disorder. A growing tissue may die if controls are lost that otherwise would keep its cells from dividing. Cancer may follow if this happens in mature tissues, as described on page 175.

We turn now to mitosis and how it can maintain the chromosome number through turn after turn of the cell cycle. Figure 7.5 only hints at the precise division mechanisms that take over as a cell leaves interphase.

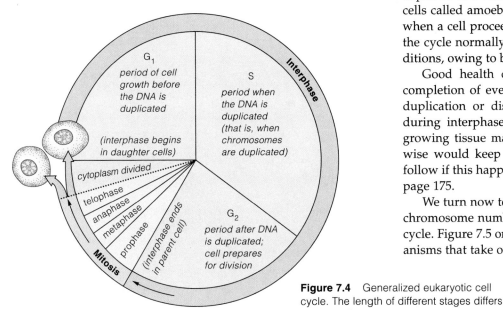

Figure 7.4 Generalized eukaryotic cell cycle. The length of different stages differs among cells.

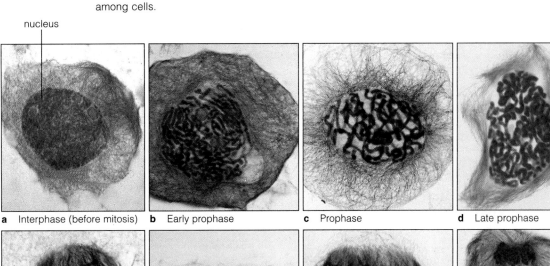

a Interphase (before mitosis) **b** Early prophase **c** Prophase **d** Late prophase

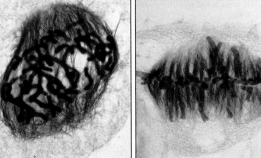

e Transition to metaphase **f** Metaphase **g** Anaphase **h** Telophase

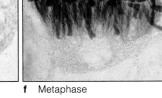

Figure 7.5 Mitosis in a cell from the African blood lily (*Haemanthus*). Chromosomes are stained blue, and microtubules that move them about are stained red.

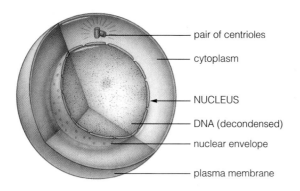

pair of centrioles

cytoplasm

NUCLEUS

DNA (decondensed)

nuclear envelope

plasma membrane

Cell at Interphase
The DNA is duplicated, then the cell prepares for division.

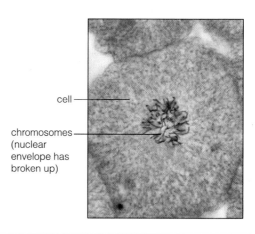

cell

chromosomes
(nuclear
envelope has
broken up)

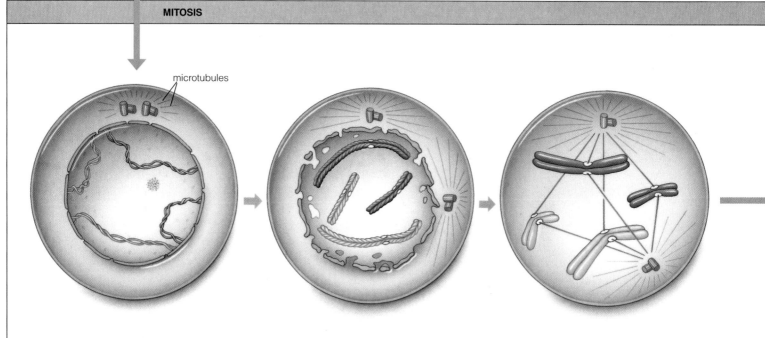

MITOSIS

microtubules

Early Prophase
The DNA and its associated proteins start to condense. The two chromosomes shaded purple were inherited from the male parent. The other two (blue) are their counterparts, inherited from the female parent.

Late Prophase
Chromosomes continue to condense. New microtubules are assembled, and they move one of two centriole pairs toward the opposite end of the cell. The nuclear envelope starts to break up.

Transition to Metaphase
Microtubules penetrate the nuclear region. Together they form a spindle apparatus. They become attached to the sister chromatids of each chromosome.

STAGES OF MITOSIS

Mitosis has four continuous stages, known as **prophase**, **metaphase**, **anaphase**, and **telophase**. Figure 7.6 shows these stages for a dividing animal cell. When you look closely at this illustration and at the preceding one, it becomes clear that the chromosomes within a cell move about dramatically during mitosis—and not on their own. A **spindle apparatus** moves them. In all cells, a fully formed spindle consists of two sets of microtubules. The microtubules extend from the spindle's two end points (poles) and overlap at its equator, which lies midway between the poles. The spindle poles establish the destinations of chromosomes during mitosis.

Prophase: Mitosis Begins

Prophase, the first stage of mitosis, is evident when chromosomes become visible in the light microscope as threadlike forms. ("Mitosis" comes from the Greek *mitos*, meaning thread.) Each chromosome was duplicated earlier, during interphase. It already consists of two sister chromatids joined at the centromere. By late prophase, all the chromosomes have condensed into thicker, rodlike forms. Part of the cell's cytoskeleton changes profoundly. Its microtubules disassemble into protein subunits—and these are put together as new microtubules for the forthcoming spindle. Meanwhile, the nuclear envelope starts to break up (Figure 7.6).

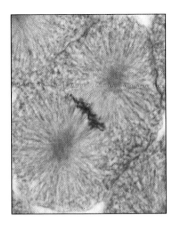
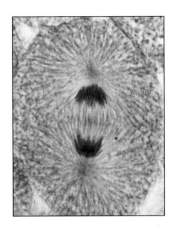
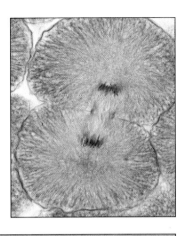

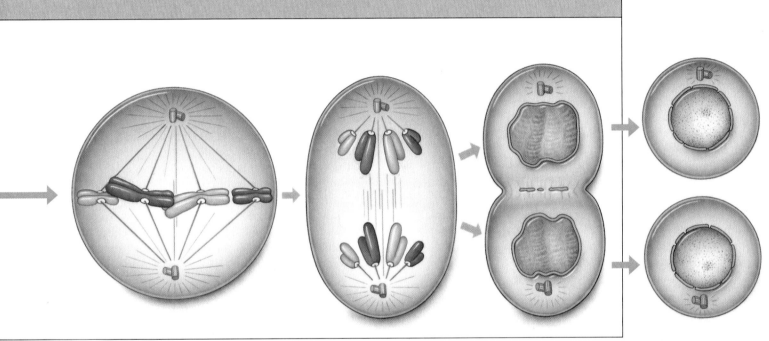

Metaphase
All chromosomes are lined up at the equator of the spindle. They are now in their most condensed form.

Anaphase
The attachment between the two sister chromatids of each chromosome breaks; the two are now chromosomes in their own right. They move to opposite spindle poles.

Telophase
Chromosomes decondense. New patches of membrane join to form nuclear envelopes around them. Most often, cytoplasmic division occurs before telophase is over.

Interphase
Two daughter cells have formed. Each is diploid, with two of each type of chromosome—just like the parent cell.

Transition to Metaphase

Major interactions occur during the transition from prophase to metaphase. (Many researchers call this transitional period "pro-metaphase.")

The nuclear envelope breaks up completely into vesicles. Microtubules are now free to interact with the chromosomes. At first they attach randomly to them. The attachments become more organized, and the spindle takes on final form. Now microtubules start pulling on each chromosome in both directions—toward both poles. The two-way yanking orients sister chromatids toward opposite poles. Meanwhile, microtubules from both poles are ratcheting past each other, and the force

Figure 7.6 Mitosis. This nuclear division mechanism assures that daughter cells will have the same chromosome number as the parent cell. For clarity, this diagram shows only two pairs of chromosomes from a diploid ($2n$) animal cell. With rare exceptions, the picture is more involved than this, as indicated by the micrographs of mitosis in a whitefish cell.

pushes the poles farther apart. All the push–pull forces are balanced when the chromosomes reach the spindle equator. When all the duplicated chromosomes are aligned here, midway between the spindle poles, we call this metaphase (*meta-* means midway between). The alignment is crucial for the next stage of mitosis.

Anaphase

During anaphase, the two chromatids of each chromosome are separated from each other and moved to opposite poles. Once they do separate, they are no longer referred to as chromatids. Each is now a chromosome in its own right:

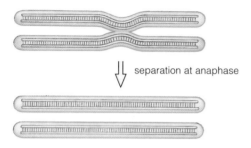

separation at anaphase

Said another way, every chromosome that was present in the parent cell now has a daughter chromosome at *both* spindle poles.

Telophase

Telophase begins once the two daughter chromosomes arrive at opposite spindle poles. The chromosomes are no longer harnessed to microtubules, and they return to threadlike form. Vesicles of the old nuclear envelope fuse together to form patches of membrane around the chromosomes. Patch joins with patch, and soon a new nuclear envelope separates each cluster of chromosomes from the cytoplasm. If the parent cell was diploid, each cluster will contain a pair of each type of chromosome. *With mitosis, each new nucleus has the same chromosome number as the parent nucleus.*

Once the two nuclei form, telophase is over—and so is mitosis.

DIVISION OF THE CYTOPLASM

The cytoplasm usually divides at some time between late anaphase and the end of telophase. **Cytoplasmic division** (or cytokinesis, as it is often called) proceeds by different mechanisms in animal and plant cells.

A layer of deposits forms around microtubules at the midsection of most animal cells. A shallow, ringlike depression appears above the layer, at the cell surface (Figure 7.7). At this depression, the **cleavage furrow**, microfilaments are at work. These components of the cytoskeleton are arranged to slide past one another. As they do, they also pull the plasma membrane inward and cut the cell in two.

The cells of land plants have fairly rigid walls, and they cannot be pinched in two. Their cytoplasm divides by the formation of a **cell plate**. Vesicles filled with wall-building material fuse with remnants from the spindle, forming a disklike structure (the "cell plate"). Here, cellulose deposits form a crosswall between the two daughter cells (Figure 7.8).

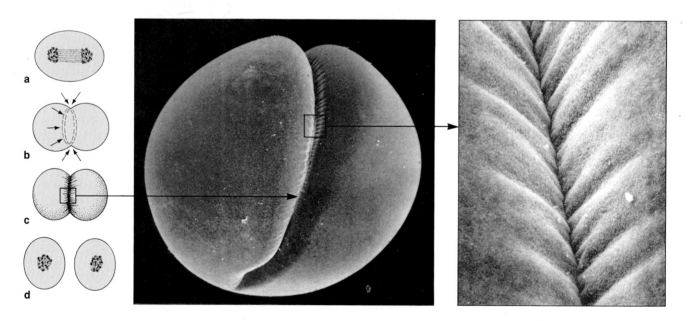

Figure 7.7 Cytoplasmic division of an animal cell. (**a**) Mitosis is complete and the spindle is disassembling. (**b**) Just beneath the plasma membrane, microfilament rings at the former spindle equator contract, like a purse string closing. (**c,d**) Continuing contractions divide the cell in two. The micrographs show how the plasma membrane sinks inward, defining the cleavage plane.

spindle equator

a As mitosis draws to a close, vesicles gather at the equator of the microtubular spindle. They contain cementing materials and starting materials for a new primary wall.

vesicles gathering

b A cell plate starts forming as the plasma membranes of vesicles fuse. Their contents become sandwiched between two membranes that form along the plane of the growing cell plate.

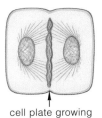

cell plate growing

c Inside the "sandwich," two walls will form as cellulose becomes deposited over both membranes. Other substances inside will form the "middle lamella" that cements the new walls together.

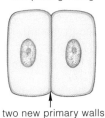

two new primary walls

d The cell plate grows at its margins until it fuses with the plasma membrane of the parent cell. During growth, when cells are expanding and walls are thin, new material also is deposited over the old primary wall.

Figure 7.8 Cytoplasmic division of a plant cell, as brought about by cell plate formation.

This concludes our picture of mitosis. Look now at your hands—and think of all the cells in your palms, thumbs, and fingers. Find a microscope and look at cells in a sample from your skin. Imagine all the divisions of all the cells that preceded them when you were developing early on, inside your mother (Figure 7.9). And be grateful for the astonishing precision that led to their formation—because the alternatives can be terrible indeed (see the *Focus* essay).

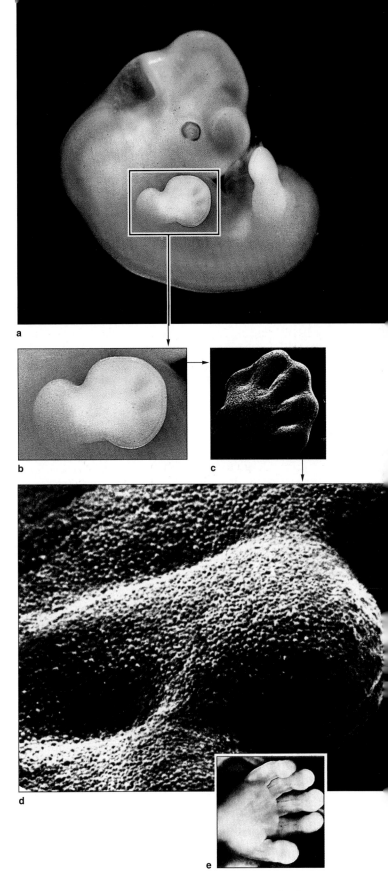

Figure 7.9 Development of the human hand by way of cell divisions and other processes. Individual cells resulting from the mitotic cell divisions are clearly visible in (**d**).

Henrietta's Immortal Cells

Each human starts out as a single fertilized egg. At birth a human body has about a trillion cells. Even in an adult, trillions of cells are still dividing. Cells in the stomach's lining divide every day. Liver cells usually don't divide—but if part of the liver becomes injured or diseased, they will divide repeatedly and produce new cells until the damaged part is replaced.

In 1951, George and Margaret Gey of Johns Hopkins University were trying to develop a way to keep human cells dividing outside the body. (Researchers could study basic life processes with such cells. They also could study cancer and other diseases, without having to experiment directly on humans.) Local physicians had provided them with normal or diseased human cells from patients. But the Geys just couldn't stop the cell lines from dying out within a few weeks.

Mary Kubicek, one of their assistants, was about to give up after dozens of failed attempts. Still, in 1951, she prepared another sample of cancer cells for culture. The sample was code-named HeLa, for the first two letters of the patient's first and last names.

The cells began to divide. And divide. And divide again. By the fourth day there were so many cells that they had to be subdivided into more tubes. As months passed, the culture continued to thrive. Unfortunately, the tumor cells inside the patient's body were just as vigorous. Six months after the patient was first diagnosed as having cancer, tumor cells had spread through her body. Two months later, Henrietta Lacks, a young woman from Baltimore, was dead.

Although Henrietta was gone, some of her cells lived on in the Geys' laboratory as the first successful human cell culture. HeLa cells were soon being shipped to other researchers, who passed cells on to others, and so HeLa cells came to live in laboratories all over the world. Some even traveled into space aboard the *Discoverer XVII* satellite. Every year, research that is described in hundreds of scientific papers is based on work with HeLa cells.

Henrietta was only thirty-one years old when runaway cell divisions killed her. Now, more than forty years later, her legacy is still benefiting humans everywhere, in cells that are still alive and dividing, day after day after day.

SUMMARY

1. Parent cells provide each daughter cell with the hereditary instructions (DNA) and cytoplasmic machinery necessary to start up its own operation.

2. Mitosis divides the hereditary instructions in the nucleus of a parent cell into two equivalent parcels. It is the basis for growth and tissue repair in multicelled eukaryotes. It also is the basis of asexual reproduction among many single-celled and multicelled eukaryotes. (Meiosis, another nuclear division mechanism, occurs only in germ cells set aside for sexual reproduction.)

3. The cytoplasm divides by different mechanisms near the end of nuclear division or at some point thereafter. (Animal cells are cleaved in two. A cell plate forms and divides plant cells.)

4. A chromosome is a single DNA molecule. In eukaryotes, many proteins are attached to it. A cell's chromosomes differ from one another in length, shape, and which portion of the hereditary instructions they carry.

Members of the same species have the same total number of chromosomes in their somatic cells.

5. "Chromosome number" is not the same thing as the total number of chromosomes. It is the number of *each type* of chromosome characteristic of a species. Mitosis keeps the chromosome number constant from one cell generation to the next. Thus, if a parent cell is diploid (with two chromosomes of each type), the daughter cells resulting from mitosis will be diploid also.

6. The cell cycle starts when a new cell forms, proceeds through interphase, and ends when the cell reproduces by mitosis and cytoplasmic division. At interphase, a cell increases in mass, doubles its number of cytoplasmic components, and duplicates all chromosomes.

7. After duplication, each chromosome consists of two DNA molecules attached at the centromere. For as long as the two stay attached, they are called sister chromatids of the chromosome.

8. Mitosis proceeds through four continuous stages:
   a. Prophase. Duplicated, threadlike chromosomes start to condense. New microtubules assemble in orga-

nized arrays near the nucleus; they will form a spindle. The nuclear envelope starts to break up.

b. Metaphase. During the *transition* to metaphase, the nuclear envelope breaks up completely. Microtubules of a forming spindle attach to the sister chromatids of each chromosome and orient the two toward opposite spindle poles. *At* metaphase, all chromosomes are aligned at the spindle equator.

c. Anaphase. Microtubules pull the sister chromatids of each chromosome away from each other, to opposite spindle poles. Every chromosome that was present in the parent cell is now represented by a daughter chromosome at both poles.

d. Telophase. The chromosomes decondense to the threadlike form. A new nuclear envelope forms around them. Each nucleus has the same chromosome number as the parent cell. Mitosis is completed.

Review Questions

1. Define the two types of nuclear division mechanisms that occur in eukaryotes. Does either one divide the cytoplasm? *101*

2. Define somatic cell and germ cell. *101*

3. What is a chromosome? What is a chromosome called in its unduplicated state? in its duplicated state (that is, with two sister chromatids)? *102*

4. Describe the spindle apparatus and its general function in nuclear division. *104*

5. Name and describe the main features of the four stages of mitosis. At what stage was the following micrograph of a human cell taken? (Fluorescent dyes stained the spindle green and the chromosomes red.) *104–106*

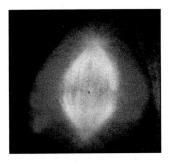

Self-Quiz *(Answers in Appendix IV)*

1. A chromosome consists of _____ .
   a. DNA only
   b. DNA plus proteins
   c. DNA plus membrane
   d. DNA plus lipids

2. A cell with pairs of the chromosomes characteristic of the species is a(n) _____ cell.
   a. diploid
   b. mitotic
   c. abnormal
   d. a and c

3. A duplicated chromosome has _____ chromatid(s).
   a. one
   b. two
   c. three
   d. four

4. The _____ is the constricted region of a duplicated chromosome, with attachment sites for microtubules.
   a. chromatid
   b. centromere
   c. cell plate
   d. cleavage furrow

5. Interphase is the stage when _____ .
   a. nothing occurs
   b. a germ cell forms its spindle apparatus
   c. a cell grows and duplicates its DNA
   d. mitosis occurs

6. After mitosis, the chromosome number of a daughter cell is _____ the parent cell's.
   a. the same as
   b. one-half
   c. rearranged compared to
   d. doubled compared to

7. Only _____ is not a stage of mitosis.
   a. prophase
   b. interphase
   c. metaphase
   d. anaphase

8. Match each stage with its key events.
   ____ metaphase
   ____ prophase
   ____ telophase
   ____ anaphase

   a. sister chromatids of each chromosome separate
   b. threadlike chromosomes start to condense
   c. chromosomes decondense, daughter nuclei form
   d. all chromosomes are aligned at spindle equator

Selected Key Terms

anaphase *104*
cell cycle *103*
cell plate *106*
centromere *102*
chromosome *102*
chromosome number *102*
cleavage furrow *106*
cytoplasmic division *106*
diploid cell *102*
germ cell *101*
interphase *103*
meiosis *101*
metaphase *104*
mitosis *101*
prophase *104*
reproduction *101*
sister chromatids *102*
somatic cell *101*
spindle apparatus *104*
telophase *104*

Readings

Gallagher, G. March 1990. "Evolution: The Mitotic Spindle." *The Journal of NIH Research.*

Murray, A., and M. Kirschner. March 1991. "What Controls the Cell Cycle?" *Scientific American* 264(3): 56–63.

Science. November 3, 1989. "Frontiers in Biology: The Cell Cycle."

Wolfe, S. 1993. *Molecular and Cellular Biology.* Belmont, California: Wadsworth.

8 MEIOSIS

Octopus Sex and Other Stories

The couple clearly are interested in each other. He caresses her with one tentacle, then another—then another and another. She reciprocates with a hug here, a squeeze there. This goes on for hours. Finally the male reaches under his mantle, removes a packet of sperm, and inserts it into a cavity under the female's mantle. For every sperm that fertilizes an egg, a new octopus may develop from this sexual encounter.

Sex for the slipper limpet is a group activity. Slipper limpets are marine relatives of land snails. They first develop into free-swimming larvae, then settle down on sand or sediments and become sexually mature adults. If one settles down alone, it will become a female. If a second larva settles down on the first, *it* will become a male—and so will any subsequent larvae. Adult slipper limpets almost always live in such piles, with the bottom one always being female (Figure 8.1*a*). The males continually release sperm. When the female finally dies, the male at the bottom of the pile becomes transformed into a female, and so it goes.

For many sexually reproducing organisms, the life cycle also has asexual episodes, based on mitotic cell divisions. Many plants, including orchids, can reproduce without sex. Flatworms can split into two roughly equivalent parts that each grow into a new flatworm. In summer, nearly all aphids are females, busily producing more females from unfertilized egg cells (Figure 8.1*b*). When autumn approaches, males finally develop and the sexual phase of the life cycle resumes.

These examples only hint at the immense variation in reproductive modes. Yet despite the variation, sexual reproduction dominates eukaryotic life cycles, and it always involves certain events. Chromosomes are duplicated in germ cells. The germ cells undergo meiosis and cytoplasmic division, and then gametes develop. When gametes join at fertilization, they form the first cell of a new individual. *Meiosis, gamete formation, and fertilization are the hallmarks of sexual reproduction.* As you will see in this chapter, these processes contribute to life's splendid diversity.

Figure 8.1 Variations in reproductive modes. (**a**) Limpets busily perpetuating the species by group participation in sexual reproduction. (**b**) Live birth of an aphid, a sexually reproducing insect that switches to an asexual mode in summer.

a

b

1. Sexual reproduction proceeds through three events: meiosis, formation of gametes (such as sperm and eggs), and fertilization.

2. Meiosis is a nuclear division mechanism that is the first step in sexual reproduction. It sorts out the chromosomes of a germ cell into four new nuclei. In most plants, spore formation and other events intervene between meiosis and gamete formation.

3. The body cells of many species are diploid, with *two* of each type of chromosome. The two function as a pair during meiosis. Most often, one is maternal (with hereditary instructions from a female parent), and the other is paternal (with comparable instructions from a male parent).

4. Meiosis divides the chromosome number in half for each forthcoming gamete. Thus, if both parents are diploid ($2n$), the union of two gametes at fertilization will restore the diploid number in the new individual ($n + n = 2n$).

5. During meiosis, each pair of chromosomes may swap segments and thereby swap hereditary instructions. Also, meiosis randomly assigns either the maternal or paternal chromosome to a gamete—and it does this for each pair. Hereditary instructions are further shuffled at fertilization. All three reproductive events lead to variations in traits among offspring.

ON ASEXUAL AND SEXUAL REPRODUCTION

What kind of offspring do orchids, flatworms, and aphids get with **asexual reproduction**? By this process, one parent alone passes on a duplicate of all its genes to each new individual. **Genes** are specific stretches of chromosomes—that is, DNA molecules. Genes are all the heritable bits of information that are required to produce new individuals. Rare mutations aside, this means that asexually produced offspring can only be clones—genetically identical copies of the parent.

Inheritance is much more interesting with **sexual reproduction**. This process commonly involves two parents, each with two genes for nearly every trait. Both parents pass on one of each gene to offspring by way of meiosis, gamete formation, and fertilization (union of two gametes). Thus the first cell of a new individual also ends up with two genes for every trait.

If instructions in every pair of genes were identical down to the last detail, then sexual reproduction would produce clones, also. Just imagine—you, everyone you know, every member of the entire human population would be clones and might end up looking exactly alike.

But a pair of genes may *not* be identical. Why not? The molecular structure of genes can change; this is what we mean by mutation. Depending on its structure, one gene of a pair may "say" slightly different things about a trait. Each different molecular form of the same gene is called an **allele**.

Such tiny differences affect thousands of traits. Figure 8.2 is an example. One allele that can occur at a certain location in a human chromosome says "put a dimple in the chin" and another says "no dimple."

This brings us to a key reason why members of any sexually reproducing species don't all look alike. *Through sexual reproduction, offspring end up with new combinations of alleles, and these lead to variations in their physical and behavioral traits.*

This chapter describes the cellular basis of sexual reproduction. More importantly, it starts us thinking about far-reaching consequences of gene shufflings at different stages of the process. The resulting variation among offspring is acted upon by agents of natural selection. Thus, *variation in traits is a foundation for evolutionary change.*

a   b

Figure 8.2 (**a**) The chin fissure, a heritable trait arising from an uncommon form of a gene. Actor Kirk Douglas received a gene that influences this trait from each of his parents. One gene called for a chin fissure and the other didn't, but one is all it takes in this case. (**b**) This photograph shows what his chin might have looked like if he had inherited two ordinary forms of the gene instead.

OVERVIEW OF MEIOSIS

Think "Homologues"

Think back on the preceding chapter, which focused on mitotic cell division. Unlike mitosis, **meiosis** divides the chromosomes in a nucleus not once but twice prior to cell division. Unlike mitosis, it is the first step leading to gamete formation. **Gametes** are sex cells, such as sperm and eggs. In multicelled organisms, gametes arise only from germ cells, which are produced in specific reproductive structures and organs (Figure 8.3).

Germ cells have the same chromosome number as the rest of the body's cells. The ones that are **diploid** (2n) have *two* of each type of chromosome, often from two parents. The two chromosomes have the same length and shape. Their genes deal with the same traits. And they line up with each other at meiosis. Think of them as **homologous chromosomes** (*hom*- means alike).

As you can deduce from Figure 7.3, there are 23 + 23 homologous chromosomes in your germ cells. After meiosis, 23 chromosomes—one of each type—end up in gametes. Said another way, meiosis halves the chromosome number, so the gametes are **haploid** (n).

Overview of the Two Divisions

Meiosis resembles mitosis in some respects, even though the outcome is different. Before interphase gives way to meiosis, a germ cell duplicates its DNA. Now each duplicated chromosome consists of two DNA molecules. These remain attached at a narrowed-down region (the centromere). For as long as the two remain attached, they are called **sister chromatids** of the chromosome:

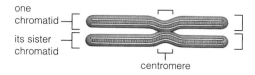

As in mitosis, the microtubules of a spindle apparatus harness the chromosomes and move them in prescribed directions (page 104).

With meiosis alone, however, *chromosomes proceed through two consecutive divisions, which end with the formation of four haploid nuclei.* The two divisions are called meiosis I and II:

	MEIOSIS I		MEIOSIS II
DNA duplication during interphase	Prophase I Metaphase I Anaphase I Telophase I	No DNA duplication between divisions	Prophase II Metaphase II Anaphase II Telophase II

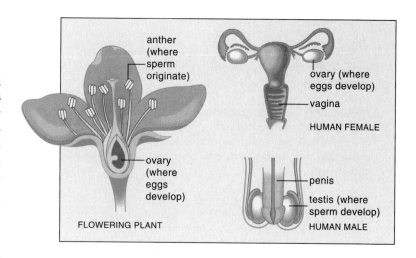

Figure 8.3 Examples of gamete-producing structures.

During meiosis I, each duplicated chromosome lines up with its partner, *homologue to homologue*; then the partners are separated from each other:

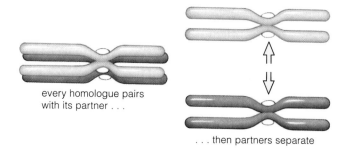

The cytoplasm typically divides after the separation of homologues. The two daughter cells are haploid, with only one of each type of chromosome. But remember, those chromosomes are still duplicated.

During meiosis II, *the sister chromatids of each chromosome are separated from each other*:

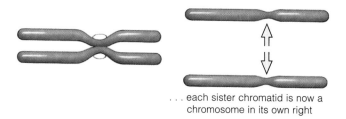

After four nuclei form, the cytoplasm typically divides once more, the outcome being four haploid cells.

Meiosis is a type of nuclear division that reduces the parental chromosome number by half—to the haploid number (n). It is the first step leading to the formation of gametes, which are required for sexual reproduction.

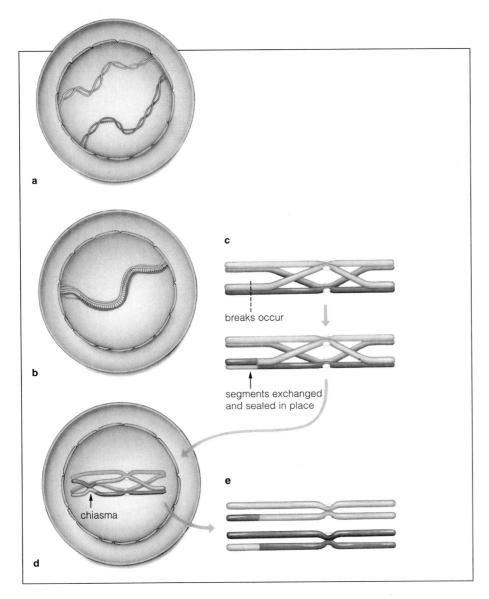

a Chromosomes become duplicated before meiosis begins. Early in prophase I, each duplicated chromosome is in threadlike form, attached at both ends to the nuclear envelope. Its two sister chromatids are so close together they look like a single thread.

b The two chromosomes become zippered together, so that all four chromatids are positioned close together.

c One or more crossovers occur at intervals along the chromosomes. In each crossover, two nonsister chromatids break at identical sites. They swap segments at the breaks, then enzymes seal the broken ends. (For clarity, the two chromosomes are shown in condensed form and pulled apart. Crossing over may seem more plausible when you realize that it occurs while chromosomes are extended like threads and tightly aligned.)

d As prophase I ends, the chromosomes continue to condense, becoming thicker, rodlike forms. They detach from the nuclear envelope and from each other—except at "chiasmata." Each chiasmata is indirect evidence that a crossover occurred at some point in the chromosomes.

e Crossing over breaks up old combinations of alleles and puts new ones together in pairs of homologous chromosomes.

Figure 8.4 Key events during prophase I, the first stage of meiosis. For clarity, only a single pair of homologous chromosomes and only one crossover event are shown. Blue signifies the paternal chromosome, and purple signifies its maternal homologue.

STAGES OF MEIOSIS

Prophase I Activities

The first stage of meiosis, prophase I, is a time of major gene shufflings. First, each chromosome is drawn close to its homologue, with little space between them. The intimate, parallel array favors **crossing over**, a molecular interaction between the chromatids of homologous chromosomes. *Nonsister* chromatids break at the same places along their length and exchange corresponding segments—that is, genes—at the break points. Figure 8.4 is a simplified picture of this interaction.

Gene swapping would be rather pointless if each type of gene never varied from one chromosome to the next. But remember, a gene can have slightly different forms—alleles. You can safely bet that all the alleles on one chromosome will *not* be identical to those on the homologue. So each crossover represents a chance to swap a *slightly different version* of the hereditary instructions for a particular trait.

We will look at the mechanism of crossing over in later chapters. For now, it is enough to know that crossing over leads to genetic recombination, which in turn leads to variation in the traits of offspring.

Crossing over is an interaction between a pair of homologous chromosomes. It breaks up old combinations of alleles and puts new ones together.

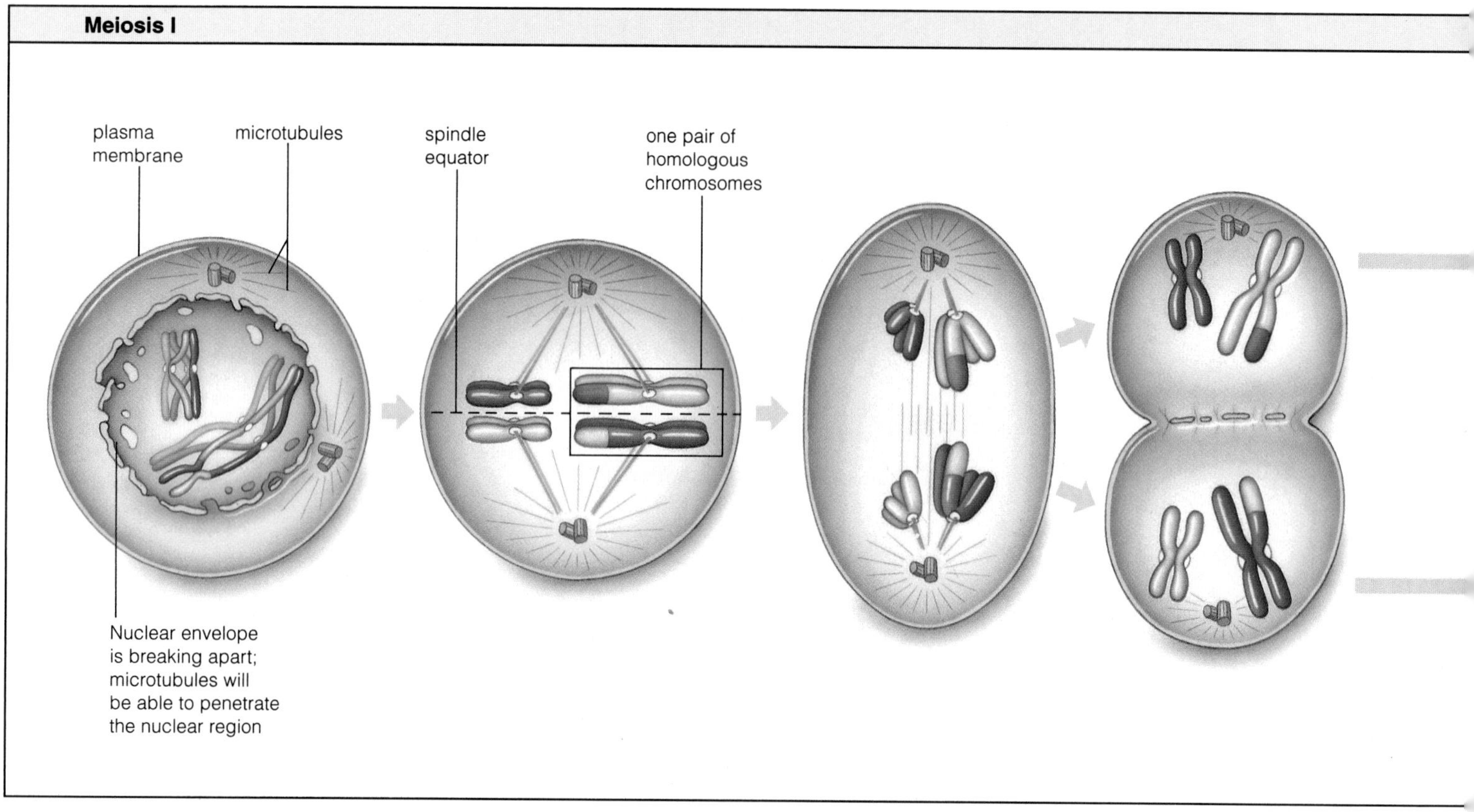

plasma membrane	microtubules	spindle equator	one pair of homologous chromosomes

Nuclear envelope is breaking apart; microtubules will be able to penetrate the nuclear region

Prophase I

By now, chromosomes are in the threadlike, duplicated form (each consists of two sister chromatids). Usually, all homologous pairs of chromosomes undergo crossing over. (For clarity, only one crossover is shown). Chromosomes start condensing to rodlike form. Each becomes attached to the microtubular spindle.

Metaphase I

All chromosomes are now positioned at the equator of the spindle.

Anaphase I

Each chromosome is separated from its homologue. The two are moved to opposite poles of the spindle.

Telophase I

When the cytoplasm divides, there are two haploid (*n*) cells; each has one chromosome of each type. Chromosomes are still in the duplicated state.

Figure 8.5 Meiosis: the nuclear division mechanism by which the parental number of chromosomes is reduced by half (to the haploid number) for forthcoming gametes. Only two pairs of homologous chromosomes are shown. Maternal chromosomes are shaded purple, and paternal ones blue.

Separating the Homologues During Meiosis I

Major shufflings of whole chromosomes begin during the transition from prophase I to metaphase I, the second stage of meiosis. Suppose the shufflings are proceeding right now in one of your germ cells. Call the twenty-three chromosomes inherited from your mother the *maternal* chromosomes and their twenty-three homologues from your father, the *paternal* ones.

As in mitosis, the spindle is completed before metaphase I (Figure 8.5). Microtubules harness and orient one chromosome of each pair toward one spindle pole and its homologue toward the other. Then they move the chromosomes toward the spindle equator. At metaphase I, all chromosomes are positioned midway between the spindle poles. Then, during anaphase I, each chromosome is pulled away from its homologue.

Will all maternal chromosomes move to one pole and all paternal chromosomes to the other? Maybe, but probably not. Their positioning at the spindle equator

There is no DNA replication between the two divisions

Prophase II

During the transition to prophase II, the two centrioles in each new cell were moved apart and a new spindle was assembled. Now, microtubules attach chromosomes to the spindle and start moving them toward the equator.

Metaphase II

All chromosomes are now positioned at the equator of the spindle.

Anaphase II

The attachment between the two chromatids of each chromosome breaks. Now the former "sister chromatids" are chromosomes in their own right and are moved to opposite poles of the spindle.

Telophase II

Four daughter nuclei form. When the cytoplasm divides, each new cell has a haploid number of chromosomes, all in the unduplicated state. One or all of the cells may develop into gametes.

and subsequent direction of movement are random. *Either chromosome of a pair may end up at either pole.*

Figure 8.6 shows how three pairs of homologues can be shuffled into any one of four possible positions at metaphase I. In this example, 2^3 or 8 combinations of maternal and paternal chromosomes are possible for the forthcoming gametes.

A human germ cell has 23 pairs of homologous chromosomes, not just three. So 2^{23} or *8,388,608 combinations* of maternal and paternal chromosomes are possible every time a germ cell gives rise to sperm or eggs! In each sperm or egg, many hundreds of alleles inherited from the mother might differ from their counterparts from the father. Are you beginning to get an idea of why such splendid mixes of traits show up even in the same family?

There is no DNA duplication between meiosis I and II. But, remember, each chromosome was duplicated earlier (during interphase). It is still in the duplicated state when meiosis II begins.

Figure 8.6 The possible outcomes of the random alignment of only three pairs of homologous chromosomes at metaphase I of meiosis. The three types of chromosomes are labeled 1, 2, and 3. The maternal chromosomes are purple; paternal ones are blue.

With merely four possible alignments, eight different combinations of maternal chromosomes and paternal chromosomes are possible in forthcoming gametes.

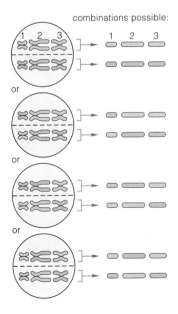

combinations possible:

Separating the Sister Chromatids During Meiosis II

Meiosis II has one overriding function: separation of the two sister chromatids of each chromosome. At metaphase II, all the duplicated chromosomes are positioned at the spindle equator. During anaphase II, sister chromatids are pulled away from each other, so that each is now a separate, daughter chromosome:

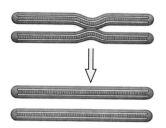

Half the parental number of chromosomes arrive at each spindle pole. But that haploid number includes one of each type of chromosome characteristic of the species. During telophase II, new nuclear membranes form around the chromosomes after they have arrived at the poles. Meiosis is completed. When the cytoplasm divides, there will be four haploid cells, one or all of which may eventually function as a gamete.

GAMETE FORMATION

The gametes that form after meiosis are not all the same. Human sperm have one tail, opossum sperm have two, and roundworm sperm have none. Crayfish sperm look like pinwheels. Most eggs are microscopic, yet ostrich eggs tucked inside a shell are as large as a softball. From appearance alone, you might not believe a plant gamete is even remotely like an animal's.

Later chapters contain details of gamete formation in the life cycles of representative organisms, including humans. Figure 8.7 and the following points may help you keep the details in perspective.

Gamete Formation in Animals. In male animals, gametes form by a process called "spermatogenesis." Inside a male reproductive system, a diploid germ cell increases in size. It becomes a large, immature cell (the primary spermatocyte) that undergoes meiosis and cytoplasmic divisions. Its four haploid daughter cells develop into immature cells called spermatids (Figure 8.8). Spermatids change in form and develop a tail. In this way, each becomes a **sperm**, a male gamete.

In female animals, gametes form by a process called "oogenesis." Each diploid germ cell develops into an immature egg (oocyte). Compared to a sperm, an oocyte accumulates more cytoplasmic components.

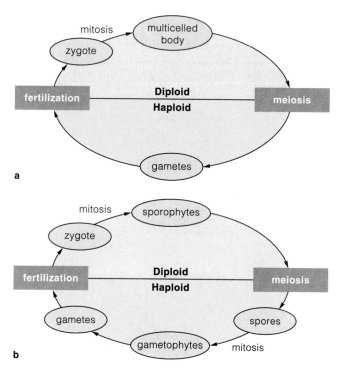

Figure 8.7 Generalized life cycles for (**a**) animals and (**b**) multicelled plants. A zygote is the first cell formed after two gametes fuse at fertilization. For plants, a sporophyte (spore-producing body) forms after fertilization and gametophytes (gamete-producing bodies) form after meiosis. A pine tree is a sporophyte. Gametophytes develop in its cones (page 283).

Also, its daughter cells differ in size and function (Figure 8.9). When the oocyte divides after meiosis I, one cell (the secondary oocyte) gets nearly all the cytoplasm. So the other cell (the first polar body) is quite small. Both may undergo meiosis II and cytoplasmic division. Here again, one cell gets most of the cytoplasm. It develops into a gamete. A mature female gamete is called an ovum or, more commonly, an **egg**.

Division of the smaller cell means there are now three polar bodies. These do not function as gametes. In effect, they are dumping grounds for chromosomes, so that the egg ends up with the required haploid number. Because polar bodies do not have much cytoplasm, they do not have much in the way of nutrients and metabolic machinery. In time they degenerate.

Gamete Formation in Plants. For pine trees, roses, and other familiar plants, certain events intervene between meiosis and gamete formation. Among other things, spores form. **Spores** are haploid cells, often with walls that allow them to resist dry periods or other adverse environmental conditions. Under favorable conditions, spores germinate and develop into a haploid body or structure that will produce gametes. Thus gamete-producing bodies *and* spore-producing bodies develop during the life cycle (Figure 8.7b).

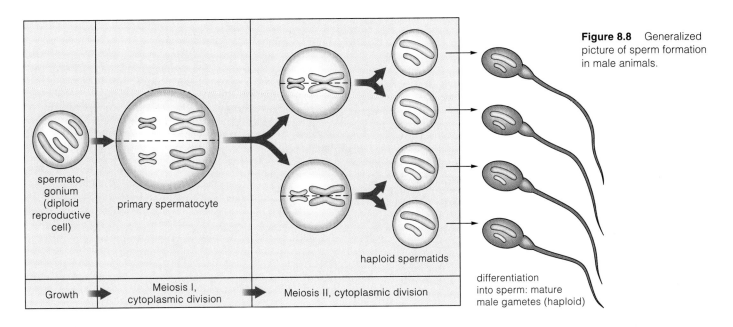

Figure 8.8 Generalized picture of sperm formation in male animals.

Growth	Meiosis I, cytoplasmic division	Meiosis II, cytoplasmic division

spermato-gonium (diploid reproductive cell)

primary spermatocyte

haploid spermatids

differentiation into sperm: mature male gametes (haploid)

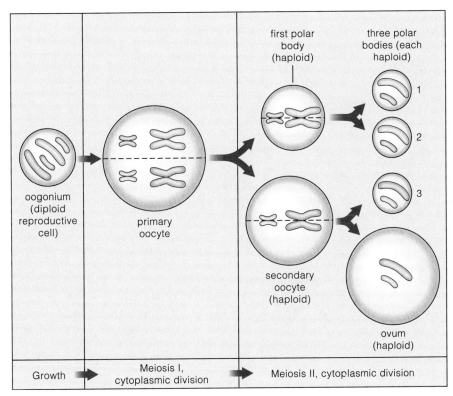

Growth	Meiosis I, cytoplasmic division	Meiosis II, cytoplasmic division

oogonium (diploid reproductive cell)

primary oocyte

first polar body (haploid)

three polar bodies (each haploid)

secondary oocyte (haploid)

ovum (haploid)

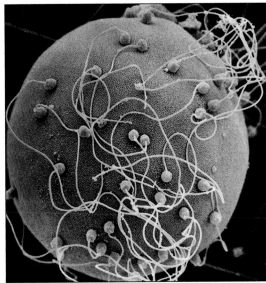

Figure 8.9 Generalized picture of egg formation in female animals. The diagram is not at the same scale as Figure 8.8. An egg is *much* larger than a sperm, as indicated by the micrograph. Also, the three polar bodies are extremely small compared to an egg.

MORE GENE SHUFFLINGS AT FERTILIZATION

The parental chromosome number is restored at **fertilization**, when the nuclei of two haploid gametes fuse. Unless meiosis precedes it, fertilization would result in a doubling of the chromosome number in every new generation.

Fertilization contributes to variation among offspring. Reflect on the possibilities for humans. During meiosis I, an average of two or three crossovers occurs in each human chromosome. Also, the random positioning of paternal and maternal chromosomes at the spindle equator results in one of 8,388,608 possible combinations of maternal and paternal chromosomes in each gamete. And of all the male and female gametes that are produced, which two actually get together is a matter of chance.

As you can see, the sheer number of combinations that can exist at fertilization is staggering!

MEIOSIS AND MITOSIS COMPARED

In this unit, our focus has been on two different mechanisms that divide the nuclear DNA of eukaryotic cells. Single-celled species use mitosis in asexual reproduction; multicelled species use mitosis during growth and tissue repair. Meiosis is the basis of gamete formation and sexual reproduction. Figure 8.10 summarizes the similarities and differences between the two mechanisms.

Keep in mind, the two mechanisms differ in a crucial way. *Mitotic cell division produces clones* (genetically identical copies of a parent cell). *Meiosis, together with fertilization, promotes variation in traits among offspring.* First, crossing over at prophase I of meiosis puts new combinations of alleles in chromosomes. Second, the movement of either member of each pair of homologous chromosomes to either spindle pole after metaphase I puts different mixes of maternal and paternal alleles into gametes. Third, different combinations of alleles are brought together by chance at fertilization. Later chapters describe how meiosis and fertilization contribute to the evolution of sexually reproducing organisms.

Figure 8.10 Summary of mitosis and meiosis, using a diploid (2*n*) animal cell as the example. The diagram is arranged to help you compare the similarities and differences between the two division mechanisms. Maternal chromosomes are shaded purple, and paternal chromosomes are shaded blue.

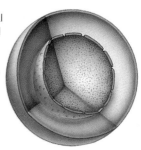

A diploid (2*n*) *somatic* cell is at interphase. DNA is replicated (all chromosomes are duplicated) before nuclear division begins.

Meiosis I

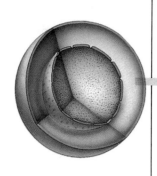  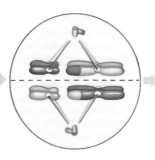 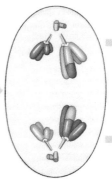 

A diploid (2*n*) *reproductive* cell is at interphase. DNA is replicated (all chromosomes are duplicated) before nuclear division begins.

Prophase I

Each duplicated chromosome (consisting of two sister chromatids) condenses to threadlike form, then rodlike form. *Crossing over* occurs. *Each chromosome separates from its homologue.* Each gets attached to the spindle during the transition to metaphase.

Metaphase I

All chromosomes are now positioned at the spindle's equator.

Anaphase I

Each chromosome is separated from its homologue. They are moved to opposite poles of the spindle.

Telophase I

When the cytoplasm divides, there are *two* cells. Each has a haploid (*n*) number of chromosomes, but these are still in the duplicated state.

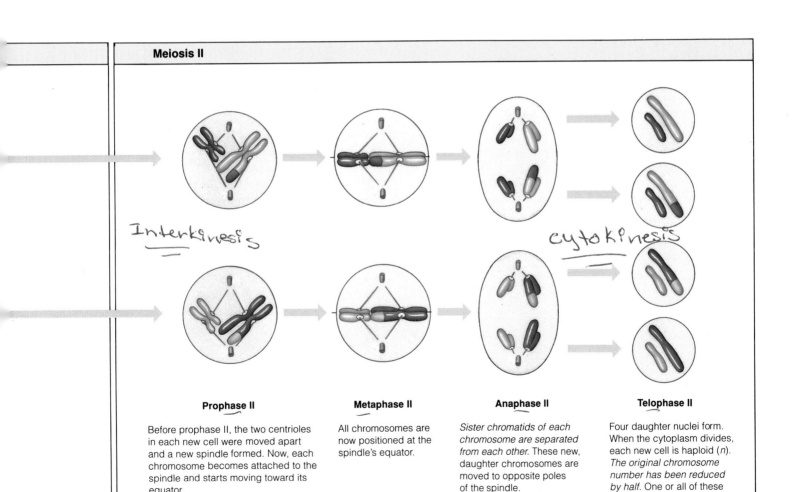

Mitosis

Prophase

Each duplicated chromosome (consisting of two sister chromatids) condenses from threadlike form to rodlike form. Each gets attached to the spindle during the transition to metaphase.

Metaphase

All chromosomes are now positioned at the spindle's equator.

Anaphase

Sister chromatids of each chromosome are separated from each other. These new, daughter chromosomes are moved to opposite poles of the spindle.

Telophase

When the cytoplasm divides, there are *two* cells. Each is diploid (2n)–*it has the same chromosome number as the parent cell.*

Meiosis II

Interkinesis

cytokinesis

Prophase II

Before prophase II, the two centrioles in each new cell were moved apart and a new spindle formed. Now, each chromosome becomes attached to the spindle and starts moving toward its equator.

Metaphase II

All chromosomes are now positioned at the spindle's equator.

Anaphase II

Sister chromatids of each chromosome are separated from each other. These new, daughter chromosomes are moved to opposite poles of the spindle.

Telophase II

Four daughter nuclei form. When the cytoplasm divides, each new cell is haploid (n). *The original chromosome number has been reduced by half.* One or all of these cells may become gametes.

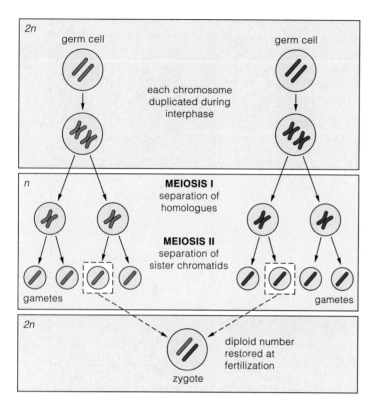

Figure 8.11 Summary of the steps required for sexual reproduction involving diploid germ cells. In each parent cell, chromosomes are duplicated during interphase. Meiosis reduces the chromosome number by half (2*n* to *n*) during two nuclear divisions. After male and female gametes form, their nuclei will fuse at fertilization and restore the chromosome number.

SUMMARY

1. The life cycle of sexually reproducing organisms includes meiosis, gamete formation, and fertilization. Meiosis reduces the chromosome number of a parent cell by half. It precedes the formation of haploid gametes (typically sperm in males, eggs in females). Fusion of a sperm nucleus and an egg nucleus at fertilization restores the chromosome number (Figure 8.11).

2. If cells of sexually reproducing organisms are diploid (2*n*), they have two of each type of chromosome characteristic of the species. Commonly, one of the two is maternal (inherited from a female parent) and the other is paternal (from a male parent).

3. Each pair of maternal and paternal chromosomes shows homology (the two are alike). Generally, the two have the same length, same shape, and same sequence of genes. They interact during meiosis.

4. Chromosomes are duplicated during interphase. So before meiosis, each consists of *two* DNA molecules that remain attached (as sister chromatids).

5. Meiosis consists of two consecutive divisions that require a spindle apparatus. In meiosis I, microtubules of the spindle move each duplicated chromosome away from its homologue (which is also duplicated). In meiosis II, they separate sister chromatids of each chromosome.

6. The following are key events of meiosis I:
   a. During prophase I, nonsister chromatids of homologous chromosomes break at corresponding sites and exchange segments. Such crossing over puts together new combinations of alleles. Alleles are slightly different molecular forms of the same gene. They code for different forms of the same trait. So new combinations of alleles lead to variations in the details of a given trait among the offspring.
   b. At metaphase I, all pairs of homologous chromosomes are positioned at the spindle equator. In each case, either the maternal chromosome or its homologue can be oriented toward either pole.
   c. During anaphase I, each maternal chromosome is separated from its homologue, and the two are moved to opposite spindle poles.

7. The following are key events of meiosis II:
   a. At metaphase II, chromosomes are still duplicated, and they are positioned at the spindle equator.
   b. During anaphase II, the sister chromatids of each chromosome are moved apart. Each is now a separate, unduplicated chromosome.
   c. During telophase II, four haploid nuclei form.

8. When the cytoplasm divides, there are four haploid cells. One or all of those cells may function as gametes (or spores, in the case of flowering plants).

1. Name the three key events of sexual reproduction. *111*

2. What is the key difference between sexual and asexual reproduction? *111*

3. Refer to the following numbers of chromosomes in the diploid body cells of a few organisms. In each case, how many chromosomes would end up gametes? *112*

Fruit fly, *Drosophila melanogaster*	8
Garden pea, *Pisum sativum*	14
Corn, *Zea mays*	20
Frog, *Rana pipiens*	26
Earthworm, *Lumbricus terrestris*	36
Human, *Homo sapiens*	46
Chimpanzee, *Pan troglodytes*	48
Amoeba, *Amoeba*	50
Horsetail, *Equisetum*	216

4. Suppose a diploid germ cell has four pairs of homologous chromosomes, designated AA, BB, CC, and DD. How would the chromosomes of the gametes be designated? *112*

5. Define meiosis and characterize its main stages. In what respects is meiosis *not* like mitosis? *112–116, 118*

6. Outline the steps involved in the formation of sperm and eggs in animals. *116*

Self-Quiz *(Answers in Appendix IV)*

1. Sexual reproduction requires _____ .
   a. meiosis
   b. gamete formation
   c. fertilization
   d. all of the above

2. Meiosis is a division mechanism that produces _____ .
   a. two cells
   b. two nuclei
   c. four cells
   d. four nuclei

3. An animal cell with two of each type of chromosome characteristic of the species is _____ .
   a. diploid
   b. haploid
   c. probably not a normal gamete
   d. both b and c

4. Meiosis _____ the parental chromosome number.
   a. doubles
   b. reduces
   c. maintains
   d. corrupts

5. Generally, a pair of homologous chromosomes _____ .
   a. carry the same genes
   b. are the same length, shape
   c. interact at meiosis
   d. all of the above

6. Generally, a gene on a paternal chromosome and its partner gene on a maternal chromosome _____ .
   a. are identical
   b. may differ slightly
   c. may be swapped
   d. b and c

7. Before the onset of meiosis, all chromosomes are _____ .
   a. condensed
   b. released from protein
   c. duplicated
   d. b and d

8. Each chromosome moves away from its homologue and ends up at the opposite spindle pole during _____ .
   a. prophase I
   b. prophase II
   c. anaphase I
   d. anaphase II

9. Sister chromatids of each chromosome move apart and end up at opposite spindle poles during _____ .
   a. prophase I
   b. prophase II
   c. anaphase I
   d. anaphase II

10. Match each term and its description.
    ____ chromosome number
    ____ alleles
    ____ metaphase I
    ____ interphase
    ____ pair of genes

    a. different molecular forms of the same gene
    b. none between meiosis I, II
    c. carry information about the same trait
    d. pairs of homologues aligned at spindle equator
    e. the number of each type of chromosome present in cell

Selected Key Terms

allele *111*
asexual reproduction *111*
crossing over *113*
diploid (chromosome number) *112*
egg (ovum) *116*
fertilization *117*
gamete *112*
gene *111*

haploid (chromosome number) *112*
homologous chromosome *112*
meiosis *112*
sexual reproduction *111*
sister chromatids *112*
sperm *116*
spore *116*

Readings

Klug, W., and M. Cummings. 1991. *Concepts of Genetics*. Third edition. New York: Macmillan.

Strickberger, M. 1985. *Genetics*. Third edition. New York: Macmillan.

Wolfe, S. 1993. *Molecular and Cellular Biology*. Belmont, California: Wadsworth.

OBSERVABLE PATTERNS OF INHERITANCE

A Smorgasbord of Ears and Other Traits

Basketball ace Charles Barkley has them. So does actor Tom Cruise. Actress Joan Chen doesn't, and neither did a monk named Gregor Mendel. To see how *you* fit in with these folks, use a mirror to check your ears. Is the fleshy lobe at the base of each ear attached to the side of your head? If so, you and Barkley and Cruise have something in common. Or is the fleshy lobe unattached, so that you can flap it back and forth? If so, you are like Chen and Mendel (Figure 9.1).

Whether a person is born with detached or attached earlobes depends on a single kind of gene. That gene comes in slightly different molecular forms—alleles. Only one form has information about detached lobes. The information is put to use while a human body is developing inside the mother. It calls for a death signal, which is sent to all the cells positioned between the newly forming lobes and the head. Without the signal, the cells don't die, and earlobes don't detach.

We all have genes for thousands of traits, such as earlobes, cheeks, lashes, and eyeballs. Most of the traits vary in their details from one person to the next. Remember, humans inherit pairs of genes, on pairs of chromosomes. In some pairings, one allele has powerful effects and overwhelms the other's contribution to a trait. The outgunned allele is said to be recessive to the dominant one. If you have *detached* earlobes, *dimpled* cheeks, *long* lashes, or *large* eyeballs, you carry at least one and possibly two dominant alleles that influence the trait in a particular way.

When both alleles of a pair are recessive, nothing masks their effect on a trait. You get *attached* earlobes with one pair of recessive alleles (and *flat* feet with another, a *straight* nose with another, and so on).

How did we discover such remarkable things about our genes? It all started with Gregor Mendel. By analyzing pea plants generation after generation, Mendel found indirect but *observable* evidence of how parents transmit units of hereditary information—genes—to offspring. This chapter focuses on Mendel's experimental methods and results. They remain a classic example of how a scientific approach can pry open important secrets about the natural world. And to this day, they serve as the foundation for modern genetics.

a Tom Cruise

b Charles Barkley

c Joan Chen

d Gregor Mendel

Figure 9.1 The attached and detached earlobes of a few representative humans. This sampling provides observable evidence of a trait that is governed by a single gene. Do you have one or the other version of this trait? It depends on whether you inherited a specific molecular form of that gene from your mother, your father, or both of your parents.

Earlobe attachment, chin dimpling, cheek dimpling, and many other single-gene traits vary from one individual to the next. As Gregor Mendel perceived, such easily observable traits can be used to identify *patterns* of inheritance that exist from one generation to the next.

1. Genes are units of information about heritable traits. Each gene has a specific location in the chromosomes of a species. But its molecular form may differ slightly from one chromosome to the next. The different molecular forms of a gene, called alleles, specify different versions of the same trait.

2. The two genes of a pair are segregated from each other during meiosis and end up in different gametes. Gregor Mendel found indirect evidence of this when he crossbred plants having observable differences in the same trait.

3. The sorting of each gene pair into different gametes tends to be independent of how the other gene pairs are sorted out. Mendel found evidence of this when he tracked plants having observable differences in *two* traits, such as flower color and height.

4. If the two genes of a pair specify different versions of a trait (if they are nonidentical alleles), one may have more pronounced effects on the trait. Besides this, two or more gene pairs often influence the same trait, and some single genes influence many traits. Finally, environmental conditions may alter gene expression.

More than a century ago, Charles Darwin explained how natural selection might bring about evolutionary change. According to his key premise, individuals vary in heritable traits. Variations that improve chances of surviving and reproducing are favored more often in each generation. Those that don't become less frequent. Thus the population changes over time—it evolves.

Darwin's theory did not fit with a prevailing idea about inheritance. It was common knowledge that sperm and eggs both transmit information about traits to offspring. But almost no one suspected the information is organized in *units* (genes). Instead, the idea was that a father's blob of information "blended" with a mother's blob at fertilization, like cream into coffee.

Carried to its logical conclusion, blending would eventually dilute hereditary information until there was only one version left of each trait. Yet why did *freckled* children keep turning up among nonfreckled generations? Why weren't all the descendants of a herd of white stallions and black mares *gray*? The blending theory scarcely explained what people could see with their own eyes. But few disputed it. Blending happened, so populations "had to be" uniform—and with uniformity in traits, evolution could not occur.

Even before Darwin presented his theory, however, someone was gathering evidence that eventually would support his key premise. A monk, Gregor Mendel, was about to prove that sperm and eggs do indeed carry "units" that deal with separate heritable traits.

MENDEL'S INSIGHTS INTO PATTERNS OF INHERITANCE

Mendel's Experimental Approach

Mendel had been raised on a farm and was well aware of agricultural principles and practices. After entering a monastery, he also studied mathematics at the University of Vienna. Shortly after his university training, he began experimenting with the garden pea plant, *Pisum sativum* (Figure 9.2).

This plant is self-fertilizing. Its flowers produce sperm *and* eggs, which get together in the same flower. Some of the plants are **true-breeding**. In other words, successive generations are just like the parents in one or more traits, as when all offspring of white-flowered parents have white flowers.

As Mendel knew, pea plants also will cross-fertilize when sperm and eggs from different plants are brought together under controlled conditions. In his experiments, he could open flower buds of a plant that bred true for a trait (say, purple flowers) and snip out the stamens. (Stamens bear pollen grains in which sperm develop.) Then he could brush the "castrated" buds with pollen from a plant that bred true for a *different* version of the trait (white flowers). Mendel hypothesized that such clearly observable differences could be used to track a trait through many generations. If there were patterns to the trait's inheritance, *those patterns might tell him something about the hereditary material itself.*

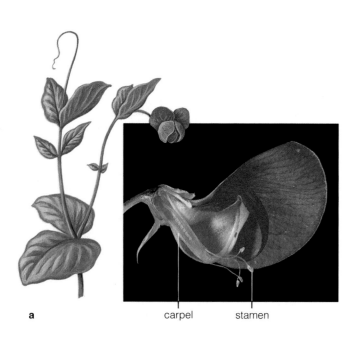

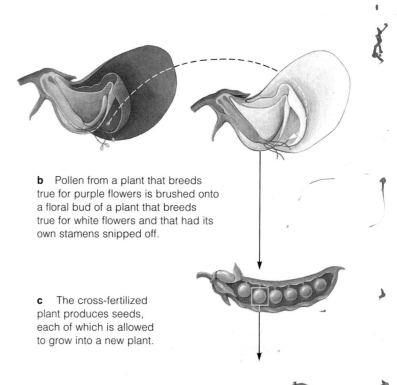

b Pollen from a plant that breeds true for purple flowers is brushed onto a floral bud of a plant that breeds true for white flowers and that had its own stamens snipped off.

c The cross-fertilized plant produces seeds, each of which is allowed to grow into a new plant.

a   carpel   stamen

Figure 9.2 The garden pea plant (*Pisum sativum*), the focus of Mendel's experiments. A flower has been sectioned to show the location of its stamens and carpel. Sperm-producing pollen grains form in stamens. Eggs develop, fertilization takes place, and seeds mature inside the carpel.

d Flower color of new plants can be used as evidence of patterns in how hereditary material is transmitted from each parent.

Some Terms Used in Genetics

Having read the chapter on meiosis, you already have insight into the mechanisms of sexual reproduction—which is more than Mendel had. He did not know about chromosomes. So he could not have known that the chromosome number is reduced by half in gametes, then restored at fertilization.

Yet Mendel sensed what was going on. As we follow his thinking, let's simplify things by substituting a few modern terms used in studies of inheritance (see also Figure 9.3):

---

1. **Genes** are units of information about specific traits, and they are passed from parents to offspring. Each gene has a specific location (locus) on a chromosome.

2. Diploid cells have a pair of genes for each trait, on a pair of homologous chromosomes.

3. Although both genes of a pair deal with the same trait, they may vary in their information about it. This happens when they have slight molecular differences, as when one gene for flower color specifies purple and another specifies white. The different molecular forms of a gene are called **alleles** of that gene.

4. If it turns out that the two alleles of a pair are the same, this is a *homozygous* condition. If different, this is a *heterozygous* condition.

5. An allele is *dominant* when its effect on a trait masks that of any *recessive* allele paired with it. We use capital letters for dominant alleles and lowercase letters for recessive ones (for instance, *A* and *a*).

6. Putting this together, a **homozygous dominant** individual has a pair of dominant alleles (*AA*) for the trait that is being studied. A **homozygous recessive** individual has a pair of recessive alleles (*aa*) for the trait. A **heterozygous** individual has a pair of nonidentical alleles (*Aa*) for the trait.

7. Two terms help keep the distinction clear between genes and the traits they specify. **Genotype** refers to the genes present in an individual. **Phenotype** refers to an individual's observable traits.

8. When tracking the inheritance of traits through generations of offspring, these abbreviations apply:

    P   parental generation
    F_1 first-generation offspring
    F_2 second-generation offspring

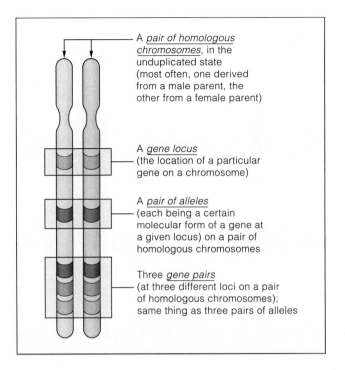

Figure 9.3 A few genetic terms illustrated. Diploid organisms have pairs of genes, on pairs of homologous chromosomes. (In your case, one chromosome of each pair was inherited from your mother, and the other, homologous chromosome was inherited from your father.)

The genes themselves may have different molecular forms, called alleles. Different alleles specify slightly different versions of the same trait. An allele at one location on a chromosome may or may not be identical to its partner on the homologous chromosome.

When the offspring of genetic crosses inherit identical alleles for a given trait, generation after generation, they are said to be a *true-breeding* lineage. When offspring of a genetic cross inherit nonidentical alleles for a trait under study, they are said to be *hybrid* offspring.

homozygous dominant parent
homozygous recessive parent
(chromosomes duplicated before meiosis)
meiosis I
meiosis II
(gametes)
(gametes)
fertilization produces heterozygous offspring

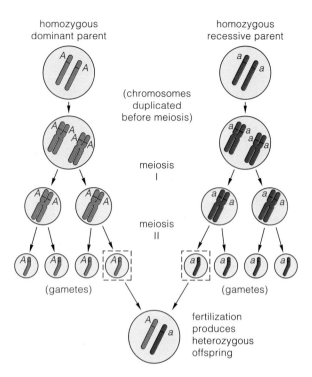

Figure 9.4 Example of a monohybrid cross, showing how one gene of a pair segregates from the other. Two parents that breed true for two different versions of a trait give rise only to heterozygous offspring.

Trait Studied
Dominant Form
Recessive Form
F_2 Dominant-to-Recessive Ratios:
seed shape
5,474 round
1,850 wrinkled
2.96:1
seed color
6,022 yellow
2,001 green
3.01:1
pod shape
882 inflated
299 wrinkled
2.95:1
pod color
428 green
152 yellow
2.82:1
flower color
705 purple
224 white
3.15:1
flower position
651 along stem
207 at tip
3.14:1
stem length
787 tall
277 dwarf
2.84:1
Average ratio for all traits studied:
3:1

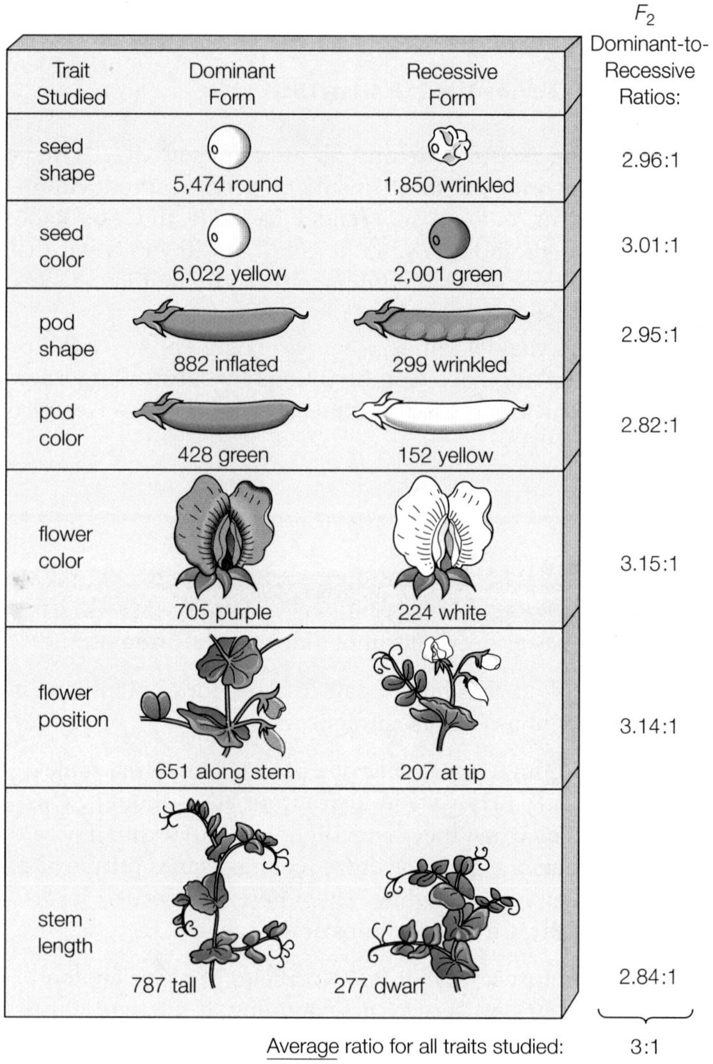

Figure 9.5 Results from Mendel's monohybrid cross experiments with the garden pea. The numbers are his counts of F_2 plants that he assumed were carrying dominant or recessive hereditary "units" (alleles) for the trait. On the average, the dominant-to-recessive ratio was 3 : 1.

The Theory of Segregation

Mendel had an idea that in every generation, a plant inherits two "units" (genes) of information for a trait, one from each parent. To test his idea, he performed what we now call **monohybrid crosses**. Offspring of such crosses are heterozygous for the one trait being studied (which is what *monohybrid* means). Their parents breed true for different versions of the trait, so the offspring inherit a pair of nonidentical alleles.

Mendel tracked many single traits through two generations. In one series of experiments, he crossed true-breeding purple-flowered plants and true-breeding white-flowered plants, then collected and planted the resulting seeds. All plants grown from the seeds had purple flowers. Then Mendel allowed those plants to self-fertilize. Some of *their* offspring had *white* flowers!

If Mendel's hypothesis were correct—if each plant had inherited two units of information about flower color—then the "purple" unit would have to be dominant, for it masked the "white" unit in F_1 plants.

Let's rephrase Mendel's thinking. Garden pea plants are diploid, with pairs of homologous chromosomes. Assume one parent is homozygous dominant (*AA*) for flower color and the other is homozygous recessive (*aa*). After meiosis, their sperm or eggs will carry only one allele for flower color (Figure 9.4). Thus, when a sperm fertilizes an egg, one outcome is possible: $A + a = Aa$.

Before we continue, you should know that Mendel crossed hundreds of plants and tracked thousands of offspring. He also *counted* and *recorded* the plants showing dominance or recessiveness. As Figure 9.5 indicates, an intriguing ratio emerged. On the average, three of every four F_2 plants were dominant for the trait and one was recessive.

This ratio suggested that fertilization is a chance event—which meant rules of probability applied to his crosses. **Probability** applies to chance events. The word

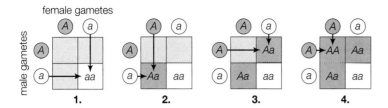

female gametes

male gametes

1.   2.   3.   4.

Figure 9.6 Punnett-square method of predicting the probable outcome of a genetic cross. Circles represent gametes. Letters on gametes represent dominant or recessive alleles. The different squares depict the different genotypes possible among offspring. In this example, gametes are from a self-fertilizing plant that is heterozygous (*Aa*) for a trait.

Figure 9.7 Results from one of Mendel's monohybrid crosses. On the average, the dominant-to-recessive ratio among F_2 plants was 3:1.

As the numerical results in Figure 9.5 show, the ratio wasn't *exactly* 3:1. Almost certainly, Mendel's reliance on a large number of crosses and his understanding of probability kept him from being confused by minor deviations from predicted results. To see why, flip a coin a few times. We all know a flipped coin is as likely to end up heads as tails. But often it ends up heads, or tails, several times in a row. When you flip the coin only a few times, the actual ratio may differ greatly from the predicted ratio (1:1). Only when you flip the coin many times will you come close to the predicted ratio.

simply refers to the most likely number of times a certain outcome will occur, divided by the total number of all possible outcomes. The **Punnett-square method**, explained in Figure 9.6, can be used to figure out the probable outcomes of Mendel's crosses.

Take a look at Figure 9.7. If half of each plant's sperm (or eggs) were *a* and half were *A*, four outcomes were possible every time a sperm fertilized an egg:

Possible event:	Probable outcome:
sperm *A* meets egg *A*	1/4 *AA* offspring
sperm *A* meets egg *a*	1/4 *Aa*
sperm *a* meets egg *A*	1/4 *Aa*
sperm *a* meets egg *a*	1/4 *aa*

By this prediction, an F_2 plant had three chances in four of getting at least one dominant allele (and purple flowers). It had one chance in four of getting two recessive alleles (and white flowers). That is a probable phenotypic ratio of three purple to one white, or 3:1.

Results from his monohybrid crosses led Mendel to formulate a theory. Stated in modern terms,

---

Mendel's theory of segregation. **Diploid cells have pairs of genes (on pairs of homologous chromosomes). During meiosis, the two genes of each pair segregate and end up in different gametes.**

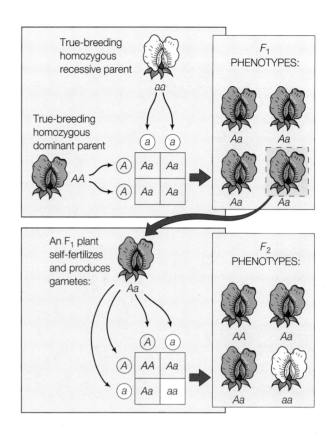

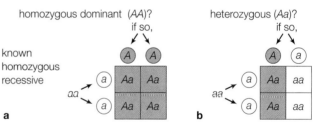

homozygous dominant (*AA*)?   heterozygous (*Aa*)?

if so,   if so,

known homozygous recessive

a   b

Figure 9.8 Punnett-square method of predicting the outcome of a testcross. A known homozygous recessive plant is crossed with a plant that may be *AA* or *Aa*. (**a**) If the plant of unknown genotype is homozygous dominant, all offspring will show the dominant form of the trait. (**b**) If heterozygous, about half the offspring will show the recessive form.

Testcrosses

Mendel gained support for his **theory of segregation** with the **testcross**. In this type of experimental cross, any organism showing the dominant form of a trait is crossed to an individual known to be homozygous recessive for that trait. Results may reveal whether the organism is homozygous dominant or heterozygous.

For example, Mendel crossed purple-flowered F_1 plants with true-breeding, white-flowered plants. He predicted there would be about as many recessive as dominant offspring from the testcross. That is exactly what happened (Figure 9.8). About half had purple flowers (*Aa*) and half had white (*aa*).

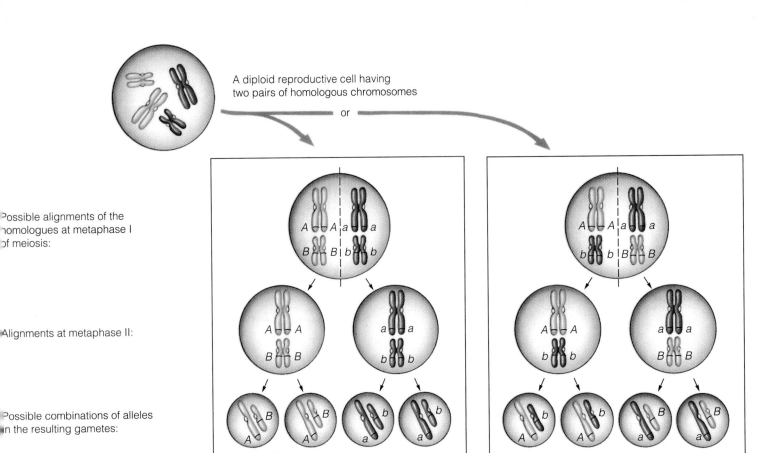

A diploid reproductive cell having two pairs of homologous chromosomes

or

Possible alignments of the homologues at metaphase I of meiosis:

Alignments at metaphase II:

Possible combinations of alleles in the resulting gametes:

1/4 AB   1/4 ab   1/4 Ab   1/4 aB

Figure 9.9 Example of independent assortment, showing just two pairs of homologous chromosomes. The allele at one location on a chromosome may or may not be identical to the allele at the same location on the homologous chromosome. In a germ cell undergoing meiosis (page 115), either chromosome of the pair can move to one spindle pole or the other. So different gametes can end up with different mixes of alleles.

The Theory of Independent Assortment

Mendel also attempted to explain how different gene pairs are sorted out into gametes. He selected true-breeding pea plants that differed in *two* traits, such as flower color and height. In such **dihybrid crosses**, the F_1 offspring inherit two gene pairs, each consisting of two nonidentical alleles.

Let's diagram one of his dihybrid crosses, using dominant alleles *A* for flower color and *B* for height, and *a* and *b* as their recessive counterparts:

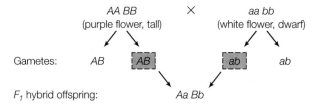

True-breeding parents:

AA BB   ×   aa bb
(purple flower, tall)   (white flower, dwarf)

Gametes: AB   AB   ab   ab

F_1 hybrid offspring: Aa Bb

As Mendel would have predicted, all the F_1 offspring are purple-flowered and tall (*Aa Bb*). How will the two gene pairs be assorted into gametes when those plants mature and reproduce?

Figure 9.10 Results from Mendel's dihybrid cross between parent plants that bred true for different versions of two traits (flower color and height). *A* and *a* represent dominant and recessive alleles for flower color. *B* and *b* represent dominant and recessive alleles for height. The probabilities of certain combinations of phenotypes among F_2 offspring occur in a 9:3:3:1 ratio, on the average.

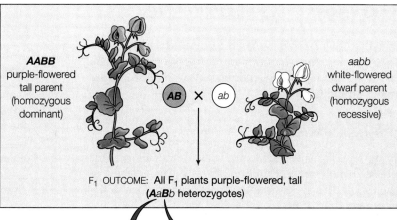

AABB
purple-flowered
tall parent
(homozygous
dominant)

AB × ab

aabb
white-flowered
dwarf parent
(homozygous
recessive)

F_1 OUTCOME: All F_1 plants purple-flowered, tall
(*AaBb* heterozygotes)

The answer partly depends on the chromosomal locations of the gene pairs. Assume the *Aa* alleles and the *Bb* alleles are located on two *different* pairs of homologous chromosomes. Next, think about how all chromosomes become positioned at the spindle equator during metaphase I of meiosis (page 114 and Figure 9.9). The chromosome with the *A* allele may be positioned to move to either spindle pole (and on into one of four gametes). The same is true for its homologue. And the same is true for the chromosomes with the *B* and *b* alleles. Therefore, after meiosis, four combinations of alleles are possible in sperm or eggs:

$$1/4 \; AB \quad 1/4 \; Ab \quad 1/4 \; aB \quad 1/4 \; ab$$

The alleles are shuffled even more at fertilization. Simple multiplication (four kinds of sperm times four kinds of eggs) tells us *sixteen* allele combinations are possible in the F_2 offspring of a dihybrid cross. Use the Punnett-square method to diagram the probabilities (Figure 9.10). Add them up, and you get 9/16 tall purple-flowered, 3/16 dwarf purple-flowered, 3/16 tall white-flowered, and 1/16 dwarf white-flowered plants. That is a probable phenotypic ratio of 9:3:3:1.

Results from all of Mendel's dihybrid F_2 crosses were close to a 9:3:3:1 ratio. Yet, without knowing that "units" of inheritance are distributed among different chromosomes, Mendel could only analyze the numbers from the dihybrid crosses. It seemed to him that the units for different traits were assorting independently into gametes. As you will read in the next chapter, we now know there are exceptions to this. So we state Mendel's theory in updated form:

Mendel's *theory of independent assortment*. During meiosis, the gene pairs of homologous chromosomes tend to be sorted into one gamete or another independently of how gene pairs on other chromosomes are sorted out.

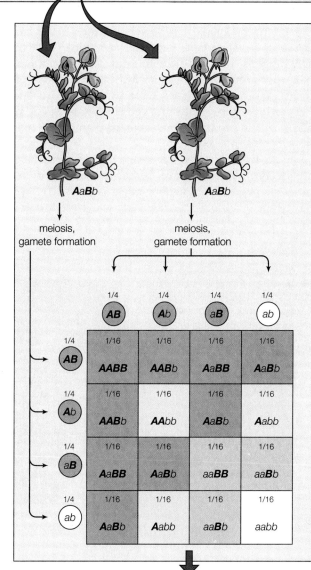

AaBb   *AaBb*

meiosis,
gamete formation

meiosis,
gamete formation

	1/4 **AB**	1/4 **Ab**	1/4 **aB**	1/4 **ab**
1/4 **AB**	1/16 **AABB**	1/16 **AABb**	1/16 **AaBB**	1/16 **AaBb**
1/4 **Ab**	1/16 **AABb**	1/16 **AAbb**	1/16 **AaBb**	1/16 **Aabb**
1/4 **aB**	1/16 **AaBB**	1/16 **AaBb**	1/16 **aaBB**	1/16 **aaBb**
1/4 **ab**	1/16 **AaBb**	1/16 **Aabb**	1/16 **aaBb**	1/16 **aabb**

ADDING UP THE F_2 COMBINATIONS POSSIBLE:

9/16 or 9 purple-flowered, tall

3/16 or 3 purple-flowered, dwarf

3/16 or 3 white-flowered, tall

1/16 or 1 white-flowered, dwarf

VARIATIONS ON MENDEL'S THEMES

Whether it was genius or good fortune, Mendel studied traits having clearly dominant or recessive forms. As the following examples will make clear, other traits are not as straightforward.

Dominance Relations

Sometimes one allele is incompletely dominant over the other. This **incomplete dominance** results in a version of the trait that is somewhere between homozygous dominant and homozygous recessive. Cross true-breeding red snapdragons with true-breeding white ones, and all F_1 offspring will have *pink* flowers (Figure 9.11). A cross between two F_1 plants produces red, pink, or white snapdragons, in a predictable ratio. Red snapdragons have two dominant alleles (which specify a red pigment). White snapdragons have two recessive alleles. Pink snapdragons are heterozygous. They have one red allele, but it produces only enough pigment to make flowers pink, not red.

In some other heterozygotes, a pair of nonidentical alleles shows **codominance**. Although they specify two different phenotypes, neither allele is masked. If you have type AB blood, for instance, you have a pair of codominant alleles that are both being expressed in your red blood cells (see the *Focus* essay).

Multiple Effects of Single Genes

A single gene may influence seemingly unrelated traits, an effect called **pleiotropy**. We see such an effect after the gene for hemoglobin (an oxygen-transporting protein in red blood cells) undergoes a certain mutation. Someone who inherits a pair of the mutated genes ends up with abnormal hemoglobin molecules—and extensive changes in phenotype. The changes are symptoms of *sickle-cell anemia*, a human genetic disorder.

Our cells use oxygen for aerobic respiration. Oxygen reaches them after diffusing out of thin-walled blood vessels called capillaries. The concentration gradient for oxygen is so steep across capillary walls, little is left inside. The near absence of oxygen causes abnormal hemoglobin molecules to stick together as long, rodlike structures. Affected cells are distorted into a shape rather like a sickle, a short-handled farm tool with a crescent-shaped blade. The cells rupture easily, and they clog and rupture capillaries. Body tissues become oxygen-starved and metabolic wastes build up. Figure 9.12 tracks the resulting phenotypic changes. Other aspects of this disorder are described in later chapters.

Figure 9.11 Example of incomplete dominance. Red-flowering and white-flowering homozygous snapdragons produce pink-flowering plants in the first generation. In heterozygotes, the red allele is only partly dominant over the white allele.

Homozygous dominant parent × Homozygous recessive parent → All F_1 offspring heterozygous

Cross between two F_1 plants

F_2 offspring show three phenotypes in a 1:2:1 ratio

ABO Blood Typing

Various membrane proteins at the surface of your cells are "self" markers—they identify the cells as being part of your own body. One kind of protein marker on red blood cells has different molecular forms. **ABO blood typing**, a method of analysis, can reveal which form a person has.

In the human population, there are three alleles of the gene that specifies this protein. They influence the protein's form in different ways. Two alleles, I^A and I^B, are codominant when paired with each other. A third allele, i, is recessive. When paired with either I^A or I^B, its effect is masked. Whenever three or more alleles of a gene exist in a population, we call this a **multiple allele system**.

Think about a cell that is making new molecules of the protein marker. Before each molecule becomes positioned at the cell surface, it takes on final form in the cytomembrane system (page 51). A carbohydrate chain is attached to it. Then an enzyme attaches a sugar unit to the chain. Alleles I^A and I^B call for two versions of that enzyme. The two attach different sugar units, and this gives the protein a special identity—either A or B.

Which two alleles for this protein do you have? With either $I^A I^A$ or $I^A i$, your blood is type A. With $I^B I^B$ or $I^B i$,

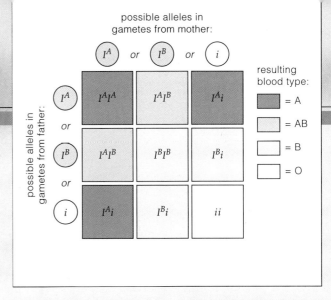

a Possible combinations of alleles that are associated with ABO blood typing.

it is type B. With codominant alleles $I^A I^B$, *both* versions of the enzyme are produced. And both attach sugar units to protein molecules. In this case, your blood is type AB. If you are homozygous recessive (*ii*), the protein markers never did get a sugar unit attached to them. Then your blood type is neither A nor B (that's what type "O" means).

When blood of two people mixes during transfusions, self markers must be compatible. Without the proper markers, red blood cells from a donor will be recognized as foreign. Antibody molecules and other protein weapons that defend the body against "nonself" will act against the foreign cells and may cause death (page 438).

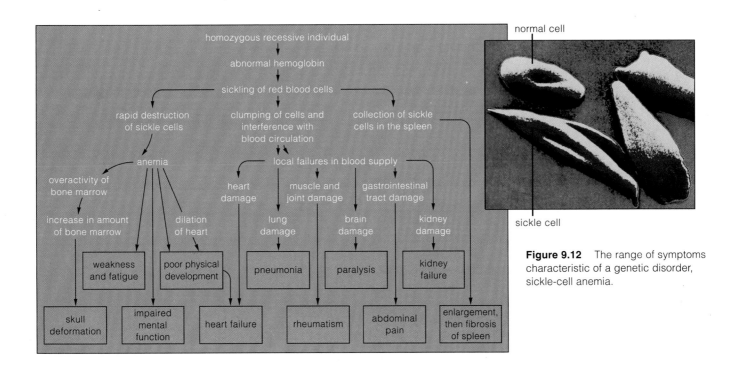

Figure 9.12 The range of symptoms characteristic of a genetic disorder, sickle-cell anemia.

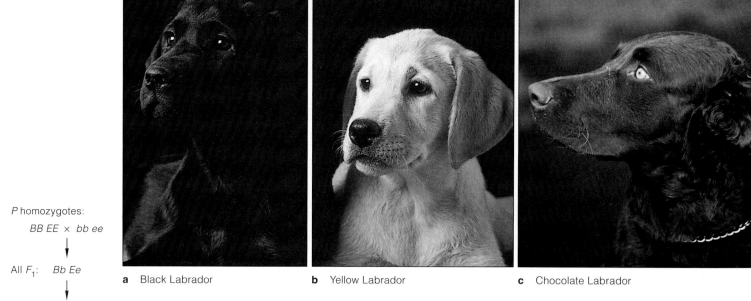

a Black Labrador

b Yellow Labrador

c Chocolate Labrador

P homozygotes:

   $BB\ EE\ \times\ bb\ ee$

All F_1: $Bb\ Ee$

F_2 combinations possible:

	BE	Be	bE	be
BE	BB EE	BB Ee	Bb EE	Bb Ee
Be	BB Ee	BB ee	BbEe	Bb ee
bE	Bb EE	Bb Ee	bb EE	bb Ee
be	Bb Ee	Bb ee	bb Ee	bb ee

RESULTING PHENOTYPES:

☐ 9/16 or 9 black

▨ 3/16 or 3 brown

☐ 4/16 or 4 yellow

Figure 9.13 How coat color of Labrador retrievers is determined by interactions among alleles of two gene pairs. Allele *B* (black) of one kind of gene involved in melanin production is dominant to allele *b* (brown). Allele *E* of a different gene allows melanin pigment to be deposited in individual hairs. Two recessive alleles (*ee*) of this gene block deposition, and a yellow coat results.

F_1 offspring of a dihybrid cross produce F_2 offspring in a 9:3:4 ratio, as the Punnett-square diagram shows. The yellow Labrador in (**b**) probably has genotype *BB ee*, because it can produce melanin but can't deposit pigment in hairs. (Looking at that photograph, can you say why?)

Interactions Between Gene Pairs

Often, a trait results from interactions among two or more gene pairs. For example, two alleles of a gene can *mask* alleles of another gene, and some expected phenotypes may not appear at all. Such interactions are called **epistasis**. They are common occurrences among the gene pairs responsible for the color of fur or skin in mammals.

Consider the black, brown, or yellow fur of Labrador retrievers (Figure 9.13). Variations in the amount and distribution of melanin, a brownish black pigment, produce the different colors. Many gene pairs affect different steps in melanin production and its deposition in certain body regions. Alleles of one gene specify an enzyme required to produce melanin. The *B* allele (black) has more pronounced effect and is dominant to *b* (brown). Alleles of a different gene control whether melanin will be deposited in a retriever's hairs. The *E* allele permits deposition, but a pair of recessive alleles (*ee*) will block it, and the retriever's fur will be yellow.

Still another gene pair codes for tyrosinase, the first enzyme required for melanin production. With one or two dominant alleles (*C*), the enzyme is produced. Two recessive alleles (*cc*) bring about the complete absence of melanin, a condition called *albinism* (Figure 9.14).

Figure 9.14 A rare albino rattlesnake. Like other animals that can't produce melanin, it has pink eyes and white surface coloration. (Eyes look pink because the absence of melanin allows red light to be reflected from blood vessels in the eyes.) In birds and mammals, surface coloration is due to pigments in feathers, fur, or skin. In fishes, amphibians, and reptiles, color-bearing cells give skin its surface coloration. Some of the cells contain melanin pigments or red to yellow pigments. Others contain crystals that reflect light and alter the effect of other pigments present.

The mutation affecting melanin production in the snake shown here had no effect on the production of yellow-to-red pigments and light-reflecting crystals. So the snake's skin appears iridescent yellow as well as white.

LESS PREDICTABLE VARIATIONS IN TRAITS

As Mendel demonstrated, the phenotypic effects of genes or gene combinations may show up in predictable ratios from one generation to the next. However, gene interactions as well as environmental factors may alter the *degree* to which a gene is actually expressed.

Differences in Environmental Conditions

Environmental effects on gene expression can be dramatic. Consider a Himalayan rabbit or Siamese cat that carries the Himalayan allele (c^h). This allele specifies a heat-sensitive version of one of the enzymes used in melanin production. At the surface of warm body regions, the enzyme is less active. There, fur grows in lighter than it does at cooler regions, including the ears and other extremities (Figures 4.11 and 9.15). Or consider a water buttercup growing half in and half out of a pond. The genes specifying leaf shape produce different phenotypes under the two environmental conditions (Figure 9.16).

Differences in Gene Interactions

Gene interactions also may introduce variability in how a gene is expressed in different individuals. Bear in mind, the path from most genes to their products (proteins) is actually a series of small metabolic steps. Genes or their products interact at many steps. If individuals differ in their combinations of alleles for those genes, their phenotypes may differ also.

This is what happens in *camptodactyly*, a human genetic disorder. Some people who carry the mutated gene have immobile, bent fingers on both hands. In others, the trait shows up on one hand only. In still others, the trait doesn't show up at all.

Figure 9.15 Effect of different environmental conditions on gene expression in animals. A Himalayan rabbit normally has black hair only on its long ears, nose, tail, and lower leg limbs. In one experiment, a patch of a rabbit's white fur was plucked clean, then an icepack was secured over the hairless patch. While the cold conditions were maintained, the hairs that grew back were black.

Himalayan rabbits are homozygous for the c^h allele of a gene that codes for tyrosinase (an enzyme needed to produce melanin). The allele specifies a heat-sensitive version of the enzyme that functions only when temperatures are below about 33°C. When hairs grow under warmer conditions, they appear light (no melanin is produced). Light fur normally covers warm body regions. Ears and other slender extremities are cooler; they tend to lose metabolic heat more rapidly.

Figure 9.16 Effect of different environmental conditions on gene expression in plants. Leaves growing from submerged stems of the water buttercup (*Ranunculus aquatilis*) are more finely divided than leaves growing in air. The variation occurs even in the same leaf if it develops half in and half out of water.

1	4	6	10	16	16	15	15	14	13	13	11	9	8	8	5	1	2

number of individuals

60 (5 feet)	61	62	63	64	65	66	67	68	69	70	71	72	73	74	75	76	77

height (inches)

a

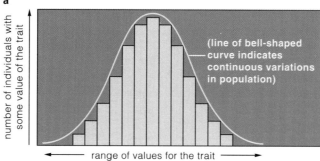

b Generalized bell-shaped curve typical of populations showing continuous variations in some trait.

(line of bell-shaped curve indicates continuous variations in population)

number of individuals with some value of the trait

← range of values for the trait →

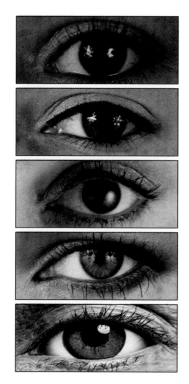

Figure 9.17 Samples from the range of continuous variation in human eye color. Different gene pairs interact to produce and deposit melanin. Among other things, this pigment helps color the eye's iris. Different combinations of alleles result in small differences in eye color. So the frequency distribution for the eye-color trait appears to be continuous over the range from black to light blue.

Figure 9.18 Continuous variation. Overall, individuals of a population show a range of small differences in most traits. Thus humans range from very short to very tall, with average heights more common than either extreme. How can you describe continuous variation of a trait within a group? First, divide the full range of different phenotypes into measurable categories. Next, count the individuals in each category. This gives you the relative frequencies of phenotypes that are distributed across the range of measurable values.

(**a**) Suppose you want to know the frequency distribution for height in a group of 168 biology students at Brigham Young University. You decide on how finely the range of possible heights should be divided. Then you measure and assign each student to the proper category. Finally, you divide the number in each category by the total number of all students in all categories. (**b**) Often a bar graph is used to depict continuous variation. Here, the proportion of students in each category is plotted against the range of measured phenotypes.

Continuous Variation in Populations

Finally, think about the color of your own eyes. That color is the cumulative result of many genes involved in the stepwise production and distribution of melanin. Black eyes have abundant melanin deposits in the iris. Dark brown eyes have less melanin, and light brown or hazel eyes have still less (Figure 9.17). Green, gray, and blue eyes don't have green, gray, or blue pigments. They have so little melanin that we readily see blue wavelengths of light being reflected from the iris.

Now think about eye color among individuals of a population. As you know, its members exhibit a *range* of small differences in the trait. In other words, they show

continuous variation. Continuous variation is especially evident in easily measurable traits, such as human height (Figure 9.18). The greater the number of gene pairs affecting a trait, the more continuous the expected distribution of all the different versions of that trait will appear to be.

The phenotypic expression of a gene may vary, by different degrees, among individuals of a population.

The degree to which a gene is expressed in different individuals depends on gene interactions. It depends also on the physical and chemical environment in which that gene or its products must function.

SUMMARY

1. A gene is a unit of information about a heritable trait. The alleles of a gene are slightly different versions of that information. Through experimental crosses with pea plants, Mendel gathered evidence that diploid organisms have two genes for each trait and that genes retain their identity when transmitted to offspring.

2. Mendel performed monohybrid crosses (between two true-breeding plants showing different versions of a single trait). The crosses provided evidence that a gene can have different molecular forms (alleles), some of which are dominant over other, recessive forms.

3. Homozygous dominant individuals have two dominant alleles (AA) for the trait being studied. Homozygous recessives have two recessive alleles (aa). Heterozygotes have two nonidentical alleles (Aa).

4. In Mendel's monohybrid crosses ($AA \times aa$), all F_1 offspring were Aa. Crosses between F_1 plants resulted in these combinations of alleles in F_2 offspring:

	A	a
A	AA	Aa
a	Aa	aa

AA (dominant)
Aa (dominant)
Aa (dominant)
aa (recessive)

This produced the expected phenotypic ratio of 3:1.

5. Armed with results from his monohybrid crosses, Mendel proposed a theory of segregation. In modern terms, diploid organisms have pairs of genes, on pairs of homologous chromosomes. The two genes of each pair segregate from each other during meiosis, such that each gamete formed ends up with one or the other.

6. Mendel also performed dihybrid crosses (between two true-breeding plants showing different versions of two traits). Results from many experiments were close to a 9:3:3:1 phenotypic ratio:

    9 dominant for both traits
    3 dominant for A, recessive for b
    3 dominant for B, recessive for a
    1 recessive for both traits

7. Mendel's dihybrid crosses led him to propose a theory of independent assortment. In modern terms, the gene pairs of two homologous chromosomes tend to be sorted into one gamete or another independently of how the gene pairs of other chromosomes are sorted out.

8. Four factors can influence gene expression. First, degrees of dominance exist between some gene pairs. Second, gene pairs can interact to produce some positive or negative effect on a trait. Third, a single gene can influence many seemingly unrelated traits. Fourth, environmental conditions can affect gene expression.

Review Questions

1. State the theory of segregation. Does segregation occur during mitosis or meiosis? 127

2. Define the difference between these terms. 125
   a. gene and allele
   b. dominant allele and recessive allele
   c. homozygote and heterozygote
   d. genotype and phenotype

3. Define true-breeding. What is a hybrid? 124, 125

4. Distinguish between monohybrid and dihybrid crosses. What is a testcross, and why is it useful in genetic analysis? 126, 128

5. State the theory of independent assortment. Does independent assortment occur during mitosis or meiosis? 129

Self-Quiz (Answers in Appendix IV)

1. Alleles are _____ .
   a. different molecular forms of a gene
   b. different molecular forms of a chromosome
   c. self-fertilizing, true-breeding homozygotes
   d. self-fertilizing, true-breeding heterozygotes

2. A heterozygote has _____ for the trait being studied.
   a. a pair of identical alleles
   b. a pair of nonidentical alleles
   c. a haploid condition, in genetic terms
   d. a and c

3. The observable traits of an organism are its _____ .
   a. phenotype   c. genotype
   b. sociobiology   d. pedigree

4. Offspring of a monohybrid cross $AA \times aa$ are _____ .
   a. all AA d. 1/2 AA and 1/2 aa
   b. all aa e. none of the above
   c. all Aa

5. Second-generation offspring from a cross are the _____ .
   a. F_1 generation   c. hybrid generation
   b. F_2 generation   d. none of the above

6. Assuming complete dominance, offspring of the cross $Aa \times Aa$ will show a phenotypic ratio of _____ .
   a. 3:1   c. 9:1
   b. 1:2:1   d. 9:3:3:1

7. Crosses between F_1 individuals resulting from the cross $AABB$ x $aabb$ lead to F_2 phenotypic ratios close to _____ .
   a. 1:2:1   c. 3:1
   b. 1:1:1:1   d. 9:3:3:1

8. Match each genetic term appropriately.
   ____ dihybrid cross   a. $AA \times aa$
   ____ monohybrid cross   b. Aa
   ____ homozygous condition   c. $AA\ BB \times aa\ bb$
   ____ heterozygous condition   d. aa

Genetics Problems (Answers in Appendix III)

1. One gene has alleles *A* and *a*. Another has alleles *B* and *b*. For each genotype listed, what type(s) of gametes will be produced? (Assume independent assortment occurs.)

   a. *AA BB*   c. *Aa bb*
   b. *Aa BB*   d. *Aa Bb*

2. Still referring to Problem 1, what will be the genotypes of offspring from the following matings? With what frequency will each genotype show up?

   a. *AA BB* × *aa BB*   c. *Aa Bb* × *aa bb*
   b. *Aa BB* × *AA Bb*   d. *Aa Bb* × *Aa Bb*

3. In one experiment, Mendel crossed a pea plant that bred true for green pods with one that bred true for yellow pods. All the F_1 plants had green pods. Which form of the trait (green or yellow pods) is recessive? Explain how you arrived at your conclusion.

4. At one gene location on a human chromosome, a dominant allele controls whether you can curl the sides of your tongue upward (*see photo*). People homozygous for the trait cannot roll their tongue. At a different gene location, a dominant allele controls whether earlobes are attached or detached (see Figure 9.1). These two gene pairs assort independently.

  Suppose a tongue-rolling woman with detached earlobes marries a man who has attached earlobes and can't roll his tongue. Their first child has attached earlobes and can't roll its tongue.

   a. What are the genotypes of the mother, father, and child?
   b. What is the probability that a second child will have detached earlobes and won't be a tongue roller?

5. Go back to Problem 1, and assume you now study a third gene having alleles *C* and *c*. For each genotype listed, what type(s) of gametes will be produced?

   a. *AA BB CC*   c. *Aa BB Cc*
   b. *Aa BB cc*   d. *Aa Bb Cc*

6. Mendel crossed a true-breeding tall, purple-flowered pea plant with a true-breeding dwarf, white-flowered plant. All F_1 plants were tall and purple-flowered. If an F_1 plant self-fertilizes, what is the probability that a randomly selected F_2 plant will be heterozygous for the genes specifying height and flower color?

7. Assume that a new gene has been identified in mice. One of its alleles specifies yellow fur color. A second allele specifies brown fur color. Suppose you are asked to determine whether the relationship between the two alleles is one of simple dominance, incomplete dominance, or codominance. What types of crosses would give you the answer? On what types of observations would you base your conclusions?

8. The ABO blood-typing system has been used to settle cases of disputed paternity. Suppose, as a geneticist, you must testify during a case in which the mother has type A blood, the child has type O blood, and the alleged father has type B blood. How would you respond to the following statements?

   a. *Attorney of the alleged father*: "The mother has type A blood, so the child's type O blood must have come from the father. Because my client has type B blood, he could not have fathered this child."
   b. *Mother's attorney*: "Further tests prove this man is heterozygous, so he must be the father."

9. As in Labrador retrievers (page 132), fur color in mice is governed by genes concerned with producing and distributing melanin. At one gene location, a dominant allele (*B*) specifies dark brown and a recessive allele (*b*) specifies light brown, or tan. At another gene location, a dominant allele (*C*) permits melanin production and a recessive allele (*c*) shuts it down and results in albinism.

   a. A homozygous *bb cc* albino mouse mates with a homozygous *BB CC* brown mouse. State the probable genotypic and phenotypic ratios for the F_1 and F_2 offspring.
   b. If an F_1 mouse from Problem 9a is backcrossed with its albino parent, what phenotypic and genotypic ratios would you expect?

Selected Key Terms

ABO blood typing *131*	monohybrid cross *126*
allele *125*	multiple allele system *131*
codominance *130*	phenotype *125*
continuous variation *134*	pleiotropy *130*
dihybrid cross *128*	probability *126*
epistasis *132*	Punnett-square method *127*
gene *125*	testcross *127*
genotype *125*	theory of independent
heterozygous *125*	assortment *129*
homozygous dominant *125*	theory of segregation *127*
homozygous recessive *125*	true-breeding *124*
incomplete dominance *130*	

Readings

Cummings, M. 1991. *Human Heredity*. Second edition. St. Paul: West Publishing Company.

Mendel, G. 1959. "Experiments in Plant Hybridization." Translation in J. Peters (editor), *Classic Papers in Genetics*. Englewood Cliffs, New Jersey: Prentice-Hall.

Orel, V. 1984. *Mendel*. New York: Oxford University Press.

Suzuki, D., et al. 1989. *An Introduction to Genetic Analysis*. Fourth edition. New York: Freeman.

10 CHROMOSOMES AND HUMAN GENETICS

Too Young To Be Old

Imagine being ten years old, and with each passing day your body becomes a bit more shriveled, frail, *old*. You are just tall enough to peer over the kitchen counter, and you weigh less than thirty-five pounds. Already you are bald, and your nose is crinkled and beaklike. Possibly you have only a few more years to live. Yet, like Mickey Hayes and Fransie Geringer (Figure 10.1), you play, laugh, and hug your friends.

Of every 8 million newborns, one is destined to grow old far too soon. One of the chromosomes inherited from its mother or father carries a mutated gene. The result will be accelerated aging and an astonishingly reduced life expectancy. This is the *Hutchinson-Gilford progeria syndrome*. There is no cure.

This particular mutation leads to disruptions in the gene interactions underlying cell division, growth, and development. Within two years, symptoms begin. Skin thins out, muscles become flabby, and limb bones soften. Hair loss is typical. Most progeriacs die in their early teens from strokes or heart attacks brought on by hardening of the arteries—a condition that is typical of advanced age.

Apparently, the gene responsible for progeria mutates spontaneously, because progeria doesn't seem to run in families. The gene isn't on a sex chromosome, because the disorder shows up in boys *and* in girls. Probably a dominant allele is involved; it is always expressed.

We began this unit of the book with a look at cell division, the starting point of inheritance. Then we started thinking about how chromosomes—and the genes they carry—are shuffled at meiosis and fertilization. In this chapter we delve more deeply into the patterns of chromosomal inheritance, with emphasis on humans. At times the methods of analysis might seem abstract. But keep in mind that we are talking about messages of inheritance in yourself and in other human individuals. When Mickey Hayes turned eighteen, he was the oldest living progeriac. Fransie was seventeen when he died.

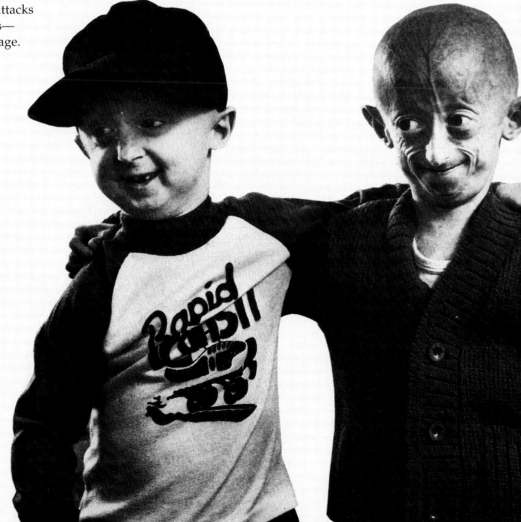

Figure 10.1 Two boys, both less than ten years old, who met during a gathering of progeriacs at Disneyland, California. Progeria is a genetic disorder characterized by accelerated aging and extremely reduced life expectancy.

KEY CONCEPTS

1. One gene follows another along the length of a chromosome, and each one has its own position in that sequence.

2. The gene sequence does not necessarily remain intact through meiosis and gamete formation. Whenever crossovers take place, some genes in the sequence are removed and replaced with genes from the homologous chromosome. Crossing over occurs more often between genes that are far apart in the sequence. It occurs less often between genes that are close together.

3. Crossing over and independent assortment during meiosis contribute to variation in traits among offspring. So does the chance union of any two gametes at fertilization.

4. A chromosome's structure can change, and so can the chromosome number. Such changes lead to variation in traits. Often they result in genetic disorders.

EARLY STUDIES OF CHROMOSOMES

Return of the Pea Plant

The year was 1884. Mendel's paper on pea plants had been gathering dust in a hundred libraries for nearly two decades, and Mendel himself had just passed away. Ironically, the experiments described in that forgotten paper were about to be devised all over again.

Improvements in microscopy had rekindled efforts to locate the hereditary material within cells. By 1882, Walther Flemming had observed threadlike bodies—chromosomes—in the nucleus of dividing cells. By 1884, a question was taking shape: Could chromosomes be the hereditary material?

Then researchers realized each gamete has half the number of chromosomes of a fertilized egg. In 1887, August Weismann proposed that a special division process must reduce the chromosome number by half before gametes form. Sure enough, in that same year meiosis was discovered. Weismann began to promote his theory of heredity: *The chromosome number is halved during meiosis, then restored at fertilization. Thus a cell's hereditary material is half paternal in origin, and half maternal*. His theory was hotly debated, and it prompted a flurry of experimental crosses—just like the ones Mendel had carried out.

Finally, in 1900, researchers came across Mendel's paper while checking literature related to their own genetic crosses. To their chagrin, their results merely confirmed what Mendel had already proposed: Diploid cells have two units (genes) for each heritable trait, and the units segregate before gametes form.

Researchers learned a great deal about chromosomes during the decades that followed. Let's start with a few high points of their work, which will serve as background for understanding human inheritance.

Autosomes and Sex Chromosomes

Generally speaking, a pair of homologous chromosomes are exactly alike in length, shape, and gene sequence. In the early 1900s, however, microscopists discovered an exception to this. Depending on the species, one type of chromosome is found in males *or* females, but never in both. Also depending on the species, that chromosome may pair up with a physically different chromosome during meiosis. For instance, human males (but not females) have a **Y chromosome**. They also have an **X chromosome** that serves as its partner during meiosis. Human females have a pair of X chromosomes.

Human X and Y chromosomes are examples of **sex chromosomes**. Inheritance of one type *or* the other governs gender—that is, whether a new individual will be male or female. All other chromosomes, which are the same in both sexes, are designated **autosomes**.

A sex chromosome is one whose presence determines a new individual's gender. An autosome is any chromosome that is the same in both sexes.

Today, microscopists routinely analyze a cell's sex chromosomes and autosomes. Chromosomes, recall, are most highly condensed at metaphase of mitosis. At that time, each has a characteristic size, length, and centromere location. Also, chromosomes of many species show banding patterns when stained in certain ways. Their more condensed regions take up more stain, so they form darker bands. Figure 7.3 is an example.

The number of metaphase chromosomes and their defining characteristics are called the **karyotype** of an individual (or species). A karyotype diagram is a cut-up, rearranged photograph of the chromosomes. The autosomes are lined up, largest to smallest, and the sex chromosomes are positioned last (Figure 10.2).

By analyzing karyotypes of human gametes, you can conclude that each egg produced by a female carries an X chromosome. Half the sperm produced by a male carry an X and half carry a Y. If an egg and sperm each carry an X chromosome and unite at fertilization, the new individual will develop into a female. Conversely, if the sperm carries a Y chromosome, the individual will develop into a male (Figure 10.3).

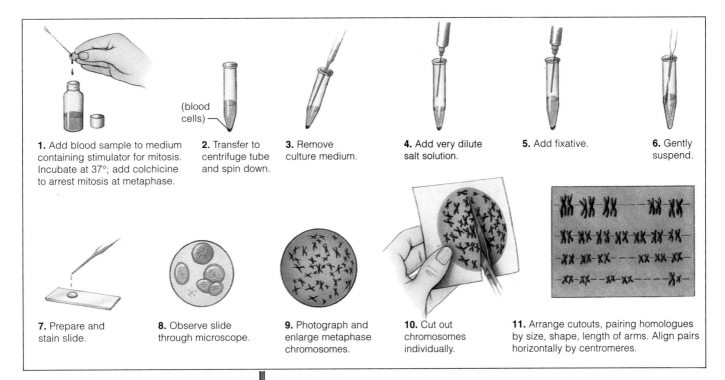

1. Add blood sample to medium containing stimulator for mitosis. Incubate at 37°; add colchicine to arrest mitosis at metaphase.

2. Transfer to centrifuge tube and spin down.

(blood cells)

3. Remove culture medium.

4. Add very dilute salt solution.

5. Add fixative.

6. Gently suspend.

7. Prepare and stain slide.

8. Observe slide through microscope.

9. Photograph and enlarge metaphase chromosomes.

10. Cut out chromosomes individually.

11. Arrange cutouts, pairing homologues by size, shape, length of arms. Align pairs horizontally by centromeres.

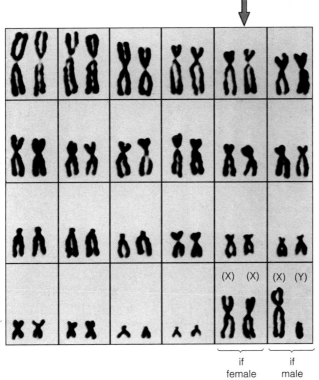

(X) (X) (X) (Y)

if female   if male

Figure 10.2 Karyotype preparation. Human cells have a diploid chromosome number of 46. The nucleus contains 22 pairs of autosomes and 1 pair of sex chromosomes (X and Y). Each chromosome in this karyotype is duplicated (it consists of two sister chromatids).

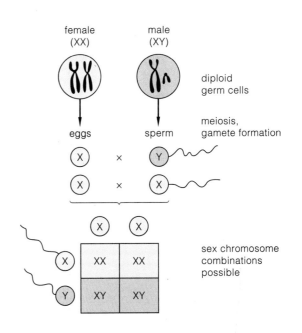

female (XX)   male (XY)

diploid germ cells

eggs   sperm

meiosis, gamete formation

X × Y

X × X

sex chromosome combinations possible

	X	X
X	XX	XX
Y	XY	XY

Figure 10.3 The pattern of sex determination in humans. This same pattern occurs in many animal species. Males transmit their Y chromosome to their sons but not to their daughters. Males receive their X chromosome only from their mother.

Girls, Boys, and the Y Chromosome

For about the first four weeks of its existence, a human embryo is neither male nor female, even though it normally carries XY or XX chromosomes. However, ducts and other structures start forming that can go either way (Figures *a* and *b*).

In XY embryos, the primary male reproductive organs (testes) start to form during the next four to six weeks. A gene region on the Y chromosome seems to govern a fork in the developmental road that can lead either to maleness or to femaleness. In XX embryos, the primary female reproductive organs (ovaries) start to form. They form automatically in the absence of a Y chromosome.

The testes start to produce testosterone and other sex hormones, which influence the development of the male reproductive system. The ovaries also start producing sex hormones that influence the development of the female reproductive system.

The master gene for sex determination is named SRY (for sex-determining region of the Y chromosome). So far, the same gene has been identified in DNA from male humans, chimpanzees, rabbits, pigs, horses, cattle, and tigers. No females that were tested had the gene. Tests with mice indicate that the gene region becomes active about the time testes start developing.

The SRY gene resembles regions of DNA that are known to specify regulatory proteins. Such proteins bind to certain parts of DNA and so turn genes on and off (page 172). Do the protein products of the SRY gene turn off genes required for female development? Maybe.

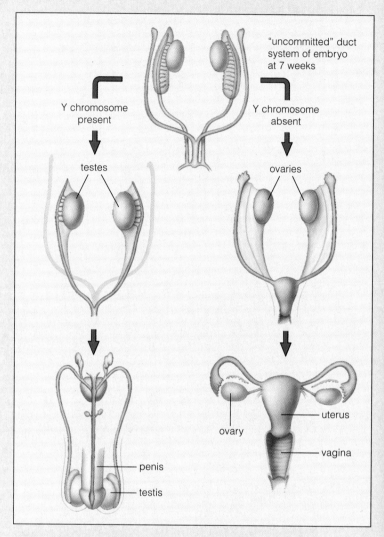

a Duct system in the early embryo that develops into a male *or* female reproductive system. Compare Figures 32.6 and 32.9.

Among the very few genes on the Y chromosome is a "male-determining gene." The expression of this gene leads to the formation of testes. In the gene's absence, ovaries form automatically, as described in the *Focus* essay. Hormones produced by the ovaries and the testes govern the development of sexual traits.

A human X chromosome probably carries more than 300 genes. Like other chromosomes, it carries some genes associated with sexual traits, such as the distribution of body fat. But most of the genes on the X chromosome deal with *nonsexual* traits, such as blood-clotting functions.

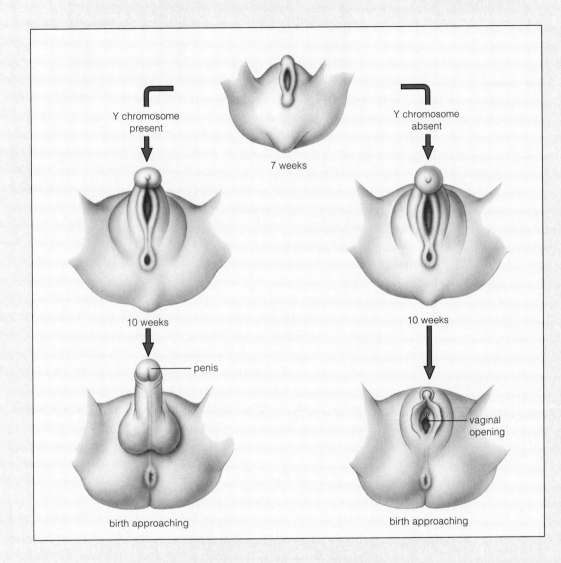

b External appearance of developing reproductive organs.

Y chromosome present

Y chromosome absent

7 weeks

10 weeks

10 weeks

penis

vaginal opening

birth approaching

birth approaching

Some time ago, Thomas Morgan and his coworkers experimented with fruit flies (*Drosophila*) in attempts to understand the nature of inheritance. Their experimental results strongly indicated that certain genes had to be located on the *Drosophila* X chromosome. This discovery helped confirm growing suspicions that *every* gene has a specific location on a specific chromosome. Figure 10.4 on the next page describes some of these landmark experiments.

For a time, genes on an X or Y chromosome were called "sex-linked genes." Today researchers use the more precise terms, **X-linked** and **Y-linked genes**.

Figure 10.4 X-linked genes: clues to inheritance patterns.

In the early 1900s, the embryologist Thomas Morgan was studying inheritance patterns. During those studies, he and his coworkers discovered an apparent genetic basis for the connection between gender and certain nonsexual traits. For example, human males and females both have blood-clotting mechanisms. Yet hemophilia (a blood-clotting disorder) shows up most often in the males, not females, of a family lineage. This gender-specific outcome was not like anything Mendel saw in his hybrid crosses between pea plants. (Either one parent plant or the other could carry a recessive allele. It made no difference *which* parent carried it; the resulting phenotype was the same.)

Morgan studied eye color and other nonsexual traits of *Drosophila melanogaster*. These fruit flies can be grown in bottles on bits of cornmeal, molasses, and agar. A female lays hundreds of eggs in a few days, and her offspring reproduce in less than two weeks. Morgan could track hereditary traits through nearly thirty generations of thousands of flies in a year's time.

At first, all the flies were wild-type for eye color; they had brick-red eyes, as in (**a**). ("Wild-type" simply means the normal or most common form of a trait in a population.) Then, through an apparent mutation in a gene controlling eye color, a *white-eyed* male appeared (**b**).

Morgan established true-breeding strains of white-eyed males and females. Then he did a series of "reciprocal crosses." These are pairs of crosses. In the first, one parent displays the trait in question. In the second, the other parent displays the trait.

White-eyed males were mated with homozygous red-eyed females. All the F_1 offspring of the cross had red eyes. But of the F_2 offspring, only some of the males had white eyes. Then white-eyed females were mated with true-breeding red-eyed males. Of the F_1 offspring of that second cross, half were red-eyed females and half were white-eyed males. Of the F_2 offspring, 1/4 were red-eyed females, 1/4 white-eyed females, 1/4 red-eyed males, and 1/4 white-eyed males!

The seemingly odd results suggested that the eye-color gene was related to gender. Probably it was located on one of the sex chromosomes. But which one? Because females (XX) could be white-eyed, the recessive allele would have to be on one of their X chromosomes. Suppose white-eyed males (XY) also carry the recessive allele on their X chromosome—*and suppose there is no corresponding eye-color allele on their Y chromosome*. Those males would have white eyes because they have no dominant allele that would mask the effect of the recessive one.

(**c**) This diagram shows the expected results when the idea of an X-linked gene is combined with Mendel's concept of segregation. By proposing that a specific gene is located on the X chromosome but not on the Y, Morgan was able to explain the outcome of his reciprocal crosses. The results of the experiments matched the predicted outcomes.

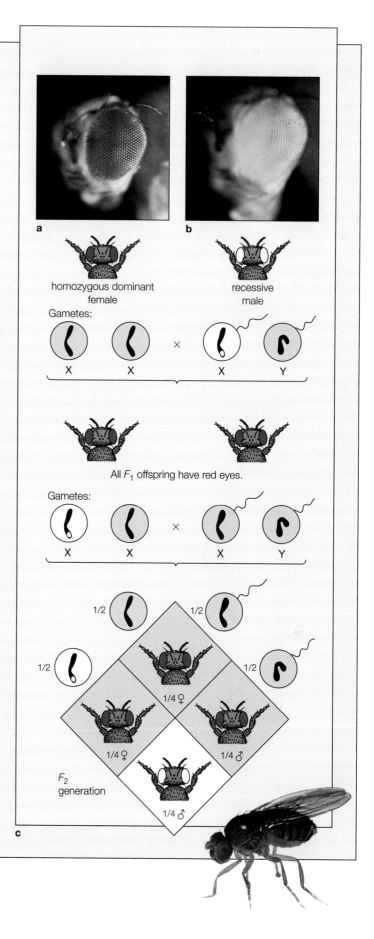

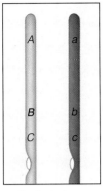

a Equivalent gene regions of a pair of homologous chromosomes at interphase, before DNA replication.

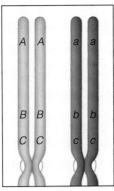

b The same regions after DNA replication at interphase. Both chromosomes are now in the duplicated state.

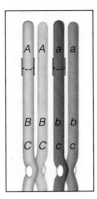

c At prophase I of meiosis, two of the nonsister chromatids break while aligned very tightly together (Figure 8.4).

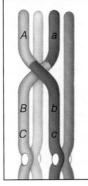

d The nonsister chromatids swap segments, then enzymes seal the broken ends.

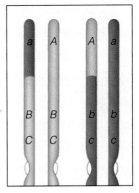

e The crossover led to genetic recombination between two of four chromatids. (Here they are shown after meiosis, as separate, unduplicated chromosomes.)

Figure 10.5 Crossing over and genetic recombination. Earlier, the diagram in Figure 8.4 started us thinking about crossing over by showing the time of its occurrence in meiosis. Only a single crossover was illustrated, so that the focus would remain on the nuclear division stages themselves. However, in humans and most other eukaryotic organisms, *every* chromosome must undergo at least one crossover. Otherwise meiosis cannot be properly completed.

Linkage and Crossing Over

Early experiments with fruit flies also suggested that genes on the same chromosome tend to end up in the same gamete. This tendency came to be called **linkage**. As we now know, linkage can be disrupted by crossing over—the breakage and exchange of segments between a pair of homologous chromosomes (Figure 10.5).

Imagine any two genes at two different locations on the same chromosome. The probability of a crossover disrupting their linkage is proportional to the distance separating them. Suppose genes A and B are twice as far apart as two other genes, C and D:

We would expect crossing over to disrupt the linkage between A and B much more often.

Two genes are very closely linked when the distance between them is small; they nearly always end up in the same gamete (Figure 10.6a). Linkage is more vulnerable to crossing over when the distance between two genes is greater (Figure 10.6b). What happens when two genes are very far apart? Crossing over disrupts their linkage so often, it is as if they assort independently.

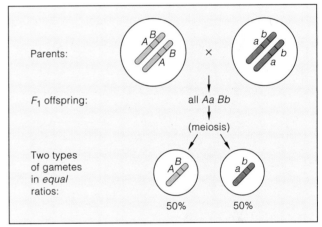

a Complete linkage (no crossing over)

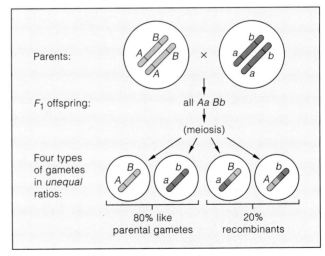

b Incomplete linkage owing to crossing over

Figure 10.6 How crossing over can affect gene linkage, using two genes on the same chromosome as the example.

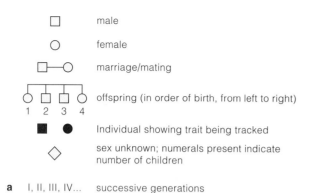

male □

female ○

marriage/mating □—○

offspring (in order of birth, from left to right)
1 2 3 4

Individual showing trait being tracked ■ ●

sex unknown; numerals present indicate number of children ◇

a I, II, III, IV... successive generations

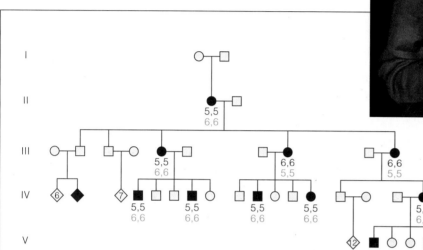

b

Figure 10.7 (**a**) Some symbols used in constructing pedigree diagrams. (**b**) This is an example of a pedigree for *polydactyly*. An individual with this condition has extra fingers, extra toes, or both. Expression of the gene governing polydactyly can vary from one individual to the next. Here, black numerals designate the number of fingers on each hand. Blue ones designate the number of toes on each foot.

HUMAN GENETICS

Pea plants and fruit flies lend themselves to genetic analysis. They grow and reproduce rapidly in small spaces, under controlled conditions. It doesn't take long to track a trait through many generations.

Humans are another story. We live under variable conditions in diverse environments. We select our own mates and reproduce if and when we want to. Human subjects live as long as the geneticists who study them, so tracking traits through generations can be rather tedious. Most human families are so small, there aren't enough offspring for easy inferences about inheritance.

To get around some of these problems, geneticists use standardized methods to construct **pedigrees**. These are charts of genetic connections among individuals. Figure 10.7 is an example. Pedigrees are analyzed to determine whether the inheritance of certain traits follows certain patterns through the generations. Such patterns are clues to the trait's genetic basis. (They may indicate whether the responsible allele is dominant or

Table 10.1	Examples of Human Genetic Disorders
Disorder or Abnormality*	Main Consequences
Autosomal Recessive Inheritance:	
Albinism (*132*)	Absence of pigmentation (melanin)
Sickle-cell anemia (*130, 170, 203*)	Severe tissue, organ damage
Galactosemia (*145*)	Brain, liver, eye damage
Phenylketonuria (*150*)	Mental retardation
Autosomal Dominant Inheritance:	
Achondroplasia (*146*)	A type of dwarfism
Camptodactyly (*133*)	Rigid, bent little fingers
Huntington's disorder (*146*)	Progressive, irreversible degeneration of nervous system
Polydactyly (*144*)	Extra digits
Progeria (*137*)	Premature aging

*Number in parentheses indicates the page(s) on which the disorder is described.

recessive and whether it occurs on an autosome or sex chromosome.) Gathering many family pedigrees increases the numerical base for analysis.

Table 10.1 lists a few traits that have been studied in detail. Some are genetic abnormalities, or deviations from the average condition. Said another way, an abnormality is simply a rare or less common version of a trait, as when a person is born with six toes on each foot instead of five. Whether we view this version of the trait as disfiguring or merely interesting is subjective. There is nothing inherently life-threatening or even ugly about it.

By contrast, a **genetic disorder** is an inherited condition that results in mild to severe medical problems.

Alleles underlying severe genetic disorders do not abound in populations, for they put individuals at great risk. They do not disappear entirely for two reasons. First, rare mutations put new copies of the alleles in the population. Second, in heterozygotes, such an allele is paired with a normal one that may cover its functions, so it can be passed to offspring.

Autosomal Inheritance

Autosomal Recessive Inheritance. When a trait results from a *recessive* allele on an autosome, two clues point to this condition. First, if both parents are heterozygous, there is a 50 percent chance each child will be heterozygous and a 25 percent chance it will be homozygous recessive (Figure 10.8). Second, if both

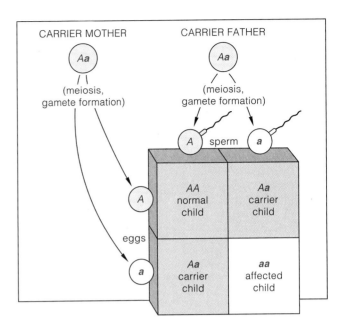

Figure 10.8 One pattern for autosomal recessive inheritance. This example shows the phenotypic outcomes possible when both parents are heterozygous carriers of the recessive allele (shown in red).

Disorder or Abnormality[*]	Main Consequences
X-Linked Inheritance:	
Hemophilia A (*146*)	Deficient blood-clotting
Testicular feminizing syndrome (*506*)	Absence of male organs, sterility
Changes in Chromosome Structure:	
Cri-du-chat (*147*)	Mental retardation, skewed larynx
Fragile X syndrome (*148*)	Mental retardation
Changes in Chromosome Number:	
Down syndrome (*148*)	Mental retardation, heart defects
Turner syndrome (*150, 153*)	Sterility, abnormal development of ovaries and sexual traits
Klinefelter syndrome (*150*)	Sterility, mental retardation
XYY condition (*150*)	Mild mental retardation in some cases; no symptoms in others

parents are homozygous recessive, any child of theirs will be, also.

Galactosemia is a genetic disorder arising from autosomal recessive inheritance. About 1 in 100,000 newborns are homozygous recessive for an enzyme that helps keep a breakdown product of lactose (milk sugar) from accumulating. Normally, lactose is converted to glucose and galactose, then to glucose-1-phosphate. Glycolysis can break down this conversion product. However, the defective enzyme produced by galactosemics blocks the full conversion:

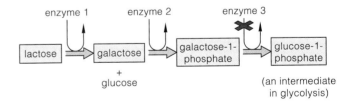

A high blood level of galactose can damage the eyes, liver, and brain. Malnutrition, diarrhea, and vomiting are early symptoms. Untreated galactosemics often die in childhood. However, the telling symptom—a high galactose level—can be detected in urine samples. With a diet that includes milk substitutes and excludes dairy products, affected individuals grow up symptom-free.

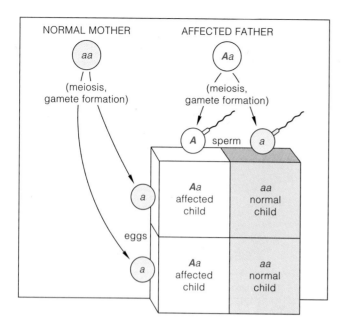

Figure 10.9 One pattern for autosomal dominant inheritance. In this example, assume the dominant allele (shown in red) is fully expressed in the carriers.

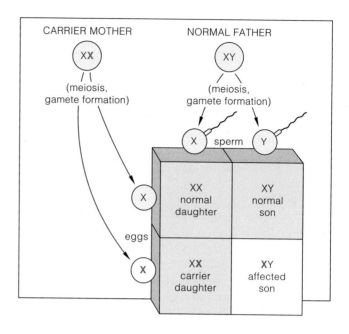

Figure 10.10 One pattern for X-linked inheritance. This example shows the phenotypic outcomes possible when the mother carries a recessive allele on one of her X chromosomes (shown in red).

Autosomal Dominant Inheritance. What clues tell us a *dominant* allele on an autosome is responsible for a trait? First, because such alleles are usually expressed (even in heterozygotes), the trait typically appears in each generation. Second, if one parent is heterozygous and the other homozygous recessive, there is a 50 percent chance that any one child of theirs will be heterozygous. Figure 10.9 shows one pattern for autosomal dominant inheritance.

A few dominant alleles that cause severe genetic disorders persist in populations. Mutation may be the reason. This is the case for progeria, the aging disorder described earlier. In other cases, *a dominant allele may not prevent reproduction, or its effect may not show up until after reproductive age.*

For example, *achondroplasia* is a type of dwarfism that affects roughly 1 in 10,000 humans. Homozygous dominants usually are stillborn. Heterozygotes cannot form cartilage properly when limb bones are growing, and this leads to abnormally short arms and legs. Adults are less than four feet, four inches tall. The dominant allele often has no other phenotypic effects on heterozygotes, who normally are able to reproduce.

As another example, *Huntington's disorder* is a progressive deterioration of the nervous system. In about half the cases, symptoms do not start until about age forty. Most people already have children by then.

X-Linked Inheritance

Certain clues may tell you that a trait is specified by an allele on the X chromosome. When the X-linked allele is *recessive*, you will detect this pattern:

1. The recessive phenotype shows up far more often in males than females. (A recessive allele can be masked in females, who may inherit a dominant allele on their other X chromosome. It cannot be masked in males, who have only one X chromosome.)

2. A son cannot inherit the recessive allele from his father. A daughter can. If she is heterozygous, there is a 50 percent chance that each son of hers will inherit the allele (Figure 10.10).

Hemophilia A is an example of X-linked recessive inheritance. In most people, a blood-clotting mechanism quickly stops bleeding from minor injuries. The reactions that lead to clot formation require the products of several genes. If any of the genes is mutated, its defective product can cause affected persons to bleed for an abnormal time.

A mutated gene for "clotting factor VIII" causes hemophilia A. It affects about 1 in 7,000 males, who run the risk of dying from untreated bruises, cuts, or internal bleeding. Blood-clotting time is more or less normal in heterozygous females.

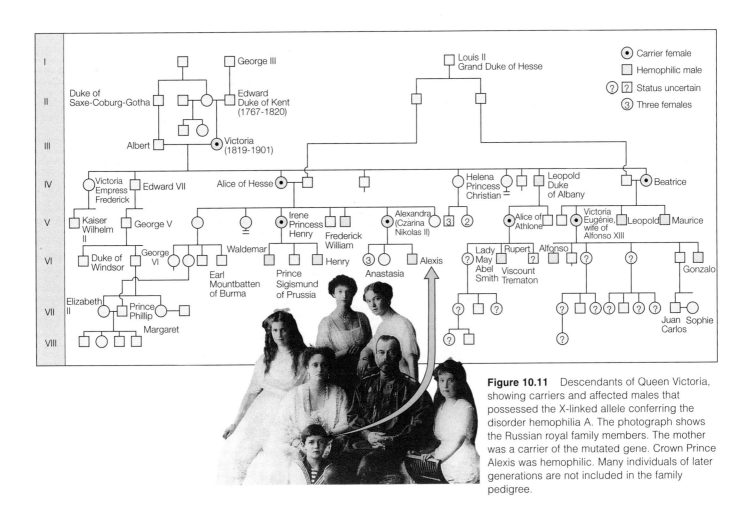

Figure 10.11 Descendants of Queen Victoria, showing carriers and affected males that possessed the X-linked allele conferring the disorder hemophilia A. The photograph shows the Russian royal family members. The mother was a carrier of the mutated gene. Crown Prince Alexis was hemophilic. Many individuals of later generations are not included in the family pedigree.

The frequency of hemophilia A was unusually high among royal families of nineteenth-century Europe, whose members often intermarried. Queen Victoria of England was a carrier. At one time, the recessive allele was present in eighteen of her sixty-nine descendants. One hemophilic great-grandchild, Crown Prince Alexis, was a focus of political intrigue that helped usher in the Russian revolution (Figure 10.11).

Changes in Chromosome Structure

On rare occasions, the structure of a chromosome may change abnormally in the following ways.

Deletion. A **deletion** is the loss of a chromosome region by irradiation, viral attack, chemical action, or other factors. One or more genes may be lost, and this nearly always causes problems. A certain deletion from human chromosome 5 causes mental retardation and an abnormally shaped larynx. When an affected infant cries, he or she produces meowing sounds. Hence the name of the disorder, *cri-du-chat* (cat-cry). Figure 10.12 shows an affected child.

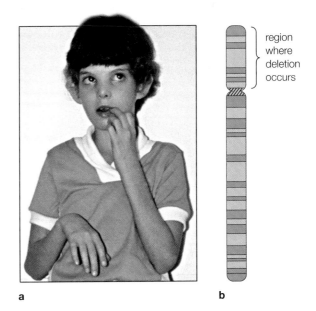

Figure 10.12 Deletion leading to cri-du-chat syndrome. Affected infants have an improperly developed larynx, and their cries sound like a cat in distress. They show severe mental retardation. A rounded face, small cranium, and misshapen ears are outward symptoms (**a**). Deletion occurs in a fragile region of chromosome 5 (**b**).

Duplication. A **duplication** is a repeated region in a chromosome. The *fragile X syndrome*, a form of mental retardation, results from an abnormally long region of repeats that may block expression of a gene in the X chromosome. In evolutionary terms, some duplications may prove beneficial, for backup copies of a gene may protect against mutation in one of them. Such copies might even evolve in separate ways. If so, they could provide individuals with related but slightly different gene products, functioning in slightly different ways.

Inversion. An **inversion** is a segment that separated from a chromosome and then was inserted at the same place—but in reverse. The reversal alters the position and order of the chromosome's genes.

Translocation. Most often, a **translocation** is part of one chromosome inserted into another (but not into its homologue). As an example, chromosome 14 may end up with a segment of chromosome 8. Controls over the segment's genes are lost at the new location, and a form of cancer results.

Long ago, the ancestors of humans and other primates took separate evolutionary paths. Structural changes in chromosomes probably contributed to the separation. Of the twenty-three pairs of human chromosomes, eighteen are nearly identical to their counterparts in chimpanzees and gorillas. Inversions and translocations introduced changes in the other five pairs. Changes also arose through duplications. For instance, each species has a set of related genes for a set of hemoglobin molecules, which have slightly different functions.

Changes in Chromosome Number

Various abnormal cellular events can put too many or too few chromosomes into gametes. New individuals end up with the wrong chromosome number. The effects range from minor physical changes to lethal disruption of organ systems. More often, affected individuals are miscarried (spontaneously aborted before birth).

Categories of Change. New individuals may end up with *one extra* or *one less* chromosome. This condition, called **aneuploidy**, is a major cause of reproductive failure. Probably it affects one of every two newly fertilized human eggs. Most of the human embryos miscarried and autopsied each year were aneuploids.

New individuals also may end up with *three or more* of each type of chromosome. This condition, called **polyploidy**, is common in the plant kingdom. (About half of all flowering plant species are polyploids.) It

also occurs among animals, including some insects and fishes. Polyploidy is lethal for humans. Probably it disrupts interactions between the genes of autosomes and sex chromosomes at key steps in the complex pathways of development and reproduction. All but 1 percent of human polyploids die before birth, and the rare newborns die within a month.

Mechanisms of Change. A chromosome number can change during mitotic or meiotic cell divisions or during the fertilization process. Suppose a cell cycle proceeds through DNA duplication and mitosis, then is arrested before the cytoplasm divides. This polyploid cell is "tetraploid," with *four* of each type of chromosome. Or suppose one or more pairs of chromosomes fail to separate properly during mitosis or meiosis. Such events are called **nondisjunction**. As Figure 10.13 shows, some or all of the resulting cells end up with too many or too few chromosomes.

Nondisjunction can be induced experimentally by exposing cells to colchicine. (This substance is derived from the autumn crocus, *Colchicum autumnale*. It must be used with caution, for it may cause cancer.) Colchicine stops microtubular spindles from forming during nuclear division, so chromosomes cannot separate. Thus a cell's chromosome number may double repeatedly, depending on the concentration and length of exposure to colchicine.

What if a gamete with an extra chromosome ($n + 1$) unites with a normal one at fertilization? The new individual will be "trisomic," with three of one type of chromosome ($2n + 1$). If the gamete is missing a chromosome, the individual will be "monosomic" ($2n - 1$).

Down Syndrome. Chromosome 21 is one of the smallest chromosomes in human cells. Someone who inherits three of them is a trisomic 21 and will show the effects of *Down syndrome*. ("Syndrome" simply means a set of symptoms that characterize a disorder.) Figure 10.14*a* shows a karyotype of an affected girl.

About 1 of every 1,000 newborns in North America is trisomic 21. Most show moderate to severe mental retardation, and about 40 percent develop heart defects. The skeleton develops more slowly than normal, and muscles are rather slack. Older children are shorter than normal and have distinctive facial features, such as a small skin fold over the inner corner of the eyelid. With special training, affected individuals often take part in normal activities, and they enjoy life to the fullest extent allowed by their condition. Down syndrome is one of many genetic disorders that can be detected by prenatal diagnosis, as described in the *Focus* essay that concludes this chapter.

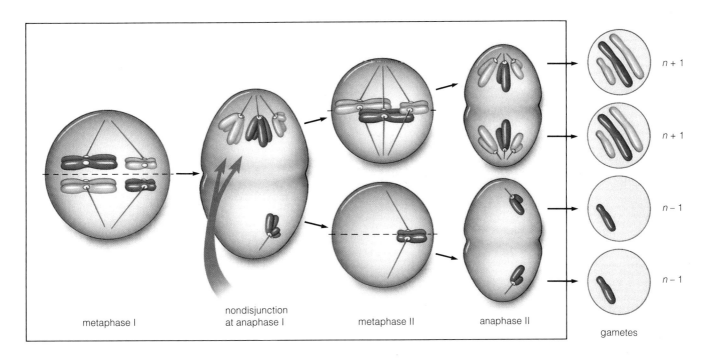

metaphase I

nondisjunction
at anaphase I

metaphase II

anaphase II

$n + 1$

$n + 1$

$n - 1$

$n - 1$

gametes

Figure 10.13 Example of nondisjunction. Here, chromosomes fail to separate during anaphase I of meiosis and so change the chromosome number in resulting gametes. As another example, sketch out a similar diagram in which nondisjunction occurs at anaphase II of meiosis. What will the chromosome numbers be in the resulting gametes?

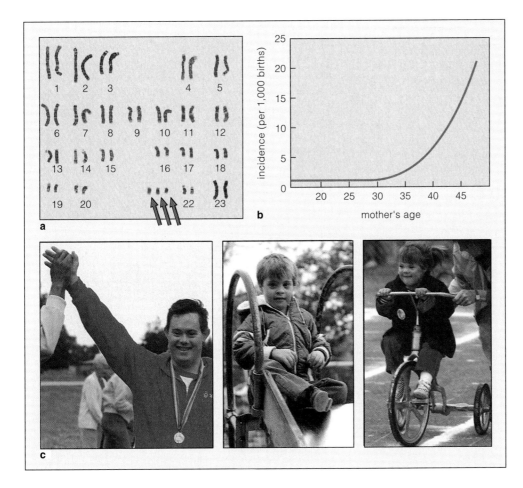

Figure 10.14 Down syndrome. (**a**) Karyotype of an affected girl. Brown arrows identify the trisomy of chromosome 21. (**b**) Relationship between the frequency of Down syndrome and the mother's age at the time her child was born. Results are from a study of 1,119 children with the disorder who were born in Victoria, Australia, between 1942 and 1957. The photographs show a few children with Down syndrome.

Sex Chromosome Abnormalities. Nondisjunction of sex chromosomes also causes several genetic disorders. For example, about 1 in every 2,500 newborn girls in North America has one X chromosome but no partner for it. This genotype leads to *Turner syndrome* (Figure 10.15).

This sex chromosome abnormality occurs less often than others, probably because most X0 embryos are miscarried early in pregnancy. Adult women who are affected are short (four feet, eight inches, on the average). Most do not have functional ovaries, which normally produce sex hormones as well as eggs. Without the hormones, sexual development cannot be completed and affected women are infertile. X0 women are within the normal range of intelligence, although they tend to show some learning disabilities.

As another example, about 1 in 500 liveborn males inherits one Y and two X chromosomes. This sex chromosome abnormality leads to *Klinefelter syndrome* (Figure 10.15). Most symptoms develop after the onset of puberty. XXY males show low fertility and some degree of mental retardation. Their testes are much smaller than normal, body hair is sparse, and there may be some breast enlargement. Injections of the hormone testosterone reverse the feminized traits but not the sterility or mental retardation.

As a final example, about 1 in every 1,000 males has one X and two Y chromosomes. With this *XYY condition*, males tend to be taller than average. Some may be mildly retarded, but most are phenotypically normal. At one time, XYY males were thought to be genetically predisposed to become criminals. We now know this is not the case.

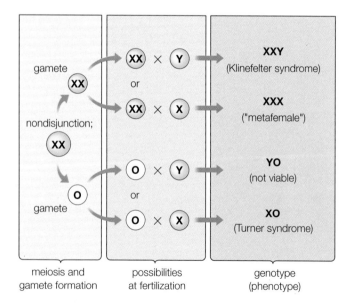

Figure 10.15 Genetic disorders resulting from nondisjunction of X chromosomes followed by fertilization involving normal sperm.

Focus on Bioethics

Prospects and Problems in Human Genetics

With the arrival of a newborn, parents typically want to know whether it is a girl or boy. Then most ask, apprehensively, "Is the baby normal?" They naturally want their baby to be free of genetic disorders, and most of the time it is. Chapter 32 describes the story of human reproduction and development when all goes well. But what are the options when it does not?

We do not approach diseases and heritable disorders the same way. "Diseases" are the outcome of infection by bacteria, viruses, and other agents from the outside. We eliminate or control infectious agents with antibiotics and other weapons. By contrast, how do we attack an "enemy" within our genes?

Do we institute regional, national, or global programs to identify people carrying harmful alleles? Do we tell them they are "defective" and run a risk of bestowing their disorder on their children? Who decides which alleles are "harmful"? Should society bear the cost of treating Down syndrome and other genetic disorders? If so, should society also have a say in whether affected embryos will be born at all, or aborted? These questions are only the tip of an ethical iceberg. And answers have not been worked out in universally acceptable ways.

Phenotypic Treatments. Genetic disorders cannot be cured. But sometimes their *symptoms* can be suppressed or minimized. For example, this can be done by controlling diets, making environmental adjustments, or intervening surgically.

Diet control works for several disorders, including galactosemia (page 145). *Phenylketonuria*, or PKU, is another case in point. A certain gene codes for an enzyme that converts one amino acid to another (phenylalanine to tyrosine). In people who are homozygous recessive for a mutated form of the gene, the first amino acid accumulates. If the excess is diverted to other pathways, phenylpyruvate and other compounds may be produced. High levels of phenylpyruvate can lead to mental retardation. But suppose affected persons restrict the amount of phenylalanine in their diet. Their body does not have to dispose of excess amounts, so they can lead normal lives.

Diet soft drinks and many other products often are artificially sweetened with aspartame, which contains phenylalanine. Such products carry warning labels directed at phenylketonurics.

Environmental adjustments help counter symptoms of some disorders. *Albinos* (page 132) can avoid direct sunlight. *Sickle-cell anemics* (page 130) can avoid strenuous activity when oxygen levels are low, as at high altitudes.

Surgical reconstructions can correct or minimize many phenotypic problems. In one form of *cleft lip*, a vertical fissure cuts through the lip midsection and often extends into the roof of the mouth. Surgery usually corrects the lip's appearance and function.

Genetic Screening. Some genetic disorders can be detected early enough to start preventive measures *before* symptoms develop. "Genetic screening" refers to large-scale programs to detect affected persons or carriers in a population. For example, most hospitals in the United States routinely screen all newborns for PKU, so it is now less common to see people with symptoms of this disorder.

Genetic Counseling. Sometimes, prospective parents suspect that they are likely to produce a severely afflicted child. (Their first child or a close relative may suffer from a genetic disorder.) Parents at risk may request information from clinical psychologists, geneticists, and social workers.

Counseling starts with accurate diagnosis of parental genotypes. This may reveal the risk of a specific disorder. Biochemical tests can be used to detect many metabolic disorders. Detailed family pedigrees may be constructed to aid the diagnosis.

For disorders showing simple Mendelian inheritance, it is possible to predict the chances of having an affected child. But not all disorders follow Mendelian patterns. Even ones that do can be influenced by other factors. Even when the risk has been defined with some confidence, prospective parents must know that the risk is the same for *each* pregnancy. For example, if a pregnancy has one chance in four of producing a child with a genetic disorder, the same odds apply to every subsequent pregnancy, also.

Prenatal Diagnosis. Suppose a woman who is forty-five years old and pregnant wants to know if her fetus is trisomic 21 and so will develop Down syndrome. Prenatal ("before birth") diagnostic methods can detect this and more than a hundred other genetic disorders. In **amniocentesis**, a fluid sample is drawn from the amnion, a sac inside the uterus that contains the fetus. Fetal cells in the sample are cultured, then analyzed by karyotyping and other tests. In a different procedure, **chorionic villi sampling** (CVS), test cells are drawn from the chorion, a sac around the amnion.

Like amniocentesis, CVS is risky. Either procedure may cause infection or puncture the fetus. Besides this, amniocentesis is performed during the fifth month of pregnancy, when a mother already feels kicks and movements inside

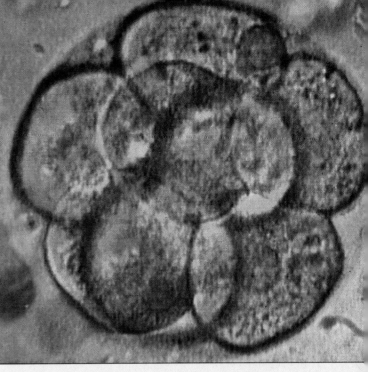

a Eight-cell stage of human development.

her. CVS can be performed earlier, between the ninth and twelfth weeks. However, even though the fetus is only about half as long as the little finger, its major organs have already started to form.

What choice do prospective parents make if either procedure reveals a devastating genetic disorder? Do they opt for **induced abortion**—an induced expulsion of the embryo from the uterus? This can be an agonizing decision. They must weigh awareness of the crushing severity of the disorder against ethical and religious beliefs. Worse, they must play out their personal tragedy on a larger stage, dominated now by a nationwide battle between fiercely vocal "pro-life" and "pro-choice" factions.

In 1992, clinical trials of **preimplantation diagnosis** proved successful. This new procedure relies on *in-vitro* fertilization. Sperm and eggs donated by prospective parents are put in an enriched medium in a petri dish. There, one or more eggs may become fertilized. Two days later, cell divisions convert each fertilized egg into a ball of eight cells (Figure *a*). The tiny ball might be considered a *prepregnancy* stage. Like unfertilized eggs flushed monthly from a woman's body, it is free-floating; it is not connected to the uterus.

All the cells in that ball have the same genes, and they are not differentiated. In other words, they are not yet committed to giving rise to specialized cells of the heart, toes, and other tissues. Researchers pluck one of the cells from each ball and analyze its genes for suspected disorders. Only a ball that is free of genetic defects is inserted back into the uterus.

At this writing, several couples who are at risk of passing on muscular dystrophy, cystic fibrosis, and other severe disorders have opted for the procedure. The procedure is still highly experimental, and it is costly. Yet five of the "test-tube" babies have been born. All are in good health—and free of the harmful genes.

SUMMARY

1. Genes, the units of instruction for heritable traits, are arranged one after the other along chromosomes.

2. Diploid (2n) cells have pairs of homologous chromosomes. Except for sex chromosomes, each pair has the same length, shape, and gene sequence, and they align with each other during meiosis.

3. Human sex chromosomes are designated X and Y. Females have a pair of X chromosomes, and males have X paired with Y. All other pairs of chromosomes are autosomes (the same in both females and males).

4. Clues to inheritance often show up in pedigrees (charts of genetic connections through lines of descent). Certain patterns are characteristic of dominant or recessive alleles on autosomes or on the X chromosome.

5. During meiosis, genes on one pair of chromosomes tend to be assorted into gametes independently of genes on the other pairs of chromosomes.

6. Genes on the *same* chromosome tend to stay together through meiosis and gamete formation. But crossing over (breakage and exchange of segments between homologues) can disrupt the linkage. The farther apart two genes are, the greater will be the frequency of crossovers between them.

7. A chromosome's structure may change on rare occasions. A segment may be deleted, duplicated, inverted, or moved to a new location.

8. The chromosome number may change on rare occasions. New individuals end up with more or fewer chromosomes than the parents.

9. Independent assortment, crossing over, and changes in chromosome number or in a chromosome's structure may influence the course of evolution. The changes in genotype (genetic makeup) lead to variations in phenotype (observable traits) among members of a population, so that evolution is possible.

1. Genes on the same chromosome tend to stay linked together during _____ and end up in the same gamete.
   a. mitosis
   b. meiosis
   c. crossing over
   d. b and c

2. The probability of a crossover between any two genes in the same chromosome is _____ .
   a. unrelated to the distance between them
   b. greater if they are close together
   c. greater if they are far apart
   d. not possible

3. A(n) _____ can alter a chromosome's structure.
   a. deletion
   b. duplication
   c. inversion
   d. translocation
   e. all of the above

4. The chromosome number may change during _____ .
   a. mitosis
   b. meiosis
   c. fertilization
   d. all of the above

5. A polyploid human embryo _____ .
   a. does not have the parental chromosome number
   b. has three or more of each type of chromosome
   c. usually dies before childbirth
   d. all of the above

6. Changes in _____ may result in a genetic disorder.
   a. genes
   b. chromosome structure
   c. chromosome number
   d. all of the above

7. _____ may contribute to variation in a population.
   a. Independent assortment
   b. Crossing over
   c. Changes in chromosome structure and number
   d. All of the above

8. Match the chromosome terms appropriately.
   ____ crossing over
   ____ deletion
   ____ nondisjunction
   ____ pedigree

   a. results in abnormal chromosome number
   b. nonhomologous chromosomes swap segments
   c. chart of genetic ties among individuals
   d. chromosome segment is lost

Genetics Problems *(Answers in Appendix III)*

1. Human females are XX and males are XY.
   a. Does a male inherit his X chromosome from his mother or father?
   b. With respect to an X-linked gene, how many different types of gametes can a male produce?
   c. If a female is homozygous for an X-linked gene, how many different types of gametes can she produce with respect to that gene?
   d. If a female is heterozygous for an X-linked gene, how many different gametes can she produce with respect to that gene?

2. One allele of a presumed Y-linked gene results in nonhairy ears in males. Another allele results in rather long hairs (hairy pinnae) (*see photo*).
   a. Why would you *not* expect females to have hairy pinnae?
   b. Any son of a hairy-eared male will be hairy-eared, but no daughter will be. Explain why.

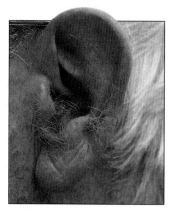

3. Suppose you are heterozygous for two linked genes with alleles *A,a* and *B,b* respectively:

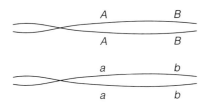

If the crossover frequency between the two genes is 0 percent, what genotypes would be expected among your gametes, and with what frequencies?

4. People with hemophilia A have a defective blood-clotting mechanism. A recessive allele of an X-linked gene gives rise to this condition. Refer to Figure 10.11. Why are the females only shown as carriers of the recessive allele?

5. Heterozygosity for a certain dominant autosomal allele nearly always results in Huntington's disorder. Neurological deterioration and other symptoms usually don't start until about age forty. The mother of a twenty-year-old woman is affected; her father is normal. What is the probability that the woman will develop the disorder? Suppose she just married someone with no family history of Huntington's disorder. What is the probability that each child of theirs will be affected?

6. Red-green color blindness is an X-linked trait. A woman heterozygous for the condition (*Gg*) marries a man with normal color vision. What is the probability that their first child will be red-green color-blind? Their second child? If they have two children only, what is the probability that both will be red-green color-blind?

7. A person with a single, unpaired X chromosome (Turner syndrome) may survive. A person with a single, unpaired Y chromosome cannot. What does this tell you about the genetic content of X and Y chromosomes?

8. Turner syndrome may result if a sperm with no sex chromosome fertilizes a normal egg. It also may result if a sperm with one X chromosome fertilizes an egg lacking a sex chromosome. Suppose a hemophilic male and a carrier (heterozygous) female have a child. The child is nonhemophilic *and* is affected by Turner syndrome. Which parent produced the gamete with the abnormal chromosome number?

9. Phenylketonuria (PKU) is due to an autosomal recessive allele. About 1 of every 50 Americans is heterozygous and symptom-free.
   a. If you select a symptom-free female at random from the population, what is the probability that she will be heterozygous?
   b. If you select a symptom-free male and a symptom-free female at random, what is the probability that both will be heterozygous?
   c. If both parents are heterozygous, what is the probability that any one of their children will have PKU?

Selected Key Terms

amniocentesis *151*
aneuploidy *148*
autosome *138*
chorionic villi sampling *151*
deletion *147*
duplication *148*
genetic disorder *145*
induced abortion *151*
inversion *148*
karyotype *138*
linkage *143*

nondisjunction *148*
pedigree *144*
polyploidy *148*
preimplantation diagnosis *151*
sex chromosome *138*
translocation *148*
X chromosome *138*
X-linked gene *141*
Y chromosome *138*
Y-linked gene *141*

Readings

Cummings, M. 1991. *Human Heredity: Principles and Issues.* St. Paul, Minnesota: West.

Holden, C. 1987. "The Genetics of Personality." *Science* 237:598–601.

Patterson, D. August 1987. "The Causes of Down Syndrome." *Scientific American* 257(2):52–60.

Suzuki, D., et al. 1989. *An Introduction to Genetic Analysis.* Fourth edition. New York: Freeman.

Weiss, R. November 1989. "Genetic Testing Possible Before Conception." *Science News* 136(21):326.

11 DNA STRUCTURE AND FUNCTION

Cardboard Atoms and Bent-Wire Bonds

Linus Pauling in 1951 did something no one had done before. Through his training in biochemistry, a talent for model building, and a few educated guesses, he deduced the structure of a protein (collagen). His discovery electrified the scientific community. If the secrets of proteins could be pried open, why not other biological molecules? Further, wouldn't structural details about those molecules provide clues to their function? And who would go down in history as having discovered the biggest prize of all—*the molecule that contains instructions for reproducing parental traits in offspring*?

Maybe hereditary instructions were encoded in the structure of some unknown class of proteins. After all, heritable traits are spectacularly diverse. Surely the molecules containing information about those traits were structurally diverse also. Proteins are put together from potentially limitless combinations of amino acid subunits, so they almost certainly could function as the sentences (genes) in each cell's book of inheritance.

Yet there was something about another substance—DNA—that excited many researchers. Among them were James Watson, a young postdoctoral student from Indiana University, and Francis Crick, an energetic researcher at Cambridge University. Could DNA, a molecule consisting of only four kinds of subunits, hold the secrets of inheritance? Watson and Crick spent long hours arguing over everything they had read about the size, shape, and bonding requirements of the subunits of DNA. They fiddled with cardboard cutouts of the subunits. They badgered chemists to identify potential bonds they might have overlooked. They assembled models from bits of metal held together with wire "bonds" bent at appropriate angles.

In 1953, they put together a model that fit all the pertinent biochemical rules and all the facts about DNA they had gleaned from other sources (Figure 11.1). They had discovered the structure of DNA. The breathtaking simplicity of that structure also enabled them to solve a long-standing riddle—*how life can show unity at the molecular level and yet give rise to so much diversity at the level of whole organisms.*

With this chapter, we turn to investigations that led to our current understanding of DNA structure and function. They are revealing of how ideas are generated in science. On the one hand, having a shot at fame and fortune quickens the pulse of men and women in any profession, and scientists are no exception. On the other hand, science proceeds as a community effort, with individuals sharing not only what they can explain but also what they do not understand. Even if an experiment "fails," it may turn up information that others can use or lead to questions that others can answer. Unexpected results, too, might be clues to something important about the natural world.

Figure 11.1 James Watson and Francis Crick posing in 1953 by their newly unveiled model of DNA structure. Behind this photograph is a recent computer-generated model. It is more sophisticated in appearance, yet basically the same as the prototype that was built nearly four decades before.

1. In living cells, DNA contains the information about heritable traits. That information is encoded in DNA's nucleotide subunits, which are strung together, one after another, in sequence. DNA has four kinds of nucleotide subunits, each with a different nitrogen-containing base. The bases are adenine, guanine, thymine, and cytosine.

2. In a DNA molecule, two strands of nucleotides twist together like a spiral stairway; they form a double helix. Hydrogen bonds connect the bases of one strand to bases of the other. As a rule, adenine pairs (hydrogen-bonds) only with thymine, and guanine only with cytosine.

3. Before a cell divides, its DNA is replicated with the help of enzymes and other proteins. Each double-stranded DNA molecule starts unwinding. A new, complementary strand is assembled on the exposed bases of each parent strand according to the base-pairing rule.

4. There is only one DNA molecule in a chromosome. Except in bacteria, great numbers of proteins are attached to the DNA and function in its structural organization.

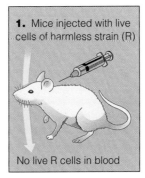
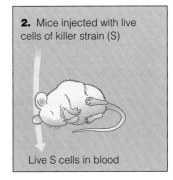
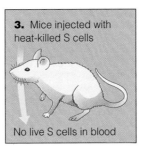
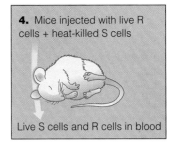

Figure 11.2 Summary of the results from Griffith's experiments with harmless (R) strains and disease-causing (S) strains of *Streptococcus pneumoniae*, as described in the text.

DISCOVERY OF DNA FUNCTION

Early Clues

In the spring of 1868, Johann Friedrich Miescher isolated a previously unknown substance in the cell nucleus. The substance became known as **deoxyribonucleic acid**, or **DNA**. For decades its role in cells remained a mystery. However, clues to its function kept popping up in different lines of investigation.

For example, in 1928 an Army medical officer, Fred Griffith, attempted to create a vaccine against a bacterium that causes a type of pneumonia. He isolated two strains of the bacterium (designated R and S). He used those strains in a series of four experiments:

1. Laboratory mice were injected with live R cells. As Figure 11.2 indicates, they did not develop pneumonia. *The R strain was harmless.*

2. Mice were injected with live S cells. The mice died. Blood samples from them teemed with live S cells. *The S strain was pathogenic* (disease-causing).

3. S cells were killed by exposure to high temperature. Mice injected with these cells did not die.

4. Live R cells were mixed with heat-killed S cells and injected into mice. The mice died—and blood samples from them teemed with *live* S cells!

What happened in the fourth experiment? Maybe heat-killed S cells in the mixture weren't really dead. But if that were true, then mice injected with heat-killed S cells alone (experiment 3) would have died. Maybe harmless R cells in the mixture mutated into the killer form. But if that were true, then mice injected with R cells alone (experiment 1) would have died.

The simplest explanation was this: Heat killed the S cells, *but it did not destroy their hereditary information, including the part on "how to cause infection."* Somehow the information had been transferred from dead S cells to living R cells—where it was put to use.

Further experiments showed that harmless cells had indeed picked up information about infection and had been permanently transformed into pathogens. Hundreds of generations of bacteria descended from the transformed cells also caused infections!

Griffith's unexpected results intrigued a microbiologist, Oswald Avery, and his colleagues. They were able to transform harmless cells with extracts of killed pathogenic cells. In 1944, they reported that the hereditary substance in the extracts was probably DNA (not proteins, as was widely believed). They had added protein-digesting enzymes to some extracts, but cells were transformed anyway. Then they added an enzyme that breaks apart DNA but not proteins—and that enzyme blocked hereditary transformation.

Figure 11.3 Example of the landmark experiments pointing to DNA as the substance of heredity. In the 1940s, Alfred Hershey and his colleague, Martha Chase, were studying the biochemical basis of inheritance. Like some other researchers of that era, they worked with bacteriophages, a class of viruses that infect bacterial cells. Compared to multicelled organisms, viruses are about as biochemically simple as you can get. They are not alive—yet they contain hereditary instructions for building new viral generations. Metabolic machinery inside a host cell carries out the instructions.

By 1952, researchers knew that some bacteriophages are constructed only of DNA and protein. Also, they knew from electron micrographs that the main part of a virus particle remains at the surface of an infected cell. Genetic material was probably injected *into* the host cell. If that were true, was the material DNA, protein, or both? To find out, Hershey and Chase designed two experiments, based on two biochemical facts. First, bacteriophage proteins incorporate sulfur (S) but not phosphorus (P). Second, DNA incorporates phosphorus but not sulfur.

(**a**) In one experiment, some bacterial cells were grown on a culture medium in which the only sulfur available for biosynthesis was the radioisotope ^{35}S. Afterward, bacteriophages were allowed to infect the cells. As Hershey and Chase had assumed, viral *proteins* built inside the cells became tagged with ^{35}S—and so did the new bacteriophage generation. Next, labeled bacteriophages were allowed to infect unlabeled bacteria suspended in fluid. Then the fluid was whirred in a kitchen blender to remove bacteriophage bodies from the cells. Labeled protein remained in the fluid—*it was part of the bacteriophage bodies*.

(**b**) In another experiment, Hershey and Chase grew bacterial cells on a culture medium in which the only phosphorus present was the radioisotope ^{32}P. Then bacteriophages infected the cells. New viral *DNA* built inside these cells became tagged. As in the first experiment, the new generation of labeled bacteriophages infected unlabeled bacteria, then their bodies were whirred off the cells. The labeled DNA remained with the cells—*it had to be the hereditary instructions for producing new bacteriophages*.

(**c**) The micrograph and sketches show one kind of bacteriophage (T4) injecting DNA into a host cell, the bacterium *Escherichia coli*.

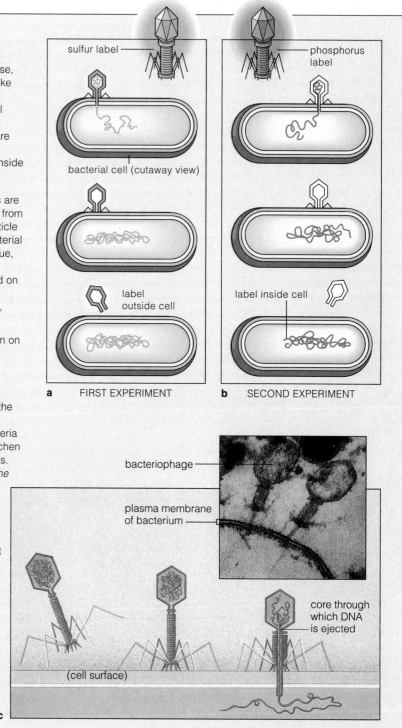

sulfur label

bacterial cell (cutaway view)

label outside cell

a FIRST EXPERIMENT

phosphorus label

label inside cell

b SECOND EXPERIMENT

bacteriophage

plasma membrane of bacterium

core through which DNA is ejected

(cell surface)

c

Confirmation of DNA Function

By the early 1950s, molecular detectives were using certain viruses as their experimental subjects. The viruses, called **bacteriophages**, infect bacterial cells such as *Escherichia coli*. Figure 11.3 describes two of the landmark experiments with bacteriophages. They provided convincing evidence that DNA, not proteins, functions as the molecule of heredity. Not long after the experimental results were reported, James Watson and Francis Crick deciphered the structure of DNA and ushered in the age of molecular genetics.

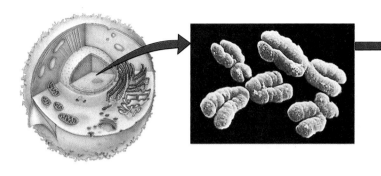

All chromosomes contain DNA. What does DNA contain? Only four kinds of nucleotides. Each nucleotide has a five-carbon sugar (shaded red). That sugar has a phosphate group attached to the fifth carbon atom of its ring structure. It also has one of four kinds of nitrogen-containing bases (shaded blue) attached to its first carbon atom. The nucleotides differ only in which base is attached to that atom:

Figure 11.4 The nucleotide subunits of DNA. The small numerals on the structural formulas identify the carbon atoms to which other parts of the molecule are attached.

DNA STRUCTURE

Components of DNA

Long before bacteriophage studies were under way, biochemists knew that DNA contains only four types of nucleotides. **Nucleotides** are the building blocks of nucleic acids. Each consists of a five-carbon sugar (deoxyribose), a phosphate group, and one of the following nitrogen-containing bases:

adenine	guanine	thymine	cytosine
(A)	**(G)**	**(T)**	**(C)**

Each type of nucleotide in DNA has its component parts joined together in much the same way as the others. But notice, in Figure 11.4, that T and C are smaller, single-ring structures (called pyrimidines). A and G are larger, double-ring structures (called purines).

The amount of adenine relative to guanine differs from species to species. But in all cases, the amount of *adenine* always equals the amount of *thymine*, and the amount of *guanine* equals the amount of *cytosine*. We may show this as A = T and G = C.

But how are nucleotides arranged in DNA? The first convincing evidence came from Maurice Wilkin's laboratory. One of Wilkin's coworkers, Rosalind Franklin, developed the best x-ray diffraction images of DNA. (DNA molecules can be spun, spooled, and pulled into gossamer fibers, like cotton candy. Such fibers scatter an x-ray beam in a regular pattern. The pattern can be used to calculate the positions of atoms in DNA.) Researchers already suspected that DNA has a uniform diameter, with nucleotides joined one after another along its length. By Franklin's calculations, the bases were positioned inside the molecule, with the sugar and phosphate components at its surface.

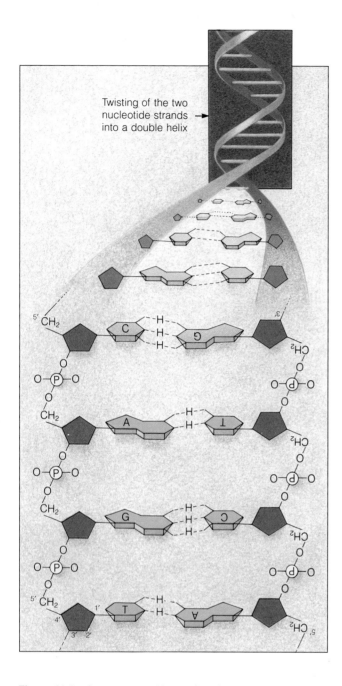

Figure 11.5 Arrangement of bases (blue) in a DNA double helix.

Patterns of Base Pairing

Using all of the available clues, Watson and Crick put together a model of DNA structure. As they alone perceived, DNA consists of *two* strands of nucleotides held together at their bases by hydrogen bonds. Those bonds form only when the two strands run in *opposing directions* and twist together into a *double helix*, much like the twisting of a circular stairway (Figure 11.5). Only two kinds of base pairings occur along the entire length of the molecule: A–T and G–C. However, the *order* of bases in a strand can vary greatly from one species to the next. In even a tiny stretch of DNA from a rose, gorilla, human, or any other organism, the sequence might be:

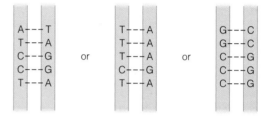

As you can see, DNA molecules show constancy *and* variation in their molecular structure. This is the molecular foundation for the unity and diversity of life.

Base pairing between the two nucleotide strands in DNA is *constant* for all species (A–T, and G–C).

The base sequence (that is, which base follows another in a nucleotide strand) is *different* from species to species.

DNA REPLICATION

The discovery of DNA structure was a turning point in studies of inheritance. Until then, no one could explain **DNA replication**—that is, how hereditary material is duplicated prior to cell division. The Watson–Crick model suggested at once how this might be done.

Hydrogen bonds hold together the two nucleotide strands making up a DNA double helix. Enzymes can readily break those weak bonds. Through enzyme action, one strand starts to unwind from the other at distinct sites, leaving the bases exposed. Cells have stockpiles of free nucleotides, and these pair with the exposed bases. Thus, each parent strand remains intact and a companion strand is assembled on each one according to the base-pairing rule (A to T, and G to C).

Each parent strand becomes twisted into a double helix with its new, partner strand. Because the parent strand is conserved, each "new" DNA molecule is really half old, half new (Figure 11.6).

> Prior to cell division, the double-stranded DNA molecule unwinds and is replicated. Each parent strand remains intact—it is conserved—and a new, complementary strand is assembled on each one.

Keep in mind that DNA does not replicate itself. Enzymes and other proteins unwind the molecule, keep the two strands separated at unwound regions, and assemble a new strand on each one. For instance, DNA polymerases govern nucleotide assembly on a parent strand. Even as one DNA region is being unwound, enzymes are winding up regions already replicated.

DNA REPAIR

DNA polymerases, DNA ligases, and other enzymes also engage in a process called **DNA repair**. If the sequence of bases in one strand of a double helix becomes altered, DNA polymerases "read" the complementary sequence on the other strand. With the aid of other repair enzymes, they restore the original sequence. Figure 11.7 gives an example of what can happen when the excision-repair function is impaired.

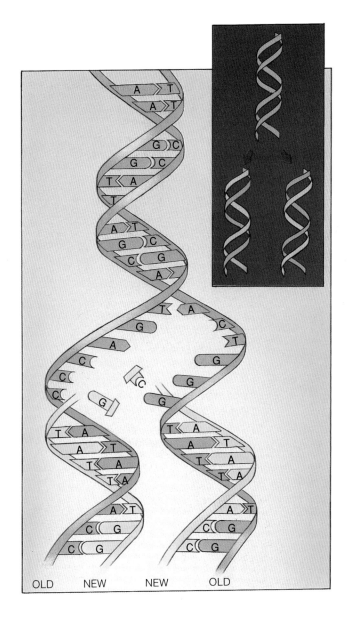

Figure 11.6 "Semiconservative" nature of DNA replication. The original two-stranded DNA molecule is shown in blue. A new strand (yellow) is assembled on each of the two parent strands.

OLD NEW NEW OLD

Figure 11.7 When DNA can't be fixed. A newborn destined to develop *xeroderma pigmentosum* faces a dim future, literally. Individuals affected by this genetic disorder cannot be exposed to sunlight, even briefly, without risking disfiguring skin tumors and early death from cancer. The disorder is a result of faulty DNA repair in the body's cells. One or more of the genes thought to govern repair processes have become damaged.

The ultraviolet wavelengths in light from the sun, tanning lamps, and other sources can cause molecular changes in DNA (Figure 5.3). Among other things, the wavelengths can promote covalent bonding between two adjacent thymine bases in a nucleotide strand. Thus the two nucleotides to which the bases belong are combined into an abnormal, bulky molecule called a "thymine dimer." Normally, a DNA repair mechanism gets rid of such bulky lesions. The mechanism requires at least seven gene products. Mutation in one or more of the required genes can skew the repair machinery. Thymine dimers can accumulate in the skin cells of individuals with such mutations. The accumulation triggers the development of serious lesions, including skin cancer. The photograph shows a common skin cancer, called *basal cell carcinoma*.

ORGANIZATION OF DNA IN CHROMOSOMES

There is one DNA molecule in each chromosome. If the DNA of all forty-six chromosomes in just *one* of your body cells were strung together end to end, the string could dangle from your head to your toes. Why isn't all that DNA a tangled mess in each tiny cell? Proteins provide the answer.

The DNA of humans and all other eukaryotes is tightly bound with many proteins, including **histones**. Some histones are like spools for winding up small stretches of DNA. Each histone-DNA spool is a **nucleosome**. Another histone stabilizes the spools (Figure 11.8).

Through interactions between histones and DNA, a chromosome can coil back on itself again and again. The coiling greatly increases its diameter. Further folding results in a series of loops. Proteins other than histones serve as a structural "scaffold" for the loops.

Apparently, scaffold regions occur *between* genes, not in regions that contain information for building proteins. Are the scaffold proteins organizing the chromosome into functional "domains"? Would that organization make it easier for DNA replication and protein synthesis to proceed? These are just two of the possibilities being probed by the new generation of molecular detectives.

SUMMARY

1. Hereditary information of organisms is encoded in DNA (deoxyribonucleic acid). DNA consists of small organic molecules called nucleotides.

2. A nucleotide has a five-carbon sugar (deoxyribose), a phosphate group, and one of four nitrogen-containing bases (adenine, thymine, guanine, or cytosine).

3. A DNA molecule consists of two nucleotide strands twisted together into a double helix. The bases of one strand pair (hydrogen-bond) with bases of the other.

4. DNA molecules show constancy in base pairing. Adenine always pairs with thymine (A–T), and guanine always pairs with cytosine (G–C).

5. DNA molecules of different species differ in the *sequence* of which base pair follows the next.

6. In DNA replication, the two strands of a double helix start unwinding from each other. A new strand of complementary sequence is assembled on each one. Two double-stranded molecules result. One strand is "old" (it is conserved) and the other is "new." Replication requires many enzymes and other proteins.

7. Eukaryotic DNA is bound tightly with many proteins. Protein-DNA interactions produce the structural organization of the metaphase chromosome.

Figure 11.8   Levels of organization of DNA in a eukaryotic chromosome.

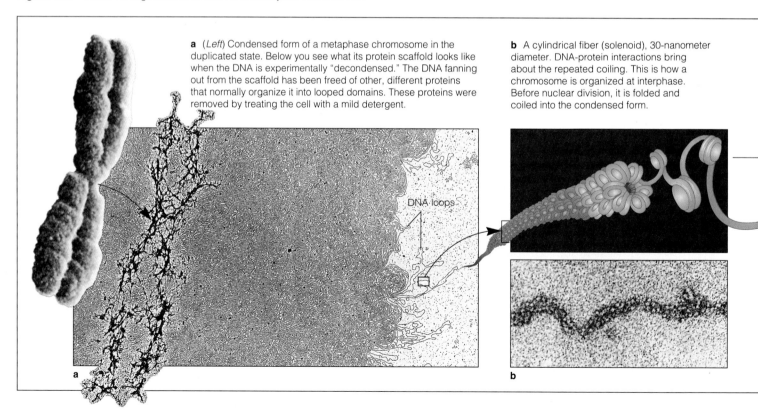

a (*Left*) Condensed form of a metaphase chromosome in the duplicated state. Below you see what its protein scaffold looks like when the DNA is experimentally "decondensed." The DNA fanning out from the scaffold has been freed of other, different proteins that normally organize it into looped domains. These proteins were removed by treating the cell with a mild detergent.

b A cylindrical fiber (solenoid), 30-nanometer diameter. DNA-protein interactions bring about the repeated coiling. This is how a chromosome is organized at interphase. Before nuclear division, it is folded and coiled into the condensed form.

DNA loops

Review Questions

1. Name the three molecular parts of a nucleotide in DNA. Name the four different kinds of nitrogen-containing bases that occur in the nucleotides of DNA. *157*

2. What kind of bond holds two DNA chains together in a double helix? Which nucleotide base-pairs with adenine? Which pairs with guanine? *157, 158*

3. The four bases in DNA may differ greatly in relative amounts from one species to the next—yet the relative amounts are always the same among members of a single species. How does base pairing explain these twin properties—the unity and diversity—of DNA molecules? *158*

4. When regions of a double helix are unwound during DNA replication, do the two unwound strands join back together again after a new DNA molecule has formed? *158–159*

Self-Quiz (Answers in Appendix IV)

1. Which is *not* a nucleotide base in DNA?
   a. adenine   c. uracil   e. guanine
   b. thymine   d. cytosine

2. What are the base-pairing rules for DNA?
   a. A–G, T–C   c. A–U, C–G
   b. A–C, T–G   d. A–T, C–G

3. A DNA strand with the sequence C–G–A–T–T–G would be complementary to the sequence _____ .
   a. C–G–A–T–T–G   c. T–A–G–C–C–T
   b. G–C–T–A–A–G   d. G–C–T–A–A–C

4. One species' DNA differs from others in its _____ .
   a. sugars   c. base sequence
   b. phosphate groups d. all of the above

5. When DNA replication begins, _____ .
   a. the two DNA strands unwind from each other
   b. the two DNA strands condense for base transfers
   c. two DNA molecules bond
   d. old strands move to find new strands

6. DNA replication results in _____ .
   a. two half-old, half-new molecules
   b. two molecules, one with old strands and one with newly assembled strands
   c. three double-stranded molecules, one with new strands and two that are discarded
   d. none of the above

7. DNA replication requires _____ .
   a. a supply of free   c. many enzymes and other
      nucleotides   proteins
   b. new hydrogen bonds   d. all of the above

8. Match the DNA concepts appropriately.
   _____ base sequence   a. two nucleotide strands
   _____ metaphase chromosome   twisted together
   _____ constancy in   b. A = T, G = C
          base pairing   c. one old strand, one new
   _____ replication   d. accounts for diversity
   _____ DNA double   among species
          helix   e. DNA with protein scaffold

Selected Key Terms

adenine (A) *157*   guanine (G) *157*
bacteriophage *156*   histone *160*
cytosine (C) *157*   nucleosome *160*
DNA (deoxyribonucleic acid) *155* nucleotide *157*
DNA repair *159*   thymine (T) *157*
DNA replication *158*

Readings

Cairns, J.; G. Stent; and J. Watson, eds. 1966. *Phage and the Origins of Molecular Biology.* Cold Spring Harbor, New York: Cold Spring Harbor Laboratories. Gives a sense of the insights, wit, humility, and personalities of the founders of and converts to molecular genetics.

Radman, M., and R. Wagner. August 1988. "The High Fidelity of DNA Duplication." *Scientific American* 259(2): 40–46.

Watson, J. 1978. *The Double Helix.* New York: Atheneum. Highly personal view of scientists and their methods, interwoven into an account of how DNA structure was discovered.

Wolfe, S. 1992. *Molecular and Cellular Biology.* Belmont, California: Wadsworth. Comprehensive, current, and accessible.

c A chromosome immersed in a salt solution loosens up to a beads-on-a-string organization. The "string" is DNA. Each "bead" is a nucleosome (**d**). Short stretches of the chromosome may look like this when DNA is being replicated or when a gene's instructions are being read. Each nucleosome consists of a double loop of DNA around a core of proteins (eight histone molecules). Another histone (H1) stabilizes the arrangement.

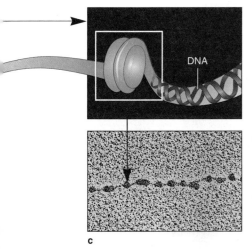

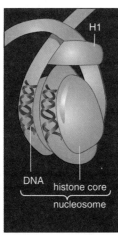

c   d

12 FROM DNA TO PROTEINS

Beyond Byssus

Picture a mussel, of the sort shown in Figure 12.1. Hard-shelled but soft of body, it is using its muscular foot to probe a wave-scoured rock. At any moment, pounding waves can whack the mussel into the water, hurl it repeatedly against the rock with shell-shattering force, and so offer up a gooey lunch for gulls.

By chance, the mussel's foot comes across a crevice in the rock. The foot moves, broomlike, and sweeps the crevice clean. Then it presses down and arches up, rather like a plumber's rubber plunger being squished down and up to unclog a drain. Beneath the foot, a vacuum-sealed chamber has been formed, and the mussel spews fluid into it. The fluid, made of keratin and other proteins, bubbles into a sticky foam. By curling its foot and pumping the foam through it, the mussel forms sticky threads about as wide as a human whisker. It varnishes the threads with another protein and ends up with an adhesive called byssus—which anchors the mussel to the rock.

Byssus is the world's premier underwater adhesive. Nothing that humans have manufactured comes close. (Sooner or later, water chemically degrades or deforms synthetic adhesives.) Byssus fascinates biochemists, dentists, and surgeons looking for better ways to do tissue grafts and to rejoin severed nerves. Genetic engineers are inserting mussel DNA into yeast cells, which reproduce in large numbers and serve as "factories" for translating mussel genes into useful amounts of proteins. This exciting work, like the mussel's own byssus building, starts with one of life's universal concepts. *Every protein is synthesized according to instructions contained in DNA.*

You are about to trace the steps leading from DNA to protein. Many enzymes are players in this pathway, as are molecules of RNA. The same steps produce *all* of the world's proteins, from mussel-inspired adhesives to the keratin in your fingernails to the insect-digesting enzymes of a Venus flytrap.

Figure 12.1 Mussels, busily demonstrating the importance of proteins for survival. When mussels come across a suitable anchoring site, they use their foot like a plumber's plunger to create a vacuum chamber. In this chamber they manufacture the world's best underwater adhesive from a mix of proteins.

1. Life cannot exist without enzymes and other proteins. Proteins consist of polypeptide chains, which consist of amino acids. The sequence of amino acids corresponds to a gene—which is a sequence of nucleotide bases in DNA.

2. The path leading from genes to proteins has two steps, called transcription and translation. In transcription, the double-stranded DNA molecule is unwound at a gene region, then an RNA molecule is assembled on the exposed bases of one of the strands. In translation, RNA directs the linkage of one amino acid after another, in the sequence required to produce a specific kind of polypeptide chain.

3. A gene's base sequence is vulnerable to permanent changes, called mutations. These changes are the source of genetic variation in populations.

4. All cells in a multicelled organism have the same genes, but they activate or suppress many of those genes in different ways. The controlled, selective use of genes leads to synthesis of the proteins that give each type of cell its distinctive structure, function, and products.

TRANSCRIPTION AND TRANSLATION: AN OVERVIEW

DNA is like a book of instructions in each cell. The alphabet used to create the book is simple enough: A, T, G, and C. But how do we get from an alphabet to a protein? The answer starts with the structure of DNA.

DNA, recall, is a double-stranded molecule (Figure 11.5). The strands consist of four types of nucleotides, which differ in only one component. That component, a nitrogen-containing base, may be adenine, thymine, guanine, or cytosine. Which base follows the next in a strand—that is, the **base sequence**—differs from one kind of organism to the next.

Before a cell divides, its DNA is replicated. During replication, the two DNA strands unwind entirely from each other. At other times in the cell's life, the two strands unwind in certain regions only, so that the cell gains access to particular genes. Those genes contain instructions for building particular proteins.

It takes two steps, **transcription** and **translation**, to carry out protein-building instructions. In eukaryotic cells, transcription occurs in the nucleus. In this step, nucleotide bases in DNA serve as a structural pattern (template) for assembling a strand of **ribonucleic acid (RNA)** from the cell's pool of free nucleotides. Afterward, the RNA is shipped to the cytoplasm, where translation occurs. In this second step, RNA directs the assembly of amino acids into polypeptide chains. Later, the chains become folded into the three-dimensional shapes of proteins.

In short, DNA guides the synthesis of RNA, then RNA guides the synthesis of proteins:

DNA ⟶ *transcription* ⟶ RNA ⟶ *translation* ⟶ protein

The new proteins will have structural and functional roles in cells. Some even will have roles in building more DNA, RNA, and proteins.

From this overview, you may have the impression that protein synthesis requires only one class of RNA molecules. Actually, it requires three. Transcription of most genes produces **messenger RNA (mRNA)**, the only class of RNA that carries *protein-building* instructions. Transcription of some other genes produces **ribosomal RNA (rRNA)**, the main components of ribosomes. As you read in Chapter 3, ribosomes are the structural workbenches on which polypeptide chains are synthesized. Transcription of still other genes produces **transfer RNA (tRNA)**. As you will see, tRNA molecules deliver amino acids one by one to the ribosome, in the order specified by mRNA.

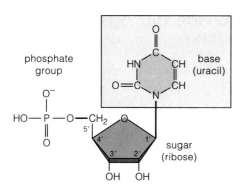

Figure 12.2 Structure of one of the four nucleotides of RNA. The other three have a different base (adenine, guanine, or cytosine instead of the uracil shown here). Compare Figure 11.4, which shows the nucleotides of DNA.

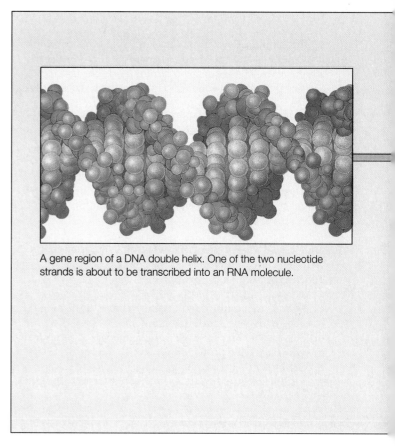

A gene region of a DNA double helix. One of the two nucleotide strands is about to be transcribed into an RNA molecule.

Figure 12.3 Transcription: the synthesis of an RNA molecule on a DNA template.

TRANSCRIPTION OF DNA INTO RNA

How RNA Is Assembled

RNA is almost, but not quite, like a strand of DNA. It contains four types of nucleotides, each consisting of a five-carbon sugar (ribose), a phosphate group, and a base. The bases are adenine, cytosine, guanine, and **uracil** (Figure 12.2). Uracil is like the thymine in DNA; it pairs with adenine. Thus a new RNA strand can be put together on a DNA region according to the base-pairing rule:

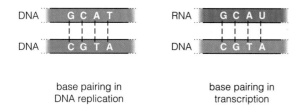

base pairing in
DNA replication

base pairing in
transcription

Transcription differs from DNA replication in three key respects. First, only one region of one DNA strand—not the whole strand—serves as the template. Second, different kinds of enzymes (RNA polymerases) are used. Third, transcription results in a *single* strand of nucleotides.

Transcription starts at a **promoter**, a base sequence that signals the start of a gene. Proteins help position an RNA polymerase on the DNA so that it binds with the promoter. The enzyme moves along the DNA, joining nucleotides together (Figure 12.3). When it reaches a base sequence that serves as a stop signal, the RNA is released as a free transcript.

Finishing Touches on mRNA Transcripts

Newly formed mRNA is an unfinished molecule that must be modified before its protein-building instructions can be used. Just as a dressmaker might snip off some threads or add bows on a dress before it leaves the shop, so does a eukaryotic cell tailor this pre-mRNA.

Very quickly, enzymes attach a cap to the start of the pre-mRNA and a tail to the end of it (Figure 12.4b). The tail, an unbroken stretch of adenine nucleotides,

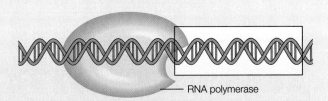

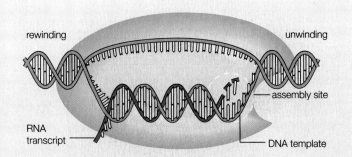

a An RNA polymerase molecule binds to a "start" site in the DNA. It will use the base sequence positioned downstream from that site as a template for linking nucleotides into a strand of RNA.

c Throughout transcription, the DNA double helix unwinds just in front of the RNA polymerase. Short lengths of the newly forming RNA strand temporarily wind up with the DNA template strand. Then they unwind from it—and the two strands of DNA wind up together again.

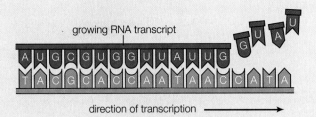

b During transcription, RNA nucleotides are based-paired, one after another, with the exposed bases on the DNA template.

d At the end of the gene region, the last stretch of the RNA is unwound from the DNA template and so is released.

becomes wound up with proteins. Later on, in the cytoplasm, the cap will promote the binding of mRNA to a ribosome. There also, enzymes will gradually destroy the wound-up tail from the tip on up, then the mRNA. Clearly, such tails "pace" enzyme access to mRNA. Perhaps they help keep protein-building messages intact for as long as the cell requires them.

Besides this, the mRNA message itself is modified. Most eukaryotic genes have one or more **introns**, which are base sequences that do *not* get translated into an amino acid sequence. Introns intervene between **exons**, the only parts of mRNA that become translated into protein. Introns are transcribed right along with exons, but enzymes snip them out before mRNA leaves the nucleus in mature form (Figure 12.4c).

Some introns may be little more than evolutionary junk, the leftovers of past mutations that led nowhere. Others are sites where instructions for building a protein can be snipped apart and spliced back together in different ways. Alternative splicing allows different cells in your body to produce modified versions of an mRNA transcript—and modified versions of the resulting protein—from the same gene.

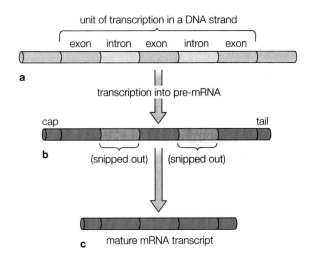

Figure 12.4 Transcription and modification of newly formed mRNA in the nucleus of eukaryotic cells. The cap simply is a nucleotide with functional groups attached. The poly(A) tail is a string of adenine nucleotides.

First Letter	Second Letter				Third Letter
	U	C	A	G	
U	phenylalanine	serine	tyrosine	cysteine	U
	phenylalanine	serine	tyrosine	cysteine	C
	leucine	serine	stop	stop	A
	leucine	serine	stop	tryptophan	G
C	leucine	proline	histidine	arginine	U
	leucine	proline	histidine	arginine	C
	leucine	proline	glutamine	arginine	A
	leucine	proline	glutamine	arginine	G
A	isoleucine	threonine	asparagine	serine	U
	isoleucine	threonine	asparagine	serine	C
	isoleucine	threonine	lysine	arginine	A
	(start) methionine	threonine	lysine	arginine	G
G	valine	alanine	aspartate	glycine	U
	valine	alanine	aspartate	glycine	C
	valine	alanine	glutamate	glycine	A
	valine	alanine	glutamate	glycine	G

Figure 12.5 The genetic code. The codons in mRNA are nucleotide bases, read in blocks of three. Sixty-one of these base triplets correspond to specific amino acids. Three others serve as signals that stop translation. In this diagram, the left column shows the first of the three nucleotides in each codon in mRNA. The middle columns show the second nucleotide; the right column shows the third. Reading from left to right, for instance, the triplet U G G corresponds to tryptophan. Both U U U and U U C correspond to phenylalanine.

TRANSLATION OF mRNA

mRNA and the Genetic Code

Like a DNA strand, mRNA is a linear sequence of nucleotides. What are the protein-building "words" encoded in its sequence? Gobind Khorana, Marshall Nirenberg, and others came up with the answer. They deduced the nature of the **genetic code**—that is, how the nucleotide sequence of DNA and mRNA corresponds to the amino acid sequence of a polypeptide chain. The genetic code is the basic language of protein synthesis in all organisms, from bacteria to humans.

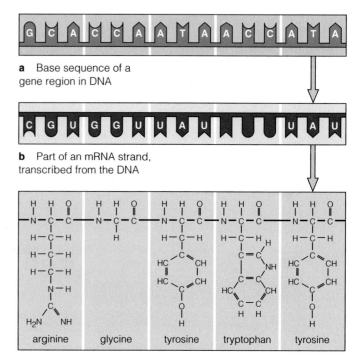

a Base sequence of a gene region in DNA

b Part of an mRNA strand, transcribed from the DNA

arginine   glycine   tyrosine   tryptophan   tyrosine

c What the amino acid sequence will be when the mRNA is translated into a polypeptide chain

Figure 12.6 Genetic code in action. (**a**) This region of a DNA double helix was unwound during transcription. (**b**) Exposed bases on one strand served as a template for assembling an mRNA strand. In the mRNA, every three nucleotide bases equaled one codon. Each codon called for one amino acid in this polypeptide chain. In (**c**), can you fill in the blank codon for tryptophan in the chain? (Refer to Figure 12.5.)

Enzymes recognize nucleotide bases three at a time, as triplets. The various combinations of base triplets in mRNA are called **codons**, and each specifies an amino acid. As you can see from Figures 12.5 and 12.6, there are sixty-four different codons. Most of the twenty kinds of amino acids correspond to more than one codon. (Glutamate, for example, corresponds to both GAA and GAG.)

One codon, AUG, also establishes the reading frame for translation. That is, enzymes start their "three-bases-at-a-time" selections at the AUG that serves as the "start" signal in an mRNA strand. Three codons (UAA, UAG, UGA) are stop signals. They prevent the addition of more amino acids to a growing polypeptide chain.

Protein-building instructions are encoded in the nucleotide sequence of both DNA and mRNA. The code words (codons) are nucleotide bases, read in blocks of three.

Sixty-one codons correspond to different amino acids. One of these also establishes the reading frame for translation. Three additional codons serve as signals to end translation.

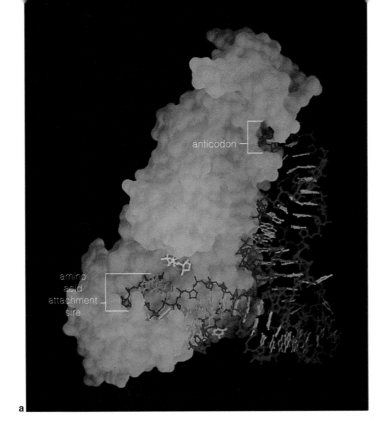

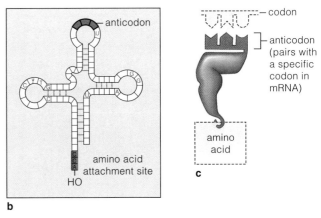

Figure 12.7 (**a**) Computer-generated, three-dimensional model of one type of tRNA molecule. The tRNA (brown) is shown attached to a bacterial enzyme (green), along with an ATP molecule (gold). This particular enzyme attaches amino acids to tRNAs. (**b**) Structural features common to all tRNAs. (**c**) Simplified model of tRNA that is used in subsequent illustrations. The "hook" at one end is the site to which a specific amino acid can become attached.

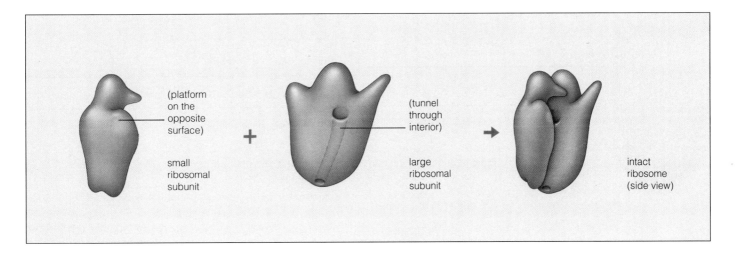

Figure 12.8 Model of eukaryotic ribosomes. Polypeptide chains are assembled on the small subunit's platform. New chains may move through the large subunit's tunnel. Notice how the platform and tunnel align in an intact ribosome.

Roles of tRNA and rRNA

Cells have pools of free amino acids and free tRNA molecules in the cytoplasm. As Figure 12.7 shows, tRNAs have a molecular "hook," an attachment site for amino acids. They also have an **anticodon**, a nucleotide triplet that can base-pair with codons. When different types of tRNAs bind to different codons, they automatically position their attached amino acids in the order specified by the mRNA sequence. Thus *tRNAs are the "translators" of mRNA.*

The tRNA molecules meet up with an mRNA strand at binding sites on the surface of ribosomes. As Figure 12.8 shows, each ribosome has two subunits, which are assembled in the nucleus from rRNA and proteins. The subunits are shipped separately into the cytoplasm. They join together during translation.

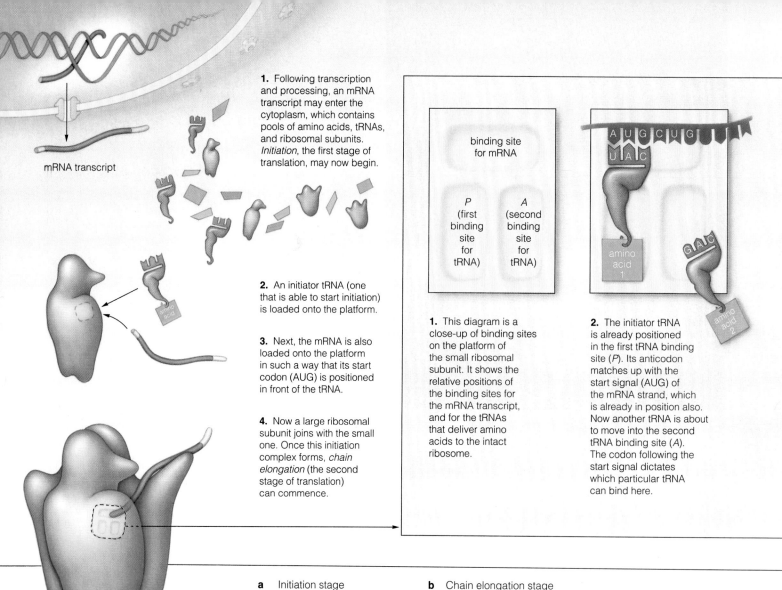

1. Following transcription and processing, an mRNA transcript may enter the cytoplasm, which contains pools of amino acids, tRNAs, and ribosomal subunits. *Initiation,* the first stage of translation, may now begin.

mRNA transcript

2. An initiator tRNA (one that is able to start initiation) is loaded onto the platform.

3. Next, the mRNA is also loaded onto the platform in such a way that its start codon (AUG) is positioned in front of the tRNA.

4. Now a large ribosomal subunit joins with the small one. Once this initiation complex forms, *chain elongation* (the second stage of translation) can commence.

binding site for mRNA

P (first binding site for tRNA) *A* (second binding site for tRNA)

1. This diagram is a close-up of binding sites on the platform of the small ribosomal subunit. It shows the relative positions of the binding sites for the mRNA transcript, and for the tRNAs that deliver amino acids to the intact ribosome.

amino acid 1

amino acid 2

2. The initiator tRNA is already positioned in the first tRNA binding site (*P*). Its anticodon matches up with the start signal (AUG) of the mRNA strand, which is already in position also. Now another tRNA is about to move into the second tRNA binding site (*A*). The codon following the start signal dictates which particular tRNA can bind here.

a Initiation stage   **b** Chain elongation stage

intact ribosome

Figure 12.9 Stages of translation, the second step of protein synthesis. Refer to Figures 12.7 and 12.8 for structural details of tRNAs and ribosomes.

Stages of Translation

Translation proceeds through three stages, called initiation, elongation, and termination.

In *initiation*, a tRNA that can start transcription *and* an mRNA transcript become loaded onto an intact ribosome. The tRNA and mRNA bind with the small ribosomal subunit. Then a large ribosomal subunit binds with the small one (Figure 12.9*a*). The next stage can begin.

In *elongation*, the mRNA strand passes between the ribosomal subunits, like a thread being moved through

the eye of a needle. Enzymes built into the ribosome join amino acids together in the sequence dictated by mRNA's codons. (They catalyze peptide bonds between amino acids.) In this way, a polypeptide chain grows (Figure 12.9*b*).

In *termination*, a stop codon is reached and there is no corresponding anticodon. Now the enzymes bind to certain proteins (release factors). The binding causes the ribosome as well as the polypeptide chain to detach from the mRNA (Figure 12.9*c*). The detached chain may join the pool of free proteins in the cytoplasm. Or it may

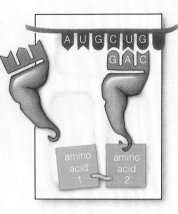
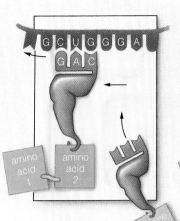

3. An enzyme breaks the bond between the initiator tRNA and the amino acid hooked to it. At the same time, a peptide bond forms between the two amino acids. After these events have occurred, the initiator tRNA will be released from the ribosome.

4. Two amino acids are now hooked onto the second tRNA. The tRNA will move into the *P* site, sliding the mRNA strand with it by one codon. With this movement, the third codon of the mRNA will become aligned above the *A* site.

5. A third tRNA is about to move into the *A* site. Its anticodon will base-pair with the third codon of the mRNA transcript. A peptide bond will form between amino acids 2 and 3.

6. Steps 3 through 5 are repeated again and again. The polypeptide chain continues to grow this way until enzymes reach a stop codon in the mRNA. Now the last stage of translation, *chain termination*, begins, as shown below in (**c**).

enter the rough ER of the cytomembrane system. As indicated on page 51, many newly formed chains take on their final form in that system before being shipped to their destinations.

Commonly, the same mRNA transcript is translated repeatedly in a given period. The transcript threads through many ribosomes, which are arranged one after another in assembly-line fashion. In this way, cells can produce many copies of a polypeptide chain from the same transcript.

1. Once a stop codon is reached, the mRNA transcript is released from the ribosome:

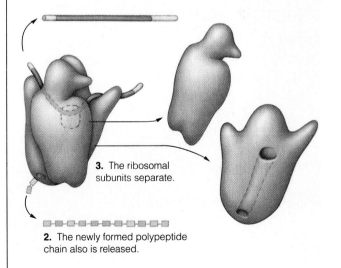

3. The ribosomal subunits separate.

2. The newly formed polypeptide chain also is released.

c Chain termination stage

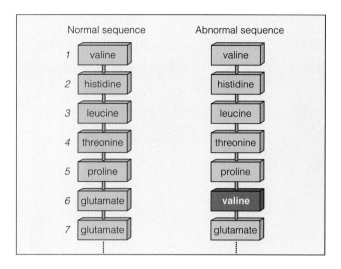

Normal sequence | Abnormal sequence
1 valine | valine
2 histidine | histidine
3 leucine | leucine
4 threonine | threonine
5 proline | proline
6 glutamate | valine
7 glutamate | glutamate

Figure 12.10 Mutation and sickle-cell anemia. The affected gene codes for one of two types of polypeptide chains in the hemoglobin molecule (page 32). A mutation has altered one base in the gene's sixth codon. During protein synthesis, that codon calls for valine instead of glutamate. The change puts a "sticky" (hydrophobic) patch on hemoglobin. When the blood level of oxygen is low, hemoglobin molecules interact at their sticky patches. They aggregate into rods and distort red blood cells, and this affects organs throughout the body.

Figure 12.11 Mutation in kernels of Indian corn. All of the kernel cells have pigment-coding genes. Yet some kernels are colorless or spottily colored. While the corn plant was growing, a transposable element in one of its cells invaded and shut down a pigment gene. Cell divisions continued, and none of the mutated cell's descendants could produce pigment. They gave rise to colorless kernel tissue. Whenever the transposable element slipped out of the pigment gene, all the descendants were able to produce pigment. They gave rise to colored kernel tissue.

MUTATION AND PROTEIN SYNTHESIS

Every so often, genes change. One base is substituted for another in the DNA sequence, an extra base is inserted, or a base is lost. These small-scale changes in the nucleotide sequence of DNA are known as **gene mutations**.

Some mutations result from "mutagens," agents that attack DNA and modify its structure. Mutagens include ultraviolet light, ionizing radiation, various chemical substances, and free radicals. Pages 60 and 176 give examples of their effects.

Other mutations occur spontaneously in cells. Among these are replication errors, as when adenine wrongly pairs with a cytosine unit on a DNA template strand. Enzymes may only detect an error, then "fix" the wrong base:

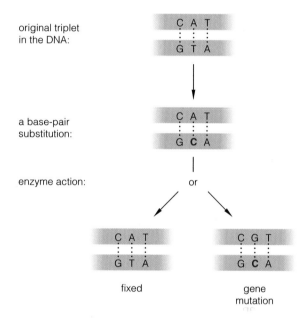

original triplet in the DNA:

C A T
G T A

a base-pair substitution:

C A T
G C A

enzyme action: or

C A T
G T A
fixed

C G T
G C A
gene mutation

Regardless of the source, base-pair substitutions may lead to substitution of one amino acid for another during protein synthesis. For example, a single base substitution in DNA results in sickle-cell anemia. As described on page 130 and in Figure 12.10, this genetic disorder has wide-ranging structural and physiological consequences for the human body.

As Barbara McClintock discovered, "jumping genes" (transposable elements) also cause spontaneous mutations. These DNA regions can move from one location to another in the same DNA molecule or in a different one. Often they inactivate genes into which they are inserted (Figure 12.11).

Let's turn now to the nature of gene control. Before doing so, review Figure 12.12, which summarizes the flow of information from genes to proteins.

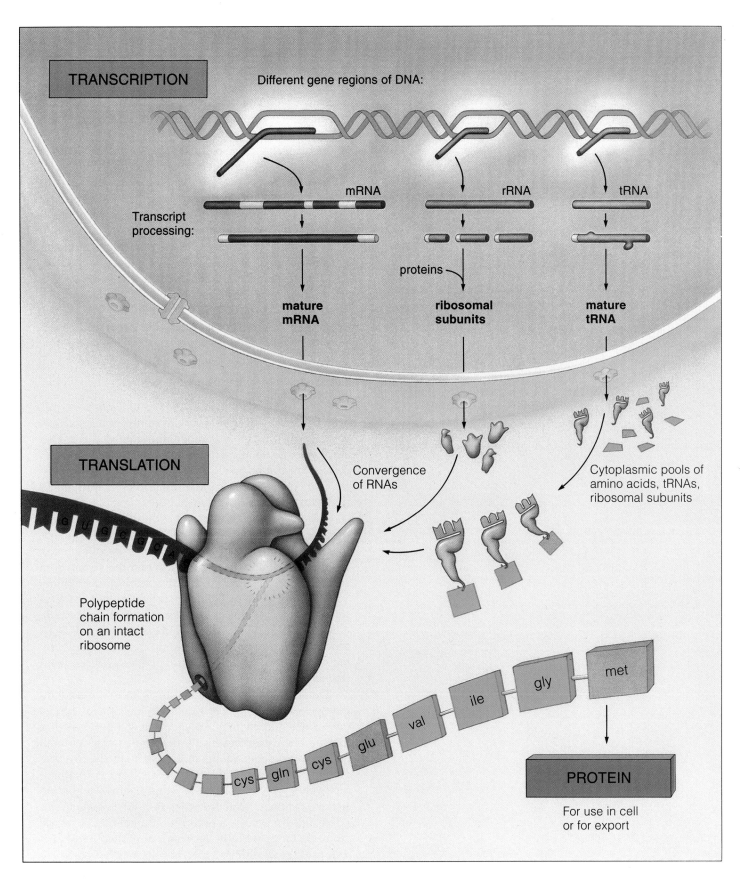

Figure 12.12 Summary of transcription and translation—the two steps leading to protein synthesis—as they occur in eukaryotic cells.

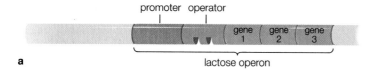

a

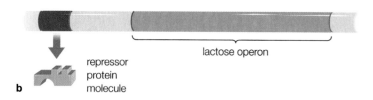

b

repressor
protein
molecule

Figure 12.13 (**a**) Components of the lactose operon in the DNA of the bacterium *E. coli.* (**b**) A gene for a repressor, a type of regulatory protein, is located elsewhere in the DNA. It interacts with the operon to afford control over the metabolism of lactose, as described in Figure 12.14.

THE NATURE OF GENE CONTROL

Your cells all carry the same genes. Most of the genes code for proteins that are basic to any cell's structure and functions. That's why the protein subunits of ribosomes are the same from one cell to the next, as are many enzymes used in metabolism.

Yet different cell types also use some of the genes in specialized ways. Only red blood cells use the genes for hemoglobin. Only certain white blood cells use the genes for the defensive weapons called antibodies. In fact, *each cell controls which genes are called into action, at what times, and in what amounts.*

Controls over genes are exerted by way of proteins, hormones, and other molecules that interact with DNA, RNA, or gene products. For instance, **regulatory proteins** enhance or suppress the rate of transcription. The next section describes how they do this for certain genes in a common prokaryote. The section after this provides a few examples of gene control in eukaryotes.

Control of Transcription in *E. coli*

You and other mammals are home to *Escherichia coli.* This bacterium lives in your intestines, where it uses some of the sugars and other nutrients you eat. One sugar, lactose, is present in milk. Probably you don't drink milk around the clock. When you do drink milk— and only then—*E. coli* rapidly transcribes its three genes for lactose-degrading enzymes.

The genes are next to each other, and a promoter precedes them. A promoter, recall, is a base sequence that signals the start of a gene. Another base sequence, an **operator**, intervenes between it and the genes. Both sequences are binding sites with roles in transcription. Such arrangements, in which a promoter and operator service more than one gene, is an **operon** (Figure 12.13). A regulatory gene at a different location in *E. coli* DNA codes for a repressor. **Repressors** are regulatory proteins that afford negative control of genes. They *prevent* transcription.

The lactose operon's repressor is able to bind with the operator *or* with a lactose molecule. When the lactose concentration in your gut is low (as when you haven't been drinking milk), the repressor binds with the operator. Being a large molecule, it *overlaps* the promoter and prevents RNA polymerase from transcribing the genes (Figure 12.14*b*). Thus lactose-degrading enzymes are not produced when they are not required.

When the lactose concentration increases, a lactose molecule binds with the repressor and alters its shape, so it can't bind with the operator. RNA polymerase can now transcribe the genes (Figure 12.14*c*). Thus lactose-degrading enzymes are produced when required.

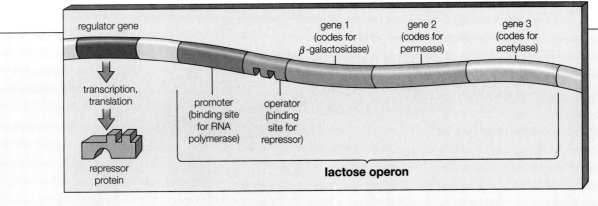

a Components of the lactose operon. Repressor proteins are produced continually and can exert negative control over the three genes of the operon by binding to the operator and inhibiting transcription.

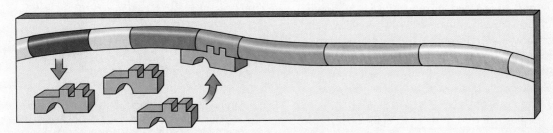

b When lactose is *absent*, a repressor protein is free to bind to the operator. The bulky repressor molecule overlaps the promoter and so prevents RNA polymerase from binding to the DNA and initiating transcription. Lactose-metabolizing enzymes (which are not needed) are not produced.

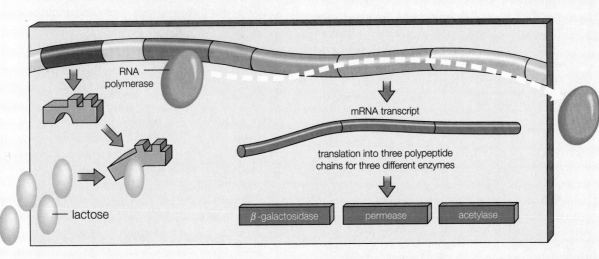

c When lactose is present, it binds to the repressor and distorts its shape. The altered shape prevents the repressor from binding to the operator. The promoter site is now exposed and RNA polymerase can start transcription. Thus lactose is an inducer of transcription of the lactose operon.

Figure 12.14 Negative control of the lactose operon. The first gene of the operon codes for an enzyme that splits lactose into two subunits (glucose and galactose). The second one codes for an enzyme that transports lactose molecules across the plasma membrane and into the cytoplasm. The third plays a role in metabolizing certain sugars.

Gene Control in Eukaryotes

Gene control is complicated business in eukaryotes. Consider how your body's cells all have the same genes (they descended from the same fertilized egg). Yet nearly all have become specialized in composition, structure, and function. This **cell differentiation** arises when cells express genes in selective ways. Different genes may be switched on once in a cell's life, left on some or all of the time, or never activated at all. *The patterns of gene activity depend on the cell type and on the control agents acting on it.*

Hormones As Control Agents. Think about the different signaling molecules called **hormones**. Most hormones are secreted from glands, picked up by the bloodstream, and distributed through the body. Some have widespread effects on gene activity in many cell types. In vertebrates, for instance, the pituitary gland secretes somatotropin (growth hormone). This hormone helps control synthesis of proteins required for cell division and, ultimately, the body's growth. Most cells have receptors for somatotropin.

Other hormones affect only certain cells at certain times. Prolactin is like this. Beginning a few days after a female mammal gives birth, prolactin activates genes in certain cells of mammary glands. Those genes alone have responsibility for milk production. Liver cells and heart cells have those genes also, but they have no means of responding to signals from prolactin.

Explaining hormonal control of gene activity is like explaining a full symphony orchestra to someone who has never seen one or heard it perform. Many separate parts must be defined before their interactions can be understood! We will return to the topic of hormonal controls in Unit VI.

X Chromosome Inactivation.

Other patterns of gene activity arise through interactions between DNA and proteins. Remember how eukaryotic chromosomes consist of DNA *and* many proteins? While a female mammal is developing as an embryo in her mother's body, certain proteins interact in ways that condense *one* of the two X chromosomes. See, for example, Figure 12.15. Which of the two condense in each cell is a matter of chance. It is a normal event, and it blocks transcription of most genes on the chromosome.

As the embryo grows through cell divisions, each daughter cell inherits the same pattern of X chromosome inactivation that occurred in its parent cell. *Either* the maternal *or* paternal X chromosome can be randomly inactivated in the cell lineage that gives rise to a given tissue region. Thus every adult female is a "mosaic" for inactivated chromosomes. She has patches

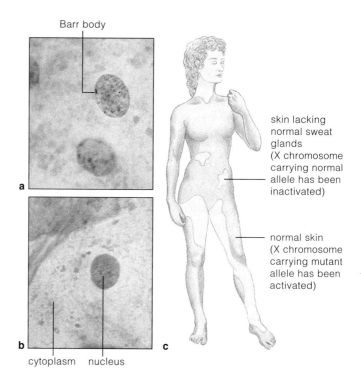

Barr body

skin lacking normal sweat glands (X chromosome carrying normal allele has been inactivated)

normal skin (X chromosome carrying mutant allele has been activated)

cytoplasm   nucleus

Figure 12.15 X chromosome inactivation. (**a**) A condensed X chromosome is visible in the interphase nucleus of a human female cell. The dark-staining spot on the inside of the nuclear envelope is called a Barr body after its discoverer, Murray Barr. (**b**) Nucleus from a male cell, which has no Barr body.

(**c**) Pattern of gene expression in a female affected by anhidrotic ectodermal dysplasia. In this disorder, patches of skin do not have normal sweat glands. Such "mosaic" tissue effects were discovered by Mary Lyon. They arise through random X chromosome inactivation.

of tissue in which an allele on the maternal *or* paternal X chromosome is being expressed.

We see evidence of the mosaic effect in females with *anhidrotic ectodermal dysplasia*. Such females are heterozygous for a certain recessive allele on the X chromosome. In this disorder, sweat glands are absent in patches of affected skin. In those patches, the X chromosome bearing the normal dominant allele has been condensed, and genes on the one bearing the mutated allele are active (Figure 12.15*c*).

When Controls Break Down

We have barely touched on gene control in eukaryotes. To gain a strong impression of their importance, take a look at the *Focus* essay, which describes what happens when cells become cancerous. Possibly more than any other example, cancerous transformations bring home the extent to which you and all other organisms depend on controls over genes and their protein products.

A Cascade of Proteins and Cancer

Every second, millions of cells in your skin, gut lining, liver, and other body regions divide and replace their worn-out, dead, and dying predecessors. They do not divide willy-nilly. They cannot divide at all unless they synthesize and stockpile cyclin, a protein. Cyclin binds with cdc2, the first of a series of enzymes that transfer phosphate from ATP to the next enzyme in line or to a final protein target. The enzyme molecules act more than once. Each activates many others, which activate many others, and so on in a growing cascade of reactions that ripple through the cell.

The enzymes switched on first replicate the cell's DNA. Others bring about nuclear division. For instance, some assemble and operate the microtubular spindle that moves chromosomes. Others orchestrate the split into two cells. As the cell divides, enzymes destroy all the cyclin, and this puts the division machinery to rest. Each daughter cell starts stockpiling cyclin. If the cell goes on to divide, cyclin will again be destroyed.

Of all proteins, only cyclin accumulates at a constant rate, disappears abruptly, then accumulates again during cell cycles (Figure *a*). But how does a cell "know" when to start and stop building cyclin? It requires signals from hormones and other regulator molecules. *Such signals lift controls that otherwise suppress the cyclin-driven engine, then put on the brakes by reinstating controls when cell division is completed.*

On rare occasions, controls over cell division are lost. It is not that affected cells divide at a horrendous rate. Rather, as long as conditions for growth stay favorable, the divisions never stop. Any tissue mass of cells that are not responding to normal controls over cell growth and division is a tumor.

When a tumor is benign, it remains in the same place in the body. Surgical removal of the tissue mass removes its threat to the functioning of the surrounding tissues. When a tumor is malignant, its cells can migrate and then grow and divide in other organs. The cells have undergone a mutation that alters or suppresses certain genes. Those genes code for the recognition proteins at the plasma membrane that allow cells to bind together in tissues and organs. When genes for those proteins are altered or suppressed, the cell can leave its proper place and travel (by way of blood or lymph). The process of invasion is called metastasis.

There are many types of malignant tumors. All are grouped into the general category of **cancer**. Cancer is not just a human affliction. It has been observed in most animal species studied to date. Similar abnormalities

interphase mitosis interphase mitosis interphase

a Chart of the controlled changes in the intracellular levels of cyclin (brown line) in normal cells. Cyclin is the protein that guides cells into mitosis (light blue bands) during the cell cycle.
b Scanning electron micrograph of a cancer cell, surrounded by some of the body's white blood cells that may or may not be able to destroy it.

cytoplasmic extension of cancer cell body

white blood cell

have even been observed in many plant species. At the minimum, all cancer cells have these characteristics:

1. *Profound changes in the plasma membrane and cytoplasm.* Membrane permeability is amplified. Some membrane proteins are lost or altered; new ones appear. The cytoskeleton shrinks, becomes disorganized, or both. Enzyme activity shifts (as in an amplified reliance on glycolysis).

2. *Abnormal growth and division.* Controls that inhibit overcrowding in tissues are lost. Cell populations increase to high densities. New proteins trigger abnormal increases in small blood vessels that service the growing cell mass.

3. *Weakened capacity for adhesion.* Cells cannot become properly anchored in the parent tissue.

4. *Lethality.* Unless cancer cells are eradicated, they will kill the individual.

Any gene with potential to induce cancerous transformations is an "oncogene." Such genes were first identified in retroviruses, a type of infectious agent (Table 17.1). But many organisms have nearly identical gene sequences that rarely trigger cancer! Those sequences are "proto-oncogenes."

Proto-oncogenes code for proteins required in normal cell function. They may become cancer-causing genes only on rare occasions, when specific mutations alter their structure or their expression. Thus the *normal* expression of proto-oncogenes is vital, even though their abnormal expression may be lethal.

Insertion of viral DNA into the DNA of a cell can skew transcription of a proto-oncogene. "Carcinogens" may do the same thing. Carcinogens can bind to DNA and introduce mutations into it. They include many natural and synthetic compounds (such as asbestos and components of cigarette smoke), x-rays, gamma rays, and ultraviolet radiation. And they can trigger cancer.

Yet cancer seems to be a multistep process, requiring mutations in more than one proto-oncogene. Look again at the listed characteristics of cancer cells. Now think about some of the products of proto-oncogenes. They include growth factors, which are signals sent by one cell to trigger growth in other cells. They include cellular receptors for growth factors. They include regulatory proteins involved in cell adhesion—and the protein signals for cell division.

SUMMARY

1. Protein-building instructions are encoded in genes. A gene is a sequence of nucleotide bases. The path leading from genes to proteins has two steps: transcription and translation.

   a. In transcription, a region of the double-stranded DNA molecule is unwound, and an RNA molecule is assembled on the exposed bases of a DNA strand.

   b. In translation, RNA molecules interact and convert the gene's message into a linear sequence of amino acids—that is, a polypeptide chain. Such chains are the structural units of proteins.

2. Protein synthesis requires three classes of RNA:

   a. rRNA (components of ribosomes, which contain the actual platform on which polypeptide chains are assembled).

   b. mRNA (the only RNA that carries the protein-building instructions from DNA to the cytoplasm).

   c. tRNA (the molecules that translate the nucleotide sequence of mRNA into a corresponding sequence of amino acids).

3. Thus DNA is used to synthesize RNA, and the RNA is used to synthesize proteins:

4. The genetic code is the correspondence between the nucleotide sequences in DNA (then mRNA) and the amino acid sequences in polypeptide chains. The code words are nucleotide bases, read in blocks of three (base triplets). Each triplet in mRNA is a codon. A complementary triplet in tRNA is an anticodon.

5. Transcription of DNA into RNA follows the same base-pairing rule that applies to DNA replication. However, uracil takes the place of thymine in an RNA strand. It pairs with adenine:

6. Translation proceeds through three stages:

   a. Initiation. A small ribosomal subunit binds with an initiator tRNA, then with an mRNA transcript. The small subunit then binds with a large ribosomal subunit to form the initiation complex.

   b. Chain elongation. tRNAs deliver amino acids to the ribosome. Their anticodons base-pair with mRNA codons. Then peptide bonds form between their amino acids. A polypeptide chain grows with each addition.

   c. Chain termination. A stop codon triggers events that cause the polypeptide chain and mRNA to detach from the ribosome.

7. Mutations are deletions, additions, or substitutions of one or more bases in the nucleotide sequence of DNA. They affect the structure of a gene and so may affect the structure and function of the gene's protein product.

8. Gene expression is controlled by many interacting elements, including regulatory proteins, enzymes, hormones, and control sites built into DNA molecules.

9. In complex eukaryotes, cell differentiation arises through selective controls over genes. This means different cell types activate and suppress some fraction of their genes in a variety of ways. The outcome is differences in appearance, composition, and function.

Review Questions

1. Define the functions of mRNA, rRNA, and tRNA. *163, 166, 167*

2. Define genetic code. *166*

3. What is a codon? an anticodon? Do they interact during transcription and translation? How? *166–169*

4. A plant, fungus, or animal is composed of diverse cell types. How might this diversity arise, given that all of the body cells in each organism inherit the *same* set of genetic instructions? *170*

5. What are the characteristics of cancer cells? *175–176*

Self-Quiz *(Answers in Appendix IV)*

1. Nucleotide bases, read _____ at a time, serve as the "code words" of genes.

2. DNA contains different genes that are transcribed into _____ .
   a. proteins
   b. mRNAs
   c. rRNAs
   d. tRNAs
   e. b, c, and d

3. The instructions of _____ are translated into proteins.
   a. DNA
   b. mRNA
   c. rRNA
   d. tRNA
   e. a and b
   f. b, c, and d

4. mRNA is produced by _____ .
   a. replication
   b. duplication
   c. transcription
   d. translation

5. _____ carries coded instructions for an amino acid sequence to the ribosome.
   a. DNA
   b. rRNA
   c. mRNA
   d. tRNA

6. tRNA _____ .
   a. delivers amino acids to ribosomes
   b. picks up genetic messages from rRNA
   c. synthesizes mRNA
   d. all of the above

7. Each codon calls for a specific _____ .
   a. protein
   b. polypeptide
   c. amino acid
   d. carbohydrate

8. The mRNA sequence CGUUUACACCGUCAC calls for how many amino acids?
   a. three
   b. five
   c. six
   d. more than six

9. Use the genetic code (Figure 12.5) to translate the mRNA sequence UAUCGCACCUCAGGAUGAGAU. Which amino acid sequence is being specified?
   a. tyr-arg-thr-ser-gly-stop-asp . . .
   b. tyr-arg-thr-ser-gly . . .
   c. tyr-arg-tyr-ser-gly-stop-asp . . .
   d. none of the above

10. Selective gene expression is the basis of _____ .
   a. transcription
   b. translation
   c. gene mutation
   d. cell differentiation

11. _____ help regulate gene expression.
   a. Controls built into DNA
   b. Regulatory proteins
   c. Enzymes
   d. Hormones
   e. All of the above

12. Match the terms related to protein building.
   ____ disrupts genetic instructions
   ____ genetic code word
   ____ transcription
   ____ translation
   ____ gene expression

   a. interacting DNA control sites, regulatory proteins, enzymes, hormones
   b. RNAs convert genetic messages into polypeptide chains
   c. base triplet for an amino acid
   d. one DNA strand is template for the process
   e. gene mutation

Selected Key Terms

anticodon *167*
base sequence *163*
cancer *175*
cell differentiation *174*
codon *166*
exon *165*
gene mutation *170*
genetic code *166*
hormone *174*
intron *165*
messenger RNA (mRNA) *163*
operator *172*
operon *172*
promoter *164*
regulatory protein *172*
repressor *172*
ribonucleic acid (RNA) *163*
ribosomal RNA (rRNA) *163*
transcription *163*
transfer RNA (tRNA) *163*
translation *163*
uracil *164*

Readings

Amato, I. January 1991. "Stuck on Mussels." *Science News* 139:8–15.

Grunstein, M. October 1992. "Histones as Regulators of Genes." *American Scientist* 267(4):68–74B.

Liotta, L. February 1992. "Cancer Cell Invasion and Metastasis." *Scientific American* 266(2): 54–63.

Murray, A., and M. Kirschner. March 1991. "What Controls the Cell Cycle?" *Scientific American* 264(3):56–63.

Wolfe, S. 1993. *Molecular and Cellular Biology.* Belmont, California: Wadsworth.

13 RECOMBINANT DNA AND GENETIC ENGINEERING

Make Way for Designer Genes

In 1990, a four-year-old girl received a historic genetic reprieve. She was born without defenses against viruses, bacteria, and other agents of disease. *She has no immune system.* Of her forty-six chromosomes, one bears a defective gene that normally would code for adenosine deaminase (ADA), an enzyme. Without the enzyme, her cells cannot properly break down excess amounts of a nucleotide (AMP), and a reaction product accumulates that is toxic to lymphoblasts in the bone marrow. Lymphoblasts give rise to white blood cells—the immune system's army. With too few of those cells (or none at all), the outcome is a devastating set of disorders—*severe combined immune deficiency* (SCID).

Bone marrow transplants help some individuals with SCID when the donated lymphoblasts go on to produce functional white blood cells. ADA injections help others. But these are not permanent cures.

Given the options, the parents of the young girl consented to the first federally approved gene therapy test for humans. Using recombinant DNA methods of the sort described in this chapter, medical researchers cultured some of her white blood cells, then inserted copies of the ADA gene into them. The modified cells were stimulated to divide. Later, the researchers inserted about a billion copies of the genetically modified cells into her bloodstream. The girl has been receiving additional cell infusions every month since then. Another ADA-deficient girl has also started treatment—and both are doing well at this writing.

In time, medical researchers would like to extract lymphoblasts from the bone marrow of these patients, insert functional ADA genes into them, and put them back in place. Only then might the girls be assured of a constant, lifelong supply of the crucial enzyme—and of functional disease fighters (Figure 13.1).

As this example suggests, recombinant DNA technology has staggering potential for medicine. It has equally staggering potential for agriculture and industry. It does not come without risks. With this chapter, we consider some basic aspects of the new technology, and we address ecological, social, and ethical questions related to its application.

Figure 13.1 White blood cells on patrol inside a blood vessel. Some people have drastically reduced numbers of these infection-fighting cells—or none at all. They are candidates for gene therapy, one of the beneficial applications of recombinant DNA research.

1. Genetic experiments have been occurring in nature for billions of years as a result of gene mutations, crossing over and recombination, and other events. Humans are now engineering genetic changes by way of recombinant DNA technology.

2. Three activities are at the heart of recombinant DNA technology. First, DNA molecules are cut into fragments. Second, the fragments are inserted into cloning tools, such as plasmids. Third, fragments containing genes of interest are copied rapidly and repeatedly. The genes, and in some cases their protein products, are produced in quantities that are large enough for research and for practical applications.

3. The new technology raises social, legal, ecological, and ethical questions regarding its benefits and risks.

Figure 13.2 A ruptured bacterial cell (*Escherichia coli*). Notice the larger bacterial chromosome and the smaller plasmids (blue arrows).

RECOMBINANT DNA TECHNOLOGY

For more than 3 billion years, nature has been conducting genetic experiments through mutation, crossing over between chromosomes, and other events. Genetic messages have changed countless times, and this is the source of life's diversity.

For many thousands of years, we humans have been changing genetically based traits of species. Through artificial selection, we coaxed modern crop plants and new breeds of cattle, birds, dogs, and cats from wild ancestral stocks. We developed meatier turkeys, sweeter oranges, seedless watermelons, and flamboyant ornamental roses and other plants. We produced the tangelo (tangerine × grapefruit) and the mule (donkey × horse).

Today we also analyze and even engineer genetic changes through **recombinant DNA technology**. With this technology, DNA from different species can be cut, spliced together, then inserted into bacteria or other types of cells that rapidly replicate their DNA and divide. The cells replicate the foreign DNA right along with their own. In short order, a huge bacterial population produces useful quantities of recombinant DNA molecules.

Also with this technology, genes can be isolated, modified, and inserted back into the same organism or into a different one. Many engineered genes are already altering specific traits of organisms.

Believe it or not, this astonishing technology had its origins in the innards of bacteria. Each bacterial cell has a single chromosome, a circular DNA molecule with all the genes needed for growth and reproduction. Many types of bacteria also have **plasmids**. Bacterial plasmids are small, circular molecules of "extra" DNA with only a few genes (Figure 13.2). They can be copied independently of the chromosome's replication. As researchers found out, portions of a plasmid can be copied independently of the bacterium.

Sometimes one bacterium transfers plasmid genes to a bacterial neighbor (page 251). Once in a great while, the transferred plasmid becomes integrated into the recipient's chromosome, forming a recombinant DNA molecule. Recombination events, including this one, depend on specific enzymes. One bacterial enzyme recognizes a short nucleotide sequence present in both the plasmid and the chromosome. It cuts the molecules at that sequence, then another enzyme splices the cut ends together.

Viruses as well as bacteria dabble in gene transfers and recombinations. So, apparently, do most eukaryotic organisms. As we turn now to the DNA recombinations in the laboratory, keep this point in mind:

Recombination following plasmid transfer is common in nature. It is made possible by specific enzymes that can make cuts in DNA molecules.

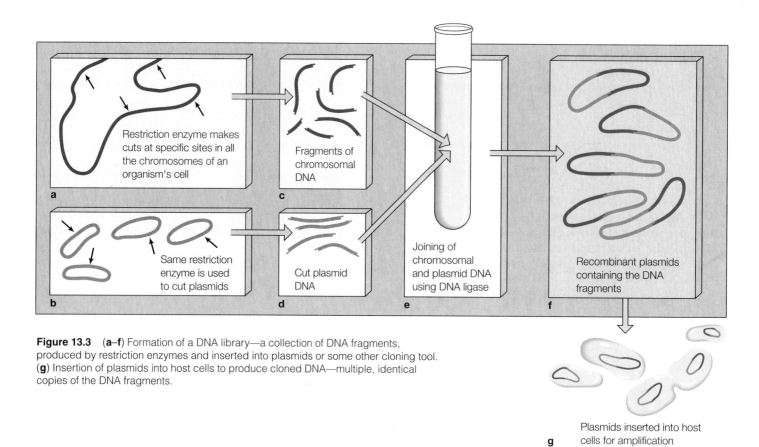

Figure 13.3 (**a–f**) Formation of a DNA library—a collection of DNA fragments, produced by restriction enzymes and inserted into plasmids or some other cloning tool. (**g**) Insertion of plasmids into host cells to produce cloned DNA—multiple, identical copies of the DNA fragments.

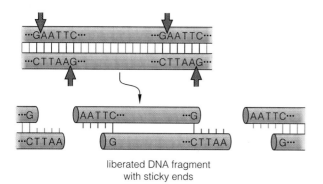

g Plasmids inserted into host cells for amplification

Producing Restriction Fragments

Bacteria are equipped with many **restriction enzymes**. These enzymes cut apart foreign DNA that has been injected into the bacterial cell, often by viruses. Each type makes a cut wherever a specific, very short nucleotide sequence occurs in the DNA. Cuts at two identical sequences in the same molecule produce a fragment. Because each cut is staggered, the fragment now is single-stranded and sticky at both ends:

liberated DNA fragment with sticky ends

By "sticky," we mean that the short, single-stranded ends of a DNA fragment can base-pair with any other DNA molecule cut by the same restriction enzyme.

Suppose you use the same restriction enzyme to cut plasmids *and* the DNA from a human cell. When you mix the cut-up molecules together, they base-pair at the cut sites. Then you add **DNA ligase** to the mixture. This enzyme seals the base-pairings, just as it does during DNA replication. In this way, you create "recombinant plasmids," which contain pieces of DNA from another organism.

You now have a **DNA library**—a collection of DNA fragments, produced by restriction enzymes, that have been incorporated into plasmids (Figure 13.3). But the library is almost vanishingly small. It must be amplified—copied again and again—into useful amounts. Bacteria, yeasts, and other cells reproduce rapidly, and they readily take up plasmids. In short order, a growing population of such cells can amplify a DNA library. Their repeated replications and cell divisions yield **cloned DNA** (multiple, identical copies of DNA fragments), inserted into plasmids.

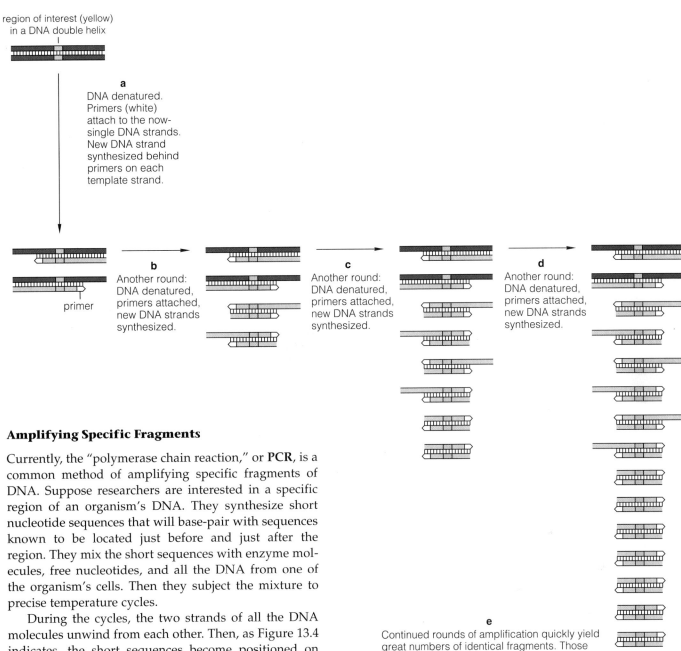

region of interest (yellow)
in a DNA double helix

a
DNA denatured.
Primers (white)
attach to the now-
single DNA strands.
New DNA strand
synthesized behind
primers on each
template strand.

primer

b
Another round:
DNA denatured,
primers attached,
new DNA strands
synthesized.

c
Another round:
DNA denatured,
primers attached,
new DNA strands
synthesized.

d
Another round:
DNA denatured,
primers attached,
new DNA strands
synthesized.

e
Continued rounds of amplification quickly yield
great numbers of identical fragments. Those
fragments contain the DNA region of interest.

Figure 13.4 Polymerase chain reaction used in gene
amplification, as described in the text.

Amplifying Specific Fragments

Currently, the "polymerase chain reaction," or **PCR**, is a
common method of amplifying specific fragments of
DNA. Suppose researchers are interested in a specific
region of an organism's DNA. They synthesize short
nucleotide sequences that will base-pair with sequences
known to be located just before and just after the
region. They mix the short sequences with enzyme molecules, free nucleotides, and all the DNA from one of
the organism's cells. Then they subject the mixture to
precise temperature cycles.

During the cycles, the two strands of all the DNA
molecules unwind from each other. Then, as Figure 13.4
indicates, the short sequences become positioned on
any exposed, complementary bases. An enzyme, **DNA
polymerase**, uses the sequences as "primers" to start
replication. The enzyme used comes from a bacterium
that thrives in hot springs and even in hot water
heaters. It is not damaged by the *elevated* temperatures
required to unwind the DNA or by the *lower* temperatures required for base-pairing.

The reactions quickly amplify a single DNA molecule to many billions of molecules. Thus PCR can
amplify samples with very little DNA, as might be
found in a single hair left at the scene of a crime.
Besides this, PCR can be used to amplify segments of
DNA found in ancient material (see the *Focus* essay on
page 182).

Applications of RFLPs

Imagine that a researcher is using restriction enzymes to cut DNA from one of your cells into fragments of specific lengths. She puts some fragments near one end of a slab of gel, then applies an electric current. Because the DNA's phosphate groups carry a charge, the fragments respond to the current and move through the gel. Large fragments can't move through it as fast as small ones do. When they separate from one another, they form different bands. DNA fragments of different lengths show up as different luminous bands after treatment with a dye. When attached to DNA and viewed under ultraviolet light, the dye fluoresces (Figure *a*).

When DNA samples from different people are cut with restriction enzymes, they may reveal slight variations in banding patterns. The variations are restriction fragment length polymorphisms. They are called **RFLPs** (pronounced RIFF-lips) for short. They arise because some base sequences in the DNA vary in molecular form from one person to the next. The molecular differences shift the number and location of sites where restriction enzymes make their cuts.

RFLP analysis has uses in basic research, such as the human genome project. Evolutionary biologists use it to decipher DNA that has been extracted from mummies, and even from mammoths and ancient humans that became preserved in glacial ice.

RFLP analysis has practical applications. Consider that mutant alleles responsible for sickle-cell anemia and other genetic disorders often have a unique restriction site closely linked to the responsible genes. This restriction site can be used to determine whether an individual carries the mutated gene. Thus RFLP analysis can be used for prenatal diagnosis (page 151).

Or consider that each person has a **DNA fingerprint**, a unique array of RFLPs inherited from each parent in a Mendelian pattern. Paternity and maternity cases can be resolved by comparing the child's DNA fingerprint with the disputed parent's. Finally, consider that murderers can be identified if they lose even a few drops of blood at the scene of the crime or if a few drops of the victim's blood are found on their clothing. The bloodstain may provide enough DNA to identify the perpetrator. Similarly, a rapist can be identified from semen recovered from the victim.

a Columns, running from left to right, in which DNA fragments (labeled with a fluorescent dye) have been separated from one another. The fragments form a luminous banding pattern that reflects differences in their length.

Locating Genetically Altered Host Cells

We study genes because we want to learn about or use their protein products. When you mix DNA with living cells, how can you find out which ones take up the DNA and contain a gene of interest? You can use **DNA probes**. These probes are short DNA sequences that are assembled from radioactively labeled nucleotides (page 19). Part of the probe must be able to base-pair with part of the gene being studied. Such base-pairing between nucleotide sequences from different sources is called **nucleic acid hybridization**.

The first step is to select those cells that have taken up the recombinant plasmids. Most plasmids contain genes that make their bacterial owners resistant to antibiotics. You put the prospective host cells on a culture medium that has the antibiotic added to it. The antibiotic prevents growth of all cells *except* the ones

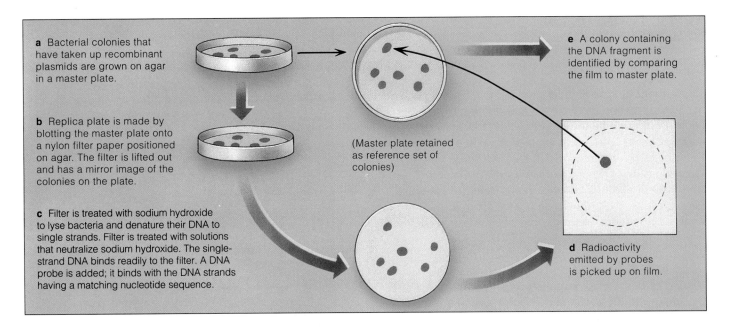

a Bacterial colonies that have taken up recombinant plasmids are grown on agar in a master plate.

b Replica plate is made by blotting the master plate onto a nylon filter paper positioned on agar. The filter is lifted out and has a mirror image of the colonies on the plate.

c Filter is treated with sodium hydroxide to lyse bacteria and denature their DNA to single strands. Filter is treated with solutions that neutralize sodium hydroxide. The single-strand DNA binds readily to the filter. A DNA probe is added; it binds with the DNA strands having a matching nucleotide sequence.

(Master plate retained as reference set of colonies)

e A colony containing the DNA fragment is identified by comparing the film to master plate.

d Radioactivity emitted by probes is picked up on film.

Figure 13.5 Use of a DNA probe to identify the colony of bacterial cells that have taken up plasmids containing a specific DNA fragment.

housing plasmids (because the plasmids carry the antibiotic-resistance genes).

The next step is to identify the particular bacterial cells having recombinant plasmids with the gene of interest. As the bacterial cells divide, they form colonies. Suppose the colonies are on agar (a solid gel) in a petri dish. You blot the agar against a nylon filter. Some cells stick to the filter in locations that mirror the locations of the original colonies (Figure 13.5). You use solutions to rupture the cells, to fix the released DNA onto the filter, and to make the double-stranded DNA molecules unwind. Then you add the DNA probes. The probe hybridizes only with the gene region having the proper base sequence. The probe-hybridized DNA emits radioactivity and allows you to tag the colonies harboring the gene of interest.

Expressing the Gene of Interest

Even if a host cell takes up a gene, it may not be able to translate it into protein. Consider human genes, which have noncoding regions (introns). Transcripts of those genes cannot be translated until introns are snipped out and coding regions (exons) are spliced together into a mature mRNA transcript. Bacterial enzymes don't recognize the splice signals, so bacterial host cells cannot always directly translate human DNA.

However, bacteria can use **cDNA**, which has been "copied" from a mature mRNA transcript for the

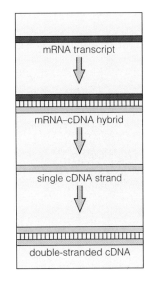

mRNA transcript

mRNA–cDNA hybrid

single cDNA strand

double-stranded cDNA

a An mRNA transcript of a desired gene is used as a template for assembling a DNA strand. An enzyme (reverse transcriptase) does the assembling.

b An mRNA-DNA hybrid molecule results.

c Enzyme action removes the mRNA and assembles a second strand of DNA on the remaining DNA strand.

d The result is double-stranded cDNA, "copied" from an mRNA template.

Figure 13.6 Formation of cDNA from an mRNA transcript.

desired gene. By a backwards process called **reverse transcription**, a matching DNA strand is assembled on mRNA. The result is a "hybrid" DNA-RNA molecule (Figure 13.6). Enzymes digest the RNA, then assemble a DNA strand on the strand remaining. This double-stranded cDNA can be inserted into a plasmid for amplification. Then the recombinant plasmids can be inserted into bacteria. The bacteria may then use the cDNA directions for making a protein.

APPLICATIONS OF THE NEW TECHNOLOGY

Uses in Basic Research

Think about Gregor Mendel growing pea plants, planting seeds, growing more plants, and forming hypotheses about what *might* be the source of their traits. Today, we study genes in cells and chromosomes. Without the new technology, we never would have discovered the gene tamperings that give rise to diseases such as cancer.

Or think about Charles Darwin, trying to convince people of the evolutionary source of diversity without knowledge of the source—gene mutations. With the new technology, we study certain DNA sequences from one organism, then see how closely or remotely they match equivalent sequences from other organisms. This is good evidence of evolutionary relationship.

Finally, think about the **human genome project**. Researchers in laboratories throughout the world are sequencing human DNA, a task that may take ten to fifteen years to complete. Some are working on specific chromosomes, others on specific genes. Of 2,000 genes mapped on chromosomes, 400 are already linked to genetic disorders. Once all of the 3 *billion* nucleotides present in haploid cells have been sequenced, we will have the ultimate reference book on human biology and genetic disorders. *About 99.9 percent of that sequence is the __same__ in every human on earth.*

Figure 13.7 Spraying an experimental strawberry patch in California with "ice-minus" bacteria. The sprayer used elaborate protective gear to meet government regulations in effect at the time.

Genetic Modification of Organisms: Some Examples

Genetic Engineering of Bacteria. Bacterial strains used in many genetic engineering experiments are harmless to begin with. They are also modified to prevent them from surviving outside the laboratory. Even so, there is concern about possible risks of introducing genetically engineered bacteria into humans or the environment.

Consider how Steven Lindow altered a bacterium that lives on many crop plants. The bacterium makes plants vulnerable to frost, because proteins at its surface promote the formation of ice crystals. Lindow excised the ice-forming gene from some cells. He planned to spray these "ice-minus bacteria" on strawberry plants in an isolated field just before a frost. He hypothesized that the bacteria would make the plants more resistant to freezing.

Here was an organism from which a harmful gene had been *deleted*, yet it triggered a bitter legal debate on the risks of releasing genetically engineered microbes in the environment. The courts finally ruled in favor of allowing the genetically engineered bacteria to be released, and researchers sprayed a small patch of strawberries (Figure 13.7). Nothing bad happened. But then, a few environmental activists had entered the patch at night and pulled up the plants.

Since then, rules governing the release of genetically engineered organisms have been clarified and are now less restrictive.

Genetic Modification of Plants. Years ago, Frederick Steward and his coworkers cultured cells of carrot plants and induced the cells to grow into small embryos. Some embryos actually grew into whole plants. Today many plant species, including major crop plants, are regenerated from cultured cells. Because the culturing methods increase mutation rates, the cultures are a source of genetic modifications. Researchers can pinpoint a useful mutation among millions of cells.

Suppose a culture medium contains a toxin that is produced by a disease agent. If a few cells have a mutated gene that confers resistance to the toxin, they will end up being the only live cells in the culture. Now

Figure 13.8 Crown gall tumors on a willow tree, as caused by a tumor-inducing plasmid from a common bacterium, *Agrobacterium tumefaciens*. Scientists now put this plasmid to work in cultures of plant cells, where it moves desirable genes into the plant chromosomes.

suppose plants are regenerated from the cells, then hybridized with other varieties. The hybrid plants may end up with disease resistance.

Researchers also can transfer genes into cultured plant cells, by using a plasmid from a bacterium that infects many flowering plants. In nature, some of the plasmid genes get integrated into a plant's DNA, then they induce tumor formation (Figure 13.8). In the laboratory, the tumor-inducing genes are removed from the plasmid and other, desired genes are inserted into it. Researchers induce bacterial cells to take up the recombinant plasmid and to infect cultured plant cells. Then they induce infected cells to grow into whole plants. The regenerated plants often express the foreign genes.

For example, one recombinant plasmid delivered a firefly gene into cultured tobacco plant cells (Figure 13.9a). The gene codes for luciferase, an enzyme of bioluminescence. Plants regenerated from the cells glow in the dark! Similarly, some genetically engineered cotton plants resist worm attacks (Figure 13.9b,c). The modification is ecologically safer than applying pesticides. Only targeted pests are killed, not "good" insects, such as ladybird beetles that prey on aphids.

a

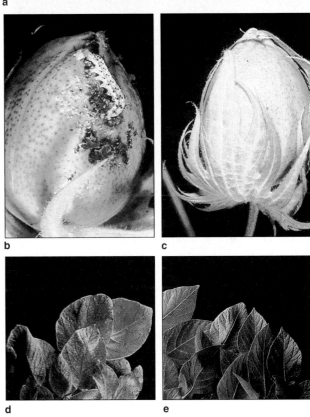

b   c

d   e

Figure 13.9 Genetically engineered plants. (**a**) This modified tobacco plant glows in the dark as a result of a gene transfer. The gene, from firefly DNA, codes for the enzyme luciferase, described on page 68. (**b**) Some worms attack buds of cotton plants, but not the modified plant shown in (**c**). (**d**) A virus damages potato plants, but the modified strain in (**e**) resists attack.

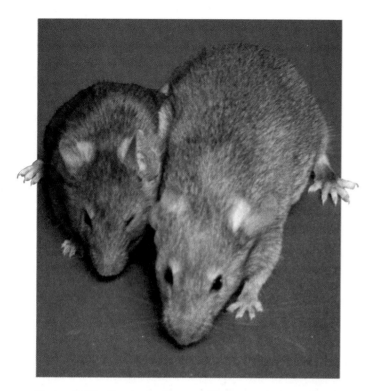

Figure 13.10 Ten-week-old mouse littermates, the one on the left weighing 29 grams, and the one on the right, 44 grams. The larger mouse grew from a fertilized egg into which the gene for human somatotropin (growth hormone) had been inserted.

Genetic Modification of Animals. In 1982, Ralph Brinster and Richard Palmiter introduced the rat gene for somatotropin (growth hormone) into fertilized mouse eggs. The gene was integrated into the mouse DNA—and the mice grew much larger than their normal littermates. Their cells had up to thirty-five copies of the gene, and the somatotropin blood level was several hundred times greater than normal. Later, the gene for human somatotropin was transferred into a mouse embryo, where it became integrated into the DNA. A "super rodent" resulted (Figure 13.10).

Today, human genes are transferred into mouse embryos during research into Alzheimer's disease. In research to develop effective vaccines, genes are being altered in viruses that each year kill millions of young children in Africa, Asia, and Latin America. In a field in North Carolina, leaves of engineered tobacco plants are producing hemoglobin, as well as an experimental AIDS drug. As part of what is fondly called "barnyard biotechnology," sheep and goats are being engineered to produce milk that will provide us with useful proteins.

Perhaps the most riveting example of genetic engineering is human **gene therapy**. Generally speaking, gene therapy refers to the transfer of one or more normal genes into the body cells of an organism to correct a genetic defect. This chapter opened with the first human application of this technology, meant to save the life of a four-year-old girl. It closes with a *Focus* essay that invites you to consider some prospects and problems associated with application of the technology.

SUMMARY

1. Genetic "experiments" have been occurring in nature for billions of years. Gene mutation, crossing over and recombination during meiosis, and other natural events have all contributed to the current diversity among organisms.

2. Humans have been manipulating the genetic character of different species for many thousands of years. The emergence of recombinant DNA technology in the past few decades has enormously expanded our capacity to modify organisms genetically.

3. Recombinant DNA technology is based on certain procedures. First, DNA molecules that contain a gene of interest are cut into fragments. Then the fragments are inserted into a cloning tool that lends itself to replication. The inserted DNA is taken up by bacteria, yeasts, or some other rapidly dividing cells. Cells of the host population replicate the inserted DNA whenever they replicate their own DNA, and so produce useful amounts of it.

Some Implications of Human Gene Therapy

Recombinant DNA technology and genetic engineering are advancing rapidly. We are only beginning to work our way through their social and ethical implications.

To most of us, human gene therapy to correct genetic abnormalities seems like a socially acceptable goal. Is it also socially acceptable to insert genes into a *normal* human individual (or sperm or egg) to alter or enhance traits? The idea of selecting desirable human traits is called eugenic engineering. Yet who decides which forms of a trait are most "desirable"? What if prospective parents could pick the sex of a child by way of genetic engineering? Three-fourths of one survey group said they would choose a boy. So what would be the long-term social implications of a drastic shortage of girls?

Would it be okay to engineer taller or blue-eyed or curlier haired individuals? If so, would it be okay to engineer "superhuman" offspring with exceptional strength or breathtaking intelligence? Suppose a person of average intelligence moved into a town composed of eight hundred Einsteins. Would the response go beyond a few mutterings of "There goes the neighborhood"?

Some say that the DNA of any organism must never be altered. Put aside the fact that nature itself alters DNA much of the time. The concern is that we don't have the wisdom to bring about beneficial changes without causing harm to ourselves or to the environment.

When it comes to manipulating human genes, one is reminded of our human tendency to leap before we look. When it comes to restricting genetic modifications of any sort, one also is reminded of an old saying: "If God had wanted us to fly, he would have given us wings." And yet, something about the human experience gave us the *capacity* to imagine wings of our own making—and that capacity has carried us to the frontiers of space.

Where are we going from here with recombinant DNA technology, this new product of our imagination? To gain perspective on the question, spend some time reading the history of our species. It is a history of survival in the face of all manner of new challenges, threats, bumblings, and sometimes disasters on a grand scale. It is also a story of our connectedness with the environment and with one another.

The questions confronting you today are these: Should we be more cautious, believing that one day the risk takers may go too far? And what do we as a species stand to lose if the risks are *not* taken?

a. Plasmids, a common cloning tool, occur in bacteria and some other organisms. Bacterial plasmids are small, circular DNA molecules that contain extra genes (in addition to those of the bacterial chromosome).

b. Researchers use restriction enzymes as well as DNA ligase to insert DNA into plasmids that have been removed from bacterial cells. Then they induce other cells to take up the recombinant plasmids.

c. In at least some host cells, a plasmid may become integrated into the host's DNA. The foreign DNA is replicated along with the host DNA before cell division.

4. A population of dividing cells produces cloned DNA—multiple, identical copies of a collection of DNA fragments that were earlier inserted into plasmids. Specific fragments of DNA can be rapidly amplified by the polymerase chain reaction (PCR).

5. Bacterial cells cannot directly translate human genes into proteins. (Human genes have introns, which bacterial enzymes don't recognize.) They can translate cDNA, which is copied from a mature mRNA transcript of a desired gene. In reverse transcription, a DNA strand is assembled on mRNA, enzymes digest the mRNA, then a matching DNA strand is assembled on the strand remaining. The double-stranded cDNA is incorporated into plasmids for amplification.

6. Genetic engineering uses recombinant DNA technology to modify genes, hence to modify traits of individuals. One application is gene therapy, which involves transferring normal genes into an individual, a sperm, or an egg to correct a genetic defect.

7. Recombinant DNA technology and genetic engineering have enormous potential for research and applications in medicine, agriculture, and home and industry. As with any new technology, potential benefits must be weighed against potential risks, including ecological and social disruptions.

1. Are restriction enzymes synthesized in the laboratory, or do they occur naturally in organisms? *180*

2. Recombinant DNA technology involves producing DNA restriction fragments, amplifying DNA, and identifying modified host cells. Briefly describe some examples of how these activities are carried out. *180–183*

3. Name two enzymes used in recombinant DNA technology and define their function. *180, 181*

4. Besides this chapter's examples, list what you believe might be some potential benefits and risks of genetic engineering.

Self-Quiz *(Answers in Appendix IV)*

1. _____ are small circles of bacterial DNA that are separate from the bacterial chromosome.

2. DNA fragments result when _____ cut DNA molecules at specific sites.
   a. DNA polymerases   c. restriction enzymes
   b. DNA probes   d. RFLPs

3. Recombinant DNA technology involves _____ .
   a. producing DNA fragments   d. all of the above
   b. making DNA libraries   e. a and c
   c. amplifying DNA

4. PCR stands for _____ .
   a. polymerase chain reaction
   b. polyploid chromosome restrictions
   c. polygraphed criminal rating

5. A _____ is a collection of DNA fragments, produced by restriction enzymes and incorporated into plasmids.
   a. DNA clone   c. DNA probe
   b. DNA library   d. gene map

6. A _____ is multiple, identical copies of a collection of DNA fragments inserted into plasmids.
   a. DNA clone   c. DNA probe
   b. DNA library   d. gene map

7. In reverse transcription, _____ is assembled on _____ .
   a. mRNA; DNA   c. DNA; enzymes
   b. DNA; mRNA   d. DNA; agar

8. _____ is the transfer of normal genes into body cells to correct a genetic defect.
   a. Reverse transcription   c. Gene mutation
   b. Nucleic acid hybridization   d. Gene therapy

9. Tobacco plant leaves that produce hemoglobin are a result of _____ .
   a. gene therapy   c. pressure on tobacco
   b. genetic engineering   growers
                   d. a and b

10. Match the terms appropriately.
   ____ DNA library   a. mutation, crossing over
   ____ plasmid   b. raises social, legal, and
   ____ nature's genetic   ethical questions
         experiments   c. rapid DNA amplification
   ____ polymerase chain   d. cut DNA fragments
         reaction   incorporated into plasmids
   ____ human gene therapy   e. extra bacterial genes

Selected Key Terms

cDNA *183*	human genome project *184*
cloned DNA *180*	nucleic acid hybridization *182*
DNA fingerprint *182*	PCR *181*
DNA library *180*	plasmid *179*
DNA ligase *180*	recombinant DNA technology *179*
DNA polymerase *181*	restriction enzyme *180*
DNA probe *182*	reverse transcription *183*
gene therapy *186*	RFLP *182*

Readings

Anderson, W. F. 1985. "Human Gene Therapy: Scientific and Ethical Considerations." *Journal of Medicine and Philosophy* 10:274–291.

Brill, W. 1985. "Safety Concerns and Genetic Engineering in Agriculture." *Science* 227:381–384.

Joyce, G. December 1992. "Directed Molecular Evolution." *Scientific American* 267(6):90–97.

Lowenstein, J. May 1992. "Whose Genome Is It, Anyway?" *Discover* 13: 28–31. Whose DNA will be immortalized by the human genome project?

White, R., and J. Lalouel. February 1988. "Chromosome Mapping with DNA Markers." *Scientific American* 258(2):40–48.

FACING PAGE: *Millions of years ago, a bony fish died, and sediments gradually buried it. Today its fossilized remains are studied as one more piece of the evolutionary puzzle.*

14 MICROEVOLUTION

Designer Dogs

We humans have tinkered rather ruthlessly with the modern descendants of a long and distinguished lineage. That lineage began some 40 million years ago with small, weasel-shaped, tree-dwelling carnivores, the forerunners of bears, raccoons, pandas, badgers—and dogs. About 10,000 years ago, we began domesticating wild dogs. No doubt the advantages of doing so were obvious. Times were tough in the days before supermarkets and police protection. Dogs could guard us and our possessions. They could eat and so dispose of rats and other unwelcome vermin in our shelters.

It didn't take us long to develop different varieties (breeds) through artificial selection. We picked dogs having the desired forms of traits and encouraged them to breed. We discarded others. Over time we shaped sheep-herding collies, badger-hunting dachshunds, wily retrievers, and snow-traversing, sled-pulling huskies. And at some point we began to delight in the odd, extraordinary dog. Imagine! In no time at all, evolutionarily speaking, we picked our way through the pool of variant dog genes and came up with such extremes as Great Danes and chihuahuas (Figure 14.1). Of course, our canine designs can exceed the limits of biological common sense. Think about how long a tiny, finicky-eating, nearly hairless, nearly defenseless chihuahua would last in the wild. Think about English bulldogs, bred for a short snout and compressed face. The roof of their mouth is now ridiculously wide and often flabby, so the dogs have trouble breathing. They sometimes get so short of air, they pass out.

Through our centuries-old fascination with artificial selection, we produced thousands of varieties of crop plants, cats, cattle, and birds as well as dogs. And now, through genetic engineering, we even mix genes of different species and get astounding new varieties, such as hemoglobin-producing tobacco plants (page 186).

So, when you hear someone wonder about whether "evolution" occurs, remind yourself that evolution simply means *change through time*. Our selective breeding practices give us abundant, tangible evidence that changes do, indeed, occur. *How* those changes occur in lines of descent is the subject of this chapter.

Figure 14.1 Two designer dogs. About 10,000 years ago, humans began domesticating wild dogs. From that ancestral stock, artificial selection produced diverse yet rather closely related breeds, such as the Great Dane (*legs, left*) and the chihuahua (*possibly fearful of being stepped on, right*).

KEY CONCEPTS

1. As Charles Darwin and Alfred Wallace perceived long ago, members of the same population share the same heritable traits—yet they don't look or act exactly the same. As we now know, variation in any heritable trait results when members differ in their alleles for the trait.

2. Within the population as a whole, one allele for a trait may become more or less common, relative to the other kinds, or even disappear. *Microevolution* refers to the changes in a population's allele frequencies over time.

3. Allele frequencies change through four processes. These are mutation, genetic drift, gene flow, and natural selection. Mutation alone produces *new* alleles. The other processes shuffle *existing* alleles into, through, or out of populations.

4. Natural selection is not an "agent," purposefully searching for the "best" individuals in a population. It simply is the *difference in survival and reproduction* among individuals that differ in one or more traits.

5. Genetic differences may build up between isolated populations of a species. When one population diverges genetically from the others, it may evolve into a new species. For sexually reproducing types, speciation has occurred when members of the separated populations no longer can interbreed and produce fertile offspring under natural conditions.

a

b

EMERGENCE OF EVOLUTIONARY THOUGHT

Early Beliefs and Confounding Discoveries

If you had lived in Europe in the fourteenth century, you would have been hard pressed to find a scholar who questioned the prevailing view of life's diversity. In this view, a great Chain of Being extended from the lowest forms of life to humans and on to spiritual beings. Each kind of being, or *species* as it was called, was a separate link in the chain. All the links were designed and forged at the same time, at the same center of creation, and had not changed since.

Then, in the sixteenth century, European explorers brought back descriptions of thousands of exotic species from Asia, Africa, the Americas, and islands of the Pacific. Many species were unique to isolated places. Others resembled species in distant lands (Figure 14.2). The biogeographical distribution was puzzling. How did so many species get from the center of creation to islands and other isolated places?

c

Figure 14.2 Examples of three species that are native to three geographically separate parts of the world. (**a**) The emu of Australia, (**b**) rhea of South America, and (**c**) ostrich of Africa. All three species of birds have features in common.

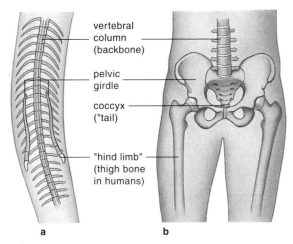

a   b

Figure 14.3 (**a**) Python bones corresponding to the pelvic girdle of other vertebrates, including humans (**b**). Small "hind limbs" protrude through skin on the underside of the snake.

Labels in figure: vertebral column (backbone); pelvic girdle; coccyx ("tail"); "hind limb" (thigh bone in humans)

In the eighteenth century, anatomists who were comparing body plans of different mammals, reptiles, and other major groups raised more questions. Consider how a human arm, whale flipper, and bat wing differ in size, shape, and function. All three body parts have similar body locations. They consist of the same tissues, arranged in the same patterns. They develop similarly in embryos. Comparative anatomists who discovered all this wondered: Why are animals as different as humans, whales, and bats so much alike?

They also discovered body parts with no apparent function. For instance, certain snakes have bones that correspond to a pelvic girdle, a set of bones to which *hind legs* attach (Figure 14.3). Snakes don't have legs, so why are the bones there? Similarly, humans have bones corresponding to a few of the *tail bones* of many other mammals. Humans don't have a tail. Why do they have parts of one?

By midcentury, geologists were adding to the confusion. They were mapping horizontal layers of sedimentary rock at sites where erosion or quarrying had cut deep into the earth. They found similar layers throughout the world. Most agreed the layers had been deposited slowly, one above the other, over time (page 211). They realized some layers contain the same kinds of **fossils**—evidence of organisms that lived in the past. For example, fossils of simple marine organisms occur in deep layers. Similar but structurally more complex fossils occur in layers above them. Fossils in uppermost layers closely resemble living marine organisms. Were fossils in the different layers somehow related?

Many scholars tried to reconcile the puzzles with the traditional view of diversity. A few thought there must have been several centers of creation, for vast oceans and impassable mountains surely would have stopped a worldwide dispersal of species from a single center. *Did species originate in more than one place?* Others wondered about the shared similarities and differences among mammals (and fossils). What if all mammals had been created according to the same plan but later were altered in different ways? *Could species become modified?* Awareness of **evolution**—of changes in lines of descent through time—was in the wind.

Voyage of the *Beagle*

In 1831, in the midst of the confusion, Charles Darwin was twenty-two years old and wondering what to do with his life. He had just earned a degree in theology from Cambridge University. Yet he was happiest when studying natural history. John Henslow, a botanist, perceived Darwin's real interests. He arranged for Darwin to become ship's naturalist aboard H.M.S. *Beagle*. Shortly thereafter, the *Beagle* set out on a five-year voyage that would take Darwin on a remarkable journey around the world (Figure 14.4).

a

Figure 14.4 (**a**) Charles Darwin and a blue-footed booby, one of the distinctive species he encountered during his five-year voyage around the world on H.M.S. *Beagle*. A replica of this ship is shown in (**b**), sailing off the coast of South America. (**c**) During stops along Argentina's coast, Darwin explored parts of the Andes. He saw fossils of marine organisms embedded in rocks 3.6 kilometers above sea level. (**d–f**) The Galápagos Islands, isolated bits of land in the ocean, far to the west of Ecuador. We now know they arose through volcanic action about 5 million years ago, so organisms could not have originated there; winds or ocean currents must have brought them. (**d**) Santa Cruz and other islands all have distinctive arrays of plants and animals.

b

c

d

e

Darwin

Wolf

Pinta

Marchena   Genovesa

EQUATOR

Santiago

Bartolomé

Seymour

Rábida   Baltra

Fernandina   Pinzón

Santa Cruz

Santa Fe

San Cristóbal

Isabela   Tortuga

Española

Floreana

Equator

Galápagos
Islands

f

Figure 14.5 (*Left*) An armadillo and (*above*) reconstruction of a glyptodont, now extinct. Their shared, unusual features and restricted geographic distribution provided Darwin with a clue that helped him develop his theory of natural selection.

Darwin's Theory Takes Form

The *Beagle* sailed first to South America, where work could be completed on mapping the coastline. During frequent stops along the coast and at various islands, Darwin had time to observe species in environments ranging from sandy shores to high mountains (Figure 14.4). He also had time to mull over a theory advanced by the geologist Charles Lyell.

According to that theory, processes now molding the earth's surface also were at work in the past. Mountain ranges rose slowly from ancient seafloors. Rain and wind eroded the mountain flanks, and sediments slowly built up in valleys and seafloors. Huge earthquakes and other catastrophic events were like abrupt punctuation points in this gradual history of geologic change.

The theory became known as **uniformitarianism**. It bothered many scholars, who believed the earth was less than 6,000 years old. Humans thought they had recorded everything that happened during those thousands of years, and in all that time, no one had ever mentioned seeing a species evolve. However, by Lyell's reckoning, it must have taken *millions* of years to mold the present landscape. Wasn't that enough time for species to evolve in splendidly different ways?

After returning to England, Darwin talked with other scholars and studied his notes for evidence of evolution. Two possibilities turned up. First, while in Argentina, he had observed fossils of glyptodonts—extinct animals that resemble living armadillos and nothing else, anywhere (Figure 14.5). If the two kinds of animals had been created at the same time, lived in the same place, and were so much alike, why is only one still living? More plausibly, the armadillos might be modified descendants of glyptodonts.

Second, as Darwin later found out, more than a dozen species of finches live on an isolated chain of islands, the Galápagos (Figure 14.6). He had observed

a

b

c

d

Figure 14.6 Examples of variation in beak shape among different finch species of the Galápagos Islands.

(**a**) *Geospiza conirostris* and (**b**) *G. scandens*, two species with a beak adapted for eating cactus flowers and fruits. Other finches have thick, strong beaks adapted for crushing cactus seeds. (**c**) *Certhidea olivacea*, a tiny tree-dwelling finch, resembles warblers in song and behavior. It uses its slender beak to probe for insects. (**d**) *Camarhynchus pallidus* feeds on wood-boring insects such as termites. It swings its small body like a woodpecker does, to hammer at bark. It does not have the woodpecker's long, probing tongue, but it has learned to break cactus spines and twigs to appropriate lengths, then hold the "tools" in its beak and use them as probes.

Figure 14.7 Alfred Wallace. Although Darwin and Wallace had worked independently, they both arrived at the same concept of natural selection. Darwin tried to insist that Wallace be credited as originator of the theory, being the first to circulate a report of his work. Wallace refused; he would not ignore the decades of work Darwin had invested in accumulating his supporting evidence.

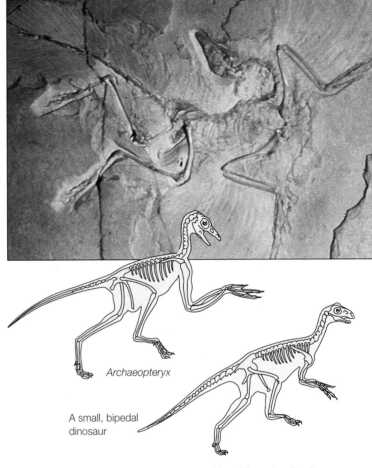

Archaeopteryx

A small, bipedal dinosaur

Figure 14.8 *Archaeopteryx*, a transitional form that lived more than 140 million years ago. It was on or very near the evolutionary road leading from reptiles to birds. The fossil might have been classified as reptilian had it not been for clear imprints of feathers in the limestone tomb.

some of them during his voyage. Each species has distinctive traits but also shares many traits with the others. Possibly they all evolved from a common ancestor.

But *how* could they evolve? A clue came from an essay by Thomas Malthus, a clergyman and economist. In Malthus's view, any population tends to outgrow its resources, and its members must compete for what is available. Darwin recalled the variations in body size, coloration, and other traits he had observed during his voyage. It dawned on him that some variations affect the ability to secure resources.

For example, one type of finch beak is suitable for crushing seeds, another type for spearing insects, and so on (Figure 14.6). Assume, as Darwin did, that members of a population compete with one another for existence. A bird born with a stronger beak can eat more hard seeds than a weak-beaked bird. If the only seeds being produced are hard, the strong-beaked bird will be more likely to survive *and reproduce*. So will strong-beaked descendants. Over time, if nature continues to "select" the trait, the population may become composed of mostly strong-beaked birds. And a population is *evolving* when traits that define it are changing.

And so Darwin recognized natural selection as a means of evolution. He worked out his evolutionary theory, then laboriously searched for flaws in his reasoning. He waited too long. In 1858 Alfred Wallace, another respected naturalist, put together the *same* the-

ory (Figure 14.7). Wallace quickly wrote up and circulated his ideas. He shared credit with Darwin when the theory was formally announced. The next year, Darwin's detailed evidence was published in book form.

Darwin's theory faced a crucial test. If organisms evolve, where were fossils of "missing links"—the transitional forms between major groups of organisms? In 1861, such a fossil was unearthed. *Archaeopteryx* appears to be a transitional form between reptiles and birds (Figure 14.8). Like modern birds, its body was covered with feathers. Like fossils of small, bipedal reptiles, it had teeth and a long, bony tail. (Bipedal means being able to walk upright, on two legs.)

Other fossils turned up during Darwin's lifetime. Even so, nearly seventy years passed before advances in genetics led to widespread acceptance of his theory of natural selection. In the meantime, his name was associated mostly with the idea that life evolves—something others had proposed before him.

Figure 14.9 Variation in shell color and banding patterns among populations of a snail species found on the Caribbean islands.

MICROEVOLUTIONARY PROCESSES

Variation in Populations

As Darwin perceived, individuals don't evolve; *populations* do. By definition, a **population** is a group of individuals of the same species occupying a given area.

Certain traits help define a population. Its members have the same overall form and appearance, as when bluejays have cobalt blue feathers, perching feet with three toes forward and one toe back, and so on. These are *morphological* traits (*morpho* means form). Also, the cells and body parts of its members operate the same way during growth, day-to-day tasks, and reproduction. These are *physiological* traits (they relate to body functions). Also, its members respond the same way to certain basic stimuli, as when humans reflexively jerk away from a spider on a wall. These are *behavioral* traits.

Yet traits vary in their *details*. Snail shells vary in patterning as well as in color (Figure 14.9). Frogs of the same population may differ in their sensitivity to win-

---

Hardy-Weinberg Rule:

In a population at genetic equilibrium, the proportions of genotypes for a locus with two alleles are

$$p^2 \; AA + 2pq \; Aa + q^2 \; aa$$

where p is the frequency of allele A, and q is the frequency of allele a. The frequencies will remain the same through the generations *if* there is no mutation, *if* the population is infinitely large and isolated from other populations, *if* mating is random, and *if* all genotypes are equally viable and fertile.

Figure 14.10 Hardy-Weinberg rule. To prove the rule stated above, let's track alleles A and a through succeeding generations of an imaginary population at genetic equilibrium. In the whole population, the frequencies of A and a must add up to 1. For example, if A occupies half of all the gene loci in the population's members and a occupies the other half, then $0.5 + 0.5 = 1$. If A occupies 90 percent of all the gene loci, then a must occupy the remaining 10 percent ($0.9 + 0.1 = 1$). No matter what the proportions of alleles A and a,

$$p + q = 1$$

When diploid organisms reproduce, each pair of alleles segregate and end up in separate gametes. Thus p is also the proportion of gametes with the A allele, and q the proportion with the a allele. To find the expected frequencies of the possible genotypes AA, Aa, and aa in the next generation, let's construct a Punnett square:

	$p\,\textcircled{A}$	$q\,\textcircled{a}$
$p\,\textcircled{A}$	$AA\ (p^2)$	$Aa\ (pq)$
$q\,\textcircled{a}$	$Aa\ (pq)$	$aa\ (q^2)$

The frequencies of genotypes add up to 1:

$$p^2 + 2pq + q^2 = 1$$

To see how the allele frequencies and genotypic frequencies remain the same through the generations, let's work through an example. We have a population of 1,000 diploid individuals that each produce two gametes:

490 AA individuals produce 980 A gametes
420 Aa individuals produce 420 A and 420 a gametes
90 aa individuals produce 180 a gametes

Notice that p, the frequency of A among the 2,000 gametes, is $(980 + 420)/2,000 = 0.7$. Also, $q = (420 + 180)/2,000 = 0.3$. These gametes combine at random to produce the next generation as given in the Punnett square, so we will have:

$$
\begin{aligned}
p^2 \quad AA &= 0.7 \times 0.7 = 0.49, \text{ or} &\quad 490\ AA \text{ individuals} \\
2pq \quad Aa &= 2 \times 0.7 \times 0.3 = 0.42, \text{ or} &\quad 420\ Aa \text{ individuals} \\
q^2 \quad aa &= 0.3 \times 0.3 = 0.09, \text{ or} &\quad 90\ aa \text{ individuals}
\end{aligned}
$$

and $p^2 + 2pq + q^2 = 0.49 + 0.42 + 0.09 = 1$.

ter cold or in their success at courtship. Pigeon feathers may be gray, brown, or blue (Figure 1.8). Human hair varies in color, texture, amount, and distribution. And these examples only hint at the immense underlying genetic variation in populations.

Sources of Variation. The members of a population generally have inherited the same number and kinds of genes. But remember, a given gene may have slightly different molecular forms, called **alleles**. Different combinations of alleles are inherited, and this leads to variations in traits. Whether your hair is black, brown, red, or blond depends on *which* alleles of certain genes you inherited from your mother and father. As described in earlier chapters, five different events contribute to that mix of alleles:

1. Gene mutation (produces new alleles)

2. Crossing over at meiosis (leads to new combinations of alleles in chromosomes)

3. Independent assortment at meiosis (leads to mixes of maternal and paternal chromosomes in gametes)

4. Fertilization (puts together combinations of alleles from two parents)

5. Changes in chromosome structure or number (lead to the loss, duplication, or alteration of alleles)

Of these events, mutation alone *creates* alleles. The rest shuffle *existing* alleles into new combinations. But what a shuffle! For example, a human gamete ends up with one of 10^{600} possible combinations of alleles. Not even 10^{10} humans are alive today. So unless you have an identical twin, it's extremely unlikely that another person with your exact genetic makeup has ever lived—or ever will.

Far more genetic variation is possible than can ever be expressed in the individuals of any population.

Notice that the allele frequencies have not changed:

$$A = \frac{2 \times 490 + 420}{2,000 \text{ alleles}} = \frac{1,400}{2,000} = 0.7 = p$$

$$a = \frac{2 \times 90 + 420}{2,000 \text{ alleles}} = \frac{600}{2,000} = 0.3 = q$$

The genotypic frequencies have not changed either, and they will stay the same over succeeding generations as long as the assumptions stated in the green box hold true. You can verify this by calculating the allele frequencies in the next generation of gametes:

F_1 genotypes: 0.49 *AA*   0.42 *Aa*   0.09 *aa*

Gametes: (A) (A) (A) (a) (a) (a)

0.49 + 0.21   0.21 + 0.09

0.7 *A*   0.3 *a*

which is back where we started from. Because the allele frequencies are exactly the same as those of the original gametes, they will yield the same frequencies of genotypes as in the second generation. You could go on with the calculations until you ran out of paper, or patience. As long as the assumptions are true for the population, you would end up with the same results stated in the green box. This is the theoretical possibility known as genetic equilibrium, or Hardy-Weinberg equilibrium.

Genetic Equilibrium. Suppose a type of snail in your garden has white or yellow shells. You count up many more yellow shells, so you assume more snails carry the "yellow" allele. With genetic analysis, you might even discover the **allele frequencies**—the abundance of each kind of allele in the whole population.

Knowing the allele frequencies allows you to track whether the population is evolving. You start with the "Hardy-Weinberg formula," which helps set up a theoretical reference point for measuring the rates of change. At that point, **genetic equilibrium**, allele frequencies for the trait you are studying would be stable from one generation to the next. The population would *not* be evolving. As Figure 14.10 makes clear, this happens only when five conditions are being met:

1. No genes are undergoing mutation.

2. The population is very, very large.

3. It is isolated from other populations of the species.

4. All members survive and reproduce (there is no natural selection).

5. Mating is random.

This situation rarely (if ever) exists, at least in nature. Mutation shifts allele frequencies over long time spans. Three other processes—gene flow, genetic drift, and natural selection—may shift them within a few generations, as you will now see.

Microevolution Defined

Microevolution refers to changes in allele frequencies brought about by mutation, gene flow, genetic drift, and natural selection. These microevolutionary processes will now be described.

Mutation

A **mutation** is a heritable change in DNA. In evolutionary terms, it is a significant event. Mutations are the source of alleles that have been accumulating in different lineages for more than 3 billion years.

In terms of individual lifetimes, a mutation is a rare event. Whether it proves to be harmful, neutral, or beneficial depends on how its product interacts with other genes and with the environment (Chapter 9).

Harmful gene mutations alter traits in such a way that an individual cannot survive or reproduce. A **lethal mutation** always leads to death. Try bending your nose. Cartilage, a tissue with many roles in the body, makes it flexible. The product of a mutated gene affects cartilage formation. Blocked nostrils and thickened ribs are among its lethal effects.

A **neutral mutation** is neither harmful nor helpful to the individual. For example, even if you have the mutated gene that leads to attached earlobes (page 122), you still should be able to make your way in the world. You might even have a mutated gene with unpleasant but not lethal effects. Who knows. If environmental conditions change, that not-great but neutral mutation might turn out to be beneficial.

Other mutations enhance prospects of surviving and reproducing. For example, a mutated regulatory gene might make a plant grow larger or faster, and so give it the best shot at sunlight and nutrients.

Mutations are the source of new alleles, hence of heritable traits. A mutation may be harmful, neutral, or beneficial, depending on how the altered gene product performs under prevailing conditions.

Gene Flow

Allele frequencies change as individuals leave a population (emigration) or new individuals enter it (immigration). This physical flow of alleles, or **gene flow**, helps keep neighboring populations genetically similar. Over time, it tends to counter the differences between populations that are brought about through mutation, genetic drift, and natural selection.

Think of the gene flow among human populations, in this age of international travel. Or think of scrubjays dispersing acorns when they store nuts for the winter.

Figure 14.11 One travel agent for gene flow among oak populations. Scrubjays hoard acorns in their home territory, but they might shop at nut-bearing trees up to a mile away. Some acorns contribute to the allele pool of an oak population some distance away from the parent tree.

Each fall they may make hundreds of round trips from acorn-bearing oak trees to soil in their home territories, which may be a mile away (Figure 14.11). The alleles flowing in with "immigrant acorns" help neutralize variations that arise among neighboring stands of oaks.

Gene flow is the physical movement of alleles into and out of populations.

Genetic Drift

With **genetic drift**, allele frequencies change randomly through the generations because of chance events alone. Often, the change is most rapid in small populations. Genetic drift may decrease variation *within* a population; it may increase variation *between* populations.

Suppose only some members of a small population carry an allele A. Whether you call it chance or bad luck, some of the carriers don't mate and the others die in accidents. It makes no difference whether A is the "best" (most advantageous) allele for a given trait. Within a few generations, it disappears (Figure 14.12). The A allele might still be around if the population had been large and hundreds carried it.

Founder effects and bottlenecks are two extreme cases of genetic drift. Both may result in severely limited genetic variation, because a population originates or is rebuilt from very few individuals.

With the **founder effect**, a few individuals leave a population and manage to establish a new one. Compared to the population left behind, the founders happen to carry fewer (or more) alleles for certain traits, such as red-tipped wings or yellow flowers. Simply by chance, the allele frequencies in the new population differ from the old. The outcome of this chance event is pronounced on isolated islands. Long ago, wind and birds carried a few seeds to the Hawaiian Islands. The resulting plant populations show limited variation.

With **bottlenecks**, disease, starvation, or some other stressful event nearly wipes out a large population. The population recovers, but abundances of its various alleles have been altered at random. Before the turn of the century, hunters killed all but twenty of a large population of northern elephant seals. The population recovered. Researchers who studied twenty-four genes found that the 30,000 members alive at the time carried the *same* alleles. The absence of variation suggests a number of alleles were lost during the bottleneck. The *Focus* essay describes what such losses may mean for species vulnerable to extinction.

With genetic drift, allele frequencies change over the generations due to chance events alone.

Figure 14.12 Genetic drift in two populations of a species. For comparative purposes, assume each population starts out with three individuals (designated type A) who bear allele *A* and three other individuals (type B) who bear allele *B*. Assume each individual who reproduces leaves two identical offspring. Finally, assume half the offspring of each generation die before they are old enough to reproduce (population size remains constant). But *which* ones live or die is random—you simply toss a coin to determine the survivors of each generation.

In (**a**), the blue numbers to the right of the diagram signify the number of individuals bearing allele *A* who survived in each generation. The relative abundances of the two types of individuals change until allele *A* disappears and allele *B* is entrenched.

Focus on the Environment

Genes of Endangered Species

Barbara Durrant and many other biologists at zoos, private conservation centers, and wildlife refuges are scrambling to buy time for the world's most **endangered species**. Populations of such species are so small, they are poised at the brink of extinction. The biologists face daunting problems, including the absence of genetic diversity.

Consider cheetahs, which must have gone through a severe bottleneck that drastically reduced their numbers and resulted in extreme genetic uniformity. These sleek, swift cats are so genetically uniform that a patch of skin from one can be successfully grafted onto another, even when their lineages are not closely related. (In other species, grafts rarely work, even among littermates.)

Today, only 20,000 or so cheetahs exist in the wild and in zoos. In the absence of genetic variation, their populations are extremely susceptible to abnormalities, diseases, and environmental changes. For example, up to 70 percent of a typical male cheetah's sperm may be abnormal, perhaps as a result of inbreeding following a bottleneck.

Intense hunting and other human activities have endangered another big cat, the Florida panther. With fewer than fifty members, the last remaining population has become highly inbred. Parents may mate with their offspring if no other mates are available. Researchers at the National Zoo retrieve DNA from unfertilized eggs of "road-killed" female panthers. With genetic engineering, the DNA may find use in captive breeding programs.

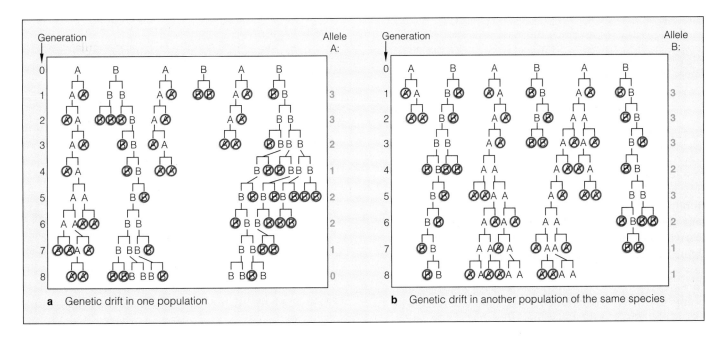

a Genetic drift in one population

b Genetic drift in another population of the same species

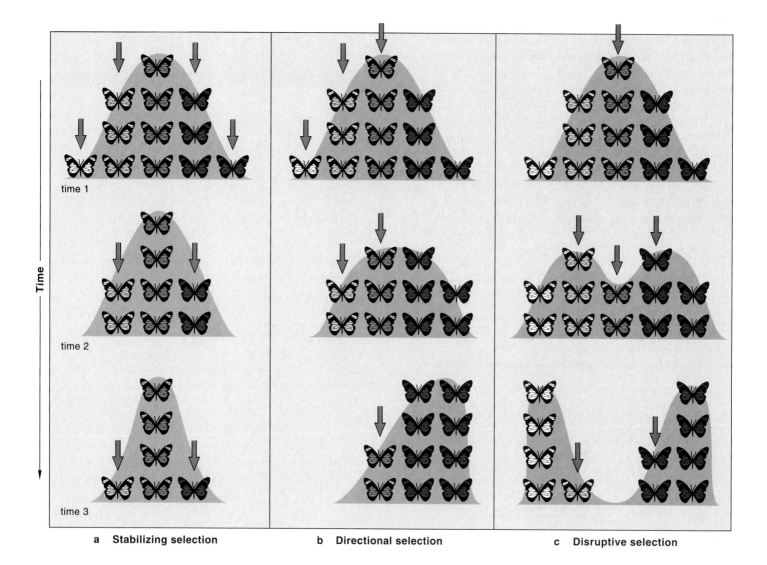

Figure 14.13 Three modes of natural selection, using the phenotypic variation of a small population of butterflies as the example. The bell-shaped curve represents the range of continuous variation in wing color. The most common forms (powder blue) are between extreme forms of the trait (white at one end of the curve, deep purple at the other). Orange arrows signify which forms are being selected against over time.

Natural Selection

Natural selection probably accounts for more changes in allele frequencies than any other microevolutionary process. Its occurrence in nature is well documented; hundreds of field studies of microorganisms, plants, and animals show that the process does indeed occur. Darwin was able to explain natural selection after correlating his understanding of inheritance with certain features of populations and the environment. Before we consider examples of the process, let's review the main points of Darwin's correlation in light of modern genetics, as expressed in the following list.

1. Individuals of a population vary in form, function, and behavior.

2. Much of the variation in a population is *heritable*. This means more than one kind of allele exists for genes that give rise to the traits.

3. Some forms of a trait are more *adaptive* than others. This means they improve chances of surviving and reproducing.

4. Natural selection is the *difference in survival and reproduction* that has occurred among individuals that differ in one or more traits.

5. A population is *evolving* when some forms of a trait are becoming more or less common, relative to the other forms. The shifts are evidence of changes in the relative abundances of alleles for that trait.

6. Life's diversity is the outcome of changes in allele frequencies in different lines of descent over time.

a Fossil　　　　**b** Living representatives of lineage, about 6 inches tall

Figure 14.14 (**a**) A fossil of the sphenophyte lineage, which extends back more than 380 million years. (**b**) The only living members of this group are horsetails (*Equisetum*).

EVIDENCE OF NATURAL SELECTION

Natural selection has different outcomes. As Figure 14.13 shows, it may *maintain* (stabilize) an existing range of traits among individuals of a population. It also may *shift* the range in some direction or *disrupt* it.

Stabilizing Selection

In **stabilizing selection**, the most common forms of a trait in a population are favored (Figure 14.13*a*). Over time, alleles for uncommon forms are eliminated. Therefore, this mode of selection tends to counter the effects of mutation, genetic drift, and gene flow.

Stabilizing selection may account for the persistence of certain forms of traits over many millions of years. We find impressive cases of this in the plant kingdom. You may have seen horsetails (*Equisetum*) growing along roadbeds. These plants retain traits of relatives that were abundant and diverse some 380 million years ago (Figure 14.14).

From studies conducted in diverse societies over the past few centuries, it appears that stabilizing selection favors human newborns who weigh about 7 pounds, on the average. Newborns weighing significantly more or less than this are more likely to die soon after birth (Figure 14.15).

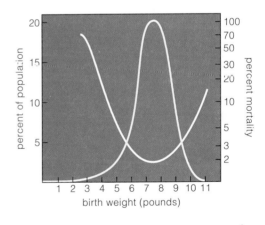

Figure 14.15 Weight distribution for 13,730 human newborns (yellow curve) correlated with their survival rate (white curve). Here, stabilizing selection favors a birth weight between 7-1/2 and 8 pounds and operates against newborns whose weight is significantly higher or lower than this.

a

b

Figure 14.16 Example of variation that is subject to directional selection in changing environments. (**a**) The light-colored and dark-colored forms of the peppered moth are resting on a lichen-covered tree trunk. (**b**) This is how they appear on a soot-covered tree trunk, which was darkened by industrial air pollution.

Directional Selection

In **directional selection**, allele frequencies shift in a steady, consistent direction. This type of shift is a response to a directional change in the environment or to new environmental conditions. Forms of traits at one end of the range of variation become more common than midrange forms (Figure 14.13*b*).

Peppered Moths. In England, directional selection has affected about a hundred moth species, including the peppered moth (*Biston betularia*). Peppered moths are active at night. They rest during the day on tree trunks, where they are vulnerable to bird predators.

Before the industrial revolution, a speckled light-gray form of the moth was common and a dark-gray form was extremely rare. Before that time, light-gray speckled lichens grew on tree trunks. Light moths resting on the lichens were hidden, but dark ones were easy prey (Figure 14.16). Then soot and other pollutants from

factories started killing the lichens and darkening the trunks. Between 1848 and 1898, dark moths became more common in the altered environment. They blended with the sooty background; light moths didn't. Now the dark coloration was the most adaptive, and allele frequencies in the populations changed.

In 1952, strict pollution controls went into effect. Lichens made comebacks, and tree trunks are largely soot-free. As you might predict, the frequency of dark moths is now declining.

Insecticide Resistance. Insect populations are getting better and better at resisting insecticides, and this is an outcome of directional selection. The first application of a chemical brew kills most of the targeted insects. But a few insects may survive. Some aspect of their structure, physiology, or behavior allows them to resist the chemical effects. If the resistance is heritable, resistant individuals will become more common in the next generation, more common in the next, and so on.

When farmers hit the resistant forms more heavily and more often, the chemicals serve as selective agents. They actually favor the resistant forms! Today, crop damage from resistant strains of insects is greater than it was before the widespread use of insecticides.

Disruptive Selection

In **disruptive selection**, forms at both ends of the range of variation are favored and intermediate forms are selected against (Figure 14.13*c*). This happened to a small population of finches on one of the Galápagos Islands. As you saw earlier, the finches differ from one another in many traits, including beak size and shape. At one end of the phenotypic range are birds with longer beaks, which are used to open cactus fruits and expose the seeds. At the other end are birds with deep, wide beaks. They can crack hard cactus seeds on the ground or strip away tree bark to get at insects. Recently, researchers observed the finches during a severe drought, when seeds and a few wood-boring insects were the only types of food available. At that time, birds with extreme beak variations survived at a greater frequency than birds in between.

Maintaining Two or More Forms of a Trait

Recall that stabilizing selection maintains the most common form of a trait, generation after generation. Stabilizing selection also can maintain *two or more* forms of a trait in fairly steady proportions over time. This balancing act is called **balanced polymorphism**. Often it exists when heterozygotes (individuals with a pair of non-identical alleles for a trait) are favored over homozygotes (with identical alleles). One allele helps the

Sickle-Cell Anemia—Lesser of Two Evils?

Sickle-cell anemia is a potentially severe human genetic disorder. It results from a mutation in the allele that codes for hemoglobin (page 32). In tropical and subtropical regions of Africa, the harmful allele (HbS) remains in the population along with the normal one (HbA). The HbS/HbS homozygotes often die in their early teens or early twenties. Yet nearly a third of the population are HbS/HbA *heterozygotes*. Notice that the harmful allele is maintained at high frequency *when it is paired with a non-identical allele* (HbS with HbA). This balancing act is an outcome of stabilizing selection. It is most pronounced in regions where malaria is most prevalent.

A parasitic protistan causes malaria (page 262). The parasite is transmitted to humans by a mosquito that is most common in the tropics and subtropics. Individuals who don't carry the HbS allele are far more likely to survive and reproduce than individuals who do—as long as

they don't get malaria. The combination of abnormal and normal hemoglobin molecules has an interesting effect. It interferes with the spread of the parasite through the body. As a result, heterozygotes are more apt to survive severe infections. In one study, there were *twice* as many severe or fatal infections among homozygotes.

Thus the persistence of a harmful allele becomes a matter of relative evils. It hurts the individual in one situation, yet has adaptive effects in a different situation.

In Central Africa, malaria has been an agent of selection for less than 2,000 years. In tropical and subtropical regions of the Middle East and Asia, it has been around much longer. Even though the sickle-cell trait occurs at high frequencies in these regions also, the symptoms are not as pronounced as they are in Africa. Alleles of other genes must be countering the serious effects of the HbS allele.

individual survive a certain environmental condition; its nonidentical partner helps the individual get through a different one. The *Focus* essay describes how a persisting combination of two different alleles helps maintain a severe genetic disorder, sickle-cell anemia, in some parts of the world.

Natural selection also works to alter or maintain differences in appearance between the males and females of a species. Such differences are called **sexual dimorphism**. Many birds and mammals are striking examples (Figure 14.17). As Darwin noticed, in some species of birds and mammals, the males are larger, more splendidly colored and patterned, and more aggressive than the females. Usually, females are the agents of selection for these traits. By choosing mates, they directly affect reproductive success. This is a form of **sexual selection**, which is based on any trait that gives an individual a competitive edge in mating and producing offspring. Chapter 38 gives additional examples.

Figure 14.17 Result of sexual selection. Male bird of paradise (*Paradisaea raggiana*) engaged in a spectacular courtship display. He caught the eye (and, perhaps, sexual interest) of a female. Males of this species compete fiercely for females, which serve as selective agents.

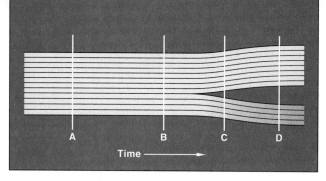

Figure 14.18 Divergence leading to speciation. Because evolution is gradual in this diagram, we cannot say at any one point in time that there are now two species rather than one. Each horizontal line represents a different population. At time A, there is only one species. At D, there are two. After time B, divergence has begun, although it is far from complete.

SPECIATION

Defining the Species

For sexually reproducing organisms, a **species** is a unit consisting of one or more populations. Individuals in each unit are able to interbreed and produce fertile off-spring under natural conditions, but they cannot do so with members of other species. No matter how diverse those individuals become, they remain members of the same species as long as they continue to interbreed successfully and share a common pool of alleles.

A species is a unit of one or more populations of individuals that can interbreed under natural conditions and produce fertile offspring, and that are reproductively isolated from other such units.

Divergence—On the Road to New Species

Imagine that some birds of a population on one island fly off to a distant island, stay there, and go about reproducing. They have become a "local" breeding unit, a small population with its own pool of alleles. After that, no birds fly one way or the other, so no genes flow either way. An absence of gene flow between populations is called **reproductive isolation**. At that point, mutation, natural selection, and genetic drift begin to operate independently in both populations. And they change the allele frequencies of both in different ways.

An accumulation of differences in allele frequencies between reproductively isolated populations is called **divergence**. It is a genetic branching that cannot be reversed. If differences in allele frequencies continue to build up, divergence may even prove to be the first step on the road to **speciation**, the evolutionary process by which species originate. For if the differences between reproductively isolated populations become great enough, their members will not be able to interbreed successfully under natural conditions. The populations will have become separate species (Figure 14.18).

Speciation Routes

Geographic Isolation. Populations of most species are not stretched out continuously, with one merging into the others. Most often they are isolated geographically to some extent, with gene flow being more of an intermittent trickle than a steady stream. But sometimes barriers form and shut off even the trickles. This can happen rapidly, as when a major earthquake changed the course of the Mississippi River in the 1800s and isolated some populations of insects that could not swim or fly. Geographic isolation also happens slowly. Millions of years ago, owing to long-term shifts in rainfall, vast forests in Africa gave way to grasslands, and sub-populations of forest-dwelling apes became isolated from one another. Regardless of how fast it happens, once geographic isolation is absolute, genetic drift or selection may lead to divergence, then to speciation. One of the subpopulations of African apes may have started the divergence that led to modern humans.

Isolation Within a Population. Speciation may follow after ecological, behavioral, or genetic barriers arise *within* the boundaries of a single population. Suppose a regulatory gene in an insect undergoes mutation and slows down its reproductive clock. The insect is able to reproduce asexually, and it produces many genetically identical offspring. The frequency of the mutated allele increases in the population—but its bearers become sexually active when their neighbors are not sexually *receptive*. They breed only with one another—and so are reproductively isolated. The stage is set for divergence and speciation.

Speciation sometimes occurs instantaneously within a population. **Polyploidy**, recall, is a condition in which offspring inherit three or more of each type of chromosome characteristic of the parental stock. This happens, for example, after chromosomes fail to separate properly during meiosis (page 148). It also happens when a reproductive cell duplicates its DNA but fails to divide, then goes on to function as a gamete.

Usually, fusion of animal gametes with mismatched chromosome numbers produces an embryo that cannot survive. Those that do survive commonly have severe genetic disorders, shorter life expectancies, or both.

By contrast, about half of all flowering plants are polyploid species. Many can self-fertilize or reproduce asexually, and they produce many offspring having the novel chromosome number. The extra chromosomes pair successfully at meiosis, and the extra genes do no harm. In such plants, speciation could have been instantaneous. Polyploidy also may result in speciation when it is followed by cross-fertilization between different plant species (Figure 14.19). However, as noted, most hybrids between species are sterile.

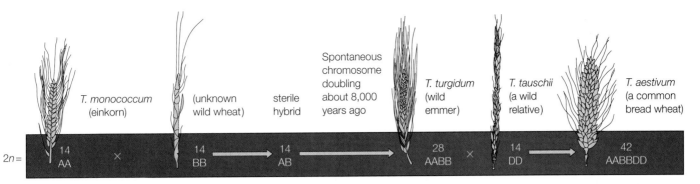

$2n =$

| 14 AA | × | 14 BB | → | 14 AB | → | 28 AABB | × | 14 DD | → | 42 AABBDD |

T. monococcum (einkorn)   (unknown wild wheat)   sterile hybrid   Spontaneous chromosome doubling about 8,000 years ago   T. turgidum (wild emmer)   T. tauschii (a wild relative)   T. aestivum (a common bread wheat)

a About 11,000 years ago, humans start cultivating wild wheats. The species *Triticum monococcum* has diploid number 14 (two sets of 7 chromosomes, shown as 14AA). It hybridizes with another species that has the same chromosome number.

b AB hybrid offspring are sterile but self-fertilizing; an interbreeding population of AB plants arises by asexual reproduction. About 8,000 years ago, by unknown events, polyploidy arises in the population. Some plants (*T. turgidum*) are tetraploid (AABB), with a chromosome number of 28 (two sets of 14). They are fertile (A chromosomes can pair with each other, and so can B chromosomes, during meiosis.)

c Later, an AABB plant hybridizes with *T. tauschii*, a wild relative with a diploid number of 14 (two sets of 7 DD). Today, populations of the hybrid descendants, *T. aestivum*, provide wheat for bread. Their chromosome number is 42 (six sets of 7 AABBDD).

Mechanisms That Prevent Interbreeding

In certain aspects of body structure, functioning, or behavior, each species is like no other. And some of its unique features are reproductive isolating mechanisms. They prevent or work against gene flow and so inhibit interbreeding between species.

Isolation of Gametes. Certain isolating mechanisms work before or during the time of fertilization. Consider two species of sea urchins. Fertilization occurs outside the body of these marine animals. Both species release eggs and sperm at the same time, in the same place, but cross-fertilization is rare. Probably the gametes of one species are not equipped to respond to molecular cues on the gametes of the other species.

Structural Isolation. Differences in body parts may prevent interbreeding. Two sage species depend on insects to carry pollen from flower to flower. The flowers of the two species are not the same size, and they are arranged differently. Each kind of flower has a "landing platform," and only one kind of pollinator can fit on it. Thus pollinators of other flowers tend not to cross-fertilize the plants (Figure 14.20).

Isolation in Time. As described earlier, differences in reproductive timing may serve as an isolating mechanism. For most animals and plants, mating or pollination is a seasonal event of relatively short duration, sometimes less than a day. Even closely related species may be isolated simply because their times of reproduction do not coincide. Consider the cicada, a type of insect that spends nearly all of the life cycle in suspended animation, underground. One species emerges and reproduces every 13 years. The other species does this every 17 years. The possibility of their meeting arises only once every 221 years!

Unworkable Hybrids. Suppose one organism manages to cross-fertilize a different kind. As mentioned above, flowering plants do this often and well. Most often, however, isolating mechanisms will kick in when

Figure 14.19 Proposed speciation in wheat by polyploidy and hybridizations. Wheat grains 11,000+ years old have been found in the Near East. Diploid wild wheats still grow there.

a   b

Figure 14.20 Reproductive isolation between two sage species, *Salvia mellifera* and *S. apiana*. (**a**) Flowers of one species have a small landing platform for small or medium-size pollinators. (**b**) Flowers of the other have a large landing platform and long pollen-bearing stamens that extend above the nectar. Although small bees can land on this larger platform, they won't brush against the stamens as larger pollinators do.

the hybrid embryo is developing. It may be that the combined genetic instructions don't work, as when certain control mechanisms, gene products, or both don't mesh. Hybrids that do live commonly are weak, and their survival rate is not good. A few types are sturdy but sterile. This is true of mules, which result from a cross between a female horse and a male donkey.

Behavioral Isolation. Differences in behavior are isolating mechanisms among related species in the same territory. For many bird species, complex courtship rituals precede mating (Chapter 38). A female typically ignores the song, wing-spreading, head-bobbing, and prancing of a male of another species. Only females of his own species are phenotypically equipped to recognize his efforts as a sexual overture.

Table 14.1	Summary of Major Microevolutionary Processes
Mutation	A heritable change in DNA
Gene flow	Change in allele frequencies as individuals leave or enter a population
Genetic drift	Random fluctuation in allele frequencies over time, due to chance occurrences alone
Natural selection	Change or stabilization of allele frequencies due to differences in survival and reproduction among variant members of a population

SUMMARY

1. Awareness of evolution—changes in lines of descent over time—emerged through the following:

a. Comparisons of body structure and patterning for major groups of animals (comparative anatomy).

b. Questions about the world distribution of plants and animals (biogeography).

c. Observations of fossils of different types (structurally simple to complex) buried in a series of horizontal layers of the earth (most ancient to more recent layers).

2. Individuals of a population share the same genes. But the genes come in different allelic forms, and this leads to variations in traits.

3. A population is evolving when some forms of a trait are becoming more or less common, relative to the other kinds. The shifts are evidence of changes in the relative abundances of alleles for that trait.

4. Allele frequencies change as a result of four microevolutionary processes: mutation, gene flow, genetic drift, and natural selection (Table 14.1).

5. Mutations (heritable changes in DNA) are the only source of *new* alleles.

6. Gene flow is a change in allele frequencies brought about by the physical movement of alleles into and out of a population (by immigration and emigration).

7. Genetic drift is a change in allele frequencies over the generations due to chance events alone.

8. Natural selection is a difference in survival and reproduction among members of a population that vary in one or more traits. (One form of a trait may be more adaptive under prevailing conditions. Its bearers tend to survive and reproduce more often, so it becomes more common than other forms of the trait.)

9. With natural selection, the range of variation for a trait may shift steadily in one direction, it may be disrupted at or near midrange, or it may be maintained over time.

10. A species is a unit of one or more populations of individuals that can interbreed and produce fertile offspring under natural conditions, and that are reproductively isolated from other populations. (This definition fits sexually reproducing species only.)

11. Divergence is a buildup of differences in allele frequencies between populations that have become reproductively isolated from one another.

12. Speciation results when the differences between reproductively isolated populations become so great that their members will not be able to interbreed successfully under natural conditions.

Review Questions

1. In what respect does mutation differ from gene flow, genetic drift, and natural selection? *197, 198–199*

2. At one time, natural selection was popularly defined as "the survival of the fittest." Explain how the process of genetic drift might skewer that definition. *198–199*

3. Define these terms:
   a. gene flow and genetic drift *198*
   b. bottleneck and founder effect *198–199*
   c. stabilizing, directional, and disruptive selection *201–202*
   d. balanced polymorphism and sexual dimorphism *202–203*
   e. divergence and speciation *204*

4. Describe three reproductive isolating mechanisms that prevent interbreeding between species. *205*

5. This photograph shows a brilliantly hued male sugarbird and a subdued-hued female. Explain this difference between the two in terms of selection theory. *205*

1. Individuals don't evolve; _____ do.

2. The allele responsible for sickle-cell anemia first appeared in tropical and subtropical regions of Asia, the Middle East, and Africa. It entered the United States population when individuals were forcibly brought over from Africa prior to the Civil War. In microevolutionary terms, this is an example of _____ .
   a. mutation
   b. genetic drift
   c. gene flow
   d. natural selection

3. Uniformitarianism is _____ .
   a. an absence of allelic diversity
   b. uniformity among observable traits
   c. an outcome of stabilizing selection
   d. a theory in geology

4. Allele frequencies change as a result of _____ .
   a. mutation
   b. gene flow
   c. genetic drift
   d. natural selection
   e. all of the above

5. The only source of new alleles is _____ .
   a. mutation
   b. gene flow
   c. genetic drift
   d. natural selection
   e. all of the above

6. Existing alleles are shuffled into, through, or out of populations by _____ .
   a. mutation
   b. gene flow
   c. genetic drift
   d. natural selection
   e. b, c, and d only

7. A bottleneck is _____ .
   a. a new population established by individuals who left an old one
   b. a chance event that wipes out nearly all members of a population
   c. an extreme case of genetic drift
   d. both b and c

8. Disruptive selection _____ .
   a. eliminates uncommon forms of alleles
   b. shifts alleles in a consistent direction
   c. doesn't favor intermediate forms of traits
   d. works against adaptive traits

9. Speciation is _____ .
   a. an extinction of one population that makes way for another
   b. a buildup of environmental factors leading to geographic isolation
   c. a process whereby species originate
   d. the means by which gene frequencies change
   e. both c and d

10. Reproductive isolation is an absence of _____ .
    a. divergence
    b. mutation
    c. gene flow
    d. genetic drift
    e. all of the above

11. Match the evolution concepts appropriately.
    _____ gene flow
    _____ sexually reproducing species
    _____ natural selection
    _____ mutation
    _____ genetic drift

    a. source of new alleles
    b. changes in a population's allele frequencies due to chance alone
    c. one or more successfully interbreeding populations
    d. immigration, emigration change allele frequencies
    e. difference in survival and reproduction among variant members of population

Selected Key Terms

allele *197*
allele frequencies *197*
balanced polymorphism *202*
bottleneck *199*
directional selection *202*
disruptive selection *202*
divergence *204*
endangered species *199*
evolution *192*
fossil *192*
founder effect *198*
gene flow *198*
genetic drift *198*
genetic equilibrium *197*

lethal mutation *198*
microevolution *198*
mutation *198*
natural selection *200*
neutral mutation *198*
polyploidy *204*
population *196*
reproductive isolation *204*
sexual dimorphism *203*
sexual selection *203*
speciation *204*
species *204*
stabilizing selection *201*
uniformitarianism *194*

Readings

de Blieu, J. *Meant to Be Wild*. 1991. Golden, Colorado: Fulcrum Publishing. Describes captive breeding programs to save endangered species.

Cook, L.; G. Mani; and M. Varley. 1986. "Postindustrial Melanism in the Peppered Moth." *Science* 231:611–613.

Moorhead, A. 1969. *Darwin and the Beagle*. New York: Harper & Row.

Stone, I. 1980. *The Origin*. New York: Doubleday.

15 LIFE'S ORIGINS AND MACROEVOLUTION

Of Floods and Fossils

About 500 years ago, Leonardo da Vinci was brooding about seashells entombed in the layered rocks of northern Italy's high mountains, hundreds of kilometers from the sea. How did they get there? If he accepted the traditional explanation, he would have to agree that turbulent floodwaters deposited shells in the mountains during the Great Deluge (Figure 15.1). But many shells were thin, fragile—and intact. Surely they would have been battered to bits if they had been swept across such distances, then up the mountains.

Da Vinci also brooded about the rock layers. They were stratified (stacked like cake layers), and some had shells but others had none. Then he remembered how large rivers deposit silt during spring floods. Had the layers slowly accumulated one atop the other, as a series of silt deposits in the past? If so, the shells in the mountains would be evidence of a series of vanished communities of organisms that had been gradually buried *in the sea!* Da Vinci did not announce his novel idea, perhaps knowing it would have been met with deafening silence, imprisonment, or worse.

By the 1700s, fossils were being accepted as evidence of past life. They were still interpreted in a traditional way, as when a Swiss naturalist excitedly unveiled the remains of a giant salamander and announced they were the skeleton of a man who drowned in the Deluge.

By midcentury, however, scholars began to question such interpretations. Extensive mining, quarrying, and canal excavations were under way. The diggers were finding similar rock layers and similar fossil sequences in distant places, such as the cliffs on both sides of the English Channel. And some scholars started analyzing the discoveries for possible connections between earth history and the history of life.

Ever since, fossils have been analyzed in more and more refined ways. Together with biochemical studies and other modern sources of information, they provide good evidence of evolution over vast spans of time—*of changes in the geologic stage, and changes in the organisms that have marched across it.*

Figure 15.1 (*Left*) From the Sistine Chapel, Michelangelo's painting of the onset of a catastrophic flood, traditionally called the Great Deluge. Puzzling events of this magnitude actually punctuated geologic time, although they have been explained in different ways throughout human history. (*Right*) A modern-day photographer captures the intricate structural pattern of fossilized ammonite shells. About 65 million years ago, all ammonites perished, along with many other groups of organisms. That mass extinction is but one piece of the macroevolutionary puzzle.

KEY CONCEPTS

1. In the evolutionary view, all species that have ever lived are related—some closely, others remotely. This is because each *new* species had to evolve from variant individuals of *existing* species, starting with the first living cells to appear on earth.

2. The fossil record, geologic record, and radioactive dating of rocks yield evidence of life's evolutionary history. Anatomical and biochemical comparisons within and between major groups of organisms provide insight into that history.

3. Lines of descent (lineages) differ from one another in terms of when their member species originated, how rapidly they evolved, and how long they have persisted. The term macroevolution refers to the patterns, trends, and rates of change among lineages over geologic time.

4. The available evidence indicates that life originated more than 3.8 billion years ago. Its origin and subsequent evolution have been linked to the physical and chemical evolution of the earth itself.

5. Many evolutionary trends ended or shifted in new directions as a result of mass extinctions and adaptive radiations. In a mass extinction, many major lineages perish abruptly and simultaneously. In an adaptive radiation, a lineage rapidly branches and fills the environment with new species.

6. Today, as in the past, species are identified and assigned names. Researchers attempt to define the evolutionary connections among species. The presumed connections are taken into account in classification schemes, which organize species into ever broader groupings, from genera on up to kingdoms.

a

b

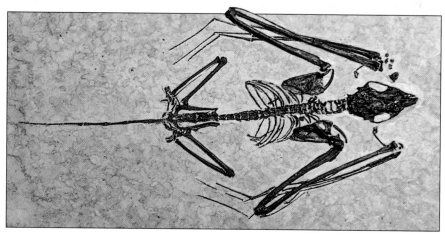

c

EVIDENCE OF MACROEVOLUTION

The history of life spans nearly 4 billion years. It is a story of how species originated, branched out through the environment, and either endured or became extinct. We call the large-scale patterns, trends, and rates of change among groups of species **macroevolution**. The groupings include all the different genera, families, phyla or divisions, and so on up to the most inclusive groups of species, the kingdoms.

Reconstructing the history of life requires knowledge of microevolutionary processes, including mutation and natural selection. This knowledge gives us confidence in a principle that can be used to make sense of scattered and sometimes puzzling scraps of evidence of past life: *New* species only evolve from variant individuals of *existing* species. With this principle to guide us, the task becomes one of identifying and sorting out lines of descent, or **lineages**, that have connected all species since the origin of life.

The fossil record and earth history yield evidence of this evolutionary connectedness. So do anatomical and biochemical comparisons.

Evolution proceeds by modification of already existing species. Therefore, a continuity of relationship exists among all species that have ever appeared on earth.

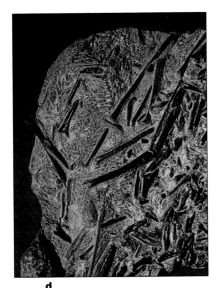

Figure 15.2 Representative fossils. (**a**) One of the most magnificent, complete fossils ever recovered—evidence of an ichthyosaur, a dolphin-like marine reptile that lived about 200 million years ago. (**b**) Fossilized leaves of an ancient tree, *Archaeopteris*. Possibly the gymnosperms (including pine trees) and angiosperms (flowering plants) are distinctly related to this genus. (**c**) The complete skeleton of a bat that lived 50 million years ago. (**d**) Another good find: the fossilized skeletons of several ducklike birds.

d

Figure 15.3 A splendid slice through time: the Grand Canyon of the American Southwest, a region that was once part of an ocean basin. Layers of sedimentary rocks were laid down gradually over hundreds of millions of years. Tectonic forces lifted them above sea level. Over time, the erosive force of rivers cut deep canyon walls.

Fossils—Evidence of Ancient Life

The Fossil Record. "Fossil" comes from a Latin word for something that has been "dug up." As generally used, a **fossil** is recognizable, physical evidence of an organism that lived long ago. Most fossils are body parts, such as bones, teeth, shells, leaves, and seeds. Some are tracks, burrows, and other telltale impressions of past life. All were preserved because they were gently buried before they could decompose or fall apart.

The best fossils come from relatively undisturbed burial sites. Few sites escaped disruption, so most fossils are broken, incomplete, crushed, or deformed. The spectacularly complete ones give more insight into the organism's functions and behavior (Figure 15.2).

The fossil record is uneven, for some organisms and environments are better represented than others. For example, fossils of hard-shelled mollusks and bony fishes are abundant. Fossils of jellyfishes and other soft-bodied animals are not. Floodplains, seafloors, swamps, and natural traps such as caves and tar pits favor fossilization. Rapidly eroding hills do not.

Besides being uneven, the fossil record is incomplete. There surely were many millions of ancient species. We have fossils of about 250,000, and no doubt we will find others. But the earth itself has changed over time, and parts of the record have been obliterated.

The completeness of the fossil record varies as a function of the kinds of organisms represented, where they lived, and the stability of their burial sites.

Dating Fossils. You can find similar layers of fossil-containing sedimentary rock over vast areas, even on different continents. Figure 15.3 shows an example. Such layers formed long ago, when silt and skeletons of tiny marine organisms gradually "rained down" onto a seafloor. Logically, then, the deepest layers formed first and the ones closest to the surface formed last.

Early naturalists did not know how old the rocks or the fossils were. But they saw some abrupt transitions in the kinds of fossils represented. They interpreted these as boundaries between four great eras in time. The most ancient fossils were from the first era, the *Proterozoic*, followed by fossils from the *Paleozoic, Mesozoic,* and finally the "modern" era, the *Cenozoic*. The boundaries between eras still serve as the basis of a **geologic time scale**. Today we know that they mark the times of mass extinctions. Thanks to radioisotope dating work (page 18), actual dates have been assigned to the time scale. The work also revealed that the Proterozoic lasted for an astoundingly immense time. The Proterozoic has since been subdivided. The first era of the geologic time scale is now called the *Archean* ("the beginning").

Figure 15.4 Forces of geologic change. (**a**) Plate tectonics. The crust is arranged in rigid plates that split apart, move about, and collide. As the plates push against a continent, they are often thrust beneath it, as shown in (**b**). The crumpling and upheavals created most major mountain ranges.

Plumes of molten material well up from the earth's interior and spread out beneath its crust. Such plumes ruptured the crust at vast ridges on the ocean floor. Here, molten material escapes, cools, and forces the seafloor away from the ridge. Seafloor spreading displaces land masses away from the ridge. Plumes also cause deep rifts in the interior of continents. In time these split apart. Such rifting is now taking place in Missouri, at Lake Baikal in Russia, and in eastern Africa. In the past, superplumes violently ruptured the crust. Volcanoes still form at the ruptured "hot spots." The Hawaiian Islands formed this way.

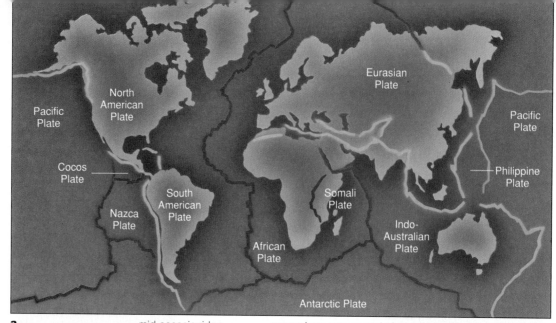

a

mid-oceanic ridge   zone where one crustal plate is being thrust under another

b

The Changing Geologic Stage

Changes in the fossil record correlate with geologic events on a grand scale. The earth's crust never has been fully stable. In the past, huge plumes of molten rocks rose from the earth's core and spread out beneath the crust or broke through it. The crust was fractured into enormous slabs (crustal plates) that still move apart and collide at their boundaries. Ocean basins opened up, then disappeared. Continents drifted about and crunched together (Figure 15.4). Besides this, objects from outer space bombarded the earth, with long-term effects on global temperature and climates.

The fossil record tells us that such changes in land masses, the oceans, and the atmosphere profoundly influenced the evolution of life. As you will see, it tells us that early life flourished in warm, shallow, isolated places along the vast shorelines of continents. When continents collided, shorelines vanished, with dire consequences for many lineages. Even as old habitats were lost, however, new ones opened up for the survivors—and evolution took off in new directions.

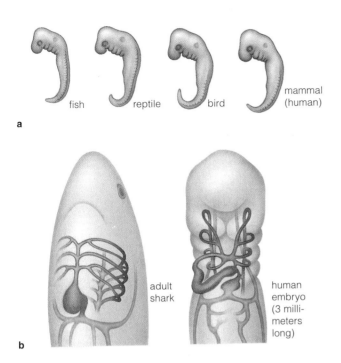

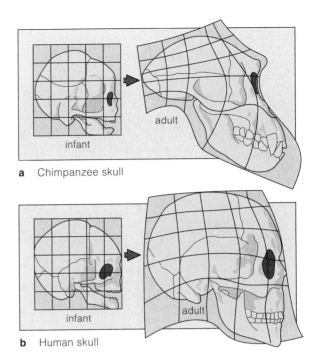

a   Chimpanzee skull

b   Human skull

Figure 15.5 Evidence of evolution from comparative embryology. (**a**) Very early embryos of vertebrates retain striking similarities. (**b**) Fishlike structures still form in early embryos of reptiles, birds, and mammals. For example, a two-chambered heart (*orange*), certain veins (*blue*), and portions of arteries called aortic arches (*red*) develop in a fish embryo and persist in adult fishes. The same structures form in an early human embryo.

Figure 15.6 Comparison of proportional changes in a chimpanzee skull (**a**) and a human skull (**b**). Both skulls are quite similar in infants. Imagine that these representations of infant skulls are paintings on a blue rubber sheet divided into a grid. Stretching the sheet deforms the grid's squares. For the adult skulls, differences in size and shape within corresponding grid sections reflect differences in growth patterns.

Comparative Morphology

Evidence of macroevolution also comes from anatomical comparisons of major lineages. This work is known as **comparative morphology**.

Stages of Development. A complex animal or plant develops in stages, from fertilized egg, through various embryonic forms, and on to the adult. Early in development, embryos of different organisms within a lineage often proceed through strikingly similar stages. Consider the examples shown in Figure 15.5. Would you know which embryo is from a fish, frog, chicken, or human?

Why do the early embryonic stages of vertebrates resemble one another so strongly? Evolution, recall, proceeds by modification of existing forms, not by starting from scratch. All vertebrate embryos inherit the same ancient plan for development. According to the plan, tissues form only when cells divide in certain patterns and interact in prescribed ways. Later on, the heart, bones, muscles, and other body parts grow in

prescribed ways, provided that each developmental step is properly completed before the next steps begin.

During vertebrate evolution, any mutations that disrupted an early stage of development would have had devastating effects on the organized interactions required for later stages. Embryos of different groups remained similar because mutations that altered early steps in development were selected against.

How, then, did the *adults* of different vertebrate groups get to be so different? At least some differences resulted from mutations that altered *rates of growth* of body parts formed during later stages. Notice, in Figure 15.6, how the proportions of chimpanzee and human skull bones are alike at birth. Thereafter, they change dramatically for chimps but only slightly for humans. Chimps and humans arose from the same ancestral stock, and their genes are nearly identical. However, at some point on the separate evolutionary road leading to humans, a regulatory gene probably mutated. Since then, instead of promoting the rapid growth required for dramatic changes in skull bones, the mutated gene has blocked it.

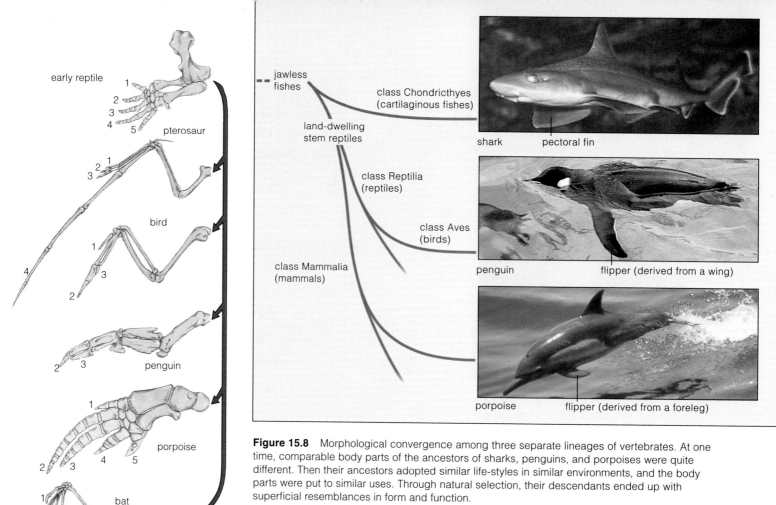

Figure 15.8 Morphological convergence among three separate lineages of vertebrates. At one time, comparable body parts of the ancestors of sharks, penguins, and porpoises were quite different. Then their ancestors adopted similar life-styles in similar environments, and the body parts were put to similar uses. Through natural selection, their descendants ended up with superficial resemblances in form and function.

Figure 15.7 Morphological divergence in the vertebrate forelimb, starting with the generalized form of ancestral early reptiles. Diverse forms evolved even while similarities in the number and position of bones were preserved.

Homologous Structures. When populations of a species branch out in different evolutionary directions, they diverge in appearance, functions, or both. Yet the related species remain alike in many ways, for their evolution proceeds by modification of a shared body plan. In such species we see **homologous structures**. These are the *same* body parts, modified in different ways in different lines of descent from a common ancestor (*homo-* means the same).

For example, the same ancestral organism probably gave rise to most land-dwelling vertebrates, which have homologous structures. Apparently their common ancestor had five-toed limbs. The limbs diverged in form and became wings in pterosaurs, birds, and bats (Figure 15.7). All these wings are homologous—they have the same parts. The five-toed limb also evolved into the flip-

pers of porpoises and the long, one-toed limbs of modern horses, stubby limbs of moles and other burrowing mammals, and pillarlike limbs of elephants. This pattern of change from a common ancestral form is an example of macroevolution. The pattern is called **morphological divergence** (*morpho* means body form).

Analogous Structures. Another macroevolutionary pattern is seen among separate lineages with large differences in form that adopted similar ways of life. Comparable body parts were put to similar uses. They became modified through natural selection and ended up resembling one another in structure and function. This pattern of change is called **morphological convergence**. Body parts that once were quite different in separate lineages but that came to resemble one another in form and function are called **analogous structures** (from *analogos*, meaning similar to one another).

For example, sharks, penguins, and porpoises are only distantly related (their lineages diverged from a common ancestor early in vertebrate evolution). All three are fast-swimming predators of the seas. As Figure 15.8 shows, penguins and porpoises have flippers that are similar in shape to the shark pectoral fin. All three structures stabilize the body in water. The shark lineage never left the water, and shark fins haven't changed much from the ancestral form. By contrast, penguins descended from flying birds that became adapted to life in water instead of air—their flippers actually are modified wings. Porpoises descended from land-dwelling, four-legged mammals that became adapted to life in the seas—their flippers are modified front legs. Thus the penguin and porpoise flippers converged on the pectoral fins of sharks.

Similar patterns of embryonic development often are evidence of evolutionary relationship.

Homologous structures are evidence of morphological divergence. They are the same body parts that became modified in different ways in different lines of descent from a common ancestor.

Analogous structures are evidence of morphological convergence. They are body parts, once quite different in separate lineages, that were put to comparable uses in the same kinds of environments. The parts ended up resembling one another in form and function.

Comparative Biochemistry

All organisms are a mix of recent and ancient traits. Whether one kind is closely or distantly related to another is reflected in the kinds and numbers of traits they share or don't share. This is true also of their biochemical traits.

For example, from morphological studies, you might already have an idea of how primates—say, monkeys, humans, and chimpanzees—are related to one another. You could test your idea by studying small differences in the amino acid sequences of certain proteins that all three primates produce. Or you could determine how closely the nucleotide sequences in their DNA match up. For reasons described next, the greatest biochemical similarities occur among the closest relatives.

Molecular Clocks. Like their primate relatives, the human species has unique traits, the outcome of unique gene mutations that accumulated after divergence from the parent stock. Humans also share many genes (and traits) with other species. The molecular structure of those shared genes has not changed much over mind-boggling spans of time. For example, the last shared ancestor of yeasts and humans lived many hundreds of millions of years ago—yet more than 90 percent of the known gene products (proteins) of certain yeasts and humans have remained the same.

Even though the overall structure of such genes has been highly conserved, neutral mutations introduced small differences in their structure in different lineages. Neutral mutations, recall, have little or no effect on survival or reproduction. They can slip past agents of selection and accumulate in the DNA.

By some calculations, neutral mutations in highly conserved genes accumulated at a regular rate. Think of the accumulation of neutral mutations in any given lineage as a series of predictable ticks of a **molecular clock**. Then turn the clock back—so that the total number of ticks "unwinds" down through the geologic time scale. Where the last tick stops, that is roughly the time of origin for the lineage.

Protein Comparisons. Now suppose two species have the same conserved gene. The amino acid sequences of the gene's product (a protein) are the same or nearly so. The absence of mutation suggests that the species are closely related. Or suppose the sequences differ. If there are many differences, many neutral mutations accumulated in the gene. And this suggests that a long time passed since the two species shared a common ancestor.

Cytochrome c, a protein of electron transport chains, is specified by a highly conserved gene. The protein is produced by organisms ranging from aerobic bacteria to corn plants to humans. Its sequence of 104 amino acids is identical in humans and chimps. It is the same except for one amino acid in rhesus monkeys. The cytochrome c of chickens differs from that of humans by 18 amino acids. The cytochrome c of turtles differs by 19 amino acids, and that of yeasts, by 56 amino acids. On the basis of such data, would you assume that humans are more closely related to a chimpanzee or a rhesus monkey? A chicken or a turtle?

Nucleic Acid Comparisons. Structural alterations caused by gene mutations typically pepper the nucleotide sequences of DNA and RNA. So the extent to which a DNA or RNA strand from one species will base-pair with a comparable strand from another species is a rough measure of evolutionary distance between them.

As an example, some nucleic acid comparisons are based on **DNA-DNA hybridization**. In this method, recall, the two strands of the double-stranded DNA from two species are induced to unwind from each other (page 182). Then the single strands are allowed to recombine into "hybrid" molecules. Heating breaks the hydrogen bonds between the two hybridized strands. The heat energy required to pull them apart is a measure of their similarity.

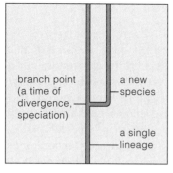

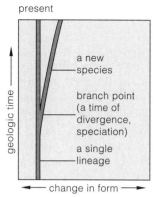

a Softly angled branching means speciation occurred through gradual changes in traits over geologic time.

b Horizontal branching means traits changed rapidly around the time of speciation. Vertical continuation of a branch means traits of the new species did not change much thereafter.

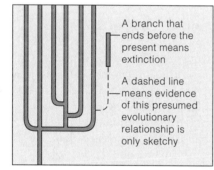

c Many branchings of the same lineage at or near the same point in geologic time means that an adaptive radiation occurred.

A branch that ends before the present means extinction

A dashed line means evidence of this presumed evolutionary relationship is only sketchy

Figure 15.9 How to read evolutionary tree diagrams. The branch points represent times of divergence and speciation, as brought about by microevolutionary processes.

Do changes from the original species form proceed slowly or rapidly? Many small changes might accumulate gradually within a lineage, over long spans of time. This is the premise of the *gradual* model of rates of change in an evolving lineage. Conversely, species might originate through rapid spurts of change over hundreds or thousands of years, then change little for the next 2 million to 6 million years. This is the premise of the *punctuation* model, which has most changes occurring about the time of speciation.

Branchings, Extinctions, and Adaptive Radiations

Evolutionary Trees. The fossil record, comparative morphology, and comparative biochemistry each provide insight into "who came from whom." Typically, information about this continuity of relationship among species is shown as **evolutionary trees**. These are tree-like diagrams in which branches represent separate lines of descent from a common ancestor. As Figure 15.9 indicates, each branch point in a tree represents a time of

Figure 15.10 How the red panda (**a**), the giant panda (**b**), and bears (**c**) are related. A giant panda resembles meat-eating bears but is a plant eater. It grasps bamboo plants with the short toes and thumblike digit of its front paws. It resembles its neighbor in China, the red panda, which eats bamboo and has no thumbs. DNA-DNA hybridization studies indicate that the giant panda is more closely related to bears. All three have a common ancestor that lived more than 40 million years ago. One divergence led to raccoons and the red panda. Much later in time, divergences led to the giant panda and to bears.

RACCOON   RED PANDA

DIVERGENCE about 40 million years ago

COMMON ANCESTOR (MEAT EATER)

divergence and speciation. Figure 15.10 shows a specific example, based on DNA-DNA hybridization studies.

The evolutionary trees for different lineages are not all alike. As you will see, lineages differ in terms of when their member species originated, how rapidly those species changed in form, and how long they persisted on the evolutionary stage. Some lineages dribbled along without much change, producing a species here, losing a species there, over tens of millions of years. Entire lineages ended abruptly during mass extinctions. Others filled the environment with the spectacular bursts of activity called adaptive radiations.

Extinctions. Inevitably, a lineage loses a number of species as local conditions change. The rather steady rate of their disappearance over time is "background extinction." By contrast, a **mass extinction** is an abrupt rise in extinction rates above the background level. It is a catastrophic, global event in which families and other major groups are wiped out simultaneously.

In the past, major groups tended to survive episodes of mass extinctions when their members were widely dispersed in different regions. Most of the hardest hit were highly specialized for life in tropical regions.

Even so, luck has a lot to do with it. What *had* been the most adaptive traits may mean absolutely nothing if an asteroid the size of Vermont hits the earth.

```
b
   GIANT              SPECTACLED   SLOTH   SUN   BLACK        POLAR   BROWN
   PANDA                 BEAR       BEAR   BEAR  BEARS        BEAR    BEAR   c


   DIVERGENCE              DIVERGENCES
   15–20 million years ago    2 million years ago
```

Adaptive Radiations. In an **adaptive radiation**, new species of a lineage fill a wide range of habitats during bursts of microevolutionary activity. Many adaptive radiations have occurred during the first few million years after a mass extinction. Figure 15.11 shows an example of this. Many others have occurred when organisms have been presented with unfilled adaptive zones. **Adaptive zones** are most easily defined as ways of life, such as "burrowing in the seafloor" or "catching insects in the air at night."

A lineage may radiate into an adaptive zone when it has physical, ecological, and evolutionary access to it. *Physical* access means that a lineage happens to be there when adaptive zones open up. At one time, mammals were all small insect eaters, distributed through rather uniform tropical regions of a single continent. Then the continent split into several land masses, so many populations of mammals became isolated from others. The landscape, climate, and resources evolved in different ways on those land masses and set the stage for independent radiations.

Evolutionary access means that modification of some structure or function will permit a lineage to exploit the environment in improved or novel ways. When the forelimbs of certain five-toed vertebrates evolved into wings, for example, this opened new adaptive zones to the ancestors of birds and bats.

Ecological access means that the lineage can enter an unoccupied adaptive zone—or outcompete the resident species. We will say more about this in Chapter 35.

Figure 15.12 sets the stage for the remainder of this chapter, which correlates milestones in the evolution of life with key events in the evolution of the earth. The figure is generalized. Only the five greatest mass extinc-

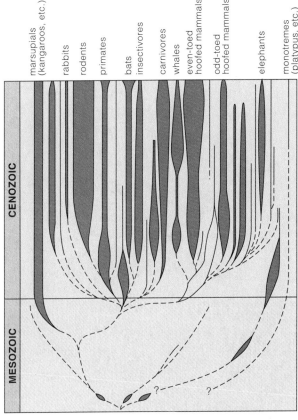

Figure 15.11 The great adaptive radiation of mammals early in the Cenozoic era. Variations in the width of a tree branch correspond to the range of diversity at different points in time. The thicker the branch, the greater the diversity.

tions are shown, along with the shrinking and expanding range of species diversity for all the major groups combined. Within that overall pattern, each major group has its own patterns of diversity, its own encounters with extinctions and radiations.

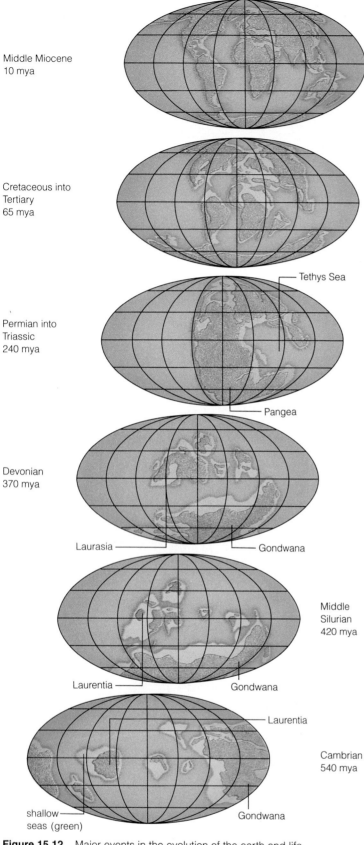

Middle Miocene
10 mya

Cretaceous into
Tertiary
65 mya

Tethys Sea

Permian into
Triassic
240 mya

Pangea

Devonian
370 mya

Laurasia — — Gondwana

Middle
Silurian
420 mya

Laurentia — — Gondwana

— Laurentia

Cambrian
540 mya

shallow
seas (green)   Gondwana

Figure 15.12   Major events in the evolution of the earth and life. Time spans of the different eras are not to scale. If they were, the Archean and Paleozoic portions of the chart would run off the page. Think of the spans as minutes on a clock. Life originated at midnight. The Paleozoic began at 10:04 A.M. The Mesozoic began at 11:09 A.M., and the Cenozoic at 11:47 A.M. The Recent epoch of the Cenozoic started during the last 0.1 second before noon.

Era	Period	Epoch	Millions of Years Ago (mya)	
CENOZOIC	Quaternary	Recent		
		Pleistocene	0.01-	
	Tertiary	Pliocene	1.65	
		Miocene	5	
		Oligocene	25	
		Eocene	38	
		Paleocene	54	
MESOZOIC	Cretaceous	Late	65	
			100	
		Early		
	Jurassic		138	
	Triassic		205	
PALEOZOIC	Permian		240	
	Carboniferous		290	
	Devonian		360	
	Silurian		410	
	Ordovician		435	
	Cambrian		505	
PROTEROZOIC			550	
ARCHEAN			2,500	

Range of Global Diversity (marine and terrestrial)	Times of Major Geologic and Biological Events

1.65 mya to present. Major glaciations. Modern humans emerge and begin what may be greatest **mass extinction** of all time on land, starting with Ice Age hunters.

65–1.65 mya. Unprecedented mountain building as continents rupture, drift, collide. Major climatic shifts; vast grasslands emerge. Major **radiations** of flowering plants, insects, birds, mammals. Origin of earliest human forms.

65 mya. Asteroid impact? **Mass extinction** of all dinosaurs and many marine organisms.

135–65 mya. Pangea breakup continues, broad inland seas form. Major **radiations** of marine invertebrates, fishes, insects, dinosaurs. Origin of angiosperms (flowering plants).

181–135 mya. Pangea breakup begins. Rich marine communities. Major **radiations** of dinosaurs.

205 mya. Asteroid impact? Mass extinction of many organisms in seas, some on land; dinosaurs, mammals survive.

240–205 mya. Recovery, **radiations** of marine invertebrates, fishes, dinosaurs. Gymnosperms the dominant land plants. Origin of mammals.

240 mya. **Mass extinction.** Nearly all species in seas and on land perish.

280–240 mya. Pangea, worldwide ocean form; shallow seas squeezed out. Major **radiations** of reptiles, gymnosperms.

360–280 mya. Tethys Sea forms. Recurring glaciations. Major **radiations** of insects, amphibians. Spore-bearing plants dominate; gymnosperms present. Origin of reptiles.

370 mya. **Mass extinction** of many marine invertebrates, most fishes.

435–360 mya. Laurasia forms, Gondwana moves north. Vast swamplands, early vascular plants. **Radiations** of fishes continue. Origin of amphibians.

435 mya. Glaciations as Gondwana crosses South Pole. **Mass extinction** of many marine organisms.

500–435 mya. Gondwana moves south. Major **radiations** of marine invertebrates, early fishes.

550–500 mya. Land masses dispersed near equator. Simple marine communities. Origin of animals with hard parts.

700–550 mya. Supercontinent Laurentia breaks up; widespread glaciations.

2,500–570 mya. Oxygen present in atmosphere. Origin of aerobic metabolism. Origin of eukaryotes: protistans, algae, fungi, animals.

3,800–2,500 mya. Origin of photosynthetic bacteria.

4,600–3,800 mya. Formation of earth's crust, early atmosphere, oceans. Chemical evolution leading to origin of life (prokaryotes; anaerobic bacteria).

4,600 mya. Origin of earth.

MACROEVOLUTION AND EARTH HISTORY

The following is a sweeping slice through time, one that cuts back to the Archean era and the chemical origins of life. This is our starting point for the next four chapters, which will take us along major lines of descent that lead to the present range of species diversity. Keep in mind that the reconstruction of life's history is incomplete, with theories and sometimes nothing more than educated guesses bridging the gaps. Even so, we have identified a principle that helps us organize separate bits of information about the past:

The origin and evolution of life have been linked to the physical and chemical evolution of the earth.

Origin of Life

Fossil evidence suggests that the first cells appeared on earth between 4 billion and 3.8 billion years ago. We have no record of the event. As far as we know, massive movements in the earth's mantle and crust, volcanic activity, and erosion obliterated all traces of it. Still, we can put together a plausible explanation of how life originated by considering three questions:

1. What physical and chemical conditions prevailed on earth at the presumed time of life's origin?

2. Based on known physical, chemical, and evolutionary principles, could proteins and other large organic molecules have formed spontaneously and then evolved into systems displaying the characteristics of life?

3. Can we devise experiments to test whether living systems could have emerged by chemical evolution?

The Early Earth and the First Atmosphere. Let's approach the first question in terms of what astronomers know about the earth's formation. Billions of years ago, explosions of dying stars ripped through our galaxy, forming a dense cloud of gases and dust that extended trillions of kilometers in space. As the cloud cooled, countless bits of matter gravitated together. By 4.6 billion years ago, the cloud had flattened out into a slowly rotating disk. Our sun was born in thermonuclear reactions at the disk's extremely dense, hot center. Farther out, the earth and other planets took form. Within 600 million years, the earth was hurtling through space as a thin-crusted inferno (Figure 15.13).

Cloudlike remnants of stars persist in space. They are mostly hydrogen gas. They also contain water, iron, silicates, hydrogen cyanide, methane, ammonia, formaldehyde, and many other simple inorganic and organic substances. Most likely, the cosmic cloud from which the earth originated was no different.

Heat and gases blanketed the earth when patches of crust were forming. This first atmosphere probably consisted of gaseous hydrogen (H_2), nitrogen (N_2), carbon monoxide (CO), and carbon dioxide (CO_2). Were gaseous oxygen (O_2) and water also present? Probably not. As we know today, rocks release very little oxygen during volcanic eruptions. Given the conditions on the early earth, small amounts of free oxygen would have reacted at once with other elements. Any water would have evaporated in the intense heat.

When the crust finally cooled and solidified, water condensed into clouds and the rains began. For millions of years, runoff from rains stripped mineral salts and other compounds from the earth's parched rocks. Salt-laden waters collected in depressions in the crust and formed the early seas.

Without an oxygen-free atmosphere, the organic compounds that started the story of life never would have formed on their own. (As described on page 60, free oxygen would have attacked them.) Without liquid water, cell membranes never would have formed. Cells, recall, are the basic units of life. Each has a capacity for independent existence.

The Synthesis of Organic Compounds. When we reduce cells to their lowest common denominator, we are left with proteins, complex carbohydrates, lipids, and nucleic acids. Today, cells assemble these molecules from small organic compounds—simple sugars, fatty acids, amino acids, and nucleotides. Energy from the environment drives the synthesis reactions. Were small organic compounds also present on the early earth? Were there sources of energy that drove their spontaneous assembly into the large molecules of life?

Mars, meteorites, the earth's moon, and the earth formed at the same time, from the same cosmic cloud. Rocks collected from Mars, meteorites, and the moon contain precursors of biological molecules—so the same precursors must have been present on the early earth. Possibly sunlight, lightning, or heat escaping from the earth's crust supplied enough energy to drive their condensation into more complex organic molecules.

In the first of many tests of this hypothesis, Stanley Miller mixed hydrogen, methane, ammonia, and water in a reaction chamber (Figure 15.14). He recirculated the mixture and kept bombarding it with a spark discharge to simulate lightning. Within a week, amino acids and other small organic compounds had formed.

In other experiments that simulated conditions on the early earth, glucose, ribose, deoxyribose, and other sugars were produced from formaldehyde. Adenine was produced from hydrogen cyanide. Adenine plus ribose occur in ATP, NAD, and other nucleotides.

Figure 15.13 Representation of the earth during its formation, when the moon's orbit presumably was much closer. If the earth had condensed into a smaller planet, its gravitational mass would not have been great enough to hold onto an atmosphere. If it had settled into an orbit closer to the sun, water would have evaporated from its surface. If the orbit had been more distant, the surface would have been too cold, locking up water as ice. Without liquid water, life never would have originated on earth.

Even if amino acids did form in the early seas, they wouldn't have lasted long. In water, the favored direction of most spontaneous reactions is toward hydrolysis, not condensation (page 27).

Maybe more lasting bonds formed at the margins of seas. By one scenario, clay in the rhythmically drained muck of tidal flats and estuaries served as templates (structural patterns) for the spontaneous assembly of proteins and other organic compounds. Clay consists of thin, stacked layers of aluminosilicates with metal ions at its surface. Clay and metal ions attract amino acids. When clay is first warmed by sunlight, then alternately dried out and moistened, it actually promotes condensation reactions that yield complex organic compounds. We know this from experiments.

Suppose proteins that formed on some clay templates had the shape and chemical behavior necessary to function as weak enzymes in hastening bonds between amino acids. If certain templates promoted such bonds, they would have selective advantage over other templates in the chemical competition for available amino acids. Perhaps selection was at work before the origin of cells, favoring the chemical evolution of enzymes.

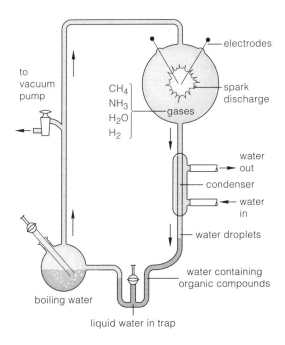

Figure 15.14 Stanley Miller's apparatus, used to study the synthesis of organic compounds under conditions that presumably existed on the early earth. The condenser cools circulating steam so that water droplets form.

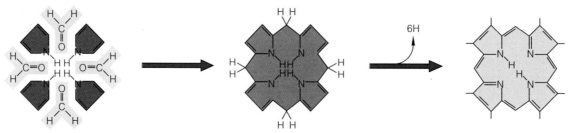

4 pyrrole rings + 4 formaldehyde molecules

6H

porphyrin ring system

Figure 15.15 How formaldehyde, a substance present on the early earth, might have undergone chemical evolution into porphyrin. Today, porphyrin is the light-trapping and electron-donating component of chlorophyll molecules. It also is a component of cytochrome, which has roles in electron transport systems of a variety of metabolic pathways.

Agents of Metabolism. A defining characteristic of life is metabolism. The word refers to all the reactions by which cells harness energy and use it to drive their activities, such as biosynthesis. During the first 600 million years of earth history, enzymes, ATP, and other molecules could have assembled spontaneously at the same locations. If so, their close association would have promoted chemical interactions—and the beginning of metabolic pathways.

Imagine an ancient sunlit estuary, rich in clay deposits. Countless aggregations of organic molecules stick to the clay. At first there are quantities of an amino acid called *D*. All over the estuary, *D* is incorporated into protein molecules—until the supply dwindles. Suppose a certain enzyme can promote the formation of *D* from a simpler substance (*C*)—which is still abundant. Aggregations having that enzyme will be favored in the chemical competition for starting materials. Now suppose *C* becomes scarce also. At that point, the advantage tilts to aggregations that can promote formation of *C* from the even simpler organic substances *B* and *A*—say, carbon dioxide and water. These substances are present in essentially unlimited amounts in the atmosphere and the sea. Selection has favored a synthetic pathway:

$$A + B \longrightarrow C \longrightarrow D$$

Suppose, finally, that some aggregations in the estuary are better than others at harnessing and using energy. What kind of molecules could give them this advantage? Think about photosynthesis, the energy-trapping pathway that now dominates the world of life. This metabolic pathway starts at light-trapping pigment molecules called chlorophylls. The part of chlorophyll that absorbs sunlight energy and gives up

one of the chlorophyll pigments in plants (chlorophyll *a*)

electrons is a porphyrin ring structure. Porphyrins also are part of cytochromes. And cytochromes are built-in parts of electron transport systems in all photosynthetic and aerobically respiring cells. As Figure 15.15 shows, porphyrins can assemble spontaneously from formaldehyde—one of the legacies of cosmic clouds. Was porphyrin the electron transporter of the first metabolic pathways? Perhaps.

Self-Replicating Systems. Another defining characteristic of life is the capacity for reproduction. Reproduction begins with DNA molecules, which are templates that encode instructions for building specific proteins. DNA is a fairly stable molecule that is easily replicated before each cell division. Enzymes and RNA carry out its encoded instructions.

Today, coenzymes assist most enzymes. Some of these small molecules have the same structure as RNA nucleotides. RNA can be assembled spontaneously from heated nucleotide precursors and short chains of phosphate groups. On the early earth, sunlight energy alone could have driven formation of those chains.

Simple, self-replicating systems of RNA, enzymes, and coenzymes have been created in the laboratory. Experiments with these systems suggest that RNA might have formed on clay templates. Did RNA later replace clay as information-storing templates for protein synthesis? Perhaps, but existing RNA is too chemically inept to serve this role. And we still don't know how DNA entered the picture. Until we identify the likely chemical ancestors of RNA and DNA, the story of life's origin will be incomplete.

The First Plasma Membranes. Experiments are more revealing of the origin of the plasma membrane, which surrounds every living cell. The membrane is a lipid bilayer, studded with proteins that carry out diverse functions. The membrane's primary function is to control which substances move into and out of the cell. Without this control, cells cannot exist.

Before cells, there must have been "proto-cells." They may have been little more than membrane sacs, protecting information-storing templates and various metabolic agents from the environment. We know that simple membrane sacs can form spontaneously. In one experiment, Sidney Fox heated amino acids until they formed protein chains, which he put in hot water. After the chains cooled, they assembled into small, stable spheres (Figure 15.16a). Like cell membranes, the spheres were selectively permeable to different substances. They also picked up lipids from the water, and a lipid-protein film formed at their surface.

In other experiments, fatty acids and glycerol combined to form long-tail lipids under conditions that simulated evaporating tidepools. The lipids self-assembled into small, water-filled sacs (Figure 15.16b). These sacs were in many ways like cell membranes.

In short, there are major gaps in the story of life's origins. But there also is good experimental evidence that chemical evolution could have given rise to the molecules and structures characteristic of life. Figure 15.17 summarizes some key events in that chemical evolution, which apparently led to the first cells.

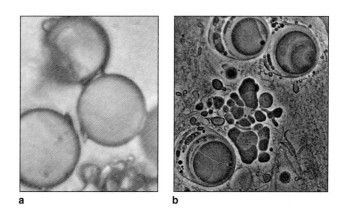

a   b

Figure 15.16 Microscopic spheres of (**a**) proteins and (**b**) lipids that self-assembled under abiotic conditions.

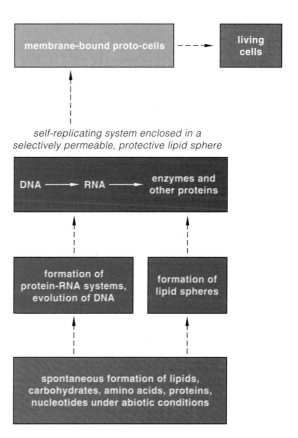

Figure 15.17 One possible sequence of events that led to the first self-replicating systems, then to the first living cells.

Figure 15.18 (**a**) One of the oldest known fossils: a strand of walled cells, 3.5 billion years old. The strand resembles certain modern-day filamentous bacteria. (**b**) From Western Australia, stromatolites that formed between 2,000 and 1,000 years ago in shallow seawater. Calcium deposits preserved their structure. They are identical to stromatolites more than 3 billion years old.

10 µm

a   b

The Archean and Proterozoic Eras

Until about 3.7 billion years ago, the earth's crust was extremely unstable. Before another 200 million years passed, the first cells had emerged. Like existing bacteria, they were prokaryotic. They probably used fermentation or similar anaerobic pathways to extract energy from organic compounds that had accumulated in the seas. Although geologic conditions were rather hellish, "food" was available, predators were absent, and biological molecules were free from oxygen attacks.

Some time ago, Carl Woese identified what may be the world's most ancient lineage. Its living representatives, the archaebacteria, include species that thrive in extremely hot, oxygen-free environments (page 252).

Between 3.5 and 3.2 billion years ago, photosynthetic bacteria emerged. They were anaerobic, but components of their electron transport systems allowed them to tap an unlimited energy source—sunlight. For nearly 2 billion years, photosynthetic cells dominated the shallow seas. Their populations formed large mats in which sediments collected. The mats slowly accumulated one atop the other and formed the structures called **stromatolites** (Figure 15.18).

At first the cyclic pathway of photosynthesis, described on page 76, prevailed. Later, the noncyclic pathway evolved in some lineages and oxygen, one of its by-products, gradually accumulated.

The resulting oxygen-rich atmosphere had two irreversible effects. First, *the spontaneous chemical origin of living cells was no longer possible in the natural world.* Except in a few anaerobic habitats, spontaneously formed organic compounds would not survive attacks by the abundant free oxygen. Second, *aerobic respiration would become the dominant energy-releasing pathway.* In some bacterial lineages, selection now favored metabolic equipment that could "neutralize" oxygen by using it as an electron acceptor. This innovation foreshadowed the evolution of large multicelled organisms

and their invasion of far-flung environments. The oxygen-rich atmosphere was an adaptive zone of global dimensions.

The first eukaryotes evolved somewhere around 1.2 billion years ago. Fossils show that green and red algae, plant spores, and fungi were already around by 900 million years ago. As you will read in Chapter 17, eukaryotes probably are descended from predatory bacteria that engulfed but could not digest other bacterial cells. The engulfed cells thrived, reproduced, and went on to evolve as organelles inside the host species.

About 800 million years ago, stromatolites began to decline dramatically. They had become an untapped food source for newly evolved, tiny, bacteria-eating animals. About the same time, small, soft-bodied animals were leaving tracks and burrows on the seafloor. They lived near the shores of a supercontinent called Laurentia. About 570 million years ago, in "Precambrian" times, their descendants may have started the first adaptive radiation of animals.

First Half of the Paleozoic Era

We divide the Paleozoic into Cambrian, Ordovician, Silurian, Devonian, Carboniferous, and Permian periods (Figure 15.12). During the transition to the Cambrian, the supercontinent broke up. The fragmented land masses straddled the equator, and warm, shallow seas lapped their margins. Global conditions restricted pronounced seasonal changes in winds, ocean currents, and the upward churning of nutrients from deep waters. Thus, nutrient supplies were stable but limited.

Early Cambrian organisms had flattened bodies, with a good surface-to-volume ratio for taking up nutrients (Figure 15.19a). Most lived on or just beneath the seafloor, where dead organisms and organic debris settled. Most of the major animal phyla evolved in Cambrian seas. They ranged from sponges to simple

Figure 15.19 (**a**,**b**) From Australia's Ediacara Hills, fossils of two Cambrian animals, about 600 million years old. (**c**) From the Burgess Shale of British Columbia, a fossilized marine worm of Cambrian times. (**d**) One of the earliest Cambrian trilobites.

vertebrates—and they were exuberantly diverse. How could so much diversity arise so early in time? Most likely, genes governing embryonic development were less intertwined than they are today, so there was not as much selection against mutated alleles and novel forms of traits among the Cambrian animals.

Also, the long, warm shorelines of the equatorial land masses offered plenty of vacant adaptive zones, with opportunities for new ways of securing food. Now we start seeing fossilized organisms with missing chunks, punctures, and healed-over wounds. These are not artifacts of fossilization; the organisms were damaged while they were still alive. Things were starting to get lively! In short order, diverse predators and prey with armorlike shells, spines, mouths, and novel feeding structures evolved.

Late in the Cambrian, temperatures in the shallow seas changed drastically. Trilobites (Figure 15.19*d*), one of the most common animals, were nearly wiped out. During the Ordovician, a continent called Gondwana drifted south, and parts became submerged in shallow seas. Vast marine environments opened up, favoring adaptive radiations. Reef organisms flourished. Fast-swimming predators called nautiloids dominated the evolutionary stage (Figure 15.20). In late Ordovician times, Gondwana straddled the South Pole. Immense glaciers formed on its surface and drained the shallow seas. This first ice age triggered the first global mass extinction. Reef life everywhere collapsed at the boundary to the Silurian epoch.

Figure 15.20 Nautiloids—tentacled, shelled animals of the phylum Mollusca. They are relatives of the now-extinct ammonites (Figure 15.1). The type in (**a**) flourished about 450 million years ago and the type in (**b**) about 400 million years ago. (**c**) An existing representative of the lineage.

Second Half of the Paleozoic Era

Gondwana drifted north during the Silurian and on into the Devonian. Reef organisms recovered, and an adaptive radiation led to formidable predators: armor-plated, massive-jawed fishes. Small, stalked plants became established along the muddy margins of land and foreshadowed a major radiation of terrestrial plants (Figure 15.21). Lobe-finned fishes ancestral to amphibians also were invading the land (Figure 15.22).

At the Devonian-Carboniferous boundary, an unknown, catastrophic event caused sea levels to swing dramatically, and another mass extinction occurred. Afterward, land plants and insects embarked on major adaptive radiations. Throughout the Carboniferous, land masses were gradually submerged and drained many times. Organic debris accumulated, became compacted, and was converted to coal, in the manner described on page 279.

In Permian times, insects, amphibians, and early reptiles flourished in vast swamp forests, composed in part of the ancestors of modern-day cycads, ginkgos, and conifers. As the Permian drew to a close, the greatest of all mass extinctions occurred. Nearly all known species on land and in the seas perished. At that time, all land masses were colliding to form Pangea. This vast supercontinent extended from pole to pole. A single world ocean lapped its margins (Figure 15.12). Changes in the distribution of water, land area, and land elevation had catastrophic effects on global temperatures and climate. Yet many species met the new challenges, including the reptilian ancestors of dinosaurs, birds, turtles, crocodiles, snakes, and lizards.

The Mesozoic Era

The Mesozoic lasted about 175 million years. It is divided into Triassic, Jurassic, and Cretaceous periods. Adaptive radiations that began during this era greatly expanded the range of global diversity (Figure 15.12). Early in the Mesozoic, divergences in a few lineages gave rise to mammals and wonderful reptilian "monsters," including the dinosaurs. At the Triassic-Jurassic

a   b

Figure 15.21 Fossils of plants that pioneered the invasion of land during Silurian times. (**a**) Stems of the oldest known plant, *Cooksonia*, less than 7 centimeters tall. (**b**) A type of Devonian plant (*Psilophyton*) that may have been among the earliest ancestors of seed-bearing plants.

Figure 15.22 Artist's reconstruction of lobe-finned fishes venturing onto a Devonian shoreline. Those fast-swimming predators probably gulped air at the surface of water that had become shallow and stagnant. The lobed fins in some lineages evolved into limbs. Such fishes were the ancestors of amphibians, reptiles, birds, and mammals.

A modern-day fish out of water—a catfish, thrashing sideways and using its fins to pull its body forward across land

Focus on the Environment

Plumes, Global Impacts, and the Dinosaurs

Rise of the Ruling Reptiles. The first dinosaurs evolved early in the Mesozoic era. They weren't much larger than a wild turkey. Many sprinted about on two legs. Most had high metabolic rates and maybe they were warmblooded. During the Triassic, they weren't the dominant lineage on land. Center stage belonged to *Lystrosaurus* and other plant-eating, mammal-like reptiles too large to be bothered by most predators.

Then *Lystrosaurus* ran out of luck, and adaptive zones opened up for the dinosaurs, possibly when an object from space struck the earth. There is a crater about the size of Rhode Island in central Quebec (Figure *a*). Even if it is not *the* impact crater, it shows how huge the impact could have been. The blast wave, firestorm, earthquakes, and lava flows would have been horrific. Atmospheric distribution of rocks and water vaporized upon impact could have darkened the skies. If so, months of acid rain followed. Animals that survived this time of extinction (and later ones) were mostly small, metabolically active, and less vulnerable to drastic swings in climate.

The surviving dinosaurs underwent a major adaptive radiation. For 140 million years, their descendants were the ruling reptiles. Some, including "ultrasaurs," reached monstrous proportions. They were 15 meters tall.

Many dinosaurs perished in a mass extinction at the end of the Jurassic, then in a pulse of extinctions during the Cretaceous. Perhaps plumes of molten material from deep in the earth ruptured the crust, releasing enough carbon dioxide in the atmosphere to change the global temperature and climate. Or perhaps major environmental

disturbances followed a swarm of bombardments from outer space. Whatever the causes, conditions changed for the dinosaurs, and not all of them lived through it. Yet some lineages recovered and new forms replaced them. By the late Cretaceous, there were perhaps a hundred different genera of dinosaurs. Duckbilled dinosaurs appeared in forests and swamps. Tanklike *Triceratops* and other plant eaters flourished in more open regions. They were prey for the agile, fearsomely toothed *Tyrannosaurus rex*.

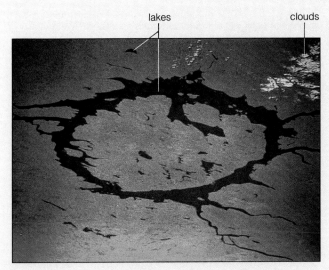

lakes   clouds

a What an impact crater looks like: Manicougan Crater, Quebec, where an object from space struck the earth.

boundary, many marine organisms perished during a mass extinction. Pangea started breaking up early in the Cretaceous, when dinosaurs ruled. About 127 million years ago, flowering plants emerged and began a spectacular radiation. Within 30 to 40 million years, they would displace the already declining conifers and related plants in nearly all environments.

About 120 million years ago, global temperatures shot up by 25 degrees. By one theory, a superplume (or a rash of them) originated deep in the earth, then broke through the crust. Plumes spreading out beneath the crust "greased" the earth's plates into moving twice as fast. Around the globe, volcanoes spewed nutrient-rich ashes. Basalt and lava poured from huge fissures. The

South Atlantic crust opened up like a zipper. Simultaneously, the plumes released huge amounts of carbon dioxide. This is one of the "greenhouse" gases that absorb heat radiating from the earth before it can escape into space. The nutrient-enriched planet warmed up—and stayed warm for 20 million years. On land and in shallow seas, photosynthetic organisms flourished. Their remains were slowly buried and converted into the world's oil reserves.

About 65 million years ago, the last of the dinosaurs and many lineages of marine organisms disappeared in a mass extinction. Their disappearance coincided with a huge asteroid impact, at a site that went on to become the northern Yucatán peninsula (see the *Focus* essay).

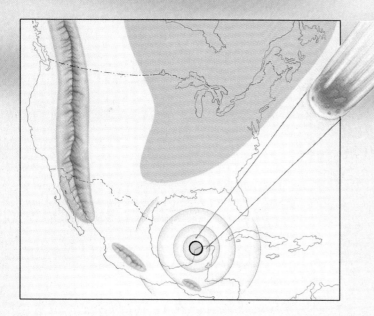

b (*Above*) Artist's interpretation of what might have happened during the last few minutes of the Cretaceous. (*Right*)If dinosaurs of this sort had not disappeared, would the then-tiny mammals ever have ventured out from under the shrubbery? Would *you* even be here today?

Horrendous End to Dominance.   The final blow came at the Cretaceous-Tertiary (K-T) boundary. Then, all (or nearly all) of the remaining dinosaurs perished when an asteroid apparently hit the earth. A thin layer of iridium-rich rock distributed around the world dates precisely to the K-T boundary. Iridium is rare on the earth's surface but common in asteroids.

By analyzing gravity maps, iridium levels in soils, and other evidence, researchers identified the impact site. Massive movements in the crust transported the site to what is now the northern Yucatán peninsula of Mexico (Figure *b*). The crater itself is 9.6 kilometers deep and 300 kilometers across—wider than the state of Connecticut. This means the asteroid hit the earth at 160,000+ kilometers (100,000 miles) per hour. At least 200,000 cubic kilometers of debris and dense gases were blasted skyward—enough to shut out sunlight for months. Monstrous, 120-meter waves raced across the oceans; the entire crust heaved with earthquakes.

Things haven't settled down much since. About 2.3 million years ago, for example, a huge object from space hit the Pacific Ocean. About the same time, vast ice sheets started forming abruptly in the Northern Hemisphere. Long-term shifts in climate may have been ushering in this most recent ice age, but a global impact might have accelerated the process. Water vaporized during the impact would have formed a cloud cover that prevented sunlight from reaching the earth's surface. As you will see in the next chapter, the early ancestors of humans were around when all of this happened. The extreme shift in climate surely put their adaptability to the test.

In short, the formation of ice sheets following the global impact is one more bit of information that compels us to look skyward, also, in our attempts to piece together the environments in which life evolved.

Figure 15.23 Representatives of the Cenozoic—the saber-tooth cat (*Smilodon*), a large bird (*Teratornis*), and a horse of the arid grasslands. This reconstruction is based on Pleistocene fossils from a pitch pool at Rancho La Brea, California.

The Cenozoic Era

The breakup of Pangea set events in motion that have continued to the present. At the dawn of the Cenozoic, land masses were on collision courses. Coastlines fractured. Intense volcanic activity and uplifting produced mountains along the margins of massive rifts and plate boundaries. The Alps, Andes, Himalayas, and Cascades were born through these upheavals. The geologic changes caused major shifts in climate that affected the further evolution of life. For example, vast, semiarid, cooler grasslands emerged. And plant-eating mammals and their predators radiated into these new adaptive zones (Figure 15.23).

Today the distribution of land masses has favored unparalleled richness in species diversity. The tropical forests of South America, Madagascar, and Southeast Asia may well be the richest ecosystems ever to appear on earth. Marine ecosystems of the island chains of the tropical Pacific are probably not far behind. Yet we are in the middle of what may turn out to be the greatest mass extinction of all time. About 50,000 years ago, early humans started following migrating herds of wild animals around the Northern Hemisphere. Within a few thousand years, major groups of large mammals disappeared. The pace of extinction has been accelerating ever since, as humans hunt animals for food, fur, feathers, or fun, and as they destroy habitats to clear land for cattle or crops. The repercussions are global in scope. We return to this topic in Chapter 37.

ORGANIZING THE EVIDENCE— CLASSIFICATION SCHEMES

We have mentioned only a tiny sample of the many millions of kinds of organisms that have originated and vanished over the past 3.5 billion years. Yet the sampling is enough to convey the challenge facing biologists who are attempting to make sense of life's diversity. The goal is to organize the known species into a classification system that is based on **phylogeny**. The word refers to evolutionary relationships, starting with the most ancestral forms and including all the branches leading to all of their descendants. In essence, a **classification system** is a useful way of retrieving information about organisms.

Classification requires the identification of each new species, which is assigned a unique name. (This work is called taxonomy.) A two-part name, devised centuries ago by Carolus Linnaeus, is still used. As Chapter 1 indicated, the first part is the generic name, or genus (plural, genera). All species that are similar to one another and distinct from others in certain traits are grouped in the same genus. The second part of the name designates the particular species within the genus.

Current classification schemes organize species into a series of ever more inclusive groupings. The following terms designate some of those groupings: kingdom, phylum (or division), class, order, family, genus, and on down to species. Table 15.1 shows how these groupings are used in the classification of four different organisms. The ranking reflects *relative* degrees of relationship. The same is true of Figure 15.24, which shows how humans are classified with respect to other primates.

In this book, we use a modified version of Robert Whittaker's phylogenetic system of classification. Major groups are assigned to five kingdoms, based on evidence of evolutionary relationship (Figure 15.25 and Appendix I). Regardless of its strengths, no classification system should be viewed as *the* system. Existing systems help summarize knowledge about the world of life. But any one of them may be modified as new evidence turns up.

Table 15.1   Classification of Four Organisms

Category (taxon)	Corn	Vanilla Orchid	Housefly	Human
Kingdom	Plantae	Plantae	Animalia	Animalia
Phylum (or Division)*	Anthophyta (flowering plants)	Anthophyta	Arthropoda	Chordata
Class	Monocotyledonae (monocots)	Monocotyledonae	Insecta	Mammalia
Order	Commelinales	Orchidales	Diptera	Primates
Family	Poaceae	Orchidaceae	Muscidae	Hominidae
Genus	*Zea*	*Vanilla*	*Musca*	*Homo*
Species	*Z. mays*	*V. planifolia*	*M. domestica*	*H. sapiens*

* Comparable taxons in different schemes.

Homo sapiens (only living species of this genus)

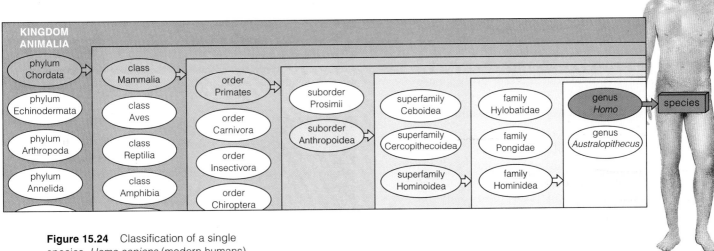

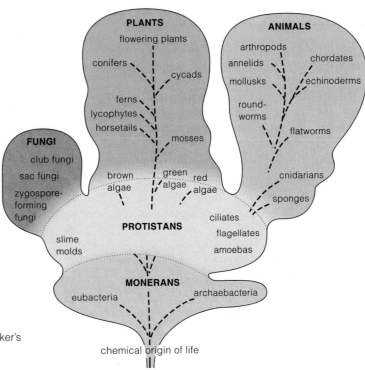

Figure 15.24   Classification of a single species, *Homo sapiens* (modern humans). Species are the only real entities in such schemes. Higher taxa (genera, families, orders, and so on) reflect our thinking about evolutionary distances among species. Not all groupings are shown, as Appendix I indicates.

Figure 15.25   Modified version of Robert Whittaker's five-kingdom system of classification.

SUMMARY

1. Evolution proceeds by modifications of already existing species. Therefore, a continuity of relationship exists among all species, past and present. The patterns, trends, and rates of change among groups of species over long spans of time are called macroevolution.

2. Evidence of evolutionary relationships comes from the fossil record and earth history, radioactive dating methods, comparative morphology, and comparative biochemistry. The evidence reveals similarities and differences in body form, functions, behavior, and biochemistry.

3. The fossil record varies as a function of the kinds of organisms represented, where they lived, and the geologic stability of the region since the time of fossilization. Together with the geologic record, it shows that the evolution of life has been linked, since the time of its origin, to the evolution of the earth.

4. Comparative morphology often reveals similarities in embryonic development that indicate evolutionary relationship. It may reveal homologous structures, shared as a result of descent from a common ancestor.

5. Comparative biochemistry relies on gene mutations that have accumulated in different species. Neutral mutations serve as a molecular clock for dating divergences from a common ancestor. Methods such as DNA-DNA hybridization reveal the degree of similarity between genes and gene products of different species.

6. Large-scale geologic events caused profound shifts in land masses, shorelines, and oceans. Also, impacts by huge objects from outer space had major effects on global temperatures and climate. These events redirected the course of biological evolution many times.

7. A mass extinction is a catastrophic event in which major lineages perish abruptly. An adaptive radiation is a burst of evolutionary activity; a lineage fills the environment with new species. Both events have changed the course of biological evolution many times.

8. An evolutionary tree shows presumed relationship by way of descent. Its branches are lines of descent (lineages). Its branch points are speciations brought about by microevolutionary processes, including mutation and natural selection. Opinions differ about the rates of change in an evolving lineage.

9. A phylogenetic classification scheme is one that reflects presumed evolutionary relationships, starting with the most ancestral forms and including all branches leading to all of the descendants.

Review Questions

1. Will the fossil record ever be complete? Why or why not? *211*

2. Explain the difference between:
   a. microevolution and macroevolution *198, 210*
   b. homologous and analogous structures *214–215*
   c. morphological divergence and convergence *214–215*

3. Define adaptive radiation and adaptive zone. *217*

4. Describe the middle-class flight from cities to suburbs and rural areas in terms of gene flow and other microevolutionary processes. Is this an example of an adaptive radiation? *198, 217*

5. What did experiments by Stanley Miller and Sidney Fox suggest with respect to the chemical origin of life? *220, 221*

6. The Atlantic Ocean is widening, and the Pacific and Indian oceans are closing. Many millions of years from now, the continents will collide and form a second Pangea. Write a short essay on what conditions might be like on that future supercontinent and what types of organisms might survive there.

7. This chapter outlined the major categories of classification schemes, but many more exist. For example, using Table 15.1, Figure 15.24, and Appendix I as guides, fill in as many groupings as you can for the following organisms:

corn   human

vanilla orchid   housefly

Superkingdom _____
Kingdom _____
Phylum/Division _____
Subphylum _____
Superclass _____
Class _____
Subclass _____
Order _____
Suborder _____
Infraorder _____
Superfamily _____
Family _____
Genus _____
Species _____

1. The fossil record is _____ and _____ in quality.
   a. complete; high
   c. incomplete; uniform
   b. complete; low
   d. incomplete; uneven

2. The geologic time scale starts with the _____ era.
   a. Paleozoic
   d. Proterozoic
   b. Mesozoic
   e. Cenozoic
   c. Archean

3. Morphological divergences may lead to _____ .
   a. analogous structures
   c. convergent structures
   b. homologous structures
   d. a and c

4. Morphological convergences may lead to _____ .
   a. analogous structures
   c. divergent structures
   b. homologous structures
   d. a and c

5. Analogous structures are evidence of _____ .
   a. descent from a common ancestor
   b. separate lineages adopting similar life-styles
   c. separate lineages adapted to similar environments
   d. b and c

6. A molecular clock is _____ .
   a. based on new technology from General Electric
   b. an internal timing mechanism in organisms
   c. an accumulation of neutral mutations
   d. an accumulation of lethal mutations

7. In an evolutionary tree, branch points are _____ .
   a. a lineage or single line of descent
   b. areas in time that were rich in fossils
   c. speciation events
   d. mass extinctions

8. Increasingly inclusive groupings in classification systems range from _____ to _____ .
   a. kingdom; species
   c. genera; kingdom
   b. kingdom; genera
   d. species; kingdom

9. A classification system based on evolutionary relationships is _____ .
   a. epigenetic
   c. Linnean
   b. phylogenetic
   d. b and c

10. Match these terms appropriately.
    _____ analogous structure
    _____ homologous structure
    _____ adaptive radiation
    _____ mass extinction
    _____ phylogeny

    a. burst of evolutionary activity
    b. descent from common ancestor is implied
    c. evolutionary relationships from most ancestral form and including all branches to all descendants
    d. wipe-out of major lineages
    e. similarities, but evolutionarily remote ancestors

Selected Key Terms

adaptive radiation *217*
adaptive zone *217*
analogous structures *214*
classification system *230*
comparative morphology *213*
DNA-DNA hybridization *215*
evolutionary tree *216*
fossil *211*
geologic time scale *211*
homologous structures *214*
lineage *210*
macroevolution *210*
mass extinction *216*
molecular clock *215*
morphological convergence *214*
morphological divergence *214*
phylogeny *230*
stromatolite *224*

Readings

Bambach, R.; C. Scotese; and A. Ziegler. 1980. "Before Pangea: The Geographies of the Paleozoic World." *American Scientist* 68(1): 26–38.

Dobb, E. February 1992. "Hot Times in the Cretaceous." *Discover* 13: 11–13.

Dott, R., Jr., and R. Batten. 1988. *Evolution of the Earth.* Fourth edition. New York: McGraw-Hill.

Horgan, J. February 1991. "Trends in Evolution: In the Beginning. . . ." *Scientific American* 264(2): 116–125.

16 HUMAN EVOLUTION

The Cave at Lascaux and the Hands of Gargas

Half a century ago, on a warm autumn day, four boys out for a romp stumbled into an intricately tunneled cave near Lascaux, a town in France. What they discovered stunned the world. Magnificent images swept out across the cave walls (Figure 16.1). Between 17,000 and 20,000 years ago, by the flickering light of crude oil lamps, artists captured the graceful, dynamic lines of bison, stags, lions, stallions, ibexes, a rhinoceros, and a heifer now named the Great Black Cow.

About 5,000 years earlier, artists put more than 150 outlines and imprints of hands on walls in the cave of Gargas, in the Pyrenees mountains. Prehistoric art also exists in other caves of France, Spain, and Africa.

Who were the people who did this? From fossils, we know they were anatomically like us. From the way they planned and executed their art, we sense a level of abstract thinking that is unique to humans.

The quality of "humanness" did not materialize out of thin air. The story of the human species began 65 million years ago, with the origin of primates in tropical forests.

In turn, the primate story began more than 200 million years ago, with the origin of mammals. The first animals emerged between 2.5 billion and 750 million years ago—and so on back in time, to the origin of the first cells.

As we poke about the branches of our family tree, keep this greater story in mind. *Our "uniquely" human traits emerged through modification of traits that had already evolved in ancestral forms.* At each branch in the family tree, mutations led to workable changes in traits, which proved useful in prevailing environments. From this perspective, "ancient" cave paintings are the legacy of individuals who departed only yesterday, so to speak. The artists of Lascaux and Gargas are not remote from us. They *are* us.

Figure 16.1 Part of the human cultural heritage—prehistoric cave paintings, a unique outcome of a long history of biological evolution.

1. Unlike other primates, humans are not restricted to a single habitat. They can remain flexible and adapt to a wide range of challenges in complex, unpredictable environments. This capacity is an outcome of trends that can be discerned in the evolutionary history of certain primate lineages. The trends involved changes in bones, muscles, teeth, sensory systems, and the brain itself.

2. Ancestral, four-legged primates underwent skeletal modifications that led to an upright stance, which in turn freed the hands for new functions.

3. Modifications in handbones and muscles led to a capacity to hold, carry, use, and make objects.

4. Reorganization of the skull and eye sockets led to increased reliance on daytime vision.

5. Teeth became less specialized, and a greater variety of food sources was tapped.

6. Brain tissue expanded and became more complex. Evolution of the brain was interlocked with cultural evolution.

THE MAMMALIAN HERITAGE

Humans belong to the class Mammalia, one of seven classes of vertebrates. Like other vertebrates, **mammals have an internal skeleton.** The skeleton's main axis is a bony column (backbone) with a cordlike bundle of nerves threading through it. Perched above the backbone is a skull. Inside the skull is a brain that is functionally divided into three regions (hindbrain, midbrain, forebrain). Also inside are sensory organs concerned with sight, sound, balance, and smell.

Of all vertebrates, only female mammals nourish their young with milk produced by mammary glands. Adult mammals feed and protect their young for an extended period and serve as models for their behavior. The young have an inborn capacity to learn and repeat a set of behaviors that have survival value. Mammals in general show behavioral flexibility. They can expand on the basics with novel forms of behavior.

Mammals also are known by their dentition (the type, number, and size of teeth). They have four types of upper and lower teeth that match up and work together to crush, grind, or cut food (Figure 16.2).

Teeth are clues to an animal's diet and life-style. **Incisors** are like flat chisels or cones. They nip or cut food. Horses and other mammals that graze in open

a

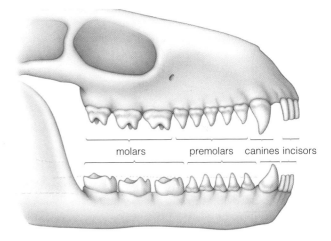

b

Figure 16.2 (**a**) Mammals descended from reptiles. Unlike mammals, reptiles have peglike upper teeth that don't match up with peglike lower ones. (**b**) Generalized picture of the type, number, size, and arrangement of mammalian teeth.

molars   premolars canines incisors

grasslands have pronounced incisors. **Canines** have piercing points. Meat-eating animals use long, sharp canines to pierce prey. **Premolars** and **molars** (the cheek teeth) are a platform for food. Their surface bumps (cusps) help crush, grind, and shear material. If a mammal has large, flat-surfaced cheek teeth, you can be fairly certain its ancestors evolved in places where fibrous plants were abundant food sources.

Teeth fossilize very well. As you will see, fragments of jaws and teeth from human ancestors give clues to their life-styles.

Mammals have an internal skeleton and a nerve cord, a three-part brain and sensory organs inside the skull, mammary glands, and distinctive teeth. Their young require an extended period of dependency and learning.

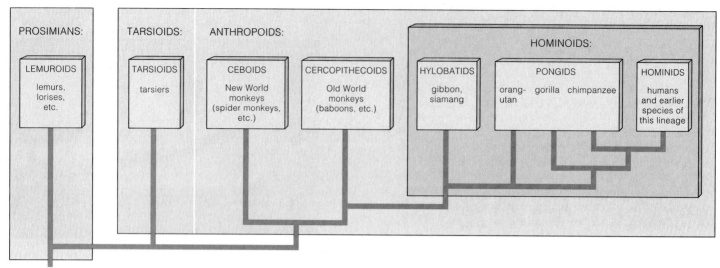

tree-dwelling, rodentlike primate of Paleocene

Figure 16.3 Simplified evolutionary tree for primates. The two highest groupings (the prosimians, and the tarsioids and anthropoids) are shown in green. Tan boxes indicate the major groups of living primates. A few representative members are named. All monkeys, apes, and humans are anthropoids. The only hominids are humans and earlier, now-extinct species of the same lineage.

Primate Classification

Figure 16.3 is an evolutionary tree for **primates,** which include prosimians, tarsioids, and anthropoids. Figure 16.4 shows a few representatives. Prosimians are the oldest primate lineage (*pro-,* before; *simian,* ape). For millions of years, prosimians dominated the trees in North America, Europe, and Asia. That was before monkeys and apes evolved and almost displaced them entirely. Tarsioids, represented by tarsiers of islands in southeastern Asia, are small primates with features that place them between prosimians and anthropoids.

Monkeys, apes, and humans are all anthropoids. In structural details and biochemistry, apes are much more similar to humans than to monkeys. That is why apes and humans (and recent human ancestors) are classified together, as **hominoids.** However, the divergence from their common ancestor began many millions of years ago. All species that appear on the separate evolutionary road leading to humans are classified as the **hominids.**

From Primate to Human: Key Evolutionary Trends

Most primates live in tropical or subtropical forests, woodlands, or savannas (open grasslands with a few stands of trees). Like their ancient ancestors, the vast majority are tree dwellers. Yet no one feature sets "the primates" apart from other mammals. Each primate lineage evolved in a distinct way and has its own defining traits. Five trends define the lineage we are concerned with here. They were set in motion when primates first started adapting to life in the trees, and they contributed to the emergence of modern humans:

1. Skeletal changes leading to upright walking, which freed the hands for new functions.

2. Changes in bones and muscles leading to refined hand movements.

3. Less reliance on a sense of smell and more on daytime vision.

4. Changes leading to fewer, less specialized teeth.

5. Brain elaboration and changes in the skull that led to speech. These developments became interlocked with each other and with cultural evolution.

Upright Walking. Of all primates, only humans can stride freely on two legs for a long time. Their habitual two-legged gait is called **bipedalism.**

By contrast, monkeys are adapted to life in the trees. Their skeleton permits rapid climbing, leaping, and running along branches. Notice, in Figure 16.5, how their armbones and legbones are about the same length. This means monkeys can run palms-down (Figure 16.4*b*). Try this yourself and see what happens.

Unlike monkeys, apes can hang onto overhead branches and use their long arms to carry some body weight. The arms often support body weight when an ape is on the ground. Because of the way their shoulder blades are positioned, apes can swivel the arms freely above the head when the body is erect or semi-erect.

Compared with monkeys and apes, humans have a shorter, S-shaped, and somewhat flexible backbone. The position and shape of their backbone, shoulder blades, and pelvic girdle are the basis of bipedalism (Figure 16.5). These skeletal traits emerged not long after the divergence that led to the hominids.

a

b

c

Figure 16.4 Representative primates. Gibbons (**a**) have limbs and a body adapted for swinging arm over arm through the trees. Monkeys are quadrupedal (four-legged) climbers, leapers, and runners, as the spider monkey in (**b**) demonstrates. Tarsiers (**c**) are vertical clingers and leapers.

Figure 16.5 Comparison of the skeletal organization and stance of monkeys, apes (the gorilla is shown here), and humans. Modifications of the basic mammalian plan have allowed three distinct modes of locomotion. The quadrupedal monkeys climb and leap, and apes climb and swing by their forelimbs. Both modes of locomotion are well suited for life in the trees. Humans are habitual two-legged walkers. The drawings are not to the same scale.

monkey

gorilla

human

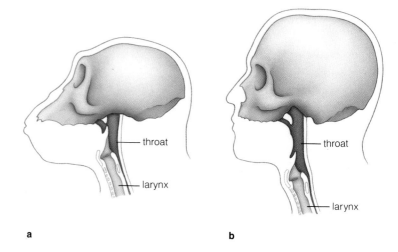

Figure 16.6 The structural basis of speech. (**a**) Early humans had a skull with a flattened base. (The same configuration exists in modern chimpanzees.) Their larynx (the tube leading to the lungs) was not that far below the skull, so the throat volume was small. (**b**) In modern humans, the base of the skull angles down sharply during development, and this moves the larynx down also. Notice the resulting increase in the volume of the throat—the area in which sounds are produced.

Precision Grips and Power Grips. The first mammals were four-legged. They spread their toes apart to help support the body as they walked or ran. Primates still spread their toes or fingers. Many also make cupping motions, as when monkeys bring food to the mouth. Two other hand movements developed among ancient tree-dwelling primates. Through alterations in handbones, fingers could be wrapped around objects (*prehensile* movements), and the thumb and the tip of each finger could now touch each other (*opposable* movements).

Hands began to be freed from load-bearing functions when early primates lived in the trees. Later, when hominids were evolving, refinements in hand movements led to the precision grip and power grip:

These hand positions gave early humans the capacity to make and use tools. They were a foundation for unique technologies and cultural development.

Enhanced Daytime Vision. Early primates had an eye on each side of the head. Later ones had forward-directed eyes. This arrangement is better for sampling shapes and movements in three dimensions. Through other modifications, the eyes were able to respond to variations in color and light intensity (dim to bright). These visual stimuli are typical of life in the trees.

Teeth for All Occasions. Monkeys have long canines and rectangular jaws. Humans have smaller teeth, of about the same length, and a bow-shaped jaw. Teeth and jaws became modified on the road from early primates to humans. There was a shift from eating insects, then fruit and leaves, and on to a mixed diet.

Better Brains, Bodacious Behavior. Living on tree branches favored shifts in reproductive and social behavior. Imagine the advantages of single births over litters, for example, or of clinging longer to the mother. In many lineages, parents started to invest more in fewer offspring. They formed strong bonds with their young, maternal care became intense, and the learning period grew longer.

Before this, the brain had been increasing in size and complexity. Now, however, brain regions concerned with encoding and processing information started to expand dramatically. New behavior stimulated development of those regions—which stimulated more new behavior. *Brain modifications and behavioral complexity became highly interlocked.*

The interlocking is most evident in the parallel evolution of human culture and the human brain. Here we define **culture** as the sum total of behavior patterns of a social group, passed between generations by learning and by symbolic behavior—especially language. The capacity for language arose among ancestral humans. And it arose through changes in the skull and expansion of parts of the brain (Figure 16.6).

PRIMATE ORIGINS

Primates evolved from ancestral mammals more than 60 million years ago, during the Paleocene. The first ones resembled small rodents or tree shrews (Figures 16.7 and 16.8). Like tree shrews, they probably had huge appetites and foraged at night for insects, seeds, buds, and eggs on the forest floor. They had a long snout and a good sense of smell, useful for detecting predators or food. They could claw their way up through the shrubbery, although not with much speed or grace.

Between 54 and 38 million years ago (the Eocene), some primates were staying in the trees. Fossils give evidence of increased brain size, a shorter snout, enhanced daytime vision, and refined grasping movements. How did these traits evolve?

Consider the trees. They offered food and safety from ground-dwelling predators. *They also were a place of uncompromising selection.* Imagine dappled sunlight, boughs swaying in the wind, colorful fruit tucked among the leaves, perhaps predatory birds. A long snout would not have been of much use—air currents disperse odors. A brain that assessed depth, shape, movement, and color would have been a definite plus. So would a brain that worked fast when its owner was running, swinging, and leaping (especially!) from branch to branch. Distance, body weight, winds, and suitability of the destination had to be estimated, and adjusting for miscalculations had to be quick.

By 35 million years ago (the dawn of the Oligocene), tree-dwelling anthropoids had evolved in tropical forests. The ancestors of monkeys, apes, and humans were among them. Some lived above swamps infested with predatory reptiles. Perhaps that's why they rarely ventured to the ground—and why it was imperative to think fast and grip strongly. Slip-ups were possible. A surprising number of primates still fall out of trees.

During the Miocene (25 million to 5 million years ago), continents began to assume their current positions (Figure 15.12). Climates were becoming cooler and drier. An adaptive radiation of apelike forms—the first hominoids—occurred during this epoch. By 13 million years ago, ape populations were scattered through Africa, Europe, and southern Asia. Among them were the chimpanzee-sized **dryopiths** (Figure 16.8). Most Miocene apes became extinct around this time. However, fossils and biochemical studies point to three divergences that occurred between 10 million and 5 million years ago. Two gave rise to the gorilla and chimpanzee lineages. The third gave rise to the early hominids—including the ancestors of humans.

Figure 16.7 A night-foraging tree shrew of Indonesia.

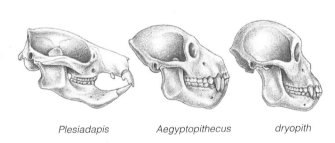

Plesiadapis Aegyptopithecus dryopith

Figure 16.8 Comparison of skull shape and teeth of some extinct primates. *Plesiadapis*, a Paleocene primate, had rodentlike teeth. *Aegyptopithecus*, an Oligocene anthropoid, probably predates the divergence leading to Old World monkeys and the apes. An adaptive radiation that began in the Miocene produced many dryopiths and other apelike forms. The drawings are not to the same scale. In size, *Plesiadapis* was like a tiny tree shrew. *Aegyptopithecus* was monkey-sized, and the dryopiths, chimpanzee-sized.

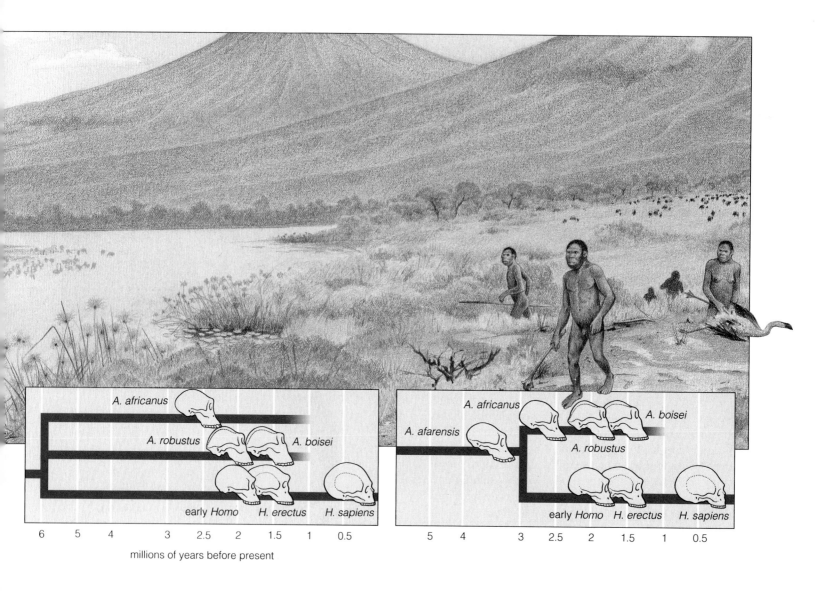

Figure 16.9 Reconstruction of one of the environments in which the first humanlike forms (early hominids) evolved. Their evolutionary connections with one another and with later hominids are not understood. Several phylogenetic trees have been proposed, including the two shown here for australopiths and for species on the road to modern humans.

THE HOMINIDS

The first hominids evolved somewhere around the Miocene-Pliocene boundary (5 million years ago). This was a "bushy" period of evolution—rapid branchings and radiations produced a variety of humanlike forms. We have only a limited number of fossil fragments, so we don't really know how they were related. Yet many had three features in common:

1. Bipedalism
2. Omnivorous behavior (varied diets)
3. Further brain expansion and elaboration

These features coincided with the emergence of cooler, drier climates. In Africa, rain forests were giving way to mixed woodlands and grasslands (Figure 16.9). Yet hominids entering the new, unpredictable environments had use of hands. They could survive on new kinds of food. And they could learn to adapt. They had three features that gave them **plasticity**—an ability to remain flexible and adapt to a wide range of demands.

All three features emerged through modifications of traits that can be observed among the other primates. *They were based on the primate heritage.*

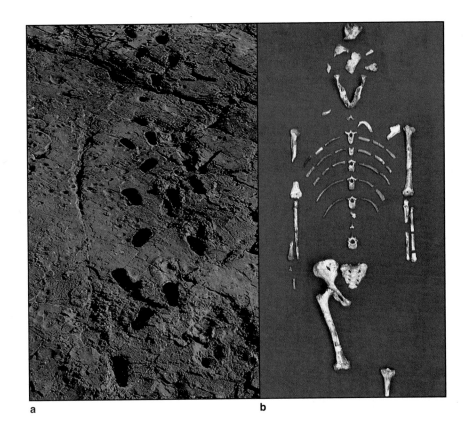

Figure 16.10 (**a**) Footprints made in soft, damp volcanic ash 3.7 million years ago at Laetoli, Tanzania, as discovered by Mary Leakey. The arch, big toe, and heel marks were made by bipedal hominids. (**b**) Fossil remains of Lucy, one of the earliest known australopiths. The density of her limb bones suggests she had strong muscles.

Australopiths

The earliest known hominids are called **australopiths**. They fall into two broad categories:

1. Gracile forms (slightly built), currently named *Australopithecus afarensis* and *A. africanus.*

2. Robust forms (muscular, heavily built), including *A. boisei* and *A. robustus.*

The australopiths were transitional forms. Like small apes, they had small brains (about 400 cubic centimeters). Like later hominids, they were bipedal. Leg-bone fragments, 4 million years old, have muscle attachment sites like those of two-legged walkers. Figures 16.10 and 16.11 show the reconstructed form and skeleton of a female dubbed Lucy. Her thighbones angled inward so that her body weight was centered directly beneath the pelvis—a sure sign of bipedalism. Thighbones angle outward in apes, which have a waddling, four-legged gait. Most telling, *the australopiths left footprints* (Figure 16.10).

Like later hominids, australopiths had slightly bowed jaws. Some apparently were vegetarians, with teeth specialized for grinding tough plant material, including seeds and nuts. Others were omnivores.

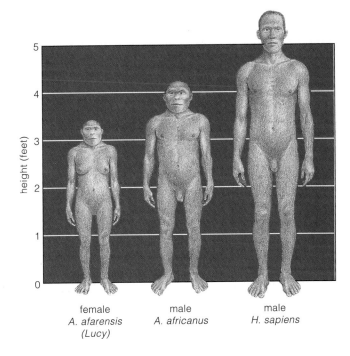

Figure 16.11 Comparison of the size and stature of australopiths and modern humans.

Figure 16.12 Comparison of skull shapes of early hominids relative to anatomically modern humans (*Homo sapiens sapiens*). The drawings are not all to the same scale. White areas of skulls are reconstructions.

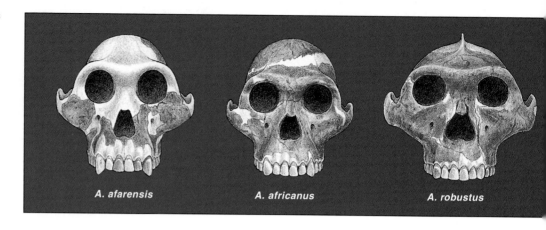

A. afarensis   A. africanus   A. robustus

Stone Tools and Early *Homo*

Early on, hominids may have used sticks and other perishable items as tools, much like apes do today. We only know of stone tools, 2.5 million years old. Skull fragments of a hominid called **early *Homo*** are dated at 2.4 million years. Early *Homo* may have made the tools. It had a smaller face, less specialized teeth, and a larger brain than australopiths (Figure 16.12). It apparently scavenged plants and small animals.

The oldest known "manufactured" tools were crudely chipped pebbles. They may have been used to dig up roots and tubers, smash bones for marrow, or poke insects out of treebark. The tools were first discovered by Mary Leakey at Olduvai Gorge (Figure 16.13). This African gorge cuts through layers of sedimentary rock. Its more recent layers contain more sophisticated tools. The sequence reflects the increasingly refined ways in which the hominids exploited a major food source—the abundant game of open grasslands.

Homo erectus

Our direct ancestors emerged between 1.5 million and 300,000 years ago, during the Pleistocene. ***Homo erectus*** had archaic traits, such as a heavy browridge (Figure 16.12). Yet in many other traits, members of this species were like modern humans. Transitional forms evolved that could easily be grouped with our own species.

H. erectus faced monumental challenges. Global temperatures were declining. More than once, glaciers advanced and retreated over vast areas of northern Europe, Asia, and North America. Some glaciers were 2 miles high! Between glacial episodes, when temperatures rose, *H. erectus* migrated out of Africa and into Southeast Asia, China, and possibly Europe—just about every place that could be reached on foot.

With long legbones and a narrow pelvis, *H. erectus* was built for travel. Its brain was put to the test during the treks into new environments. Increases in brain complexity were probably interlocked with rapid cultural evolution. Over time, *H. erectus* became better at toolmaking and learned how to control fire.

Homo sapiens

Somewhere between 300,000 and 200,000 years ago, *Homo sapiens* (modern humans) evolved in Europe, the Near East, and China. Various forms coexisted in time with *H. erectus*, which then gradually disappeared.

Compared to their ancestors, early *H. sapiens* had smaller teeth and jaws. Many individuals had a chin. They had thinner facial bones, a larger brain, and a rounder, higher skull. About 1.5 million years ago, the skull's base began to change in some groups. It gradually angled downward, forcing the larynx down. This change led to complex human language (Figure 16.6).

The capacity for language may not have been well developed among the Neandertals. Fossils of this group of early humans were first discovered in Germany's Neander Valley. The group arose 130,000 years ago. Their skull base was flatter than that of *H. erectus*. Although large brained, they had heavy facial bones and often large browridges.

Neandertal populations existed in southern France, central Europe, and possibly Israel and Iraq. They lived at the edge of forests, in caves, rock shelters, and open-air camps. Apparently the males hunted or gathered food that passed by their doorstep, so to speak. They did not learn how to exploit the abundant game herds of the world's vast grasslands.

About 35,000 or 40,000 years ago, Neandertals disappeared. Anatomically modern humans (*H. sapiens sapiens*) arose at that time. These groups learned to store and share food among themselves. They learned to anticipate the seasonal migrations of grazing animals in the grasslands—and to plan community hunts. They developed spectacularly rich cultures. They were the creators of exquisite tools and artistic treasures at Lascaux and elsewhere.

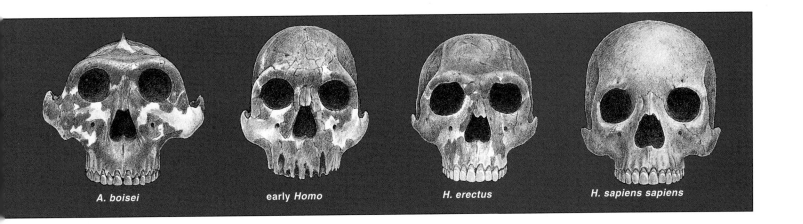

A. boisei   early *Homo*   H. erectus   H. sapiens sapiens

Figure 16.13 A sampling of the 37,000+ stone tools from Olduvai Gorge. *Upper row, left to right:* A crude chopper. Two more advanced forms having a joint as well as a sharp edge. The stone ball may reflect a transition to using tools aggressively against animals. It resembles Argentine bolas, which are strung on lengths of hide and thrown at animals to entangle the legs and bring them down. *Lower row:* A hand ax and a cleaver.

By one theory, *H. sapiens* evolved from different hominid stocks in each major geographic region. Maybe each geographic group remained isolated, or maybe gene flow knitted them together over time. By another theory, *H. sapiens* evolved in Africa from a single ancestral stock, then radiated into different parts of the world over the past 100,000 years. If so, did they displace native hominid populations entirely, or were there instances of gene flow?

At this writing we simply don't have the answers. We can only look forward to new discoveries that may help clarify the picture.

From 40,000 years ago to the present, human evolution has been almost entirely cultural rather than biological, and so we leave the story. From the biological perspective, however, we can make these concluding remarks: Humans spread throughout the world by rapidly devising the cultural means to deal with a broad range of environments. Compared with their predecessors, modern humans developed rich and varied cultures, moving from "stone-age" technology to the age of "high tech." Yet hunters and gatherers persist in parts of the world, attesting to the great plasticity and depth of human adaptations.

SUMMARY

1. Like other mammals, humans and other primates have an internal skeleton, a complex brain and sensory organs within a skull, mammary glands (in females), and distinctive teeth. Adults nourish, protect, and serve as behavioral models for the young.

2. Primates include prosimians (lemurs and related forms) and anthropoids (including tarsioids, monkeys, apes, and humans). Only apes and humans are hominoids. Only modern humans (*H. sapiens*) and their most recent ancestors (from *A. afarensis* to *H. erectus*) are further classified as hominids.

3. Unlike most primates, humans are not restrictively specialized; they remain flexible and adapt to a wide range of challenges in complex, unpredictable environments. This capacity resulted from five evolutionary trends that occurred among certain primate lineages:

   a. From a four-legged gait to bipedalism.

   b. Increased manipulative skills owing to changes in the hands, which began to be freed from load-bearing functions among tree-dwelling primates.

   c. Less reliance on the sense of smell and more on enhanced daytime vision.

   d. From specialized to omnivorous eating habits.

   e. Increased brain complexity, together with skull changes that led to language.

4. The first primates—small, rodentlike mammals—evolved more than 60 million years ago. Anthropoids, including the ancestors of monkeys, apes, and humans, had evolved by 35 million years ago. The first hominids (australopiths) arose about 5 million years ago.

5. About 2.4 million years ago, early *Homo* evolved. It may have "manufactured" the first stone tools. Our direct ancestor, *Homo erectus*, existed between 1.5 million and 300,000 years ago. It was larger brained than earlier hominids and adapted to a wide range of habitats. Modern humans (*H. sapiens*) arose between 300,000 and 200,000 years ago. Starting about 40,000 years ago, cultural evolution outstripped biological evolution of the human form.

Selected Key Terms

australopith *241*	hominid *236*	molar *235*
bipedalism *236*	hominoid *236*	plasticity
canine *235*	*Homo erectus 242*	(behavioral) *240*
culture *238*	*Homo sapiens 242*	premolar *235*
dryopith *239*	incisor *235*	primate *236*
early *Homo 242*	mammal *235*	

Review Questions

1. What are the general evolutionary trends that occurred among the primates as a group? What way of life apparently was the foundation for these trends? *236*

2. What conditions seem to have been responsible for the great adaptive radiation of apelike forms during the Miocene? *239*

3. What is the difference between "hominoid" and "hominid"? Are we hominoids, hominids, or both? *236*

Readings

Stringer, C. December 1990. "The Emergence of Modern Humans." *Scientific American* 263: 98–104.

Weiss, M., and A. Mann. 1990. *Human Biology and Behavior*. Fifth edition. New York: Harper Collins.

FACING PAGE: *Patterns of diversity in nature, here represented by different species of plants and fungi.*

17 VIRUSES, BACTERIA, AND PROTISTANS

The Unseen Multitudes

Did a friend ever mention that you are nearly 1/1,000 of a mile tall? Probably not. What would be the point of measuring people in units as big as miles? Yet we think this way, in reverse, when we measure **microorganisms**. These are mostly single-celled organisms too small to be seen without a microscope. The bacteria in Figure 17.1 are a case in point. To measure them, you'd have to divide a meter into a thousand units (millimeters). Then you'd have to divide one millimeter into a thousand smaller units (micrometers). One millimeter would be as small as the dot of this "i." A *thousand* bacteria would fit side by side across the dot!

With this chapter we turn to the great spectrum of life, starting with its tiniest members—bacteria and protistans. To be sure, viruses are smaller. We measure them in nanometers (one billionth of a meter). But viruses are not living things. We consider them here because they infect just about every known organism.

Microbes vastly outnumber all other organisms combined. Their reproductive potential is staggering.

Under ideal conditions, some bacteria can divide about every twenty minutes. At that rate, a single bacterium could have nearly a billion descendants in ten hours! So why don't microbes take over the world? Sooner or later, their activities typically ruin the conditions favoring their reproduction. Also, they eat one another, other organisms attack them, and viruses and seasonal changes help keep them in check.

Many microbes are **pathogens** (infectious, disease-causing agents). They invade and multiply inside other organisms, and disease follows when their activities damage tissues and interfere with normal body functions. Yet we couldn't do without other kinds of microbes. Think of the vast numbers of photosynthetic cells in the sea (page 81). Those cells provide food and oxygen for entire communities. They also have major roles in the global cycling of carbon. Still other microbes decompose organic debris and so help cycle the nutrients that sustain nearly all other organisms on earth.

Figure 17.1 How small are bacteria? Shown here, *Bacillus* cells on the tip of a pin. The cells in (**c**) are magnified 14,000 times.

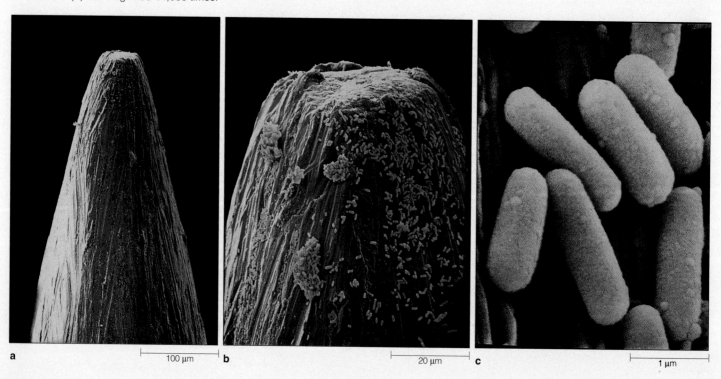

a   100 µm   b   20 µm   c   1 µm

KEY CONCEPTS

1. All living organisms have the metabolic means to maintain themselves, grow, and reproduce. Viruses are nonliving, infectious particles. It takes the metabolic machinery of a living host cell to assemble new ones.

2. Bacteria are single-celled, prokaryotic organisms. They do not have the membrane-bound nucleus and other organelles seen in eukaryotes. Yet they show great metabolic diversity, and many make complex behavioral responses to the environment.

3. Virtually all single-celled eukaryotes belong to the protistan kingdom. Some protistan lineages also extend continuously from single-celled forms into the kingdoms of animals, plants, and fungi. Unlike the multicelled members of these other kingdoms, protistans do not have specialized tissues.

4. Some protistans may resemble the first eukaryotes, which arose more than a billion years ago from prokaryotic ancestors. According to the theory of endosymbiosis, certain key features of eukaryotes resulted from accidental partnerships between prokaryotic species, which became permanently interdependent.

VIRUSES

In ancient Rome, *virus* meant "poison" or "venomous secretion." In the late 1800s, this rather nasty word was bestowed on newly discovered pathogens, smaller than the bacteria being studied by Louis Pasteur and others. Many viruses deserve the name. They attack humans, cats, cattle, insects, crop plants, fungi, bacteria, even other viruses. You name it, and there probably are viruses that can infect it.

Today we define a **virus** as a noncellular infectious agent having two characteristics. First, a virus consists of a nucleic acid core (its genetic material) inside a protective protein coat. Second, a virus cannot reproduce itself. It can be reproduced only after its genetic material enters a host cell and subverts the cell's biosynthetic machinery. Thus a virus is no more alive than a chromosome is alive.

The genetic material of a virus is DNA *or* RNA. The viral coats are not all alike. Some coats even become wrapped in membrane remnants from the previously infected cell. The virus particle shown in Figure 17.2c is an example of this.

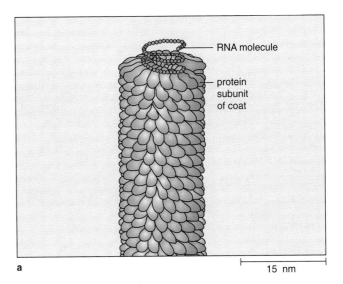

a   15 nm

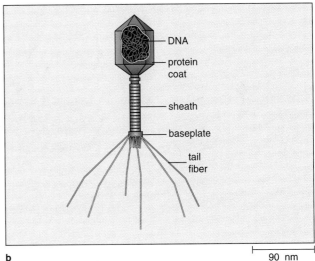

b   90 nm

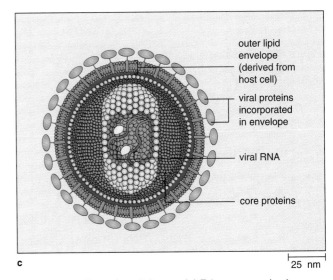

c   25 nm

Figure 17.2 Examples of viruses. (**a**) Tobacco mosaic virus, with its rod-shaped coat, is about 300 nanometers long. (**b**) T4 bacteriophage. (**c**) HIV, with its lipid envelope. Compare page 448.

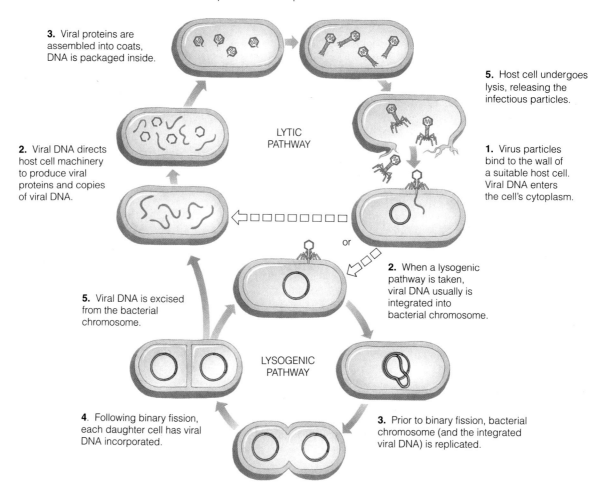

4. Tail fibers, other components are added to coats. New viral particles are completed.

3. Viral proteins are assembled into coats, DNA is packaged inside.

5. Host cell undergoes lysis, releasing the infectious particles.

2. Viral DNA directs host cell machinery to produce viral proteins and copies of viral DNA.

LYTIC PATHWAY

1. Virus particles bind to the wall of a suitable host cell. Viral DNA enters the cell's cytoplasm.

or

2. When a lysogenic pathway is taken, viral DNA usually is integrated into bacterial chromosome.

5. Viral DNA is excised from the bacterial chromosome.

LYSOGENIC PATHWAY

4. Following binary fission, each daughter cell has viral DNA incorporated.

3. Prior to binary fission, bacterial chromosome (and the integrated viral DNA) is replicated.

Figure 17.3 Examples of viral replication pathways. (The lysogenic route is one of the temperate pathways.)

Viral Replication

Viruses replicate in a variety of ways. But nearly all of the replication pathways include five basic steps:

1. The virus attaches to a host cell. Any cell is a suitable "host" if a virus can chemically recognize and lock onto specific molecular groups at its surface.

2. The whole virus or its genetic material alone enters the cell's cytoplasm.

3. In an act of molecular piracy, the viral DNA or RNA directs the host cell into producing many copies of viral nucleic acids and proteins, including enzymes.

4. The viral nucleic acids and proteins are put together to form new virus particles.

5. The newly formed virus particles are released from the infected cell.

Two kinds of pathways are common. In a *lytic* pathway, steps 1 through 4 proceed rapidly, and virus particles are released by lysis. "Lysis" means the host cell's plasma membrane is damaged and its cytoplasm leaks out. Cell death is quick (Figure 17.3).

In *temperate* pathways, a viral infection enters a "latent" period, meaning the host cells aren't killed outright. Sometimes genetic recombination occurs. In the Figure 17.3 example, a viral enzyme cuts the host chromosome and integrates viral genes into it. This recombinant DNA is replicated and passed on to all of the infected cell's descendants. Later, if the viral genes move out of the chromosome, a new cycle of infection will begin.

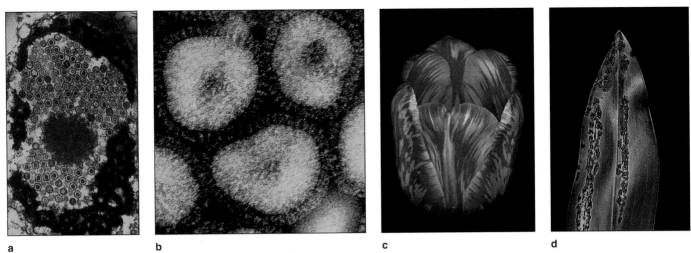

Figure 17.4 (**a**) Particles of a DNA virus (*Herpes*) in an infected cell. (**b**) Particles of the RNA virus that causes influenza. Visible effects of two plant viruses: (**c**) A streaked tulip blossom. A relatively harmless virus infected pigment-forming cells in the colorless parts. (**d**) An orchid leaf infected by a rhabdovirus.

Examples of Viruses

Bacteria-infecting viruses, or **bacteriophages**, follow lytic or temperate pathways. Some types were used in early experiments that revealed DNA as the genetic material (Figure 13.4). Bacteriophages are still being used as research tools in genetic engineering.

Animal viruses cause diseases as varied as warts, chicken pox, the common cold, influenza, and several forms of cancer (Table 17.1). HIV, the human immuno-deficiency virus, causes AIDS. Like many viruses, HIV enters cells by endocytosis, then departs by exocytosis, as shown on page 448. Because it is an RNA virus, its replication requires an extra step. A viral enzyme puts together a strand of DNA on the viral RNA. This DNA transcript becomes integrated into the host chromosome.

HIV is one of the viruses that enters a long latent period. Another is *Herpes simplex* virus (Figure 17.4a). Type I *Herpes simplex*, which causes recurring cold sores, is latent in just about everybody.

Plant viruses must breach plant cell walls to cause diseases. They typically hitch rides on the piercing or sucking devices of insects that feed on plant juices. Viral diseases severely damage potatoes, barley, tomatoes, and other major crop plants. Figures 17.4c and d show the visible effects of some infections.

Some pathogens are even more stripped down than viruses. Prions, which are protein particles, cause rare, fatal degeneration of the nervous system in humans, sheep, and other animals. Viroids are strands or circles of RNA, smaller than any viral DNA or RNA molecule, and they have no protein coat. Viroids are plant pathogens that can destroy entire fields of citrus, potatoes, and other crop plants.

Bear in mind, viruses, viroids, and prions are not living organisms. Only when we turn to the simplest cells do we cross the threshold into the living world.

Table 17.1 Classification of Animal Viruses

DNA Viruses	Some Diseases
Adenoviruses	Respiratory infections
Hepatitis virus	Liver diseases
Herpesviruses:	
H. simplex type I	Oral herpes, cold sores
H. simplex type II	Genital herpes
Varicella-zoster	Chicken pox, shingles
Epstein-Barr	Infectious mononucleosis, implicated in some cancers
Papovaviruses	Benign and malignant warts
Parvoviruses	Roseola (fever, rash) in small children; aggravation of sickle-cell anemia
Poxviruses	Smallpox, cowpox

RNA Viruses	Some Diseases
Picornaviruses:	
Enteroviruses	Polio, hemorrhagic eye disease; hepatitis A (infectious hepatitis)
Rhinoviruses	Common cold
Togaviruses	Encephalitis, yellow fever, dengue fever
Paramyxoviruses	Measles, mumps
Rhabdoviruses	Rabies
Coronaviruses	Respiratory infections
Orthomyxoviruses	Influenza
Arenaviruses	Hemorrhagic fevers
Reoviruses	Respiratory, intestinal infections
Retroviruses:	
HTLV I, II	Associated with cancer (leukemia)
HIV	AIDS

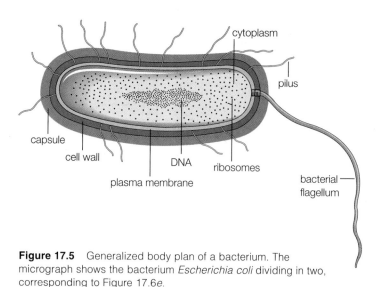

Figure 17.5 Generalized body plan of a bacterium. The micrograph shows the bacterium *Escherichia coli* dividing in two, corresponding to Figure 17.6e.

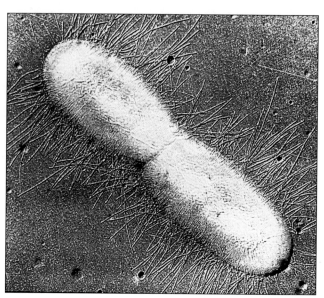

BACTERIA

Of all organisms, bacteria are the most abundant and far-flung. Different kinds live in places ranging from deserts to hot springs, snow, and deep oceans. Tens of billions may exist in a handful of rich soil. The ones in your gut and on your skin outnumber your body cells! Your cells are larger, however, so you are only "a few percent bacterial" by weight.

Characteristics of Bacteria

Metabolic Diversity. Compared to other organisms, bacteria show the greatest metabolic diversity. Some are *photosynthetic autotrophs*. Like plants, they make their own biological molecules from simpler compounds, using sunlight as an energy source. Unlike plants, these bacteria use a variety of photosynthetic pathways. Other bacteria are *chemosynthetic autotrophs*. They make their own biological molecules, using a variety of inorganic substances as the energy source. The vast majority of bacteria are *heterotrophs*. They derive organic compounds from the tissues, products, or wastes and remains of other organisms (page 71).

Structural Features. Bacteria alone are prokaryotic. They don't have a nucleus or the other organelles seen in eukaryotic cells (Table 17.2). Photosynthesis and many other reactions occur at the plasma membrane. Protein synthesis occurs at ribosomes attached to the membrane or distributed through the cytoplasm.

Bacteria generally have a cell wall (Figure 17.5). Usually it is a mesh of peptidoglycan (polysaccharide strands, linked together by modified amino acids). Such

---

> **Table 17.2 Characteristics of Bacterial Cells**
>
> 1. Bacterial cells are prokaryotic (they have no membrane-bound nucleus or other organelles in the cytoplasm).
>
> 2. Bacterial cells have a single chromosome (a circular DNA molecule); many species also have plasmids (page 179).
>
> 3. Most bacteria have a cell wall composed of peptidoglycan.
>
> 4. Most bacteria reproduce by binary fission.
>
> 5. Collectively, bacteria show great diversity in their modes of metabolism.

walls are strong, semirigid, and help maintain bacterial cells in one of three shapes:

coccus (plural, cocci);
spherical

rod or bacillus (plural, bacilli);
cylindrical

spirillum (plural, spirilla);
helical

Polysaccharides form a cover (glycocalyx) around the cell wall of many bacterial species. They help the bacterial cell attach to teeth, assorted membranes, rocks in streambeds, and other surfaces. In some types, they form a jellylike capsule that deters a host's infection-fighting cells.

Some bacteria have one or more motile structures called bacterial flagella (singular, flagellum). These don't have the same structure as eukaryotic flagella,

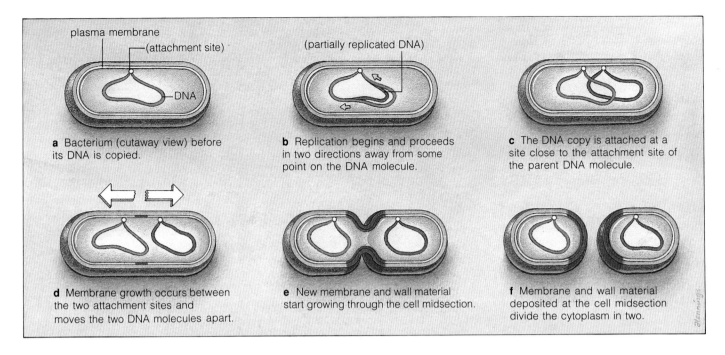

plasma membrane
(attachment site)

DNA

a Bacterium (cutaway view) before its DNA is copied.

(partially replicated DNA)

b Replication begins and proceeds in two directions away from some point on the DNA molecule.

c The DNA copy is attached at a site close to the attachment site of the parent DNA molecule.

d Membrane growth occurs between the two attachment sites and moves the two DNA molecules apart.

e New membrane and wall material start growing through the cell midsection.

f Membrane and wall material deposited at the cell midsection divide the cytoplasm in two.

Figure 17.6 Bacterial reproduction by binary fission, a cell division mechanism.

and they don't operate the same way. They move the cell by rotating like a propeller. Many bacteria have pili (singular, pilus). These are short, filamentous proteins that project above the cell wall. Some help cells adhere to surfaces. Others help them attach to one another as a prelude to conjugation, described next.

Bacterial Reproduction

Compared to eukaryotes, bacteria have a simple body plan and only a single chromosome (a circular DNA molecule). Most reproduce by a cell division mechanism called **binary fission**. A parent cell divides into two genetically identical daughter cells right after DNA replication (Figure 17.6b). Sometimes the daughter cells remain stuck together in pairs, clusters, or chains.

In many species, daughter cells also inherit one or more plasmids. A plasmid, recall, is a small, self-replicating circle of extra DNA with only a few genes. Usually the genes confer a survival advantage, as when they code for an enzyme that allows a cell to synthesize or use some nutrient. Some plasmid genes confer resistance to antibiotics, substances that kill or inhibit the growth of bacteria (see the *Focus* essay). Others confer the means to engage in **bacterial conjugation**. By this process, one bacterial cell transfers plasmid DNA to another (Figure 17.7). Sometimes the transfer is called sexual reproduction between "male" and "female" bacterial cells, although comparing this to eukaryotic sex requires a rather breathtaking leap of the imagination.

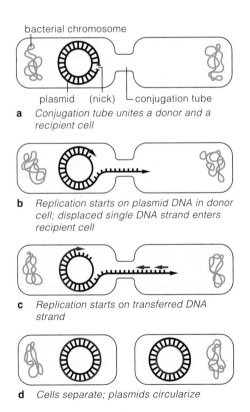

bacterial chromosome

plasmid   (nick)   conjugation tube

a *Conjugation tube unites a donor and a recipient cell*

b *Replication starts on plasmid DNA in donor cell; displaced single DNA strand enters recipient cell*

c *Replication starts on transferred DNA strand*

d *Cells separate; plasmids circularize*

Figure 17.7 Bacterial conjugation—the transfer of a plasmid from a donor to a recipient cell. A long pilus brings the two cells into close contact so that a conjugation tube can form between them. For clarity, the plasmid in the diagrams is enormously enlarged, compared to the bacterial chromosome.

Antibiotics

When your grandparents were children, bacterial agents of tuberculosis, pneumonia, and scarlet fever may have caused 25 percent of all deaths each year in the United States. Bacterial agents of dysentery, diphtheria, and whooping cough also were common killers. Bacterial infections during childbirth killed or maimed women by the thousands. From the 1940s onward, we started treating these and other diseases with antibiotics.

An **antibiotic** is a normal metabolic product of certain microorganisms, and it kills or inhibits the growth of other microorganisms. For example, streptomycins block protein synthesis in their targets. Penicillins disrupt formation of covalent bonds that hold bacterial cell walls together. Penicillin derivatives cause the wall to weaken until it ruptures. The known antibiotics don't have comparable effects on protein coats of viruses. If you have a viral infection, antibiotics won't help.

Antibiotics must be carefully prescribed. Besides performing their intended function of counterattacking pathogens, some can disrupt the normal populations of bacteria in the intestines and of yeast cells in the vaginal canal. Such disruptions can lead to secondary infections.

Overprescribed antibiotics have lost their punch. Over time, they destroyed the most susceptible cells of target populations—and favored their replacement by more resistant ones. Antibiotic-resistant strains have made tuberculosis, typhoid, gonorrhea, "staph" infections, and other bacterial diseases more difficult to treat. In a few cases, "superbugs" that cause tuberculosis cannot be treated successfully.

We are now exploring the seas for weapons against pathogens. Certain jellyfishes, lobsters, and shrimps carry fungus-fighting bacteria. So do shrimps living near hydrothermal vents. Unknown microbes abound in mud samples brought up by deep-sea drilling projects. One produces an antibiotic that may work against colon tumor cells—even against the AIDS-causing virus.

Classification of Bacteria

Traditionally, the many thousands of known bacterial species have been classified on the basis of cell shape, mode of nutrition, metabolic patterns, how their cell wall takes up certain stains, and other traits. About twenty years ago, Carl Woese and others started to classify bacteria by comparing traits that more closely measured evolutionary relationship. Through comparative biochemical studies, they traced three major bacterial lineages back to the common ancestor of all cells. As you will see, one of those lineages led to modern-day eukaryotes. The other two gave rise to two great subkingdoms of bacteria, the **archaebacteria** and **eubacteria**.

Archaebacteria

In many respects, the cell structure, metabolism, and nucleic acid sequences of archaebacteria are unique. These organisms differ as much from other bacteria as they do from eukaryotes. They probably resemble the first cells on earth. Life apparently arose in extremely hot, acidic, and salty places—and such places are where we find archaebacteria. Hence their name (*archae-* means beginning).

As Table 17.3 shows, methanogens, halophiles, and thermophiles belong to this domain. **Methanogens** ("methane-makers") live only in oxygen-free habitats, such as swamp muck, sewage, and the animal gut. They make ATP by converting carbon dioxide and hydrogen gases to methane (CH_4). Pungent stockyard fumes testify to their presence. So does the "marsh gas" of swamps and sewage treatment plants. As a group, methanogens produce about 2 billion tons of methane gas each year. They affect the global cycling of carbon and carbon dioxide levels in the atmosphere.

Halophiles ("salt-lovers") live in brackish seas, Utah's Great Salt Lake, and other high-salinity habitats (Figure 17.8). Some thrive in (and spoil) salted fish. Most species are heterotrophs, and most form ATP by aerobic pathways. When oxygen levels drop, some strains can produce ATP by photosynthesis. They have a unique type of light-trapping pigment called bacteriorhodopsin.

Thermophiles ("heat-lovers") live in hot springs, in very acidic soils, and near hydrothermal vents, which are fissures in the floor of deep oceans. The vent species, which use hydrogen sulfide to make ATP, are the basis of remarkable food webs (page 605).

Table 17.3 Some Major Groups of Bacteria

Group	Main Habitats	Characteristics	Representatives
Archaebacteria			
Methanogens	Anaerobic sediments of lakes, swamps; also animal gut	Chemosynthetic; methane producers; used in sewage treatment facilities	*Methanobacterium*
Halophiles	Brines (extremely salty water)	Heterotrophic; also have photosynthetic machinery of a unique sort	*Halobacterium*
Thermophiles	Acidic soil, hot springs, hydrothermal vents on seafloor	Heterotrophic or chemosynthetic; use inorganic substances such as sulfur as a source of electrons for ATP formation	*Sulfolobus, Thermoplasma*
Photosynthetic eubacteria			
Cyanobacteria	Mostly lakes, ponds; some marine, terrestrial	In photosynthesis, water is electron donor, oxygen a by-product; some fix nitrogen	*Anabaena, Nostoc*
Purple sulfur bacteria	Anaerobic, organically rich muddy soils, and sediments of aquatic habitats	In photosynthesis, reduced sulfur compounds are electron donors	*Rhodospirillum, Chlorobium*
Chemosynthetic eubacteria			
Nitrifying bacteria	Soil; freshwater, marine habitats	Major ecological role (nitrogen cycle)	*Nitrosomonas, Nitrobacter*
Heterotrophic eubacteria			
Spirochetes	Aquatic habitats; parasites of animals	Helically coiled, motile; free-living and parasitic species; some major pathogens	*Spirochaeta, Treponema*
Gram-negative, aerobic rods and cocci	Soil, aquatic habitats; parasites of animals, plants	Some major pathogens; some (e.g., *Rhizobium*) fix nitrogen	*Pseudomonas, Neisseria, Rhizobium, Agrobacterium*
Gram-negative, facultative anaerobic rods	Soil, plants, animal gut	Many are major pathogens; one (*Photobacterium*) is bioluminescent	*Samonella, Proteus, Escherichia, Photobacterium*
Rickettsias and chlamydias	Host cells of insects, other animals	Intracellular parasites; many pathogens	*Rickettsia, Chlamydia*
Myxobacteria	Decaying plant, animal matter; bark of living trees	Gliding, rod-shaped; aggregate and migrate together	*Myxococcus*
Gram-positive cocci	Soil; skin and mucous membranes of animals	Some major pathogens	*Staphylococcus, Streptococcus*
Endospore-forming rods and cocci	Soil; animal gut	Some major pathogens	*Bacillus, Clostridium*
Gram-positive nonsporulating rods	Fermenting plant, animal material; gut, vaginal tract	Some important in dairy industry, others major contaminators of milk, cheese	*Lactobacillus, Listeria*
Actinomycetes	Soil; some aquatic habitats	Include anaerobes and strict aerobes; major producers of antibiotics	*Actinomyces, Streptomyces*

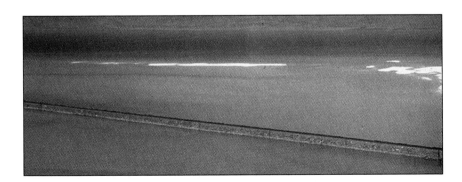

Figure 17.8 Great Salt Lake, Utah. Halophilic bacteria and algae growing in this vast, saline lake impart a pink cast to the water. The diagonal strip across the photograph is a raised bed for a railroad track.

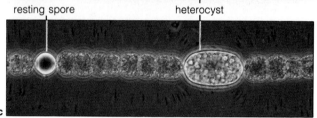

resting spore   heterocyst

Figure 17.9 (**a**) A population of cyanobacteria near the surface of a nutrient-enriched pond. (**b,c**) Resting spores form when conditions do not favor growth. When they germinate, they give rise to a new chain of cells. A nitrogen-fixing heterocyst also is shown in (**c**).

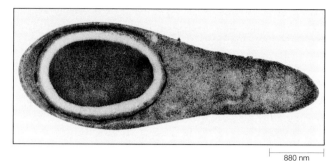

880 nm

Figure 17.10 An endospore developing in *Clostridium tetani*.

Eubacteria

The *eu-* in eubacteria implies "typical." Table 17.3 lists representative eubacteria, which are far more common than archaebacteria. Let's look at a few types, using modes of nutrition as a conceptual framework.

Photosynthetic Eubacteria. Freshwater ponds are home to most cyanobacteria (also called blue-green algae). These are the most common photosynthetic eubacteria. Many species grow as chains of cells, sheathed in mucus. The chains form dense, slimy mats near the surface of nutrient-enriched water (Figure 17.9). *Anabaena* and other species also convert nitrogen in the air to ammonia, a nitrogen source for biosynthesis. When nitrogen-containing compounds are in short supply, some cells develop into heterocysts. These modified cells alone can make a nitrogen-fixing enzyme. Heterocysts share nitrogen compounds with the photosynthetic cells and get carbohydrates in return. Substances move freely through junctions between the cytoplasm of neighboring cells.

Chemosynthetic Eubacteria. The members of this group use inorganic substances as an energy source. Many affect the global cycling of nitrogen, sulfur, phosphorus, and other nutrients. Consider nitrogen, a building block for amino acids and proteins. Without it, there would be no life. Nitrifying bacteria in soil strip electrons from ammonia when they make ATP. Plants can use the end product, nitrate, as a nitrogen source.

Heterotrophic Eubacteria. Most bacteria are heterotrophs. Most of these benefit the world of life; they are major decomposers in nearly all communities. Other beneficial members include *Lactobacillus* species, which help us make cheese, sour cream, yogurt, and other fermented milk products. Some of the soil bacteria called actinomycetes provide us with antibiotics. *Escherichia coli*, a gut dweller, produces vitamin K and compounds useful in fat digestion. Its activities keep many food-borne pathogens from colonizing the gut. It also helps newborns digest milk.

Yet some *E. coli* strains are among the pathogenic heterotrophs. They cause a form of diarrhea that is the leading cause of infant death in developing countries. Other pathogens use insects to taxi from one host to another. *Borrelia burgdorferi* travels inside blood-sucking ticks. Tick bites transmit it from deer, field mice, and some other wild animals to humans, who develop *Lyme disease*. The first sign is a circular "bulls-eye" rash around the tick bite, then severe headaches, backaches, chills, and fatigue. Without prompt treatment, serious problems can develop.

As part of their life cycle, some heterotrophs form an endospore around their DNA and a bit of cytoplasm. Endospores are structures that resist heat, drying, boiling, and radiation. New bacterial cells form after the endospore germinates. Figure 17.10 shows *Clostridium tetani*, an endospore former that causes the disease tetanus (page 479). A relative, *C. botulinum*, can taint fermented grain as well as food in improperly sterilized or sealed cans and jars. Its toxins cause *botulism*, a form of poisoning that can lead to respiratory failure and death.

About the "Simple" Bacteria

Bacteria are small. Their insides are not elaborate. *But bacteria are not simple.* A brief look at their behavior will reinforce this point. Bacteria move toward regions with more nutrients. Aerobes move toward oxygen, anaerobes move away from it. Photosynthetic species move toward more intense light (or away from light if it is too intense). Many species tumble away from toxins. Their behavior depends on membrane receptors, which change shape when they absorb light or encounter chemical compounds. When a bacterium changes direction, its receptors are stimulated in a different way. This triggers a fleeting "memory," a changing biochemical condition that can be compared against that of the immediate past.

Some species show collective behavior. Millions of cells of *Myxococcus xanthus* form "predatory" colonies. Their enzyme secretions digest "prey" (cyanobacteria and other microbes) that become stuck to the colony. The cells absorb the breakdown products. What's more, migrating *M. xanthus* cells change direction and *move as a single unit* toward what may be food!

PROTISTANS

Classification of Protistans

Like bacteria, the protistans show splendid diversity. Here we define **protistans** as single-celled eukaryotes. Yet many protistan lineages appear to be continuous with multicelled species that commonly are placed in the kingdoms of plants, fungi, and animals. Ongoing studies are shedding light on the evolutionary connections. In time, classification schemes probably will be restructured to reflect the new evidence. Meanwhile, at the very least, the protistan kingdom includes the slime molds, euglenoids, chrysophytes, dinoflagellates, sporozoans, and the amoeboid, flagellated, and ciliated protozoans (Table 17.4). Compared to bacteria, most of these protistans are larger and structurally more complex, as the representative in Figure 17.11 suggests.

The amoeboid protozoans are the simplest protistans we know about. They have no hard parts, so they vanish quickly after they die. If the eukaryotic ancestors of protistans were soft-bodied also, we may never find fossils of them. Even so, evidence from comparative biochemistry suggests ancestors of the eukaryotic lineage arose not long after the first cells appeared on earth. Two billion years later, cells of the eukaryotic lineage probably were larger than bacteria, they had a nucleus, and they had mitochondria, chloroplasts, or both. How did they get that way? The theories described in the *Commentary* give us interesting things to think about.

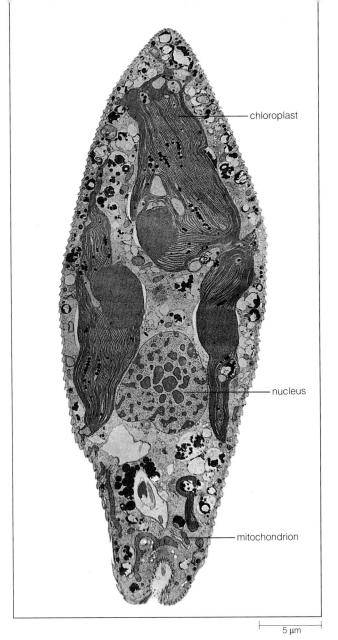

Figure 17.11 A longitudinal slice through the cell body of a single-celled eukaryote (*Euglena*). A profusion of diverse organelles is typical of most members of this kingdom.

Table 17.4 Classification of Major Groups of Protistans	
Phylum (Division)	Common Name
Myxomycota	Plasmodial slime molds
Gymnomycota	Cellular slime molds
Euglenophyta	Euglenoids
Chrysophyta	Yellow-green algae, golden algae, diatoms
Pyrrhophyta	Dinoflagellates
Mastigophora	Flagellated protozoans
Sarcodina	Amoeboid protozoans (amoebas, foraminiferans, radiolarians, heliozoans)
Ciliophora	Ciliated protozoans
	Sporozoans*

*The name has no formal taxonomic status.

The Rise of Eukaryotic Cells

Somewhere between 3.8 million and 2.5 billion years ago, divergences split a prokaryotic lineage into three great evolutionary roads. Those divergences led to the first archaebacteria, the first eubacteria—and the bacterial ancestor of eukaryotes (Figure *a*).

Today, the key defining traits of eukaryotes are the nucleus and other organelles. Many of these organelles evolved as a result of mutations in the ancestral cells' DNA. Other organelles apparently arose through a remarkable evolutionary event called endosymbiosis.

Origin of the Nucleus and ER.   Eukaryotic organelles are membrane-bound compartments in the cytoplasm. Prokaryotes don't have such organelles, but some species have infoldings of the plasma membrane (Figure *b*). That plasma membrane is peppered with enzymes and other metabolic agents. Long ago, some infoldings in prokaryotic cells may have extended far enough into the cell interior to serve as a channel to the cell surface. Perhaps some evolved into ER channels and into an envelope around the DNA.

What would be the advantage of such membranous enclosures? Maybe they protected genes and protein products from "foreigners." Remember how bacterial species can transfer plasmid DNA among themselves? Yeasts, which are simple eukaryotic cells, do the same thing. Some yeast species have up to fifty plasmids. Long ago, a nuclear envelope may have been favored because it got the cell's genes, replication enzymes, and transcription enzymes out of the cytoplasm. It would have allowed vital genetic messages to be produced free of metabolic competition from an unmanageable hodgepodge of foreign genes. Similarly, ER channels might have kept certain proteins and other substances away from metabolically hungry "guests"—foreign cells that became permanent residents inside the host cell.

Endosymbiosis.   Accidental partnerships between different prokaryotic species probably formed countless times on the evolutionary road to eukaryotes. And some partnerships produced mitochondria, chloroplasts, and other organelles. This is a theory of **endosymbiosis**, developed in greatest detail by Lynn Margulis. *Endo-* means within, and *symbiosis* means living together. In endosymbiosis, one cell (a guest species) lives permanently inside another kind of cell (the host species), and the interaction benefits both. By this theory, eukaryotes arose after the noncyclic pathway of photosynthesis emerged and oxygen accumulated to significant levels in the atmosphere.

a   An evolutionary tree of life that reflects current thinking about the connections among major groups of organisms. The diagram incorporates a few ideas about the endosymbiotic origins of eukaryotic cells.

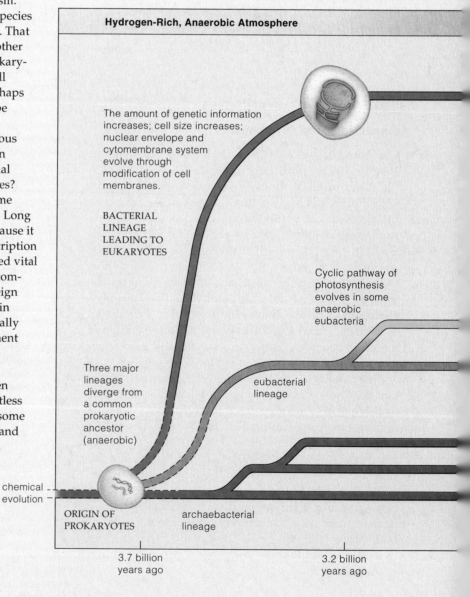

Hydrogen-Rich, Anaerobic Atmosphere

The amount of genetic information increases; cell size increases; nuclear envelope and cytomembrane system evolve through modification of cell membranes.

BACTERIAL LINEAGE LEADING TO EUKARYOTES

Cyclic pathway of photosynthesis evolves in some anaerobic eubacteria

Three major lineages diverge from a common prokaryotic ancestor (anaerobic)

eubacterial lineage

chemical evolution

ORIGIN OF PROKARYOTES

archaebacterial lineage

3.7 billion years ago

3.2 billion years ago

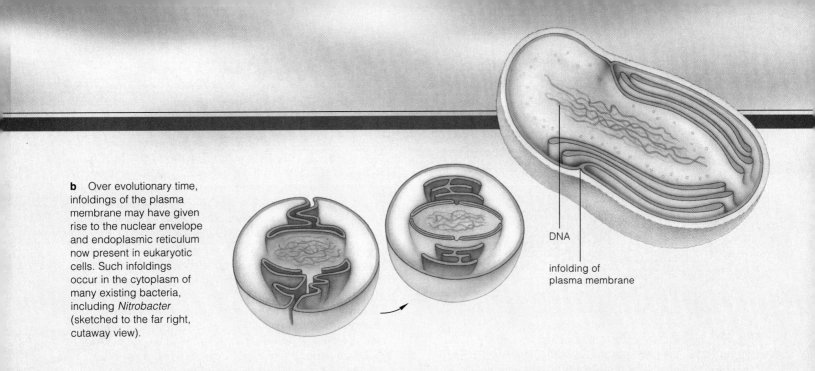

b Over evolutionary time, infoldings of the plasma membrane may have given rise to the nuclear envelope and endoplasmic reticulum now present in eukaryotic cells. Such infoldings occur in the cytoplasm of many existing bacteria, including *Nitrobacter* (sketched to the far right, cutaway view).

DNA

infolding of plasma membrane

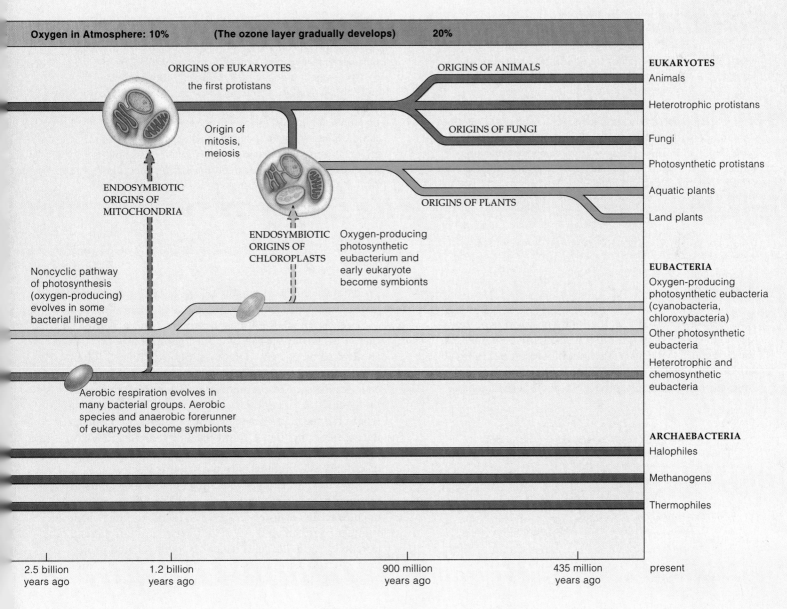

Oxygen in Atmosphere: 10%　　　(The ozone layer gradually develops)　　　20%

ORIGINS OF EUKARYOTES
the first protistans

ORIGINS OF ANIMALS

Origin of mitosis, meiosis

ORIGINS OF FUNGI

ENDOSYMBIOTIC ORIGINS OF MITOCHONDRIA

ENDOSYMBIOTIC ORIGINS OF CHLOROPLASTS

Oxygen-producing photosynthetic eubacterium and early eukaryote become symbionts

ORIGINS OF PLANTS

Noncyclic pathway of photosynthesis (oxygen-producing) evolves in some bacterial lineage

Aerobic respiration evolves in many bacterial groups. Aerobic species and anaerobic forerunner of eukaryotes become symbionts.

EUKARYOTES
Animals

Heterotrophic protistans

Fungi

Photosynthetic protistans

Aquatic plants

Land plants

EUBACTERIA
Oxygen-producing photosynthetic eubacteria (cyanobacteria, chloroxybacteria)

Other photosynthetic eubacteria

Heterotrophic and chemosynthetic eubacteria

ARCHAEBACTERIA
Halophiles

Methanogens

Thermophiles

| 2.5 billion years ago | 1.2 billion years ago | 900 million years ago | 435 million years ago | present |

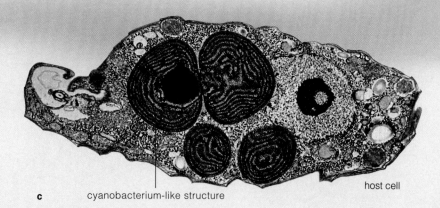

c   cyanobacterium-like structure   host cell

In some bacterial groups, electron transport systems had already expanded to include "extra" cytochromes. As it happened, those cytochromes were able to donate electrons to oxygen. Aerobic respiration had emerged.

By 1.2 billion years ago, and possibly much earlier, our prokaryotic ancestors were engulfing aerobic bacteria. They probably were similar to existing amoebalike cells that weakly tolerate free oxygen. Like amoebas, they trapped food with their cytoplasmic extensions, then formed endocytic vesicles around it for delivery and digestion in the cytoplasm. But some aerobic bacteria resisted digestion. They actually thrived in the new, protected, nutrient-rich environment. In time they were releasing extra ATP, which the hosts came to depend on for growth, greater activity, and assembly of more structures, such as hard body parts. The guests were no longer duplicating metabolic functions that the hosts were performing for them. The anaerobic and aerobic cells were now incapable of independent existence. The guests had become mitochondria, supreme suppliers of ATP.

Strong evidence supports Margulis's theory. For example, mitochondria are like bacteria in size and structure. They divide independently of the host cell's division. The inner mitochondrial membrane is like a bacterial plasma membrane. And mitochondria have their own self-replicating DNA. Their DNA contains instructions for building some proteins required for specialized mitochondrial tasks. A few of the genetic code words have uniquely mitochondrial meanings. That is, the "mitochondrial code" has a few slight differences, compared to the genetic code of cells.

Also consider those food-producing factories called chloroplasts. Chloroplasts may be descended from oxygen-producing photosynthetic cells that also became endosymbionts. In their metabolism and overall nucleic acid sequence, chloroplasts resemble some eubacteria. Their DNA is self-replicating, and they divide independently of the cell's division. Chloroplasts vary in shape and in their light-absorbing pigments, just as different photosynthetic eubacteria do. They may have originated a number of times, in a number of different lineages.

And so new kinds of cells, housing mitochondria, chloroplasts, or both, emerged on the evolutionary road. They were the first protistans. With their efficient metabolic strategies, the early protistans underwent rapid divergences and adaptive radiations. In no time at all, evolutionarily speaking, some of their descendants gave rise to the kingdoms of animals, fungi, and plants. And within those kingdoms we find plenty of examples that nature continues to tinker with endosymbionts (Figure c).

Euglenoids

The *Commentary* hints at the kinds of novel evolutionary events that may have given rise to diverse lineages of protistans. With this essay in mind, look once more at the *Euglena* cell in Figure 17.11. It is one of the **euglenoids** (phylum Euglenophyta). Most of these flagellated

protistans live in stagnant or freshwater ponds and lakes. Most are photosynthetic; many are heterotrophs.

Euglena is 40,000 times larger and more complex than a bacterium. It can live on organic compounds in sunless settings—yet it also is photosynthetic. Light readily passes through its translucent outer layer (pellicle). An eyespot of pigment granules partly shields its light-sensitive receptor. *Euglena* takes up positions where the receptor is best exposed to sunlight. Perhaps its ancestors acquired chloroplasts by endosymbiosis. Perhaps it also is related to green algae and land plants; its chlorophyll pigments are the same as theirs.

Euglena lives in hypotonic places, so water tends to move into its cytoplasm by osmosis. The excess collects in **contractile vacuoles** (Figure 17.12). When filled, each vacuole contracts and water is forced through a small pore to the outside. As is true of most flagellated protistans, the cell reproduces by longitudinal fission. After duplicating its organelles, the cell divides lengthwise.

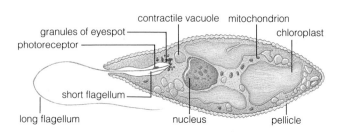

contractile vacuole mitochondrion

granules of eyespot

photoreceptor

chloroplast

short flagellum

long flagellum   nucleus   pellicle

Figure 17.12 Some components of a *Euglena* cell.

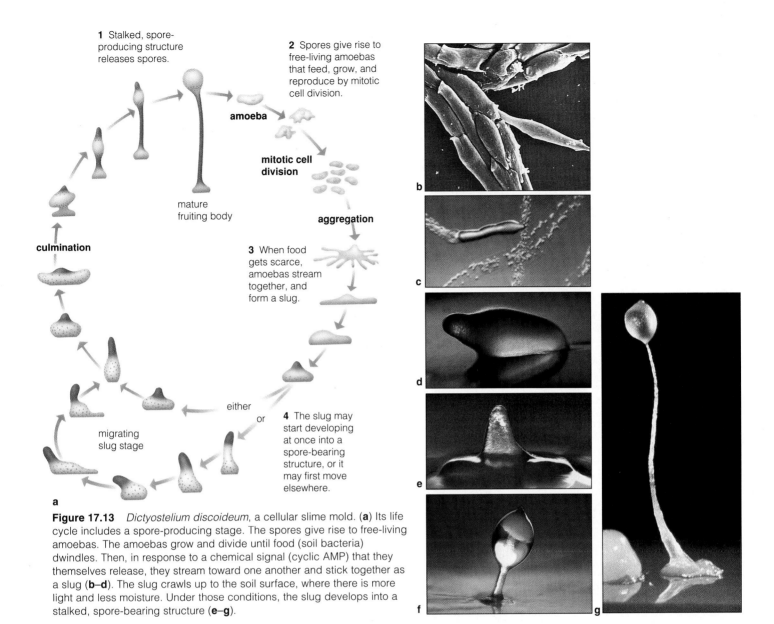

1 Stalked, spore-producing structure releases spores.

2 Spores give rise to free-living amoebas that feed, grow, and reproduce by mitotic cell division.

amoeba

mitotic cell division

mature fruiting body

culmination

aggregation

3 When food gets scarce, amoebas stream together, and form a slug.

either

or

migrating slug stage

4 The slug may start developing at once into a spore-bearing structure, or it may first move elsewhere.

a

b

c

d

e

f

g

Figure 17.13 *Dictyostelium discoideum*, a cellular slime mold. (**a**) Its life cycle includes a spore-producing stage. The spores give rise to free-living amoebas. The amoebas grow and divide until food (soil bacteria) dwindles. Then, in response to a chemical signal (cyclic AMP) that they themselves release, they stream toward one another and stick together as a slug (**b–d**). The slug crawls up to the soil surface, where there is more light and less moisture. Under those conditions, the slug develops into a stalked, spore-bearing structure (**e–g**).

Slime Molds

Call someone a slimy scum and you may get punched in the nose. Yet that is an apt description of **slime molds**, heterotrophic protistans of obscure ancestry. During a slime mold life cycle, free-living amoebalike cells congregate on rotting plant parts, such as decaying leaves and bark. In each wet, scummy patch, cells busily engulf bacteria, spores, and organic compounds. Later in the cycle, the amoebalike cells differentiate and form stalked, spore-bearing reproductive structures.

Figure 17.13 shows one of the cellular slime molds (phylum Gymnomycota). Figure 17.14 shows a congregation of cells of one of the "plasmodial" slime molds (phylum Myxomycota). No plasma membranes or walls separate their cytoplasm. Flows of cytoplasm distribute nutrients and oxygen throughout the mass of cells.

Figure 17.14 A plasmodial slime mold (*Physarum*), migrating along a rotting log. Some populations spread over a square meter.

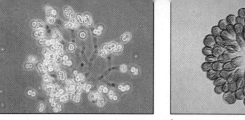

a   b

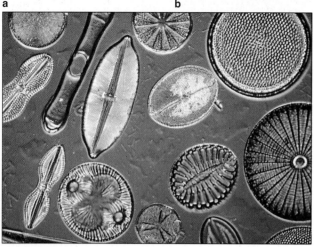

c

Figure 17.15   A few chrysophytes. (**a**) *Mischococcus*, a yellow-green alga that grows in colonies in plankton. (**b**) *Synura*, a golden alga with a fishy odor. It also grows in colonies in plankton. (**c**) Diverse shells of diatoms.

a

b

Figure 17.16   (**a**) Source of red tides along Florida's coast—the dinoflagellate *Gymnodinium breve*. (**b**) Part of a fish kill resulting from a dinoflagellate "bloom."

Chrysophytes

The **chrysophytes** (phylum Chrysophyta) include yellow-green algae, golden algae, and golden-brown diatoms (Figure 17.15). Most are photosynthetic. Their coloration results from xanthophylls and beta carotene, which mask their chlorophyll. Chrysophytes are major oxygen suppliers and food producers in freshwater and marine communities. Water pollution is devastating to chrysophytes and other community members.

Diatoms have shells (two perforated, glasslike structures that overlap like a pillbox). Many ocean sediments contain deposits of finely crumbled diatom shells, which we use in abrasives, filters, and insulating materials. More than 270,000 metric tons of this crumbly material are quarried annually near Lompoc, California.

Dinoflagellates

Floating or weakly swimming organisms, mostly microscopic, make up the aquatic communities that are called **plankton** (after *planktos*, meaning to wander). The photosynthetic species serve as a food base for other members of the aquatic community. The **dinoflagellates** (phylum Pyrrhophyta) are important in the feeding relationships in plankton. Some are photosynthesizers, others are heterotrophs. Certain types have flagella in grooves between cellulose plates at the body surface (Figure 17.16).

Various pigments color the different dinoflagellates blue, yellow-green, green, brown, or red. In nutrient-enriched seawater, some red or brown species may undergo a population explosion. Their sheer numbers tint the water red or brown, a condition called a "red tide." Some of these species produce a toxin, and hundreds of thousands of plankton-eating fish may be poisoned during a red tide (Figure 17.16). The toxin doesn't kill clams, oysters, and other edible mollusks, but it accumulates in their tissues. The accumulated toxin can kill humans who eat the mollusks.

Protozoans

More than 65,000 kinds of protistans have predatory or parasitic habits. They are named **protozoans** ("first animals"), because they may resemble the single-celled heterotrophs that gave rise to animals. Fewer than two dozen kinds cause diseases in humans, but there are no effective vaccines against them. In any year, hundreds of millions of humans suffer from protozoan infections.

By one scheme, protozoans are classified in four major groups. These are the flagellated, amoeboid, and ciliated protozoans and the sporozoans (Table 17.4).

Flagellated Protozoans. Among the **flagellated protozoans** (phylum Mastigophora) are free-living and parasitic species. *Trypanosoma brucei* is one of the parasitic trypanosomes (Figure 17.17*a*). It causes *African sleeping sickness*, a terrible disease spread by bites from tsetse flies. Without treatment, the nervous system is damaged and death follows. *T. cruzi*, common in Mexico and South America, causes *Chagas disease*. Bugs transfer this parasite to humans and other mammals. Scratches or abrasions invite infection. The liver and spleen enlarge, the face swells, then the brain and heart are damaged. There is no treatment or cure.

Giardia lamblia infects wild mammals and foraging cattle. Largely because of increased camping in remote areas, it also has infected about 10 percent of the American population. This parasite causes mild intestinal disturbances. It can kill a few susceptible people. *G. lamblia* forms **cysts** (walled, resting structures) that leave the body in feces. Ingesting feces-contaminated water or food leads to infection. Once cysts enter a new host, the parasite reproduces and damages tissues. Even water from remote mountain streams may contain cysts and should be boiled before drinking.

Trichomonads are in this group (Figure 17.17*b*). One type, *Trichomonas vaginalis*, can be transmitted during sexual intercourse. Untreated infections damage the urinary and reproductive tracts.

Amoeboid Protozoans. The **amoeboid protozoans** (phylum Sarcodina) are distributed worldwide. They include amoebas, foraminiferans, heliozoans, and radiolarians (Figure 17.18). These single-celled heterotrophs are among the simplest protistans. Most feed on algae, bacteria, and other protozoans. They reproduce asexually, simply by dividing in two. Predatory types capture prey with temporary extensions of the cell body (pseudopods). Parasitic types are transmitted from host to host.

The amoebas are "naked," soft-bodied cells that live in freshwater, seawater, and soil. They include *Amoeba proteus* of biology laboratory fame. *Entamoeba histolytica*, a cyst-forming parasite, causes *amoebic dysentery*, a severe intestinal disorder. It is transmitted by contact with feces, which may contaminate water and soil in regions where sewage treatment is inadequate or nonexistent.

Foraminiferans are shelled species that live mostly in the seas. Their pseudopods extend through holes in their hardened shells. Most radiolarians are shelled members of marine plankton. Some species form colonies, with many cells cemented together. The needlelike pseudopods of heliozoans ("sun animals") radiate like sun rays. Most types live in freshwater habitats. The fossil record for shelled amoeboid protozoans extends back into the Paleozoic.

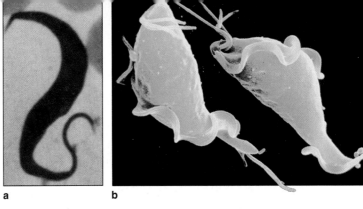

a   b

Figure 17.17 Two of the flagellated protozoans that infect humans. (**a**) *Trypanosoma brucei*, which causes African sleeping sickness. (**b**) *Trichomonas vaginalis*, which causes a sexually transmitted disease (trichomoniasis).

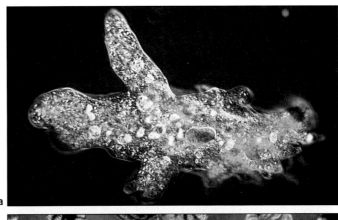

a

b

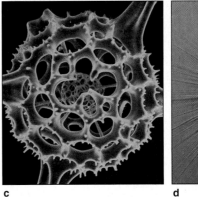

c   d

Figure 17.18 Amoeboid protozoans. (**a**) *Amoeba proteus*, one of the "naked" amoeboid protozoans. Other members of the group are shelled: (**b**) foraminiferan shells, (**c**) a radiolarian shell, and (**d**) a living heliozoan, with needlelike pseudopods. Fossilized shells that date back to the early Paleozoic are evidence of the group's ancient ancestry. The very first eukaryotes may have been soft-bodied, like *Amoeba*.

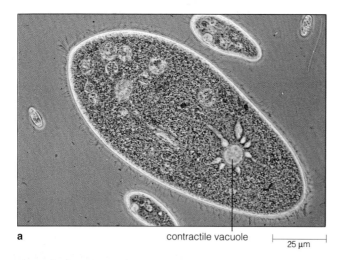

a   contractile vacuole   ⊢——⊣ 25 μm

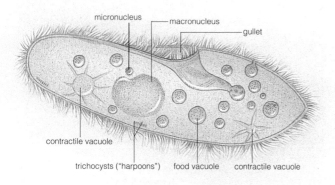

micronucleus — macronucleus — gullet

contractile vacuole

trichocysts ("harpoons")   food vacuole   contractile vacuole

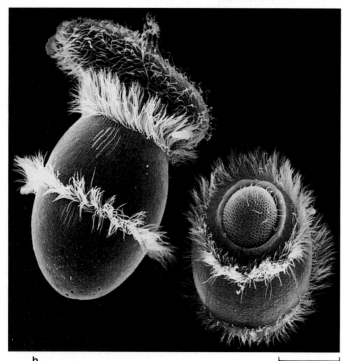

b   ⊢——⊣ 50 μm

Figure 17.19   **(a)** *Paramecium*, a ciliated protozoan. **(b)** Mealtime for big-mouthed *Didinium*, another ciliated protozoan. Dinner in this case is a *Paramecium* cell, poised at the mouth (*left*), then swallowed (*right*).

Ciliated Protozoans. **Ciliated protozoans** abound in freshwater and marine habitats, where they feed on bacteria, algae, and one another. Often the cell surface bristles with cilia (motile structures) and trichocysts (poison-charged "harpoons" that may be used against predators and prey). *Paramecium*, a typical member of the group, has a gullet (a cavity leading into the cell). When its rows of cilia beat in synchrony, water laden with bacteria and food particles is swept into the gullet. Once inside, the particles become enclosed in enzyme-filled vesicles and are digested. Wastes are moved to an "anal pore" and eliminated. Like the amoeboid protozoans, *Paramecium* relies on contractile vacuoles to rid the cell of excess water (Figure 17.19).

Sporozoans. The **sporozoans** are diverse parasites. All produce sporelike infectious agents (sporozoites), which insects often transmit to new hosts. One type, *Plasmodium*, causes *malaria* (Figure 17.20). Mosquitoes transmit its sporozoites to bird or human hosts. About 200 million people have the long-lasting disease, mostly in tropical and subtropical regions, and 1 to 2 million die annually. Travelers in countries with high rates of malaria are advised to use antimalarial drugs such as chloroquine. But drug resistance is common, and a vaccine has been difficult to develop. Vaccines induce the body to respond to a specific pathogen. But they are not equally effective against all of the different stages that develop during sporozoan life cycles. This is generally true of parasites with complex life cycles.

    Toxoplasma, another sporozoan, completes the sexual phase of its life cycle in cats and asexual phases in humans as well as cattle, pigs, and other animals. It can reach new hosts in infected meat, raw or undercooked. Sporozoite-containing cysts in the feces of infected cats are spread by houseflies, cockroaches, and other insects. The resulting disease, *toxoplasmosis*, is not prevalent in the population but is a major cause of birth defects. A woman who becomes infected during pregnancy may transmit the disease to her fetus, which may suffer brain damage or die. A pregnant woman should never empty litterboxes or otherwise clean up after any cat.

Summing Up—Eukaryotes and Prokaryotes Compared

This concludes our survey of single-celled prokaryotes and eukaryotes. We have touched on the features and possible evolutionary connections of these two kinds of organisms in many parts of this book. Table 17.5 summarizes and compares those features.

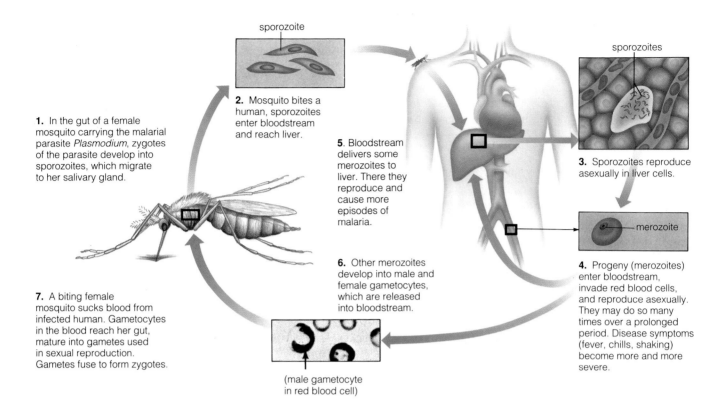

1. In the gut of a female mosquito carrying the malarial parasite *Plasmodium*, zygotes of the parasite develop into sporozoites, which migrate to her salivary gland.

2. Mosquito bites a human, sporozoites enter bloodstream and reach liver.

3. Sporozoites reproduce asexually in liver cells.

4. Progeny (merozoites) enter bloodstream, invade red blood cells, and reproduce asexually. They may do so many times over a prolonged period. Disease symptoms (fever, chills, shaking) become more and more severe.

5. Bloodstream delivers some merozoites to liver. There they reproduce and cause more episodes of malaria.

6. Other merozoites develop into male and female gametocytes, which are released into bloodstream.

7. A biting female mosquito sucks blood from infected human. Gametocytes in the blood reach her gut, mature into gametes used in sexual reproduction. Gametes fuse to form zygotes.

sporozoite

sporozoites

merozoite

(male gametocyte in red blood cell)

Figure 17.20 *Plasmodium*, a sporozoan that causes malaria. Its life cycle requires a human host and an insect intermediate host.

Table 17.5	Comparison of Prokaryotes With Eukaryotes	
	Prokaryotes	Eukaryotes
Organisms represented:	Bacteria only	Protistans, fungi, plants, and animals
Ancestry:	Two lineages (archaebacteria and eubacteria) that began more than 3.5 billion years ago	Equally ancient prokaryotic ancestors gave rise to the forerunners of eukaryotes, which emerged more than 1.2 billion years ago
Level of organization:	Single-celled	Mostly multicelled, often with division of labor among differentiated cells and tissues
Typical cell size:	Small (1–10 micrometers)	Large (10–100 micrometers)
Cell wall:	Mostly distinctive sugars and peptides	Cellulose or chitin; none in animal cells
Membrane-bound organelles:	Absent	Present
Modes of metabolism:	Both anaerobic and aerobic prevalent	Aerobic modes predominate
Genetic material:	Single bacterial chromosome (and sometimes plasmids)	Complex chromosomes (DNA and many associated proteins) within a nucleus
Mode of cell division:	Binary fission, mostly	Mitosis, meiosis, or both

SUMMARY

1. A virus is a nonliving, noncellular infectious agent. It consists of a nucleic acid core and a protein coat that sometimes is enclosed in a lipid envelope. A virus cannot reproduce itself. Its genetic material must enter a host cell and direct the cellular machinery to synthesize the materials necessary to produce new virus particles.

2. Divergences from a common prokaryotic ancestor apparently gave rise to three lineages: archaebacteria, eubacteria, and the forerunners of eukaryotes.

3. Archaebacteria (methanogens, halophiles, and thermophiles) live in extreme environments, like those in which life probably originated. All other bacteria are eubacteria. These more common types differ from archaebacteria in wall structure and other features.

4. Protistans are single-celled eukaryotes; they have a membrane-bound nucleus. We have classified protistans as these groups:

   a. Slime molds. These heterotrophic cells feed independently for part of the life cycle, migrate as a coordinated body, then form a spore-bearing structure.

   b. Euglenoids, chrysophytes, and dinoflagellates. Most are photosynthetic. Chrysophytes include yellow-green algae, golden algae, and diatoms.

   c. Flagellated, amoeboid, and ciliated protozoans. These are motile predators or parasites.

   d. Sporozoans. These parasites have a sporelike infectious stage.

Review Questions

1. What is a virus? Why is a virus considered to be no more alive than a chromosome? *247*

2. Describe the key characteristics of a bacterium. What are some differences between archaebacteria and eubacteria? *250–251, 252–254*

3. Name a few photosynthetic, chemosynthetic, and heterotrophic eubacteria. Describe some that are likely to give you the most trouble medically or recreationally (if you enjoy water sports). *253, 254*

4. Name the main categories of protistans. Think about where most of them live, then correlate some of their structural features with environmental conditions. *255, 258–262*

Self-Quiz *(Answers in Appendix IV)*

1. Viruses are _____ .
   a. the simplest living organisms
   b. infectious particles
   c. nonliving
   d. a and b
   e. b and c

2. The defining features of viruses are _____ .
   a. DNA core; protein coat
   b. nucleic acid core; plasma membrane
   c. DNA-containing nucleus; lipid envelope
   d. nucleic acid core; protein coat

3. Nondividing bacteria have _____ chromosome(s) and may have extra circles of _____ called plasmids.
   a. one; RNA
   b. two; RNA
   c. one; DNA
   d. two; DNA

4. Biochemical studies point to _____ primary lines of descent leading to all existing organisms.
   a. one
   b. two
   c. three
   d. four

5. _____ live in extreme environments, much like those of the early earth.
   a. Cyanobacteria
   b. Eubacteria
   c. Archaebacteria
   d. Protozoans

6. Most bacterial species are _____ .
   a. photosynthetic
   b. chemosynthetic
   c. heterotrophic
   d. omnivorous

7. Eukaryotes apparently arose through _____ .
   a. exocytosis
   b. endocytosis
   c. endosymbiosis
   d. sheer force of will

8. Euglenoids and chrysophytes are mostly _____ .
   a. photosynthetic
   b. chemosynthetic
   c. heterotrophic
   d. omnivorous

9. Match the groups to their descriptions.
   ____ viruses
   ____ bacteria
   ____ protistans
   ____ protozoans
   ____ archaebacteria

   a. one chromosome, maybe plasmids
   b. single-celled eukaryotes
   c. predators and parasites
   d. nonliving infectious particle
   e. most like first cells on earth

Selected Key Terms

amoeboid protozoan *261*
antibiotic *252*
archaebacteria *252*
bacterial conjugation *251*
bacteriophage *249*
binary fission *251*
chrysophyte *260*
ciliated protozoan *262*
contractile vacuole *258*
cyst *261*
dinoflagellate *260*
endosymbiosis *256*
eubacteria *252*
euglenoid *258*
flagellated protozoan *261*
halophile *252*
methanogen *252*
microorganism *246*
pathogen *246*
plankton *260*
protistan *255*
protozoan *260*
slime mold *259*
sporozoan *262*
thermophile *252*
virus *247*

Readings

Fraenkel-Conrat, H., P. Kimball, and J. Levy. 1988. *Virology*. Second edition. Englewood Cliffs, New Jersey: Prentice-Hall.

Margulis, L. 1993. *Symbiosis in Cell Evolution*. Second edition. New York: Freeman. Paperback.

Margulis, L., and K. Schwartz. 1992. *Five Kingdoms*. Second edition. New York: Freeman. Paperback.

Stanier, R., et al. 1986. *The Microbial World*. Fifth edition. Englewood Cliffs, New Jersey: Prentice-Hall.

18 FUNGI AND PLANTS

Pioneers in a New World

Seven hundred million years ago, no shorebirds stirred and noisily announced the dawn of a new day. No crabs clacked their tiny claws together and skittered off to burrows. The only sounds were the rhythmic muffled thuds of waves in the distance, at the outer limits of another low tide. Nearly 3 billion years before, life had its beginnings in the nearshore waters—and now, quietly, the invasion of the land was under way.

Astronomical numbers of photosynthetic cells had come and gone, and oxygen-producing ones had slowly changed the atmosphere. High above the earth, the sun's energy had converted much of the oxygen into a dense layer of ozone. This became a shield against lethal doses of ultraviolet radiation—which had kept early organisms from poking above the water's surface.

By one scenario, cyanobacteria were the first to adapt to intertidal zones, where the land dried out with each retreating tide. They were the first to move into shallow, freshwater streams meandering down to the coasts. Later, certain green algae and fungi made the same journey together. Every land plant around you is a descendant of green algae that either lived near the water's edge or made it onto land. And fungi are still associated with nearly all of them.

We have some tantalizing fossils of the pioneers. We also are learning about them through comparative biochemistry and studies of existing species. Today, as in Precambrian times, cyanobacteria and green algae grow in mats in nearshore waters and on the banks of freshwater streams (Figure 18.1). When a volcanic eruption or some other event exposes rocks for the first time, cyanobacteria colonize them. Mutually beneficial associations of green algae and fungi follow. Through their activities, these organisms actually create soils in which mosses and other plants can take hold.

With this chapter, we turn to the existing fungi and plants. The vast majority reside on land. Be glad their ancient ancestors left the water. Without them, we humans and other land-dwelling animals never would have made it onto the evolutionary stage.

Figure 18.1 Filaments of a green alga, massed in a shallow stream. More than 400 million years ago its ancestors may have lived in a similar stream that meandered down to the shore of an early continent.

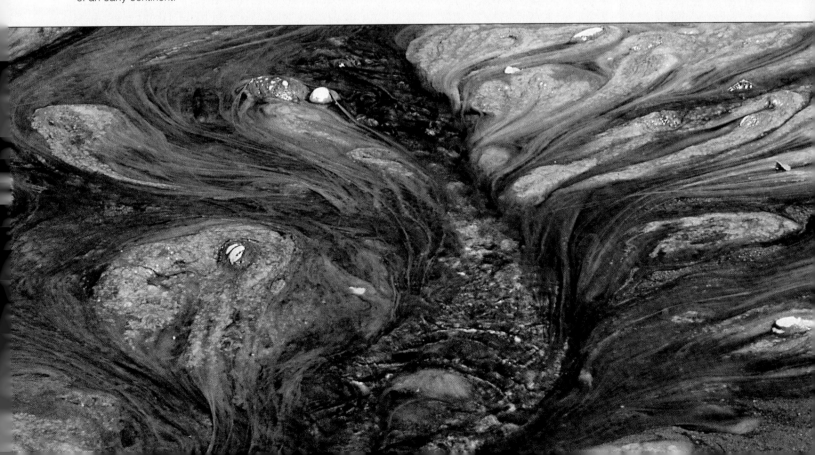

a Scarlet hood (*Hygrophorus*)

b Frost's bolete (*Boletus*)

c Yellow coral fungus (*Clavaria*)

d Trumpet chanterelle (*Craterellus*)

e Big laughing mushroom (*Gymnophilus*)

Figure 18.2 Fungal species from southeastern Virginia. This small sampling from a single group (club fungi) merely hints at the rich diversity within the boundaries of the kingdom Fungi.

1. Fungi are heterotrophs. Their cells secrete digestive enzymes in or on organic material, then absorb the breakdown products. Most fungi are decomposers with vital roles in cycling nutrients in nature.

2. The vast majority of fungi are multicelled. Like their early ancestors, many fungi are permanently linked with plants in symbiotic relationships.

3. With few exceptions, plants are multicelled, photosynthetic autotrophs. Plants are the primary producers for nearly all communities of life.

4. Long ago, when plants invaded the land, they became structurally and functionally adapted to new, harsh conditions. In many groups, the adaptations included a waxy cuticle, root and shoot systems, and tissues for conducting water and solutes. They included special means of nourishing, protecting, and dispersing the gametes as well as the new generation of plants.

PART I. KINGDOM OF FUNGI

For many of us, our thoughts about fungi are limited to deciding between whole or sliced brown mushrooms in grocery stores. How many people know those drab mushrooms are produced by a fungus that belongs to a huge group of diverse species? Figure 18.2 shows just a few of its relatives. These don't begin to do justice to the 80,000 fungal species we know about. And there may be at least a million more species we *don't* know about!

General Characteristics

Nutritional Modes. All fungi are heterotrophs; they feed on organic compounds synthesized by other organisms. Most are **saprobes**; they get nutrients from nonliving organic matter and so cause its decay. Saprobic fungi are major decomposers in nature. Other fungi are **parasites**; they get nutrients directly from tissues of a living host and may cause disease. All fungi digest dinner out on the table, so to speak. As their cells grow in or on organic matter, they secrete digestive enzymes, then absorb breakdown products. This "extracellular digestion" also benefits plants, which absorb some of the nutrients being released.

Think of all the wastes and remains of plants, insects, birds, and other animals. In autumn, one elm tree alone may shed 400 pounds of withered leaves.

Figure 18.3 An example of the filaments (hyphae) of a mycelium, the food-absorbing portion of many fungi.

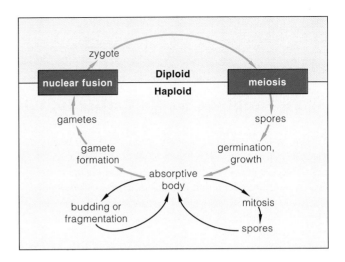

Figure 18.4 Generalized life cycle for many fungi. Asexual reproductive events (black arrows) dominate. Sexual reproduction (blue arrows) is less frequent.

Without saprobic fungi and other decomposers, natural communities would become buried in their own garbage, nutrients would not be recycled, and life could not go on.

Fungal Body Plans. The vast majority of fungi are multicelled, land-dwelling species. The food-absorbing part of their body consists of a mesh of branching filaments (Figure 18.3). This mesh, the **mycelium** (plural, mycelia), rapidly grows over or into organic matter. It has a good surface-to-volume ratio for absorbing nutrients. Each filament in a mycelium is a **hypha** (plural, hyphae). The cell walls of most hyphae are reinforced with chitin, a nitrogen-containing polysaccharide.

Reproductive Modes. During the life cycle of all multicelled fungi, some portion of the body becomes modified to produce spores. Such body parts are called sporangia (singular, sporangium). A **spore** is a walled cell or multicelled structure that germinates at some point after dispersal from the parent body. As Figure 18.4 shows, spores are produced asexually or sexually. Occasionally, gametes also form in special reproductive structures that develop during the life cycle. The gamete-producing parts of these structures are called gametangia (singular, gametangium).

Major Groups of Fungi

Table 18.1 lists the main groups of fungi. These are the water molds, chytrids, zygospore-formers, sac fungi, and club fungi.

Table 18.1 Classification of Fungi

Group	Common Name	Typical Habitats
Oomycetes	Water molds, related forms	Aquatic; some parasitic
Chytridiomycetes	Chytrids	Aquatic (mud, decaying plants or animals); some parasitic
Zygomycetes	Zygospore-forming fungi	Soil, decaying plants; most saprobic, a few parasitic
Ascomycetes	Sac fungi	Soil, decaying plants; many pathogens of plants, some of animals
Basidiomycetes	Club fungi	Soil, decaying plants; symbionts or pathogens of plants; many saprobic
	Imperfect fungi*	Diverse (damp grain, humans, etc.)

*These fungi are of undetermined affiliation; the name has no formal taxonomic status.

Commentary

A Few Fungi We Would Rather Do Without

You know that you are a serious student of biology when you view organisms objectively in terms of their place in nature, not in terms of their impact on humans generally and yourself in particular. As a student you can indeed respect saprobic fungi as vital decomposers and salute parasitic fungi that keep populations of destructive insects and weeds in check.

The true test is when you open the refrigerator for a dish of high-priced raspberries and discover that a fungus beat you to them or when another fungus starts feeding on warm, damp tissues between your toes. Who among us praises a fungus that makes our skin reddened, scaly, and cracked? Which home gardener waxes poetic about black spot or powdery mildew on roses? Which farmers happily lose millions of dollars each year to rust or smut invaders of their crops?

And who denies that fungi have influenced the very course of human history? Consider how potatoes used to be *the* main food crop of Irish peasants. Between 1845 and 1860, growing seasons were cool and damp—and perfect for growth of the water mold *Phytophthora infestans*. This fungus rots potatoes and tomatoes, a disease called *late blight*. For fifteen growing seasons, its abundant spores spread unimpeded through the watery films on potato plants. The destruction was appalling. A third of Ireland's population starved to death, died in the outbreak of typhoid fever that followed as a secondary effect, or fled to the United States and other countries.

Or consider *Claviceps purpurea*, a fungal parasite of rye and other grains. We use its metabolic by-products (alkaloids) to treat migraine headaches and to shrink the uterus (to prevent hemorrhaging) after childbirth. However, large amounts of the alkaloids are toxic. Eat a lot of bread made with contaminated rye flour and you end up with *ergotism*. Symptoms of this disease include hysteria, hallucinations, convulsions, vomiting, diarrhea, dehydration, and gangrenous limbs. Severe cases are fatal.

Some Pathogenic and Toxic Fungi	
Water molds	
Phytophthora infestans	Late blight of potato, tomato
Plasmopara viticola	Downy mildew of grapes (*above*)
Zygospore-forming fungi	
Rhizopus	Food spoilage
Sac fungi	
Ophiostoma ulmi	Dutch elm disease
Cryphonectria parasitica	Chestnut blight
Venturia inaequalis	Apple scab
Claviceps purpurea	Ergot of rye, ergotism
Monilinia fructicola	Brown rot of stone fruits
Club fungi	
Puccinia graminis	Black stem wheat rust
Ustilago maydis	Smut of corn
Amanita (some)	Severe mushroom poisoning
Imperfect fungi	
Verticillium	Plant wilt
Microsporum, Trichophyton, Epidermophyton	Various species cause ringworms, including athlete's foot
Candida albicans	Infection of mucous membranes

Ergotism epidemics were common in Europe in the Middle Ages, when rye was a major crop. Ergotism thwarted Peter the Great, the Russian czar who was obsessed with conquering ports along the Black Sea for his vast and nearly landlocked empire. Soldiers laying siege to the ports ate mostly rye bread and fed rye to their horses. The former went into convulsions and the latter, into "blind staggers." Quite possibly, outbreaks of ergotism were an excuse to launch the Salem witch-hunts in colonial Massachusetts.

Water Molds and Chytrids. The **water molds** and their relatives, including downy mildews, belong to a group called the Oomycetes. Some species are single celled; others have profuse hyphae. Most water molds are saprobes or parasites in fresh water or moist soil. The cottony growths you may have observed on goldfish or tropical fish are hyphae of a parasitic water mold, *Saprolegnia*. Other water molds destroy many crop plants, including potatoes and grapes (see the *Commentary*).

Muddy or aquatic habitats are home to the single-celled **chytrids** (Chytridiomycetes). Most species feed on decaying plants. Some are parasites of corn and other plants. Like most fungi, some chytrids have chitin-reinforced cell walls. They may resemble the eukaryotic cells that gave rise to the fungal kingdom.

Both chytrids and water molds produce flagellated spores that actively swim to new sources of food. In this respect they are distinct from the other fungal groups.

slice of bread

a

b

500μm

Figure 18.5 Zygospore-forming fungi. (**a**) Sporangia of *Rhizopus stolonifer*, the black bread mold. (**b**) *Pilobolus*, which grows on animal feces. Dark sacs above the swollen portion of each stalk contain spores. Water pressure builds up inside the swollen portion so that the spore sac can be blasted 2 meters away—a remarkable feat, considering the stalk is less than 10 millimeters tall!

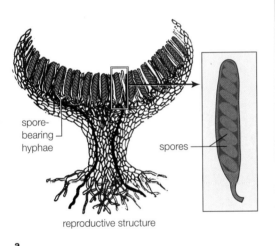

spore-bearing hyphae

spores

reproductive structure

a

b

c

Figure 18.6 Sac fungi. (**a**) Diagram and (**b**) photograph of the reproductive structure of the scarlet cup fungus. Spores are produced inside sacs (pouches). (**c**) One of the edible morels. It has a poisonous relative.

Zygospore-Forming Fungi. When members of this group reproduce sexually, a thick spore wall forms around the zygote. When this resting spore germinates, it gives rise to stalked, spore-bearing structures. Hence the name, **zygospore-forming fungi** (Zygomycetes).

The parasitic species live on houseflies and other insects. Saprobic species live in soil, decaying plant matter, and stored food. For example, the black bread mold (*Rhizopus stolonifer*) spoils baked goods. It can produce so many spores, moldy bread looks black (Figure 18.5*a*). Winds have dispersed its lightweight, dry spores just about everywhere, including the North Pole. Another saprobe, *Pilobolus*, disperses its spores in a wonderfully blastful way (Figure 18.5*b*).

Sac Fungi. Single-celled yeasts are the simplest **sac fungi** (Ascomycetes). They live in flower nectar and on fruits and leaves. One yeast species, shown earlier in Figure 6.8, produces the carbon dioxide that makes bread rise and the ethanol in wine, beer, and other alcoholic beverages.

However, most sac fungi are multicelled. When they reproduce sexually, spores develop inside large cells. These cells form in reproductive structures that are shaped like globes, flasks, or dishes (Figure 18.6). Certain truffles, morels, and other multicelled species are edible. Many others are pathogens (see *Commentary*). One *Neurospora* species ruins baked goods; another is used in genetic research.

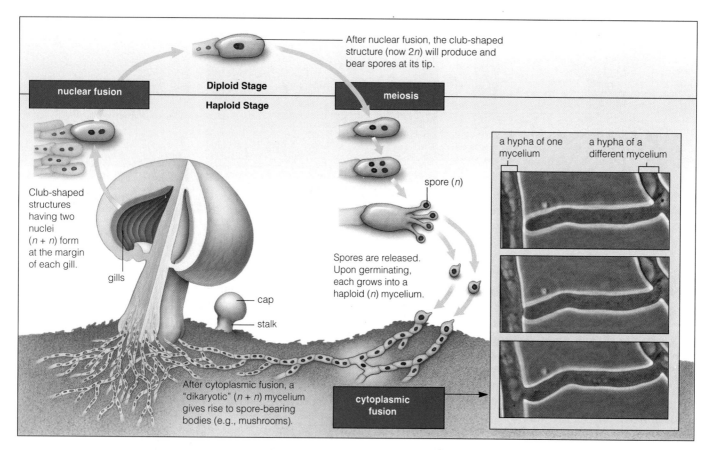

Figure 18.7 More club fungi. (**a**) Light-red coral fungus (*Ramaria*). (**b**) Shelf fungus (*Polyporus*) growing on a rotting log. (**c**) Fly agaric mushroom (*Amanita muscaria*), which causes hallucinations when eaten. Some other *Amanita* species are usually fatal when eaten.

After nuclear fusion, the club-shaped structure (now 2*n*) will produce and bear spores at its tip.

nuclear fusion

Diploid Stage

Haploid Stage

meiosis

Club-shaped structures having two nuclei (*n* + *n*) form at the margin of each gill.

gills

a hypha of one mycelium

a hypha of a different mycelium

spore (*n*)

Spores are released. Upon germinating, each grows into a haploid (*n*) mycelium.

cap

stalk

After cytoplasmic fusion, a "dikaryotic" (*n* + *n*) mycelium gives rise to spore-bearing bodies (e.g., mushrooms).

cytoplasmic fusion

Figure 18.8 Generalized life cycle for many club fungi. When two compatible mating strains grow together, the cytoplasm (but not the nuclei) of two hyphal cells may fuse. Cell divisions produce a "dikaryotic" mycelium (its cells have two nuclei). When mushrooms develop, club-shaped structures form over the surface of the gills. Each structure contains two nuclei, which fuse to form a diploid zygote.

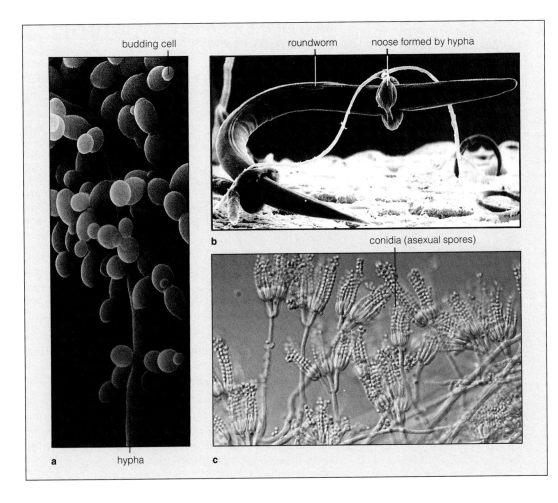

budding cell

roundworm   noose formed by hypha

b

conidia (asexual spores)

a   hypha   **c**

Figure 18.9 Imperfect fungi. (**a**) *Candida albicans*, cause of "yeast infections" of the vagina and mouth. (**b**) Hyphae of *Arthrobotrys dactyloides*, a predatory fungus, form nooselike rings that swell rapidly with incoming water when stimulated. The "hole" in the noose shrinks and captures this worm. (**c**) Rows of conidia (asexual spores) of *Penicillium*.

Club Fungi.   Figure 18.2 only hints at the diversity among the **club fungi** (Basidiomycetes). Figure 18.7 shows a few more members. Reproductive structures of this group have microscopically small, club-shaped cells that produce and bear spores on their surface. Figure 18.8 shows the location of these cells in the reproductive structures of the common mushroom (*Agaricus brunnescens*). The common mushroom is the basis of a multimillion-dollar business.

Some club fungi are major decomposers of plant material. Rusts, smuts, and other pathogenic types can destroy entire fields of wheat, corn, and other crops. Ingest *Amanita muscaria*, the fly agaric mushroom, and you will hallucinate (Figure 18.7). This mushroom was used ritualistically in ancient societies in Central America, Russia, and India. Other varieties in the United States make you vomit, and some can quickly kill you. There is no easy way to distinguish harmless from deadly mushrooms. *No one should eat mushrooms gathered in the wild unless they have been accurately identified as edible.*

The club fungus *Armillaria bulbosa* is among the oldest and largest organisms on earth. The mycelium of one specimen, discovered in a northern Michigan forest, extends through at least fifteen hectares of soil. By one estimate, it weighs more than 10,000 kilograms and has been growing for more than 1,500 years.

"Imperfect" Fungi.   Some fungi are not easily classified. If a species has no sexual phase in its life cycle, it is called an "imperfect" fungus. If researchers later detect a sexual phase, the fungus is assigned to a recognized group.

Some imperfect fungi cause human diseases. *Candida albicans* is a notorious cause of vaginal infections (see Figure 18.9a and the *Commentary*). Other, predatory types ensnare tiny worms (Figure 18.9b). *Aspergillus* produces citric acid for candies and soft drinks, and it ferments soybeans for soy sauce. Certain *Penicillium* species "flavor" Camembert and Roquefort cheeses; and others produce penicillins, which we use as antibiotics.

Figure 18.10 Lichens. Upper photograph, *Usnea*, commonly called old man's beard. Lower photograph, *Cladonia rangiferina*, sometimes called reindeer moss.

BENEFICIAL ASSOCIATIONS BETWEEN FUNGI AND PLANTS

Many fungi and plants form mutually beneficial associations. And many communities depend on their interactions.

Lichens

Thousands of sac fungi and club fungi can enter into mutually beneficial interactions with cyanobacteria, green algae, or both. These "mutualistic" associations between a fungus and photosynthetic cells are called **lichens** (Figure 18.10).

Lichens form after a fungal hypha penetrates a host cell and starts absorbing carbohydrates from it. If the cell survives, both it and the fungus multiply together into crusty formations (Figure 18.11). The fungus continues to draw nutrients from the cell's descendants. The photosynthetic species suffers in terms of its own growth, but it also benefits from the lichen's sheltering effect.

Sheltering is sometimes vital. Lichens colonize places that are too hostile for other organisms. They grow (slowly) on bare rocks in deserts and mountains, on tree bark and fence posts. Some types live close to the South Pole! They endure by suspending activities when it becomes extremely hot, cold, or bright. Then, their crusty part thickens and blocks sunlight, so photosynthesis stops. They become active only when moistened with raindrops, fog, or dew. It is easy to imagine that ancient lichens were the first invaders of the land.

Lichens absorb minerals from rock and nitrogen from the air. Soils slowly form through their activities, and this sets the stage for colonization by different species. Lichens also signal environmental deterioration. They absorb but cannot rid themselves of toxins. When lichens die around cities, air pollution is getting bad. We know this from studies in industrialized regions of England and in New York City.

fungal hyphae

algal layers (dark cells)

fungal hyphae

asexual reproductive body

Figure 18.11 One type of lichen in cross-section.

Where Have All the Fungi Gone?

Throughout European forests, wild mushrooms are getting smaller, and species are declining at an alarming rate. Mushroom gatherers aren't to blame; toxic as well as edible species may be on the road to extinction. Also throughout Europe, driving cars, burning coal, prodding crop plants with nitrogen fertilizers, and other human practices are pumping ozone, nitrogen oxides, and sulfur oxides into the air. The decline of fungi is correlated with the rise in air pollution.

Forest trees and assorted mycorrhizal fungi are symbiotic partners. The fungi promote enhanced uptake of water and nutrients, and the trees provide the fungi with carbohydrates. Normally, as a tree ages, one fungal species gradually supplants another in predictable patterns. If fungi are indeed disappearing, trees will lose their support system and will become highly vulnerable to severe frost and drought. Entire forests may already be at risk. In Europe, collectors have been recording information about the wild mushroom populations since the 1900s. Similarly extensive records have not been kept in the United States, but comparable environmental conditions suggest that North American forests also are at risk.

Mycorrhizae

The small roots of at least 80 percent of all vascular plants associate with fungi in a mutually beneficial way. Such associations are **mycorrhizae**, or "fungus-roots." A mycorrhiza is an example of **symbiosis**—the most permanent and intimate of all mutually beneficial interactions between species. (Symbiosis means "living together.")

In some cases, fungal hyphae penetrate the plant cells, as they do in lichens. In other cases, they form a velvety wrapping over a small root, then surround but do not penetrate the living cells inside. Figure 18.12*a* shows hyphae threading densely around the young roots of a forest tree.

A mycorrhizal fungus absorbs carbohydrates from the host plant, which absorbs mineral ions from the fungus. Collectively, fungal hyphae have an enormous surface area for taking up water and dissolved ions. The fungus takes up ions when they are abundant in soil and releases them to the plant when ions are scarce. Many plants cannot grow as efficiently when mycorrhizae are not available to help them absorb phosphorus and other crucial ions (Figure 18.12*b*).

As the *Focus* essay suggests, air pollution damages mycorrhizae, and this is affecting the world's forests. We will return to this topic in Chapter 37.

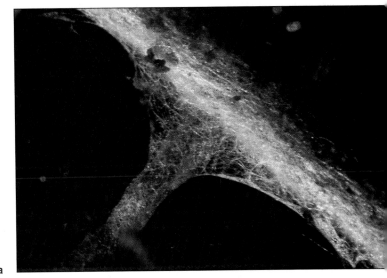

a

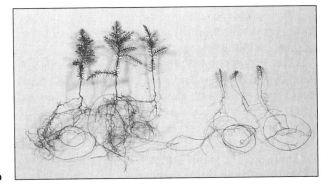

b

Figure 18.12 (**a**) Mycorrhiza of a hemlock tree. White threads are hyphal strands around a small root. (**b**) The juniper seedlings are six months old. The three on the left were grown in sterilized, phosphorus-poor soil with a mycorrhizal fungus. The three seedlings to the right were grown without the fungus.

PART II. KINGDOM OF PLANTS

General Characteristics

We turn now to the kingdom of plants. Most plants are multicelled, photosynthetic autotrophs. This means they produce food from water and dissolved minerals, using sunlight as the energy source. Together with photosynthetic bacteria and protistans, plants are the primary producers for nearly all communities.

More than 280,000 species live in water, on land, even high above forest floors, attached to other plants. In size alone, they range from microscopic algae to giant redwoods. Most are **vascular plants**, with tissues that conduct and distribute water and solutes through well-developed roots, stems, and leaves.

There are about 13,000 kinds of **red**, **brown**, and **green algae**. Nearly all are multicelled, aquatic species. There are fewer than 16,000 species of **bryophytes**, which are "nonvascular" land plants. Bryophytes don't have true roots, stems, and leaves, which are partly defined by internal vascular tissues.

Evolutionary Trends Among Plants

From the time their protistanlike ancestors arose until about 435 million years ago, plants evolved in the seas. Evolutionarily speaking, the pace picked up after that (Table 18.2). Stalked species evolved along coasts and streams, in partnership with fungi. Within 60 million years, diverse species cloaked much of the land. Some long-term changes in structure and reproductive events help explain how the diversity came about.

Evolution of Roots, Stems, and Leaves. Underground structures started evolving in early land plants. On the evolutionary road leading to vascular plants, these developed into root systems that could take up water and scarce mineral ions from a large volume of soil. Other structures evolved into aboveground systems of stems and leaves that could exploit abundant resources—sunlight energy and carbon dioxide in the air.

Roots, stems, and leaves never would have evolved without the development of internal pipelines. The pipelines evolved as components of two vascular tissues: xylem and phloem. Xylem distributed water and dissolved ions through plant parts. Phloem distributed sugars and other photosynthetic products.

Also, extensive growth of stems and branches became possible when lignin evolved. This organic compound strengthens cell walls. Today, lignin-reinforced tissues structurally support plant parts that display leaves and help increase the surface area exposed to sunlight.

Finally, water-conserving structures evolved by way of natural selection in land habitats. Stems and leaves

Table 18.2		Milestones in the Evolution of Plants and Fungi
Mesozoic	Cretaceous	*138–65 mya.* * Origin and rise to dominance of flowering plants.
	Jurassic	*205–138 mya.* Worldwide dominance of conifers, cycads.
	Triassic	*240–205 mya.* Origin of cycads, ginkgos.
Paleozoic	Permian	*290–240 mya.* Extinction of most groups of land plants, origin of conifers.
	Carboniferous	*360–290 mya.* Forests, coal swamps with diverse fungi, lycopods, bryophytes, ferns, horsetails, and progymnosperms.
	Devonian	*410–360 mya.* Rise of early vascular plants (*Rhynia*, lycopods, horsetails, ferns, progymnosperms). Origin of seed plants by end of Devonian.
	Silurian	*435–410 mya.* Earliest known vascular land plants (*Cooksonia*) by mid-Silurian.
	Cambrian to Ordovician	*550–435 mya.* Red, brown, green algae diversify in the seas. Green algae invade the land.
	Late Precambrian	*900–700 mya.* Multicelled algae and fungi in warm, shallow seas.
	Early Precambrian	*1,700–900 mya?* Single-celled, eukaryotic algae evolve from photosynthetic, protistanlike ancestors (page 257).

*Millions of years ago.

became protected by a waxy cuticle that reduced water loss on hot, dry days. Tiny openings (stomata) across stem and leaf surfaces evolved. Carbon dioxide uptake and water loss could be controlled at these openings. The next unit describes these tissue specializations.

From Haploid to Diploid Dominance. As early plants moved into higher, drier parts of the world, adaptations accumulated in their life cycles. Think about aquatic green algae, which spend most of their time producing and releasing gametes into the surrounding water. Actually, their gametes can't get together *except* in liquid water. This also was true of ancestral green algae. Figure 18.13*a* shows how you might diagram the life cycle of a green alga. The cycle has two phases. The haploid (*n*) phase, which starts with meiosis and gamete formation, dominates. A diploid (2*n*) phase starts when two gametes fuse at fertilization.

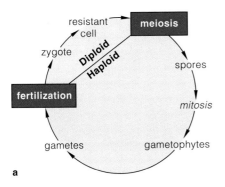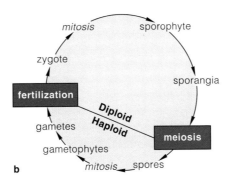

Figure 18.13 Typical life cycle of (**a**) some algae and of (**b**) vascular plants.

Now think about the life cycles of land plants. As Figure 18.13*b* suggests, the *diploid* phase dominates. This is a key adaptation to life in land habitats, most of which do not have limitless supplies of water and nutrients. Long ago, natural selection favored the evolution of complex **sporophytes**. These multicelled plant bodies grow during the diploid phase of the life cycle, and they produce spores. A pine tree is an example. Complex sporophytes have good root systems. In addition, with their mycorrhizal fungi, they can obtain water and scarce nutrients—even in seasonally dry regions.

The haploid phase starts in reproductive parts of the sporophyte. There, haploid spores form by meiosis. Such spores divide by mitosis and produce multicelled, haploid **gametophytes**. These structures produce the gametes. *Unlike algae, land plants nourish and protect the forthcoming generation.* They hold onto spores and gametophytes until environmental conditions favor dispersal and fertilization.

Evolution of Pollen and Seeds. Gymnosperms (such as pine trees and other conifers) and angiosperms (flowering plants) have radiated into nearly all high and dry habitats. The evolution of pollen grains and seeds contributed to their successful radiations.

Both groups of plants produce not one but *two* kinds of spores. One kind of spore develops into **pollen grains**. These become sperm-bearing *male* gametophytes. The other kind of spore develops into *female* gametophytes, the plant parts where eggs form and fertilization occurs. Pollen grains hitch rides on air currents or insects, birds, and so on. *They don't require liquid water to meet up with the eggs.*

Also, in both groups of plants, the parent sporophyte holds onto its female gametophytes. It nourishes and protects fertilized eggs as they develop into young embryos. The nutritive tissues, protective tissues, and the embryo itself constitute a **seed**. It's no coincidence that the dominant seed plants arose during the extreme

Table 18.3	Comparison of Trends Among Plants				
Green Algae	Bryophytes	Ferns		Gymnosperms	Angiosperms
Nonvascular ──────→ Vascular ──────────────────→					
Haploid dominance ────── Diploid dominance ───────────					
Spores of one type ──────────→ Spores of two types ─────					
Motile gametes ──────────────────────── Nonmotile gametes* ──					
Seedless ─────────────────→ Seeds ──────→					

*Require pollination by wind, insects, etc.

climates of Permian times. *Seeds are packages that endure hostile conditions.* Putting this all together,

1. During the evolution of complex land plants, vascular tissues and other structural adaptations to dry conditions emerged.

2. Sporophytes with well-developed roots and shoots came to dominate the life cycle of complex land plants. They had the means to nourish and protect spores and gametophytes through unfavorable conditions.

3. Some plants started producing two types of spores instead of one. This led to the evolution of male gametes specialized for dispersal without liquid water. It led to the evolution of seeds—embryos packaged with protective and nutritive tissues.

Table 18.3 summarizes some of the key evolutionary trends among plants. With these trends in mind, let's now turn to the spectrum of plant diversity.

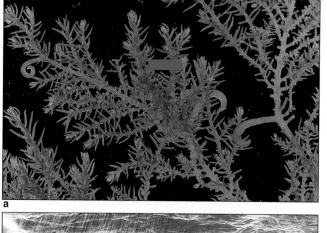

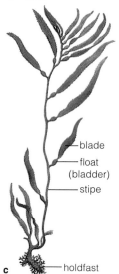

Figure 18.14 (**a**) Red alga showing finely branched growth. (**b**) Red alga having sheetlike growth. (**c**) Components of many brown algae. (**d**) Sea palms, a brown alga of intertidal zones. (**e**) A kelp that stays submerged even at low tide.

blade
float (bladder)
stipe

c — holdfast   d   e

Red, Brown, and Green Algae

Algae is a term that originally was used to define simple aquatic "plants." It no longer has formal meaning in classification schemes. The organisms once grouped under the term are now assigned to different kingdoms. In this book, recall, we classify "blue-green algae" (cyanobacteria) as eubacteria and "golden algae" as protistans. Although we classify red, brown, and green algae as plants, keep in mind that they have strong evolutionary links with protistans.

> Most red, brown, and green algae are multicelled aquatic plants with strong evolutionary links to protistan groups.

Red Algae. When you eat moist cookies or gelatin desserts, the red algae (Rhodophyta) probably don't pop into your mind. Yet commercial baked goods and gelatins incorporate agar as a moisture-fixing or setting agent. This inert, gelatinous substance comes from the cell walls of some species of red algae.

Red algae appear green, red, purple, or greenish black, depending on the types and amounts of pigments besides chlorophylls (Figure 18.14). Their phycobilin pigments trap the blue wavelengths that often

penetrate far below the water's surface. There some red algae have an edge in competition for nutrients.

Most red algae live in marine habitats. They abound in tropical seas, often at surprising depths (more than 200 meters) when the water is clear. The ones with stonelike cell walls help build coral reefs (page 607).

Brown Algae. If you enjoy ice cream, pudding, salad dressing, canned and frozen foods, jellybeans, or beer, if you use cough syrup, toothpaste, cosmetics, paper, or floor polish, thank the brown algae (Phaeophyta). The cell walls of some species contain algin, which is used as a thickening, emulsifying, and suspension agent.

Xanthophyll pigments help color various species olive-green, golden, or dark brown. Structurally, the simplest species have a branching, filamentous body. Complex species have leaflike blades, stemlike stipes, and rootlike holdfasts (Figure 18.14c).

Most brown algae live just offshore or in intertidal zones. Giant kelps (*Macrocystis*) often are 50 meters tall, forming underwater forests that are among the most productive ecosystems. They have hollow, gas-filled bladders that help keep blades near the sunlit surface waters. The brown alga *Sargassum* floats as extensive masses through the vast Sargasso Sea, which lies between the Azores and the Bahamas.

Figure 18.15 Life cycle of a single-celled species of *Chlamydomonas*, one of the most common green algae of freshwater habitats. *Chlamydomonas* reproduces asexually most often. It reproduces sexually under certain environmental conditions.

Zygote (cross-section)

A thick-walled resistant zygote develops.

Meiosis and Germination

Nuclear Fusion

Diploid Stage

Haploid Stage

haploid cell (+ strain)

haploid cell (− strain)

Mitosis occurs. Whether the resulting cells develop into spores or gametes will depend on environmental conditions.

Cytoplasmic Fusion

SEXUAL REPRODUCTION: Mainly when nitrogen levels are low and light is of a certain quality and intensity, the cells develop into gametes.

ASEXUAL REPRODUCTION: More spores are produced.

ASEXUAL REPRODUCTION: More spores are produced.

Gametes of different mating types meet.

+

−

+

−

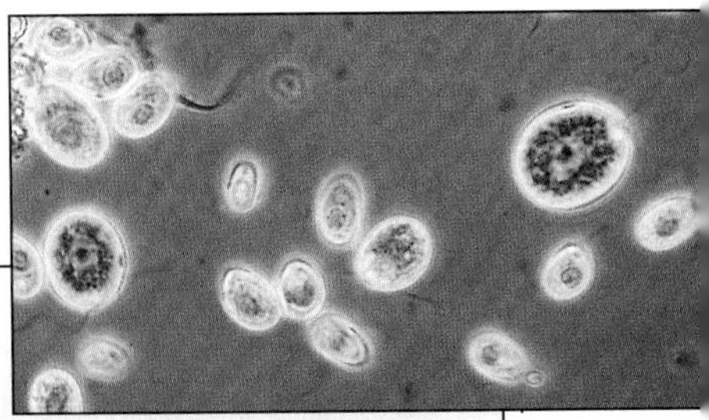

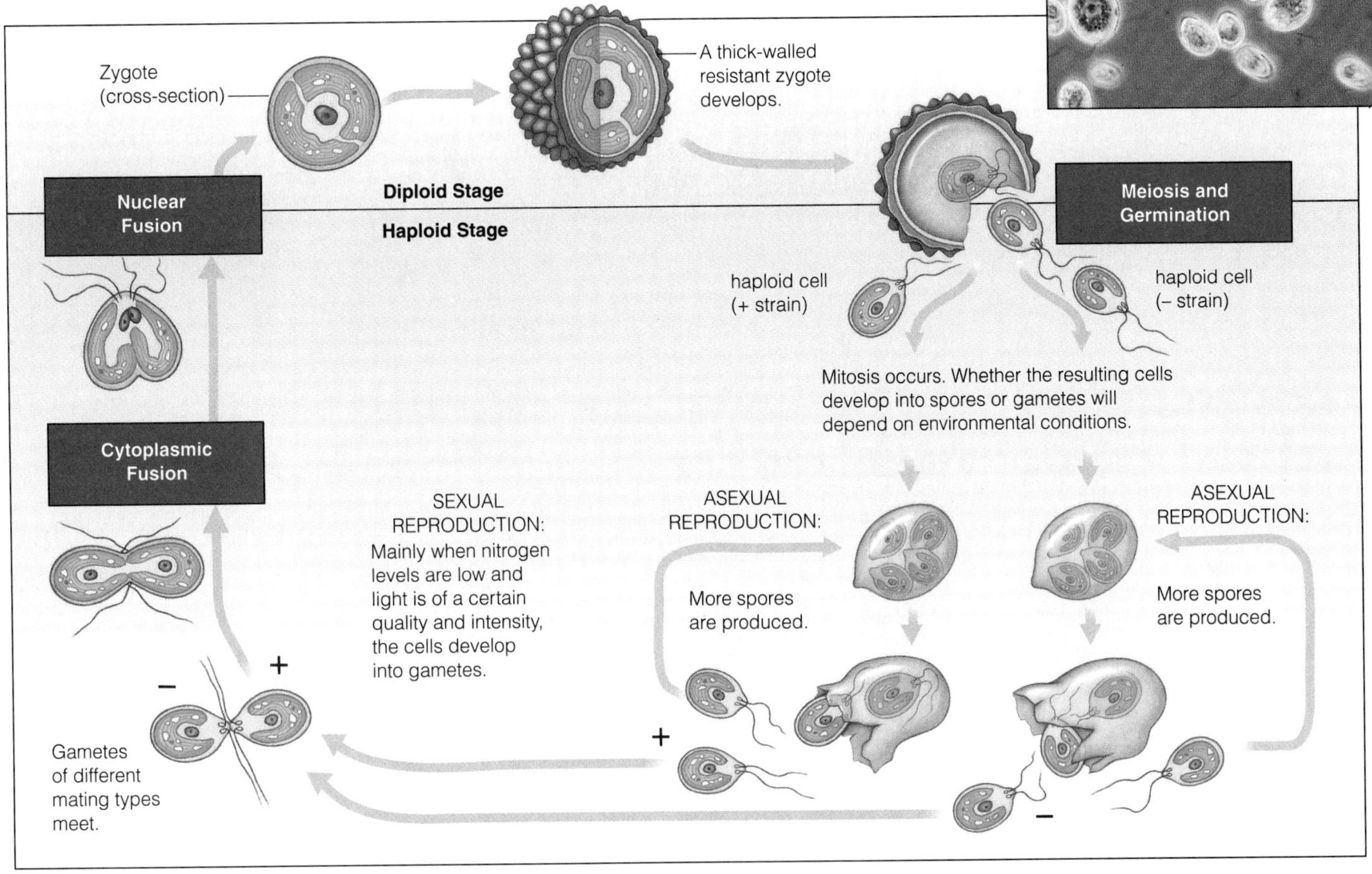

Green Algae. Think of astronauts spending years in a space station. Where would they get oxygen? Maybe from green algae (Chlorophyta), which would require only light, carbon dioxide, and some minerals and could grow in small spaces. The algae also would take up carbon dioxide exhaled by the aerobically respiring crew.

Of all aquatic plants, green algae bear the greatest resemblance to land plants. They, too, accumulate starch in chloroplasts. They, too, have an abundance of chlorophylls *a* and *b* as well as carotenoids and xanthophylls, and most have cell walls with cellulose. These and other similarities support the theory that land plants evolved from ancestral green algae.

Green algae grow in fresh water, on the surface of open oceans, just below the surface of soil and marine sediments, on rocks, on tree bark as well as on other organisms, even on snow. Their reproductive modes are diverse. Figures 18.15 and 18.16 are just two examples.

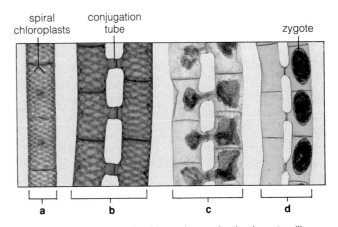

spiral chloroplasts   conjugation tube   zygote

a   b   c   d

Figure 18.16 One mode of sexual reproduction in watersilk (*Spirogyra*), a green alga with ribbonlike, spiral chloroplasts (**a**). A conjugation tube forms between cells of adjacent haploid filaments of different mating strains (**b**). The cellular contents of one strain pass through the tubes into cells of the other strain, where zygotes form (**c,d**). The zygotes will develop thick walls. They will undergo meiosis as they germinate and give rise to new haploid filaments.

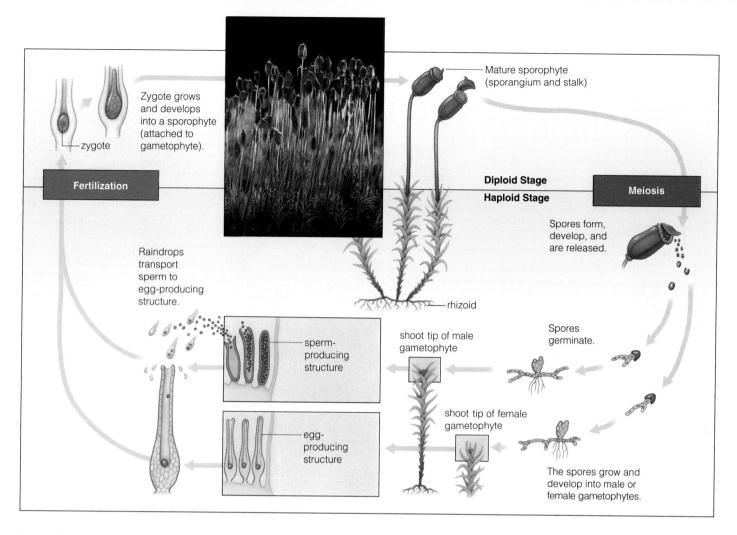

Figure 18.17 Life cycle of a moss, a representative bryophyte. The sporophyte remains attached to and depends on the gametophyte for water and nutrients.

Within the figure:

Zygote grows and develops into a sporophyte (attached to gametophyte).

zygote

Fertilization

Mature sporophyte (sporangium and stalk)

Diploid Stage
Haploid Stage

Meiosis

Spores form, develop, and are released.

Raindrops transport sperm to egg-producing structure.

rhizoid

sperm-producing structure

shoot tip of male gametophyte

Spores germinate.

Spores

egg-producing structure

shoot tip of female gametophyte

The spores grow and develop into male or female gametophytes.

Bryophytes

The nonvascular plants called bryophytes grow in fully or seasonally moist habitats. They have three adaptations that evolved in early land plants. First, a cuticle prevents water loss from aboveground parts. Second, a jacket of cells around their sperm-producing and egg-producing parts holds in moisture. Third, the embryo sporophyte starts developing *inside* the female gametophyte, which supplies it with nutrients and water.

Mosses, liverworts, and hornworts are bryophytes. All are small plants, generally less than 20 centimeters (8 inches) tall. Although they have leaflike, stemlike, and rootlike parts, they don't have xylem or phloem. Like lichens and some algae, they can dry out, then revive by absorbing moisture. Most species have rhizoids, elongated cells or threads that attach the gametophytes to soil. Rhizoids also absorb water and dissolved mineral ions.

Of all land plants, bryophytes alone have "free-living" gametophytes that hold onto sporophytes. Look at the moss life cycle in Figure 18.17. The green "moss plants" are the gametophytes. Eggs and sperm develop in jacketed structures at shoot tips. Sperm reach eggs by swimming through a film of water on the plant parts. After fertilization, sporophytes develop on the tips, drawing nutrients and water from them.

Bryophytes require liquid water for fertilization. Their sporophytes stay attached to the gametophytes, which provide them with nutrients and water.

Lycophytes, Horsetails, and Ferns

Lycophytes, **horsetails**, and **ferns** are seedless vascular plants. One of their earliest ancestors (*Cooksonia*) was leafless. In time, this lineage dominated the land (see the *Focus* essay). The ancestors of gymnosperms also may have belonged to this lineage. Figure 15.2 shows fossilized leaves from one of these progymnosperms.

Focus on the Environment

Ancient Carbon Treasures

a Reconstruction of a Carboniferous forest.

Between 360 and 280 million years ago, during the Carboniferous, swamp forests carpeted the wet lowlands of continents. Among the diverse species were the ancient ancestors of lycophytes, horsetails, ferns, and possibly the seed-bearing plants (Figure *a*). This was a period when sea levels rose and fell fifty times. When the seas moved out, swamp forests flourished. When the seas moved in, forest plants were submerged and became buried in sediments that protected them from decay. Gradually, the sediments compressed the saturated, undecayed remains into what we now call **peat**. As more sediments accumulated,

increased heat and pressure made the peat even more compact. It became **coal** (Figure *b*).

Coal has a high percentage of carbon; it is energy-rich. It is one of our premier "fossil fuels." It took a fantastic amount of photosynthesis, burial, and compaction to form each major seam of coal in the earth. It has taken us only a few centuries to deplete much of the known coal deposits. Often you will hear about our annual "production rates" for coal or some other fossil fuel. But how much do we really produce each year? None. We simply *extract* it from the earth. Coal is a nonrenewable source of energy.

Seedless vascular plants differ from bryophytes in three major ways. First, the sporophyte doesn't remain attached to the gametophyte. Second, it has complex vascular tissues. Third, it is the larger, longer lived phase of the life cycle.

Although their sporophytes are adapted to land, seedless vascular plants are confined largely to wet, humid regions. Their gametophytes have no vascular tissues for water transport. Also, the male gametes must

have water to reach the eggs. The few species living in extreme environments (such as deserts) can reproduce sexually only when adequate water is available.

Lycophytes, horsetails, and ferns are seedless vascular plants. All require liquid water for fertilization. Their sporophytes dominate the life cycle.

Chapter 18 Fungi and Plants **279**

a

Figure 18.18 Sporophytes of two seedless vascular plants. (**a**) A lycophyte (*Lycopodium*). (**b**) A horsetail (*Equisetum*).

b

Lycophytes.   The most familiar lycophytes are tiny club mosses growing on forest floors. Their sporophyte has true roots, stems, and small leaves with a strand of vascular tissue. Often it has conelike leaf clusters, which bear the spore sacs (Figure 18.18*a*). Spores give rise to small, free-living gametophytes.

Figure 18.19   Fern life cycle. The large photograph shows ferns in a moist habitat in Indiana.

The sporophyte (still attached to the gametophyte) grows, develops

zygote

fertilization

Diploid Stage

Haploid Stage

meiosis

rhizome

spores develop

Spores are released from a sporangium.

egg

egg-producing structure

sperm

sperm-producing structure

mature gametophyte (underside)

A spore germinates and grows into a gametophyte.

Horsetails. Only one genus of horsetails (*Equisetum*) has survived to the present. Horsetails grow in moist soil along streams and in disturbed habitats, such as roadsides. Their sporophytes have underground stems and scalelike leaves on aboveground stems (Figures 14.14 and 18.18*b*). Pioneers of the American West used horsetails to scrub cooking pots. The silica-containing walls of stem cells worked like sandpaper.

Ferns. Most ferns are native to tropical and temperate regions. Except for tropical tree ferns, the stems are mostly underground. Fern leaves (fronds) are usually finely divided, like feathers. You may have noticed rust-colored patches on the lower surface of many fern fronds. Each patch is a sorus (plural, sori), a cluster of sporangia. When the sporangium snaps open, spores catapult through the air. A spore develops into a small gametophyte, such as the green, heart-shaped type shown in Figure 18.19.

Most ferns are adapted to moist, dimly lit conditions on forest floors. They are being hit hard in the tropics, where forests are rapidly disappearing (page 620).

We now leave the seedless vascular plants. Take a moment to review Table 18.4, which compares their characteristics with those of the remaining major plant groups.

Table 18.4	Comparison of Major Plant Groups*
Structurally simple, mostly aquatic plants. Mostly haploid dominance.	
Red algae	4,000 species. Abundant phycobilin pigments. Mostly marine but some freshwater habitats.
Brown algae	1,500 species. Abundant xanthophyll pigments. Nearly all marine habitats, especially cool waters.
Green algae	7,000 species. Presumed ancestors of land plants. Mostly freshwater but also marine land habitats.
Nonvascular land plants. Fertilization requires free water. Haploid dominance. Cuticle, stomata present in some.	
Bryophytes	16,000 species. Moist, humid habitats.
Seedless vascular plants. Fertilization requires free water. Diploid dominance. Cuticle, stomata present.	
Lycophytes	1,000 species with simple leaves. Mostly wet or shady habitats.
Horsetails	15 species of single genus. Swamps, disturbed habitats.
Ferns	12,000 species. Wet, humid habitats in mostly tropical, temperate regions.
Vascular plants with "naked seeds" (gymnosperms). Diploid dominance. Cuticle, stomata present.	
Conifers	550 species, mostly evergreen, woody trees and shrubs having pollen- and seed-bearing cones. Widespread distribution.
Cycads	100 slow-growing species. Tropics, subtropics.
Ginkgos	1 species, a tree with fleshy-coated seeds.
Gnetophytes	70 species, limited distribution in deserts and tropics.
Vascular plants with flowers and protected seeds (angiosperms). Diploid dominance. Cuticle, stomata present.	
Flowering plants:	
Monocots	65,000 species. Floral parts often arranged in threes or multiples of three; one seed leaf; parallel leaf veins common.
Dicots	Nearly 200,000 species. Floral parts often arranged in fours, fives, or multiples of these; two seed leaves; net-veined leaves common.

*More than 307,200 known species total.

Figure 18.20 Lesser-known gymnosperms. (**a**) A cycad and its cone. (**b**) Ginkgo trees and their seeds with fleshy coats. (**c**) A gnetophyte (*Welwitschia*).

Gymnosperms

In sheer numbers and distribution, the seed-bearing plants (gymnosperms and angiosperms) are the most successful plants. Their seeds are mature **ovules**. An ovule is a structure that contains the egg-producing female gametophyte, surrounded by tissues and a jacket of cell layers. As a fertilized egg develops into an embryo sporophyte, the outer layers become a seed coat.

Gymnosperm is a word derived from the Greek *gymnos* (meaning naked) and *sperma* (which is taken to mean seed). As the name implies, gymnosperm seeds aren't covered. They are perched on scales of cones that are attached to the sporophyte. Gymnosperms include conifers as well as cycads, ginkgos, and gnetophytes.

Cycads. Cycads (Cycadophyta) flourished along with the dinosaurs. Existing species are confined to the tropics and subtropics. Cycads have massive, cone-shaped structures that bear either pollen or ovules (Figure 18.20*a*). Air currents or insects transfer pollen from "male" plants to developing seeds on "female" plants. In parts of Asia, people eat cycad seeds and a starchy flour made from cycad trunks, but only after they rinse out the plant's poisonous alkaloids.

Ginkgos. Ginkgos (Ginkgophyta) were diverse during dinosaur times, but only a single species survived to the present. Several thousand years ago, ginkgo trees were planted in cultivated grounds around temples in China. Then the natural populations almost became extinct. This is puzzling, for ginkgos seem hardier than many other trees. Perhaps, as human populations grew, ginkgos were cut for firewood. Male ginkgo trees are now planted in cities because of their attractive, fan-shaped leaves (Figure 18.20*b*) and their resistance to insects, disease, and air pollutants. Fleshy-coated seeds of female trees produce an awful stench when stepped on.

Gnetophytes. This gymnosperm group (Gnetophyta) includes *Welwitschia* of south and west African deserts. This plant is mostly a deep root. Its exposed part is a woody stem with cones and two strap-shaped leaves. As the plant matures, the leaves repeatedly split lengthwise into a scraggly pile (Figure 18.20*c*).

Conifers. Most conifers (Coniferophyta) are woody trees and shrubs with needlelike or scalelike leaves. They include pines, spruces, firs, hemlocks, junipers, cypresses, and redwoods. Most species are evergreen. Although they shed old leaves throughout the year, they retain enough leaves to distinguish them from deciduous species. Conifer cones are clusters of modified leaves in which spore-producing structures develop. Seeds develop on shelflike scales of the female cones.

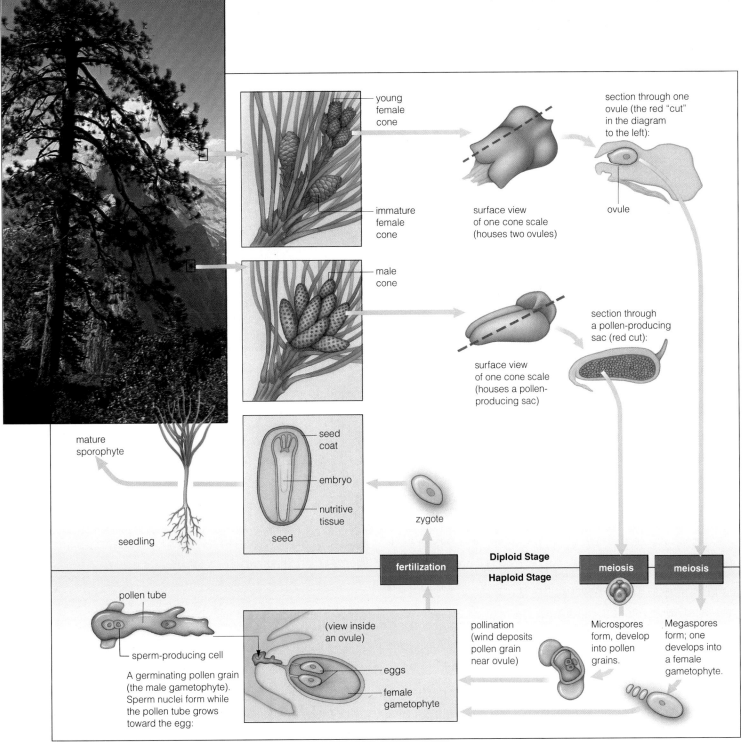

Figure 18.21 Life cycle of ponderosa pine, a conifer.

Figure 18.21 shows a pine life cycle. Pine trees produce two kinds of spores. Microspores, which develop into pollen grains, are produced in male cones. Megaspores are produced in female cones, and they develop into female gametophytes.

Each spring, millions of pollen grains drift off male cones, and some land on ovules. The arrival of pollen on female reproductive parts is called pollination. Now pollen grains germinate and develop into pollen tubes, which transport sperm toward female gametophytes.

Pollen tubes are the mature male gametophytes. Often, fertilization occurs months to a year after pollination.

Conifers radiated into many land environments during the Mesozoic. Compared to flowering plants, their reproductive pace was slow, and this put them at a competitive disadvantage when angiosperms began their radiations. But conifers still dominate many regions, including the far north, high altitudes, and parts of the Southern Hemisphere. Various species provide lumber, furniture, paper, and many other products.

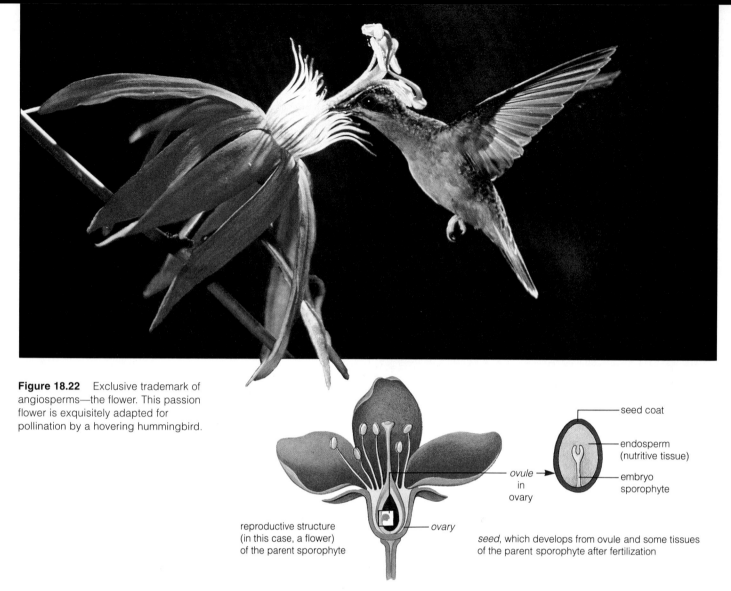

Figure 18.22 Exclusive trademark of angiosperms—the flower. This passion flower is exquisitely adapted for pollination by a hovering hummingbird.

seed coat

endosperm (nutritive tissue)

ovule in ovary

embryo sporophyte

reproductive structure (in this case, a flower) of the parent sporophyte

ovary

seed, which develops from ovule and some tissues of the parent sporophyte after fertilization

Angiosperms—The Flowering Plants

Angiosperm is a word derived from the Greek *angeion* (a vessel) and *sperma*. The "vessel" refers to tissues that surround and protect ovules and, later, developing seeds. Of all the plants, angiosperms alone have flowers (Figures 18.22 and 18.23). They have dominated the land for more than 100 million years. Today there are about 265,000 known species. New ones are discovered almost daily in previously unexplored regions of the tropics. Unless tropical deforestation is brought under control, untold numbers may become extinct before we learn about them.

Angiosperms are also the most diverse plants. They live on dry land and in wetlands, fresh water, and seawater. They range in size from tiny duckweeds (about a millimeter long) to *Eucalyptus* trees more than 100 meters tall. Most angiosperms are free-living and photosynthetic. A few nonphotosynthetic types depend on mycorrhizal fungi that are symbionts with a photosynthetic plant. Some, including mistletoe, are parasites on other plants.

There are two classes of angiosperms, the **monocots** and **dicots**. (The formal names are Monocotyledonae and Dicotyledonae.) Palms, lilies, and orchids are among the 65,000 species of monocots. So are the grasses, including our main crop plants (wheat, corn, rice, rye, sugarcane, and barley). With nearly 200,000 known species, dicots are the most diverse of all plants. They include most shrubs and trees, most nonwoody (herbaceous) plants, cacti, and water lilies. Figure 18.23 shows a monocot life cycle. A typical dicot life cycle is described in detail in the next unit, which focuses on the structure and function of flowering plants.

Many factors contributed to the adaptive success of angiosperms. As with other seed plants, a large diploid sporophyte dominates the life cycle, and it retains and nourishes the gametophytes. Angiosperm seeds alone contain a nutritive tissue called endosperm. The seeds are packaged in fruits, which help protect and disperse them. And above all, angiosperms alone have the reproductive structures called flowers.

As you will see in the next unit, many diverse floral structures coevolved with animal pollinators—insects, bats, rodents, and birds (Figure 18.22). This innovation probably figured in the rise of angiosperms and the gradual decline of gymnosperms in many regions over the past 100 million years.

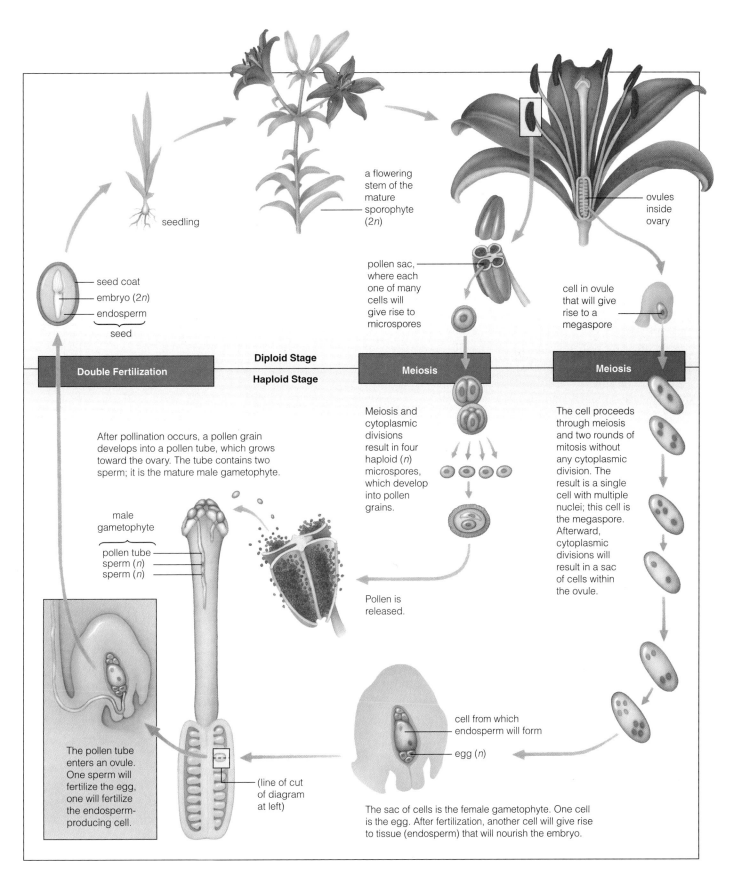

seedling

a flowering stem of the mature sporophyte (2n)

ovules inside ovary

seed coat
embryo (2n)
endosperm
seed

pollen sac, where each one of many cells will give rise to microspores

cell in ovule that will give rise to a megaspore

Diploid Stage

Double Fertilization

Haploid Stage

Meiosis

Meiosis

After pollination occurs, a pollen grain develops into a pollen tube, which grows toward the ovary. The tube contains two sperm; it is the mature male gametophyte.

Meiosis and cytoplasmic divisions result in four haploid (n) microspores, which develop into pollen grains.

The cell proceeds through meiosis and two rounds of mitosis without any cytoplasmic division. The result is a single cell with multiple nuclei; this cell is the megaspore. Afterward, cytoplasmic divisions will result in a sac of cells within the ovule.

male gametophyte

pollen tube
sperm (n)
sperm (n)

Pollen is released.

The pollen tube enters an ovule. One sperm will fertilize the egg, one will fertilize the endosperm-producing cell.

(line of cut of diagram at left)

cell from which endosperm will form

egg (n)

The sac of cells is the female gametophyte. One cell is the egg. After fertilization, another cell will give rise to tissue (endosperm) that will nourish the embryo.

Figure 18.23 Life cycle of a monocot (*Lilium*). "Double" fertilization occurs in flowering plants. The male gametophyte delivers two sperm to an ovule. One sperm fertilizes the egg, and the other fertilizes a cell that gives rise to a tissue (endosperm) that will nourish the forthcoming embryo. Figure 22.5 provides a closer look at flowering plant life cycles, using a dicot as the example.

SUMMARY

1. Fungi are heterotrophs. Many are major decomposers of organic matter. Saprobic types feed on nonliving organic matter. Parasitic types obtain nutrients from living organisms. The cells of all fungi secrete digestive enzymes that break down food into small molecules, which are absorbed across the plasma membrane.

2. Most fungi are multicelled. Their food-absorbing part (mycelium) is a mesh of filaments (hyphae). These often interweave into aboveground reproductive structures. Fungi reproduce asexually and sexually. Many interact with plants in mutually beneficial ways. Mycorrhizae are examples.

3. The plant kingdom includes aquatic and terrestrial species. Nearly all are photosynthetic autotrophs. Land plants apparently evolved from aquatic green algae.

4. The following trends emerged during the evolution of certain lineages of land plants:

   a. The evolution of vascular tissues and other structural adaptations to dry conditions.

   b. A shift from haploid to diploid dominance. Complex sporophytes evolved that could hold onto, nourish, and protect spores and gametophytes.

   c. A shift from one to two kinds of spores. Among gymnosperms and angiosperms, this led to the evolution of pollen grains and seeds.

5. Red, brown, and green algae are mostly multicelled aquatic plants. Bryophytes do not have vascular tissues. Seedless vascular plants include lycophytes, horsetails, and ferns. Existing seed-bearing vascular plants are gymnosperms and angiosperms.

6. The distribution of bryophytes and seedless vascular plants is limited by water availability. All require liquid water at the time of fertilization.

7. The evolution of pollen grains freed gymnosperms and angiosperms from dependence on free water for fertilization. Their seeds are efficient means of dispersing the new generation and helping it through hostile conditions. Pollen grains and seeds were key adaptations underlying the move into high, dry habitats.

Review Questions

1. Describe the fungal mode of nutrition. *266*

2. How does a mycorrhiza differ from a lichen? *272, 273*

3. How do bryophytes and vascular plants differ? *279*

4. Choose a garden plant, crop plant, or weed that grows in your neighborhood. Make a diagram of it, labeling its parts. Attempt to correlate some of its structures with seasonal variations in temperature, moisture, and other local environmental conditions.

Self-Quiz *(Answers in Appendix IV)*

1. A "mushroom" is _____ .
   a. the food-absorbing part of a fungal body
   b. a nonhyphal part of the fungal body
   c. a reproductive structure
   d. a nonessential part of the fungus

2. New mycelia form after _____ germinate.
   a. hyphae   c. mycelia
   b. spores   d. mushrooms

3. The _____ stage dominates vascular plant life cycles.
   a. diploid   c. sporophyte
   b. haploid   d. both a and c

4. Which is *not* a feature of gymnosperms and angiosperms?
   a. vascular tissues   c. two kinds of spores
   b. diploid dominance   d. single spore type

5. Which statement is *not* true?
   a. Red, brown, and green algae are mostly aquatic plants.
   b. Bryophytes are nonvascular plants.
   c. Lycophytes, horsetails, ferns, gymnosperms, and angiosperms are vascular plants.
   d. Gymnosperms are the simplest vascular plants.

6. A seed is _____ .
   a. a female gametophyte   c. a mature pollen tube
   b. a mature ovule   d. an immature embryo

7. Match the terms appropriately.
   ____ mycelium   a. gymnosperm
   ____ flower   b. angiosperm
   ____ "naked" seeds   c. bryophyte
   ____ nonvascular land plant   d. fungus

Selected Key Terms

angiosperm 284	horsetail 278	red alga 274
brown alga 274	hypha 267	sac fungus 269
bryophyte 274	lichen 272	saprobe 266
chytrid 268	lycophyte 278	seed 275
club fungus 271	monocot 284	spore 267
coal 279	mycelium 267	sporophyte 275
dicot 284	mycorrhiza 273	symbiosis 273
fern 278	ovule 282	vascular plant 274
gametophyte 275	parasite 266	water mold 268
green alga 274	peat 279	zygospore-forming
gymnosperm 282	pollen grain 275	fungus 269

Readings

Gifford, E., and A. Foster. 1989. *Morphology and Evolution of Vascular Plants*. Third edition. New York: Freeman. Outstanding reference.

Gray, J., and W. Shear. September–October 1992. "Early Life on Land." *American Scientist* (80): 444–456.

Moore-Landecker, E. 1990. *Fundamentals of the Fungi*. Third edition. Englewood Cliffs, New Jersey: Prentice-Hall.

19 ANIMALS

Making Do (Rather Well) With What You've Got

It has taken the platypus nearly two centuries to earn a little respect. In 1798, skeptical naturalists at the British Museum poked and probed a specimen, looking for signs that a prankster had stitched the bill of an over-sized duck onto the pelt of a small furry mammal. No wonder. At first blush the platypus looks like nature's practical joke (Figure 19.1).

The platypus is a mammal, about half the size of a housecat. Like other mammals, it has mammary glands and fur. Yet, like birds and most reptiles, it lays eggs! Its flat, furry tail serves as a rudder in water, a blanket to keep warm inside burrows, and a storehouse for energy-rich fats. When flared out, its webbed feet are underwater paddles. Its strong, clawed hind feet are great for digging. Its "duckbill" does look like an evolutionary accident. However, with 800,000 built-in sensory receptors, the bill detects even faint electrical fields or pressure waves. A shrimp-hunting platypus can plunge into murky water, snap shut its eyes and ears, and zero in on a meal with awesome precision.

About 100 million years ago, the ancestors of the platypus coexisted with dinosaurs in Gondwanaland. When that supercontinent broke up, the ancestral forms happened to be on the huge fragment that eventually became Australia. There, the lineage has endured in splendidly isolated habitats. With that track record, who are we to chuckle over the collection of platypus traits? As is the case for all other animals, the traits meet nature's most important test. They work.

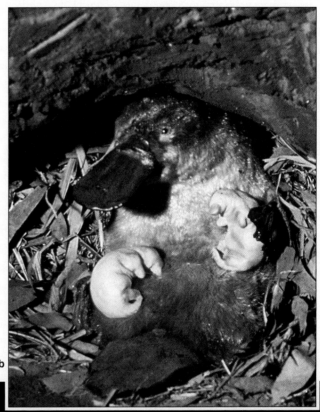

b

Figure 19.1 One of evolution's success stories—the platypus, underwater (**a**) and in its burrow with offspring (**b**).

a

KEY CONCEPTS

1. Animals are multicelled, motile heterotrophs that pass through embryonic stages during their life cycle.

2. There are probably more than 2 million existing species of animals. Fewer than 50,000 species are vertebrates (animals with a backbone). The rest are invertebrates (animals with no backbone).

3. Several trends occurred during the evolution of certain animal lineages. These are evident through comparisons of body plans of existing animals, in conjunction with the fossil record.

4. The most revealing aspects of an animal's body plan are its type of symmetry, gut, and cavity (if any) between the gut and body wall; whether it has a distinct head end; and whether it is divided into a series of segments.

Table 19.1 Animal Phyla Described in This Chapter

Phylum	Some Representatives	Number of Known Species
Porifera (poriferans)	Sponges	8,000
Cnidaria (cnidarians)	Hydrozoans, jellyfishes, corals, sea anemones	11,000
Platyhelminthes (flatworms)	Turbellarians, flukes, tapeworms	15,000
Nematoda (roundworms)	Pinworms, hookworms	20,000
Rotifera (rotifers)	Species with crown of cilia	1,800
Mollusca (mollusks)	Snails, slugs, clams, squids, octopuses	110,000
Annelida (segmented worms)	Leeches, earthworms, polychaetes	15,000
Arthropoda (arthropods)	Crustaceans, spiders, insects	1,000,000 +
Echinodermata (echinoderms)	Sea stars, sea urchins	6,000
Chordata (chordates)	Invertebrate chordates: Tunicates, lancelets	2,100
	Vertebrates:	
	Fishes	21,000
	Amphibians	3,900
	Reptiles	7,000
	Birds	8,600
	Mammals	4,500

OVERVIEW OF THE ANIMAL KINGDOM

General Characteristics of Animals

Generally speaking, an **animal** is defined by the following characteristics:

1. Animals are multicelled. In most groups, cells form tissues, which are arranged as organs and organ systems.

2. Animals are heterotrophs; they cannot build organic compounds from simple inorganic substances. They eat other organisms or absorb nutrients from them.

3. Animals require oxygen (for aerobic respiration).

4. Animals reproduce sexually and, in many cases, asexually. Most are diploid organisms.

5. Animal life cycles include a period of embryonic development. In brief, cell divisions transform the zygote into a multicelled embryo. The embryo's cells become arranged as "primary" tissue layers: ectoderm, endoderm, and, in most species, mesoderm. These give rise to the adult's tissues and organs (Chapter 32).

6. Most animals are motile during at least part of the life cycle.

Diversity in Body Plans

Table 19.1 lists the groups of animals described in this chapter. Their shared characteristics arose early in time, before divergences from a common ancestor gave rise to separate lineages. Later, morphological differences accumulated among the lineages, and they were the foundation for today's bewildering diversity. How can we get a conceptual handle on animals as different as flatworms, hummingbirds, platypuses, humans, and giraffes? We can compare their similarities and differences with respect to five basic features. These are *body symmetry, cephalization, type of gut, type of body cavity,* and *segmentation.*

Body Symmetry and Cephalization. Nearly all animals are radial or bilateral. Animals with **radial symmetry** have body parts arranged regularly around a central axis, like spokes of a bike wheel. Thus a cut down the center of a hydra (Figure 19.2*a*) divides it into equal halves; another cut at right angles to the first divides it into equal quarters. Radial animals are aquatic. With their body plan, they can respond to food suspended in the water all around them.

Animals with **bilateral symmetry** have right and left halves that are mirror images of each other. Most of

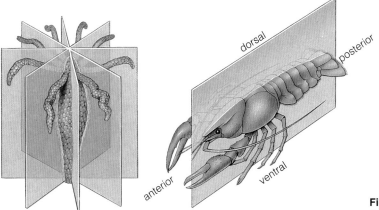

a   b

Figure 19.2 (**a**) Radial symmetry of a hydra. (**b**) Bilateral symmetry of a crayfish.

these animals have an *anterior* end (head) and an opposite, *posterior* end. They have a *dorsal* surface (a back) and an opposite, *ventral* surface. Figure 19.2*b* shows this body plan, which evolved after forward-creeping animals emerged. Most often, their forward end was the first to encounter food and other stimuli. So we can imagine there was strong selection for **cephalization**. By this evolutionary process, sensory structures and nerve cells became concentrated in a head. Bilateral plans and cephalization evolved together. Their joint evolution resulted in *pairs* of muscles, sensory structures, nerves, and brain regions.

Type of Gut. A **gut** is a body region where food is digested and then absorbed. Saclike guts have only one opening (a mouth) for taking in food and expelling residues. Other guts are part of tubelike systems with openings at both ends (mouth and anus). These "complete" digestive systems are more efficient. Their different regions have specialized functions, such as preparing, digesting, and storing food. The evolution of such systems helped pave the way for increases in body size and activity.

Body Cavities. A body cavity separates the gut and body wall of most bilateral animals (Figure 19.3). One type of cavity, a **coelom**, has a unique lining called a peritoneum. The lining also covers organs in the coelom and helps hold them in place. You have a coelom. A sheetlike muscle divides it into an upper, *thoracic* cavity and a lower, *abdominal* cavity. Your thoracic cavity holds a heart and lungs; your abdominal cavity holds a stomach, intestines, and other organs.

Some types of worms don't have a body cavity. The region between their gut and body wall is packed with tissues. Other types have a "false coelom"—a body cavity but no peritoneum. A true coelom was a major step in the evolution of animals larger and more complex than worms. By cushioning and protecting organs, it favored increases in their size and activity.

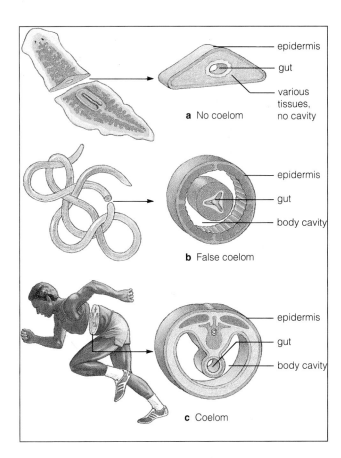

a No coelom
epidermis
gut
various tissues, no cavity

b False coelom
epidermis
gut
body cavity

c Coelom
epidermis
gut
body cavity

Figure 19.3 Type of body cavity (if any) in animals.

Segmentation. "Segmented" animals consist of a series of body units that may or may not be similar to one another. Most earthworm segments look alike on the outside. Insect segments are grouped into three body parts (head, thorax, and abdomen), and they differ greatly. Insects are an example of how segmentation favored the evolution of specialized headparts, legs, wings, and other appendages.

An Evolutionary Framework

Animals arose from protistanlike ancestors during the 200-million-year span before the Cambrian era. Most likely, the first animals were soft-bodied, so we may never find evidence of them. But the fossil record is clear on one point. By 560 million years ago, those ancestral forms had given rise to all major animal phyla, including the ones named in Figure 19.4.

There are now thirty animal phyla, some wildly successful, others obscure. We are most familiar with **vertebrates** (animals with a backbone). But the vast majority are **invertebrates**—they have no backbone.

Suppose we select just ten phyla—the sponges, cnidarians, flatworms, roundworms, rotifers, mollusks, annelids, arthropods, echinoderms, and chordates (Table 19.1). Suppose we compare them in terms of body symmetry, cephalization, type of gut and body cavity, and segmentation. By doing so, we can identify certain trends in animal evolution. Keep in mind, these trends did *not* occur in every line of descent. But their absence doesn't mean an animal is "primitive" or evolutionarily stunted. As you will see, even "simple" sponges are exquisitely adapted to their environment.

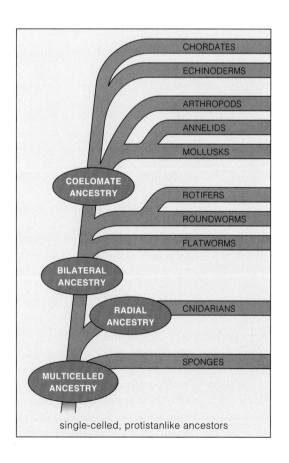

single-celled, protistanlike ancestors

PART I. THE INVERTEBRATES

Sponges

Sponges (Porifera) are one of nature's success stories. Since Cambrian times, they have abounded in the seas, including coastal waters and coral reefs. A few species live in freshwater habitats. Some are fingernail size, others are big enough to sit in. Different sponges are sprawling, flattened, compact, lobed, branching, tubular, cuplike, or vaselike (Figure 19.5).

A sponge body has no symmetry and no organs. It has a framework of glasslike fibers, needles, rods, or other structures, often cemented together. Within this framework are passages for the inflow and outflow of water. The outer surface and portions of cavities in the body wall have sheets of flattened cells, but these aren't the same as the tissue linings of other animals.

Many flagellated cells with "collars" of microvilli (absorptive structures) are arranged around canals or chambers in the body wall. As the flagella beat, they help keep water moving past, then out through larger openings (Figure 19.6). The microvilli trap bacteria and other waterborne edibles. Some food gets transferred to the amoebalike cells for further breakdown, storage, and distribution.

Sponges reproduce sexually when collar cells give rise to eggs and sperm. The young sponges pass through a larval stage. A **larva** (plural, larvae) is a sexually immature form that precedes the adult stage in many animal life cycles. Some sponges also reproduce asexually. For example, small fragments that break away from the body can grow into new sponges.

Figure 19.4 Evolutionary relationships among major groups of animals. Take a moment to study this family tree. We will use it repeatedly as our road map through discussions of each group.

Figure 19.5 Representative sponges. (**a**) A basket sponge of the Caribbean, releasing a cloud of sperm into the water. (**b**) A sprawling, red-orange sponge, one of many types that encrust underwater ledges in temperate seas. (**c**) Interior view of a vase-shaped sponge.

a

b

c

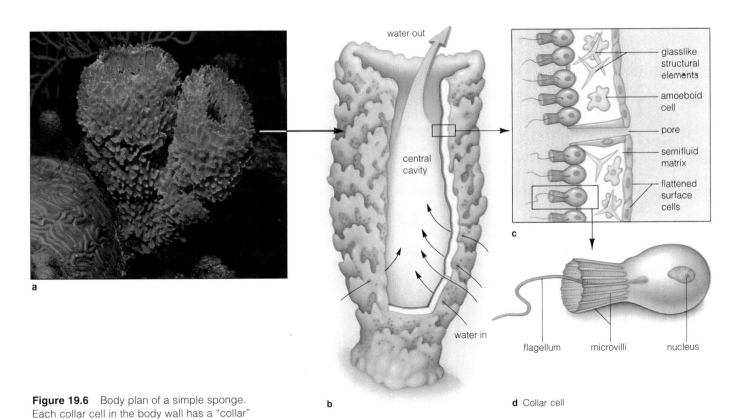

a

water out

central
cavity

water in

b

glasslike
structural
elements

amoeboid
cell

pore

semifluid
matrix

flattened
surface
cells

c

flagellum   microvilli   nucleus

d Collar cell

Figure 19.6 Body plan of a simple sponge. Each collar cell in the body wall has a "collar" of food-trapping microvilli.

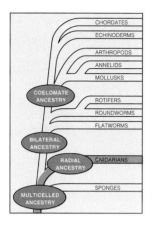

Cnidarians

Most **cnidarians** (Cnidaria) live in the seas. They include scyphozoans (jellyfishes), anthozoans (such as corals and sea anemones), and hydrozoans (such as *Hydra*). Figures 19.7 and 19.8 show examples. Two forms are common among these radial animals. As Figure 19.7*a* suggests, one form is a bell-shaped medusa (plural, medusae). The other, a tubelike polyp, attaches to firm substrates. Polyps and medusae both have tentacles, a mouth, a saclike gut, and tissues. A jellylike tissue called mesoglea is sandwiched between two sheetlike **epithelia** (singular, epithelium)—a tissue common to most animals. **Nerve cells** form a "nerve net," a simple nervous system that coordinates responses to stimuli. Together with sensory and contractile cells, they control movement and shape changes. Cnidarians alone have nematocysts—capsules that discharge prey-capturing threads (Figure 19.8*b*).

Most of the corals also have calcium-reinforced external skeletons—the main material of reefs. Their most extravagant work, the Great Barrier Reef, parallels Australia's east coast for 1,600 kilometers. Often the reef-forming hard corals are colonial; they are composed of many interconnected polyps.

Cnidarians reproduce sexually and asexually. Their larvae, called planulas, swim or creep about before settling down and growing into adults. Figure 19.9 shows one of these larval forms.

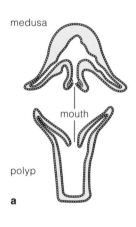

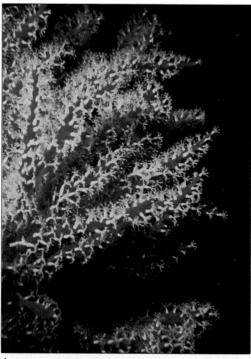

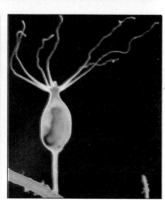

Figure 19.7 (**a**) The two main body plans of cnidarians, midsection. (**b**) Medusa of the sea nettle (*Chrysophora*), one of the jellyfishes. (**c**) A hydrozoan polyp (*Hydra*), capturing and then digesting a small crustacean. Two reef-building corals: (**d**) a "soft" coral (*Telesto*) with small individual polyps and (**e**) a "hard" coral (*Tubastrea*) with an external calcium-containing skeleton.

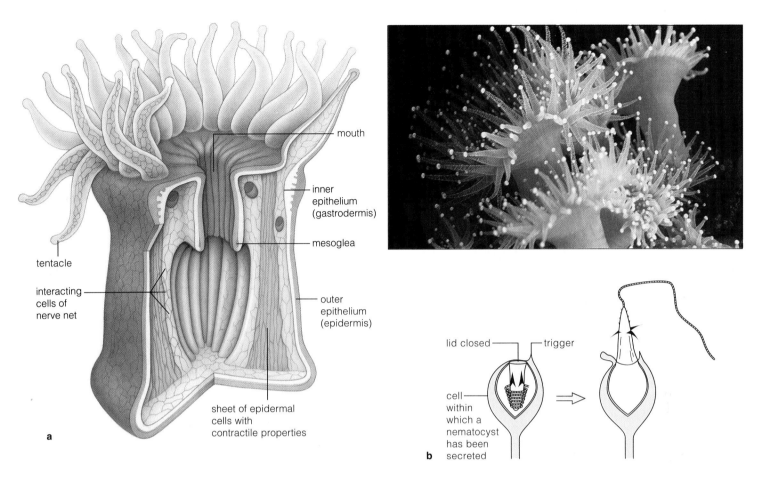

Figure 19.8 (**a**) Body plan of a sea anemone, showing its tissue organization. (**b**) As in other cnidarians, nematocysts are embedded in the tentacles. These capsules contain an inverted tubular thread. The one shown has a bristlelike trigger. When prey touch it, the capsule becomes more "leaky" to water. Water diffuses inward, pressure inside the capsule builds up, and the thread is forced to turn inside out. The thread's tip may penetrate prey and release a toxin.

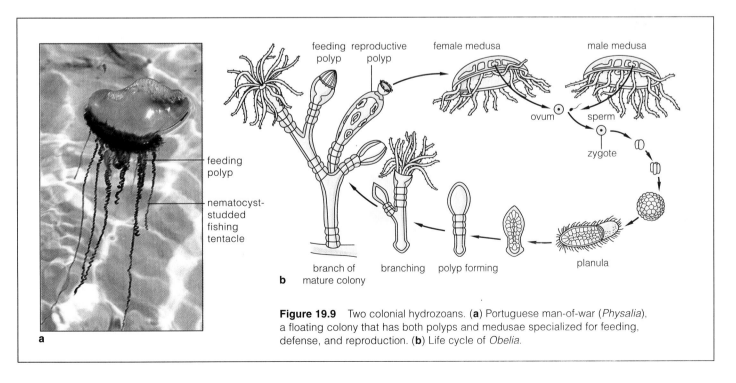

Figure 19.9 Two colonial hydrozoans. (**a**) Portuguese man-of-war (*Physalia*), a floating colony that has both polyps and medusae specialized for feeding, defense, and reproduction. (**b**) Life cycle of *Obelia*.

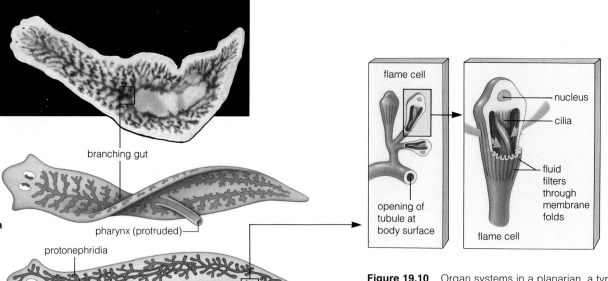

a

branching gut

pharynx (protruded)

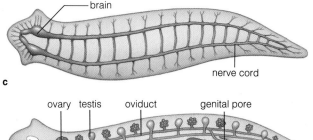

protonephridia

b

brain

nerve cord

c

ovary   testis   oviduct   genital pore

penis

d

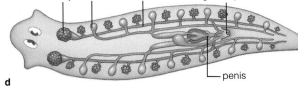

flame cell

nucleus

cilia

fluid filters through membrane folds

flame cell

opening of tubule at body surface

Figure 19.10 Organ systems in a planarian, a type of flatworm. (**a**) A pharynx opens to the gut and projects to the outside. It retracts into a chamber between feedings. (**b**) Water-regulating system, (**c**) nervous system, and (**d**) reproductive system.

Flatworms

Most **flatworms** (Platyhelminthes) have more or less flattened bodies, hence the name. The turbellarians, flukes, and tapeworms are in this phylum. Unlike sponges or cnidarians, they all are bilateral, cephalized animals with organ systems. Food typically enters a **pharynx**, a tube leading to a gut.

More important to our story, three germ tissue layers form in flatworm embryos. The midlayer, **mesoderm**, gives rise to muscles and reproductive parts. It was pivotal in the evolution of complex animals. *Mesoderm allowed contractile cells to evolve independently of the two other layers. It also became the embryonic source of blood, bones, and many other tissues.*

Turbellarians. Most turbellarians live in the seas. Some eat diatoms, but many more feed on whole small animals or suck tissues from dead or wounded ones.

Planarians are among the few freshwater species. One of their organ systems helps control the volume and composition of body fluids. It has one or more small, branched tubes called protonephridia. These extend from pores at the body surface to bulb-shaped cells in body tissues (Figure 19.10). These so-called flame cells have an internal tuft of cilia that "flickers" and drives excess fluid down the tubes to the outside.

Planarians often reproduce asexually. They divide in half, and each half regenerates the missing portion. However, most flatworms reproduce sexually in an uncommon way. Each flatworm is a hermaphrodite, with female *and* male organs. Two partners exchange sperm and fertilize each other.

Flukes. Adult flukes (trematodes) are **parasites**, a type of heterotroph that lives on or in another living organism (the host). Parasites take up nutrients from a host's tissues and may or may not end up killing it. The vast majority of adult parasitic flukes live in the gut, liver, lungs, bladder, or blood vessels of vertebrates.

Fluke life cycles have sexual and asexual phases, and they require at least two kinds of hosts. Like many other parasites, flukes require a *primary* host in which they reach maturity and sexually reproduce. Their eggs develop into swimming larvae, which infect an *intermediate host*. There they grow and may reproduce asexually, giving rise to many small, motile juveniles. These travel to a new primary host, and the cycle begins again. As you will read shortly, humans are primary hosts for flukes that cause some awful diseases.

CHORDATES
ECHINODERMS
ARTHROPODS
ANNELIDS
MOLLUSKS
COELOMATE ANCESTRY
ROTIFERS
ROUNDWORMS
FLATWORMS
BILATERAL ANCESTRY
RADIAL ANCESTRY
CNIDARIANS
SPONGES
MULTICELLED ANCESTRY

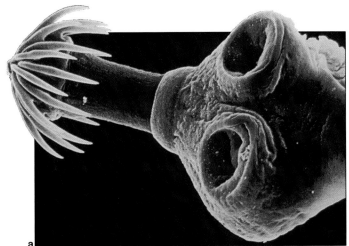

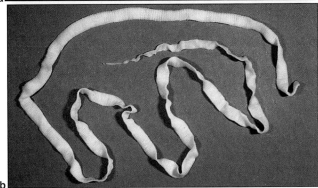

Figure 19.11 (**a**) Hooks and suckers of a scolex, the part of a tapeworm that attaches to a primary host (in this case, a shorebird). (**b**) A sheep tapeworm.

Tapeworms. As adults, parasitic tapeworms (cestodes) live in the intestines of vertebrates. Existing types have no gut; they simply absorb soluble nutrients. Possibly the gut became useless as ancestral tapeworms evolved in the intestines, which are chockful of food that has already been digested into absorbable bits.

Tapeworms attach to the intestinal wall by a scolex, a structure with suckers, hooks, or both (Figure 19.11). Behind the scolex, tapeworm body units (proglottids) form by budding. Page 296 shows such units, which are hermaphroditic—they mate and transfer sperm. Fertilized eggs accumulate in older proglottids. These are farthest from the scolex and first to leave the body, in feces. Out in the world, the fertilized eggs may be ingested by an intermediate host, then become encysted "future tapeworms" in its tissues. When humans eat undercooked, infected beef or mutton, they may become the next primary host of certain tapeworms.

Simple flatworms, fluke larvae, and tapeworm larvae happen to resemble jellyfish planulas. The resemblance has inspired speculation that complex, bilateral animals evolved from planula-like forms.

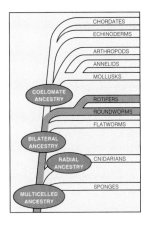

Roundworms

Roundworms (nematodes) are parasites of living animals and plants or free-living scavengers of dead ones. They live just about everywhere, yet most people have never seen one. A handful of rich soil has thousands of them. Humans alone are infected by about thirty species. One long roundworm lives just under skin and lifts it in thin, serpentlike ridges. Thousands of years ago, healers removed the "serpents" by winding them out slowly, around a stick. The symbol of our medical profession continues to be a serpent wound around a staff.

Roundworms have a bilateral, cylindrical body, usually tapered at both ends (Figure 19.12). A tough, flexible covering—a type of cuticle—covers the surface. The body cavity (a false coelom) contains reproductive organs and a fluid that circulates nutrients. It also contains a complete digestive system. In this respect, roundworms are like the more complex animals, which emerged on a different evolutionary road.

We turn now to a *Focus* essay on some diseases caused by a few of the notorious flatworms and roundworms.

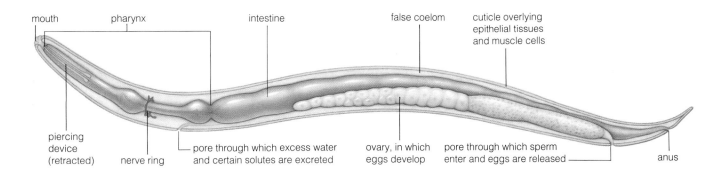

mouth   pharynx   intestine   false coelom   cuticle overlying epithelial tissues and muscle cells

piercing device (retracted)   nerve ring   pore through which excess water and certain solutes are excreted   ovary, in which eggs develop   pore through which sperm enter and eggs are released   anus

Figure 19.12 Body plan of a roundworm (*Paratylenchus*) that parasitizes the roots of certain plants. The worm shown is female.

Focus on Health

A Rogues' Gallery of Parasitic Worms

To our enormous discomfort, many parasitic flatworms and roundworms call the human body home. In a given year, about 200 million people house blood flukes responsible for *schistosomiasis*. Figure *a* shows the life cycle of a Southeast Asian blood fluke (*Schistosoma japonicum*). The cycle requires a human primary host, standing water through which larvae can swim, and an aquatic snail that serves as intermediate host. Flukes reproduce sexually, and their eggs mature in the human body (1). Eggs leave the body in feces, then hatch into ciliated, swimming larvae (2). These burrow into a snail and multiply asexually (3). In time many fork-tailed larvae develop (4). These leave the snail and swim until they encounter human skin (5). They bore inward and migrate to thin-walled intestinal veins, and the cycle begins anew. In infected humans, white blood cells that defend the body attack the masses of fluke eggs, and grainy masses form in tissues. In time, the liver, spleen, bladder, and kidneys deteriorate.

Tapeworms also infect humans. One kind uses pigs as intermediate hosts; another uses cattle. Humans become infected when they eat insufficiently cooked pork or beef (Figure *b*). Larvae of another tapeworm are eaten by copepods (tiny crustaceans). The larvae avoid digestion, then develop further in fishes that eat the copepods. Humans who eat infected fishes that are raw, improperly pickled, or insufficiently cooked can become hosts for adult tapeworms.

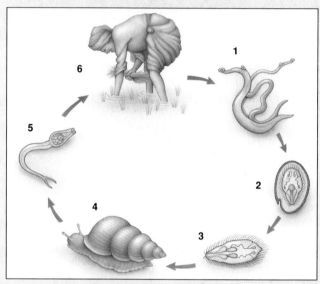

a Life cycle of a dangerous blood fluke, *Schistosoma japonicum*.

b Life cycle of a beef tapeworm, *Taena saginata*.

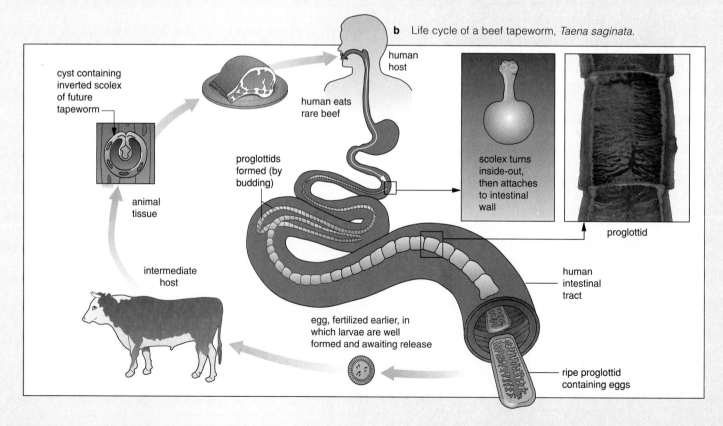

cyst containing inverted scolex of future tapeworm

animal tissue

intermediate host

proglottids formed (by budding)

egg, fertilized earlier, in which larvae are well formed and awaiting release

human host

human eats rare beef

scolex turns inside-out, then attaches to intestinal wall

proglottid

human intestinal tract

ripe proglottid containing eggs

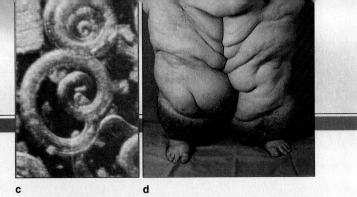

(c) Juveniles of a roundworm, *Trichinella spiralis*, inside muscle tissue. **(d)** Elephantiasis in a woman, caused by the roundworm *Wuchereria bancrofti*.

Then there are the types of roundworms called pinworms and hookworms. *Enterobius vermicularis*, a pinworm, parasitizes humans in temperate regions. It lives in the large intestine. At night, centimeter-long female pinworms migrate to the anal region and lay eggs. Their presence causes itching, and scratchings made in response will transfer some eggs to other objects. Newly laid eggs contain embryos, but within a few hours they are juveniles and ready to hatch if another human inadvertently ingests them.

Hookworms are a serious problem in impoverished parts of the tropics and subtropics. Adult hookworms live in the small intestine. They feed on blood and other tissues after teeth or sharp ridges bordering their mouth cut into the intestinal wall. Adult females, about a centimeter long, can release a thousand eggs daily. These leave the body in feces, then hatch into juveniles. A juvenile may penetrate the skin of a barefoot person. Inside a host, the parasite travels the bloodstream to the lungs, where it works its way into the air spaces. After moving up the windpipe, the parasite is swallowed. Soon it is in the small intestine, where it may mature and live for several years.

Another roundworm, *Trichinella spiralis*, causes painful, sometimes fatal symptoms. Adults live in the lining of the small intestine. Female worms release juveniles (Figure *c*), and these work their way into blood vessels and travel to muscles. There they become encysted (they produce a covering around themselves and enter a resting stage). Humans usually become infected by eating insufficiently cooked meat from pigs or certain game animals. The presence of encysted juveniles cannot easily be detected when fresh meat is examined, even in a slaughterhouse.

Figure *d* shows the results of prolonged, repeated infections by *Wuchereria bancrofti*, a roundworm. Adult worms live in the lymph nodes, where they can obstruct the flow of a fluid (lymph) that normally is returned to the bloodstream. The obstruction causes fluid to accumulate in legs and other body regions. These undergo grotesque enlargement, a condition called *elephantiasis*. A mosquito is the intermediate host. Female *Wuchereria* produce active young that travel the bloodstream at night. If a mosquito sucks blood from an infected human, the juveniles may enter the insect's tissues. In time they move near the insect's sucking device, where they are ready to enter another human host when the mosquito draws blood again.

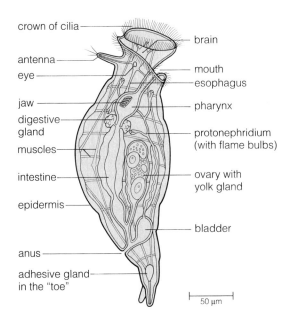

crown of cilia — brain
antenna
eye — mouth
— esophagus
jaw — pharynx
digestive gland
— protonephridium (with flame bulbs)
muscles
intestine — ovary with yolk gland
epidermis
— bladder
anus
adhesive gland in the "toe"
50 µm

Figure 19.13 A rotifer. Many species produce few males or none at all (diploid eggs can develop into diploid females). The males are dwarfed and short-lived.

Rotifers

Most **rotifers** (Rotifera) are common in food webs of lakes and ponds. They eat bacteria, and other animals eat them. Most are tinier than a millimeter, yet they illustrate the complexity possible in animals with a false coelom (Figure 19.13). Rotifers probably are an evolutionary side road. It is among coelomate animals that complexity has developed on a spectacular scale.

Two Main Evolutionary Roads

Bilateral animals not much more complex than flatworms emerged during Cambrian times. Not long afterward, a major divergence occurred. One line of descent led to mollusks, annelids, and arthropods. These are called "protostomes." Another led to the "deuterostomes"—the echinoderms and chordates.

Animals of both lineages usually have a coelom and a complete digestive system. But their embryos develop in different ways. For example, in deuterostomes, cell divisions divide the fertilized egg as you might divide an apple, by cutting it into four wedges from top to bottom. Each wedge is then cut in half at its midsection, then each piece is cut at *its* midsection. In protostomes, the "wedges" are cut at diagonal angles.

Another difference arises at surface "dents" that form on the early embryos. In deuterostomes, a first indentation opens and becomes the anus; a second one becomes the mouth. In protostomes, a mouth forms at the first indentation, and the anus forms elsewhere. As a final example, a deuterostome's coelom starts to form at places where the gut wall pouches outward. In protostomes, it starts to open up within solid tissue masses near the sides of the gut.

Mollusks

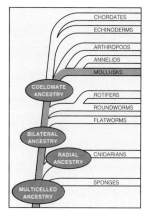

Snails, clams, and octopuses are familiar **mollusks** (Mollusca, or "soft-bodied animal"). All have a mantle, a tissue fold draped around the soft, fleshy body (Figure 19.14). Most mollusks have a shell and a muscular foot.

Gastropods. Snails and slugs are not nature's ballerinas. When they move, their foot spreads out under the body; hence the name, gastropod ("belly-footed"). Snails have coiled or cone-shaped shells (Figures 14.9 and 19.14). These balance the mass of organs above the foot, like a backpack. Sea slugs and land slugs have no shells, as Figures 19.15 and 38.2 suggest.

Gastropods that graze on algae and other photosynthetic organisms have a radula, an organ rather like a tongue with teeth on it. A radula shreds food destined for the gut. Predatory gastropods may have biting jaws. A few parasitic types have piercing devices.

Generally, water-dwelling gastropods have various kinds of **gills**, and land-dwellers have **lungs**. Gills and lungs are organs of respiration, which enhance gas exchange. Oxygen dissolved in water or in the air moves into them, then diffuses into blood. Aerobically respiring cells take up oxygen from the blood. At the same time, the cells give up carbon dioxide, which is moved to the gills or lungs and out from the body.

As most gastropods develop, some body parts grow at different rates and become realigned inside the body. By this process, called torsion, the rear of the mantle cavity twists forward and becomes a space into which the head may withdraw in times of danger.

Figure 19.15 Sea slugs, a kind of soft-bodied gastropod. These two "Mexican dancers" are engaged in mating behavior.

Bivalves. A bivalve's shell opens and closes like a hinged double door (which is what "bivalve" means). Humans have been prying open the doors and eating scallops, oysters, mussels, clams, and other bivalves since prehistoric times. They also have valued bivalve secretions called pearls and the iridescent mother-of-pearl that lines the interior of certain shells.

Some bivalves are only a millimeter across. A few giant clams of the South Pacific are more than a meter across and weigh 225 kilograms (close to 500 pounds). Bivalves are filter-feeders; they strain bacteria and other food from the water (Figure 19.16). Their head is not much to speak of, but their large, muscular foot is good for slow movements and digging burrows. Oysters, mussels, and some other bivalves glue themselves to substrates (page 261).

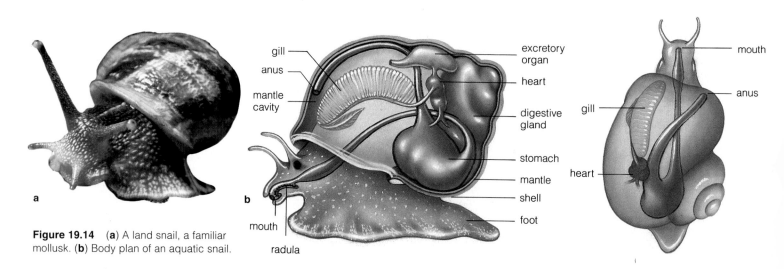

Figure 19.14 (**a**) A land snail, a familiar mollusk. (**b**) Body plan of an aquatic snail.

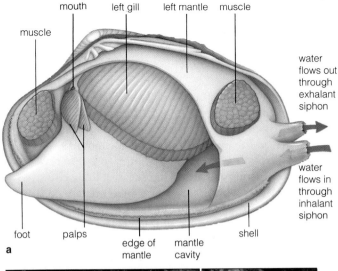

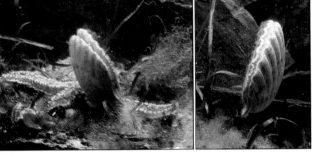

b

Figure 19.16 (**a**) Body plan of a clam, with half of its shell (the left valve) removed for this diagram. Palps sort food that gets trapped in mucus on the gills. Cilia on the palps sweep suitable tidbits to the mouth. (**b**) By clapping their valves together, scallops create a jet of water that propels them for a distance.

a

Cephalopods. The fried calamari served in many restaurants are what's left of squids. Like their octopus, cuttlefish, and nautilus relatives, squids are cephalopods. Some types are a few centimeters long. If you start measuring from the tip of its longest tentacle, the giant squid is 18 meters (about 60 feet) long. It is the largest known invertebrate.

Cephalopods are footless predators of the oceans. Long ago, the foot became modified into muscular arms or tentacles equipped with suckers. Cephalopods now move by jet propulsion. Muscles in their mantle relax, and water is drawn into the mantle cavity. The muscles contract and the mantle's free edge closes down, and water is shot out in a jet through a siphon. When cephalopods catch up with prey, their tentacles help ensnare it (Figure 19.17). Venom paralyzes the prey, beaklike jaws bite or crush it, and a radula draws it into the mouth.

Cephalopods have gills. Being so active, they also have great demands for oxygen. By contracting their mantle, they speed the flow of water (and dissolved oxygen) into the mantle cavity and over the gills. They also benefit from a "closed" circulatory system. In this system, blood is contained inside muscular pumps (hearts) and blood vessels, including vast networks of capillaries where blood and tissue fluid exchange oxygen and carbon dioxide. Cephalopods are the only mollusks with such a system.

Because they move rapidly, cephalopods make good use of well-developed eyes. They also have a splendid nervous system with a large brain. Giant nerve fibers connect the brain with muscles used in jet propulsion and coordinate quick responses to food or danger. Besides this, cephalopods can learn. For example, give an octopus a mild electric shock after it sees an object with a distinctive shape, and it will thereafter avoid the object. In terms of memory and learning, cephalopods are the world's most complex invertebrates.

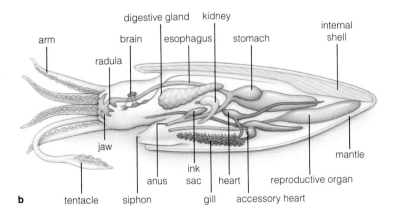

b

Figure 19.17 (**a**) An octopus, one of the cephalopods. (**b**) General body plan of a cuttlefish. The tentacles are more slender than the arms and are specialized for prey capture.

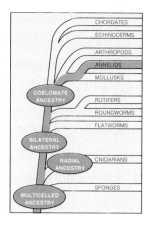

Annelids

Annelid Groups. There are three groups of **annelids**, which are segmented worms (Annelida). These are the oligochaetes, leeches, and polychaetes. Figures 19.18 through 19.20 show a few representatives.

Earthworms are familiar oligochaetes. They burrow in moist soil or mud, where they scavenge decomposing plant material and other organic matter. By doing so, they aerate soil and bring nutrients up to the surface, and so benefit many plants.

Leeches live in freshwater, seawater, and moist land habitats. They inch about on suckers at both ends of their muscular body. Most swallow small animals or kill them and suck their juices. The leeches most people hear about feed on vertebrates (Figure 19.20).

Polychaetes, which live mostly in marine habitats, are the most diverse annelids. Some excavate burrows in sand or mud, swim freely, or live in tubes that they construct. Different types feed on small invertebrates, algae, or organic matter present in sediments.

Annelid Adaptations. Let's use the earthworm as our representative annelid (Figure 19.19). Like most annelids (but not leeches), it has chitin-reinforced bristles called setae in nearly all segments. Setae can extend from and retract into the body. They provide traction for crawling and burrowing. A thin, flexible cuticle covers the segments. It is thin enough to permit bending as well as gas exchange. Inside the worm body, the coelom is partitioned into chambers. A gut threads through all of them, from mouth to anus. So does a double **nerve cord**, which carries information to and away from a brain. There are a well-developed circulatory system and nephridia (singular, nephridium). The nephridia regulate the volume and composition of body fluids.

The coelomic chambers serve as a hydrostatic skeleton that works with the earthworm's muscles. Some muscles encircle each body segment, others span several segments. When longitudinal muscles contract and the circular ones relax, segments shorten and fatten. When the circular muscles contract and the longitudinal ones relax, the segments elongate—then setae protrude and hold the body in its newly gained position. When longitudinal muscles contract in the next segments down the line, those segments move forward. Alternating contractions and elongations move the worm forward.

a

worm inside self-constructed tube

Figure 19.18 (**a**) A tube-dwelling polychaete, with featherlike structures coated with food-trapping mucus. (**b**) A marine polychaete inside its burrow. Setae (reinforced bristles) along the length of its body help this marine worm crawl about and grip the burrow wall.

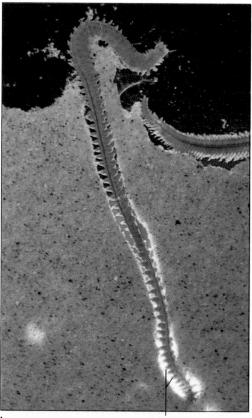

b

setae

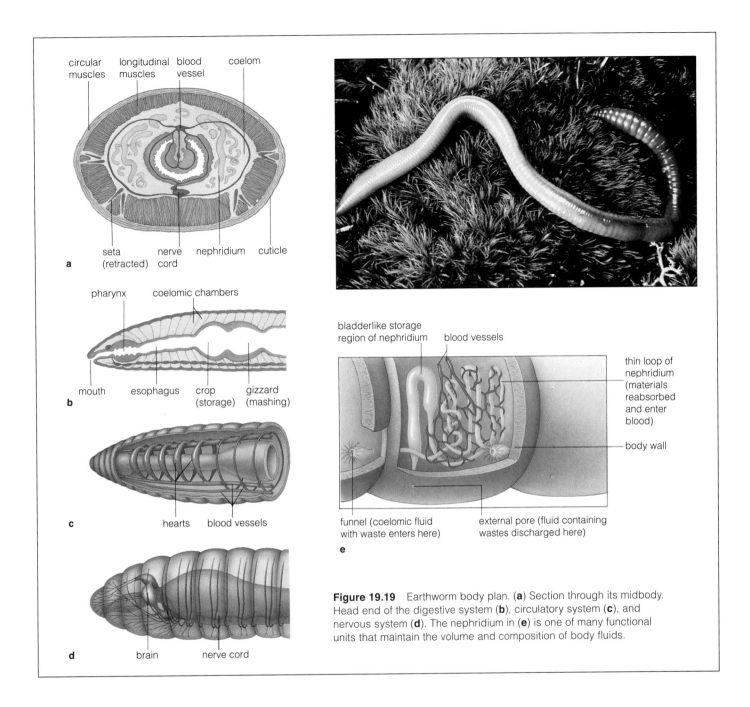

Figure 19.19 Earthworm body plan. (**a**) Section through its midbody. Head end of the digestive system (**b**), circulatory system (**c**), and nervous system (**d**). The nephridium in (**e**) is one of many functional units that maintain the volume and composition of body fluids.

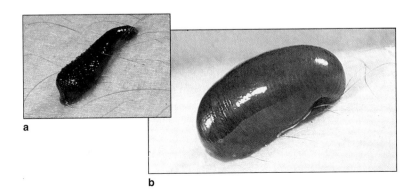

Figure 19.20 A leech (**a**) before and (**b**) after gorging itself with human blood. For 2,000 years, medical practitioners used leeches as a blood-letting tool to treat problems ranging from nosebleeds to obesity. Leeches are still used, but more selectively. For example, they remove pooled blood from a severed ear, lip, or fingertip that's been surgically reattached. The body can't do this on its own until it reestablishes blood circulation routes.

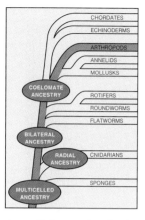

Insects and Other Arthropods

By far, **arthropods** (Arthropoda) are the most diverse animals on earth. Biologists have named more than a million species—most of which are insects—and new ones are discovered weekly. The arthropods called trilobites are extinct (page 225). The major existing groups are chelicerates (such as spiders, ticks, and mites), crustaceans (such as crabs and lobsters), and uniramians (the centipedes, millipedes, and insects).

Arthropod Adaptations. Arthropods are truly abundant, they have enormously different life-styles, and they are successful in almost all habitats. Six adaptations contributed to the success of the group in general and insects in particular:

1. A hardened exoskeleton

2. Specialized segments

3. Jointed appendages

4. Specialized respiratory structures

5. Efficient nervous system and sensory organs

6. Division of labor in the life cycle

Hardened exoskeletons: The arthropod cuticle is actually an exoskeleton (external skeleton). Although it may be hardened, its protein and chitin components make it

Figure 19.21 Molting of a centipede exoskeleton.

light, flexible—and protective. Think of the calcium-stiffened exoskeleton of a lobster or crab. It's like armor plating. Probably exoskeletons evolved as defenses against predators. They took on added functions when arthropods first invaded the land. An exoskeleton restricts evaporative water loss, and it can support a body deprived of water's buoyancy. It does restrict increases in size, but arthropods grow in spurts, by **molting**. At different stages in the life cycle, they grow a new, soft exoskeleton under the old one, which they shed (Figure 19.21). Before this new one hardens, aquatic arthropods swell with water; land-dwelling types swell with air.

Specialized segments: As arthropods evolved, body segments became more specialized, fewer in number, and grouped or fused in diverse ways. For example, in spiders, segments fused and become modified into a forebody and hindbody.

Jointed appendages: Arthropods have highly specialized, jointed appendages. (Arthropod means "jointed foot.") For example, paired appendages on the head are used in feeding and in detecting stimuli. Other appendages are used in locomotion, transferring sperm to females, or spinning silk.

Respiratory structures: Water-dwelling arthropods rely on a variety of gills for gas exchange. Among the land-dwellers, the evolution of tracheas helped bring about diversity in insects especially. Insect tracheas begin as pores on the body surface and branch into narrow tubes that deliver oxygen directly to body tissues (page 454). They help support such energy-consuming activities as insect flight.

Specialized sensory structures: Intricate eyes and other sensory organs contributed to arthropod success. Many species have a wide angle of vision, and they can process visual information from many directions.

Division of labor: Many insects have a division of labor among different stages of their life cycle. They first develop as sexually immature larvae that molt and change as they grow. Then the larvae enter a pupal stage, which involves massive reorganization and remodeling of body tissues. We saw an example of this in Figure 1.5, which showed the emergence of an adult moth. Growth and major transformation of a larva into the adult form is called **metamorphosis**.

Moths, butterflies, beetles, and flies are among the metamorphosing insects. Their larval stages specialize in *feeding* and *growth*, whereas the adult is concerned mostly with *dispersal* and *reproduction*. This division of labor is highly adaptive to seasonal changes in the environment, including variations in food sources.

With this overview of adaptations in mind, we turn now to the spectrum of diversity among the existing arthropods.

a

Figure 19.22 Horseshoe crab, one of the few sea-going chelicerates alive today. Flip its hard, shieldlike cover over and you will see five pairs of legs, a feature that is characteristic of the group.

b

c

Figure 19.23 A few land-dwelling chelicerates. (**a**) Wolf spider. Like most spiders, it is harmless to humans and helps keep insect populations in check. (**b**) Female black widow spider, with a red hourglass-shaped mark on its hindbody. Its bite is painful, sometimes dangerous. (**c**) Brown recluse spider, with a violin-shaped mark on its forebody. Its bite can be severe to fatal for humans.

Chelicerates. Chelicerates evolved in the shallow seas of the early Paleozoic era. The only surviving types that still live in the seas are the so-called horseshoe crabs (Figure 19.22) and sea spiders. Of the familiar existing chelicerates—the spiders, scorpions, ticks, and mites— we might say this: Never have so many been loved by so few.

Spiders and scorpions are predators. Many keep insect pests in check. Yet humans are bitten enough times by the venomous types that the whole group has a bad name. Ticks are blood-sucking parasites. Often they transmit pathogens to human hosts and so help spread Lyme disease, Rocky Mountain spotted fever, and other diseases. Most mites are free-living scavengers; a number are agricultural pests.

The spider forebody has several good eyes and four pairs of legs (Figure 19.23). Another pair of appendages inflicts wounds, discharges venom, and otherwise handles prey. (Spiders pump digestive enzymes into wounds, then suck up liquefied remains.) Still another pair may be used in mating rituals. Hindbody appendages (if any) spin silk for webs and egg cases.

Crustaceans. Shrimps, crayfishes, lobsters, crabs, and pillbugs are familiar crustaceans. Most live in the seas; some live in fresh water or on land. All have roles in food webs, and humans harvest many edible types.

Crabs and lobsters are examples of how crustacean segments are highly modified. They have strong claws for shredding seaweed, attacking prey, or intimidating other animals. A shieldlike cover obscures some or all of the segments. Their thorax has five pairs of legs and other specialized appendages (Figure 19.24).

Barnacles and copepods are common crustaceans, especially in marine habitats. An adult barnacle lives attached to a substrate (Figure 19.24c). Its calcified shell (exoskeleton) resists predators, drying out, and the force of waves. Most copepods are free-swimming members of plankton (Figure 19.24d). Huge numbers feed on tiny photosynthetic organisms. By converting this food into their own tissues, they become food for other organisms, and so on up through vital food webs.

a

b

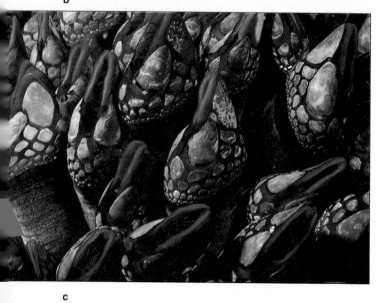
c

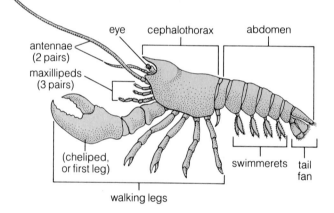

d

Figure 19.24 Representative crustaceans. (**a**) A crab and (**b**) a lobster, displaying their splendid appendages. (**c**) Stalked barnacles that grow in intertidal zones of the Pacific coast. (**d**) A marine copepod.

Millipedes and Centipedes. Closely related to insects are the millipedes and centipedes, which typically live under damp rocks, logs, or forest litter. As Figure 19.25 shows, both kinds of arthropods have a long body with many legs. Most millipedes are slow-moving, nonaggressive scavengers of decaying plant material. Centipedes are fast-moving, aggressive carnivores, complete with fangs and venom glands. They prey mostly on insects, earthworms, and snails, although some tropical species can subdue small lizards and toads. A few species can inflict painful bites.

Insect Body Plans. As a group, insects share the adaptations listed on page 302. Here we add a few more details to the list. All insects have a head, thorax, and abdomen. The head has a pair of sensory antennae and paired mouthparts, specialized for biting, chewing, sucking, or puncturing (Figure 19.26). The thorax has three pairs of legs and usually two pairs of wings. For most insects, the only abdominal appendages are reproductive structures, such as egg-laying devices.

Insects dispose of by-products of protein metabolism through small tubes (Malpighian tubules) that connect with the gut. The by-products diffuse from blood into the tubes, where they are converted into harmless crystals of uric acid that are eliminated with feces. This system allows land-dwelling insects to get rid of potentially toxic wastes without losing precious water.

a

b

Figure 19.25 (**a**) A mild-mannered millipede that scavenges on decaying plant parts. (**b**) An aggressive centipede of Southeast Asia that preys on small frogs and lizards.

Figure 19.26 Specialized arthropod appendages. These examples are insect headparts, adapted for feeding.

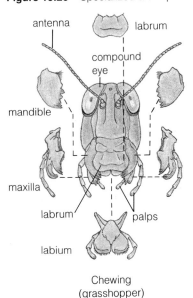

antenna
labrum
compound eye
mandible
maxilla
labrum
palps
labium

Chewing
(grasshopper)

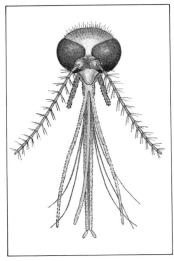

Piercing and sucking
(mosquito)

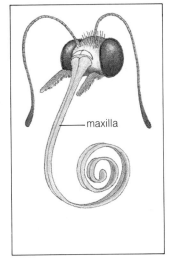

maxilla

Siphoning tube
(butterfly)

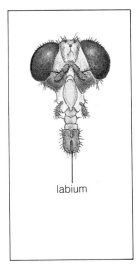

labium

Sponging
(housefly)

Figure 19.27 Representative insects.

(**a**) Off with the old, on with the new. On the side of a tree trunk, a green cicada (order Homoptera) wriggling out of its outgrown cuticle during a molting cycle.

(**b**) A male praying mantid (order Mantodea) mating with a larger female. He may be eaten during or immediately after mating with her.

(**c**) Mediterranean fruit fly (order Diptera). Its larvae destroy citrus fruit and other valuable crops.

(**d**) A honeybee (order Hymenoptera) attracting its hive mates with a dance, as described in Chapter 38.

(**e**) Flea (order Siphonaptera), with big strong legs for jumping onto and off animal hosts. (**f**) Duck louse (order Mallophaga). It eats feather particles and bits of skin.

(**g**) European earwig (order Dermaptera), a common household pest.

(**h**) Stinkbugs (order Hemiptera), newly hatched.

(**i**) Ladybird beetles (order Coleoptera) swarming. These beetles are raised commercially and released in great numbers as biological controls of aphids and other pests. Also in this order, the scarab beetle (**j**). With more than 300,000 named species, Coleoptera is the largest order in the animal kingdom.

(**k**) Luna moth (order Lepidoptera), a flying insect of North America. Like most other moths and butterflies, its wings and body are covered with microscopic scales.

(**l**) Dragonfly (order Odonata). This swift aerialist captures and eats other insects in midflight.

a

b

c

d

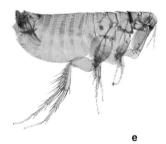

e   f   g

Insect Diversity. Figure 19.27 shows members of major orders of insects. In sheer numbers and distribution, the most successful types are small. Many grow and reproduce in great numbers on a single plant that might be only an appetizer for another animal. The insect reproductive capacity often is staggering. (By one estimate, if all the progeny of a single female fly were to survive and reproduce through six more generations, that fly would have more than 5 trillion descendants!) The most successful types also rely on metamorphosis. This allows them to use different resources at different times. And the most successful types are winged; they can fly among widely scattered food sources. They are the only winged invertebrates. Their capacity for flight contributed greatly to their widespread distribution on land.

The factors that contribute to their success also make insects our most aggressive competitors. They destroy vegetable crops, stored food, wool, paper, and timber. They draw blood from us and our pets and transmit diseases. On the bright side, many insects pollinate flowering plants in general and crop plants in particular. And many "good" insects attack or parasitize the ones we would rather do without.

tube feet of ray

Figure 19.28 Some echinoderms. (**a**) Sea star with five rays (arms), and (**b**) a closer view of its tube feet. (**c**) Sea cucumber. (**d**) Feather star, with finely branched food-gathering appendages. (**e**) Brittle stars, which move by rapid, snakelike action of their rays. (**f**) Sea urchin, which moves about on spines and tube feet.

tube feet   spine

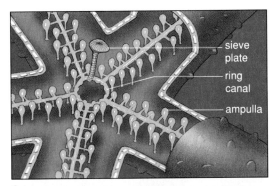

sieve
plate

ring
canal

ampulla

a

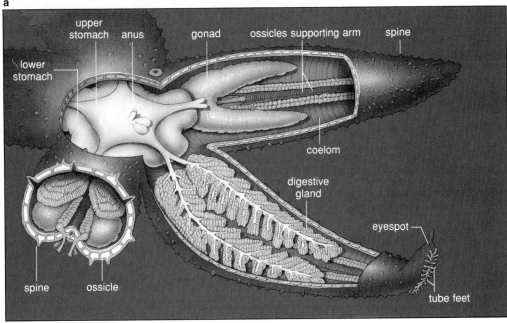

upper
stomach   anus   gonad   ossicles supporting arm   spine

lower
stomach

coelom

digestive
gland

eyespot

spine   ossicle

tube feet

b

Figure 19.29 (**a**) Water-vascular system of sea stars. Together with tube feet, this radially arranged system is the basis for movement. (**b**) Some of the internal organs of sea stars.

Echinoderms

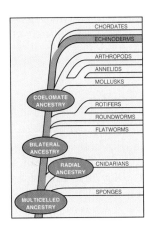

We turn now to the deuterostome branch of the animal family tree. **Echinoderms** (Echinodermata) are invertebrates that have calcified spines, needles, or plates in their body wall. (Hence the name, which means "spiny-skinned.") They include sea stars, sea cucumbers, feather stars, brittle stars, and sea urchins. Figure 19.28 shows only a small sampling of the diversity that exists within these groups.

Adult echinoderms have a curious body plan. They are radial with some bilateral features. (Some ancient echinoderms were bilateral; so are many echinoderm larvae.) The nervous system is decentralized; there is no brain. This allows, say, any one of a sea star's several arms to head off in a new direction, with the rest of the body following suit.

Turn a sea star over and you see tube feet. These fluid-filled, muscular structures have suckerlike disks that help sea stars walk, burrow, or grip rocks or prey. Tube feet are part of a "water-vascular system" unique to echinoderms (Figure 19.29). In sea stars, each arm contains a water canal. Short canals extending from it deliver water to tube feet. Each tube foot has a fluid-filled, muscular structure (ampulla), much like the rubber bulb on a medicine dropper. When it contracts, the foot lengthens with the force of incoming fluid. Tube feet change shape when muscles redistribute fluid through the system. Each swings forward, reattaches to the substrate, then swings backward and is released before swinging forward again.

Although you may consider this bad table manners, many sea stars push part of their stomach outside the mouth and around prey, then start digesting it even before swallowing.

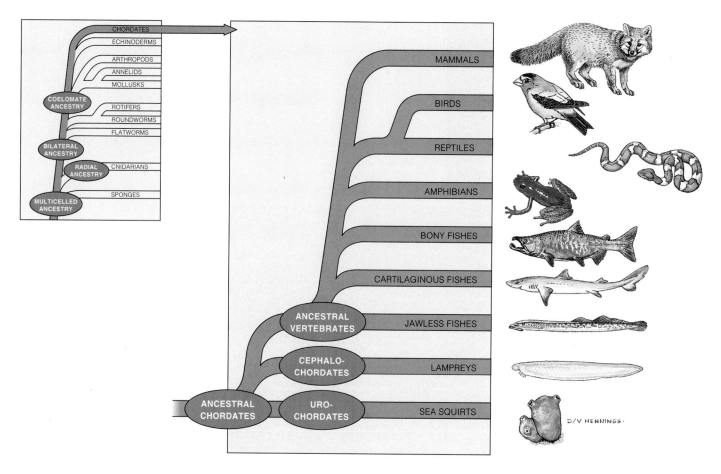

Figure 19.30 Family tree for the chordates.

PART II. VERTEBRATES AND THEIR KIN

The Chordate Heritage

Like echinoderms, the **chordates** (Chordata) are on the deuterostome branch of the animal family tree (Figure 19.30). The "vertebrate chordates" have a backbone of cartilage or bone, and they have skull bones that protect the brain. They greatly outnumber the "invertebrate chordates," which do not have a backbone.

Four features are evident in chordate embryos, and in many species these persist into adulthood. First, a *notochord*, a long rod of stiffened tissue (not cartilage or bone), helps support the body. Second, a hollow, *dorsal nerve cord* runs above and parallel to the notochord and gut. Third, a *pharynx* with slits in its wall functions in feeding, respiration, or both. Fourth, a *tail* extends past the anus.

Chordates are classified as sea squirts and kin, lancelets, and vertebrates. Vertebrates are further classified as jawless fishes, jawed armored fishes, cartilagi-

nous fishes, bony fishes, amphibians, reptiles, birds, and mammals. Appendix I includes a classification scheme for these groups. Unit VI describes their structure and function. Here we survey trends that occurred during vertebrate evolution. We find clues to those trends among existing invertebrate chordates.

Invertebrate Chordates

Sea squirts (urochordates) live mostly in the seas. They are so named because they squirt water through a siphon when they are irritated. They also are called tunicates, a name alluding to the jellylike or leathery "tunic" that adults secrete around themselves.

Sea squirt larvae are bilateral, free-swimming forms that look like tadpoles (Figure 19.31). They have a notochord. When muscles on one side or the other of their tail contract, the notochord bends. When muscles relax, the notochord springs back. The tail's strong, side-to-side motion propels the animal forward—a propulsive motion also used by most fishes. After swimming about

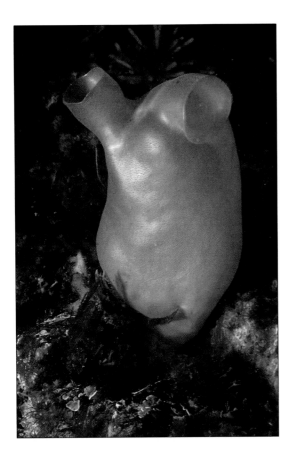

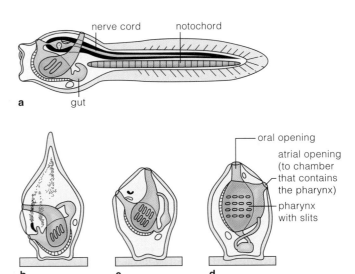

Figure 19.31 Body plan of a sea squirt (tunicate), an invertebrate chordate. The tadpole-like larva (**a**) swims a bit, then attaches its head to a substrate (**b**). It metamorphoses into the adult (**c,d**). The tail, notochord, and most of the nervous system are resorbed (recycled to form new tissues). The pharynx wall gains more slits. Organs rotate until the openings through which water enters and leaves point away from the substrate.

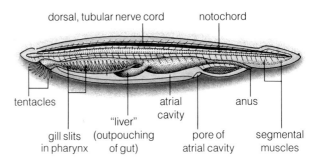

Figure 19.32 Photograph and cutaway view of a lancelet, an invertebrate chordate.

briefly, the larva settles down and undergoes metamorphosis into the adult.

Ancient sea squirts were not on the evolutionary road leading to vertebrates, but they may resemble animals that were. In some ways, they are like echinoderms. And echinoderm larvae bear resemblances to fishlike invertebrate chordates, the lancelets.

Lancelets (cephalochordates) have a body that tapers sharply at both ends (Figure 19.32). They have segmented muscles on both sides of a full-length notochord. Lancelets use the notochord and muscles for swimming. But they spend much of the time buried in sediments almost up to their mouth, which filters food particles out of the water. Probably as a result of convergent evolution, lancelets and the larvae of certain fishes (lampreys) are similar in structure and feeding behavior. Lancelets are not in the vertebrate lineage, but they are closer to it than sea squirts.

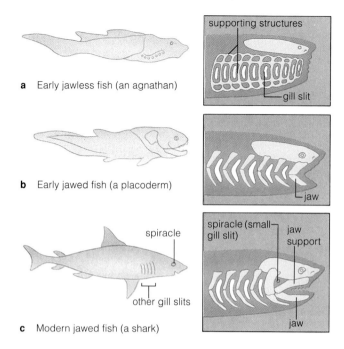

a Early jawless fish (an agnathan)

supporting structures

gill slit

b Early jawed fish (a placoderm)

jaw

spiracle

other gill slits

c Modern jawed fish (a shark)

spiracle (small gill slit)

jaw support

jaw

Figure 19.33 Evolution of jaws from structural elements that supported the gills of early jawless fishes. Fossils of early fishes, both jawless (**a**) and jawed (**b**), suggest these modifications occurred and led to the more formidable structures seen in modern jawed fishes (**c**).

Evolutionary Trends Among the Vertebrates

How do we get from sea squirts to vertebrates? Let's start with a broad evolutionary picture. One evolutionary trend involved a shift away from the notochord to reliance on a skeletal column of separate, hard segments. We call the skeletal elements **vertebrae** (singular, vertebra). The vertebral column proved to be a strong internal skeleton (endoskeleton) for muscles to work against. *The vertebral column was the foundation for fast-moving predators—some of which were ancestral to all other vertebrate animals.*

In a related trend, part of the nerve cord expanded and developed into a complex brain. The expansion began after **jaws** evolved. In early fishes, the first jaws were modified gill-supporting structures (Figure 19.33). Jaws led to new feeding possibilities—and to intensified competition among predators. Fishes able to recognize food or predators *from a distance* were favored. Over time, they developed better senses of smell and vision. The brain became better at processing information (Figure 19.34). *The trend toward complex sensory organs and nervous systems began in fishes and continued among land vertebrates.*

Another trend began when paired fins evolved. **Fins** are appendages that help propel, stabilize, and guide

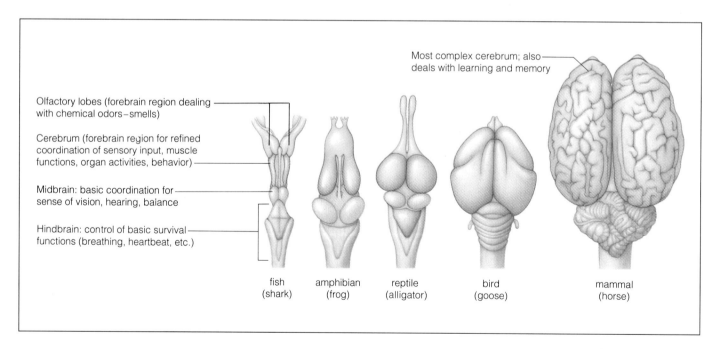

Most complex cerebrum; also deals with learning and memory

Olfactory lobes (forebrain region dealing with chemical odors–smells)

Cerebrum (forebrain region for refined coordination of sensory input, muscle functions, organ activities, behavior)

Midbrain: basic coordination for sense of vision, hearing, balance

Hindbrain: control of basic survival functions (breathing, heartbeat, etc.)

fish (shark)   amphibian (frog)   reptile (alligator)   bird (goose)   mammal (horse)

Figure 19.34 Evolutionary trend toward an expanded, more complex brain, as suggested by comparisons of existing vertebrates. These are dorsal views (looking down on the top of the head). Think about the head size of a frog and a horse, and you know these drawings are not to the same scale.

the body through water. Among some fishes, ventral fins became fleshy and equipped with skeletal supports—the forerunners of limbs. *Paired, fleshy fins were the starting point for the legs, arms, and wings seen among amphibians, reptiles, birds, and mammals.*

Another trend involved gas exchange structures. As fishes became larger and more active, oxygen uptake and distribution improved. Gills became more efficient in aquatic lineages. But gills cannot work out of water.

(They stick together unless water flows through them and keeps them moist.) In fishes ancestral to land vertebrates, pouches developed on the gut wall. These evolved into lungs—internally moistened sacs for gas exchange. In a related trend, modifications to the heart enhanced the pumping of oxygen and carbon dioxide through the body (Figure 19.35). *Ancestors of land vertebrates relied less on gills and more on lungs. More efficient circulatory systems accompanied the evolution of lungs.*

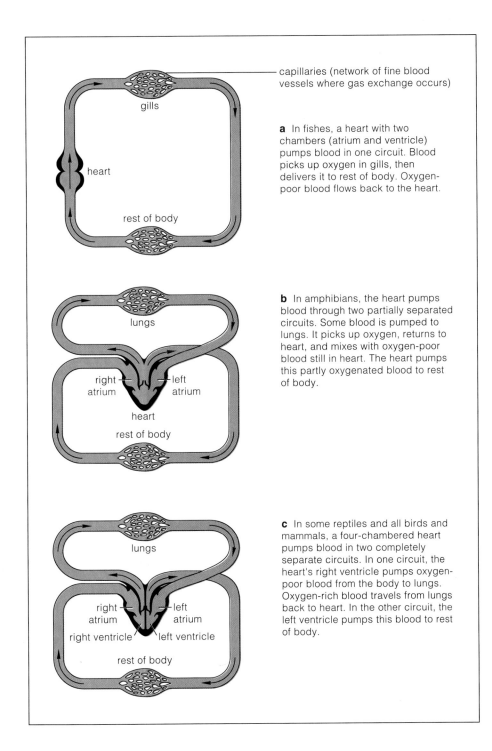

capillaries (network of fine blood vessels where gas exchange occurs)

a In fishes, a heart with two chambers (atrium and ventricle) pumps blood in one circuit. Blood picks up oxygen in gills, then delivers it to rest of body. Oxygen-poor blood flows back to the heart.

b In amphibians, the heart pumps blood through two partially separated circuits. Some blood is pumped to lungs. It picks up oxygen, returns to heart, and mixes with oxygen-poor blood still in heart. The heart pumps this partly oxygenated blood to rest of body.

c In some reptiles and all birds and mammals, a four-chambered heart pumps blood in two completely separate circuits. In one circuit, the heart's right ventricle pumps oxygen-poor blood from the body to lungs. Oxygen-rich blood travels from lungs back to heart. In the other circuit, the left ventricle pumps this blood to rest of body.

Figure 19.35 Circulatory systems of vertebrates. All vertebrates have a "closed" circulatory system, in which blood is confined within blood vessels and a heart. Some of these systems are more efficient than others at providing body tissues with oxygen.

Figure 19.36 A lamprey, pressing its toothed oral disk to the glass wall of an aquarium.

a

b

c

Figure 19.37 Examples of cartilaginous fishes: (**a**) shark, (**b**) blue-spotted reef ray, and (**c**) chimaera (ratfish).

Fishes

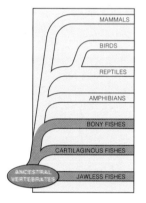

Collectively, living **fishes** are the most numerous and diverse vertebrates. Their body form tells us about their watery world. Being about 800 times denser than air, water resists rapid movements. Predatory fishes of the oceans are streamlined for pursuit, with a long, trim body that reduces friction. Their tail muscles are organized for propulsive force and forward motion. Bottom-dwelling fishes spend most of their lives hiding from predators or prey. Their flattened body is easy to conceal, but the shape tells us these fishes are sluggish, not Corvettes of the deep.

Finally, a motionless trout suspended in shallow water is a great example of an adaptation to water's density. Like many fishes, a trout can maintain neutral buoyancy with a swim bladder. This is an adjustable flotation device that exchanges gases with the blood. When a trout gulps air at the water's surface, it's adjusting the gas volume in its swim bladder.

The First Vertebrates. The first free-swimming vertebrates arose during the Cambrian. They gave rise to two kinds of fishes—those without and those with jaws. One or the other kind probably gave rise to all vertebrate lineages that followed.

Ostracoderms were among the earliest jawless fishes (Agnatha). The armorlike plates of these small bottom feeders protected them from sea scorpions but weren't much good against jaws. Ostracoderms disappeared when jawed fishes began their adaptive radiations.

Among the first fishes with jaws and paired fins were the placoderms. But these bottom-dwelling scavengers and predators were extinct by the end of the Paleozoic. Cartilaginous and bony fishes replaced them in the seas.

Existing Jawless Fishes. Lampreys and hagfishes are the only living jawless fishes. They have cylindrical, eel-like bodies and a skeleton of cartilage. Lampreys are specialized predators, almost parasites. Their suckerlike oral disk has horny, toothlike parts that rasp flesh from prey (Figure 19.36). Some types latch onto salmon, trout, and other commercially valuable fishes, then suck out juices and tissues. In fisheries of the Great Lakes and other regions, the water is treated with a chemical that poisons lamprey larvae. But the battle goes on.

Imagine a large worm with feelers around the mouth and you know what the scavenging hagfishes look like. They are not the favorite of fishermen. They

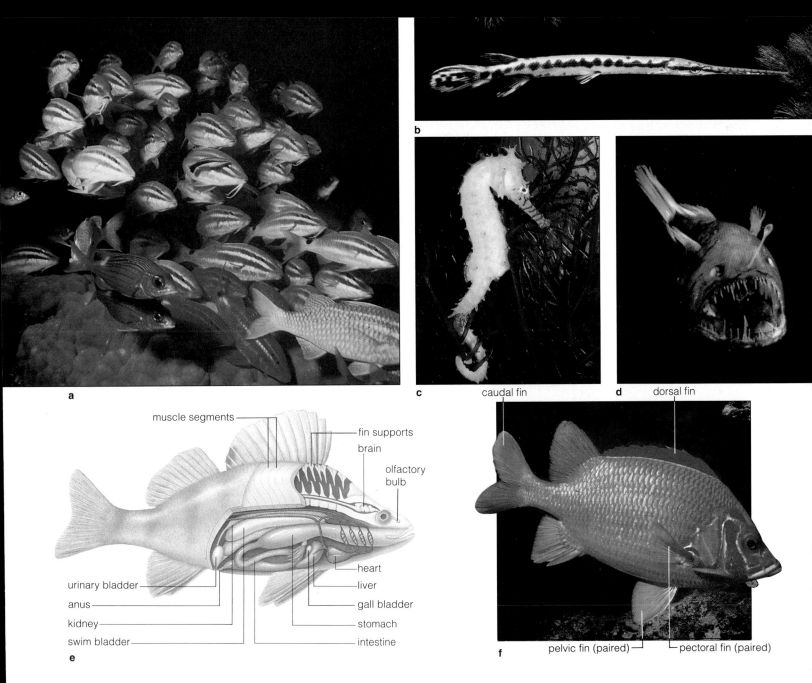

Figure 19.38 A few of the diverse ray-finned fishes: (**a**) goatfish demonstrating schooling behavior, (**b**) long-nose gar, (**c**) seahorse, and (**d**) deep-sea angler fish. Most ray-finned fishes have the same general body plan, represented here by a perch (**e**) and soldierfish (**f**).

burrow into fish trapped by setlines or nets. The ones brought on deck secrete copious amounts of sticky, slimy mucus.

Cartilaginous Fishes. This group (Chondrichthyes) includes rays, skates, sharks, and chimaeras (Figure 19.37). All have a skeleton of cartilage. Most are marine predators. Skates and rays are mostly bottom dwellers that eat hard-shelled invertebrates. Both have distinctive, enlarged fins extending onto the side of the head. Sharks have formidable jaws and can detect traces of blood in water. A few types are longer than two pickup trucks parked end to end! Some have sharp, triangular teeth that rip chunks of flesh from large fishes and marine mammals. Sharks continually shed and replace their teeth. Chimaeras are also called ratfishes. Most use their modified jaws to crush hard-shelled mollusks.

Bony Fishes. Bony fishes (Osteichthyes) have skeletons of bone. There are two major lineages: the *ray-finned* fishes and the *lobe-finned* fishes. Figure 19.38 shows examples of the spectacularly diverse ray-finned fishes. Figure 19.38*e* shows how their paired fins have supporting rays (skin extensions). The teleosts of this group are the most abundant fishes. They include salmon, tuna, rockfish, catfish, minnows, and eels.

Figure 19.39 (**a**) Coelacanth (*Latimeria*), a "living fossil" that resembles early lobe-finned fishes. (**b,c**) Proposed evolution of the skeletal elements inside the lobed fins of certain fishes into the limb bones of early amphibians.

a

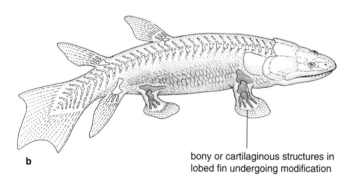

b

bony or cartilaginous structures in lobed fin undergoing modification

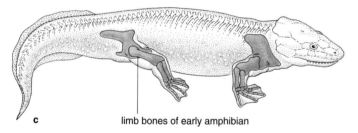

c

limb bones of early amphibian

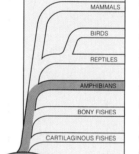

MAMMALS

BIRDS

REPTILES

AMPHIBIANS

BONY FISHES

CARTILAGINOUS FISHES

ANCESTRAL VERTEBRATES

JAWLESS FISHES

About Those Lobed Fins . . . . Coelacanths and lungfishes are existing lobe-finned fishes (Figure 19.39). Their paired fins incorporate fleshy extensions of the body. The ancestors of these fishes arose during the Devonian, when swamp forests flooded and drained repeatedly (page 226). Those enormously challenged animals used their fins to pull themselves from disappearing pond to pond. And their lungs proved most adaptive during the pond-to-pond lurchings.

Amphibians

Challenges of Life on Land. The lobe-finned fishes of the Devonian traveled across land simply as a way of reaching more water. Yet their travels favored the evolution of more efficient lungs and stronger fins. Among those evolving fishes were the ancestors of the first amphibians.

An **amphibian** is a vertebrate somewhere between fishes and reptiles in body plan and reproductive mode. For early amphibians, the land was dangerous—and promising. Temperatures fluctuated more on land than in water, air didn't support the body as water did, and water was not always available. But air has far more oxygen. Lungs evolved and enhanced the uptake of oxygen. Circulatory systems evolved and became better at distributing oxygen. Both modifications increased the energy base for more active life-styles.

The land also was rich with sensory information. Swamp forests abounded with tasty insects and other invertebrate prey. Animals with good vision, hearing, and balance were favored. Brain regions concerned with those senses expanded dramatically.

There are now three groups of amphibians—the salamanders, frogs and toads, and caecilians. Figure 19.40 shows representatives. None has escaped water entirely. They all have thin skin that dries out easily. Most species shed eggs into water or lay them in moist places, and their larvae are aquatic.

Figure 19.40 Representative amphibians. (**a**) Terrestrial stage in the life cycle of the red-spotted salamander. (**b**) A frog, splendidly jumping. (**c**) American toad. (**d**) A caecilian.

Salamanders. Like fishes and the first amphibians, salamanders bend from side to side when they walk. The adults of some species retain many larval features. Also, the larvae of some groups are sexually precocious; they can breed. This is true of the Mexican axolotl. It retains the larval tail and external gills, and its tooth and bone development is arrested.

Frogs and Toads. With more than 3,000 species, frogs and toads are the most successful amphibians. With their long hindlimbs and powerful muscles, they can catapult into the air or propel themselves forcefully through water. Their sticky-tipped tongue is usually attached at the front of the mouth and flips out to capture prey. Skin glands of some species produce toxins.

Poisonous types often have bright coloration that advertises their inedibility.

Frogs have a closed circulatory system with a three-chambered heart, in which oxygen-rich and oxygen-poor blood mix in the third chamber (Figure 19.35). This may seem inefficient, compared with your own four-chambered heart, but it keeps some oxygen circulating when these animals are underwater and cannot breathe. At that time, some gas exchange also is accomplished at the skin and pharynx.

Caecilians. Ancestors of caecilians lost their limbs and gave rise to decidedly worm-shaped amphibians. Nearly all caecilians burrow through soft, moist soil in pursuit of insects and earthworms.

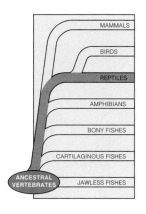

MAMMALS
BIRDS
REPTILES
AMPHIBIANS
BONY FISHES
CARTILAGINOUS FISHES
JAWLESS FISHES
ANCESTRAL VERTEBRATES

Reptiles

The Rise of Reptiles. In the Late Carboniferous, insects radiated into lush habitats on land. About the same time, reptiles evolved from amphibians. Like modern species, those amphibians almost certainly were carnivores. The rapidly expanding quantities and choices of insects surely represented a major, untapped food source. Modifications of the brain and limb bones of the early reptiles allowed them to pursue prey with greater cunning and speed. Modifications of teeth and jaws allowed them to feed more efficiently on diverse insects—and some fellow vertebrates.

By definition, **reptiles** were the first vertebrates to escape dependency on standing water. They did so by two adaptations: internal fertilization and a new kind of egg. (Internal fertilization means sperm do not have to swim through water to reach eggs; they are deposited in the female parent's body.) Inside an **amniote egg**, the embryo develops to an advanced stage before being hatched or born into dry habitats. These eggs have specialized membranes and often a leathery or calcified shell (Figure 19.41). As Chapter 32 describes, these membranes retain water, protect the embryo, and support the embryo metabolically.

Reptiles underwent a major adaptive radiation in the Mesozoic. Dinosaurs and related forms emerged and ruled the land for the next 125 million years. Their domination ended when the Cretaceous drew to a close. The essay on page 227 describes their dramatic story.

Today's reptiles include turtles, crocodilians, lizards and snakes, and tuataras (Figure 19.42). All have scaly skin that resists drying out. All can adjust their internal body temperature by changes in physiology and behavior, as when they seek and absorb heat by basking on sun-warmed rocks. Most reptiles lay eggs; some give birth to fully formed offspring.

Turtles. Turtles live in a mobile home—a shell that protects them from predators. Most can pull their head and limbs into the shell. When this fails, some use powerful jaws and fierce dispositions to keep predators at bay. All turtles lay eggs on land, then leave them. Predators eat most of the eggs, so few hatch. Prospects are not good for sea turtles. Humans value their shell and meat, and have put them on the brink of extinction.

Lizards and Snakes. Nearly all living reptiles are lizards and snakes. Most of these distant relatives of dinosaurs are small. But the Komodo monitor lizard is

Figure 19.41 Eastern hognose snakes, emerging from leathery-shelled amniote eggs. Shelled or not, this type of egg contributed to the successful colonization of land habitats by reptiles and, later, birds and mammals.

large enough to hunt young water buffalo. The longest snake would stretch across ten yards of a football field.

Most lizards grab insect prey with small, peglike teeth. Chameleons zap them with a long tongue. When grabbed by a predator, many lizards give up their tail. The disconnected tail wriggles for a bit and may be distracting enough to allow time for a getaway. Some lizards intimidate predators (and rivals) by flaring their throat fan.

Long ago, short-legged, long-bodied lizards gave rise to snakes. Although limbless, a few modern snakes have remnants of hindlimbs. Most snakes move in S-shaped waves. "Sidewinders" move in J-shaped waves across unstable sediments and desert sands. Snakes eat other vertebrates, mostly. Their jaws are wonderfully movable—some swallow animals wider than they are. Pythons and boas coil tightly around prey and suffocate it. Others, including rattlesnakes, use venom to subdue prey. Generally, snakes are not aggressive toward humans, but each year as many as 40,000 people die from bites by the venomous ones.

Tuataras. Existing tuataras are restricted to small islands near New Zealand. Their body plan hasn't changed much since the age of dinosaurs, about 150 million years ago. Tuataras don't engage in sex until they are twenty years old. This works well because tuataras, like turtles, may live for sixty years or more.

Figure 19.42 Representative reptiles. (**a**) One type of Galápagos tortoise that impressed Charles Darwin during his five-year voyage. (**b**) A frilled lizard, flaring a ruff of neck skin in a defensive display. (**c**) A rattlesnake of the American Southwest. (**d**) Tuatara, a "living fossil" that hasn't changed much since the age of dinosaurs. (**e**) An African crocodile sunning itself.

Crocodilians. Crocodiles and alligators live in or near water. All have similar body plans, including a long snout (Figure 16.2). Although they look like big lizards, they are evolutionarily closer to birds. Crocodilians show complex social behavior, as when male and female parents assist hatchlings in their move out of the egg and into the water. Young alligators of the Gulf Coast swamps remain with the mother for two years after hatching. At one time, saltwater crocodiles were up to seven meters long. Their belly skin is prized as leather, however. They are now hunted so intensively that few live long enough to grow to impressive size.

a

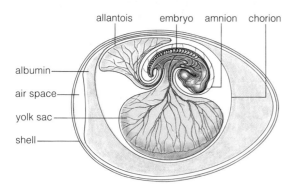

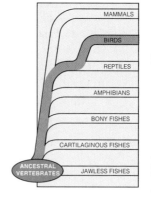

c

allantois   embryo   amnion   chorion

albumin

air space

yolk sac

shell

b

Figure 19.43 Some characteristics of birds. (**a**) Of all living vertebrates, only birds and bats fly by flapping their wings. (**b**) Speckled eggs of a magpie. All birds lay hard-shelled eggs of the sort shown in the generalized diagram. (**c**) Canada geese, one of the types of birds that show migratory behavior, at their wintering grounds in New Mexico. (**d**) A male pheasant with flamboyant plumage, an outcome of sexual selection. This native of the Himalaya Mountains of India is an endangered species. As is the case for many birds, its jewel-colored feathers end up adorning humans—in this case, on the caps of native tribespeople.

MAMMALS

BIRDS

REPTILES

AMPHIBIANS

BONY FISHES

CARTILAGINOUS FISHES

ANCESTRAL VERTEBRATES   JAWLESS FISHES

Birds

Judging from *Archaeopteryx* fossils, birds descended from bipedal reptiles that lived some 160 million years ago (page 195). Birds still resemble reptiles in many internal structures, their horny beaks and scaly legs, and their egg-laying habit (Figure 19.43). Today, many biologists actually view birds as a branch of the reptilian lineage.

By definition, **birds** alone are animals with feathers—lightweight structures with roles in flight, insulation, or both. Birds vary tremendously in size, proportions, coloration, and capacity for flight. Some flightless ones are long-legged sprinters (Figure 14.2). Warblers and other perching birds especially show pro-

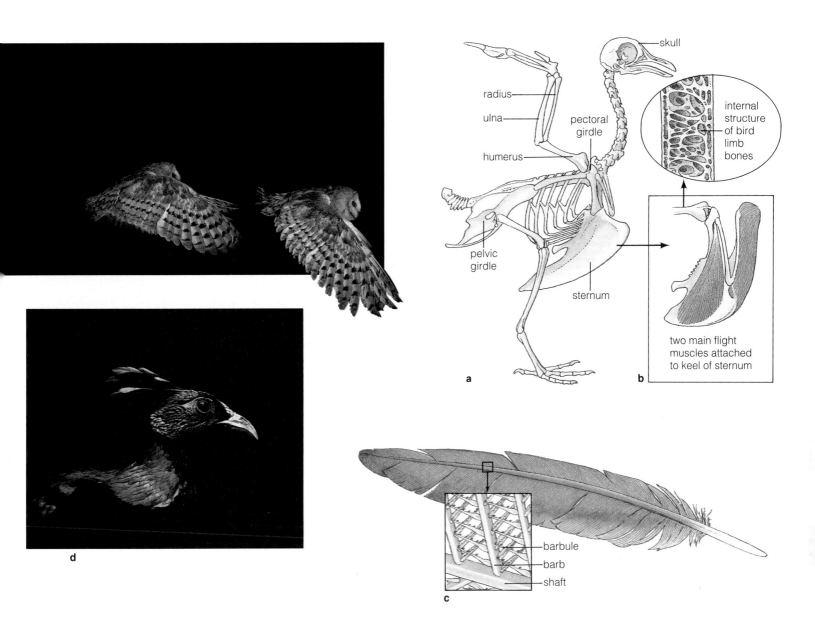

Figure 19.44 Bird flight. Bird flight requires an airstream and a powerful downstroke that will provide *lift*, a force at right angles to the airstream. (**a,b**) Body plan of birds, showing the large, keeled breastbone (sternum) to which flight muscles attach. (**c**) The bird wing is a system of lightweight bones and feathers. Feathers gain strength from a hollow central shaft and from tiny barbules interlocked in a latticelike array. With its long flight feathers, the bird wing serves as an airfoil. Usually, feathers are spread out on the downstroke; this increases the size of the surface pushing against air. On the upstroke, feathers fold somewhat, so the wing presents the least possible profile against the air.

nounced differences in coloration of their feathers and in their territorial songs. Bird songs and other social behaviors are splendidly complex. These are topics of later chapters.

Flight demands high metabolic rates, which require a good deal of oxygen. A unique respiratory system greatly enhances oxygen uptake in birds (page 455). Flight also demands low weight and high power. The bird wing (a forelimb) consists of feathers, powerful muscles, and lightweight bones. Flight muscles attach to an enlarged breastbone and the upper limb bones adjacent to it (Figure 19.44). When they contract, they produce the powerful downstroke for flight. Bird bones are strong and yet weigh very little because of air cavities in the bone tissue. The skeleton of a frigate bird, which has a seven-foot wingspan, weighs only four ounces. That's less than the feathers weigh!

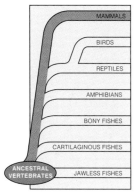

Mammals

Chapter 16 introduced some general characteristics of mammals. Here we round out the picture. The ancestors of mammals arose during the Carboniferous. Mammal-like reptiles called therapsids wandered about during the late Mesozoic. If not the direct ancestors of mammals, they were close to them.

By definition, **mammals** alone nourish their young with milk secreted from the female's mammary glands. They alone have hair, although this trait was lost in most whale lineages. Unlike reptiles, which generally swallow prey whole, most mammals use specialized kinds of teeth to kill, cut, and sometimes chew food before swallowing it (page 235). Mammals also can adjust their body temperature. This is one reason why early mammals were able to radiate across continents and through oceans. Their hair or thick skin helps retain metabolic heat.

A great mammalian radiation began at the dawn of the Cenozoic. Then, the mammalian brain began to reveal its true potential. Especially among primates, interconnected masses of information-encoding and information-processing cells expanded dramatically. This expansion was the foundation for our own capacity for memory, learning, and conscious thought.

Today there are three mammalian groups: the egg-laying, pouched, and placental mammals. Figure 19.45 shows a few representatives.

Egg-Laying Mammals. The platypus described at the start of this chapter is one egg-laying mammal (prototherian) that has survived to the present. Another is the spiny anteater, a burrowing animal of Australia and New Guinea (Figure 19.45a). Both kinds lost most or all of their teeth over evolutionary time. The loss correlates with specialized diets. Platypuses feed mostly on small aquatic invertebrates. Spiny anteaters eat termites and ants, which they capture with their long, sticky tongue.

Pouched Mammals. Pouched mammals (metatherians) also are called marsupials. Nearly all live in Australia and nearby islands. Marsupial young are born alive but not quite "finished"—they are tiny, blind, and hairless. The newborns have a good sense of smell and strong forelimbs, which they use to locate and reach the mother's pouch. They are suckled in the pouch, where they complete early development.

For at least 50 million years, Australian marsupials evolved in relative isolation from placental mammals. Their ancestors had crossed a narrow sea that opened

a

b

c

up between two land masses, but placental mammals stayed behind. In the absence of competition, the marsupials radiated freely into adaptive zones in their own land mass, which became Australia. Kangaroos and wallabies now occupy adaptive zones comparable to the ones filled by deer, antelope, and other herbivores on other continents. Siberian and North American forests have wolves; Australia has (or had) the Tasmanian "wolf," a marsupial that is now probably extinct. Other marsupials glide like flying squirrels, and some climb like monkeys. In Australia, humans have introduced cattle, sheep, horses, and other placental mammals. Many native marsupials are being threatened with displacement.

d

e

SUMMARY

1. Animals are multicelled, motile heterotrophs that reproduce sexually and, often, asexually. Embryonic development occurs during their life cycle. Except for sponges, animals have true tissues. Most have organs and organ systems.

2. Sponges, cnidarians, flatworms, roundworms, rotifers, mollusks, annelids, arthropods, and echinoderms are invertebrates. Sea squirts and lancelets are invertebrate chordates. Fishes, amphibians, reptiles, birds, and mammals are vertebrate chordates.

3. Certain aspects of animal body plans reflect major trends in animal evolution. These are the body's type of symmetry, gut, and cavity (if any) between the gut and body wall; whether it has a distinct head end; and whether the body is partitioned into segments.

4. These trends occurred among certain lineages:

a. Bilateral, cephalized body plans led to *pairs* of body parts, including nerves, brain regions, muscles, sensory structures (such as eyes), and appendages (such as fins and legs).

b. A saclike gut evolved into a complete digestive system, with mouth, anus, and specialized regions for processing food.

c. Animals with tissue-packed bodies gave rise to coelomate animals. A coelom is a cavity between the gut and body wall, lined with peritoneum. Organs in this cavity increased in size and complexity.

d. Unsegmented animals gave rise to segmented ones. Movements became more complex. Appendages became specialized for different tasks in different segments.

Placental Mammals. One or more kinds of placental mammals (eutherians) live in virtually every kind of aquatic and terrestrial environment. Their "placenta" is a spongy tissue that lines the mother's uterus (page 522). It consists of maternal tissue and embryonic membranes. It is the means by which an embryo receives nutrients and oxygen and gets rid of metabolic wastes. Placental mammals grow faster inside their mother than marsupials do in a pouch. At birth, many are fully developed and move about almost at once. However, they remain with adults through an extended period of dependency and learning.

As Appendix I reveals, placental mammals include exotic as well as familiar species. Rats, mice, squirrels, and other rodents are the most diverse, followed by bats. More familiar placental mammals are the dogs, bears, cats, walruses, and dolphins (all meat eaters, or carnivores), as well as the horses, camels, deer, and elephants (plant eaters, or herbivores). The structure and functions of many of these animals will occupy our attention in chapters to follow.

5. Arthropods, especially insects, are the most abundant, widely distributed animals. Five adaptations underlie their success: a hardened exoskeleton, jointed appendages, specialized respiratory structures, well-developed nervous and sensory systems, and division of labor in the life cycle.

6. The following trends occurred during the evolution of certain vertebrate lineages:

a. A vertebral column supplanted the notochord as the structural element against which muscles act. The vertebral column foreshadowed the evolution of fast-moving, predatory animals.

b. Jaws evolved and led to increased predator-prey competition. This in turn favored the evolution of more efficient nervous systems and sensory organs.

c. Fins evolved into fleshy lobes with internal structural elements—the forerunners of limbs.

d. Lungs evolved and proved adaptive in the invasion of land. The circulatory system became better at distributing oxygen as a related development.

Review Questions

1. In what ways do invertebrates and vertebrates differ? *290, 312–313*

2. Describe the features of one group of invertebrates. *290ff.*

3. Name some of your paired body parts that evolved in your bilateral, cephalized ancestors. *289*

4. Name some animals with a saclike gut. Evolutionarily, what advantages does a complete gut afford? *289, 292, 294*

5. What is a coelom? Why was it important in the evolution of certain animal lineages? *289*

6. Choose an insect that thrives in your neighborhood and describe some of the adaptations that underlie its success. *302, 307*

Self-Quiz *(Answers in Appendix IV)*

1. Five body features that help us identify trends in animal evolution are _____ , _____ , _____ , _____ , and _____ .

2. Animals are _____ .
   a. autotrophs
   b. heterotrophs
   c. chemotrophs
   d. saprobes

3. Which is *not* characteristic of the animal kingdom?
   a. multicellularity; cells form tissues, organs
   b. exclusive reliance on sexual reproduction
   c. motility at some stage of the life cycle
   d. embryonic development during life cycle

4. Which animals have radial *and* bilateral features?
   a. cnidarians
   b. mollusks
   c. arthropods
   d. echinoderms
   e. This is a trick question; no animal shows both.

5. In sheer numbers and distribution, _____ are the most successful animals.
   a. arthropods
   b. sponges
   c. snails and clams
   d. sea stars
   e. vertebrates

6. Jointed appendages and a hardened exoskeleton occur among the _____ .
   a. arthropods
   b. sponges
   c. snails and clams
   d. sea stars
   e. vertebrates

7. Which animals cause serious diseases in humans?
   a. cnidarians
   b. flatworms, roundworms
   c. segmented worms
   d. chordates

8. More complex animals have a _____ between the gut and body wall.
   a. pharynx
   b. peritoneum
   c. coelom
   d. archenteron

9. Match the terms with the appropriate groups.
   _____ sponges
   _____ cnidarians
   _____ flatworms
   _____ roundworms
   _____ rotifers
   _____ mollusks
   _____ annelids
   _____ arthropods
   _____ echinoderms
   _____ chordates
   a. spiny-skinned
   b. vertebrates and kin
   c. flukes and tapeworms
   d. no tissue organization
   e. no males for some
   f. nematocysts, radial symmetry
   g. hookworms, elephantiasis
   h. jointed appendages
   i. "belly-foots" and kin
   j. segmented worms

Selected Key Terms

amniote egg *318*	fish *314*	molting *302*
amphibian *316*	flatworm *294*	nerve cell *292*
animal *288*	gill *298*	nerve cord *300*
annelid *300*	gut *289*	parasite *294*
arthropod *302*	invertebrate *290*	pharynx *294*
bilateral symmetry *288*	jaw *312*	radial symmetry *288*
bird *320*	lancelet *311*	reptile *318*
cephalization *289*	larva *290*	rotifer *297*
chordate *310*	lung *298*	roundworm *295*
cnidarian *292*	mammal *322*	sea squirt *310*
coelom *289*	mesoderm *294*	sponge *290*
echinoderm *309*	metamorphosis *302*	vertebra *312*
epithelium *292*	mollusk *298*	vertebrate *290*
fin *312*		

Readings

Hickman, C. P., and L. S. Roberts. 1988. *Integrated Principles of Zoology*. Eighth edition. St. Louis: Mosby.

Kozloff, E. 1990. *Invertebrates*. Philadelphia: Saunders.

Pough, F. H.; J. Heiser; and W. McFarland. 1989. *Vertebrate Life*. Third edition. New York: Macmillan.

Welty, J., and L. Baptista. 1988. *The Life of Birds*. Fourth edition. New York: Saunders.

FACING PAGE: *A flowering plant* (Prunus) *busily doing what it does best: producing flowers for the fine art of reproduction.*

20 PLANT TISSUES

The Greening of the Volcano

In the spring of 1980, Mount St. Helens in southwestern Washington exploded, and 540 million tons of ash blew skyward. Within minutes, the shock waves blew down or incinerated hundreds of thousands of mature trees near the volcano's northern flank. Rivers of ash and fire surged down the slopes faster than a hundred miles an hour. Twenty billion gallons of water, released when the intense heat melted snow and glacial ice, turned the nightmarish rivers into torrents of cementlike mud. In one brief moment, nearly 100,000 acres of magnificent forests became a barren sweep of land (Figure 20.1). It was a mind-numbing picture of what our world would be like without plants.

With this chapter we open a unit dedicated to the most successful plants on earth—the flowering plants. No other kind of plant surpasses them in diversity or distribution. They have distinctive patterns of structural organization and splendid modes of reproduction, growth, and development. Their adaptations help them survive hostile conditions on land—even momentary takeovers by volcanoes.

a

Figure 20.1 (**a**) Eruption of Mount St. Helens in 1980. (**b**) Nothing remained of the forest that had cloaked the volcano's flanks. (**c**) Less than a decade later, plants were making a comeback.

b

c

1. Flowering plants dominate the plant kingdom. They have aboveground shoots (which include stems, leaves, and flowers) and descending roots that typically grow downward and outward through soil.

2. Flowering plant tissues are organized into three main kinds of tissue systems. Ground tissue systems make up most of the plant body. Vascular tissue systems distribute water, minerals, and products of photosynthesis through roots, stems, and leaves. Dermal tissues cover and protect plant surfaces.

3. A simple plant tissue (parenchyma, sclerenchyma, or collenchyma) consists of one type of cell. Complex tissues have two or more types of cells. They include xylem and phloem (vascular tissues) and epidermis (a dermal tissue).

4. Plants grow by cell divisions and cell enlargements at meristems, which are localized regions of self-perpetuating embryonic cells. Their primary growth (lengthening of stems and roots) originates at apical meristems of stem and root tips. Their secondary growth (increases in diameter) originates at lateral meristems inside stems and roots.

5. Many plants show primary and secondary growth, year after year. Wood is the outcome of this secondary growth.

THE PLANT BODY: AN OVERVIEW

Chapter 18 sprints through the spectrum of nearly 310,000 known plant species. Even that fleeting run through plant diversity reveals why no one species can be used as a typical example of plant body plans. However, when we hear the word "plant," we mostly think of gymnosperms (such as pine trees) and angiosperms (such as roses, apple trees, and corn).

Angiosperms, recall, are the only plants that produce flowers. *And with more than 265,000 species, they dominate the plant kingdom.* We devote this chapter to the tissues and body plans of flowering plants. Chapter 21 explains how these plants take up water and nutrients, restrict water loss, and distribute organic substances through the plant body. Chapter 22 looks at key aspects of their growth, development, and reproduction.

Shoots and Roots

Many flowering plants have a body plan like that shown in Figure 20.2. The aboveground parts are **shoots**. These are stems and their branchings, leaves,

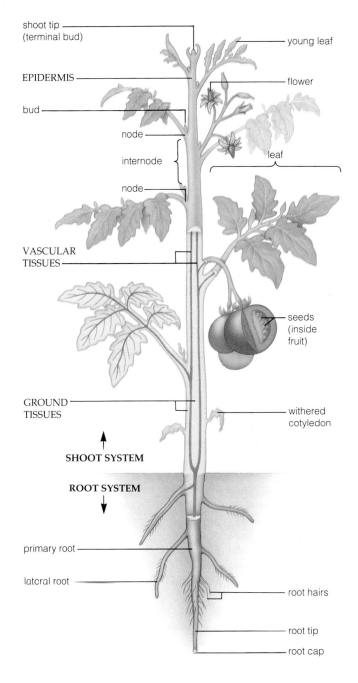

Figure 20.2 Body plan for one type of flowering plant. Vascular tissues (purple) conduct water, nutrients, and organic substances through the plant. They thread through ground tissues, which make up most of the plant body. Dermal tissues (epidermis) cover the surfaces of the root and shoot systems.

flowers, and other components. Stems are like structural beams for upright growth. Upright growth favorably exposes photosynthetic cells in young stems and leaves to light. It also displays the flowers for pollinators. The plant's descending parts, or **roots**, typically penetrate downward and spread out through soil, absorbing water and dissolved nutrients. Roots also anchor aboveground parts. Most store food as well, releasing it as required to root cells or for transport to aboveground parts.

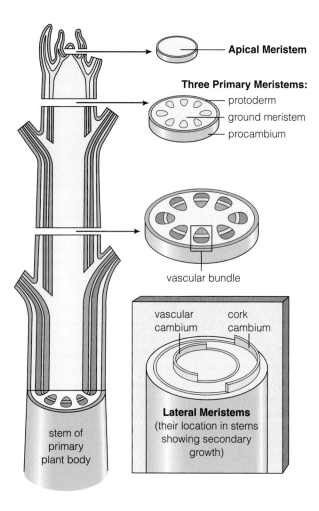

Apical Meristem

Three Primary Meristems:
- protoderm
- ground meristem
- procambium

vascular bundle

vascular cambium   cork cambium

Lateral Meristems
(their location in stems showing secondary growth)

stem of primary plant body

Figure 20.3 Approximate locations of primary meristems (yellow) and lateral meristems (purple) in a plant showing both primary and secondary growth.

Meristems

Flowering plants grow at meristems (localized regions of self-perpetuating, embryonic cells). As they grow, stems and shoots lengthen and thicken. Increases in *length* originate at **apical meristems** located inside the dome-shaped tips of stems and roots. There, cell divisions and enlargements give rise to three *primary* meristems, which go on to produce specialized tissue systems (Figure 20.3). In all, a stem or root lengthens through growth that originates at apical meristems and at meristematic tissues derived from them. This growth produces the *primary* tissues of the plant body.

Increases in diameter originate at *lateral* meristems, which form inside a stem or root (Figure 20.3). There are two kinds of lateral meristems: **vascular cambium** and **cork cambium**. Growth originating here produces *secondary* tissues of the plant body. Each spring, for example, primary growth lengthens a maple tree's stems, branches, and roots; and secondary growth

thickens them. Some of the new cells that form during primary and secondary growth perpetuate the meristems. Others differentiate and become part of the tree's specialized tissues.

Primary growth (lengthening of stems and roots) originates at apical meristems and at meristematic tissues derived from them.

Secondary growth (thickening of stems and roots) originates at lateral meristems.

Three Plant Tissue Systems

The stems, branches, leaves, and roots of a flowering plant are similar in a key respect. Their main tissues are grouped into three systems. The **ground tissue system** is the most extensive; its tissues make up the bulk of the plant body. A **vascular tissue system** contains two kinds of conducting tissues that distribute water and solutes through the plant body. The **dermal tissue system** covers and protects the plant's surfaces.

As you will now see, some tissues in these systems are simple, in that they contain one type of cell only. Parenchyma, collenchyma, and sclerenchyma fall in this category. Other tissues are complex, with highly organized arrays of two or more types of cells. Xylem, phloem, and epidermis are like this.

Simple Tissues

Parenchyma is the most common tissue in ground tissue systems. Its cells are typically thin-walled, pliable, and many-sided (Figure 20.4a). The cells are metabolically active at maturity. They retain the capacity to divide, so they often can heal wounded plant parts.

Parenchyma tissues make up most of the soft, moist, primary tissues of roots, stems, leaves, flowers, and fruits. One type specializes in photosynthesis. Carbon dioxide and oxygen diffuse through abundant air spaces between its cells. Other types specialize in storage, secretion, and other tasks. Strands of parenchyma also thread through vascular tissue systems.

Near the surface of lengthening stems are patches or cylinders of **collenchyma**, a tissue that gives flexible support for primary tissues (Figure 20.4b). Collenchyma also forms pliable ribs or strings in many leaf stalks. Collenchyma cells are mostly elongated, with unevenly thickened walls. Pectin in the cell walls imparts pliability to this tissue.

Sclerenchyma mechanically supports mature plant parts, and it often protects seeds. Generally, its cells have thick, lignin-impregnated walls (Figure 20.4c).

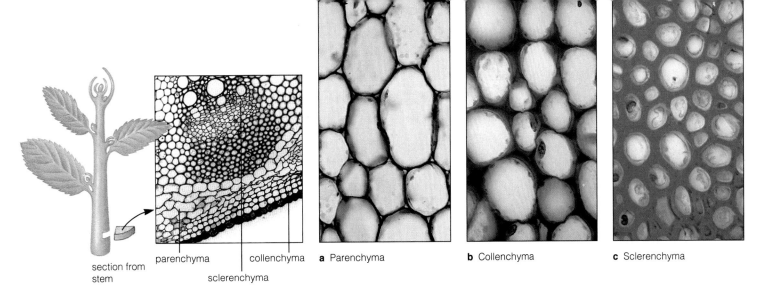

section from
stem

parenchyma

collenchyma

sclerenchyma

a Parenchyma

b Collenchyma

c Sclerenchyma

Figure 20.4 Examples of ground tissues from the stem of a sunflower plant (*Helianthus*).

Lignin strengthens and waterproofs cell walls. Without it, land plants would not have evolved to new heights.

Sclerenchyma tissues consist of fibers or sclereids. Fibers are in the vascular tissue systems of some stems and leaves. They are long, tapered cells that flex and twist without stretching. Fibers from flax plants are used to make rope, paper, cloth, and thread. Sclereids are shorter cells. They form clusters in pears and some other fruits and give them a gritty texture. They also form coconut shells, peach pits, and other seed coats.

Complex Tissues

Vascular Tissues. Two kinds of vascular tissues, called xylem and phloem, distribute substances through the plant. Often their conducting cells are bundled in a sheath composed of fibers and parenchyma cells.

Xylem conducts soil water and dissolved minerals, and it mechanically supports the plant. Its conducting cells (vessel members and tracheids) are not alive at maturity. The lignified *walls* of the dead cells connect to form tubes and strengthen the plant. Water flows between adjacent cells through pits in the side walls.

Phloem transports sugars and other solutes. Its main conducting cells (sieve-tube members) are alive at maturity. Adjacent cells interconnect at perforations in their side walls and at open or perforated end walls. In leaves, sieve-tube members become loaded with sugars from photosynthetic cells, often with the help of specialized, living parenchyma cells (companion cells). The sugars are unloaded in regions where cells are growing or storing food. The next chapter has more to say about the functions of xylem and phloem.

Dermal Tissues. All surfaces of primary plant parts are covered and protected by a dermal tissue system

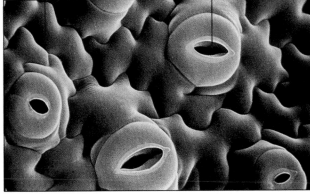

cuticle-coated
epidermal cell (surface view)

stoma (passageway
across epidermis)

Figure 20.5 Scanning electron micrograph of a leaf surface, showing some cuticle-covered epidermal cells and stomata.

called **epidermis**. In most plants, epidermis is mainly a single layer of unspecialized cells. Waxes and cutin (a fatty substance) coat the outermost cell walls. The surface coating, called a cuticle, restricts water loss and resists attacks by some microorganisms.

Stem and leaf epidermis also contains specialized cells, including pairs of guard cells (Figure 20.5). Guard cells can change shape. When they do, a gap called a stoma (plural, stomata) opens up between them or closes. In the chapter to follow, you will see how the movement of water vapor, carbon dioxide, and oxygen across the epidermis is controlled at stomata.

Periderm replaces the epidermis of stems and roots showing secondary growth. Most of this protective cover is composed of the walls of cork cells that are no longer alive. The walls are heavily impregnated with suberin, a fatty substance.

a Typical monocot flower

b Typical dicot flower

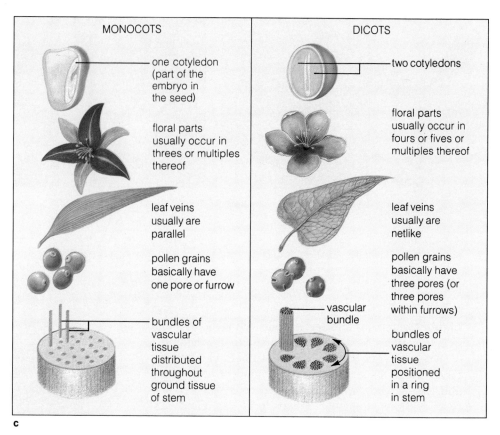

MONOCOTS

one cotyledon (part of the embryo in the seed)

floral parts usually occur in threes or multiples thereof

leaf veins usually are parallel

pollen grains basically have one pore or furrow

bundles of vascular tissue distributed throughout ground tissue of stem

DICOTS

two cotyledons

floral parts usually occur in fours or fives or multiples thereof

leaf veins usually are netlike

pollen grains basically have three pores (or three pores within furrows)

vascular bundle

bundles of vascular tissue positioned in a ring in stem

c

Figure 20.6 Comparison of the main features that distinguish monocots from dicots. The photographs show the flower of a wild iris (*Iris*, a monocot) and of St. John's wort (*Hypericum*, a dicot).

Monocots and Dicots Compared

Recall that we informally refer to the two classes of flowering plants as **monocots** and **dicots** (page 284). Grasses, lilies, orchids, irises, cattails, and palms are familiar monocots. Dicots include nearly all of the well-known trees and shrubs, other than conifers.

Monocots resemble dicots in structure and function, but they differ in some distinctive ways. For example, monocot seeds have one cotyledon and dicot seeds have two. A "cotyledon" is a leaflike structure that forms in the seed, as part of a plant embryo, and stores or absorbs food for it. It withers after the seed germinates and the leaves start functioning. Figure 20.6 shows other differences between monocots and dicots.

PRIMARY STRUCTURE OF SHOOTS

Picture yourself scanning a menu in a wood-framed restaurant, slouched comfortably in an old wicker chair. Grasscloth panels the walls. On the carved oak table, a woven basket holds bread, and a vase holds delicately stemmed flowers. Nearby, a waiter tosses a salad in a wooden bowl. Count the ways in which you are being pampered by stems and leaves (or what's left of them). Let's consider what stems and leaves of this sort look like before someone carves, weaves, snips, or chops them or mashes, moistens, and presses them flat.

Internal Structure of Stems

Recall that when a stem is lengthening, primary meristems give rise to its ground, vascular, and dermal tissues. Commonly, the vascular tissues (primary xylem and phloem) develop as **vascular bundles**. These are multistranded, sheathed cords that thread lengthwise through the ground tissue system. Inside each bundle, the phloem is often positioned near the side of the sheath facing the stem surface, and the xylem is positioned near the side facing the stem center.

Examine the primary stems of flowering plants, and you see that two patterns predominate. Vascular bundles are scattered through the ground tissue system of most monocots and some dicots. They are organized in a ring within most dicot stems. The ring divides the ground tissue system into two zones: an outer *cortex* and inner *pith*. Figures 20.7 and 20.8 give examples of the two patterns.

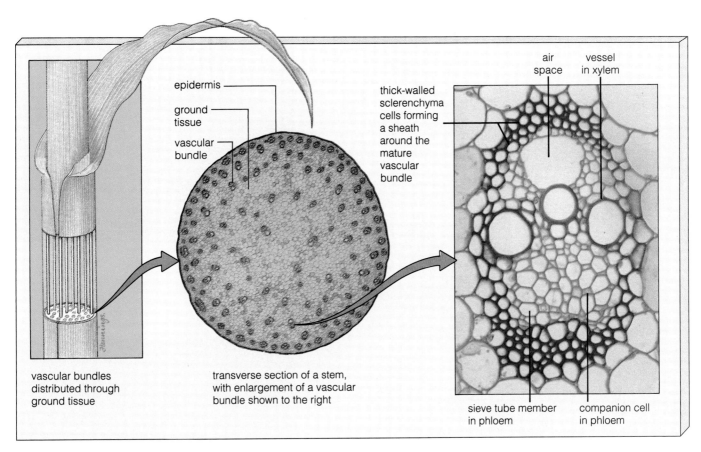

Figure 20.7 Stem structure of corn, a monocot.

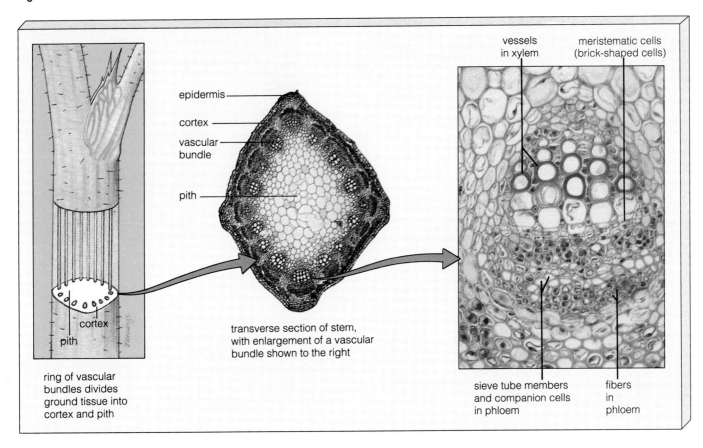

Figure 20.8 Stem structure of alfalfa, a dicot.

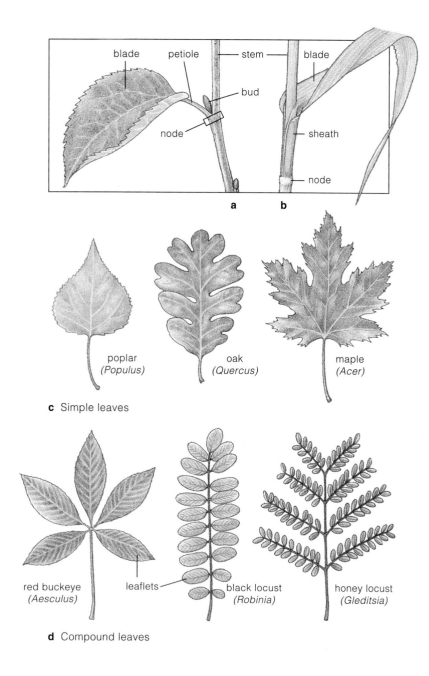

c Simple leaves

d Compound leaves

Figure 20.9 Common leaf forms among (**a**) dicots and (**b**) monocots. (**c,d**) Examples of simple and compound leaves.

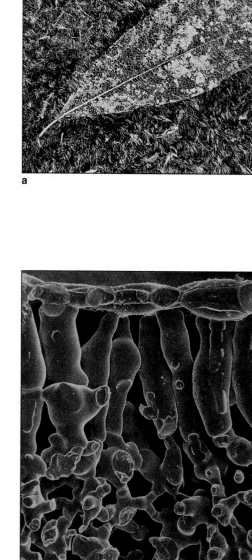

50 μm

Leaf Structure

By definition, **leaves** are metabolic factories equipped with food-producing, photosynthetic cells.

Leaf Shapes. The leaves of ryegrass, corn, and most other monocots are flat surfaced, like a knife blade. The base of the blade encircles and sheathes the stem. The leaves of dicots are diverse. Many have one broad blade, attached to the stem by a stalk (petiole). Often the blade is variously lobed. Some dicots have "compound" leaves, with several stalked leaflets. Figure 20.9 shows some leaf plans. Many leaves also have surface specializations, including hairs, scales, spines, and hooks (page 342).

Leaves of birches and other "deciduous" species drop away from the stem as winter approaches. Leaves of camellias and other "evergreen" species also drop, but they don't all drop at the same time.

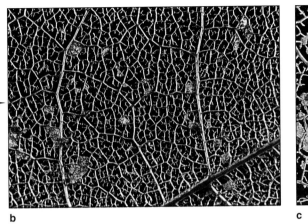

b

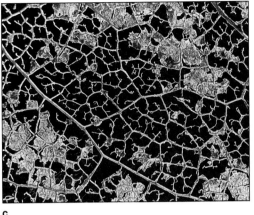

c

Figure 20.10 Leaf from a magnolia tree, showing veins at increasing magnifications. Networks of minor veins thread through photosynthetic tissue.

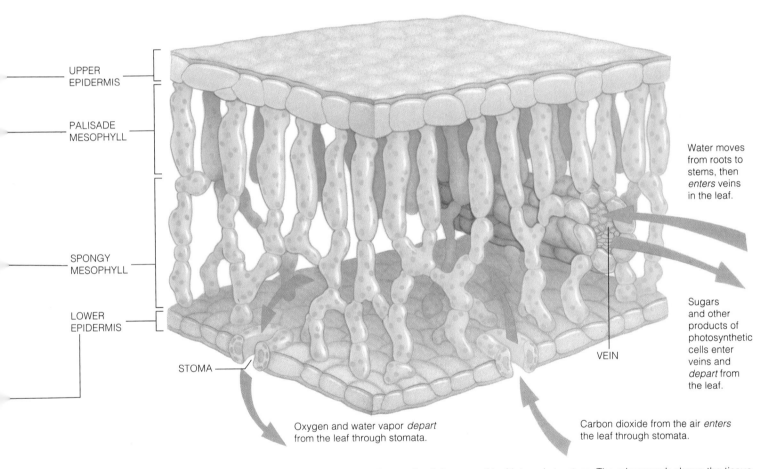

UPPER EPIDERMIS

PALISADE MESOPHYLL

SPONGY MESOPHYLL

LOWER EPIDERMIS

STOMA

VEIN

Water moves from roots to stems, then *enters* veins in the leaf.

Sugars and other products of photosynthetic cells enter veins and *depart* from the leaf.

Oxygen and water vapor *depart* from the leaf through stomata.

Carbon dioxide from the air *enters* the leaf through stomata.

Figure 20.11 Generalized diagram of leaf internal structure. The micrograph shows the tissue layers of a leaf from the kidney bean plant (*Phaseolus*).

Leaf Internal Structure. The leaves just described have a large surface area exposed to sunlight and carbon dioxide. Their upper surface tissue is cuticle-covered epidermis. Next comes **mesophyll**, a type of parenchyma tissue with photosynthetic cells and plenty of air spaces. The spaces enhance diffusion of carbon dioxide into the cells and oxygen out of them.

Distinct vascular bundles called **veins** thread lacily through the interior of the leaf (Figure 20.10). They move water and nutrients to the cells and carry sugars and other substances away from them. Below the mesophyll is another cuticle-covered epidermis, the leaf's lower surface tissue. This often contains most of the stomata (Figure 20.11).

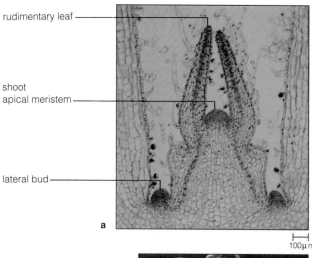

rudimentary leaf

shoot
apical meristem

lateral bud

a

100μm

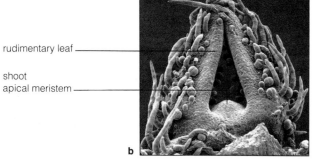

rudimentary leaf

shoot
apical meristem

b

100μm

Figure 20.12 (**a**) New leaves forming at the shoot tip of *Coleus*, in thin section. (**b**) Scanning electron micrograph of the same type of shoot tip.

Origin of Leaves

How do leaves originate? The ones described so far form at the tip of a main stem or its youngest branches. As Figure 20.12 indicates, each starts out as a bulge on the flanks of the apical meristem and enlarges into a rudimentary leaf. The bulges that form are close together at first. But as primary growth continues, the leaves that form from the bulges become spaced at intervals along the lengthening stem. Each stem region where one or more leaves are attached is a "node." The stem region between two successive nodes is an "internode." Figure 20.2 shows an example of this. At each node, vascular bundles continuous with the stem's vascular system develop within the forming leaf.

Look at a twig of a walnut tree in winter, after all its leaves have dropped (Figure 20.13). At the shoot tip is a **bud**: an undeveloped shoot of mostly meristematic tissue, often covered and protected by scales (modified leaves). Besides this *terminal* bud, the stem has other buds spaced at intervals along its length. Each of these *lateral* buds forms in the upper angle where a leaf was attached to the stem. Some buds give rise to new stems. Others give rise to leaves, flowers, or both. Vascular bundles diverge from the main stem and develop in the new plant parts, also.

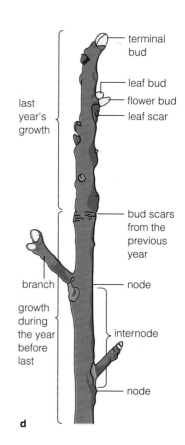

terminal bud

leaf bud
flower bud
leaf scar

last year's growth

bud scars from the previous year

branch

node

growth during the year before last

internode

node

a   **b**   **c**   **d**

Figure 20.13 (**a–c**) Leaves forming at a terminal bud of a dogwood tree. (**d**) Arrangement of buds on a twig from a walnut tree.

Uses and Abuses of Leaves

Imagine a leaf-deprived grocery cart. Out go the lettuces, parsley, spinach, and chard. No thyme, basil, oregano, and rosemary for your secret spaghetti sauce. No more mint, darjeeling, and other teas.

Beyond the table, oils from leaves of lavender scent perfumes and soaps. Oils of eucalyptus and camphor find their way into medicine chests. Leaves of nightshade plants (belladonna) provide extracts to treat Parkinson's disease. Digitalis from foxglove leaves helps regulate heartbeat and blood circulation. Juices from the leaves of *Aloe vera* soothe sun-damaged skin. Rosy periwinkle leaves provide alkaloids that treat Hodgkin's disease.

Landscape architects and imaginative homeowners use leaves of various trees and shrubs much as an artist uses colors and textures. They plant shrubs next to buildings and so help keep them from heating up in summer and

losing heat in winter. We get twine and rope from century plant leaves (*Agave*), cords and textiles from leaf fibers of Manila hemp, hats from Panamanian palm fronds, and thatched roofs from leaves of palms and grasses.

We kill cockroaches, fleas, lice, and flies with insecticides derived from Mexican cockroach plants. We kill more than a hundred kinds of insects, mites, and nematodes with extracts of neem tree leaves without killing off natural predators of those common pests.

Some people also smoke, chew, or tuck into their mouth the leaves of tobacco plants—and so become candidates for the hundreds of thousands of annual deaths from lung, mouth, and throat cancers. Cocaine, derived from coca leaves, is used medicinally and abused by increasing numbers of individuals—with devastating social and economic effects, as described in the *Focus* essay on page 473.

PRIMARY STRUCTURE OF ROOTS

Unless tree roots start to buckle a sidewalk or choke off a sewer line, most of us don't pay much attention to the elaborate root systems of flowering plants. For most species, roots extend to a depth of 2 to 5 meters. But in hot deserts, where water is scarce, one hardy mesquite shrub is known to have sent roots down 53.4 meters (175 feet) near a streambed. The shallow roots of some cacti may radiate out for 15 meters. Someone once measured the roots of a young rye plant that had been growing for four months in only 6 liters of soil. If the surface area of that root system were laid out as a single sheet, it would extend more than 600 square meters!

Taproot and Fibrous Root Systems

The first root to poke through the coat of a germinating seed is the "primary" root. In most dicot seedlings, it increases in diameter and grows downward. Later, **lateral roots** start forming in internal tissues and erupt through the epidermis. The youngest of these lateral roots are closest to the root tip. A primary root and its lateral branchings are a **taproot system**. Carrot and dandelion plants have a taproot system (Figure 20.14*a*). So does an oak tree.

Grasses and other monocots generally have short-lived primary roots. In their place, *adventitious* roots

arise from the young plant's stem, then lateral roots branch from them. ("Adventitious" refers to a structure arising at an uncommon location, such as roots arising from stems or leaves.) These lateral roots, which are similar in diameter and length, form a **fibrous root system** (Figure 20.14*b*). The rye plant mentioned above has a fibrous root system.

a   **b**

Figure 20.14 (**a**) Taproot system of a dandelion. (**b**) Fibrous root system of a grass plant.

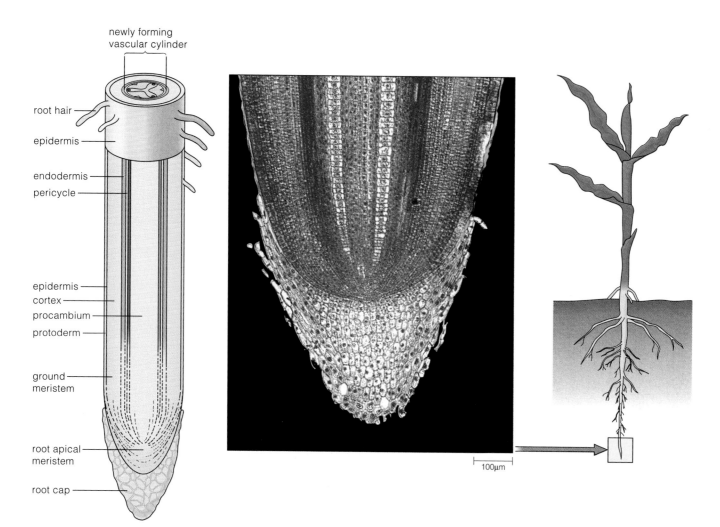

Figure 20.15 Generalized root tip, sliced lengthwise. The micrograph shows a root tip from corn (*Zea mays*).

- newly forming vascular cylinder
- root hair
- epidermis
- endodermis
- pericycle
- epidermis
- cortex
- procambium
- protoderm
- ground meristem
- root apical meristem
- root cap

100μm

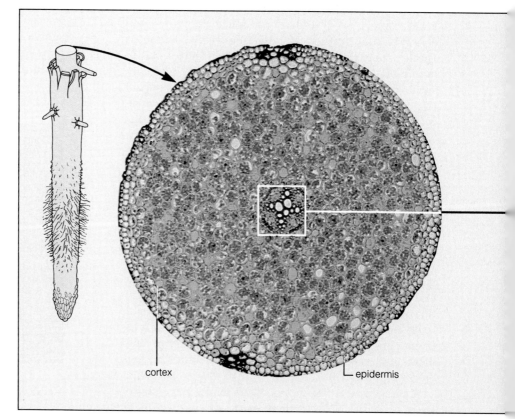

Figure 20.16 Section through a young root from a buttercup (*Ranunculus*). The close-up shows details of its vascular cylinder.

- cortex
- epidermis

Internal Structure of Roots

Figure 20.15 shows the location of the apical meristem in a root tip. Behind this, cellular descendants of the apical meristem divide, enlarge, and become the specialized cells of the root's tissue systems.

A root tip has a dome-shaped cell mass called a root cap. Root apical meristem produces the cap and in turn is protected by it. As the root grows, it pushes the root cap forward against soil particles, so that some cells tear loose. The slippery remnants lubricate the cap as it moves through the soil.

A root's epidermis is the absorptive interface with soil. Some epidermal cells send out long extensions called **root hairs** (Figure 20.15). Root hairs greatly increase the surface area available for taking up water and solutes. Gardeners learn never to yank a plant from the ground when transplanting it. Too much of the fragile absorptive surface would be torn off.

Vascular tissues form a **vascular cylinder**, a central column inside the root. Ground tissues surrounding the cylinder are called the root cortex (Figure 20.16). An abundance of spaces in the cortex allows dissolved oxygen to reach living root cells. (Like other cells in the plant, the cells require oxygen for aerobic respiration.) Water entering the root moves from cell to cell until it reaches the endodermis, the innermost part of the root cortex. Endodermis is a sheetlike layer, one cell thick, around the vascular cylinder. Abutting walls of its cells are waterproof, so they force incoming water to pass through the cytoplasm of endodermal cells. As described in Chapter 21, this arrangement helps control the movement of water and dissolved substances into the vascular cylinder.

Just inside the endodermis is the pericycle. This part of the vascular cylinder has one or more layers of cells that give rise to lateral roots, which erupt through the cortex and epidermis (Figure 20.17).

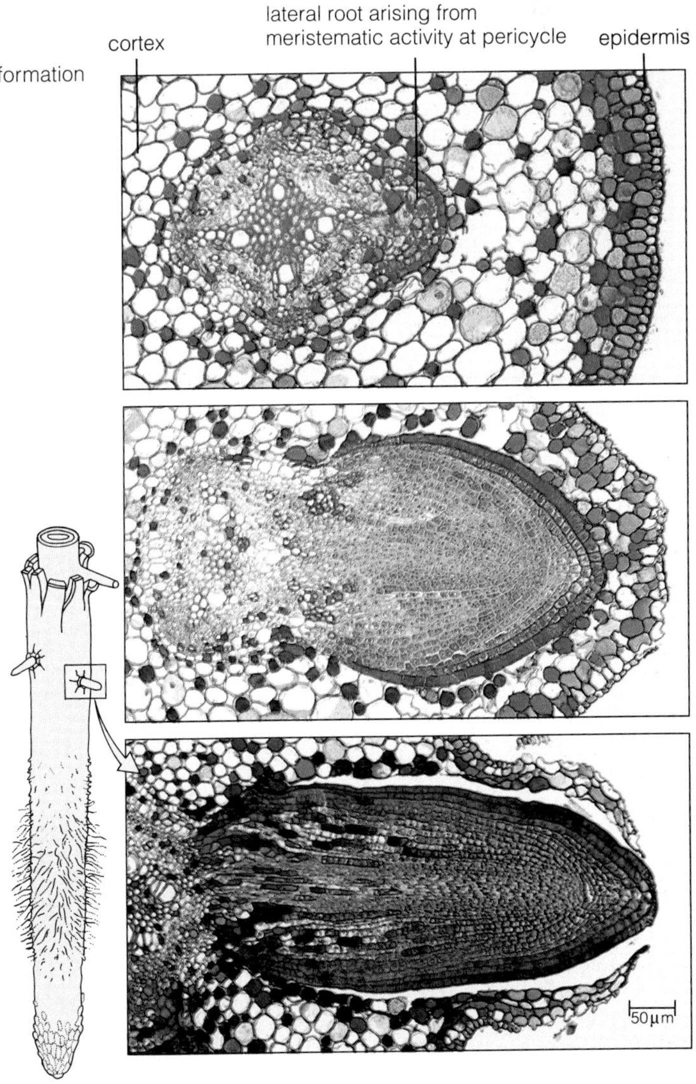

Figure 20.17 Lateral root formation in a willow (*Salix*).

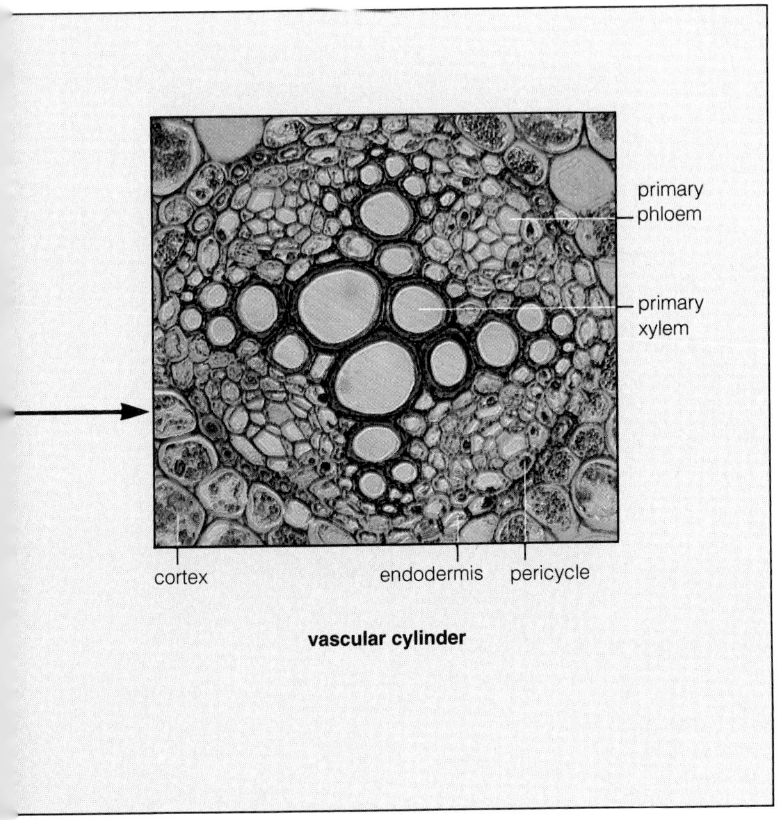

primary phloem

primary xylem

cortex   endodermis   pericycle

vascular cylinder

cortex   lateral root arising from meristematic activity at pericycle   epidermis

50 µm

WOODY PLANTS

Nonwoody and Woody Plants Compared

The life cycles of flowering plants extend from seed germination to seed formation, then eventual death. In that time, most monocots and some dicots show little or no secondary growth; they are nonwoody (herbaceous) plants. By contrast, many dicots (and all gymnosperms) show secondary growth during two or more growing seasons; they are woody plants. In this respect, flowering plants fall into three categories:

annuals   Life cycle is completed in one growing season. There is little or no secondary growth. Examples: corn, marigolds.

biennials   Life cycle is completed in two growing seasons. Roots, stems, and leaves form in the first season; the plant forms seeds and flowers, then dies in the second season. Example: carrots.

perennials   Vegetative growth and seed formation continue year after year. Some have secondary tissues, others do not. Examples: nonwoody cacti, woody shrubs (roses), vines (ivy, grape), and trees (elms, magnolias).

Figure 20.18 shows the structure of a tree trunk—an older, woody stem with extensive secondary growth. Notice how living phloem cells are restricted to a thin zone just beneath the corky surface. Stripping off a band of phloem all the way around a tree's circumference (an activity called girdling) cuts all the vertical phloem pipelines. Photosynthetically derived food can't reach roots, which will die—and so, in time, will the tree.

Growth at Lateral Meristems

How do older stems and roots of plants get massive and woody? Each growing season, new tissues originate at their **lateral meristems**, especially vascular

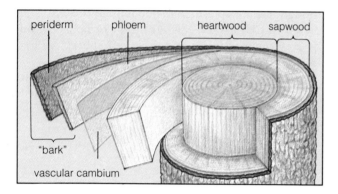

Figure 20.18 Structure of a woody stem showing extensive secondary growth. Heartwood, the mature tree's core, has no living cells. Sapwood is the cylindrical zone of xylem between heartwood and vascular cambium. It contains some living parenchyma cells among nonliving conducting cells of xylem. Everything outside the vascular cambium is often called bark. Everything inside it is called wood.

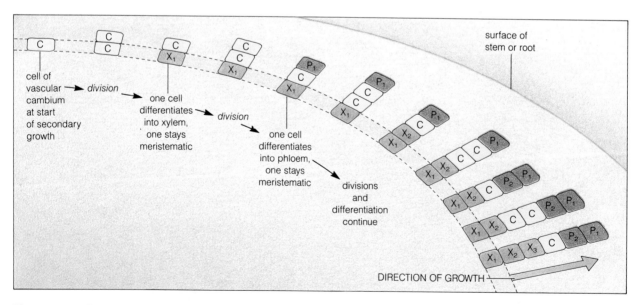

Figure 20.19 Relationship between the vascular cambium and the cells (secondary xylem and phloem) originating from it. This composite drawing depicts growth in a stem through successive seasons. Ongoing divisions displace the vascular cambium, moving its cells steadily outward as the enlarging core of xylem makes the stem or root thicker.

cambium (Figure 20.3). Fully developed vascular cambium is like a cylinder, one cell (or a few cells) thick. Its cells give rise to secondary xylem and phloem that conduct water and dissolved substances up, down, and sideways through the enlarging stem or root. Xylem forms on the inner face of the vascular cambium, and phloem forms on the outer face (Figure 20.19).

The layers of xylem increase season after season. They usually crush the thin-walled phloem cells from the preceding growth period. New phloem cells are added each year, outside the growing core of xylem.

In time, the pressure exerted by the tissues accumulating inside a stem or root ruptures the cortex and outer phloem. Parts of the cortex split away, carrying epidermis with them. But cork cambium forms, and its cells produce periderm—a corky replacement for the lost epidermis. "Cork" is not the same as "bark," which refers to all living and nonliving tissues external to the vascular cambium (Figure 20.18).

Early and Late Wood

In regions having prolonged dry spells or cool winters, vascular cambium becomes inactive during parts of the year. The first xylem cells produced at the start of the growing season tend to have large diameters and thin walls. They form early wood (Figure 20.20). As the season progresses, cell diameters become smaller and their walls thicker. These cells form late wood.

In a full-diameter slice from an old tree trunk, early and late wood reflect light differently. You can identify them as alternating light and dark bands. The alternating bands represent annual growth layers, or "tree rings" (Figure 20.21).

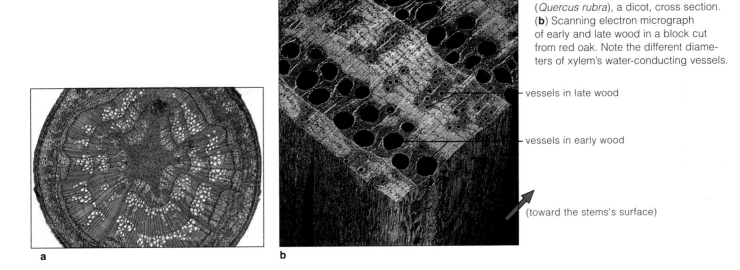

a

b

— vessels in late wood

— vessels in early wood

(toward the stems's surface)

Figure 20.20 (**a**) Stem of red oak (*Quercus rubra*), a dicot, cross section. (**b**) Scanning electron micrograph of early and late wood in a block cut from red oak. Note the different diameters of xylem's water-conducting vessels.

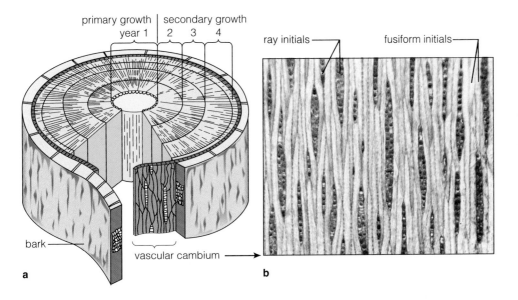

primary growth | secondary growth
year 1 | 2 3 4

ray initials ——   fusiform initials ——

bark

vascular cambium ——

a

b

Figure 20.21 (**a**) Location of vascular cambium in an older stem showing secondary growth. This one has three annual rings, corresponding to secondary growth in years one through three. Some cells of the vascular cambium (the fusiform initials) produce the secondary xylem and phloem that conduct water and food vertically through the stem. Other cells (ray initials) produce parenchyma and other cells that serve as lateral channels for water and food, and as food storage centers. (**b**) Vascular cambium from the trunk of an apple tree (*Malus*).

SUMMARY

1. Monocots and dicots are two classes of flowering plants (angiosperms). Their shoots (stems and leaves) and roots are composed mainly of dermal, ground, and vascular tissue systems.

2. Plant growth originates at meristems, localized regions of self-perpetuating, embryonic cells.

    a. Primary growth (lengthening of stems and roots) originates at apical meristems in root and shoot tips.

    b. Secondary growth (increases in diameter) originates at lateral meristems (vascular cambium and cork cambium) inside stems and roots.

3. Parenchyma, sclerenchyma, and collenchyma are simple tissues that each consist of only one type of cell (Table 20.1).

    a. Parenchyma cells are alive and metabolically active at maturity. Various types of parenchyma function in photosynthesis and other tasks, and they make up the bulk of ground tissue systems. Strands of parenchyma cells also thread through vascular tissues.

    b. Sclerenchyma mechanically supports growing plant parts. Collenchyma supports mature plant parts.

4. Vascular tissues (xylem and phloem) and dermal tissues (epidermis and periderm) are complex tissues, each consisting of two or more cell types (Table 20.1).

    a. Vascular tissues distribute water and dissolved substances through the plant. Typically, strands of conducting cells are bundled together inside sheaths of sclerenchyma and parenchyma. Many such vascular bundles thread through the ground tissues.

    b. Xylem's water-conducting cells are dead at maturity. Their lignified, pitted walls interconnect as pipelines for water and dissolved minerals.

    c. Phloem's conducting cells are alive at maturity. The cytoplasm of adjacent cells interconnects across perforated end walls and side walls. Dissolved sugars and other products of photosynthesis are loaded into these cells in leaves and unloaded in regions where cells are growing or storing food.

    d. Epidermis covers and protects all surfaces of primary plant parts. Periderm replaces epidermis on plants with secondary growth.

5. Stems favorably position photosynthetic cells of leaves to intercept light, and they display flowers. Their vascular bundles distribute substances to and from all plant parts. Most monocot stems have vascular bundles distributed throughout the ground tissue. Most dicot stems have a ring of bundles that divides the ground tissue into cortex and pith.

6. Leaves contain veins and mesophyll (photosynthetic parenchyma) between the upper and lower epidermis.

Table 20.1 Summary of Flowering Plant Tissues and Their Component Cells

Simple Tissues:

Parenchyma	Parenchyma cells
Collenchyma	Collenchyma cells
Sclerenchyma	Fibers or sclereids

Complex Tissues:

Xylem	Conducting cells (tracheids and vessel members); parenchyma cells; sclerenchyma cells
Phloem	Conducting cells (sieve-tube members); parenchyma cells; sclerenchyma cells
Epidermis	Mostly undifferentiated cells; also guard cells, other specialized cells
Periderm	Cork cells; cork cambium cells; parenchyma cells; sclerenchyma cells

Air spaces around the photosynthetic cells enhance gas exchange. Water vapor and gases move across the epidermis through tiny openings (stomata).

7. Roots absorb water and dissolved minerals from the surroundings for distribution to aboveground parts. They anchor the plant. Most store food, and some help support the plant body.

Review Questions

1. Choose a flowering plant and list some functions of its roots and shoots. *327*

2. Name and define the basic functions of a flowering plant's three main tissue systems. *328–329*

3. Describe the differences between:
    a. apical and lateral meristems *328*
    b. parenchyma and sclerenchyma *328–329*
    c. xylem and phloem *329*
    d. epidermis and periderm *329*

4. Which of the following stem sections is typical of most monocots? of most dicots? Label the main tissue regions of each. *330*

5. Label the component parts of this three-year-old tree section. Correctly label the annual growth layers: *339*

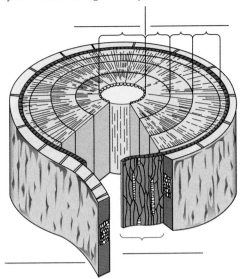

6. What would happen to a tree if you stripped away a band of phloem all the way around its circumference? *338*

Self-Quiz *(Answers in Appendix IV)*

1. _____ and _____ are two classes of flowering plants.
   a. Angiosperms; gymnosperms
   b. Monocots; dicots
   c. Shrubs; trees
   d. Herbs; shrubs

2. Fleshy plant parts consist mostly of _____ cells.
   a. parenchyma
   b. sclerenchyma
   c. collenchyma
   d. epidermal

3. Xylem and phloem are _____ tissues.
   a. ground
   b. vascular
   c. dermal
   d. both b and c

4. _____ conducts water and minerals; _____ conducts food.
   a. Phloem; xylem
   b. Cambium; phloem
   c. Xylem; phloem
   d. Xylem; cambium

5. Roots and shoots lengthen through activity at _____.
   a. apical meristems
   b. lateral meristems
   c. vascular cambium
   d. cork cambium

6. Older roots and stems thicken through activity at _____.
   a. apical meristems
   b. lateral meristems
   c. vascular cambium
   d. both b and c

7. Buds give rise to _____.
   a. leaves
   b. flowers
   c. stems
   d. all of the above

8. Mesophyll consists of _____.
   a. waxes and cutin
   b. lignified cell walls
   c. cork but not bark
   d. photosynthetic parenchyma cells
   e. sclerenchyma cells

9. Early wood forms from cells having _____ diameters and _____ walls.
   a. small; thick
   b. small; thin
   c. large; thick
   d. large; thin

10. Match the terms with the appropriate plant parts.
    _____ apical meristem
    _____ lateral meristem
    _____ xylem, phloem
    _____ periderm
    _____ vascular bundle
    _____ wood

    a. masses of xylem
    b. stems, roots lengthen
    c. corky covering
    d. older roots, stems thicken
    e. conduct water, food
    f. central column in roots

Selected Key Terms

annual *338*
apical meristem *328*
biennial *338*
bud *334*
collenchyma *328*
cork cambium *328*
dermal tissue system *328*
dicot *330*
epidermis *329*
fibrous root system *335*
ground tissue system *328*
lateral meristem *338*
lateral root *335*
leaf *332*
lignin *329*
mesophyll *333*

monocot *330*
parenchyma *328*
perennial *338*
periderm *329*
phloem *329*
root *327*
root hair *337*
sclerenchyma *328*
shoot *327*
taproot system *335*
vascular bundle *330*
vascular cambium *328*
vascular cylinder *337*
vascular tissue system *328*
vein *333*
xylem *329*

Readings

Raven, P.; R. Evert; and S. Eichhorn. 1986. *Biology of Plants.* Fourth edition. New York: Worth. Exquisite color micrographs and illustrations.

Rost, T., et al. 1984. *Botany: An Introduction to Plant Biology.* Second edition. New York: Wiley.

Stehlin, D. October 1990. "Harvesting Drugs from Plants." *FDA Consumer* 24: 20–23.

Stern, K. 1991. *Introductory Plant Biology.* Fifth edition. Dubuque, Iowa: Brown. Beautifully illustrated, accessible. Paperback.

21 PLANT NUTRITION AND TRANSPORT

Flies for Dinner

How often do we think that plants actually do anything impressive? Being mobile, intelligent, and emotional, we tend to be fascinated more with ourselves than with immobile, expressionless plants. Yet plants don't just stand around soaking up sunlight. Consider a flytrap that evolved long before humans ever built one. The Venus flytrap, a native of North and South Carolina, has two-lobed, spine-fringed leaves that open and close much like a steel trap. Like other organisms, this plant can't grow without nitrogen and other essential minerals. Certain minerals happen to be scarce in the soggy soils where the Venus flytrap lives. Ah, but insects flying in from nearby areas are abundant.

Sticky, sugary substances ooze out of leaf glands. The ooze is part bait, part snare. Insects enticed to land on it brush against tiny hairlike projections—triggers for the trap. If an insect touches two hairs at the same time, or even the same hair twice in rapid succession, the leaf snaps shut (Figure 21.1). Then digestive juices pour out from leaf cells, pool around the insect—and dissolve its minerals. The plant, in short, makes its own mineral-rich water!

With this chapter we turn to adaptations by which land plants function in the environment. As you know

already, most plants are photosynthetic autotrophs. Using only sunlight, water, carbon dioxide, and some minerals, they nourish themselves. Yet plants, like people, don't have unlimited supplies of required resources. Of every million molecules of air, only 350 are carbon dioxide. Unlike the soggy home of Venus flytraps, most soils are frequently dry. Nowhere except in overfertilized gardens does soil water hold lavish amounts of dissolved minerals. And this brings us to a key concept: *Many aspects of plant structure and function are responses to low environmental concentrations of essential resources.*

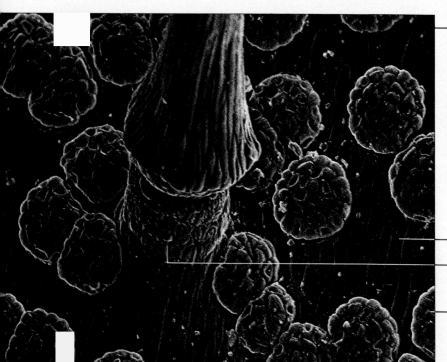

surface of leaf

base of one of the hairlike projections— the triggers for the trap

one of the secretory glands at the leaf surface

Figure 21.1 Venus flytrap (*Diondea muscipula*)—a "carnivorous" plant that turns the dinner table on animals.

NUTRITIONAL REQUIREMENTS

1. Many aspects of plant structure and function are adaptive responses to low concentrations of water, minerals, and other environmental resources.

2. Leaves have passageways (stomata) across their epidermis. Plants lose water during the day, when stomata remain open and so allow carbon dioxide (used in photosynthesis) to move into the leaves. Most plants conserve water by closing the stomata at night.

3. Plants have mechanisms for water transport. The force of dry air causes evaporation from aboveground plant parts (transpiration). This pulls continuous columns of water molecules (which are hydrogen-bonded to one another) from roots to aboveground parts.

4. Sucrose and other organic compounds are distributed throughout the plant by translocation. In this energy-requiring process, organic compounds are loaded into conducting cells of phloem. They are unloaded at actively growing regions or storage regions of the plant body.

It took you eighteen years or so to grow to your present height. A corn plant can grow more than that in three months! Of course, if you or it were deprived of certain elements, neither of you would be all that vigorous.

Sixteen elements keep plants alive and well. Three of these essential elements—oxygen, carbon, and hydrogen—are the main components of carbohydrates, lipids, proteins, and nucleic acids. Plants get them from water (H_2O), and from gaseous oxygen (O_2) and carbon dioxide (CO_2) in the air.

The other essential elements become available to plants as salts dissolved in water; they are **mineral ions**. As Table 21.1 indicates, six are macronutrients, meaning they are present in easily detectable concentrations in plant tissues. Seven are micronutrients; only small traces occur in plant tissues. All mineral ions have vital roles in metabolic events such as photosynthesis and aerobic respiration. Taken as a whole, they also contribute to the solute concentration gradients that help move substances into and out of cells.

Table 21.1 Role of Mineral Elements in Plant Function

Macronutrient	Some Known Functions	Some Deficiency Symptoms	Micronutrient	Some Known Functions	Some Deficiency Symptoms
Nitrogen	Component of proteins, nucleic acids, coenzymes, chlorophylls	Stunted growth; light green older leaves; older leaves yellow and die (chlorosis)	Chlorine	Role in root, shoot growth; role in photolysis	Wilting; chlorosis; some leaves die
Potassium	Activation of enzymes, role in maintaining water-solute balance* and thus affecting osmosis	Reduced growth; curled, mottled, or spotted older leaves; burned leaf margins; weakened roots and stems	Iron	Roles in chlorophyll synthesis, electron transport	Chlorosis; yellow and green striping in grasses
Calcium	Roles in cementing cell walls, regulation of many cell functions	Leaves deformed; terminal buds die; reduced root growth	Boron	Roles in flowering, germination, fruiting, cell division, nitrogen metabolism	Terminal buds, lateral branches die; leaves thicken, curl, become brittle
Magnesium	Component of chlorophylls; activation of enzymes	Chlorosis; drooped leaves	Manganese	Role in chlorophyll synthesis; coenzyme activity	Light green leaves with green major veins; leaves whiten and fall off
Phosphorus	Component of nucleic acids, phospholipids, ATP	Purplish veins in older leaves; fewer seeds and fruits; stunted growth	Zinc	Role in formation of auxin, chloroplasts, and starch; enzyme component	Chlorosis; mottled or bronzed leaves; abnormal roots
Sulfur	Component of most proteins, two vitamins	Light green or yellow leaves; reduced growth	Copper	Component of several enzymes	Chlorosis; dead spots in leaves; stunted growth; terminal buds die
			Molybdenum	Component of enzyme used in nitrogen metabolism	Possible nitrogen deficiency; pale green, rolled or cupped leaves

*All mineral elements contribute to the water-solute balance, but potassium is notable because there is so much of it.

a

b

Figure 21.2 (**a**) Root nodules on a soybean plant. (**b**) Soybean plants growing in nitrogen-poor soil. The seeds that produced the plants on the right were inoculated with symbiotic bacterial cells (*Rhizobium*) and developed root nodules.

UPTAKE OF WATER AND NUTRIENTS

The availability of water and dissolved mineral ions profoundly affects root development—hence the growth of the entire plant. Roots branch out through some parts of the surrounding soil; then they are replaced by roots that branch into different parts as conditions change. It is not that roots "explore" the soil for resources. Rather, areas that have greater concentrations of water and mineral ions stimulate outward growth.

Specialized Absorptive Structures

Certain organisms help many flowering plants take up nutrients—and they get something in return. Often the organisms and the plants spend their entire lives together, locked in intimate dependency in which benefits flow both ways. This mutually beneficial type of interaction is called **symbiosis**.

Consider how many crops suffer from nitrogen scarcity. There actually is plenty of gaseous nitrogen (N≡N) in air. But plants don't have the metabolic means to break apart the three covalent bonds in each molecule, so they can't attach the nitrogen to organic compounds. High crop yields depend on applications of nitrogen compounds in fertilizers or on the activity of "nitrogen-fixing" bacteria in the soil. These bacteria convert gaseous nitrogen to forms that they—and their plant neighbors—can use.

String beans, peas, alfalfa, clover, and other legumes have an advantage in this respect. Nitrogen-fixing bacteria actually reside in **root nodules**, which are localized swellings on plant roots (Figure 21.2). The bacteria withdraw some organic compounds from the plant. But they also provide the plant with usable nitrogen.

Mycorrhizae (singular, mycorrhiza) also are forms of symbiosis. As Chapter 18 described, they are mutually beneficial associations between young roots and a

fungus (hence the name, which means "fungus-root"). In some cases, the fungal filaments (hyphae) coat the roots. Collectively, the hyphae have a large surface area for absorbing mineral ions from a large volume of soil. The fungus gets sugars and nitrogen-containing compounds from the roots. The roots get some scarce minerals that the fungus is better able to absorb.

Finally, the plants themselves have numerous **root hairs** that greatly increase their surface area for absorbing water and nutrients. Root hairs are slender extensions of specialized epidermal cells (Figure 21.3 and page 337). A single root system might develop millions or billions of root hairs.

Root nodules, mycorrhizae, and root hairs help many plants absorb water and scarce mineral ions.

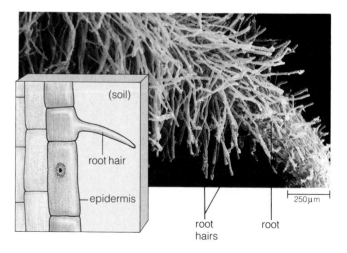

Figure 21.3 Scanning electron micrograph of root hairs on part of a small plant root.

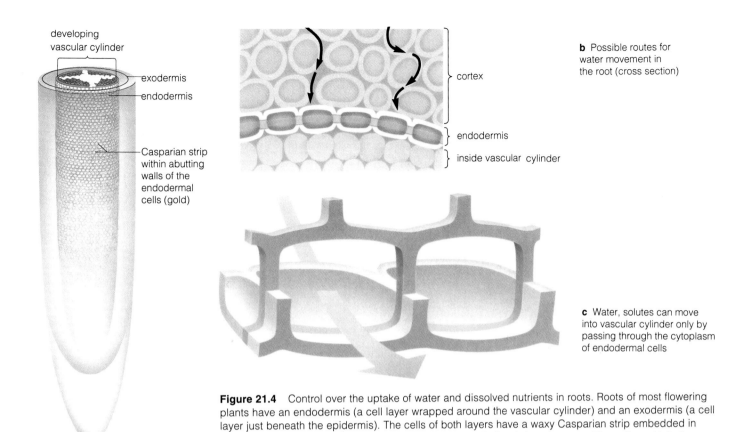

developing
vascular cylinder

exodermis

endodermis

Casparian strip within abutting walls of the endodermal cells (gold)

b Possible routes for water movement in the root (cross section)

cortex

endodermis

inside vascular cylinder

c Water, solutes can move into vascular cylinder only by passing through the cytoplasm of endodermal cells

a Older root

Figure 21.4 Control over the uptake of water and dissolved nutrients in roots. Roots of most flowering plants have an endodermis (a cell layer wrapped around the vascular cylinder) and an exodermis (a cell layer just beneath the epidermis). The cells of both layers have a waxy Casparian strip embedded in their abutting walls. The strip keeps water from moving indiscriminately *around* the cells and into the vascular column. It makes water move *through* the cells. In this way, transport proteins that span the plasma membrane of those cells can selectively control the uptake of water and particular nutrients.

Controlling Nutrient Uptake

Roots don't absorb mineral ions willy-nilly. For example, salts and heavy metals are toxic. They would quickly kill the plant if they were absorbed at the concentrations found in some soils.

Look at Figure 20.16, which shows a root's **vascular cylinder**—a central column of vascular tissue. Wrapped around the column is a sheetlike layer of cells called the **endodermis**. At this layer, a type of waxy band makes abutting walls of neighboring cells impenetrable to water and dissolved nutrients. The band, a **Casparian strip**, stops water from trickling *around* the cells—it has to go *through* their cytoplasm. As Figure 21.4 shows, water moves into the vascular cylinder only by crossing the plasma membrane of endodermal cells, diffusing through their cytoplasm, then crossing the plasma membrane on the other side. Plasma membranes, recall, permit the movement of some solutes but not others across the lipid bilayer. *Transport proteins and other membrane components help control the types of absorbed solutes that will become distributed through the plant.*

Recently, botanists discovered an **exodermis**, a layer of cells just inside the roots of most flowering plants. This layer also has a Casparian strip that functions like the one next to the vascular cylinder.

Once nutrients enter the vascular cylinder, they are distributed to different tissues in coordinated ways that affect growth. Here are the interrelated processes:

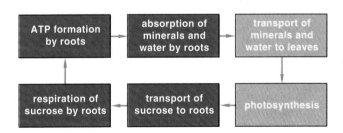

Living cells expend ATP energy and actively transport nutrients at membrane proteins (page 43). In photosynthetic cells, the required ATP forms during photosynthesis and aerobic respiration. In other cells, ATP forms mostly during aerobic respiration.

Plants control the uptake and distribution of nutrients at a cell layer near the root surface, at another layer around the root vascular cylinder, then at the plasma membranes of living cells throughout the plant body.

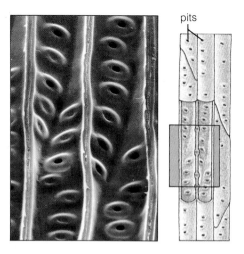

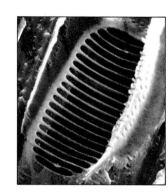

a Part of two tracheids, cut lengthwise. Tracheids have tapering, unperforated end walls. Water does not flow rapidly through the small pits that match up between neighboring tracheids. But air bubbles that might otherwise obstruct water flow through the tubes can be confined to individual tracheids.

b Part of one type of vessel, showing how thick-walled cells (vessel members) interconnect to form a water-conducting tube.

c A different type of vessel member, showing its end wall perforations. In the vessels of phloem, water flows unimpeded from one vessel member to the next. So can air bubbles that might collect and obstruct the flow. This may be why natural selection has favored the retention of tracheids along with vessel members in the same plant.

Figure 21.5 Examples of tracheids and vessel members. These cells are dead at maturity. Their walls remain interconnected and form the water-conducting tubes of xylem. Tracheids probably evolved before vessel members, but both are present in nearly all flowering plants.

WATER TRANSPORT AND CONSERVATION

Transpiration

Let's turn now to the means by which water—and the nutrients dissolved in it—moves from roots to stems, then into leaves. Plants use a fraction of the water in growth and metabolism, but most evaporates into air by the mechanism described on page 23. Water evaporation from stems, leaves, and other plant parts is called **transpiration**.

How does water get all the way to the top of plants, including trees that are 100 meters tall? As Chapter 20 indicated, water travels through conducting pipelines in the vascular tissue called **xylem**. The pipelines are formed by conducting cells called **tracheids** and **vessel members** (Figure 21.5). These cells are dead at maturity and only their walls remain—so they aren't pulling the water "uphill." Instead, *water is pulled up by the drying power of air, which creates continuous negative pressures (tensions) that extend downward from leaves to roots.* That's the idea behind the **cohesion–tension theory of water transport**.

The key points of this theory are illustrated in Figure 21.6 and summarized here:

1. The drying power of air causes transpiration, the evaporation of water from plant parts exposed to air.

2. Transpiration puts the water in xylem in a state of tension that extends from veins in leaves, down through the stems, to roots.

3. As long as water molecules continue to escape from the plant, the continuous tension in the xylem permits more molecules to be pulled up to replace them.

4. Unbroken, fluid columns of water show cohesion; they resist rupturing as they are pulled upward under tension. The collective strength of hydrogen bonds between water molecules, which are confined in the narrow, tubular xylem cells, imparts this cohesion.

5. Hydrogen bonds are enough to hold water molecules together in the xylem. But they are not strong enough to prevent the molecules from breaking away from each other during transpiration and then escaping from leaves.

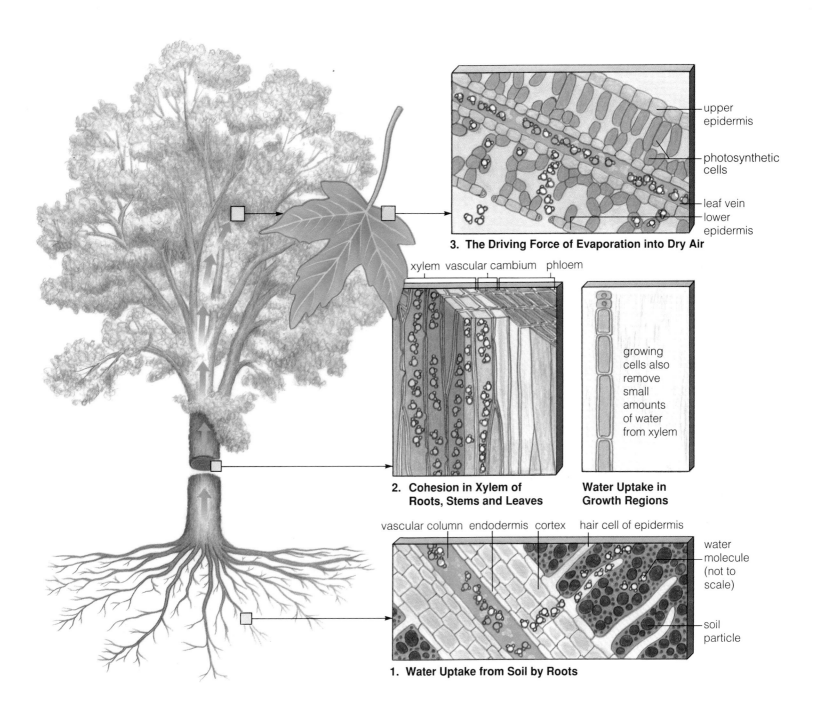

upper
epidermis

photosynthetic
cells

leaf vein

lower
epidermis

3. The Driving Force of Evaporation into Dry Air

xylem vascular cambium phloem

growing
cells also
remove
small
amounts
of water
from xylem

**2. Cohesion in Xylem of
Roots, Stems and Leaves**

**Water Uptake in
Growth Regions**

vascular column endodermis cortex hair cell of epidermis

water
molecule
(not to
scale)

soil
particle

1. Water Uptake from Soil by Roots

Figure 21.6 Cohesion-tension theory of water transport. Tensions in water within the xylem
extend from leaf to root. The tensions result mostly from transpiration (the evaporation of water
from plant parts). They allow columns of water molecules that are hydrogen-bonded to one
another to be pulled upward. Hydrogen bonding gives water its cohesive properties (page 24).

Figure 21.7 Micrograph showing the thick cuticle on the upper epidermis of a kaffir lily leaf.

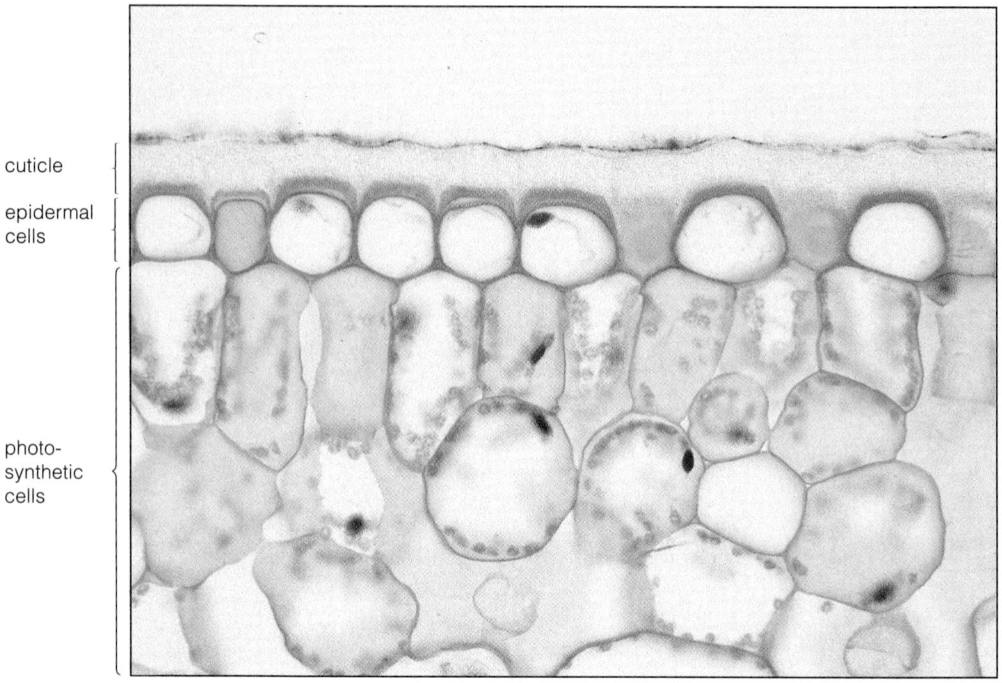

cuticle

epidermal cells

photo-synthetic cells

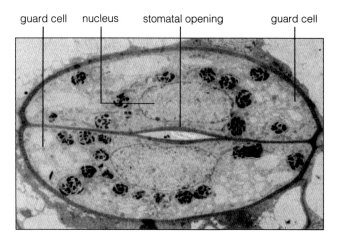

guard cell   nucleus   stomatal opening   guard cell

Figure 21.8 Transmission electron micrograph of the structure of guard cells from the stem of a beavertail cactus (*Opuntia*), thin section.

Control of Water Loss

Of the water moving into a leaf, 90 percent or more is lost by transpiration. Of the water conserved inside it, only 2 percent is used in photosynthesis, membrane functions, and other activities. Yet that tiny amount is vital. When evaporation exceeds water uptake by the roots for very long, plant tissues wilt and water-dependent activities are seriously disrupted.

Even mildly stressed plants would rapidly wilt and die without their **cuticle**. All epidermal cell walls exposed to air have this waxy covering (Figure 21.7). A cuticle reduces the rate of water loss from leaves. It also limits the diffusion rate of carbon dioxide into the leaf.

Still, the cuticle-covered epidermis is peppered with tiny openings called **stomata** (singular, stoma). These are the sites where almost all of the water departs from the plant and carbon dioxide moves in. A pair of **guard cells** defines each opening (Figures 21.8 and 21.9). When the pair swell with water uptake, the force of internal fluid pressure distorts their cell walls. As the guard cells bend under the pressure, a gap (the stoma) forms between them. When they lose water and internal fluid pressure drops, the guard cells collapse against each other and so close the gap.

In most plants, stomata stay open during the day, when photosynthesis takes place. They lose water—but they gain carbon dioxide for the reactions. Stomata stay closed at night. Then, plants conserve water—and precious carbon dioxide accumulates inside the leaf as cells busily engage in aerobic respiration.

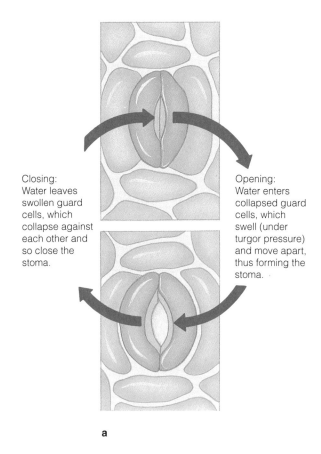

Closing: Water leaves swollen guard cells, which collapse against each other and so close the stoma.

Opening: Water enters collapsed guard cells, which swell (under turgor pressure) and move apart, thus forming the stoma.

a

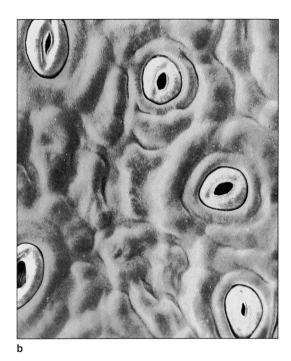

b

c

Whether a stoma opens or closes depends on how much water and carbon dioxide are in the two guard cells. Photosynthesis starts when the sun comes up. As the morning progresses, carbon dioxide levels drop in cells—including guard cells. The decrease helps trigger active transport of potassium ions into the guard cells. So do blue wavelengths of light, which penetrate the atmosphere better as the sun arcs higher in the sky. Water follows potassium into the cells (by osmosis). The inward movement of water provides the fluid pressure to open the stoma.

Carbon dioxide levels rise when the sun goes down and photosynthesis stops. Potassium, then water, moves out of the guard cells—and the stoma closes.

CAM plants, which include most cacti, conserve water in a different way. They open stomata at night, when they fix carbon dioxide by a special metabolic C4 pathway (compare page 82). The cells of CAM plants use carbon dioxide in photosynthesis the following day, when their stomata close.

Figure 21.9 Stomata, small openings that permit water vapor and gases to move across leaf and stem epidermis. (**a**) A stoma remains open or closes, depending on shape changes in the two guard cells that define the opening. (**b**) Surface view of holly leaf stomata. (**c**) What the leaf looks like when a holly plant grows in industrialized regions. Gritty airborne pollutants clog stomata and prevent much of the sun's rays from reaching photosynthetic cells inside the leaf.

By opening and closing stomata at different times, plants control carbon dioxide uptake and water loss.

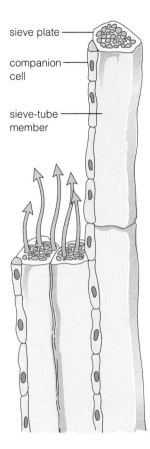

sieve plate

companion cell

sieve-tube member

Figure 21.10 Examples of cells (sieve-tube members) that interconnect to form the conducting tubes of phloem. Companion cells expend energy to load sugars and other organic compounds into the sieve-tube members.

Figure 21.11 Honeydew droplet exuding from the tail end of an aphid feeding on sugars flowing through the phloem.

TRANSPORT OF ORGANIC SUBSTANCES

Storage and Transport Forms of Organic Compounds

What happens to sucrose and other organic products of photosynthesis? Leaf cells use some; the rest end up in roots, stems, buds, flowers, and fruits. Carbohydrates become stored as starch in most plant cells. Proteins and fats become stored in many seeds. Some fruits (avocados especially) also accumulate fats.

Starch molecules are too large for transport across the plasma membrane of cells that produce them. They also are too insoluble for transport to other plant parts. Fats (which are mostly insoluble) and protein reserves can't be transported away from storage sites. Cells convert storage forms of organic compounds to smaller solutes that are more readily transported. For instance, hydrolysis of starch liberates glucose units, which combine with fructose to form sucrose. Sucrose is the main transport form for sugars.

Storage starch, fats, and proteins are converted to smaller units that are soluble and transportable through the plant.

Translocation

Translocation distributes organic compounds through plants in the vascular tissue called **phloem**. Like xylem, this tissue consists of conducting tubes, fibers, and strands of parenchyma cells. Unlike xylem, the tubes are composed of living cells positioned side by side and end to end. These cells are **sieve-tube members**. Dissolved organic compounds flow rapidly through large pores in their abutting end walls (Figure 21.10). Specialized parenchyma cells, called **companion cells**, adjacent to the sieve tubes have supporting roles in translocation.

The pressure that forces sucrose and other organic compounds to flow through sieve tubes is high—often five times as high as in automobile tires. The small insects called aphids demonstrate this when they force a mouthpart into sieve tubes and feed on the dissolved sugars inside. Pressure can force the fluid through the aphid gut and out the other end as "honeydew" (Figure 21.11). Park your car under trees being attacked by aphids and it might get a spattering of sticky honeydew droplets, thanks to the high fluid pressure.

Pressure Flow Theory

Organic compounds flow along gradients of decreasing solute concentration and pressure. The flow's source is any region where organic compounds are being loaded into sieve tubes. Photosynthetic tissues in leaves are a common source. The flow ends at a "sink"—any region where organic compounds are being unloaded and used or stored. Young, growing flowers or developing fruits are common sink regions.

Why do organic compounds flow from a source to a sink? According to the **pressure flow theory**, pressure builds at the source end of a sieve tube system and *pushes* the solute-rich solution toward a sink, where solutes are removed.

Figure 21.12 follows what happens after sucrose moves from photosynthetic cells into small veins in leaves. There, companion cells expend energy to load sucrose into adjacent sieve-tube members. As the sucrose concentration increases in the tubes, water also moves in, by osmosis. More internal fluid pressure is exerted against the tube walls. As the pressure increases, it pushes the sucrose-laden fluid out of the leaf, into the stem, and on to the sink.

Organic compounds are distributed through the plant body in response to concentration and pressure gradients in the sieve-tube system of phloem.

The gradients exist as long as companion cells supply energy to load compounds into sieve tubes at sources (such as mature leaves) and as long as compounds are removed at sinks (such as roots).

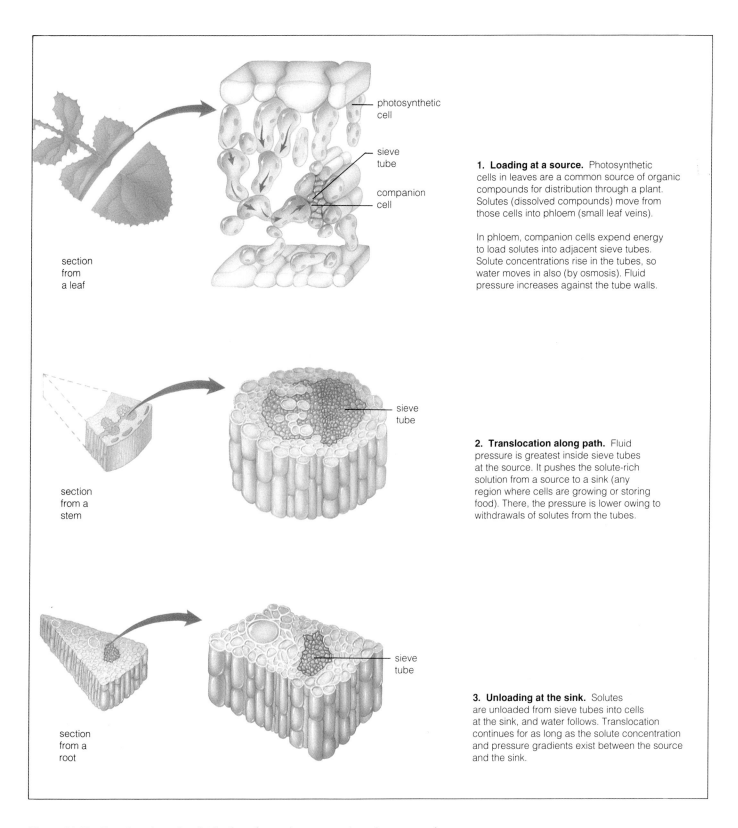

photosynthetic cell

sieve tube

companion cell

section from a leaf

1. Loading at a source. Photosynthetic cells in leaves are a common source of organic compounds for distribution through a plant. Solutes (dissolved compounds) move from those cells into phloem (small leaf veins).

In phloem, companion cells expend energy to load solutes into adjacent sieve tubes. Solute concentrations rise in the tubes, so water moves in also (by osmosis). Fluid pressure increases against the tube walls.

sieve tube

section from a stem

2. Translocation along path. Fluid pressure is greatest inside sieve tubes at the source. It pushes the solute-rich solution from a source to a sink (any region where cells are growing or storing food). There, the pressure is lower owing to withdrawals of solutes from the tubes.

sieve tube

section from a root

3. Unloading at the sink. Solutes are unloaded from sieve tubes into cells at the sink, and water follows. Translocation continues for as long as the solute concentration and pressure gradients exist between the source and the sink.

Figure 21.12 Translocation—the distribution of organic compounds under pressure in phloem. The plant parts come from *Sonchus*, commonly called sow thistle. Compare these illustrations with Figure 5.2 to see how translocation is linked with the process of photosynthesis.

SUMMARY

1. Plants require oxygen, carbon, hydrogen, and thirteen essential nutrients.

2. The large surface area of root systems favors the uptake of water and nutrients, which often are present in low concentrations in the surrounding soil.

3. These are the key points of the cohesion-tension theory of water transport:

   a. In water-conducting cells of xylem, continuous negative pressures (tensions) extend from leaves to roots. Transpiration (evaporation of water from leaves and other parts exposed to air) causes the tension.

   b. When water molecules escape from leaves, replacements are pulled under tension to the sites.

   c. The collective strength of hydrogen bonds between water molecules imparts cohesion that allows the water to be pulled upward as continuous fluid columns.

4. Water cannot cross a plant's waxy cuticle, which covers most aboveground parts. Plants transpire and take up carbon dioxide mostly at stomata. These are small openings across leaf (and stem) epidermis.

5. Stomata commonly are open during the day. Then, plants lose water but take in carbon dioxide for photosynthesis. Stomata close at night, conserving water and carbon dioxide produced by aerobically respiring cells. The controlled opening and closing of stomata balance carbon dioxide uptake with water conservation.

6. Translocation distributes organic compounds through plants; it occurs in sieve tubes of phloem.

7. According to the pressure flow theory, translocation is driven by differences in solute concentrations and pressure between a source (where solutes are loaded into sieve tubes) and a sink (where they are unloaded from them). The differences exist as long as companion cells expend energy to load solutes into the sieve tubes and as long as solutes are removed at the sink.

Review Questions

1. Refer to Table 21.1. What are some signs that a plant suffers a deficiency in certain mineral elements? *343*

2. When moving a plant from one place to another, it helps to include some native soil around the roots. Explain why, given what you know about mycorrhizae and root hairs. *344*

3. What is the function of a Casparian strip in roots? *345*

4. Most plants would die if their stomata remained open all the time or closed all the time. Explain why. *348*

5. Describe the pressure flow theory as a way of explaining translocation. *351*

Self-Quiz (Answers in Appendix IV)

1. Water can be pulled up through a plant due to the cumulative strength of _____ between water molecules.

2. Water leaves and carbon dioxide enters a plant through _____ (tiny openings across leaf epidermis).

3. In plants, mineral ions _____ .
   a. have roles in metabolism
   b. help establish gradients across membranes
   c. influence water movement into cells
   d. maintain cell shape and growth
   e. all of the above

4. Water evaporation from plant parts is called _____ .
   a. translocation   c. transpiration
   b. expiration   d. tension

5. During the day, most plants lose _____ and take up _____ .
   a. carbon dioxide; water   c. oxygen; water
   b. water; oxygen   d. water; carbon dioxide

6. At night, plants conserve _____ and restrict _____ intake.
   a. carbon dioxide; water   c. oxygen; water
   b. water; oxygen   d. water; carbon dioxide

7. In phloem, organic compounds flow through _____ .
   a. companion cells   c. xylem
   b. sieve tubes   d. both a and b

8. Match the concepts of plant nutrition and transport.
   _____ stomata   a. evaporation from plant parts
   _____ sink   b. balancing water loss with
   _____ root system   carbon dioxide requirements
   _____ hydrogen bonds   c. cohesion in water transport
   _____ transpiration   d. sugars unloaded from sieve tubes
   _____ translocation   e. distribution of organic
                              compounds through the plant
                           f. response to scarce soil nutrients

Selected Key Terms

Casparian strip *345*
cohesion-tension theory
   of water transport *346*
companion cell *350*
cuticle *348*
endodermis *345*
exodermis *345*
guard cell *348*

mineral ions *343*
mycorrhiza *344*
phloem *350*
pressure flow
   theory *352*
root hair *344*
root nodule *344*
sieve-tube
   member *350*

stoma (stomata) *348*
symbiosis *344*
tracheid *346*
translocation *350*
transpiration *346*
vascular cylinder *345*
vessel member *346*
xylem *346*

Readings

Epstein, E. 1973. "Roots." *Scientific American* 228(5):48–58.

Galston, A.; P. Davies; and R. Satter. 1980. *The Life of a Green Plant.* Englewood Cliffs, New Jersey: Prentice-Hall.

Salisbury, F., and C. Ross. 1991. *Plant Physiology.* Fourth edition. Belmont, California: Wadsworth.

22 PLANT REPRODUCTION AND DEVELOPMENT

Chocolate From the Tree's Point of View

When tax collectors came around in ancient Mexico, they liked to be paid in seeds from the cacao tree (*Theobroma cacao*). Tax payments went straight to the king, Montezuma, and from him to the royal kitchen. Fermented, roasted, and ground to a paste, the seeds were the basis of chocolatl—a rich drink spiced with hot chili peppers, cinnamon, and vanilla. Montezuma sipped chocolatl from gold goblets after meals, during religious rites, and reportedly before visits to his harem. Later the goblets were ceremonially tossed into a lake. In 1519 Hernando Cortez dethroned Montezuma, drained the lake, and recovered the goblets. These he shipped to Spain, along with three chests filled with cacao seeds—and so began a worldwide fascination with chocolate.

T. cacao evolved in Central America, where wild animals ate its fruits and dispersed its seeds. Today its trees flourish on vast plantations in Central America,

the West Indies, and West Africa. Growers harvest its seeds, or "cacao beans." These are processed into cocoa butter and essences, which end up in chocolate products (Figure 22.1). Unknown interactions among the 1,000 or so compounds in chocolate exert compelling (some say addictive) effects on the human brain. Each year, for instance, the average American feels compelled to buy 8 to 10 pounds of the stuff. With that kind of demand, growers do their very best to keep cacao trees growing and reproducing.

And so, through interactions with humans, *T. cacao* is like orange trees, corn, and other domesticated crop plants that now reproduce in great numbers, in many different parts of the world. In this respect it is like flowering plants in the wild that interact with bees, birds, beetles, bats, and other animals in ways that promote reproductive success. And in evolutionary terms, *reproductive success* is what life is all about.

Figure 22.1 A cacao tree (*Theobroma cacao*) in its native habitat. Unlike most flowering trees, its flowers grow from buds on the trunk. After eggs in the female floral parts are fertilized, seed-filled fruits develop. Cravers of chocolate—a product derived from the seeds—have indirectly contributed to this plant's widespread distribution, far beyond what it could have accomplished on its own.

1. Sexual reproduction of flowering plants requires spores and gametes, which develop in flowers. Commonly it depends on animals that help pollinate the flowers and disperse the seeds.

2. Male reproductive parts of flowers produce microspores. These develop into pollen grains, which give rise to sperm-bearing male gametophytes. Pollen grains are released, then carried by wind, water, or animals to the female reproductive parts of flowers.

3. Female reproductive parts of flowers produce megaspores. These give rise to egg-producing female gametophytes. The female gametophytes remain embedded within and are nourished by tissues of the parent plant.

4. After sperm fertilize the eggs, seeds develop. Each seed is an embryo sporophyte along with tissues that function in its nutrition, protection, and dispersal.

5. From the time a seed germinates, hormones influence a plant's growth and development. Seasonal changes and other environmental cues trigger hormone secretions from its cells.

REPRODUCTIVE MODES

Although you probably don't think about this very often, flowering plants engage in sex. Like humans, they have splendid reproductive systems that produce, protect, and nourish sperm and eggs. Like human females, flowering plants house embryos during early development. Flowers serve as invitations to third parties—pollinators that help get sperm and egg together. Long before humans ever thought of it, flowering plants were using tantalizing colors and fragrances that improve the odds for sexual success.

Most plants also do something humans cannot do, at least not yet. They can reproduce asexually. In *sexual* reproduction, a fertilized egg has two sets of genetic instructions (from two gametes). *Asexual* reproduction occurs by way of mitosis, so all the new plants are genetically identical to the parent (they form a clone).

When we hear the word "plant," we often think of something like a cherry tree (Figure 22.2). This is an example of a **sporophyte**, a vegetative body that grows (by mitosis) from a fertilized egg. Sporophytes bear reproductive shoots called **flowers**. Flowers produce haploid spores, which develop into tiny, haploid bodies called **gametophytes**. Sperm form in male gametophytes, and eggs form in female gametophytes.

Sexual reproduction dominates most flowering plant life cycles, and it will be our focus here. But bear in mind, sporophytes also reproduce asexually—as

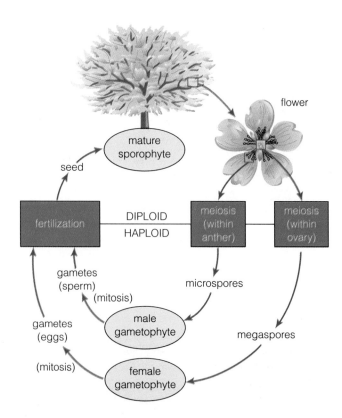

Figure 22.2 Overview of flowering plant life cycles.

when new plants grow along runners (aboveground stems) of strawberry plants or from underground stems of onions, lilies, and Bermuda grass. Many crop plants also are propagated asexually. For example, orchard pear trees are grown from cuttings or buds of a parent tree. Southern California's navel-orange industry is a huge clone from a single tree that grew in Riverside.

GAMETE FORMATION IN FLOWERS

Floral Structure

As a flower develops at a shoot tip, it differentiates into nonfertile components (sepals and petals) and fertile components (stamens and carpels). Directly or indirectly, all of these are attached to a receptacle, the modified base of the floral shoot (Figure 22.3).

Like leaves, the sepals and petals have ground tissues, a vascular system, and epidermis. What makes many flowers sweet smelling? Epidermal cells of their petals contain fragrant oils. What gives petals their color and shimmer? Various cells of their ground tissues contain pigments, such as carotenoids (yellow to red-orange) and anthocyanins (red to blue), and tiny, light-refracting crystals. In what ways do the fragrances, colors, patterns, and arrangements of petals function? As you will see, they attract pollinators.

Peel open a rosebud. Notice how its sepals enclose petals, which in turn enclose its fertile components. This is a common arrangement. Closest to the petals are **stamens**, the male reproductive parts. Nearly all stamens consist of an anther and a single-veined stalk (filament). An anther is internally divided into pollen sacs, the chambers in which walled spores develop into haploid, sperm-producing bodies called **pollen grains** (Figure 22.4). Wind, water currents, or animals carry the pollen grains to the female reproductive parts.

Female reproductive parts are at the center of a flower. These parts are sometimes called pistils, but their more recent name is **carpels**. Some flowers have a single carpel. Others have more, often fused together as a compound structure. The lower portion of single or fused carpels is the **ovary**, where eggs develop, fertilization takes place, and seeds mature. The upper portion is the **stigma**, a sticky or hairy surface tissue that "captures" pollen grains and favors their germination. Commonly, the stigma is elevated by a slender, upward extension of the ovary's wall (a style).

Not all flowers have stamens *and* carpels. The ones with both male and female parts are called "perfect" flowers. But many plants produce "imperfect" flowers, with male *or* female parts. In some species, such as oaks, the same plant bears male and female flowers. In willows and other species, they are on separate plants.

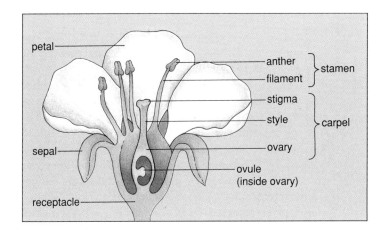

Figure 22.3 Structure of a cherry (*Prunus*) blossom. Like the flowers of many plants, it has a single carpel (a female reproductive part) and stamens (male reproductive parts). Flowers of other plants have two or more carpels, often united as a single structure. Single or fused carpels consist of an ovary and a stigma. Often the ovary extends upward as a slender column (style).

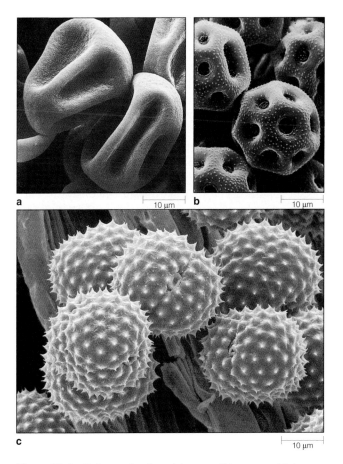

Figure 22.4 Pollen grains from (**a**) grass, (**b**) rose, and (**c**) ragweed plants. Pollen grains of most families of plants differ in size, wall sculpturing, and number of wall pores.

From Microspores to Pollen Grains

We turn now to the formation of pollen grains, of the sort shown in Figure 22.4. While anthers grow, four masses of spore-producing cells form by mitotic cell divisions, and walls develop around them. The anther now has four chambers, called pollen sacs. As shown in Figure 22.5a, cells in each pollen sac undergo meiosis and cytoplasmic division. This results in haploid spores, called **microspores**, each of which becomes encased in a sculpted wall. The walled microspores divide once or twice by mitosis and become pollen grains, which enter a period of arrested growth. In time, they will be released from the anther, and their wall components will protect them from decomposer organisms and assorted chemicals in the environment.

As soon as they form, many types of pollen grains produce sperm nuclei, which are the male gametes of flowering plants. Other types don't do this until after they travel to a carpel and start growing toward its ovule. Thus, a pollen grain is a mature or immature male gametophyte, depending on the plant species.

From Megaspores to Eggs

Meanwhile, one or more dome-shaped cell masses have been developing on the inner wall of a flower's ovary. Each mass is the start of an **ovule** that, if all goes well, will become a seed. As each domed mass grows, a tissue forms inside it, and one or two protective layers (integuments) form around it. Inside the mass, a cell divides by meiosis. Four haploid spores form. Spores that form in flowering plant ovaries are called **megaspores**.

Commonly, all but one megaspore disintegrate. The one remaining undergoes mitosis three times *without* cytoplasmic division. So at first it's a cell with eight nuclei (Figure 22.5b). Its cytoplasm divides after each nucleus migrates to a specific location. The result is a seven-celled embryo sac, the female gametophyte. One of those cells, the "endosperm mother cell," has two nuclei. It will help form **endosperm**, a nutritive tissue for the forthcoming embryo. Another cell is the egg.

FROM POLLINATION TO FERTILIZATION

Each spring, flowering plants release pollen. You are acutely aware of this reproductive event if you are one of the millions of people who suffer from *hay fever*. This is an allergic reaction to wall proteins of pollen grains released by ragweed and many other plants.

Pollination refers to the transfer of pollen grains to a receptive stigma. Air currents, water currents, insects, birds, or other pollinating agents make the transfer. The relationship between flowering plants and their pollina-

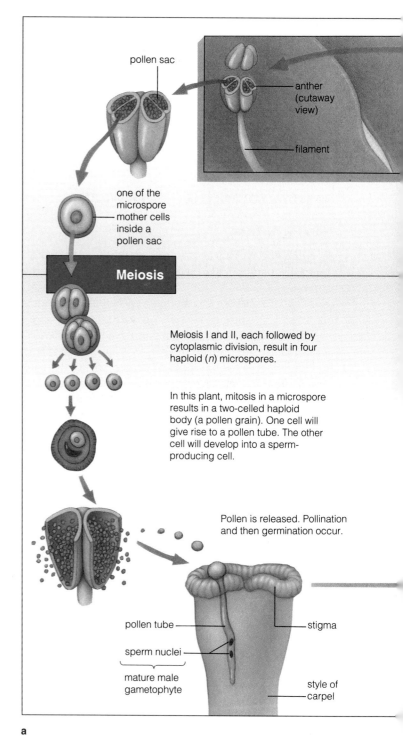

a

tors is one of the most intriguing of all evolutionary stories. It is the topic of the *Commentary* on page 358.

Once a pollen grain lands on a receptive stigma, it germinates. In this case, germination means that the pollen grain resumes growth and develops into a multicelled, tubular structure. This "pollen tube" starts burrowing through tissues of the ovary, carrying sperm nuclei with it (Figure 22.5b). Chemical and molecular cues guide a pollen tube's growth through tissues of the ovary, toward the egg chamber and sexual destiny.

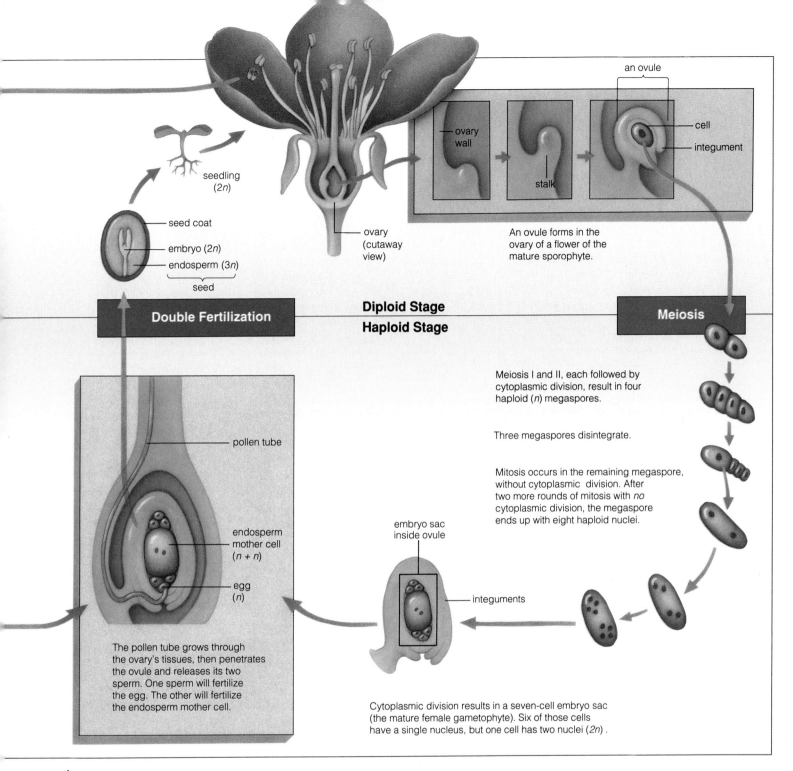

seedling
(2n)

seed coat
embryo (2n)
endosperm (3n)
seed

ovary
(cutaway
view)

an ovule

ovary
wall

stalk

cell
integument

An ovule forms in the
ovary of a flower of the
mature sporophyte.

Double Fertilization

Diploid Stage

Haploid Stage

Meiosis

pollen tube

endosperm
mother cell
(n + n)

egg
(n)

The pollen tube grows through
the ovary's tissues, then penetrates
the ovule and releases its two
sperm. One sperm will fertilize
the egg. The other will fertilize
the endosperm mother cell.

embryo sac
inside ovule

integuments

Meiosis I and II, each followed by
cytoplasmic division, result in four
haploid (n) megaspores.

Three megaspores disintegrate.

Mitosis occurs in the remaining megaspore,
without cytoplasmic division. After
two more rounds of mitosis with *no*
cytoplasmic division, the megaspore
ends up with eight haploid nuclei.

Cytoplasmic division results in a seven-cell embryo sac
(the mature female gametophyte). Six of those cells
have a single nucleus, but one cell has two nuclei (2n).

b

When the tube reaches an ovule, it penetrates the embryo
sac, its tip ruptures, and the two sperm are released.

"Fertilization" generally means the fusion of a
sperm nucleus and an egg nucleus. **Double fertilization** occurs in flowering plants alone. In diploid
species, one sperm nucleus fuses with that of the egg,
producing a diploid (2n) zygote. Meanwhile, the other
sperm nucleus fuses with both nuclei of the endosperm
mother cell, forming a cell with a triploid (3n) nucleus.
That 3n cell gives rise to endosperm, a nutritive tissue.

Figure 22.5 Life cycle of cherry (*Prunus*), one of the flowering
plants classified as dicots. (**a**) This part of the diagram shows
how pollen grains (the male gametophytes) develop and
germinate. (**b**) This part shows what goes on inside the single
ovule in the cherry flower's ovary. An egg forms within an embryo
sac (the female gametophyte). After double fertilization, an
embryo sporophyte and nutritive tissue develop and become
encased in a seed coat. Figure 18.23 is a life cycle for a
monocot (*Lilium*).

Flowering Plants and Pollinators—A Coevolutionary Tale

Water lilies. Arctic lupine. A towering saguaro cactus. Wonderfully different flowering plants live almost everywhere, from icy tundra to deserts, ponds, even in the seas. Few organisms match their distribution and diversity. How did the plants manage it?

We can piece together a plausible answer, based on the fossil record and observations of existing species. About 500 million years ago, when plants were first invading the land, insects that feasted on decaying plant parts and spores were probably right behind them. In no time at all, taller plants with cuticle-covered, lignified stems appeared, as did cuticle-covered insects adapted to drier life-styles. The smorgasbord of lofty, tough-stemmed plants seemed to have favored natural selection of winged insects with an amazing assortment of sucking, piercing, and chewing mouthparts.

By 390 million years ago, the first *seed*-bearing plants were flourishing in humid coastal forests. The ancestors of existing gymnosperms and flowering plants were among them. Many species had separate conelike structures for pollen sacs and ovules. Pollen may have reached the ovules simply by drifting on air currents.

Pollen happens to be rich in proteins. At some point, insects made the connection between "cone" and "food source." *They started serving as pollinating agents.* The plants lost some dustlike pollen to insect appetites—but they gained a major reproductive advantage when pollen-dusted insects brushed up against ovules. To be sure, insects clambering over the cones weren't precision pollinators. But they were more accurate at delivering pollen grains than air currents alone. Fewer pollen grains could be produced and yet still be delivered right to the door, so to speak. The tastier the pollen, the more home deliveries, and the more seeds formed. *And the greater the number of seeds formed, the greater the chance of reproductive success.*

What we are describing is **coevolution**. The word refers to two (or more) species jointly evolving as an outcome of their close ecological interactions. When one species evolves, the change affects selection pressures operating between the two, so the other also evolves.

In our coevolutionary tale, there was selection of more accurate pollen-delivering insects—and of plant structures attractive to them. Pollinators quick to recognize and locate particular plants had a competitive edge. So did plants with come-hither advertisements—fragrances and flowers.

Advertising also attracted pollen-eating beetles, which chewed on the ovules they pollinated. Chewing behavior, too, may have been a selective force in the evolution of flowers.

Today we can correlate many floral features with specific pollinators. Red petals of the flower in Figure *a* form a tube for a tempting, sugar-rich fluid—nectar. Beetles and honeybees don't respond to red colors, and they also can drown in the nectar of flowers with large, deep tubes. Birds do respond to flowers with red as well as yellow components. And certain birds have a bill as long as the floral tube. This flower's anthers and stigma are located where birds brush against them. The birds visit nectar cups of many plants of this species and so promote cross-pollination.

Bird-pollinated plants don't spend energy producing potent perfumes (birds have a poor sense of smell). Flowers pollinated by beetles (and flies) smell like decaying meat or moist dung. Their odors resemble the

a Bahama woodstar, sipping nectar from a hibiscus blossom. Like other hummingbirds, it can forage for nectar in midflight. Its long, narrow bill coevolved with long, narrow floral tubes.

smells of decaying matter in forest litter, where beetles first evolved.

Flowers commonly pollinated by bees have strong, sweet odors and bright yellow, blue, and purple parts. They also contain pigments that absorb ultraviolet light. The distribution of these floral pigments creates patterns of reflected ultraviolet light that honeybees find alluring. Unlike us, bees can see ultraviolet light (Figure *b*).

A daisy's yellow center serves another function. This tight, flattened clump of tiny flowers (florets) is a landing platform for pollen-gathering insects. As the insects bustle about, they transfer pollen from floret to floret. Landing platforms of some flowers strategically position bee bodies so they brush against pollen-laden anthers (Figure *c*).

Butterflies forage by day. They are attracted to fragrant flowers with distinctive patterns and shapes, such as upright daisies with horizontal landing platforms. Some also are attracted to flowers with red and orange components. Most moths forage by night. They pollinate strong, sweet-smelling flowers with white or pale-colored petals, which are more visible in the dark (Figure *d*). Butterflies and moths have long, narrow mouthparts, corresponding to narrow floral tubes or spurs. The uncoiled mouthpart of a Madagascar hawkmoth is 22 centimeters long—the same length as the floral tube of an orchid (*Angraecum sesquipedale*)! Like hummingbirds, hawkmoths don't use landing platforms; they hover near the floral tubes as they draw nectar from them.

b To human eyes, a marsh marigold flower is solid yellow (*left*). To bee eyes, the flower looks something like the photograph to the right, taken with film that is sensitive to ultraviolet light.

d Stephanotis, a night-flowering plant, has no color pattern. Its strong scent and white petals (which reflect more light at night than colors do) attract moth pollinators.

c The landing platform of scotch broom corresponds to the size and shape of bees. A bee's weight forces the platform's petals apart. Pollen-laden stamens, positioned to whap the bee, are released and dust the bee with pollen. Before returning to its hive, the bee grooms itself and packs pollen (the orange mass) inside "baskets" of leg hairs.

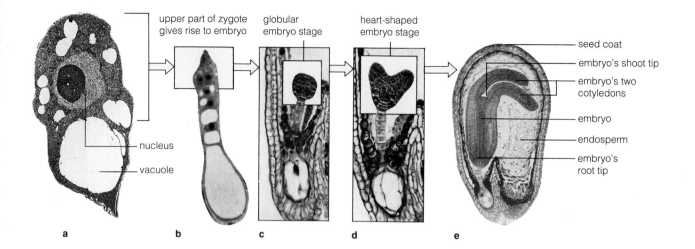

upper part of zygote gives rise to embryo

globular embryo stage

heart-shaped embryo stage

nucleus

vacuole

seed coat

embryo's shoot tip

embryo's two cotyledons

embryo

endosperm

embryo's root tip

a   b   c   d   e

EARLY DEVELOPMENT

Formation of the Embryo Sporophyte

Following fertilization, the newly formed zygote embarks on a course of mitotic cell divisions that will lead to a mature embryo sporophyte. Take a look at Figure 22.6, which shows how an embryo forms in *Capsella*, a dicot. Early on, cell divisions produce a single row of cells. The lower cells develop into a stalklike structure that anchors the embryo and absorbs nutrients from the endosperm for it. The cells above it develop into the embryo proper. By the time the embryo reaches the stage shown in Figure 22.6*e*, its primary meristems have formed and two cotyledons have started to develop from two tissue lobes.

Cotyledons, or "seed leaves," develop as part of all flowering plant embryos. Dicot embryos have two; monocot embryos have one. In many plants, including peas and beans, large cotyledons absorb the endosperm and function in food storage. In wheat, corn, and other plants, thin cotyledons may produce enzymes for transferring stored food from the endosperm to the germinating seedling.

Seeds and Fruits

From the time a zygote forms until a mature embryo has developed, the parent plant transfers nutrients to tissues of the ovule. Food reserves accumulate in the expanding endosperm or cotyledons. Eventually, the connection between the ovule and the ovary wall separates. The ovule's integuments thicken and harden into a seed coat. The embryo and food reserves are now a self-contained package. The ovule has become a **seed**.

While seeds are forming, changes also occur in other parts of the flower. The ovary itself expands in size and its tissues become modified. What we call a **fruit** is a mature ovary, with or without other floral structures

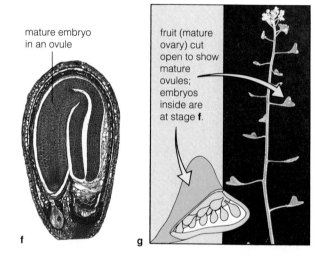

mature embryo in an ovule

fruit (mature ovary) cut open to show mature ovules; embryos inside are at stage **f**.

f   g

Figure 22.6 Stages in the development of shepherd's purse (*Capsella*), a dicot. The micrographs are not to the same scale. (**a**) The single-celled zygote. (**b–d**) The early embryo is identified by the yellow boxes. The row of cells below it transfers nutrients to the developing embryo from the parent plant. The embryo in (**e**) is well developed; the one in (**f**) is mature.

that have become incorporated into it. Different types of fruit are listed in Table 22.1, and Figure 22.7 shows some examples. Many fruits, including apples, are juicy and fleshy. Others, such as grains and nuts, are dry. Still others, including pineapples, are a multiple fruit, formed from clusters of many flowers.

Fruits promote seed dispersal from the parent plant. Like flowering plants and their pollinators, fruits have coevolved with wind, water, and animals that assist their dispersal in specific environments. Look at the winglike extensions of maple fruits in Figure 22.7*e*. When the fruit drops, its wings spin it sideways. The wings may whirl the fruit far enough away that its enclosed embryo won't have to compete with the parent plant for soil water, minerals, and sunlight. Other

a Apple blossoms

b Petals fallen away

d floral remnants — ovaries

e seed (in carpel) — wing

c Enlarging ovaries — ovary

Figure 22.7 A few splendid fruits. (**a–c**) Fruit formation on an apple (*Malus*) tree. Petals dropping from the flower usually mean fertilization has occurred. After this, the ovary and receptacle expand. The sepals and stamens are still visible on the immature fruit. (**d**) Multiple fruit of a pineapple plant. (**e**) Dry fruits of a maple tree.

fruits taxi to new locations by adhering to feathers or fur with hooks, spines, hairs, and sticky surfaces. Embryos inside strawberries, cherries, and other fleshy fruits survive being eaten and assaulted by digestive enzymes in the animal gut. Besides digesting a fruit's flesh, the enzymes digest some of the seed coat. This will help the embryo break through the hard coat after seeds are expelled from the animal's body.

A mature ovule, which encases an embryo sporophyte and food reserves in a hardened coat, is a seed.

A mature ovary, with or without additional floral parts that have become incorporated into it, is a fruit.

Table 22.1 Kinds of Fruits of Some Flowering Plants

Type	Characteristics	Some Examples
Simple (formed from single carpel, or two or more united carpels of one flower)	1. Fruit wall *dry; split* at maturity	Pea, magnolia, tulip, mustard
	2. Fruit wall *dry; intact* at maturity	Sunflower, wheat, rice, maple
	3. Fruit wall *fleshy*, sometimes with leathery skin	Grape, banana, lemon, cherry, orange
Aggregate (formed from numerous but separate carpels of single flower)	*Aggregate* (cluster) of matured ovaries (fruits), all attached to receptacle (modified stem end)	Blackberry, raspberry
Multiple (formed from carpels of several associated flowers)	*Multiple* matured ovaries, grown together into a mass; may include accessory structures (such as receptacle, sepal, and petal bases)	Pineapple, fig, mulberry
Accessory (formed from one or more ovaries *plus* receptacle tissue that becomes fleshy)	1. *Simple:* a single ovary surrounded by receptacle tissue	Apple, pear
	2. *Aggregate:* swollen, fleshy receptacle with dry fruits on its surface	Strawberry

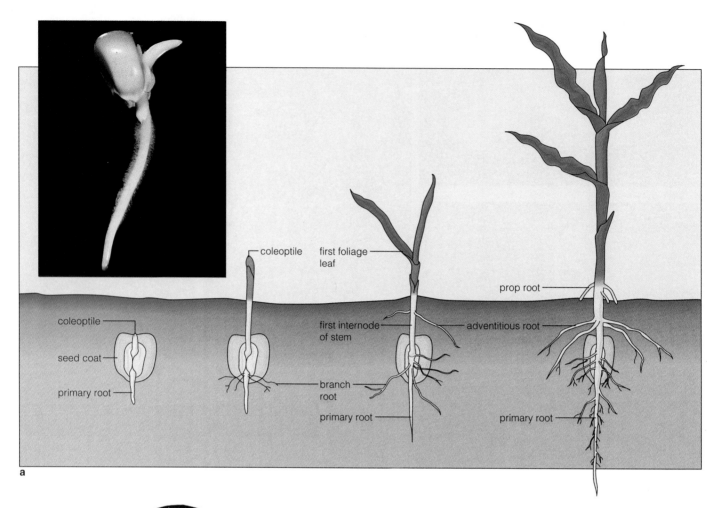

a

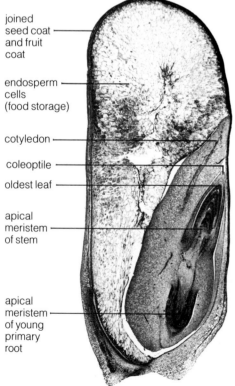

joined
seed coat
and fruit
coat

endosperm
cells
(food storage)

cotyledon

coleoptile

oldest leaf

apical
meristem
of stem

apical
meristem
of young
primary
root

b

Figure 22.8 (**a**) Stages in the development of a corn plant (*Zea mays*), a monocot. (**b**) The organization of an embryo sporophyte within a corn grain, sliced lengthwise through the middle. This corn grain is oriented the same way as the one shown already germinated in the above photograph. Following germination, the coleoptile protects young leaves when the new seedling grows through soil. Adventitious roots develop at the coleoptile's base.

PATTERNS OF GROWTH
AND DEVELOPMENT

After its dispersal from the parent plant, a flowering plant embryo grows into a seedling, which develops into a mature sporophyte. The sporophyte produces flowers, fruits, and seeds; it may drop leaves in autumn. What mechanisms govern its growth and development? Does the environment affect those mechanisms? Let's consider some answers (and best guesses).

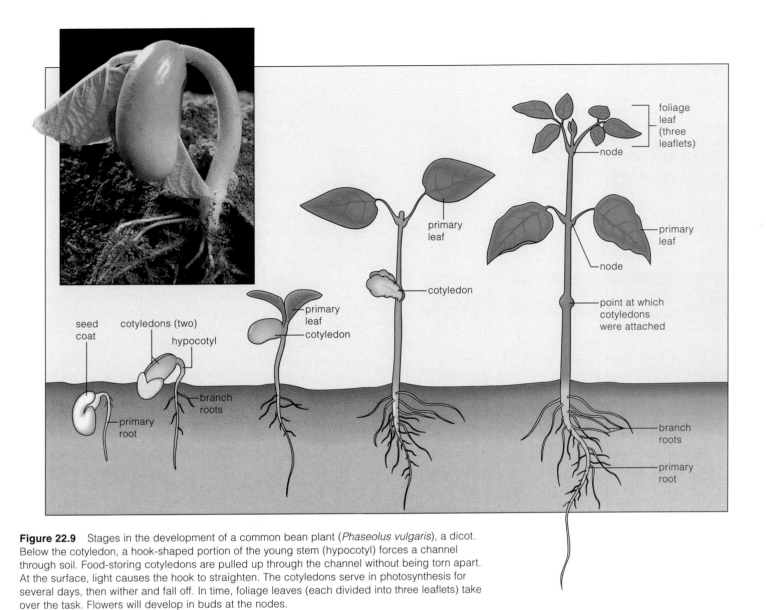

Figure 22.9 Stages in the development of a common bean plant (*Phaseolus vulgaris*), a dicot. Below the cotyledon, a hook-shaped portion of the young stem (hypocotyl) forces a channel through soil. Food-storing cotyledons are pulled up through the channel without being torn apart. At the surface, light causes the hook to straighten. The cotyledons serve in photosynthesis for several days, then wither and fall off. In time, foliage leaves (each divided into three leaflets) take over the task. Flowers will develop in buds at the nodes.

Seed Germination and Early Growth

Before or after seed dispersal, the embryo's growth idles. When the seed germinates, the embryo absorbs water, resumes growth, and breaks through the seed coat. Seed germination depends on water, oxygen, temperature, light, daylength, and other environmental factors. Usually it coincides with the return of spring rains. As more and more water molecules move inside it, the seed swells and its coat ruptures. More oxygen now reaches the embryo, and aerobic respiration moves into high gear. Cells of the embryo rapidly grow and divide, become specialized, and so produce the seedling sporophyte.

Figures 22.8 and 22.9 show the patterns of growth and development for a monocot and a dicot. The basic growth patterns are heritable; they are dictated by the plant's genes. They also can be adjusted in response to the environment. Suppose a seed germinates in a vacant lot and a heavy paper bag blows on top of it. Its cells will grow in such a way that the first shoot bends and grows out from under the bag, toward sunlight. Enzymes and other proteins in those cells carry out the growth responses. And hormones stimulate them to act.

Effects of Plant Hormones

In flowering plants, as in animals, a **hormone** is a signaling molecule released from one cell that changes the activity of target cells. Any cell is a target if it has receptors that can bind the signaling molecule.

Table 22.2 lists the five types of known plant hormones. They are the auxins, gibberellins, cytokinins, abscisic acid, and ethylene. There may be other types, including one tentatively named florigen.

Auxins make stems lengthen and may influence growth responses to light and gravity (Figure 22.10). Orchardists spray trees and other crops with various auxins to thin out overcrowded seedlings in spring. They also use auxins to prevent premature fruit drop, which cuts farm labor costs—all the fruit can be picked at the same time. As the *Focus* essay indicates, some synthetic auxins are used as herbicides.

Gibberellins make stems lengthen (Figure 22.11a,b). They induce dormant seeds, buds, and maybe flowers to grow. Certain doses of a gibberellin produce radishes as big as beach balls and cabbage shoots as tall as you are. Gibberellins make celery longer and crispier, they keep navel orange skins pliable, and they give seedless grapes market appeal (Figure 22.11c).

Cytokinins promote cell division and leaf expansion, and retard leaf aging. Natural and synthetic versions are used to prolong the shelf life of cut flowers as well as lettuces, mushrooms, and other vegetables.

Abscisic acid (ABA) inhibits cell growth, promotes bud dormancy, and keeps seeds from germinating prematurely. It also causes stomata to close and so helps conserve water when the plant is water stressed. Often ABA is applied to nursery stock about to be shipped; dormant plants aren't damaged as much.

Ethylene promotes fruit ripening and the dropping of flowers, fruits, and leaves from trees. Humans have been using it to ripen picked fruit for centuries. Ancient Chinese burned incense to make fruit ripen faster, although they didn't know the smoke contained ethylene. Food

Table 22.2	Main Plant Hormones and Some Known (or Suspected) Effects
Auxins	Promote cell elongation in coleoptiles and stems; involved in phototropism and gravitropism
Gibberellins	Promote stem elongation; might help break dormancy of seeds and buds; stimulate breakdown of starch
Cytokinins	Promote cell division; promote leaf expansion and retard leaf aging
Abscisic acid	Promotes stomatal closure; promotes bud and seed dormancy
Ethylene	Promotes fruit ripening; promotes abscission of leaves, flowers, and fruits
Florigen (?)	Arbitrary designation for as-yet unidentified hormone (or hormones) thought to cause flowering

Figure 22.10 Effects of auxin, a plant hormone. (**a**) A cutting from a gardenia plant (*left*), four weeks after an auxin was applied to its base. The other cutting (*right*) was untreated.

(**b**) Experiment showing that an auxin (IAA) in a coleoptile's tip stimulates the cells below it to lengthen. (1) First, cut off the tip of an oat coleoptile. The stump of the cut coleoptile does not elongate much, compared to a normal oat coleoptile (2) used as a control. (3) Next, place a tiny block of agar under the cut-off tip and leave it for several hours. During that time, IAA diffuses into the agar. (4) Place the agar on another de-tipped coleoptile. You see that elongation proceeds about as rapidly as in an intact coleoptile growing alongside it (5).

a   treated with auxin   untreated

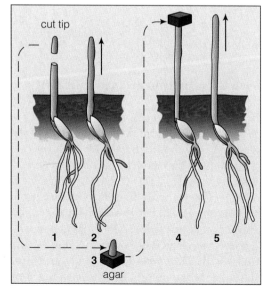

b

About Those Herbicides

An herbicide is any compound that, at proper concentration, kills plants. "Round-Up" is an example. Many useful herbicides kill some plants but not others. Among the widely used, selective herbicides are some synthetic auxins. One is 2,4-D (short for 2,4-dichlorophenoxyacetic acid). Farmers use it to prevent broadleaf weeds from growing in fields of cereal crops. Before the development of synthetic herbicides, farmers and horticulturalists controlled weeds in crops and gardens with backbreaking hand labor or extremely poisonous chemicals.

After exhaustive testing, it appears that 2,4-D does not adversely affect animals, including humans. This is not the case with a compound present in a related herbicide, 2,4,5-T. When mixed in equal proportions, the two compounds produce Agent Orange. The United States Armed Forces used that herbicide to defoliate thickly forested and nearly impenetrable war zones during the Vietnam conflict. Later, experiments suggested that minute traces of dioxin, a substance in 2,4,5-T, might cause miscarriages, birth defects, leukemia, and disorders of the liver and lungs in laboratory animals. Consequently, 2,4,5-T is now banned from most uses in the United States.

a   b

c

Figure 22.11 Hormonal effect on stem growth. (**a**) This California poppy was left alone; (**b**) gibberellin was applied to this plant. (**c**) Seedless grapes, radiating market appeal. Gibberellin made their stems lengthen, which improved air circulation around the grapes and gave them more growing room. Grapes get larger and weigh more. This delights growers; grapes are sold by the pound.

distributors now use ethylene to ripen green fruit after shipment. Fruit that is picked green doesn't bruise or deteriorate as fast.

We have evidence of other hormones besides the five types mentioned. Root and leaf cells produce their own hormones. An unidentified hormone (florigen) may trigger the flowering process. An unknown hormone in shoot tips blocks the growth of lateral buds, an effect called "apical dominance." By pinching off shoot tips, gardeners keep the hormone from reaching lateral buds. The buds branch out, and gardeners get bushier plants.

bean   pea   oat

Figure 22.12 Phototropism in seedlings. The plants were grown in darkness, then exposed to light from the right side for a few hours before being photographed.

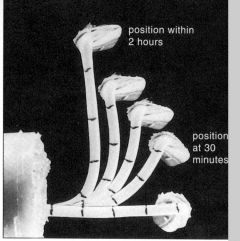

position within 2 hours

position at 30 minutes

Figure 22.13 Composite time-lapse photograph of gravitropism by a sunflower seedling grown in the dark. This 5-day-old plant was turned on its side and marked at 0.5-centimeter intervals.

Plant Tropisms

Directional growth made in response to sunlight or some other environmental stimulus is a "tropism." No one fully understands how tropisms work.

Phototropism. Put a pot of bean sprouts next to a window, where sunlight is more intense on one side of their stems. The stems will start curving toward the light (Figure 22.12). Similarly, many leaves will turn until their flat surface faces the light. When a plant adjusts its direction and rate of growth in response to light, we call this **phototropism**.

In the late 1800s, Charles Darwin noticed that a coleoptile makes a directional response toward light. (A coleoptile is a cylinder around a grass seedling's leaves; it keeps them from being torn apart as the seedling grows up through soil.) Darwin reported that the coleoptile grew toward a light source when light struck its tip from one side. Today we know that plants make the strongest phototrophic response to light of blue wavelengths. The response probably operates through flavoprotein, a pigment molecule that absorbs blue wavelengths.

Gravitropism. A growth response to the earth's gravitational force is called **gravitropism**. Turn a potted seedling on its side in a dark room. Cells on the stem's upward-facing side won't lengthen much—but those on the downside will (Figure 22.13). The stem curves up, even in the absence of light. It may be that the downward-facing cells have become more sensitive to auxin (IAA) and are growing faster than those on top. By contrast, turn a root on its side and it will curve downward, possibly under the prodding of auxin from the root cap. Slice off the cap and the root won't curve down, but it will do so if you put the cap back on.

Thigmotropism. Peas, beans, and many other plants with long, slender stems are climbing vines; they don't usually grow upright without physical support. They show **thigmotropism**—unequal growth after contacting solid objects in the surroundings. Think of a vine, which is a type of stem too slender or soft to grow upright without support. When one side of a vine grows against a fence post, cells on the "contact" side stop elongating, and within minutes the vine starts curling around the post, maybe several times. Then cells on both sides start growing at the same rate again. The same response is made by tendrils (coiling, modified leaves or young stems that help support the plant). Auxin and ethylene may be involved in this response.

Biological Clocks and Their Effects

Like all other organisms, plants have biological clocks (internal time-measuring mechanisms). The clocks have roles in adjusting daily activities. They also function in seasonal adjustments to the plant's patterns of growth, development, and reproduction.

Circadian Rhythms. Some plant activities occur regularly in cycles of about twenty-four hours, even when outside conditions are held constant. These are *circadian* rhythms (meaning "about a day"). Think of a plant that positions its leaves horizontally during the day but folds them closer to the stem at night (Figure 22.14). Keep the plant in constant light or darkness for a few days, and it still folds its leaves into the "sleep" position! It's measuring time *without* light-on (sunrise) and light-off (sunset) signals.

Photoperiodism. There are more hours of daylight and warmer weather in summer than in winter. Wherever seasons change, biological clocks are reset and

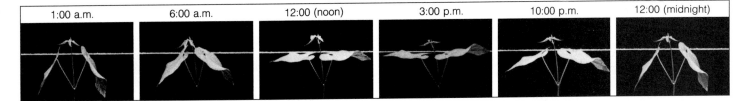

Figure 22.14 Rhythmic leaf movements. This bean plant was kept in constant darkness for twenty-one hours. Its leaf movements continued independently of sunrise (6 A.M.) and sunset (6 P.M.). Leaves folded close to the stem may prevent phytochrome from being activated by bright moonlight, which could interrupt the dark period that triggers flowering. Or perhaps folding helps slow heat loss from leaves otherwise exposed to the cold night air.

adjustments are made in plant growth, development, and reproductive activities.

Any biological response to a *change* in the relative length of daylight and darkness in a cycle of twenty-four hours is an example of **photoperiodism**. In plants, a blue-green pigment molecule called **phytochrome** is often the alarm button for a biological clock. This pigment absorbs light of red or far-red wavelengths, with different results. It converts to active form (Pfr) at sunrise, when red wavelengths dominate the sky. It reverts to inactive form (Pr) at sunset, at night, even in shade, where far-red wavelengths dominate (Figure 22.15). Pfr may influence which enzymes are being produced in particular cells—*and different enzymes are required for different growth responses*. Those cells take part in seed germination, stem lengthening and branching, leaf expansion, and formation of flowers, fruits, and seeds.

The Flowering Process. As a flowering plant matures, it channels more energy and nutrients into producing flowers, seeds, and fruits. Flowering is often a photoperiodic response. In fact, most flowering plants respond in exquisitely predictable ways to changes in daylength through the year and to resulting changes in environmental conditions. "Long-day" plants flower in spring, when daylength becomes *longer* than some critical value. "Short-day" plants flower in late summer or early autumn, when daylength becomes *shorter* than a critical value. "Day-neutral" plants flower when they are mature enough to do so.

Figure 22.16 shows what happens when spinach (a long-day plant) grows under short-day conditions and long-day conditions. A spinach plant won't flower and produce seeds unless it is exposed to fourteen hours of light each day for two weeks. These conditions don't exist in the tropics—which is why spinach is a poor choice for a seed farm there.

Consider how sensitively the cocklebur, a short-day plant, measures time. It flowers after a single night that's longer than 8-1/2 hours. But if artificial light

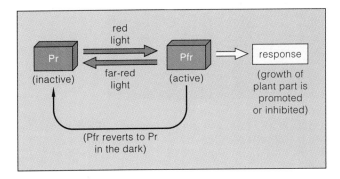

Figure 22.15 Interconversion of phytochrome from active form (Pfr) to inactive form (Pr). This pigment molecule is part of a switching mechanism that can promote or inhibit growth of different plant parts.

Figure 22.16 Effect of daylength on spinach, a long-day plant. The plant on the left was grown under short-day conditions; the one on the right, under long-day conditions.

interrupts that dark period for even a minute or two, cockleburs won't flower! Or consider why short-day poinsettias planted along California's interstate highways never did flower. Headlights of cars and trucks blocked the flowering response.

Probably at least one hormone, as yet unidentified, interacts with phytochrome to influence flowering and other growth responses. Evidence suggests that the hormone may be produced in leaves and transported to newly forming buds. Trim all but one leaf from a cocklebur plant, cover the remaining leaf with black paper for 8-1/2 hours, and the plant flowers. Cut off the leaf immediately after the dark period and the plant will *not* flower.

experimental plant
(pods removed)

control plant
(pods not removed)

Figure 22.17 Results of an experiment to demonstrate how the removal of seed pods from a soybean plant can delay its senescence.

Senescence. Plants make major investments in reproduction. They actually withdraw nutrients from leaves, roots, and stems and distribute them to newly forming flowers, fruits, and seeds. Most plants end up with dead leaves. Deciduous species (which shed leaves at the end of each growing season) transport nutrients to storage sites in twigs, stems, and roots before the leaves wither and fall off.

The dropping of leaves, flowers, fruits, or other plant parts is called "abscission." Ethylene formed in cells near the break points may trigger the process. Abscisic acid may cause cells near the break point to produce the required ethylene.

The sum total of processes leading to the death of plant parts or the whole plant is called **senescence**. The funneling of nutrients into reproductive parts may be a cue for senescence of leaves, stems, and roots. Stop the drain of nutrients by removing each newly emerging flower or seed pod, and a plant's leaves and stems will stay green and vigorous much longer (Figure 22.17). Gardeners routinely remove flower buds from many plants to maintain vegetative growth.

Dormancy. As autumn approaches and days grow shorter, growth slows or stops in many perennial and biennial plants. It stops even if temperatures are still mild, the sky is bright, and water is plentiful. When a plant stops growing under conditions that seem (to us) quite suitable for growth, it has entered a state of **dormancy**. Ordinarily, its buds will not resume growth until early spring. Short days and long nights are strong cues for dormancy. It may be that less Pfr can form when daylength shortens and nights lengthen in late summer. Also, cold nights, dry soil, and nitrogen deficiency seem to promote dormancy.

The Rise and Fall of a Giant

In this unit, you surveyed the spectrum of flowering plant tissues. You skimmed through the mechanisms and outcomes of plant growth, development, and reproduction. To pull all of this into a coherent picture, imagine walking near the central California coast, where the land rolls inward as steep, rounded hills. Those sandstone hills started forming 65 million years ago as a jagged new coastal range. Over time, rains and winds softened the stark contours of the inland hills and sent mineral-laden sediments into canyons. Grasses took hold, and their organic remains accumulated and enriched the soil. More than 10 million years ago, this is where the coast live oak (*Quercus agrifolia*) began to evolve.

As you walk, you come across a giant tree that is 300 years old. You stop briefly under trees more than 100 feet tall, with evergreen branches that spread even wider. It's early spring, so you see golden catkins—clusters of male flowers—among the leaves. A light wind is dispersing pollen from catkins to female flowers near the branch tips of the same or neighboring trees. Long after your walk, the sperm-bearing pollen tubes will still be growing toward the plant's ovaries. In time, newly formed zygotes will undergo the first of millions of cell divisions. Seed coats will form and ovarian walls will become shells. By early fall, giant trees will shed their seeds as hard-shelled acorns.

Three centuries ago, long before Gaspar de Portola sent landing parties ashore to found colonies throughout upper California, oaks were shedding the seeds of a new generation. By chance, one particular acorn escaped the attention of foraging squirrels and bluejays. The next spring, it germinated and embarked on a journey of continued growth.

Cells differentiated and developed into the cortex, epidermis, and a vascular cylinder through which water and ions would flow. Lateral roots emerged. As new roots grew

SUMMARY

1. In flowering plant life cycles, a multicelled spore-producing body (sporophyte) alternates with a small, haploid, gamete-producing body (gametophyte). Most of the sporophytes are vegetative bodies that have roots, stems, and leaves and that produce flowers on specialized floral shoots.

2. Sepals and petals are the outermost, nonfertile parts of a flower. Stamens and carpels are the innermost, fertile parts. The stalked anthers of stamens contain pollen

oak seed (acorn) germinating

longer, their absorptive surfaces increased. When the first shoot began its upward surge, vascular bundles began forming. In time they would form a continuous cylinder of secondary xylem and phloem.

The seedling's roots, stems, and leaves demanded more water and dissolved nutrients. Fungi formed velvety coverings around the roots. And these fungus-roots—mycorrhizae—enhanced the uptake of water and mineral ions. Leaves started conserving precious water with the development of stomata. As the oak seedling grew, its parts became interconnected by vascular pipelines. Through xylem, water and minerals moved from roots to stems and leaves. Through phloem, sugars were shuttled to cells actively growing or storing food.

By chance, the seed had sprouted in a well-drained, sunlit basin at the foot of a canyon. Each winter, rainwater accumulated, and it kept the soil moist enough to encourage the plant's growth through spring and the dry summers. The sun's red wavelengths activated phytochrome, triggering hormonal events that encouraged stem branching and leaf expansion. Other hormone-mediated responses were made to the winds, the sun, the tug of gravity, the changing seasons.

And so the oak flourished every season. Century after century, roots snaked through a huge volume of the moist soil. Branches continued to spread beneath the sun. Leaves proliferated, and their photosynthetic cells put together food with sunlight energy, water, carbon dioxide, and the few simple minerals being mined from the soil.

In 1849, prospectors on their way to the gold fields rested in the shade of the oak's immense canopy. The great earthquake of 1906 scarcely disturbed the giant, anchored as it was by a root system extending 80 feet through the

soil. The few small brush fires that swept through the canyon did no serious damage to the tree. Fungi that could have rotted its roots never took hold in the well-drained soil. Canyon birds and the tree's own protective chemicals kept insects in check.

Then, in the 1960s, the once-wild hills began to give way to suburban housing. A developer turned his tractors into the canyon but spared the giant oak. Death came later.

The new homeowners were ignorant of the ancient, delicate relationships between the giant tree and the land that sustained it. They graded the soil between the trunk and the drip line of the overhanging canopy. They mounded flower beds against the trunk and planted lawns beneath the branches, then kept the sprinklers busy. Overwatering in summer created standing water next to the great trunk. Before then, the giant had successfully resisted the oak root fungus (*Armillaria*). In the changed environment, the fungus took hold. With its roots rotting away, the oak began to suffer massive disruptions in the feedback relations among its roots, stems, and leaves. Eventually it had to be cut down. In their fifth winter, in their red brick fireplace, the homeowners began burning three centuries of firewood.

sacs. In these chambers, cells divide by meiosis, and a wall develops around the resulting haploid cells. The walled cells are microspores. These develop into pollen grains (male gametophytes). At some point, a pollen grain produces two sperm nuclei.

3. A carpel is a female reproductive part. A single carpel (or two or more carpels fused together) has an ovary, where eggs develop, fertilization takes place, and seeds mature. Above this is a stigma, a sticky or hairy surface tissue that captures pollen grains and promotes their germination. Often the upper portion of the ovary is modified into a slender column (style).

4. One or more ovules form on the inner wall of an ovary. Before fertilization, an ovule consists of a female gametophyte with egg cell, a surrounding tissue, and one or two protective layers (integuments). After the egg is fertilized, the ovule matures into a seed, and the ovary expands and matures into a fruit.

5. After pollination (transfer of pollen grains to a receptive stigma), a pollen grain germinates. It becomes a pollen tube that grows down through tissues of the ovary, carrying the two sperm nuclei with it.

6. Only flowering plants undergo double fertilization. By this process, one sperm nucleus fuses with one egg

nucleus to form a diploid (2*n*) zygote. The other sperm nucleus fuses with both nuclei of an endosperm mother cell to form a cell that gives rise to 3*n* endosperm, a nutritive tissue for the forthcoming embryo.

7. Following dispersal from the parent plant, seeds germinate and develop into seedlings. These young sporophytes increase in volume and mass, and develop new tissues and organs. At maturity, they produce flowers, fruits, and seeds; then older leaves drop off.

8. Five types of plant hormones have been identified. Auxins and gibberellins promote stem elongation. Cytokinins stimulate cell division, promote leaf expansion, and retard leaf aging. Abscisic acid promotes bud and seed dormancy, and it limits water loss (by triggering stomatal closure). Ethylene promotes fruit ripening and abscission.

9. Plant hormones interact with one another to produce patterns of growth and development. They help adjust those patterns in response to environmental rhythms, including seasonal changes in daylength and temperature. They also adjust the patterns to the amount of sunlight, shade, and other environmental circumstances in which a plant finds itself.

Review Questions

1. Label and define the functions of the floral parts in this diagram: *355*

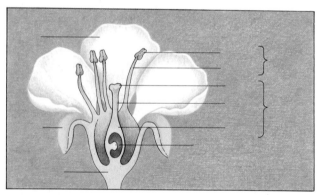

2. Distinguish between these terms:
   a. megaspore and microspore *356*
   b. pollination and fertilization *356–357*
   c. pollen grain and pollen tube *356*
   d. ovule and female gametophyte *356*

Self-Quiz *(Answers in Appendix IV)*

1. In the fertile parts of flowers, _____ form.
   a. spores
   b. gametes
   c. gametophytes
   d. embryo sporophytes
   e. all of the above

2. Seeds are mature _____ ; fruits are mature _____ .
   a. ovaries; ovules
   b. ovules; stamens
   c. ovules; ovaries
   d. stamens; ovaries

3. A _____ contains an ovary in which eggs develop, fertilization occurs, and seeds mature.
   a. pollen sac
   b. carpel
   c. receptacle
   d. sepal

4. After meiosis in pollen sacs, _____ develop.
   a. megaspores
   b. sperm
   c. microspores
   d. pollen grains

5. After meiosis in carpels, _____ develop.
   a. megaspores
   b. sperm
   c. microspores
   d. pollen grains

6. In flowering plants alone, _____ occur(s).
   a. seed formation
   b. double fertilization
   c. triple fertilization
   d. both a and b

7. _____ as well as gibberellins promote stem lengthening.
   a. Auxins
   b. Cytokinins
   c. Abscisic acid
   d. Ethylene

8. In flowering plants, _____ serve as a switching mechanism for biological clocks.
   a. auxins
   b. gibberellins
   c. phytochromes
   d. dormant periods

9. Match the descriptions with the proper concepts.
   _____ pollination
   _____ seed formation
   _____ double fertilization
   _____ phototropism
   _____ senescence

   a. zygote and the endosperm mother cell form
   b. death of plant or parts of it
   c. growth response to light
   d. pollen lands on stigma
   e. ovule matures after egg is fertilized

Selected Key Terms

Readings

Proctor, M., and P. Yeo. 1973. *The Pollination of Flowers.* New York: Taplinger.

Salisbury, F., and C. Ross. 1992. *Plant Physiology.* Fourth edition. Belmont, California: Wadsworth.

FACING PAGE: *How many and what kinds of body parts does it take to function as a lizard in a tropical forest? Make a list of what comes to mind as you start reading Unit VI, then see how resplendent the list can become at the unit's end.*

23 TISSUES, ORGAN SYSTEMS, AND HOMEOSTASIS

Meerkats, Humans, It's All the Same

After a cold night in Africa's Kalahari Desert, animals small enough to fit in a coat pocket emerge stiffly from their burrows. These "meerkats" stand on their hind legs and face east, exposing their chilled bodies to the warm rays of the morning sun (Figure 23.1). Meerkats don't know it, but sunning behavior helps their enzymes. If their body temperature were to fall below or exceed a tolerable range, enzyme activity would drop sharply. With such a change in enzyme activity, metabolism would suffer.

Once meerkats warm up, they fan out from the burrows, looking for food. Into the meerkat gut go insects and an occasional lizard. These are pummeled, dissolved, and digested into glucose and other nutritious tidbits small enough to move across the gut wall, into the bloodstream, and on to cells throughout the body. In cells, aerobic machinery cracks nutrients apart and so releases vital energy. A respiratory system and the bloodstream supply the machinery with oxygen

and take away its carbon dioxide leftovers. All of this activity changes the composition and volume of the "internal environment"—the bloodstream and fluids bathing the body's cells. Drastic changes in that fluid would kill cells, but a urinary system works to keep this from happening. Governing this system and all others are the body's command posts—a nervous system and an endocrine system. They mobilize the body as a whole for everything from simple housekeeping tasks to heart-thumping flights from predators.

And so meerkats start us thinking about this unit's central topics: how an animal body is structurally put together (its *anatomy*) and how the body functions (its *physiology*). This chapter provides us with an overview of the animal tissues and organ systems that we will be considering. It also introduces the important concept of "homeostasis"—stability in the internal environment, brought about by the coordinated activity of the body's cells, tissues, organs, and organ systems.

1. The cells of most animals interact at three levels of organization—in *tissues*, many of which are combined in *organs*, which are components of *organ systems*.

2. Most animals are constructed of only four types of tissues: epithelial, connective, nervous, and muscle tissues.

3. Each cell engages in basic metabolic activities that assure its own survival. At the same time, cells of a tissue or organ perform activities that contribute to the survival of the animal as a whole.

4. The combined contributions of cells, tissues, organs, and organ systems help maintain a stable internal environment, which is required for individual cell survival. This concept helps us understand the functions of any organ system, regardless of its complexity.

Figure 23.1 In the Kalahari Desert, meerkats line up and face the warming rays of the morning sun, just as they do every morning. This simple behavior helps maintain internal body temperature. How animals function in their environment is the subject of this unit.

ANIMAL STRUCTURE AND FUNCTION: AN OVERVIEW

Regardless of whether it is a flatworm or salmon, a meerkat or human being, each animal is structurally and physiologically adapted to perform these tasks:

1. Maintain internal operating conditions within a tolerable range even as external conditions change.

2. Locate and take in nutrients and other materials, distribute them through the body, and dispose of wastes.

3. Protect itself against injury or attack from viruses, bacteria, and other disease-causing agents.

4. Reproduce, and often help nourish and protect the new individuals during their early development.

Your body, and those of other complex animals, has only four basic types of tissues. These are epithelial, connective, muscle, and nervous tissues. A **tissue** is an interacting group of cells and intercellular substances that take part in one or more of the tasks listed above. An **organ**, such as a heart, consists of different tissues organized in specific proportions and patterns. An **organ system** consists of two or more organs that interact physically, chemically, or both in a common task, such as blood circulation. Cells, tissues, organs, and organ systems split up the work, so to speak, in ways that contribute to the survival of the animal as a whole. This is sometimes called a division of labor.

How Tissues Form

To get a general sense of how animal tissues form, let's start with sperm and eggs. These develop from germ cells, which are immature reproductive cells. The rest of the body's cells are said to be "somatic," after the Greek word for body. After a sperm fertilizes an egg, the resulting zygote embarks on a course of mitotic cell divisions that produce the early animal embryo. In vertebrate embryos, cells soon become arranged as three "primary" tissues—ectoderm, mesoderm, and endoderm. These are the embryonic forerunners of all tissues in the adult. **Ectoderm** gives rise to the skin's outer layer and to tissues of the nervous system. **Mesoderm** gives rise to the tissues of muscle, bone, and most of the circulatory, reproductive, and urinary systems. **Endoderm** gives rise to the gut's lining and organs derived from it.

With these points in mind, let's review the main features of specialized tissues in the adult animal.

Epithelial Tissue

An epithelial tissue is also called **epithelium** (plural, epithelia). One side of this tissue faces some kind of body fluid or the outside environment (Figure 23.2). Its cells are close together, with little intervening material. Specialized junctions provide structural and functional links between the cells, which absorb, synthesize, and secrete a variety of substances (Figure 23.3).

Simple epithelium, with a single layer of cells, functions as a lining for body cavities, ducts, and tubes. Figure 23.4 shows examples. *Stratified* epithelium, with two

Figure 23.2 Characteristics of epithelium. All epithelia have a free surface, and a basement membrane is interposed between the opposite surface and an underlying connective tissue. The diagram shows this arrangement in simple epithelium, which consists of a single layer of cells. The micrograph shows the upper portion of stratified epithelium, which has more than one layer of somewhat flattened cells.

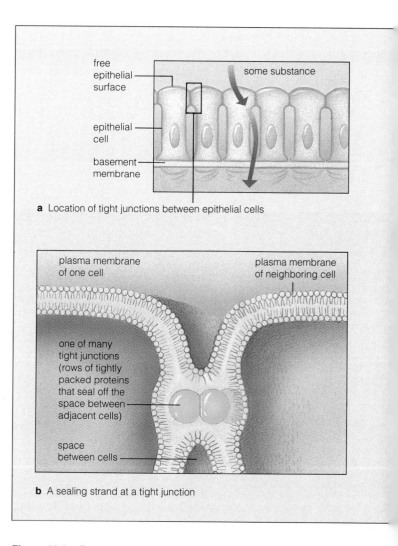

a Location of tight junctions between epithelial cells

b A sealing strand at a tight junction

Figure 23.3 Examples of cell junctions. (**a**,**b**) In some epithelia, protein strands form tight seals that ring each cell and seal it to its neighbors. The seals prevent substances from leaking across the free epithelial surface. The only way that substances can reach the tissues below is to pass *through* the epithelial cells—which have built-in mechanisms that can control their passage. (**c**) Adhesion junctions are like spot welds that "cement" cells of

or more cell layers, typically functions in protection, as it does in skin.

Glands are secretory cells or multicelled structures derived from epithelium and often connected to it. Various *exocrine* glands secrete mucus, saliva, earwax, oil, milk, digestive enzymes, and other products. These products usually are released onto a free epithelial surface through ducts or tubes.

By contrast, *endocrine* glands have no ducts. Their products—hormones—are secreted directly into the fluid bathing the gland, then are picked up and distributed by the bloodstream.

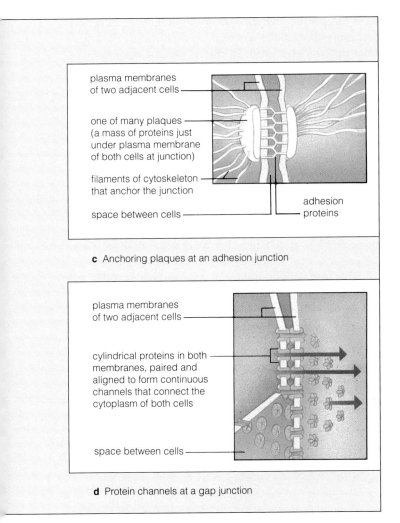

plasma membranes of two adjacent cells

one of many plaques (a mass of proteins just under plasma membrane of both cells at junction)

filaments of cytoskeleton that anchor the junction

space between cells

adhesion proteins

c Anchoring plaques at an adhesion junction

plasma membranes of two adjacent cells

cylindrical proteins in both membranes, paired and aligned to form continuous channels that connect the cytoplasm of both cells

space between cells

d Protein channels at a gap junction

epithelium (and all other tissues) together so that they function as a unit. They are abundant in the skin's surface layer and other tissues subjected to abrasion. (**d**) Gap junctions promote diffusion of ions and small molecules from cell to cell. They are abundant in the heart, liver, and other organs in which cell activities must be rapidly coordinated.

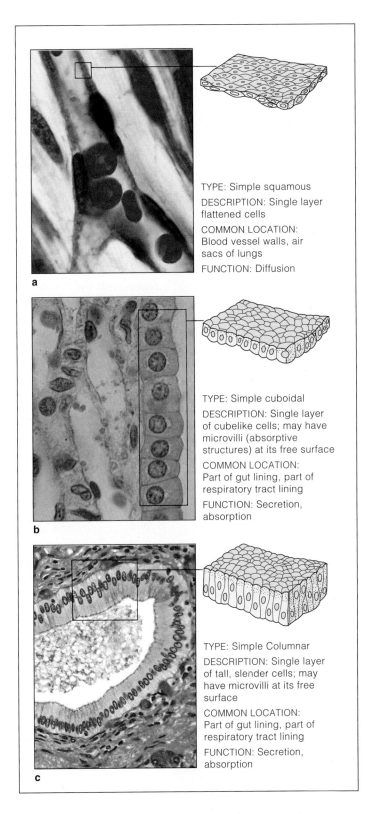

a

TYPE: Simple squamous
DESCRIPTION: Single layer flattened cells
COMMON LOCATION: Blood vessel walls, air sacs of lungs
FUNCTION: Diffusion

b

TYPE: Simple cuboidal
DESCRIPTION: Single layer of cubelike cells; may have microvilli (absorptive structures) at its free surface
COMMON LOCATION: Part of gut lining, part of respiratory tract lining
FUNCTION: Secretion, absorption

c

TYPE: Simple Columnar
DESCRIPTION: Single layer of tall, slender cells; may have microvilli at its free surface
COMMON LOCATION: Part of gut lining, part of respiratory tract lining
FUNCTION: Secretion, absorption

Figure 23.4 Examples of simple epithelium, showing the three basic cell shapes in this type of tissue.

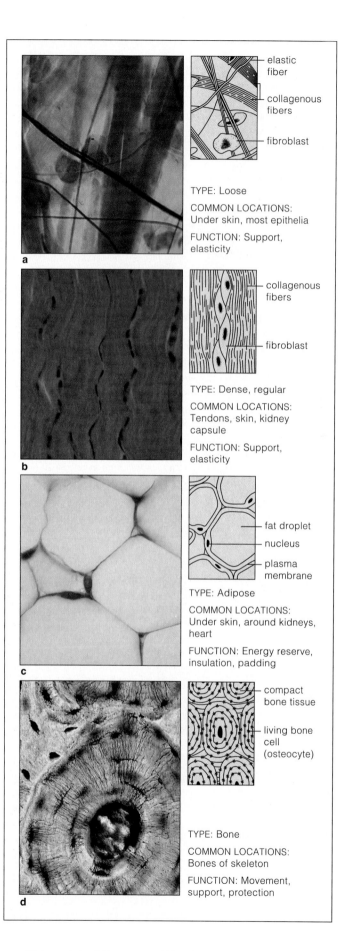

TYPE: Loose

COMMON LOCATIONS: Under skin, most epithelia

FUNCTION: Support, elasticity

a

TYPE: Dense, regular

COMMON LOCATIONS: Tendons, skin, kidney capsule

FUNCTION: Support, elasticity

b

TYPE: Adipose

COMMON LOCATIONS: Under skin, around kidneys, heart

FUNCTION: Energy reserve, insulation, padding

c

TYPE: Bone

COMMON LOCATIONS: Bones of skeleton

FUNCTION: Movement, support, protection

d

Connective Tissue

These tissues range from **connective tissue proper** to specialized types, including **cartilage**, **bone**, **adipose tissue**, and **blood** (Figure 23.5 and Table 23.1). In all types except blood, fibroblasts and certain other cells secrete fibers of collagen or elastin. These are structural proteins. (Plastic surgeons know about collagen. They use it to plump wrinkled skin, sunken acne scars, even lips.) Fibroblasts secrete modified polysaccharides as well. This material accumulates between the cells and fibers and so becomes the tissue's "ground substance."

Connective Tissue Proper. Tissues in this category have mostly the same components, but in different proportions. *Loose* connective tissue has more cells and fewer, thinner fibers (Figure 23.5*a*). Often it functions as a supporting framework for epithelium. Infection-fighting cells wander through loose connective tissue or take up residence in it. In the tissue just beneath skin, these cells combat bacteria and other disease agents entering the body through cuts or other wounds.

Dense, irregular connective tissue has fewer cells and more fibers, which are thick and irregularly arranged. This tissue forms protective capsules around organs that are not stretched much.

Specialized Connective Tissue. Many collagen fibers, bundled together in parallel, impart strength to *dense, regular* connective tissue (Figure 23.5*b*). They do so in ligaments (which attach bones together) and tendons (which attach muscles to bones). Both of these organs resist being torn apart.

The intercellular material of cartilage is solid yet somewhat pliable, like a piece of solid rubber, so it resists compression. Vertebrate embryos start out with limbs of cartilage. These serve as structural models for

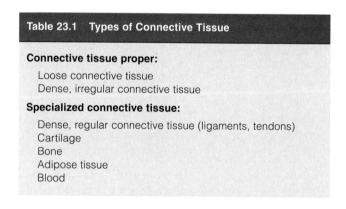

Table 23.1   Types of Connective Tissue
Connective tissue proper:
Loose connective tissue
Dense, irregular connective tissue
Specialized connective tissue:
Dense, regular connective tissue (ligaments, tendons)
Cartilage
Bone
Adipose tissue
Blood

Figure 23.5 Examples of connective tissue proper and of specialized connective tissue.

the bones that replace them. Cartilage also maintains the shape of the nose, outer ear, and other body parts. It cushions joints between adjacent bones of the vertebral column, limbs, hands, and elsewhere.

Adipose tissue is so chockful of large fat cells, it no longer looks like the connective tissue that it is (Figure 23.5c). The body's excess carbohydrates and proteins are converted to storage fats and tucked away in this tissue. Adipose tissue has a rich supply of blood, which serves as an immediately accessible "highway" for moving fats to and from the tissue's individual cells.

Bone (Figure 23.5d) is the weight-bearing tissue of the skeleton, which supports or protects softer tissues and organs. This connective tissue is hardened with minerals; its collagen fibers and ground substance are loaded with calcium salts. Limb bones, such as the ones in your arms and legs, interact with attached muscles to bring about movements. Certain tissues within some bones also function in red blood cell production.

Blood is a fluid derived from connective tissue and so is categorized here. Blood transports oxygen to cells and carries wastes away from them. It transports hormones and enzymes. Some of its components protect against blood loss through clotting mechanisms, and others defend against disease-causing agents. Chapter 26 describes this tissue and its complex functions.

Muscle Tissue

In **muscle tissue** alone, cells contract (shorten) in response to stimulation, then lengthen and so return to their original state. Muscle tissue helps move the body and its individual parts. There are three types, called skeletal, smooth, and cardiac muscle tissues.

Skeletal muscle tissue has many long, cylindrical cells (Figure 23.6a). Typically, these cells are bundled together. Then several bundles of the muscle cells are enclosed in a tough connective tissue sheath to form "a muscle," such as a biceps. The structure and function of skeletal muscle tissue are topics of the next chapter.

The contractile cells of *smooth* muscle tissue are tapered at both ends, and cell junctions hold them together (Figure 23.6b). They, too, are enclosed in connective tissue. The walls of blood vessels, the stomach, and other internal organs contain smooth muscle. We humans think of smooth muscle action as "involuntary," because we usually can't *make* it contract (as we can do with skeletal muscle).

Cardiac muscle tissue is the heart's contractile tissue (Figure 23.6c). The plasma membranes of adjacent cardiac muscle cells are fused together by cell junctions. Certain junctions at these fusion points allow the cells to contract as a unit. When one muscle cell receives a signal to contract, its neighbors are also stimulated into contracting.

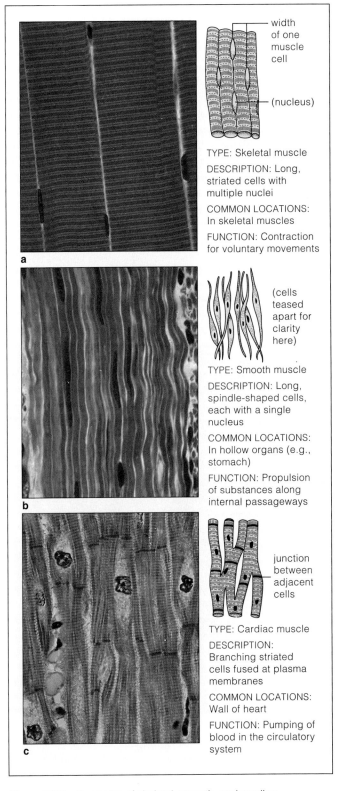

TYPE: Skeletal muscle
DESCRIPTION: Long, striated cells with multiple nuclei
COMMON LOCATIONS: In skeletal muscles
FUNCTION: Contraction for voluntary movements

TYPE: Smooth muscle
DESCRIPTION: Long, spindle-shaped cells, each with a single nucleus
COMMON LOCATIONS: In hollow organs (e.g., stomach)
FUNCTION: Propulsion of substances along internal passageways

TYPE: Cardiac muscle
DESCRIPTION: Branching striated cells fused at plasma membranes
COMMON LOCATIONS: Wall of heart
FUNCTION: Pumping of blood in the circulatory system

Figure 23.6 Examples of skeletal, smooth, and cardiac muscle tissues.

Nervous Tissue

Of all tissues, **nervous tissue** is the one with greatest control over the body's responsiveness to changing conditions. Its cells, called neurons, are organized as communication lines that extend through the body. Some neurons detect specific changes in environmental conditions. Others coordinate the body's immediate and long-term responses to change. Neurons of the sort shown in Figure 23.7 relay signals to muscles and glands that can carry out appropriate responses. How these remarkable cells function is a topic of Chapter 30.

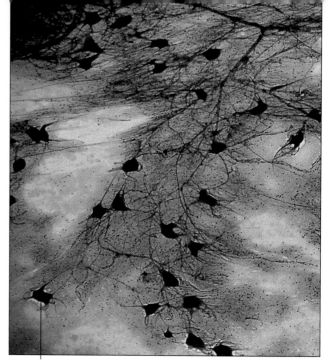

Figure 23.7 A sampling of the millions of neurons that form communication lines within and between different regions of the human body. These motor neurons relay signals from the brain or spinal cord to muscles and glands. Collectively, these and other neurons sense environmental change, integrate signals about those changes, and initiate appropriate responses.

cell body of one of the motor neurons in this nervous tissue sample

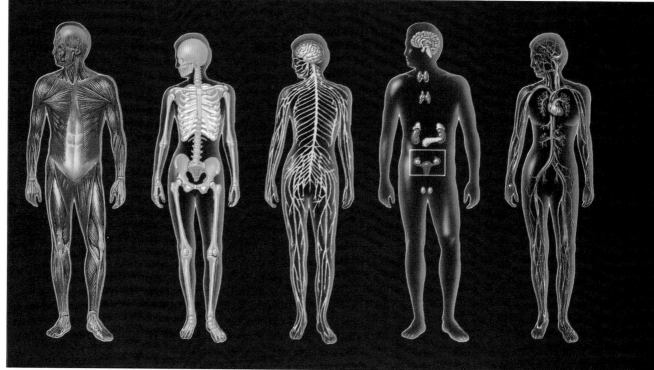

INTEGUMENTARY SYSTEM	MUSCULAR SYSTEM	SKELETAL SYSTEM	NERVOUS SYSTEM	ENDOCRINE SYSTEM	CIRCULATORY SYSTEM
Protection from injury and dehydration; body temperature control; excretion of some wastes; reception of external stimuli; defense against microbes.	Movement of internal body parts; movement of whole body; maintenance of posture; heat production.	Support, protection of body parts; sites for muscle attachment, blood cell production, and calcium and phosphate storage.	Detection of external and internal stimuli; control and coordination of responses to stimuli; integration of activities of all organ systems.	Hormonal control of body functioning; works with nervous system in integrative tasks.	Rapid internal transport of many materials to and from cells; helps stabilize internal temperature and pH.

Figure 23.8 Organ systems of the human body. All vertebrates have the same types of systems, serving similar functions.

MAJOR ORGAN SYSTEMS

The tissues just described form organs and organ systems that are much the same in all vertebrates. Figure 23.8 shows the organ systems of humans. Figure 23.9 shows major body cavities in which they are located.

You might think we are stretching things a bit when we say that each one of those systems contributes to the survival of all living cells in the body. After all, what could the body's bones and muscles have to do with the life of a tiny cell? Yet interactions between the skeletal and muscular systems allow us to move about—toward sources of nutrients and water, for example. Parts of those systems assist blood circulation, as when contractions of leg muscles help move blood in veins back to the heart. The bloodstream transports nutrients and other substances to individual cells, and transports secreted products and wastes away from them.

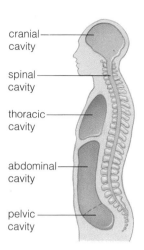

Figure 23.9 Major cavities in the human body.

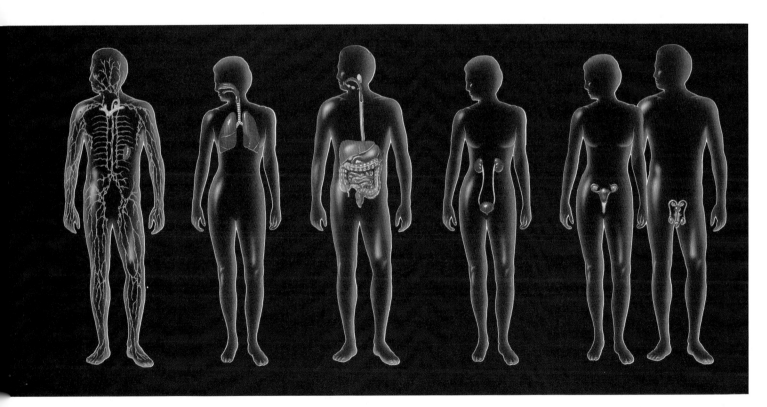

LYMPHATIC SYSTEM	RESPIRATORY SYSTEM	DIGESTIVE SYSTEM	URINARY SYSTEM	REPRODUCTIVE SYSTEM
Return of some tissue fluid to blood; roles in immunity (defense against specific invaders of the body).	Delivery of oxygen to cells; removal of carbon dioxide wastes produced by cells; pH regulation.	Ingestion of food, water; preparation of food molecules for absorption; elimination of food residues from the body.	Maintenance of the volume and composition of extracellular fluid. Excretion of blood-borne wastes.	Male: production and transfer of sperm to the female. Female: production of eggs; provision of a protected, nutritive environment for developing embryo and fetus. Both systems have hormonal influences on other organ systems.

HOMEOSTASIS AND SYSTEMS CONTROL

The Internal Environment

To stay alive, your cells must be continually bathed in a fluid that supplies them with nutrients and carries away metabolic wastes. In this they are no different from an amoeba or any other free-living, single-celled organism. However, many *trillions* of cells crowd together in your body—and they all must draw nutrients from and dump wastes into the same fifteen liters of fluid. That is less than sixteen quarts.

The fluid *not* inside cells is called **extracellular fluid**. Much of it is *interstitial*, meaning it occupies spaces between cells and tissues. The rest is *plasma*, the fluid portion of blood. Interstitial fluid exchanges substances with blood and with the cells it bathes.

In functional terms, extracellular fluid is continuous with the fluid inside cells. That's why drastic changes in its composition and volume have drastic effects on cell activities. Its ion concentrations are especially important in this regard. They must be maintained at levels that are compatible with cell survival. Otherwise, the animal itself cannot survive.

It makes no difference whether an animal is simple or complex. *The component parts of any animal work together to maintain the stable fluid environment required by its living cells.* This concept is absolutely central to understanding the structure and function of animals, and it may be summarized this way:

---

1. Each cell of the animal body engages in basic metabolic activities that ensure its own survival.

2. Concurrently, the cells of a given tissue perform one or more activities that contribute to the survival of the whole organism.

3. The combined contributions of individual cells, organs, and organ systems help maintain the stable internal environment—that is, the extracellular fluid—required for individual cell survival.

Mechanisms of Homeostasis

Homeostasis refers to stable operating conditions in the internal environment. Three components, called sensory receptors, integrators, and effectors, interact to maintain this state. **Sensory receptors** are cells or cell parts that can detect a **stimulus**, which is a specific change in the environment. When someone kisses you, for example, there is a change in pressure on your lips. Receptors in the skin of your lips translate the stimulus into a signal that can be sent to the brain. Your brain is

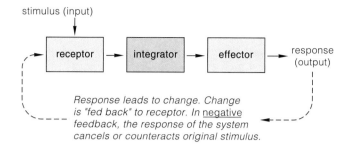

Figure 23.10 Components necessary for negative feedback at the organ level.

an **integrator**, a control point where different bits of information are pulled together in the selection of a response. It can send signals to your muscles or glands (or both). Muscles and glands are **effectors**—they carry out the response. In this case, the response might include flushing with pleasure and kissing the person back. Of course, you cannot engage in a kiss indefinitely, for this would prevent you from eating and performing other tasks that maintain your body's operating conditions.

How does your brain reverse the physiological changes induced by the kiss? Receptors only provide it with information about how things *are* operating. The brain also receives information about how things *should be* operating—that is, information from "set points." When physical or chemical conditions deviate sharply from a set point, the brain functions to bring them back to an effective operating range. It does this by way of signals that cause specific muscles and glands to increase or decrease their activity.

Feedback mechanisms are among the controls that help keep physical and chemical aspects of the body within tolerable ranges. In a **negative feedback mechanism**, an activity alters a condition in the internal environment, and this triggers a response that reverses the altered condition (Figure 23.10). Think of a furnace with a thermostat. A thermostat senses the air temperature and "compares" it to a preset point on a thermometer built into the furnace control system. When the temperature falls below the preset point, the thermostat signals a switching mechanism that turns on the heating unit. When the air becomes heated enough to match the prescribed level, the thermostat signals the switching mechanism, which shuts off the heat.

Similarly, feedback mechanisms help keep the body temperature of meerkats, humans, huskies, and many other animals near 37°C (98.6°F), even during hot or cold weather. Imagine a husky running around on a hot summer day. Its body gets hot—and receptors trigger events that slow down the whole dog *and* its cells. The husky flops down and rests in the shade of a tree. Mois-

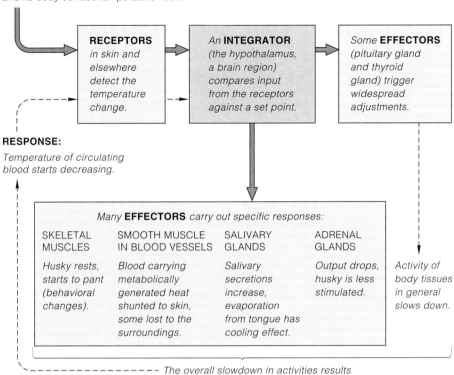

STIMULUS:

The husky is overactive on a hot, dry day and its body surface temperature rises.

RECEPTORS in skin and elsewhere detect the temperature change.

An **INTEGRATOR** (the hypothalamus, a brain region) compares input from the receptors against a set point.

Some **EFFECTORS** (pituitary gland and thyroid gland) trigger widespread adjustments.

RESPONSE:

Temperature of circulating blood starts decreasing.

Many **EFFECTORS** carry out specific responses:

SKELETAL MUSCLES	SMOOTH MUSCLE IN BLOOD VESSELS	SALIVARY GLANDS	ADRENAL GLANDS	
Husky rests, starts to pant (behavioral changes).	*Blood carrying metabolically generated heat shunted to skin, some lost to the surroundings.*	*Salivary secretions increase, evaporation from tongue has cooling effect.*	*Output drops, husky is less stimulated.*	*Activity of body tissues in general slows down.*

The overall slowdown in activities results in less metabolically generated heat.

ture from its respiratory system evaporates from the tongue and carries some body heat with it (Figure 23.11). These and other control mechanisms counter overheating by curbing the body's heat-generating activities and by giving up excess heat to the surroundings.

Sometimes **positive feedback mechanisms** operate. These set in motion a chain of events that *intensify* a change from an original condition—and after a limited time, the intensification reverses the change. Positive feedback is associated with instability in a system. For example, during sexual intercourse, signals from a female's nervous system trigger intense physiological responses to her sexual partner. Those responses stimulate changes in her partner that further stimulate the female, and so on until an explosive, climax level is reached. As another example, during childbirth, a fetus exerts pressure on the wall of its mother's uterus. This stimulates production and secretion of oxytocin, a hormone. Oxytocin causes wall muscles to contract and exert pressure on the fetus, which exerts more pressure on the wall, and so on until the fetus is expelled.

Homeostatic control mechanisms help maintain physical and chemical aspects of the body's internal environment within ranges that are most favorable for cell activities.

Figure 23.11 Homeostatic controls over the internal temperature of a husky's body. Blue arrows indicate the main control pathways. Dashed line shows how a feedback loop is completed.

What we have been describing here is a general pattern of monitoring and responding to a constant flow of information about an animal's internal and external environments. During this activity, organ systems operate together in coordinated fashion. Throughout this unit, we will be asking the following questions about their operation:

1. What physical or chemical aspect of the internal environment are organ systems working to maintain as conditions change?

2. By what means are organ systems kept informed of change?

3. By what means do they process incoming information?

4. What mechanisms are set in motion in response?

As you will see in chapters to follow, the operation of all organ systems is under neural and endocrine control.

SUMMARY

1. Cells and intercellular substances of a tissue work together in a common task. In organs, combinations of tissues function in a common task. In organ systems, organs interact chemically, physically, or both in ways that contribute to the survival of the body as a whole.

2. Epithelial tissues cover external body surfaces and line internal cavities and tubes. They have one free surface exposed to body fluids or the environment.

3. Connective tissue proper and specialized connective tissues (such as bone and blood) bind together other tissues or give them structural or metabolic support.

4. Muscle tissues contract and so help move the body or parts of it. Nervous tissue detects and integrates information about internal and external conditions, and it governs the body's responses to change.

5. Tissues, organs, and organ systems work together to maintain the stable internal environment (the extracellular fluid) required for individual cell survival. At homeostasis, conditions in the internal environment are most favorable for cell activities.

6. Feedback controls help maintain internal conditions. In negative feedback, a change in some condition triggers a response that reverses the change.

7. Homeostasis depends on receptors, integrators, and effectors. Receptors detect stimuli (specific changes in the environment). Integrating centers (such as a brain) process the information and direct muscles and glands (the body's effectors) to carry out responses.

Review Questions

1. Identify and describe the following tissues: *374–377*

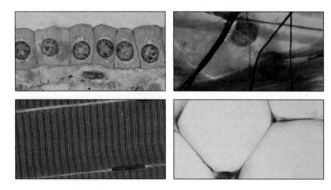

2. Define an animal tissue, organ, and organ system. List and define the functions of the human body's major organ systems. *378–379*

3. Define extracellular fluid and interstitial fluid. *380*

4. Define homeostasis. *380*

Selected Key Terms

adipose tissue *376*
blood *376*
bone *376*
cartilage *376*
connective tissue proper *376*
ectoderm *373*
effector *380*
endoderm *373*

epithelium *374*
extracellular fluid *380*
gland *375*
homeostasis *380*
integrator *380*
mesoderm *373*
muscle tissue *377*
negative feedback mechanism *380*

nervous tissue *378*
organ *373*
organ system *373*
positive feedback mechanism *381*
sensory receptor *380*
stimulus *380*
tissue *373*

Readings

Bloom, W., and D. W. Fawcett. 1986. *A Textbook of Histology.* Eleventh edition. Philadelphia: Saunders.

Ross, M., and E. Reith. 1985. *Histology: A Text and Atlas.* New York: Harper & Row.

24 PROTECTION, SUPPORT, AND MOVEMENT

"All Right—GO!"

On that command, Susan Butcher's trained huskies leap forward for another race along Alaska's Iditarod Trail. For the next eleven days and nights, they will tow a 90-kilogram sled—that's 200 pounds—across 1,860 kilometers of snow, ice, and treacherous rivers between Anchorage and Nome. Three times in the past ten years, Butcher has mushed her team to victory.

Huskies have astounding strength and endurance, an outcome of artificial selection practices. Their leg bones are sturdy, yet lightweight. Their forelegs move freely, thanks to a rib cage that is deep but not too broad. Their hind legs have massive muscles. These are not the muscles of a sprinting greyhound or cheetah. They are the muscles of a load-pulling, long-distance runner. Huskies also have thick foot pads—cushions against sharp ice and frozen rock. And they have an insulating layer of thick, fine hairs next to the skin. Above this is a slightly oily layer of tougher, longer hairs that keep out biting winds and near-freezing moisture.

We humans don't even approach a huskie's stamina and built-in protection against the elements. Long before each Iditarod race begins, Butcher must start a marathoner's regimen of diet and exercise to put her arm and leg muscles in peak condition for the extraordinary effort that lies ahead. Lacking the fur of mammals native to the Far North, she must acquire clothing that will insulate and protect her during a race without restricting her movements (Figure 24.1).

In this chapter, our focus will be on human systems of protection, support, and movement. Human skin, which did not evolve in arctic environments, cannot withstand bitter cold. Muscles and bones in human legs are suitable for walking and striding, not for long-distance, load-pulling runs. From this perspective, it is mainly human ingenuity that keeps Butcher and others mushing down the Iditarod Trail—in the company of huskies supremely bred for it.

Figure 24.1 Susan Butcher and her Alaskan huskies, superbly illustrating systems of support, movement, and protection along the Iditarod Trail.

1. Nearly all animals have an integument (an outer covering, such as skin), muscles, and a skeleton.

2. Skin protects the body from abrasion, ultraviolet radiation, bacterial attack, and other environmental insults. It also contributes to overall body functioning, as when it helps control moisture loss.

3. Smooth muscle and cardiac muscle help move internal organs. Skeletal muscle helps move the body's limbs and other structural elements. When properly stimulated, cells of all three types of muscle tissue contract (shorten) in controlled ways, then return to the resting position.

4. Skeletal systems interact with muscles to bring about movement. They protect and support soft organs and store minerals. Blood cells are produced in tissues of some bones.

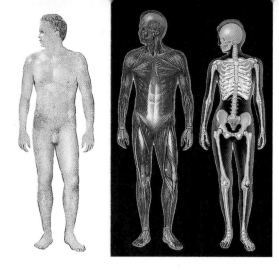

Figure 24.2 Overview of the human integumentary system (skin and its derivatives), muscle system, and skeletal system.

You have three systems to thank for your superficial features, shape, and movements. These are the integumentary, skeletal, and muscular systems (Figure 24.2).

INTEGUMENTARY SYSTEM

Animals ranging from worms to humans have an outer covering, or **integument** (after a Latin word meaning "to cover"). For you and other vertebrates, this is skin and components derived from it. Figure 24.3 shows an example. Skin's derivatives range from feathers, hairs, nails, scales, beaks, hooves, and horns to assorted glands and a lathering of slime.

Functions of Skin

No garment ever made comes close to skin's qualities. What besides skin holds its shape after repeated stretchings and washings, is waterproof, holds in moisture, kills many bacteria on contact, blocks harmful rays from the sun, repairs small cuts and burns on its own, and with a little care will last as long as you do?

Skin does more than protect against dehydration and various environmental insults. It helps stabilize the body's internal temperature. Its many small blood vessels are a reservoir for blood that can be shunted to metabolically active regions, such as legs running along the Iditarod Trail. Skin also produces the vitamin D required for calcium metabolism. And signals from skin's sensory receptors help the brain assess what's going on in the outside world.

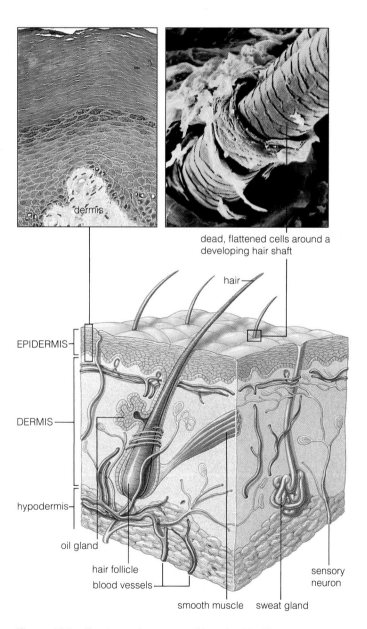

dead, flattened cells around a developing hair shaft

hair

EPIDERMIS

DERMIS

hypodermis

oil gland

hair follicle

blood vessels

smooth muscle   sweat gland

sensory neuron

Figure 24.3 Two-layered structure of human skin. The hypodermis is a layer beneath the skin, not part of it.

Focus on Health

Sunlight and Skin

Are you someone who tans by rotating beneath the sun's rays or tanning lamps, like a chicken in an oven broiler? If so, remember what a broiler does to the chicken.

Exposure to the sun's ultraviolet light stimulates your melanin-producing cells. Continued exposure increases the amount of melanin in light skin, and so visibly darkens it (the sought-after tan). Tanning provides some protection against ultraviolet radiation. But prolonged exposure causes elastin fibers to clump together in the dermis, so skin loses its resiliency. In time it starts to look like old shoe leather.

Prolonged exposure to ultraviolet radiation also suppresses the immune system. For instance, infection-fighting phagocytes in the epidermis combat viruses and bacteria. Sunburns interfere with their functioning. This may be why sunburns can trigger small, painful blisters (*cold sores*) that announce the recurrence of a *Herpes simplex* infection. Nearly everyone harbors this virus. It remains hidden in the face, inside a ganglion (a cluster of neuron cell bodies). Stress factors, including sunburn, can activate the virus. When that happens, virus particles move down the neurons to their endings near the skin. There they infect epithelial cells and cause the painful skin eruptions.

Ultraviolet radiation also can activate proto-oncogenes that can trigger cancerous transformation of skin cells. *Epidermal skin cancers* start out as scaly, reddened bumps. They grow rapidly and can spread to adjacent lymph nodes unless they are surgically removed. *Basal cell carcinomas* start out as small, shiny bumps and slowly grow into ulcers with beaded margins (page 159). Their threat ceases if they are surgically removed in time.

Structure of Skin

Your skin weighs about four kilograms (nine pounds), and most is thin as a paper towel. Skin has two distinct regions—an outermost **epidermis** and an underlying **dermis** (Figure 24.3). Below this, a tissue (hypodermis) anchors skin yet allows it to move a bit. Fat stored in the hypodermis insulates the body and cushions some body parts.

Epidermis. Epidermis, like puff pastry, consists of stacked sheets (epithelia). Cells of the uppermost sheets are dead, with flattened plasma membranes. Most are filled with **keratin**, a water-insoluble protein they made while still alive. All these keratin-filled bags form a tough, waterproof covering. Millions wear off daily, but rapid, ongoing cell divisions in the epidermis push up replacements and also help skin mend fast after cuts or burns. Deep in the epidermis, other cells (melanocytes) synthesize **melanin**, a brownish black pigment. Skin color and sensitivity to ultraviolet radiation depend on melanocyte distribution and activity. Hemoglobin and carotene pigments also contribute to skin color.

Dermis. The dermis is mostly dense connective tissue that resists everyday stretching, although chronic abrasion can separate it from the epidermis and cause blisters. Blood vessels and tips of sensory receptors thread through it. Sweat glands, oil glands, and husklike cavities (hair follicles) develop from epidermal cells but reside mostly in the dermis. Your 2-1/2 million sweat glands secrete sweat, which helps dissipate excess body heat. Oil gland secretions soften and lubricate hair and skin. They also kill surface bacteria. In *acne*, bacteria have managed to infect oil gland ducts.

An average scalp has 100,000 hairs—flexible, keratinized shafts rooted in skin. The number depends on genes, hormones, and nutrition. In a shaft's outer layer, flattened cells overlap like roof shingles (Figure 24.3). These have frizzed out when hair has "split ends."

As we age, epidermal cells divide less often, so our skin becomes thinner. Secretions that kept it soft and moistened start dwindling. Collagen and elastin fibers in the dermis break down and become sparser, so skin loses elasticity and wrinkles deepen. Skin ages faster with tobacco smoking, prolonged exposure to dry winds, and excessive tanning (see the *Focus* essay).

With its multiple layers of keratinized, melanin-shielded epidermal cells, skin helps the body conserve water, avoid damage by ultraviolet radiation, and resist mechanical stress.

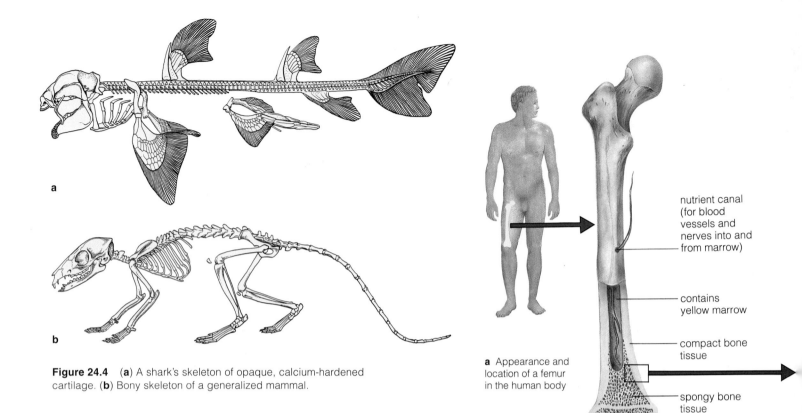

Figure 24.4 (**a**) A shark's skeleton of opaque, calcium-hardened cartilage. (**b**) Bony skeleton of a generalized mammal.

a Appearance and location of a femur in the human body

nutrient canal (for blood vessels and nerves into and from marrow)

contains yellow marrow

compact bone tissue

spongy bone tissue

Figure 24.5 Structure of a femur (thighbone), one of the long bones of mammals.

SKELETAL SYSTEM

Humans and other vertebrates have an internal skeleton (an endoskeleton). Sharks have one of the simplest kinds; it consists of opaque, calcium-hardened cartilage (Figure 24.4). In you and most other adult vertebrates, the skeleton is primarily bone.

Characteristics of Bone

Functions of Bone.   Just as skin is more than a baglike covering, so is your skeletal system more than a frame to hang muscles on. Its major parts, **bones**, are complex organs that serve these functions:

***

1. *Movement*. By interacting with skeletal muscles, bones maintain or change the position of body parts.

2. *Protection*. Bones are hard compartments that enclose and protect the brain, lungs, and other organs.

3. *Support*. Bones support and anchor muscles.

4. *Mineral storage*. Bone tissue is a bank for depositing and withdrawing mineral ions, and so helps maintain body fluids and support metabolic activities.

5. *Blood cell formation*. Blood cells are produced in some bones.

***

Bone Structure.   In size, human bones range from tiny earbones to clublike thighbones. In shape, bones are long, short (or cubelike), flat, and irregular. All incorporate epithelial and connective tissues as well as bone tissue. The calcium-hardened bone tissue consists of living cells and collagen fibers in a ground substance.

Figure 24.5 shows the organization of a thighbone. *Compact* bone tissue in its shaft and at its two ends withstands mechanical shocks. The tissue's thin, dense layers are arranged as multiple cylinders around small, interconnected canals. Blood vessels and nerves in these "Haversian canals" service the living bone cells.

Within the bone ends and shaft, *spongy* bone tissue imparts strength without adding too much weight. With its abundant spaces, the tissue has a spongelike appearance, but its flattened parts are quite firm. **Red marrow** fills the spaces in some bones, including the breastbone. Red marrow is a major site of blood cell formation. Cavities in most mature bones contain **yellow marrow**. Although yellow marrow is mostly fat, it converts to red marrow and produces red blood cells when blood loss from the body is severe.

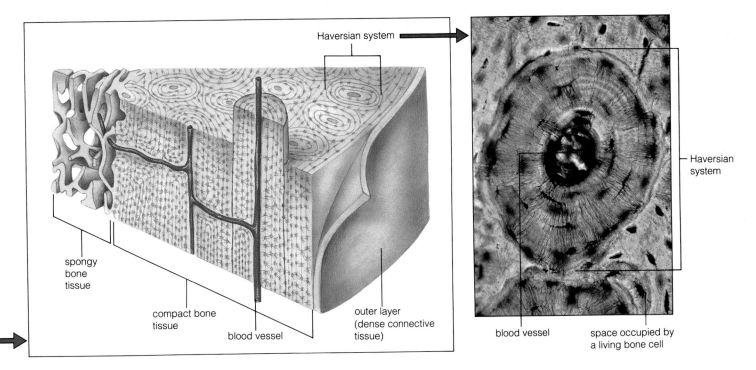

b Appearance of spongy bone tissue and compact bone tissue in a femur. The thin, dense layers of compact bone tissue are arranged as cylinders around interconnecting canals, which contain blood vessels and nerves. Each cylinder is a Haversian system.

c Micrograph of a Haversian system. The blood vessel services osteocytes, living bone cells within small spaces in bone tissue. Small tunnels connect neighboring spaces.

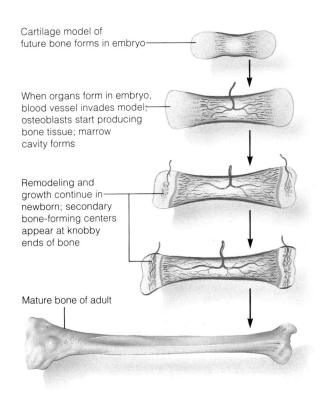

Figure 24.6 Long bone formation, starting with osteoblast activity in a cartilage model (here, already formed in the embryo). Bone-forming cells are active first in the shaft region, then at the knobby ends. In time, the only cartilage left is at the ends.

How Bones Develop. Before a long bone forms in a growing embryo, a cartilage model for it forms. Later, bone-forming cells (osteoblasts) secrete material inside the model's shaft and on its surface. A marrow cavity develops as cartilage breaks down inside the shaft (Figure 24.6). When bone-forming cells are finally surrounded by their own secretions, we call them osteocytes (living bone cells). Through their metabolic activities, these cells maintain mature bones.

Bone Tissue Turnover. Minerals are continually deposited in bone tissue and withdrawn from it. Exercising stimulates mineral deposition in adult bones and generally increases bone density. Stress or injury triggers withdrawals of mineral ions. As children grow, their bones are remodeled, and this involves tissue turnovers. For instance, a thighbone is made thicker and stronger as bone cells *deposit* minerals at the surface of its shaft. At the same time, the bone is being made less heavy as other bone cells *destroy* bone tissue inside the shaft.

Bone turnover is coordinated to maintain calcium levels. Bone cells secrete enzymes that break down bone tissue, releasing calcium and other minerals required for metabolism into interstitial fluid. The bloodstream distributes the calcium through the body.

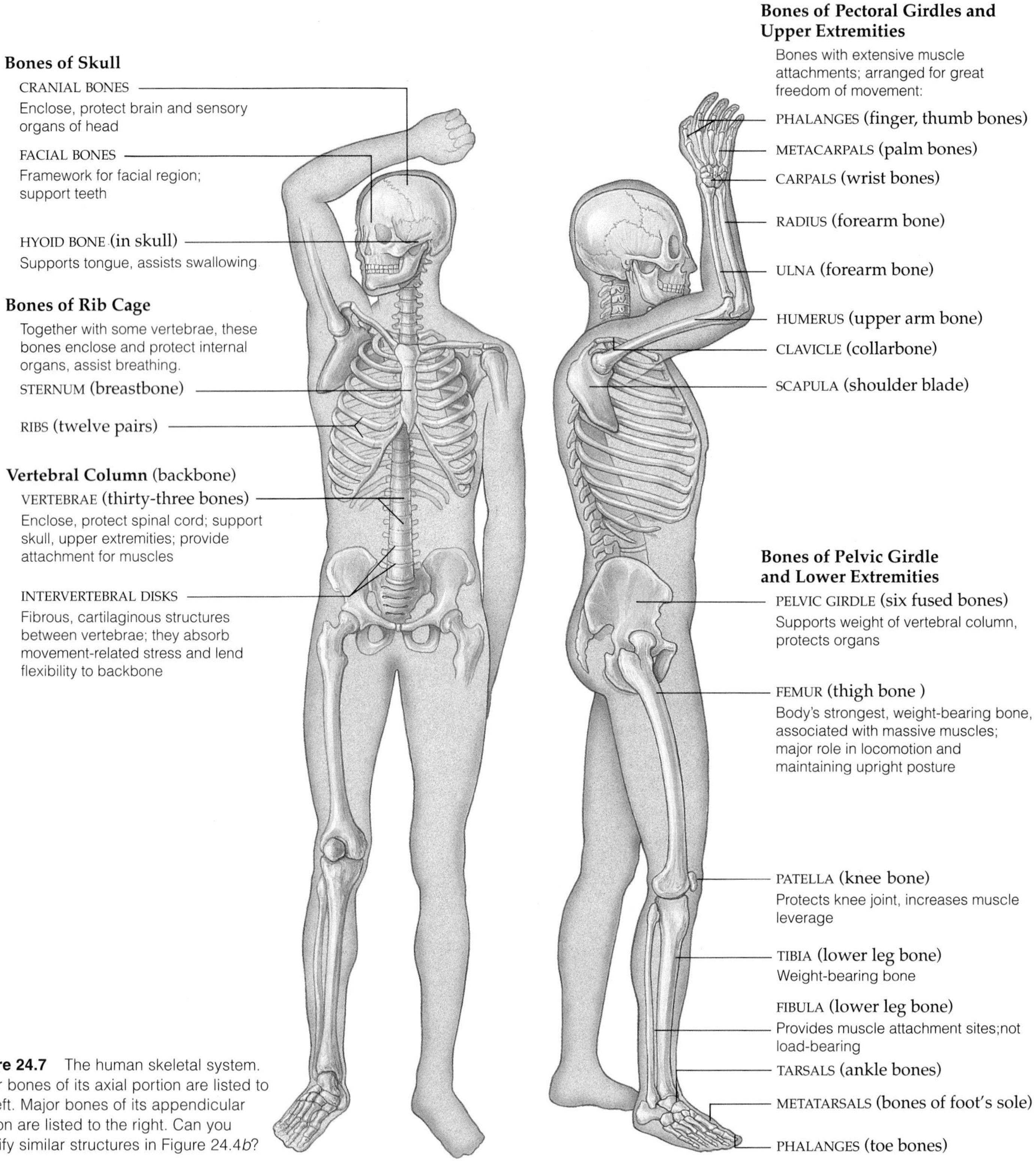

Bones of Skull

CRANIAL BONES
Enclose, protect brain and sensory organs of head

FACIAL BONES
Framework for facial region; support teeth

HYOID BONE (in skull)
Supports tongue, assists swallowing

Bones of Rib Cage

Together with some vertebrae, these bones enclose and protect internal organs, assist breathing.

STERNUM (breastbone)

RIBS (twelve pairs)

Vertebral Column (backbone)

VERTEBRAE (thirty-three bones)
Enclose, protect spinal cord; support skull, upper extremities; provide attachment for muscles

INTERVERTEBRAL DISKS
Fibrous, cartilaginous structures between vertebrae; they absorb movement-related stress and lend flexibility to backbone

Bones of Pectoral Girdles and Upper Extremities

Bones with extensive muscle attachments; arranged for great freedom of movement:

PHALANGES (finger, thumb bones)

METACARPALS (palm bones)

CARPALS (wrist bones)

RADIUS (forearm bone)

ULNA (forearm bone)

HUMERUS (upper arm bone)

CLAVICLE (collarbone)

SCAPULA (shoulder blade)

Bones of Pelvic Girdle and Lower Extremities

PELVIC GIRDLE (six fused bones)
Supports weight of vertebral column, protects organs

FEMUR (thigh bone)
Body's strongest, weight-bearing bone, associated with massive muscles; major role in locomotion and maintaining upright posture

PATELLA (knee bone)
Protects knee joint, increases muscle leverage

TIBIA (lower leg bone)
Weight-bearing bone

FIBULA (lower leg bone)
Provides muscle attachment sites;not load-bearing

TARSALS (ankle bones)

METATARSALS (bones of foot's sole)

PHALANGES (toe bones)

Figure 24.7 The human skeletal system. Major bones of its axial portion are listed to the left. Major bones of its appendicular portion are listed to the right. Can you identify similar structures in Figure 24.4*b*?

Skeletal Structure

Our distant four-legged ancestors started walking upright about 4 million years ago and we haven't stopped since. As Figure 16.5 suggests, an upright posture puts the backbone into an S-shaped curve. This evolutionary experiment is not a perfectly workable modification of the four-legged plan. The older we get, the longer we have resisted gravity in a compromised way—and the more lower back pain we suffer.

Figure 24.7 lists a few of the 206 human bones and their functions. Some belong to the *axial* portion of the skeleton. They include the skull, backbone (vertebral column), ribs, and breastbone. The vertebral column

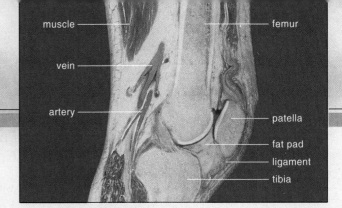

a Longitudinal section through the knee joint

Focus on Health

The Battered Knee

A knee joint is a wonderful thing. It lets you swing, bend, and twist the long bones below it. When you run, a knee joint absorbs the full force of your weight each time a foot hits the ground. But this freely movable joint can withstand only so much stress.

Knee joints are self-lubricating. Ligaments, tendons, and fibers strap its bones and muscles together. A small bone (patella) embedded in one ligament stabilizes and protects the joint. Where one bone touches another, cartilage cushions them and absorbs shocks.

Suddenly stretch a knee joint too far or twist it in an odd way and you may *strain* it. Tear its ligaments or tendons and you *sprain* it. Move the wrong way and you may dislocate its bones. Overexert your legs, and tendons can become stiff and sore (*tendonitis*). Besides this, blows to the knee during football and other collision sports can sever an entire ligament. Surgically repairing the connective tissue fibers of a severed ligament is like sewing two hairbrushes together. The sewing must be done within ten days of the injury. Wandering about in the joint's lubricating fluid are phagocytic cells that clean up after everyday wear and tear. Present them with torn ligaments and they will indiscriminately turn the tissue to mush.

has cartilage-containing disks that serve as shock absorbers and flex points. With severe or rapid shocks, a disk may slip out of place and possibly rupture. Such *herniated disks* may protrude into nerves or the spinal cord and cause excruciating pain.

Other bones belong to the *appendicular* portion of the skeleton. They include the pectoral girdles (at the shoulders), arms, hands, pelvic girdle (at the hips), legs, and feet. Slender collarbones and flat shoulder blades are rather flimsily arranged parts of the pectoral girdles. Fall on an outstretched arm and you might dislocate your shoulder or fracture a collarbone, which is the bone most frequently broken.

Tendons (cords or straps of dense connective tissue) attach muscle to bones. **Ligaments** (straps of dense connective tissue) bridge **joints**, which are areas of contact or near-contact between bones. Some joints are freely movable, and this can be bad as well as good (see the *Focus* essay). As people age, cartilage covering the bone ends of freely movable joints may wear away, a condition called *osteoarthritis*. By contrast, *rheumatoid arthritis* (page 447) is a degenerative disorder.

At certain joints, cartilage intervenes between bones and limits movement. This is true of joints in the backbone. At other joints, fibrous tissue connects and entirely fills the space between bones. Fibrous joints loosely connect the flat skull bones of newborns. Together with "soft spots," which are membranous areas between bones, these joints allow the bones to slide over each other a bit during childbirth and so prevent skull fractures. In time, the joints and soft spots harden and the bones become fused into a single unit.

With increasing age, the backbone, hip bones, and other bones decrease in mass, especially in women. This progressive bone deterioration is called *osteoporosis* (Figure 24.8). In time, a weakened backbone may collapse and curve. This lowers the rib cage, which causes problems for internal organs. Decreasing osteoblast activity, calcium loss, sex hormone deficiencies, excessive protein intake, and decreased physical activity may contribute to the disorder.

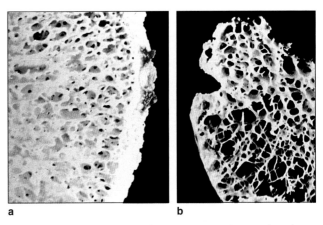

a   b

Figure 24.8 Osteoporosis. (**a**) In normal bone tissue, mineral deposits continually replace mineral withdrawals. (**b**) After the onset of osteoporosis, replacements cannot keep pace with withdrawals. In time the tissue erodes, and bones become hollow and brittle.

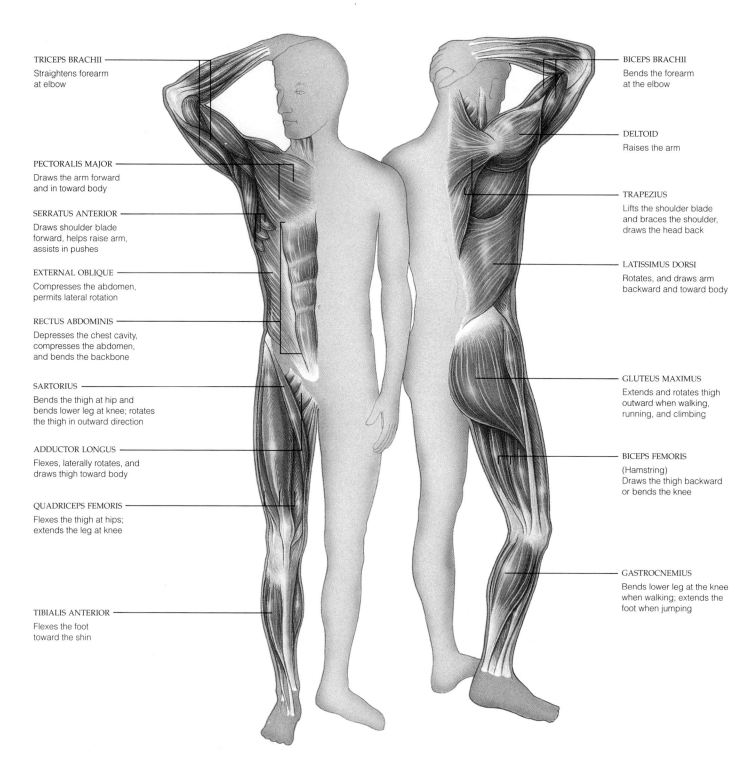

TRICEPS BRACHII

Straightens forearm
at elbow

PECTORALIS MAJOR

Draws the arm forward
and in toward body

SERRATUS ANTERIOR

Draws shoulder blade
forward, helps raise arm,
assists in pushes

EXTERNAL OBLIQUE

Compresses the abdomen,
permits lateral rotation

RECTUS ABDOMINIS

Depresses the chest cavity,
compresses the abdomen,
and bends the backbone

SARTORIUS

Bends the thigh at hip and
bends lower leg at knee; rotates
the thigh in outward direction

ADDUCTOR LONGUS

Flexes, laterally rotates, and
draws thigh toward body

QUADRICEPS FEMORIS

Flexes the thigh at hips;
extends the leg at knee

TIBIALIS ANTERIOR

Flexes the foot
toward the shin

BICEPS BRACHII

Bends the forearm
at the elbow

DELTOID

Raises the arm

TRAPEZIUS

Lifts the shoulder blade
and braces the shoulder,
draws the head back

LATISSIMUS DORSI

Rotates, and draws arm
backward and toward body

GLUTEUS MAXIMUS

Extends and rotates thigh
outward when walking,
running, and climbing

BICEPS FEMORIS

(Hamstring)
Draws the thigh backward
or bends the knee

GASTROCNEMIUS

Bends lower leg at the knee
when walking; extends the
foot when jumping

Figure 24.9   The human muscular system, showing some of the outermost skeletal muscles.

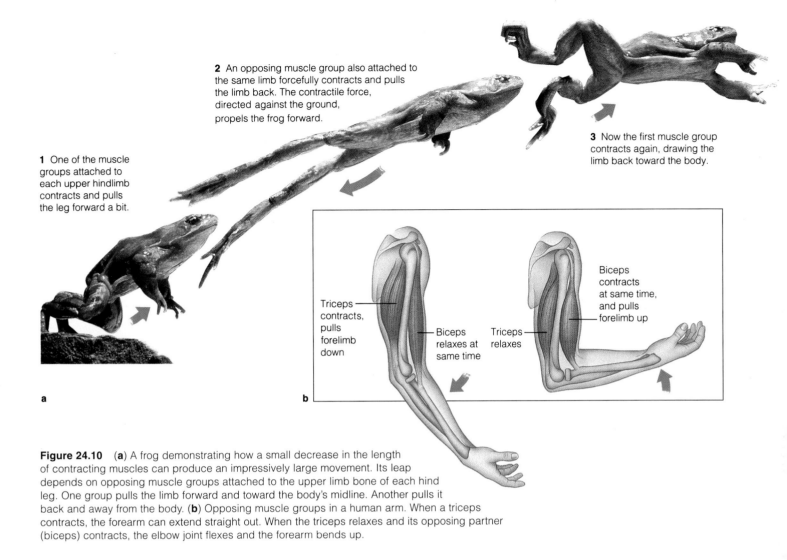

2 An opposing muscle group also attached to the same limb forcefully contracts and pulls the limb back. The contractile force, directed against the ground, propels the frog forward.

1 One of the muscle groups attached to each upper hindlimb contracts and pulls the leg forward a bit.

3 Now the first muscle group contracts again, drawing the limb back toward the body.

Triceps contracts, pulls forelimb down

Biceps relaxes at same time

Triceps relaxes

Biceps contracts at same time, and pulls forelimb up

a

b

Figure 24.10 (**a**) A frog demonstrating how a small decrease in the length of contracting muscles can produce an impressively large movement. Its leap depends on opposing muscle groups attached to the upper limb bone of each hind leg. One group pulls the limb forward and toward the body's midline. Another pulls it back and away from the body. (**b**) Opposing muscle groups in a human arm. When a triceps contracts, the forearm can extend straight out. When the triceps relaxes and its opposing partner (biceps) contracts, the elbow joint flexes and the forearm bends up.

MUSCULAR SYSTEM

How Muscles and Bones Interact

We turn now to skeletal muscle. In muscle tissue, recall, cells contract (shorten) in response to stimulation, then lengthen and so return to their original state. When you dance, scribble notes, or tilt your head skyward, contracting muscle cells are helping to move your body or to change the positions of its parts.

Your body is equipped with more than 600 skeletal muscles. Figure 24.9 shows a few and lists their functions. By definition, a **skeletal muscle** contains hundreds to many thousands of muscle cells, which look like long, striped fibers. The fiberlike cells are arranged in bundles. Connective tissue surrounds the bundles and extends beyond them to form tendons, which attach both ends of the muscle to bone.

Most attachments between skeletal muscle and bones are rather like the gearshift in a car. They are a lever system, in which a rigid rod is attached to a fixed point but able to move about at it. The muscles connect with bones (rigid rods) near a joint (fixed point). When the muscles contract, they transmit force to the bones and make them move. Take a look at Figure 24.10*b*, then extend your right arm. Now place your left hand over the upper arm's biceps and slowly "bend your elbow," so that you can feel the biceps contract. Even if your biceps contracts only a bit, it can produce a large movement in the forearm bone connected to it. This is true of most leverlike arrangements.

Skeletal muscles interact with one another as well as with bones. Some are arranged as pairs or groups that work together to promote the same movement. Others, like those in Figure 24.10, work in opposition; the action of one opposes or reverses the action of another.

Only skeletal muscle is the functional partner of bone. As described on page 377, smooth muscle is present mostly in walls of internal organs, such as the stomach and intestines. Cardiac muscle occurs only in the heart wall, and its action is described in a later chapter.

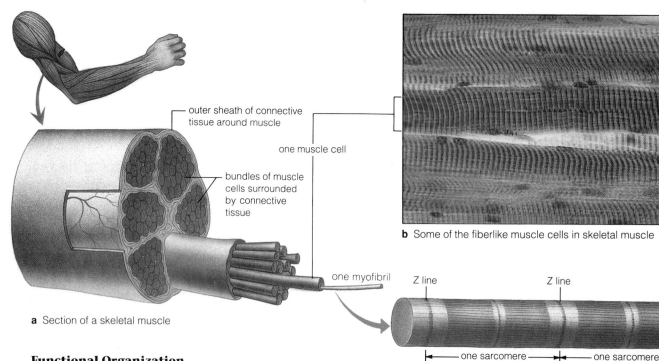

outer sheath of connective tissue around muscle

one muscle cell

bundles of muscle cells surrounded by connective tissue

one myofibril

a Section of a skeletal muscle

b Some of the fiberlike muscle cells in skeletal muscle

Z line   Z line   Z line

one sarcomere   one sarcomere

c Two of the myofibrils inside a muscle cell. In each myofibril, the contractile units called sarcomeres are arranged one after the other. The oval-shaped organelle is a mitochondrion.

Figure 24.11 (**a**) Components of a skeletal muscle. (**b**) Light micrograph of skeletal muscle cells. (**c**) Each muscle cell contains myofibrils. Each myofibril is functionally divided into sarcomeres, the basic units of contraction. Dark "bands" (Z lines) define the two ends of each sarcomere.

Functional Organization of a Skeletal Muscle

Bones move—they are pulled in some direction—when the skeletal muscles that are attached to them shorten. When a skeletal muscle shortens, its component muscle cells are shortening. When a muscle cell shortens, many units of contraction within that cell are shortening. The basic units of contraction are called **sarcomeres**.

Take a look at Figure 24.11. It shows how the bundles of cells in a skeletal muscle run parallel with the muscle itself. It also shows that each muscle cell contains **myofibrils**, threadlike structures packed together in parallel array. Every one of those myofibrils is functionally divided into sarcomeres, which are arranged one after another along its length.

Now look closely at the micrograph in Figure 24.11*c*, and you see that a sarcomere contains many filaments, side by side in parallel array. Some of the filaments are thin, others are thick. Each *thin* filament is actually two beaded strands, twisted together. The "beads" are ball-shaped molecules of the protein **actin**:

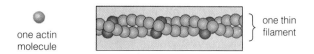

one actin molecule

one thin filament

Each *thick* filament consists of molecules of **myosin**, a protein with a head and long tail. The myosin tails are packed together in parallel. The heads, which look like double-headed golf clubs, stick out to the sides:

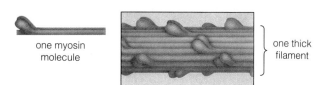

one myosin molecule

one thick filament

The orderly arrays of actin and myosin filaments in the sarcomeres give skeletal muscle (and cardiac muscle) its striped appearance.

Think of the parallel orientation of all the myofibrils, muscle cells, and muscle bundles within a skeletal muscle. The orientation of a skeletal muscle's component parts focuses the force of contraction onto the bone in a particular direction.

A skeletal muscle shortens through the combined decreases in length of its sarcomeres, the basic units of contraction.

The parallel orientation of a muscle's component parts directs the force of contraction toward a bone that must be pulled in some direction.

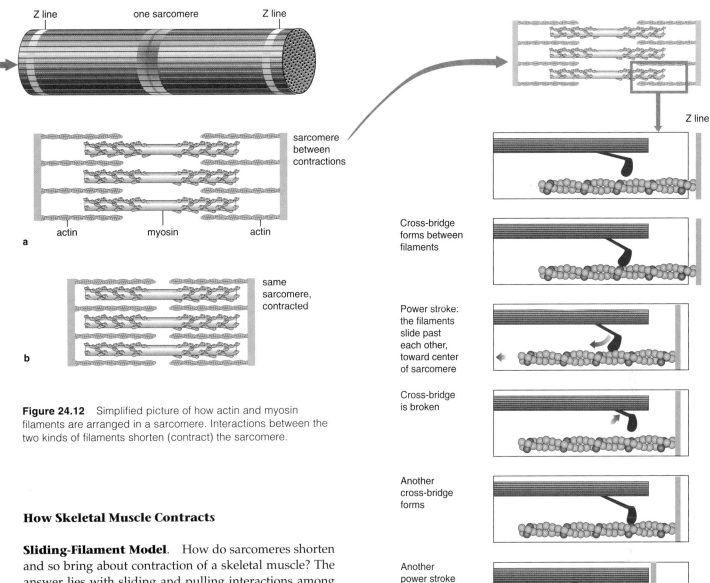

Z line   one sarcomere   Z line

sarcomere
between
contractions

actin   myosin   actin

a

same
sarcomere,
contracted

b

Figure 24.12 Simplified picture of how actin and myosin filaments are arranged in a sarcomere. Interactions between the two kinds of filaments shorten (contract) the sarcomere.

Z line

Cross-bridge
forms between
filaments

Power stroke:
the filaments
slide past
each other,
toward center
of sarcomere

Cross-bridge
is broken

Another
cross-bridge
forms

Another
power stroke
toward
center of
sarcomere

Figure 24.13 Sliding-filament model of contraction in the sarcomeres of muscle cells. For simplicity, the action of only one myosin head is shown.

How Skeletal Muscle Contracts

Sliding-Filament Model. How do sarcomeres shorten and so bring about contraction of a skeletal muscle? The answer lies with sliding and pulling interactions among the sarcomere's filaments. As Figure 24.12 shows, there are two sets of actin filaments, attached to opposite sides of the sarcomere and extending partway to its center. There is one set of myosin filaments. These don't extend all the way to the sides, but they partially overlap the actin filaments. During contraction, myosin filaments physically slide along and *pull* the two sets of actin filaments toward the center of the sarcomere, which thereby shortens. This is the key premise of the **sliding-filament model** of muscle contraction.

The interactions between myosin and actin filaments occur through **cross-bridge formation**. As indicated by Figure 24.13, each cross-bridge is an attachment between a myosin "head" and a binding site on actin. When a muscle cell is stimulated, myosin heads are energized. They attach to an adjacent actin filament and tilt in a short power stroke toward the sarcomere's center. An input of energy from ATP drives the power stroke. During the stroke, the heads pull the

actin filament along with them. Then a new energy input makes the heads let go, attach to another region of the actin filament, tilt in another power stroke, and so on down the line. A single contraction of a sarcomere takes a whole series of power strokes.

Energy-driven interactions between myosin and actin filaments shorten the many sarcomeres of a muscle cell and collectively account for its contraction.

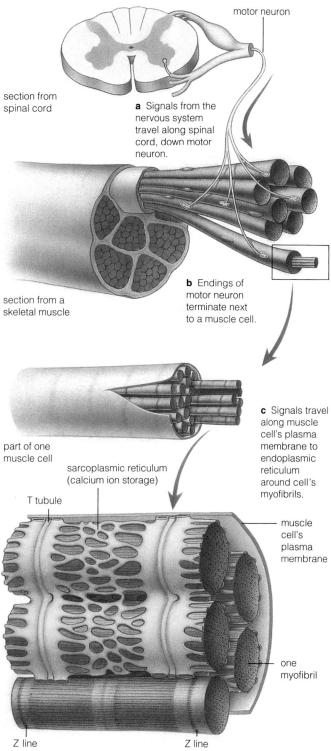

a Signals from the nervous system travel along spinal cord, down motor neuron.

section from spinal cord

motor neuron

section from a skeletal muscle

b Endings of motor neuron terminate next to a muscle cell.

part of one muscle cell

sarcoplasmic reticulum (calcium ion storage)

c Signals travel along muscle cell's plasma membrane to endoplasmic reticulum around cell's myofibrils.

T tubule

muscle cell's plasma membrane

one myofibril

Z line   Z line

d Signals trigger calcium release from sarcoplasmic reticulum lacing around the myofibrils. Calcium allows actin and myosin filaments inside the myofibrils to interact and bring about contraction.

Figure 24.14 Pathway for signals from the nervous system that stimulate contraction of skeletal muscle. The plasma membrane of each muscle cell surrounds myofibrils and connects with inward-threading tubes (T tubules). The membrane-bound tubes are close to the sarcoplasmic reticulum, a calcium-storing system that functions in the control of contraction.

Control of Contraction. Skeletal muscles contract to help move the body and its assorted parts at certain times, in certain ways—and they do so in response to commands from the nervous system. The nervous system communicates with muscles by way of **motor neurons**, which deliver signals that can stimulate or inhibit contraction of muscle cells. We will consider these signals in Chapter 30. For now, the following brief description will give you an idea of what goes on.

Like neurons, muscle cells are "excitable." As is true of all cells, they show a difference in electric charge across their plasma membrane. (The cytoplasm just beneath the membrane is a bit more negative than the fluid outside the membrane.) But in excitable cells only, the difference in charge *reverses* suddenly and briefly in response to adequate stimulation. This sudden reversal in charge, an **action potential**, results from a flow of charged ions across the membrane. The commotion spreads without diminishing along the membrane, away from the point of stimulation.

Figure 24.14 shows what happens after commands from the nervous system stimulate action potentials in a muscle cell. Action potentials spread from the point of stimulation and rapidly reach small, tubular extensions of the plasma membrane. The small tubes connect with a system of membrane-bound chambers that thread lacily around the cell's myofibrils. That system—called the **sarcoplasmic reticulum**—takes up, stores, and releases calcium ions in controlled ways.

When the signals reach the sarcoplasmic reticulum, they trigger the outward flow of calcium ions from it. The released calcium ions diffuse into the myofibrils and reach actin filaments. Before this, the muscle was resting; the actin binding sites were blocked and myosin could not form cross-bridges with them. The arrival of calcium ions clears these sites—so contraction can proceed. Afterward, calcium ions are actively transported back into the membrane storage system.

Commands from the nervous system initiate action potentials in muscle cells. These are the signals for cross-bridge formation—and contraction.

Muscle Tension. Collectively, the cross-bridges that form during contraction exert muscle **tension**. This is a mechanical force that resists gravity, the weight of packages or dumbbells or other objects being lifted, and other opposing forces. When muscle tension is greater than the forces opposing it, contracting muscle cells *shorten*. When opposing forces are stronger, muscle cells *lengthen*. People who engage in isometric exercises consciously maintain contraction of muscles at a constant length for a brief time.

Exercise Levels and Energy for Contraction

A resting muscle cell has a rather tiny supply of ATP. When called upon to contract, however, its demands for ATP skyrocket. At that time, the cell plucks phosphate from creatine phosphate (an organic compound) and attaches it to ADP. This ATP-forming reaction is simple, fast, and good for a few seconds' worth of contraction before the cell depletes its limited store of creatine phosphate. It also buys time for other, slower ATP-forming pathways to kick in.

During prolonged, moderate exercise, the oxygen-requiring reactions of aerobic respiration provide most of the ATP required for contraction. At first, the glucose required for the reactions comes from glycogen stores inside muscle cells. However, when contraction must be sustained, the cells rapidly switch to glucose and fatty acids delivered to them by the bloodstream. (Here you may wish to review Figure 6.10.)

Suppose the level of exercise is so intense, it exceeds the capacity of the body's respiratory and circulatory systems to deliver oxygen to muscles. Then, muscle cells produce more and more ATP by lactate fermentation. In this anaerobic pathway, recall, glucose is broken down to lactate. The ATP yield is small, and lactate builds up in the cell.

Contracting skeletal muscle cells form ATP by creatine phosphate breakdown, aerobic respiration, and fermentation.

Muscle Strength and Muscle Fatigue

Arnold Schwarzenegger, with his bulging muscles, is Hollywood's icon of physical strength. His strength depends on how forcefully his muscles can contract. That, in turn, depends on (1) muscle size, (2) how many muscle cells are contracting, and (3) how rapidly the nervous system is stimulating them.

Look closely at Figure 24.14, and you see that the endings of a motor neuron form junctions with more than one muscle cell. Each motor neuron and the muscle cells that receive its messages comprise a motor unit.

Physiologists can record the responses of motor units to stimulation. They create "artificial" action potentials (electrical impulses), then record any changes in muscle contraction. With a single, brief stimulus, a muscle contracts briefly, then relaxes. This response is a **muscle twitch** (Figure 24.15a). A second stimulus applied before the response is over makes the muscle twitch again. A **tetanus** is a large contraction in which a motor unit is stimulated repeatedly, so that twitches mechanically run together. (In a disease by the same name, toxins prevent muscle relaxation.)

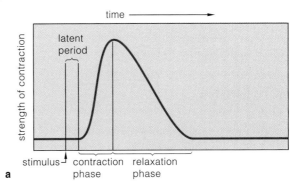

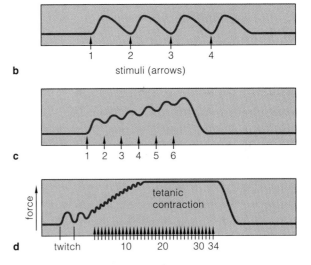

Figure 24.15 Recordings of twitches in artificially stimulated muscles. (**a**) A single twitch. (**b**) Two stimulations per second cause a series of twitches. (**c**) Six per second cause a summation of twitches. (**d**) About twenty per second cause tetanic contraction.

Suppose continuous, high-frequency stimulation keeps a muscle cell in a state of tetanic contraction. In time, the result will be **muscle fatigue**—a decline in the tension resulting from its cross-bridges. After a few minutes of rest, a fatigued muscle will contract again in response to stimulation. But the extent of its recovery depends on how long and how frequently it was stimulated before. Muscles associated with brief, intense exercise fatigue fast but recover rapidly. This is true of weightlifting. Muscles associated with prolonged, moderate exercise fatigue more slowly but take longer to recover—often by as much as 24 hours. This is true of hiking.

With regular exercise, muscles may become larger, have a higher metabolic capacity, and become more resistant to fatigue. Among other things, exercising improves circulation and increases the number of mitochondria in muscle cells. The *Focus* essay on the next page takes a look at what can happen when the desire for stronger muscles is carried to an extreme.

Muscle Mania

Some call it a will-to-win gone bonkers, a desire for muscles larger than life. Others say it's a modern necessity in athletic competition. Either way, many athletes illegally use performance-enhancing drugs, mainly stanazol and other anabolic steroids. Ten athletes were disqualified from the 1988 Olympics for using the banned drugs. Others dropped out rather than submit to stringent drug tests.

Each year in the United States alone, about 1 million athletes use anabolic steroids. Maybe 85 percent of all professional football players use or have used them. Even adolescent boys use them to gain a winning edge in wrestling, football, and weight-lifting tournaments.

What Anabolic Steroids Are.　Anabolic steroids are synthetic hormones, developed in the 1930s as therapeutic drugs that could mimic a sex hormone, testosterone. Secondary sexual traits, among other things, depend on testosterone. Under its influence, boys get a deeper voice; more hair on their facial, underarm, and pubic skin; more sweat gland secretions; and more massive muscles in the arms, legs, shoulders, and elsewhere. Testosterone also stimulates the more aggressive behavior often associated with maleness. Anabolic steroids can do these things as well—but at a physical and psychological price.

What Anabolic Steroids Do.　Twenty or so varieties of anabolic steroids stimulate the synthesis of protein molecules, including muscle proteins. Supposedly, they induce rapid gains in muscle mass and strength during weight-training exercise programs. The claim is disputed (results of most studies are based on too few subjects). Even so, testimonials pour in from weightlifters, football players, and other athletes who specialize in events involving "brute power." Users commonly combine daily oral doses with a single hefty injection each month.

Yet athletes who use steroids suffer minor and major side effects. In men, acne, baldness, shrinking testes, and infertility are early signs of toxicity. These symptoms arise when high blood levels of anabolic steroids trigger a sharp drop in normal testosterone production. The drugs also may cause early heart disease. Even brief or occasional use may damage kidneys or set the stage for cancer of the liver, testes, and prostate gland. In women, anabolic steroids deepen the voice and produce pronounced facial hair. Menstrual cycles become irregular. Breasts may shrink and the clitoris may become grossly enlarged.

'Roid Rage.　Not all steroid users have developed severe physical side effects. Far more common are *mental* difficul-

SUMMARY

1. Skin and other integumentary systems cover the body's surface. In humans, skin protects against abrasion, bacterial attack, ultraviolet radiation, and dehydration. It helps control internal temperature and serves as a blood reservoir. Its receptors detect specific stimuli in the external environment.

2. Bones are structural elements of vertebrate skeletons. They function in movement (by interacting with skeletal muscles), protection and support of other body parts, mineral storage, and blood cell formation.

3. A human skeleton has an axial portion (skull, backbone, ribs, breastbone) and an appendicular portion (including limb bones, pelvic girdle, pectoral girdles).

4. Bones and skeletal muscles work together like a system of levers, with rigid rods (bones) moving at fixed points (joints). Pairs and groups of muscles work together or in opposition to bring about movement of the body or positional changes in its parts.

5. In response to action potentials (commands from the nervous system), cells of smooth, cardiac, and skeletal muscle tissues contract (shorten).

6. Each skeletal muscle cell contains many threadlike myofibrils, which contain actin and myosin filaments. The filaments are organized in orderly, parallel arrays in sarcomeres, the basic units of contraction.

7. Sarcomeres contract when action potentials trigger the release of calcium ions from a membrane system (sarcoplasmic reticulum) around myofibrils. Calcium binds to actin filaments and induces changes at the binding sites that allow heads of adjacent myosin filaments to form cross-bridges. Cross-bridges form during repeated, ATP-driven power strokes. The repeated strokes make actin filaments slide past myosin filaments and so shorten the sarcomere.

3. Which is *not* a function of skin?
   a. resist abrasion   c. produce movement
   b. restrict dehydration   d. act as blood reservoir

4. _____ are shock pads and flex points.
   a. Vertebrae   c. Marrow cavities
   b. Femurs   d. Intervertebral disks

5. Blood cells form in _____ .
   a. red marrow   c. many bones
   b. all bones   d. a and c

6. In skeletal muscle, the _____ is the basic unit of contraction.
   a. myofibril   c. muscle fiber
   b. sarcomere   d. myosin filament

7. Muscle contraction requires _____ .
   a. calcium ions   c. action potential arrival
   b. ATP   d. all of the above

8. _____ provide(s) ATP for muscle contraction.
   a. Aerobic respiration   d. a and b only
   b. Lactate fermentation   e. a and c only
   c. Creatine phosphate   f. a, b, and c
      breakdown

9. Match the M words with their defining feature.
   _____ muscle   a. actin's partner
   _____ muscle twitch   b. all in the hands
   _____ muscle tension   c. blood cell production
   _____ melanin   d. tension decline
   _____ myosin   e. pigment
   _____ marrow   f. motor unit response
   _____ metacarpals   g. force exerted by cross-bridges
   _____ myofibrils   h. muscle cells bundled in
   _____ muscle fatigue   connective tissue
                           i. threadlike structures inside
                              muscle cell

Selected Key Terms

actin 392	ligament 389	sarcoplasmic
action potential 394	melanin 385	reticulum 394
bone 386	motor neuron 394	skeletal muscle 391
cross-bridge	muscle fatigue 395	sliding-filament
formation 393	muscle twitch 395	model 393
dermis 385	myofibril 392	tendon 389
epidermis 385	myosin 392	tension 394
integument 384	red marrow 386	tetanus 395
joint 389	sarcomere 392	yellow marrow 386
keratin 385		

ties, called 'roid rage or body-builder's psychosis. Some men become irritable and increasingly aggressive. Others become uncontrollably aggressive, delusional, and wildly manic. One steroid user deliberately drove his rapidly moving car into a tree.

Given the suspected dangers, why would people jeopardize themselves and their future? Possibly not everyone believes the drugs do enough damage to outweigh the "edge" they give in competition. What should a competitor do when "winning" accords athletes wealth and hero status but relegates others to the pile of also-rans? What would *you* do?

Review Questions

1. List some functions of skin. *384, 385*

2. What are some of the functions of bone tissue? *386*

3. In what respect are smooth, cardiac, and skeletal muscle tissues alike? How do they differ? *377, 391*

4. Explain how actin and myosin interact to shorten a sarcomere. Mention the steps at which ATP and calcium come into play. *393*

Self-Quiz *(Answers in Appendix IV)*

1. In upper layers of epidermis, most cells are _____ .
   a. dead   c. keratinized
   b. flattened   d. all of the above

2. Melanin is produced in the _____ .
   a. epidermis   c. hypodermis
   b. dermis   d. a and b

Readings

Eckert, R.; D. Randall; and G. Augustine. 1988. *Animal Physiology: Mechanisms and Adaptations.* Third edition. New York: Freeman.

Weeks, O. December 1989. "Vertebrate Skeletal Muscle: Power Source for Locomotion." *BioScience* 39(11):791–798.

25 DIGESTION AND HUMAN NUTRITION

Sorry, Have To Eat and Run

Fall through winter, along mountain ridges from central Canada down into northern Mexico, pronghorn antelope browse on wild sage. Come spring they move down to open grasslands and deserts, where they browse on new growth (Figure 25.1*a*). And can they eat and run! With a top speed of 95 kilometers per hour, these antelope leave coyotes and other predators in the dust.

The teeth of an antelope or any other mammal are clues to its life-style. Think of your own cheek teeth, each with a flattened crown that serves as a grinding platform. The crown of an antelope's cheek teeth dwarfs yours (Figure 25.1*b*). You probably don't rub your mouth on the ground while eating. An antelope does, and abrasive bits of soil enter its mouth. Its teeth wear down more rapidly than yours do—and natural selection has favored more crown to wear down.

After bedding down, antelope digest their cellulose-packed meals. The breakdown of cellulose—a tough, fibrous substance—starts in the first two of *four* stomach sacs. There, symbiotic bacteria produce cellulose-

digesting enzymes to the antelope's benefit as well as their own. As the enzymes do their thing, the antelope regurgitates, rechews the contents of the first two stomach sacs, then swallows again. (This is what "chewing cud" means.) Pummeling the material more than once exposes more surface area to the enzymes and gives them more time to act.

We humans, too, do amazing things with food. Somewhere in the world, one person might eat only raw whale blubber in a given day, another might eat only rice, and still another might eat only pizza and bananas, or couscous or snake meat, or drink dandelion wine. Through its metabolic magic, the human body converts these and a dizzying variety of other substances into usable energy and tissues of its own.

And with this interesting thought in mind, we start our tour of **nutrition**. The word encompasses processes by which food is ingested, digested, absorbed, and later converted to the body's own carbohydrates, lipids, proteins, and nucleic acids.

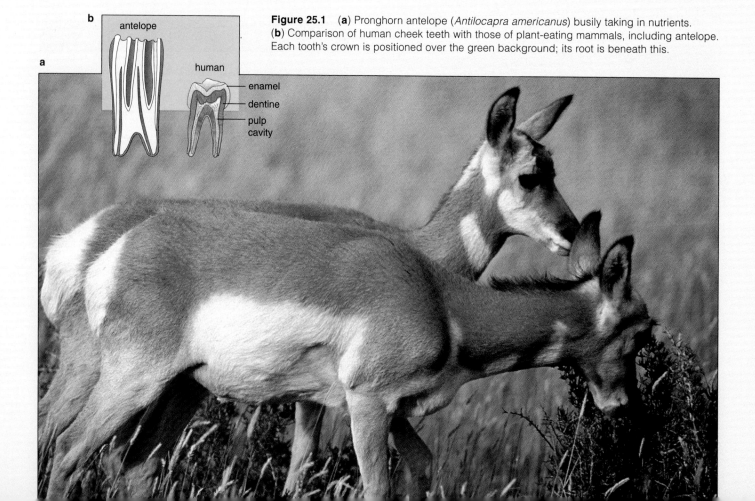

Figure 25.1 (**a**) Pronghorn antelope (*Antilocapra americanus*) busily taking in nutrients. (**b**) Comparison of human cheek teeth with those of plant-eating mammals, including antelope. Each tooth's crown is positioned over the green background; its root is beneath this.

1. Interactions among the digestive, circulatory, respiratory, and urinary systems supply the body's cells with raw materials, dispose of wastes, and maintain the volume and composition of extracellular fluid.

2. Most digestive systems have specialized parts for food transport, processing, and storage. Different parts mechanically break apart and chemically break down food, absorb breakdown products, and eliminate the unabsorbed residues.

3. To maintain an acceptable body weight and overall health, energy intake must balance energy output (by way of metabolic activity, physical exertion, and so on). Complex carbohydrates provide the most glucose, which typically is the body's main energy source.

4. Nutrition involves the intake of vitamins, minerals, and certain amino acids and fatty acids that the body cannot produce itself.

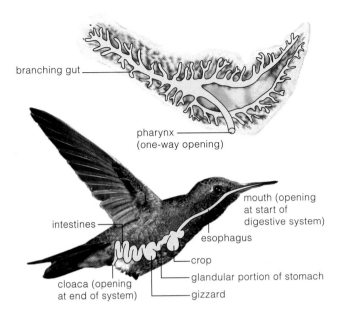

Figure 25.2 (**a**) Incomplete digestive system of a flatworm, with two-way traffic of food and undigested material through one opening. The branched gut cavity serves both digestive and circulatory functions. (**b**) Complete digestive system of a bird—basically a tube with regional specializations and an opening at each end.

FUNCTIONS OF DIGESTIVE SYSTEMS

A **digestive system** is a body cavity or tube in which food is reduced to particles, then to molecules small enough to be absorbed into the internal environment.

Some invertebrates have an *incomplete* digestive system—a saclike gut with one opening (Figure 25.2). What goes in but cannot be digested goes out the same way. Most animals have a *complete* digestive system. This is a tube with an opening at one end for food intake and an opening at the other end for elimination of unabsorbed residues. The tube also has specialized regions for food transport, processing, and storage.

Digestive systems correlate with feeding behavior. For example, meals may be few and far between for coyotes. When they do capture prey, they gorge on it. Their digestive system stores food being gulped down faster than it can be digested and absorbed. The digestive system of antelope and other ruminants (hoofed mammals with multiple stomach chambers) can accept a steady flow of tough plant parts, then slowly release nutrients from them.

This chapter focuses primarily on the human digestive system and its nutritional needs. As Figure 25.3 suggests, the system does not act alone. After nutrients are absorbed, a circulatory system distributes them to cells throughout the body. A respiratory system helps cells use nutrients by supplying them with oxygen (for

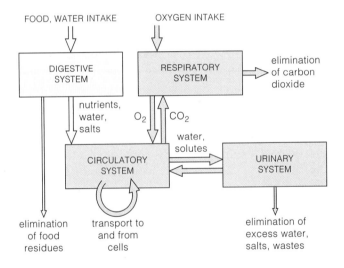

Figure 25.3 Links between the digestive, respiratory, circulatory, and urinary systems. These organ systems work together to supply the body's cells with raw materials and eliminate wastes. This chapter focuses on the digestive system; subsequent chapters will address the other systems shown here.

aerobic respiration) and taking away carbon dioxide wastes. A urinary system counters variations in the kinds and amounts of nutrients being absorbed. It helps maintain the composition and volume of the internal environment for cells.

HUMAN DIGESTIVE SYSTEM

Overview of the Components

Figure 25.4 shows the complete digestive system of an adult human and lists the functions of its component parts. Stretched out, the system would extend 6.5 to 9 meters (21 to 30 feet). Its main regions are the mouth, pharynx, esophagus, and **gut** (gastrointestinal tract). The gut is subdivided into a stomach, small intestine, large intestine (colon), rectum, and anus. Salivary glands and a gallbladder, liver, and pancreas secrete enzymes and other substances into different parts of the system. The overall functions of the system may be summarized this way:

---

1. *Motility.* Muscular movement of the system's tubular walls. The movement leads to the mechanical reduction, mixing, and passage of food material, then elimination of undigested and unabsorbed residues.

2. *Secretion.* Release of digestive enzymes, fluids, and other substances into the lumen (the space inside the digestive tube).

3. *Digestion.* Breakdown of food into particles, then into nutrient molecules small enough to be absorbed.

4. *Absorption.* Passage of digested nutrients, fluid, and ions across the tube wall and into the blood or lymph, which will distribute them through the body.

Controls Over the System

Recall that homeostatic controls respond to changes in the internal environment. By contrast, controls over digestion come into play *before* food is absorbed into the internal environment. Changes in the volume and composition of food in the stomach provide an example.

Suppose food distends the stomach wall. Local networks of neurons *in the wall itself* trigger mixing and wavelike contractions. For example, they direct smooth muscle in the stomach wall to contract behind food and relax in front of it. Contraction forces the food onward, this distends the next wall region, and so on. (Such movement is called peristalsis.) Sphincters help keep the food moving forward and prevent backflow. Sphincters are muscle rings, circularly arranged, at the start and end of the stomach and some other regions.

Suppose the stomach lumen contains fragments of proteins. This causes endocrine cells *in the stomach wall* to secrete gastrin—and this hormone activates cells with roles in protein digestion. Similarly, endocrine cells in the intestinal wall secrete hormones (secretin, CCK, and GIP) that help control digestive processes.

Control of the digestive system is more than just a local event, however. Later chapters will explain how the nervous and endocrine systems monitor the digestive system and coordinate it with other events.

Into the Mouth, Down the Tube

Food is chewed and polysaccharide breakdown begins in the oral cavity, or **mouth**. Normally, adult humans tackle food with thirty-two teeth, which are arranged this way along the upper and lower jaws:

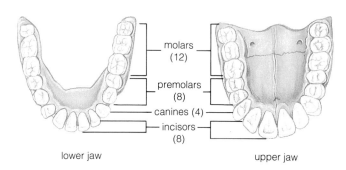

As suggested by Figure 25.1, each **tooth** has an enamel coat (hardened calcium deposits), dentine (a thick bone-like layer), and an inner pulp (with nerves and blood vessels). It is an engineering marvel, able to withstand years of chemical insults and mechanical stress. Recall that the chisel-shaped incisors shear off chunks of food (page 235). Cone-shaped canines tear it. The premolars and molars, with their broad crowns and rounded cusps, are good at grinding and crushing food.

Chewing mixes food with saliva. Saliva contains an enzyme (salivary amylase), a buffer (bicarbonate, or HCO_3^-), mucins, and water. **Salivary glands** produce and secrete this fluid through ducts beneath and in back of the tongue. The enzyme breaks down starch. Bicarbonate helps maintain your mouth's pH when you eat acidic foods such as tomatoes. Mucins (modified proteins) help form the mucus that binds food into a softened, lubricated ball.

When you swallow, tongue muscles contract and force softened balls of food into your **pharynx**. This is the entrance for the tubular part of the digestive tract. Contractions in the pharynx wall propel food into the **esophagus**, a sphinctered tube that leads to the stomach. Contractions in the esophagus propel food forward, into the stomach.

The pharynx also is the entrance to the trachea, an airway leading to the lungs. You normally don't choke on food because a flaplike valve (epiglottis) closes off the airway and keeps you from breathing while food is moving into the esophagus (page 457).

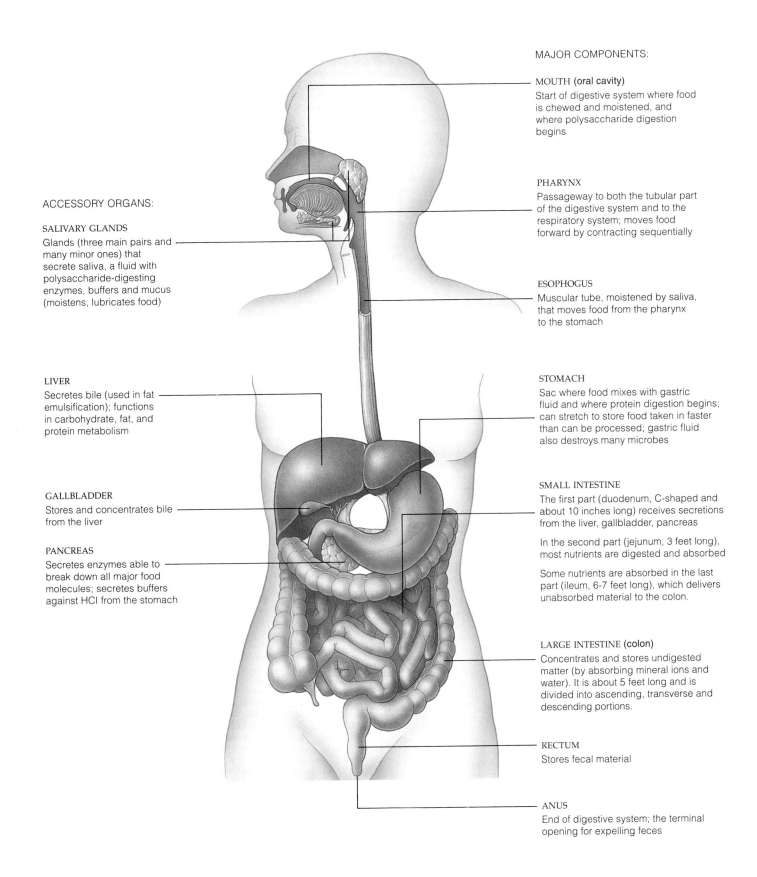

MAJOR COMPONENTS:

MOUTH (oral cavity)
Start of digestive system where food is chewed and moistened, and where polysaccharide digestion begins

PHARYNX
Passageway to both the tubular part of the digestive system and to the respiratory system; moves food forward by contracting sequentially

ESOPHOGUS
Muscular tube, moistened by saliva, that moves food from the pharynx to the stomach

STOMACH
Sac where food mixes with gastric fluid and where protein digestion begins; can stretch to store food taken in faster than can be processed; gastric fluid also destroys many microbes

SMALL INTESTINE
The first part (duodenum, C-shaped and about 10 inches long) receives secretions from the liver, gallbladder, pancreas

In the second part (jejunum, 3 feet long), most nutrients are digested and absorbed

Some nutrients are absorbed in the last part (ileum, 6-7 feet long), which delivers unabsorbed material to the colon.

LARGE INTESTINE (colon)
Concentrates and stores undigested matter (by absorbing mineral ions and water). It is about 5 feet long and is divided into ascending, transverse and descending portions.

RECTUM
Stores fecal material

ANUS
End of digestive system; the terminal opening for expelling feces

ACCESSORY ORGANS:

SALIVARY GLANDS
Glands (three main pairs and many minor ones) that secrete saliva, a fluid with polysaccharide-digesting enzymes, buffers and mucus (moistens, lubricates food)

LIVER
Secretes bile (used in fat emulsification); functions in carbohydrate, fat, and protein metabolism

GALLBLADDER
Stores and concentrates bile from the liver

PANCREAS
Secretes enzymes able to break down all major food molecules; secretes buffers against HCl from the stomach

Figure 25.4 Major components of the human digestive system and their functions. Organs with accessory roles in digestion are also listed.

The Stomach

The **stomach**, a muscular, stretchable sac, has three main functions. First, it stores and mixes food. Second, its secretions help dissolve and degrade food. Third, it helps control passage of food into the small intestine. Figure 25.5 compares the structure of the human stomach with the more elaborate stomach of antelope.

Stomach Acidity. Epithelium loaded with glands lines the inside of the stomach wall. Each day, cells in that lining secrete about two liters of hydrochloric acid (HCl), pepsinogens, mucus, and other substances. The substances make up gastric fluid (that is, the stomach's own fluid). Stomach acidity increases when the HCl separates into H^+ and Cl^-, and this helps dissolve bits of food to form a liquid mixture called chyme. The acidity kills many microorganisms present in food. It may cause *heartburn* when gastric fluid from the stomach backs up into the esophagus.

In response to the sight, aroma, and taste of food, the brain signals the secretory cells to step up activity. Also, when food enters the stomach, it distends the wall and activates receptors in the lining. This also leads to increased secretory activity.

Protein digestion starts in the stomach. The high acidity resulting from HCl secretions alters protein structure and exposes peptide bonds. It also converts pepsinogens to active enzymes (pepsins) that break the peptide bonds. Protein fragments start accumulating, and this triggers secretion of gastrin—a hormone that stimulates the pepsinogen- and HCl-secreting cells even more.

Mucus and buffers (bicarbonate ions especially) keep HCl and pepsin from acting on proteins in the stomach wall. When something interferes with controls over their secretion, HCl is free to diffuse into the stomach's lining. There, it triggers the release of a chemical (histamine) from tissue cells. Among other things, histamine stimulates more HCl secretion. A positive feedback loop is set up, and stomach tissues become damaged. If the tissue damage leads to bleeding in the stomach, the condition is called a *peptic ulcer*. Think about this if you are sensitive to aspirin. Aspirin promotes inflammation of the stomach—largely by suppressing mucus and bicarbonate secretions.

Stomach Emptying. In the stomach, waves of contractions mix the chyme and build up force as they approach a sphincter at the start of the small intestine

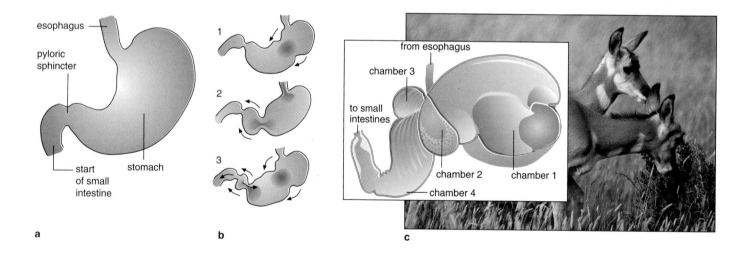

Figure 25.5 (**a**) Human stomach. (**b**) Peristaltic wave down the stomach, produced by alternating contraction and relaxation of muscles in the stomach wall.

(**c**) Multiple-chambered stomach of antelopes and other ruminants. The first chamber is a large pouch. The second is a smaller chamber and has a honeycombed inner surface (this is what cooks call tripe). In both of these, food is mixed with fluid, kneaded, and exposed to fermentation activities of symbiotic bacteria and protozoans. (Some of the symbionts break down cellulose. Others synthesize organic compounds, fatty acids, and vitamins, some of which their host uses.) The kneaded food is regurgitated into the mouth, chewed further, then swallowed again. This time it enters the third stomach chamber, where it is pummeled again before entering the final chamber, the true stomach.

(Figure 25.5*b*). The arrival of a strong contraction closes the sphincter, so most of the chyme is squeezed back. But a small amount moves into the small intestine.

Chyme's volume and composition affect how fast the stomach empties. For example, large meals activate more stomach receptors, these call for more forceful contraction, and emptying speeds up. As another example, increased acidity (or fat content) in the small intestine stimulates hormone secretions that cause a slowdown in stomach emptying—so food isn't moved along faster than it can be processed. Fear, depression, and other emotional upsets also trigger slowdowns.

Only alcohol and a few other substances are absorbed across the stomach wall. Most substances move onward, to the small intestine.

The Small Intestine

Table 25.1 lists the sites where carbohydrates, lipids (including fats), proteins, and nucleic acids are digested by specific enzymes, then absorbed into the internal environment. As you can see, digestion is completed and most nutrients are absorbed in the **small intestine**. Secretions from the **pancreas** and **liver** assist in these tasks. Each day, the small intestine receives about nine liters of fluid from the stomach, liver, and pancreas. At least 95 percent is absorbed across an epithelial lining that is exposed to the intestinal lumen.

Processes of Digestion. Pancreatic enzymes are essential for digestion. They break down carbohydrates, proteins, nucleic acids, and fats. The pancreas also secretes bicarbonate, a buffer that helps neutralize HCl arriving from the stomach. (You may have heard of two other pancreatic secretions, insulin and glucagon, but these have roles in nutrition, not digestion.)

Fat digestion depends on more than enzyme action. It depends also on bile, which is secreted continually by the liver. Bile contains bile salts, bile pigments, cholesterol, and lecithin (a phospholipid). It flows through ducts from the liver to the small intestine. When food isn't moving through the gut, a sphincter closes off the main duct. Bile backs up into the saclike **gallbladder**, where it is stored and concentrated.

By a process called **emulsification**, bile salts speed up fat digestion. Most fats in our diet are triglycerides. These are insoluble in water and tend to aggregate into large fat globules in the chyme. When movements of the intestinal wall agitate chyme, fat globules break up into small droplets, which become coated with bile salts. Because bile salts carry negative charges, the coated droplets repel each other and stay separated. This suspension of fat droplets is the "emulsion."

Compared to fat globules, emulsion droplets give fat-digesting enzymes a much greater surface area to act upon. Thus triglycerides can be broken down much more rapidly to fatty acids and monoglycerides.

Table 25.1 Major Enzymes of Digestion

Enzyme	Source	Where Active	Substrate	Main Breakdown Products*
Carbohydrate Digestion:				
Salivary amylase	Salivary glands	Mouth	Polysaccharides	Disaccharides
Pancreatic amylase	Pancreas	Small intestine	Polysaccharides	Disaccharides
Disaccharidases	Intestinal lining	Small intestine	Disaccharides	Monosaccharides (e.g., glucose)
Protein Digestion:				
Pepsins	Stomach lining	Stomach	Proteins	Protein fragments
Trypsin and chymotrypsin	Pancreas	Small intestine	Proteins	Protein fragments
Carboxypeptidase	Pancreas	Small intestine	Protein fragments	Amino acids
Aminopeptidase	Intestinal lining	Small intestine	Protein fragments	Amino acids
Fat Digestion:				
Lipase	Pancreas	Small intestine	Triglycerides	Free fatty acids, monoglycerides
Nucleic Acid Digestion:				
Pancreatic nucleases	Pancreas	Small intestine	DNA, RNA	Nucleotides
Intestinal nucleases	Intestinal lining	Small intestine	Nucleotides	Nucleotide bases, monosaccharides

*Yellow parts of table identify breakdown products that can be absorbed into the internal environment.

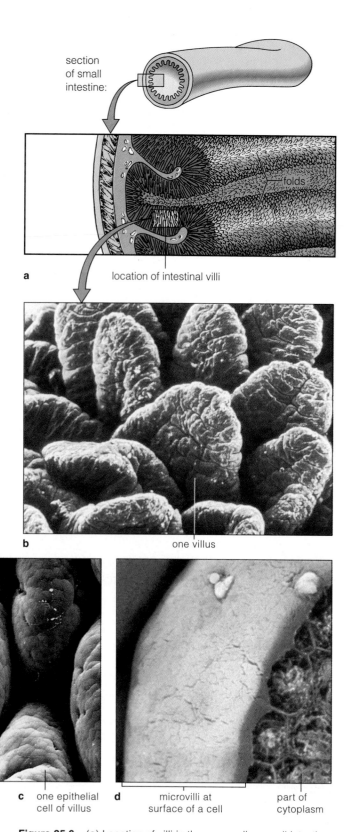

section of small intestine:

folds

a   location of intestinal villi

b   one villus

c   one epithelial cell of villus

d   microvilli at surface of a cell   part of cytoplasm

Figure 25.6 (**a**) Location of villi in the mammalian small intestine. (**b**) Surface view of the deep, permanent folds of the inner layer of the intestinal tube. (**c**) Some of the fingerlike projections (villi) that cover the inner layer. The villi are so dense and numerous, they give the surface a velvety appearance. Individual epithelial cells are visible. (**d**) The dense crown of microvilli at the surface of a single cell.

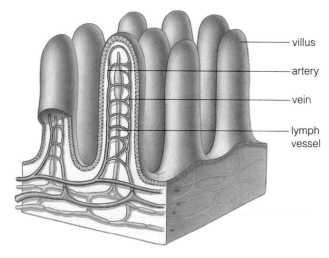

villus

artery

vein

lymph vessel

Figure 25.7 Blood and lymph vessels in intestinal villi. Monosaccharides and most amino acids that have crossed the intestinal lining enter the blood vessels. Fats enter the lymph vessels.

Processes of Absorption.   The intestinal lining is one of the body's marvels. Figure 25.6 shows its dense folds, which are absorptive structures called **villi** (singular, villus). Villi increase the surface area available for interactions with chyme. Epithelial cells line these structures, and each cell has a crown of **microvilli**. Each microvillus is a threadlike projection of the plasma membrane. Collectively, they greatly increase the surface area available for absorption.

By the time carbohydrate, protein, and lipid molecules are halfway through this elaborate tube, most have been broken down and digested. Some of the product, including monosaccharides and amino acids, are absorbed in straightforward fashion. So are water and mineral ions. Transport proteins in the plasma membrane of epithelial cells actively shunt them across the intestinal lining. By contrast, **bile** salts assist the absorption of fatty acids and monoglycerides. These products of fat digestion can diffuse across the lipid bilayer of the plasma membrane of epithelial cells.

By a process called **micelle formation**, bile salts combine with the products of fat digestion, forming tiny droplets (micelles). Product molecules in the micelles continuously exchange places with product molecules that are dissolved in the chyme. When concentration gradients favor it, molecules diffuse out of micelles, out of the chyme, and into epithelial cells. There, fatty acids and monoglycerides recombine as triglycerides. Then the triglycerides combine with proteins into particles (chylomicrons) that leave the cell by exocytosis and enter the internal environment.

Once absorbed into the internal environment, glucose and amino acids enter blood vessels. The triglycerides enter lymph vessels, which drain into blood vessels of the circulatory system (Figure 25.7). Figure 25.8 summarizes the absorption routes just described.

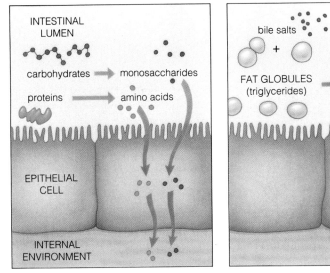

1 Digestion of carbohydrates to monosaccharides and proteins to amino acids completed by pancreatic enzymes.

2 Active transport of monosaccharides, amino acids across plasma membrane of cells, then out of cells, into internal environment.

1 Emulsification. Wall movements break up fat globules into small droplets. Bile salts keep globules from re-forming. Pancreatic enzymes digest droplets to fatty acids and monoglycerides.

2 Formation of micelles (as bile salts combine with digested products and phospholipids). Products readily slip into and out of micelles.

3 Concentration of monosaccharides, fatty acids in micelles enhance gradients that lead to their diffusion across lipid bilayer of cells' plasma membranes.

4 Reassemby of products into triglycerides inside cells. These join with proteins, then are expelled (by exocytosis) into internal environment.

Figure 25.8 Summary of digestion and absorption processes in the small intestine.

The Large Intestine

Material not absorbed in the small intestine moves into the large intestine, or **colon**. The colon concentrates and stores feces, a mixture of undigested and unabsorbed material, water, and bacteria. The mixture becomes concentrated as water moves across the colon's lining. Cells in the lining actively transport sodium ions out of the lumen. As the ion concentration in the lumen drops, the water concentration increases—so water also moves out of the lumen, by osmosis.

As the sketch below indicates, the colon starts out as a cup-shaped pouch known as the cecum. The **appendix** is a slender projection from that pouch, as the following sketch indicates:

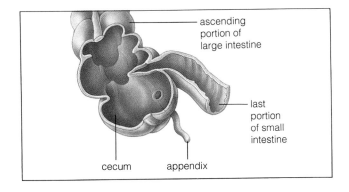

The appendix has no known digestive functions. It may have roles in the body's defense against infections.

The colon ascends on the right side of the abdominal cavity, continues across to the other side, then descends and connects with a short tube, the rectum (Figure 25.4). Distension of the rectal wall triggers a reflex pathway that leads to expulsion of feces from the body. The nervous system controls the expulsion. It stimulates or inhibits contractions of a muscle sphincter at the anus, the terminal opening of the gut.

Bulk refers to a volume of fiber and other undigested material that cannot be decreased by absorption in the colon. As bulk increases, it puts more pressure on the colon wall. Without adequate bulk, it takes longer for fecal material to move through the colon, with irritating and perhaps carcinogenic effects. People of rural Africa and India rarely suffer colon cancer or appendicitis. (*Appendicitis* is an infection of the appendix. If the infected appendix ruptures, bacteria normally living in the colon can enter the abdominal cavity and cause severe problems.) Most people who live in those areas cannot afford to eat much more than whole grains, which are high in fiber. When they move to wealthier nations and leave behind their fiber-rich diet, they are *more* likely to suffer colon cancer and appendicitis. This example brings us to a survey of what constitutes good and bad nutrition.

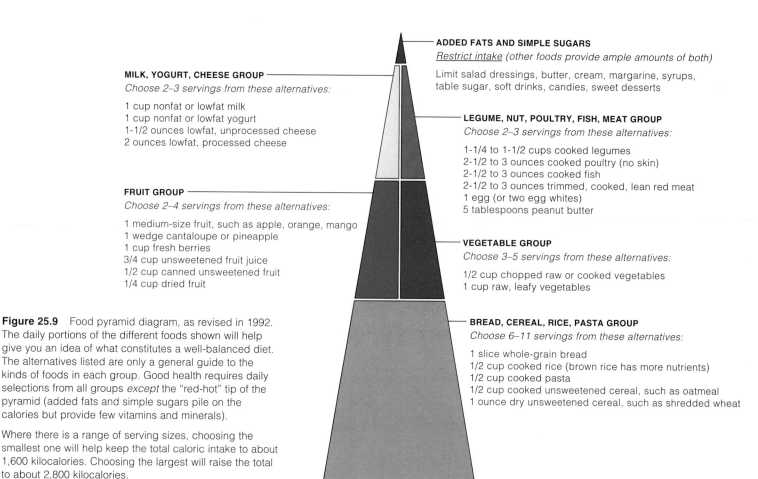

MILK, YOGURT, CHEESE GROUP

Choose 2–3 servings from these alternatives:

1 cup nonfat or lowfat milk
1 cup nonfat or lowfat yogurt
1-1/2 ounces lowfat, unprocessed cheese
2 ounces lowfat, processed cheese

FRUIT GROUP

Choose 2–4 servings from these alternatives:

1 medium-size fruit, such as apple, orange, mango
1 wedge cantaloupe or pineapple
1 cup fresh berries
3/4 cup unsweetened fruit juice
1/2 cup canned unsweetened fruit
1/4 cup dried fruit

ADDED FATS AND SIMPLE SUGARS

Restrict intake (other foods provide ample amounts of both)

Limit salad dressings, butter, cream, margarine, syrups, table sugar, soft drinks, candies, sweet desserts

LEGUME, NUT, POULTRY, FISH, MEAT GROUP

Choose 2–3 servings from these alternatives:

1-1/4 to 1-1/2 cups cooked legumes
2-1/2 to 3 ounces cooked poultry (no skin)
2-1/2 to 3 ounces cooked fish
2-1/2 to 3 ounces trimmed, cooked, lean red meat
1 egg (or two egg whites)
5 tablespoons peanut butter

VEGETABLE GROUP

Choose 3–5 servings from these alternatives:

1/2 cup chopped raw or cooked vegetables
1 cup raw, leafy vegetables

BREAD, CEREAL, RICE, PASTA GROUP

Choose 6–11 servings from these alternatives:

1 slice whole-grain bread
1/2 cup cooked rice (brown rice has more nutrients)
1/2 cup cooked pasta
1/2 cup cooked unsweetened cereal, such as oatmeal
1 ounce dry unsweetened cereal, such as shredded wheat

Figure 25.9 Food pyramid diagram, as revised in 1992. The daily portions of the different foods shown will help give you an idea of what constitutes a well-balanced diet. The alternatives listed are only a general guide to the kinds of foods in each group. Good health requires daily selections from all groups *except* the "red-hot" tip of the pyramid (added fats and simple sugars pile on the calories but provide few vitamins and minerals).

Where there is a range of serving sizes, choosing the smallest one will help keep the total caloric intake to about 1,600 kilocalories. Choosing the largest will raise the total to about 2,800 kilocalories.

HUMAN NUTRITIONAL REQUIREMENTS

A few million years ago, our hominid ancestors dined mostly on fresh fruits and other fibrous plant material. From this nutritional beginning, humans in many parts of the world have become habituated to poor diets. Consider Americans. On the whole, they are among the world's best-fed people, yet they suffer a high incidence of digestive disorders. Along with affluence, bad eating habits are rampant. Americans skip meals, eat too much and too fast when they sit down at the table, and generally give their gut erratic workouts. Worse, their diet tends to be loaded with cholesterol, sugar, and salt—and short on bulk. Remember, too little bulk means delayed expulsion of feces, with irritating and possibly carcinogenic consequences. Or, as epidemiologist Denis Burkitt put it, if you pass a small volume of feces, you have large hospitals.

The New, Improved Food Pyramid

From elementary school onward, you probably have encountered "food pyramid" charts that supposedly showed you what is meant by a well-balanced diet.

Every so often, these charts are revised to incorporate new findings of nutritional studies. As of the summer of 1993, many nutritionists are now arguing that food in the proportions shown in Figure 25.9 and listed below constitutes a well-balanced diet:

Complex carbohydrates:	60–80%
Proteins:	9% or less
Fats and other lipids:	10–30%

Of course, not all carbohydrates, fats, and proteins are equal in nutritional value, as you will now read.

Carbohydrates

Starch and, to a lesser extent, glycogen should be the main carbohydrates in our diet. These complex carbohydrates are easily broken down to glucose units, which are the body's main energy source (Chapter 6). Starch is abundant in fleshy fruits, cereal grains, and legumes, including beans and peas.

The carbohydrates called simple sugars don't have the fiber of complex carbohydrates or the vitamins and minerals of whole foods. Each week, the average Amer-

No Limiting Amino Acid	Low in Lysine	Low in Methionine, Other Sulfur-Containing Amino Acids	Low in Tryptophan

legumes:
  soybean
  tofu
  soy milk
 	
cereal grains:
  wheat germ
 	
nuts:
 	
milk
cheeses (except
  cream cheese)
yogurt
eggs
meats
	legumes:
	  peanuts
	
	cereal grains:
	  barley
	  buckwheat
	  corn meal
	  oats
	  rice
	  rye
	  wheat
	
	nuts, seeds:
	  almonds
	  cashews
	  coconut
	  English walnuts
	  hazelnuts
	  pecans
	  pumpkin seeds
	  sunflower seeds
		legumes:
		  beans (dried)
		  black-eyed peas
		  garbanzos
		  lentils
		  lima beans
		  mung beans
		  peanuts
		
		nuts:
		  hazelnuts
		
		fresh vegetables:
		  asparagus
		  broccoli
		  green peas
		  mushrooms
		  parsley
		  potatoes
		  soybeans
		  Swiss chard
			legumes:
			  beans (dried)
			  garbanzos
			  lima beans
			  mung beans
			  peanuts
			
			cereal grains:
			  corn meal
			
			nuts:
			  almonds
			  English walnuts
			
			fresh vegetables:
			  corn
			  green peas
			  mushrooms
			  Swiss chard

Figure 25.10 Essential amino acids—a small portion of the total protein intake. All eight must be available at the same time, in certain amounts, if cells are to build their own proteins. Milk and eggs have high amounts of all eight in proportions that humans require; they are among the complete proteins.

Nearly all plant proteins are incomplete, so vegetarians should construct their diet carefully to avoid protein deficiency. For example, they can combine different foods from the three columns of incomplete proteins shown to the left. Also, vegetarians who avoid dairy products and eggs should take vitamin B_{12} and B_2 (riboflavin) supplements. Animal protein is a luxury in most traditional societies. Yet their cuisines include good combinations of plant proteins, including rice/beans, chili/cornbread, tofu/rice, and lentils/wheat bread.

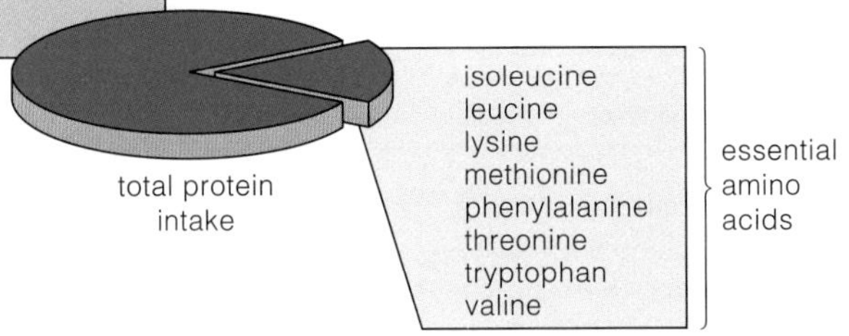

total protein intake

isoleucine
leucine
lysine
methionine
phenylalanine
threonine
tryptophan
valine

essential amino acids

ican eats as much as two pounds of refined sugar (sucrose). You may think this a far-fetched statement, but start looking at the ingredients listed on packages of cereal, frozen dinners, soft drinks, and other prepared foods. Many sugars are "hidden" as corn syrup, corn sweeteners, dextrose, and so on.

Lipids

The body can't function without fats and other lipids. For instance, the phospholipid lecithin is a required component of cell membranes. Fats serve as energy reserves, they cushion many organs, and they provide insulation beneath the skin. Dietary fats also help the body store fat-soluble vitamins.

Yet the body can synthesize most of its own fats, including cholesterol, from protein and carbohydrates. The ones it cannot produce are called **essential fatty acids**, but whole foods provide plenty of them. Linoleic acid is an example. One teaspoon a day of corn oil, olive oil, or some other polyunsaturated fat in food provides enough of it.

Today, butter and other fats make up 40 percent of the average diet in the United States. Most of the medical community agrees the proportion should be less than 30 percent. Like other animal fats, butter is a saturated fat that tends to raise the level of cholesterol in the blood. Cholesterol is used in the synthesis of bile acids and steroid hormones, and it is a required component of our cell membranes. However, too much cholesterol may damage the circulatory system (page 427). Work is under way to develop edible but nondigestible oils, such as sucrose polyester, for those who have difficulty cutting back on the fats.

Proteins

The amino acid components of dietary proteins are required for the body's own protein-building programs. Of twenty common types, eight are **essential amino acids**. Our cells cannot synthesize them, so we must obtain them from food. The eight are methionine (or an equivalent called cysteine), isoleucine, leucine, lysine, phenylalanine (or tyrosine), threonine, tryptophan, and valine.

Most animal proteins are "complete," meaning their ratios of amino acids match human nutritional needs. Plant proteins are "incomplete," meaning they lack one or more of the essential amino acids. To get enough protein, vegetarians must eat certain combinations of different plants (Figure 25.10).

A measure called net protein utilization (NPU) is used to compare proteins from different sources. NPU values range from 100 (all essential amino acids present in ideal proportions) to 0 (one or more absent; when eaten alone, the protein can't satisfy nutritional needs).

Because enzymes and other proteins are vital for the body's structure and function, protein-deficient diets may have severe consequences. Protein deficiency is most damaging among the young, for the brain grows and develops rapidly early in life. Unless enough protein is taken in just before and just after birth, irreversible mental retardation occurs. Even mild protein starvation can retard growth and affect mental and physical performance.

Table 25.2 Vitamins: Sources, Functions, and Effects of Deficiencies or Excesses*

Vitamin	Common Sources	Main Functions	Signs of Severe Long-Term Deficiency	Signs of Extreme Excess
Fat-Soluble Vitamins:				
A	Its precursor comes from beta-carotene in yellow fruits, yellow or green leafy vegetables; also in fortified milk, egg yolk, fish liver	Used in synthesis of visual pigments, bone, teeth; maintains epithelia	Dry, scaly skin; lowered resistance to infections; night blindness; permanent blindness	Malformed fetuses; hair loss; changes in skin; liver and bone damage; bone pain
D	D_3 formed in skin and in fish liver oils, egg yolk, fortified milk; converted to active form elsewhere	Promotes bone growth and mineralization; enhances calcium absorption	Bone deformities (rickets) in children; bone softening in adults	Retarded growth; kidney damage; calcium deposits in soft tissues
E	Whole grains, dark-green vegetables, vegetable oils	Possibly inhibits effects of free radicals; helps maintain cell membranes; blocks breakdown of vitamins A and C in gut	Lysis of red blood cells; nerve damage	Muscle weakness, fatigue, headaches, nausea
K	Colon bacteria form most of it; also in green leafy vegetables, cabbage	Blood clotting; ATP formation via electron transport	Abnormal blood clotting; severe bleeding (hemorrhaging)	Anemia; liver damage and jaundice
Water-Soluble Vitamins:				
B_1 (thiamin)	Whole grains, green leafy vegetables, legumes, lean meats, eggs	Connective tissue formation; folate utilization; coenzyme action	Water retention in tissues; tingling sensations; heart changes; poor coordination	None reported from food; possible shock reaction from repeated injections
B_2 (riboflavin)	Whole grains, poultry, fish, egg white, milk	Coenzyme action	Skin lesions	None reported
Niacin	Green leafy vegetables, potatoes, peanuts, poultry, fish, pork, beef	Coenzyme action	Contributes to pellegra (damage to skin, gut, nervous system, etc.)	Skin flushing; possible liver damage
B_6	Spinach, tomatoes, potatoes, meats	Coenzyme in amino acid metabolism	Skin, muscle, and nerve damage; anemia	Impaired coordination; numbness in feet
Pantothenic acid	In many foods (meats, yeast, egg yolk especially)	Coenzyme in glucose metabolism, fatty acid and steroid synthesis	Fatigue, tingling in hands, headaches, nausea	None reported; may cause diarrhea occasionally
Folate (folic acid)	Dark green vegetables, whole grains, yeast, lean meats; colon bacteria produce some folate	Coenzyme in nucleic acid and amino acid metabolism	A type of anemia; inflamed tongue; diarrhea; impaired growth; mental disorders	Masks vitamin B_{12} deficiency
B_{12}	Poultry, fish, red meat, dairy foods (not butter)	Coenzyme in nucleic acid metabolism	A type of anemia; impaired nerve function	None reported
Biotin	Legumes, egg yolk; colon bacteria produce some	Coenzyme in fat, glycogen formation and in amino acid metabolism	Scaly skin (dermatitis), sore tongue, depression, anemia	None reported
C (ascorbic acid)	Fruits and vegetables, especially citrus, berries, cantaloupe, cabbage, broccoli, green pepper	Collagen synthesis; possibly inhibits effects of free radicals; structural role in bone, cartilage, and teeth; role in carbohydrate metabolism	Scurvy, poor wound healing, impaired immunity	Diarrhea, other digestive upsets; may alter results of some diagnostic tests

*The guidelines for appropriate daily intakes are being worked out by the Food and Drug Administration.

Vitamins and Minerals

Metabolic activity depends on small amounts of more than a dozen organic substances called **vitamins**. Most plants synthesize all of these substances. Animals generally have lost the ability to do so and must obtain vitamins from food. Human cells need at least thirteen different vitamins, each with specific metabolic roles. Many metabolic activities require several vitamins, so the absence of one affects the functions of others. Metabolic activities also require certain inorganic substances called **minerals**. For example, most of your cells use calcium and magnesium in many reactions. All cells require iron in electron transport chains. Red blood cells can't function without iron in hemoglobin, the oxygen-carrying pigment in blood. Neurons can't function without sodium and potassium.

Table 25.3 Major Minerals: Sources, Functions, and Effects of Deficiencies or Excesses*

Mineral	Common Sources	Main Functions	Signs of Severe Long-Term Deficiency	Signs of Extreme Excess
Calcium	Dairy products, dark green vegetables, dried legumes	Bone, tooth formation; blood clotting; neural and muscle action	Stunted growth; possibly diminished bone mass (osteoporosis)	Impaired absorption of other minerals; kidney stones in susceptible people
Chloride	Table salt (usually too much in diet)	HCl formation in stomach; contributes to body's acid-base balance; neural action	Muscle cramps; impaired growth; poor appetite	Contributes to high blood pressure in susceptible people
Copper	Nuts, legumes, seafood, drinking water	Used in synthesis of melanin, hemoglobin, and some transport chain components	Anemia, changes in bone and blood vessels	Nausea, liver damage
Fluorine	Fluoridated water, tea, seafood	Bone, tooth maintenance	Tooth decay	Digestive upsets; mottled teeth and deformed skeleton in chronic cases
Iodine	Marine fish, shellfish, iodized salt, dairy products	Thyroid hormone formation	Enlarged thyroid (goiter), with metabolic disorders	Goiter
Iron	Whole grains, green leafy vegetables, legumes, nuts, eggs, lean meat, molasses, dried fruit, shellfish	Formation of hemoglobin and cytochrome (transport chain component)	Iron-deficiency anemia, impaired immune function	Liver damage, shock, heart failure
Magnesium	Whole grains, legumes, nuts, dairy products	Coenzyme role in ATP-ADP cycle; roles in muscle, nerve function	Weak, sore muscles; impaired neural function	Impaired neural function
Phosphorus	Whole grains, poultry, red meat	Component of bone, teeth, nucleic acids, ATP, phospholipids	Muscular weakness; loss of minerals from bone	Impaired absorption of minerals into bone
Potassium	Diet provides ample amounts	Muscle and neural function; roles in protein synthesis and body's acid-base balance	Muscular weakness	Muscular weakness, paralysis, heart failure
Sodium	Table salt; diet provides ample to excessive amounts	Key role in body's acid-base balance; roles in muscle and neural function	Muscle cramps	High blood pressure in susceptible people
Sulfur	Proteins in diet	Component of body proteins	None reported	None likely
Zinc	Whole grains, legumes, nuts, meats, seafood	Component of digestive enzymes; roles in normal growth, wound healing, sperm formation, and taste and smell	Impaired growth, scaly skin, impaired immune function	Nausea, vomiting, diarrhea; impaired immune function and anemia

*The guidelines for appropriate daily intakes are being worked out by the Food and Drug Administration.

People who are in good health can get all the vitamins and minerals they need from a balanced diet of whole foods. Generally, specific vitamin and mineral supplements are necessary only for strict vegetarians, the elderly, and individuals suffering from a chronic illness or taking medication that affects the body's use of specific nutrients. For example, in one comprehensive study, vitamin K supplements helped older women retain calcium and so diminish bone loss (osteoporosis) by 30 percent. Preliminary research suggests two vitamins (C and E) and beta-carotene (the precursor for vitamin A) may inactivate free radicals—those rogue metabolic fragments that may contribute to aging and may impair immune function (page 60).

No one should take massive doses of any vitamin or mineral supplement except under medical supervision.

As Tables 25.2 and 25.3 indicate, excess amounts of many vitamins and minerals are harmful. For example, large doses of at least two vitamins (A and D) can actually damage the body. Like all fat-soluble vitamins, excess amounts can accumulate in tissues and interfere with normal metabolic function. Similarly, sodium is present in plant and animal tissues, and it is a component of table salt. Sodium has roles in the body's salt–water balance, muscle activity, and nerve function. Yet prolonged, excessive intake of sodium may contribute to high blood pressure.

Severe shortages or self-prescribed, massive excesses of vitamins and minerals can disturb the delicate balances in body function that promote health.

Man's Height	Size of Frame		
	Small	Medium	Large
5' 2"	128–134	131–141	138–150
5' 3"	130–136	133–143	140–153
5' 4"	132–138	135–145	142–156
5' 5"	134–140	137–148	144–160
5' 6"	136–142	139–151	146–164
5' 7"	138–145	142–154	149–168
5' 8"	140–148	145–157	152–172
5' 9"	142–151	148–160	155–176
5'10"	144–154	151–163	158–180
5'11"	146–157	154–166	161–184
6' 0"	149–160	157–170	164–188
6' 1"	152–164	160–174	168–192
6' 2"	155–168	164–178	172–197
6' 3"	158–172	167–182	176–202
6' 4"	162–176	171–187	181–207

Woman's Height	Size of Frame		
	Small	Medium	Large
4'10"	102–111	109–121	118–131
4'11"	103–113	111–123	120–134
5' 0"	104–115	113–126	122–137
5' 1"	106–118	115–129	125–140
5' 2"	108–121	118–132	128–143
5' 3"	111–124	121–135	131–147
5' 4"	114–127	124–138	134–151
5' 5"	117–130	127–141	137–155
5' 6"	120–133	130–144	140–159
5' 7"	123–136	133–147	143–163
5' 8"	126–139	136–150	146–167
5' 9"	129–142	139–153	149–170
5'10"	132–145	142–158	152–173
5'11"	135–148	145–159	155–176
6' 0"	138–151	148–162	158–179

Figure 25.11 "Ideal" weights for adults according to one insurance company in 1983. Values shown are for people twenty-five to fifty-nine years old wearing shoes with one-inch heels and three pounds of clothing (for women) or five pounds (for men). Extreme obesity puts severe strain on the circulatory system. The body produces many more small blood vessels (capillaries) to service the increased tissue masses. The heart becomes more stressed. It must pump harder to keep blood circulating.

Energy Needs and Body Weight

We grow and maintain ourselves when we supply the body with energy and materials from foods of certain types, in certain amounts. Nutritionists measure energy in units called **kilocalories**. Each unit is 1,000 calories of heat energy (that is, the amount needed to raise the temperature of one kilogram of water by 1°C).

To maintain an acceptable weight and keep the body functioning normally, caloric intake must be balanced with energy output. The output varies from one person to another because of differences in physical activity, basic rate of metabolism, age, sex, hormone activity, and emotional state. Some of these factors are influenced by a person's social environment. Others have a genetic basis. For instance, researchers have studied many identical twins (with identical genes) who, owing to family problems, were separated at birth and raised apart, in different households. At adulthood, the body weights of the separated twins were still similar.

When caloric input balances the output, body weight doesn't change dramatically over long periods. As any dieter knows, the body behaves as if it has a set point for what that weight is going to be—and it works to counteract deviations from its set point.

How many kilocalories should you take in each day to maintain "acceptable" body weight? First, multiply the desired weight (in pounds) by 10 if you are not very active physically, by 15 if you are moderately active, and by 20 if you are quite active. Then, depending on your age, subtract the following amount from the value obtained from the first step:

Age:	Subtract:
25–34	0
35–44	100
45–54	200
55–64	300
Over 65	400

For example, if you want to weigh 120 pounds and are very active, $120 \times 20 = 2,400$ kilocalories. If you are thirty-five years old, you would take in $(2,400 - 100)$ or 2,300 kilocalories a day. Such calculations provide a rough estimate of caloric intake. Other factors, such as height, must be considered also. An active person who is 5 feet, 2 inches tall doesn't need as much energy as an active person who weighs the same but is 6 feet tall.

To maintain an acceptable body weight, energy input (caloric intake) must be balanced with energy output (as through metabolic activity and exercise).

By definition, **obesity** is an excess of fat in the body's adipose tissues, most often caused by imbalances between caloric intake and energy output. Yet what is too fat or too thin? What is an "ideal weight"? Insurance companies have developed many charts (Figure 25.11), mostly to identify overweight people who might be insurance risks. Such charts factor in height.

Focus on Health

When Does Dieting Become a Disorder?

By current standards, how much of our total tissue mass should be fat? The proportion of fat should be no more than 18 to 24 percent for a female who is less than thirty years old. For a male, it should be no more than 12 to 18 percent. Yet an estimated 34 million Americans are overweight to the point of obesity. Many sincerely attempt to shed the excess fat by dieting. Yet most eventually regain what they lost—and sometimes put on more.

Dieting and Exercise. Why is it so difficult to lose weight permanently? The difficulty is partly a result of our evolutionary heritage. The fat-storing cells of adipose tissue are an adaptation for survival. Collectively, they represent an energy reserve that may carry an animal through times when food just isn't available. Remember this when you start to put on unwanted fat. Once a fat-storing cell is added to your body, food intake may affect how empty or full it gets—but apparently that cell is in your body to stay. Dieting *can* decrease the fat stores. But research now suggests that the body interprets dieting as "starvation."

Dieting triggers changes in metabolism. With these changes, food is used more conservatively—so that *fewer* calories are burned. Meanwhile, the dieter's appetite surges, and "starved" fat cells quickly refill when a diet ends. In "yo-yo dieting," a person repeatedly gains and loses weight. This may alter cell metabolism in ways that make it *more* difficult to shed weight with each new round of dieting. Besides this, frequent changes in body weight may also increase the risk of heart disorders.

Some people opt to shed pounds by exercising, only to discover the going is slow indeed. Losing just a single pound of fat requires an energy expenditure of about 3,500 kilocalories. You can do this by, say, jogging for four hours or playing tennis for nearly eight hours straight. A more feasible way to keep off the fat is to combine a *moderate* reduction in caloric intake with increased physical activity. This approach might minimize the "starvation" response and increase the rate at which the body uses energy. Exercise also increases muscle mass. Even a resting muscle burns more calories than other types of tissues. Once you lose fat, the "starvation" response kicks in. To keep off unwanted weight, you have to continue to eat in moderation and maintain a program of regular exercise.

Anorexia Nervosa. When dieting becomes obsessive, one result can be a potentially fatal eating disorder called *anorexia nervosa* (Figure *a*). The disorder is most common among women in their teens and early twenties.

People with anorexia nervosa have an abnormal perception of their body weight. Their fear of being fat and hungry becomes overwhelming. They start starving themselves and often overexercise. Emotional factors contribute to the disorder. Some anorexic people fear growing up in general and maturing sexually in particular. Others have irrational measures of what to expect of themselves and of what they might accomplish.

Bulimia. At least 20 percent of college-age women now suffer to varying degrees from *bulimia* ("an oxlike appetite"). Outwardly they may look healthy, but their food intake is out of control. During an hour-long eating binge, a bulimic may take in more than 50,000 kilocalories—then vomit or use laxatives to purge the body. The binge-purge routine may occur once a month; it may occur several times a day. Repeated purgings can damage the gut. Repeated vomiting, which brings gastric fluid into the mouth, can erode teeth to stubs. Extreme bulimics can die of heart failure, stomach rupturing, or kidney failure.

Some women start the routine because it seems like a simple way to lose weight. Others have emotional problems. Often bulimics are well-educated and accomplished, but they strive for perfection and may have problems with family members who exert control over them. Eating may actually be unpleasant for bulimics, but the purging (which they themselves control) relieves them of anger and frustration.

Treatment for both anorexia nervosa and bulimia usually includes long-term psychotherapy and diet counseling. Severe cases may require hospitalization.

People who are 25 percent heavier than "ideal" are viewed as obese.

Some researchers who study causes of death suspect that the "ideal" may be ten to fifteen pounds heavier than the charts indicate. Some dieticians are convinced the chart values should be less. Whatever the ideal range may be, serious disorders do arise with extremes at either end of that range (see the *Focus* essay).

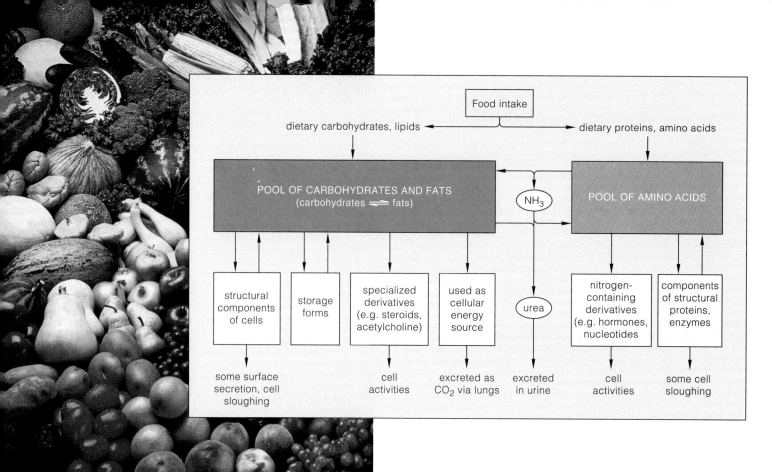

Figure 25.12 Summary of major pathways of organic metabolism. Carbohydrates, fats, and proteins are continually broken down and synthesized. Urea forms mainly in the liver.

NUTRITION AND METABOLISM

Figure 25.12 outlines the main routes by which organic molecules are shuffled and reshuffled after the body absorbs them. Your body continually breaks down most of its own carbohydrates, lipids, and proteins, then picks up their component parts, such as glucose, for use in new molecules. At the molecular level, the body undergoes massive and sometimes rapid turnover, using conversion pathways such as the ones shown earlier in Figure 6.10.

When you eat, the body builds up its pools of organic molecules. Excess carbohydrates and other dietary molecules are transformed mostly into fats, which are stored in adipose tissue. Some are also converted to glycogen in the liver and in muscle tissue. While organic molecules are being absorbed and stored, most cells use glucose as their main energy source. There is no net breakdown of protein in muscle or other tissues during this period.

Between meals, the body taps into its fat stores for energy. Fats stored mainly in adipose tissue are broken

down to glycerol and fatty acids, which are released into blood. The glycerol is converted to glucose in the liver. Cells can take up the circulating fatty acids and use them for ATP production.

As described on page 504, the nervous system and endocrine system interact to control these and other aspects of organic metabolism.

During a meal, glucose moves into cells, where it can be used for energy and where the excess can be stored.

Between meals, the body metabolizes fat molecules as the main energy source.

This brief overview merely hints at the central role that the liver plays in storing and interconverting organic molecules. The liver serves other functions as well. As Table 25.4 suggests, it helps maintain the blood's concentrations of organic substances and removes many toxic substances from it. The liver inactivates most hormone molecules and sends them to the

Table 25.4	Some Activities That Depend on Liver Functioning

1. Carbohydrate metabolism

2. Role in controlling synthesis of proteins dissolved in blood

3. Assembly and disassembly of certain proteins

4. Urea formation from nitrogen-containing wastes

5. Assembly and storage of some fats

6. Fat digestion (bile is formed by the liver)

7. Inactivation of many chemicals (such as hormones and some drugs)

8. Detoxification of many poisons

9. Degradation of worn-out red blood cells

10. Immune response (removal of some foreign particles)

11. Red blood cell formation (liver absorbs, stores factors needed for red blood cell maturation)

kidneys for excretion, in urine. Also, ammonia (NH_3) is produced when cells break down amino acids, and it can be toxic to cells. The circulatory system carries ammonia to the liver, and there it is converted to urea. Urea is a much less toxic waste product, and it leaves the body by way of the kidneys, in urine.

SUMMARY

1. Nutrition includes all the processes by which the body takes in, digests, absorbs, and uses food.

2. Mammals have a complete digestive system, with two openings (mouth and anus). The system breaks down food molecules by mechanical and enzymatic means. At certain regions, breakdown products as well as water and mineral ions are absorbed into the internal environment. At the terminal opening, unabsorbed residues are eliminated.

3. The human digestive system includes the mouth, pharynx, esophagus, stomach, small intestine, large intestine (colon), rectum, and anus. The salivary glands, liver, gallbladder, and pancreas have accessory roles in digestion.

4. The nervous system, endocrine system, and local networks of neurons in the gut wall interact to govern activities of the digestive system. Some controls operate in response to the volume and composition of food passing through the stomach and intestines. The controls trigger changes in muscle activity and in the rate at which hormones and enzymes are secreted.

5. Starch digestion begins in the mouth and protein digestion in the stomach. But most digestion occurs and nearly all nutrients are absorbed in the small intestine. The pancreas produces and secretes the main digestive enzymes. Bile from the liver assists in fat digestion.

6. During absorption, glucose and most amino acids are actively transported out of the intestinal lumen by cells making up the intestinal lining. Fatty acids and monoglycerides diffuse across the lipid bilayer of those cells. There they are recombined as triglycerides, then are released, by exocytosis, into interstitial fluid.

7. To maintain acceptable weight and overall health, caloric intake must balance energy output. A well-balanced daily diet of whole foods (60–80% complex carbohydrates, 10–30% fats and other lipids, and 9% proteins) normally provides all required vitamins and minerals.

Review Questions

1. Name the organs of the human digestive system, as well as glandular organs with accessory roles in its functioning: *401*

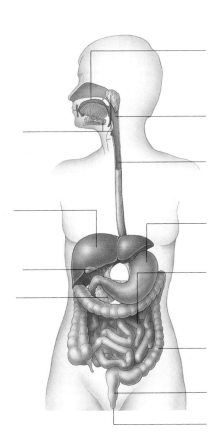

2. What are the main functions of the stomach? the small intestine? the large intestine (colon)? *402, 403, 405*

3. Name four kinds of breakdown products that are small enough to be absorbed across the intestinal lining and into the internal environment. *403*

4. A glass of whole milk contains lactose, proteins, butterfat (mostly triglycerides), vitamins, and minerals. Explain what happens to each type of component when it passes through your digestive tract. *403–404, 405*

5. As a person ages, the number of body cells steadily decreases, and energy needs decline. If you were planning an older person's diet, what foods would you emphasize, and why? Which ones would you deemphasize? *406–410*

Self-Quiz *(Answers in Appendix IV)*

1. The _____ provides cells with a tolerable internal environment, raw materials, and waste disposal.
   a. digestive system   d. urinary system
   b. circulatory system   e. interaction of all of the
   c. respiratory system   systems listed

2. Most digestive systems have regions for _____ food.
   a. transporting   c. storing
   b. processing   d. all of the above

3. Maintaining good health and normal body weight requires that _____ intake be balanced by _____ output.

4. Most of our caloric intake should come from _____ .
   a. complex carbohydrates   c. proteins
   b. simple carbohydrates   d. lipids

5. The human body cannot produce all of its own _____ .
   a. vitamins and minerals   d. a through c
   b. essential fatty acids   e. a and c
   c. essential amino acids

6. Secretions from the _____ do *not* assist in digestion and absorption.
   a. salivary glands   c. liver
   b. thymus gland   d. pancreas

7. Digestion is completed and most nutrients are absorbed in the _____ .
   a. mouth   c. small intestine
   b. stomach   d. colon

8. Glucose and most amino acids are absorbed across the gut lining _____ .
   a. by active transport   c. at lymph vessels
   b. by diffusion   d. as fat droplets

9. Bile has roles in _____ digestion and absorption.
   a. carbohydrate   c. protein
   b. fat   d. amino acid

10. Match the organ with its key digestive function(s).
    _____ gallbladder   a. secrete bile and bicarbonate
    _____ stomach   b. digest, absorb most nutrients
    _____ colon   c. store, mix, dissolve food; start
    _____ pancreas   protein breakdown
    _____ salivary gland   d. store, concentrate bile
    _____ small intestine   e. concentrate undigested
    _____ liver   matter
                            f. secrete substances that
                            moisten food, start
                            polysaccharide breakdown
                            g. secrete digestive enzymes,
                            bicarbonate

Selected Key Terms

appendix *405*
bile *404*
bulk *405*
colon *405*
digestive system *399*
emulsification *403*
esophagus *400*
essential amino acid *407*
essential fatty acid *407*
gallbladder *403*
gut *400*
kilocalorie *410*
liver *403*
micelle formation *404*

microvillus *404*
mineral *408*
mouth *400*
nutrition *398*
obesity *410*
pancreas *403*
pharynx *400*
salivary gland *400*
small intestine *403*
stomach *402*
tooth *400*
villus *404*
vitamin *408*

Readings

Campbell-Platt, G. May 1988. "The Food We Eat." *New Scientist* 19:1–4.

Cohen, L. 1987. "Diet and Cancer." *Scientific American* 257(5):42–68.

Wardlaw, G.; P. Insel; and M. Seyler. 1992. *Contemporary Nutrition: Issues and Insights*. St. Louis: Mosby.

Weiss, P. September 1, 1990. "Fat and Fiction." *Science News* 138(9):138–139.

26 CIRCULATION

Heartworks

For Augustus Waller, Jimmie the bulldog was no ordinary pooch. Connected to wires and soaked to his ankles in buckets of salty water, Jimmie was a four-footed window into the workings of the heart.

Feel the repeated thumpings of your heart at the chest wall. Waller and other physiologists of the nineteenth century wondered if every heartbeat might produce a pattern of electrical currents that could be recorded painlessly at the body surface. That is where Jimmie and the buckets came in. Saltwater conducts electricity so efficiently, it carried faint signals from Jimmie's beating heart, through the skin of his legs, to a crude monitoring device. With this device, Waller made one of the world's first graphic recordings of a beating heart—an electrocardiogram (Figure 26.1).

A graph of your own heart's activity would look much the same. The pattern emerged a few weeks after you started growing as an embryo inside your mother. Patches of cardiac muscle started to contract. One patch took the lead and has been the pacemaker ever since.

It sets, adjusts, and resets the rate at which blood is pumped from your heart, through a vast network of blood vessels, then back to the heart. When you rest, the rate is moderate, around 70 beats a minute. When you jog, your muscles demand more blood-borne oxygen and glucose. Then, your heart may start pounding 150 times a minute to deliver sufficient blood to them.

We have come a long way from Jimmie in monitoring the heart. Sensors now detect the faintest signals of an impending heart attack. Computers are used to analyze a patient's beating heart and build images of it on a video screen. Surgeons substitute battery-powered pacemakers for malfunctioning natural ones.

With this chapter, we turn to the circulatory system—the means by which substances are rapidly moved to and from the body's living cells. As you will see, the system is absolutely central to the body's ability to maintain stable operating conditions in the internal environment—a state we call homeostasis.

Figure 26.1 History in the making—Dr. Augustus Waller's pet bulldog, Jimmie, taking part in a painless experiment (**a**) that yielded one of the world's first electrocardiograms (**b**).

a   BUCKET

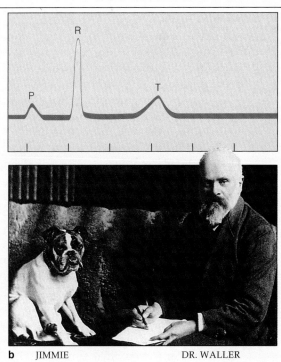

b   JIMMIE   DR. WALLER

1. Cells survive by exchanging substances with their surroundings. In most animals, substances move rapidly to and from living cells by way of a closed circulatory system. Blood, a fluid connective tissue confined within the heart and blood vessels, is the transport medium.

2. Human blood flows in two circuits. In the pulmonary circuit, the heart pumps oxygen-poor blood to the lungs, where it picks up oxygen; then blood flows back to the heart. In the systemic circuit, the heart pumps oxygen-enriched blood to all body regions. After giving up oxygen in those regions, blood flows back to the heart. Blood also transports carbon dioxide, plasma proteins, vitamins, hormones, lipids, and other solutes.

3. Arteries and veins are large-diameter transport tubes. Capillaries and to some extent venules are fine-diameter tubes for diffusion. Arterioles have adjustable diameters. They are control points for distributing different volumes of blood flow to different regions.

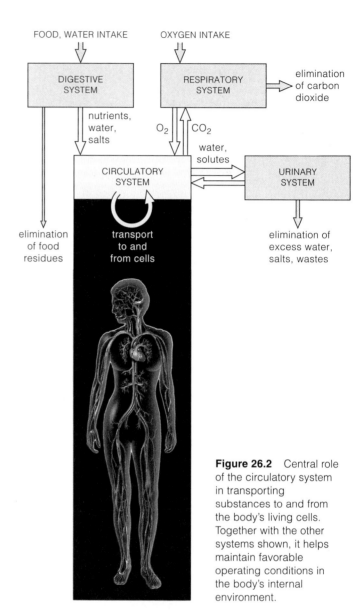

Figure 26.2 Central role of the circulatory system in transporting substances to and from the body's living cells. Together with the other systems shown, it helps maintain favorable operating conditions in the body's internal environment.

CIRCULATORY SYSTEMS: AN OVERVIEW

General Characteristics

Imagine that an earthquake has closed off the highways around your neighborhood. Grocery trucks can't enter and waste-disposal trucks can't leave, so food supplies dwindle and garbage piles up. Cells would face similar predicaments if your body's highways were disrupted. The highways are part of a **circulatory system**, which functions in the rapid internal transport of substances to and from cells.

A circulatory system helps maintain favorable neighborhood conditions, so to speak (Figure 26.2). This is important. Your body's differentiated cells, which perform specialized tasks, cannot fend for themselves. Different types must interact in coordinated ways to maintain the composition, volume, and temperature of the tissue fluid surrounding them, the **interstitial fluid**. A circulating connective tissue—**blood**—interacts with that fluid, making continual deliveries and pickups that help keep conditions tolerable for cell activities. (Together, blood and interstitial fluid are the body's internal environment.) The blood flows through blood vessels, which are tubes that differ in wall thickness and diameter. A muscular pump, the **heart**, generates the pressure that keeps blood flowing.

Like most animals, you have a *closed* circulatory system—blood flow is confined within the heart and blood vessels that have continuously connected walls. This is not true of the *open* circulatory systems of arthropods and most mollusks. In such systems, blood is pumped into tubes that open onto a space in the body's tissues. It mingles with tissue fluids, then moves into open-ended tubes leading back to the heart. Figure 26.3*a* is a diagram of this arrangement.

Think about the overall "design" of a closed system. The heart pumps incessantly, so the volume of flow through the entire system equals the volume returned to the heart. Yet the rate and volume have to be *adjusted* along the route. As Figure 26.3*b* suggests, blood flows rapidly through large-diameter vessels. Elsewhere in

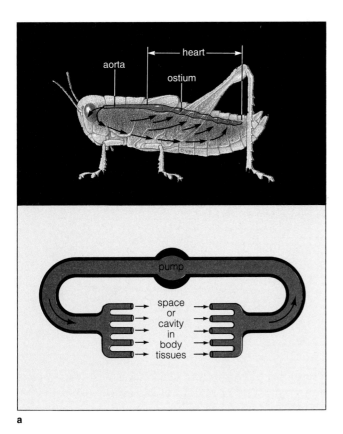

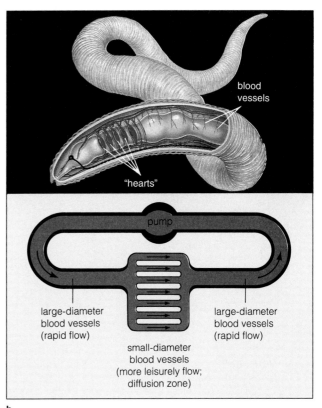

a

b

Figure 26.3 Examples of fluid flow in open and closed circulatory systems. (**a**) In a grasshopper's open system, a "heart" pumps blood through a vessel (aorta), which dumps the blood into body tissues. Blood diffuses through the tissues, then through openings that lead back to the heart. (**b**) In an earthworm's closed system, blood vessels lead away from and back to several muscular "hearts" near its head end. Walls of the hearts and blood vessels interconnect. (**c**) Example of a blood capillary typical of a closed circulatory system.

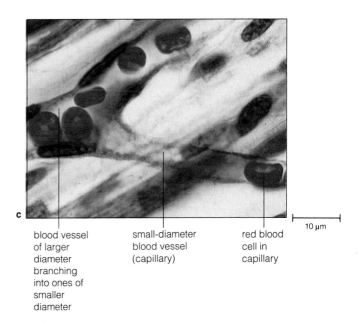

c

blood vessel of larger diameter branching into ones of smaller diameter

small-diameter blood vessel (capillary)

red blood cell in capillary

the system, it must flow slowly, so that there is time enough for substances to be exchanged with interstitial fluid. As you will see, the required slowdown occurs at **capillary beds**, where the flow fans out through vast numbers of small-diameter blood vessels called capillaries. By dividing up the blood flow, capillaries handle the same total volume of flow as the large-diameter vessels—but at a more leisurely pace.

Functional Links With the Lymphatic System

The heart's pumping action puts pressure on blood flowing through the circulatory system. Partly because of this pressure, small amounts of water and a few of the proteins dissolved in blood are forced out of the capillaries and become part of interstitial fluid. How-

ever, an elaborate network of drainage vessels picks up excess interstitial fluid and reclaimable solutes and returns them to the circulatory system. This network is part of the **lymphatic system**. Later in the chapter, you will see how other parts of the lymphatic system help cleanse bacteria and other disease agents from fluid being returned to the blood.

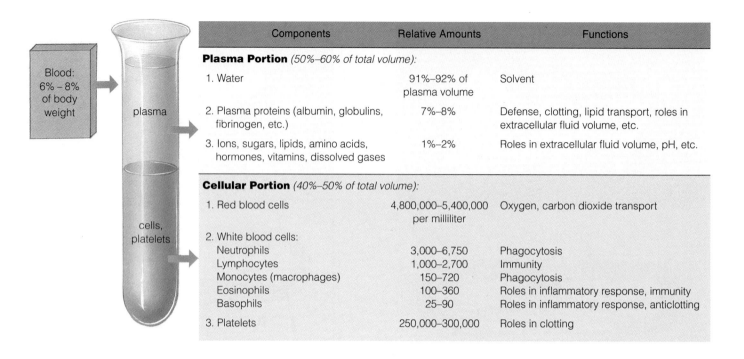

Components	Relative Amounts	Functions
Plasma Portion *(50%–60% of total volume):*		
1. Water	91%–92% of plasma volume	Solvent
2. Plasma proteins (albumin, globulins, fibrinogen, etc.)	7%–8%	Defense, clotting, lipid transport, roles in extracellular fluid volume, etc.
3. Ions, sugars, lipids, amino acids, hormones, vitamins, dissolved gases	1%–2%	Roles in extracellular fluid volume, pH, etc.
Cellular Portion *(40%–50% of total volume):*		
1. Red blood cells	4,800,000–5,400,000 per milliliter	Oxygen, carbon dioxide transport
2. White blood cells:		
Neutrophils	3,000–6,750	Phagocytosis
Lymphocytes	1,000–2,700	Immunity
Monocytes (macrophages)	150–720	Phagocytosis
Eosinophils	100–360	Roles in inflammatory response, immunity
Basophils	25–90	Roles in inflammatory response, anticlotting
3. Platelets	250,000–300,000	Roles in clotting

Blood: 6%–8% of body weight

plasma

cells, platelets

Figure 26.4 Components of blood. If a blood sample placed in a test tube is kept from clotting, it will separate into a layer of straw-colored liquid (the plasma) that floats over the red-colored cellular portion of blood.

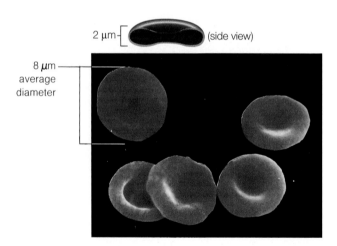

2 μm (side view)

8 μm average diameter

Figure 26.5 Size and shape of red blood cells.

CHARACTERISTICS OF BLOOD

Functions of Blood

Blood is a connective tissue with multiple functions. It transports oxygen, nutrients, and other solutes to cells. It carries away their secretions (including hormones) and metabolic wastes. Blood helps stabilize internal pH, and it serves as a highway for phagocytic cells that scavenge tissue debris and fight infections. In birds and mammals, blood helps equalize body temperature. It does this by carrying excess heat from skeletal muscles and other regions of high metabolic activity to the skin, where heat can be dissipated.

Blood carries substances to cells, carries away products and wastes from them, and helps maintain an internal environment that is favorable for cell activities.

Blood Volume and Composition

The volume of blood depends on body size and on the concentrations of water and solutes. Blood volume for average-size adult humans is about 6 to 8 percent of the body weight. That amounts to about 4 or 5 quarts.

As for all vertebrates, human blood is a sticky fluid, thicker than water and slower flowing. **Plasma**, **red blood cells**, **white blood cells**, and **platelets** are its components (Figures 26.4 and 26.5). Plasma normally accounts for 50 to 60 percent of the total blood volume.

Plasma. Plasma is mostly water. It serves as a transport medium for blood cells and platelets. It also serves as a solvent for ions and molecules, including hundreds of different plasma proteins. Some plasma proteins transport lipids and fat-soluble vitamins. Others help clot blood or defend against disease agents. The movement of water between blood and interstitial fluid—hence the blood's fluid volume—is influenced by the concentrations of plasma proteins. Ions, glucose and other simple sugars, lipids, amino acids, vitamins, and hormones also are dissolved in plasma. So are oxygen, carbon dioxide, and nitrogen.

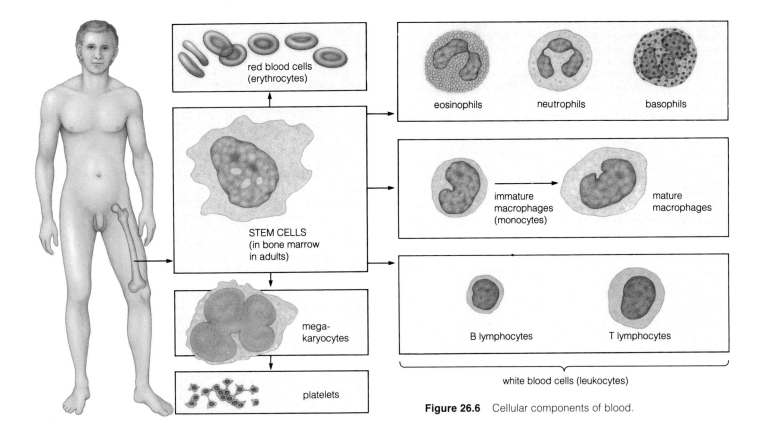

Figure 26.6 Cellular components of blood.

Labels in figure: red blood cells (erythrocytes); STEM CELLS (in bone marrow in adults); mega-karyocytes; platelets; eosinophils; neutrophils; basophils; immature macrophages (monocytes); mature macrophages; B lymphocytes; T lymphocytes; white blood cells (leukocytes)

Red Blood Cells. Erythrocytes, or red blood cells, are biconcave disks—rather like doughnuts with a squashed-in center instead of a hole (Figure 26.5). They transport oxygen (required for aerobic respiration) and carry away some carbon dioxide wastes. When oxygen diffuses into blood, it binds with hemoglobin (page 32), an iron-containing pigment that gives red blood cells their color. Oxygenated blood is bright red. Poorly oxygenated blood is darker red but appears blue inside blood vessel walls near the body surface.

As Figure 26.6 shows, red blood cells are derived from stem cells in red bone marrow. **Stem cells** are unspecialized cells that replace themselves by ongoing mitotic divisions. Portions of their daughter cells also divide, then differentiate into specialized cells.

Mature red blood cells no longer have their nucleus, nor do they require it. They have enough hemoglobin, enzymes, and other proteins to function for about 120 days. Phagocytes continually engulf the oldest or already dead cells, but ongoing replacements keep the cell count fairly stable. A **cell count** is the number of cells of a given type in a microliter of blood. The average number of red blood cells is 5.4 million in males and 4.8 million in females.

White Blood Cells. Leukocytes, or white blood cells, function in day-to-day housekeeping and defense. Some engulf old, damaged, or dead cells and anything that is chemically perceived as foreign to the body. Others target or destroy specific bacteria, viruses, and other agents of disease. All types arise from stem cells in bone marrow. They travel the circulation highways, but most go to work after they squeeze out of capillaries and enter the surrounding tissues.

White blood cells differ in size, nuclear shape, and staining traits. There are five types: neutrophils, eosinophils, basophils, monocytes, and lymphocytes (Figure 26.4). Their cell counts vary, depending on whether an individual is highly active, healthy, or under siege, as described in the next chapter. Neutrophils and monocytes are "search-and-destroy" cells. The monocytes follow chemical trails to inflamed tissues. There they develop into wandering macrophages ("big eaters") that engulf invaders and debris. Two classes of lymphocytes (B cells and T cells) respond to specific invaders. Life for most white blood cells is challenging and short—typically measured in days or, during a major battle, a few hours.

Platelets. Some stem cells in bone marrow give rise to "giant" cells (megakaryocytes). The cells shed fragments of cytoplasm, which become enclosed in a bit of plasma membrane. The membrane-bound fragments are platelets. Each lasts five to nine days, but hundreds of thousands are always circulating in blood. Substances released from platelets initiate blood clotting.

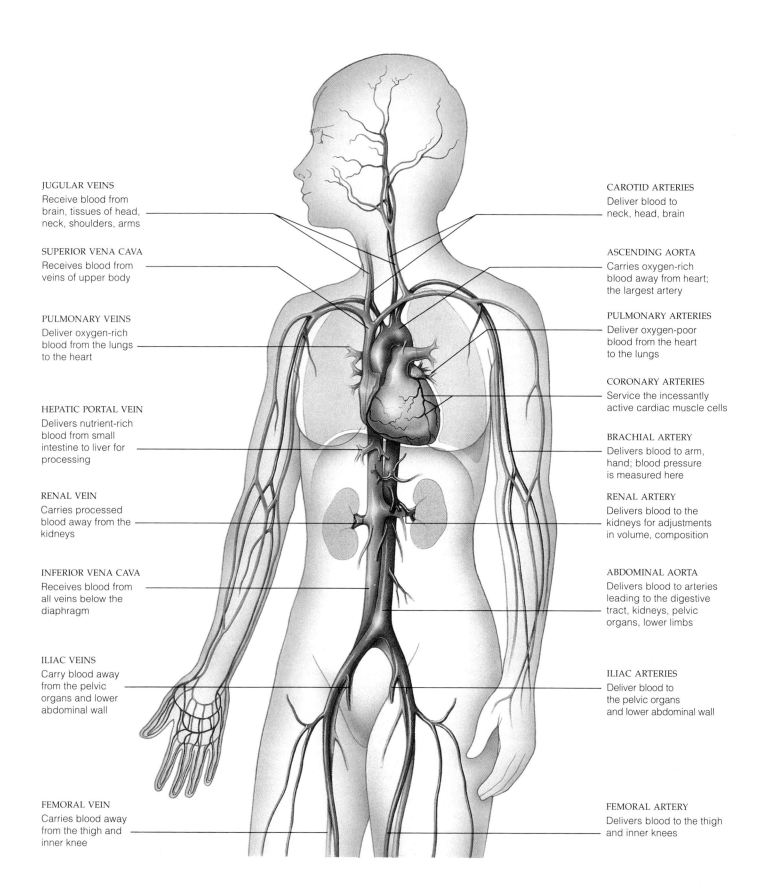

JUGULAR VEINS
Receive blood from brain, tissues of head, neck, shoulders, arms

SUPERIOR VENA CAVA
Receives blood from veins of upper body

PULMONARY VEINS
Deliver oxygen-rich blood from the lungs to the heart

HEPATIC PORTAL VEIN
Delivers nutrient-rich blood from small intestine to liver for processing

RENAL VEIN
Carries processed blood away from the kidneys

INFERIOR VENA CAVA
Receives blood from all veins below the diaphragm

ILIAC VEINS
Carry blood away from the pelvic organs and lower abdominal wall

FEMORAL VEIN
Carries blood away from the thigh and inner knee

CAROTID ARTERIES
Deliver blood to neck, head, brain

ASCENDING AORTA
Carries oxygen-rich blood away from heart; the largest artery

PULMONARY ARTERIES
Deliver oxygen-poor blood from the heart to the lungs

CORONARY ARTERIES
Service the incessantly active cardiac muscle cells

BRACHIAL ARTERY
Delivers blood to arm, hand; blood pressure is measured here

RENAL ARTERY
Delivers blood to the kidneys for adjustments in volume, composition

ABDOMINAL AORTA
Delivers blood to arteries leading to the digestive tract, kidneys, pelvic organs, lower limbs

ILIAC ARTERIES
Deliver blood to the pelvic organs and lower abdominal wall

FEMORAL ARTERY
Delivers blood to the thigh and inner knees

Figure 26.7 Human circulatory system, showing some of the major blood vessels. Arteries are shaded red. Veins are shaded blue.

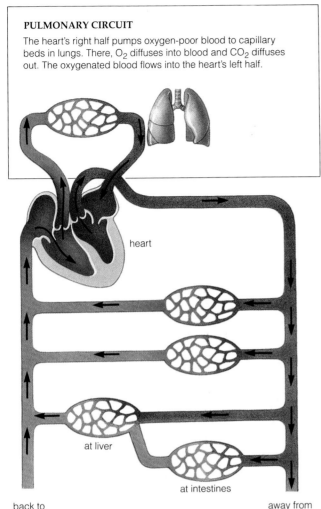

Figure 26.8   Diagram of the pulmonary and systemic circuits for blood flow through the human circulatory system.

HUMAN CARDIOVASCULAR SYSTEM

"Cardiovascular" comes from the Greek *kardia* (heart) and the Latin *vasculum* (vessel). Figure 26.7 shows the human cardiovascular system. In this system, the heart pumps blood into large-diameter **arteries**. From there, blood flows into small, muscular **arterioles**, which branch into small-diameter **capillaries**. Blood flows from capillaries into small **venules**, then into large-diameter **veins** that return blood to the heart.

Blood Circulation Routes

A muscular partition divides the human heart into two halves. The partition is the basis of two cardiovascular circuits (Figure 26.8). Each circuit has its own set of arteries, arterioles, capillaries, venules, and veins.

In the **pulmonary circuit**, the heart's *right* half pumps blood to the lungs. There, blood picks up oxygen and gives up carbon dioxide. From the lungs, the freshly oxygenated blood flows to the heart's left half.

In the **systemic circuit**, the heart's *left* half pumps the oxygenated blood to all tissue regions, where oxygen is used and carbon dioxide is produced. Then the oxygen-poor blood flows to the heart's right half.

You might think that a given volume of blood passes through only one capillary bed during a trip away from and back to the heart. This is not always the case. For example, part of the systemic circuit threads around the intestines. There, blood flows through one capillary bed and picks up glucose and other absorbed substances. Then the blood moves on through another capillary bed, in the liver—an organ with a key role in nutrition. The second bed gives the liver time to process absorbed substances before blood is circulated further.

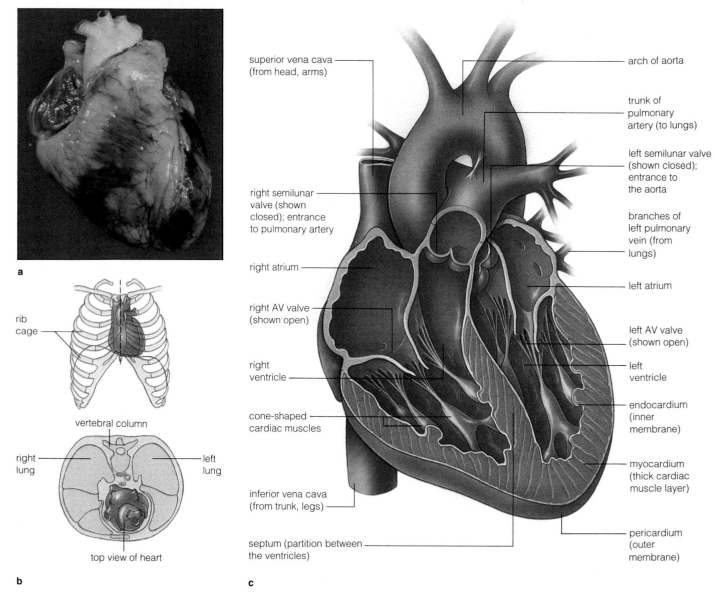

superior vena cava
(from head, arms)

arch of aorta

trunk of
pulmonary
artery (to lungs)

right semilunar
valve (shown
closed); entrance
to pulmonary artery

left semilunar valve
(shown closed);
entrance to
the aorta

branches of
left pulmonary
vein (from
lungs)

right atrium

left atrium

right AV valve
(shown open)

left AV valve
(shown open)

right
ventricle

left
ventricle

cone-shaped
cardiac muscles

endocardium
(inner
membrane)

inferior vena cava
(from trunk, legs)

myocardium
(thick cardiac
muscle layer)

septum (partition between
the ventricles)

pericardium
(outer
membrane)

a

rib
cage

vertebral column

right
lung

left
lung

top view of heart

b

c

Figure 26.9 (**a**) The human heart and (**b**) its location. (**c**) Cutaway view showing the heart's internal organization.

The Human Heart

Heart Structure. During a seventy-year life span, the human heart beats some 2.5 billion times and rests only briefly between heartbeats. Its structure reflects its role as a durable pump. The heart is mostly cardiac muscle tissue protected by a tough outer membrane. Its inner chambers are lined with connective tissue and endothelium. Endothelium is a layer of epithelial cells found only in the heart and blood vessels.

Each half of the heart has two chambers—an **atrium** (plural, atria) located above a **ventricle**. Membrane flaps separate the two chambers and serve as a one-way valve. As Figure 26.9 shows, these flaps are called an AV valve (short for atrioventricular). Each half of the heart also has a semilunar valve between the ventricle

and the arteries leading away from it. During a heart-beat, both types of valves open and close in ways that keep blood moving in one direction.

The heart has its own "coronary circulation." Two coronary arteries lead into a capillary bed that services most of its cardiac muscle cells (Figure 26.7). They branch off the **aorta**, the major artery carrying oxygen-enriched blood away from the heart.

Cardiac Cycle. Each time the heart beats, its four chambers go through phases of contraction (systole) and relaxation (diastole). The sequence of contraction and relaxation is a **cardiac cycle** (Figure 26.10).

When the relaxed atria are filling, fluid pressure increases inside them and forces the AV valves open. Blood flows into the ventricles, which fill completely

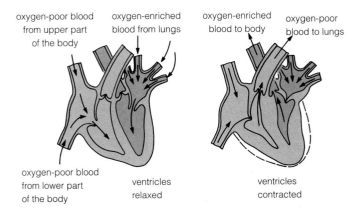

oxygen-poor blood from upper part of the body

oxygen-enriched blood from lungs

oxygen-enriched blood to body

oxygen-poor blood to lungs

oxygen-poor blood from lower part of the body

ventricles relaxed

ventricles contracted

Figure 26.10 Blood flow through the heart during part of a cardiac cycle. The blood and heart movements generate vibrations, producing a "lub-dup" sound that can be heard at the chest wall. At each "lub," AV valves are closing as the ventricles contract. At each "dup," semilunar valves are closing as the ventricles relax.

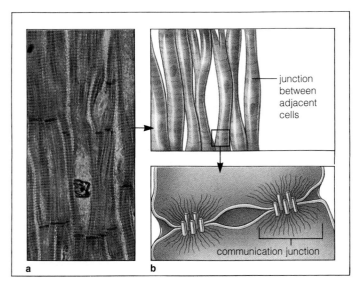

junction between adjacent cells

communication junction

a   b

Figure 26.11 Communication junctions at the ends of abutting cardiac muscle cells. Signals travel rapidly across the junctions and cause cells to contract nearly in unison.

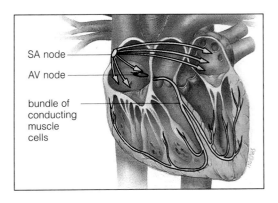

SA node

AV node

bundle of conducting muscle cells

Figure 26.12 Location of cardiac muscle cells that conduct signals for contraction through the heart.

when the atria contract. When the filled ventricles start to contract, fluid pressure inside them increases and forces the AV valves to shut. Their continued contraction causes the pressure to rise sharply above that in blood vessels leading away from the heart. The pressure forces the semilunar valves open, and blood flows out of the heart. Now the ventricles relax, the semilunar valves close—and the already filling atria are ready to repeat the cycle.

During a cardiac cycle, atrial contraction simply helps fill the ventricles. Contraction of the *ventricles* is the driving force for blood circulation.

Mechanisms of Contraction. In skeletal muscle tissue, the ends of the fiberlike cells are attached to bones. In *cardiac* muscle tissue, the ends of cells branch, then connect with one another. Where the plasma membranes of abutting cells are joined together, there are communication junctions (Figure 26.11). Signals for contraction spread across these junctions. With each heartbeat, the signals spread so rapidly that cardiac muscle cells contract together, almost as if they were a single unit.

Cardiac and skeletal muscle cells differ in another way. The nervous system makes skeletal muscle contract. But cardiac muscle contracts spontaneously—the nervous system can only *adjust* the rate and strength of its contraction. Even if all nerves leading to the heart

are severed, the heart will keep on beating! How? Some self-excitatory cardiac muscle cells continue to produce and conduct signals for contraction.

The cell bodies of these pacemaking cells are clustered mainly in a region of the right atrium's wall (Figure 26.12). This region is the SA node (short for sinoatrial). It generates waves of excitation, usually seventy or eighty times a minute. Each wave spreads over both atria, causes them to contract, then reaches the AV node (for atrioventricular). The wave spreads more slowly here. The delay allows the atria to finish contracting before a wave of excitation spreads over the ventricles. Thus the SA node is the **cardiac pacemaker**. Its spontaneous, rhythmic signals are the basis for the normal rate of heartbeat.

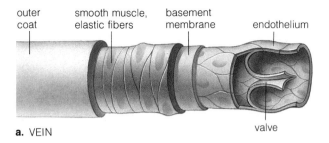

a. VEIN

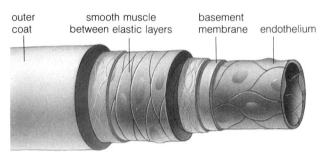

b. ARTERY

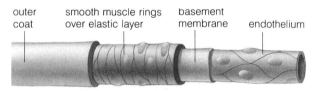

c. ARTERIOLE

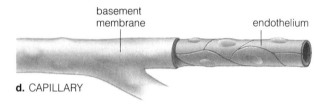

d. CAPILLARY

Figure 26.13 Structure of blood vessels. The basement membranes are noncellular layers, rich with proteins and polysaccharides, that intervene between the endothelium and another tissue layer.

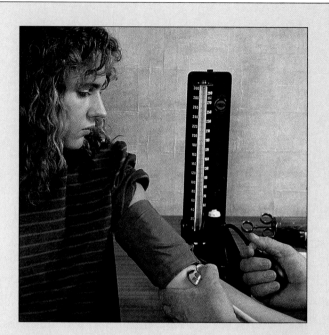

Figure 26.14 Measuring blood pressure. A hollow cuff attached to a pressure gauge is wrapped around the upper arm. Then it is inflated with air to a pressure above the highest pressure of the cardiac cycle (at systole, when the ventricles contract). Above this pressure, no sounds are heard through a stethoscope positioned above the artery (because no blood is flowing through it).

Air in the cuff is slowly released, so some blood flows into the artery. The turbulent flow causes soft tapping sounds, and when this first occurs, the value on the gauge is the systolic pressure—about 120 mm mercury (Hg) in young adults at rest. (This means the measured pressure would force mercury to move upward 120 millimeters in a narrow glass column.) More air is released from the cuff. Just after the sounds become dull and muffled, blood flows continuously. So the turbulence and tapping sounds stop. The silence corresponds to the diastolic pressure (at the end of a cardiac cycle, just before the heart pumps out blood). Generally the reading is about 80 mm Hg. In this example, the *pulse* pressure (the difference between the highest and lowest pressure readings) is 120 − 80, or 40 mm Hg.

Blood Pressure in the Vascular System

Blood pressure is the fluid pressure generated by heart contractions. It does not stay the same throughout the systemic or pulmonary circuit. For example, pressure normally is high in the aorta, then drops along the circuit away from and back to the heart. Why does the pressure drop? As blood passes through different kinds of blood vessels, energy in the form of pressure is lost as it overcomes resistance to the flow of blood. The structure of the blood vessels themselves, shown in Figure 26.13, is a factor in this loss of energy.

Arterial Blood Pressure. Arteries are pressure reservoirs that can "smooth out" pulsations in blood pressure. Such pulsations are generated during each cardiac cycle.

Figure 26.13*b* shows the elastic, muscular wall of an artery. The wall bulges under the pressure surge caused by ventricular contraction, then it recoils and forces blood onward. With their large diameters, arteries present little resistance to flow, so pressure does not drop much in the arterial portion of the blood circuits. Figure 26.14 shows how blood pressure is measured at large arteries of the upper arms.

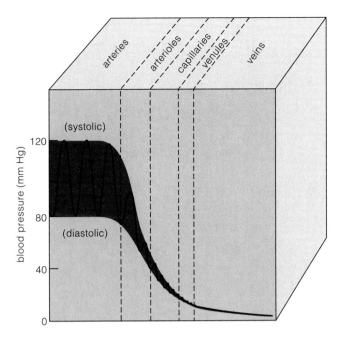

Figure 26.15 Blood pressure. This diagram shows the drop in fluid pressure for a volume of blood making a trip along the systemic circuit.

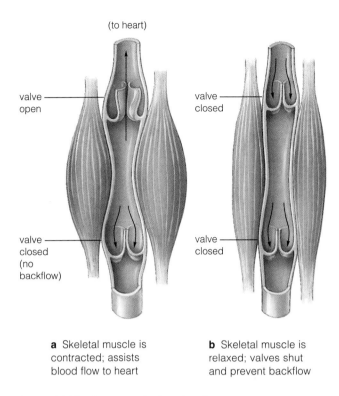

a Skeletal muscle is contracted; assists blood flow to heart

b Skeletal muscle is relaxed; valves shut and prevent backflow

Figure 26.16 Structure of valves in veins.

Resistance at Arterioles. Arteries branch into vessels of smaller diameter, the arterioles (Figure 26.13c). The greatest pressure drop occurs in arterioles, which offer the greatest resistance to blood flow (Figure 26.15).

The diameter of an arteriole shrinks when smooth muscle in its wall contracts. It enlarges when the smooth muscle relaxes. Such changes are initiated by signals from the nervous and endocrine systems, as well as by changes in local chemical conditions. And they allow adjustments in the distribution of blood flow. Blood is directed to a region of great metabolic activity when the diameters of arterioles in those regions enlarge. Blood is directed away from a less active region when arteriole diameters constrict.

Capillary Function. Capillary beds are diffusion zones for exchanges between blood and interstitial fluid. Of all blood vessels, capillaries have the thinnest wall—a single layer of flat endothelial cells, separated from one another by narrow clefts (Figure 26.13d). The capillary lumen is so small, blood cells must squeeze through it single file.

A single capillary presents high resistance to flow. Yet there are so many in a capillary bed, the combined diameters are greater than the diameters of arterioles.

Thus a capillary bed presents less *total* resistance to flow than the arterioles leading into it, and the total drop in blood pressure is not great.

Venous Pressure. Capillaries merge into "little veins," or venules. Some diffusion occurs across the venule wall, which is only a little thicker than that of a capillary. Venules merge into large-diameter veins (Figure 26.13a). Veins are transport tubes leading back to the heart, and they are equipped with valves that prevent backflow. When gravity beckons, blood in veins tends to reverse direction—and pushes the valves shut.

Veins also are blood volume reservoirs; they contain 50 to 60 percent of the total blood volume. A vein wall is thin enough to bulge greatly under pressure. It also contains smooth muscle. When blood must circulate faster (as during exercise), the smooth muscle contracts. The wall stiffens, the vein doesn't bulge as much—and venous pressure rises. This drives more blood to the heart, so a larger volume of blood can be pumped to active tissue regions. Skeletal muscles bulging against adjacent veins also help drive blood back to the heart (Figure 26.16).

We turn now to a *Focus* essay that describes the major cardiovascular disorders.

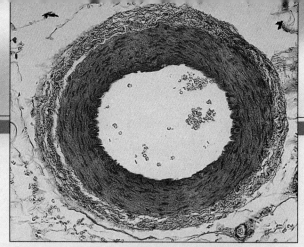

Cardiovascular Disorders

Cardiovascular disorders are the leading cause of death in the United States. They affect at least 40 million people and kill about a million each year. The most common disorders are *hypertension* (sustained high blood pressure) and *atherosclerosis* (progressive thickening of the arterial wall and narrowing of the arterial lumen). Both affect blood circulation and so cause most *heart attacks* (damage or death of heart muscle) and *strokes* (brain damage).

In most heart attacks, a "crushing" pain behind the breastbone lasts a half hour or more. Mild to severe pain may radiate into the left arm, shoulder, or neck. Sweating, nausea, vomiting, dizziness, or loss of consciousness may accompany an attack.

Risk Factors. The following risk factors have been linked to cardiovascular disorders:

1. High level of cholesterol in the blood
2. High blood pressure
3. Obesity (page 410)
4. Lack of regular exercise
5. Smoking (page 460)
6. Diabetes mellitus (page 505)
7. Genetic predisposition to heart failure
8. Age (the older you get, the greater the risk)
9. Gender (until age fifty, males are at much greater risk than females)

Factors 1 through 5 are controllable by eating properly, exercising, and not smoking.

For example, the fatter you become, the more blood capillaries you get (they service the increased adipose tissue masses). So your heart has to work harder to pump blood through the increasingly divided vascular circuit. As another example, smoking can damage your heart. Nicotine in tobacco makes the adrenal glands secrete epinephrine, a hormone that constricts blood vessels and so triggers an accelerated heartbeat and a rise in blood pressure. Carbon monoxide in cigarette smoke outcompetes carbon dioxide for the binding sites on hemoglobin—so the heart has to pump harder to rid the body of carbon dioxide.

The next sections describe some of the tissue damage that results from cardiovascular disorders.

Hypertension. Hypertension results from gradual increases in the resistance to blood flow through small arteries. In time, blood pressure remains elevated, even when a person is resting. Heredity may be a factor; the disorder tends to run in families. But so is diet. For instance, high salt intake can raise blood pressure in

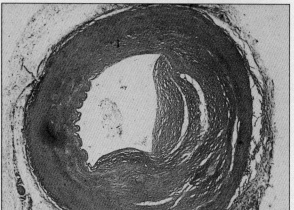

a Cross-section of a normal artery (*above*) and a partially obstructed one (*below*).

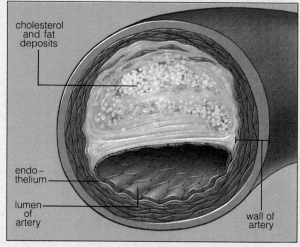

cholesterol and fat deposits

endothelium

lumen of artery

wall of artery

b Diagram of an atherosclerotic plaque.

susceptible people and increase the heart's workload. Eventually the heart may enlarge and fail to pump blood effectively. High blood pressure also may contribute to the "hardening" of arterial walls that hampers delivery of oxygen to the brain, heart, and other vital organs.

Hypertension is called the "silent killer" because affected persons may have no outward symptoms. Even when they know their blood pressure is high, some tend to resist medication, changes in diet, and regular exercise. Of 23 million hypertensive Americans, most don't undergo treatment. About 180,000 die each year.

Atherosclerosis.   Arteries may thicken and lose elasticity (a condition called arteriosclerosis). In atherosclerosis, this condition worsens as cholesterol and other lipids build up in the wall of arteries and cause the lumen to narrow.

Normally, the liver produces enough cholesterol to satisfy the body's needs. Together with the liver's output, cholesterol from the diet also ends up circulating in the blood. If you habitually eat cholesterol-rich food, you may end up with a high blood level of cholesterol. If you have a certain heritable (genetic) disorder, the same thing might happen no matter what kinds of food you eat.

When circulating in blood, cholesterol is bound to proteins as *low-density lipoproteins* (LDLs). These can bind to receptors on cells throughout the body. Cells take up LDLs and their cholesterol cargo for use in cell activities. Excess cholesterol is attached to proteins as *high-density lipoproteins* (HDLs) and transported back to the liver, where it can be metabolized.

In some people, cells may not have enough LDL receptors, so not enough LDL is removed from the blood. As the blood level of LDL increases, so does the risk of atherosclerosis. LDLs—with their bound cholesterol—can *infiltrate* arterial walls. Abnormal smooth muscle cells multiply and connective tissue components increase in arterial walls. Cholesterol accumulates in cells and extracellular spaces of the wall's endothelial lining. Calcium salts are deposited on top of the lipids, and a fibrous net forms over the mass. This *atherosclerotic plaque* sticks out into the arterial lumen (Figures *a,b*).

When platelets get caught on a plaque's rough edges, they secrete chemicals that initiate clot formation. Growth of the plaque and clot can narrow or block the artery. Blood flow to the tissues serviced by the artery may decrease to a trickle or stop entirely. A clot that stays in place is a *thrombus*. If it becomes dislodged and travels the bloodstream, it is an *embolus*.

Coronary arteries and their branches have narrow diameters. They are extremely vulnerable to clogging by a plaque or clot. When they narrow to one-quarter of their former diameter, the outcome ranges from mild chest pains (angina pectoris) to a heart attack.

Atherosclerosis involving coronary arteries can be diagnosed through stress electrocardiograms, or ECGs. These are recordings of the electrical activity of the cardiac cycle while a person is exercising on a treadmill. It also can be diagnosed by *angiography*. This procedure involves injections of a dye that is opaque to x-rays. Severe blockage may require surgery. In *coronary bypass surgery*, a section of an artery from the chest is stitched to the aorta and to the coronary artery below the narrowed or blocked region (Figure *c*). In *laser angioplasty*, laser beams vaporize

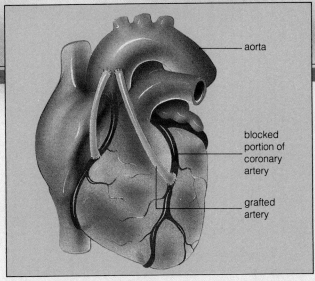

aorta

blocked portion of coronary artery

grafted artery

c Two coronary bypasses (green).

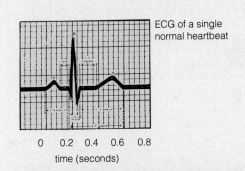

ECG of a single normal heartbeat

0   0.2   0.4   0.6   0.8
time (seconds)

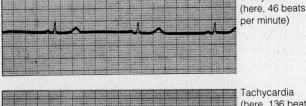

Bradycardia (here, 46 beats per minute)

Tachycardia (here, 136 beats per minute)

d Examples of ECG readings.

the atherosclerotic plaques. In *balloon angioplasty*, a small balloon is inflated within a blocked artery to break up plaques.

Arrhythmias.   Irregular heart rhythms, called *arrhythmias*, can be detected by an ECG (Figure *d*). Some arrhythmias are normal. For example, many endurance athletes have a lower-than-average resting cardiac rate, a condition called *bradycardia*. Their nervous system has adjusted the cardiac pacemaker's rate of contraction downward. This is an adaptive response to ongoing strenuous exercise. Also, during exercise or times of stress, more than 100 heartbeats a minute are common (*tachycardia*).

427

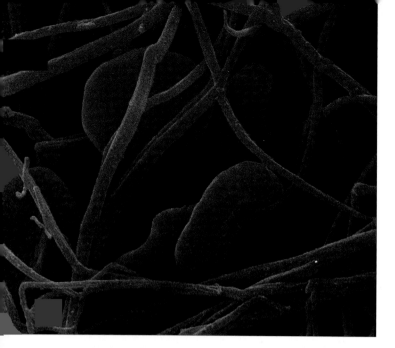

Figure 26.17 Red blood cells trapped during clot formation. Blood coagulates when damage exposes collagen fibers in the wall of small blood vessels. This initiates reactions that cause rod-shaped plasma proteins to stick together as long, insoluble threads. The threads stick to the exposed collagen, forming a net that traps blood cells and platelets. The entire mass is a blood clot.

Hemostasis

The smaller blood vessels just described are vulnerable to ruptures, cuts, and other damage. A process called **hemostasis** repairs the damage and so prevents blood loss. The process includes blood vessel spasm, platelet plug formation, and blood coagulation.

First, smooth muscle in a damaged wall contracts in an automatic response called a spasm. The blood vessel constricts, so blood flow through it temporarily stops. Second, platelets clump together as a temporary plug in the damaged wall. They also release substances that help prolong the spasm and attract more platelets. Third, blood coagulates (converts to a gel) and forms a clot (Figure 26.17). Finally, the clot retracts into a compact mass, and the breach in the wall is sealed.

Blood Typing

Your body protects itself in another way, through "self markers" on every one of its cells. These are proteins that identify the cells as belonging to you. Your body also produces **antibodies**. These proteins ignore *your* self markers, but they bind to a *foreign* marker and so target its bearer for destruction by the immune system (page 438). Bacteria, viruses, and anything else that isn't one of your own normal cells carry foreign markers.

During a transfusion (when blood from two people mixes), red blood cells bearing the "wrong" marker will be recognized as foreign and destroyed, with serious consequences. The same thing will happen during pregnancy, if antibodies diffuse from the mother's circulatory system into that of her unborn child.

Based on an understanding of cell surface markers and antibodies, scientists have devised ways to analyze what forms of self markers are present on a person's red blood cells. Blood typing is based on these analyses.

ABO Blood Typing. Molecular variations in one kind of self marker on red blood cells are analyzed in **ABO blood typing**. (The genetic basis of this variation was described earlier, on page 131.) People with one form of the marker are said to have type A blood, and those with another form have type B blood. Many people have *both* forms of the marker on their red blood cells; they have type AB blood. Others have *neither* form of the marker; they have type O blood.

If you are type A, your antibodies ignore A markers but will act against B markers. If you are type B, your antibodies ignore B markers but will act against A markers. If you are type AB, your antibodies ignore both forms of the marker, so you can tolerate donations of type A, B, or AB blood. However, if you are type O, you have antibodies *against* both forms of the marker, so your options are limited to type O donations.

Figure 26.18 shows what happens when blood from incompatible donors and recipients intermingles. In a response called agglutination, antibodies act against the "foreign" cells and cause them to clump. The clumps can clog small blood vessels. They may lead to tissue damage and death.

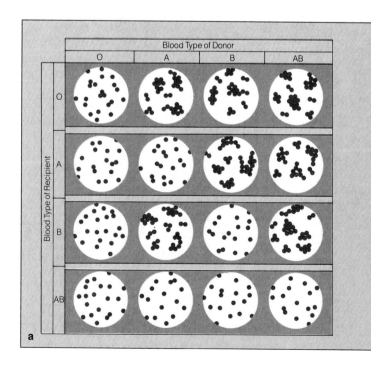

Rh Blood Typing. Other kinds of surface markers on red blood cells also can cause agglutination responses. For example, **Rh blood typing** is based on the presence or absence of an Rh marker (so named because it was first identified in blood samples of *rh*esus monkeys). If you are type Rh⁺, your blood cells bear this marker. If you are type Rh⁻, they don't. Ordinarily, people don't have antibodies against Rh markers. However, a recipient of an Rh⁺ blood transfusion will produce antibodies against the marker, and these will continue circulating in the blood.

If an Rh⁻ woman becomes pregnant by an Rh⁺ man, there is a chance the unborn child will be Rh⁺. During pregnancy or childbirth, some red blood cells of the fetus may leak into the mother's bloodstream. If they do, her body will produce antibodies against Rh (Figure 26.19). If she becomes pregnant *again*, Rh antibodies will enter the bloodstream of this new fetus. If its blood is type Rh⁺, the antibodies will cause red blood cells to swell, rupture, and release hemoglobin.

In *erythroblastosis fetalis*, an extreme case of this disorder, too many cells are destroyed and the fetus dies. If diagnosed before birth or if delivered alive, it can survive by having its blood slowly replaced with transfusions that are free of Rh antibodies. Currently, a known Rh⁻ woman can be treated right after her first pregnancy with an anti-Rh gamma globulin (RhoGam) that will protect her next fetus. The drug will inactivate any Rh⁺ fetal blood cells circulating in the mother's bloodstream before she can become sensitized and begin producing the potentially dangerous antibodies.

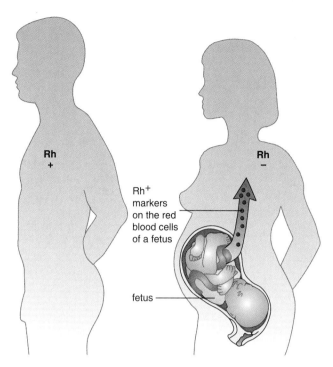

a A forthcoming child of an Rh⁻ woman and Rh⁺ man inherits the genetic instructions for the Rh⁺ marker. The placenta, a complex tissue that forms during pregnancy, allows the mother's blood to intermingle with that of the growing fetus. So fetal blood cells bearing the Rh⁺ markers enter her bloodstream.

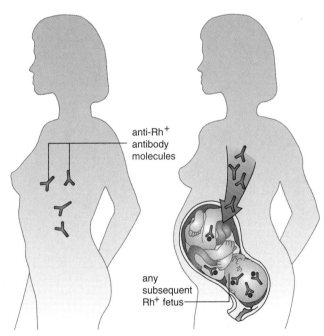

b The foreign markers in the mother's body stimulate production of antibody molecules. Suppose the woman becomes pregnant again. If the new fetus (or any other one) inherits instructions for the Rh⁺ marker, the anti-Rh⁺ antibodies will act against them.

Figure 26.19 Production of antibodies in response to Rh⁺ markers on the red blood cells of a developing fetus.

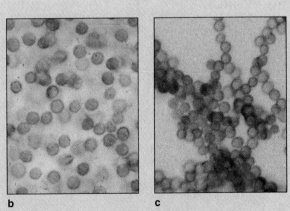

Figure 26.18 (**a**) Agglutination responses in blood types A, B, AB, and O when mixed with blood samples of the same and different types. Micrographs showing the absence of agglutination in a mixture of two different but compatible types (**b**) and agglutination in a mixture of incompatible blood types (**c**).

LYMPHATIC SYSTEM

We conclude this chapter with a section on the manner in which the lymphatic system supplements blood circulation. But think of this section as a bridge to the next chapter, on immunity, for the lymphatic system also helps defend the body against injury and attack. As Figure 26.20 shows, the system consists of drainage vessels, lymphoid organs, and lymphoid tissues. Tissue fluid that has moved into the vessels is called **lymph**.

Lymph Vascular System

The vascular portion of the lymphatic system, which includes **lymph capillaries** and **lymph vessels**, has three functions. First, it takes up water and plasma proteins that have leaked out at capillary beds and returns them to the blood. Second, it transports fats absorbed from the small intestine (page 404). Third, it transports

Figure 26.20 The human lymphatic system, which includes the lymph vascular network and the lymphoid organs and tissues. Green dots show some of the major lymph nodes. Patches of lymphoid tissue in the small intestine and appendix also are part of the system.

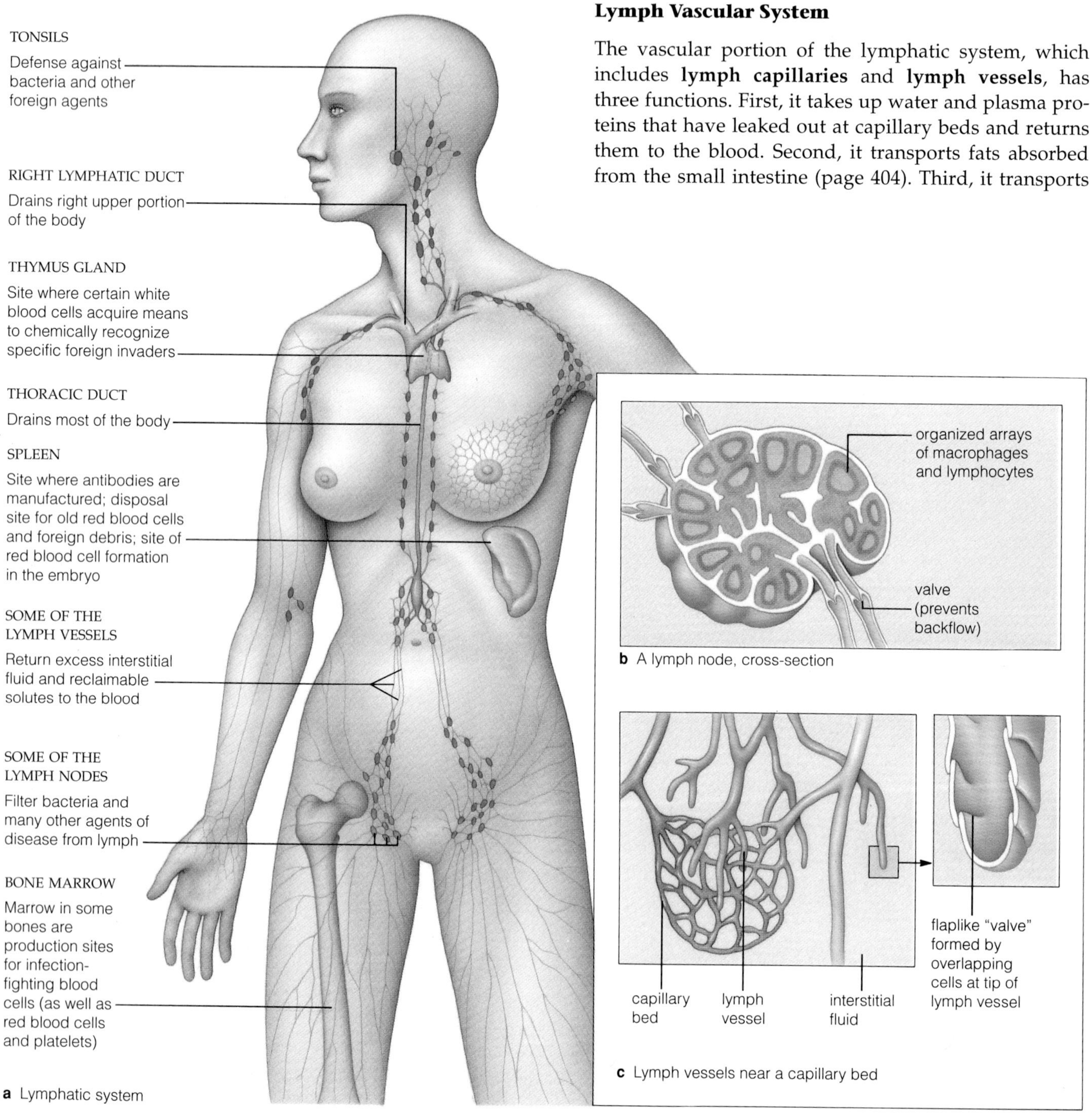

TONSILS

Defense against bacteria and other foreign agents

RIGHT LYMPHATIC DUCT

Drains right upper portion of the body

THYMUS GLAND

Site where certain white blood cells acquire means to chemically recognize specific foreign invaders

THORACIC DUCT

Drains most of the body

SPLEEN

Site where antibodies are manufactured; disposal site for old red blood cells and foreign debris; site of red blood cell formation in the embryo

SOME OF THE LYMPH VESSELS

Return excess interstitial fluid and reclaimable solutes to the blood

SOME OF THE LYMPH NODES

Filter bacteria and many other agents of disease from lymph

BONE MARROW

Marrow in some bones are production sites for infection-fighting blood cells (as well as red blood cells and platelets)

a Lymphatic system

organized arrays of macrophages and lymphocytes

valve (prevents backflow)

b A lymph node, cross-section

flaplike "valve" formed by overlapping cells at tip of lymph vessel

capillary bed   lymph vessel   interstitial fluid

c Lymph vessels near a capillary bed

foreign particles and cellular debris to disposal centers (**lymph nodes**).

The lymph vascular system starts at capillary beds. Tissue fluid enters lymph capillaries. These have no pronounced entrance. Instead, their tips contain regions of overlapping endothelial cells (Figure 26.20). Water and solutes move inward at these flaplike "valves." Lymph capillaries merge with lymph vessels. These have a larger diameter, smooth muscle in their wall, and flaplike valves that prevent backflow. Lymph vessels converge into collecting ducts, which drain into veins in the lower neck.

Lymphoid Organs and Tissues

The defense portion of the lymphatic system includes the lymph nodes, spleen, thymus, tonsils, adenoids, and patches of tissue in the small intestine and appendix. These lymphoid organs and tissues contain production centers for the white blood cells called lymphocytes. Some also serve as battlegrounds.

After lymphocytes are produced in bone marrow, they enter the blood and travel to lymphoid organs, where they take up residence. When they detect a foreign agent, they rapidly divide and produce vast armies that can destroy it.

Lymph nodes are located at intervals along lymph vessels (Figure 26.20). Lymph trickles through at least one node before entering the bloodstream. Chambers inside a node are packed with lymphocytes and macrophages.

The largest lymphoid organ, the **spleen**, is a filtering station for blood and a holding station for lymphocytes. The spleen has inner chambers filled with red and white "pulp." The red pulp is a large reservoir of red blood cells and the phagocytic cells called macrophages. In human embryos, the spleen produces red blood cells.

In the **thymus**, hormones are produced that help govern lymphocyte activity. Here also, lymphocytes multiply, differentiate, and mature into fighters of specific disease agents. The thymus is central to immunity, the focus of the chapter to follow.

SUMMARY

1. Humans and other vertebrates have a closed circulatory system. It consists of a heart (a muscular pump), blood vessels (arteries, arterioles, capillaries, venules, and veins), and blood. The system functions in rapid internal transport of substances to and from cells.

2. Blood, a fluid connective tissue, helps maintain favorable conditions for cells. It delivers oxygen and other substances to the interstitial fluid around cells. It also picks up cell products and wastes from that fluid.

3. Blood consists of plasma, red and white blood cells, and platelets.

   a. Plasma, the liquid component of blood, is a transport medium for blood cells and platelets. It is a solvent for plasma proteins, simple sugars, lipids, amino acids, mineral ions, vitamins, hormones, and several gases.

   b. Red blood cells transport oxygen from the lungs to interstitial fluid. They are packed with hemoglobin, an iron-containing pigment that binds reversibly with oxygen. These cells also transport some carbon dioxide from interstitial fluid to the lungs.

   c. Certain phagocytic white blood cells cleanse tissues of dead cells, cellular debris, and anything detected as not belonging to the body. Other white blood cells (lymphocytes) form armies that destroy specific bacteria, viruses, and other disease agents.

4. An internal partition divides the human heart into two halves, each with two chambers (an atrium and a ventricle). The partition separates the blood flow into two circuits, one pulmonary and the other systemic.

   a. In the pulmonary circuit, oxygen-poor blood in the heart's *right* half is pumped to capillary beds in the lungs. The blood picks up oxygen, then flows to the heart's left half.

   b. In the systemic circuit, oxygenated blood in the heart's *left* half is pumped to all body tissues. There, cells take up oxygen for aerobic respiration and give up carbon dioxide. The blood, now oxygen-poor, flows to the heart's right half.

5. Ventricular contraction drives blood through both circuits. Blood pressure is high in contracting ventricles, then drops in arteries, arterioles, capillaries, venules, and veins. It is lowest in relaxed atria.

   a. Arteries are an elastic pressure reservoir. They smooth out pressure changes resulting from heartbeats and so smooth out blood flow through capillaries.

   b. Arterioles are control points for distributing different volumes of blood to different regions.

   c. Capillary beds are diffusion zones where blood and interstitial fluid exchange substances.

   d. Venules overlap capillaries and veins somewhat in function.

   e. Veins are a blood volume reservoir that can be tapped to adjust the volume of flow back to the heart.

6. The lymphatic system has these functions:

   a. Its vascular portion takes up water and plasma proteins that seep out of blood capillaries, then returns them to the blood circulation. It transports absorbed fats and delivers agents of disease to disposal centers. Lymph capillaries and lymph vessels are components.

   b. Its lymphoid organs and tissues contain production centers for lymphocytes as well as battlegrounds.

1. Describe the cellular components of blood. Describe the plasma portion of blood. *418–419*

2. Define the functions of the circulatory system and the lymphatic system. *416–417*

3. Distinguish between:
   a. blood and interstitial fluid *416*
   b. systemic and pulmonary circuits *421*
   c. atrium and ventricle *422*

4. State the main functions of arteries, arterioles, capillaries, veins, and lymph vessels. *424–425, 430–431*

5. Label the heart's components: *422*

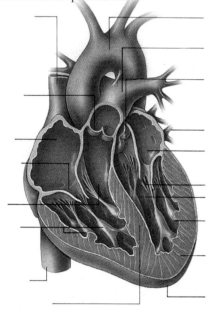

1. Cells directly exchange substances with _____ .
   a. blood vessels   c. interstitial fluid
   b. lymph vessels   d. both a and b

2. Which are *not* components of blood?
   a. plasma
   b. blood cells and platelets
   c. gases and other dissolved substances
   d. all of the above

3. The _____ produces red blood cells, which transport _____ and some _____ .
   a. liver; oxygen; mineral ions
   b. liver; oxygen; carbon dioxide
   c. bone marrow; oxygen; hormones
   d. bone marrow; oxygen; carbon dioxide

4. The _____ produces white blood cells, which function in _____ and _____ .
   a. liver; oxygen transport; defense
   b. lymph nodes; oxygen transport; pH stabilization
   c. bone marrow; housekeeping; defense
   d. bone marrow; pH stabilization; defense

5. In the pulmonary circuit, the heart's _____ half pumps blood to lungs, then _____ blood flows to the heart.
   a. right; oxygen-poor   c. right; oxygen-rich
   b. left; oxygen-poor   d. left; oxygen-rich

6. In the systemic circuit, the heart's _____ half pumps _____ blood to all body regions.
   a. right; oxygen-poor   c. right; oxygen-rich
   b. left; oxygen-poor   d. left; oxygen-rich

7. Blood pressure is high in _____ and lowest in _____ .
   a. arteries; veins   c. arteries; ventricles
   b. arteries; relaxed atria   d. arterioles; veins

8. Match the blood vessels with their main functions.
   ____ artery   a. overlaps capillary, vein functions
   ____ arteriole   b. rapid transport; pressure reservoir
   ____ capillary   c. transport; blood volume reservoir
   ____ venule   d. transport; blood volume distribution
   ____ vein   e. leisurely diffusion

9. Match the components with their descriptions.
   ____ heart   a. red blood cell, lymphocyte reservoir
   ____ ventricle   b. self-replacing, undifferentiated cell;
   ____ atrium   source of specialized cells
   ____ SA node   c. muscular pump
   ____ stem cell   d. contractions drive blood circulation
   ____ lymph node   e. receives blood from veins
   ____ spleen   f. heart's own blood supply
   ____ coronary   g. cleansing station for lymph;
        circulation   lymphocytes, macrophages
                       h. cardiac pacemaker

ABO blood
  typing *428*
antibody *428*
aorta *422*
arteriole *421*
artery *421*
atrium *422*
blood *416*
blood pressure *424*
capillary *421*
capillary bed *417*
cardiac cycle *422*
cardiac
  pacemaker *423*

cell count *419*
circulatory
  system *416*
heart *416*
hemostasis *428*
interstitial
  fluid *416*
lymph *430*
lymph capillary *430*
lymph node *431*
lymph vessel *430*
lymphatic
  system *417*
plasma *418*

platelet *418*
pulmonary
  circuit *421*
red blood cell *418*
RH blood
  typing *429*
spleen *431*
stem cell *419*
systemic circuit *421*
thymus *431*
vein *421*
ventricle *422*
venule *421*
white blood
  cell *418*

Golde, D. W., and J. C. Gasson. July 1988. "Hormones That Stimulate the Growth of Blood Cells." *Scientific American* 259(1):62–70.

Raloff, J., September 1989. "Do You Know Your HDL?" *Science News* 136(11):171–173.

Robinson, T. F., et al. June 1986. "The Heart as a Suction Pump." *Scientific American* 254(6):84–91.

27 IMMUNITY

Russian Roulette, Immunological Style

Until about a century ago, smallpox epidemics swept repeatedly through the world's cities. Some outbreaks were so severe, only half of those stricken survived. Survivors had permanent scars on the face, neck, shoulders, and arms—but they seldom contracted the disease again. They were "immune" to smallpox.

No one knew what caused smallpox. But the idea of acquiring immunity was dreadfully appealing. In twelfth-century China, healthy people were gambling with deliberate infections. They sought out survivors of mild cases of smallpox (who were only mildly scarred), then removed crusts from the scars, ground them up, and inhaled the powder. By the seventeenth century, some Europeans were risking inoculations of crust material. Others soaked threads in fluid from the sores, then poked the threads into scratches on the body. Those who survived these practices acquired immunity to smallpox—but many came down with raging infections.

While this immunological version of Russian roulette was going on, Edward Jenner was growing up in the English countryside. At the time, it was known that people who contracted cowpox never got smallpox. (Cowpox is a mild disease that can be transmitted from cattle to humans.) No one thought much about this until 1796, when Jenner, by now a physician, injected material from a cowpox sore into a boy's arm. After the reaction subsided, Jenner injected fluid from *smallpox* sores into

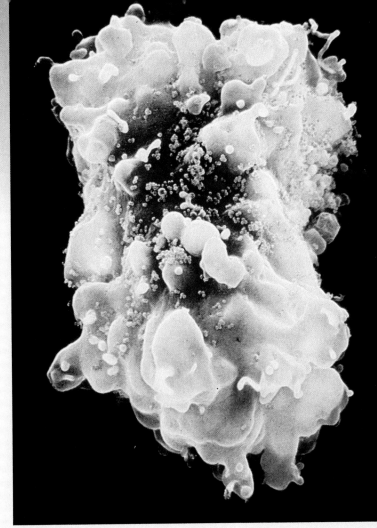

b

the boy (Figure 27.1). He hypothesized that the earlier injection might provoke immunity to smallpox—and he was right. The boy remained free of smallpox. The French mocked the procedure, calling it "vaccination" (which translates as "encowment"). Much later a French chemist, Louis Pasteur, devised similar procedures for other diseases, and the term became respectable.

By Pasteur's time, improved microscopes were revealing diverse bacteria, fungal spores, and other previously invisible forms of life. Were some of these forms responsible for diseases? In the late 1870s Robert Koch, a German physician, linked one of the microorganisms to a disease—anthrax. Koch had injected blood from infected animals into uninfected ones. The recipients ended up with blood teeming with cells of a bacterium (*Bacillus anthracis*)—and they developed anthrax. Even more convincing, injections of bacterial cells that were cultured outside the body also caused the disease!

Thus, by the beginning of the twentieth century, the promise of understanding the basis of infectious disease and immunity loomed large. The battles against those diseases were about to begin in earnest. Those battles are the focus of this chapter.

Figure 27.1 (**a**) Statue honoring Edward Jenner's development of an immunization procedure against smallpox, one of the most dreaded diseases in human history. (**b**) Micrograph of a white blood cell being attacked by the virus (blue particles) that causes AIDS. Immunologists are working to develop weapons against this modern-day scourge.

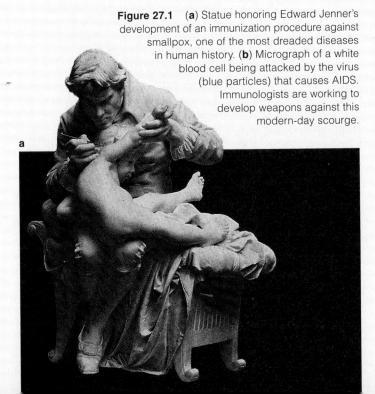

a

1. The vertebrate body has physical, chemical, and cellular defenses against invasion by viruses, bacteria, and other agents of disease.

2. During early stages of an invasion, white blood cells and plasma proteins take part in an inflammatory response. During this rapid, nonspecific counterattack, phagocytic white blood cells ingest invaders. Plasma proteins promote phagocytosis, and some also destroy invaders directly.

3. If the invasion persists, some white blood cells make immune responses. These cells can recognize distinct configurations on molecules that are abnormal or foreign to the body, such as those on bacteria and viruses. If the foreign or abnormal molecule triggers an immune response, it is called an antigen.

4. In one type of immune response, some of the white blood cells produce antibodies in huge amounts. Antibodies are molecules that bind to a specific antigen and tag it for destruction. In another type of immune response, executioner cells directly destroy body cells that have become abnormal, as by infection or by a tumor-producing process.

Table 27.1   The Vertebrate Body's Three Lines of Defense Against Pathogens

Barriers at Body Surfaces (*nonspecific* targets):

1. Intact skin; mucous membranes at other body surfaces
2. Infection-fighting substances in tears, saliva, etc.
3. Normally harmless bacterial inhabitants of body surfaces that outcompete pathogenic visitors
4. Flushing effect of tears, urination, and diarrhea

Nonspecific Responses (*nonspecific* targets):

1. Inflammation
   a. Fast-acting white blood cells (neutrophils, eosinophils, basophils)
   b. Macrophages (also take part in immune responses)
   c. Complement proteins, blood-clotting proteins, other infection-fighting substances
2. Organs with phagocytic functions (e.g., lymph nodes, spleen)

Immune Responses (*specific* targets):

1. White blood cells (macrophages, T cells, B cells)
2. Communication signals (e.g., interleukins) and chemical weapons (e.g., antibodies, complement proteins)

You may not be aware of it, but throughout your life, you are attacked by an amazing assortment of pathogens. **Pathogens** include viruses, bacteria, fungi, protozoans, and parasitic worms that cause disease. You and other vertebrates coevolved with most of them, so you need not lose sleep over this. Physical and chemical barriers, as listed in Table 27.1, exist at the body's surface. When pathogens do breach the barriers, white blood cells and chemical weapons destroy most of them. If all else fails, armies of elite white blood cells can make highly focused counterattacks.

FIRST LINE OF DEFENSE—SURFACE BARRIERS TO INVASION

Most often, pathogens cannot even get past skin or the membranes that line other body surfaces. Think of the conditions on your skin—low moisture, low pH, and thick layers of dead cells. Dense populations of normally harmless bacteria are adapted to one another and to life at the skin's surface. Few pathogens can compete with established types unless conditions change. For instance, pathogenic fungi that lurk in gym locker rooms can thrive between warm, damp toes encased in warm, damp shoes. These fungi can penetrate sodden, weakened tissue and give you *athlete's foot.*

Similarly, the gut's mucous lining and its resident bacteria help keep pathogens in check. *Lactobacillus* populations on the vagina's lining do the same. Lactate, one of their metabolic by-products, helps maintain a low vaginal pH that most bacteria and fungi cannot tolerate.

Other barriers stop airborne pathogens that enter the tubular airways leading to your lungs. As air rushes past the turns and branchings of the mucus-coated tube walls, it becomes turbulent and flings the bacteria against the sticky sides. The mucus contains protective substances. One of these substances, **lysozyme**, is an enzyme that digests bacterial cell walls and so contributes to the pathogen's death. In a final bit of housekeeping, broomlike cilia lining the airways sweep out the trapped and enzymatically whapped pathogens.

Still more protection comes from lysozyme and other enzymes in tears, saliva, gastric fluid, and intestinal fluid. Tears give eyes a sterile washing. Urine, with its low pH and flushing action, helps keep pathogens from

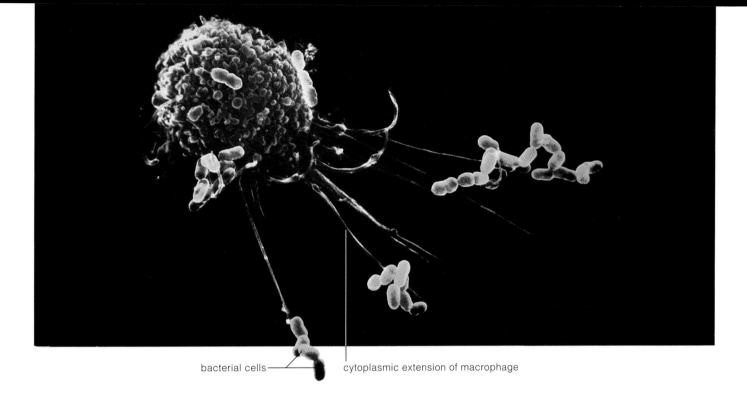

bacterial cells ——— cytoplasmic extension of macrophage

Figure 27.2 Scanning electron micrograph of a macrophage, probing its surroundings with cytoplasmic extensions. The macrophage engulfs bacterial cells that come in contact with it.

moving into the urinary tract. The flushing action of diarrhea can rid the gut of irritating pathogens. Although controlling diarrhea in young children helps them avoid dehydration, stopping its flushing action in adults can actually prolong infection.

Skin and the mucous membranes that line body surfaces are the first line of defense against pathogens.

By their population densities and metabolic activities, resident bacteria on body surfaces also help deflect invasion.

SECOND LINE OF DEFENSE— NONSPECIFIC RESPONSES

Natural defenses against a great variety of pathogens are already in place inside your body—even if you have never before encountered them. The defenders are certain types of white blood cells (leukocytes) and plasma proteins. Both are built according to instructions encoded in your genes. Both are adapted to repel invasions in general, not one particular pathogen or another. Their "nonspecific" responses are usually triggered by tissue damage. By contrast, "specific" responses are

triggered by a molecular configuration that is present on only one kind of bacterium or some other pathogen, and they are made whether tissues are damaged or not.

First on the Front Lines

Phagocytes and Kin. Recall that white blood cells arise from stem cells in bone marrow. Many circulate in blood and lymph. Many others take up stations in the lymph nodes, spleen, liver, kidneys, lungs, and brain. (Here you may wish to refer to Figures 26.4 and 26.20.)

Three kinds of white blood cells are like SWAT teams—they act swiftly against danger in general but are not adapted for sustained battles. **Neutrophils**, the most abundant kind, phagocytize bacteria. They ingest, kill, and digest bacterial cells to simple molecular bits. **Eosinophils** secrete enzymes that punch holes in the surface of parasitic worms. **Basophils** secrete histamine and other substances that help keep inflammation going after it starts.

Although slower to act, **macrophages** are the "big eaters." Figure 27.2 shows one of them. A macrophage engulfs and digests just about any foreign agent. It also helps clean up damaged tissues. Immature macrophages circulating in blood are called monocytes.

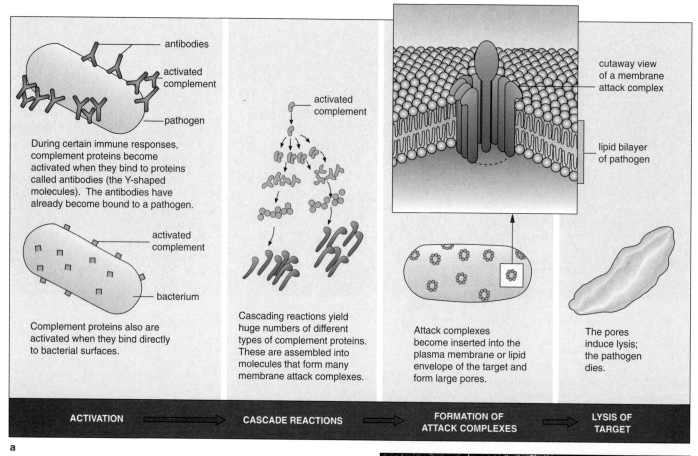

During certain immune responses, complement proteins become activated when they bind to proteins called antibodies (the Y-shaped molecules). The antibodies have already become bound to a pathogen.

- antibodies
- activated complement
- pathogen

Complement proteins also are activated when they bind directly to bacterial surfaces.

- activated complement
- bacterium

Cascading reactions yield huge numbers of different types of complement proteins. These are assembled into molecules that form many membrane attack complexes.

- activated complement

Attack complexes become inserted into the plasma membrane or lipid envelope of the target and form large pores.

- cutaway view of a membrane attack complex
- lipid bilayer of pathogen

The pores induce lysis; the pathogen dies.

ACTIVATION ⇒ **CASCADE REACTIONS** ⇒ **FORMATION OF ATTACK COMPLEXES** ⇒ **LYSIS OF TARGET**

a

Figure 27.3 How complement proteins form membrane attack complexes. (**a**) One reaction pathway begins when complement binds to bacterial surfaces. Another operates during immune responses to specific invaders, as described later in the chapter.

Both pathways produce membrane pore complexes that induce lysis in the target pathogen. They protect against many bacteria, some parasitic protistans, as well as viruses with lipid envelopes —which are derived from the plasma membrane of a previously infected host cell. (**b**) What membrane attack complexes look like.

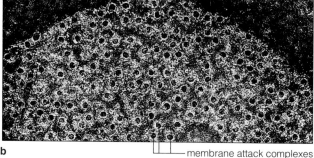

b — membrane attack complexes

Complement Proteins. A set of plasma proteins has roles in both nonspecific *and* specific defenses. Together, these proteins are the **complement system**.

About twenty kinds of complement proteins circulate in blood in inactive form. If even a few molecules of one kind are activated, they trigger a huge "cascade" of reactions. They activate many molecules of another kind of complement protein. Each of these activates many molecules of another kind of protein at the next reaction step, and so on. Thus great numbers of molecules are deployed, with the following effects.

Some complement proteins join together to form pore complexes. As Figure 27.3 shows, these are molecular structures with an interior channel. The pore complexes become inserted into the plasma membrane of many pathogens and induce lysis. **Lysis**, recall, refers to a gross structural disruption of a plasma membrane, and it leads to cell death. Pore complexes also become inserted into the lipid coat surrounding some bacteria. Lysozyme molecules diffuse through the pores, and when they reach the bacterial cell wall, they destroy it.

Some activated complement proteins promote inflammation. With their huge cascades, these proteins create concentration gradients that attract phagocytes to an irritated or damaged tissue. They also encourage the phagocytes to dine. The surface of many invaders has binding sites for the complement proteins. The invader ends up with a complement "coat." The coat adheres to the phagocytes, and this is rather like putting a basted turkey on a dinner table.

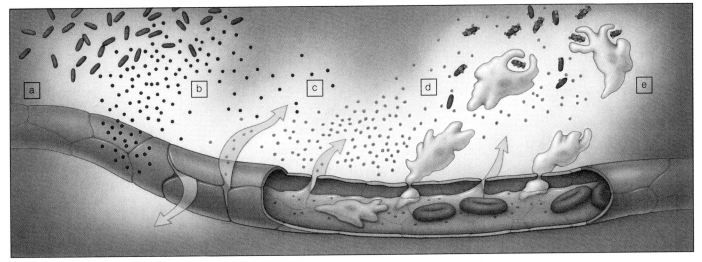

a Bacteria invade a tissue. They kill cells or release harmful metabolic by-products.

b The substances released by bacteria and by damaged or killed body cells accumulate in the tissue.

c The substances make the tissue's small blood vessels more permeable. Plasma fluid and various plasma proteins escape into the tissue.

d Some plasma proteins attack bacteria. Others create chemical gradients that facilitate migration of phagocytes to the tissue. Still others repair tissue damage (as by clotting mechanisms).

e Phagocytic white blood cells engulf bacteria.

Figure 27.4 Acute inflammation. In this example, a bacterial invasion induces chemical changes in a local tissue that trigger increased blood flow to the region. Small blood vessels become more permeable. Plasma fluid, certain plasma proteins, and phagocytic white blood cells leave the blood and enter the tissue. The invaders are killed and the tissue can be repaired.

Inflammation

By a mechanism called **acute inflammation**, fast-acting phagocytes, complement proteins, and other plasma proteins can escape from the bloodstream and enter a besieged tissue. Localized redness, swelling, heat, and pain are signs of acute inflammation. The signs are an outcome of changes in capillaries and other small blood vessels that thread through the tissue.

An inflammatory response develops when a tissue's cells are damaged or killed, as happens during an infection. Histamine and some other substances that promote dilation of small blood vessels are released from mast cells, which are a kind of tissue-dwelling basophil. The vessels become engorged with blood, so the tissue reddens and becomes warmer. (Blood, remember, carries metabolic heat.) Dilation also causes cells that make up the blood vessel walls to pull apart slightly and become "leaky" to plasma. Plasma, with its infection-fighting weapons, enters the tissue. Swelling and pain follow. Voluntary movements aggravate the pain and tend to be avoided—and this promotes tissue repair.

Within hours, neutrophils squeeze out through gaps in the blood vessel walls and swiftly go to work (Figure 27.4). Monocytes arrive later, differentiate into macrophages, and engage in more sustained action.

While macrophages are busily engulfing pathogens, they secrete several substances. Among the secretions are **interleukins**, the communication signals among white blood cells. One of these (interleukin-1) also signals a brain region that controls body temperature—and induces it to raise the "set point" on the body's thermostat. A **fever** is a body temperature that has climbed to the higher set point. A fever of about 39°C (100°F) is not a bad thing. It promotes an increase in host defense activities. It also increases body temperature to a level that is "too hot" for the functioning of most pathogens.

Besides this, interleukin-1 induces drowsiness during a fever. Drowsiness reduces the body's demands for energy, so that more can be diverted to defense and repair tasks. Macrophages also take part in cleanup and repair operations. So do blood-clotting proteins that repair blood vessels, in the manner described on page 428.

In acute inflammation, phagocytes and many plasma proteins (including complement proteins) leave the bloodstream, then defend and help repair a besieged tissue.

THIRD LINE OF DEFENSE— THE IMMUNE SYSTEM

Overview of Immune Responses

Characteristics of the System. Sometimes physical barriers and inflammation are not enough to overwhelm an invader, and an infection becomes well established. When that happens, armies of white blood cells called **B** and **T lymphocytes** form and join the battle. Lymphocytes are central to the body's third line of defense—the **immune system**. The immune system has two defining features. The first is immunological *specificity*, whereby lymphocytes can zero in on specific pathogens and eliminate them. The second feature is immunological *memory*, whereby some of the lymphocytes that form during a first-time confrontation are set aside for a future battle with the same pathogen.

Keep in mind, each kind of pathogenic virus, bacterium, or any other disease-causing agent bears various molecular markers that give it a unique identity. So do the cells of each host organism. A host's lymphocytes can recognize the *self* markers (on the body's own cells), which they normally ignore. They also can recognize *nonself* (foreign) markers. When they encounter a nonself marker, B and T lymphocytes are stimulated to divide repeatedly by way of mitosis, and huge populations form. While the divisions are proceeding, different subpopulations of the new cells become specialized to respond to the invader in different ways. Some subpopulations are *effector* cells. These fully differentiated cells engage and destroy the enemy. Other subpopulations are *memory* cells. These enter a resting phase. But they will "remember" the invader and undertake a larger, more rapid response if it ever shows up again.

Thus, immunological specificity and memory involve three events: *recognition* of a specific invader, *repeated cell divisions* that form huge populations of lymphocytes, and *differentiation* into subpopulations of specialized effector and memory cells.

Any nonself marker that triggers the formation of lymphocyte armies is called an **antigen**. Many antigens are protein molecules at the surface of infectious agents or tumor cells, and each has a unique configuration. As you will see, receptor molecules at the surface of lymphocytes can bind to these configurations. This is how lymphocytes "recognize" their targets. Besides this, the kind of antigen-binding receptors called **antibodies** can be secreted by lymphocytes that produce them. Secreted antibodies are freely circulating weapons.

An antigen is any unique molecular configuration that triggers an immune response. Lymphocytes recognize antigens by way of special receptors, such as antibodies.

Overview of the Defenders. During all immune responses, the same kinds of white blood cells are called into action. The following list gives their names and functions:

1. **Macrophages.** Besides engulfing anything detected as foreign, these phagocytes also inform T lymphocytes that a specific antigen is present.

2. **Helper T cells.** When activated, these T lymphocytes produce and secrete chemicals that promote formation of large effector and memory cell populations.

3. **Cytotoxic T cells.** These T lymphocytes eliminate infected body cells and tumor cells by a lethal hit.

4. **B cells.** These lymphocytes produce antibodies, then position them at their surface or secrete them.

Immune Functions of Macrophages

Consider your own macrophages. As is true of the rest of your cells, their plasma membrane incorporates various proteins. Among the proteins are **MHC markers** (named after the genes coding for them). Some MHC markers are common to all of your body cells. Others are unique to your macrophages and lymphocytes.

Suppose a wood splinter sinks into one of your fingers and bacteria enter the punctured tissue. They multiply, inflammation follows, and then phagocytes start engulfing them. Some of the phagocytes are macrophages that also can do interesting things with antigen. They destroy the bacterial cells but *not* the antigen. Digestive enzymes cleave the antigen molecules into fragments—which become attached to MHC molecules. Then the antigen-MHC complexes are carted off to the macrophage surface, where they are displayed. Any cell that processes and displays antigen with an appropriate MHC molecule is an **antigen-presenting cell**.

When antigen fragments and a certain MHC marker are displayed together, lymphocytes take notice (Figure 27.5). This is the antigen recognition that promotes the cell divisions leading to huge numbers of lymphocytes.

Macrophages can function as antigen-presenting cells. They combine antigen fragments with special MHC molecules, then display the complex at their surface. Such complexes trigger an immune response.

Let's turn now to the functions of the immune system's two fighting branches, as outlined in Figure 27.6.

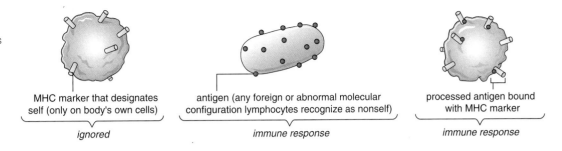

Figure 27.5 Molecular cues that stimulate lymphocytes to make immune responses.

MHC marker that designates self (only on body's own cells)

ignored

antigen (any foreign or abnormal molecular configuration lymphocytes recognize as nonself)

immune response

processed antigen bound with MHC marker

immune response

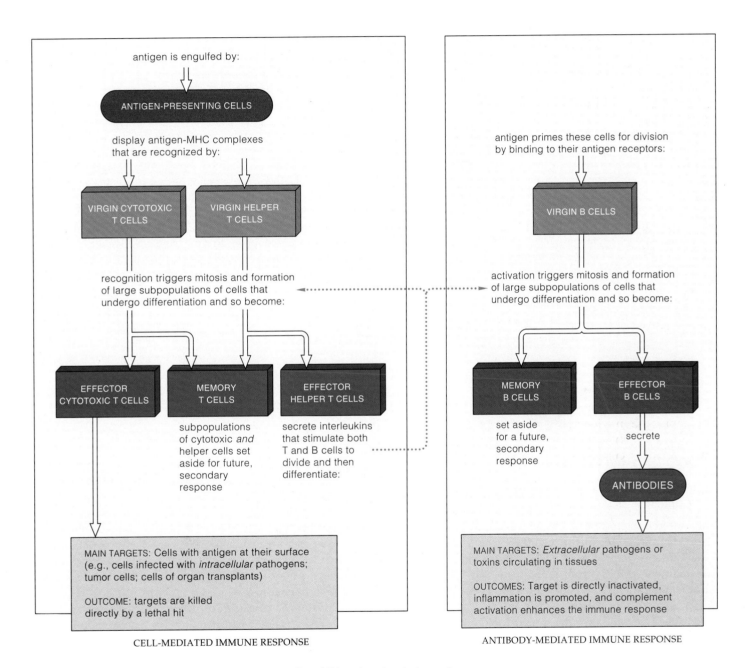

antigen is engulfed by:

ANTIGEN-PRESENTING CELLS

display antigen-MHC complexes that are recognized by:

VIRGIN CYTOTOXIC T CELLS

VIRGIN HELPER T CELLS

recognition triggers mitosis and formation of large subpopulations of cells that undergo differentiation and so become:

EFFECTOR CYTOTOXIC T CELLS

MEMORY T CELLS

EFFECTOR HELPER T CELLS

subpopulations of cytotoxic *and* helper cells set aside for future, secondary response

secrete interleukins that stimulate both T and B cells to divide and then differentiate:

MAIN TARGETS: Cells with antigen at their surface (e.g., cells infected with *intracellular* pathogens; tumor cells; cells of organ transplants)

OUTCOME: targets are killed directly by a lethal hit

CELL-MEDIATED IMMUNE RESPONSE

antigen primes these cells for division by binding to their antigen receptors:

VIRGIN B CELLS

activation triggers mitosis and formation of large subpopulations of cells that undergo differentiation and so become:

MEMORY B CELLS

EFFECTOR B CELLS

set aside for a future, secondary response

secrete

ANTIBODIES

MAIN TARGETS: *Extracellular* pathogens or toxins circulating in tissues

OUTCOMES: Target is directly inactivated, inflammation is promoted, and complement activation enhances the immune response

ANTIBODY-MEDIATED IMMUNE RESPONSE

Figure 27.6 Overview of the key interactions among B and T lymphocytes during an immune response. Most often, both types of white blood cells are activated when an antigen has been detected. An antigen is any large molecule that lymphocytes recognize as not being "self" (normal body molecules). A first-time encounter with antigen elicits a *primary* response. A subsequent encounter with the same type of antigen elicits a *secondary* immune response. This response is larger and more rapid. Memory cells that formed but were not used during the first battle can immediately engage in the second one.

Cancer and Immunotherapy

Carcinomas, sarcomas, leukemia—these chilling words refer to malignant tumors in skin, muscle, bone, and other tissues. Such tumors arise when viral attack, chemical change, or irradiation alters genes and cells turn cancerous (page 175). The transformed cells divide again and again. Unless they are destroyed or surgically removed, they kill the individual.

Often a transformed cell bears abnormal proteins at its surface. The nonspecific and specific responses to transformed cells can make the tumor regress, but sometimes the responses may be inadequate. Also, some tumors release many copies of the abnormal proteins. If these saturate antigen receptors on the defenders, the tumor escapes detection. If the tumor hides long enough and reaches a certain critical mass, it may overwhelm the immune system's capacity for an effective response.

Researchers are working to develop various procedures that will deliberately enhance immunological defenses against tumors as well as against certain pathogens. This prospect is called *immunotherapy.*

Suppose antibodies against tumor-specific antigens could be produced in quantity and injected into a cancer patient. Normal antibody-secreting B cells won't live long enough in culture to mass-produce pure antibody. Also, being end cells, they cannot reproduce. But Cesar Milstein and Georges Kohler showed how to make antibody "factories." They injected an antigen into a mouse. The mouse's B cells produced antibodies against it. Later, they extracted the antibody-producing mouse B cells and fused them with cells from B cell tumors. Some descendants of the hybrid cells divided nonstop, and they, too, produced the antibody. Clones of these proliferating hybrid cells are now being maintained indefinitely—and they make identical copies of antibodies in useful amounts. We call their products *monoclonal antibodies.*

A few cancer patients have been inoculated with monoclonal antibodies that are expected to home in on the malignant tumors. In some cases, cell-killing chemicals have been artificially attached to such antibodies. At this writing, success is limited.

Helper T Cells and Cytotoxic T Cells

T cells, recall, arise from stem cells in bone marrow (page 419). There, they start down a pathway toward a fully differentiated state. Each cell travels to the thymus gland (Figure 26.20). In this gland, helper T and cytotoxic T cells develop. There, each acquires receptors for self markers (MHC) and antigen-specific receptors. After leaving the thymus, they circulate in the blood as untouched, "virgin" T cells.

Virgin T cells ignore unadorned MHC markers on the body's own cells. They ignore free antigen. But they recognize and bind with antigen-MHC complexes on antigen-presenting cells. This stimulates them to divide and give rise to clones. (A clone is a population of genetically identical cells.) Their descendants differentiate into subpopulations of effector and memory cells—all with receptors for the antigens.

The effector helper T cells secrete interleukins. These secretions fan the cell divisions and differentiation outlined in Figure 27.6.

The effector cytotoxic T cells recognize antigen combined with a particular MHC marker. The combination forms on body cells infected by intracellular pathogens, such as viruses, and on tumor cells (see the *Focus* essay). And it serves as a "double signal" to destroy the cells that bear it.

Cytotoxic effectors destroy infected cells with a lethal hit. They secrete **perforins**, proteins that form doughnut-shaped pores in a target's plasma membrane, similar to those shown in Figure 27.3*b*. They secrete toxins that disrupt organelles and wreak havoc on nuclear DNA. Having made its hit, a cytotoxic effector quickly disengages and moves on to new targets (Figure 27.7).

Cytotoxic T cells also are players in the rejection of tissue and organ transplants. Parts of MHC markers on

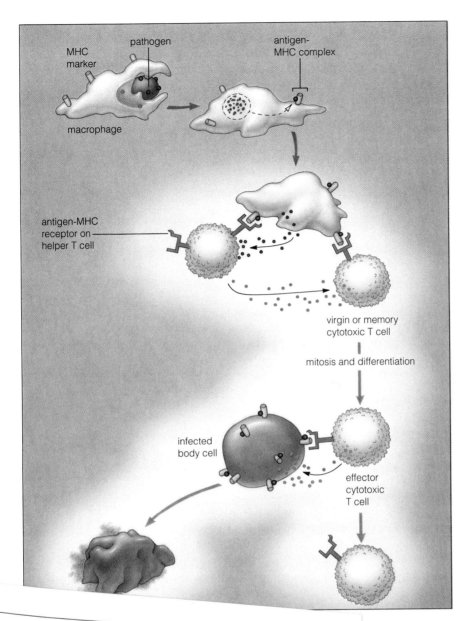

a A macrophage engulfs and digests a pathogen, then cleaves its antigen into fragments that bind to MHC markers. The macrophage becomes an antigen-presenting cell; it displays processed antigen-MHC complexes at its surface.

b Receptors on T cells bind to the complexes. Binding stimulates the macrophage to secrete interleukin-1 (pink dots). This stimulates helper T cells to secrete other interleukins (blue dots). These stimulate virgin or memory cytotoxic T cells to divide and differentiate into large populations of effector and memory cells. Only the effector cytotoxic T cells have cell-killing abilities.

c An effector encounters a target: an infected body cell that has the processed antigens bound with MHC markers at its surface. The effector delivers a lethal hit. It releases perforins and toxic substances (green dots) onto its target and so programs it for death.

d The effector disengages from the doomed cell and reconnoiters for new targets. Meanwhile, perforins make holes in the target's plasma membrane. Toxins move into the cell, disrupt organelles, and make the DNA disassemble. The infected cell dies.

out by activated cytotoxic
arts this immune response.

must be bearing odd molecular configurations at their surfaces.

Effector helper T cells secrete interleukins that promote the cell divisions and differentiations required for an immune response.

Effector cytotoxic T cells destroy infected cells, tumor cells, or foreign cells with a lethal hit.

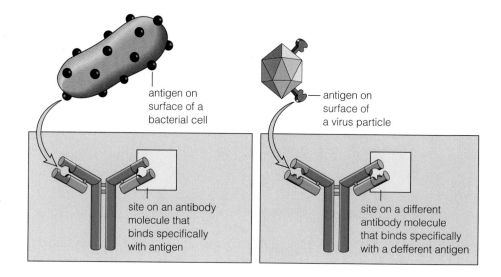

antigen on surface of a bacterial cell

antigen on surface of a virus particle

site on an antibody molecule that binds specifically with antigen

site on a different antibody molecule that binds specifically with a defferent antigen

Figure 27.8 Location of binding sites for two different antigens on two different antibody molecules.

B Cells and Antibodies

Antibody-Mediated Responses. Like T cells, the B cells also arise from stem cells in bone marrow and start down a pathway that will culminate in full differentiation. Along *their* pathway, B cells start synthesizing many copies of a single kind of antibody molecule.

All antibodies are proteins, but each kind has binding sites that match up to only a single antigen. Many of the molecules are more or less Y-shaped. They have a tail as well as two arms that bear identical antigen receptors (Figure 27.8). After they are synthesized, antibody molecules move to the plasma membrane of a maturing B cell. There, the tail of each one becomes embedded in the lipid bilayer, and the two arms stick out above it. Bristling with antigen receptors (its bound antibodies), the B cell enters the bloodstream as a virgin cell.

Suppose the receptors encounter and lock onto antigen. When that happens, the B cell is primed for division. But it will not divide without the proper signals. As Figure 27.9 shows, the signal must come from a helper T cell *already activated* by an antigen-presenting cell. In the presence of helper T cell secretions, the primed B cell and its descendants undergo repeated cell divisions.

The resulting clonal B cell population differentiates into effector and memory B cells. The effectors (formerly called plasma cells) produce and secrete staggering numbers of antibody molecules. When freely circulating antibody binds antigen, it tags the invader for destruction, as by phagocytes and complement proteins (Figure 27.3a).

The main targets of antibody-mediated responses are *extracellular* pathogens and toxins, which are freely circulating in tissues or body fluids. Antibodies cannot bind to pathogens or toxins hidden inside a host cell.

Antibodies that are secreted by B cells bind to antigens of extracellular pathogens or toxins and tag them for disposal, as by phagocytes and complement proteins.

The Immunoglobulins. Four classes of antibodies are produced with each immune response. Collectively, the four classes are known as **immunoglobulins** (Igs). They all have binding sites for antigen—but the ones in each class also have other sites that permit them to function in a specialized way. The different classes are the protein products of gene shufflings, which take place during the cell divisions that give rise to subpopulations of effector and memory B cells.

IgM antibodies are the first to be secreted during immune responses. After binding antigen, they set the complement cascade in motion. They also can bind invaders together in clumps, which are more handily eliminated by phagocytes.

IgG antibodies activate complement proteins and neutralize many toxins. The IgGs are long lasting. They are the only antibodies to cross the placenta during pregnancy, and so protect a fetus with the mother's acquired immunities. They also are secreted into the milk produced early on by mammary glands, then are absorbed into the suckling newborn's bloodstream.

IgA antibodies enter mucus-coated body surfaces, including those in the respiratory, digestive, and reproductive tracts. There they neutralize infectious agents.

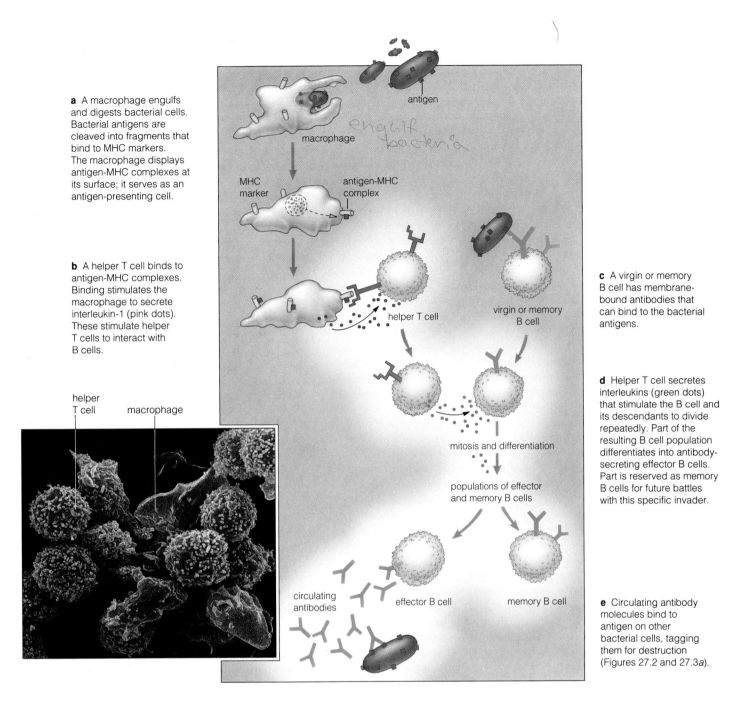

a A macrophage engulfs and digests bacterial cells. Bacterial antigens are cleaved into fragments that bind to MHC markers. The macrophage displays antigen-MHC complexes at its surface; it serves as an antigen-presenting cell.

b A helper T cell binds to antigen-MHC complexes. Binding stimulates the macrophage to secrete interleukin-1 (pink dots). These stimulate helper T cells to interact with B cells.

c A virgin or memory B cell has membrane-bound antibodies that can bind to the bacterial antigens.

d Helper T cell secretes interleukins (green dots) that stimulate the B cell and its descendants to divide repeatedly. Part of the resulting B cell population differentiates into antibody-secreting effector B cells. Part is reserved as memory B cells for future battles with this specific invader.

e Circulating antibody molecules bind to antigen on other bacterial cells, tagging them for destruction (Figures 27.2 and 27.3a).

Figure 27.9 Example of an antibody-mediated immune response to a bacterial invasion. The micrograph shows helper T cells interacting with an antigen-presenting macrophage.

A mother's milk delivers them to the mucous lining of her newborn's gut.

IgE triggers inflammation during attacks by parasitic worms. As described later, it also figures in allergies. The tails of IgE antibodies bind to basophils and mast cells, and the antigen receptors face outward. IgE antibodies are like traps. When antigen springs the trap, the basophils and mast cells release substances that promote inflammation.

Control of Immune Responses

Antigen provokes an immune response—and removal of antigen stops it. For instance, when the tide of battle turns, effector cells have destroyed most of the antigen-bearing pathogens in the body. With fewer antigen molecules around, stimulation of the response declines and finally stops. Inhibitory signals from cells with suppressor functions also help shut down the response.

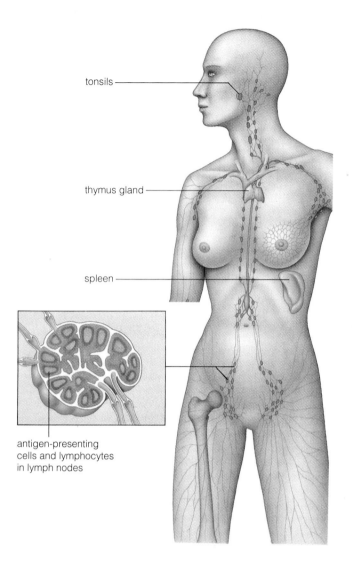

tonsils

thymus gland

spleen

antigen-presenting
cells and lymphocytes
in lymph nodes

Figure 27.10 Lymphoid organs—battlegrounds for the immune system.

Battlegrounds for the Immune System

So far, we have described some key battles—but where are the battlegrounds? The antigen-presenting cells and lymphocytes interact in organs that actually promote immune responses. Here we are referring to **lymphoid organs**, as shown in Figure 27.10. Think of the location of tonsils and most other lymph nodules just beneath the mucous membranes of the respiratory, digestive, and reproductive systems. By their very location, the organs allow antigen-presenting cells and lymphocytes to intercept invaders that have just penetrated the surface barriers. Or think of antigen present in tissue fluid that has entered lymphatic vessels. This eventually drains into the expressways for blood transport—which could distribute antigen to every tissue in the body.

However, before antigen can reach those vital expressways, it must trickle through lymph nodes, which are packed with defending cells. Even in the few cases where antigen manages to enter the blood, immune system cells in the spleen intercept it.

Inside the lymphoid organs, cells are organized for maximum effect. Antigen-presenting cells are arrayed at the front line. There, antigen is engulfed and presented, and cell divisions produce populations of effector and memory lymphocytes. As lymph drains through an organ, it moves the effector activities to the back of the organ and beyond. All the while, virgin and memory cells circulate through the organ, reconnoitering at the front line for antigen and antigen-presenting cells.

Immune responses are carried out within lymph nodes and other lymphoid organs, where antigen-presenting cells and lymphocytes are organized to do battle.

The Basis of Specificity and Memory

Immunological Specificity. People, food, water, air, soil, pets, and just about everything else in your surroundings can house a mind-boggling variety of pathogens, each with unique antigens. How do your lymphocytes make the millions of different antigen-specific receptors required to detect those threats?

All of your T and B cells inherit the same genes. But while they are maturing, their DNA undergoes recombination in the manner described in Chapter 13. *Different regions of the genes that code for antigen receptors are shuffled at random into one of millions of possible combinations.* This random process produces a nucleotide sequence that codes for one of millions of possible antigen receptors in each T or B cell.

For example, remember those Y-shaped antibodies? As shown in Figure 27.11, each arm of an antibody molecule consists of two polypeptide chains. In each B cell, the established gene sequence became translated into an amino acid sequence, which dictated the folding of those chains into grooves and bumps having a particular charge distribution. This is an antigen binding site. Only antigen having the appropriate shape, charge distribution, and so on will be able to bind with it.

DNA recombinations help explain the "clonal selection theory," first proposed by Macfarlane Burnet. All the clonal descendants of a virgin lymphocyte have the same randomly shuffled gene sequence. That sequence codes for the one receptor configuration that can bind to the specific antigen that "selected" the parent cell (Figure 27.12a).

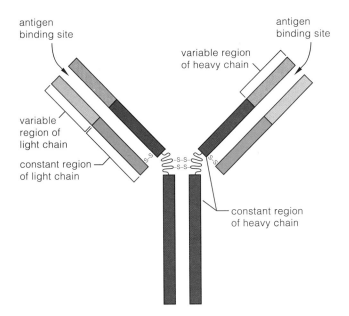

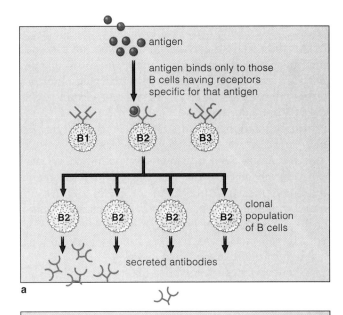

a

Figure 27.11 Structure of antibodies. An antibody molecule has four polypeptide chains joined into a Y-shaped structure. Some regions are almost the same in all antibody molecules. But the molecular configuration is unique in one region of each kind of antibody; this is the binding site for one kind of antigen. The antigen fits both into the grooves and onto the protrusions of the binding site.

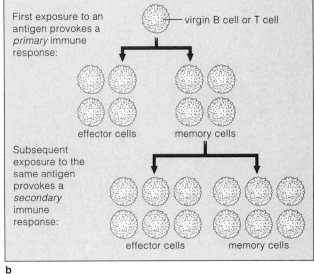

b

Figure 27.12 (a) Clonal selection of a B cell that produced the specific antibody that can combine with a specific antigen. Only antigen-selected B cells (and T cells) are activated and give rise to a population of immunologically identical clones.

(b) Immunological memory. Not all B and T cells are used in a primary immune response to an antigen. Many continue to circulate as memory cells, which become activated during a secondary immune response.

Immunological Memory. The clonal selection theory helps explain how a person can have "immunological memory" of a first-time encounter with an antigen. The term refers to the body's capacity to make a *secondary* immune response to any subsequent encounter with the same antigen (Figure 27.13).

Memory cells that form during a primary response do not engage in battle. They continue to circulate for years, sometimes for decades. Compared to the virgin cells that initiate a primary response, the patrolling battalions have far more cells, so antigen is intercepted much sooner. Effector cells form sooner, in greater numbers, so the infection is terminated before the host gets sick. Hence the advantage of immunity. And even greater numbers of memory cells form during the secondary response (Figures 27.6 and 27.12*b*).

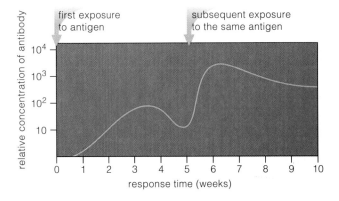

Figure 27.13 Differences in magnitude and duration between a primary and a secondary immune response to the same antigen. (In this example, the secondary response starts at week 5.)

IMMUNIZATION

Jenner didn't know why his cowpox vaccine provided immunity against smallpox. Today we know that the viruses causing the two diseases are related, and they bear similar antigens at their surface. Let's express what goes on in modern terms.

Immunization refers to various processes that promote increased immunity against specific diseases. In *active* immunization, an antigen-containing preparation called a **vaccine** is either taken orally or injected into the body (Figure 27.14). A first injection elicits a primary immune response. Another injection (booster) elicits a secondary response, with formation of more effector cells *and* memory cells that can provide long-lasting protection against the disease.

Many vaccines are manufactured from weakened or killed pathogens. (For example, weakened polio virus particles are used for the Sabin polio vaccine.) Other vaccines are based on inactivated forms of natural toxins, such as the bacterial toxin that causes tetanus. Still others are made with the help of harmless, genetically engineered viruses. These incorporate genes from three or more different viruses in their genetic material. After an engineered virus is introduced into a host, the incorporated genes are expressed, antigens are produced—and immunity is established.

Passive immunization often helps people already infected with pathogens, including the ones that cause diphtheria, tetanus, measles, and hepatitis B. A person receives injections of antibody molecules purified from some other source. The best source is another person who already has produced a lot of the required antibody. The effects are not lasting; the person's own B cells are not producing antibodies. However, injected antibody molecules may help counter the immediate attack.

Figure 27.14 Immunization schedule for children in the United States. Pediatricians routinely immunize infants and children during office visits. Low-cost or free vaccinations also are available at many community clinics.

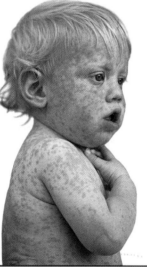

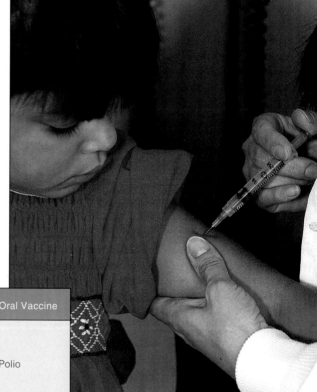

Age	Combined Injected Vaccines	Oral Vaccine
1–2 months	Hepatitis B (two injections, one shortly after birth, another before two months old)	
2 months	DPT (diphtheria, whooping cough, tetanus),* *Hemophilus influenzae*	Polio
4 months	DPT, *H.influenzae*	
6 months	DPT, *H.influenzae*	
12–18 months	Measles, mumps, rubella (German measles)	
15 months	Hepatitis B, *H.influenzae*	
18 months	DPT	Polio
4–6 years	DPT	
11–12 years	Measles, mumps, rubella	

* Also called tetramune.

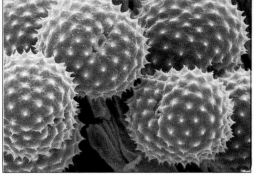

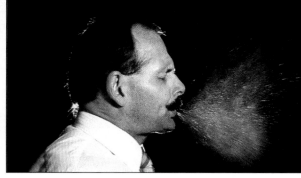

a Grass pollen   **b** Ragweed pollen   **c**

Figure 27.15 (**a,b**) Two common allergens and (**c**) a demonstration of one of their effects in sensitive people.

ABNORMAL OR DEFICIENT IMMUNE RESPONSES

Allergies

In many millions of people, immune responses are made against normally harmless substances. Substances that provoke the abnormal responses are known as allergens, and the response itself is called an **allergy**. Common allergens are pollen (Figure 27.15), a variety of drugs and foods, dust mites, fungal spores, insect venom, and cosmetics. Some responses to allergens start within minutes. Others are delayed. Either way, the allergens provoke inflammation.

Some people are genetically inclined to allergies. Besides this, infections, emotional stress, or changes in air temperature may trigger reactions that otherwise might not occur. Upon exposure to certain antigens, IgE antibodies are secreted, and these bind to mast cells. When IgE on mast cells binds to antigen, the mast cells are stimulated to secrete histamine, prostaglandins, and other substances. These substances fan the inflammatory response. They also stimulate secretion of mucus and cause airways to constrict. In *asthma* and *hay fever*, congestion, sneezing, a drippy nose, and labored breathing are symptoms of the allergic response.

In some cases, inflammatory reactions proceed throughout the body and trigger a life-threatening condition called *anaphylactic shock*. For example, a person who is allergic to wasp or bee venom can die within minutes of a single sting. Air passages leading to the lungs constrict massively. Fluid escapes too rapidly from dilated, grossly permeable capillaries. Blood pressure plummets and may lead to circulatory collapse. Antihistamines (anti-inflammatory drugs) are often used to relieve the short-term symptoms of allergies. Desensitization programs may help over the long term. In such programs, skin tests are used to identify the particular allergens that bother a patient. Inflammatory responses to some of them can be blocked if the patient's body can be stimulated to make IgG instead of IgE. Over an extended period, larger and larger doses of specific allergens are administered. Each time, the body produces more circulating IgG molecules and IgG memory cells. IgG binds with the allergen and blocks its attachment to IgE—and thus blocks inflammation.

Autoimmune Disorders

In an **autoimmune response**, the powerful weapons of the immune system are unleashed against normal body cells. Consider *Grave's disorder*, in which the body over-produces thyroid hormones. These hormones influence the development of many tissues and overall metabolic rates, and delicate feedback mechanisms control their production (page 502). Affected persons produce antibodies that unfortunately bind to receptors on the hormone-producing cells. The antibodies, which do not respond to feedback controls, stimulate the cells to produce hormones in excess. Typical disease symptoms include elevated metabolic rates, heart fibrillations, nervousness, excessive sweating, and weight loss.

Or consider *myasthenia gravis*. Here, antibodies are directed against acetylcholine receptors on skeletal muscle cells, and progressive muscular weakness follows.

Finally, consider *rheumatoid arthritis*, in which skeletal joints are chronically inflamed. Affected persons have a genetic (inherited) predisposition to the disorder. Macrophages, T cells, and B cells become activated by antigens associated with the joints. Immune responses are made against the body's own collagen and apparently against antibody that is bound to an unknown antigen. Complement activation and inflammation further damage joint tissues. So do skewed repair mechanisms. Over time, the joint becomes filled with synovial membrane cells and is eventually immobilized.

Deficient Immune Responses

When the body has inadequate numbers of lymphocytes, immune responses are not effective. This condition can be inherited or acquired. Either way, a person who makes deficient immune responses becomes highly vulnerable to infections that are not life-threatening to the general population. This is what happens in *AIDS* (acquired immunodeficiency syndrome). AIDS is caused by the human immunodeficiency virus, or HIV. The *Focus* essay that follows describes how the virus replicates inside certain lymphocytes and destroys the capacity to fight infections. We return to this topic on page 530.

Focus on Health

AIDS—The Immune System Compromised

AIDS is a constellation of disorders that follow infection by a type of virus designated HIV. The virus cripples the immune system, and the body becomes highly susceptible to usually harmless infections and some otherwise rare forms of cancer. At present there is no vaccine against the known forms of this virus (HIV-1 and HIV-2). There is no cure for those already infected.

About a million Americans are now infected with HIV. In the past decade, more than 200,000 cases of AIDS were diagnosed, and nearly 65 percent of those affected have already died. Worldwide, an estimated 8 million to 10 million people are HIV infected.

How HIV Replicates. HIV infects macrophages, antigen-presenting cells, and helper T cells (also called CD4 lymphocytes). HIV is one of the retroviruses. It has a protein coat and an inner protein core that contains RNA and several copies of an enzyme (reverse transcriptase). Its outermost lipid envelope is a bit of plasma membrane that surrounded the virus particle when it departed from a previously infected cell. HIV proteins are incorporated in the envelope.

Once inside a host cell, the viral enzyme uses the RNA as a template for making DNA, which then is inserted into a host chromosome (Figure *a*). The inserted viral DNA may remain dormant in some cells for months to years. Or it may be activated. Then, transcription yields copies of

viral RNA, which are translated into viral proteins. New virus particles are put together. They bud from the plasma membrane of the host cell and are released (Figures *b–d*). With each round of infection, more macrophages, more antigen-presenting cells, and more helper T cells are impaired or destroyed.

The host immune system produces antibodies in response to HIV antigenic proteins. (The antibodies are the basis of diagnostic tests to identify HIV infection.) However, antibodies do not eliminate the virus. Over the next decade or more, key lymphocytes are lost—and so, eventually, is the ability to mount immune responses.

At first an infected person might appear to be in good health, suffering no more than a bout of "the flu." Then he or she starts displaying symptoms that foreshadow AIDS. These include persistent weight loss, fever, fatigue, bed-drenching night sweats, and enlarged lymph nodes. In time, the diseases that follow certain opportunistic infections are signs of AIDS itself. Such diseases are rare in the population at large. Among these are widespread yeast infections and a form of pneumonia caused by *Pneumocystis carinii*. Spots that resemble bruises may appear, especially on the legs and feet. These are signs of Kaposi's sarcoma, a form of cancer that develops from endothelial cells of blood vessels. The immune system cannot control the infections or cancers, which end up killing the person with AIDS.

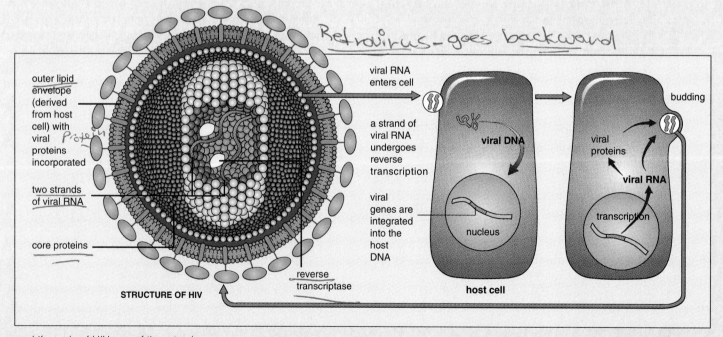

a Life cycle of HIV, one of the retroviruses.

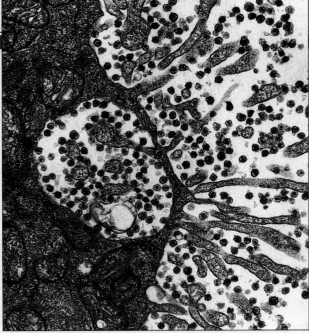

b

497 nm

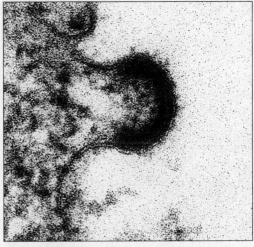

c

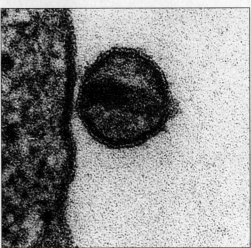

45 nm

d

(**b**) Transmission electron micrograph of HIV particles (black specks) escaping from an infected cell. (**c, d**) Closer views of a virus particle budding from the host cell's plasma membrane.

How HIV Is Transmitted.   Like any human virus, HIV requires a medium that allows it to leave one host, survive in the environment into which it is released, then enter another host.

HIV is transmitted when body fluids of an infected person enter another person's tissues. Initially in the United States, transmission occurred most often between males during homosexual activities, especially anal intercourse. It has spread among intravenous drug abusers who share blood-contaminated syringes and needles. It also has spread among the heterosexual population, increasingly by vaginal intercourse. HIV also has traveled from infected mothers to offspring during pregnancy, birth, and breast-feeding. Contaminated blood supplies accounted for some cases before screening was implemented in 1985. Tissue transplants caused four infections in 1991. In several developing countries, health care providers have spread HIV by way of contaminated transfusions and reuse of unsterile needles and syringes.

The molecular structure of HIV is not stable outside the human body, so transmission must be quite direct. At this time, there is no evidence that HIV can be effectively transmitted by way of food, air, water, casual contact, or insect bites. The virus has been isolated from blood, semen, vaginal secretions, saliva, tears, breast milk, amniotic fluid, cerebrospinal fluid, and urine. It is likely to be present in other fluids, secretions, and excretions. However, only infected blood, semen, vaginal secretions, and breast milk contain the virus in concentrations that seem to be high enough for successful transmission.

Prospects for Treatment.   Developing effective drugs or vaccines against HIV is a formidable challenge. HIV mutates rapidly, so it will be difficult to produce something that works against all of its mutated forms. In the meantime, drugs such as AZT (azidothymidine) and ddI (dideoxyinousine) are being used to slow down HIV replication and possibly to extend the life expectancy of men, women, and children.

Until effective vaccines and treatments are developed, checking the spread of HIV depends absolutely on persuading people to avoid or modify social behaviors that put them at risk. We return to this topic on page 530.

Table 27.2	Summary of Major White Blood Cells and Their Roles in Defense
Cell Type	**Main Characteristics**
Macrophage	Phagocyte; takes part in nonspecific defense responses; presents antigen to T cells; cleans up and helps repair tissue damage
Neutrophil	Fast-acting phagocyte; takes part in inflammation but not in sustained responses
Eosinophil	Secretes enzymes that attack parasitic worms
Basophil and mast cell	Secrete histamines, prostaglandins, other substances that act on small blood vessels, producing inflammation; also roles in allergies
Lymphocytes:	(All take part in most immune responses; following antigen recognition, all form clonal populations of effector cells and memory cells.)
1. B cell	Effectors secrete four classes of antibodies (IgA, IgE, IgG, and IgM) that protect the host in specialized ways
2. Helper T cell	Effectors secrete interleukins that stimulate rapid divisions and differentiation of both B cells and T cells
3. Cytotoxic T cell	Effectors kill infected cells, tumor cells, and foreign cells by way of a lethal hit
Natural killer (NK) cell	Cytotoxic cell of undetermined affiliation (may be a kind of lymphocyte); kills infected cells and tumor cells by way of a lethal hit

SUMMARY

1. Vertebrates are protected from many pathogens (infection-causing agents) by physical and chemical barriers at body surfaces. They also are protected by the nonspecific and specific responses of the white blood cells listed in Table 27.2.

a. Nonspecific responses to tissue irritation or damage include inflammation and involve organs with phagocytic functions, such as the liver and spleen.

b. Immune responses are made against specific pathogens, foreign cells, or abnormal body cells.

2. Intact skin and mucous membranes lining various body surfaces are physical barriers to infection. Glandular secretions (as in tears, saliva, and gastric fluid) are examples of chemical barriers. So are metabolic products of resident bacteria on body surfaces.

3. Tissue redness, warmth, swelling, and pain are signs of inflammation. The inflammatory response begins with changes in blood flow to a besieged tissue:

a. Substances released by pathogens and by dead or damaged body cells make small blood vessels leaky. Circulating white blood cells enter the tissue and destroy invaders.

b. Plasma proteins also enter the tissue. The complement proteins induce lysis of pathogens and attract phagocytes. Blood-clotting proteins help repair damaged blood vessels.

4. Immune responses have these characteristics:

a. Each response is triggered by antigen, a unique molecular configuration that lymphocytes recognize as foreign (nonself). An example is a protein at the surface of a particular virus.

b. Each immune response has specificity, meaning it is directed against one antigen alone.

c. Each immune response has memory. A subsequent encounter with the same antigen will trigger a more rapid, secondary response, of greater magnitude.

d. An immune response normally is not made against the body's own self marker proteins.

5. Macrophages and other antigen-presenting cells process and display fragments of antigen with their own MHC markers. Lymphocytes have receptors that can bind to antigen-MHC complexes. Binding is the start signal for an immune response.

6. An immune response begins with recognition of antigen. It proceeds through repeated cell divisions that form clonal populations of lymphocytes, which differentiate into subpopulations of effector and memory cells. Interleukins and other signals among white blood cells drive the responses. Effector helper T and cytotoxic T cells, as well as effector B cells and antibodies, carry out the response. Memory cells are set aside for secondary responses.

7. Effector cytotoxic T cells reconnoiter for any cell bearing antigen-MHC complexes. They directly destroy cells infected with intracellular pathogens, tumor cells, and cells of tissue or organ transplants.

8. Antibodies are Y-shaped protein molecules, each with binding sites for one kind of antigen. Only B cells produce them. When antibody binds to antigen, toxins may become neutralized, pathogens may be tagged for destruction, or the attachment of pathogens to body cells may be prevented.

9. In active immunization, vaccines provoke an immune response, with production of effector and memory cells. In passive immunization, injections of purified antibodies help patients through an infection.

10. An allergy is an immune response to a generally harmless substance. An autoimmune response is an attack by lymphocytes on the body's own cells. An immune deficiency is a weakened or nonexistent capacity to mount an immune response.

1. While jogging barefoot along a seashore, your toes accidentally land on a jellyfish. Soon the bottoms of your toes are swollen, red, and warm to the touch. Describe the events that result in these signs of inflammation. *437*

2. Distinguish between:
   a. neutrophil and macrophage *435, 438*
   b. cytotoxic T cell and natural killer cell *440, 441*
   c. effector cell and memory cell *438*
   d. antigen and antibody *438*

3. Describe antigen processing. *438*

4. HIV doesn't kill its host. Why are so many people dying of AIDS? *448–449*

5. Write a short essay on how the immune response contributes to homeostasis. (You may wish to refer to the discussion of homeostasis on page 380.)

6. Before each influenza season starts, you get an influenza vaccination. But this year you come down with "the flu" anyway. What do you suppose happened? (There are at least three explanations.)

Self-Quiz (Answers in Appendix IV)

1. _____ are barriers to pathogens at body surfaces.
   a. Intact skin and mucous membranes
   b. Tears, saliva, and gastric fluid
   c. Resident bacteria
   d. Urine
   e. All of the above

2. Macrophages are derived from white blood cell precursors called _____ .
   a. lymphocytes   d. monocytes
   b. basophils   e. eosinophils
   c. neutrophils

3. Activated complement functions in defense by _____ .
   a. neutralizing toxins   c. promoting inflammation
   b. enhancing resident   d. forming holes in memory
   bacteria   lymphocyte membranes

4. _____ are large molecules that lymphocytes recognize as foreign and that elicit an immune response.
   a. Interleukins   d. Antigens
   b. Antibodies   e. Histamines
   c. Immunoglobulins

5. Immunoglobulins designated _____ increase antimicrobial activity in mucus.
   a. IgA   c. IgG   e. IgZ
   b. IgE   d. IgM

6. Antibody-mediated responses work best against _____ .
   a. intracellular pathogens   d. b and c
   b. extracellular pathogens   e. all of the above
   c. toxins

7. Antigens (nonself molecular markers) are _____ .
   a. nucleotides   c. steroids
   b. triglycerides   d. proteins

8. _____ would be a target of an effector cytotoxic T cell.
   a. Extracellular virus particles in blood
   b. Cervical tumor cells
   c. Parasitic flukes in the liver
   d. Bacterial cells in pus
   e. Pollen grains in nasal mucus

9. Development of a secondary immune response is based on populations of _____ .
   a. memory cells   d. effector cytotoxic T cells
   b. circulating antibodies   e. mast cells
   c. effector B cells

10. Match the immunity concepts.
    ____ inflammation   a. neutrophil
    ____ antibody secretion   b. effector B cell
    ____ phagocyte   c. nonspecific response
    ____ immune memory   d. deliberately provoking
    ____ vaccination   memory cell production
    ____ allergy   e. basis of secondary
                                  immune response
                               f. nonprotective immune
                                  response

Selected Key Terms

acute inflammation *437*	immunoglobulin *442*
allergy *447*	interleukin *437*
antibody *438*	lymphoid organ *444*
antigen *438*	lysis *436*
antigen-presenting cell *438*	lysozyme *434*
autoimmune response *447*	macrophage *435*
B lymphocyte (B cell) *438*	memory cell *445*
basophil *435*	MHC marker *438*
complement system *436*	natural killer cell *441*
cytotoxic T cell *438*	neutrophil *435*
eosinophil *435*	pathogen *434*
fever *437*	perforin *440*
helper T cell *438*	T lymphocyte (T cell) *438*
immune system *438*	vaccine *446*
immunization *446*	

Readings

Edelson, R., and J. Fink. June 1985. "The Immunologic Function of Skin." *Scientific American* 252(6):46–53.

Golub, E., and D. Green. 1991. *Immunology: A Synthesis*. Second edition. Sunderland, Massachusetts: Sinauer Associates.

Kimball, J. 1990. *Introduction to Immunology*. Third edition. New York: Macmillan.

Tizard, I. 1992. *Immunology: An Introduction*. Third edition. Philadelphia: Saunders.

Tonegawa, S. October 1985. "The Molecules of the Immune System." *Scientific American* 253(4):122–131.

28 RESPIRATION

Conquering Chomolungma

To experienced climbers, Chomolungma is the ultimate challenge. The summit of this Himalayan mountain, also known as Everest, is 9,700 meters (29,108 feet) above sea level (Figure 28.1). It is the highest place on earth.

Chomolungma's challenge is not merely iced-over vertical rock, driving winds, blinding blizzards, and heart-stopping avalanches. It is the extreme danger that oxygen-poor air poses to the brain.

Most of us live at low elevations. Of the air we breathe, one molecule in five is oxygen. In mountains higher than 3,300 meters (10,000 feet), the breathing game changes. Gravity's pull is not as great and gas molecules spread out—so breathing the way we do at sea level will not deliver enough oxygen to our lungs. The oxygen deficit can cause headaches, shortness of breath, heart palpitations, even loss of appetite, nausea, and vomiting. We call this "altitude sickness."

At Chomolungma's base camp, climbers are 6,300 meters (19,000 feet) above sea level. More than half of the atmosphere's oxygen is below them. At 7,000 meters (23,000 feet), oxygen and other gaseous molecules are extremely diffuse. The very low air pressure and oxygen scarcity combine to make blood vessels leaky. Brain and lung tissues swell with plasma fluid. When the edema continues, climbers become comatose and die.

During bad moments at high altitudes, climbers inhale bottled oxygen. Chomolungma's casualties have even been zipped inside airtight bags. Oxygen is pumped in and carbon dioxide removed until "air" in the bag is closer to the air at 6,000 to 9,000 feet.

Few of us will ever find ourselves near the peak of Chomolungma, pushing our reliance on oxygen to the limits. Here in the lowlands, disease, smoking, and other environmental insults push it in more ordinary ways. The risks can be just as great, as you will see by this chapter's end.

Figure 28.1 A climber approaching the summit of Chomolungma, where oxygen is brutally scarce.

Figure 28.3 A flatworm, an animal that uses integumentary exchange. Oxygen reaches individual cells and carbon dioxide is disposed of simply by diffusing across the body surface.

1. Of all organisms, multicelled animals require the most energy. The energy comes mainly from aerobic metabolism, which requires oxygen and produces carbon dioxide wastes. In a process called respiration, animals move oxygen into their internal environment and give up carbon dioxide to the external environment.

2. Oxygen diffuses into the body as a result of a pressure gradient. The pressure of this gas is higher in air than it is in metabolically active tissues, where cells rapidly use oxygen. Carbon dioxide follows a gradient in the other direction. Its pressure is higher in tissues (where it is a by-product of metabolism) than it is in the air.

3. In most respiratory systems, oxygen and carbon dioxide diffuse across a respiratory surface, such as the thin, moist epithelium in the human lungs. Blood flowing through the body's circulatory system picks up oxygen and gives up carbon dioxide at this respiratory surface.

THE NATURE OF RESPIRATION

High in the mountains, in underground burrows, in deep and shallow waters, animals are engaged in aerobic respiration. This oxygen-requiring metabolic pathway yields energy and produces carbon dioxide leftovers. Most animals take in oxygen and rid themselves of carbon dioxide by a process called "respiration." And most have a **respiratory system** that functions in this exchange of gases. By interacting with

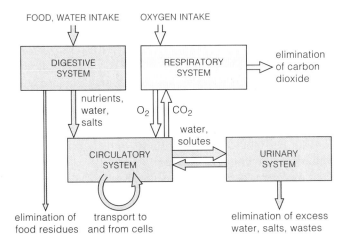

Figure 28.2 Interactions between the respiratory system and other organ systems in complex animals.

other organ systems, the respiratory system helps maintain stable operating conditions for the body's cells (Figure 28.2).

Why Oxygen and Carbon Dioxide Move Into and Out of the Body

Partial Pressure Gradients. Oxygen or carbon dioxide tends to diffuse down its concentration gradient—or, as we say for gases, its pressure gradient. When molecules of either gas are more concentrated in the surrounding air or water, more pressure is available to drive them into the body. When they are more concentrated inside, more pressure is available to drive them outside.

This is not to say both gases exert the *same* pressure. Pump air into a flat tire near a beach in San Diego. You will be filling it with about 78 percent nitrogen, 21 percent oxygen, 0.04 percent carbon dioxide, and 0.96 percent of some other gases. (This is true of dry air anywhere at sea level.) These numbers tell you that oxygen or carbon dioxide can only exert *part* of the total pressure on the tire wall. And the "partial pressure" of oxygen is obviously greater than that of carbon dioxide.

Typically, gases enter and leave the body by crossing a **respiratory surface**, such as a thin epithelial layer. The more extensive the surface area and the larger the gradient, the faster a gas diffuses across it. The respiratory surface is kept moist, for gases cannot diffuse across it unless they are dissolved in some fluid.

Surface-to-Volume Ratio. Despite their diversity, all animal body plans are adapted to promote favorable rates of inward diffusion of oxygen and outward diffusion of carbon dioxide. For example, the requirement for favorable rates of gas exchange is one reason why animals without respiratory organs are flattened, tubelike, or tiny. Imagine a flatworm growing in all directions, like an inflating balloon. Its surface area does not increase at the same rate as its volume (page 37). Once its girth exceeds a single millimeter, the diffusion distance between the surface and internal cells is so great, the worm will die.

Ventilation. Gas exchange requirements are high for large-bodied, active animals and cannot be maintained by diffusion alone. For example, a trout must move flaps located above its respiratory organs (gills). The movement stirs the surrounding water, bringing in more dissolved oxygen and moving carbon dioxide away from the body. Your own circulatory system must rapidly deliver oxygen to cells and transport carbon dioxide to your lungs for disposal. Similarly, breathing ventilates your lungs.

Transport Pigments. Animals boost gas exchange rates with respiratory pigments, mainly **hemoglobin**, that circulate in blood. Hemoglobin rapidly transports oxygen away from a respiratory surface. In human lungs, each hemoglobin molecule binds loosely with as many as four oxygen molecules, then releases them in oxygen-poor tissues. Its action helps maintain a steep partial pressure gradient for oxygen between the internal environment and the surroundings.

Environmental Constraints. Aquatic animals expend energy to maintain pressure gradients across respiratory surfaces. Whereas a trout may devote 20 percent of its energy output to stirring water around its gills, a massive buffalo staring out over the plains may devote a mere 2 percent to breathing. Why? Even when water is saturated with dissolved oxygen, it holds only 5 percent of what air holds. Salty or warm water holds even less. The same is true of dimly lit waters, where you will not find an abundance of oxygen-producing photosynthesizers.

Living on land poses a different threat to respiration. When respiratory membranes dry out, they stick together and gases cannot diffuse across them. Land animals keep respiratory surfaces moist by bathing them in secretions.

By a process called respiration, animals take in oxygen for an energy-releasing pathway (aerobic respiration) and remove the pathway's carbon dioxide wastes.

This gas exchange requires steep partial pressure gradients between an organism's internal environment and the surroundings.

Modes of Respiration

If an animal is not massive and if its life-style does not require high rates of metabolism, its demands for respiration are not great. Flatworms are like this. Their demands are met by **integumentary exchange**, in which gases diffuse directly across the body covering (integument). Earthworms rely on integumentary exchange. They also transport gases to and from the body surface by way of blood vessels.

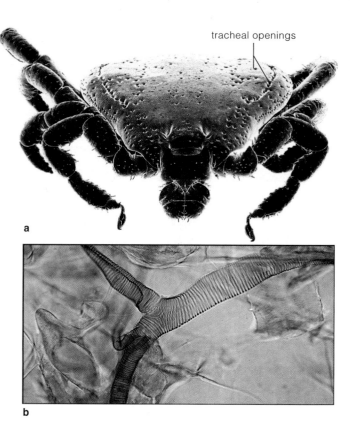

tracheal openings

a

b

Figure 28.4 (**a**) Scanning electron micrograph of a tick. The openings across the integument are the start of airways used in tracheal respiration. (**b**) A closer view of an insect trachea. This branching tube is reinforced with rings of chitin.

The integument of animals living in dry habitats helps conserve precious water. But it is usually too thick, hardened, small, or sparsely supplied with blood vessels to be a good respiratory surface. This is true of spiders and insects that rely on **tracheal respiration**. Figure 28.4 shows how small openings perforate the integument of an insect. Each opening is the start of a tube that branches repeatedly inside the body. The last branchings dead-end at a fluid-filled tip, where gases diffuse directly into tissues. Tips abound in muscle and other tissues with high oxygen demands.

Other animals exchange gases at **gills**—respiratory organs with a moist, thin, vascularized epidermis. Some insects, a few fish larvae, and a few amphibians have external gills that project into the surrounding water. Adult fishes have a pair of internal gills. These are rows of slits or pockets at the back of the mouth that extend to the body surface. Water flows into the mouth and pharynx, then over filaments in the gills.

As Figure 28.5 suggests, respiratory surfaces in a gill filament are richly endowed with blood vessels. When water flows past, it first exchanges gases with a vessel carrying blood *out* of the filament. This blood has less oxygen than the water does, so oxygen diffuses into it. Then the water flows over a vessel carrying blood *into* the filament. The water has already given up oxygen,

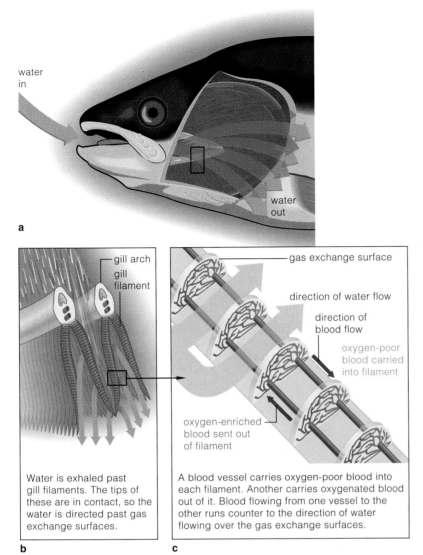

a

water in

water out

b

gill arch
gill filament

Water is exhaled past gill filaments. The tips of these are in contact, so the water is directed past gas exchange surfaces.

c

gas exchange surface

direction of water flow

direction of blood flow

oxygen-poor blood carried into filament

oxygen-enriched blood sent out of filament

A blood vessel carries oxygen-poor blood into each filament. Another carries oxygenated blood out of it. Blood flowing from one vessel to the other runs counter to the direction of water flowing over the gas exchange surfaces.

Figure 28.5 Respiratory system of many fishes. (**a**) One of a pair of gills, each located under a bony lid (removed for this sketch). Water drawn into the fish mouth flows past the gills. (**b**) Each gill has filaments that contain richly vascularized respiratory surfaces. (**c**) Here, blood flow runs counter to the direction of water flow. The arrangement favors the movement of oxygen (down its partial pressure gradient) into the blood.

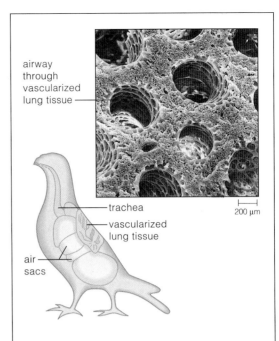

airway through vascularized lung tissue

trachea

vascularized lung tissue

air sacs

200 µm

Figure 28.6 Respiratory system of birds. Typically, five air sacs are attached to each lung, which is small and inelastic. When the bird inhales, air is drawn into air sacs through tubes, open at both ends, that thread through the vascularized lung tissue. This tissue is the respiratory surface, where gases are exchanged. When the bird exhales, air is blown out of the sacs, through the small tubes, and out of the trachea. Thus, air is not merely drawn into the bird lungs. It is continuously drawn *through* them and across the respiratory surface. This unique ventilating system supports the high metabolic rates that birds require for flight and other energy-intensive activities.

but it still has more than blood inside this vessel does— so more oxygen diffuses into the filament.

Movement of two fluids in opposing directions is called countercurrent flow. With their flow mechanism, fish gills extract about 80 to 90 percent of the oxygen in water that flows past them. That is far more than they would get from a one-way flow mechanism, at far less energy cost.

Lungs are internal respiratory surfaces in the shape of a cavity or sac. They evolved in some fishes more than 450 million years ago (page 313). Evolutionarily ancient lungfishes have gills, but they also have lungs that supplement respiration in oxygen-poor habitats. In fact, they drown if kept from gulping air at the water surface. With their low metabolic rates, amphibians rely mostly on integumentary exchange, but they, too, use a pair of small lungs. Reptiles, birds, and mammals use paired lungs as *the* major respiratory surfaces. In all of these lungs, *airways* carry gas molecules to and from one side of the respiratory surface, and *blood vessels* carry gas molecules to and from the other side:

1. Air moves by bulk flow into and out of the lungs, and new air is delivered to the respiratory surface.

2. Gases diffuse across the respiratory surface.

3. Blood flow to and from the lungs enhances the diffusion of gases into and out of lung capillaries.

4. In tissues, blood exchanges gases with interstitial fluid, which exchanges gases with cells.

We turn next to the human respiratory system. Its operating principles apply to most vertebrates. Birds are a notable exception. As Figure 28.6 shows, a unique system of air sacs helps ventilate bird lungs.

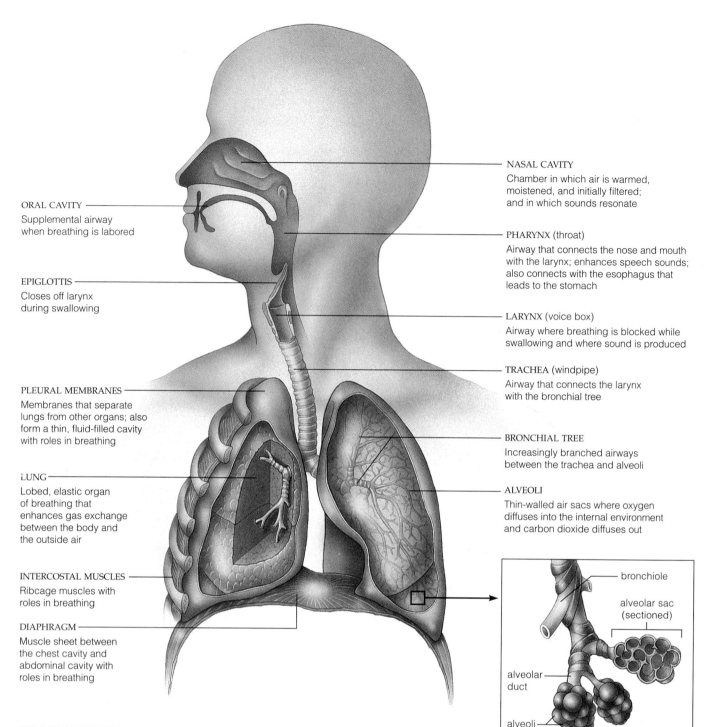

ORAL CAVITY
Supplemental airway
when breathing is labored

EPIGLOTTIS
Closes off larynx
during swallowing

PLEURAL MEMBRANES
Membranes that separate
lungs from other organs; also
form a thin, fluid-filled cavity
with roles in breathing

LUNG
Lobed, elastic organ
of breathing that
enhances gas exchange
between the body and
the outside air

INTERCOSTAL MUSCLES
Ribcage muscles with
roles in breathing

DIAPHRAGM
Muscle sheet between
the chest cavity and
abdominal cavity with
roles in breathing

NASAL CAVITY
Chamber in which air is warmed,
moistened, and initially filtered;
and in which sounds resonate

PHARYNX (throat)
Airway that connects the nose and mouth
with the larynx; enhances speech sounds;
also connects with the esophagus that
leads to the stomach

LARYNX (voice box)
Airway where breathing is blocked while
swallowing and where sound is produced

TRACHEA (windpipe)
Airway that connects the larynx
with the bronchial tree

BRONCHIAL TREE
Increasingly branched airways
between the trachea and alveoli

ALVEOLI
Thin-walled air sacs where oxygen
diffuses into the internal environment
and carbon dioxide diffuses out

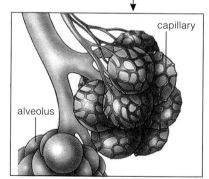

bronchiole

alveolar sac
(sectioned)

alveolar
duct

alveoli

capillary

alveolus

HUMAN RESPIRATORY SYSTEM

Airways to the Lungs

It will take at least 300 million breaths to get you to age
seventy-five. You may find yourself going without food
for a few hours or days. But stop breathing even for five
minutes and normal brain function is over.

Take a deep breath, then look at Figure 28.7 to get an
idea of where the air is going. Unless you are out of
breath and panting heavily, the air has just entered two
nasal cavities, not your mouth. There it is warmed and
picks up moisture from mucus. Ciliated epithelium lining the nasal cavities filters dust and particles from it.
So do those unseemly but functional nose hairs.

Figure 28.7
Components of the
human respiratory
system
and their functions.
Also shown are the
diaphragm and other
structures with
secondary roles in
respiration. The
structure of alveoli,
the main sites of gas
exchange, is shown
in the boxes.

Sites of Gas Exchange in the Lungs

The two bronchi carry inhaled air into the organs of gas exchange—the lungs. Humans have two elastic, cone-shaped lungs inside the rib cage, to the left and right of the heart. They are positioned above the **diaphragm**, a muscular partition between the chest cavity and abdominal cavity. A thin membrane lines the outer surface of the lungs *and* the inner surface of the chest cavity wall. This is the pleural membrane. Visualize the lungs as two baseballs pushed into a partly inflated balloon. The balls take up so much space, they press the balloon's opposing sides together. Similarly, the pleural membrane is saclike, with the chest wall and the lungs pressing its opposing surfaces together. Only a thin film of lubricating fluid separates the two membrane surfaces, and it prevents friction between them.

During the respiratory disorder called *pleurisy*, the pleural membrane becomes inflamed and swollen, friction follows, and breathing can be painful.

Inside each lung, air moves through finer and finer branchings of a "bronchial tree" (Figure 28.7*a*). These airways are **bronchioles**. They end with *respiratory* bronchioles, which have cup-shaped outpouchings from their walls. Each outpouching is an **alveolus** (plural, alveoli). Most often, alveoli are clustered together, forming a larger pouch called an alveolar sac (Figure 28.7*b*). These sacs are the major sites of gas exchange.

Blood capillaries surround 150 million alveoli in each lung. Together, the alveoli provide a tremendous surface area for exchanging gases with blood. If your alveoli were stretched out as a single layer, they would cover the floor of a racquet ball court!

Now the air is poised at the **pharynx**, or throat. This is the entrance to both the **larynx**, an airway, and the esophagus (the tube leading to the stomach).

Later, on its way out, the air may interact with two **vocal cords**—thickened, muscular folds of the larynx wall that help produce sound waves for speech. You make sounds by forcing air through the glottis, the gap between the cords. When you increase the pressure, you make louder sounds. When you increase muscle tension in the vocal cords, you make high-pitched sounds or squeaks.

Right now the **epiglottis**, a flaplike structure at the start of the larynx, is pointing up so the air can move into your **trachea**, or windpipe. This branches into two airways, one leading to each lung. Each airway is a **bronchus** (plural, bronchi). Its epithelial lining has an abundance of cilia and mucus-secreting cells (Figure 28.8). The lining serves as a barrier to infection. Bacteria and airborne particles stick in the mucus, then cilia sweep the debris-laden mucus toward the mouth. Where it goes from there really is up to you, but possibly the sidewalk is not a suitable destination.

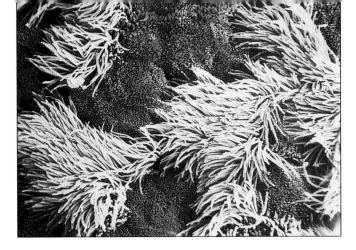

Figure 28.8 Color-enhanced scanning electron micrograph of cilia (gold) and mucus-secreting cells (rust-colored) in the respiratory tract.

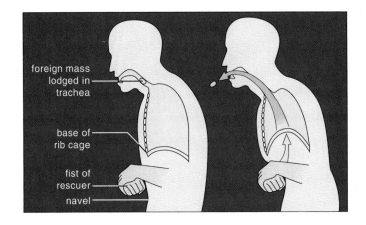

Figure 28.9 The Heimlich maneuver. Each year, several thousand people choke to death when food enters the trachea instead of the esophagus (compare Figure 25.4). Their air flow is blocked for as little as four or five minutes. The Heimlich maneuver, *an emergency procedure only*, often can dislodge the food. The idea is to elevate the diaphragm forcibly, causing a sharp decrease in the chest cavity volume and a sudden increase in alveolar pressure. The increased pressure forces air up the trachea and may be enough to dislodge the obstruction.

To perform the Heimlich maneuver, stand behind the victim, make a fist with one hand, then position the fist, thumb-side in, against the victim's abdomen. The fist must be slightly above the navel and well below the rib cage. Next, press the fist into the abdomen with a sudden upward thrust. Repeat the thrust several times if needed. The maneuver can be performed on someone who is standing, sitting, or lying down. Once the obstacle is dislodged, a physician must see the person at once, for an inexperienced rescuer can inadvertently cause internal injuries or crack a rib. It could be argued that the risk is worth taking, given that the alternative is death.

Bear in mind, when you swallow food, muscle contractions force the epiglottis down, to its closed position. This prevents food from going down the wrong tube and disrupting gas exchange. Such misdirected chunks of food can cause strangulation, as Figure 28.9 indicates.

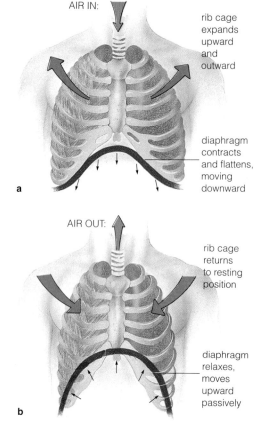

AIR IN:

rib cage
expands
upward
and
outward

diaphragm
contracts
and flattens,
moving
downward

a

AIR OUT:

rib cage
returns
to resting
position

diaphragm
relaxes,
moves
upward
passively

b

Figure 28.10 Changes in the size of the chest cavity during breathing. The blue line indicates the diaphragm's position when you (**a**) inhale and (**b**) exhale.

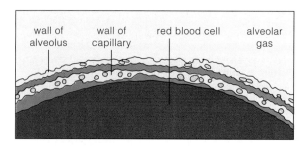

| wall of alveolus | wall of capillary | red blood cell | alveolar gas |

Figure 28.11 What a section through an alveolus and an adjacent lung capillary would look like. Compared to the red blood cell's diameter, the diffusion distance across the capillary wall, the interstitial fluid, and the alveolar wall is small.

How Breathing Changes Air Pressure in the Lungs

When you breathe, you are ventilating your lungs. Air is inhaled (drawn into the airways), then exhaled (expelled from them). The chest cavity's volume increases and decreases in a rhythmic way. Each time you change the volume, you reverse pressure gradients between the lungs and air outside your body. Gases in the respiratory tract follow those gradients.

As you start to inhale, the dome-shaped diaphragm contracts and flattens, and muscles lift the ribs upward and outward (Figure 28.10). Then, as the chest cavity expands, the rib cage moves away slightly from the lung surface. Pressure is lowered in the potential space between each lung and the pleural membrane. This pressure is transmitted to air spaces inside the alveoli, and it is lower than the atmospheric pressure at the mouth. Fresh air follows the pressure gradient. It flows down the airways and causes the lungs to expand.

When you start to exhale, elastic tissue in the lungs and chest wall recoils passively, so the chest cavity's volume shrinks. This compresses the air in the alveolar sacs. The air pressure in the sacs is now greater than the atmospheric pressure, so air follows the gradient—out of the lungs. When oxygen demands increase, as during exercise, muscles of the rib cage and abdomen are recruited to hasten the airflow.

GAS EXCHANGE AND TRANSPORT

Gas Exchange in Alveoli

Each alveolus is only a single layer of epithelial cells, surrounded by a thin basement membrane. Lung capillaries also have thin walls, and only a thin film of interstitial fluid separates them from the alveoli. Thus gases diffuse rapidly in both directions (Figure 28.11).

Figure 28.12 shows the partial pressure gradients for oxygen and carbon dioxide through the human respiratory system. Passive diffusion alone moves oxygen across the respiratory surface and into the bloodstream. And it moves carbon dioxide in the reverse direction.

Driven by its partial pressure gradient, oxygen diffuses from alveolar air spaces, through interstitial fluid, and into lung capillaries.

Carbon dioxide, driven by its partial pressure gradient, diffuses in the opposite direction.

Gas Transport To and From Metabolically Active Tissues

Blood can carry only so much dissolved oxygen and carbon dioxide. The transport of both gases must be assisted to meet the energy requirements of humans and other large animals. In humans, the hemoglobin in red blood cells binds and transports both. This pigment increases oxygen transport by seventy times. It also increases carbon dioxide transport away from tissues by seventeen times.

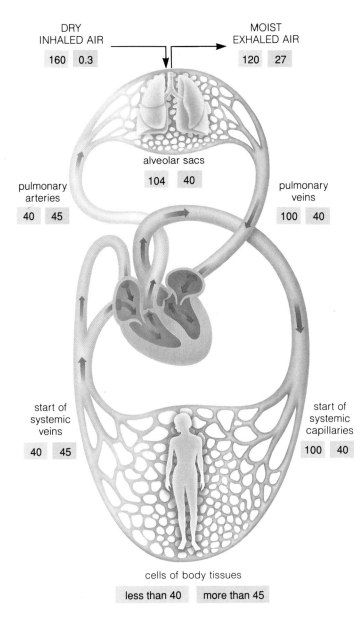

DRY INHALED AIR
160 0.3

MOIST EXHALED AIR
120 27

alveolar sacs
104 40

pulmonary arteries
40 45

pulmonary veins
100 40

start of systemic veins
40 45

start of systemic capillaries
100 40

cells of body tissues
less than 40   more than 45

Figure 28.12 Partial pressure gradients for oxygen (blue boxes) and carbon dioxide (pink boxes) through the respiratory tract.

This is the point to remember about the values shown: *Each gas moves from regions of higher to lower partial pressure.* That is why, for example, you become light-headed when you first visit places at high altitudes. The partial pressure of oxygen decreases with altitude, and your body does not function as well when the pressure gradient between the surrounding air and your lungs is lower than what you normally encounter.

Oxygen Transport. Think about the inhaled air that reaches the alveoli. It has plenty of oxygen and not much carbon dioxide—and the opposite is true of blood in lung capillaries. Thus, in the lungs, oxygen diffuses into the plasma portion of blood and then into red blood cells, where it rapidly binds with hemoglobin.

Hemoglobin molecules hold onto the oxygen rather weakly. They give it up in any tissue where the partial pressure of oxygen is low. They give it up even faster in tissues where the blood is warmer, the pH lower, and the partial pressure of carbon dioxide high. Such conditions exist in contracting muscle and other metabolically whipped-up tissues.

Carbon Dioxide Transport. Now think about a tissue where carbon dioxide's partial pressure is higher than in the blood flowing by in capillaries. Carbon dioxide diffuses into the capillaries, which transport it toward the lungs. Some carbon dioxide remains dissolved in plasma or binds with hemoglobin. But most of it—about 70 percent—is transported in the form of bicarbonate.

The bicarbonate is produced when carbon dioxide combines with water to form carbonic acid. The carbonic acid separates into bicarbonate and hydrogen ions:

$$CO_2 + H_2O \rightleftharpoons \underset{\text{carbonic acid}}{H_2CO_3} \rightleftharpoons \underset{\text{bicarbonate}}{HCO_3^-} + H^+$$

The reactions don't amount to much in blood plasma, where only 1 of every 1,000 carbon dioxide molecules is converted. It's a different story in red blood cells. They contain an enzyme (carbonic anhydrase) that increases the reaction rate by 250 times. In these cells, most of the carbon dioxide *not* bound to hemoglobin is converted to carbonic acid. The enzyme-mediated reactions make the blood level of carbon dioxide drop swiftly. This helps maintain the gradient that keeps carbon dioxide diffusing from interstitial fluid into the bloodstream.

What happens to the bicarbonate ions formed during the reaction? They tend to diffuse out of the red blood cells and into the blood plasma. What about the hydrogen ions? Hemoglobin acts as a buffer for them and keeps the blood from becoming too acidic. A buffer, recall, is a molecule that combines with or releases hydrogen ions in response to changes in cellular pH.

The reactions are reversed in the alveoli. There the partial pressure of carbon dioxide is lower than it is in the surrounding capillaries. Water and carbon dioxide formed during the reactions diffuse into alveolar sacs. From there, carbon dioxide is exhaled from the body.

Hemoglobin in red blood cells greatly enhances the oxygen-carrying capacity of blood.

Most carbon dioxide is transported in blood as bicarbonate, nearly all of which forms through enzyme-mediated reactions in red blood cells.

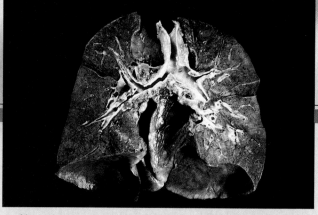

a   Normal appearance of tissues in the paired human lungs.

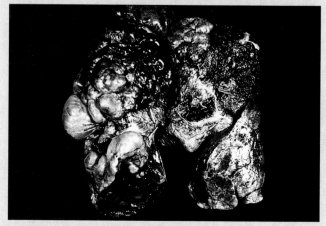

b   Lungs from someone affected by emphysema.

When the Lungs Break Down

In large cities, in certain workplaces, even near a cigarette smoker, airborne particles and certain gases are present in abnormal amounts, and they put extra work loads on the respiratory system.

Bronchitis.   The ciliated, mucous epithelium that lines your bronchioles helps protect you from respiratory infections. Toxins in cigarette smoke and other airborne pollutants irritate the lining and may lead to *bronchitis*. In this respiratory ailment, excess mucus is secreted when epithelial cells that line the airways become irritated. As mucus accumulates, so do bacteria and other particles stuck in it. Coughing brings up some of the gunk. If the irritation persists, so does the coughing. Initial attacks of bronchitis are treatable. When the aggravation is allowed to continue, bronchioles become inflamed. Bacteria, chemical agents, or both attack cells of the bronchiole walls. Ciliated cells are destroyed and mucus-secreting cells multiply. Fibrous scar tissue forms; in time it may narrow or obstruct airways.

Emphysema.   When bronchitis persists, airways become clogged with mucus. Tissue-destroying bacterial enzymes attack the thin, stretchable walls of alveoli. As the walls break down, inelastic fibrous tissue surrounds them. Gas exchange now depends on fewer alveoli—which become enlarged. In time, the balance between air flow and blood flow becomes permanently compromised because the lungs remain distended and inelastic. Compare Figures *a* and *b* to get an idea of what goes on. Running, walking, even exhaling become difficult. These are symptoms of *emphysema*, a respiratory disease that affects about 1.3 million people in the United States alone.

   A few people are genetically predisposed to emphysema. They do not have a functional gene for antitrypsin, an enzyme that can inhibit bacterial attack on the alveoli.

Also, poor diet and chronic (persistent or recurring) colds and other respiratory infections make people susceptible to emphysema later in life. But smoking is the major cause of the disease. Emphysema can develop slowly, over twenty or thirty years. By the time it is detected, the damage to lung tissue cannot be repaired.

Effects of Cigarette Smoke.   Every thirteen seconds, one of 50 million cigarette-puffing Americans dies of emphysema, chronic bronchitis, or heart disease. For every eight of them, one *nonsmoker* dies of ailments brought on by prolonged exposure to tobacco smoke in the surroundings. Children who breathe secondhand smoke can expect to suffer more allergies and lung ailments. Smoking is the major cause of lung cancer, which is now the leading cause

Control of Air Flow and Blood Flow

Gas exchange is most efficient when the rate of air flow matches the rate of blood flow. The rates can be brought into balance by adjustments in local tissues as well as in tissues through the body as a whole.

   Some local controls operate at alveoli in the lungs. Suppose your heart is pounding and you don't breathe deeply enough. Blood flows too fast and the air too

sluggishly for the efficient disposal of carbon dioxide. An increase in the blood level of carbon dioxide affects smooth muscle in bronchiole walls. The wall diameter widens, enhancing the air flow.

   Local controls also work on blood vessels in the lungs. When air flow is too great relative to blood flow, oxygen levels rise in some parts of the lungs. The increase affects smooth muscle in blood vessel walls. The wall diameter widens, so blood flow to the regions

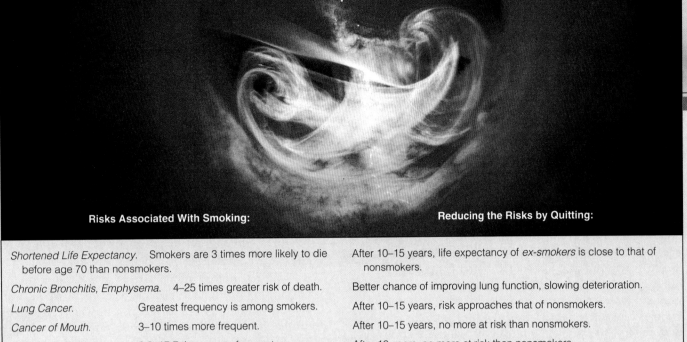

Risks Associated With Smoking: **Reducing the Risks by Quitting:**

Risks Associated With Smoking:		Reducing the Risks by Quitting:
Shortened Life Expectancy.	Smokers are 3 times more likely to die before age 70 than nonsmokers.	After 10–15 years, life expectancy of *ex-smokers* is close to that of nonsmokers.
Chronic Bronchitis, Emphysema.	4–25 times greater risk of death.	Better chance of improving lung function, slowing deterioration.
Lung Cancer.	Greatest frequency is among smokers.	After 10–15 years, risk approaches that of nonsmokers.
Cancer of Mouth.	3–10 times more frequent.	After 10–15 years, no more at risk than nonsmokers.
Cancer of Larynx.	2.9–17.7 times more frequent.	After 10 years, no more at risk than nonsmokers.
Cancer of Esophagus.	2–9 times greater risk of death.	Depends on amount smoked; quitting should lower risk.
Cancer of Pancreas.	2–5 times greater risk of death.	Depends on amount smoked; quitting should lower risk.
Cancer of Bladder.	7–10 times more frequent.	After 7 years, no more at risk than nonsmokers.
Coronary Heart Disease.	Smoking is a major factor.	After 1 year, risk drops sharply; after 10, no more risk than nonsmokers.
Effects on Offspring.	Smoking during pregnancy increases risk of stillbirths or underweight liveborns more vulnerable to disease.	No smoking after fourth month of pregnancy eliminates risk.
Impaired Immune System Function.	Increased allergies, death of macrophages that protect respiratory tract.	Avoidable by not smoking.

c   Risks incurred by smoking and the benefits of quitting. The photograph shows swirls of cigarette smoke poised at the entrance to the two bronchi that lead into the lungs.

of death among women. Yet every day, 3,000 to 5,000 Americans light a cigarette for the first time. Children spend a billion dollars a year on cigarettes. The economy loses 22 billion a year because of direct medical costs of treating smoke-induced respiratory disorders.

Consider how cigarette smoke affects the lungs. Noxious particles in the smoke from just one cigarette immobilize cilia in the bronchioles for several hours. The particles also stimulate mucus secretions that in time clog the airways. They can kill infection-fighting macrophages in the respiratory tract. What starts as "smoker's cough"

can end in bronchitis and emphysema. Marijuana smoke also can cause extensive lung damage.

Or consider how cigarette smoke contributes to lung cancer. Inside the body, certain compounds in coal tar and cigarette smoke are converted to highly reactive intermediates. These are the carcinogens. They provoke the uncontrolled cell divisions in lung tissues. On the average, 90 of every 100 smokers who develop lung cancer will die from it. If you now smoke, or if you are thinking about starting or quitting, you may wish to consider the information presented in Figure *c*.

increases. If the volume of air flow is too small, the diameter shrinks, and blood flow decreases.

The nervous system controls oxygen and carbon dioxide levels in arterial blood for the entire body. Some sensory receptors notify the brain of rising carbon dioxide levels in the blood. Others notify the brain of decreases in the partial pressure of oxygen dissolved in arterial blood. The brain sends signals to muscles that are located in the diaphragm and chest wall. Through changes in muscle

activity, adjustments are made in the rate and depth of breathing. With these adjustments, more oxygen is delivered to affected tissues and more carbon dioxide removed from them.

The control mechanisms that govern gas exchange are like a fine Swiss watch—intricately coordinated and smooth in their operation. The *Focus* essay describes a few respiratory ailments, some serious enough to stop the watch from ticking.

RESPIRATION IN UNUSUAL ENVIRONMENTS

Decompression Sickness

As professional divers know, water pressure increases greatly with depth. When diving in deep water, they take along tanks of compressed air (air under pressure). They also ascend from the depths with utmost caution. Why? Because of the increased pressure, more gaseous nitrogen (N_2) than usual has become dissolved in their body tissues. When the pressure decreases, N_2 moves back out of the tissues and into the bloodstream. If an ascent is too rapid, N_2 enters the blood faster than the lungs can dispose of it. When that happens, bubbles of nitrogen may form in the blood and tissues. Too many bubbles cause pain, especially at joints. Hence the common name, "the bends," for what is otherwise known as *decompression sickness*. If bubbles obstruct blood flow to the brain, deafness, impaired vision, and paralysis may result.

At depths of about 150 meters, N_2 poses another threat. At high partial pressures it produces feelings of euphoria, even drunkenness. Divers have been known to offer the mouthpiece of their airtank to a fish.

Hypoxia

We conclude this chapter by coming full circle, back to the example used to introduce it. The partial pressure of oxygen decreases with increasing altitude. Unlike the occasional climbers of Chomolungma, humans, llamas, and other animals accustomed to living at high altitudes have permanent adaptations to the thinner air. While they were growing up, more air sacs and blood vessels developed in their lungs. The ventricles in their heart became enlarged enough to pump larger volumes of blood. Llamas have an additional advantage. Compared to human hemoglobin, llama hemoglobin has a greater affinity for oxygen. It picks up oxygen more efficiently at the lower pressures characteristic of high altitudes.

Visitors who have not had time to adapt to the thinner air at high altitudes can suffer *hypoxia*, or cellular oxygen deficiency. Generally, at 2,400 meters (about 8,000 feet) above sea level, respiratory centers work to compensate for the oxygen deficiency by triggering *hyperventilation* (breathing much faster and more deeply than normal). At 3,300 meters (10,000 feet) or so, altitude sickness sets in.

Hypoxia also occurs when the partial pressure of oxygen in arterial blood falls because of *carbon monoxide poisoning*. Carbon monoxide, a colorless, odorless gas, is present in automobile exhaust fumes and in smoke from tobacco, coal, or wood burning. It binds to hemoglobin at least 200 times more strongly than oxygen does. Even very small amounts can tie up half of the body's hemoglobin and so impair oxygen delivery to tissues.

SUMMARY

1. Animal cells rely mainly on aerobic respiration, a metabolic pathway that provides enough energy for active life-styles. This pathway requires oxygen and produces carbon dioxide. The process by which the animal body as a whole acquires oxygen and disposes of carbon dioxide is called respiration.

2. Air is a mixture of oxygen, carbon dioxide, and other gases, each exerting a partial pressure. Each gas tends to move from areas of higher to lower partial pressure. Respiratory systems make use of this tendency.

3. In all respiratory systems, oxygen and carbon dioxide diffuse across a respiratory surface (a moist, thin layer of epithelium). In vertebrates, airways carry gases to and from one side of the respiratory surface, and blood vessels carry gases to and from the other side.

4. The airways of the human respiratory system are the nasal cavities, pharynx, larynx, trachea, bronchi, and bronchioles. Alveoli at the end of the terminal bronchioles are the main sites of gas exchange.

5. During inhalation, the chest cavity expands, lung pressure falls below atmospheric pressure, and air flows into the lungs. During normal exhalation, these events are reversed.

6. Driven by its partial pressure gradient, oxygen in the lungs diffuses from alveolar air spaces into the lung capillaries. Then it diffuses into red blood cells and binds weakly with hemoglobin. In tissues where cells are metabolically active, hemoglobin gives up oxygen, which diffuses out of the capillaries, across interstitial fluid, and into the cells.

7. Driven by its partial pressure gradient, carbon dioxide diffuses from cells, across interstitial fluid, and into the bloodstream. Most reacts with water to form bicarbonate, but the reactions are reversed in the lungs. There, carbon dioxide diffuses from lung capillaries into the air spaces of the alveoli, then is exhaled.

Review Questions

1. A few of your friends who have not taken a biology course ask you what insect lungs look like. What is your answer (assuming your instructor is listening)? *454*

2. Describe the features of the respiratory surface that are common to all respiratory systems. *453*

3. Distinguish between:
   a. aerobic respiration and respiration *453*
   b. pharynx and larynx *456, 457*
   c. bronchiole and bronchus *457*
   d. pleural sac and alveolar sac *457*

4. Explain why humans, including the female shown above, cannot survive on their own for very long in underwater environments. *453–454*

5. Label the components of the human respiratory system and the structures that enclose it: *456*

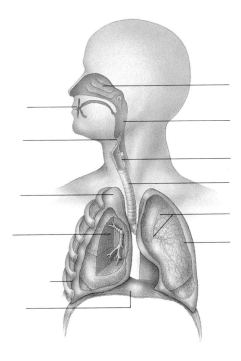

Self-Quiz *(Answers in Appendix IV)*

1. A partial pressure gradient of oxygen exists between _____ .
   a. air and lungs
   b. lungs and metabolically active tissues
   c. air at sea level and air at high altitudes
   d. all of the above

2. In humans, oxygen in the air must diffuse across _____ as it follows its partial pressure gradient into the internal environment.
   a. pleural sacs   c. a respiratory surface
   b. alveolar sacs   d. both b and c

3. Each human lung encloses a _____ .
   a. diaphragm   c. pleural sac
   b. bronchial tree   d. both b and c

4. Gas exchange occurs at the _____ .
   a. two bronchi   c. alveolar sacs
   b. pleural sacs   d. both b and c

5. The Heimlich maneuver may help people _____ .
   a. with altitude sickness   c. choking on food
   b. with the bends   d. quit smoking

6. Breathing _____ .
   a. ventilates the lungs   d. causes reversals
   b. draws air into airways   in pressure gradients
   c. expels air from airways   e. all of the above

7. When you inhale, the diaphragm _____ .
   a. curves upward   c. relaxes
   b. flattens   d. remains stationary

8. After oxygen diffuses into lung capillaries, it diffuses into _____ and binds with _____ .
   a. interstitial fluid; red blood cells
   b. interstitial fluid; carbon dioxide
   c. red blood cells; hemoglobin
   d. red blood cells; carbon dioxide

9. Most carbon dioxide in blood is in the form of _____ .
   a. carbon dioxide   c. carbonic acid
   b. carbon monoxide   d. bicarbonate

10. Match the components with their descriptions.
   ___ trachea   a. airway leading into a lung
   ___ pharynx   b. throat
   ___ alveolus   c. fine bronchial tree
   ___ hemoglobin   branchings
   ___ bronchus   d. windpipe
   ___ bronchiole   e. respiratory pigment
                      f. site of gas exchange

Selected Key Terms

alveolus *457*   lung *455*
bronchiole *457*   nasal cavity *456*
bronchus *457*   pharynx *457*
diaphragm *457*   respiratory surface *453*
epiglottis *457*   respiratory system *453*
gill (of fish) *454*   trachea *457*
hemoglobin *454*   tracheal respiration *454*
integumentary exchange *454*   vocal cord *457*
larynx *457*

Readings

American Cancer Society. 1980. *Dangers of Smoking; Benefits of Quitting and Relative Risks of Reduced Exposure.* Revised edition. New York: American Cancer Society.

Vander, A.; J. Sherman; and D. Luciano. 1990. *Human Physiology: The Mechanisms of Body Function.* Fifth edition. New York: McGraw-Hill. Clear introduction to the respiratory system.

West, J. 1989. *Respiratory Physiology: The Essentials.* Fourth edition. Baltimore: Williams & Wilkins. Excellent, brief introduction to respiratory functions. Paperback.

29 WATER-SOLUTE BALANCE

Tale of the Desert Rat

About 375 million years ago, some animals left the seas for life on land, and they didn't consult their cells about the move. Their cells were adapted to life in a *salty fluid*, for goodness' sake! The cells survived anyway. The animals brought salty fluid along with them, as an internal environment (blood and tissue fluid). Still, it wasn't easy. On land they found intense sunlight, dry winds, water of dubious salt content, and sometimes no water at all. How could they keep from drying out? How could they replace the water and specific salts lost through everyday activities? How could they maintain the volume and composition of that precious internal environment and so prevent cellular anarchy?

Any of their existing descendants provides you with answers. Think of a kangaroo rat in an isolated desert of New Mexico (Figure 29.1). After a brief rainy season, the sun bakes the sand for months. The only obvious water is imported, sloshing in canteens of a researcher or tourist. Yet with nary a sip of free water, this tiny animal counters threats to its internal environment.

The kangaroo rat waits out the daytime heat in burrows, then forages in the cool of night for dry seeds and maybe a succulent. It is not sluggish about this. It hops rapidly and far, searching for seeds and fleeing from coyotes and snakes. All that hopping requires ATP energy and water. Seeds, chockful of energy-rich carbo-

hydrates, supply both. Metabolic reactions that release energy from carbohydrates and other organic compounds also yield water. Each day, this "metabolic water" represents about 12 percent of your own total water intake. It represents a whopping 90 percent of the total for a kangaroo rat.

Inside its cool burrow, the kangaroo rat conserves and recycles water. As the animal breathes cool air into its warm lungs, water vapor condenses on the epithelial lining inside its nose—and some of it diffuses back into the body. Also, after a night of foraging, the animal empties its cheek pouches of seeds—which soak up water vapor that does escape from the nose. Eating the seeds recycles the water.

The kangaroo rat can't lose water by perspiring; it has no sweat glands. It can lose water by piddling—but specialized kidneys don't let it piddle away much. Kidneys filter the blood's water and dissolved salts (solutes). They adjust *how much water* and *which solutes* return to the blood or leave the body as urine. Overall, the kangaroo rat and every other animal takes in enough water and solutes to replace the daily losses (Figure 29.1c). Let's turn now to the ways in which this balancing act is accomplished.

b

Figure 29.1 A kangaroo rat (**a**), master of water conservation in the desert (**b**).

a

c

	Kangaroo Rat*	Human**
Water Gain (milliliters):		
by ingesting solids	6.0	850
by ingesting liquids	0	1,400
by way of metabolism	54.0	350
	60.0	2,600
Water Loss (milliliters):		
in urine	13.5	1,500
in feces	2.6	200
by evaporation	43.9	900
	60.0	2,600

*Measured over 4 weeks.
**Measured daily.

1. Animals continually gain and lose water, nutrients, and ions. They continually produce metabolic wastes. Even with all these inputs and outputs, the body's extracellular fluid does not change much in its overall volume and composition.

2. In humans and other mammals, a urinary system helps balance the intake and output of water and dissolved substances (solutes). This system eliminates excess water and solutes as a fluid called urine.

3. Urine forms in tubelike nephrons. These structures are packed inside a pair of kidneys, which are blood-filtering organs. Nephrons receive water and solutes from a set of blood capillaries. They return most of the filtrate to a second set of capillaries; the rest leaves the body as urine.

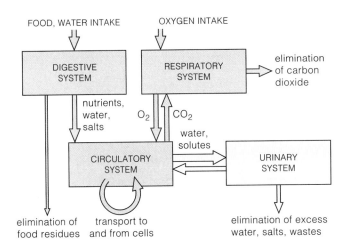

Figure 29.2 Organ systems that interact to maintain homeostasis, or favorable operating conditions in the body.

MAINTAINING EXTRACELLULAR FLUID

In most animals, recall, interstitial fluid fills the spaces between living cells and other components of the body's tissues. Another fluid, blood, circulates inside blood vessels. Taken together, interstitial fluid and blood are the body's **extracellular fluid**.

In humans, as in all mammals, a well-developed urinary system helps keep the volume and composition of extracellular fluid within tolerable ranges. It does this by balancing the daily losses of water and solutes with daily intakes. Other organ systems, especially those shown in Figure 29.2, assist in this homeostatic task.

Water Gains and Losses

Ordinarily, and on a daily basis, humans and other mammals take in as much water as they give up (Figure 29.1c). The body *gains* water by two processes:

1. Absorption from solid food and liquids
2. Metabolism

Water, recall, is absorbed from solids and liquids in the gut. A thirst mechanism, described in this chapter, affects how much water enters the gut in the first place. Also recall that water forms as a by-product of many metabolic reactions, such as carbohydrate breakdown.

The body *loses* water by the following processes:

1. Urinary excretion
2. Evaporation from lungs and through skin
3. Sweating
4. Elimination (in feces)

Of these, **urinary excretion** affords the greatest control over water loss. This process, to be described shortly, eliminates excess water and excess or harmful solutes in the form of urine. Some water is always evaporating from the respiratory surface in lungs. It evaporates from sweaty skin. Normally, most of the water traveling through the gut is absorbed; very little leaves in feces.

Solute Gains and Losses

The body's extracellular fluid *gains* solutes mainly as a result of four processes:

1. Absorption from solid food and liquids
2. Secretion
3. Respiration
4. Metabolism

Meals provide a variety of nutrients (including glucose) and mineral ions (such as potassium and sodium ions). These are absorbed from the gut, as are drugs and food additives. Cells throughout the body secrete substances into interstitial fluid and the blood. The respiratory system puts oxygen into the blood, and aerobically respiring cells put carbon dioxide into it.

Extracellular fluid *loses* mineral ions and metabolic wastes in these ways:

1. Urinary excretion
2. Respiration
3. Sweating

Mineral ions depart in sweat. The most abundant metabolic waste, carbon dioxide, is exhaled in respiration. Other metabolic wastes, such as ammonia, urea, uric acid, phosphoric acid, and sulfuric acid, are excreted. Ammonia, formed during amino acid breakdown, is toxic in excess amounts. Urea, the main by-product of protein breakdown, forms in the liver when two ammonia molecules join to form carbon dioxide.

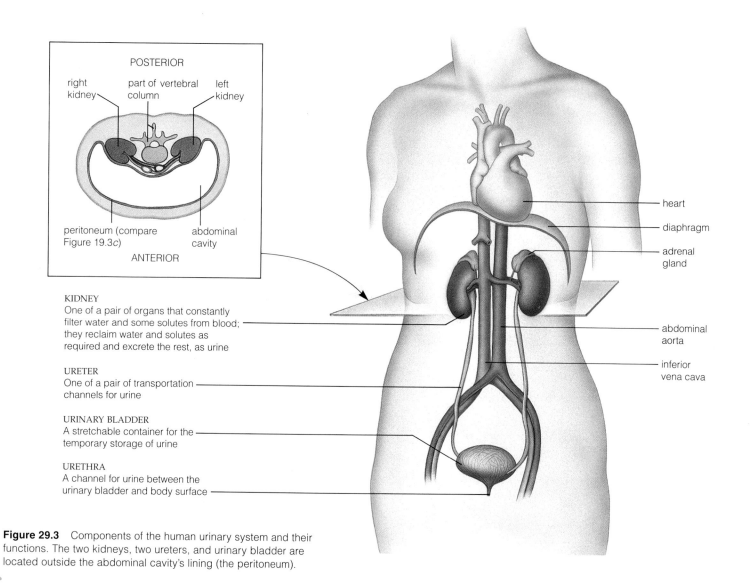

right kidney

part of vertebral column

left kidney

peritoneum (compare Figure 19.3*c*)

abdominal cavity

ANTERIOR

KIDNEY
One of a pair of organs that constantly filter water and some solutes from blood; they reclaim water and solutes as required and excrete the rest, as urine

URETER
One of a pair of transportation channels for urine

URINARY BLADDER
A stretchable container for the temporary storage of urine

URETHRA
A channel for urine between the urinary bladder and body surface

heart

diaphragm

adrenal gland

abdominal aorta

inferior vena cava

Figure 29.3 Components of the human urinary system and their functions. The two kidneys, two ureters, and urinary bladder are located outside the abdominal cavity's lining (the peritoneum).

URINARY SYSTEM OF MAMMALS

Different solid foods and fluids intermittently hit your gut. Afterward, different amounts of water and diverse ions, nutrients, and other substances move into blood, then into interstitial fluid throughout your body. Any unwanted shifts in this extracellular fluid are dealt with swiftly by a pair of organs called **kidneys**. Kidneys are the key components of the **urinary system**, as shown in Figures 29.3 and 29.4.

Kidneys filter water, mineral ions, organic wastes, and other substances out of blood. They adjust the filtrate's composition, then return all but about 1 percent of it to the blood. This tiny portion of water and solutes is urine.

By definition, **urine** is a fluid that rids the body of water and solutes in excess of the amounts required to maintain extracellular fluid. Urine flows from each kidney into a **ureter**, a tubular channel. The ureters lead to

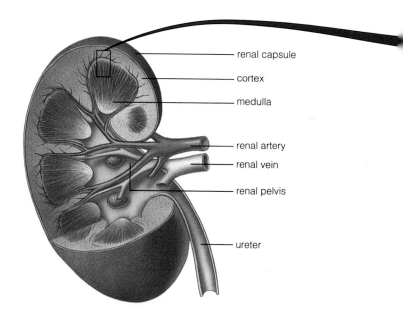

renal capsule

cortex

medulla

renal artery

renal vein

renal pelvis

ureter

Figure 29.4 Internal structure of the kidney and the major blood vessels leading into and out of it.

the **urinary bladder**. Here urine is temporarily stored before moving out through the **urethra**, a channel that opens at the body surface. All mammals have this urinary system of two kidneys, two ureters, a urinary bladder, and a urethra.

By adjusting blood's volume and composition, kidneys help maintain conditions in the extracellular fluid.

Kidney Structure

As Figure 29.4 suggests, a kidney has a tough coat of connective tissue (the renal capsule). Beneath the coat are two regions, first the cortex, then the medulla. Notice how several lobes extend from the cortex down through the medulla. Each lobe contains blood vessels and numerous slender tubes called **nephrons**. Nephrons filter water and solutes from blood. Most of the filtrate is reclaimed from them. The rest (urine) enters tubelike collecting ducts. These lead to a central cavity (renal pelvis) and the entrance to a ureter.

Nephron Structure

More than a million nephrons are packed inside each fist-sized human kidney. The nephron wall is a single layer of epithelial cells, but the cells and junctions between them are not all the same. As you will see, the differences prevent or assist the movement of water and solutes across different parts of the nephron wall.

As Figure 29.5 shows, the nephron starts as part of a **glomerulus**, a blood-filtering unit. Here the nephron wall forms a cup around a set of blood capillaries. The cupped wall region, **Bowman's capsule**, receives water and solutes filtered from blood. The rest of the nephron is tubelike. The **proximal tubule** (closest to Bowman's capsule) is followed by a hairpin-shaped **loop of Henle**. The **distal tubule** (most distant from Bowman's capsule) delivers urine to a collecting duct.

Blood entering a glomerulus doesn't give up *all* of its water and solutes to the nephron. The unfiltered portion flows into a second set of capillaries that thread around the nephron's tubular parts (Figure 29.5c). There blood reclaims water and solutes, then flows into veins and back to the general circulation.

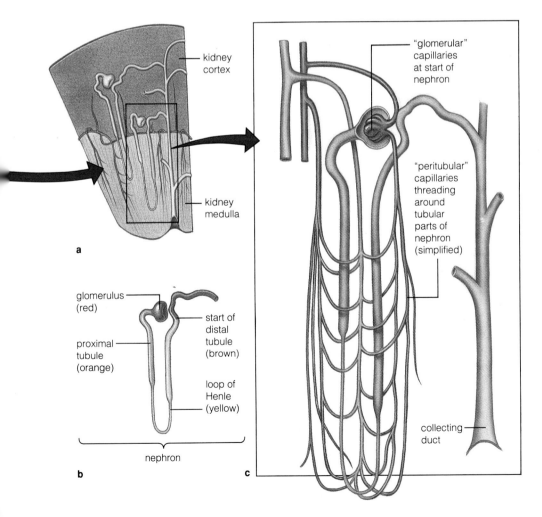

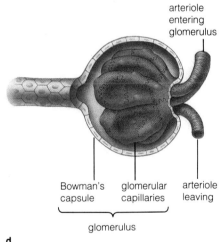

Figure 29.5 Simplified diagrams of a nephron and its association with two connected sets of blood capillaries. (**a**) Orientation of nephrons relative to the kidney's cortex and medulla. (**b**) Functional regions of a nephron. (**c**) The first set of capillaries (glomerular capillaries) is clustered inside a capsule at the start of the nephron. The second set (peritubular capillaries) threads around all tubular parts of the nephron. (**d**) A closer look at the glomerulus, the nephron's blood-filtering unit.

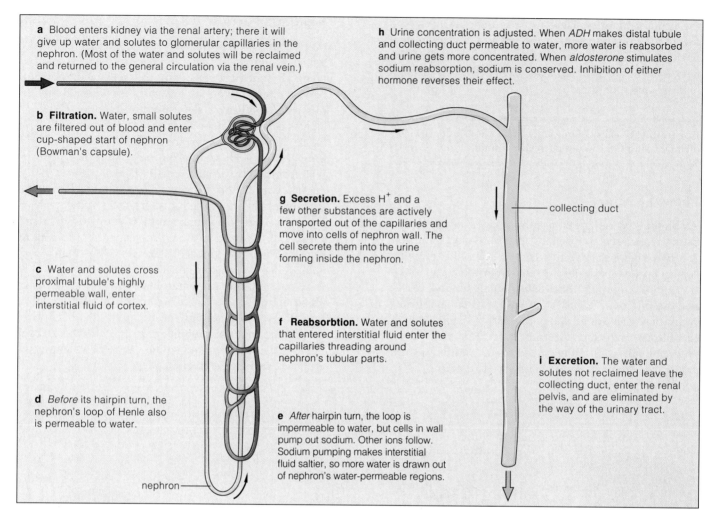

a Blood enters kidney via the renal artery; there it will give up water and solutes to glomerular capillaries in the nephron. (Most of the water and solutes will be reclaimed and returned to the general circulation via the renal vein.)

b Filtration. Water, small solutes are filtered out of blood and enter cup-shaped start of nephron (Bowman's capsule).

c Water and solutes cross proximal tubule's highly permeable wall, enter interstitial fluid of cortex.

d *Before* its hairpin turn, the nephron's loop of Henle also is permeable to water.

nephron

g Secretion. Excess H$^+$ and a few other substances are actively transported out of the capillaries and move into cells of nephron wall. The cell secrete them into the urine forming inside the nephron.

f Reabsorbtion. Water and solutes that entered interstitial fluid enter the capillaries threading around nephron's tubular parts.

e *After* hairpin turn, the loop is impermeable to water, but cells in wall pump out sodium. Other ions follow. Sodium pumping makes interstitial fluid saltier, so more water is drawn out of nephron's water-permeable regions.

h Urine concentration is adjusted. When *ADH* makes distal tubule and collecting duct permeable to water, more water is reabsorbed and urine gets more concentrated. When *aldosterone* stimulates sodium reabsorption, sodium is conserved. Inhibition of either hormone reverses their effect.

collecting duct

i Excretion. The water and solutes not reclaimed leave the collecting duct, enter the renal pelvis, and are eliminated by the way of the urinary tract.

Figure 29.6 Processes involved in urine formation and excretion.

URINE FORMATION

Urine-Forming Processes

Urine forms through three processes, called filtration, tubular reabsorption, and secretion (Figure 29.6).

Filtration starts and ends at the glomerulus. Blood pressure, generated by heart contractions, drives the process. It "filters" blood by forcing water and small solutes out of the glomerular capillaries, leaving blood cells, proteins, and other large solutes behind. After this occurs, the filtrate moves from the cupped part of the nephron into the proximal tubule.

Reabsorption proceeds along the nephron's tubular parts. Most of the filtrate's water and solutes move out of the nephron, then into adjacent blood capillaries. Table 29.1 lists some daily reabsorption values.

Secretion also proceeds at tubular wall regions *but in the opposite direction*. Substances move *out* of the capillaries and into cells of the nephron wall. Then the substances move across the cells, which secrete them into the forming urine. Among other things, this highly con-

Table 29.1	Average Daily Reabsorption Values for a Few Substances		
	Filtered	Excreted	Reabsorbed
Water	180 liters	1.8 liters	99%
Glucose	180 grams	None, normally	100%
Sodium ions	630 grams	3.2 grams	99.5%
Urea	54 grams	30 grams	44%

trolled process rids blood of excess hydrogen ions (H$^+$) and potassium ions. It also prevents some metabolic wastes (such as uric acid) and foreign substances (such as penicillin) from accumulating in blood.

A concentrated or dilute urine forms through the processes of filtration, reabsorption, and secretion.

Factors Influencing Filtration

About 1.5 liters (1-1/2 quarts) of blood flow through an adult's kidneys every minute. From this enormous flow volume, filtration diverts 120 milliliters per minute—*180 liters per day*—into the nephron. Two mechanisms are responsible for this high rate of filtration. First, blood pressure in the glomerular capillaries is high. (Arterioles delivering blood to the glomerulus have a wider diameter and less resistance to flow than most arterioles.) Second, these capillaries are 10 to 100 times more permeable than others in the body.

The volume of blood flow to the kidneys affects the rate of filtration. When you run a race or dance until dawn, for instance, your nervous system diverts an above-normal volume of blood away from the kidneys, toward your heart and skeletal muscles.

Reabsorption of Water and Sodium

Reabsorption Mechanism. Your kidneys precisely adjust how much water and sodium your body excretes or conserves. It makes no difference if you drink too much or too little water at lunch, or wolf down salty potato chips, or lose too much sodium by sweating.

Adjustments start promptly, when filtrate enters the nephron's proximal tubule. Cells of this tubule wall actively transport some sodium ions out of the filtrate into interstitial fluid, and other ions follow. Then water follows the ions, by osmosis. Because this wall region is highly permeable to water, the body reabsorbs most of the filtrate's water here (Figure 29.6c–f).

Now the filtrate enters the loop of Henle, which plunges into the kidney medulla. Where the loop makes its hairpin turn, the surrounding interstitial fluid is saltiest. Water moves out by osmosis before the turn. The fluid left behind gets saltier, until it matches the interstitial fluid. After the turn, no more water can cross the tubule wall—but sodium is pumped out. As the surroundings become even saltier, more water is drawn out of fluid just now entering the loop.

When fluid remaining in the nephron reaches the distal tubule in the kidney cortex, it is quite dilute. The stage is set for excretion of highly dilute or highly concentrated urine—or anywhere in between.

Hormone-Induced Adjustments. Receptors for two hormones—ADH and aldosterone—occur on cells of the nephron's distal tubule and on cells of collecting ducts.

When the body must conserve water, the action of **ADH** (antidiuretic hormone) results in a small volume of concentrated urine. ADH secretion is stimulated by the hypothalamus. This hormone makes the walls of distal tubules and collecting ducts more permeable to water (Figure 29.6h). More water is reabsorbed, so the urine becomes more concentrated.

When the body must lose excess water, ADH secretion is inhibited. Then, distal tubules and collecting ducts remain impermeable to water, less water is reabsorbed, and the urine becomes dilute.

Aldosterone enhances sodium reabsorption. When extracellular fluid loses too much sodium, its volume decreases. Sensory receptors in blood vessels and the heart detect the decrease, and certain kidney cells are called into action (Figure 29.7). The cells secrete renin, an enzyme that lops off part of a protein circulating in blood. Additional reactions convert the protein to a hormone. This hormone stimulates aldosterone-secreting cells of the **adrenal cortex**, the outer portion of a gland perched on top of each kidney (Figure 29.3). Aldosterone makes cells of the distal tubules and collecting ducts reabsorb sodium at a faster pace, so less sodium is excreted in the urine. Conversely, when the body contains too much sodium, aldosterone secretion is inhibited. Less sodium is reabsorbed, and more is excreted.

ADH conserves water by enhancing its reabsorption at distal tubules and collecting ducts. Excess water is excreted when ADH secretion is inhibited.

Aldosterone conserves sodium by enhancing its reabsorption at distal tubules and collecting ducts. When aldosterone secretion is inhibited, the body excretes excess sodium.

Thirst Behavior. Finally, bear in mind that when the solute concentration in extracellular fluid rises, the hypothalamus does more than stimulate ADH secretion. Some of its cells act as a "thirst center." Signals from these cells inhibit saliva production. Your brain interprets the resulting sensation of dryness in your mouth as "thirst" and leads you to seek out drinking fluids. Actually, a cottony mouth is an early sign that your body is becoming dehydrated.

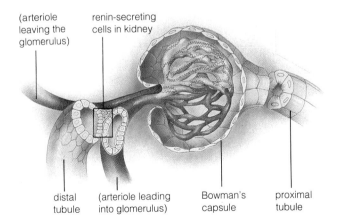

(arteriole leaving the glomerulus) renin-secreting cells in kidney

distal tubule (arteriole leading into glomerulus) Bowman's capsule proximal tubule

Figure 29.7 Location of renin-secreting cells that play a role in sodium reabsorption.

Kidney Failure and Dialysis

Through illness or accident, about 13 million people in the United States alone suffer from kidney disorders. Kidney malfunctions disrupt the controls over the volume and composition of extracellular fluid. Among other things, toxic by-products of protein breakdown may accumulate in the bloodstream. Nausea, fatigue, and loss of memory are outcomes. In advanced cases, death may follow. A *kidney dialysis machine* can restore the proper solute balances. Like the kidney, this machine helps maintain extracellular fluid by selectively removing solutes from blood and adding solutes to it. "Dialysis" refers to an exchange of substances across a membrane between solutions of differing compositions. In *hemodialysis*, the machine is connected to an artery or a vein, then blood is pumped through tubes made of a material similar to sausage casing or cellophane.

The tubes are submerged in a warm-water bath. The precise mix of salts, glucose, and other substances in the bath sets up the correct gradients with blood. In *peritoneal dialysis*, fluid of the proper composition is put into the patient's abdominal cavity, left in place for a specific length of time, then drained out. The cavity's lining (peritoneum) serves as the dialysis membrane.

Hemodialysis is performed three times a week. Each time, the procedure takes about four hours, for blood must circulate repeatedly to improve the solute concentrations in the body. It is used as a temporary measure in reversible kidney disorders. In chronic cases, the procedure must be used for the rest of the patient's life or until a functional kidney is transplanted. With treatment and controlled diets, many patients resume fairly normal activity.

Kidney Malfunctions

Good health depends absolutely on normal kidney function. For example, when something interferes with sodium excretion, the body's sodium content rises. So does the volume of extracellular fluid. This leads to a rise in blood pressure. Abnormally high blood pressure, or *hypertension*, can adversely affect the kidneys, brain, and vascular system. Restricted intake of sodium chloride—table salt—lessens the problem. As another example, uric acid, calcium salts, and other wastes can collect in the renal pelvis as *kidney stones* that may lodge in the ureter (or urethra), then disrupt urine flow. They usually are passed naturally (in urine). If not, they must be removed by medical or surgical procedures. Advanced cases of kidney malfunctioning often can be treated by procedures described in the *Focus* essay.

ACID-BASE BALANCE

So far, we've focused on how kidneys maintain the volume and composition of extracellular fluid. They also help keep it from becoming too acidic or too basic. The overall acid-base balance is maintained through control of H^+ and other dissolved ions. Buffer systems, respiration, and urinary excretion provide the control.

Normally, the extracellular pH of humans must be maintained between 7.37 and 7.45. As you know, acids lower the pH and bases raise it. A variety of acidic and basic substances enter the blood (by absorption from the gut and by metabolism). Typically, cell activities produce an excess of acids, which separate into H^+ and other fragments. This lowers the pH. The effect is minimized when excess H^+ reacts with buffer molecules. The bicarbonate–carbon dioxide buffer system, introduced earlier on page 25, is an example:

$$H^+ + \underset{\text{bicarbonate}}{HCO_3^-} \rightleftharpoons \underset{\text{carbonic acid}}{H_2CO_3} \rightleftharpoons H_2O + CO_2$$

Here, the H^+ is neutralized, and the carbon dioxide that forms during the reactions is exhaled from lungs. Like other buffer systems, this one has a temporary effect; it doesn't eliminate acid. *Only the urinary system eliminates excess H^+ and restores buffers.*

The same reactions proceed in reverse in cells of the nephron's tubular walls. After forming in those cells, the HCO_3^- moves into interstitial fluid, then into capillaries around the nephron, then into the general circulation, where it buffers excess acid. The H^+ that forms in the cells is secreted into the nephron. There, it may combine with bicarbonate ions to form CO_2—which is returned to the blood and exhaled by the lungs. Or it may combine with phosphate ions or ammonia (NH_3), then leave the body in urine. In such ways, hydrogen ions are *permanently* removed from extracellular fluid.

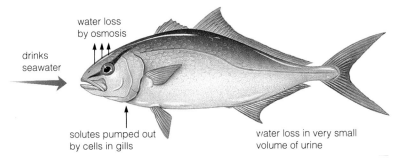

water loss
by osmosis

drinks
seawater

solutes pumped out
by cells in gills

water loss in very small
volume of urine

a Marine bony fish (body fluids less salty than the surroundings).

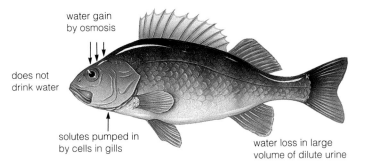

water gain
by osmosis

does not
drink water

solutes pumped in
by cells in gills

water loss in large
volume of dilute urine

b Freshwater bony fish (body fluids much saltier than the surroundings).

Figure 29.8 Water-solute balancing acts in fishes.

ON FISH, FROGS, AND KANGAROO RATS

Now that you have an idea of how your own body maintains water and solute levels, consider what goes on in some other vertebrates, including that kangaroo rat hopping about at the start of the chapter.

Body fluids of herring, snapper, and other marine fishes are about three times less salty than seawater. These fishes lose water by osmosis, drink replacements, and excrete ingested solutes against concentration gradients (Figure 29.8). Fish kidneys don't have loops of Henle, so urine can't ever be saltier than body fluids. Cells in fish gills actively pump out most of the excess solutes in blood. In fresh water, bony fishes and amphibians gain water and lose solutes. They don't drink. Water moves in by osmosis, through thin gill membranes or, in adult amphibians, across skin. Excess water leaves as dilute urine, formed in kidneys. Solute losses are balanced by solutes gained from food and by the inward pumping of sodium by cells of the gills.

And about that kangaroo rat! Its nephrons have extremely long loops of Henle. The solute concentration in interstitial fluid around the loops becomes very high. The osmotic gradient between this fluid and the urine is so steep, nearly all the water that reaches the equally long collecting ducts is reabsorbed. Kangaroo rats give up only a tiny volume of urine, and it's three to five times more concentrated than that of humans.

SUMMARY

1. Inside the animal body, the cellular environment consists of certain types and amounts of substances dissolved in water. This extracellular fluid fills tissue spaces and blood vessels. Its volume and composition are maintained only when the animal's daily intake and output of water and solutes are in balance. In mammals, the following processes maintain the balance:

a. Water is gained by absorption (from the gut) and metabolism. It is lost by urinary excretion, evaporation from lungs and skin, sweating, and elimination of feces.

b. Solutes are gained by absorption from the gut, secretion, respiration, and metabolism. They are lost by excretion, respiration, and sweating.

c. Losses of water and solutes are controlled mainly by adjusting the volume and composition of urine.

2. The human urinary system consists of two kidneys, two ureters, a urinary bladder, and a urethra.

3. Blood is filtered and urine forms in many nephrons in the kidneys. Each nephron interacts closely with two sets of blood capillaries (glomerual and peritubular).

a. A nephron has a cup-shaped beginning (Bowman's capsule), then three tubelike regions (proximal tubule, loop of Henle, and distal tubule, which empties into a collecting duct).

b. Bowman's capsule surrounds a set of highly permeable capillaries. Together, they are a blood-filtering unit (glomerulus).

c. Blood pressure forces water and small solutes out of the capillaries, into the cup. Most of the filtrate is reabsorbed by the tubules and returned to the blood. A portion is excreted as urine.

4. Urine forms in the nephron by three processes:

a. Filtration of blood at the glomerulus, which puts water and small solutes into the nephron.

b. Reabsorption. Water and solutes to be retained leave the nephron's tubular parts and enter capillaries that thread around them. A small volume of water and solutes remains in the nephron.

c. Secretion. A few substances leave peritubular capillaries and enter the nephron, for disposal in urine.

5. Urine becomes more or less concentrated by the action of two hormones (ADH and aldosterone). These act on cells of distal tubules and collecting ducts:

a. ADH conserves water by enhancing reabsorption across the nephron wall. Inhibition of ADH allows more water to be excreted.

b. Aldosterone conserves sodium by enhancing reabsorption. Inhibition of aldosterone allows more sodium to be excreted.

6. Together with the respiratory system and other mechanisms, the kidneys also help maintain the body's overall acid-base balance.

1. How does urine formation help maintain the body's internal environment? *465*

2. Define the components of the human urinary system. Then label the kidney's component parts: *466–467*

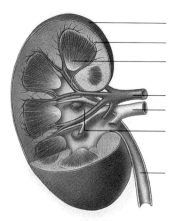

3. Label the regions of the nephron, and identify where filtration, reabsorption, and secretion occur: *467, 468*

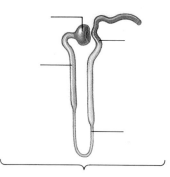

4. Which hormone influences water reabsorption and conservation? sodium reabsorption and conservation? *469*

5. Fatty tissue holds the kidneys in place. Extremely rapid weight loss may cause the tissue to shrink, so that the kidneys slip from their normal position. On rare occasions, the slippage may put a kink in one or both ureters and block urine flow. Speculate on what might happen to the kidneys if this happens. *470*

1. In mammals, water intake depends on _____ .
   a. absorption from gut
   b. metabolism
   c. a thirst mechanism
   d. all of the above

2. In mammals, water is lost by way of the _____ .
   a. skin
   b. respiratory system
   c. digestive system
   d. urinary system
   e. c and d
   f. all of the above

3. Water and small solutes return to blood during _____ .
   a. filtration
   b. reabsorption
   c. secretion
   d. both a and b

4. Water and small solutes leave blood during _____ .
   a. filtration
   b. reabsorption
   c. secretion
   d. both a and c

5. A few substances move out of the capillaries threading around tubular parts of the nephron. These substances are moved into the nephron during _____ .
   a. filtration
   b. reabsorption
   c. secretion
   d. both a and c

6. A nephron's reabsorption mechanism depends on _____ .
   a. osmosis across nephron wall
   b. active transport of sodium across nephron wall
   c. a steep solute concentration gradient starting at kidney cortex and descending into medulla
   d. all of the above

7. Water is conserved and urine becomes more concentrated by the action of _____ .
   a. ADH
   b. renin
   c. aldosterone
   d. both b and c

8. Sodium is conserved and urine becomes more dilute by the action of _____ .
   a. ADH
   b. renin
   c. aldosterone
   d. both b and c

9. Match the term with the appropriate description.
   _____ glomerulus
   _____ distal tubule
   _____ loop of Henle
   _____ acid-base balance
   _____ kangaroo rat
   a. surrounded by saltiest fluid
   b. extra-long loops of Henle
   c. involves buffer systems
   d. blood-filtering unit
   e. ADH, aldosterone act here

ADH *469*
adrenal cortex *469*
aldosterone *469*
Bowman's capsule *467*
distal tubule *467*
extracellular fluid *465*
filtration *468*
glomerulus *467*
kidney *466*
loop of Henle *467*

nephron *467*
proximal tubule *467*
reabsorption *468*
secretion *468*
ureter *466*
urethra *467*
urinary bladder *467*
urinary excretion *465*
urinary system *466*
urine *466*

Flieger, K. March 1990. "Kidney Disease: When Those Fabulous Filters Are Foiled." *FDA Consumer* 24: 26–29.

Smith, H. 1961. *From Fish to Philosopher*. New York: Doubleday. Paperback.

Vander, A.; J. Sherman; and D. Luciano. 1990. "The Kidneys and Regulation of Water and Inorganic Ions" in *Human Physiology*. Fifth edition. New York: McGraw-Hill, Chapter 15.

30 NEURAL CONTROL AND THE SENSES

Why Crack the System?

Suppose your biology instructor asks you to volunteer for an experiment. A microchip will have to be implanted in your brain. It will make you feel *really* good. But it may mess up your health, lop ten years off your life, and destroy a good part of your brain. Your behavior will change for the worse, so you may have trouble completing school, getting or keeping a job, even having a normal family life.

The longer the chip is implanted, the less you will want to give it up. You won't get paid. *You* pay the experimenter—first at bargain rates, then a little more each week. The chip is illegal. If you get caught using it, you and the experimenter will go to jail.

Sometimes Jim Kalat, a professor at North Carolina State University, proposes this experiment (which of course is hypothetical). Hardly any students volunteer. Then he substitutes *drug* for "microchip" and *dealer* for "experimenter"—and an amazing number come forward! Like 30 million other Americans, the "volunteers" seem ready to engage in self-destructive uses of drugs that alter emotional and behavioral states.

The destruction shows up in unexpected places. Each year, for instance, about 300,000 newborns are already addicted to crack—thanks to their addicted mothers. *Crack* is a cheap form of cocaine. It causes relentless stimulation of brain regions that govern our sense of "pleasure." It dampens normal urges to eat and sleep, and blood pressure rises. Elation and sexual desire intensify. In time, brain cells that produce the stimulatory chemicals can't keep up with the incessant demands. The chemical vacuum makes crack users frantic, then profoundly depressed. Only crack makes them "feel good" again.

Each of us possesses a body of great complexity. Its architecture, its functioning are legacies of millions of years of evolution. Its nervous system is unparalleled in the living world. One of its most astonishing products is language—the encoding of shared experiences of groups of individuals in time and space. Through the evolution of our nervous system, the sense of history was born, and the sense of destiny. Through this system we can ask how we have come to be what we are, and where we are headed from here. Perhaps the sorriest consequence of drug abuse is its implicit denial of this legacy—the denial of self when we cease to ask, and cease to care.

Figure 30.1 Owners of an evolutionary treasure—a complex brain that is the foundation for our memory and reasoning, and our future.

1. Nerve cells, or neurons, are the basic units of communication of nervous systems. Collectively, they detect and integrate information about external and internal conditions, then select or control the body's muscles and glands in ways that produce appropriate responses.

2. The distribution of ions across the plasma membrane of a neuron is unequal, with the inside of the neuron being slightly more negative than the outside. When the neuron is stimulated, this polarity of charge across the membrane may briefly and abruptly reverse. Such reversals—action potentials—are the basis of messages sent through the nervous system.

3. Action potentials are self-propagating along a neuron. But they can't cross small gaps that exist between neurons. Chemical signals bridge the gaps and help stimulate or inhibit the adjoining neuron.

4. Nervous systems coevolved with sensory organs, such as eyes, and motor structures, such as legs and wings. The coevolution of nervous, sensory, and motor systems paved the way for increasingly intricate behavior.

CELLS OF THE NERVOUS SYSTEM

Carefully observe almost any animal, and you see that it senses and actively responds to the environment in complex ways. If its life-style is elaborate, you can safely bet the animal has a well-developed **nervous system**. In such systems, cells called **neurons** monitor conditions and issue commands for responsive actions that benefit the body as a whole. They are the basic units of communication of the brain, spinal cord, and nerves.

There are three classes of neurons. *Sensory* neurons are specialized to respond to specific kinds of stimuli (light, pressure, or some other form of energy). They relay information about the stimulus to the spinal cord and brain. Within the spinal cord and brain, *inter*neurons receive sensory input, integrate it with other information, then influence the activity of other neurons. *Motor* neurons relay information *away* from the brain and spinal cord to muscles or glands. Muscles and glands, the body's effectors, carry out responses:

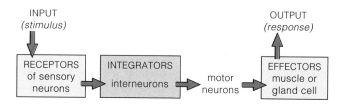

Neurons make up less than half the volume of vertebrate nervous systems. The rest (neuroglia) consists of a variety of specialized cells that protect, structurally support, and functionally assist the neurons.

Structure of the Neuron

All neurons have a cell body, and most have slender cytoplasmic extensions, although these differ greatly in number and length. The motor neuron shown in Figure 30.2 has many **dendrites**—short, slender extensions of the cell body. It also has one **axon**, a long, cylindrical extension. An axon of a motor neuron has finely branched endings that adjoin muscle cells. Think of dendrites and the cell body as "input zones," where a neuron receives signals about changing conditions. Axon endings are "output zones," where it sends messages to other cells.

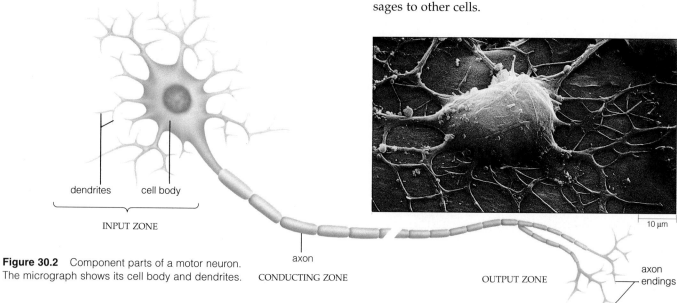

Figure 30.2 Component parts of a motor neuron. The micrograph shows its cell body and dendrites.

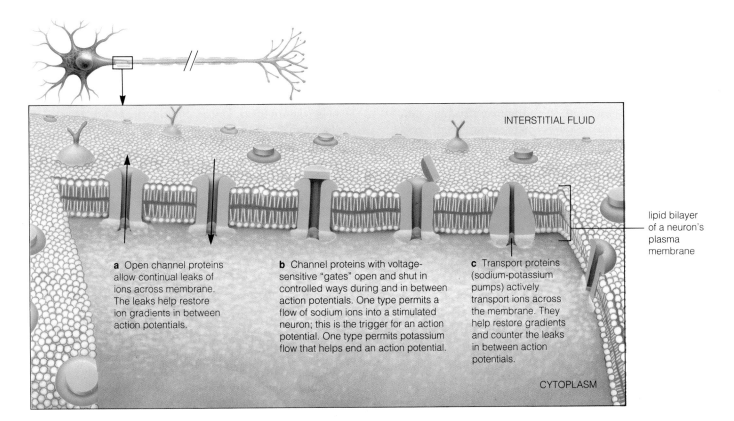

a Open channel proteins allow continual leaks of ions across membrane. The leaks help restore ion gradients in between action potentials.

b Channel proteins with voltage-sensitive "gates" open and shut in controlled ways during and in between action potentials. One type permits a flow of sodium ions into a stimulated neuron; this is the trigger for an action potential. One type permits potassium flow that helps end an action potential.

c Transport proteins (sodium-potassium pumps) actively transport ions across the membrane. They help restore gradients and counter the leaks in between action potentials.

lipid bilayer of a neuron's plasma membrane

INTERSTITIAL FLUID

CYTOPLASM

How a Neuron Responds to Stimulation

Different types of signals travel through a nervous system. Let's start with one that begins and ends on the same neuron, and is not transferred to another cell.

When a neuron is not being bothered, it works to maintain a voltage difference (a difference in electric charge) across the plasma membrane. It keeps the cytoplasmic fluid next to the membrane negatively charged, compared to the interstitial fluid outside. The amount of energy inherent in this steady voltage difference is the **resting membrane potential**.

Suppose a weak signal reaches a patch of membrane in the neuron's input zone. The voltage difference across the patch changes only slightly, if at all. By contrast, a strong signal might trigger an **action potential**—an abrupt, short-lived *reversal* in the voltage difference across the plasma membrane. For a fraction of a second, the inside becomes positive with respect to the outside. The reversal triggers another action potential at the adjoining patch of membrane, this triggers another at the next patch, and so on away from the input zone.

In short, *stimulation of a neuron disturbs the distribution of electric charge across its plasma membrane.*

In an undisturbed neuron, a difference in electric charge across the plasma membrane has been established and is being maintained. An action potential is an abrupt, short-lived reversal in that difference.

Figure 30.3 Pathways for ions across the plasma membrane of a neuron.

Neurons "At Rest." In between action potentials, the neuron restores and maintains the voltage difference across each patch of membrane. It can do so because of two membrane properties. First, the membrane's lipid bilayer bars the passage of potassium ions (K^+), sodium ions (Na^+), and other charged substances. Thus the neuron can build up differences in ion concentrations across the membrane. Second, ions can flow from one side to the other in controllable ways—through the interior of proteins that span the bilayer.

Suppose a motor neuron has 15 sodium ions inside the membrane for every 150 outside. Suppose it also has 150 potassium ions inside for every 5 on the outside. We can depict each ion's concentration gradient in this way (from the large to the small letter):

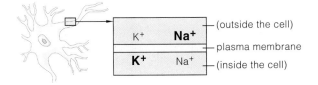

Such gradients dictate the direction in which sodium and potassium ions will diffuse across the membrane, through the interior of channel proteins. Some channels never shut, so ions leak (diffuse) through them all the time. Other channels are "gated" (Figure 30.3). The gates open when a neuron is adequately stimulated.

Leaks and the Sodium-Potassium Pump. Imagine being small enough to stand on an input zone of a motor neuron that is not being disturbed. Many bumps—the tops of channel proteins—spread out before you. Sodium channels are shut, so sodium isn't rushing into the neuron. Some potassium is leaking out through a few open potassium channels, making the neuron a bit more negative inside—so some potassium is attracted back in.

When the inward pull of electric charge balances the outward force of diffusion, there is no more *net* movement of potassium across the membrane. As you will see, the concentration and electric gradients that now exist are the basis of the neuron's responsiveness to

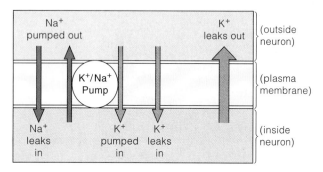

Figure 30.4 Pumping and leaking processes that maintain the distribution of sodium and potassium ions across the plasma membrane of a neuron at rest. Arrow widths indicate the magnitude of the movements. Notice how the total inward and outward movements for each kind of ion are balanced.

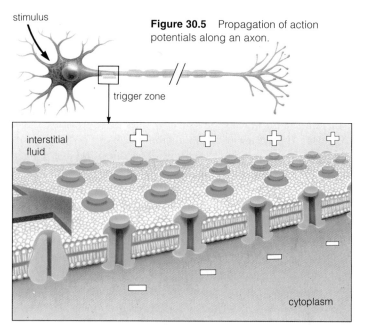

Figure 30.5 Propagation of action potentials along an axon.

a Membrane at rest (inside negative with respect to the outside). An electrical disturbance (red arrow) spreads from an input zone to an adjacent trigger region of the membrane, which has many gated sodium channels.

stimulation. The neuron must expend energy to maintain these gradients. Why? A tiny fraction of the potassium that leaks out is not attracted back in. Also, a tiny fraction of sodium leaks in, through a *few* open channels (Figure 30.4). Unless the tiny leaks are countered, the gradients would gradually disappear.

An active transport mechanism, of the sort described on page 43, counters the leaks. Spanning the membrane are transport proteins called **sodium-potassium pumps**. When they get an energy boost from ATP, they pump potassium in and sodium out at the same time.

In short, *ion leaks and ion pumps keep the neuron in a state of readiness between action potentials.*

Approaching Threshold. Stimulate a neuron at its input zone and you disturb the ion balance across the membrane—but not much. Say you trip over a cat and so put a bit of pressure on its skin. Sensory neurons have receptors (input zones) in the tissue beneath the skin's epidermis. The pressure deforms membrane patches on the receptors, and this allows some ions to flow across the membrane. The voltage difference changes slightly, producing a type of graded, local signal.

"Graded" means the signals at an input zone can vary in magnitude. They can be large or small; it depends on the intensity and duration of the stimulus.

"Local" means the signals don't spread far from the point of stimulation. It takes certain ion channels to propagate a signal along the membrane, and input zones simply don't have them.

However, when a stimulus is intense or long-lasting, graded signals can spread out of the input zone and into an adjacent "trigger zone." At this membrane site, the voltage difference across the membrane can change by a certain minimum amount—a *threshold* level. That

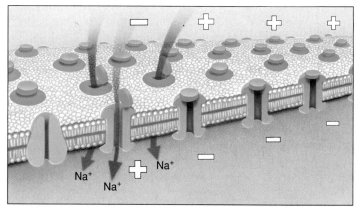

b A strong disturbance initiates an action potential. Sodium gates open, the inflow decreases the negativity inside; this causes more gates to open, and so on, until threshold is reached and the voltage difference across the membrane reverses.

amount is what it takes to trigger an action potential. Threshold can be reached at any membrane patch that has voltage-sensitive, gated channels for sodium ions. As Figure 30.5 suggests, a stimulus causes sodium ions to flow into the neuron. With the influx of these positively charged ions, the cytoplasmic side of the membrane becomes less negative. This causes more gates to open, more sodium to enter, and so on. The ever increasing inward flow of sodium is an example of positive feedback, whereby an event intensifies as a result of its own occurrence:

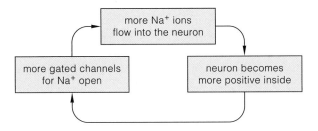

At threshold, the opening of more sodium gates no longer depends on the strength of the stimulus. Because the positive-feedback cycle is now under way, the inward-rushing sodium itself is enough to cause more sodium gates to open.

Figure 30.6 is a recording of the voltage difference across the membrane before, during, and after an action potential. Notice how the membrane potential spikes once threshold is reached. All action potentials in a given neuron spike to the same level above threshold as an *all-or-nothing* event. Once the positive-feedback cycle starts, nothing can stop the full spiking.

If threshold is not reached, however, the membrane disturbance will subside when the stimulus is removed.

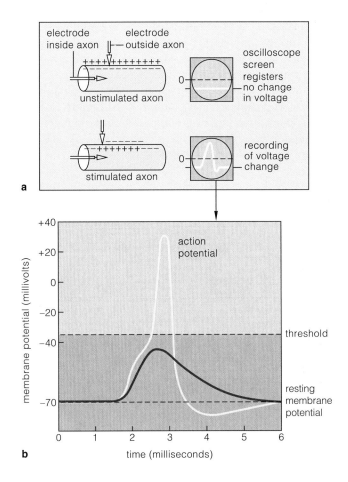

Figure 30.6 Action potentials. (**a**) Electrodes placed inside and outside an axon can be used to detect voltage changes when the axon is stimulated. Changes show up as deflections in a beam of light across the screen of an oscilloscope. (**b**) The yellow line is a typical waveform for an action potential. The red line represents a local signal that did not reach the threshold of an action potential, so spiking did not occur.

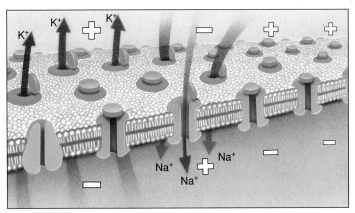

c The reversal causes sodium gates to shut and potassium gates to open (at purple arrows). Potassium follows its gradient (out of the neuron). Voltage is restored. The disturbance produced by the action potential triggers another action potential at the adjacent membrane site, and so on, away from the point of stimulation.

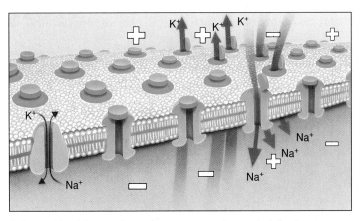

d The inside of the membrane becomes negative again following each action potential, but the sodium and potassium concentration gradients are not yet fully restored. Active transport at sodium-potassium pumps restores the gradients.

Duration of Action Potentials. An action potential lasts only a millisecond. At the membrane site where the charge reversal occurred, gated sodium channels closed and shut off the inflow of sodium ions. About halfway through the reversal, the gates of potassium channels opened, so many more potassium ions flowed out and restored the original voltage difference across the membrane. And sodium-potassium pumps restored the tiny changes in the ion gradients.

The inside of a neuron at rest is negative with respect to the outside.

During an action potential, the inside is more positive than the outside at a disturbed patch of membrane.

Following an action potential, resting conditions are restored at the membrane patch.

Propagation of Action Potentials. The membrane disturbances underlying an action potential are self-propagating, and they don't diminish in magnitude. When they spread to an adjacent membrane patch, the opening of just as many gated channels is repeated. The disturbance causes gated channels to open in the next patch, and so on.

For a brief period after the disturbance, each membrane patch remains insensitive to stimulation because its sodium gates cannot open. That is why action potentials do not spread back to the trigger zone, but rather propagate themselves away from it.

Action potentials are propagated rapidly along the axons of many sensory and motor neurons. As Figure 30.7 shows, the axons are wrapped in a myelin sheath. The sheath consists of the plasma membranes of neuroglial cells called Schwann cells. Each cell is separated from adjacent ones by a small, exposed gap (node) where the axon membrane is loaded with voltage-sensitive gated sodium channels. In a manner of speaking, the action potentials jump from node to node.

Chemical Synapses

When action potentials reach the output zone of a neuron, they usually don't proceed further. But their arrival can make the neuron release one or more **neurotransmitters**. These are signaling molecules that diffuse across junctions called **chemical synapses**. The junctions are no more than gaps between the output zone of one neuron and an input zone of an adjacent cell (Figure 30.8). Some occur between two neurons, others between a neuron and a muscle cell or gland cell.

At any chemical synapse, one of the two cells stores a neurotransmitter in vesicles. Think of it as the *pre*-synaptic cell. When the neurotransmitter binds to receptors on a *post*synaptic cell, channels open up across the cell's plasma membrane. Depending on the type of channels being opened up, a neurotransmitter may *excite* or *inhibit* the membrane (help drive it toward or away from threshold). After neurotransmitter molecules exert their effect, the presynaptic cell takes them up or enzymes inactivate them.

Acetylcholine (ACh) is an example of a neurotransmitter. It has excitatory and inhibitory effects on various cells in muscles, glands, the brain, and the spinal cord. Think of a junction between a motor neuron and a muscle cell, as shown in Figure 24.14. The motor neuron releases ACh when an action potential spreads down to each axon ending. The ACh diffuses across the cleft and binds to receptors on the muscle cell membrane. In this case, its effect is excitatory. It triggers action potentials, which in turn initiate muscle contraction.

Membrane responses to neurotransmitters are often enhanced or reduced by neuromodulators, another kind of signaling molecule. These include endorphins, peptide molecules that may affect memory, learning, body temperature, emotional states, and behavior.

Synaptic Integration

At any time, excitatory and inhibitory signals are arriving at a postsynaptic cell. Some types drive the membrane closer to threshold; others maintain the resting level or drive it away from threshold. All these signals compete for control of the membrane.

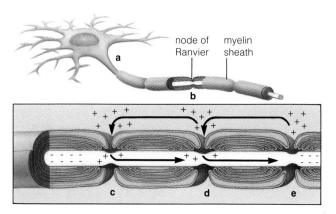

Figure 30.7 Propagation of an action potential along a motor neuron having a myelin sheath. (**a**) The sheath is a series of Schwann cells, each wrapped like a jellyroll around the axon. Each "jellyroll" blocks ion movements across the membrane. But ions can cross it at nodes *between* Schwann cells (**b**). The unsheathed nodes have dense arrays of gated sodium channels. (**c**,**d**) A disturbance caused by an action potential spreads down the axon. When it reaches a node, sodium gates open, sodium ions rush inward, and another action potential results. (**e**) This new disturbance spreads rapidly to the next node and triggers another action potential, and so on down the line.

node of Ranvier   myelin sheath

Deadly Imbalances at Synapses

Clostridium botulinum, an anaerobic bacterium, lives in soil. It is an endospore former. When its endospores wind up in improperly stored, preserved, or canned food, they germinate and produce a dangerous toxin—which can be ingested and absorbed. The toxin ends up binding to neurons that synapse with muscle cells and blocks their release of acetylcholine (ACh). Once this happens, muscles cannot contract. They undergo progressive, flaccid paralysis. Death may follow within ten days, usually from respiratory and cardiac failure. Recovery is possible when antitoxins are quickly administered. Extended use of an artificial respirator may also be required.

A related bacterium, *C. tetani*, lives in the gut of grazing animals and causes the disease *tetanus*. Its endospores survive in soil. They can enter the human body through a deep puncture or cut, then germinate in dead (anaerobic) tissues. A toxin produced by *C. tetani* acts on interneurons that help control motor neurons in the spinal cord.

a

It stops them from releasing inhibitory neurotransmitters (GABA and glycine) and so frees the motor neurons from normal inhibitory control. After four to ten days of overstimulation, muscles cannot be released from contraction and undergo prolonged, spastic paralysis. Fists and jaws may stay clenched (the disease is also called lockjaw). The back may arch permanently. Muscle spasms may cause bones to break. When respiratory and cardiac muscles also become paralyzed, death nearly always follows.

Vaccines now prevent nearly all cases of tetanus in the United States. They weren't available for soldiers of early wars, when battlefields were littered with bodies of (and manure from) cavalry horses. Figure *a* is a painting of a young victim of a contaminated battle wound, as he lay dying in a military hospital.

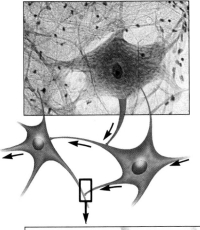

Figure 30.8 Chemical synapse (junction) between a neuron and another cell. Arrival of an action potential at the neuron's output zone (axon endings) triggers the release of neurotransmitter molecules. These molecules diffuse across the junction and bind to receptors on the adjoining cell.

In a process called **synaptic integration**, competing signals at an input zone of a neuron are summed. This process is the means by which signals arriving at any neuron in the body are reinforced or dampened, sent on or suppressed. When something interferes with it, the outcome can be lethal, as the *Focus* essay suggests.

Synaptic integration is the moment-by-moment combining of excitatory and inhibitory signals acting on a neuron.

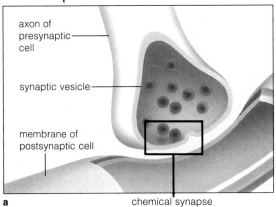

axon of presynaptic cell

synaptic vesicle

membrane of postsynaptic cell

a   chemical synapse

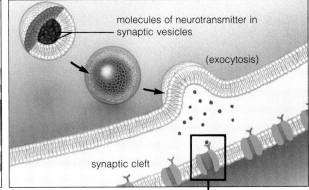

molecules of neurotransmitter in synaptic vesicles

(exocytosis)

synaptic cleft

b   receptor on the postsynaptic membrane

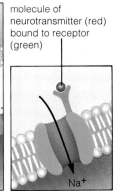

molecule of neurotransmitter (red) bound to receptor (green)

Na^+

c

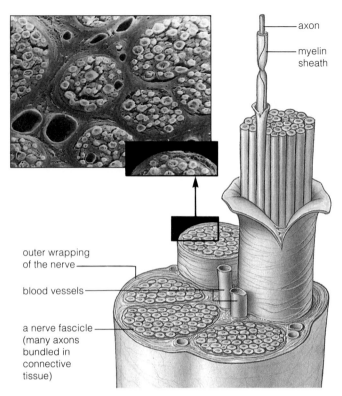

- axon
- myelin sheath

outer wrapping of the nerve

blood vessels

a nerve fascicle (many axons bundled in connective tissue)

Figure 30.9 Structure of a nerve. Axons in the nerve are bundled together inside wrappings of connective tissue.

Nerves

We are about to consider how neurons are organized in the body. As you will see, many signals are confined to "local circuits" of neurons, such as those making up a brain center. Others travel along **nerves**. These lead into and out from the spinal cord or brain and connect it with other body regions. In each nerve, bundled-together axons of sensory neurons, motor neurons, or both serve as the interstate highways for information flow (Figure 30.9). They are distinct from nerve tracts (bundled-together axons of interneurons inside the spinal cord and brain).

Synaptic integration occurs every time neurons chatter in a local circuit. It occurs every time neurons send information long-distance, on a nerve.

Figure 30.10 Reflex response to muscle stretching. Inside a skeletal muscle, stretch-sensitive receptors of a sensory neuron are located in muscle spindles. Stretching disturbs them and generates action potentials that travel down to the axon endings. These synapse with a motor neuron that has axons leading back to the stretched muscle. The motor neuron can initiate muscle contraction.

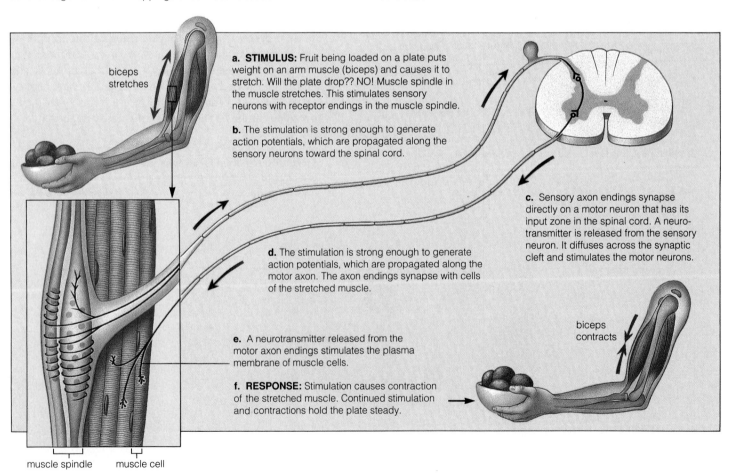

biceps stretches

a. STIMULUS: Fruit being loaded on a plate puts weight on an arm muscle (biceps) and causes it to stretch. Will the plate drop?? NO! Muscle spindle in the muscle stretches. This stimulates sensory neurons with receptor endings in the muscle spindle.

b. The stimulation is strong enough to generate action potentials, which are propagated along the sensory neurons toward the spinal cord.

c. Sensory axon endings synapse directly on a motor neuron that has its input zone in the spinal cord. A neurotransmitter is released from the sensory neuron. It diffuses across the synaptic cleft and stimulates the motor neurons.

d. The stimulation is strong enough to generate action potentials, which are propagated along the motor axon. The axon endings synapse with cells of the stretched muscle.

e. A neurotransmitter released from the motor axon endings stimulates the plasma membrane of muscle cells.

f. RESPONSE: Stimulation causes contraction of the stretched muscle. Continued stimulation and contractions hold the plate steady.

biceps contracts

muscle spindle   muscle cell

VERTEBRATE NERVOUS SYSTEMS

Suppose a sensory neuron that is tripped into action synapses on a single motor neuron, which stimulates a muscle to contract. No other neurons alter the flow of signals from reception of the stimulus to the response. In this case, sensory stimulation has directly caused a simple, stereotyped movement—a reflex. In nervous systems, this is the simplest **reflex pathway** (Figure 30.10).

When nervous systems first evolved, signals flowed through reflex pathways. The oldest parts of the vertebrate brain still deal with reflex coordination of breathing and other vital functions. Now, however, interneurons synapse on the sensory and motor neurons of ancient pathways. They form intricate synapses in the most recent additions to the brain. There, interneurons store and compare information about experiences. They come up with possibilities for brand-new responses. They give *you* the capacity to reason, remember, and learn.

Billions of neurons interact in the human nervous system, which is shown in Figure 30.11*a*. Think of the neurons as belonging to central or peripheral regions. Interneurons are confined to the **central nervous system**, the spinal cord and brain. Sensory and motor neurons belong to the **peripheral nervous system**, the communication lines that thread through the rest of the body and carry signals into and out of the central nervous system (Figure 30.11*b*).

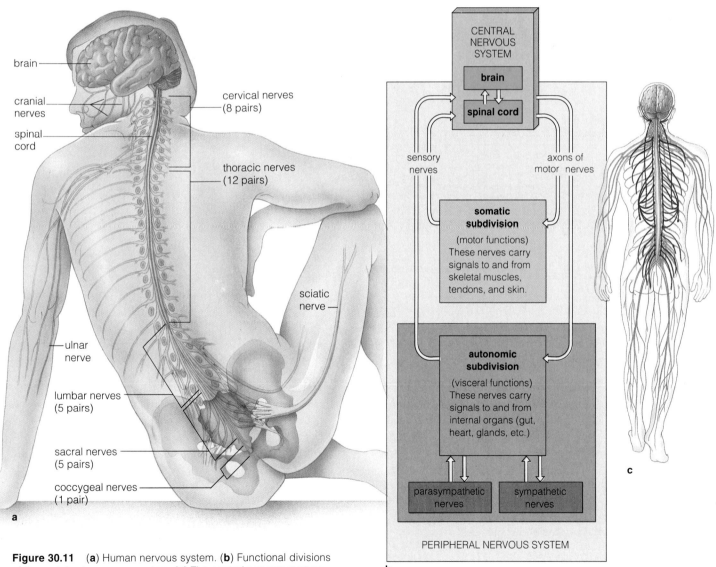

Figure 30.11 (**a**) Human nervous system. (**b**) Functional divisions of the vertebrate nervous system. (**c**) The central nervous system is coded blue. For the peripheral nervous system, somatic nerves are coded green, and autonomic nerves are coded red.

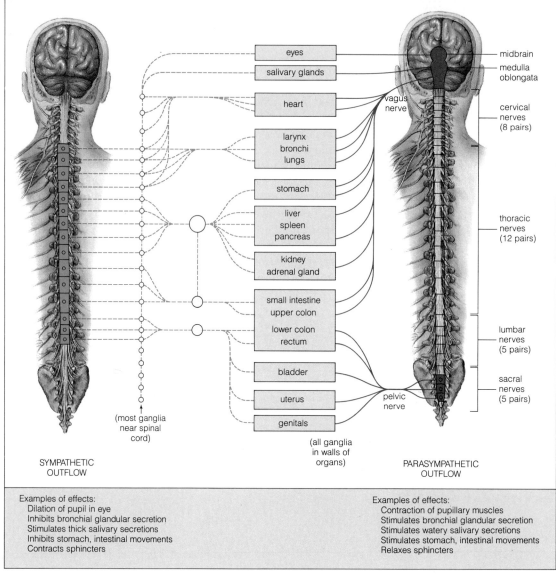

Figure 30.12 Autonomic nervous system. Its pairs of sympathetic and parasympathetic nerves extend from the central nervous system to internal organs. Ganglia (singular, ganglion) are clusters of cell bodies of the neurons that are bundled together in nerves.

eyes
salivary glands
heart
larynx
bronchi
lungs
stomach
liver
spleen
pancreas
kidney
adrenal gland
small intestine
upper colon
lower colon
rectum
bladder
uterus
genitals

midbrain
medulla oblongata
vagus nerve
cervical nerves (8 pairs)
thoracic nerves (12 pairs)
lumbar nerves (5 pairs)
sacral nerves (5 pairs)
pelvic nerve

(most ganglia near spinal cord)

(all ganglia in walls of organs)

SYMPATHETIC OUTFLOW

PARASYMPATHETIC OUTFLOW

Examples of effects:
Dilation of pupil in eye
Inhibits bronchial glandular secretion
Stimulates thick salivary secretions
Inhibits stomach, intestinal movements
Contracts sphincters

Examples of effects:
Contraction of pupillary muscles
Stimulates bronchial glandular secretion
Stimulates watery salivary secretions
Stimulates stomach, intestinal movements
Relaxes sphincters

Peripheral Nervous System

The human peripheral nervous system includes thirty-one pairs of spinal nerves, which connect with the spinal cord. It also includes twelve pairs of cranial nerves, which connect directly with the brain.

Somatic and Autonomic Subdivisions. Cranial and spinal nerves are classified according to function. The ones that help you move your head, trunk, and limbs are *somatic* nerves (Figure 30.11). Some relay signals from receptors in skeletal muscles, tendons, and skin to the brain. Others relay signals to those muscles. By contrast, spinal and cranial nerves dealing with smooth muscle, cardiac muscle, and gland cells are the *autonomic* nerves. They deal with the gut, heart, and many other internal organs and structures.

Sympathetic and Parasympathetic Nerves. There are two categories of autonomic nerves. We call them parasympathetic and sympathetic. Both types carry excitatory and inhibitory signals—often to the same organ, where they compete for control. In such cases, synaptic integration leads to minor adjustments in the organ's level of activity.

Parasympathetic nerves dominate when the body is not receiving much outside stimulation. They tend to slow down the body overall and divert energy to basic "housekeeping" tasks, such as digestion (Figure 30.12).

Sympathetic nerves dominate during heightened awareness, excitement, or danger. They tend to shelve housekeeping tasks. At the same time, they prepare the animal to fight or escape (when threatened) or to frolic (as in play and sexual behavior).

As you read this, sympathetic nerves are asking your heart to beat a bit faster, and parasympathetic nerves are asking it to beat a bit slower. The heart rate is influenced by the integration of these opposing signals. If something scares or excites you, parasympathetic input to your heart will drop. Sympathetic nerves will cause the release of a hormone (epinephrine). This hormone makes your heart beat faster. It makes you

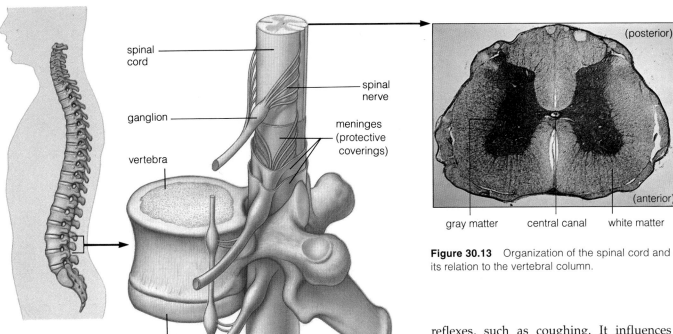

spinal cord

ganglion

vertebra

spinal nerve

meninges (protective coverings)

(location of intervertebral disk)

(posterior)

(anterior)

gray matter   central canal   white matter

Figure 30.13 Organization of the spinal cord and its relation to the vertebral column.

breathe faster and sweat. In this state of intense arousal, you are primed to fight (or play) hard or to get away fast. Hence the name, "fight-flight response."

Central Nervous System

We turn now to the brain and spinal cord. Both are protected by bones and membranes. And both are bathed in cerebrospinal fluid. This clear extracellular fluid cushions them from sudden, jarring movements.

The Spinal Cord. By definition, the **spinal cord** is a vital expressway for signals between the peripheral nervous system and the brain. It also is a center for controlling some reflexes. The cord itself threads through a canal formed by bones of the vertebral column. Figure 30.13 shows its "white matter," axons bundled in glistening sheaths. The "gray matter" consists of dendrites, cell bodies of neurons, and neuroglial cells. The cord's gray matter deals mainly with reflexes for limb movements (such as walking) and organ activity (such as bladder emptying).

Divisions of the Brain. The spinal cord merges with the **brain**, the body's master control panel. The brain receives, integrates, stores, and retrieves information. It also coordinates responses by intricately adjusting activities throughout the body. Its three divisions are called the hindbrain, midbrain, and forebrain.

The *hind*brain consists of the medulla oblongata, cerebellum, and pons. The medulla oblongata has reflex centers for respiration, blood circulation, and other vital tasks. It also coordinates motor responses and complex

reflexes, such as coughing. It influences other brain regions that help you sleep or wake up.

The cerebellum helps maintain the body's posture and coordinate its movements. It integrates signals from eyes, muscle spindles, skin, and elsewhere. The pons (meaning bridge) has bands of axons extending into each side of the cerebellum. It is a major traffic center for information passing between the cerebellum and cerebral cortex.

The *mid*brain coordinates reflex responses to sights and sounds. In most vertebrates—but not in mammals—the midbrain contains optic lobes, which deal with sensory input from the eyes. Its roof of gray matter (tectum) is the major center for coordinating nearly all sensory input and initiating motor responses. (A frog surgically deprived of its highest brain center but with its tectum intact still does just about everything frogs do.) In mammals, the tectum has been reduced to a reflex center; it swiftly relays sensory signals to higher integrating centers.

Collectively, the midbrain, pons, and medulla oblongata are the brain's "stem." Inside the brain stem's core is a major network of interneurons, the reticular formation. This network extends through the length of the brain stem, connects with the forebrain, and helps govern activities of the nervous system as a whole.

At one time, the *fore*brain of all vertebrates was dominated by an olfactory lobe, concerned mostly with odors emanating from prey and predators (Figure 19.34). A brain center called the cerebrum integrated olfactory input and selected motor responses to it. The thalamus was the coordinating center for sensory input and the relay station for signals flowing to and from the cerebrum. As it does today, the hypothalamus monitored internal organs and influenced behavior related to organ activities, such as thirst, hunger, and sex. During the evolution of mammals especially, the forebrain evolved in spectacularly intricate ways.

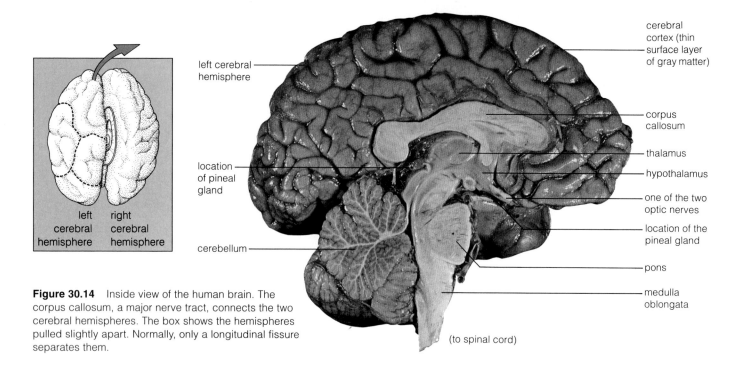

Figure 30.14 Inside view of the human brain. The corpus callosum, a major nerve tract, connects the two cerebral hemispheres. The box shows the hemispheres pulled slightly apart. Normally, only a longitudinal fissure separates them.

Table 30.1	Regional Divisions of the Human Brain	
Divisions	Main Components	Key Functions
Forebrain	Cerebrum (including cerebral cortex) Olfactory lobes Limbic system Thalamus Hypothalamus Pituitary gland Pineal gland	Most complex centers for receiving and processing sensory information and motor responses; memory; abstract thought; neural-endocrine control of growth, metabolism, behavior
Midbrain	Tectum	Coordinating, relaying sensory input to forebrain
Hindbrain	Pons Cerebellum Medulla oblongata	Reflex centers for basic tasks (e.g., breathing); coordinating motor activity for movement, positioning, orientation

Case Study—The Human Brain

Figure 30.14 shows what the inside of your brain looks like. If your brain is average in size, it weighs about 1,300 grams (3 pounds) and contains about a hundred billion neurons. Table 30.1 lists the components of its three regional divisions.

The Cerebrum. The neurons that contribute most to your humanness reside in the cerebrum. The cerebrum, with its shell of hard skull bones, resembles the much-folded nut in a hard walnut shell. A deep fissure divides it into two parts, the left and right cerebral

hemispheres. Major nerve tracts (white matter) functionally link the halves with each other and with the rest of the body.

Each half of the cerebrum receives, processes, and coordinates responses to information mostly from the opposite side of the body. (For example, signals from your *right* foot travel to your *left* cerebral hemisphere.) Most brain regions dealing with speech reside in the left half of the cerebrum, in which mathematics and analytical skills also are localized. Most regions dealing with nonverbal (abstract) skills, such as spatial abilities and music, reside in the right half.

The **cerebral cortex** is a thin layer of gray matter at the surface of the cerebrum. Figure 30.15 shows some of its centers for receiving, encoding, and processing information. Among these are motor centers that coordinate instructions for motor responses. Thumb and tongue muscles get much of the brain's attention, which gives you an idea of how much control is required for hand movements and verbal expression. Compared to the odor-interpreting centers of early vertebrates, our visual and auditory centers have become more important. The record of human evolution (Unit III) suggests why the shift occurred.

Memory. Even before birth, the brain starts to store and retrieve information about previous experiences. It shows **memory**, which underlies the capacity to learn. The treasure houses of experience are constructed in many regions, including those shown in Figure 30.16.

According to a commonly accepted theory, the brain stores information in stages. Short-term storage is a fleeting stage of neural excitation, lasting a few seconds to a few hours. It seems to be limited to a few bits of

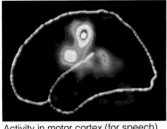

Activity in motor cortex (for speech)

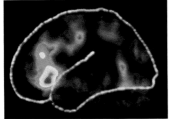

Activity in auditory cortex (hearing)

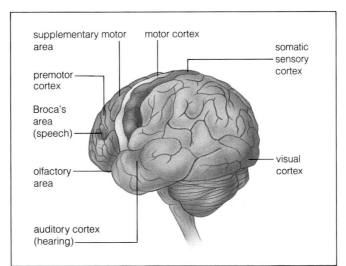

supplementary motor area

premotor cortex

Broca's area (speech)

olfactory area

auditory cortex (hearing)

motor cortex

somatic sensory cortex

visual cortex

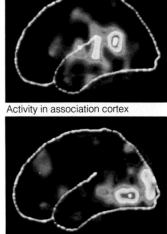

Activity in association cortex

Activity in visual cortex (vision)

Figure 30.15 Primary receiving and association areas for the human cerebral cortex. Signals from receptors near the body's surface enter primary cortical areas. Association centers coordinate and process sensory input from receptors. The PET scans (page 19) show which brain regions are active during the performance of specific tasks (speaking, hearing, generating, and observing words).

information—numbers, words of a sentence, and so on. By contrast, long-term storage apparently occurs through chemical and structural changes in the brain. Seemingly limitless amounts of information are stored on a more or less permanent basis.

The theory helps explain *retrograde amnesia*. This ailment occurs after a person receives a severe blow to the head and loses consciousness. There is a loss of memory of what happened during the time (variable in length) that preceded the blow. Yet memories of events *before* then often remain intact.

Long-term memory storage probably resides in neurons that have changed, chemically and physically, as a result of synaptic connections. We have micrographs of synapses that fell into disuse, so the functional link between neurons was weakened or broken. Such withered synapses have been observed in the visual cortex of mice raised without visual stimulation. We also have evidence that intensely stimulated synapses form stronger connections, grow in size, or sprout buds or spines to form more connections!

Emotional States. You are an emotional animal. For this you can thank your cerebral cortex and **limbic system**, which consists of several brain regions (Figure 30.16). The limbic system is distantly related to the olfactory lobes that were so important in vertebrate evolution. It still retains connections with the sense of smell. That's why you might recall a delicious odor of a grandmother's kitchen, or the scent of a cologne when you have a pleasant memory of the person who wore it.

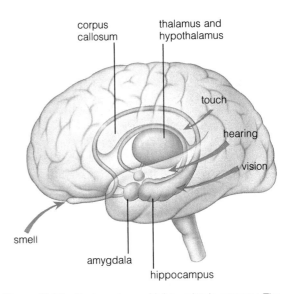

corpus callosum

thalamus and hypothalamus

touch

hearing

vision

smell

amygdala

hippocampus

Figure 30.16 Brain regions with key roles in memory. The cerebral cortex processes sensory input, then sends it to parts of the limbic system and forebrain. The limbic system, our "emotional brain," includes regions called the thalamus, hypothalamus, amygdala, and hippocampus.

The hypothalamus still monitors internal organs, as it did in the past. It also is gatekeeper of the limbic system. Many connections from the cerebral cortex and from ancient brain centers pass through it. Through these connections, the hypothalamus correlates organ activities with self-gratifying behavior, such as eating and sex. Also through these connections, the reasoning possible in the cerebral cortex can dampen rage, hatred, and other "gut reactions."

Think of the signals passing through the hypothalamus when your heart and stomach are on fire—either with passion or indigestion. Also think of how many parts of your mind and body are disrupted by psychoactive drugs, as described in the *Focus* essay on the next page.

Drugs, the Brain, and Behavior

Broadly speaking, a drug is any substance introduced into the body to provoke a specific physiological response. Some drugs help a person cope with illness or stress. Others act on the brain, artificially fanning the pleasure we associate with sex and other self-gratifying behaviors.

Many drugs are habit-forming. Even if the body functions well without them, they continue to be used for the real or imagined relief they afford. Often the body develops tolerance of such drugs (it takes larger or more frequent doses to produce the same effect). Habituation and tolerance are signs of **drug addiction**—chemical dependence on a drug. *The drug has taken on an "essential" biochemical role in the body.* Abruptly deprive addicts of the drug, and they suffer physical pain and mental anguish. The entire body suffers major biochemical upheavals. Stimulants, depressants, hypnotics, narcotic analgesics, hallucinogens, and psychedelics have such effects.

Stimulants. Caffeine, nicotine, amphetamines, cocaine, and other stimulants increase alertness and body activity. Then they cause depression.

Coffee, tea, chocolate, and many soft drinks contain caffeine, a widely used stimulant. Low doses stimulate the cerebral cortex first, leading to increased alertness and restlessness. Higher doses acting at the medulla oblongata disrupt motor coordination and mental coherence. Nicotine, a component of tobacco, has powerful effects throughout the nervous system. It mimics acetylcholine and can directly stimulate a variety of sensory receptors. Its short-term effects include water retention, irritability, increased heart rate and blood pressure, and gastric upsets.

Amphetamines (including "speed") resemble two neurotransmitters, dopamine and norepinephrine. In time, the brain produces less and less of these and depends more on artificial stimulation. Cocaine produces a rush of pleasure by *blocking* reabsorption of dopamine, norepinephrine, and other neurotransmitters. These accumulate in synaptic clefts and incessantly stimulate receptor cells for an extended period. Heart rate and blood pressure rise; sexual appetite increases. Then neurotransmitters diffuse away. Replacements can't be made fast enough to counter the loss. The sense of pleasure evaporates as the now-hypersensitive receptor cells demand stimulation. After prolonged, heavy use of cocaine, "pleasure" cannot be experienced. Addicts become anxious and depressed. They lose weight and cannot sleep properly. Their immune system weakens, and heart abnormalities set in.

Granular cocaine, which is inhaled (snorted), has been around for some time. Crack cocaine is burned and the smoke inhaled. As suggested at the start of this chapter, crack is extremely addictive. Its highs are higher, but the crashes are more devastating.

Depressants, Hypnotics.   These drugs lower activity in nerves and parts of the brain. Some act at synapses in the reticular formation system and in the thalamus. Depending on the dosage, responses range from emotional relief, drowsiness, sleep, anesthesia, and coma to death. Low doses act more on inhibitory synapses, so the person feels excited or euphoric at first. Increased doses suppress excitatory synapses as well, leading to depression. Depressants and hypnotics have additive effects. One amplifies another, as when alcohol plus barbiturates increases the depression.

Alcohol (ethyl alcohol) acts directly on plasma membranes to alter cell function. Some people mistakenly think of it as a harmless stimulant (it produces an initial "high"). But alcohol is one of the most powerful psychoactive drugs and a major cause of death. Small doses even over the short term produce disorientation, uncoordinated motor functions, and diminished judgment. Long-term addiction destroys nerve cells and causes permanent brain damage. It can permanently damage the liver (cirrhosis).

Analgesics.   When stress causes physical or emotional pain, the brain produces natural analgesics (pain relievers). Endorphins and enkephalins are examples. They act on many parts of the nervous system, including brain centers that deal with emotions and pain. Narcotic analgesics, including codeine and heroin, sedate the body and relieve pain. They are extremely addictive. Deprivation following massive doses of heroin leads to fever, chills, hyperactivity and anxiety, violent vomiting, cramping, and diarrhea.

Psychedelics, Hallucinogens.   These drugs alter sensory perception by interfering with acetylcholine, norepinephrine, or serotonin activity. LSD (lysergic acid diethylamide) affects the action of serotonin, a neurotransmitter with roles in inducing sleep, in temperature regulation—and in sensory perception. Even in small doses, LSD warps perceptions. For example, some users "perceived" that they could fly and "flew" off buildings.

Marijuana, another hallucinogen, is made from crushed leaves, flowers, and stems of the plant *Cannabis*. In low doses marijuana is like a depressant. It slows down but does not impair motor activity; it relaxes the body and elicits mild euphoria. It also can produce disorientation, increased anxiety bordering on panic, delusions (including paranoia), and hallucinations.

Like alcohol, marijuana affects the performance of complex tasks, such as driving a car. In one study, pilots showed a marked deterioration in instrument-flying ability for more than two hours after smoking marijuana. Over time, marijuana smoking can suppress the immune system and impair mental functions.

SENSORY SYSTEMS

We turn now to sensory systems, the front doors of the nervous system. **Sensory systems** receive and notify the brain of specific changes that are going on outside and inside the body. Each consists of (1) sensory receptors, (2) nerve pathways leading from the receptors to the brain, and (3) brain regions where sensory information is processed and translated into sensation.

Sensation is the conscious awareness of a stimulus. It is not the same thing as perception (understanding what the sensation means). The reactions of a live lobster placed in a cookpot tell us that a stimulus (boiling water) has indeed registered. But the lobster reacts the same way to any strong stimulus, as when someone grips it firmly. Most likely, the lobster doesn't "understand" its predicament or "feel" pain as humans do.

A sensory system has sensory receptors for specific stimuli, nerve pathways that conduct information from receptors to the brain, and brain regions where the information is processed.

Types of Sensory Receptors

There are five major types of sensory receptors, based on the type of stimulus energy they detect:

1. **Chemoreceptors** detect chemical energy of specific substances dissolved in the fluid surrounding them.

2. **Mechanoreceptors** detect forms of mechanical energy (changes in pressure, position, or acceleration).

3. **Photoreceptors** detect visible and ultraviolet light.

4. **Thermoreceptors** detect infrared energy (heat).

5. **Nociceptors** (pain receptors) detect tissue damage.

Depending on the kinds and numbers of their sensory receptors, animals sample the environment in different ways and differ in their awareness of it. Unlike bees, you have no receptors for ultraviolet light and don't see many flowers the way they do (page 359). Unlike many bats, you and bees have no receptors for ultrasound. Unlike pythons, you, bees, and bats have no receptors to detect warm-blooded prey in the dark (Figure 30.17).

a

b

Figure 30.17 Examples of diversity in sensory reception. (**a**) A python has thermoreceptors in pits above and below its mouth. The receptors detect body heat (infrared energy) of prey—small, night-foraging mammals that are warmer than the night air. The snake brain assesses signals about prey location with near-perfect accuracy. (**b**) Some bats listen to echoes of their own high-frequency sounds. The bat brain deciphers echoes bouncing back from prey and other objects in the environment as a "sound map." The map helps the bat capture mosquitoes and moths in midair, without even seeing them.

Table 30.2 Receptors Associated With the Major Senses

Category of Receptor	Examples	Stimulus
Chemoreceptors:		
Internal chemical senses	Carotid bodies in blood vessel wall	Substances (CO_2, etc.) dissolved in extracellular fluid
Taste	Taste receptors of tongue	Substances dissolved in saliva, etc.
Smell	Olfactory receptors of nose	Odors in air, water
Mechanoreceptors:		
Touch, pressure	Pacinian corpuscles in skin	Mechanical pressure against body surface
Stretch	Muscle spindle in skeletal muscle	Stretching
Auditory	Hair cells within ear	Vibrations (sound or ultrasound waves)
Balance	Hair cells within ear	Fluid movement
Photoreceptors:		
Visual	Rods, cones of eye	Wavelengths of light
Thermoreceptors	Cold or warm receptors in skin; central thermoreceptors in hypothalamus, etc.	Presence of or change in radiant energy (heat)
Nociceptors	Free nerve endings in skin	Any stimulus that causes tissue damage and leads to sensation of pain

Sensory Pathways

Sensory neurons carry signals from receptors to the cerebral cortex. Different parts of the cortex are laid out like maps corresponding to the body surface. Larger maps represent functionally important regions and receive signals from more receptors. That's why you have a large "visual cortex," for example.

Sensory receptors convert stimulus energy to action potentials. But action potentials aren't like a wailing ambulance siren; they don't vary in amplitude. What do they "tell" the brain about the stimulus? In effect:

The brain senses a stimulus based on which nerve pathways carry the incoming signals, the frequency of signals traveling along each axon of that pathway, and the number of axons that have been recruited.

First, the genetically prescribed network of neurons in each animal's brain can interpret action potentials only in certain ways. That's why you "see stars" when your eye is poked, even in the dark. Photoreceptors in the eye were mechanically disturbed enough to trigger signals that traveled along an optic nerve to your brain. Your brain always interprets signals arriving from an optic nerve as "light."

Second, when a stimulus is stronger, receptors fire action potentials faster. The same receptor detects the sounds of a throaty whisper and a wild screech. The brain senses the difference through frequency variations in the signals that the receptor sends to it.

Third, strong stimulation recruits more receptors. Tap a spot of skin on your arm and you activate some receptors. Press hard on the same area and you activate many more in a larger area. The increased disturbance sets off action potentials in many sensory axons at the same time. The brain interprets the combined activity as an increase in stimulus intensity.

Let's focus on a few of the sensory receptors listed in Table 30.2. Some have more than one location; they contribute to somatic sensations. Others are restricted to certain locations, such as the eyes. They contribute to the special senses.

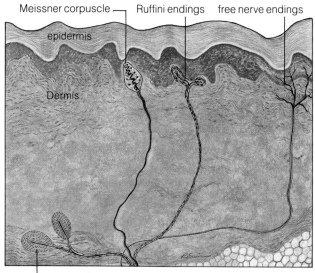

Figure 30.18 Tactile receptors in human skin. Free nerve endings detect changes in temperature, light pressure, and pain. Pacinian corpuscles detect vibration. Meissner corpuscles signal the onset and end of sustained pressure. Ruffini endings react continually to ongoing stimulation.

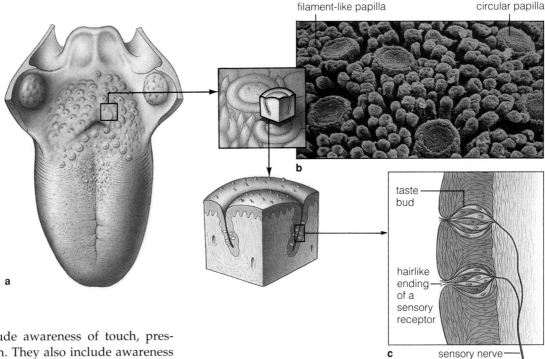

Figure 30.19 Location of taste receptors inside taste buds in the human tongue. The peglike projections don't contribute to taste (they help move food).

filament-like papilla

circular papilla

a

b

taste bud

hairlike ending of a sensory receptor

c

sensory nerve

Somatic Sensations

Somatic sensations include awareness of touch, pressure, heat, cold, and pain. They also include awareness of movements and positions in space. Such sensations start with receptor endings in surface tissues, in skeletal muscles, and in walls of internal organs.

Touch and Pressure. Mechanoreceptors are abundant in many parts near the body surface, including the tongue tip and fingertips. Figure 30.18 shows some skin receptors, which easily detect deformation caused by pressure. Types that signal the onset and removal of a stimulus make you aware of touch, even light tickles. Other types signal throughout the stimulation; they make you aware of pressure.

Temperature Changes. When temperatures near the body surface remain much the same, thermoreceptors in skin fire a steady barrage of signals to the brain. With increases in temperature, the firing frequency increases. Free nerve endings serve as heat receptors. Cold receptors haven't been identified yet.

Pain. Perception of injury to some body region, or pain, begins with signals from nociceptors. Among these are free nerve endings, which detect any stimulus strong enough to damage the surrounding tissue. They respond to strong mechanical stimulation, intense heat or cold, and chemical irritation.

Responses to pain often depend on the brain's ability to identify the affected tissue. Get hit in the face with a snowball and you "feel" the contact on facial skin. For reasons not fully understood, the brain sometimes associates pain with a tissue located some distance away from an *internal* injury. This is "referred pain." For example, a heart attack can be felt as pain in skin above the heart and along the left shoulder and arm.

Muscle Sense. Sensing limb motions and the body's position in space requires mechanoreceptors in skeletal muscle, joints, tendons, ligaments, and skin. Among these are stretch receptors in skeletal muscle (Figure 30.10). Their responses to stimulation depends on how much and how fast the muscle stretches.

The Special Senses

Taste and Smell. Different animals taste substances with their mouth, antennae, legs, tentacles, or fins. It all depends on where chemoreceptors called *taste* receptors are distributed. These detect differences in substances that become dissolved in fluid next to some body surface. The ones on animal tongues often occur in sensory organs called taste buds (Figure 30.19).

Animals smell substances with *olfactory* receptors. These provide clues or advance warning of predators, food, or anything else that gives off chemicals able to diffuse through water or air. Humans have about 10 million olfactory receptors in the nose. A bloodhound nose has more than 200 million.

Potential mates and rivals give off pheromones. These signaling molecules help members of the same species communicate with each other, as described in Chapter 38. Here, simply consider a male silk moth's olfactory receptors. They respond to bombykol, a sex-attracting pheromone. Contact with merely one bombykol molecule per second sends action potentials to the moth brain. They help a male locate a female in the dark, even more than a kilometer upwind from him.

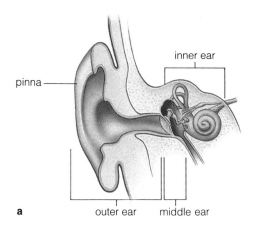

pinna
inner ear
outer ear   middle ear

a

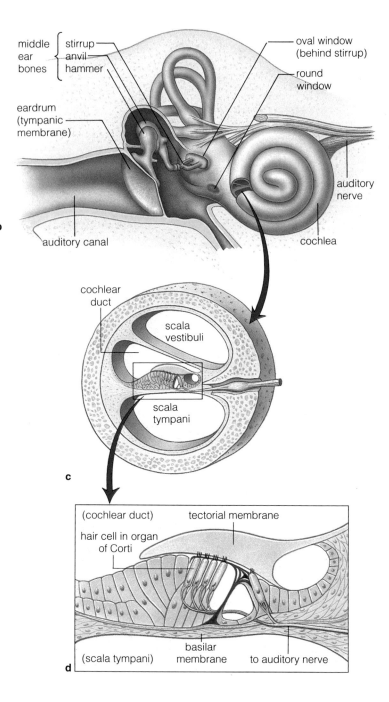

middle ear bones { stirrup, anvil, hammer }
oval window (behind stirrup)
round window
eardrum (tympanic membrane)
auditory nerve
auditory canal
cochlea

b

cochlear duct
scala vestibuli
scala tympani

c

(cochlear duct)   tectorial membrane
hair cell in organ of Corti
(scala tympani)   basilar membrane   to auditory nerve

d

Figure 30.20 (**a**) The human ear. (**b**) The *outer ear's* external flaps collect sound waves, which move through a canal to an eardrum (tympanic membrane). The waves cause the eardrum to vibrate. Small bones of the *middle ear* pick up the vibrations.

Middle earbones amplify the stimulus: they transmit the force of pressure waves to a smaller surface (oval window). This is an elastic membrane over the entrance to the coiled *inner ear*. (**c**) It bows in and out, producing fluid pressure waves in two ducts (scala vestibuli and scala tympani). The waves reach a membrane (round window). When this membrane bulges under pressure, fluid moves back and forth in the inner ear.

(**d**) Pressure waves are sorted out at a third duct (cochlear duct) in the coiled inner ear. A duct membrane (basilar membrane) starts out narrow and stiff. It is broader and flexible deep in the coil. At different points, the membrane vibrates more strongly to sounds of different frequencies. When mechanoreceptors (hair cells) on the membrane are stimulated, action potentials are sent along an auditory nerve to the brain.

Hearing. Hearing requires mechanoreceptors that detect vibrations. A vibration is a wavelike form of mechanical energy. For example, clapping produces waves of compressed air. Each time your hands clap together, molecules are forced outward and a low-pressure state is created in the region they vacated. The pressure variations can be depicted as a wave form, and the amplitude of its peaks corresponds to loudness:

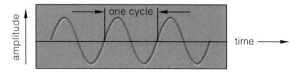

one cycle
amplitude
time →

The frequency of a sound is the number of wave cycles per second. Each cycle extends from the start of one wave peak to the start of the next. The more cycles per second, the higher the frequency, and the higher the perceived pitch of the sound.

Like all land-dwelling mammals, humans have a pair of ears. Each ear has regions that receive, amplify, and sort out sound waves. Figure 30.20 shows the location of their hair cells, mechanoreceptors that respond to disturbances created by sounds. Prolonged exposure to some intense sounds permanently damages the inner ears (Figure 30.21). The hair cells are not adapted to amplified music, jet planes, and other recent developments.

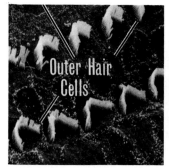

Outer Hair Cells
'Scars'

Figure 30.21 Intense sound and the inner ear. (**a**) Normal organ of Corti from a guinea pig, showing two rows of outer hair cells. (**b**) Same organ after 24 hours of noise levels comparable to extremely loud music (2,000 cycles/second at 120 decibels).

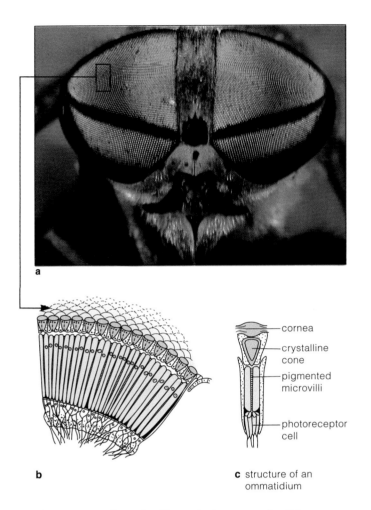

a

b

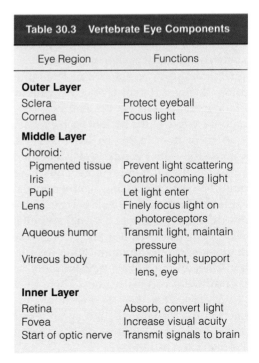

cornea

crystalline cone

pigmented microvilli

photoreceptor cell

c structure of an ommatidium

Figure 30.22 Example of invertebrate eyes: a deerfly's compound eyes. Crystal-like cones of light-sensitive units (ommatidia) focus light on pigmented photoreceptor cells.

Vision. All organisms, whether they see or not, are sensitive to light. Even a single-celled amoeba stops abruptly when you shine a light on it. Vision, however, requires a *system* of photoreceptors that encode information about a visual stimulus, including its position, shape, brightness, distance, and motion. Vision also requires a brain that can interpret patterns of signals arriving from different parts of the system.

Many invertebrates have eyespots, clusters of photosensitive cells in a cuplike depression at the body surface. Others have eyes—photoreceptor organs that contribute to image formation. Insects and crustaceans have compound eyes, with many closely packed photosensitive units (Figure 30.22). Octopus eyes are as well developed as vertebrate eyes, described next.

Structure of Vertebrate Eyes. Figure 30.23 shows the structure of the vertebrate eye. Think of it as having three layers, as listed in Table 30.3. The outer layer consists of a white *sclera* (a dense, fibrous cover that protects most of the eye) and a curved *cornea* (a transparent cover that bends light rays so that they converge at the back of the eyeball). The inner layer includes the *retina*, a thin, light-sensitive tissue.

In the eye's middle layer, an adjustable, transparent *lens* focuses incoming light onto the retina. A clear fluid (aqueous humor) bathes both sides of the lens. A jellylike substance (vitreous body) fills a chamber behind the lens. The *choroid*, a pigmented tissue, keeps light from scattering inside the eyeball. Suspended beneath the cornea is an iris, a ring of pigmented, contractile tissue that can be adjusted to admit more or less light. The ring's center is the *pupil*.

Table 30.3	Vertebrate Eye Components
Eye Region	Functions
Outer Layer	
Sclera	Protect eyeball
Cornea	Focus light
Middle Layer	
Choroid:	
Pigmented tissue	Prevent light scattering
Iris	Control incoming light
Pupil	Let light enter
Lens	Finely focus light on photoreceptors
Aqueous humor	Transmit light, maintain pressure
Vitreous body	Transmit light, support lens, eye
Inner Layer	
Retina	Absorb, convert light
Fovea	Increase visual acuity
Start of optic nerve	Transmit signals to brain

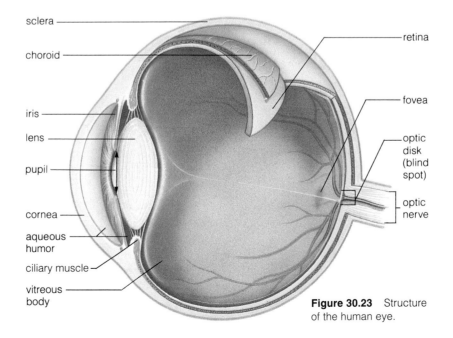

sclera

choroid

iris

lens

pupil

cornea

aqueous humor

ciliary muscle

vitreous body

retina

fovea

optic disk (blind spot)

optic nerve

Figure 30.23 Structure of the human eye.

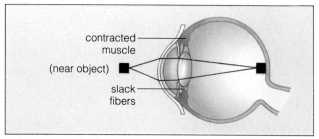

a Accommodation for near objects (lens bulges)

contracted
muscle

(near object)

slack
fibers

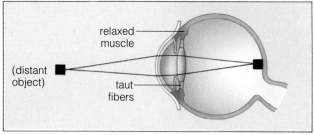

b Accommodation for distant objects (lens flattens)

relaxed
muscle

(distant
object)

taut
fibers

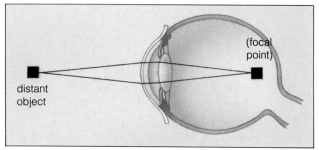

c Focal point in nearsighted vision.

(focal
point)

distant
object

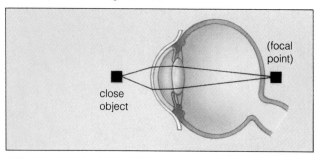

d Focal point in farsighted vision.

(focal
point)

close
object

Figure 30.24 Focusing light on the retina by adjusting the lens. A muscle ringing the lens attaches to it by fiberlike ligaments. (**a**) Close objects are brought into focus when the muscle contracts and makes the lens bulge, so the focal point moves closer. (**b**) Distant objects are brought into focus when the muscle relaxes and makes the lens flatten, so the focal point moves farther back. (**c**) In one form of *nearsightedness*, the eyeball is longer than wide, so images of distant objects are focused in front of the retina. (**d**) In *farsightedness*, the eyeball is wider than long, so close images are focused behind the retina.

Focusing Light Onto the Retina. In birds and mammals, a circular muscle adjusts the shape of the lens, so incoming light is focused onto the retina. Figure 30.24 shows what happens when focusing is precise—and not so precise.

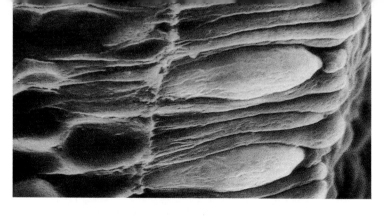

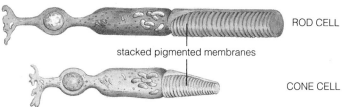

ROD CELL

stacked pigmented membranes

CONE CELL

Figure 30.25 Structure of rods and cones, the photoreceptors of vertebrate eyes.

Vertebrate Photoreception. Birds and mammals have a well-developed retina. Its basement layer, a pigmented epithelium, covers the choroid. Resting on this are densely packed photoreceptors, called *rod* cells and *cone* cells (Figure 30.25). Rod cells are sensitive to very dim light. They contribute to coarse perception of movement by detecting changes in light intensity across the field of vision. Typically they are abundant in the periphery of the retina. Cones respond to bright light. They contribute to sharp daytime vision and color perception. Pigments in different cone cells respond to wavelengths of red, green, or blue light. Cones of human eyes are densely packed in the *fovea*, a funnel-shaped depression near the retina's center. These contribute the most to visual acuity—that is, to the precise discrimination between adjacent points in space.

How are signals from photoreceptors translated into the sense of vision? Many different signals activate or inhibit adjacent neurons. Axons of the neurons converge to form an optic nerve at the back of each eyeball. Both optic nerves relay information to the thalamus, which sends it on to the visual cortex in the forebrain (Figure 30.26). Information becomes more and more organized through levels of synapsing neurons in the retina, then in different parts of the visual cortex.

Signals about different aspects of a visual stimulus (form, movement, depth, color, texture, and so on) seem to be processed along separate communication lines to the visual cortex. Then all the signals travel rapidly, at the same time, to different cortical regions. There they are processed to discern what the stimulus might be, where it is located in the visual field, and so on. Final integration of the signals produces organized electrical activity—and the sensation of sight.

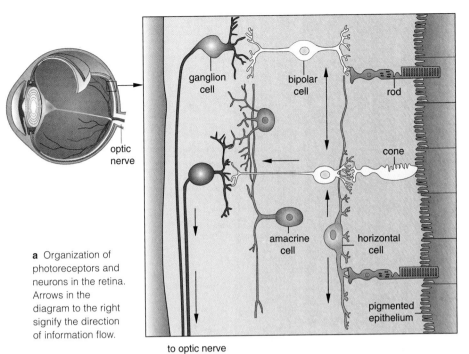

a Organization of photoreceptors and neurons in the retina. Arrows in the diagram to the right signify the direction of information flow.

to optic nerve

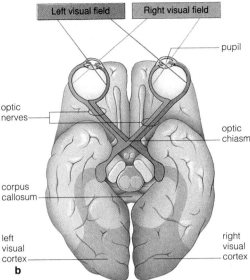

Figure 30.26 (**a**) Neurons and photoreceptors of the retina and (**b**) the sensory pathway to the visual cortex.

SUMMARY

1. The nervous system senses, interprets, and issues commands for responses to specific aspects of the environment. In nearly all animals, its communication lines consist of sensory neurons, interneurons, and motor neurons (which activate muscle and gland cells).

2. A stimulus is a form of energy that the body detects by specific receptors of sensory neurons. Information about a stimulus travels as electrical and chemical signals through the nervous system.

3. Dendrites and the cell body are the input zones of a neuron. Arriving signals may spread to a trigger zone (the start of an axon, most often). They may give rise to an action potential, which travels to the neuron's output zone (axon endings). Action potentials are the basis of messages sent through the nervous system.

   a. In an undisturbed neuron, concentration gradients for sodium and potassium ions hold steady across the plasma membrane. They are the basis of the resting membrane potential—a steady difference in electric charge (a voltage difference) across the membrane.

   b. Between action potentials, the gradients are *established* as some ions leak across the membrane, through channel proteins. The gradients are *restored* and *maintained* by sodium-potassium pumps.

   c. When a neuron is adequately stimulated, an ever increasing number of gated sodium channels open until an action potential occurs: the difference in charge across the resting membrane briefly and abruptly reverses. The accelerated ion flow is self-propagating; it triggers an action potential in patch after patch of membrane.

   d. At a neuron's output zone, action potentials trigger the release of neurotransmitters. These diffuse across chemical synapses—junctions between a neuron and another cell. A neurotransmitter may excite that cell's membrane (drive it closer to threshold) or inhibit it (drive it away from threshold).

4. Integration is the moment-by-moment combining of all signals, excitatory and inhibitory, acting at all synapses on a neuron. It is the means by which information is played down or reinforced, suppressed or sent on to other neurons of the nervous system.

5. The central nervous system consists of the brain and spinal cord. It is functionally connected to all other regions by nerves of the peripheral nervous system.

6. Reflexes are simple, stereotyped movements in response to stimuli. Some brain regions deal with reflex coordination of sensory inputs and motor outputs beyond that afforded by the spinal cord. More recently evolved regions are the basis of memory, learning, and reasoning.

7. The hypothalamus monitors internal organs and influences thirst, hunger, sexual activity, and other behaviors related to organ function. It also relays signals to and from the limbic system, which has roles in learning, memory, and emotional behavior.

8. Specific parts of the cerebral cortex (such as the visual cortex) receive signals from sensory receptors, process and integrate new information with memories of past events, and coordinate motor responses.

9. Somatic sensations include touch, pressure, temperature, pain, and muscle sense. Special senses include taste, smell, hearing, and vision.

1. Label the input and output zones of this motor neuron: 474

2. What is a chemical synapse? What is synaptic integration, and why is it important for information flow through the nervous system? 478–479

3. Distinguish between the following:
   a. neuron and nerve 474, 480
   b. central and peripheral nervous systems 481
   c. somatic and autonomic nerves 482
   d. parasympathetic and sympathetic nerves 482
   e. stimulus and sensation 487

4. Label the major parts of the human brain: 484

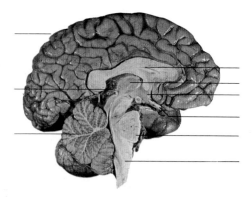

1. An action potential is _____ across the plasma membrane of a neuron.
   a. a steady voltage difference
   b. an abrupt reversal in the voltage difference
   c. a short-lived reversal in the voltage difference
   d. both b and c

2. Action potentials occur when _____ .
   a. a neuron is adequately stimulated
   b. sodium gates open in an accelerating way
   c. sodium-potassium pumps kick into action
   d. both a and b

3. Compared to its surroundings, an undisturbed neuron carries a slight _____ charge inside the plasma membrane.
   a. positive   c. graded, local
   b. negative   d. both b and c

4. The resting membrane potential is maintained by _____ .
   a. ion leaks   c. neurotransmitters
   b. ion pumps   d. both a and b

5. Neurotransmitters diffuse across a _____ .
   a. chemical synapse   c. myelin sheath
   b. channel protein   d. both a and b

6. Parasympathetic and sympathetic nerves play off one another to refine _____ .
   a. limb movements   c. organ activities
   b. motion of whole body d. both a and b

7. The _____ includes the visual cortex, motor cortex, and other centers that are most highly developed for receiving, encoding, and processing sensory information.
   a. cerebellum   c. tectum
   b. cerebrum   d. medulla oblongata

8. A _____ is conscious awareness of a stimulus.
   a. perception   c. reflexive response
   b. sensation   d. both a and b

9. Match the term with its description.
   ____ motor neuron   a. detects tissue injury
   ____ sensory neuron   b. reflex responses to sights, sounds
   ____ hindbrain
   ____ midbrain   c. our emotional brain
   ____ forebrain   d. photoreceptors in retina
   ____ limbic system   e. reflex centers for vital tasks
   ____ nociceptors   f. detects specific stimulus
   ____ rods and cones   g. activates muscle, gland cells
   ____ interneuron   h. in brain and spinal cord only
                              i. most recent brain centers

action potential 475
axon 474
brain 483
central nervous system 481
cerebral cortex 484
chemical synapse 478
chemoreceptor 487
dendrite 474
drug addiction 486
limbic system 485
mechanoreceptor 487
memory 484
nerve 480
nervous system 474
neuron 474
neurotransmitter 478

nociceptor 487
parasympathetic nerve 482
peripheral nervous system 481
photoreceptor 487
reflex pathway 481
resting membrane potential 475
sensation 487
sensory system 487
sodium-potassium pump 476
spinal cord 483
sympathetic nerve 482
synaptic integration 479
thermoreceptor 487

Fischbach, G. D. September 1992. "Mind and Brain." *Scientific American* 267(3): 48–57. This issue is devoted to the development, function, and certain disorders of the brain.

Julien, R. 1985. *A Primer of Drug Action.* Fourth edition. New York: Freeman. Paperback.

31 ENDOCRINE CONTROL

Hormone Jamboree

In the 1960s, at her forest camp by the shore of Lake Tanganyika in Kenya, the primatologist Jane Goodall let it be known that bananas were available. One of the first chimpanzees attracted to the food was a female—Flo, as she came to be called—who was devoted to her offspring. Three years later, motherhood gave way to a preoccupation with sex. Male chimps followed Flo to the camp, and stayed for more than the bananas.

Sex, Goodall discovered, is the premier force in the social life of chimps. These primates don't mate for life as eagles do, or wolves. Before the rainy season, mature females entering a fertile cycle become sexually active. Blood levels of sex hormones change. External sexual organs become swollen and pink—a visual signal to males. The swellings are the flags of sexual jamborees, of great gatherings of highly stimulated chimps.

The gathering of many flag-waving females draws together individuals that otherwise forage alone or in small family groups. It reestablishes bonds that unite their community. As infants and juveniles play with one another and with adults, their aggressive and submissive jostlings help map out future dominance hierarchies. Flo and other females sexually allied with high-ranking males aggressively help their male offspring win confrontations with other young males.

Intriguingly, sex hormones even induce swelling when a female is not fertile or after she is pregnant. Sexual selection has no doubt favored the prolonged swelling. Males groom a sexually attractive female more often, protect her, give her more food, and let her tag along to new foraging sites. The more that males accept a female, the higher she goes in the social hierarchy—and the more her individual offspring benefit.

Hormones shape the growth, development, and reproductive cycles of animals, from worms to chimps to humans. They influence minute-by-minute and day-to-day metabolic functions. Hormonal interplays with the nervous system affect physical appearance, well being, and behavior. How animals behave affects whether they survive, either on their own or in social groups.

This chapter focuses mainly on hormones—their sources, targets, and interactions as well as mechanisms involved in their secretion. If the details start to seem remote, remember: This is the stuff of life. Hormones underwrote Flo's appearance, behavior, and rise through the chimpanzee social hierarchy—and just imagine what they have been doing for you.

Figure 31.1 (**a**) Jane Goodall in Gombe National Park, near the shores of Lake Tanganyika, scouting for chimpanzees. (**b**) Flo and three of her offspring.

a

b

1. Hormones and other signaling molecules help integrate cell activities in ways that benefit the whole body. Some hormones help the body adjust to short-term changes in diet and levels of activity. Others help induce long-term adjustments in growth, development, and reproduction.

2. The hypothalamus and pituitary gland coordinate the activities of many endocrine glands. Their interactions control many of the body's functions.

3. Neural signals, hormonal signals, chemical changes, and environmental cues trigger hormone secretions. Only cells with receptors for a specific hormone are its targets. Protein hormones alter enzyme activity, and steroid hormones induce gene activation and protein synthesis in target cells.

"THE ENDOCRINE SYSTEM"

The word "hormone" dates back to the early 1900s. W. Bayliss and E. Starling were trying to figure out what causes pancreatic secretions when food travels through the canine gut. They already knew that acids mix with food in the stomach, and that the pancreas secretes an alkaline solution after the acidic mixture moves into the small intestine. Was the nervous system or something else stimulating the pancreatic response?

To find the answer, Bayliss and Starling blocked nerves—but not blood vessels—leading to a laboratory animal's upper small intestine. Later, when acidic food entered the intestine, the pancreas still responded to it. More telling, extracts of cells from the intestinal lining also induced its response. Glandular cells in the lining had to be the source of a pancreas-stimulating substance.

The substance came to be called secretin. Proof of its existence and mode of action confirmed a centuries-old idea: *Organ activities are influenced by internal secretions that have been picked up by the bloodstream.* Starling coined the word hormone for such internal glandular secretions (after *hormon*, meaning to set in motion).

Later, researchers identified other hormones and their sources. Figure 31.2 shows the locations of the following sources, which are typical of vertebrates:

Pituitary gland
Adrenal glands (*two*)
Thyroid gland
Parathyroid glands (*four in humans*)
Gonads (*two*)
Pancreatic islets
Thymus gland
Pineal gland

Besides these major sources, glandular cells of the liver, placenta, and other organs also secrete hormones.

The hormone sources in an animal came to be viewed as the **endocrine system**. The name implies there is a separate control system for the body, apart from the nervous system. (*Endon* means within; *krinein* means separate.) However, both biochemical studies and electron microscopy now show that endocrine sources and the nervous system function in intricately connected ways.

HORMONES AND OTHER SIGNALING MOLECULES

Cells respond to changing conditions by taking up and releasing chemical substances. In vertebrates, the responses of millions to many billions of cells must be integrated in ways that benefit the whole body. *Signaling* molecules make integration possible. These include hormones, neurotransmitters, local signaling molecules, and pheromones. All act on **target cells**, which are any cells that have receptors for a given type of signaling molecule and that may alter their behavior in response to it. A target may or may not be next to the cell sending the signal.

By definition, **hormones** are secretions from endocrine glands, endocrine cells, and some neurons that the bloodstream distributes to *nonadjacent* target cells. They are this chapter's focus.

By contrast, neurotransmitters, released from neurons, act swiftly on abutting target cells (page 478). Local signaling molecules, released by many cell types, alter conditions in local tissues. Prostaglandins, for instance, target smooth muscle cells in bronchiole walls, which then constrict or dilate and so alter air flow in lungs. Some types influence blood flow through inflamed tissues and from the uterine lining during menstrual cycles. Pheromones, secreted by some exocrine glands, diffuse through water or air to targets outside the body. They act on cells of other animals of the same species and so help integrate social behavior (pages 489 and 634).

Hormones are signaling molecules secreted by endocrine glands, endocrine cells, and some neurons. The bloodstream distributes them to nonadjacent target cells.

Figure 31.2 On the facing page, an overview of some components of the human endocrine system and the primary effects of their major hormones. The system also includes endocrine cells of many organs, including the liver, kidneys, heart, and small intestine. The hypothalamus, a major component of the brain, also secretes some hormones.

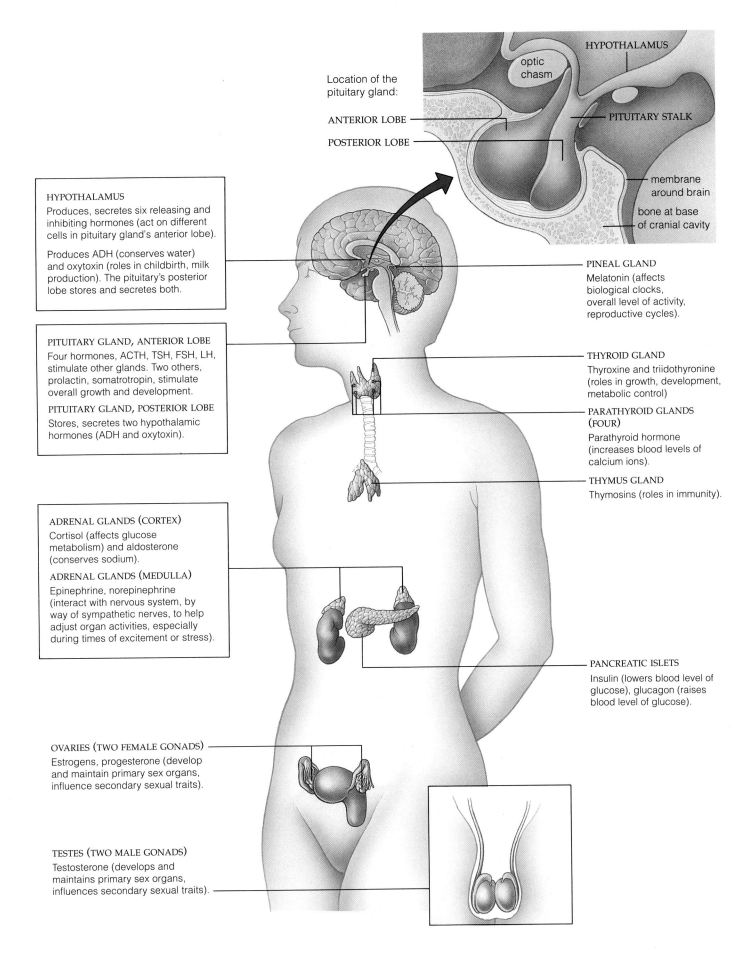

Location of the pituitary gland:

HYPOTHALAMUS

optic chasm

ANTERIOR LOBE

POSTERIOR LOBE

PITUITARY STALK

membrane around brain

bone at base of cranial cavity

HYPOTHALAMUS
Produces, secretes six releasing and inhibiting hormones (act on different cells in pituitary gland's anterior lobe).

Produces ADH (conserves water) and oxytoxin (roles in childbirth, milk production). The pituitary's posterior lobe stores and secretes both.

PITUITARY GLAND, ANTERIOR LOBE
Four hormones, ACTH, TSH, FSH, LH, stimulate other glands. Two others, prolactin, somatrotropin, stimulate overall growth and development.

PITUITARY GLAND, POSTERIOR LOBE
Stores, secretes two hypothalamic hormones (ADH and oxytoxin).

ADRENAL GLANDS (CORTEX)
Cortisol (affects glucose metabolism) and aldosterone (conserves sodium).

ADRENAL GLANDS (MEDULLA)
Epinephrine, norepinephrine (interact with nervous system, by way of sympathetic nerves, to help adjust organ activities, especially during times of excitement or stress).

OVARIES (TWO FEMALE GONADS)
Estrogens, progesterone (develop and maintain primary sex organs, influence secondary sexual traits).

TESTES (TWO MALE GONADS)
Testosterone (develops and maintains primary sex organs, influences secondary sexual traits).

PINEAL GLAND
Melatonin (affects biological clocks, overall level of activity, reproductive cycles).

THYROID GLAND
Thyroxine and triidothyronine (roles in growth, development, metabolic control)

PARATHYROID GLANDS (FOUR)
Parathyroid hormone (increases blood levels of calcium ions).

THYMUS GLAND
Thymosins (roles in immunity).

PANCREATIC ISLETS
Insulin (lowers blood level of glucose), glucagon (raises blood level of glucose).

Table 31.1 Hormones Released From the Mammalian Pituitary Gland

Pituitary Lobe	Secretions	Designation	Main Targets	Primary Actions
Posterior Nervous tissue (extension of hypothalamus)	Antidiuretic hormone	ADH	Kidneys	Induces water conservation required in control of extracellular fluid volume (and, indirectly, solute concentrations)
	Oxytocin		Mammary glands	Induces milk movement into secretory ducts
			Uterus	Induces uterine contractions
Anterior Mostly glandular tissue	Corticotropin	ACTH	Adrenal cortex	Stimulates release of adrenal steroid hormones
	Thyrotropin	TSH	Thyroid gland	Stimulates release of thyroid hormones
	Gonadotropins: Follicle-stimulating hormone	FSH	Ovaries, testes	In females, stimulates egg formation; in males, helps stimulate sperm formation
	Luteinizing hormone	LH	Ovaries, testes	In females, stimulates ovulation, corpus luteum formation; in males, promotes testosterone secretion, sperm release
	Prolactin	PRL	Mammary glands	Stimulates and sustains milk production
	Somatotropin (also called growth hormone)	STH (GH)	Most cells	Promotes growth in young; induces protein synthesis, cell division; roles in glucose, protein metabolism in adults
Intermediate* Glandular tissue, mostly	Melanocyte-stimulating hormone	MSH	Pigmented cells in skin, other surface coverings	Induces color changes in response to external stimuli; affects behavior

*Present in most vertebrates (not adult humans). MSH is associated with the anterior lobe in humans.

THE HYPOTHALAMUS AND PITUITARY GLAND

Deep in the brain is the **hypothalamus**. This brain region monitors internal organs and activities related to their functioning, such as eating and sexual behavior. It also secretes some hormones. Suspended from its base by a slender stalk is a pea-size lobed gland. The hypothalamus and this **pituitary gland** interact as a major neural-endocrine control center.

The *posterior* lobe of the pituitary stores and secretes two of the hormones that are synthesized in the hypothalamus. The *anterior* lobe produces and secretes its own hormones, most of which govern the release of hormones from other endocrine glands (Table 31.1). The pituitary of many vertebrates (but not humans) also has an intermediate lobe. In many cases, this lobe secretes a hormone that governs reversible changes in skin or fur color.

Posterior Lobe Secretions

Figure 31.3*a* shows the cell bodies of certain neurons in the hypothalamus. Their axons extend down into the posterior lobe, then terminate next to a capillary bed. The neurons produce antidiuretic hormone (ADH) and oxytocin, then store them in the axon endings. When either hormone is released, it diffuses through interstitial fluid and enters capillaries, then travels the bloodstream to its targets.

ADH acts on cells of nephrons and collecting ducts in the kidneys. Kidneys, recall, filter blood and rid the body of excess water and salts, in urine. ADH promotes water reabsorption when the body must conserve water. Oxytocin has reproductive roles. It triggers contractions in the uterus during labor and causes milk release when offspring are being nursed.

Anterior Lobe Secretions

Anterior Pituitary Hormones. Other hypothalamic hormones enter a capillary bed in the pituitary stalk, which delivers them to a second capillary bed in the anterior lobe. There, the hormones leave the bloodstream and act on target cells. As Figure 31.4 shows, different cells of the anterior pituitary secrete six hormones that they themselves produce:

Corticotropin	ACTH
Thyrotropin	TSH
Follicle-stimulating hormone	FSH
Luteinizing hormone	LH
Prolactin	PRL
Somatotropin (or growth hormone)	STH (or GH)

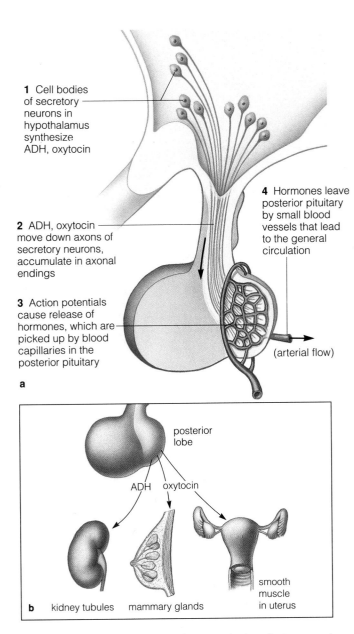

1 Cell bodies of secretory neurons in hypothalamus synthesize ADH, oxytocin

2 ADH, oxytocin move down axons of secretory neurons, accumulate in axonal endings

3 Action potentials cause release of hormones, which are picked up by blood capillaries in the posterior pituitary

4 Hormones leave posterior pituitary by small blood vessels that lead to the general circulation

(arterial flow)

a

posterior lobe

ADH   oxytocin

kidney tubules   mammary glands   smooth muscle in uterus

b

Figure 31.3 (**a**) Functional links between the hypothalamus and the posterior lobe of the pituitary. (**b**) Main targets of the posterior lobe secretions.

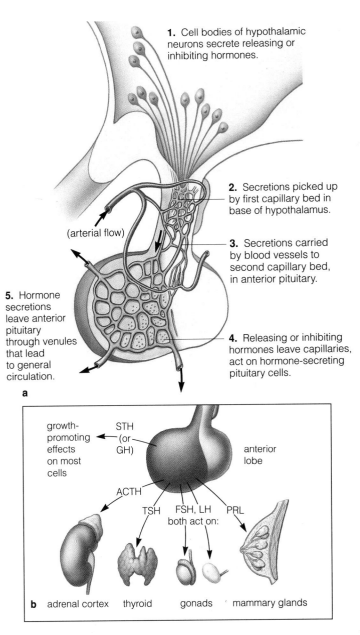

1. Cell bodies of hypothalamic neurons secrete releasing or inhibiting hormones.

2. Secretions picked up by first capillary bed in base of hypothalamus.

(arterial flow)

3. Secretions carried by blood vessels to second capillary bed, in anterior pituitary.

5. Hormone secretions leave anterior pituitary through venules that lead to general circulation.

4. Releasing or inhibiting hormones leave capillaries, act on hormone-secreting pituitary cells.

a

growth-promoting effects on most cells

STH (or GH)

anterior lobe

ACTH

TSH   FSH, LH both act on: PRL

adrenal cortex   thyroid   gonads   mammary glands

b

Figure 31.4 (**a**) Functional links between the hypothalamus and the anterior lobe of the pituitary. (**b**) Main targets of the anterior lobe secretions.

ACTH acts on adrenal glands, and TSH on the thyroid gland, as described shortly. FSH and LH have roles in reproduction, the central topic of Chapter 32.

Somatotropin affects body tissues in general (Table 31.1 and Figure 31.4). It works by stimulating protein synthesis and cell division and so has profound influence over growth, especially of cartilage and bone.

Prolactin also has general effects. But it is better known for its role in stimulating and sustaining milk production in mammary glands, after other hormones have primed the tissues. Prolactin also affects hormone production in ovaries.

About the Hypothalamic Triggers. Most hypothalamic hormones acting in the anterior lobe of the pituitary are *releasers*; they stimulate secretion from their target cells. For example, the one called GnRH (short for gonadotropin-releasing hormone) brings about secretion of FSH and LH—both of which are classified as gonadotropins. Similarly, the one called TRH stimulates the secretion of thyrotropin.

Some hypothalamic hormones are *inhibitors* of secretion from their targets in the anterior pituitary. For instance, the one called somatostatin brings about a decrease in somatotropin and thyrotropin secretion.

a

b

age nine   sixteen

thirty-three   fifty-two

Figure 31.5 (**a**) Manute Bol, an NBA center, is 7 feet 6-3/4 inches tall owing to excess somatotropin production during childhood. (**b**) Somatotropin's effect on overall growth. Gigantism (person at the center) results from overproduction of the hormone during childhood. Pituitary dwarfism (person to the right) results from insufficient production. The person at the left is average in size.

Figure 31.6 Acromegaly, which resulted from excess production of somatotropin during adulthood. Before this female reached maturity, she was symptom-free.

Examples of Abnormal Pituitary Output

The body does not produce enormous numbers of hormone molecules. Roger Guilleman and Andrew Schally realized this when they isolated the first known releasing hormone. After four years of dissecting 500 tons of sheep brains, then 7 tons of hypothalamic tissue, they finally extracted a single milligram of TRH.

Yet normal body function depends on those tiny hormonal outputs. Consider what happens when anterior pituitary cells produce too much somatotropin during childhood. *Gigantism* results. With this condition, an adult is similar in proportion to a normal person but much larger (Figure 31.5*a*). Or consider what happens when not enough somatotropin is produced during childhood. *Pituitary dwarfism* results. With this condition, an adult is similar in proportion to a normal person but much smaller (Figure 31.5*b*). What if somatotropin output becomes excessive during adulthood, when long bones no longer can lengthen? *Acromegaly* results. Bone, cartilage, and other connective tissues in hands, feet, and jaws thicken abnormally, as do epithelia of the skin, nose, eyelids, lips, and tongue (Figure 31.6).

SELECTED EXAMPLES OF HORMONAL CONTROL

Table 31.2 lists hormones from endocrine glands other than the pituitary. Let's look at a few examples to get an idea of how hormonal controls work. The examples will make more sense if you keep three points in mind.

First, hormones often interact with one another. One or more hormones may oppose, add to, or prime target cells for another hormone's effects. Second, the hypothalamus and pituitary govern many responses through **homeostatic feedback loops**. (One or both detect a change in a hormone's concentration in some region, then respond by inhibiting or stimulating its secretion.) Third, responses vary, depending on the hormone's concentration at any given time and on the nature of receptors on the target cell.

Responses to hormones may be influenced by homeostatic feedback loops to the hypothalamus and pituitary, variations in hormone concentrations, and the nature of receptors on a target cell.

Table 31.2 Hormone Sources Other Than the Mammalian Hypothalamus and Pituitary

Source	Secretion(s)	Main Targets	Primary Actions
Adrenal cortex	Glucocorticoids (including cortisol)	Most cells	Promote protein breakdown and conversion to glucose
	Mineralocorticoids (including aldosterone)	Kidney	Promote sodium reabsorption; control salt–water balance
Adrenal medulla	Epinephrine (adrenalin)	Liver, muscle, adipose tissue	Raises blood level of sugar, fatty acids; increases heart rate, force of contraction
	Norepinephrine	Smooth muscle of blood vessels	Promotes constriction or dilation of blood vessel diameter
Thyroid	Triiodothyronine, thyroxine	Most cells	Regulate metabolism; have roles in growth, development
	Calcitonin	Bone	Lowers calcium levels in blood
Parathyroids	Parathyroid hormone	Bone, kidney	Elevates calcium levels in blood
Gonads:			
Testes (in males)	Androgens (including testosterone)	General	Required in sperm formation, development of genitals, maintenance of sexual traits; influence growth, development
Ovaries (in females)	Estrogens	General	Required in egg maturation and release; prepare uterine lining for pregnancy; required in development of genitals, maintenance of sexual traits; influence growth, development
	Progesterone	Uterus, breasts	Prepares, maintains uterine lining for pregnancy; stimulates breast development
Pancreatic islets	Insulin	Muscle, adipose tissue	Lowers blood sugar level
	Glucagon	Liver	Raises blood sugar level
	Somatostatin	Insulin-secreting cells of pancreas	Influences carbohydrate metabolism
Endocrine cells of stomach, gut	Gastrin, secretin, etc.	Stomach, pancreas, gallbladder	Stimulate activity of stomach, pancreas, liver, gallbladder
Liver	Somatomedins	Most cells	Stimulate cell growth and development
Kidneys	Erythropoietin*	Bone marrow	Stimulates red blood cell production
	Angiotensin*	Adrenal cortex, arterioles	Helps control blood pressure, aldosterone secretion
	Vitamin D$_3$*	Bone, gut	Enhances calcium resorption and uptake
Heart	Atrial natriuretic hormone	Kidney, blood vessels	Increases sodium excretion; lowers blood pressure
Thymus	Thymosins, etc.	Lymphocytes	Have roles in immune responses
Pineal	Melatonin	Gonads (indirectly)	Influences daily biorhythms, seasonal sexual activity

*These hormones are not produced in the kidneys but are formed when *enzymes* produced in kidneys activate specific substances in the blood.

Adrenal Glands

Adrenal Cortex. Humans have a pair of adrenal glands, one above each kidney (Figure 31.7). The outer portion of each gland is the **adrenal cortex**. Some of its cells secrete glucocorticoids and other hormones. Homeostatic feedback loops to the hypothalamus and pituitary govern their secretion.

Glucocorticoids help maintain blood levels of glucose, and they also suppress inflammatory responses. Cortisol, for instance, blocks the uptake and use of glucose by muscle cells. It also stimulates liver cells to form glucose from amino acids.

In *hypoglycemia*, blood levels of glucose fall below a set point, and a negative feedback mechanism operates. The hypothalamus detects the decrease and secretes CRH in response (Figure 31.7). This releasing hormone makes the anterior pituitary secrete corticotropin (ACTH). ACTH makes the adrenal cortex secrete cortisol—which helps raise the blood glucose level by preventing muscle cells from taking up more glucose from the blood. These cells are major glucose users.

During severe stress, painful injury, or prolonged illness, the nervous system overrides feedback control of cortisol secretion. It initiates a *stress response* in which secretions of cortisol help suppress inflammation. If left unchecked, prolonged inflammation could damage tissues. Cortisol-like drugs, such as cortisone, counter asthma and other chronic inflammatory disorders.

Adrenal Medulla. The **adrenal medulla** is the inner portion of the adrenal gland (Figure 31.7). It contains neurons that secrete epinephrine and norepinephrine. These substances are hormones in some contexts and neurotransmitters in others. Sympathetic nerves carry stimulatory and inhibitory signals to them from the hypothalamus and other brain regions.

Both substances help adjust blood circulation and carbohydrate metabolism when the body is excited or stressed. They increase heart rate, dilate arterioles in some body regions and constrict them in others, and dilate bronchioles. Thus more blood volume is shunted to heart and muscle cells from other regions, and more oxygen flows to energy-demanding cells throughout the body. These are features of the "fight-flight" response described on pages 482 and 483.

Thyroid Gland

The human **thyroid gland** is located at the base of the neck in front of the trachea, commonly known as the windpipe (Figure 31.8). Its main hormones are thyroxine and triiodothyronine. These thyroid hormones have

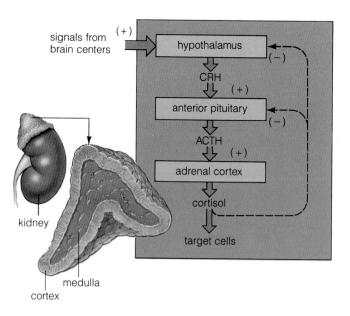

Figure 31.7 Negative feedback loop that governs cortisol secretion.

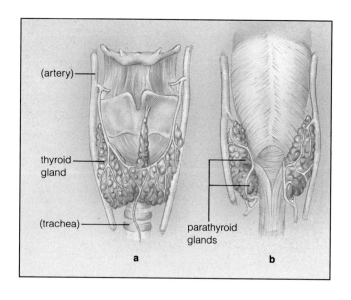

Figure 31.8 (**a**) Anterior view of the human thyroid gland. (**b**) Posterior view, showing four parathyroid glands next to it.

widespread effects on most body cells. They are critical for the normal development of many tissues, and they control the overall metabolic rate in humans and other warm-blooded animals. Their importance is brought into sharp focus by cases of abnormal thyroid secretion.

The synthesis of thyroid hormones requires iodine, which is obtained from the diet. In the absence of iodine, blood levels of these hormones decrease. The anterior pituitary responds by secreting a thyroid-stimulating hormone (TSH). Because the thyroid hormones cannot be synthesized, the feedback signal continues—and so does TSH secretion. Excess TSH overstimulates the thyroid gland and causes it to enlarge. The enlargement is a form of *goiter* (Figure 31.9). Goiter caused by iodine deficiency is no longer common in countries where iodized salt is used.

Hypothyroidism results from insufficient concentrations of thyroid hormones. Hypothyroid adults often are overweight, sluggish, dry-skinned, intolerant of cold, and sometimes confused and depressed. Affected women often have menstrual disturbances. *Hyperthyroidism* results from excess concentrations of thyroid hormones. Hyperthyroid adults show an increased heart rate, elevated blood pressure, weight loss despite normal caloric intake, intolerance of heat, and profuse sweating. Typically they are nervous, agitated, and have trouble sleeping.

Parathyroid Glands

Instead of responding directly to other hormones or nerves, some endocrine glands respond homeostatically to chemical change in the immediate surroundings. The **parathyroid glands** are like this. Four such glands are positioned next to the back of the human thyroid (Figure 31.8). They secrete parathyroid hormone (PTH) when the concentration of calcium ions in their surroundings decreases. Their action affects how much calcium is available for enzyme activation, muscle contraction, blood clotting, and many other tasks.

PTH stimulates bone cells to release calcium and phosphate, and the kidneys to conserve it. PTH also helps activate vitamin D. The activated form, a hormone, enhances calcium absorption from food in the gut. In vitamin D deficiency, not enough calcium and phosphorus is absorbed and bones don't develop properly. This ailment is called *rickets* (Figure 31.10).

Thymus Gland

The lobed **thymus gland** is located behind the breastbone, between the lungs (Figure 31.2). White blood cells multiply, differentiate, and mature in this gland, which secretes hormones collectively called thymosins. These have roles in immunity (Chapter 27).

Figure 31.9 A mild case of goiter, as displayed by Maria de Medici in the year 1625. During the late Renaissance, a rounded neck was a sign of beauty. It occurred regularly in parts of the world where iodine supplies were insufficient for normal thyroid function.

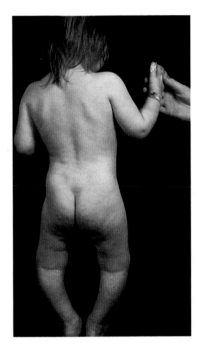

Figure 31.10 A child with rickets. Bowed legs are typical of the disorder.

Gonads

Homeostatic feedback loops also govern the function of **gonads**—the primary reproductive organs. Male gonads are testes (singular, testis) and female gonads, ovaries. Gonads produce gametes. They also secrete estrogens, progesterone, and androgens (including testosterone). These sex hormones control reproductive function and secondary sexual traits, as they did for Flo and others in the chimpanzee community described at the start of this chapter. Chapter 32 describes their action.

Figure 31.11 Some homeostatic controls over glucose metabolism.

Following a meal, glucose enters the bloodstream faster than cells can use it. The blood glucose level rises, and pancreatic beta cells are stimulated to secrete insulin. Insulin's targets (mainly liver, fat, and muscle cells) use glucose or store it as glycogen.

Between meals, the blood glucose level decreases. Pancreatic alpha cells are stimulated to secrete glucagon. This hormone's target cells convert glycogen back to glucose, which enters the blood. Also, the hypothalamus prods the adrenal medulla to secrete other hormones that slow down conversion of glucose to glycogen in liver, fat, and muscle cells.

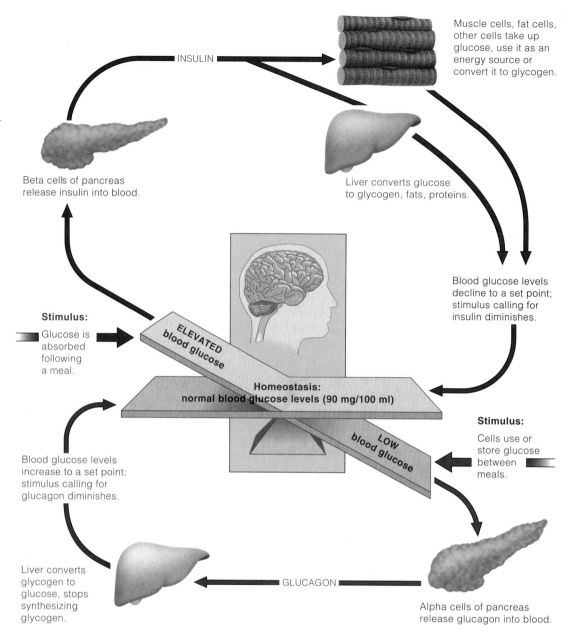

Muscle cells, fat cells, other cells take up glucose, use it as an energy source or convert it to glycogen.

INSULIN

Beta cells of pancreas release insulin into blood.

Liver converts glucose to glycogen, fats, proteins.

Blood glucose levels decline to a set point; stimulus calling for insulin diminishes.

Stimulus: Glucose is absorbed following a meal.

ELEVATED blood glucose

Homeostasis: normal blood glucose levels (90 mg/100 ml)

LOW blood glucose

Stimulus: Cells use or store glucose between meals.

Blood glucose levels increase to a set point; stimulus calling for glucagon diminishes.

Liver converts glycogen to glucose, stops synthesizing glycogen.

GLUCAGON

Alpha cells of pancreas release glucagon into blood.

Pancreatic Islets

The pancreas has exocrine and endocrine functions. (Its exocrine cells secrete digestive enzymes). About 2 million clusters of endocrine cells also are scattered through the gland. Each small cluster, a **pancreatic islet**, contains three types of hormone-secreting cells:

1. *Alpha* cells secrete the hormone glucagon. Between meals, cells throughout the body use the glucose delivered to them by the bloodstream. The blood glucose level decreases. Then, glucagon secretion causes glycogen (a storage polysaccharide) and amino acids to be converted to glucose in the liver. In such ways, *glucagon raises the glucose level in the blood.*

2. *Beta* cells secrete the hormone insulin. After meals, when the blood glucose level is high, insulin stimulates uptake of glucose by liver, muscle, and adipose cells especially. It also promotes synthesis of proteins and fats, and inhibits protein conversion to glucose. Thus, *insulin lowers the glucose level in the blood.*

3. *Delta* cells secrete somatostatin, a hormone that helps control food digestion. It also can block secretion of insulin and glucagon.

Figure 31.11 shows how pancreatic hormones interact to maintain blood glucose levels even though the times and amounts of food intake vary. Bear in mind, insulin is the only hormone that stimulates cells to take up and store glucose in forms that can be rapidly tapped when glucose blood levels fall. Its central role in carbohydrate, protein, and fat metabolism becomes clear when we observe people who can't produce enough insulin or who lack body cells that can respond to it.

Insulin deficiency can lead to *diabetes mellitus*. Blood glucose levels rise, and glucose accumulates in urine. This promotes excessive urination, which disrupts the body's water-solute balance. Affected people become dehydrated and abnormally thirsty. Without a steady supply of glucose, their cells start breaking down proteins and fats for energy. This leads to weight loss. Ketones, which are normal acidic products of fat breakdown, accumulate in the blood and urine. They promote excessive water loss. The imbalance disrupts brain function. In extreme cases, death may follow.

In "type 1 diabetes," the body mounts an autoimmune response against its own insulin-secreting beta cells and destroys them. Genetic susceptibility and environmental triggers combine to produce the disorder, which is the less common but more immediately dangerous of the two types of diabetes. Symptoms usually emerge during childhood and adolescence (the disorder also is called juvenile-onset diabetes). Type 1 diabetic patients survive with insulin injections.

In "type 2 diabetes," insulin levels are close to or above normal, but target cells can't respond to insulin. As affected persons grow older, their beta cells produce less and less insulin. Type 2 diabetes usually occurs in middle age. It is less dramatically dangerous than the other type. Affected persons lead normal lives by controlling their diet, controlling weight, and sometimes taking drugs to enhance insulin action or secretion.

Pineal Gland

Hormonal responses to the external environment influence growth, development, and reproduction. The **pineal gland**, a photosensitive organ located in the brain, provides a good example of this (Figure 31.2). In the absence of light, this gland secretes the hormone melatonin. Thus blood levels of melatonin vary from day to night, and they vary with the changing seasons. The variations are known to influence the development of gonads, reproductive cycles, and reproductive behavior in a variety of species.

Consider the hamster. In winter, when nights are longest, melatonin levels are high, and sexual activity is suppressed. In summer, when days are longer, melatonin levels are low—and hamster sex peaks. Or consider a male white-throated sparrow (Figure 31.12). Melatonin suppresses growth of the bird's gonads until spring, when days start to lengthen. Then, stepped-up gonad activity leads to the production of hormones that influence singing behavior. With his distinctive song, the sparrow defines his territory and may hold the interest of a mate.

Does melatonin also influence human behavior? Perhaps. It could be that decreased melatonin secretion

Focus on the Environment

Rhythms and Blues

At sunset, when light is waning, a bump of tissue in the brain is stimulated to secrete a hormone. The bump is the pineal gland. The hormone, melatonin, acts on certain neurons that lower your body temperature and make you sleep. At sunrise, when melatonin secretion slows, body temperature increases and you wake up and become active.

An internal, biological clock governs the cycle of sleep and arousal. It seems to tick in synchrony with daylength. Think of the night worker who tries to sleep in the morning but ends up staring groggily at sunbeams on the ceiling. Two hours past midnight they are sitting up in bed, wondering where the coffee and croissants are. Two hours past noon they are ready for bed. They will shift to a new routine when melatonin's signals start arriving at their target neurons on Paris time.

In winter, some people get depressed, go on carbohydrate binges, and have an overwhelming desire to sleep. Their "winter blues" may be symptoms of a biological clock out of synch with the season's shorter daylengths. Doses of melatonin make their seasonal symptoms worse. Exposure to intense light—which shuts down pineal activity—leads to dramatic improvements.

triggers *puberty*, the age at which human reproductive organs and structures start to mature. In cases where disease has destroyed the pineal gland, puberty begins prematurely. The *Focus* essay mentions some other interesting possibilities.

Figure 31.12 A male white-throated sparrow, belting out a song that began, indirectly, with an environmentally induced decline in melatonin secretion.

Table 31.3	Two Main Categories of Hormones
Type of Hormone	Examples
Steroid	Estrogens, testosterone, aldosterone, cortisol
Nonsteroid:	
Amines	Norepinephrine, epinephrine
Peptides	ADH, oxytocin, TRH
Proteins	Insulin, somatotrophin, prolactin
Glycoproteins	FSH, LH, TSH

SIGNALING MECHANISMS

Hormones and other signaling molecules have diverse effects. They induce target cells to take up substances. They make cells alter rates of protein synthesis, tinker with existing proteins, change cell shape, and modify internal structures. Two factors influence responses to hormonal signals. First, different hormones activate different cellular mechanisms. Second, not all cells *can* respond to a given signal. Many have receptors for cortisol, for instance, so this hormone has widespread effects. Only a few cell types have receptors for hormones that have highly directed effects.

Let's consider the effect of two different types of hormone molecules on target cells. They are categorized as steroid and nonsteroid hormones (Table 31.3).

Steroid Hormone Action

Steroid hormones are synthesized from cholesterol. Being lipid-soluble, they diffuse directly across the lipid bilayer of a target cell's plasma membrane. Once inside, a steroid hormone molecule usually moves into the nucleus and binds to a receptor for it (Figure 31.13). In some cases, the molecule binds to a receptor that is in the cytoplasm, then the hormone-receptor complex moves into the nucleus. There, the molecular configuration of the complex allows it to interact with DNA. The complex stimulates (or inhibits) transcription of certain gene regions into mRNA. Enzymes and other proteins are produced by mRNA translation. These proteins carry out the cellular response to the hormonal signal.

Steroid hormones stimulate or inhibit protein synthesis by entering the nucleus of target cells and switching certain genes on or off.

Testosterone, a steroid hormone, affects development of male sexual traits. Development proceeds normally when target cells have functional receptors for testosterone. In *testicular feminization syndrome*, the

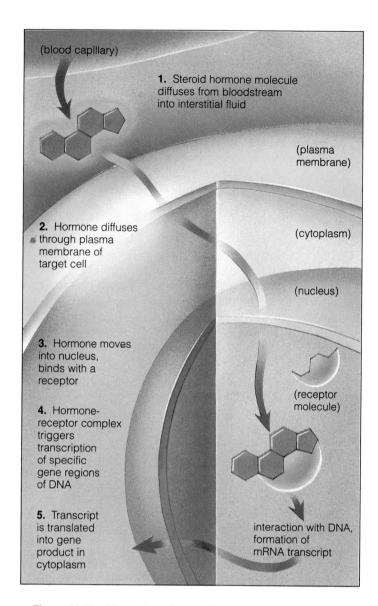

Figure 31.13 Mechanism of steroid hormone action on a target cell. This same type of mechanism is also thought to occur for thyroid hormones, with one qualification. Transport proteins assist thyroid hormones across the plasma membrane.

1. Steroid hormone molecule diffuses from bloodstream into interstitial fluid

2. Hormone diffuses through plasma membrane of target cell

3. Hormone moves into nucleus, binds with a receptor

4. Hormone-receptor complex triggers transcription of specific gene regions of DNA

5. Transcript is translated into gene product in cytoplasm

interaction with DNA, formation of mRNA transcript

receptors are defective. Genetically, the affected individual is male (XY); he has functional testes that secrete testosterone. But none of the target cells can respond to the hormone, so the secondary sexual traits that do develop are like those of females.

Nonsteroid Hormone Action

Nonsteroid hormones include amines, peptides, proteins, and glycoproteins. Like other water-soluble signaling molecules, they cannot cross the plasma membrane of target cells without assistance.

For example, protein hormones bind to receptors at the plasma membrane. In some cases, the hormone-receptor complex moves into the cytoplasm by way of

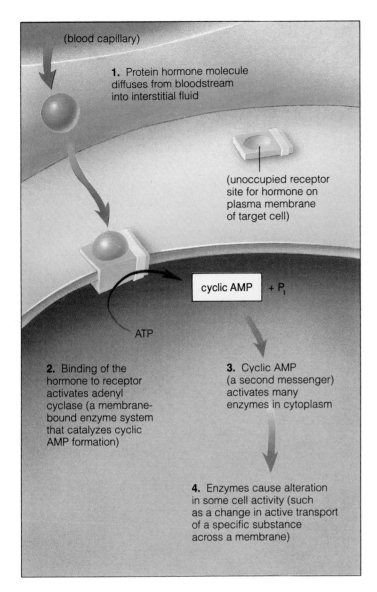

1. Protein hormone molecule diffuses from bloodstream into interstitial fluid

(blood capillary)

(unoccupied receptor site for hormone on plasma membrane of target cell)

cyclic AMP + P$_i$

ATP

2. Binding of the hormone to receptor activates adenyl cyclase (a membrane-bound enzyme system that catalyzes cyclic AMP formation)

3. Cyclic AMP (a second messenger) activates many enzymes in cytoplasm

4. Enzymes cause alteration in some cell activity (such as a change in active transport of a specific substance across a membrane)

Figure 31.14 Mechanism of one protein hormone's action on a target cell. A second messenger inside the cell (in this case, cyclic AMP) amplifies the cellular response to the hormone.

endocytosis (Figure 3.10), then further action occurs inside the cell. In other cases, a hormone simply binds to receptors at the membrane surface. The hormone-receptor complex activates transport proteins or triggers the opening of channel proteins across the membrane. In either case, specific ions or other substances move inside and their cytoplasmic concentrations change. The changes influence specific cell activities.

Often, nonsteroid hormones activate **second messengers**. These are molecules inside the cell that mediate the response to a hormone. Cyclic AMP (cyclic adenosine monophosphate) is an example. Suppose a protein hormone binds to a receptor on a target cell (Figure 31.14). Binding triggers activity at a membrane-bound enzyme system. Now an enzyme (adenyl

cyclase) speeds the conversion of ATP to cyclic AMP. The hormone-receptor complex activates many molecules of the enzyme, in a cascade of reactions. Each molecule speeds the conversion of many ATP molecules to cyclic AMP. Each cyclic AMP molecule switches on many enzyme molecules. These convert many substrate molecules into activated enzymes, and so on. Soon the number of molecules representing the final cellular response to the hormonal signal is enormous.

Some protein hormones enter cells by receptor-mediated endocytosis. Others bind to receptors and activate transport or channel proteins. Still others bind to receptors and so activate second messengers, some of which trigger an amplified response to the hormone.

SUMMARY

1. Cells continually take up and release substances. In complex animals, myriad withdrawals and secretions must be integrated in ways that ensure cell survival through the whole body.

2. Integration requires signaling molecules: chemical secretions by one cell that adjust the behavior of other, target cells. Any cell with receptors for the signal is the target. It may or may not be next to the signaling cell. Hormones, neurotransmitters, local signaling molecules, and pheromones are all signaling molecules.

3. The hypothalamus and pituitary gland interact to integrate many body activities.

   a. ADH and oxytocin, two hypothalamic hormones, are stored in and released from the posterior lobe of the pituitary. ADH influences extracellular fluid volume. Oxytocin has roles in reproduction and other events.

   b. Six other hypothalamic hormones (releasing and inhibiting hormones) control secretion by different cells of the anterior lobe of the pituitary.

   c. Of the six hormones produced in the anterior lobe, two (prolactin and somatotropin, or growth hormone) have general effects on body tissues. Four (ACTH, TSH, FSH, and LH) act on specific endocrine glands.

4. Responses to hormones may be influenced by hormone interactions, homeostatic feedback loops to the hypothalamus and pituitary, variations in hormone concentrations, and the number and kind of receptors on a target cell.

5. Steroid (and thyroid) hormones trigger gene activation and protein synthesis. Most nonsteroid hormones alter activity of existing proteins in target cells. The cellular responses help maintain the internal environment or contribute to the developmental or reproductive program.

Review Questions

1. Name the main endocrine glands and state where each is located in the human body. *496–497, 502–505*

2. Distinguish among hormones, neurotransmitters, local signaling molecules, and pheromones. *496*

3. A hormone molecule binds to a receptor on a cell membrane. It doesn't enter the cell; rather, the binding activates a second messenger inside the cell that triggers an amplified response to the hormonal signal. Is the signaling molecule a steroid or protein hormone? *506–507*

4. Which secretions of the posterior and anterior lobes of the pituitary gland have the targets indicated? *(Fill in the blanks; see pages 498 and 499.)*

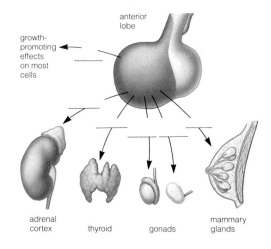

anterior lobe

growth-promoting effects on most cells

adrenal cortex   thyroid   gonads   mammary glands

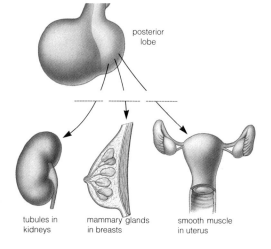

posterior lobe

tubules in kidneys   mammary glands in breasts   smooth muscle in uterus

Self-Quiz *(Answers in Appendix IV)*

1. _____ are molecules released from a signaling cell that have effects on target cells.
   a. hormones   d. pheromones
   b. neurotransmitters   e. a and b
   c. local signaling molecules   f. all of the above

2. Hormones are products of _____ .
   a. endocrine glands and cells   d. a and b
   b. some neurons   e. a and c
   c. exocrine cells   f. a, b, and c

3. ADH and oxytocin are hypothalamic hormones secreted from the pituitary's _____ lobe.
   a. anterior   c. intermediate
   b. posterior   d. secondary

4. _____ and _____ have effects on body tissues in general.
   a. ACTH; FSH   c. LH; prolactin
   b. ACTH; TSH   d. Somatotropin; prolactin

5. Which do *not* stimulate hormone secretions?
   a. neural signals   d. environment cues
   b. local chemical changes e. all of the above can
   c. hormonal signals   stimulate hormone secretion.

6. _____ lowers blood sugar levels; _____ raises it.
   a. Glucagon; insulin   c. Gastrin; insulin
   b. Insulin; glucagon   d. Gastrin; glucagon

7. The pituitary detects a rising hormone concentration in some region and inhibits the gland secreting the hormone. This is a _____ feedback loop.
   a. positive   c. long-term
   b. negative   d. b and c

8. Second messengers assist _____ .
   a. steroid hormones   c. releasing hormones
   b. protein hormones   d. both a and b

9. Match the term with its closest description.
   ___ adrenal cortex   a. affected by light levels
   ___ adrenal medulla   b. cortisol source
   ___ thyroid gland   c. roles in immunity
   ___ parathyroids   d. raise blood calcium level
   ___ pancreatic islets   e. epinephrine source
   ___ pineal gland   f. insulin, glucagon
   ___ thymus gland   g. hormones require iodine

Selected Key Terms

adrenal cortex *502*   parathyroid glands *503*
adrenal medulla *502*   pineal gland *505*
endocrine system *496*   pituitary gland *498*
gonads *503*   second messenger *507*
homeostatic feedback loop *500* steroid hormone *506*
hormone *496*   target cell *496*
hypothalamus *498*   thymus gland *503*
nonsteroid hormone *506*   thyroid gland *502*
pancreatic islet *504*

Readings

Goodall, J. 1986. *The Chimpanzees of Gombe.* Cambridge, Massachusetts: Belknap Press of Harvard University Press.

Hadley, M. 1992. *Endocrinology.* Third edition. Englewood Cliffs, New Jersey: Prentice-Hall.

Snyder, S. October 1985. "The Molecular Basis of Communication Between Cells." *Scientific American* 253(4):132–141.

32 REPRODUCTION AND DEVELOPMENT

From Frog to Frog and Other Mysteries

With a quavering, low-pitched call that only a female of its kind could find seductive, a male frog proclaims the onset of warm spring rains, of ponds, of sex in the night. There, in the darkened waters, he attracts a hormone-primed female. He clamps his forelegs about her in a prolonged, tight squeeze that forces a ribbon of hundreds of eggs from her swollen abdomen. These he blankets with a milky cloud of sperm. Soon, tiny fertilized eggs—zygotes—are suspended in the water.

For the leopard frog, *Rana pipiens*, a drama now begins to unfold that has been reenacted each spring, with only minor variations, for many millions of years. Within a few hours after fertilization, each zygote divides into two cells, then four, then many more to produce an early embryo. In less than twenty hours, the embryo is a ball of cells, no larger than the zygote.

Soon those cells are signaling one another, dividing and growing and changing shape, and migrating about within the embryo! Internal tissue layers form, then embryonic organs. In less than a week the embryo has become a tadpole—a swimming, algae-eating larva (Figure 32.1). Several months pass. Legs start to grow.

The tail shortens, then disappears. The mouth develops jaws and now snaps shut on insects and worms. Eventually the transformations lead to an adult frog. With luck the frog will avoid predators, disease, and other threats in the months ahead. In time it may even call out quaveringly across a moonlit pond, and the cycle will begin again.

How does the single-celled zygote of a frog or any other complex animal become transformed into all the specialized cells and structures of the adult? With this question we turn to one of life's greatest dramas—the development of offspring in the image of their parents.

Figure 32.1 Development of the leopard frog, *Rana pipiens*. (**a**) A male clasping a female. When she releases eggs into the water, he releases sperm over the eggs. (**b**) Cell divisions of a fertilized egg (zygote) help produce a many-celled embryo, which develops further into a larval form, the tadpole. (**c**) A tadpole on the way to becoming an adult. The photographs only hint at the complex cell divisions and changes in cell size, shape, position, and function that transform an animal zygote into the sexually mature adult.

Zygote →

KEY CONCEPTS

1. Biologically, the cost of sexual reproduction is high. Separation into sexes requires elaborate reproductive structures and special controls to synchronize the timing of male and female reproductive activities. Also, energy that might otherwise be spent on individual survival is typically spent on courtship and parental behavior. The cost is offset by an evolutionary advantage—variation in traits among offspring.

2. The human reproductive system consists of a pair of primary reproductive organs (testes in males, ovaries in females), accessory glands, and ducts. Testes produce sperm, ovaries produce eggs, and both release sex hormones in response to signals from the hypothalamus and pituitary gland.

3. Human males continually produce sperm from puberty onward. The hormones testosterone, LH, and FSH control male reproductive functions.

4. Human females are fertile on a cyclic basis. Each month during their reproductive years, an egg is released from an ovary, and the lining of the uterus is prepared for pregnancy. The hormones estrogen, progesterone, FSH, and LH control this cyclic activity.

5. In humans and other vertebrates, development proceeds through six stages: gamete formation, fertilization, cleavage, gastrulation, organ formation, and growth and tissue specialization.

of asexual reproduction, offspring are genetically the same as parents, or nearly so. This absence of variation is useful when gene-encoded traits are highly adaptive to a limited and rather consistent set of environmental conditions.

Most animals live in places where opportunities, resources, and danger vary in complicated ways. Such animals rely mainly on sexual reproduction, with offspring getting different mixes of alleles from male and female parents. The resulting variation in traits improves the odds that some offspring, at least, will be able to survive and reproduce (Chapter 14).

Separation into sexes comes at a cost. Sex means building special reproductive structures and colorful feathers or other attractive parts. It means engaging in behavior that promotes courtship and fertilization. It means having control mechanisms to synchronize the timing of sperm and egg production, sexual readiness, even parental behavior. Assuring survival of offspring is also costly. Aquatic animals that release eggs into the water for fertilization often must produce them in great numbers—so at least some will survive egg-eating predators. Your own mother put great demands on her body to protect and nourish you through many months of early development, inside her body.

Separation into male and female sexes requires special reproductive structures, control mechanisms, and behaviors. The biological cost is offset by a selective advantage: variation in traits among the offspring.

THE BEGINNING: REPRODUCTIVE MODES

In earlier chapters, you read about the cellular basis of sexual reproduction, in which offspring are produced by way of meiosis, gamete formation, and fertilization. You also read about asexual reproduction, in which offspring are produced by means other than gamete formation. Consider now a few structural, behavioral, and ecological aspects of these two different reproductive modes.

Think of a new sponge budding from a parent sponge (page 290). Or think of flatworm fission. As certain flatworms glide along, their body starts to constrict below the midsection. The part behind the constricting region grips a substrate and initiates a tug-of-war with the part in front of it. After several hours it splits off, both parts go their separate ways—and both regenerate the missing part to become a whole worm. In such cases

BASIC PATTERNS OF DEVELOPMENT

Stages in Development

You don't look like a frog. You didn't look like one when the two of you were **embryos**, either. Embryos are transitional forms on the road from a fertilized egg to the adult. Yet despite the differences, it is possible to identify patterns in the way animals develop. These patterns can serve as a framework for your reading.

Figure 32.2 lists the stages of animal development. In the first stage, **gamete formation**, eggs or sperm form inside a parent. The second stage, **fertilization**, starts when a sperm and egg fuse. It ends when the egg nucleus and sperm nucleus fuse to form a zygote, the first cell of a new individual. Next comes **cleavage**—mitotic cell divisions that convert the zygote into a multicelled embryo. In many animals, cleavage produces a blastula, a ball of cells no wider in diameter than the zygote was.

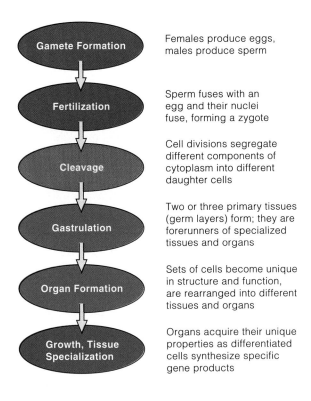

Gamete Formation	Females produce eggs, males produce sperm
Fertilization	Sperm fuses with an egg and their nuclei fuse, forming a zygote
Cleavage	Cell divisions segregate different components of cytoplasm into different daughter cells
Gastrulation	Two or three primary tissues (germ layers) form; they are forerunners of specialized tissues and organs
Organ Formation	Sets of cells become unique in structure and function, are rearranged into different tissues and organs
Growth, Tissue Specialization	Organs acquire their unique properties as differentiated cells synthesize specific gene products

Figure 32.2 Overview of the stages of animal development.

As cleavage draws to a close, the pace of cell division slackens. The embryo enters **gastrulation**, a stage of major reorganization. Now cells become arranged into two or three primary tissues, or "germ layers," which will give rise to all tissues of the adult:

endoderm inner germ layer; forerunner of the gut's inner lining and organs derived from it

mesoderm intermediate layer; forerunner of muscle, of circulatory, reproductive, and excretory organs, of most of the skeleton, and of connective tissue layers of the gut and integument

ectoderm surface germ layer; forerunner of nervous tissues and the integument's outer layer

All three form in the embryos of most animal species.

Next, germ layers split into subpopulations of cells. This marks the onset of **organ formation**. Different sets of cells become unique in structure and function, and they give rise to different tissues and organs. The time line in Figure 32.3 hints at how rapidly this developmental stage can proceed. Finally, during **growth and tissue specialization**, organs expand and take on specialized properties. This stage continues into adulthood.

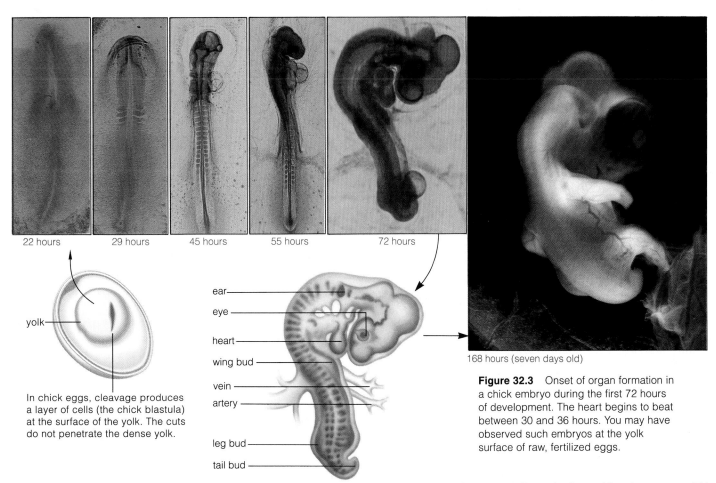

22 hours 29 hours 45 hours 55 hours 72 hours

168 hours (seven days old)

yolk

In chick eggs, cleavage produces a layer of cells (the chick blastula) at the surface of the yolk. The cuts do not penetrate the dense yolk.

ear
eye
heart
wing bud
vein
artery
leg bud
tail bud

Figure 32.3 Onset of organ formation in a chick embryo during the first 72 hours of development. The heart begins to beat between 30 and 36 hours. You may have observed such embryos at the yolk surface of raw, fertilized eggs.

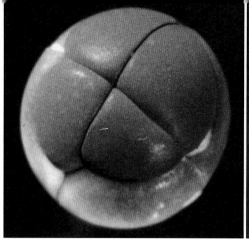
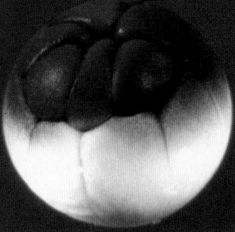
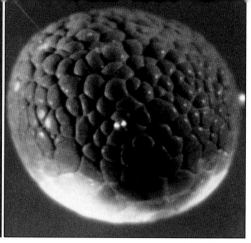

Figure 32.4 Two experiments showing how regional differences in the cytoplasm of a fertilized egg influence the fate of cells in a developing embryo. Frog eggs contain dark pigment granules in their "cortex" (the plasma membrane and the cytoplasm just below it). Granules are concentrated near one pole of the egg; yolk is concentrated near the other pole. At fertilization, a portion of the granule-containing cortex shifts away from the yolk. This exposes lighter-colored cytoplasm in a crescent-shaped gray area:

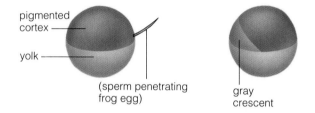

pigmented cortex

yolk

(sperm penetrating frog egg)

gray crescent

Usually the first cleavage gives both daughter cells a portion of the gray crescent. (**a**) If the two cells are experimentally split apart, each may still give rise to a whole tadpole. (**b**) A fertilized egg can be manipulated so the first cleavage plane misses the gray crescent. The daughter cell that gets the gray crescent develops into a normal tadpole. The daughter cell deprived of substances in the gray crescent gives rise to a ball of undifferentiated cells. Notice, in the photographs above, how cleavage segregates different cytoplasmic regions.

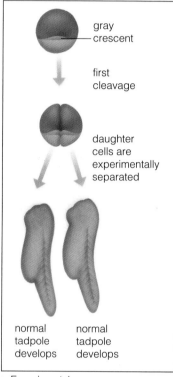

gray crescent

first cleavage

daughter cells are experimentally separated

normal tadpole develops

normal tadpole develops

a Experiment 1

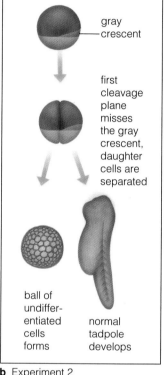

gray crescent

first cleavage plane misses the gray crescent, daughter cells are separated

ball of undiffer-entiated cells forms

normal tadpole develops

b Experiment 2

Mechanisms of Development

Early Marching Orders. You probably don't have an arm attached to your nose or toenails growing from your navel. Your parts generally are arranged in predictable patterns. For this you can thank (1) the cytoplasm of an immature egg and (2) interactions among embryonic cells.

Before a maturing egg is even fertilized, it stockpiles "maternal instructions" that will influence the shape and arrangement of future body parts. Such instructions are enzymes, mRNA, and other components that become distributed in prescribed ways in different regions of the egg cytoplasm. Then, after fer-

tilization, cleavage produces daughter cells at prescribed locations in the embryo. Simply by virtue of their location, cells end up with different marching orders. Only some cells get the type of molecule that can activate a certain gene, for example.

The experiments shown in Figure 32.4 will give you a sense of how localized differences in an egg's cytoplasm help seal the fate of embryonic cells.

Following cleavage, groups of cells start to interact physically and chemically. Together with the segregation of maternal instructions, the interactions lead to the formation of specialized tissues and organs. They are the basis of two key processes of development—cell differentiation and morphogenesis.

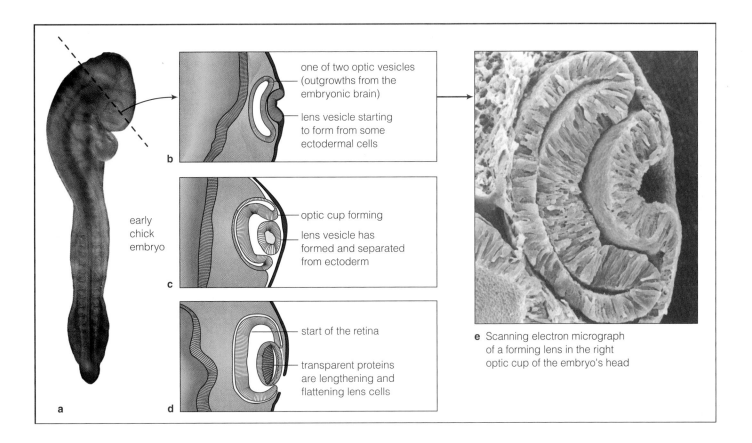

b one of two optic vesicles (outgrowths from the embryonic brain)

lens vesicle starting to form from some ectodermal cells

early chick embryo

c optic cup forming

lens vesicle has formed and separated from ectoderm

d start of the retina

transparent proteins are lengthening and flattening lens cells

a

e Scanning electron micrograph of a forming lens in the right optic cup of the embryo's head

Cell Differentiation. All cells of an animal embryo descend from the same zygote. They generally have the same number and kind of genes. Yet from gastrulation onward, different cells and their descendants start to use a fraction of the genes in different ways. In **cell differentiation**, a cell selectively uses certain genes and synthesizes proteins that are not found in other cell types. The specialized proteins are the basis of distinct cell structures, products, and functions.

Thus, when your eyes were developing, some cells started making crystallin proteins for transparent lens fibers. Only these cells could activate the required genes. The fibers made lens cells lengthen and flatten out, and so gave each lens its unique optical properties. Crystallin-producing cells are only one of 150 or so differentiated cell types now present in your body.

Morphogenesis. As an embryo continues to develop, the size, shape, and proportion of tissues and organs change, and they become organized in patterns. This process, called **morphogenesis**, involves localized cell divisions and tissue growth. It involves movements of cells and entire tissues from one site to another. And it involves the folding of sheetlike tissues as well as the controlled death of certain cells. For example, your hands were paddle shaped before you were born. But skin cells between lobes in the paddles died on cue, leaving separate fingers (Figure 7.9).

During this process, embryonic cells sense their position relative to other cells, and they respond by

Figure 32.5 Eye formation in a chick embryo. (**a**) Each eye starts forming at a fluid-filled pouch (optic vesicle) beneath ectoderm of the head. (**b**) When ectoderm cells come into contact with cells of the optic vesicle, they are induced to elongate, fold inward, and form a lens vesicle. (**c**) At the same time, the contact induces cells of the optic vesicle to sink inward, forming an optic cup from which most of the eyeball will form. (**d**) The cup's inner layer will produce the retina.

forming ordered arrangements of specialized tissues. For example, the lens of a vertebrate eye develops from ectoderm directly above a newly forming retina (Figure 32.5). Hans Spemann, a pioneer in developmental biology, surgically removed retina-forming tissue from a salamander embryo and inserted it into belly ectoderm. Afterward, a lens developed on the embryo's belly! Spemann correctly concluded that, during normal eye formation, a signal emanating from retinal cells induces ectodermal cells to develop into a lens.

Embryonic development occurs through cell differentiation and morphogenesis. Both processes depend on the segregation of cytoplasm during cleavage and on interactions among embryonic cells.

With these concepts in mind, let's look more closely at the events of sexual reproduction and development, using humans as our example.

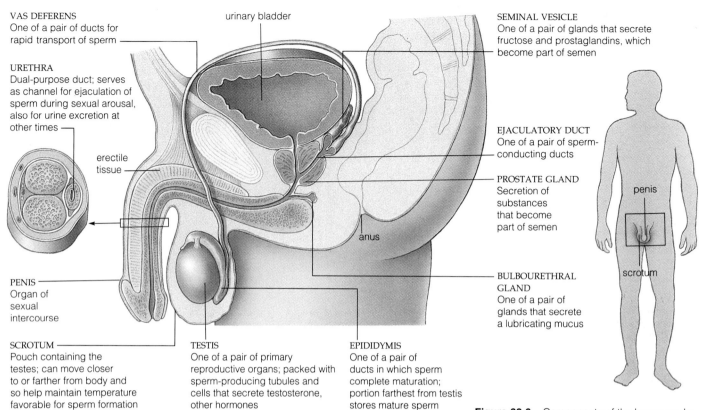

VAS DEFERENS
One of a pair of ducts for rapid transport of sperm

urinary bladder

SEMINAL VESICLE
One of a pair of glands that secrete fructose and prostaglandins, which become part of semen

URETHRA
Dual-purpose duct; serves as channel for ejaculation of sperm during sexual arousal, also for urine excretion at other times

erectile tissue

EJACULATORY DUCT
One of a pair of sperm-conducting ducts

PROSTATE GLAND
Secretion of substances that become part of semen

penis

anus

PENIS
Organ of sexual intercourse

BULBOURETHRAL GLAND
One of a pair of glands that secrete a lubricating mucus

scrotum

SCROTUM
Pouch containing the testes; can move closer to or farther from body and so help maintain temperature favorable for sperm formation

TESTIS
One of a pair of primary reproductive organs; packed with sperm-producing tubules and cells that secrete testosterone, other hormones

EPIDIDYMIS
One of a pair of ducts in which sperm complete maturation; portion farthest from testis stores mature sperm

Figure 32.6 Components of the human male reproductive system and their functions.

HUMAN REPRODUCTIVE SYSTEM

The reproductive system of both men and women consists of a pair of gonads (primary reproductive organs), accessory glands, and ducts. Female gonads are egg-producing **ovaries**. Male gonads are sperm-producing **testes** (singular, testis). Testes and ovaries also secrete sex hormones that influence reproduction and the development of **secondary sexual traits**. Such traits are associated with maleness and femaleness but don't play a direct role in reproduction. Examples are the amount and distribution of body fat, hair, and skeletal muscle.

Gonads look the same in all early human embryos. After seven weeks, genes on the sex chromosomes and hormone secretions induce their development into testes *or* ovaries (page 140). By birth, gonads and accessory organs are already formed. They reach full size and become functional ten to sixteen years later.

Male Reproductive Organs

Where Sperm Form. Figure 32.6 shows the organs of the male reproductive system and lists their functions. In an embryo, testes form on the abdominal cavity wall. Before birth, they descend into the scrotum, an outpouching of skin below the pelvic region. Sperm develop properly when the inside of the pouch is a few degrees cooler than normal body temperature.

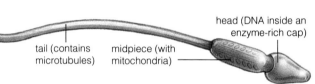

head (DNA inside an enzyme-rich cap)

tail (contains microtubules)

midpiece (with mitochondria)

Figure 32.7 A mature human sperm.

Many wedge-shaped lobes partition the interior of a testis. Coiled inside each lobe are two or three tubes, the seminiferous tubules. Just inside the tube walls are undifferentiated diploid cells that give rise to **sperm**, the male gametes (Figure 32.7). From puberty onward, males continuously produce sperm, with many millions in different stages of development on a given day. A mature sperm has a tail, a midpiece, and a head with a DNA-packed nucleus. Its mitochondria supply energy for whiplike movement. Its cap (acrosome) contains enzymes that assist in egg penetration.

How Semen Forms. Mammalian sperm are not quite mature when they leave a testis. They enter a long, coiled duct (epididymis), and secretions from the duct wall trigger the finishing touches. Mature sperm are stored in the last stretch of the epididymis. When a male is sexually aroused, muscle contractions propel the sperm down a thick-walled tube (vas deferens), ejaculatory ducts, and then the urethra, which also functions in urine excretion. This last tube threads through the penis (the male sex organ) and opens at its tip.

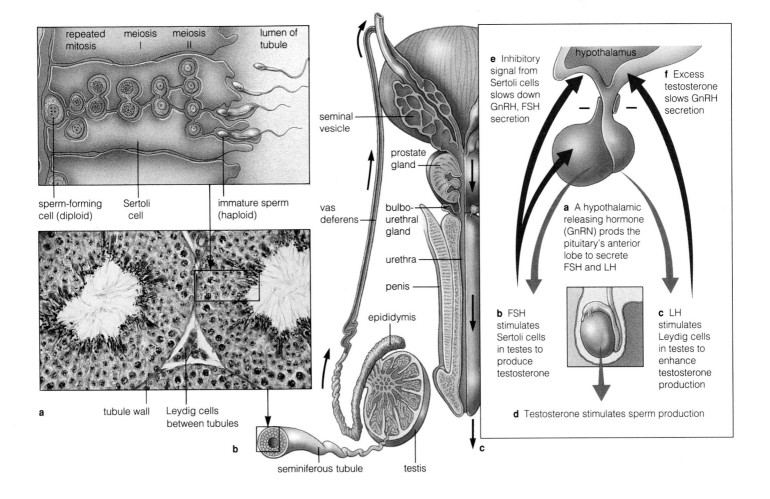

a | repeated mitosis | meiosis I | meiosis II | lumen of tubule

sperm-forming cell (diploid) | Sertoli cell | immature sperm (haploid)

tubule wall | Leydig cells between tubules

b | seminiferous tubule | testis

seminal vesicle

prostate gland

vas deferens | bulbo-urethral gland

urethra

penis

epididymis

e Inhibitory signal from Sertoli cells slows down GnRH, FSH secretion

hypothalamus

f Excess testosterone slows GnRH secretion

a A hypothalamic releasing hormone (GnRN) prods the pituitary's anterior lobe to secrete FSH and LH

b FSH stimulates Sertoli cells in testes to produce testosterone

c LH stimulates Leydig cells in testes to enhance testosterone production

d Testosterone stimulates sperm production

During the trip to the urethra, glandular secretions become mixed with sperm. The result is semen, a thick fluid that is expelled from the penis. Seminal vesicles secrete sperm-nourishing fructose and prostaglandins (these induce contractions in the female reproductive tract and assist sperm movement). Prostate gland secretions help buffer vaginal pH, which is about 3.5–4.0. Sperm motility improves at pH 6. Bulbourethral glands secrete mucus that improves sperm motility.

Hormonal Control of Male Reproductive Functions

Male reproduction depends on the hormones testosterone, LH, and FSH. **Testosterone** is secreted by Leydig cells, located in tissue between lobes in the testes. It governs the growth, form, and functions of the male reproductive tract. It stimulates sexual and aggressive behavior. It also promotes secondary sexual traits, including facial hair growth and deepening of the voice at puberty.

LH and **FSH**, recall, are secreted by the anterior lobe of the pituitary gland. The hypothalamus governs interactions among testosterone, LH, and FSH and so controls sperm formation (Figure 32.8). It secretes GnRH when the blood level of testosterone decreases. This

Figure 32.8 (**a**) Sperm formation, starting with a diploid germ cell in a seminiferous tubule. Repeated mitotic cell divisions, then meiosis, produce a clone of haploid cells that differentiate into sperm. Leydig cells outside the tubule and Sertoli cells inside interact to produce testosterone, which promotes sperm formation. (**b**) Posterior view of the male reproductive tract, showing the route that mature sperm follow before ejaculation from the penis of a sexually aroused male.

(**c**) Negative feedback loops to the hypothalamus and pituitary gland from the testes. With these loops, excess testosterone production shuts off the mechanisms leading to its production. This helps maintain the testosterone level in amounts required for sperm formation.

releasing hormone stimulates the pituitary to release LH and FSH, which the bloodstream distributes to the testes. There, LH stimulates Leydig cells to secrete testosterone, which enters the tubes where sperm form. FSH enters the same tubes and diffuses into Sertoli cells. It improves testosterone uptake by those cells, in ways that enhance sperm production.

Over the long term, the testosterone level doesn't fluctuate much. Whenever it increases past a set point, feedback loops to the hypothalamus slow testosterone secretion. The hormone balancing act assures that sperm will be available when the female becomes receptive.

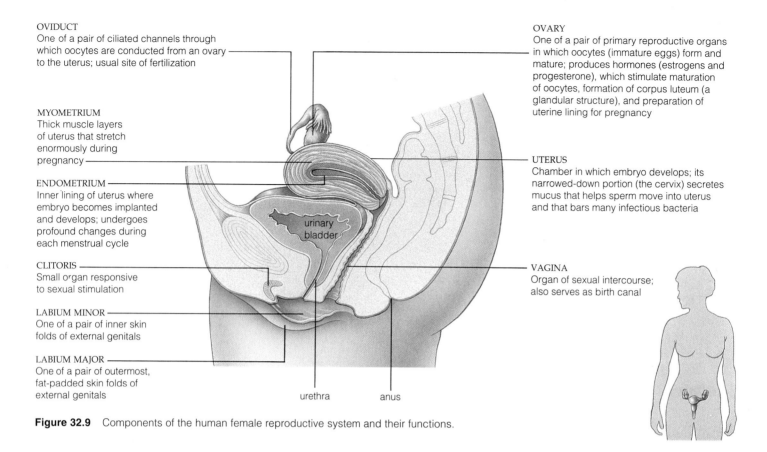

OVIDUCT
One of a pair of ciliated channels through
which oocytes are conducted from an ovary
to the uterus; usual site of fertilization

MYOMETRIUM
Thick muscle layers
of uterus that stretch
enormously during
pregnancy

ENDOMETRIUM
Inner lining of uterus where
embryo becomes implanted
and develops; undergoes
profound changes during
each menstrual cycle

CLITORIS
Small organ responsive
to sexual stimulation

LABIUM MINOR
One of a pair of inner skin
folds of external genitals

LABIUM MAJOR
One of a pair of outermost,
fat-padded skin folds of
external genitals

urinary
bladder

urethra   anus

OVARY
One of a pair of primary reproductive organs
in which oocytes (immature eggs) form and
mature; produces hormones (estrogens and
progesterone), which stimulate maturation
of oocytes, formation of corpus luteum (a
glandular structure), and preparation of
uterine lining for pregnancy

UTERUS
Chamber in which embryo develops; its
narrowed-down portion (the cervix) secretes
mucus that helps sperm move into uterus
and that bars many infectious bacteria

VAGINA
Organ of sexual intercourse;
also serves as birth canal

Figure 32.9   Components of the human female reproductive system and their functions.

Female Reproductive Organs

Figure 32.9 shows the main components of the female reproductive system and lists their functions. Eggs are produced in the paired ovaries. After an immature egg, or **oocyte**, is released from an ovary, it enters a nearby oviduct. The paired oviducts are channels to the uterus, a hollow, pear-shaped organ where the embryo grows and develops. The uterus has a thick layer of smooth muscle (myometrium), lined inside with connective tissue, glands, and blood vessels. This lining is called the **endometrium**. The narrowed part of the uterus is the cervix. A muscular tube, the vagina, extends from the cervix to the body surface. This tube receives sperm and functions as part of the birth canal.

At the body surface are external genitals (vulva) that include organs for sexual stimulation. Outermost are a pair of fat-padded skin folds, the labia majora. They enclose a smaller pair of skin folds (labia minora) that are highly vascularized but have no fatty tissue. The smaller folds partly enclose the clitoris, an organ sensitive to stimulation. The urethra's opening is about midway between the clitoris and the vaginal opening.

Menstrual Cycle

Overview of the Cycle.   Most mammalian females follow an estrous cycle. Only at certain times of year are they fertile *and* in heat (sexually receptive to males). By

contrast, female primates, including humans, follow a **menstrual cycle**. They are fertile intermittently, on a cyclic basis. The times of heat and fertility are not synchronized. Females of reproductive age can get pregnant at certain times only, but they may be receptive to sex at any time.

During a menstrual cycle, an oocyte matures and is released from an ovary. As it matures, hormones prime the endometrium to receive and nourish an embryo, *if* fertilization occurs. Blood-rich fluid, about four to six tablespoons total, starts flowing out through the vaginal canal on the first day of the cycle. It means there is no embryo at this time. The uterine "nest" is being sloughed off, and it is about to be constructed all over again.

The menstrual cycle operates through feedback loops to the hypothalamus and pituitary gland from the ovaries. As you will see, FSH and LH promote cyclic changes in ovaries. They also stimulate the ovaries to secrete **estrogens** and **progesterone**. These hormones promote the cyclic changes in the endometrium.

A human female's menstrual cycles start between ages ten and sixteen. Each cycle lasts about twenty-eight days, but this is simply the average. It runs longer for some women and shorter for others. The cycles continue until a woman is in her late forties or early fifties, when her egg supply is dwindling and hormone secretions slow down. This is the onset of menopause, the twilight of reproductive capacity.

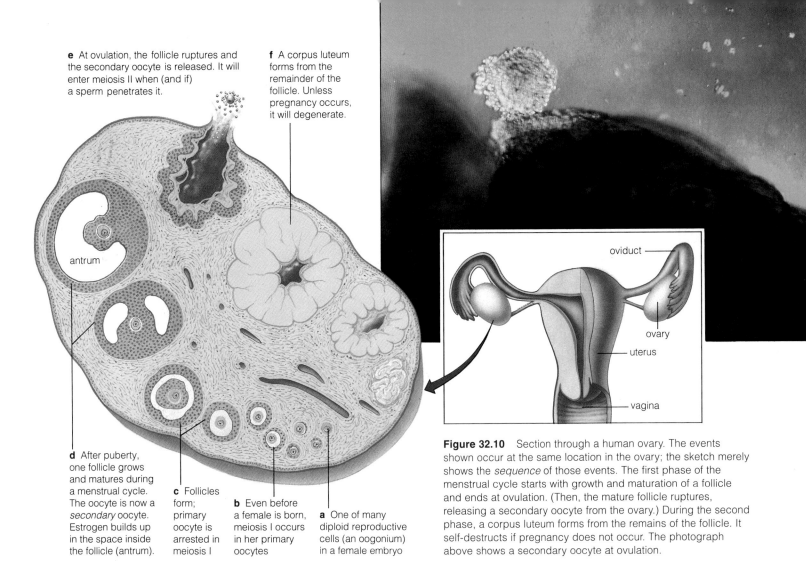

e At ovulation, the follicle ruptures and the secondary oocyte is released. It will enter meiosis II when (and if) a sperm penetrates it.

f A corpus luteum forms from the remainder of the follicle. Unless pregnancy occurs, it will degenerate.

antrum

d After puberty, one follicle grows and matures during a menstrual cycle. The oocyte is now a *secondary* oocyte. Estrogen builds up in the space inside the follicle (antrum).

c Follicles form; primary oocyte is arrested in meiosis I

b Even before a female is born, meiosis I occurs in her primary oocytes

a One of many diploid reproductive cells (an oogonium) in a female embryo

oviduct

ovary

uterus

vagina

Figure 32.10 Section through a human ovary. The events shown occur at the same location in the ovary; the sketch merely shows the *sequence* of those events. The first phase of the menstrual cycle starts with growth and maturation of a follicle and ends at ovulation. (Then, the mature follicle ruptures, releasing a secondary oocyte from the ovary.) During the second phase, a corpus luteum forms from the remains of the follicle. It self-destructs if pregnancy does not occur. The photograph above shows a secondary oocyte at ovulation.

Ovarian Function. Take a moment to review Figure 8.9, the generalized picture of how meiosis proceeds in an oocyte. A normal baby girl has about 2 million *primary* oocytes in her ovaries. By the time she is seven years old, about 300,000 remain (her body has resorbed the rest). These oocytes have entered meiosis I, but the division process has been arrested. Meiosis will resume in one oocyte at a time, starting with the first menstrual cycle. Only about 400 or 500 will be released during the reproductive years.

A primary oocyte is located near the surface of an ovary. A layer of cells (granulosa cells) surrounds and nourishes it. Together, the primary oocyte and the cell layer are a **follicle**. Figure 32.10 shows one of these.

At the start of a menstrual cycle, the hypothalamus secretes GnRH, which stimulates the anterior pituitary to release FSH and LH. These hormones stimulate a follicle to grow. (Hence the name FSH, for "follicle-stimulating hormone." Soon the meaning of LH will be revealed to you.) The oocyte increases in size, and more cell layers form around it. Glycoprotein deposits build up between the oocyte and the cell layers, widening the space between them. In time they form a noncellular coating, the zona pellucida, around the oocyte.

In response to FSH and LH, cells outside the zona pellucida secrete several estrogens. Estrogen-containing fluid starts to accumulate inside the follicle. And estrogen levels in the blood start to rise.

About eight to ten hours before being released from the ovary, the oocyte completes meiosis I. Then the cytoplasm divides, so there are now two cells inside the follicle. One cell, the *secondary* oocyte, ends up with nearly all the cytoplasm. The other cell is the first **polar body**. Chromosomes are distributed between the two cells so that the secondary oocyte will be haploid—the chromosome number required of gametes.

About midway through the cycle, the pituitary gland detects the rising estrogen levels and responds with a brief outpouring of LH. The LH triggers cellular contractions that make the fluid-filled follicle balloon outward, then rupture. The fluid escapes, carrying the secondary oocyte with it (Figure 32.10). The midcycle surge of LH has triggered **ovulation**—the release of a secondary oocyte from the ovary.

Uterine Function. The estrogens released during the early phase of the menstrual cycle also contribute to changes that prepare the uterus for pregnancy. They stimulate the growth of the endometrium and its glands. Then, just before the midcycle surge of LH, cells of the zona pellucida also start secreting progesterone as well as estrogens. The progesterone causes blood vessels to grow rapidly in the thickened endometrium.

At ovulation, estrogens also act on the cervix, the narrow opening to the uterus from the vagina. The cervix now secretes large amounts of a thin, clear mucus, an ideal medium for sperm travel.

Another hormone dominates events after ovulation. Granulosa cells left behind in the follicle differentiate into a yellowish glandular structure, the **corpus luteum**. The name means "yellow body." Formation of the corpus *luteum* results from the midcycle surge of LH (hence the name, *luteinizing* hormone).

The corpus luteum secretes progesterone as well as some estrogen. Progesterone prepares the reproductive tract for the arrival of an embryo. For example, it causes mucus in the cervix to become thick and sticky. The mucus can serve as a barrier against vaginal bacteria that might enter the uterus through the cervix and endanger the embryo. Progesterone also maintains the endometrium during a pregnancy.

The corpus luteum persists for about twelve days. During that time, the hypothalamus calls for a decrease in FSH secretion. This prevents other follicles from developing. However, if an embryo doesn't arrive and burrow into the endometrium, the corpus luteum self-destructs during the last days of the menstrual cycle. It secretes prostaglandins, which apparently disrupt its own functioning.

After this, progesterone and estrogen levels fall rapidly, so the endometrium starts to break down. Deprived of oxygen and nutrients, its blood vessels constrict, and its tissues die. Blood escapes from the ruptured walls of weakened capillaries. The blood and sloughed endometrial tissues make up the menstrual flow, which continues for three to six days. Then the cycle begins anew, and rising estrogen levels stimulate the repair and growth of the endometrium.

Each year, as many as 10 million American women are affected by *endometriosis*, the spread and growth of endometrial tissue outside the uterus. Estrogen acts on endometrial tissue wherever it occurs. Its action may cause sensations of pain during menstruation, sex, or urination. Also, endometrial scar tissue may form on the ovaries or oviducts and lead to infertility. The ailment may arise when some menstrual flow backs up through the oviducts and spills into the pelvic cavity. Or perhaps some embryonic cells were positioned in the wrong place before birth and are stimulated to grow at puberty, when sex hormones become active.

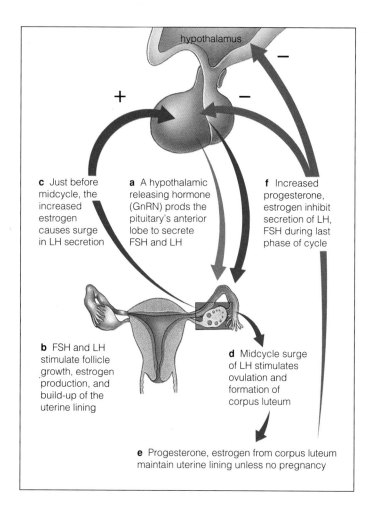

Figure 32.11 Feedback loops to the hypothalamus and pituitary gland from the ovaries during a menstrual cycle. A positive feedback loop causes a surge in LH secretion that triggers ovulation. Negative feedback loops after ovulation inhibit FSH and LH secretion to prevent another follicle from maturing until the cycle is completed.

Summing Up. By now, you probably sense that the menstrual cycle isn't a simple tune on a biological banjo. It is a full-blown hormonal symphony! Figures 32.11 and 32.12 will help you correlate the cycle's changing hormone levels with the changes they bring about in the ovary and uterus.

Sexual Intercourse

Suppose a secondary oocyte is on its way down an oviduct when a female and male are engaged in sexual intercourse (coitus). Within seconds of sexual arousal, the penis undergoes changes that will help it penetrate into the vaginal canal. As Figure 32.6 shows, a penis contains three cylinders of spongy tissue. One of these has a mushroom-shaped tip (glans penis) loaded with sensory receptors that are activated by friction. In sexually unaroused males, blood vessels leading into the three cylinders are constricted and the penis is limp. In

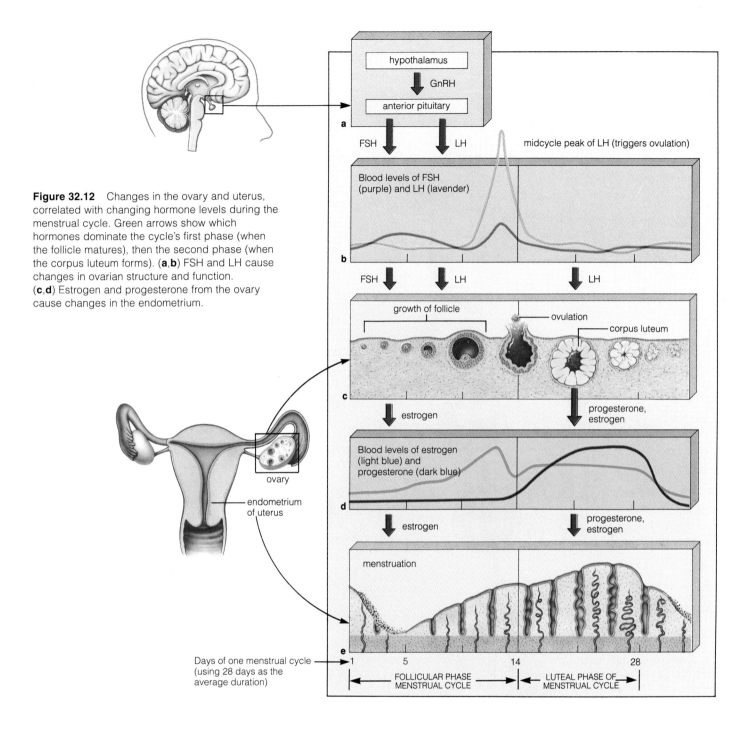

Figure 32.12 Changes in the ovary and uterus, correlated with changing hormone levels during the menstrual cycle. Green arrows show which hormones dominate the cycle's first phase (when the follicle matures), then the second phase (when the corpus luteum forms). (**a**,**b**) FSH and LH cause changes in ovarian structure and function. (**c**,**d**) Estrogen and progesterone from the ovary cause changes in the endometrium.

aroused males, blood flows into the cylinders faster than it flows out. As blood collects in the spongy tissue, the penis lengthens and stiffens.

During coitus, pelvic thrusts stimulate the penis as well as the female's vaginal walls and clitoral region. Mechanical stimulation causes rhythmic, involuntary contractions in the male reproductive tract. Sperm are moved rapidly out of the epididymis. The contents of seminal vesicles and the prostate gland are forced into the urethra. And semen is ejaculated into the vagina.

(During ejaculation, a sphincter closes and prevents urine from being excreted from the bladder.)

Together, the muscular contractions, ejaculation, and associated sensations of release, warmth, and relaxation are called "orgasm." Similar events occur during female orgasm. They include an intense vaginal awareness, involuntary uterine and vaginal contractions, and sensations of relaxation and warmth. Even if the female does not reach this state of excitation, she can still get pregnant.

Figure 32.13 Fertilization and implantation in the uterus.

a. A sperm penetrates the zona pellucida of a secondary oocyte. Only the sperm's nucleus and centrioles enter the oocyte's cytoplasm. This triggers meiosis II in the first polar body and in the oocyte.

b. The sperm nucleus fuses with the egg nucleus at fertilization, producing the zygote. With the first cleavage, the fertilized egg enters the two-cell stage.

c. The second cleavage produces the four-cell stage.

d. Successive cleavages produce a solid ball of cells, the morula.

e. Fluid enters the ball and lifts some cells, forming a cavity. This produces the blastocyst, a ball of cells having a surface layer and an inner cell mass.

surface layer of cells

inner cell mass

blastocyst

endometrium

(uterine cavity)

f. Implantation begins when the blastocyst attaches to and invades the endometrium. During the second week after fertilization, it slowly embeds itself in the endometrium.

FROM FERTILIZATION TO BIRTH

Fertilization

So now sperm are in the vagina. A single ejaculation can put 150 million to 350 million there. If it is a few days before or after ovulation, or any time in between, fertilization may be the outcome. Within thirty minutes after ejaculation, muscle contractions move sperm deeper into the female reproductive tract. Only a few hundred sperm actually reach the upper portion of the oviduct, where fertilization most often occurs.

The stunning micrograph on page 99 shows living sperm at the surface of a secondary oocyte. Upon contacting an oocyte, sperm release digestive enzymes from the cap on their head. The enzymes clear a path through the zona pellucida. Although many sperm might get this far, usually only one fuses with the oocyte. Only its nucleus and centrioles actually enter the oocyte's cytoplasm. The secondary oocyte and the first polar body are stimulated to complete meiosis II (Figure 32.13). There are now three polar bodies and a

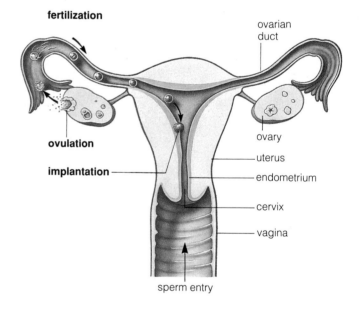

fertilization

ovarian duct

ovulation

ovary

implantation

uterus

endometrium

cervix

vagina

sperm entry

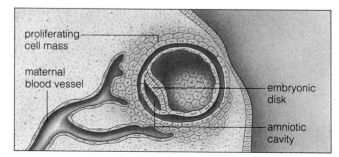

proliferating cell mass

maternal blood vessel

embryonic disk

amniotic cavity

g. During implantation, a slitlike cavity forms between the inner cell mass and the surface layer. The inner cell mass is transformed into a flattened, embryonic disk from which the embryo develops.

mature egg, or ovum (plural, ova). The sperm nucleus and egg nucleus fuse. Their chromosomes intermingle and so restore the diploid number for the zygote.

Late in a woman's life, eggs have been waiting in the wings for a long time. Mistakes are more likely to occur when they finally enter meiosis II. For example, their chromosomes may fail to separate properly, leading to disorders such as Down syndrome (page 148).

Implantation

Three or four days after fertilization, the zygote is moving through the oviduct, and cleavage has begun. By the time the cluster of dividing cells reaches the uterus, it is a solid ball (morula). A fluid-filled cavity forms in it, producing a blastocyst. A human blastocyst has a surface layer of cells and an inner cell mass (Figure 32.13e). **Implantation** begins five or six days after fertilization. By this process, the blastocyst adheres to the uterine lining, some of its cells send out projections that invade the mother's tissues, and connections start forming that will transfer nutrients and otherwise sustain the developing embryo through the many months ahead. While the invasion is proceeding, the inner cell mass will be transformed into two cell layers having a flattened, somewhat circular shape. The embryo proper will develop from this "embryonic disk" (Figure 32.14).

Membranes Around the Embryo

As implantation progresses, a fluid-filled cavity starts forming between the embryonic disk and the blastocyst surface. In time, the embryo will be surrounded by its membranous lining, the **amnion**. The amniotic fluid will be its buoyant cradle, where the embryo can grow, move freely, and be protected from sudden temperature shifts and impacts. Meanwhile, the blastocyst cavity becomes lined with another membrane, the **yolk sac**, which is part of the evolutionary heritage of land vertebrates (page 317). In most shelled eggs, the yolk sac holds nutritive yolk. In human development, however, part becomes a site of blood cell formation, and some of its cells give rise to germ cells, the forerunners of gametes.

Before the blastocyst is fully implanted, spaces open in maternal tissues and fill with blood seeping from ruptured capillaries. In the blastocyst itself, a membrane-lined cavity opens up around the amnion and yolk sac. Fingerlike projections start forming on the membranous lining, the **chorion**. The chorion will become part of a spongy, blood-engorged tissue called the placenta. Two weeks after a blastocyst is fully implanted, a membrane outpouching, the **allantois**, will appear on the yolk sac. For reptiles, birds, and some mammals, the allantois functions in respiration and storage of metabolic wastes. In human development, it functions in early blood formation and in the formation of the urinary bladder.

In time the blastocyst secretes a hormone, **HCG** (human gonadotropin), which makes the corpus luteum keep on secreting estrogen and progesterone. Thus the blastocyst prevents menstrual flow and avoids being sloughed off. By the third week of pregnancy, HCG can be detected in the mother's blood or urine. At-home pregnancy tests use a "dip-stick" that changes color when HCG is present in urine.

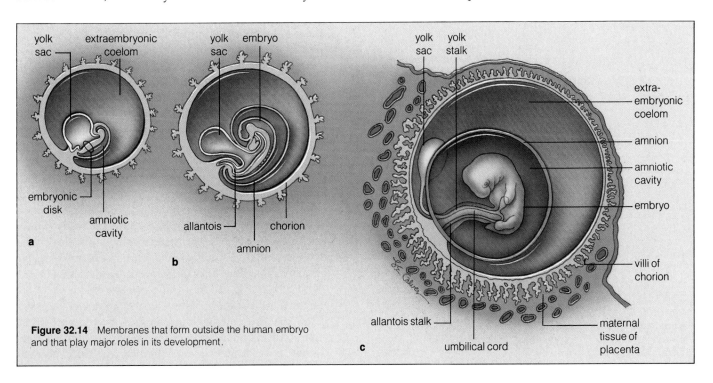

Figure 32.14 Membranes that form outside the human embryo and that play major roles in its development.

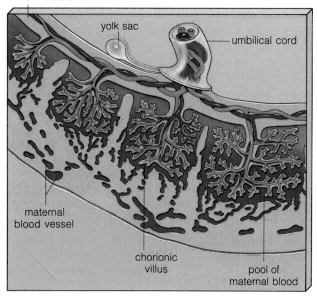

Figure 32.15 Relationship between fetal and maternal tissues in the placenta.

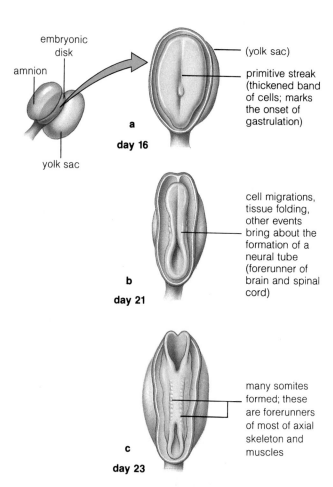

a
day 16

embryonic disk

amnion

yolk sac

(yolk sac)

primitive streak (thickened band of cells; marks the onset of gastrulation)

b
day 21

cell migrations, tissue folding, other events bring about the formation of a neural tube (forerunner of brain and spinal cord)

c
day 23

many somites formed; these are forerunners of most of axial skeleton and muscles

Figure 32.16 Transformation of an embryonic disk into the early human embryo. These are three dorsal views (of the embryo's back).

The Placenta

While the blastocyst is implanting itself, bumps on its surface start to become chorionic villi, which project into the endometrium. Small blood vessels migrate into each villus, and blood-filled spaces form around them. In time, one-fourth of the inner surface of the uterus will be a spongy, blood-engorged tissue—the **placenta**. It will form from endometrium and extraembryonic membranes, the chorion especially (Figure 32.15).

The placenta is the body's way of sustaining an embryo while allowing its blood vessels to develop apart from the mother's. Oxygen and nutrients diffuse out of the maternal blood vessels, across the placenta's blood-filled spaces, then into the embryonic blood vessels. So do antibodies, prescription drugs, cocaine, and alcohol. Carbon dioxide and other wastes diffuse in the opposite direction, and they are disposed of quickly by the mother's lungs and kidneys.

After the third month of pregnancy, the placenta will secrete progesterone and estrogens—and so take over the task of maintaining the uterine lining.

Embryonic and Fetal Development

First Trimester. Pregnancy lasts about nine months. Its "first trimester" extends from fertilization to the end of the third month.

During the week after implantation, the embryonic disk gives rise to three primary tissue layers—and the embryo proper. At the disk's surface, cell migrations foreshadow the formation of the nervous system (Figure 32.16). Inward-migrating cells form mesoderm—the start of the heart, muscles, bone, and other internal organs. Later, endoderm will give rise to parts of the respiratory and digestive systems. After the third week, an early, tubelike heart is already beating.

By the end of the fourth week, the embryo has grown 500 times its original size. Its growth spurt gives way to four weeks in which the main organs develop more slowly. The nerve cord and the four heart chambers form. Respiratory organs form but are not yet functional.

A segmentation pattern appears in all vertebrate embryos, and it shows up during the first trimester. Some of the paired segments, or somites, are forerunners of connective tissues, bones, and muscles (Figure 32.16). Arms, legs, fingers, and toes now develop, along with the tail that emerges in all vertebrate embryos. The human tail slowly disappears after the eighth week.

Second Trimester. The "second trimester" extends from the start of the fourth month to the end of the sixth. All major organs have formed, and the growing individual is now called a **fetus**. Figure 32.17*d* shows a

gill arches   somites

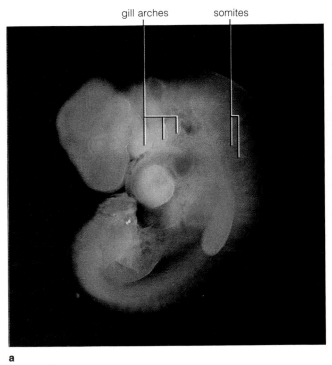

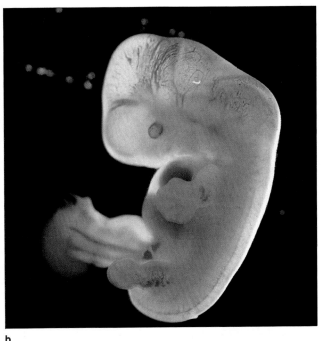

a

b

Figure 32.17 (**a**) The embryo at four weeks, about 7 millimeters (0.3 inch) long. Its tail and gill arches are embryonic features of all vertebrates. Arm and leg buds are visible. (**b**) Embryo at the end of five weeks, about 12 millimeters long. The head starts to enlarge and the trunk starts to straighten. Finger rays appear in the paddlelike forelimbs. A pigmented retina outlines the forming eyes. (**c**) An embryo near the end of the first trimester, suspended in fluid within the amniotic sac. The chorion, which covers the amniotic sac, has been opened and pulled aside. (**d**) The fetus at sixteen weeks.

umbilical cord   amniotic sac

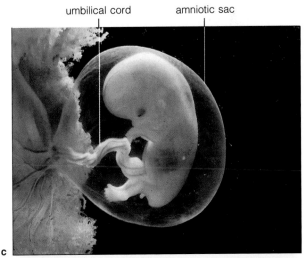

c

fetus at this stage of pregnancy. Movements of facial muscles produce frowns and squints. The sucking reflex, shown in the *Focus* essay on the next page, also is evident. Before the second trimester draws to a close, the mother can sense movements of the fetal arms and legs.

When the fetus is five months old, its heart can be heard through a stethoscope on the mother's abdomen. Soft, fuzzy hair (the lanugo) covers its body. Its skin is wrinkled, reddish, and protected from abrasion by a thick, cheesy coating. In the sixth month, eyelids and eyelashes form. In the seventh month, the eyes open.

Third Trimester. The "third trimester" extends from the seventh month until birth. After the middle of the third trimester, the fetus can survive on its own if born prematurely or removed surgically from the uterus. However, with medical support, fetuses only 23–25 weeks old have survived early delivery. Although development might appear to be relatively complete by the seventh month, few fetuses are able to breathe normally or maintain a normal body temperature then, even with the best medical care. By the ninth month, survival chances increase to about 95 percent.

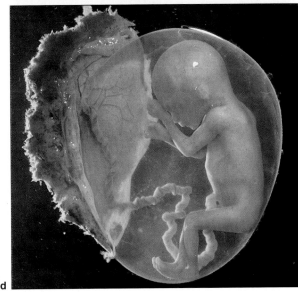

d

Mother as Protector, Provider, Potential Threat

A woman who decides to become pregnant is committing a large part of her body's resources and functions to the development of a new individual. From the time of fertilization to birth, her future child will be at the mercy of her diet, health habits, and life-style (Figure *a*).

Some Nutritional Considerations. How does a pregnant woman best provide nutrients for the embryo, and then the fetus? Normally, the same balanced diet of whole foods that is good for her provides it with all the required carbohydrates, lipids, and proteins (pages 406–407). Her own need for vitamins and minerals increases, but the placenta absorbs enough of these from the bloodstream, even at her expense. If her diet is marginally poor, the addition of nutritious, protein-rich food will do her and the fetus more good than vitamin pills and other food supplements.

A pregnant woman also must eat adequate amounts of food, so that she gains between twenty and twenty-five pounds, on the average. If she gains a great deal less than this, she is stacking the deck against her future child. Compared to newborns of normal weight, the significantly underweight ones go through more postdelivery complications. They also are more at risk of being mentally impaired later in life.

As birth approaches, the fetus makes greater nutritional demands of the mother, and her diet profoundly influences the remaining developmental events. As is true of most other fetal organs, the brain is especially vulnerable in the weeks just before and after birth, when it undergoes its greatest expansion. All of its neurons are formed; that is why poor nutrition now will have repercussions on intelligence and other brain functions later in life.

Risk of Infections. Antibodies circulating in a pregnant woman's bloodstream continually cross the placenta. They help protect a developing individual from all but the most severe bacterial infections. But certain viral diseases may be dangerous during the first six weeks after fertilization—the critical time of organ formation. Suppose a woman contracts rubella (German measles) during that period. There is a 50 percent chance some organs won't form properly. For example, if embryonic ears are forming, her child might be born deaf. (Vaccination *before* pregnancy can prevent German measles.) If the disease is contracted during or after the fourth month, it has no discernible effect.

Effects of Prescription Drugs. During the first trimester, an embryo is highly sensitive to drugs. Consider what happened after *thalidomide* started being prescribed in Europe. Women who had used this tranquilizer during the first trimester gave birth to infants with missing or severely deformed arms and legs. Thalidomide was withdrawn from the market after the deformities were traced to it. However, other tranquilizers, sedatives, and barbiturates are still prescribed. They may cause similar, although less severe, damage. Even certain anti-acne drugs increase the risk of facial and cranial deformities. Tetracycline, a commonly prescribed antibiotic, yellows the teeth. Strepto-

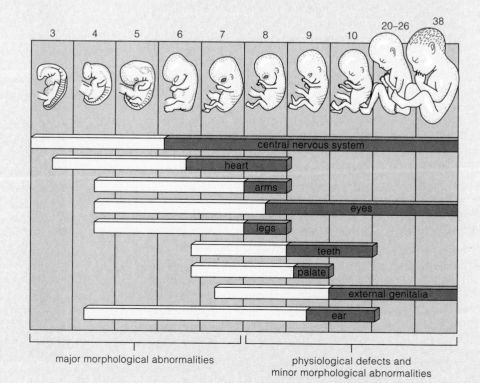

a Critical periods of embryonic and fetal development. Red indicates periods in which organs are most sensitive to damage from cigarette smoke, alcohol, viral infection, and so on. Numbers signify the week of development.

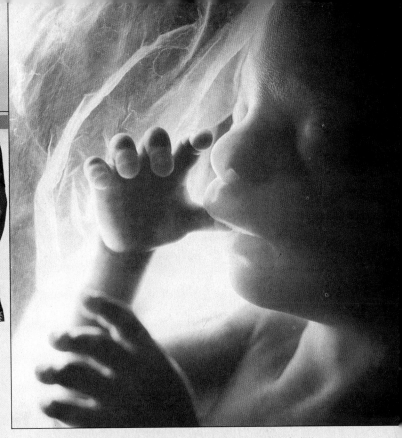

mycin causes hearing problems and may affect the nervous system.

At no stage of development is the embryo impervious to drugs. A pregnant woman should not take any drugs at all unless they are prescribed by a good physician.

Effects of Alcohol. As a fetus grows, its physiology becomes increasingly like the mother's. Alcohol passes freely across the placenta and has the same effect on the fetus as it has on her. *Fetal alcohol syndrome* (FAS) is a set of deformities that result from excessive alcohol intake during pregnancy. FAS is the third most common cause of mental retardation in the United States, and the number of reported cases is increasing. Newborns have facial deformities, poor coordination and, sometimes, heart defects (Figure *b*). Between 60 and 70 percent of the newborns of alcoholic women have FAS. Some researchers suspect that any alcohol at all during pregnancy may be dangerous for the fetus. Increasingly, physicians are urging total or near-abstinence during pregnancy.

c The fetus at eighteen weeks, about 18 centimeters (a little more than 7 inches) long. The sucking reflex begins in the earliest fetal stage, as soon as nerves establish connections with muscles. Legs kick, arms wave, fingers make grasping motions—all reflexes that will be vital skills in the world outside the uterus.

Effects of Cocaine. A mother who uses cocaine, particularly crack, disrupts the nervous system of her future child as well as her own. The consequences extend beyond birth. Here you may wish to read again the introduction to Chapter 30.

Effects of Cigarette Smoke. Cigarette smoking impairs fetal growth and development. Newborns of women who smoked every day during pregnancy have a low birth weight. They do even if the woman's weight, nutritional status, and all other relevant variables are identical with those of pregnant nonsmokers. Smoking has other effects. In Great Britain, all infants born during a certain week were tracked for seven years. Besides being smaller, newborns of smokers had a 30 percent greater incidence of postdelivery deaths and a 50 percent greater incidence of heart abnormalities. At age seven, their average "reading age" was nearly half a year behind that of the children of nonsmokers.

In this last study, newborns of women who stopped smoking by the middle of the second trimester were indistinguishable from those born to nonsmokers. The mechanisms by which smoking affects the fetus are not known. But its demonstrated effects are further evidence that the placenta, marvelous structure that it is, cannot prevent all assaults on the fetus that the human mind can dream up.

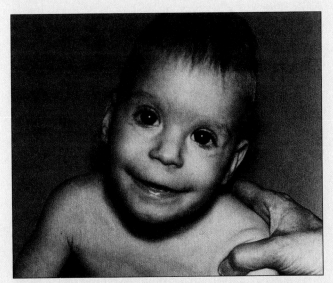

b An infant affected by FAS. Symptoms include a small head, low and prominent ears, poorly developed cheekbones, and a long, smooth upper lip. The child can expect to encounter growth problems and abnormalities of the nervous system. About 1 in 750 newborns in the United States is affected by this disorder.

525

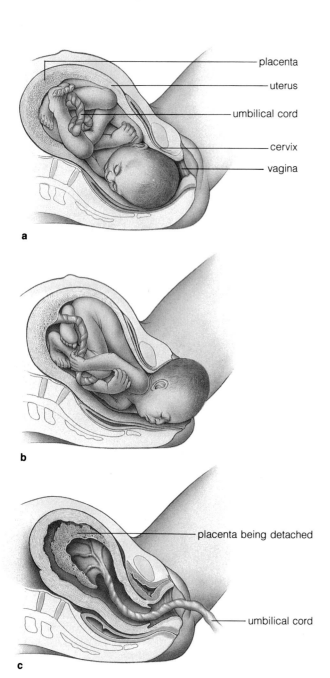

Figure 32.18 Expulsion of the fetus during the birth process. The placenta, fluid, and blood are expelled shortly afterward (this is the "afterbirth").

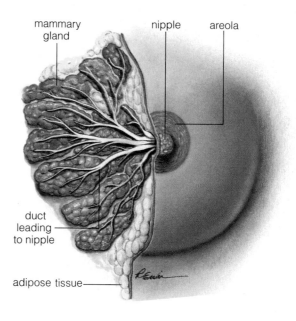

Figure 32.19 Breast of a lactating female. This cutaway view shows the mammary glands and ducts.

Birth and Lactation

Birth takes place thirty-nine weeks after fertilization, give or take a few weeks. It starts when the uterus begins to contract. For the next two to eighteen hours, contractions become stronger and more frequent. Inside the cervix, the canal through which the fetus will pass dilates fully. Just before birth, the amnion usually ruptures, and "water" (amniotic fluid) gushes out from the vagina.

Birth typically occurs less than an hour after full dilation. Immediately afterward, uterine contractions force fluid, blood, and the placenta from the body (Figure 32.18). The umbilical cord—the lifeline to the mother—is now severed, and the newborn embarks on its nurtured existence in the outside world.

During **lactation**, hormone-primed glands produce milk. Estrogen and progesterone stimulated the growth of mammary glands and ducts in the mother's breasts during the pregnancy (Figure 32.19). For the first few days following birth, the glands produce a fluid rich in proteins and lactose. Then prolactin secreted by the pituitary stimulates milk production (page 499). When a newborn suckles, the pituitary also releases oxytocin. This hormone causes breast tissue to contract and force milk into the ducts. It also triggers uterine contractions that shrink the uterus back to normal size.

Table 32.1	Stages of Human Development: A Summary

Prenatal Period:

1.	Zygote	Single cell resulting from fusion of sperm nucleus and egg nucleus at fertilization
2.	Morula	Solid ball of cells produced by cleavages
3.	Blastocyst	Ball of cells with surface layer and inner cell mass
4.	Embryo	All developmental stages from two weeks after fertilization until end of eighth week
5.	Fetus	All developmental stages from the ninth week until birth (about thirty-nine weeks after fertilization)

Postnatal Period:

6.	Newborn	Individual during the first two weeks after birth
7.	Infant	Individual from two weeks to about fifteen months after birth
8.	Child	Individual from infancy to about ten or twelve years
9.	Pubescent	Individual at puberty, when secondary sexual traits develop; girls between ten and fifteen years, boys between twelve and sixteen years
10.	Adolescent	Individual from puberty until about three or four years later; physical, mental, emotional maturation
11.	Adult	Early adulthood (between eighteen and twenty-five years); bone formation and growth completed. Changes proceed very slowly afterward.
12.	Old age	Aging follows late in life.

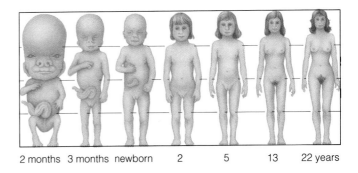

2 months   3 months   newborn   2   5   13   22 years

Figure 32.20 Changing proportions of the human body during prenatal and postnatal growth.

POSTNATAL DEVELOPMENT AND AGING PROCESSES

Following birth, the new individual follows a prescribed course of further growth and development that leads to the adult, the mature form of the species. Table 32.1 summarizes all the prenatal and postnatal stages. (Prenatal means before birth; postnatal means after birth.) Figure 32.20 shows how the human body changes in proportions as the course is followed.

Late in life, the body gradually deteriorates through processes collectively called **aging**. Cells start to break down, and this leads to structural changes and gradual loss of body functions. All organisms with extensively differentiated cells undergo aging.

No one knows what causes aging, although research gives us interesting things to think about. More than two decades ago, Paul Moorhead and Leonard Hayflick cultured normal human embryonic cells. All of the cell lines proceeded to divide about fifty times, then the entire population died off. Hayflick took some cultured cells that were partway through the series of divisions and froze them for several years. Then he thawed the cells and placed them in a culture medium. The cells proceeded to complete the cycle of fifty doublings—whereupon they all died on schedule.

Such experiments suggest that normal cell types have a limited division potential. That is, mitosis may be genetically programmed to decline at a certain time in the life cycle. But does the change in mitosis *cause* aging or is it a *result* of the aging process? Think of neurons, which stop dividing after birth. They still deteriorate over time. Maybe age-related changes also affect dividing cells throughout the body.

It may be that cells gradually lose the capacity for DNA self-repair as a result of an accumulation of environmental insults. Over time, DNA mutations could thwart the production of enzymes and other proteins required for proper cell functioning. Consider how cells depend on smooth exchanges of materials between the cytoplasm and the extracellular fluid. Collagen is present in extracellular spaces throughout the body. If deteriorating DNA regions code for defective collagen molecules, it is conceivable that the movement of oxygen, nutrients, hormones, and so forth to and from cells could be hampered. Repercussions could ripple through the entire body.

Finally, consider what might happen if genes coding for membrane proteins deteriorate. What if the proteins serve as self markers? If they become altered, does the immune system then perceive the body's own cells as "foreign" and attack them? According to one theory, such autoimmune responses might increase over time. And so would the increased vulnerability to disease and stress associated with aging.

CONTROL OF HUMAN FERTILITY

Some Ethical Considerations

The transformation of a zygote into an intricately detailed adult raises profound questions. *When does development begin?* As we have seen, key developmental events occur even before fertilization. *When does life begin?* During her lifetime, a human female can produce as many as 400 eggs, all of which are alive. During one ejaculation, a human male can release a quarter of a billion sperm, which also are alive. Even before sperm and egg merge by chance and establish the genetic makeup of a new individual, they are as much alive as any other form of life. It is scarcely tenable, then, to say "life begins" when they fuse. *Life began billions of years ago; and each gamete, each zygote, each mature individual is only a fleeting stage in the continuation of that beginning.*

This fact cannot diminish the meaning of conception. It is no small thing to entrust a new individual with the gift of life, wrapped in the unique evolutionary threads of our species and handed down through an immense sweep of time.

Yet how can we reconcile the marvel of individual birth with growing awareness of the astounding birth rate for our whole species? While this book is being written, an average of 10,700 newborns enter the world every single hour. By the time you go to bed tonight, there will be 257,000 more people on earth than there were last night at that hour. Within a week, the number will reach 1,800,000—about as many people as there are now in the entire state of Massachusetts. *Within one week.* Worldwide population growth has outstripped resources, and each year millions face the horrors of starvation. Living as we do on one of the most productive continents on earth, few of us can know what it means to give birth to a child, to give it the gift of life, and have no food to keep it alive.

And how can we reconcile the marvel of birth with the confusion surrounding unwanted pregnancies? Even highly developed countries have inadequate educational programs concerning fertility. And a great number of people are not inclined to exercise control. Each year in the United States alone, there are more than 100,000 "shotgun" marriages, about 200,000 unwed teenage mothers, and perhaps 1,500,000 abortions. Many parents encourage early boy-girl relationships, ignoring the risk of premarital intercourse and unplanned pregnancy. Advice is often condensed to a terse "Don't do it. But if you do it, be careful!"

The motivation to engage in sex has been evolving for more than 500 million years. A few centuries of moral and ecological reasoning that call for its suppression have not prevented unwanted pregnancies. And complex social factors have contributed to a population growth rate that is out of control.

How will we reconcile our biological past and the need for a stabilized cultural present? Whether and how fertility is to be controlled is one of the most volatile issues of our time. We will return to this issue in the next chapter, in the context of principles governing the growth and stability of populations. Here, we briefly consider some control options.

Birth Control Options

The most effective method of birth control is complete *abstinence*, no sexual intercourse whatsoever. It is unrealistic to expect many people to practice it.

A modified form of abstinence is the *rhythm method*. The idea is to avoid intercourse during the woman's fertile period, starting a few days before ovulation and ending a few days after. The woman identifies and tracks her fertile period. She keeps records of the length of her menstrual cycles, takes her temperature each morning when she wakes up, or both. (Body temperature rises by one-half to one degree just before the fertile period.) But ovulation may not be regular, and miscalculations are frequent. Also, sperm deposited in the vaginal canal a few days before ovulation may survive until ovulation. The method is inexpensive; it costs nothing after buying a thermometer. It doesn't require fittings and periodic checkups by a physician. But its practitioners do run a large risk of pregnancy, as Figure 32.21 indicates.

Withdrawal, or removing the penis from the vagina before ejaculation, dates back at least to biblical times. But withdrawal requires very strong willpower, and the method may fail anyway. Fluid released from the penis just before ejaculation may contain some sperm.

Douching, or rinsing out the vagina with a chemical right after intercourse, is next to useless. Sperm can move past the cervix and out of reach of the douche within ninety seconds after ejaculation.

Controlling fertility by surgical intervention is less chancy. In *vasectomy*, a tiny incision is made in a man's scrotum, and each vas deferens is severed and tied off. The simple operation can be performed in twenty minutes in a physician's office, with only a local anesthetic. After vasectomy, sperm cannot leave the testes and so will not be present in semen. So far there is no firm evidence that vasectomy disrupts hormonal interactions in males. There seems to be no difference in sexual activity. Vasectomies can be reversed. But half of those who submit to surgery later develop antibodies against sperm and may not be able to regain fertility.

For females, surgical intervention includes *tubal ligation*, in which the oviducts are cauterized or cut and tied off. Tubal ligation is usually performed in a hospi-

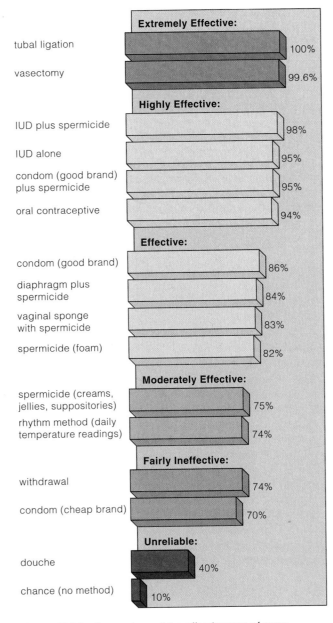

Extremely Effective:
tubal ligation — 100%
vasectomy — 99.6%

Highly Effective:
IUD plus spermicide — 98%
IUD alone — 95%
condom (good brand) plus spermicide — 95%
oral contraceptive — 94%

Effective:
condom (good brand) — 86%
diaphragm plus spermicide — 84%
vaginal sponge with spermicide — 83%
spermicide (foam) — 82%

Moderately Effective:
spermicide (creams, jellies, suppositories) — 75%
rhythm method (daily temperature readings) — 74%

Fairly Ineffective:
withdrawal — 74%
condom (cheap brand) — 70%

Unreliable:
douche — 40%
chance (no method) — 10%

Figure 32.21 Comparison of the effectiveness of some contraceptive methods in the United States. Percentages shown are based on the number of unplanned pregnancies per 100 couples who used the method as the only form of birth control for a year. For example, "94% effectiveness" for oral contraceptives (the Pill) means that 6 of every 100 women will become pregnant, on average.

tal. A small number of women who have had the operation suffer recurring bouts of inflammation and pain in the pelvic region where surgery was performed. The operation sometimes can be reversed.

Other, less drastic methods of controlling fertility involve physical or chemical barriers to prevent sperm from entering the uterus and moving to the ovarian ducts. *Spermicidal foam* and *spermicidal jelly* are toxic to sperm. They are transferred from an applicator into the vagina just before intercourse. These products are not always reliable unless used with another device, such as a diaphragm or condom.

A *diaphragm* is a flexible, dome-shaped device, inserted into the vagina and positioned over the cervix before intercourse. A diaphragm is relatively effective when fitted initially by a doctor, used with foam or jelly before each sexual contact, inserted correctly each time, and left in place for a prescribed length of time.

Condoms are thin, tight-fitting sheaths worn over the penis during intercourse. They are about 85 to 93 percent reliable. Only the kinds made of latex provide protection against sexually transmitted diseases, described in the *Focus* essay on the page that follows. However, condoms can tear and leak, at which time they become absolutely useless.

The most widely used method of fertility control is *the Pill*, an oral contraceptive of synthetic estrogens and progesterone. It suppresses the normal release of these hormones from the pituitary and so stops eggs from maturing and being released at ovulation. The Pill is a prescription drug. Formulations vary and are selected to cause the least possible disruption to each patient's hormone balance. That is why one woman shouldn't borrow oral contraceptives from another.

When a woman does not forget to take her daily dosage, the Pill is one of the most reliable methods of controlling fertility. It does not interrupt sexual intercourse, and the method is easy to follow. Often the Pill corrects erratic menstrual cycles and decreases cramping. However, the Pill causes some side effects in some women. In the first month or so of use, it may cause nausea, weight gain, tissue swelling, and headaches. Its continued use may lead to blood clotting in the veins of a few women (3 out of every 10,000) who are predisposed to this disorder. Some cases of elevated blood pressure, gallbladder disorders, and possibly breast cancer may be related to use of the Pill.

A pregnancy test does not register positive until after implantation. It has been argued that a woman is not pregnant until that time. From this perspective, *RU-486*, the "morning-after pill," intercepts pregnancy. It interferes with hormonal signals that govern events between ovulation and implantation. Three pills, taken within seventy-two hours after sexual intercourse, either block fertilization or prevent a blastocyst from burrowing into the uterine lining. RU-486 has side effects, mainly nausea, vomiting, and breast tenderness. These are temporary, and not all women experience them. But RU-486 disrupts complex hormonal interactions and should only be taken under medical supervision. As is true of oral contraceptives, it may trigger elevated blood pressure, blood clots, or breast cancer in some women. At this writing, RU-486 is available in Europe. In the United States, the Federal Drug Administration has just approved clinical trials.

Termination of Pregnancy

Once implantation has occurred, the only way to terminate a pregnancy is *abortion*, the dislodging and removal of the embryo from the uterus. At one time, abortions were generally forbidden by law in the United States unless the pregnancy endangered the mother's life. The Supreme Court has ruled that the government does not have the right to forbid abortions during the early stages of pregnancy (typically up to five months). Before this ruling, there were dangerous, traumatic, and often fatal attempts to abort embryos, either by pregnant women themselves or by quacks.

Vacuum suctioning and other methods make abortion relatively rapid, painless, and free of complications when performed during the first trimester. Abortions in the second and third trimesters will probably remain extremely controversial unless the mother's life is threatened. For both medical and humanitarian reasons, it is generally agreed in this country that the preferred route to birth control is not through abortion. Rather, it is through sexually responsible behavior that prevents unwanted pregnancy in the first place.

In Vitro Fertilization

Control of fertility also extends in the other direction—to help childless couples who are desperate to conceive a child. In the United States, about 15 percent of all couples cannot conceive because of sterility or infertility. For example, hormonal imbalances may prevent ovulation in females, or the sperm count in the male may be too low to assure fertilization.

With "in vitro fertilization," external conception is possible, provided sperm and oocytes obtained from the couple are normal. A hormone is administered that prepares the ovaries for ovulation. Then a physician locates and removes the preovulatory oocyte with a suction device. Before the oocyte is removed, sperm from the male are placed in a solution that simulates the fluid in oviducts. When the suctioned oocyte is placed with the sperm, fertilization may occur a few hours later. About twelve hours later, the newly dividing zygote is transferred to a solution that will support further development. About two to four days after that, the blastocyst is transferred to the female's uterus. Implantation is successful in about 20 percent of the cases. Each attempt costs several thousand dollars.

Sexually Transmitted Diseases

Sexually transmitted diseases (STDs) have reached epidemic proportions in the United States. Urban poverty, prostitution, intravenous drug abuse, and sex-for-drugs are fanning the epidemic. STDs are rampant even on college campuses, where students who are not poor, not promiscuous, and not uninformed still think, "It can't happen to me." Ominously, antibiotic-resistant strains of bacteria are on the rise, and some viral diseases cannot be cured. At this writing, an estimated 57 million Americans have some form of sexually transmitted disease. Two-thirds of those infected are under age twenty-five; one-fourth are teenagers. Women and children are hardest hit.

The economics of this health problem are staggering. In 1993, the Centers for Disease Control informed us that the annual cost of treating the most prevalent STDs is as follows: *Herpes*, $759 million; gonorrhea, $1 billion; chlamydial infection, $2.4 billion; and pelvic inflammatory disease, $4.2 billion. This does not include the accelerating cost of treating AIDS patients. In many developing countries, AIDS alone threatens to overwhelm health-care delivery systems and to unravel decades of economic progress. The social consequences are sobering. Mothers bestow a chlamydial infection on 1 of every 20 newborns in the United States. They bestow Type II *Herpes* virus on 1 of every 10,000 newborns; one-half of the infected babies die, and one-fourth have severe neurological defects. Each year 1 million female Americans contract pelvic inflammatory disease. Between 100,000 and 150,000 become sterile.

AIDS. Someone may become infected by HIV, the human immunodeficiency virus, and not even know it. At first there may be no outward symptoms. But five to ten years later, a set of chronic disorders develops. Collectively, these are called *AIDS* (acquired immune deficiency syndrome). The virus has slowly crippled the immune system, in the manner described in Chapter 27. It has opened the door to "opportunistic" infection by the body's resident and normally harmless bacteria. These are positioned to be the first to take advantage of the lowered resistance. Dangerous pathogens also take their toll. In time, infections simply overwhelm the immune-compromised person.

Most commonly, HIV spreads by sexual intercourse (vaginal, anal, and oral) and by IV drug users. Most infections occur through the transfer of blood, semen, urine, or vaginal secretions between people. HIV enters the internal environment through cuts or abrasions on the penis, vagina, or rectum, and maybe in mucous membranes in the mouth. Once inside, it locks onto cells in which it can be replicated. Targets include helper T cells (T4 lymphocytes), macrophages, and epithelial cells of the cervix.

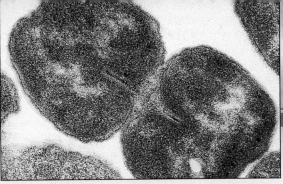

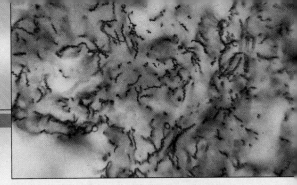

a   Bacterial agents of gonorrhea.

b   Bacterial agents of syphilis.

At present there is no vaccine against HIV and no effective way to treat AIDS. If you get it, you die. *There is no cure.*

HIV wasn't identified until 1981, although it seems to have been present in some parts of Central Africa for at least several decades. In the 1970s and early 1980s, it spread to the United States and other developed countries. Most of those initially infected were male homosexuals. Today, a significant portion of the heterosexual population is infected or at risk.

Worldwide, 8 million to 10 million may now be HIV infected. By early 1992, AIDS had become the second leading cause of death among men and the fifth leading cause among women between ages 25 and 44. The risk is growing among teenagers. In one poll of American high schools, two-thirds of the students said they don't use condoms. More than 40 percent of the teenagers polled reported having two or more sex partners.

Free or low-cost, confidential testing for exposure to HIV is available through public health facilities and many physicians' offices. People who suspect they have put themselves at risk should know there may be a time lag, from a few weeks to six months or more, until detectable antibodies form in response to infection. The presence of antibodies only indicates exposure to the virus; by itself, it doesn't mean AIDS will develop. Even so, anyone who tests positive for HIV should be considered capable of spreading the virus.

Public education programs are under way to stop the spread of HIV. Most advocate safe sex. Yet there is confusion about what "safe" means. High-quality latex condoms, used with a spermicide containing nonoxynol-9, are assumed to be effective in preventing transmission, but there still is a small risk of infection. Open-mouthed, intimate kissing with someone who tests positive for HIV should be avoided. Caressing carries no risk, *if* there are no lesions or cuts through which body fluids containing the virus can enter the body. Certain lesions are symptoms of other sexually transmitted diseases. They may increase susceptibility to HIV infection.

In sum, AIDS reached epidemic proportions mainly for three reasons. First, it took a while to discover that the virus is transmitted by semen, blood, and vaginal fluid and that *behavioral* controls can limit its spread. Second, it took time to develop ways to test symptom-free carriers, who can infect others. Third, many still don't understand that the medical, economic, and social consequences of AIDS affect everyone.

Gonorrhea.   During sexual intercourse, *Neisseria gonorrhoeae* can enter the body at mucous membranes of the urethra, cervix, and anal canal. This bacterium, shown in Figure *a*, causes *gonorrhea*. An infected female may only notice a slight vaginal discharge or a burning sensation while urinating. If the bacterium spreads to her oviducts, it may induce severe cramps, fever, vomiting, and scar tissue formation, which may cause sterility. Symptoms are more noticeable in males. Within a week after infection, the male's penis discharges yellow pus. Urinating is more frequent and may be painful.

This STD is rampant even though prompt treatment quickly cures it. Why? There are no troubling symptoms among women in early stages. Also, infection doesn't confer immunity to the bacterium, perhaps because there are sixteen or more different strains of it. Contrary to common belief, someone can get gonorrhea over and over again. Finally, oral contraceptives, which are widely used, encourage infection by altering vaginal pH. Resident bacterial populations decline, and *N. gonorrhoeae* can move in.

Syphilis.   *Syphilis*, a dangerous STD, is caused by a spirochete, *Treponema pallidum* (Figure *b*). Sex with an infected partner puts the motile, spiral bacterium onto the surface of genitals or into the cervix, vagina, or oral cavity. It can enter the body through tiny tears in the epidermis. One to eight weeks later, new treponemes are twisting about in a flattened, painless *chancre* (local ulcer). This first chancre is a symptom of the primary stage of syphilis. By then, treponemes are in the blood. Treponemes can cross the placenta of an infected, pregnant woman. The result will be a miscarriage, stillbirth, or syphilitic newborn.

Usually the chancre heals, but treponemes are now multiplying in mucous membranes, joints, bones, and the eyes, spinal cord, and brain. More chancres and a skin rash develop in this highly infectious, secondary stage of syphilis. Then symptoms subside. In about 25 percent of the cases, the immune system cures the disease. Another 25 percent remain infected but symptom-free. In the rest, mild to major lesions and scars appear in the skin, liver, bones, aorta, and other internal organs. Few treponemes are present during this tertiary stage, but the immune system is hypersensitive to them. Chronic immune reactions can severely damage the brain and spinal cord and bring about general paralysis.

Probably because the symptoms are so alarming, more people seek early treatment for syphilis than for gonorrhea. Later stages require prolonged treatment.

Pelvic Inflammatory Disease. *Pelvic inflammatory disease* (PID) is one of the most serious complications of gonorrhea, chlamydial infections, and other STDs. It also can arise when bacteria that normally inhabit the vagina ascend into the pelvic region. Most often, the uterus, oviducts, and ovaries are affected. There is bleeding and vaginal discharge. Pain in the lower abdomen may be so severe, infected women often think they are having an attack of acute appendicitis. The oviducts may become scarred, and scarring can lead to abnormal pregnancies as well as to sterility.

Genital Herpes. The type II *Herpes simplex* virus has already infected about 25 million people in the United States. It causes *genital herpes*. Infection requires direct contact with active *Herpes* viruses or sores that contain them. Mucous membranes of the mouth or genital area are highly susceptible to invasion. Symptoms often are mild or absent. Among infected women, small, painful blisters may appear on the vulva, cervix, urethra, or anal tissues. Among men, blisters form on the penis and anal tissues. Within three weeks, the virus enters latency, and the sores crust over and heal.

The virus is reactivated sporadically. Each time, it causes new, painful sores at or near the original site of infection. Sexual intercourse, menstruation, emotional stress, or other

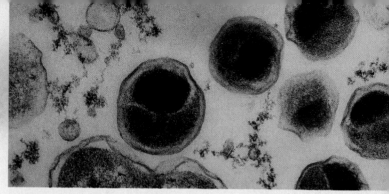

c Bacterial agents of chlamydia.

infections can trigger new *Herpes* infections. Acyclovir, an antiviral drug, decreases the healing time and often decreases the pain and viral shedding.

Chlamydial Infection. *Chlamydia trachomatis* is a parasitic bacterium (Figure *c*). It must spend part of its life cycle inside cells of the genital and urinary tracts. It causes several diseases, including *NGU* (chlamydial nongonococcal urethritis). NGU is far more common than syphilis or gonorrhea. Often, *N. gonorrhoeae* and *C. trachomatis* are transmitted at the same time. Prompt penicillin treatment cures the gonorrhea—but not NGU, which requires tetracycline and sulfonamide treatment.

NGU leads to inflammation of the cervix and, in both sexes, the urethra. An infected person may experience a

SUMMARY

1. Most animals reproduce sexually. Specialized reproductive structures and forms of behavior assist fertilization and nutritionally support the young.

2. Human males produce sperm in a pair of testes, and human females produce eggs in a pair of ovaries. These primary reproductive organs also produce sex hormones, which control reproductive functions and influence secondary sexual traits, including hair and fat distribution.

3. The hypothalamus governs FSH and LH secretion from the anterior lobe of the pituitary. In males, these hormones stimulate testosterone secretion, which stimulates sperm formation. Feedback loops from the testes back to the hypothalamus and anterior pituitary maintain blood levels of FSH and LH.

4. Human females of reproductive age are fertile on a monthly basis, as part of a menstrual cycle (Table 32.2).

   a. FSH and LH govern the menstrual cycle. They stimulate growth of follicles (primary oocytes and a surrounding cell layer), estrogen production, and preparation of the uterine lining (endometrium) for pregnancy.

   b. A midcycle surge of LH causes ovulation

(release of a secondary oocyte from the ovary). Follicle remnants develop into a corpus luteum, a glandular structure that secretes progesterone and some estrogen. Its secretions maintain the endometrium for the last phase of the cycle. The corpus luteum self-destructs if pregnancy does not occur.

   c. Feedback loops from the ovaries back to the hypothalamus and anterior pituitary govern the blood levels of FSH and LH.

Table 32.2	Summary of Events of the Menstrual Cycle	
Phase	Events	Days of the Cycle*
Follicular phase	Menstruation; endometrium breaks down	1–5
	Follicle matures in ovary; endometrium rebuilds	6–13
Ovulation	Secondary oocyte released from ovary	14
Luteal phase	Corpus luteum forms; endometrium thickens and develops	15–28

*Assuming a 28-day cycle.

burning sensation while urinating. But sometimes there are no symptoms, treatment is not sought, and serious complications develop. In males, lymph vessels become blocked, so testes and prostate become swollen and inflamed. In females, the infection may spread into the uterus and oviducts to cause pelvic inflammatory disease.

Symptom-free infected persons are unwittingly spreading destructive chlamydial infections through all ethnic groups, among poor and affluent, among dropouts and college students alike.

Genital Warts. More than sixty types of the human papillomaviruses (HPV) have been identified. A few cause benign, bumplike growths called *genital warts*. HPV infection of the genitals and anus has become the most prevalent STD in the United States. The increase of Type 16 HPV among sexually active men and women is especially alarming.

Type 16 HPVdoes not usually cause obvious warts, but it is strongly associated with precancerous sores and cancers of the cervix, vagina, vulva, penis, and anus. In one clinical study in Seattle, Washington, 22 percent of the female college students who were examined tested positive for this virus. So did 44 percent of the female patients who were examined for other STDs.

5. As is true of animals generally, human embryonic development proceeds through six stages: gamete formation, fertilization, cleavage, gastrulation, organ formation, and growth and tissue specialization.

6. Embryos develop by way of cell differentiation and morphogenesis. Both processes depend on the segregation of cytoplasm during cleavage and on interactions among embryonic cells.

    a. In cell differentiation, a cell selectively uses certain genes and synthesizes proteins not found in other cell types. Specialized proteins are the basis of distinct cell structures, products, and functions.

    b. In morphogenesis, the size, shape, and proportion of tissues and organs change, and they become organized in patterns that in time produce the adult.

7. All tissues and organs in the human embryo arise from three primary tissues (germ layers): endoderm, ectoderm, and mesoderm.

8. As is true of vertebrates generally, four extraembryonic membranes form during human embryonic development.

    a. Amnion: a fluid-filled sac; it surrounds and protects the embryo from mechanical shocks, abrupt temperature changes, and drying out.

    b. Yolk sac: it stores nutritive yolk in most shelled eggs; but in humans, part becomes a major site of blood formation and some of its cells give rise to germ cells.

    c. Chorion: it becomes a protective membrane around the embryo and the other membranes. It also becomes a major component of the placenta.

    d. Allantois: functions in early blood formation and in development of the urinary bladder.

9. The placenta is a blood-engorged organ, composed of endometrium and extraembryonic membranes. It keeps the maternal and embryonic circulatory systems separate while allowing oxygen, nutrients, and wastes to diffuse across it. The placenta cannot protect the embryo and fetus against a mother's harmful behavior, such as poor nutrition, and use of cigarettes, alcohol, or illegal or unprescribed drugs.

10. At delivery, uterine contractions dilate the cervical canal and expel the fetus and afterbirth. Estrogen and progesterone stimulate growth in the mammary glands. After delivery, nursing stimulates the release of hormones that in turn stimulate milk production and release.

11. Control of human sexuality and fertility raises important ethical questions. These questions extend to physical, chemical, surgical, and behavioral interventions used to prevent or end pregnancies.

Review Questions

1. Distinguish between:
    a. Sertoli cell and Leydig cell *515*
    b. sperm and semen *514–515*
    c. primary oocyte and ovum *517, 520*
    d. follicle and corpus luteum *517, 518*
    e. ovulation and implantation *517, 520–521*

2. Label the components of the human male and female reproductive systems and state their functions. *514, 516*

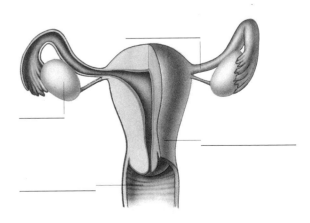

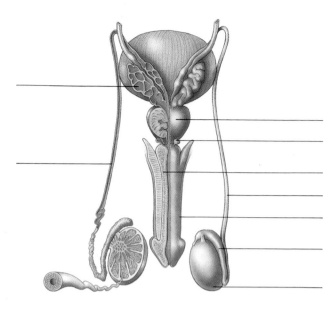

1. _____ is the forerunner of muscle and of circulatory, reproductive, and excretory organs, most of the skeleton, parts of the gut, and parts of the skin.
   a. Ectoderm   c. Endoderm
   b. Mesoderm   d. Plasmoderm

2. A fertilized egg undergoes _____ , which segregates different cytoplasmic components into daughter cells.
   a. cleavage   c. cell differentiation
   b. gastrulation   d. morphogenesis

3. In _____ , a cell selectively uses certain genes and makes proteins not found in other cell types.
   a. cleavage   c. cell differentiation
   b. gastrulation   d. morphogenesis

4. During _____ , organs change in size, shape, and proportion and become organized in patterns.
   a. cleavage   c. cell differentiation
   b. gastrulation   d. morphogenesis

5. _____ directly stimulate(s) reproductive cells that give rise to sperm.
   a. FSH   c. Testosterone
   b. LH   d. Both a and b

6. During a menstrual cycle, _____ and _____ directly stimulate a follicle to mature.
   a. FSH; LH   c. estrogens; progesterone
   b. FSH; testosterone   d. FSH; progesterone

7. During the first phase of a menstrual cycle, _____ and _____ prime the endometrium for pregnancy.
   a. FSH; LH   c. estrogens; progesterone
   b. FSH; testosterone   d. estrogens only

8. About midway through a menstrual cycle, a surge of _____ triggers ovulation.
   a. FSH   c. estrogens
   b. LH   d. progesterone

9. During the last phase of a menstrual cycle, _____ and _____ from the corpus luteum maintain the endometrium.
   a. FSH; LH   c. estrogens; progesterone
   b. FSH; testosterone   d. estrogens only

10. Is this statement true or false? A placenta consists of embryonic and maternal blood vessels.

11. Match the term with its description.
   ____ seminiferous tubule   a. sperm penetration has triggered meiosis II
   ____ corpus luteum   b. a burrowing blastocyst
   ____ allantois   c. glandular remnant of follicle
   ____ implantation   d. sperm form here
   ____ amnion   e. secretes HCG
   ____ chorion   f. oxygen, nutrient transport
   ____ follicle   g. primary oocyte plus cell layer
   ____ ovum   h. absorbs shocks, keeps embryo from drying out

aging *527*
allantois *521*
amnion *521*
cell differentiation *513*
chorion *521*
cleavage *510*
corpus luteum *518*
ectoderm *511*
embryo *510*
endoderm *511*
endometrium *516*
estrogen *516*
fertilization *510*
fetus *522*
follicle *517*
FSH *515*
gamete formation *510*
gastrulation *511*
growth, tissue specialization *511*
HCG *521*

implantation *521*
lactation *526*
LH *515*
mature egg (ovum) *521*
menstrual cycle *516*
mesoderm *511*
morphogenesis *513*
oocyte *516*
organ formation *511*
ovary *514*
ovulation *517*
placenta *522*
polar body *517*
progesterone *516*
secondary sexual trait *514*
sperm *514*
testis *514*
testosterone *515*
yolk sac *521*

Aral, S., and K. Holmes. February 1991. "Sexually Transmitted Diseases in the AIDS Era." *Scientific American* 2(264): 62–69.

Caldwell, M. November 1992. "How Does a Single Cell Become a Whole Body?" *Discover* 13(11): 86–93.

Gilbert, S. 1991. *Developmental Biology*. Third edition. Sunderland, Massachusetts: Sinauer.

Larsen, W. 1993. *Human Embryology*. New York: Churchill Livingston.

Nilsson, L., et al. 1986. *A Child Is Born*. New York: Delacorte Press/Seymour Lawrence.

Zack, B. July 1981. "Abortion and the Limitations of Science." *Science* 213: 291.

FACING PAGE: *Two organisms—a fox in the shadows cast by a snow-dusted spruce tree. What are the nature and consequences of their interactions with each other, with other organisms, and with their environment? By the end of this last unit, you possibly will see worlds within worlds in such photographs.*

33 POPULATION ECOLOGY

A Tale of Nightmare Numbers

Suppose this year the United States Congress passes legislation to control population growth. By law, there can be no more than three children per family. After the third child is born, the father must be sterilized, with or without his consent. *It would never happen here*, you may be thinking. Such an invasion of privacy would never be tolerated in our society. Besides, family size is not much of an issue in North America, where rates of food production and standards of hygiene and medical care are among the world's highest.

Elsewhere in the world, especially where living conditions are already marginal, many populations are growing at alarming rates. Consider India. Its population, which already surpasses that of North and South America combined, grows by about 2 percent annually. Most people there do not have adequate food, shelter, or medical care. Each *week*, 100,000 enter the job market, with little hope for gainful employment. Each *day*, 100 acres of croplands that provide food for the population are removed permanently from agriculture. Why? Too many salts have built up in the intensively irrigated soil; India does not get enough rain to flush them out.

Birth control programs sponsored by the government of India have not worked well. Administering the programs has been difficult, for many people live in remote villages. Information must be conveyed by word of mouth because of widespread illiteracy. Many children die of disease and starvation, so villagers often resist limiting family size. Without a large family, they ask, who will survive and help a father tend fields? Who will go to the cities and earn money to send back home? Who will care for parents when they are too old to work? How can a father otherwise know he will be survived by a son? By Hindu tradition, a son must conduct the last rites so the soul of his dead father will rest in peace.

In 1976, out of desperation, government officials subjected some men to compulsory vasectomies. Public outrage over the policy contributed to the eventual downfall of Indira Gandhi's government, and the law

Figure 33.1 Bathers crowding the banks of the Ganges River in India—a tiny sampling of the more than 5.5 billion humans on earth. In this chapter we turn to the principles governing the growth and sustainability of populations, including our own.

was rescinded. By the end of 1993, there were more than 5.5 billion people on earth. Of those, 901 million were living in India.

Is there a way out of such dilemmas? Should the wealthier, less densely populated nations that now use most of the world's resources learn to get by more efficiently, on less? Should they donate surplus food to less fortunate nations? Would donations help, or would they encourage dependency and further population growth? What would happen if the benefactor nations suffered severe droughts year after year and had trouble meeting their own resource requirements?

Whether we consider humans or any other kind of organism, *certain ecological principles govern the growth and sustainability of populations over time*. This chapter describes these principles, then shows how they apply to the past, present, and future growth of the human population.

KEY CONCEPTS

1. Ecological principles govern the growth and sustainability of all populations, including our own.

2. A population may display an exponential growth pattern, in which it increases in size by ever larger amounts over increments of time. Alternatively, it may display a logistic growth pattern. By this pattern, a low-density population rapidly increases in numbers, then levels off in size as resource scarcity limits its further increase or triggers a decline in numbers. Finally, some populations display large fluctuations in numbers that are not easily explained.

3. All populations face limits to growth, for no environment can indefinitely sustain a continuously increasing number of individuals. Competition for resources, disease, predation, and other factors act as controls over population growth. The controls vary in their relative effects on populations of different species, and they vary over time.

FROM POPULATIONS TO THE BIOSPHERE

With this unit, we turn to the ways in which organisms interact with one another and with the physical and chemical features of their environment. The study of these interactions, called **ecology**, is undertaken at the levels of populations, communities, ecosystems, and the biosphere.

We define a **population** as a group of individuals of the same species occupying a given area. The actual place where a population (or individual) lives is its *habitat*. The populations of all species that occupy a habitat make up a **community**. Ecologists also use the term for evolutionarily related or ecologically similar groups of organisms in a habitat, such as a community of birds or of animals that feed on grasses. An **ecosystem** is a community *and* its environment. It has a biotic component (all of its living organisms) and abiotic (nonliving) components, such as nutrients, temperature, and rainfall. The **biosphere** is the sum total of all places in which organisms live—from the lower atmosphere and the waters of the earth to the surface rocks, soils, and sediments of the earth's crust.

In Units V and VI, we considered ways in which individuals are physiologically adapted to their environments. We turn now to relationships that influence the size, structure, and distribution of the populations to which they belong. Communities, ecosystems, and the biosphere are the focus in chapters to follow.

Figure 33.2 Generalized population distribution patterns. (**a**) Most commonly, individuals are distributed in clumps that correspond to patchy conditions in the habitat. (Some parts offer more shade, more water, better hiding places, more hunting possibilities, and so on.) Many animals, such as the baboons shown here, form social groups. Also, seeds, larvae, and other forms of offspring may not travel far; they may simply settle near the parent organisms.

(**b**) Under some circumstances, individuals are evenly spaced in a habitat. This pattern, called uniform spacing, seems to be caused by competition among individuals within the population. Creosote plants near Death Valley, California, are an example. Large, mature bushes deplete the soil around them of water. In addition, seed-eating ants and rodents are most active near mature plants. They tend to eat seeds that fall on the open ground between plants.

(**c**) Random dispersion may result when conditions are fairly uniform and individuals of a population neither attract nor repel each other. Spiders that actively search for prey on a forest floor may be an example.

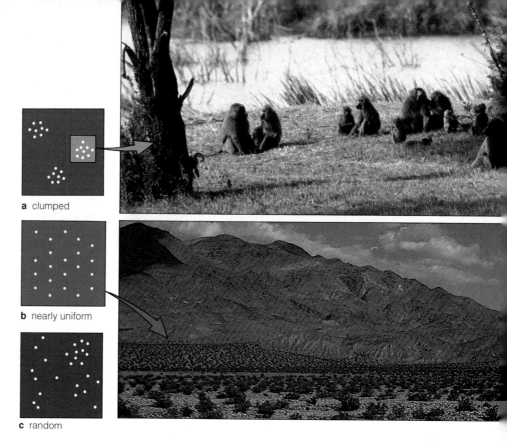

a clumped

b nearly uniform

c random

POPULATION DYNAMICS

Characteristics of Populations

Each population has certain characteristics, such as size, density, and distribution. **Population size** is the number of individuals that make up the gene pool. **Population density** is the number of individuals living in a specified area or volume, such as the number of guppies in each liter of water in a stream. **Distribution** refers to the general pattern of dispersion through the habitat. Most commonly, individuals clump together at specific sites. Sometimes they are randomly or uniformly dispersed through the habitat (Figure 33.2).

We further characterize a population in terms of **age structure**—the number of individuals in each of several to many age categories. For example, we may divide a population into pre-reproductive, reproductive, and post-reproductive ages. The total number in the middle category is the population's "reproductive base."

Population Size and Patterns of Growth

Over a specified time interval, any change in the size of a population depends on how many individuals enter and leave. Population size increases through births and **immigration** (individuals enter and take up residence). It decreases through deaths and **emigration** (individuals leave and take up residence elsewhere). If we assume that immigration and emigration are balanced, we can ignore their effects on population size. Given this assumption, there will be **zero population growth** if the number of births is balanced by the number of deaths over a specified period. When births and deaths are so balanced, population size is stabilized.

Exponential Growth. Births, deaths, and other variables that affect populations can be measured in terms of rates per individual ("per capita" rates). Think of 2,000 mice living in a cornfield. If 1,000 are born each month, the birth rate is 1,000/2,000 = 0.5 per mouse per month. If 800 of the 2,000 die during the same period, the death rate is 800/2,000 = 0.4 per mouse per month.

For clarity, let's not think about how much or how little the mouse population grows. Let's simply assume the birth and death rates remain constant, so that we can combine them into a single variable. We call this variable the **net population growth rate**, or r for short. The combined rate for the mice would be 0.5 − 0.4 = 0.1 per mouse per month. With this example, we have a way to represent population growth rate:

$$\begin{pmatrix} \text{population} \\ \text{growth} \\ \text{rate} \end{pmatrix} = \begin{pmatrix} \text{net population} \\ \text{growth rate} \\ \text{per individual} \end{pmatrix} \times \begin{pmatrix} \text{number of} \\ \text{individuals} \end{pmatrix}$$

or, more simply, $G = rN$. As long as r remains constant, any population will show **exponential growth**.

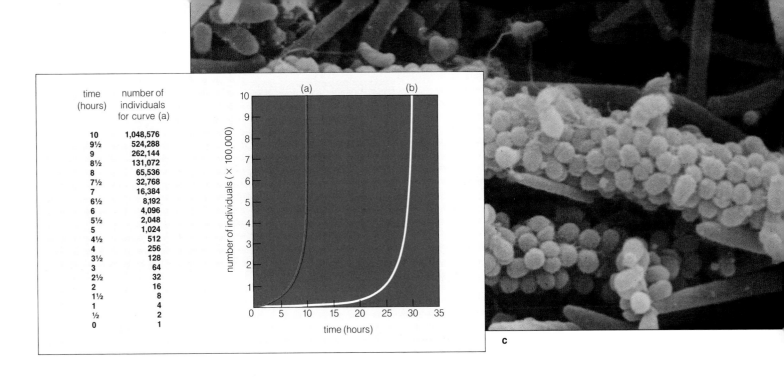

time (hours)	number of individuals for curve (a)
10	1,048,576
9½	524,288
9	262,144
8½	131,072
8	65,536
7½	32,768
7	16,384
6½	8,192
6	4,096
5½	2,048
5	1,024
4½	512
4	256
3½	128
3	64
2½	32
2	16
1½	8
1	4
½	2
0	1

We can observe exponential growth by supplying a lone bacterium in a culture flask with all the nutrients that it requires for growth. Thirty minutes later, the cell divides into two new cells. Each of its descendants also divides every thirty minutes. No cells die between divisions, so every thirty minutes the population size doubles—from 1 to 2, then 4, 8, 16, 32, 64, and so on. The length of time it takes for a population to double in size is called (appropriately) its **doubling time**.

The larger the bacterial population becomes, the more cells there are to divide. As a result, the number added to the population in each thirty-minute interval increases through time. After only 9-1/2 hours (nineteen doublings), the population exceeds 500,000. After 10 hours (twenty doublings), it exceeds 1 million!

When we plot these doublings in population size against time, we get curve *a* in Figure 33.3. This is a **J-shaped curve**, which is characteristic of populations undergoing unrestricted, exponential growth.

What if some bacteria in the culture flask die? To examine the effect of deaths on the growth rate, let's start over with a single bacterium. This time, assume that 25 percent of the population dies during each thirty-minute interval. With this death rate, it takes almost two hours (not thirty minutes) to double population size. So it takes thirty hours (not ten) to arrive at a million bacteria. *But only the time scale changes*. We still end up with a J-shaped curve that signifies exponential growth (curve *b* in Figure 33.3).

As long as the per capita birth rate remains even slightly above the per capita death rate, a population will grow. If both rates remain constant, it will grow exponentially.

Figure 33.3 J-shaped curves characteristic of exponential growth. (**a**) Growth curve for a population of bacterial cells that are reproducing every half hour. (**b**) Growth curve when cells divide every half hour but when 25 percent die between divisions. Although deaths slow the rate of increase, they do not stop exponential growth.

Do growth curves seem far removed from your personal experience? Think of bacteria growing in, say, your mouth (**c**) when they are presented with a buffet of nutrients in candy or some other sugar-laden food.

Biotic Potential. Suppose members of a population have plenty of food, living space, and other resources. They do not have to compete with other organisms for resources. And their habitat is free of predators and pathogens. Under such ideal conditions, a population will show a *maximum* rate of increase per individual. That rate is its **biotic potential**.

Species differ greatly in biotic potential. For many bacteria, the maximum rate of increase is 100 percent every half hour. For humans and other large mammals, it is between 2 and 5 percent per year. The rate depends on the age at which each new generation starts reproducing, how many times each individual reproduces, and how many offspring are born each time.

Even when a population is not growing at its full biotic potential, its growth still may be exponential. For example, each human female is biologically capable of bearing twenty or more children, but many women do not reproduce at all. Even so, as you will see shortly, the human population has been growing exponentially since the mid-eighteenth century.

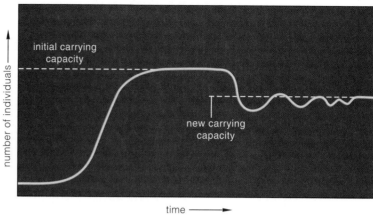

Figure 33.4 Idealized S-shaped curve characteristic of logistic growth. After a rapid growth phase, growth slows and the curve flattens out as the carrying capacity is reached. Variations can occur in S-shaped growth curves, as when changed environmental conditions bring about a decrease in the carrying capacity. This happened to the human population of Ireland before 1900, when a disease destroyed potato crops that were the mainstay of the diet (page 268).

Limits on Population Growth. Given their exponential growth, what will keep humans from filling up the planet? For that matter, given that each female sea star can produce 2,500,000 eggs a year, what keeps sea stars from filling the oceans? Most often, *environmental circumstances* prevent populations from reaching their full biotic potential.

In nature, where organisms interact in complex ways, it is not easy to identify factors that limit population growth. To get a sense of what some of those factors might be, start again with a lone bacterium in a culture flask, where you can control the variables. First you enrich the culture medium with glucose and the other nutrients required for bacterial growth. Then you allow bacterial cells to reproduce for many generations. At first the growth pattern appears to be exponential. Then growth slows, and population size remains rather stable. After the stable period, the population size starts to decline rapidly—and all the bacteria die.

What happened? As the population expanded faster and faster, the growing number of individual cells used more and more nutrients. When the supply of nutrients dwindled, cell divisions decreased. When the supply was exhausted, the bacterial cells starved to death.

Any essential resource that is in short supply is a **limiting factor** on population growth. Food, micronutrients, refuge from predators, living space, and a pollution-free environment are examples. The number of such factors can be huge, and their effects can vary. Even so, usually one factor alone is putting the brakes on a population's growth at any particular moment. Suppose you kept on freshening the supply of *all* required nutrients for the growing bacterial culture. After an episode of exponential growth, the population would still crash. Like all organisms, bacteria produce metabolic wastes. Concentrations of wastes produced by such a huge population of bacterial cells would be so great, they would drastically alter the conditions in the culture. By its own activities, the population would pollute its environment and put a stop to growth.

Collectively, all the limiting factors acting on a population represent the environmental resistance to its growth. Environmental resistance affects the number of individuals of a given species that can be sustained indefinitely in a particular area.

Carrying Capacity and Logistic Growth. Imagine a small population, with individuals dispersed through the habitat. As it increases in size, more and more individuals must share the nutrients, living quarters, and other resources their habitat offers. As the share available to each one shrinks, fewer individuals may be born and more may die through starvation or lack of nutrients. The population's rate of growth will decline until births are balanced or outnumbered by deaths. The final population size will largely depend on the sustainable supply of resources.

Biologists have a name for the maximum number of individuals of a population (or species) that can be sustained indefinitely by a given environment. They call it **carrying capacity**. We can identify the effect of carrying capacity in the pattern called **logistic population growth**. By this pattern, a low-density population at first grows slowly in size, then it proceeds through a rapid growth phase, and finally its size levels off once the carrying capacity is reached. We can represent logistic growth in this simplified way:

$$\begin{pmatrix} \text{population} \\ \text{growth} \\ \text{rate} \end{pmatrix} = \begin{pmatrix} \text{maximum net} \\ \text{reproduction} \\ \text{per individual} \end{pmatrix} \times \begin{pmatrix} \text{number} \\ \text{of indi-} \\ \text{viduals} \end{pmatrix} \times \begin{pmatrix} \text{proportion} \\ \text{of resources} \\ \text{not yet used} \end{pmatrix}$$

or $G = r_{max} N [(K - N)/K]$. Here, K represents the carrying capacity. The term inside the parentheses is close to 1 when a population is small. It approaches zero when population size is close to carrying capacity.

A plot of logistic growth gives us an **S-shaped curve**, as shown in Figure 33.4. Such curves are only a simple approximation of what goes on in nature. For example, there are times when a rapidly growing population temporarily exceeds its carrying capacity. In time, high death rates and low birth rates push the number of individuals down to the carrying capacity—or lower.

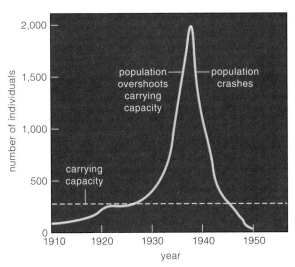

Figure 33.5 Carrying capacity and a reindeer herd. In 1910, four male and twenty-two female reindeer were introduced on one of Alaska's Pribilof Islands. In less than thirty years, the herd increased to 2,000. Its members had to compete for dwindling vegetation, and overgrazing destroyed most of it. In 1950 the herd plummeted to eight members. The growth pattern reflects how population size overshot the carrying capacity, then crashed.

Natural controls may keep a population below the carrying capacity set by the resources that it uses. Such controls include predation, disease, storms, lightning-triggered fires, and other disturbances that kill individuals even when they face no shortage of resources.

The carrying capacity itself may decrease when a burgeoning population damages or depletes its own resource supply. When mountain lions, wolves, and other natural predators are killed off, the size of deer populations often increases dramatically. The deer forage more intensively and damage the plants that sustain them. Severe damage may decrease their resource supply and so dictate a new, lower carrying capacity. The same thing may happen when resource levels decline for other reasons (Figure 33.5). Will the activities of the human population bring about a similar decrease in carrying capacity? You may wish to return to this question after reading Chapter 37, which describes what we are doing to the soil, air, water, and forests that sustain us.

Carrying capacity is the maximum number of individuals of a population that can be sustained indefinitely by resources in a given environment.

Resource availability can change with time, and such changes introduce variations in the carrying capacity and the size of most populations.

Checks on Population Growth

Density-Dependent Control. The logistic growth equation describes *density-dependent* increases in population size. When population density is low, a population grows rapidly. When density is high, the population grows very slowly or even declines in size. High density and overcrowding also put individuals at greater risk of being killed by predators, colonized by parasites, or infected by pathogenic microorganisms that cause contagious diseases. Predators, parasites, and pathogens exert density-dependent controls on population growth. With increases in population size, these interactions intensify and so bring about a decline in the population's density. Once the size has been reduced, the density-dependent interactions are less common, and the population may grow once again.

A classic example is the *bubonic plague* that swept through Europe in the fourteenth century. *Yersinia pestis*, the bacterium that causes the disease, normally lives in wild rodents. Hungry fleas transmit it to new hosts. *Y. pestis* multiplies in the flea gut and blocks digestion, the fleas attempt to feed more and more often, and so the disease spreads. It spread like wildfire through cities, where humans were crowded together, sanitary conditions were poor, and rats were abundant. By the plague's end, the populations in European cities had declined by 25 million.

Density-Independent Control. Some events cause more deaths or fewer births regardless of whether the members of a population are crowded or not. As an example of *density-independent* control, think of the freak summer snowstorms in the Colorado Rockies. Every so often, they wipe out butterfly populations. Similarly, spraying pesticides in your backyard may kill most insects, mice, cats, birds, and other animals regardless of how dense their populations are.

Table 33.1 Life Table for the United States Human Population, 1989*

Age Interval (category for individuals between the two ages listed)	Survivorship (number alive at start of age interval, per 100,000 individuals)	Mortality (number dying during the age interval)	Life Expectancy (average lifetime remaining at start of age interval)	Number of Reported Live Births for Total U.S. Population
0 – 1	100,000	896	75.3	
1 – 5	99,104	192	75.0	
5 – 10	98,912	117	71.1	
10 – 15	98,795	132	66.2	11,486
15 – 20	98,663	429	61.3	506,503
20 – 25	98,234	551	56.6	1,077,598
25 – 30	97,683	606	51.9	1,263,098
30 – 35	97,077	737	47.2	842,395
35 – 40	96,340	936	42.5	293,878
40 – 45	95,404	1,220	37.9	44,401
45 – 50	94,184	1,766	33.4	1,599
50 – 55	92,418	2,727	28.9	
55 – 60	89,691	4,334	24.7	
60 – 65	85,357	6,211	20.8	
65 – 70	79,146	8,477	17.2	
70 – 75	70,669	11,470	13.9	
75 – 80	59,199	14,598	10.9	
80 – 85	44,601	17,448	8.3	
85+	27,153	27,153	6.2	
				Total: 4,040,958

* Compiled by Marion Hansen, based on data from US Bureau of the Census, *Statistical Abstract of the United States*, 1992 (edition 112).

LIFE HISTORY PATTERNS

So far, we have thought about populations as if each were made up of identical individuals. Yet members of most species pass through many stages of development, with each new stage having its own perils and rewards. Let's take a look at a few examples of age-specific, life history patterns.

Life Tables

Each species has a characteristic life span, but few individuals reach the maximum age possible. Death is more probable at some ages and less so at others. Also, individuals tend to reproduce or leave a population at certain ages, which vary from one species to the next.

Such age-specific patterns first aroused the interest of life insurance and health insurance companies, but they also interest ecologists. Typically, researchers keep track of a group of individuals from the time of birth until the last one dies. Such a group is called a cohort. They also track the number of offspring born to individuals at each age interval. Computing the age-specific birth and death rates produces birth and death "sched-

ules." A death schedule can be transformed to a "survivorship" schedule, which more cheerily lists the number of individuals that live long enough to reach age x. Such age-specific patterns of birth and death can be summarized in **life tables**, as in Table 33.1.

To ecologists, dividing a population into age classes and assigning birth rates and mortality risks to each class have practical applications. Unlike a crude census (head count), such data can be the basis for informed policy decisions on conservation of endangered species, pest management, social planning for human populations, and other concerns. For example, birth and death schedules for the spotted owl figured in the court decisions that halted logging in old-growth forests, the owl's habitat.

Survivorship Curves and Reproductive Patterns

Plots of the age-specific survival of a cohort in a given environment are called **survivorship curves**. Three types are common in nature. *Type I* curves reflect high

Figure 33.6 Three generalized types of survivorship curves. For Type I populations, there is high survivorship until some age, then high mortality. Type II populations show a fairly constant death rate at all ages. For Type III populations, survivorship is low early in life.

survivorship until fairly late in life, then a large increase in deaths. Figure 33.6 shows an example of this type of curve. They are typical of elephants and other large mammals that produce only one or a few large offspring at each reproduction and provide them with extended parental care. Female elephants produce only four or five calves in a lifetime, and they devote several years of parental care to each one.

Type I curves also are typical of human populations that have access to good health care services. Historically, and where health care is poor today, infant deaths cause a sharp drop at the start of the curve. Following the drop, the curve then levels off from childhood to early adulthood.

Type II curves reflect a fairly constant death rate at all ages. They are typical of organisms that are just as likely to be killed or die of disease at any age. This is true of some songbirds, lizards, and small mammals.

Type III curves reflect a high death rate early in life. They are typical of animals that produce many small offspring, with little or no parental care. Figure 33.6c shows how the curve plummets for sea stars. Although sea stars produce mind-boggling numbers of tiny offspring, their young must feed and grow rapidly on their own, without support, protection, or guidance from parents. Most offspring are quickly eaten by corals and other animals. Plummeting survivorship curves are typical of many other marine invertebrates, most insects, and many fishes, plants, and fungi.

At one time, ecologists thought that selection processes favored either the early, rapid production of many small offspring or the late production of only a few large ones. These two patterns are now known to be extremes, at opposite ends of a range of possible life histories. Also, both patterns (and intermediate ones) can occur in different populations of the same species.

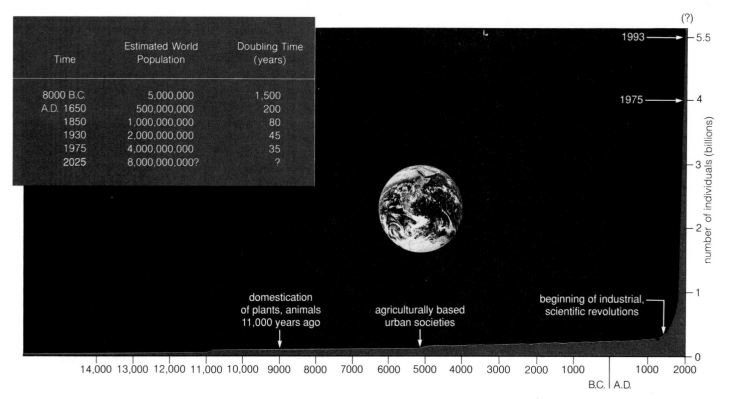

Time	Estimated World Population	Doubling Time (years)
8000 B.C.	5,000,000	1,500
A.D. 1650	500,000,000	200
1850	1,000,000,000	80
1930	2,000,000,000	45
1975	4,000,000,000	35
2025	8,000,000,000?	?

Figure 33.7 Growth curve for the human population. The diagram's vertical axis represents world population, in billions. (The slight dip between the years 1347 and 1351 is the time when 25 million people died in Europe as a result of bubonic plague.) The growth pattern over the past two centuries has been exponential, sustained by agricultural revolutions, industrialization, and improvements in health care. The list in the blue box tells us how long it took for the human population to double in size at different times in its history.

HUMAN POPULATION GROWTH

In 1993, the human population reached 5.5 billion (Figure 33.7). In that year alone, 94 million were added to it. All of these individuals are not dispersed uniformly through the environment. About 4.5 billion people are clumped together on only 10 percent of the land's surface. About 3 billion crowd together within 300 miles of the seas.

Our staggering population growth continues while at least a billion of the humans already on the planet are malnourished or starving, without clean drinking water, and without adequate shelter. Growth continues when 1.5 billion already go without the benefits of health care delivery and sewage treatment facilities.

Suppose it were possible, by monumental efforts, to double the food supply to keep pace with growth. We would do little more than maintain marginal living conditions for most people, and deaths from starvation could still be 20 million to 40 million a year. Even this would come at great cost, for we are drastically modifying the very environment that must sustain us.

Salted-out cropland, desertification, deforestation, pollution—these are some of the consequences you will read about in Chapter 37, and they do not bode well for our future.

For a while, it would be like the Red Queen's garden in Lewis Carroll's *Through the Looking Glass*, where one is forced to run as fast as one can to remain in the same place. But what happens when the human population doubles again? Can you brush this picture aside as being too far in the future to warrant your concern? It is no farther removed from you than the sons and daughters of the next generation.

How We Began Sidestepping Controls

How did we get into this predicament? Our population growth has been slow for most of human history. But in the past two centuries, there have been astounding increases in the growth rate. There are three possible reasons for this:

1. We steadily developed the capacity to expand into new habitats and new climate zones.

2. We increased carrying capacities in existing habitats.

3. We sidestepped several limiting factors.

The human population did all three of these things. Consider the first point. Early humans lived mostly in savannas. They were vegetarians who added scavenged bits of meat to their diet. By 200,000 years ago, small bands of hunters and gatherers had emerged. By 40,000 years ago, hunter-gatherers had spread through much

Populations and Resource Consumption

This chapter began with a brief look at the appalling conditions confronting most of the people of India, with its whopping 16 percent of the human population. By comparison, the United States has merely 4.7 percent. Yet which country is the most "overpopulated"—not in terms of numbers, but rather in terms of resource consumption and environmental damage?

In his book *Living in the Environment*, G. Tyler Miller considers reports that the average American consumes fifty times as much as the average person in India. The United States produces 21 percent of all goods and services. It uses about 25 percent of the available processed minerals and nonrenewable energy resources. It also produces at least 25 percent of the global pollution and trash. By contrast, India produces 1 percent of all goods and services. It uses about 3 percent of the available minerals and nonrenewable energy resources. And India produces about 3 percent of the global pollution and trash. Extrapolating from these numbers, Miller writes that it would take 12.9 billion impoverished people in India to match the present impact on the environment by 258 million average Americans. Think about it.

of the world. Most other species could not have expanded into such a broad range of habitats. Humans did so by relying on their very complex brains. They applied learning and memory to problems such as how to build fires, assemble shelters, create clothing and tools, and plan community hunts to exploit huge game herds. Learned experiences were not confined to individuals but spread quickly from one band to another because of language—the ability for cultural communication. Thus, *the human population expanded into diverse new environments in an extremely short time span, compared with the geographic spread of other organisms.*

What about the second possibility? About 11,000 years ago, humans began to shift from the hunting and gathering way of life to agriculture. They stopped following the game herds and moving about to harvest ripe fruits and grains. They settled down and developed a more dependable basis for existence in more favorable settings. A milestone was the domestication of wild grasses, including species ancestral to modern wheats and rice. Seeds were harvested, stored, and planted in one place. Animals were domesticated and kept close to home for food and pulling plows. Water was diverted into hand-dug ditches to irrigate crops.

The emerging agricultural practices increased productivity. With larger, more dependable food supplies, population growth rates increased. As towns and cities developed, a social hierarchy emerged that provided a labor base for more intensive agriculture. Much later, food supplies increased again through the use of fertilizers and pesticides. Transportation for food distribution improved. *Even at its simplest, managing food supplies through agriculture increased the carrying capacity for the human population.*

What about the third possibility—sidestepping several limiting factors? Consider what happened when medical practices and sanitary conditions improved. Until about 300 years ago, malnutrition, contagious diseases, and poor hygiene kept death rates high enough to more or less balance birth rates. Contagious diseases (density-dependent factors) swept through crowded settlements and cities. Without proper hygiene and sewage disposal, and plagued with such disease carriers as fleas and rats, the human population increased only slowly in size. Then plumbing and sewage treatment methods were developed. Bacteria and viruses were recognized as disease agents. Vaccines, antitoxins, and antibiotics were developed. The result was a sharp drop in the death rate. With births now exceeding deaths, the human population grew rapidly.

Finally, consider what happened when humans discovered how to harness the energy stored in fossil fuels, starting with coal. This discovery occurred in the mid-eighteenth century. Within a few decades, large industrialized societies emerged in western Europe and North America. More efficient technologies developed after World War I. Cars, tractors, and other affordable goods were mass-produced in factories. Machines replaced many of the farmers needed to produce food, and fewer farmers could support a larger population.

Thus, *by bringing many disease agents under control and by tapping into concentrated, existing stores of energy, humans sidestepped certain factors that had previously limited their population growth.*

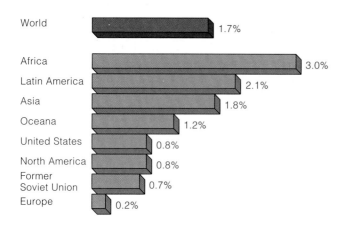

Figure 33.8 Average annual population growth rate in various groups of countries in 1993.

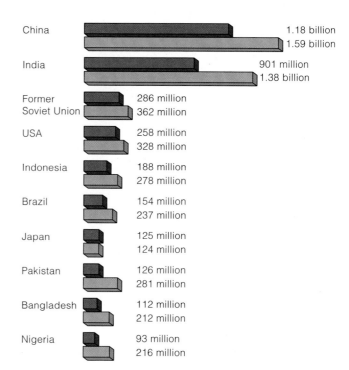

Figure 33.9 The ten most populous countries in 1993 (red bars). The blue bars indicate their population size projected for 2025.

Present and Future Growth

Where have our farflung migrations and advances in agriculture, industrialization, and health care taken us? It took *2 million years* for the human population to reach 1 billion. It took only 130 years to reach the second billion, 30 years to reach the third, 15 years to reach the fourth—*and only 12 years to reach the fifth!*

Figure 33.8 shows the rates at which populations in different regions grew during 1993. The annual growth rate for the entire human population averaged out at 1.7 percent. If that rate is maintained, the human population may approach *8 billion* just thirty years from now (Figure 33.9). It is mind-numbing to think about the resources that will be required to sustain 8 billion people. We will have to increase food production, supplies of drinkable water, energy reserves, and all the wood, steel, and other materials we use to meet everyone's basic needs—something we are not even doing now. Besides this, the massive manipulation of resources is likely to intensify pollution—which will adversely affect the atmosphere, water supplies, and productivity on land and in the seas (Chapter 37).

From what we know of the principles governing population growth—and unless there are technological breakthroughs that will once again increase the carrying capacity—we should expect a dramatic increase in death rates. *Although the stupendously accelerated growth of the human population continues, it cannot be sustained indefinitely.*

Besides having adverse effects on resource supplies, our skyrocketing numbers invite the return of severe density-dependent controls. Consider that the largest *cholera* epidemic of this century is now sweeping through South Asia. Like six others before it, the epidemic began in India and may spread into Africa, the Middle East, and on into Mediterranean countries before peaking, two to three years from now. It may claim 5 million lives.

Cholera is caused by a bacterium, *Vibrio cholerae*. Existing vaccines are useless against its most recently mutated form. People get infected by drinking water or eating food contaminated with raw sewage. As the bacterium multiplies in the small intestine, it produces a toxin that causes massive loss of body fluids (by diarrhea). Within two to seven days, untreated people can die of extreme dehydration. For many centuries, *V. cholerae* has thrived and mutated in the sewage-rich waters of the Ganges and other rivers in India (Figure 33.1). In the slums of Calcutta, where clean water is nowhere to be found, urban poor by the millions are forced to bathe in stagnant ponds and polluted waterways. In 1992, cholera struck tens of thousands in that city alone.

Controlling Population Growth

Today, there is growing awareness of the links between population growth, resource depletion, pollution, and the quality of life. Many governments attempt to control growth rates by restricting immigration. Only the United States, Canada, Australia, and a few other countries allow large annual increases. In 1991 alone, the United States admitted 1.8 million legal immigrants.

Some countries attempt to reduce population pressures by encouraging emigration. Most focus on cutting the birth rates through economic development and family planning. The first involves providing more economic security and educational programs. Such programs may ease the pressure on individuals to have large numbers of children to help them survive. Family planning involves educating individuals about choosing how many children they will have, and when.

Control Through Economics. Changes in population growth can be correlated with changes that unfold in four stages of economic development. This is the key point of the **demographic transition model** (Figure 33.10). During the *preindustrial* stage, living conditions are harsh. Birth rates are high, but so are death rates, so population growth is slow. During the *transitional stage*, industrialization begins, food production rises, and health care improves. Although death rates drop, birth rates remain fairly high, so the population grows rapidly. Growth continues at high rates (2.5 to 3 per-cent, on the average) over a long period. Then growth starts to level off as living conditions improve and birth rates begin to decline.

Population growth slows dramatically during the *industrial stage*, when industrialization is in full swing. The slowdown occurs mostly because people move from the countryside to cities, and urban couples control family size. Many decide that raising more than a few children is expensive and puts them at an economic disadvantage. In the *postindustrial stage*, zero population growth is reached. The birth rate falls below the death rate, and population size slowly decreases.

Today, the United States, Canada, Australia, Japan, the Soviet Union, and most countries of western Europe are in the industrial stage, and their growth rate is slowly decreasing. In Germany, Bulgaria, Hungary, and some other countries, birth rates are lower than death rates, and the populations are getting smaller.

Mexico and other less-developed countries are in the transitional stage. They do not have enough skilled workers to complete the transition to a fully industrial economy. Fossil fuels and other resources that drive industrialization are being used up there as well as in the industrialized countries. Fuel costs may become prohibitive for countries at the bottom of the economic ladder before they can enter the industrial stage. If population growth keeps outpacing economic growth, death rates will increase. Thus many countries may now be stuck in the transitional stage. Some may return to the harsh conditions of the preceding stage.

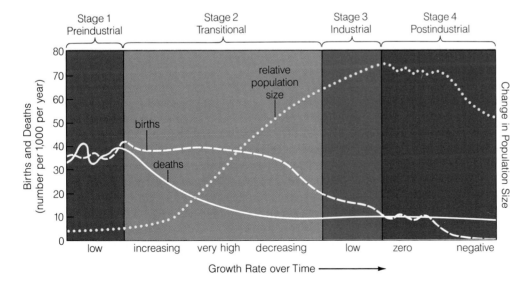

Figure 33.10 Demographic transition model of changes in the growth characteristics and size of populations, as correlated with changes in economic development. The model explains changes that have taken place in western Europe and other industrialized regions.

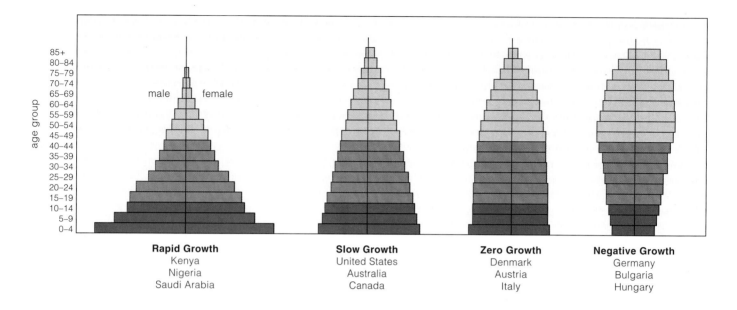

Figure 33.11 Age structure diagrams for countries with rapid, slow, zero, and negative population growth rates. Dark green indicates the pre-reproductive base. Purple indicates reproductive years; light blue, the post-reproductive years. The portion of the population to the left of the vertical axis in each diagram represents males; the portion to the right represents females. The width of each bar indicates the proportion of males and females within each age group.

Control Through Family Planning.   When thoughtfully developed and carefully administered, family planning programs may bring about a faster decline in birth rates, at less cost, than economic development alone. Such programs vary from country to country, but all provide information on methods of birth control, as described on page 528.

Suppose family planning programs were successful beyond our wildest imagination, so that each couple decided to have only two children. (Actually, the average "replacement rate" for zero population growth is slightly higher, for some female children die before reaching reproductive age. It is about 2.5 children per woman in less-developed countries, and 2.1 in more-developed countries.) Even if the replacement rate for zero population growth were achieved globally, the human population would keep on growing for at least another sixty years! Why? An immense number of *existing* children will eventually be reproducing.

Take a look at Figure 33.11, which shows age structure diagrams for populations growing at different rates. In these diagrams, 15–44 is the average range of childbearing years. The diagram for Kenya and other rapidly growing populations has a broad base. The central portion of each diagram includes women and men of reproductive age. The lower portion includes an even larger number of children who will move into the reproductive category during the next fifteen years. Today, *more than a third of the world population falls in the*

broad pre-reproductive base. This gives you an idea of the magnitude of the effort it will take to control population growth.

One way to slow the birth rate is to bear children in the early thirties, rather than in the mid-teens or early twenties. Delayed reproduction slows the tendency toward growth and lowers the average number of children in each family. In China, for example, the government has established the most extensive family planning program in the world. Premarital sex is strongly discouraged, and couples are strongly urged to postpone the age at which they marry. Married couples have ready access to free contraceptives, abortion, and sterilization. Paramedics and mobile units ensure access in remote rural areas. Couples who pledge not to have more than one child are given extra food, better housing, free medical care, and salary bonuses. Their child will be granted free tuition and preferential treatment when he or she enters the job market. Those who break the pledge forego benefits and sometimes pay penalties.

Are these outrageous measures? Think about this: Family planning has been China's way of avoiding mass starvation. Between 1958 and 1962 alone, an estimated 30 million Chinese died because of famine. Even so, the population time bomb has not stopped ticking. China's population now numbers 1.18 billion—and 340 million of its young women are moving into the reproductive age category. By the year 2025, China is projected to have a population of 1.59 billion people.

Questions About Zero Population Growth

For us, as for all species, the biological implications of extremely rapid growth are staggering. Yet so are the social implications of achieving and maintaining zero population growth or of a population decline.

For instance, as you have seen, most individuals of an actively growing population fall in younger age brackets. Under conditions of constant growth, the age distribution guarantees the availability of a future work force. A large work force is capable of supporting older, nonproductive people with social security, low-cost housing, health care, and other programs. If zero population growth is maintained, a larger proportion of the population will fall in the older age brackets. Will the nonproductive members be provided with goods and services if productive ones are asked to carry a greater and greater share of the burden? These are not abstract questions. Put them to yourself. How much are you willing to bear for the sake of your parents? Your grandparents? How much will your children be able to bear for you?

We have arrived at a major turning point, not only in our biological evolution but in our cultural evolution as well. The decisions awaiting us are among the most difficult we will ever have to make, yet it is clear that they must be made, and soon.

All species face limits to growth. We may think we are different from the rest, for our unique ability to undergo cultural evolution has allowed us to postpone the action of most of the factors limiting growth. But the key word here is "postpone." *No amount of cultural intervention can hold back the ultimate check of limited resources and a damaged environment.*

We have sidestepped a number of the smaller laws of nature. In doing so, we have become more vulnerable to those laws which cannot be repealed. Today there may be only two options available. Either we make a global effort to limit population growth in accordance with environmental carrying capacity, or we wait until the environment does it for us.

SUMMARY

1. A population's growth rate depends on the rates of birth, death, immigration, and emigration. Putting aside the effects of immigration and emigration, if the birth rate per individual exceeds the death rate per individual by a constant amount, a population will grow exponentially.

2. Carrying capacity is the maximum number of individuals in a population that can be sustained indefi-

nitely by the resources that are available in a given environment.

3. Population size is determined by the availability of sustainable resources, predation, competition, and other factors that limit population growth. Limiting factors vary in their relative effects and they vary over time, so population size also changes over time.

4. Limiting factors such as competition for resources, disease, and predation are density dependent. Other factors tend to increase the death rate or decrease the death rate more or less independently of population density.

5. Patterns of reproduction, death, and migration vary over the life span characteristic of a species. Environmental variables help shape its life history (age-specific) patterns.

6. Currently, human population growth varies from below zero in a few developed countries to more than 4 percent per year in some less-developed countries. In 1993 the annual growth rate for the entire human population was 1.7 percent.

7. Rapid growth of the human population during the past two centuries was possible largely because of a capacity to expand into new environments, and because of agricultural and technological developments that increased the carrying capacity. Ultimately, we must confront the reality of limits to our population growth.

Review Questions

1. Why do populations that are not restricted in some way tend to grow exponentially? *538–539*

2. If the birth rate equals the death rate, what happens to the growth rate of a population? If the birth rate remains slightly higher than the death rate, what happens? *538–539*

3. What defines the carrying capacity for a particular environment? Diagram what happens when a low-density population shows a logistic growth pattern. *540–541*

4. At present growth rates, how long will it take to add another billion to the human population? *546*

5. How did human populations develop the means to expand steadily into new environments? How did they increase the carrying capacity of their habitats? How have they avoided some of the limiting factors on population growth? Is the avoidance an illusion? *544–545*

6. If a third of the world population is now below age fifteen, what effect will this age distribution have on the future growth rate of the human population? What sorts of humane recommendations would you make that would encourage individuals of this age group to limit the number of children they plan to have? What are some of the social, economic, and environmental factors that might keep them from following those recommendations? (*Reflect upon pages 536, 545, 547, and 548.*)

7. Write a short essay about a hypothetical population that shows one of the following age structures. Describe what might happen to younger and older age groups when individuals move into new categories. *548*

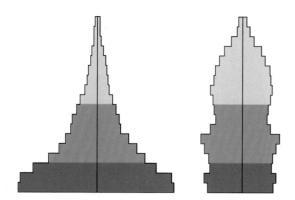

1. _____ is the study of how organisms interact with one another as well as with their physical and chemical environment.

2. A _____ is a group of individuals of the same species that occupy a certain area.

3. The rate at which a population grows or declines depends upon the rate of _____ .
   a. births
   b. deaths
   c. immigration
   d. emigration
   e. all of the above

4. Populations grow exponentially when _____ .
   a. birth rate exceeds death rate and neither changes
   b. death rate remains above birth rate
   c. immigration and emigration rates are equal
   d. emigration rates exceed immigration rates
   e. both a and c

5. The maximum number of individuals in a population that can be sustained indefinitely by the resources in a given environment is the _____ .
   a. biotic potential
   b. carrying capacity
   c. environmental resistance
   d. density control

6. Which of the following factors does not affect sustainable population size?
   a. predation
   b. competition
   c. resources
   d. pollution
   e. All of the above can affect population size.

7. Resource competition, disease, and predation are _____ controls on population growth rates.
   a. density-independent
   b. population-sustaining
   c. age-specific
   d. density-dependent

8. In 1993, the average annual growth rate for the human population was _____ percent.
   a. 0
   b. 1.0
   c. 1.5
   d. 1.7
   e. 2.7
   f. 4.0

9. During the past two centuries, _____ contributed to the rapid growth of the human population.
   a. worldwide increased birth rate
   b. worldwide increased death rate
   c. carrying capacity reduction
   d. carrying capacity expansion

10. Match the population ecology terms with the appropriate description.
    _____ carrying capacity
    _____ exponential growth
    _____ population growth rate
    _____ density-dependent controls
    _____ population

    a. disease, predation, etc.
    b. group of individuals of same species occupying a given area
    c. depends on rates of birth, death, immigration, emigration
    d. maximum number of individuals that can be sustained indefinitely by a given environment's resources
    e. increases in population size by ever larger amounts per unit time

age structure *538*
biosphere *537*
biotic potential *539*
carrying capacity *540*
community *537*
demographic transitional model *547*
distribution *538*
doubling time *539*
ecology *537*
ecosystem *537*
emigration *538*
exponential growth *538*

immigration *538*
J-shaped curve *539*
life table *542*
limiting factor *540*
logistic population growth *540*
net population growth rate (*r*) *538*
population *537*
population density *538*
population size *538*
S-shaped curve *540*
survivorship curve *542*
zero population growth *538*

Krebs, C. 1985. *Ecology*. Third edition. New York: Harper & Row.

Miller, G. T. 1994. *Living in the Environment*. Eighth edition. Belmont, California: Wadsworth. This author consistently pulls together information on human population growth into a coherent picture.

Polgar, S. 1972. "Population History and Population Policies From an Anthropological Perspective." *Current Anthropology* 13(2):203–241. Analyzes often-ignored cultural barriers to programs for population control.

Ricklefs, R. 1990. *Ecology*. Third edition. New York: Freeman. Chapters 15–19.

34 COMMUNITY INTERACTIONS

No Pigeon Is an Island

Flying through the rain forests of New Guinea is an extraordinary pigeon, with cobalt blue feathers and plumes on its head. It is about as big as a turkey, and it flaps so slowly and noisily that its flight sounds like an idling truck. Like eight species of smaller pigeons living in the same forest, it perches on branches to eat fruit.

Why are there *nine* species of large and small fruit-eating pigeons in the same forest? Why hasn't competition for food left one species the winner? In fact, within that forest, each species lives, grows, and reproduces in a particular way, as defined by its relationships with other organisms and with the surroundings. Big pigeons perch on sturdy branches when they feed, and they eat big fruit. Smaller ones eat fruit hanging from branches too thin to support the weight of a turkey-size pigeon, and their bill is too small to open big fruit. Also, forest trees differ in the thickness of their fruit-bearing branches and fruit size, so they attract different pigeons. In such ways, the fruit supply is *partitioned* among the nine pigeon species.

And what about the forest's trees? Their pigeon-enticing fruits contain seeds with tough coats that resist digestion. While the seeds are passing through the pigeon gut, the pigeons are flying about and are likely to disperse them to new locations. Dispersal improves the odds that seedlings won't have to compete directly with parent trees for sunlight, water, and nutrients. With their extensive roots and leafy crowns, the parent trees might well outcompete their offspring for resources.

Leaf-eating, fruit-munching, and bud-nipping insects interact with other organisms and the surroundings in certain ways, as do nectar-drinking bats, birds, and insects that pollinate trees. The same is true of diverse decomposers, which busily extract energy from the remains and wastes of other organisms on the forest floor—and thereby cycle nutrients back to the trees.

Like humans, then, no pigeon is an island, isolated from the rest of the living world. The nine species of New Guinea pigeons eat fruit of different sizes, so they disperse seeds of different sorts. Dispersal influences where different trees grow and where decomposers flourish—and the tree distribution and the decomposition activities influence how the entire forest community is organized. *Directly or indirectly, populations of each kind of organism are affected by interactions with neighboring populations.* With this chapter, we turn to community interactions that influence populations over time and in the space of their environment.

Figure 34.1 The turkey-size Victoria crowned pigeon, one of nine pigeon species in the same tropical rain forest of New Guinea. Within this habitat, each species has its own niche, defined by the full range of conditions in its habitat and its relations with other organisms.

1. A habitat is the type of place where individuals of a particular species normally live. A community is an association of all the populations of all species that occupy the same habitat. Its members directly or indirectly interact with one another, as by predation and competition for resources.

2. In any community, each species has its own niche, defined by the full range of the physical and biological conditions under which its members live and reproduce.

3. Community stability depends on a balancing of forces, although it is often an uneasy balance. Partitioning of resources among species that have similar niches helps maintain the balance. So do predation, disease, parasitism, and physical disturbances to the habitat.

4. Through a process of succession, species in a habitat are replaced by others in reasonably orderly progression until there is a climax community—a more or less stable array of species under prevailing conditions. Even in stable climax communities, fires and other disturbances introduce patches of different successional stages.

5. The overall pattern and actual history of changing population sizes, the arrival and disappearances of species, and physical disturbances to the habitat.

The factors just listed combine to influence community structure in three ways. First, they help determine the number of species at different "feeding levels," starting with producers and continuing on up through assorted levels of consumers. Second, they help determine the overall number of species. (For example, the high solar radiation, high temperatures, and high humidity of tropical habitats favor the growth of many kinds of plants, which in turn support many kinds of animals. Conditions in arctic habitats do not favor such great numbers of species.) Finally, the five factors influence how many individuals of each species are present.

The next two chapters address the flow of energy through feeding levels and the geographic factors that shape community structure. Here we begin with interactions among species, using the niche concept as our guide.

CHARACTERISTICS OF COMMUNITIES

Factors That Shape Community Structure

Think of a damselfish darting above a coral reef, a maple tree growing on a hillside in Vermont, or a mole burrowing about under your lawn. The type of place where you normally will find a damselfish, maple, or mole is its **habitat**. The habitat of any organism is characterized by physical and chemical features, as well as by certain other species living in the same place.

In any given habitat, the populations of all species directly or indirectly associate with one another as a **community**. The structure of that community is shaped by five factors:

1. Interactions between climate and topography that help dictate temperatures, rainfall, soil composition, and other physical conditions in the habitat.

2. The kinds and amounts of food and other resources that are available through the year.

3. The adaptive traits that allow members of a species to survive physical and chemical conditions of the habitat and to exploit specific resources.

4. Interactions between species, including competition, predation, and mutually beneficial activities.

The Niche

In any particular habitat, certain combinations of environmental conditions and resources allow individuals of a species to engage fully in the business of surviving and reproducing—as long as other species do not interfere with their necessary activities. For a predatory fish, some of the conditions might include water of suitable temperature and salinity, prey of acceptable size and nutritional value, and an absence of fishermen. The full range of environmental and biological conditions under which its members can live, grow, and reproduce is called the **niche** of that species. As you will see, conditions that define a niche are not static. They shift in large and small ways over time, creating a mosaic of changes to which members of the species must respond.

For each species in a particular habitat, certain combinations of environmental and biological conditions represent the limits within which its members can survive and reproduce. The full range of those conditions is the species' niche.

Types of Species Interactions

Even simple communities consist of dozens or hundreds of species engaged in a bewildering number of interactions. Yet it is possible to define certain *types* of interactions between any two species (see Table 34.1). Often, the species do not interact directly, and one has a *neutral* effect on the other. At most, interactions with other species provide only indirect links between the two.

Table 34.1	Types of Interactions Between Two Species	
	Direct Effect of Interaction*	
Type of Interaction	Species 1	Species 2
Neutral	0	0
Commensalism	+	0
Mutualism	+	+
Interspecific competition	–	–
Predation	+	–
Parasitism	+	–

*0 indicates no direct effect on population growth, + indicates positive effect, – indicates negative effect.

Eagles and meadow grasses, for example, have neutral effects on each other. By preying on grass-eating rabbits, eagles help control the size of rabbit populations and so help the grasses; and grasses help eagles by fattening their prey. But these are both indirect interactions.

Commensalism is a rather lopsided type of interaction. It directly benefits one species but does not harm or help the other much, if at all. This is true of birds that use a tree as their roosting site. The tree is not harmed by the interaction, and it gets nothing in return.

Mutualism is an interaction from which both species benefit. (Don't think of it as cozy cooperativeness. The two-way benefits really are an outcome of reciprocal exploitation.) By contrast, **interspecific competition** has adverse effects on both of the interacting species. **Predation** and **parasitism** directly benefit one species (the predator or parasite) and directly harm the other (the prey or host). Let's take a look at some examples of these four categories of interactions.

Figure 34.2 A mutualistic interaction in Colorado's high desert. Each species of yucca plant (**a**) is pollinated exclusively by one species of yucca moth (**b**), an insect that cannot complete its embryonic development in any other plant. The adult stage of the moth life cycle coincides with blossoming of yucca flowers. Using specialized mouthparts, a female moth gathers sticky pollen and rolls it into a ball. Then she flies to another flower. She pierces the wall of the flower's ovary, where seeds will develop, and lays eggs inside. As she crawls out of the flower, she pushes a ball of pollen onto a pollen-receiving surface. The pollen grains germinate and grow down through the ovary tissues, carrying sperm to the flower's eggs. After fertilization, seeds develop. Meanwhile, the moth eggs develop to the larval stage. When larvae emerge (**c**), they eat a few seeds, then gnaw their way out of the ovary. Uneaten seeds give rise to new yucca plants.

MUTUALLY BENEFICIAL INTERACTIONS

Mutualistic interactions, in which positive benefits flow both ways, abound in nature. The chapter introduction described how trees provide New Guinea pigeons with fruit. Such interactions exist between many kinds of plants and animals that help them by dispersing seeds to new germination sites. Most flowering plants and their pollinators are mutualists. Consider the yucca plant and yucca moth. Each yucca plant species can be pollinated by only one yucca moth species. Yucca moth larvae can grow nowhere except in yucca plants; they eat only yucca seeds (Figure 34.2).

Symbiosis is a special form of mutualism. The word refers to organisms of different species that cannot grow and reproduce unless they spend their entire lives together, in intimate dependency. This is true of the mycorrhiza, an intermeshing of two kinds of absorptive structures (fungal hyphae and young roots). Hyphae of a fungal species penetrate the roots of certain plants and become virtual extensions of them. In fact, the plant cannot take up enough vital nutrients without its fungal symbiont (page 273). In turn, the fungus depends on sugar molecules produced by its photosynthetic partner. When photosynthesis stops, the fungus stops producing spores. As a final example, think back on the presumed endosymbiotic origins of eukaryotes (page 256). If cells of different bacterial species had not evolved in such intimate, mutually beneficial ways, you and all other eukaryotic organisms would not even be around today.

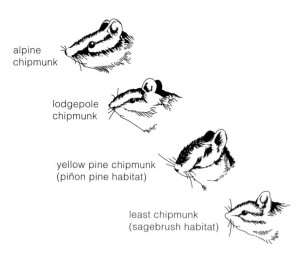

alpine
chipmunk

lodgepole
chipmunk

yellow pine chipmunk
(piñon pine habitat)

least chipmunk
(sagebrush habitat)

Figure 34.3 Competition in nature. On the eastern slopes of the Sierra Nevada, different chipmunk species occupy different habitats. The alpine habitat is at the highest elevation. Below this are the lodgepole pine, piñon pine, and finally the sagebrush habitats. The least chipmunk lives in sagebrush at the base of the mountains. It could move up into the piñon pine habitat, but the aggressively competitive behavior of the yellow pine chipmunk in that habitat stops it. (Dietary habits keep the yellow pine chipmunk out of the sagebrush habitat.)

a

b

Figure 34.4 From the time spring wildflowers start blooming in the Rocky Mountains until late summer, a male broadtailed hummingbird (**a**) will aggressively chase other males and females of its species out of his blossom-rich feeding territory. Then, in August, male rufous hummingbirds (**b**) migrate from their breeding grounds in the Pacific Northwest, and they cross through the high Rockies on their way to wintering grounds in Mexico. Rufous males are even more competitive. They evict resident broadtails from their territories all along their migratory route!

COMPETITIVE INTERACTIONS

Categories of Competition

Competition among individuals of a population can be fierce, as the preceding chapter indicated. In such *intraspecific* competition, some individuals suffer when others deplete or restrict access to a resource. By contrast, *interspecific* competition (between populations of different species) usually is less intense. Requirements of two species might be similar, but they never will be as close as they are for members of the same species.

Competitive interactions within or between species take two forms. In some cases, all individuals have equal access to a required resource, but some are better than others at exploiting it. This interaction tends to reduce the common supply of a shared resource—unless the resource is abundant. (If you and a friend were sharing a 10-gallon milkshake, each drinking from your own straw, you wouldn't care how fast your friend drank.) In other cases, certain individuals control access to a resource. They partially or wholly prevent others from using it, regardless of its scarcity or abundance. (Even if you shared a 10-gallon milkshake, you would object if your friend pinched your straw.) Figures 34.3 and 34.4 give two examples of this "interference" competition.

Competitive Exclusion

To a greater or lesser extent, any two species differ in their adaptations for securing food or avoiding enemies, so one usually competes more effectively for scarce resources. The two are less likely to coexist in the same habitat when they use resources in very similar ways.

G. Gause demonstrated this by growing two species of *Paramecium* separately, then together (Figure 34.5). Both species exploited the same food (bacterial cells) and competed intensely for it. The test results suggested that two species that require identical resources cannot coexist indefinitely, a concept now known as **competitive exclusion**.

In other experiments, Gause used two other species of *Paramecium* that did not overlap as much in their requirements. When grown together, one species tended to feed on bacteria suspended in the liquid in the culture tube. The other tended to feed on yeast cells at the bottom of the tube. The rate of population growth was slower for both species, but not enough for one to exclude the other. They continued to coexist.

The more two species in the same habitat differ in their use of scarce resources, the more likely they are to coexist.

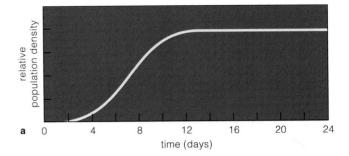

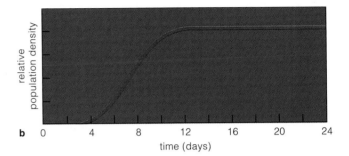

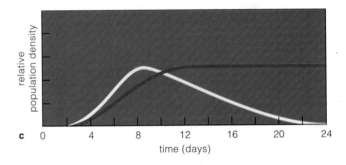

Figure 34.5 Competitive exclusion, as demonstrated by an experiment with two bacterial species that compete for the same food resource.

First, *Paramecium caudatum* (**a**) and *P. aurelia* (**b**) were grown apart, in separate culture tubes. They both established stable populations, as indicated by the S-shaped growth curves in (**a**) and (**b**). Next, populations of the two species were grown together in the same culture tube. *P. aurelia* (the red growth curve in **c**) drove the other species toward extinction (as indicated by the yellow growth curve in **c**).

From this experiment and others, it appears that two species cannot coexist indefinitely in the same habitat if they require identical resources. If their requirements do not overlap as much, one might influence the population growth rate of the other, but both may still coexist.

Figure 34.6 Idealized cycling of abundances of predator and prey. (The scale exaggerates predator density; predators usually are less common than their prey at all points in the cycle.) The pattern arises through time lags in predator responses to changes in prey abundance. Starting at time *a*, prey density is low, so predators have more difficulty capturing food and their population is declining. In response to the predator decline, prey start increasing, but predators do not start increasing until reproduction gets under way (time *b*). Both populations grow until the increased number of predators causes the prey population to decline (time *c*). Predators continue to increase and take out more prey, but the lower prey density leads to starvation among predators and their growth rate slows (time *d*). At time *e*, a new cycle begins.

CONSUMER-VICTIM INTERACTIONS

"Predator" Versus "Parasite"

Of all community interactions, predation is the most riveting of our attention—as well as the prey's. But what, exactly, is a predator-prey interaction? Take a look at the leopard in Figure 34.6 as it closes in for the kill. A goat pulling up a thistle plant for breakfast, although less dramatic, is also a predator. Its prey is a living organism, killed for food. Can the same be said of a horse grazing on but not killing plants? What about a

mosquito taking blood from your arm before it flies off? What about ticks or fleas taking blood for long periods before they get off one host and lay their eggs elsewhere? What about tapeworms, mistletoe, and other parasites that live in or on other organisms?

For simplicity, let's use two broad definitions for all interactions between consumers and their victims. A **predator** is any animal that feeds on other living organisms (its prey) but does not take up residence on or in them. Prey may or may not die as a result of the interaction. A **parasite** takes up residence in or on other living organisms (its hosts) and feeds on their tissues. It does this for a good part of its life cycle, and host organisms may or may not die as a result of the interaction.

Dynamics of Predator-Prey Interactions

Some predator and prey populations coexist at more or less steady population levels. Others undergo recurring cycles of abundance and population crashes, erratic population cycles, or prey extinction. Three factors influence the outcome of predator-prey interactions:

1. Carrying capacity of the prey population. Carrying capacity, recall, is the maximum number of individuals

Figure 34.7 Predator-prey interactions between the Canadian lynx and snowshoe hare. Abundances of both populations, shown in the diagram, are based on counts of pelts that trappers sold to Hudson's Bay Company over a ninety-year period. The dashed line represents the abundance of lynx, and the solid line, the abundance of hares. This figure is a good test of how willing you are to accept conclusions without questioning their scientific basis. (Remember the discussion of scientific methods in Chapter 1?) What other factors may have influenced the relative abundances of lynx and hare? Did weather vary greatly, with more rigorous winters imposing greater demand for food (required to keep warm) and higher death rates? Did competition between lynx and other predators (owls, goshawks, coyotes, foxes) complicate the lynx cycle? Did predators turn to alternate prey species during low points of the hare cycle? Did trapping increase with rising fur prices in Europe, and did it decrease as pelt supply outstripped the demand?

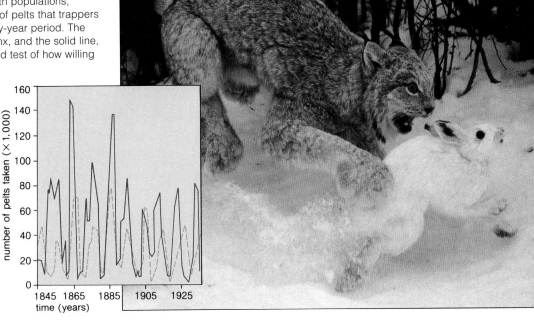

that can be maintained indefinitely by the resources available in a given environment.

2. Reproductive rates of the predator and prey (predators rarely reproduce as fast as prey).

3. How individual predators respond to increases in prey density (for example, they might eat more or move to areas where prey are more abundant).

When predation keeps the prey population from overshooting its carrying capacity, the two populations tend to coexist at relatively steady levels. This happens when predators reproduce promptly and eat more when there are more prey organisms around to eat.

By contrast, population densities tend to fluctuate when predators do not reproduce as fast as the prey, when they can eat only so many prey organisms in a given time interval, and when the carrying capacity for the prey is high.

The graph lines in Figure 34.6 represent an idealized case of cyclic changes brought about by time lags in a predator's response to changes in prey abundance. In nature, we sometimes find such a correspondence between the rise and fall of predator and prey populations. Other factors also contribute to these cyclic changes.

For example, long-term studies of the snowshoe hare population in Canada show that the hare population density changes every nine to ten years. The cycle

tends to be synchronized across much of Canada and Alaska. Records of pelts gathered by trappers and sold to the Hudson's Bay Company seem to suggest a similar rise and fall in Canadian lynx populations (Figure 34.7). However, lynx are not the only predators in the cycle. When hares are near their peak abundance, they also are feasted upon by great horned owls, goshawks, foxes, coyotes, and other predators. After the hares decline, predator populations recover only when the density of hares begins to increase once again.

To see if food scarcity causes the declines in the hare population, researchers provided extra food when hares were at peak densities. The hare population declined anyway. Most likely, the quality and quantity of the hare's food supply affects hare population density. But most hares die of predation, not starvation. As food becomes scarce, hares are forced to take more risks to reach the remaining edible plants—and so expose themselves to increasing risk of predation.

When predation keeps a prey population from overshooting its carrying capacity, the predator and prey populations tend to coexist at relatively stable levels.

When there are time lags in a predator's response to changes in prey abundance, the populations tend to undergo cyclic or irregular fluctuations in density.

a

b

c

d

Figure 34.8 The fine art of camouflage by prey and by predators. (**a**) Find the plants (*Lithops*) that hide in the open from herbivores; they have the form, pattern, and coloring of stones. (**b**) Find the scorpion fish, a dangerous predator that lies motionless and camouflaged on the sea bottom, the better to surprise unsuspecting prey. (**c**) This caterpillar uses coloration and body positions to mimic an unappetizing bird dropping. (**d**) What bird??? With the approach of a predator, the least bittern stretches its neck (colored much like the surrounding withered reeds), thrusts its beak upward, and sways gently like reeds in the wind.

Prey Defenses

Predators and prey exert continual selection pressure on each other. When some new, heritable means of defense appears in a prey population, predators not equipped to counter the defense won't eat. *When the prey evolves, the predator also evolves to some extent because the change affects selection pressures operating between the two.* This is an example of **coevolution**. The word refers to the joint evolution of two (or more) species that are interacting in close ecological fashion. Let's take a look at some of the outcomes for consumers and their victims.

Camouflage. Predation pressure has favored the evolution of prey that can **camouflage** themselves—that is, hide in the open. Adaptations in form, patterning, color, or behavior help an organism blend with its surroundings and escape detection. One desert plant looks like a small rock (Figure 34.8). It flowers only during a brief rainy season, when other plants and water are available for plant eaters. Camouflage also works for stealthy predators. Think of polar bears against snow, tigers against tall-stalked and golden grasses, and pastel-colored spiders on pastel-colored flower petals.

Warning Coloration and Mimicry. Predation pressure has favored the evolution of prey species that are bad-tasting, toxic, or able to inflict pain (as by stingers) on their attackers. Many toxic types have bold patterns

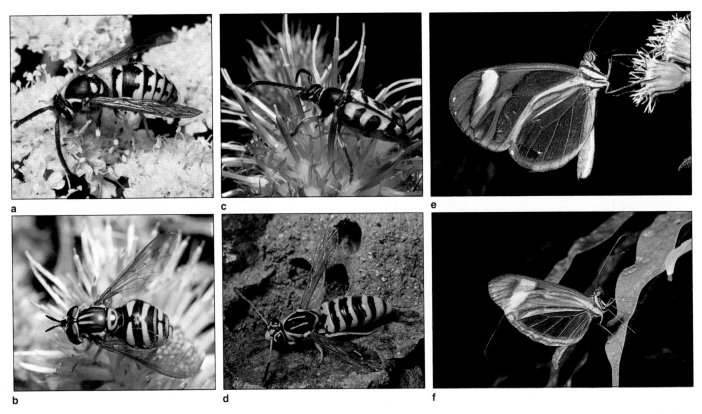

a

c

e

b

d

f

and conspicuous colors that serve as warning signals to predators. Inexperienced predators may attack a striped skunk, yellow-banded wasp, or bright-orange monarch butterfly—once. They quickly learn to associate the colors and patterning with pain or digestive upsets.

Weaponless prey species often have warning colors very similar to those of bad-tasting, toxic, or dangerous ones. Figure 34.9 gives examples. The resemblance of an edible species to a relatively inedible one is called **mimicry**. They are splendid cases of adaptive evolution.

Moment-of-Truth Defenses. When cornered, some prey animals startle or intimidate the predator with display behavior. For example, a baboon on the run that turns and displays its formidable canines may startle the leopard chasing it (Figure 34.6). The interaction may create a moment of confusion, and a moment may be all it takes for a getaway.

Other cornered animals release chemicals as warning odors, repellants, and outright poisons. Earwigs, skunks, and stink beetles produce awful odors. Several beetles take aim and let loose with a noxious spray (Figure 34.10). Similarly, many plants emit chemicals that repel predators. Tannins in the foliage and seeds of certain plants taste bitter and make the plant tissues hard to digest. Nibble on the delicate yellow petals of a buttercup (*Ranunculus*) and you will badly irritate the lining of your mouth.

Figure 34.9 Mimicry. Many animals avoid predation by having a bad taste, obnoxious secretion, or painful bite or sting. Each young predator learns this the hard way, by unpleasant taste trials. If dangerous or unpalatable prey animals were not easy to recognize and remember, many would be lost while inexperienced predators were learning their lessons. Many avoid being eaten by displaying bright colors and bold markings. Many do not even bother hiding.

Many edible, unrelated species avoid predation by mimicking the appearance and behavior of the dangerous or unpalatable species. The aggressive yellow jacket shown in (**a**) is the probable model for similar-looking but edible flies (**b**), beetles (**c**), and wasps (**d**). The unpalatable butterfly in (**e**) is a model for the palatable mimic *Dismorphia* (**f**).

Figure 34.10 Moment-of-truth defensive behavior. As a last resort, some beetles spray noxious chemicals at their attackers, which works some of the time but not all of the time. Grasshopper mice plunge the chemical-spraying tail end of the beetle into the ground and feast on the head end.

Parasitic Interactions

Parasites. Remember those flukes and tapeworms described in Chapter 19? They are prime examples of parasites—consumers that live on or in a host organism and exploit its tissues for nutrients. Sometimes parasites indirectly cause death, as when a weakened host dies from secondary infections. This is not advantageous for the parasite. After all, hosts able to live longer can spread far more of their offspring. Thus parasites tend to coevolve with their hosts, in ways that produce less-than-fatal effects. Generally, death results only when a parasite attacks a novel host (one that has no coevolved defenses against it) or when many individual parasites infect a single host organism.

Some parasites do not consume host tissues. Rather, they complete their life cycle by exploiting the social behavior of another species. Such animals are known as social parasites. Consider the North American brown-headed cowbird. It never builds a nest, incubates its eggs, or cares for its offspring. The cowbird removes an egg from the nest of another kind of bird and lays one as a "replacement." Some birds can't tell the difference, so they end up hatching the egg and raising a young cowbird. The large, aggressive cowbird often pushes the rightful occupants out of the nest or gets most of the food.

Parasitoids. Somewhere between predators and parasites are the **parasitoids**, insect larvae that always kill what they eat. While parasitoids are growing up, they completely consume the soft tissues of their hosts. This sounds horrendous, but fortunately their hosts are not humans but the larvae or pupae of other insect species. Actually, parasitoids serve as natural controls over other insects. Many species are raised commercially and released as an alternative to chemical pesticides.

COMMUNITY ORGANIZATION, DEVELOPMENT, AND DIVERSITY

Community stability is an outcome of forces that have come into balance—sometimes an uneasy balance. Resources are sustained, as long as populations do not flirt dangerously with the carrying capacity. Predators and prey coexist, as long as neither wins. Competitors have no sense of fair play. Even mutualists are stingy. A flower produces as little nectar as necessary to attract a pollinator, and the pollinator takes as much nectar as it can for the least effort. Let's take a look at some community patterns arising from these conflicting forces.

Resource Partitioning

Those nine species of fruit-eating pigeons in the same forest are a good example of **resource partitioning**. In this community pattern, similar species generally share the same kind of resource in different ways, in different areas, or at different times. Figure 34.11 gives another example of this. In large part, resource partitioning is an outcome of natural selection, which promotes differences among the community's competing, established populations. The pattern of resource partitioning may change when novel species immigrate to the community.

Community resources may become partitioned when competition promotes the evolution of differences among species or when novel species join the community.

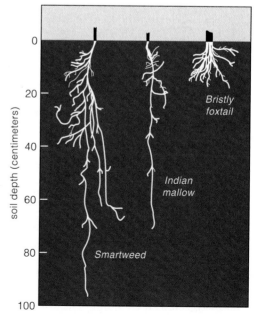

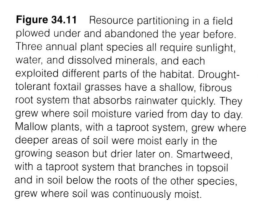

Figure 34.11 Resource partitioning in a field plowed under and abandoned the year before. Three annual plant species all require sunlight, water, and dissolved minerals, and each exploited different parts of the habitat. Drought-tolerant foxtail grasses have a shallow, fibrous root system that absorbs rainwater quickly. They grew where soil moisture varied from day to day. Mallow plants, with a taproot system, grew where deeper areas of soil were moist early in the growing season but drier later on. Smartweed, with a taproot system that branches in topsoil and in soil below the roots of the other species, grew where soil was continuously moist.

a   Periwinkles in a tidepool

b   *Enteromorpha*, a filamentous green alga

Figure 34.12   Effect of grazing by periwinkles (*Littorina littorea*) on the number of algal species in tidepools. Jane Lubchenco's studies showed that the number of algal species is greatest in tidepools that have an intermediate number of algae-eating periwinkles. In tidepools having only a few periwinkles, the competitively dominant alga (*Enteromorpha*) eliminates all other algal species. In tidepools with high densities of periwinkles, the grazing pressure eliminates all algal species—*except* the tough, unpalatable *Chondrus*.

c   *Chondrus*, another alga

How Predation and Disturbances Influence Competition

Predation reduces the density of prey populations. By doing so, it also can reduce competition between prey species and promote their coexistence. For several years, Robert Paine kept sea stars out of experimental plots in a rocky intertidal zone. He left sea stars and their prey—fifteen invertebrate species—in control plots. In the predator-free plots, almost half of the prey species were crowded out by mussels. Mussels are the main prey of sea stars—yet the strongest competitors in their absence. In this community, predation helps maintain diversity of prey species by preventing competitive exclusion.

Storms, fires, and other disturbances also contribute to diversity among competing species. Communities subject to moderate predation or disturbances often show greater diversity than undisturbed or severely disturbed ones, as the example in Figure 34.12 suggests.

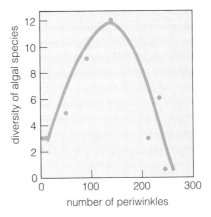

d

Figure 34.13 Primary succession in Alaska's Glacier Bay region (**a**), where changes in newly deglaciated regions have been carefully documented. Comparison of maps from 1794 onward shows that ice has been retreating at annual rates ranging from 8 meters at the glacier's sides to 600 meters at its tip over bays. (**b**) When a glacier retreats, a constant flow of meltwater tends to leach the newly exposed soil of minerals, including nitrogen. Less than ten years ago, the soil here was still buried below ice. (**c**,**d**) The first invaders of these nutrient-poor sites are feathery seeds of mountain avens (*Dryas*), drifting on the winds. Mountain avens is a pioneer species that benefits from the nitrogen-fixing activities of symbiotic microbes. It grows and spreads rapidly over glacial till.

(**e**) Within twenty years, young alders take hold. These deciduous shrubs are symbiotic with nitrogen-fixing microbes. Young cottonwood and willows also become established (**f**). In time, alders form dense thickets (**g**). As the thickets mature, cottonwood and hemlock trees grow rapidly. So do a few evergreen spruce trees. (**h**) By eighty years, spruce crowd out the mature alders. (**i**) In areas deglaciated for more than a century, dense forests of Sitka spruce and western hemlock dominate. By this time, nitrogen reserves are depleted, and much of the biomass is tied up in *peat*—excessively moist, compressed organic matter that resists decomposition and forms a thick mat on the forest floor.

Succession

How do communities come to exist in the first place? New ones may arise in barren habitats. They also may arise in inhabited areas that have been disturbed, such as abandoned pastures. In a process called **succession**, the first species in a habitat thrive, then they are replaced by other species. These are replaced by others in reasonably orderly progression until the composition of species becomes relatively steady under prevailing conditions. The most persistent array of species in a habitat is the **climax community**.

Primary Succession. In **primary succession**, the changes begin when pioneer species colonize a barren habitat. A new volcanic island is such a habitat; so is land exposed by the retreat of a glacier that had kept it buried for many thousands of years (Figure 34.13). **Pioneer species** are adapted to growing in habitats that cannot support most species. These include exposed areas with intense sunlight, wide swings in temperature, and nutrient-deficient soil. Pioneers typically are small plants with short life cycles. Each year they produce an abundance of small seeds, which are quickly dispersed.

Once pioneers are established, they improve living conditions for other species. In doing so, they commonly set the stage for their own replacement. Many pioneer plants can grow in nitrogen-poor soil because of their mutualistic interactions with nitrogen-fixing bacteria. The gradual accumulation of their wastes and remains adds volume to the soil and enriches it with nutrients that allow other species to take hold. Also, the pioneers form low-growing mats. These shelter the seeds of later

species, yet they cannot shade out the seedlings. And so the later successional species crowd out the pioneers, whose seeds travel as fugitives on the wind or water—destined, perhaps, for a new but temporary habitat.

Secondary Succession. In **secondary succession**, a patch of habitat or a community that has been disturbed is once again changing in a direction toward the climax state. This successional pattern is typical of ponds, shallow lakes, and abandoned fields. It is typical of patches of established forests when falling trees and other disturbances open up part of the continuous canopy of leaves. Sunlight reaches seeds and seedlings already on the forest floor and spurs their growth.

During secondary succession on land, both early and late species often grow together under prevailing conditions. The later species simply are growing more slowly, although in time they will exclude the others through competition. Also, the early species may be inhibiting growth of later species, which prevail only if some disturbance removes the established competitors.

e

f

g

h

i

The Role of Disturbance. Winds, fires, overgrazing, and insect infestations are among the disturbances that give rise to successional patches in even the most stable climax communities. They, too, modify and shape the direction of succession by encouraging some species and eliminating others in different parts of the habitat. In fact, many communities persist *because* of episodes of disturbance that permit regeneration of dominant species.

Giant sequoia trees grow in isolated groves in the Sierra Nevada in California. Talk about persistent! Some trees of this type of climax community are more than 4,000 years old. They endure partly because of brush fires that hit the forests every so often. Sequoia seeds germinate only in the absence of smaller, shade-tolerant plant species. Too much litter on the forest floor inhibits their germination. Modest fires eliminate trees and shrubs that compete with the young sequoias but do not damage the sequoias themselves. Mature sequoias have very thick bark, which burns poorly and insulates them against modest heat damage.

Fires once were prevented in many sequoia groves in national and state parks—not just accidental fires from campsites and discarded cigarettes, but also natural fires touched off by lightning. Litter builds up when small fires are stopped, fire-susceptible species take hold, and dense underbrush forms. Sequoia seeds cannot germinate in underbrush—which is fuel for hotter fires that damage the giants. Controlled fires are now set to eliminate underbrush and promote conditions necessary for cyclic replacements.

A climax community is the most persistent array of species in a habitat.

Episodes of disturbance, which permit the regeneration of dominant species, often maintain climax communities.

Hello Lake Victoria, Goodbye Cichlids

Given the exponential growth of the human population, it seems imperative that we find better ways to manage our food supply. Such efforts are well intentioned, but they can have disastrous consequences when ecological principles are not taken into account.

Consider what happened several years ago, when someone thought it would be a great idea to introduce the Nile perch into Lake Victoria in East Africa. People had been using simple, traditional methods of fishing there for thousands of years, but now they were taking too many fish. Soon there would be too few fish to feed the local populations and no excess catches to sell for profit. But Lake Victoria is a very big lake, and the Nile perch is a very big fish (more than 2 meters long). This seemed an ideal combination to attract commercial fishermen from the outside, with their big, elaborate nets—right? Wrong.

Native fishermen had been harvesting native fishes called cichlids, which eat mostly detritus and aquatic plants. But the Nile perch eats other fish—including cichlids. Having had no prior evolutionary experience with the Nile perch, the cichlids that were native to the lake simply had no defenses against it.

And so the Nile perch ate its way through cichlid populations and destroyed the natural fishery. Dozens of cichlid species found nowhere else are now extinct. By wiping out its own food source, the Nile perch destroyed the basis of its own population growth. It ceased to be a potentially large, exploitable food source for people living around the lake.

As if that weren't enough, the Nile perch is an oily fish. Unlike cichlids, which can be sun-dried, it must be preserved by smoking—and smoking requires firewood. And so the people started cutting down more trees in the local forests, and trees are not rapidly renewable resources. To add insult to injury, people living near Lake Victoria never liked to eat Nile perch anyway; they preferred the flavor and texture of cichlids.

What is the lesson? A little knowledge and some simple experiments in a contained setting could have prevented the whole mess at Lake Victoria.

Species Introductions

Many species have been introduced to new geographic regions, some on purpose and others by accident. Nearly all the species we know about are agriculturally useful or are pests. The useful ones include rice, wheat, corn, and potatoes grown far from their place of origin. The pests include water hyacinths.

In the 1880s, the water hyacinth from South America was displayed at the New Orleans Cotton Exposition. Flower fanciers from Florida and Louisiana took home clippings of the blue-flowered plants and set them out as ornamental additions to ponds and streams. Unchecked by their natural predators, the fast-growing hyacinths spread through the nutrient-rich waters and displaced many native species. In time they choked off ponds and streams, then rivers and canals. Now they thrive as far west as San Francisco—and they are still bringing river traffic in many areas to a halt. The *Focus* essay describes the awful results of a more recent species introduction.

Introduced species don't always lead to wholesale disasters, but most have ecological impact on climax communities. "Imported" honeybees displaced native bees in many parts of North America. As another example, remember the African bees that were accidentally released in South America? Their Africanized, "killer bee" descendants have already spread hundreds of miles beyond the border between Mexico and the United States. At this writing, they have killed several hundred people in Latin America. They have attacked 140 Americans, one of whom died from the multiple stings. Table 34.2 lists other examples of introduced species.

Table 34.2 Detrimental Effects of Species Introduced Into the United States

Species Introduced	Origin	Mode of Introduction	Outcome
Water hyacinth	South America	Intentionally introduced (1884)	Clogged waterways; shading out of other vegetation
Dutch elm disease: *Ophiostoma ulmi* (the fungal pathogen)	Europe	Accidentally imported on infected elm timber (1930)	Destruction of millions of elms; great disruption of forest ecology
Bark beetle (the disease carrier)		Accidentally imported on unbarked elm timber (1909)	
Chestnut blight fungus	Asia	Accidentally imported on nursery plants (1900)	Destruction of nearly all eastern American chestnuts; disruption of forest ecology
Argentine fire ant	Argentina	In coffee shipments from Brazil? (1891)	Crop damage; destruction of native ant communities; mortality of ground-nesting birds
Camphor scale insect	Japan	Accidentally imported on nursery stock (1920s)	Damage to nearly 200 species of plants in Louisiana, Texas, and Alabama
Japanese beetle	Japan	Accidentally imported on irises or azaleas (1911)	Defoliation of more than 250 plant species, including commercially important species such as citrus
Carp	Germany	Intentionally released (1887)	Displacement of native fish; uprooting of water plants with loss of waterfowl populations
Sea lamprey	North Atlantic Ocean	Through Erie Canal (1860s), then through Welland Canal (1921)	Destruction of lake trout and lake whitefish in Great Lakes
European starling	Europe	Released intentionally in New York City (1890)	Competition with native songbirds; crop damage; transmission of swine diseases; airport runway interference; noisy and messy in large flocks
House sparrow	England	Released intentionally (1853)	Crop damage; displacement of native songbirds; transmission of some diseases
European wild boar	Russia	Intentionally imported (1912); escaped captivity	Destruction of habitat by rooting; crop damage
Nutria (large rodent)	Argentina	Intentionally imported (1940); escaped captivity	Alteration of marsh ecology; damage to earth dams and levees; crop destruction

After David W. Ehrenfeld, *Biological Conservation*, 1970, Holt, Rinehart and Winston, and *Conserving Life on Earth*, 1972, Oxford University Press.

Figure 34.14 Surtsey, a volcanic island, at the time of its formation. Such islands are natural laboratories for ecologists.

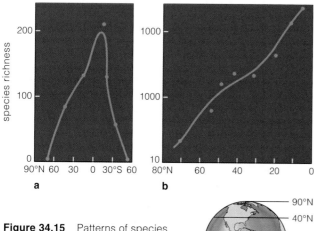

a   b

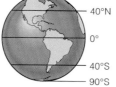

90°N
40°N
0°
40°S
90°S

Figure 34.15 Patterns of species diversity corresponding to latitude, as represented by (**a**) ants and (**b**) breeding birds of North and Central America.

Patterns of Species Diversity

Island Patterns. In 1965, a volcanic eruption formed a new island southwest of Iceland. Within six months, bacteria, fungi, seeds, flies, and some seabirds were established on it. A vascular plant appeared after two years, and a moss two years after that (Figure 34.14). As soils improved, the number of plant species increased. All species were colonists from Iceland. None originated on the island itself, which was named Surtsey. The number of new species will not increase indefinitely. Why is this so? Studies of community patterns on islands around the world provide us with two ideas.

First, islands far from a source of potential colonists receive few colonizing species; the few that do arrive are adapted for long-distance dispersal. This is called the *distance* effect. Second, larger islands tend to sup-

port more species than smaller islands at equivalent distances from source areas. This is the *area* effect. The larger islands tend to be physically more complex and often higher above sea level, and so offer more habitats that favor species diversity. Also, being bigger targets, large islands may "intercept" more colonists.

Most importantly, extinctions keep species diversity lower on small islands. The smaller populations are more vulnerable to storms, volcanic eruptions, diseases, and random shifts in birth and death rates.

The number of species on an island is a balance between the immigration rate for new species and the extinction rate for established ones.

Mainland and Marine Patterns. The most striking patterns of species diversity on land and in the seas relate to distance from the equator. As Figure 34.15 suggests, the number of coexisting species is highest in the tropics. Why is this so? First, tropical latitudes intercept more sunlight of consistently greater intensity, rainfall is high, and the growing season is long. Thus resource availability tends to be higher and more reliable. Tropical trees provide new leaves, flowers, and fruit all year in wet tropical forests. Year in and year out, they support many diverse herbivores, nectar foragers, and fruit consumers. In temperate and arctic regions, such specializations would never evolve.

Second, species diversity is self-reinforcing. The diversity of trees in tropical forests is far greater than in comparable areas of forests at higher latitudes. When more plant species coexist, more species of herbivores evolve, partly because no one herbivore can overcome the chemical defenses of all plants. More predators and parasites evolve in response to the diversity of prey and hosts. The same is true of diversity on tropical reefs.

Third, the speciation rate in the tropics historically has exceeded the background extinction rate. At higher latitudes, episodes of mass extinction have helped keep diversity low. But keep in mind that millions of species in tropical forests may disappear in the next decade, for reasons that will be described in chapters to follow.

SUMMARY

1. A community is an association of populations of different species in the same habitat. Mutualism, competition, predation, and other species interactions directly or indirectly link its populations.

2. Each species has its own niche, as defined by the full range of conditions in its habitat and its relations with other organisms. A habitat is the type of place where a particular species (or individual) lives.

3. Two species that require the same limited resource are likely to compete, either by using up the resource as rapidly or efficiently as possible or by preventing the other from using it. Species are more likely to coexist when their niches are not too similar.

4. Prey populations often stay below their carrying capacity because of predation. Time lags in a predator's response to changes in prey density may lead to cyclic fluctuations in the abundance of both populations.

5. Evolutionary adaptations for capture and escape include threat displays, chemical weapons, mimicry, and camouflaging.

6. Parasites may help control the populations of their hosts. Coevolution tends to favor resistant hosts and only moderately harmful parasites.

7. Coexistence in a community is often an uneasy balance, maintained partly by resource partitioning and partly by predation among competing species.

8. Succession is the replacement of early, pioneer species by later species in an orderly progression until the climax community is established. This is the most persistent state for a particular habitat. Primary succession occurs in barren habitats; secondary succession occurs in disturbed parts of established communities.

9. The number of species in a given community depends on the size of a region, the colonization rate, and extinction rates. It depends also on the level and pattern of resource availability.

Review Questions

1. What is the difference between the habitat and the niche of a species? Why do you suppose it might be difficult to define "the human habitat"? *552*

2. Define mutualism and give a few examples. *553*

3. Describe competitive exclusion. How might two species that compete for the same resource coexist? Can you think of some possible examples besides the ones used in the chapter? For example, consider some of the animals and plants living in your own neighborhood. *555*

4. Define primary and secondary succession. *562*

Self-Quiz *(Answers in Appendix IV)*

1. _____ are associations of populations of all species in the same habitat.

2. Each species has its own _____ , defined by the full range of conditions in its habitat and its relations with other organisms.

3. An organism's _____ is the type of place where it normally lives.

4. Community structure is shaped by _____ .
 a. geography and climate
 b. the amount and kinds of available resources
 c. adaptations of individuals for exploiting resources
 d. interspecific competition
 e. all of the above

5. _____ occurs when two species require the same limited resource.
 a. Mutualism   c. Parasitism
 b. Predation   d. Competition

6. Species competing for resources are most likely to coexist when their niches are _____ .
 a. identical   c. overlapping almost
 b. very similar but not   completely
 identical   d. not too similar

7. Chemical weapons, mimicry, and camouflage _____ .
 a. suggest coevolution   d. are strictly for defensive
 b. enhance mate selection   behavior
 c. are used by predators and   e. all of the above
 prey for capture and escape

8. Coexistence is possible because _____ .
 a. species differ in their means of using resources
 b. species partition available resources
 c. predators slow competition among prey species
 d. all of the above

9. The most stable state for a habitat is _____ .
 a. the pioneer community   c. the climax community
 b. the bare community   d. any early successional stage

10. Match the community interaction terms with the closest description.
 _____ community
 _____ parasitism
 _____ mutualism
 _____ succession
 _____ commensalism
 a. orderly progression of change to most stable state of community
 b. associations of populations of all species in a habitat
 c. one species benefits, one is neither helped nor harmed
 d. one species benefits, one is harmed
 e. two-way flow of benefits

Selected Key Terms

camouflage *558*	interspecific	pioneer species *562*
climax community *562*	competition *553*	predation *553*
coevolution *558*	mimicry *559*	predator *556*
commensalism *553*	mutualism *553*	primary succession *562*
community *552*	niche *552*	resource partitioning *560*
competitive	parasite *556*	secondary succession *562*
exclusion *555*	parasitism *553*	succession *562*
habitat *552*	parasitoid *560*	symbiosis *553*

Readings

Begon, M.; J. Harper; and C. Townsend. 1990. *Ecology: Individuals, Populations, and Communities*. Second edition. Sunderland, Massachusetts: Sinauer.

Moore, P. 1987. "What Makes a Forest Rich?" *Nature* 329:292.

Smith, R. 1992. *Elements of Ecology*. Third edition. New York: Harper-Collins.

35 ECOSYSTEMS

Crêpes for Breakfast, Pancake Ice for Dessert

Think of Antarctica, and you think of ice. Mile-thick slabs of the stuff hide all but a small fraction of a continent whipped by fierce winds and kept permanently frozen by murderously low temperatures, on the order of −100°F. Yet on patches of exposed rocky soil and on nearby islands, mosses and lichens grow. There, during the breeding season, penguins as well as seals form great noisy congregations, reproduce, and raise offspring. They cruise offshore or venture out in the open ocean in pursuit of food—krill, fishes, and squids.

In 1961, thirty-eight nations signed a treaty to set aside Antarctica as a reserve for scientific research. This was the start of scientific outposts—and of the trashing of Antarctica. Onto the ice or into the water went oil drums, old tires, and used equipment. Just offshore, marine life became acquainted with sewage and chemical wastes.

Antarctica also became a destination of cruise ships. Each summer thousands of tourists, fortified by three sumptuous meals a day plus snacks, are ferried from ship to land across channels glistening with pancake ice (Figure 35.1). They trample vegetation and bob around penguins, cameras clicking. At the communal breeding ground at Cape Royds alone, the penguin population has declined by half.

Meanwhile, nations looking for new sources of food started licking their chops over Antarctica's krill. Word got out about potentially rich deposits of uranium, oil, and gold—and by 1988, treaty nations were poised to authorize digs and drillings. Of course, oil spills or krill harvesting *could* destroy the fragile ecosystem. For example, the tiny, shrimplike krill are all that Adélie penguins eat. They are a key food for baleen whales and other marine animals that are, in turn, food for still others.

In 1991, the treaty nations thought about all of this and imposed a fifty-year ban on mineral exploration. Research stations started burying or incinerating wastes, treating sewage, and taking other measures to curb pollution. Tour operators promised to supervise tourists. Krill are not being harvested on a massive scale, yet.

Because Antarctica seems so remote from the rest of the world, it is easier to see how life is interconnected there. We can ask whether harm to one species or one habitat will lead to collapse of the whole, and be fairly sure of the answer. What about places not as sharply defined? Does interconnectedness prevail there, also? Are other places as vulnerable to disturbance or more resilient? These topics will now occupy our attention.

Figure 35.1   A boatload of tourists crossing a channel of "pancake ice" off the coast of Antarctica.

1. An ecosystem is an association of organisms and their physical environment, linked by a flow of energy and a cycling of materials. It is an open system, with inputs and outputs of both energy and nutrients.

2. Energy flows in one direction through an ecosystem. Most often the flow begins when photosynthetic autotrophs harness sunlight energy and convert it to forms that they and other organisms of the ecosystem can use. These "primary producer" organisms directly or indirectly nourish an array of consumers, decomposers, and detritivores, which interact as part of grazing or detrital food webs.

3. Water and major nutrients move from the physical environment, through organisms, then back to the environment. Their movements, called biogeochemical cycles, are global in scale. Human activities are disrupting these natural cycles and so are endangering ecosystems.

CHARACTERISTICS OF ECOSYSTEMS

Overview of the Participants

Diverse types of natural systems abound on the earth's surface. In climate, soils, vegetation, and animal life, deserts differ from hardwood forests, which differ from tropical rain forests, prairies, and arctic tundra. In their physical properties and arrays of organisms, open oceans differ from coral reefs, which differ from lakes and rivers. *Yet despite the differences, such systems are alike in many aspects of their structure and function.*

With few exceptions, each system runs on energy from the sun. Plants and other photosynthetic autotrophs described in earlier chapters are the most common "self-feeders." They capture sunlight energy and convert it to forms they can use to synthesize organic compounds from simple inorganic substances. By securing energy from the physical environment, autotrophs serve as the **primary producers** for the entire system.

All other organisms in the system are heterotrophs, not self-feeders. They depend directly or indirectly on energy stored in tissues of primary producers. Some heterotrophs are **consumers**, which feed on the tissues of other organisms. The ones called *herbivores* eat plants, *carnivores* eat animals, *parasites* reside in or on living hosts and extract energy from them, and *omnivores* par-

take of a variety of edibles. Other heterotrophs, the **detritivores**, consume dead or decomposing particles of organic matter. Crabs, nematodes, and earthworms are examples. Still other heterotrophs, the **decomposers**, include fungi and bacteria that extract energy from the remains or products of organisms (Figure 35.2).

Autotrophs also secure nutrients as well as energy for the entire system. During growth, they take up water and carbon dioxide (as sources of oxygen, carbon, and hydrogen) along with dissolved minerals, including nitrogen and phosphorus. Such materials are building blocks for carbohydrates, lipids, proteins, and nucleic acids. When decomposers and detritivores get their turn at consuming this organic matter, they break it down to small inorganic molecules. If the molecules are not washed away or otherwise removed from the system, autotrophs can use them again as nutrients.

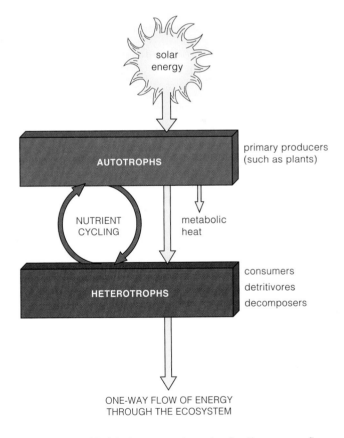

Figure 35.2 Model of an ecosystem, showing the one-way flow of energy through it and the cycling of nutrients between its autotrophic and heterotrophic organisms.

Ecosystem Defined

What we have just described in broad outline is an ecosystem. An **ecosystem** is an array of organisms *and* their physical environment, all interacting through (1) a flow of energy and (2) a cycling of materials.

Keep this important point in mind: Ecosystems are *open* systems and so are not self-sustaining. They require an *energy input* (as from the sun) and often *nutrient inputs* (as from minerals carried by erosion into a lake). Ecosystems also have *energy output* and *nutrient outputs*. Energy cannot be recycled; in time, most of the energy originally fixed by autotrophs is lost to the environment in the form of metabolic heat. Nutrients typically are cycled, but some are still lost (as through soil leaching). This chapter deals with the inputs, internal transfers, and outputs of ecosystems.

An ecosystem is an entire complex of producers, consumers, decomposers, and detritivores and their physical environment, all interacting through energy flow and materials cycling.

Table 35.1	Examples of Trophic Levels	
Members of Trophic Levels	Energy Source	Representative Organisms
Primary producers:		
Photosynthetic autotrophs	Sunlight energy	Grasses, diatoms, cyanobacteria
Chemosynthetic autotrophs	Oxidation of (stripping electrons from) inorganic substances	Nitrifying bacteria
Primary consumers:		
Herbivores	Primary producers	Grasshoppers, deer, krill
Secondary consumers:		
Primary carnivores	Herbivores	Spiders, foxes, small squids
Tertiary consumers:		
Secondary carnivores	Primary carnivores	Emperor penguins

STRUCTURE OF ECOSYSTEMS

Trophic Levels

In all ecosystems, feeding relationships are structured in much the same way. Their members fit somewhere in a hierarchy of energy transfers called **trophic levels** (from *troph*, meaning nourishment). "Who eats whom?" we can ask. When organism B eats organism A, energy is transferred to B from A. All organisms at a particular trophic level are the same number of transfer steps away from the energy input into the system.

Primary producers are closest to the initial source of energy, so they make up the first trophic level. Photosynthetic autotrophs in a lake, including cyanobacteria and aquatic plants, are examples. Rotifers, snails, and other herbivores feeding directly on producers are at the next trophic level. Birds and other primary carnivores that prey directly on the herbivores form a third level.

Many organisms, including decomposers and humans, can extract energy from more than one source, so we cannot assign them to a single trophic level. They feed at several levels and must be partitioned among them or considered as a separate group. Even so, the categories in Table 35.1 are a useful starting point for understanding feeding relationships.

Food Webs

A straight-line sequence of who eats whom in an ecosystem is sometimes called a **food chain**. However, you will have a hard time finding such a simple, isolated sequence. Why? Most often, the same food resource is part of more than one chain. This is especially true of resources at low trophic levels. It is more accurate to view food chains as *cross-connecting* with one another—as **food webs**. Figure 35.3 shows some of the participants in a food web in the seas that surround Antarctica.

A food web is a network of crossing, interlinked food chains, encompassing primary producers and an array of consumers and decomposers.

A bad-luck story about a fisherman can clarify the difference between a food chain and food web. The fisherman nets some fish that were feeding on algae near the ocean's surface. Come lunchtime, he cooks some of the catch but later loses his footing and falls into the water, where other carnivores lurk. You might think this is a simple food chain:

algae ⟶ fish ⟶ fisherman ⟶ shark

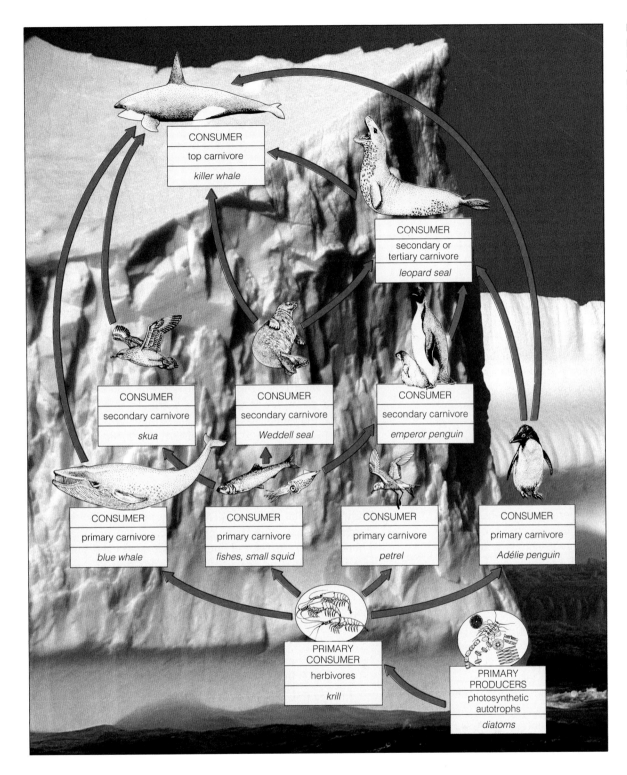

Figure 35.3 Simplified picture of a food web in the waters of the Antarctic. There are many more participants, including an array of decomposer organisms.

CONSUMER
top carnivore
killer whale

CONSUMER
secondary or tertiary carnivore
leopard seal

CONSUMER
secondary carnivore
skua

CONSUMER
secondary carnivore
Weddell seal

CONSUMER
secondary carnivore
emperor penguin

CONSUMER
primary carnivore
blue whale

CONSUMER
primary carnivore
fishes, small squid

CONSUMER
primary carnivore
petrel

CONSUMER
primary carnivore
Adélie penguin

PRIMARY CONSUMER
herbivores
krill

PRIMARY PRODUCERS
photosynthetic autotrophs
diatoms

Yet the chain excludes alternate feeding relationships. Crustaceans also were grazing on the algae, small squids and mid-sized fishes were feeding on crustaceans, and some larger fishes were feeding on the smaller ones. The sharks may have been moving in to feed on the large and mid-sized fishes. The fisherman, who ate a sandwich of cooked fish, lettuce, and bread, shifted back and forth between herbivore and carnivore. He became even more omnivorous by accompanying the sandwich with wine, an alcoholic product of decomposers (the yeasts whose fermentation activities yield wine from crushed grapes).

ENERGY FLOW THROUGH ECOSYSTEMS

Primary Productivity

To get an idea of how energy flow is studied, consider one type of land ecosystem, for which multicelled plants are the primary producers. The plants trap light energy and convert it into the chemical energy of organic compounds. The rate at which the ecosystem's producers capture and store a given amount of energy in a given length of time is its **primary productivity**. How much energy actually gets stored depends on (1) how many plants are present and (2) the balance between photosynthesis and aerobic respiration in the plants.

Gross primary productivity is the total rate of photosynthesis for an ecosystem during a specified interval.

Net primary productivity is the rate of energy storage in plant tissues in excess of the rate of aerobic respiration by the plants themselves.

Heterotrophic consumption affects the rate of energy storage.

Other factors influence the amount of net primary production, its seasonal patterns, and its distribution through the habitat. The size and form of the primary producers are major factors. So are the availability of mineral ions, the range of temperature, and the amount of sunlight and rainfall during the growing season. The harsher the environment, the fewer shoots will be produced—and the lower the productivity.

Major Pathways of Energy Flow

What is the direction of energy flow through ecosystems on land? Only a small part of the energy from sunlight becomes fixed in plants. The plants themselves use as much as half of what they fix (they lose metabolically generated heat). Other organisms tap into the energy that is conserved in plant tissues, remains, or wastes. They, too, lose heat to the environment. *All of these heat losses represent a one-way flow of energy out of the ecosystem.*

Figures 35.4 and 35.5 diagram this one-way flow of energy. Notice how it flows through two kinds of food webs. In **grazing food webs**, the energy flows from plants to herbivores, then through an array of carnivores. In **detrital food webs**, it flows mainly from plants through detritivores and decomposers. Usually the two kinds cross-connect, as when a small crab of a detrital food web wanders in front of a herring gull of a grazing food web.

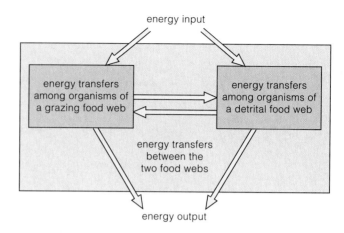

Figure 35.4 Overview of the one-way flow of energy through two kinds of cross-connected food webs in ecosystems.

The amount of energy moving through the two kinds of food webs differs from one ecosystem to the next and often varies with the seasons. In most cases, however, the greatest portion of net primary production passes through detrital food webs. You may be skeptical about this statement. After all, when cattle graze heavily on pasture plants, about half the net primary production enters a grazing food web. But cattle don't *use* all the energy contained in plant parts. Quantities of undigested plant parts and feces become available for decomposers and detritivores. Or consider marshes. There, most of the stored energy is not even used until plant parts die and become available for detrital food webs.

In short, these are the points to remember about major pathways of energy flow through ecosystems on land:

1. Energy flows into ecosystems from an outside source, which in most cases is the sun.

2. Energy flows through the food webs of ecosystems. Living tissues of photosynthesizers are the basis of grazing food webs. The remains of photosynthesizers and consumers are the basis of detrital food webs.

3. Energy leaves ecosystems mainly by loss of metabolic heat, which each organism generates.

Ecological Pyramids

Biomass Versus Energy Pyramids. Often, the trophic structure of an ecosystem is represented as an "ecological pyramid." In such pyramids, producers form a base for

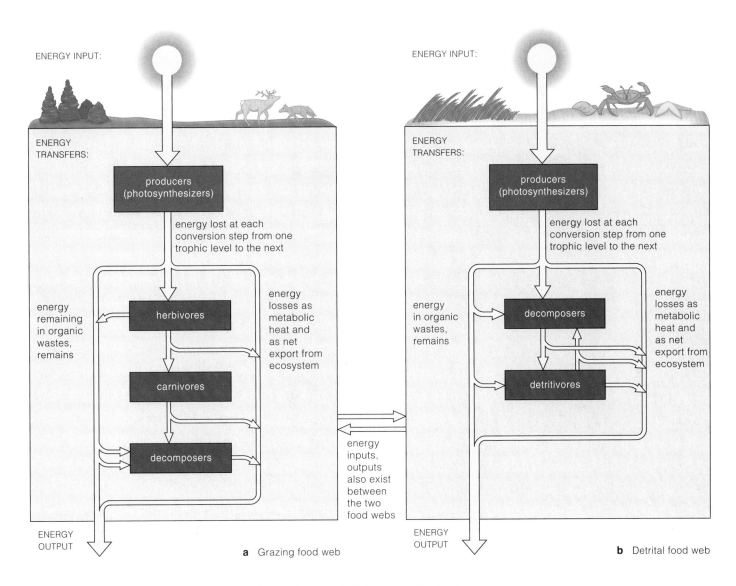

ENERGY INPUT:

ENERGY TRANSFERS:

producers (photosynthesizers)

energy lost at each conversion step from one trophic level to the next

energy remaining in organic wastes, remains

herbivores

carnivores

decomposers

energy losses as metabolic heat and as net export from ecosystem

ENERGY OUTPUT

a Grazing food web

energy inputs, outputs also exist between the two food webs

ENERGY INPUT:

ENERGY TRANSFERS:

producers (photosynthesizers)

energy lost at each conversion step from one trophic level to the next

energy in organic wastes, remains

decomposers

detritivores

energy losses as metabolic heat and as net export from ecosystem

ENERGY OUTPUT

b Detrital food web

Figure 35.5 Models of the energy input, transfers, and outputs for (**a**) a grazing food web and (**b**) a detrital food web.

successive tiers of consumers above them. Some pyramids are based on "biomass," as determined by the weight of all the members at each trophic level. Here is a pyramid of biomass (grams/square meter) for the aquatic ecosystem at Silver Springs, Florida:

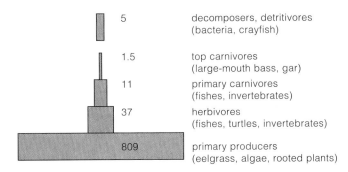

5   decomposers, detritivores (bacteria, crayfish)

1.5   top carnivores (large-mouth bass, gar)

11   primary carnivores (fishes, invertebrates)

37   herbivores (fishes, turtles, invertebrates)

809   primary producers (eelgrass, algae, rooted plants)

Some pyramids of biomass are "upside down," with the smallest tier on the bottom. Think of phytoplankton (including diatoms) and zooplankton (including rotifers) in a pond. A small biomass of rapidly growing and reproducing phytoplankton may support a greater biomass of larger-bodied zooplankton that grow more slowly than they do and consume less energy per unit of weight.

An **energy pyramid** is a more useful representation of an ecosystem's trophic structure. Such pyramids illustrate the energy losses at each transfer to a different trophic level. They start with a large energy base at the bottom and so are always "right-side up." They give a more accurate picture of ever-diminishing amounts of energy flowing through successive trophic levels of the ecosystem.

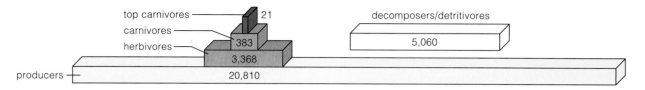

Figure 35.6 Pyramid of energy flow during one year at an aquatic ecosystem, Silver Springs, Florida.

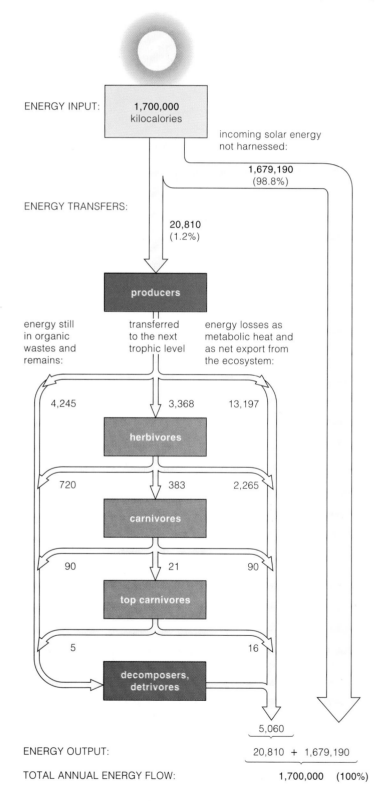

ENERGY INPUT:

incoming solar energy not harnessed:

1,679,190 (98.8%)

ENERGY TRANSFERS:

20,810 (1.2%)

energy still in organic wastes and remains:

transferred to the next trophic level

energy losses as metabolic heat and as net export from the ecosystem:

ENERGY OUTPUT: 20,810 + 1,679,190

TOTAL ANNUAL ENERGY FLOW: 1,700,000 (100%)

Figure 35.7 Annual energy flow, in kilocalories/square meter/year, for Silver Springs, Florida. Producers in this small spring are mostly aquatic plants. Carnivores are insects and small fishes; top carnivores are larger fishes. The energy source (sunlight) is available all year. Detritivores and decomposers cycle organic compounds from other trophic levels.

The producers trap 1.2 percent of incoming solar energy, and only a little more than a third of this amount is fixed in new plant biomass (4,245 + 3,368). The producers use more than 63 percent of the fixed energy for their own metabolism. About 16 percent of it is transferred to herbivores, and most of this is used for metabolism or transferred to detritivores and decomposers. Only 11.4 percent of the energy transferred to herbivores reaches the next trophic level (carnivores), and the carnivores use all but about 5.5 percent, which is transferred to top carnivores. In time, all of the 5,060 kilocalories transferred through the system appears as metabolically generated heat. This diagram is oversimplified, for no community is isolated from others. Organisms and materials constantly drop into the springs. Organisms and materials are slowly lost by way of a stream that leaves the springs.

Example of an Energy Pyramid. To gather data for an energy pyramid, researchers measure the energy that each type of individual in the ecosystem takes in, burns during metabolism, and stores in body tissues. They measure the energy remaining in its waste products. They also multiply the energy per individual by population size. Energy inputs and outputs are calculated so that energy flow can be expressed per unit of land (or water) per unit of time.

For example, the pyramid in Figure 35.6 is based on a long-term study of a grazing food web in an aquatic ecosystem in Florida. Figure 35.7 shows some of the calculations used to construct it. Such studies tell us that, given the metabolic demands of organisms and the amount of energy shunted into organic wastes, *only about 6 to 16 percent of the energy entering one trophic level becomes available to organisms at the next level.* Because the efficiency of these energy transfers is so low, ecosystems generally have no more than four consumer trophic levels.

The amount of useful energy flowing through consumer trophic levels declines at each energy transfer. It also declines as metabolically generated heat is lost and as food energy becomes shunted into organic wastes.

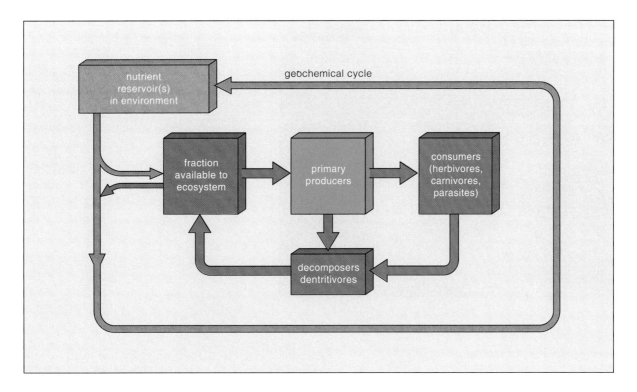

Figure 35.8 Generalized model of nutrient flow through a land ecosystem. The overall movement of nutrients from the physical environment, through organisms, and back to the environment constitutes a biogeochemical cycle.

BIOGEOCHEMICAL CYCLES

Availability of nutrients as well as energy profoundly influences the structure of ecosystems. Photosynthetic producers require carbon, oxygen, and hydrogen, which they get from water and air. They also require nitrogen, phosphorus, and other minerals (Table 21.1). Because mineral deficiencies adversely affect primary productivity, they adversely affect the ecosystem at large.

Nutrients move in **biogeochemical cycles**. In such cycles, the nutrient is transferred from the environment to organisms, then back to the environment—part of which serves as a reservoir for it. Generally, a nutrient moves slowly through the reservoir, compared to its rapid exchange between organisms and the environment.

Figure 35.8 is a model of the relationship between geochemical cycles and most ecosystems. This model is based on four factors:

1. Elements used as nutrients usually become available to producer organisms as mineral ions, such as ammonium (NH_4^+).

2. Inputs from the physical environment and the cycling activities of decomposers and detritivores maintain an ecosystem's nutrient reserves.

3. The amount of a nutrient being cycled through most major ecosystems is greater than the amount entering or leaving in a given year.

4. Inputs to an ecosystem's nutrient reserves occur by rainfall or snowfall, metabolism (such as nitrogen fixation), and weathering of rocks. Outputs for land ecosystems include losses by runoff.

There are three types of biogeochemical cycles. In the *water* cycle (or hydrologic cycle), oxygen and hydrogen move in the form of water molecules. In *atmospheric* cycles, a large portion of a nutrient is in the form of an atmospheric gas. This is true of carbon and nitrogen, for example. In *sedimentary* cycles, the nutrient doesn't exist in gaseous form. Rather, it moves from land to the seafloor and only "returns" to land through geological uplifting, which may take millions of years. The earth's crust is the main storehouse for such nutrients, which include phosphorus. Let's consider what goes on in some examples of biogeochemical cycles.

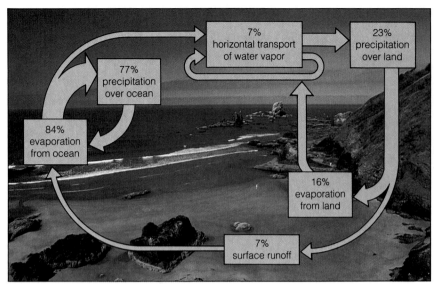

	Annual Volume $(10^{18}$ grams)
ocean	1,380,000
sedimentary layers	210,000
evaporation from ocean	319
precipitation over ocean	283
precipitation over land	95
evaporation from land	59
runoff and groundwater	36
atmospheric water vapor	13

b

a

Figure 35.9 (**a**) Simplified picture of the water cycle. Values in boxes indicate the total quantities of water present at a given time in the atmosphere, in the oceans, and on land. Values next to the arrows show an annual net rate of transfer of 37.3 x 103 cubic kilometers of water from the atmosphere to the land. This is balanced by a comparable net loss from the oceans to the atmosphere and a net gain by oceans of that amount (through runoff).

(**b**) Global water budget. Percentages show the annual movement of water into and out of the atmosphere.

Water Cycle

Driven by solar energy, the waters of the earth move slowly and on a vast scale through the atmosphere, on or through the surface layers of land masses, to the oceans, and back again. Water that evaporates into the atmosphere remains aloft as vapor, clouds, and ice crystals. It falls to earth as precipitation—rain and snow, mostly. Ocean currents and wind patterns play roles in this global **water cycle**, which is shown in Figure 35.9*a*.

The turnaround time for airborne water molecules is rapid; on average, they stay aloft for no more than ten days. Water reaching the land as rain or snow stays there for about 10 to 120 days, depending on the season and the location, then evaporates or flows to the seas. Most evaporation is from the oceans (Figure 35.9*b*).

Water in itself is vital for all ecosystems. *But water is also an important medium by which nutrients move into and out of ecosystems.* This became clear through studies in **watersheds**, where all the precipitation in a specified region becomes funneled into a single stream or river. Watersheds can be any size. The Mississippi River

watershed extends across roughly one-third of the United States. Watersheds at Hubbard Brook Valley in the White Mountains of New Hampshire average about 14.6 hectares (36 acres).

As Figure 35.10 suggests, most of the water entering a watershed seeps into the soil or becomes surface runoff, which moves into a stream. Plants take up water (and dissolved minerals) from the soil, then lose it by transpiration. Measurements of such inputs and outputs have practical application. Cities that depend on surface water supplies in watersheds are aware of their vulnerability to variations in the water cycle.

Other watershed studies have revealed that plants greatly influence the movement of nutrients through the ecosystem phase of biogeochemical cycles. For example, you might think that water draining a watershed would rapidly leach away calcium ions and other minerals. Yet in studies of young, undisturbed forests in Hubbard Brook watersheds, each hectare lost only about 8 kilograms of calcium, and rainfall and the weathering of rocks brought in calcium replacements. Tree roots were also "mining" the soil, so calcium was being stored in a growing biomass of tree tissues.

Nutrient outputs changed when experimental watersheds were stripped of vegetation. The soil was left undisturbed, and herbicides were applied for three years to prevent regrowth. Compared to undisturbed watersheds, *six times* as much calcium was lost by way of stream outflow (Figure 35.11*c*).

Because calcium and other nutrients move so slowly through geochemical cycles, *deforestation may have disruptive effects on nutrient availability for an entire ecosystem.* This is true of forests that cannot regenerate themselves over the short term. Coniferous forests of the Pacific Northwest are like this.

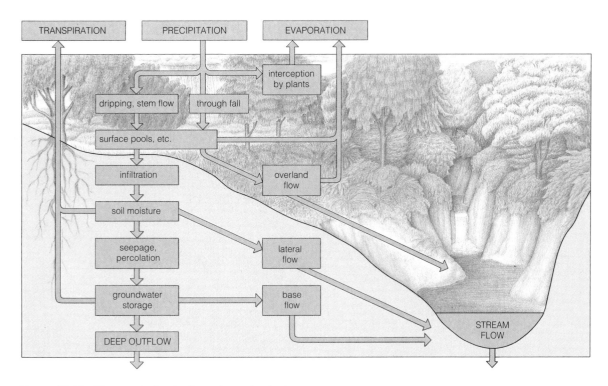

Figure 35.10 Movement of water through a watershed.

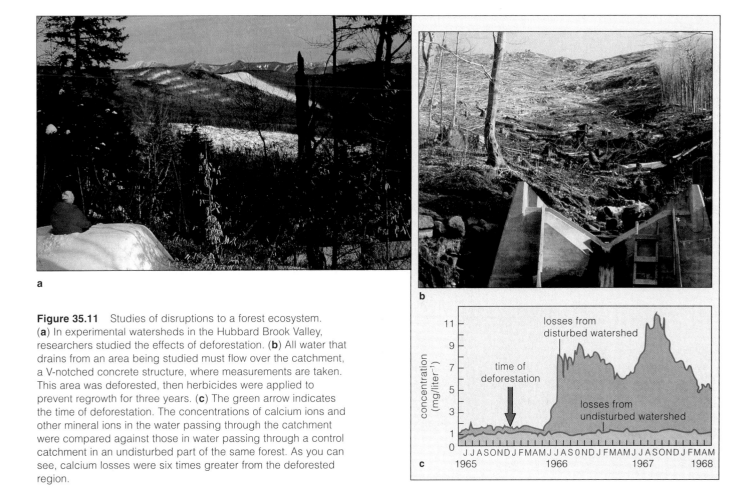

a

b

c

Figure 35.11 Studies of disruptions to a forest ecosystem. (**a**) In experimental watersheds in the Hubbard Brook Valley, researchers studied the effects of deforestation. (**b**) All water that drains from an area being studied must flow over the catchment, a V-notched concrete structure, where measurements are taken. This area was deforested, then herbicides were applied to prevent regrowth for three years. (**c**) The green arrow indicates the time of deforestation. The concentrations of calcium ions and other mineral ions in the water passing through the catchment were compared against those in water passing through a control catchment in an undisturbed part of the same forest. As you can see, calcium losses were six times greater from the deforested region.

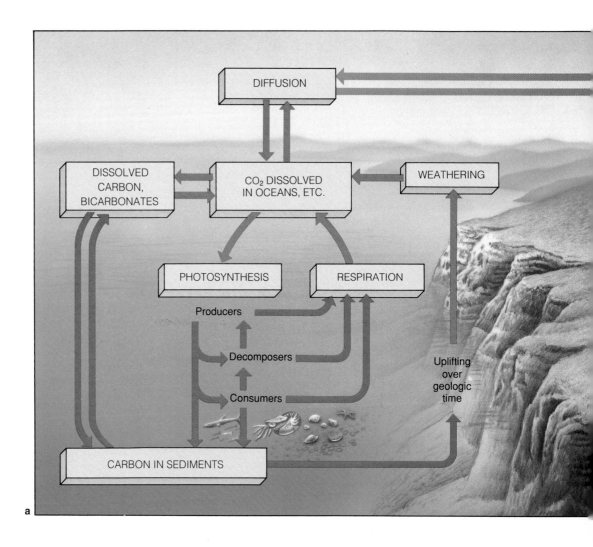

Figure 35.12 (**a**) Global carbon cycle. To the left, the movement of carbon through marine ecosystems. To the right, its movement through terrestrial ecosystems. (**b**) Fossil fuel burning is one of the ways in which humans are adding carbon to the atmosphere.

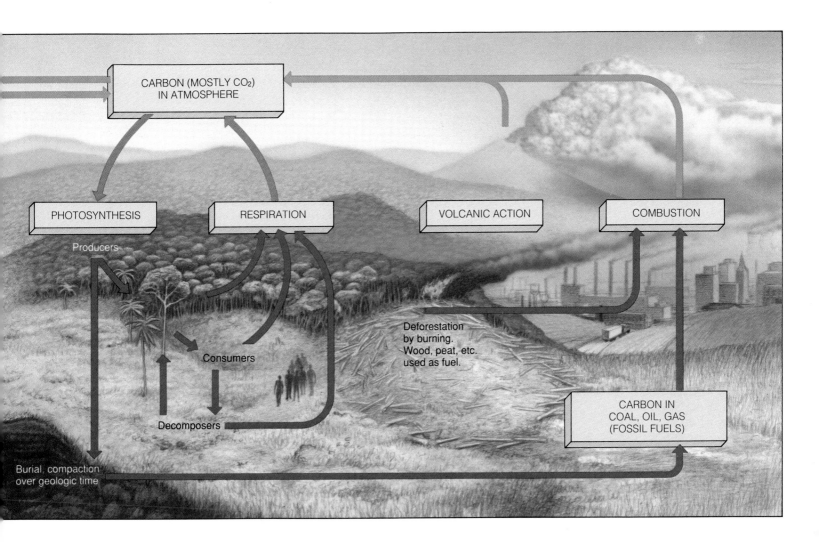

Carbon Cycle

In one of the most vital of all atmospheric cycles, carbon moves from reservoirs in the atmosphere and oceans, through organisms, then back to reservoirs. This is the **carbon cycle**.

As Figure 35.12 shows, carbon enters the atmosphere through aerobic respiration, fossil fuel burning, and volcanic eruptions that release carbon from rocks deep in the earth's crust. Most of the carbon in the atmosphere is in the form of carbon dioxide (CO_2). Each year, about half of all the carbon that enters the atmosphere moves into two large "holding stations"—the oceans and the accumulation of plant biomass.

Through carbon dioxide fixation, photosynthesizers incorporate billions of metric tons of carbon atoms into organic compounds on an annual basis. However, the average length of time that a carbon atom is held in any given ecosystem varies greatly.

In tropical forests, for example, leaves decompose rapidly, so not much carbon is tied up in litter on the soil surface. In marshes, bogs, and other anaerobic settings, organic compounds cannot be broken down completely, so carbon slowly accumulates in compressed organic matter, such as peat (refer to Figure 34.13i).

In aquatic food webs, carbon is incorporated into shells and other hard parts. When shelled organisms die, they sink and become buried in sediments at different depths. Carbon in deep sediments may stay buried for millions of years, until geologic movements bring it to the surface. Still more carbon is slowly converted to long-standing reserves of gas, petroleum, and coal deep in the earth—reserves we tap for use as fossil fuels.

Human activities, including the worldwide burning of fossil fuels, are putting more carbon into the atmosphere than can be cycled to the global holding stations (oceans and plant biomass). This adds to the greenhouse effect and so may help bring on global warming. The *Focus* essay on the next page describes this effect and some possible outcomes of increases in it.

From Greenhouse Gases to a Warmer Planet?

Near the earth's surface, atmospheric concentrations of gaseous molecules play a profound role in shaping the average global temperature, which in turn has enormous effect on the global climate. Molecules of carbon dioxide, water, ozone, methane, nitrous oxide, and chlorofluorocarbons are the key players. Collectively, they act somewhat like a pane of glass in a greenhouse (hence their name, "greenhouse gases"). They let wavelengths of visible light reach the earth's surface, but they impede the escape of longer, infrared wavelengths—that is, heat—from the earth into space. They absorb infrared wavelengths, and much of the energy inherent in those wavelengths is reradiated back toward the earth (Figure *a*). In short, greenhouse gases cause heat to build up in the lower atmosphere, a warming action known as the **greenhouse effect**.

Without greenhouse gases, the earth would be cold and lifeless. But there can be too much of a good thing. Largely as a result of human activities, greenhouse gases are building up to higher levels in the atmosphere (Figure *b*). The increase may be contributing to an alarming trend toward global warming.

What is so alarming about a warmer planet? Suppose the temperature of the lower atmosphere were to rise by only 4°C (7°F). Sea levels could rise by about 2 feet, or 0.6

a The greenhouse effect

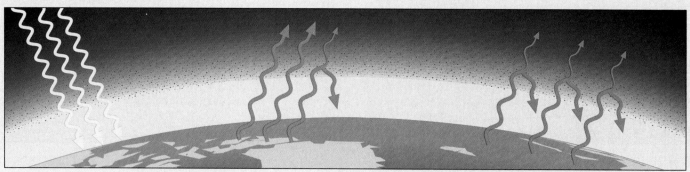

1. Sunlight penetrates the atmosphere and warms the earth's surface.

2. The earth's surface radiates heat (infrared wavelengths) to the atmosphere. Some heat escapes into space. Greenhouse gases and water vapor absorb some infrared wavelengths and reradiate a portion back toward the earth.

3. When greenhouse gases build up in the atmosphere, more heat is trapped near the earth's surface. Ocean surface temperatures rise, more water vapor enters the atmosphere, and the earth's surface temperature rises.

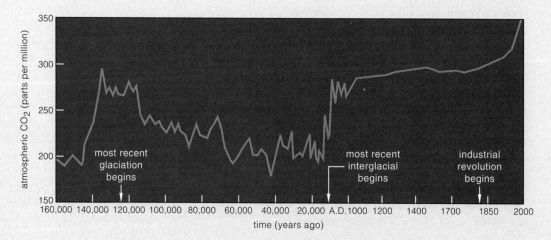

b Shifts in atmospheric concentrations of carbon dioxide, correlated with the most recent glaciation and interglacial period during the past 160,000 years.

meter. Why? The warming would increase ocean surface temperatures—and water expands when heated. Global warming also could make glaciers and the Antarctic ice sheet melt faster, so low coastal regions could flood.

Think of a long-term rise in sea level, combined with high tides and storm waves. Waterfronts of Vancouver, Boston, San Diego, Galveston, and other coastal cities would be submerged. So would agricultural lowlands and deltas in India, China, and Bangladesh, where much of the world's rice is grown. Huge tracts of Florida and Louisiana would face saltwater intrusions. Besides this, global warming could disturb regional patterns of precipitation and temperature. Crop yields would decline in currently productive regions, including parts of Canada and the United States, and increase in others.

In the late 1950s, researchers on a mountaintop in the Hawaiian Islands started measuring concentrations of different greenhouse gases, and their monitoring activities are still going on. They chose the remote site because it was free of local contamination and reflected average conditions for the Northern Hemisphere.

Consider what they found out about carbon dioxide alone. Atmospheric levels of carbon dioxide follow the annual cycle of plant growth in the Northern Hemisphere. They are lower in summer, when photosynthesis rates are highest. They are higher in winter, when aerobic respiration continues and photosynthesis slows. In Figure c (part 1),

the peaks and troughs around the graph line represent the highs and lows. For the first time, scientists saw the integrated effects of the carbon balances for the land and water ecosystems of an entire hemisphere. The midline of the peaks and troughs in the cycle has steadily increased. Many scientists take this as evidence of a buildup of carbon dioxide that may intensify the greenhouse effect over the next century.

The global burning of fossil fuels is probably contributing most to increasing carbon dioxide levels. Deforestation adds to it; carbon is released when wood burns. Today, vast tracts of tropical forests are being cleared and burned at a rapid rate (refer to Figure 37.10). More importantly, the plant biomass is plummeting—and this affects global absorption of carbon dioxide in photosynthesis.

Many scientists wonder whether atmospheric levels of greenhouse gases will continue to increase until the middle of the twenty-first century—and whether the global temperature will rise by several degrees. If this is indeed a trend already in motion, we will not be able to reverse it now by stopping fossil fuel burning and deforestation. So there is widespread agreement that we should begin preparing for the consequences. For example, we might step up genetic engineering studies to develop drought-resistant and salt-resistant plants. Such plants may prove crucial in regions of saltwater intrusions and climatic change.

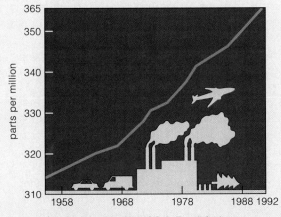

1. Carbon Dioxide (CO_2). Fossil fuel burning, factory emissions, car exhaust, and deforestation are contributing to the increased atmospheric concentration.

c Recently documented increases in atmospheric concentrations of four greenhouse gases.

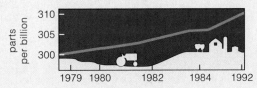

2. Chlorofluorocarbons (CFCs). These are used in plastic foams, air conditioners, refrigerators, and industrial solvents (page 615).

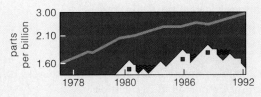

3. Methane (CH_4). This is produced by anaerobic bacteria in swamps, landfills, and termite activities. It is produced also by bacteria in the digestive tract of cattle and other ruminants.

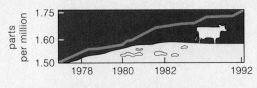

4. Nitrous Oxide (N_2O). This is a natural by-product of denitrifying bacteria. It also is released in great amounts from fertilizers and animal wastes, as in livestock feedlots.

Nitrogen Cycle

Since the time of life's origin, the atmosphere and oceans have contained nitrogen. This component of all proteins and nucleic acids moves in an atmospheric cycle called the **nitrogen cycle**. Gaseous nitrogen (N_2) makes up about 80 percent of the atmosphere—the largest nitrogen reservoir. Triple covalent bonds hold its two atoms together ($N\equiv N$), and few organisms can break the bonds. Only certain bacteria, volcanic action, and lightning can convert N_2 into forms that can enter food webs.

Nitrogen is often the scarcest of all nutrients required for plant growth. Today, nearly all nitrogen in soils has been put there by nitrogen-fixing organisms. Ecosystems lose it through the activities of bacteria that "unfix" the fixed nitrogen. Land ecosystems lose more through leaching of soils, although this is the basis of nitrogen inputs to aquatic ecosystems such as streams, lakes, and the oceans (Figure 35.13).

Cycling Processes. Let's follow nitrogen atoms through the ecosystem part of the nitrogen cycle. They make the rounds among organisms by processes called nitrogen fixation, assimilation and biosynthesis, decomposition, ammonification, and nitrification.

In **nitrogen fixation**, a few kinds of bacteria convert N_2 to ammonia (NH_3), which dissolves quickly in the cytoplasm to form ammonium (NH_4^+). Cyanobacteria, such as *Anabaena* and *Nostoc*, are nitrogen fixers of aquatic ecosystems. *Rhizobium* and *Azotobacter* are nitrogen fixers of many land ecosystems. Although small in size, the nitrogen fixers are mighty in number. Collectively they fix about 200 million metric tons of nitrogen each year! This is the nitrogen that plants will assimilate and use in the biosynthesis of amino acids, proteins, and nucleic acids. Plant tissues will serve as the only nitrogen source for animals, which feed directly or indirectly on plants.

Through **decomposition** and **ammonification**, bacteria and fungi break down nitrogen-containing wastes and remains of organisms in the ecosystem. The decomposers use a portion of the released proteins and amino acids for their own metabolism. But most of the nitrogen is still in the decay products, in the form of ammonia or ammonium, which plants take up. Ammonia or ammonium in soil also gets the attention of nitrifying bacteria. In **nitrification**, a chemosynthetic process, these compounds are stripped of electrons, and nitrite (NO_2^-) is the result. Then other bacteria use the nitrite in metabolism and produce nitrate (NO_3^-), which plants take up.

Certain plants have an advantage in securing nitrogen. For example, peas, beans, clover, and other legumes have nitrogen-fixing bacterial symbionts living in their roots. Fungal symbionts with plant roots (mycorrhizae) also enhance nitrogen uptake, as described on page 273.

Nitrogen Scarcity. You'd think that land plants are assured of enough nitrogen, given the cycling processes. However, the ammonium, nitrite, and nitrate that form during the cycle are highly vulnerable to leaching and runoff. With **leaching**, soil water moves out of an area, which thereby loses the nutrients dissolved in it.

In addition, some of the fixed nitrogen is lost to the air by **denitrification**. By this process, bacteria convert nitrate or nitrite to N_2 and a bit of nitrous oxide (N_2O). Ordinarily, most denitrifying bacteria rely on aerobic respiration. When the soil is waterlogged and poorly aerated, they switch to anaerobic pathways and use nitrate, nitrite, or nitrous oxide as the final electron acceptor instead of oxygen. (Chapter 6 describes this type of metabolic pathway.) In these reactions, the fixed nitrogen is converted to N_2, much of which escapes into the atmosphere.

Besides this, nitrogen fixation comes at high metabolic cost to plants that are symbionts of nitrogen-fixing bacteria. In exchange for nitrogen, the plants give up sugars and other photosynthetic products that can only be assembled with large ATP and NADPH investments. Such plants do have the competitive edge in nitrogen-poor soil. In nitrogen-rich soil, however, species that do not have to pay the metabolic price often displace them.

Finally, losses are enormous in agricultural regions. Nitrogen in the tissues of harvested plants departs from the fields. Soil erosion and leaching remove more. In Europe and North America, farmers traditionally have rotated crops, as when they alternate wheat with legumes. In combination with other conservation practices, crop rotation has helped keep soils stable and productive, sometimes for thousands of years.

Today, intensive agriculture is based on nitrogen-rich fertilizers. Plant varieties are bred for their ability to take up fertilizers, and crop yields per hectare have doubled and even quadrupled over the past forty years. Whether pest control and soil management technologies can sustain high yields indefinitely remains uncertain, for reasons that will become apparent in Chapter 37.

Another catch is this: We can't get something for nothing. Fertilizer production requires huge amounts of energy. The energy comes from fossil fuels, not the free, unending stream of sunlight. At one time, few thought about fossil fuel supplies running out, so there was little concern about fertilizer costs. And it is still a common practice to pour more energy into the soil (in the form of fuels, fertilizers, and other chemicals) than we get out of it (in the form of food). Yet as any hungry person will tell you, food calories are more basic to survival than

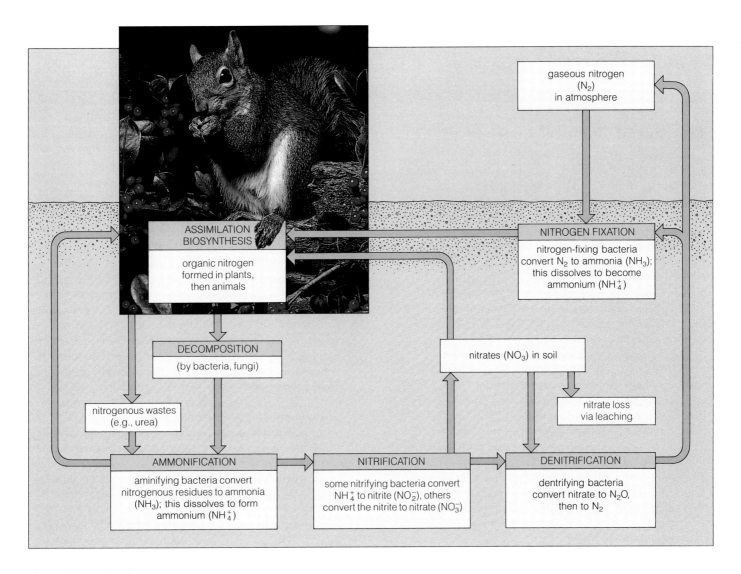

Figure 35.13 The nitrogen cycle.

gasoline calories or perhaps, even, than a car. As long as the human population continues to grow exponentially, farmers will be engaged in a constant race to grow as much food as they can for as many people as possible. Soil enrichment with nitrogen-containing fertilizers is part of the race, as it is now being run.

Prospects for Ecosystem Modeling

We now come full circle to the story that opened this chapter, which conveyed how disturbances to one part of an ecosystem can have unexpected effects on other, seemingly unrelated parts.

A recent approach to predicting unforeseen effects is through **ecosystem modeling**. With this method, crucial bits of information about different ecosystem components are identified. Then computer programs and models are used to combine the information, and the

resulting data are used to predict the outcome of the next disturbance. For example, an analysis of which species feed on which others in the food web shown in Figure 35.3 can be turned into a series of equations that describe how much of each population is consumed. The equations can be used to predict what the effect would be, say, of overharvesting whales or of greatly expanding the harvest of krill.

As we attempt to deal with larger and more complex ecosystems, it becomes more difficult and expensive to run experiments in the field. The temptation is to run them instead on the computer. This is a valid exercise if the computer model adequately represents the system. The danger is that we may not have identified all of the key relationships in the ecosystem and incorporated them accurately into the model. The most important fact may be one that we do not yet know. The *Focus* essay on the next page, which concludes this chapter, reinforces this point.

Transfer of Harmful Compounds Through Ecosystems

DDT, the first of the synthetic organic pesticides, was first used during World War II. In mosquito-infested regions of the tropical Pacific, people were vulnerable to a dangerous disease, malaria. DDT helped control the mosquitoes that were transmitting the sporozoan disease agents (*Plasmodium japonicum*). In war-ravaged cities of Europe, people were suffering from the crushing headaches, fevers, and rashes associated with typhus. DDT helped control the body lice that were transmitting *Rickettsia rickettsii*, the bacterial agent of this terrible disease. After the war, it seemed like a good idea to use DDT against insects that were agricultural or forest pests, transmitters of pathogens, or merely nuisances in homes and gardens.

DDT is a relatively stable hydrocarbon compound. It is nearly insoluble in water, so you might think that it would stay put and act only where applied. But winds can carry DDT in vapor form; water can transport fine particles of it. DDT also is highly soluble in fats, so it accumulates in the tissues of organisms. Thus, as we now know, DDT can show **biological magnification**. The term refers to an increase in concentration of a nondegradable (or slowly degradable) substance in organisms as it is passed along food chains. Most of the DDT from all the organisms that a consumer eats during its lifetime will become concentrated in its tissues. Besides this, many organisms have the means to partially metabolize DDT to DDE and other modified compounds with different but still disruptive effects. Both DDT and the modified compounds are toxic or physiologically disruptive to *many* aquatic and terrestrial animals.

After the war, DDT began to move through the global environment, infiltrate food webs, and affect organisms in ways that no one had predicted. In cities where DDT was sprayed to control Dutch elm disease, songbirds started dying. In streams flowing through forests where DDT was sprayed to control spruce budworms, salmon started dying. In croplands sprayed to control one kind of pest, new kinds of pests moved in. DDT was indiscriminately killing off the natural predators that had been keeping pest populations in check! It took no great leap of the imagination to make the connection. All of those organisms were dying at the same time and place as the DDT applications.

Then side effects of biological magnification started showing up in places far removed from the areas of DDT application—and much later in time. Most devastated were species at the end of food chains, including bald eagles, peregrine falcons, ospreys, and brown pelicans. One product of DDT breakdown interferes with physiological processes. As one consequence, birds produced eggs with brittle shells—and many of the chick embryos didn't make it to hatching time. Some species were at the brink of extinction.

Since the 1970s, DDT has been banned in the United States, except for restricted applications where public health is endangered. Many hard-hit species have partially recovered in numbers. Even today, however, some birds lay thin-shelled eggs. They pick up DDT at their winter ranges in Latin America. As recently as 1990, the California State Department of Health recommended that a fishery off the coast of Los Angeles be closed. DDT from industrial waste discharges that ended twenty years before is still in that ecosystem.

SUMMARY

1. An ecosystem is an entire complex of producers, consumers, detritivores, and decomposers and their physical environment, all interacting through a flow of energy and a cycling of materials.

2. Ecosystems are open systems, with inputs and outputs of energy and nutrients. With few exceptions, photosynthetic autotrophs are the primary producers. They secure energy from sunlight and assimilate much of the nutrients required by all other members of the system. Ecosystems generally are most open for inputs and outputs of water, carbon, and energy. Most nutrients are cycled within a natural ecosystem.

3. Energy fixed by photosynthesizers passes through grazing and detrital food webs. Both types of food webs typically are interconnected in the same ecosystem. Both lose energy (as heat) through metabolism.

4. In biogeochemical cycles, water, nutrients, and other elements and compounds move from the physical environment to organisms, then back to the environment.

5. Ecosystems on land have predictable rates of nutrient losses that generally increase when the land is cleared or otherwise disturbed.

6. Fossil fuel burning and conversion of natural ecosystems to cropland or grazing land are contributing to increased atmospheric concentrations of carbon dioxide. The increase may be contributing to a global warming trend.

7. Nitrogen availability is often a limiting factor for the total net primary productivity of land ecosystems. Gaseous nitrogen is abundant in the atmosphere, but it must be converted to ammonia and to nitrates that primary producers can use. Some bacteria as well as volcanic action and lightning can cause the conversion.

Review Questions

1. Define ecosystem, and name the central roles that autotrophs play in all ecosystems. 569, 570

2. Define and give examples of trophic levels in ecosystems. 570

3. Distinguish between food chain and food web. Can you imagine an extreme situation whereby you would be a participant in a food chain? 570–571

4. If you were growing a vegetable garden, what variables might affect its net primary production? 572

5. Characterize grazing and detrital food webs. Indicate how energy leaves each one and how the two are interconnected. 572

6. Describe the greenhouse effect. Make a list of agricultural products and manufactured goods that you depend on. Are any implicated in the amplification of the greenhouse effect? 580–581

7. Describe the reservoirs and organisms involved in one of the biogeochemical cycles. 578–583

8. Define nitrogen fixation, nitrification, ammonification, and denitrification. 582

Self-Quiz (Answers in Appendix IV)

1. _____ is an ecosystem.
   a. A freshwater spring
   b. Antarctica
   c. A city
   d. all of the above

2. Ecosystems have _____ .
   a. energy inputs and outputs
   b. nutrient cycling but not outputs
   c. one trophic level
   d. a and b

3. The _____ is an atmospheric cycle.
   a. water cycle
   b. carbon cycle
   c. nitrogen cycle
   d. b and c

4. Trophic levels can be described as _____ .
   a. structured feeding relationships
   b. who eats whom in an ecosystem
   c. a hierarchy of energy transfers
   d. all of the above

5. A feeding relationship that proceeds from algae to a fish, then to a fisherman and then to a shark is _____ .
   a. a food chain
   b. a food web
   c. bad luck for the fisherman
   d. a and c

6. Primary productivity is affected by _____ .
   a. photosynthesis and plant respiration
   b. how many plants are neither eaten nor decomposed
   c. rainfall
   d. temperatures
   e. all of the above

7. Deforestation of watersheds _____ most nutrient outputs.
   a. lessens
   b. equalizes
   c. increases
   d. stabilizes

8. Match the ecosystem terms with the correct description.
   _____ primary producers
   _____ consumers
   _____ decomposers
   _____ detritivores

   a. herbivores, carnivores, omnivores, parasites
   b. feed on partly decomposed organic particles
   c. break down remains or products of other organisms
   d. photosynthetic autotrophs

9. Match the ecosystem terms with the appropriate description.
   _____ nitrogen availability
   _____ ecosystem components
   _____ phosphorus
   _____ ammonium
   _____ biogeochemical cycle

   a. movement of water or nutrients from environment, to organisms, and back
   b. form of nitrogen plants can take up
   c. limiting factor for net primary production in land ecosystems
   d. producers, consumers, detritivores, decomposers
   e. moves through a sedimentary cycle

Selected Key Terms

ammonification 582
biogeochemical cycle 575
biological magnification 584
carbon cycle 579
consumer 569
decomposer 569
decomposition 582
denitrification 582
detrital food web 572
detritivore 569
ecosystem 570
ecosystem modeling 583
energy pyramid 573
food chain 570
food web 570
grazing food web 572
greenhouse effect 580
leaching 582
nitrification 582
nitrogen cycle 582
nitrogen fixation 582
primary producer 569
primary productivity 572
trophic level 570
water cycle 576
watershed 576

Readings

Begon, M.; J. Harper; and C. Townsend. 1990. *Ecology: Individuals, Populations, and Communities*. Second edition. Sunderland, Massachusetts: Sinauer.

Botkin, D. B. 1990. *Discordant Harmonies: A New Ecology for the Twenty-first Century*. New York: Oxford University Press.

Post, W., et al. 1990. "The Global Carbon Cycle." *American Scientist* 78:310–326.

Rambler, M.; L. Margulis; and R. Fester. 1989. *Global Ecology: Towards a Science of the Biosphere*. San Diego, California: Academic Press.

36 THE BIOSPHERE

Does a Cactus Grow in Brooklyn?

Suppose you live in the American Southwest but find yourself touring the deserts of Africa (Figure 36.1). There you come across flowering plants with spines, tiny leaves, and columnlike, fleshy stems—just like some cactus plants back home. Or suppose you live in the coastal hills of California and decide to tour the Mediterranean coast, Africa's southern tip, or even central Chile. There you come across woody, many-branched plants—very much like the chaparral plants back home.

In both cases, the plants are separated by enormous geographic and evolutionary distances. Why are they so much alike? The question intrigues you, so you decide to compare their locations on a global map. As you quickly discover, American and African desert plants grow about the same distance from the equator. Chaparral plants and their distant look-alikes grow along western or southern coasts of continents between latitudes 30° and 40°. As Charles Darwin and other naturalists did before, you have stumbled onto one of many patterns in the world distribution of species.

In part, "accidents of history" put many species in particular places. Go back to Pangea's colossal breakup, more than 100 million years ago. When vast chunks of that supercontinent drifted apart, many species became isolated from others. Among them were the ancestors of eucalyptus trees, wombats, and kangaroos on the chunk that became Australia. But species also owe their distribution to topography, climate, and species interactions. With diligence, you might grow a cactus under artificial lights in a heated room in Brooklyn or some other New York City borough. Plant that cactus outside and it won't last one winter.

This last example reminds us that we humans tinker with the distribution of species. Not all of our tinkering is as harmless as growing a cactus in Brooklyn. Think of our predatory effects on the world's fisheries or the effects of our pesticide battles with insect competitors for food. Earlier chapters provided you with a general picture of predation, competition, and other species interactions. Consider now the physical forces shaping the biosphere itself. This will serve as a foundation for addressing the impact of the human species on the biosphere—the topic of the chapter to follow.

a

b

Figure 36.1 Convergent evolution in the form of plants that are geographically and evolutionarily distant. (**a**) *Echinocereus*, of the cactus family, grows in deserts of the American Southwest. (**b**) *Euphorbia*, of the spurge family (Euphorbiaceae), grows in deserts of southwestern Africa. Although the plants appear to be similar, both groups evolved from leafy plants that are not related.

1. Besides being the main energy source for ecosystems, solar radiation influences their global distribution. It provides heat energy that warms the atmosphere and drives the earth's weather systems.

2. Global air circulation patterns, ocean currents, and topographic features interact to produce regional variations in patterns of temperature and rainfall. These patterns influence the composition of soils and sediments, the growth and distribution of primary producers and, through them, ecosystem distribution.

3. A biome is a large, regional unit of land that can be characterized mainly by the climax vegetation of the ecosystems it encompasses. Deserts, shrublands, woodlands, forests, and tundra are examples. Their distribution corresponds roughly with regional variations in climate, topography, and soil type.

4. The water provinces cover more than 70 percent of the earth's surface. They include oceans, inland seas, and bodies of standing freshwater and moving freshwater. Each freshwater and marine ecosystem has gradients in light availability, temperature, and dissolved gases. The gradients, which vary daily and seasonally, influence primary productivity and the composition of species.

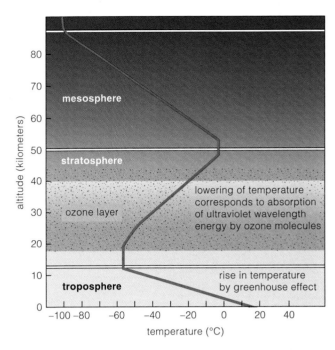

Figure 36.2 Earth's atmosphere. Most global air circulation occurs in the troposphere, where temperatures decrease rapidly with altitude. Ultraviolet wavelengths from the sun are absorbed in the upper atmosphere mainly at the ozone layer, between 17 and 25 kilometers above sea level.

CHARACTERISTICS OF THE BIOSPHERE

Biosphere Defined

The **biosphere** is the sum total of all places in which organisms live. The biosphere encompasses the waters of the earth, the lower atmosphere, and the crust's surface rocks, soils, and sediments. Collectively, its oceans, polar ice caps, and all other forms of liquid or frozen water constitute the **hydrosphere**. Its envelope of gases and airborne particles is the **atmosphere**. About 80 percent of the atmosphere's molecular components are distributed within 17 kilometers of the earth's surface.

Global Patterns of Climate

The biosphere's ecosystems range from continent-wide forests to tiny pools. Except for a few ecosystems in deep oceans, all are partly shaped by climate. **Climate** refers to prevailing weather conditions, such as temperature, humidity, wind speed, cloud cover, and rainfall.

Climate is an outcome of many factors. The major ones are (1) variations in the amount of incoming solar radiation, (2) the earth's daily rotation and its path around the sun, (3) world distribution of continents and oceans, and (4) elevations of land masses. These factors interact to produce prevailing winds and ocean currents that influence global patterns of climate. Climate affects the development of soils and sediments. Together, climate and the composition of soils and sediments affect growth of primary producers—hence the distribution of entire ecosystems.

The Atmosphere's Mediating Effects. Of the solar radiation that reaches the outer atmosphere, only about half gets to the earth's surface. Atmospheric molecules of ozone (O_3) and oxygen (O_2) absorb most of the ultraviolet wavelengths, which are lethal for most forms of life. Absorption is greatest in the **ozone layer**, a region between 17 and 25 kilometers above sea level where ozone is most concentrated (Figure 36.2). Clouds, dust, and water vapor absorb a portion of the other types of wavelengths or reflect them back into space.

Radiation that gets through the atmosphere warms the earth's surface, which gives up heat by radiation or evaporation. Molecules of the lower atmosphere absorb some heat, then reradiate part of it toward the earth. The effect is a bit like heat retention in a greenhouse, which lets in the sun's rays while retaining heat being lost from the plants and soil inside (page 580).

Why is this important? Heat energy derived from the sun warms the atmosphere—*and that energy drives the earth's weather systems.*

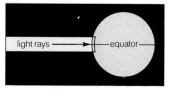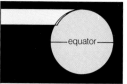

a

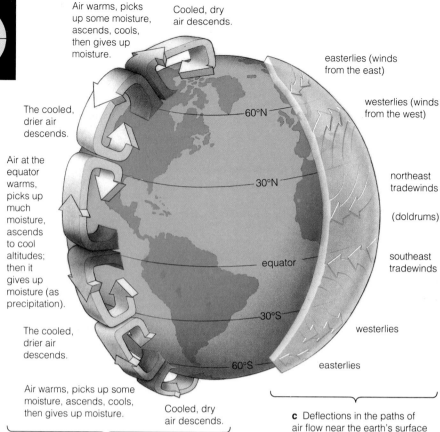

Air warms, picks up some moisture, ascends, cools, then gives up moisture.

Cooled, dry air descends.

The cooled, drier air descends.

Air at the equator warms, picks up much moisture, ascends to cool altitudes; then it gives up moisture (as precipitation).

The cooled, drier air descends.

Air warms, picks up some moisture, ascends, cools, then gives up moisture.

Cooled, dry air descends.

easterlies (winds from the east)

60°N

westerlies (winds from the west)

30°N

northeast tradewinds

(doldrums)

equator

southeast tradewinds

30°S

westerlies

60°S

easterlies

c Deflections in the paths of air flow near the earth's surface

b Initial pattern of air circulation

Figure 36.3 Global air circulation patterns, brought about by three interrelated factors. *First*, the sun's rays are less spread out in equatorial regions than in polar regions (**a**). Warm equatorial air rises and spreads northward and southward to produce the initial pattern of air circulation (**b**).

Second, the nonuniform distribution of land masses creates variations in air pressure. Land absorbs and gives up heat faster than oceans do, so some parcels of air rise (or sink) faster than others. Air pressure decreases where warm air rises (and increases where it sinks). The differences give rise to winds, which disrupt the overall air movement from the equator to the poles.

Third, the earth's rotation and overall shape introduce easterly and westerly deflections in wind directions. Each time the ball-shaped earth makes a full rotation, its surface turns faster beneath air masses at the equator and slower beneath those at the poles. Thus, a rising air mass can't really move "straight north" or "straight south." It is deflected to the east or west. This is the source of prevailing east and west winds (**c**).

Air Currents. The sun's rays have different heating effects at different latitudes. Its rays are more concentrated at the equator than at the poles, so air is heated more at the equator. The global pattern of air circulation begins when warm equatorial air rises and spreads northward and southward. The earth's rotation modifies the circulation into worldwide belts of prevailing east and west winds (Figure 36.3). Together, the differences in solar heating at different latitudes and the modified air circulation patterns define the earth's major temperature zones (Figure 36.4).

Global air circulation patterns also cause differences in rainfall at different latitudes. Consider that warm air can hold more moisture than cool air. As air heats up at the equator, it picks up moisture and rises to cooler altitudes, then it gives up moisture as rain. That rain supports luxuriant tropical forests. The now-drier air moves away from the equator, then descends at latitudes of about 30°. As it descends, the air gets warmer and drier—and deserts tend to form at these latitudes. Air even farther from the equator picks up moisture, ascends to high altitudes, then creates another moist belt at latitudes of about 60°. Finally, air descends in polar regions, where low temperatures and almost nonexistent precipitation create cold, dry, polar deserts. Thus, *sunlight and global air circulation produce latitudinal*

Figure 36.4 World temperature zones.

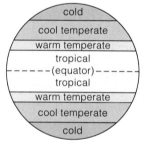

| cold |
| cool temperate |
| warm temperate |
| tropical |
| ----(equator)---- |
| tropical |
| warm temperate |
| cool temperate |
| cold |

belts of temperature and rainfall, which influence the locations of different ecosystems.

Seasonal Variations in Climate. The amount of solar radiation reaching the earth varies through the year in the Northern and Southern Hemispheres. The variation leads to seasonal changes in climate (Figure 36.5). Many biological rhythms coincide with the seasons. Plants undergo cycles of leaf formation, flowering, fruiting, and leaf drop. Caribou, many birds, many butterflies, and other animals follow breeding and migration cycles. Sea turtles, seals, and whales are migrants that follow the seasonal bursts of primary productivity in the oceans, as described on page 81. In temperate regions, organisms respond mainly to seasonal changes in daylength and temperature. In deserts and tropical forests, they respond more to seasonal changes in rainfall.

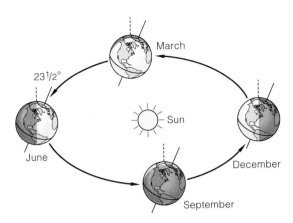

Figure 36.5 Annual variation in the amount of incoming solar radiation. The northern end of the earth's fixed axis tilts toward the sun in June and away from it in December. Thus the position of the equator relative to the boundary of illumination between day and night varies annually. Such variations in the intensity and duration of daylight lead to seasonal variations in temperature in the two hemispheres. Seasonal change becomes more pronounced with distance from the equator. It is greatest in the central regions of continents, where the moderating effects of oceans are minimal.

Ocean Currents. Surface currents and drifts in the oceans are created by the earth's rotation, prevailing surface winds, and variations in water temperature. The water movements tend to run parallel with the equator, but land masses intervene and modify the flows in the Atlantic, Pacific, and Indian oceans. Immense circular movements in each major ocean basin drive warm equatorial waters north and south (Figure 36.6).

Along the western coasts of Africa, North America, and South America, cold currents from the poles move toward the equator and help shape coastal climates. People living in the Pacific Northwest have the California current to thank for their cool, wet, fog-shrouded coasts. Where cool air masses that have formed over cold currents can move inland, temperatures are moderate. In such ways, *atmospheric and oceanic circulation patterns influence regional temperatures.*

Topography. Mountains, valleys, and other land formations influence regional climates. Imagine a warm air mass picking up moisture off California's western coast. After moving inland, it reaches the Sierra Nevada, a mountain range that runs parallel with the coast. As the warm, moist air ascends to higher altitudes, it cools and loses moisture (Figure 36.7).

Vegetation belts at different elevations correspond to the changes in air temperature and moisture. The

Figure 36.6 Ocean currents and surface drifts in January. Solid arrows indicate warm-water movements; dashed arrows indicate cold-water movements, as by the Humboldt current.

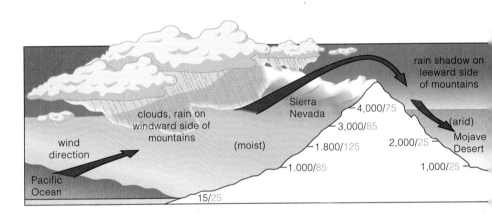

Figure 36.7 Rain shadow effect, a reduction of rainfall on the side of high mountains facing away from the prevailing wind. Only plants adapted to arid or semiarid conditions grow in such places. Blue numbers are the average yearly precipitation (in centimeters) measured at different locations on both sides of the mountain range. Black numbers signify the elevation (in meters).

western base of the range supports plants that are adapted to semiarid conditions. At higher elevations, moisture levels and cool temperatures support forests of deciduous and evergreen trees. The subalpine belt higher up supports only evergreen trees adapted to the rigors of a cold climate. Only low herbaceous plants and dwarfed shrubs can grow above the subalpine belt.

After air flows over the mountain crests and starts to descend, it becomes warmer and can hold more moisture. The air now draws moisture out of plants and the soil rather than giving it up as rain. Only plants that are adapted to arid or semiarid conditions grow in this **rain shadow**. The term refers to the reduction in rainfall on the leeward side of high mountains. ("Leeward" is the direction not facing a wind; "windward" is the direction from which the wind is blowing.) The high mountain ranges of Europe, the Himalayas of Asia, and the Andes of South America also create rain shadows.

THE WORLD'S BIOMES

By now, it should be clear that complex interactions between the atmosphere, oceans, and landforms shape the climates of different regions. As you will now see, the resulting variations in temperature and moisture give us an idea of what kinds of organisms we can expect to find in those regions.

Prevailing climates are a key reason why grasslands, deserts, forests, and tundra exist in some places but not others. They help explain the global distribution of species, each of which is adapted to regional conditions. This also is true of the similar-looking but distantly related species of plants mentioned in the chapter introduction; they are classic examples of convergent evolution. Their ancestors were quite different, but they became adapted to similar environments. Body parts used in comparable ways came to resemble each other.

From the broadest perspective, there are six distinct realms on land, each with distinguishing types of plants and animals. They retain their identity partly because of climate—and partly because oceans, mountain ranges, and other major barriers tend to restrict gene flow and so keep their component species isolated. Long ago, W. Sclater and then Alfred Wallace named these areas **biogeographic realms**. Figure 36.8 shows the six-realm scheme, which is still widely accepted.

We can subdivide each biogeographic realm into biomes. A **biome** is a large region of land that is characterized mainly by the climax vegetation of the ecosystems within its boundaries. Figure 36.9 shows a few familiar examples. Figure 36.10 shows the world distribution of the major biomes.

Effect of Latitude and Elevation. As Figure 36.9 suggests, we typically find certain biomes at certain latitudes and elevations. Thus, short plant species often prevail in dry regions, at high elevations, and at high latitudes. Biomes dominated by tall, leafy plant species prevail at tropical latitudes, temperate latitudes, and low elevations with warm temperatures and high rainfall. Tropical rain forests are an example.

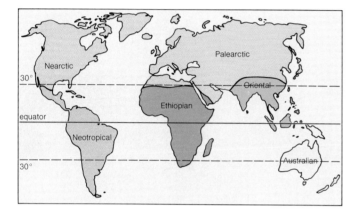

Figure 36.8 Major biogeographic realms.

Figure 36.9 Changes in plant form along environmental gradients for North America. Shown here, gradients in water availability and elevation that influence primary productivity. Mean annual temperature, soil drainage, and other factors also have effects.

arctic tundra   boreal coniferous forest   temperate deciduous forest

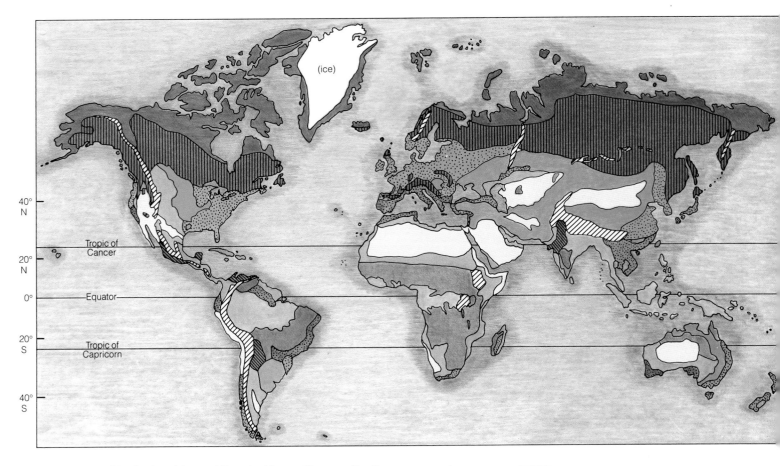

Figure 36.10 Distribution of the world's major biomes. The overall pattern corresponds roughly with distribution patterns for climate and soil type. Compare this map with the two-page photograph of the earth's surface that precedes Chapter 1.

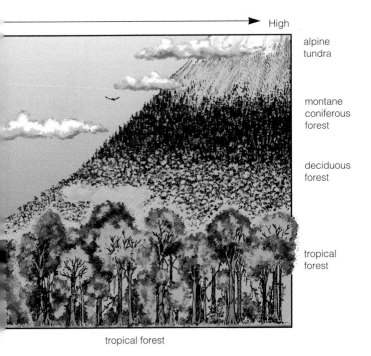

tropical forest

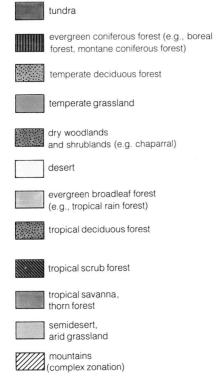

tundra

evergreen coniferous forest (e.g., boreal forest, montane coniferous forest)

temperate deciduous forest

temperate grassland

dry woodlands and shrublands (e.g. chaparral)

desert

evergreen broadleaf forest (e.g., tropical rain forest)

tropical deciduous forest

tropical scrub forest

tropical savanna, thorn forest

semidesert, arid grassland

mountains (complex zonation)

Figure 36.11 Soil characteristics that influence the distribution of ecosystems on land. Soil is a mix of rock, mineral ions, and organic matter in a state of physical and chemical breakdown. The rocks range from coarse-grained gravel to sand, silt, and fine-grained clay.

Humus (partly decomposed organic matter) helps retain water-soluble ions. Loam, a mixture of sand, silt, clay, and humus, retains nutrients and is well aerated. Most plants suffer in poorly aerated, poorly draining soils.

In gravelly or sandy soils, leaching depletes mineral ions. Clay soils with fine, closely packed particles are poorly aerated and do not drain well. Few plants can grow in waterlogged clay soils.

Topsoil, the most fertile soil layer, is just below any surface litter. It may be less than a centimeter deep on steep slopes to more than a meter deep in grasslands. Most crops are grown in former grasslands. When tropical rain forests are cleared for agriculture, the heavy seasonal rains leach most nutrients from the exposed topsoil.

Not enough rain falls in deserts to support crops. In California's Imperial Valley and some other deserts, crops are grown only by soil management and extensive irrigation. They become unproductive from waterlogging and salt buildup if irrigated without proper drainage.

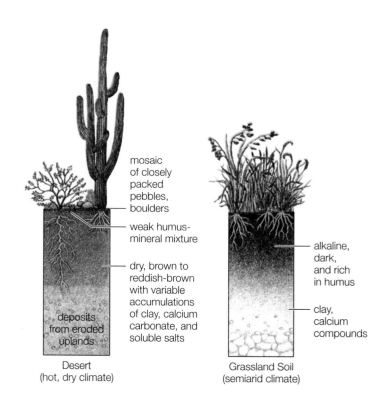

mosaic of closely packed pebbles, boulders

weak humus-mineral mixture

dry, brown to reddish-brown with variable accumulations of clay, calcium carbonate, and soluble salts

deposits from eroded uplands

Desert (hot, dry climate)

alkaline, dark, and rich in humus

clay, calcium compounds

Grassland Soil (semiarid climate)

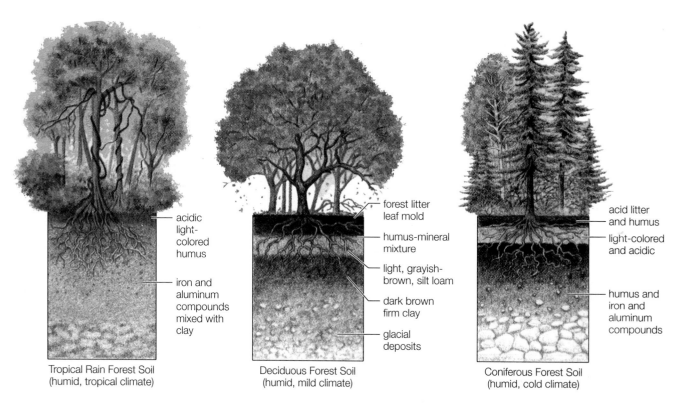

acidic light-colored humus

iron and aluminum compounds mixed with clay

Tropical Rain Forest Soil (humid, tropical climate)

forest litter leaf mold

humus-mineral mixture

light, grayish-brown, silt loam

dark brown firm clay

glacial deposits

Deciduous Forest Soil (humid, mild climate)

acid litter and humus

light-colored and acidic

humus and iron and aluminum compounds

Coniferous Forest Soil (humid, cold climate)

Soils of Major Biomes. What we call **soil** is a mixture of rock, mineral ions, and organic matter in some state of physical and chemical breakdown. Water, air, and organisms are mixed in with its components.

A soil that is rich with decomposed organic matter is good at retaining soluble mineral ions. Such ions are the micronutrients on which plant growth depends. The primary productivity of a biome depends largely on whether its soils are rich or poor in nutrients, and on whether they drain rapidly or are waterlogged and poorly aerated. Figure 36.11 shows the vertical structure, or soil profile, typical of some major biomes.

Figure 36.12 Warm desert near Tucson, Arizona. Among the plants shown are creosote bushes, multistemmed ocotillo, columnlike saguaro cacti, and prickly pear cacti with rounded pads.

Let's turn now to the major biomes—the deserts, shrublands, woodlands, grasslands, forests, and tundra. Bear in mind, no biome is uniform throughout. Within its borders, local climates, landforms, and other physical features favor patches of distinct communities. Regardless of the variation, this key concept holds true:

Atmospheric circulation patterns, ocean currents, and landforms interact in ways that influence regional temperatures and moisture levels—which affect the distribution and dominant features of biomes.

Deserts

Deserts exist where the potential for evaporation greatly exceeds rainfall. Such conditions prevail at latitudes of about 30° north and south. There we find the deserts of the American Southwest, northern Chile, Australia, northern and southern Africa, and Arabia. Farther north are the vast Gobi of Asia, the Kyzyl-Kum east of the Caspian Sea, and the high deserts of eastern Oregon. These northern deserts result largely from the rain shadow effect of extensive mountain ranges.

Vegetation cover is limited in deserts. The infrequent rains often fall in heavy, brief pulses that erode the unprotected soil. Humidity is so low, solar rays easily penetrate the air and quickly heat the ground surface. The surface radiates heat and cools quickly at night.

Although arid or semiarid conditions do not favor large, leafy plants, deserts show plenty of diversity. A patch of Arizona desert may have deep-rooted evergreen shrubs, including creosote, and fleshy-stemmed, shallow-rooted cacti (Figure 36.12). It may have tall saguaro, short prickly pear, and ocotillo—which drops leaves more than once a year, then grows new ones after a rain. Its annuals and perennials flower spectacularly but briefly after spring rains. Mesquite, cottonwood, and other deep-rooted species may grow near streambeds with a permanent underground water supply.

Of all land surfaces, more than one-third is arid or semiarid. Alarmingly, many parts of the world are undergoing **desertification**—the wholesale conversion of grasslands and other productive biomes to dry wastelands. This trend is described on page 619.

Figure 36.14 Natural tallgrass prairie in eastern Kansas.

Figure 36.13 (*Above*) Chaparral-covered foothills in California. What appear to be dirt "roads" are firebreaks. (*Below*) Closer view of a chaparral plant community.

Dry Shrublands and Woodlands

Western or southern coastal regions of continents between latitudes 30° and 40° have a semiarid climate, like that around the Mediterranean Sea. They get more rain than deserts, but not much more. Rain falls mostly during mild winter months, and the summers are long, hot, and dry. Dry shrublands and woodlands prevail in these regions and other areas with related climates. Their dominant plants often have hard, tough, evergreen leaves.

Dry shrublands dominate when annual rainfall is less than 25 to 60 centimeters. These biomes have exotic local names, such as fynbos and chaparral. California alone has 2.4 million hectares (6 million acres) of chaparral-covered hills, where dominant plants may form a nearly impenetrable vegetation cover. These plants are woody, multi-branched, and a few meters tall at most (Figure 36.13). Every so often, lightning-sparked firestorms sweep through these biomes. Many of the shrubs have highly flammable leaves, and their aboveground parts burn rapidly. Yet they are exquisitely adapted to episodes of fire and quickly resprout from their root crowns. Trees (and suburban housing developments) do not fare as well in firestorms. The shrubs—which feed the fires—have the competitive edge.

Dry woodlands dominate when annual rainfall is about 40 to 100 centimeters. The dominant trees can be tall, but they do not form a dense, continuous canopy. Eucalyptus woodlands of southwestern Australia and oak woodlands of the Pacific states are like this.

Grasslands and Savannas

Grasslands sweep across parts of southern Africa, Australia, South America, and midcontinental regions of North America and the Soviet Union. The main types are *shortgrass prairie, tallgrass prairie,* and *tropical savannas.* Usually the land is flat or rolling. Warm temperatures prevail in summer in temperate zones and throughout the year in the tropics. The 25 to 100 centimeters of annual rainfall prevents deserts from forming but is not enough to support forests. The dominant animals are grazing and burrowing types. Grazing and periodic fires keep shrublands and forests from encroaching on the fringes of many grasslands.

Tallgrass prairie (Figure 36.14) once extended west from the temperate deciduous forests of North America. Daisies, sunflowers, and other composites as well as legumes also were abundant. Most tallgrass prairie has been converted to farmland. Large areas are now being restored in several locations. Shortgrass prairie prevails when winds are strong, rainfall light and infrequent, and evaporation rapid (Figure 36.15). Plant roots above the permanently dry subsoil soak up the brief, seasonal rain. In the 1930s, the shortgrass prairie of the American Great Plains was largely overgrazed and plowed under for wheat, which requires more moisture than the region sometimes provides. Strong prevailing winds, drought, and poor farming practices turned much of the prairie into a Dust Bowl (Figure 37.11). Steinbeck's *The Grapes of Wrath* and Michener's *Centennial,* two historical novels, speak eloquently of the disruption of this biome and its consequences.

Tropical savannas cover broad belts of Africa, South America, and Australia. Where rainfall is low, rapidly growing grasses dominate (Figure 36.16). Scattered patches of acacia and other shrubs exist where there is slightly more moisture. Where more rain falls, savannas grade into tropical woodlands with tall, coarse grasses.

Monsoon grasslands thrive in parts of southern Asia. "Monsoon" refers to a season of heavy rain that corresponds to a shift in prevailing winds over the Indian Ocean. It alternates with a pronounced dry season. The climate favors dense stands of tall, coarse grasses. These die back and often burn during the dry season.

Figure 36.15 Rolling shortgrass prairie to the east of the Rocky Mountains. Bison—60 million at one time—were the dominant large herbivore of North American grasslands.

Figure 36.16 African savanna—warm grasslands with scattered stands of shrubs and trees. More large ungulates (hooved, plant-eating mammals) live here than anywhere else. They include migratory wildebeests (shown), giraffes, Cape buffalo, zebras, and impalas. Major predators are lions and cheetahs; scavengers include vultures, hyenas, and jackals.

bromeliad

Figure 36.17 Tropical forests, with their spectacular biodiversity. Among the millions of known species are jaguars and *Rafflesia*, a leafless, foul-smelling plant with a fly-pollinated flower that can be 3 meters across. Bromeliads, orchids, and other plants grow on tree branches, obtaining minerals from organic remains of leaves, insects, and other litter that dissolved in tiny pockets of water.

Forests

In the world's major **forests**, tall trees grow together closely enough to form a fairly continuous canopy over a broad region. The trees fall into three general categories. Which category prevails in a region depends partly on distance from the equator. "Evergreen broadleafs" dominate between latitudes 20° north and south. "Deciduous broadleafs" are most common at moist, temperate latitudes where winters are not severe. "Evergreen conifers" are most common at higher, colder latitudes and in the mountains of temperate zones.

Evergreen Broadleaf Forests. These biomes sweep across tropical zones of Africa, the East Indies, southeastern Asia, the Malay Archipelago, South America, and Central America. Annual rainfall can exceed 200 centimeters and is never less than 130 centimeters.

Highly productive *tropical rain forests* exist where rainfall is regular and heavy, the annual mean temperature is 25°C, and the humidity is 80 percent or more (Figure 36.17). Evergreen trees produce new leaves and shed old ones through the year, so tropical rain forests

Spring

Summer

Autumn

Winter

produce more litter than other forest biomes. But decomposition and mineral cycling are rapid in the hot, humid climate. The soils are highly weathered, humus-poor, and not good nutrient reservoirs. We take a closer look at these forests in the next chapter.

Deciduous Broadleaf Forests. Leaving tropical rain forests, we enter regions where temperatures stay mild but rainfall dwindles during part of the year. Here we find *tropical deciduous forests*, in which many trees drop some or all of their leaves during a pronounced dry season. The *monsoon forests* of India and southeastern Asia also have such trees. Farther north, in the temperate zone, rainfall is even lower and temperatures become cold during the winter. Here we find *temperate deciduous forests* (Figure 36.18). Decomposition is not as

Figure 36.18 The changing character of a temperate deciduous forest in spring, summer, autumn, and winter. The one shown here is south of Nashville, Tennessee.

rapid as in the humid tropics, and nutrients are conserved in accumulated litter on the forest floor.

At one time, forests of ash, beech, birch, chestnut, elm, and deciduous oaks stretched across northeastern North America, Europe, and eastern Asia. They shrank drastically when land was cleared for farming. In North America, diseases brought in by introduced species wiped out nearly all chestnuts and many elms (Table 34.2). Now, maple and beech forests predominate in the Northeast. Farther west, oak-hickory forests prevail, then oak woodlands and the tallgrass prairie.

Evergreen Coniferous Forests. Evergreen conifers are primary producers of boreal forests, montane coniferous forests, temperate rain forests, pine barrens, and many other forest biomes. Conifers are cone-bearing trees, and most have needle-shaped leaves adapted to arid conditions. The needles have a thick cuticle and recessed stomata that help the trees conserve water.

Expanses of coniferous trees in northern Europe, Asia, and North America are called *boreal forests*, or *taiga* (meaning "swamp forest"). Most are in glaciated regions with cold lakes and streams (Figure 36.19a). It rains mostly in summer, and evaporation is low in the cool summer air. The cold, dry winters are more severe in eastern parts of the continents than in the west (where oceanic winds moderate climate). Spruce and balsam fir dominate North America's boreal forests.

Montane coniferous forests exist in the great mountain ranges of North America and other continents. Spruce and fir dominate in the north and at higher elevations. They give way to fir and pines in the south and at lower elevations (Figure 36.19b). In the late 1800s and early 1900s, logging destroyed most of the coniferous forests that once flourished around the northern Great Lakes.

Coniferous forests grow in some temperate lowlands. A *temperate rain forest* parallels the coast from Alaska into northern California. It includes some of the world's tallest trees—Sitka spruce to the north and redwoods to the south (Figure 1.7d). Sandy, nutrient-poor soil of New Jersey's coastal plain supports *pine barrens*—scrub forests in which grasses and low shrubs grow beneath stands of pitch pine and oak trees. The forests recover quickly from fires, which are frequent. Managed pine plantations have replaced most of the fire-maintained pine forests that once cloaked the coastal plains of North Carolina, South Carolina, Georgia, and Florida.

Tundra

Alpine tundra exists at high elevations in mountains throughout the world (Figure 36.20a). The dominant plants often form cushions and mats that withstand the buffeting of strong winds. Primary productivity is low, given the low temperatures and nutrient-poor soils.

Actually, the word tundra is derived from the Finnish *tuntura*, which means treeless plain. This type of biome is **arctic tundra**. It lies between the polar ice cap and belts of boreal forests in North America, Europe, and Asia. Much of Alaska's arctic tundra is flat, windswept, and wet (Figure 36.20b). Temperatures are cool in summer and below freezing in winter, so there is little evaporation. Little rain or snow falls. In summer's nearly continuous sunlight, short plants grow and flower profusely; seeds ripen fast.

a

a

Snow does not cloak arctic tundra all year long, but summers are too short to thaw much more than surface soil. Just beneath the surface is a permanently frozen layer, the **permafrost**. It is more than 500 meters thick in some places. The soil above it cannot drain and remains waterlogged. Anaerobic conditions and low temperatures limit nutrient cycling. Organic matter decomposes so slowly, it accumulates in soggy masses (peat). All but about 5 percent of the carbon in the arctic tundra is locked up in peat.

b

Figure 36.19 (**a**) Example of a boreal forest. Spruce trees dominate this one. (**b**) The montane coniferous forest of Yosemite Valley in the Sierra Nevada of California.

b

Figure 36.20 (**a**) Short, hardy plants typical of alpine tundra. (**b**) Arctic tundra, where rain and snowmelt cannot soak downward because of the permafrost.

Figure 36.21 (*Above*) In the Canadian Rockies, a lake basin formed by glacial action during the last ice age. (*Below*) In Oregon, Crater Lake, a collapsed volcanic cone now filled with water from rains and melting snow. Like other volcanoes of the Cascade Range, it started forming at the dawn of the Cenozoic, when crustal plates were undergoing major reorganization.

THE WATER PROVINCES

The water provinces are far more extensive than the earth's biomes. They include the world's lakes, rivers, ponds, estuaries, and wetlands. They include rocky and sandy shores, coral reefs, parts of the open ocean, even hydrothermal vents on the ocean floor. "Typical" examples are hard to find. Some ponds can be waded across; Lake Baikal in Siberia is more than 1.7 kilometers deep. All aquatic ecosystems have gradients in light penetration, temperature, and dissolved gases, but these differ greatly from one to the next. All we can do here is sample the diversity.

Lake Ecosystems

The topography, climate, and geologic history of a lake dictate the kinds and numbers of its residents, how they are distributed, and how nutrients are cycled among them. All lakes form in land basins (Figure 36.21). In time, most fill with sediments or drain when erosion deepens their outlets.

Lake Zones. A **lake** is a body of fresh water with littoral, limnetic, and profundal zones (Figure 36.22). The littoral is a shallow, usually well-lit zone extending all around the shore to the depth at which rooted aquatic plants stop growing. Diversity is greatest in the littoral, the home of assorted plants, decomposers, and consumers, including snails and frogs. The limnetic zone is the open, sunlit water beyond the littoral, down to a depth where photosynthesis is insignificant. The limnetic has **plankton** (communities of floating or weakly swimming organisms, mostly microscopic). Cyanobacteria, diatoms, green algae, and other photosynthetic autotrophs make up the *phyto*plankton. Rotifers, copepods, and other heterotrophs make up *zoo*plankton.

Figure 36.22 Lake zonation. The littoral extends all around the lake and from the shore to the depth where aquatic plants stop rooting. The profundal includes areas below the depth of light penetration. Above the profundal are the open, sunlit waters of the limnetic zone.

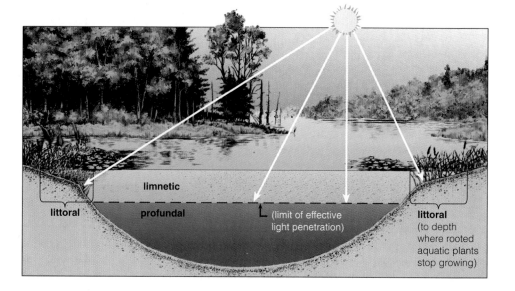

littoral

limnetic

profundal

(limit of effective light penetration)

littoral (to depth where rooted aquatic plants stop growing)

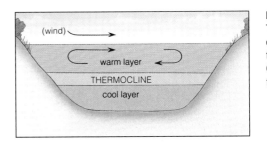

Figure 36.23 Thermal layering of the water in a temperate lake in Connecticut during the summer.

Figure 36.24 Experiment demonstrating eutrophication of a lake in Ontario, Canada. Researchers stretched a plastic curtain across a narrow channel between two basins of the same lake. They added phosphorus, carbon, and nitrogen to one basin (in the background) and carbon and nitrogen only to the other (in the foreground). Within two months, the phosphorus-enriched basin showed accelerated eutrophication, with a dense algal bloom turning the water green.

A lake's profundal zone is all the open water below the depth at which wavelengths suitable for photosynthesis can penetrate. Detritus sinks through the profundal to bottom sediments, which contain communities of bacterial decomposers. Decomposition activities release nutrients into the water.

Seasonal Changes in Lakes. In temperate regions that have warm summers and cold winters, lakes show seasonal changes in density and temperature from the surface to the bottom. A layer of ice forms over many of the lakes in midwinter. Water near the freezing point is the least dense and accumulates beneath the ice. Water at 4°C is densest; it accumulates in deeper layers, which are a bit warmer than the surface layer in midwinter.

In spring, daylength increases and the air is less cold. Ice on a lake melts, and the surface water slowly warms to 4°C so that temperatures become uniform throughout. Winds blowing across the lake surface now cause a **spring overturn**. In such overturns, strong vertical movements carry dissolved oxygen from the lake's surface layer to its depths, and nutrients released by decomposition are brought from the bottom sediments to the surface layer.

By midsummer, the surface layer is well above 4°C. Now the lake has a "thermocline," a middle layer that changes abruptly in temperature and that prevents vertical mixing (Figure 36.23). Being warmer and less dense, the surface water floats on the thermocline. Decomposers may deplete the water below of dissolved oxygen. In autumn, the lake's upper layer cools, becomes denser, and sinks. The thermocline vanishes. During this **fall overturn**, lake water mixes vertically, and once again dissolved oxygen moves down and nutrients move up.

Primary productivity shifts with the seasons. After a spring overturn, the longer daylengths and cycled nutrients support increased primary productivity. Then, phytoplankton and rooted aquatic plants rapidly take up phosphorus, nitrogen, and other nutrients. As the growing season progresses, the thermocline cuts off vertical mixing. All the while, nutrients tied up in the remains of microscopic producers and consumers sink into deeper waters. Nutrients dwindle in the upper waters, and primary productivity declines. By late summer, nutrient shortages are limiting photosynthesis.

After a fall overturn, cycled nutrients drive another burst of primary productivity. However, with autumn's shorter daylight hours and declining temperatures, this burst does not last long. Primary productivity will not increase again until spring.

Trophic Nature of Lakes. Geologic processes create lakes, as when the grinding action of advancing glaciers carves basins in the earth. After the glaciers retreat, the exposed basins fill with water. Over time, erosion, sedimentation, and other processes change a lake's dimensions. Soils of the basin and the surrounding regions contribute to the type and amount of nutrients available to support the lake's primary producers.

Interactions among soils, basin shape, and climate lead to conditions ranging from oligotrophy to eutrophy. *Oligotrophic* lakes are often deep, poor in nutrients, and low in primary productivity. *Eutrophic* lakes are often shallow, rich in nutrients, and high in primary productivity. As lake sediments accumulate, conditions may progress from oligotrophy to eutrophy, then on to a final successional stage with a filled-in basin.

Human activities can disrupt the geologic, climatic, and biological interactions that dictate the trophic nature of lakes. For example, dumping sewage into a lake or logging over the surrounding land can lead to **eutrophication**. The term refers to nutrient enrichment of a lake or some other body of water, and it typically results in reduced transparency and a phytoplankton-dominated community. The field experiment shown in Figure 36.24 nicely demonstrates this outcome.

Figure 36.25 The type of estuary called a New England salt marsh. A marsh grass (*Spartina*) is the major producer, with its microbe-enriched litter providing food for consumers in the creeks and sound.

SALT MARSH (estuary)

open ocean

sound

shallow bay

creek

tidal river

Stream Ecosystems

The flowing-water ecosystems called **streams** start out as freshwater springs or seeps. They grow and merge as they flow downslope, then often combine to form a river. Between the headwaters and the river's end, we find three kinds of habitats—riffles, pools, and runs. Riffles are shallow, turbulent stretches where water flows swiftly over a rough bottom of sand and rock (Figure 7.1). Pools have deep water flowing slowly over a smooth, sandy, or muddy bottom. Runs are smooth-surfaced but fast-flowing stretches over bedrock or rock and sand.

A stream's average flow volume and temperature depend on rainfall, snowmelt, geography, altitude, even the shade cast by plants. Its solute concentrations are influenced by the streambed's composition as well as by agricultural, industrial, and urban wastes.

Especially in forested regions, streams import most of the organic matter that supports their food webs. Where trees cast shade, photosynthesis is limited, and most of the organic matter is litter, which enters detrital food webs. Aquatic organisms continually take up and release nutrients as water flows downstream. (Nutrients move upstream only as components of migrating fishes and other animals.) Think of nutrients as spiraling between water and aquatic organisms as the stream flows on its one-way course, usually to the sea.

Ever since cities formed, streams have been sewers for industrial and municipal wastes. The wastes, as well as other pollutants from poorly managed farmlands, have left many streams choked with sediment and poisoned by chemicals. Streams are resilient, however, and show impressive recovery when pollution is controlled.

Marine Ecosystems

Like freshwater ecosystems, oceans and seas differ in their physical and chemical properties, including light penetration and water temperature, depth, and salinity. Here we focus on estuaries, then ecosystems of the intertidal zone and open ocean.

Estuaries. An **estuary** is a partly enclosed coastal ecosystem where seawater mixes with nutrient-rich freshwater from rivers, streams, and runoff from the land. The confined conditions, slow mixing of water, and tidal action combine to trap the nutrient-enriched water being drained from the land. This is the reason why estuaries are the most productive aquatic ecosystems on earth.

Chesapeake Bay, Mobile Bay, and San Francisco Bay are broad, shallow estuaries. Estuaries in Alaska and British Columbia are narrow and deep; so are Norway's fjords. In Texas and Florida, they lie behind long spits of sand and mud. The estuary in Figure 36.25 is on the New England coast, where salt marshes are common.

Primary producers of estuaries include salt-tolerant grasses that withstand submergences at high tides. They include phytoplankton as well as algae that grow on surfaces of plants and mud. Much of the primary production enters detrital food webs in which bacterial and fungal decomposers are the first to feed.

The estuary is home to a great diversity of invertebrates and the rest stop for many ducks, geese, and other migratory birds. Many of these highly productive ecosystems are being polluted by sewage, agricultural runoff, urban and industrial wastes, and diversion of fresh water for human use.

upper littoral

mid-littoral

sea anemone

sea star

lower littoral

Figure 36.26 Vertical zonation in the intertidal zone of a rocky shore in the Pacific Northwest. The vertical difference between high and low tides is about 3 meters. Elsewhere, it varies from a few centimeters (in the Mediterranean Sea) to more than 15 meters (in the Bay of Fundy, next to Nova Scotia).

Figure 36.27 Typical tide pool residents in the Pacific Northwest.

Life Along the Coasts. Along rocky and sandy coastlines, we find ecosystems of the **intertidal zone**, which is not exactly renowned for creature comforts. The resident organisms are battered by waves, fiercely so during storms. They are alternately submerged and exposed by tides. The higher up they are on the shore, the more they might dry out, freeze in winter, or bake in summer, and the less food comes their way. The lower they are, the more they must compete in the limited space. At low tides, birds, rats, and raccoons move in to feed on them. High tides bring the predatory fishes.

Generalizing about life along the coasts is difficult, because waves and tides constantly resculpt a dizzying array of habitats. Vertical zonation is about the only feature that rocky and sandy shores have in common. *Rocky shores* often have three zones (Figure 36.26). The highest (upper littoral), which is submerged only during the highest tide of the lunar cycle, is sparsely populated. Each day, the midlittoral is submerged during the

highest regular tide, then exposed during the lowest. In its tide pools we find red, brown, and green algae, small invertebrates such as hermit crabs and nudibranchs, and small fishes (Figure 36.27). Diversity is greatest in the lower littoral, which is exposed only during the lowest tide of the lunar cycle. As is true of the other two zones, erosion prevents detritus from accumulating, so grazing food webs tend to predominate here.

Sandy and *muddy shores* are stretches of loose sediments, constantly rearranged by waves and currents. Few large plants grow in these unstable places. Detrital food webs start with imported bits of organic matter. Vertical zonation is less obvious. Along temperate coasts, blue crabs and sea cucumbers live below the low tide mark. Marine worms, crabs, and other invertebrates live between the high and low tide marks. At night, at the high tide mark, beach hoppers and ghost crabs leave their burrows and bound or lurch about the beach, seeking food.

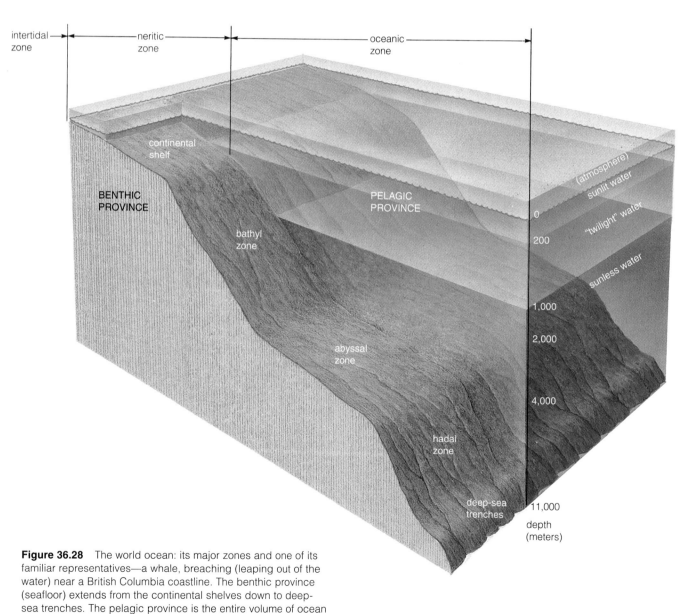

Figure 36.28 The world ocean: its major zones and one of its familiar representatives—a whale, breaching (leaping out of the water) near a British Columbia coastline. The benthic province (seafloor) extends from the continental shelves down to deep-sea trenches. The pelagic province is the entire volume of ocean water. The dimensions of each zone are not to scale.

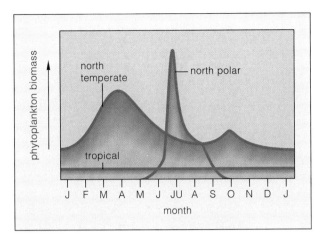

Figure 36.29 Seasonal variations in primary production in the ocean corresponding to latitude. The strongest peak in the graph line for north polar and temperate seas corresponds to phytoplankton blooms, brought about by a seasonal increase in daylength. The smaller peak corresponds to an increase in nutrient availability, something like the fall overturn in lakes. In most tropical seas, daylength and nutrient availability do not vary much, and neither does primary production (compare Figure 36.5).

Figure 36.30 Hydrothermal vent ecosystems. In 1977, biologists discovered a distinct type of ecosystem near the Galápagos Islands. In the Galápagos Rift, a volcanically active boundary between two of the earth's crustal plates, near-freezing seawater seeps into fissures, becomes heated, and is spewed out through vents at high temperatures. The hydrothermal outpouring results in deposits of zinc, iron, and copper sulfides as well as calcium and magnesium sulfates. All are leached from rocks as pressure forces the heated water upward.

In contrast to most of the deep ocean floor, these nutrient-rich warm "oases" support marine communities. Chemosynthetic bacteria use hydrogen sulfide as an energy source. They are primary producers in a food web that includes tube worms (**a**), crustaceans (**b**), clams, and fishes.

Other hydrothermal vent ecosystems have been located in the South Pacific, near Easter Island; the Gulf of California, about 150 miles south of the tip of Baja California, Mexico; and the Atlantic. In 1990, a team of United States and Russian scientists located another in Lake Baikal, the world's deepest lake. This tectonically formed lake basin seems to be splitting apart (hence the hydrothermal vents) and may mark the beginning of a new world ocean.

The Open Ocean. Beyond the intertidal are two vast provinces of the open ocean (Figure 36.28). The **benthic province** includes all sediments and rocks of the ocean bottom. It starts with the continental shelf and extends down to deep-sea trenches. The **pelagic province** is the entire volume of ocean water. Its *neritic* zone consists of all the water above the continental shelves. Its *oceanic* zone is the water filling the ocean basins.

Photosynthesis is restricted to the ocean's upper surface waters, and it varies seasonally (Figure 36.29). Phytoplankton drifting with the currents are the start of food webs that include zooplankton, copepods, shrimp-like krill, whales, squids, and fishes. Organic remains and wastes from these communities sink down to become the basis of detrital food webs for most communities in the benthic province.

At the bottom of the ocean, in the abyssal zone, are communities that are *not* based on photosynthesis. There, at **hydrothermal vents**, heated water is spewed from fissures between two crustal plates (Figure 36.30). Chemosynthetic bacteria are the primary producers of these unique communities. They obtain the energy required for biosynthesis by way of reactions involving hydrogen sulfide.

sea anemone

crown-of-thorns sea star

moray eel

pillar coral

green tube coral

a

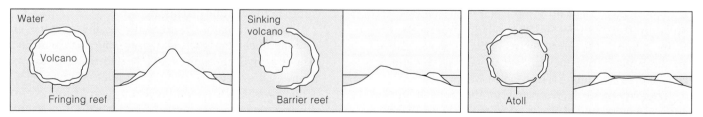

Water

Volcano

Fringing reef

Sinking volcano

Barrier reef

Atoll

b   Types of coral reefs

coral reef   island   lagoon   open ocean

Coral Reefs and Banks. The wave-resistant formations called **coral reefs** are the accumulated remains of countless corals and other organisms. They include fringing reefs, barrier reefs, and atolls, one of which is shown in Figure 36.31a. Most coral reefs are located in clear, warm waters between latitudes 25° north and south. These are renowned for their diversity of marine life. Figure 36.31 includes a small sampling of reef organisms. Figures 19.5 through 19.9 show others.

Farther north and south, **coral banks** have formed at the edges of continental shelves. Among these are the smooth, vertical banks in cold, deep waters near Japan, California, Norway, England, and New Zealand.

Figure 36.31 (**a**) A sampling of tropical reef diversity.

Long ago, corals began to grow and reproduce in the warm, nearshore waters off the island in the photograph. The skeletons they left behind were a substrate for more corals to grow upon. Skeletons and residues accumulated, the reef grew, and tides and currents carved ledges and caverns. Today, the reef's spine may be decked out with as many as 750 species of corals. Red algae encrust the coral foundation. In shallow waters behind the reef, red algae give way to blue-green forms. Many small, transparent animals feed on algae and other plants and in turn are food for predators, including fishes, sea stars, and sea anemones. Fishes and other animals are food for moray eels.

(**b**) The three types of coral reefs.

El Niño and Seesaws in the World's Climates

The winter of 1982–1983 was one for the books. Record amounts of rain caused massive mudslides along the California coast. Month after month, storms caused major floods along the normally arid and semiarid coasts of Ecuador and Peru. In India, the life-sustaining monsoon rains hardly materialized. Droughts devastated natural ecosystems and farms in Australia and the Hawaiian Islands. An ongoing drought in Africa intensified, with consequent and appalling amplification of human starvation and death.

What caused the massive dislocations in the global patterns of rainfall? The answer lies with interactions among sea surface temperature, air circulation patterns, and drought-related conditions on land. Every three to seven years, the interactions produce a climatic event of global proportions. Meteorologists call it the *El Niño Southern Oscillation* (ENSO).

The "Southern Oscillation" part of the name refers to a seesaw in atmospheric pressure in the western equatorial Pacific. This area is the world's largest reservoir of warm water (Figure *a*). More warm, moisture-laden air rises here than anywhere else. Rainfall is heavy, and it releases much of the heat energy that drives the world's air circulation system.

The Southern Oscillation may be triggered by pulses of heat from the earth, possibly at clusters of a thousand or more active volcanoes recently discovered on the ocean floor. Whatever the cause, heat moves upward and warms the surface waters.

Normally, the warm reservoir and the heavy rainfall associated with it move westward. But now they move to the east (Figure *b*). This causes prevailing surface winds in the western equatorial Pacific to pick up speed. The stronger winds have a more pronounced effect on "dragging" the ocean surface waters to the east. Upper ocean currents are affected to the extent that the westward transport of water slows down and the eastward transport increases. *More* warm water in the vast reservoir moves east—and so on in a feedback loop between the ocean and the atmosphere.

The reversal in the usual westward flow of air and water displaces the cold, deep Humboldt current—and so stops the upwelling of nutrients along the western coast of South America. Peruvian fishermen refer to the warm, nutrient-poor current from the east as "El Niño." Usually it reaches their coast around Christmas (hence the name, meaning "the Christ child").

Numerical models are used to study these and other episodes of climatic change a few seasons in advance. More reliable forecasting from this ecosystem modeling should follow when more and better observations are made of the interrelated systems of the ocean, land, and atmosphere.

Upwelling. Wherever currents stir ocean water and circulate nutrients to the surface, primary productivity increases. A case in point is **upwelling**, an upward movement of deep, nutrient-rich water along the margins of continents. Upwelling occurs when winds force surface waters to move away from a coastline and deep water moves in vertically to replace it.

Consider what happens off Peru's west coast. Prevailing winds from the south and southeast force surface water away from shore, and cold, deeper water brought to the continental shelf by the Humboldt current moves toward the surface. It pulls up tremendous amounts of nitrate and phosphate. Phytoplankton using the nutrients are the basis of one of the richest fisheries; the populations of anchoveta and other fishes are huge. Every three to seven years, however, warm surface waters of the western equatorial Pacific move eastward. The massive displacement of warm water influences the prevailing winds, which accelerate the eastward flow. The flow is enough to displace the cooler waters of the Humboldt current and prevent upwelling. This phenomenon, which local fishermen named **El Niño**, has a catastrophic effect on productivity, on anchoveta-eating birds, and on the anchoveta industry. The *Focus* essay takes a closer look at the El Niño phenomenon. With it, we return to a concept presented at the start of this chapter—that interactions among the atmosphere, oceans, and land profoundly influence the world of life.

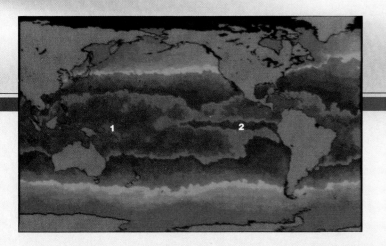

a Satellite images of the distribution of ocean surface temperatures in normal years, with the warmest waters found in the western equatorial Pacific (1), and a tongue of relatively cold water extending westward along the equator from South America (2).

b Distribution of ocean surface temperatures that were associated with the 1982–1983 ENSO episode.

SUMMARY

1. The biosphere encompasses the earth's waters, lower atmosphere, and uppermost portions of its crust in which organisms live. It contains diverse ecosystems that are influenced by the flow of energy and the movement of materials on a global scale.

2. Earth history, climate, topography, and species interactions shaped the world distribution of species.

3. Climate refers to prevailing weather conditions, including temperature, humidity, wind velocity, cloud cover, and rainfall. It is an outcome of differences in the amount of solar radiation reaching equatorial and polar regions, the earth's daily rotation and its annual path around the sun, the distribution of continents and oceans, and the elevation of land masses.

4. Interacting climatic factors produce prevailing winds and ocean currents, which influence global weather patterns. Weather affects soil composition, sedimentation, and water availability, which influence the growth and distribution of primary producers. Through these interactions, climate influences ecosystems.

5. Land masses are classified as six major biogeographic realms. Each realm has a characteristic array of plant and animal species, and each is more or less isolated by oceans, mountain ranges, or desert barriers.

6. Biomes are categories of ecosystems shaped by regional variations in climate, landforms, and soil composition. Plant species adapted to a certain set of conditions dominate each one. Deserts, dry shrublands and woodlands, grasslands, forests, and tundra are major types.

7. Water provinces cover more than 70 percent of the earth's surface. They include standing fresh water (such as lakes), running fresh water (such as streams), and oceans and seas. All of their aquatic ecosystems have gradients in light penetration, temperature, salinity, and dissolved gases. These vary daily and seasonally, and they influence primary productivity.

8. Estuaries, intertidal zones, rocky and sandy shores, tropical reefs, and scattered ecosystems of the open ocean are major marine ecosystems. Of these, photosynthetic activity is greatest in shallow coastal waters and in regions of upwelling along the margins of continents. Upwelling is an upward movement of deep, cooler water that carries nutrients to the surface.

9. The interrelatedness of ocean surface temperatures, the atmosphere, and the land is evident in studies of the El Niño Southern Oscillation. This recurring phenomenon is accompanied by abnormal drought conditions in many parts of the world.

Review Questions

1. Define climate, then list the major interacting factors that influence climate. *587*

2. List a few ways in which air currents, ocean currents, or both probably influence the biome in which you live. *588–589*

3. How does the composition of regional soils affect ecosystem distribution? *591, 592*

4. List some features that all freshwater and marine ecosystems have in common. *600*

5. Spend some time outdoors observing the land or any freshwater or marine ecosystem near you. Write a short essay on some of its features.

1. Solar radiation drives the distribution of weather systems and so influences the distribution of _____ .
   a. every single ecosystem
   b. ecosystems on land only
   c. all ecosystems except those at hydrothermal vents

2. The _____ is like a shield against harmful ultraviolet wavelengths from the sun.
   a. upper atmosphere   c. ozone layer
   b. lower atmosphere   d. greenhouse effect

3. Regional variations in patterns of rainfall and temperature depend on _____ .
   a. global air circulation   c. topography
   b. ocean currents   d. all of the above

4. A rain shadow is a reduction in rainfall on the _____ of a mountain range.
   a. windward side   c. highest elevation
   b. leeward side   d. lowest elevation

5. Biogeographic realms are _____ .
   a. land and water provinces   c. divided into biomes
   b. six major land provinces   d. b and c

6. Biome distribution corresponds roughly with regional variations in _____ .
   a. climate   c. topography
   b. soils   d. all of the above

7. _____ are highly adapted to episodes of fire.
   a. Dry shrublands   c. Pine barrens
   b. Grasslands   d. All of the above

8. During _____, strong vertical movements transport dissolved oxygen in the surface layer of a body of water down to decomposers in its depths, and transport nutrients from the depths up to photosynthesizers.
   a. upwelling   c. fall overturns
   b. spring overturns   d. b and c

9. Match the concepts appropriately.
   ____ taiga
   ____ permafrost
   ____ chaparral
   ____ desert
   ____ deciduous broadleaf forest
   ____ tropical rain forest

   a. high productivity, poor mineral cycling
   b. swamp forest
   c. common at moist, temperate latitudes with mild winters
   d. evaporation greatly exceeds infrequent rainfall
   e. a type of dry shrubland
   f. feature of arctic tundra

10. Match the concepts appropriately.
    ____ plankton
    ____ upwelling
    ____ eutrophication
    ____ estuary
    ____ benthic province

    a. deep, nutrient-rich water moves up along coast
    b. sediments, rocky formations of ocean bottom
    c. seawater, freshwater slowly mix
    d. nutrient enrichment of body of water, resulting in reduced transparency, phytoplankton blooms
    e. community of mostly microscopic floating or weakly swimming organisms

alpine tundra *598*	fall overturn *601*
arctic tundra *598*	forest *596*
atmosphere *587*	grassland *594*
benthic province *605*	hydrosphere *587*
biogeographic realm *590*	hydrothermal vent *605*
biome *590*	intertidal zone *603*
biosphere *587*	lake *600*
climate *587*	ozone layer *587*
coral bank *607*	pelagic province *605*
coral reef *607*	permafrost *598*
desert *593*	plankton *600*
desertification *593*	rain shadow *589*
dry shrubland *594*	soil *592*
dry woodland *594*	spring overturn *601*
El Niño *608*	stream *602*
estuary *602*	upwelling *608*
eutrophication *601*	

Colinvaux, P. May 1989. "The Past and Future Amazon." *Scientific American* 260(3):102-108.

Garrison, T. 1993. *Oceanography: An Invitation to Marine Science*. Belmont, California: Wadsworth.

Gibbons, B. September 1984. "Do We Treat Our Soil Like Dirt?" *National Geographic* 166(3):350-388.

Smith, R. 1989. *Ecology and Field Biology*. Fourth edition. New York: Harper & Row.

37 HUMAN IMPACT ON THE BIOSPHERE

Tropical Forests—Disappearing Biomes?

On the first night of an expedition into a tropical forest, a young graduate student fell asleep under a cotton mosquito net. He woke up before dawn with sudden, gut-wrenching awareness that things were crawling all over him. During the hot, humid night, a platoon of voracious insects had devoured most of the drenched net.

Tropical forests contain diverse hungry insects and the world's biggest ones. They are home to the greatest variety of birds and plants with the largest flowers (Figure 36.17). They support tapirs, monkeys, and jaguars in South America and apes, okapi, and leopards in Africa. Entire communities of insects, spiders, and amphibians live, breed, and die in small pools that collect in the leaves of orchids, mosses, and lichens.

Developing countries in Latin America, Africa, and Southeast Asia have growing populations and limited food, fuel, and lumber. They have been clearing their tropical forests on a massive scale. Most of the forests may disappear within your lifetime.

Why does it matter? For purely ethical reasons, many condemn the obliteration of a major chunk of the biosphere. For practical reasons, the destruction will affect your own life. Our food supply is based on very few crop and livestock species. By developing new or hybrid crop plants, we can make the supply less vulnerable. With genetic engineering and tissue culturing, we can tap tropical rain forests for genetic "resources." We can tap them to develop new antibiotics and vaccines.

Many tropical plants already give us alkaloids for treating cardiovascular disorders and cancer. Aspirin, the most widely used drug in the world, is based on a chemical "blueprint" of a compound extracted from tropical willow leaves. Coffee, bananas, cinnamon, cocoa, sweeteners, Brazil nuts, and many other spices and foods we take for granted originated in the tropics. So did many ornamental plants. So did the latex, gums, resins, dyes, waxes, and many oils used in ice cream, toothpaste, shampoo, condoms, cosmetics, perfumes, compact discs, tires, shoes, and other diverse products.

If aspirin or condoms don't catch your attention, think about this: Wholesale destruction of tropical forests is changing the atmosphere. Burning trees releases carbon dioxide, carbon monoxide, hydrocarbons, nitric oxide, and nitrogen dioxide. Increasing atmospheric concentrations of these gases are contributing to acid rain, smog, and maybe global warming.

As an individual, you might choose to cherish, brood about, or ignore any aspect of the world of life. Whatever the choice, the bottom line is that you and all other organisms are in this together. Our lives interconnect, to degrees that we are only now starting to comprehend.

Figure 37.1 El Yunque rain forest, Puerto Rico.

1. The human population has been growing exponentially since the mid-eighteenth century. Today we have the population size, technology, and cultural inclination to use energy and modify the environment at astonishing rates.

2. The world of life ultimately depends on energy from the sun. That energy drives the complex interactions among the atmosphere, ocean, and land. The increasing accumulation of human-generated pollutants in the atmosphere especially may be disrupting the interactions and may have serious consequences in the near future.

3. We as a species must come to terms with the principles of energy flow and resource utilization that govern all systems of life on earth.

ENVIRONMENTAL EFFECTS OF HUMAN POPULATION GROWTH

Of all the concepts introduced in the preceding chapter, the one that should be foremost in your mind is this: Interactions among the atmosphere, oceans, and land are the engines of the biosphere. Driven by energy from the sun, they create the global temperatures and circulation patterns on which life ultimately depends. With this chapter, we turn to a related concept of equal importance. Simply put, the growing human population has been straining the global engines without fully comprehending how the engines work.

To gain perspective on what is happening, think of something we take for granted—the air around us. The composition of the present atmosphere is the outcome of geologic and metabolic events, including photosynthesis, that began billions of years ago. A few million years ago, the first humans emerged. Like us, they breathed in oxygen from an atmosphere of ancient origins. Like us, they were protected from harmful ultraviolet radiation by an ozone shield in the stratosphere. Their population sizes were not much to speak of, and their interactions with the biosphere were trivial. About 10,000 years ago, however, agriculture began in earnest, and it laid the foundation for rapid population growth. With agriculture, and with the medical and industrial revolutions that followed, human population growth became exponential in a mere blip of evolutionary time (Figure 33.7).

Today, our burgeoning population may be demanding more than the biosphere can sustain. As we take energy and resources from it, we put back monumental amounts of wastes. In the process, we threaten the sta-

bility of ecosystems on land, taint the hydrosphere, and change the composition of the atmosphere. Our carbon dioxide wastes alone are helping to amplify the "greenhouse effect," described on page 580, and this may have unforeseen effects on the global climate.

In a few developed countries of Europe, population growth has more or less stabilized. Rates of increase in the resource use per individual have slowed somewhat. But the resource utilization levels in these countries are already high. At the same time, population growth and demands for resources are increasing rapidly in Central America, South America, Asia, Africa, and elsewhere—even though millions in these developing countries are already starving to death and hundreds of millions more suffer from malnutrition and inadequate health care.

Many of the problems sketched out in this chapter are not going to go away tomorrow. It will take decades, even centuries, to reverse some of the trends already in motion, and not everyone is ready to make the effort. A few enlightened individuals in Michigan or Alberta or New South Wales can make good attempts at resource conservation and at minimizing pollution—but scattered attempts will not be enough. Individuals of all nations will unite to reverse global trends only when they perceive that the dangers of *not* doing so outweigh the personal benefits of ignoring them.

Does this seem pessimistic? Think of the exhaust fumes released into the atmosphere each time you drive a car. Think of the oil refineries, paper mills, and food-processing plants that supply you with goods—and also release chemical wastes into the nation's waterways. Think of Mexico and other developing countries that produce cheap food by using an unskilled labor force and dangerous pesticides. Uncontrolled pesticide applications poison people who work the land, the land itself, and perhaps those who purchase the exported produce. Who changes behavior first? We have no answer to the question. We can suggest, however, that a strained biosphere can rapidly impose an answer upon us.

CHANGES IN THE ATMOSPHERE

If you were to compare the earth to an apple from a supermarket, the atmosphere would be no thicker than the layer of shiny wax applied to that apple. Yet this thin, finite wrapping of air around the earth receives more than 700,000 metric tons of pollutants each day in the United States alone. **Pollutants** are substances with which ecosystems have had no prior evolutionary experience, in terms of kinds or amounts, so ecosystems have no mechanisms for dealing with them. From the human perspective, pollutants are substances that adversely affect our health, activities, or survival.

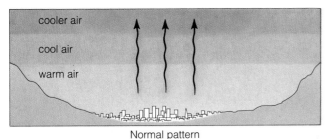

Normal pattern

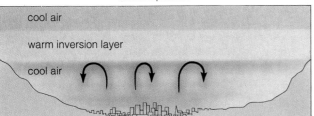

Thermal inversion

Figure 37.2 Mexico City on an otherwise bright, sunny morning, under its self-generated blanket of smog. Topography, staggering numbers of people and motor vehicles, and industrialization combine to make its air some of the world's dirtiest. Breathing here is like smoking two packs of cigarettes a day. The diagrams show how a thermal inversion layer traps airborne pollutants.

Table 37.1	Major Classes of Air Pollutants
Carbon oxides	Carbon monoxide (CO), carbon dioxide (CO_2)
Sulfur oxides	Sulfur dioxide (SO_2), sulfur trioxide (SO_3)
Nitrogen oxides	Nitric oxide (NO), nitrogen dioxide (NO_2), nitrous oxide (N_2O)
Volatile organic compounds	Methane (CH_4), benzene (C_6H_6), chlorofluorocarbons (CFCs)
Photochemical oxidants	Ozone (O_3), peroxyacyl nitrates (PANs), hydrogen peroxide (H_2O_2)
Suspended particles	Solid particles (dust, soot, asbestos, lead, etc.), liquid droplets (sulfuric acid, oils, dioxins, pesticides)

Table 37.1 lists the main classes of air pollutants. Among them are carbon dioxide, sulfur oxides, nitrogen oxides, and chlorofluorocarbons. Also among them are photochemical oxidants, formed by interactions between sunlight and many of the chemicals that we release to the atmosphere.

Local Air Pollution

Air pollutants can be dispersed through the atmosphere or concentrated at their source. What happens to them in a given stretch of time depends on local climate and topography. Consider what happens during a **thermal inversion**, when weather conditions trap a layer of cool, dense air under a layer of warm air (Figure 37.2). Pollutants cannot be dispersed by winds or rise higher in the atmosphere, so they can accumulate to dangerous levels

near the ground. By intensifying a phenomenon called smog, thermal inversions have contributed to some of the worst local air pollution disasters.

Two types of smog (gray air and brown air) form in major cities. Where winters are cold and wet, **industrial smog** develops as a gray haze over industrialized cities that burn coal and other fossil fuels for heating, manufacturing, and generating electric power. The burning releases airborne pollutants, including dust, smoke, soot, ashes, asbestos, oil, bits of lead and other heavy metals, and sulfur oxides. When pollutants are not dispersed by winds and rain, they may reach lethal concentrations. Industrial smog was the cause of London's 1952 air pollution disaster, in which 4,000 people died. New York, Pittsburgh, and Chicago, like London, were gray air cities until coal burning was restricted. Today, most industrial smog forms in cities of China, India, and other developing countries, as well as in Hungary, Poland, and other coal-dependent countries of eastern Europe.

In warm climates, **photochemical smog** develops as a brown, smelly haze over large cities. Where the surrounding land forms a natural basin, as it does around Los Angeles and Mexico City, photochemical smog reaches harmful levels. The main culprit is nitric oxide, produced mainly by cars and other vehicles with internal combustion engines. When nitric oxide reacts with oxygen in the air, nitrogen dioxide forms. When exposed to sunlight, nitrogen dioxide reacts with hydrocarbons to form photochemical oxidants. (Most of the hydrocarbons come from spilled or partly burned gasoline.) The main oxidants in smog are ozone and PANs (short for *peroxyacyl nitrates*). PANs are similar to tear gas. Even traces can sting the eyes, irritate the lungs, and damage crops.

Figure 37.3 1984 average acidities of precipitation and soil sensitivities to acid deposition, for regions of North America. Prolonged exposure to multiple air pollutants is contributing to the rapid destruction of trees in New York's Whiteface Mountains (*left photograph*), Germany (*right photograph*), and elsewhere. Often, pollutants harm mycorrhizal fungi on which trees depend. Weakened trees are more vulnerable to drought, disease, and insect attack.

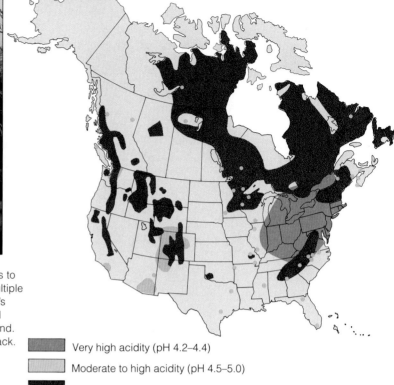

Very high acidity (pH 4.2–4.4)

Moderate to high acidity (pH 4.5–5.0)

Areas sensitive to acid deposition

Acid Deposition

Oxides of sulfur and nitrogen are among the most dangerous air pollutants. (Coal-burning power plants, factories, and metal smelters are major sources of sulfur dioxides. Motor vehicles, power plants that burn fossil fuels, and nitrogen fertilizers are sources of nitrogen oxides.) Depending on climatic conditions, oxides may stay airborne for a while as tiny particles, then fall to earth as dry acid deposition. Most sulfur and nitrogen dioxides dissolve in atmospheric water to form a weak solution of sulfuric and nitric acids. Winds may distribute them over great distances before they fall to earth in rain and snow. This is wet acid deposition, or **acid rain**.

Normal rainwater has a pH of about 5. Acid rain can be 10 to 100 times more acidic than this. Sometimes it becomes as acidic as lemon juice. The acids chemically attack marble buildings, metals, mortar, rubber, plastic, even nylon stockings. They also can disrupt the physiology of organisms and the chemistry of ecosystems.

Depending on their soils and vegetation, some regions are more sensitive than others to acid deposition (Figure 37.3). Highly alkaline soils neutralize the acids before they enter the streams and lakes of watersheds. Water with a high carbonate content also neutralizes acids. However, in watersheds throughout much of northern Europe and southeastern Canada and in scattered regions of the United States, thin soils overlie solid granite—and cannot buffer the acids much.

Rain in much of eastern North America is thirty to forty times more acidic than it was several decades ago, and croplands and forests are suffering (Figure 37.3). Fish populations have vanished from more than 200 lakes in the Adirondack Mountains of New York. Some Canadian biologists predict that fish will vanish from 48,000 lakes in Ontario within the next two decades.

Large forests in northern Europe are under attack by airborne acids and other pollutants from industrial regions of England and Germany. High in the Appalachian Mountains, such pollutants may be killing fungi that are mycorrhizal symbionts of spruce and fir—and so are destroying the trees (page 273). They also are emerging as a serious problem in heavily industrialized parts of Asia, Latin America, and Africa.

Researchers confirmed years ago that power plants, factories, and vehicles are the main sources of acid pollutants, and that acid deposition does indeed adversely affect the environment. Sadly, some responses to local air pollution problems that were recognized decades ago actually contributed to the current problems. For example, tall smokestacks were added to power plants and smelters. The idea was to dump the smoke high in the atmosphere so winds could distribute it elsewhere—which winds readily did. The winds helped transform local air pollution into regional acid deposition. Today the world's tallest smokestack, in Sudbury, Ontario, accounts for about 1 percent by weight of the annual worldwide emissions of sulfur dioxide.

Canada cannot be singled out in this issue. Canada presently receives more acid deposition from industrialized regions of the midwestern United States than it sends across its southern border. Most acidic pollutants in Finland, Norway, Sweden, the Netherlands, Austria, and Switzerland arise in industrialized parts of western and eastern Europe. Prevailing winds do not stop at national boundaries; the problem is global.

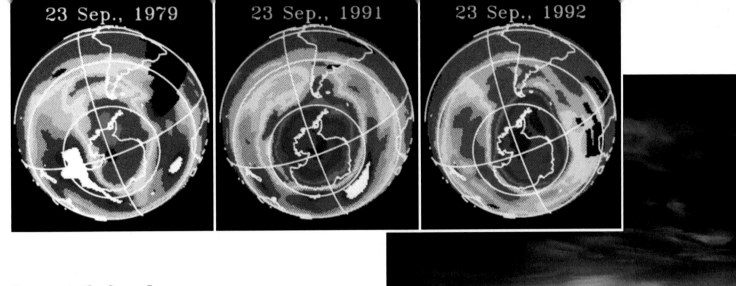

23 Sep., 1979 — 23 Sep., 1991 — 23 Sep., 1992

Figure 37.4 Expansion of the seasonal ozone hole over Antarctica, as recorded by special high-altitude planes. The lowest ozone values (the "hole") are indicated by magenta and purple. Droplets of sulfuric acid, released during the 1991 volcanic eruption of Mount Pinatubo, contributed to the thinning. The photograph shows the ice clouds over Antarctica that have a role in the ozone thinning each spring.

Damage to the Ozone Layer

The ozone layer in the lower stratosphere is almost twice as far above sea level as the summit of Mount Everest, the highest place on earth. Each year, from September through mid-October, it thins down by as much as *half* above Antarctica. The pronounced seasonal thinning is called an **ozone hole**. In 1992, it spanned an area three times the size of the continental United States (Figure 37.4). In 1993, the ozone level was the lowest ever recorded. Within sixty years, it may decrease by as much as 30 percent above heavily populated regions of North America, Europe, and Asia.

Why is the ozone reduction so alarming? It lets more ultraviolet radiation reach the earth's surface. A dramatic rise in skin cancers is one consequence (page 81). Cataracts, an eye disorder, are becoming more common. Besides this, ultraviolet radiation weakens the immune system and our ability to fight many infections. It harms photosynthesizers, and so may alter the atmosphere. Collectively, phytoplankton alone take up huge amounts of carbon dioxide and release huge amounts of oxygen on an annual basis (page 385).

More than any other factor, **chlorofluorocarbons** (CFCs) are bringing about the ozone reduction. CFCs are odorless, invisible, and otherwise harmless compounds of chlorine, fluorine, and carbon. They have been widely used as propellants in aerosol spray cans, coolants in refrigerators and air conditioners, industrial solvents, and plastic foams. CFCs enter the atmosphere slowly and resist breakdown. When a free CFC molecule absorbs ultraviolet light, it gives up a chlorine atom. Chlorine can react with ozone, forming an oxygen molecule and a chlorine monoxide molecule. When chlorine monoxide reacts with a free oxygen atom, another chlorine atom is released that can attack another ozone molecule. Each chlorine atom released in the reactions can convert as many as 10,000 molecules of ozone to oxygen!

Chlorine monoxide levels above Antarctica are 100 to 500 times higher than at mid-latitudes. Why? At high altitudes, ice clouds form there during the winter. Winds that rotate around the South Pole for most of the winter months keep the ice clouds separate from other latitudes (Figure 37.4). The same thing happens, on a lesser scale, in the Arctic. Ice provides a surface that promotes the breakdown of chlorine compounds. So chlorine is free to destroy ozone when the Antarctic air warms somewhat in the Antarctic spring. Hence the ozone hole.

By some estimates, nearly all of the CFCs released between 1955 and 1990 are still making their way up to the stratosphere. Even if we stop using CFCs today, it may be a century before ozone levels return to pre-1985 levels. Since 1978, CFCs in aerosol spray cans have been banned in the United States, Canada, and most Scandinavian countries but not in western Europe. Nonaerosol uses of CFCs continue throughout the world. In 1992, the bad news about the ozone thinning prodded more than half of the world's nations to agree to phase out CFC production except for a few essential uses. Such agreements are steps in the right direction, although some call them too little, too late. CFCs already in the air will be there for over a century, before natural processes neutralize them. You, your children, and your grandchildren will be living with their destructive effects. Are you part of the problem? Think of how you might take an active role in finding solutions instead (see the *Focus* essay on the page that follows).

Ray Turner's Refrigerator

Ray Turner was rummaging around in his refrigerator, wondering whether the bottle of vinegar or the lemons held promise for the ozone layer. Ray was looking for a harmless alternative to the CFC-containing solvents being used to clean the electronic circuit boards being manufactured by his employer, Hughes Aircraft in California. He put a few drops of vinegar on a well-traveled penny, but this didn't remove its film of oxidation. He tried pulverized lemon peel. No luck. Then he put drops of lemon juice on the penny—and the rest is history. Today, Hughes uses citrus-based solvents that are inexpensive and CFC-free. Following their lead, AT&T cleans computer chips and circuit boards with an acid extracted from peaches, plums, and cantaloupes. Ray's refrigerator gave him what he needed to give the ozone layer a break.

Think about it. Maybe solutions to an environmental problem are no farther away than your backyard, your basement, or your refrigerator.

Figure 37.5 Irrigation-dependent crops growing in the Sahara Desert, in Algeria.

CHANGES IN THE HYDROSPHERE

There is a tremendous amount of water in the world, but most is too salty for human consumption or agriculture. For every million liters of water, only about 6 liters are in a readily usable form.

Consequences of Large-Scale Irrigation

Our expanding food-producing capability has helped fan the human population's exponential growth (page 544). How long can expansion continue? Today, we produce about one-third of our food supply on irrigated land (Figure 37.5). We must pipe water into agricultural fields from groundwater or from lakes, reservoirs, and other sources of surface waters. But irrigation often changes the land's suitability for agriculture. Irrigation water often is loaded with mineral salts. In regions with poorly draining soils, evaporation may cause **salination** (salt buildup). Salination can stunt growth, decrease yields, and in time kill crop plants.

Besides this, improperly drained, irrigated lands can become waterlogged. Water accumulating underground can gradually raise the **water table** (the upper limit at which the ground is fully saturated with water). When the water table is close to the ground's surface, soil becomes saturated with saline water that can damage plant roots. Salinity and waterlogging can be corrected with proper management of the water-soil system. The economic cost of doing so is high.

We use groundwater for many purposes, but irrigation is often paramount. Consider the American farmers who withdraw water from the vast Ogallala aquifer (Figure 37.6). Their annual *over*draft—the amount that

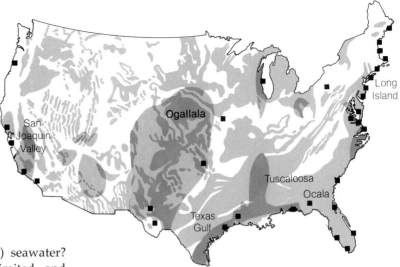

Figure 37.6 Underground aquifers (blue areas) that contain 95 percent of all fresh water in the United States. Gold shading indicates regions where aquifers are being depleted, mainly for agriculture. Black boxes indicate aquifers being contaminated by saltwater intrusion.

nature does not replenish—is nearly equal to the annual flow of the Colorado River! Water tables that were low to begin with are dropping rapidly, and stream and underground spring flows are drying up. Where will the water come from when the aquifer is depleted?

Why not desalinate (remove salt from) seawater? The supply of seawater is essentially unlimited, and desalination technologies already exist. However, it takes fuel energy to distill seawater or force it through membranes (a method called reverse osmosis). Desalination may be feasible in Saudi Arabia and a few other countries with limited population sizes, large energy reserves, and cash to spare. In some situations it may be the only alternative to running out of water, as Santa Barbara and some other California cities nearly did during a severe, prolonged drought. Even so, such efforts cannot solve the core problem. *Desalination may never be cost-effective for large-scale agriculture—which accounts for almost two-thirds of the human population's annual use of fresh water.*

Maintaining Water Quality

The problem of water scarcity is compounded by water pollution. Human sewage, animal wastes, and toxic chemicals make water unfit to drink, even to swim in. Pollutants can encourage contamination by pathogenic microbes. Agricultural runoff pollutes water with sediments, pesticides, and plant nutrients. Power plants and factories pollute water with chemicals, radioactive materials, and excess heat (thermal pollution).

Pollutants accumulate in lakes, rivers, and bays before reaching the oceans. Many cities throughout the world dump untreated sewage into coastal waters. Cities along rivers and harbors maintain shipping channels by dredging the polluted muck and barging it out to sea. They also barge out sewage sludge—coarse, settled solids that contain bacteria, viruses, and toxic metals. Sometimes black sludge washes ashore after storms.

In the United States, about 15,000 facilities treat the liquid wastes from about 70 percent of the population and 87,000 industries. The remaining wastes are mostly from suburban and rural populations. They are treated in lagoons or septic tanks, or they are directly discharged—untreated—into waterways.

There are three levels of wastewater treatment. In *primary* treatment, screens and settling tanks remove sludge, which is dried, burned, dumped in landfills, or treated further. Chlorine often is used to kill pathogens in the water, but it does not kill them all. Also, chlorine can react with some industrial chemicals. Some of the resulting chlorinated organic compounds are carcinogens.

In *secondary* treatment, microbial populations are used to break down organic matter after primary treatment but before chlorination. The wastewater is either (1) sprayed and trickled through microbe-containing gravel beds or (2) aerated in tanks and seeded with microbes. Toxic solutes sometimes poison the microbial helpers. Then, the facilities are shut down until the microbial populations are reestablished.

Primary and secondary treatment combined remove most suspended solids and oxygen-demanding wastes. They do not remove all of the nitrogen, phosphorus, and toxic substances, including heavy metals, pesticides, and industrial chemicals. Usually the water is chlorinated before being released into the waterways.

Tertiary treatment may adequately reduce pollution levels but is largely experimental and expensive. It is applied to only 5 percent of the nation's wastewater.

In short, most wastewater is not being treated adequately. A pattern is repeated thousands of times along our waterways. Water for drinking is removed *upstream* from a city, and wastes from industry and sewage treatment are discharged *downstream*. It takes no great leap of the imagination to see that pollution intensifies as rivers flow toward the oceans. In Louisiana, waters drained from the central states flow toward the Gulf of Mexico. Its pollution levels are high enough to threaten public health. Water destined for drinking does get treated to remove pathogens, but this does not remove toxic wastes dumped by numerous factories upstream. *You may find it illuminating to investigate where your own city's supply of water comes from and where it has been.*

CHANGES IN THE LAND

Solid Wastes

Many resources are scarce in developing countries, so very few materials are discarded. In the United States, as in some other developed countries, a "throwaway" mentality prevails. People use something once, discard it, then buy another. Each year, millions of metric tons of solid wastes are dumped, burned, and buried. Paper products make up half the total volume, which also includes 50 billion nonreturnable cans and bottles.

It takes more than 500,000 trees to supply the Sunday paper to Americans. Every week. If everyone recycled even one out of ten newspapers, 25 million trees a year could be left standing. Using recycled paper also reduces by 95 percent the air pollution that results from paper manufacture. Also, it takes 30 to 50 percent less energy to recycle paper than to make new paper.

Accompanying our throwaway mentality is a unique problem in the world of life—what to do with the solid wastes. Unlike natural ecosystems, cities don't cycle materials. They bury them in landfills or burn them in incinerators, which produce toxic air pollutants as well as toxic ashes that must be disposed of safely. Besides this, all landfills eventually "leak" and pose a threat to groundwater supplies.

A transition from a throwaway mentality to one based on recycling and reuse is feasible. It is affordable, and we have the technology to do it. Consumers can influence manufacturers by refusing to buy goods that are lavishly wrapped and boxed, packaged in indestructible containers, and designed for one-time use. Individuals can find out from the local post office how to reduce their daily flow of junk mail. Unsolicited mail wastes an astounding amount of paper, time, and energy—and means higher postage rates for everyone.

Individuals also can ask local governments for well-designed, large-scale resource recovery centers. Such a plant operates in Saugus, Massachusetts.

Conversion of Marginal Lands for Agriculture

Today, agriculture is expanding onto marginally productive land. Almost 21 percent of the earth's land surfaces are already used for cropland or grazing land. Another 28 percent is said to be potentially suitable for these practices, but productivity would be so low that conversion may not be worth the cost (Figure 37.7).

Asia and some other heavily populated regions have severe food shortages, yet more than 80 percent of their productive land is already cultivated. Valiant efforts have been made to improve crop production on existing land. Under the banner of the **green revolu-**

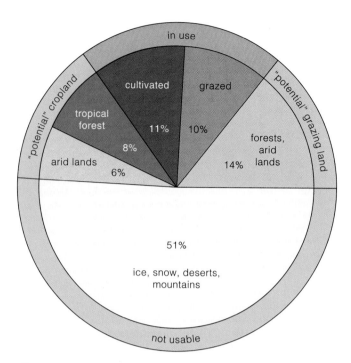

Figure 37.7 Classification of the earth's land. Theoretically, the world's cropland could be more than doubled by clearing tropical forests and irrigating arid lands. But converting this marginal land to cropland would destroy valuable forest resources, cause serious environmental problems, and possibly cost more than it is worth.

tion, research has been directed toward (1) improving crop plants for higher yields and (2) exporting modern agricultural practices and equipment to developing countries. Many developing countries rely on *subsistence* agriculture, which runs on energy inputs from sunlight and human labor. They also rely on *animal-assisted* agriculture, with energy inputs from oxen and other draft animals.

By contrast, mechanized agriculture requires massive inputs of fertilizers, pesticides, and ample irrigation to sustain high-yield crops. It requires fossil fuel energy to drive farm machines. Crop yields *are* four times as high. But the modern practices use up a hundred times more energy and mineral resources.

Many farmers in developing countries cannot afford to take widespread advantage of the new, high-yield crop strains. Of necessity, costs of fertilizers and machinery are reflected in market food prices—which are too high for much of the country's own population. Farmers who make the investment end up depending on industrialized producers of fertilizers and machinery that often must be imported. Such importations can add to the foreign debt of developing countries.

Pressures for increased food production are greatest in areas of Central and South America, Asia, the Middle East, and Africa where human populations are rapidly expanding into marginal lands. As described next, the repercussions extend beyond national boundaries.

Figure 37.8 Dust storm approaching Prowers County, Colorado, in 1934. The Great Plains of the American Midwest are dry, windy, and subject to severe recurring droughts. Their conversion to agriculture started in the 1870s. Overgrazing left the ground bare through vast tracts of the natural grassland. In May 1934, a cloud of topsoil that blew off the land blanketed the entire eastern portion of the United States, giving the Great Plains a new name—the Dust Bowl. About 3.6 million hectares (9 million acres) of cropland were destroyed. Today, without massive irrigation and intensive conservation farming, desertlike conditions could prevail in this region.

Figure 37.9 Desertification in the Sahel, a region of West Africa that forms a belt between the dry Sahara Desert and tropical forests. This savanna country is rapidly undergoing desertification as a result of overgrazing and overfarming.

Desertification

Desertification refers to the conversion of large tracts of grasslands, rain-fed cropland, or irrigated cropland to a more desertlike state, with a 10 percent or greater drop in agricultural productivity. In the past fifty years, 9 million square kilometers worldwide have become desertified. At least 200,000 square kilometers are being converted annually. Prolonged drought can accelerate the process, as it did in the American Great Plains years ago (Figure 37.8). Today, overgrazing on marginal lands is the main cause of large-scale desertification.

In Africa, for example, there are too many cattle in the wrong places. Cattle require more water than the wild herbivores that are native to the region. This means the cattle have to move back and forth between grazing areas and watering holes. As they do, they trample grasses and compact the soil surface (Figure 37.9). By contrast, gazelles and other native herbivores get most (if not all) of their water from plants. They also are better water conservers; they lose little in feces, compared to cattle.

In 1978 a biologist, David Holpcraft, formed a ranch composed of antelope, zebras, giraffes, ostriches, and other native herbivores. He raised cattle as "control groups" in order to compare costs and meat yields on the same land. His initial results exceeded expectations. Native herds increased steadily and yielded meat. Range conditions improved rather than deteriorating. Vexing problems remained. African tribes have their own idea of what constitutes "good" meat, and some tribes view cattle as the symbols of wealth in their society.

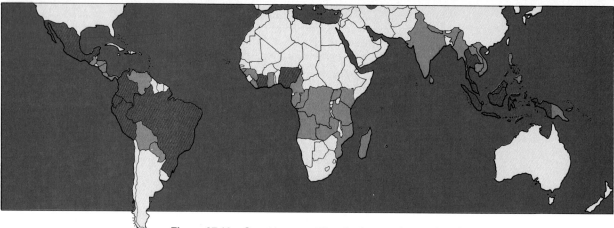

Figure 37.10 Countries permitting the largest destruction of tropical forests. Red shading indicates where 2,000 to 14,800 square kilometers are deforested annually. Orange indicates "moderate" deforestation (100 to 1,900 square kilometers).

Deforestation

The world's great forests have major influence on the biosphere. Forested watersheds are like giant sponges that absorb, hold, and gradually release water. By intervening in the downstream flow of water, forests help control soil erosion, flooding, and sediment buildup in rivers, lakes, and reservoirs.

Deforestation is the removal of all trees from large tracts of land. (The photograph in Figure 37.10 shows this happening in the Amazon basin in South America.) It leads to loss of fragile soils and disrupts watersheds, especially in steeply sloped regions.

In the tropics, the soil that is lost when forests are cleared for agriculture leads to a long-term loss in productivity. The irony is that tropical forests are one of the worst places to grow crops and raise pasture animals. The high temperatures and heavy, frequent rainfall favor decomposition. In intact forests, organic remains and wastes decompose too fast for litter to build up. The forest trees and other plants absorb and assimilate nutrients as they become available.

Shifting cultivation (once called slash-and-burn agriculture) disrupts the forest ecosystem. Trees are cut and burned, then the ashes tilled into the soil. Then crops are grown for one to several seasons. However, nutrients are quickly leached from the exposed soils, so cleared plots soon become infertile and are abandoned. Forest ecosystems are probably not damaged much by shifting cultivation on small, widely scattered plots. Their fertility plummets when larger areas are cleared and when plots are cleared again at shorter intervals.

Rates of evaporation, transpiration, and runoff shift after deforestation, so regional patterns of rainfall can be disrupted. Between 50 and 80 percent of the water vapor above tropical forests alone is released from the trees. Without trees, annual precipitation declines, and the rain that does fall rapidly runs off the bare soil. As the region gets hotter and drier, soil fertility and moisture decline even more. In time, sparse grassland or even desertlike conditions can prevail where there had once been a rich tropical forest.

Figure 37.11 The vast Amazon River Basin of South America in September 1988. Its features were completely obscured by smoke from fires that had been set to clear tropical forests, pasturelands, and croplands during the dry season.

Smoke extends to the Andes Mountains on the western horizon, about 650 miles (1,046 kilometers) away. The smoke cover was the largest that astronauts had ever before observed. It extended almost 175 million square kilometers (1,044,000 square miles).

The smoke plume near the center of the photograph alone covered an area comparable to the huge forest fire in Yellowstone National Park in that same year.

Widespread tropical forest destruction may have global repercussions. These forests absorb much of the sunlight reaching equatorial regions of the earth's surface. When they are cleared, the land becomes shinier, so to speak, and reflects more incoming energy back into space. Also, the trees of these vast forests help maintain the global cycling of carbon and oxygen by their photosynthetic activities. When they are harvested or burned, carbon stored in their biomass is released to the atmosphere in the form of carbon dioxide—and this may be amplifying the greenhouse effect.

Almost half the world's tropical forests have been destroyed for cropland, grazing land, timber, and fuelwood. Deforestation is greatest in Brazil, Indonesia, Colombia, and Mexico (Figures 37.10 and 37.11). If clearing and destruction continue at present rates, only Brazil and Zaire will have large tropical forests in the year 2010. By 2035, most of their forests will be gone.

A QUESTION OF ENERGY INPUTS

Paralleling the J-shaped curve of human population growth is a steep rise in total and per capita energy consumption. The rise is due to increased numbers of energy users and to extravagant consumption and waste. For example, in one of the most pleasant of all climates, a major university constructed seven- and eight-story buildings with narrow, sealed windows. The windows can't be opened to catch the prevailing ocean breezes. The buildings and windows were not designed or aligned to use the abundant sunlight for passive solar heating and breezes for passive cooling. Massive energy-demanding cooling and heating systems were installed.

When you hear talk of abundant energy supplies, keep in mind that there is a huge difference between the total supply and the net amount available. **Net energy** is that left over after subtracting the energy used to locate, extract, transport, store, and deliver energy to consumers. Some sources of energy, such as direct solar energy, are renewable. Others, such as coal and petroleum, are not. Currently, 83 percent of the energy stores being tapped falls in the second category (Figure 37.12).

Fossil Fuels

Fossil fuels are the carbon-containing remains of plants that lived hundreds of millions of years ago (page 279). The plants were buried and compressed in sediments, then gradually transformed into coal, petroleum (oil), and natural gas. They are nonrenewable resources.

Even with strict conservation efforts, known petroleum and natural gas reserves may be used up during the next century. As petroleum and natural gas deposits in easily accessible areas are depleted, we seek new deposits. Our explorations have taken us to wilderness areas in Alaska and to other fragile environments, such as the continental shelves. The *net* energy decreases as costs of extraction and transportation to and from remote areas increase. Environmental costs of extraction and transportation escalate. The long-term impact of the 11-million-gallon spill from the tanker *Valdez* in Alaska's coastal waters is still not understood.

What about coal? In theory, world reserves can meet the energy needs of the entire human population for at least several centuries. But coal burning has been the largest single source of air pollution. Most coal reserves contain low-quality, high-sulfur material. Unless sulfur is removed before or after burning, sulfur dioxides are released into the air. They add to the global problem of acid deposition. Fossil fuel burning also releases carbon dioxide and so adds to the greenhouse effect.

Extensive strip mining of coal reserves close to the earth's surface carries its own problems. It removes

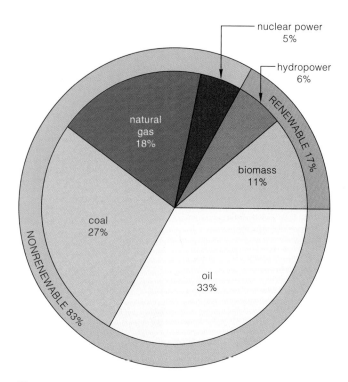

Figure 37.12 World consumption of nonrenewable and renewable energy sources in 1991.

land from agriculture, grazing, and wildlife. Restoration is difficult and expensive in arid and semiarid lands, where much strip mining is proceeding.

Nuclear Energy

Nuclear Reactors. As Hiroshima burned in 1945, the world recoiled in horror from the destructive force of nuclear energy. By the 1950s, nuclear energy was being viewed as an instrument of progress. Many energy-poor industrialized nations, France included, now depend heavily on nuclear power. Yet new nuclear plants have been delayed or cancelled in most other countries. Since 1970, in the United States alone, plans for 117 nuclear power plants have been cancelled. Other plants were abandoned before completion. A few are being converted at great cost to fossil fuel burning.

Questions surround nuclear energy's costs, efficiency, safety record, and environmental impact. By 1990 in the United States, it cost slightly more to generate electricity by nuclear energy than by using coal. Electricity-generating methods now have average costs below those of new nuclear power plants. By the year 2000, solar energy with natural gas backup also should cost less.

What about safety? Radioactivity and carbon dioxide escaping from nuclear plants during normal opera-

Focus on the Environment

Incident at Chernobyl

On April 26, 1986, errors in judgment during a routine test led to runaway reactions at the Chernobyl power plant in the Ukraine. Explosions destroyed the plant's nuclear core and ripped open its 1,000-metric-ton concrete roof.

Thirty-one people died instantly; others died of radiation sickness during the following weeks. Inhabitants of entire villages were relocated; their former homes were bulldozed under.

After the accident, the number of people opposed to nuclear plants rose sharply, even in France. Figure *a* gives us an indication of why opposition increased. The radioactive fallout put an estimated 300 million to 400 million people at risk of leukemia and other radiation-induced disorders. Besides this, hundreds of millions of dollars were lost throughout eastern and western Europe when the fallout made crops and livestock unfit for consumption.

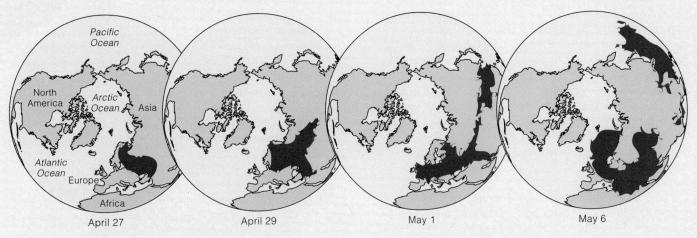

a Global distribution of radioactive fallout after explosions ripped through the Chernobyl nuclear plant on April 26, 1986.

tion are actually less than the amount released from coal-burning plants of the same capacity. And they emit no sulfur dioxide. The danger lies with the potential for **meltdown**. As nuclear fuel breaks down, it releases considerable heat, which typically is absorbed by water that is circulated over the fuel. Heated water produces the steam that drives electricity-generating turbines. Should a leak develop in the circulating water system, water levels might plummet around the fuel—which might then heat rapidly, past its melting point. On the floor of the generator, melting fuel would instantly convert the remaining water to steam. Together with other chemical reactions, the formation of enough steam could blow the system apart and release radioactive material into the environment. Besides this, an overheated reactor core could melt through its thick concrete containment slab and contaminate groundwater.

Is this scenario far-fetched? The *Focus* essay suggests not, but this is a controversial issue.

Nuclear Waste Disposal. Unlike coal, nuclear fuel cannot be burned to harmless ashes. A reactor's fuel elements are spent after about three years, but they still contain uranium fuel and hundreds of new radioisotopes produced by the reactions. Taken together, the wastes are extremely radioactive and dangerous. As they undergo radioactive decay, they become extremely hot. They are plunged at once into water-filled pools. The water cools them and keeps radioactive material from escaping. After being stored for several months, the remaining isotopes are still lethal. Some must be isolated for at least 10,000 years. If one kind of plutonium isotope (^{239}Pu) is not removed, the wastes must be kept isolated for a quarter of a million years!

After nearly fifty years of research, scientists still cannot agree on the best way to store high-level radioactive wastes. Even if they could do so, there is no politically acceptable solution. No one wants radioactive wastes anywhere near where they live.

Biological Principles and the Human Imperative

Molecules, cells, tissues, organs, organ systems, multicelled organisms, populations, communities, ecosystems, the biosphere. These are architectural systems of life, assembled in increasingly complex ways over the past 3.5 billion years. We are latecomers to this immense biological building program. Yet, within the relatively short span of 10,000 years, we have been changing the character of the land, oceans, and atmosphere—even the genetic character of species.

It would be presumptuous to think we alone have had major effects on the world of life. As long ago as the Proterozoic era, photosynthetic organisms irrevocably changed the course of biological evolution by gradually enriching the atmosphere with oxygen. In the past as well as the present, competitive adaptations assured the rise of some groups, whose dominance assured the decline of others. Thus change is nothing new to this biological building program. What is new is the accelerated, potentially cataclysmic change accompanying the stupendous growth of the human population. We now have the population size, the technology, and the cultural inclination to use energy and modify the environment at frightening rates.

Where will rampant, accelerated change lead us? Will feedback controls begin to operate as they do, for example, when population growth exceeds the carrying capacity of the environment? In other words, will negative feedback controls come into play and keep things from getting too far out of hand?

Feedback control will not be enough, for it operates only when deviation already exists. Our population growth rates and patterns of resource consumption are founded on an illusion of unlimited resources and a forgiving environment. A prolonged, global shortage of food or the passing of a critical threshold for the global engines can come too fast to be corrected. At some point, deviations may have too great an impact to be reversed.

What about feedforward mechanisms, which might serve as early warning systems? For example, skin receptors sense a drop in outside air temperature. Each sends messages to the nervous system, which responds by triggering mechanisms that raise the core temperature before the body itself becomes dangerously chilled. If we could develop feedforward control mechanisms, maybe we could begin corrective measures before we alter the environment too significantly.

By themselves, feedforward controls won't work, for they start operating when change is already under way. Think of the DEW line—the Distant Early Warning system. It is like a sensory receptor, one that detects intercontinental ballistic missiles that may be launched against North America. By the time the system detects what it is designed

On the Drawing Boards. Breeder reactors "breed" fuel by converting an abundant isotope of uranium into a fissionable isotope of plutonium. Unlike a conventional nuclear reactor, which can't explode like an atomic bomb, breeder reactors could produce a small nuclear explosion. Also, it may cost several times more to put them into operation. It may take 100 to 200 years to produce enough plutonium to fuel them.

In theory, fusion power might provide energy. The idea is to fuse hydrogen atoms to form helium atoms, with the release of considerable energy. The process is analogous to reactions that produce heat energy in the sun. The scientific, technological, and economic obstacles are great. Without major breakthroughs, fusion power is not expected to produce electricity on a commercial basis until the last half of the next century, if ever.

A Solar-Hydrogen Future?

For more than a billion years, photosynthetic cells have stayed alive by harnessing sunlight energy and splitting water molecules for their building programs. Imagine a future in which we do the same thing.

We have a virtually limitless supply of water; it covers nearly three-fourths of the earth's surface. We know that its molecules can be split into hydrogen gas (H_2) and oxygen by passing an electric current through it. Hydrogen gas released by this process can be used as an energy source. Suppose we develop low-cost solar cells that can harness sunlight energy, then use that energy to generate the electricity required to produce hydrogen gas. Then, the human population will be assured of an affordable and renewable energy source—and the cleansing of the environment can begin in earnest. Why? In the words of G. Tyler Miller, Jr., an environmental scientist, "If we can make the transition to an energy-to detect, it may be too late to stop wide-spread destruction.

It would be naive to assume we can ever reverse who we are at this point in evolutionary time, to de-evolve ourselves culturally and biologically into becoming less complex in the hope of averting disaster. Yet there is reason to believe we can avert disaster by using a third kind of control mechanism, one that is uniquely our own. We have the capacity to anticipate events before they happen. We are not locked into responding only after irreversible change has begun. We have the capacity to anticipate the future—it is the essence of our visions of utopia or of nightmarish hell. *We all have the capacity to adapt to a future that we can partly shape.* We can, for example, learn to live with less. Far from being a return to primitive simplicity, it would be one of the most complex and intelligent behaviors of which we are capable.

Having that capacity and using it are not the same thing. We have already put the world of life on dangerous ground because we have not yet mobilized ourselves as a species to work toward self-control.

Our survival depends on predicting possible futures. It depends on designing and constructing ecosystems that are in harmony with our definition of basic human values and with the biological models available to us. Human values can change; our expectations can and must be adapted to biological reality. *For the principles of energy flow and resource utilization, which govern the survival of all systems of life, do not change.*

It is our biological and cultural imperative that we come to terms with these principles, and ask ourselves what our long-term contribution will be to the world of life.

efficient solar-hydrogen age, we can say goodbye to smog, oil spills, acid rain, and nuclear energy, and per-haps to global warming. The reason is simple. When hydrogen burns in air, it reacts with oxygen gas to produce water vapor—not a bad thing to have coming out of our tailpipes, chimneys, and smokestacks. If clean solar-hydrogen technology comes on line soon enough, it could help heavily populated developing countries raise their living standards without severely disrupting the earth's life support systems."

Researchers in Germany, Japan, and elsewhere have been working intensively to develop the required tech-nologies for hydrogen-fueled vehicles, factories, and homes. Prototype **photovoltaic cells**, which convert sunlight energy into electricity, are in place in thou-sands of villages in southeast Asia, 30,000 homes around the world, even drilling platforms in the Persian Gulf. Solar-hydrogen systems that will meet the heat-ing, cooling, and other electrical needs of homes may be

on the production line in Germany before the 1990s are over. Mercedes, BMW, and Mazda now have hydrogen-powered prototype vehicles on the road.

With serious commitment, we could expect solar cells to supply as much energy as nuclear power plants by the year 2010. By 2050, they could be satisfying half the energy needs of the United States. They could do so at lower cost and at much lower risk than the existing alternatives. Cuts in the federal budget have hampered research efforts and applications in the United States.

In sum, what are our best options? For now and the immediate future, we must reduce consumption by using more energy-efficient vehicles, architectural designs, heating and cooling systems, and appliances. For the long term, some argue in favor of hydrogen fuel systems based on energy from the sun—or from wind, flowing water, heated steam being vented from the earth, or some other renewable resource. The *Focus* essay gives other arguments.

SUMMARY

1. Accompanying the exponential growth of the human population are increased energy demands and environmental pollution.

2. Pollutants are substances with which ecosystems have had no prior evolutionary experience (in terms of kinds and amounts) and so have no mechanisms for absorbing or cycling them. Many pollutants are an outcome of human activities, and they adversely affect the health, activities, or survival of human populations.

3. Industrial smog, photochemical smog, and acid rain are examples of regional air pollution. Thinning of the ozone layer and possible enhancement of the greenhouse effect are examples of global air pollution.

4. Human population growth presently depends on the expansion of agriculture, made possible by large-scale irrigation and heavy applications of fertilizers and pesticides. Global freshwater supplies are limited, yet they are being polluted by agricultural runoff (which includes sediments as well as pesticides and fertilizers), industrial wastes, and human sewage.

5. Human populations are damaging land surfaces by the mind-boggling accumulation of solid wastes and by conversion of marginal lands for agriculture. Wide-spread desertification and the destruction of tropical forest biomes may be altering regional soils and patterns of rainfall.

6. Energy supplies in the form of fossil fuels are non-renewable, dwindling, and environmentally costly to extract and use. Nuclear energy in itself is less polluting, but the costs and risks associated with fuel containment and with storing radioactive wastes are enormous. The challenge is to develop affordable alternatives based on renewable resources, such as solar-hydrogen energy.

Review Questions

1. What are CFCs? Describe how they have helped form the ozone hole above Antarctica. *615*

2. Make a list of advantages you personally enjoy as a member of an affluent, industrialized society. List some of the drawbacks. Do you believe the benefits outweigh the costs? This is not a trick question. *612*

3. List six activities you pursue each day that harm the environment. List six ways in which you might reduce the harmful effects.

4. After reading this chapter, write your own caption about the flow of energy and materials into and out of the human ecosystem shown in the photograph below.

1. Since the mid-eighteenth century, human population growth has been _____ .
   a. leveling off
   b. growing slowly
   c. growing exponentially
   d. not much to speak of

2. Pollutants disrupt ecosystems because _____ .
   a. their component elements differ from those of natural substances
   b. only humans have uses for them
   c. there are no evolved, established mechanisms that can deal with them
   d. their only effect is on ecosystems but not humans

3. During a thermal inversion, weather conditions trap a layer of _____ air under a layer of _____ air.
   a. warm; cool
   b. cool; warm
   c. warm; sooty
   d. cool; sooty

4. _____ is (are) a case of local air pollution.
   a. Smog
   b. Acid deposition
   c. Ozone layer thinning
   d. a and b

5. _____ is (are) a case of air pollution with global effects.
   a. Smog
   b. Acid deposition
   c. Ozone layer thinning
   d. b and c

6. Two-thirds of the water used annually by the human population goes to _____ .
   a. urban centers
   b. agriculture
   c. treatment facilities
   d. a and c

7. The upper limit at which the ground is fully saturated with water is called _____ .
   a. groundwater
   b. an aquifer
   c. the water table
   d. the salination limit

8. Energy from fossil fuels is _____; their extraction and use come at _____ cost to the environment.
   a. renewable; low
   b. nonrenewable; low
   c. renewable; high
   d. nonrenewable; high

9. Nuclear energy normally pollutes _____ than fossil fuels; it poses _____ dangers than fossil fuels.
   a. less; lesser
   b. more; greater
   c. more; lesser
   d. less; greater

10. Match each term with its appropriate description.
    _____ desertification
    _____ deforestation
    _____ green revolution
    _____ pollutant
    _____ solar-hydrogen power

    a. substance with adverse effects on environment, humans, or both
    b. one of our best options
    c. soil loss, watershed loss, altered rainfall patterns follow
    d. attempt to improve crop production on existing land
    e. conversion of marginal land for agriculture, with at least 10 percent drop in productivity

acid rain *614*
chlorofluorocarbon (CFC) *615*
deforestation *620*
desertification *619*
fossil fuel *622*
green revolution *618*
industrial smog *613*
meltdown *623*
net energy *622*
ozone hole *615*
photochemical smog *613*
photovoltaic cell *625*
pollutant *612*
salination *616*
shifting cultivation *620*
thermal inversion *613*
water table *616*

Collins, M. 1990. *The Last Rain Forests*. New York: Oxford University Press.

Gruber, D. 1989. "Biological Monitoring and Our Water Resources." *Endeavour* 13(3):135–140.

Miller, G. T., Jr. 1994. *Living in the Environment*. Eighth edition. Belmont, California: Wadsworth. Chapter 17 is especially valuable reading. This author consistently puts information from numerous sources into a current, accessible survey of the present and future state of the environment.

Mohnen, V. August 1988. "The Challenge of Acid Rain." *Scientific American* 259(2):30–38.

Western, D., and M. Pearl. 1989. *Conservation for the Twenty-First Century*. New York: Oxford University Press.

Wilson, E. 1988. *Biodiversity*. Washington, D.C.: National Academy of Sciences.

38 ANIMAL BEHAVIOR

Deck the Nest With Sprigs of Green Stuff

About a century ago, the European starling hitched a boat ride to North America. Ever since, starlings have been multiplying and evicting great numbers of native birds from scarce nest sites.

Curiously, starlings decorate the usurped nests with many sprigs of wild carrot, freshly plucked. Why do they do this? Does greenery camouflage the nests from predators? Not really. The nests are already concealed, in the dark cavities of tree holes. Does the greenery insulate eggs from the cold? Actually, moist, green plant parts promote heat loss. Well, then, does the greenery combat parasites? After all, mites parasitize birds and infest the nest holes (Figure 38.1). Even a few tiny mites can rapidly produce thousands of descendants. In large numbers, the mites can suck enough blood from a nestling to weaken it and interfere with its growth.

A bit of wild carrot does indeed fumigate a nest—and it measurably increases a starling's chance of producing healthy, surviving offspring. Wild carrot contains a highly aromatic steroid compound. Most likely, the compound functions in the plant's chemical defense against herbivorous animals. By coincidence, it also happens to arrest immature bird mites in their developmental tracks.

As biologists Larry Clark and Russell Mason discovered, fresh wild carrot sprigs prevent mites from becoming sexually mature—and therefore prevent a mite population explosion in the nest. They constructed experimental nests with and without carrot sprigs, and these were occupied by starlings. In time, by Clark's and Mason's count, an average of 750,000 mites occupied undecorated nests. A mere 8,000 or so occupied the green-sprigged ones.

And so starlings lead us into the world of behavioral research. As you will see, some behavioral studies focus on structural and functional mechanisms that enable individuals to behave as they do. Others focus on the adaptive value of some behavior to an individual's reproductive success.

Figure 38.1 (**a**) A most excellent fumigator in nature—the European starling (*Sturnus vulgaris*), which combats infestations of mites by decorating previously owned nests with fresh sprigs of wild carrot (**b**).

a

b

1. "Behavior" refers to the observable, coordinated responses an animal makes to stimuli. The responses are instinctive, learned, or a combination of both.

2. Forms of behavior have a genetic basis. Certain genes contain instructions that govern the development of the nervous and endocrine systems. These are the systems by which an animal detects, processes, and issues commands for behavioral responses to stimuli.

3. Like other traits having a genetic basis, forms of behavior have evolved by way of natural selection—the measure of which is reproductive success. Thus sexual selection and other evolutionary processes have favored behavioral mechanisms that enhance the ability of the individual to pass on his or her genes to offspring.

4. Social behavior may be explained in terms of natural selection. It has costs and benefits that can be measured in terms of the reproductive success of the individual.

MECHANISMS UNDERLYING BEHAVIOR

Genetic Effects on Behavior

An animal's nervous system detects and processes information about conditions outside and inside the body, then it commands muscles and glands to make appropriate responses. The observable responses that animals make to stimuli are what we call **animal behavior**.

Genes contribute in an indirect yet major way to behavior. Consider Stevan Arnold's studies of the feeding preferences of different populations of a garter snake species. For snake populations living along the California coast, the food of choice is the banana slug (Figure 38.2). Farther inland, tadpoles and small fishes are preferred. Offer inland snakes a banana slug and they ignore it.

In one set of experiments, Arnold offered captive newborn garter snakes a chunk of slug as the first meal. Offspring of coastal parents usually ate it and even flicked their tongue at cotton swabs drenched in essence of slug. (Snakes "smell" by tongue-flicking, which draws chemical odors into the mouth.) Offspring of inland parents ignored the swabs and only rarely ate the slug meat.

Here was a clear difference between captive baby snakes that had no prior, direct experience with slugs. It suggested that the snakes were programmed to accept

Figure 38.2 (**a**) Banana slug, food for (**b**) a grown-up garter snake of coastal California. (**c**) A newborn garter snake from a coastal population, tongue-flicking at a cotton swab drenched with banana slug fluids.

or reject slugs before they were born; they didn't learn to do so through "taste trials." To test this hypothesis, Arnold crossed coastal male snakes with inland female snakes. He also crossed coastal females with inland males. If the difference between the populations has a genetic basis, then "hybrid" offspring might make an intermediate response to slug chunks and odors.

Arnold's observations matched the predicted results. Compared with typical newborn inland snakes, many baby snakes of mixed parentage tongue-flicked *more* often at slug-scented cotton swabs—but *less* often than typical newborn coastal snakes. Thus, the difference in feeding behavior could stem from differences in genes that affect how odor-detecting mechanisms are put together in the snake embryo during its development.

By influencing the development of the nervous system, genes contribute in an indirect yet major way to behavior.

Hormonal Effects on Behavior

The gene products called hormones contribute directly and indirectly to behavior. Imagine it is early spring in a Canadian forest. A male white-throated sparrow whistles a song that sounds rather like "Sam Peabody, Peabody, Peabody." He repeats it thousands of times, clearly and consistently, and you wonder how he does it. For white-throated sparrows and most other songbirds, most of the vocalists are males. In a roundabout way, their singing behavior starts with melatonin, a hormone secreted by the pineal gland (page 505). As one of its effects, melatonin suppresses the growth and function of gonads. In spring, the increase in daylength inhibits melatonin secretion. Then, gonads grow and step up secretion of estrogen and testosterone—two hormones that *directly* influence singing.

The males and females differ in the structure and size of a **sound system**, which consists of brain regions that govern muscles of the vocal organ. Before male birds hatch, a high estrogen level triggers development of a masculinized brain. Later, at the start of the breeding season, their enlarged gonads step up their secretion of testosterone. This hormone acts on cells in the sound system and prepares the male bird to sing when properly stimulated.

Hormones influence the organization and activation of mechanisms required for particular forms of behavior.

INSTINCTIVE AND LEARNED BEHAVIOR

Instinct and Learning Defined

As the preceding examples suggest, built-in mechanisms control behavioral responses to environmental cues. Traditionally, such responses have been split into two categories: instinctive and learned.

In **instinctive behavior**, components of the nervous system allow an animal to carry out complex, stereotyped responses to certain environmental cues, which are often simple. Consider the cuckoo, a social parasite. The adult females lay eggs in the nests of other bird species. Young cuckoos eliminate the natural-born offspring, then receive the undivided attention of their unsuspecting foster parents. Newly hatched cuckoos are blind, but when they contact an egg, they maneuver it onto their back, then push it out of the nest (Figure 38.3). This is an instinctive response, triggered by a well-defined, simple stimulus—and once it is set in motion, it is performed in its entirety. We call it a **fixed action pattern**.

Figure 38.3 (**a**) A newly hatched cuckoo making a complex, innate behavioral response to an environmental cue—spherical objects in the nest. The European cuckoo lays eggs in the nests of other species. Even before a cuckoo hatchling opens its eyes, it responds to the shape of the host's eggs by shoving them out of the nest. (**b**) The foster parents keep feeding the usurper, even when it has grown larger than they are.

With their tongue-flicking, orientation, and strikes, newborn garter snakes give us more examples of instinctive behavior. So do human infants. When they are only two or three weeks old, they smile instinctively when an adult's face comes close. The infants will make the same smiling response to a simple stimulus—a flat, face-size mask with two dark spots where eyes would be on a real human face. A mask with one "eye" will not do the trick. Older infants also show instinctive behaviors (Figure 38.4).

Yet animals also incorporate and process information that has been gained from specific experiences, then use the information to vary or change their responses to stimuli. Such responses are called **learned behavior**.

Think of a young toad, with special wiring in its brain that commands it to flip its sticky-tipped tongue instinctively at any dark object moving across its field of vision. In the toad world, such objects are usually edible insects. But suppose one object turns out to be a bumblebee that stings the tongue. The lesson of this experience might be "Bumblebee-size objects with black and yellow bands can sting." The toad learns to leave the bumblebees alone. Figure 38.5 gives other examples of learning.

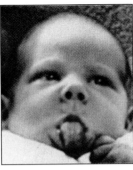

Figure 38.4 Examples of instinctive responses in humans. Older infants have the instinctive capacity to imitate facial expressions of adults.

a No one can tell these imprinted baby geese that Konrad Lorenz is not Mother Goose.

b An imprinted rooster wading out to meet the objects of his affections.

Figure 38.5 Imprinting and a few other traditional categories of learned behavior.

Imprinting. For many animals, this is a time-dependent form of learning that requires exposure to key stimuli, usually early in development. (**a**) As an example, baby geese formed an attachment to the ethologist Konrad Lorenz—or to any moving object—if they were separated from their mother shortly after hatching. But they did this only after being exposed to a moving object during a short, sensitive period early in life. At that time only, they were primed to learn a valuable bit of information—the identity of the individual they would follow in the months ahead. In nature, that individual is normally the mother or father.

Imprinting has later consequences for the sexual behavior of various birds. For example, many birds direct sexual behavior toward members of whatever species they had been sexually imprinted upon in their youth (normally their own, because they encounter their mother soon after hatching). The rooster shown in (**b**) had been imprinted on a mallard duck early in life. He courted ducks instead of hens of its own species.

Classical conditioning. Ivan Pavlov's experiments with dogs are an example. Dogs salivate just before eating. Pavlov's dogs were conditioned to salivate even in the absence of food. They did so in response to the sound of a bell or a flash of light that was initially associated with the presentation of food. In this case, the animals learned to associate an automatic, unconditioned response with a novel stimulus that does not normally trigger the response.

Operant conditioning. An animal learns to associate a voluntary activity with its consequences, as when a toad learns to avoid bad-tasting insects after voluntary attempts to eat them.

Habituation. Through experience, an animal learns not to respond to a situation if the response has no positive or negative consequences. Many of the birds living in cities learn not to flee from people who pose no real threat to them.

Spatial or latent learning. Through inspection of its environment, an animal acquires a mental map of a region, often by learning the position of local landmarks. Bluejays, for instance, can store information about the position of dozens, if not hundreds, of places where they have stashed food.

Insight learning. An animal abruptly solves a problem without trial-and-error attempts at the solution. Chimpanzees often exhibit insight learning in captivity when they suddenly solve a novel problem created for them by their captors. For example, some abruptly stacked together several boxes and used a stick to reach a bunch of bananas suspended out of their reach.

Genes and the Environment

Don't fall into the trap of separating instinctive and learned behavior, as if one is determined by genes and the other by the environment. Behavior is influenced by genes *and* the environment. For instance, male white-crowned sparrows in different habitats belt out variations ("dialects") of their species song. As Peter Marler found out, males acquire the full song by hearing it ten to fifty days after hatching. During this sensitive period, a learning mechanism is primed to receive a certain bit of information from the environment.

As Marler also discovered, these birds learn parts of the song by picking up acoustical cues from other males. He raised male nestlings to maturity in sound-proof chambers so they would not hear adult males singing. At maturity, the captives produced a song with none of the detailed structure of a typical adult's song. Marler also exposed isolated, captive males to recordings of white-crowned sparrows *and* song sparrows. At maturity, the captive males sang only the white-crown song (they even mimicked the dialects). Evidently, they must learn some—but not all—acoustical cues.

In a different experiment, young, hand-reared white-crowns were allowed to interact with a "social tutor," as opposed to listening to taped songs. They tended to learn the song of the alien species. Therefore, their learning mechanisms must be primed to be influenced *by social experience* as well as by acoustical cues.

Whether instinctive or learned, behavior is influenced by genes *and* by the environment.

THE ADAPTIVE VALUE OF BEHAVIOR

We turn now to evolutionary mechanisms by which diverse forms of behavior come about. We begin by defining a few terms used in animal behavior studies:

---

1. **Reproductive success** refers to the survival and production of the offspring of an individual.

2. **Adaptive behavior** is a behavior that promotes the propagation of an individual's genes and tends to occur at increased frequency in successive generations.

3. **Selfish behavior** is a behavior by which an individual protects or increases its own chance of producing offspring, regardless of the consequences for the group to which it belongs.

4. **Altruistic behavior** is self-sacrificing behavior. The individual behaves in a way that helps others but, by doing so, decreases its chance to produce its own offspring.

5. **Natural selection** is a measure of the difference in survival and reproduction among individuals that differ from one another in their heritable traits.

When behavioral biologists speak of a "selfish" or an "altruistic" individual, they don't mean the individual is consciously aware of what it is doing or of the reproductive goal of its behavior. A lion doesn't have to know that eating zebras is good for reproductive success. Its nervous system simply calls for hunting behavior when the lion is hungry and sees a vulnerable zebra.

Whether selfish or altruistic, a behavior persists in a population because the genes that are responsible for it are persisting also. Think about Norwegian lemmings, which disperse when population density skyrockets and food becomes scarce. As lemmings move out of an overcrowded habitat, many die. Are they committing suicide to help their species? Or are they merely dying by starvation, predation, and accidental drowning? A wonderfully instructive cartoon by Gary Larson shows lemmings plunging into water, presumably in the act of suicide—but one has an inflated inner tube about its waist! If lemmings tended to be that altruistic, only the "selfish" ones would reproduce—and genes underlying the altruistic behavior would disappear from the population. A more likely explanation is this: Overcrowded lemmings disperse to less crowded places, where *individuals* have a better chance to survive and reproduce.

In sum, when studying behavior, you usually will find it more profitable to look for evidence of natural (individual) selection than for something that benefits the species. A few examples of reproductive and feeding behavior will reinforce this point.

Selection Theory and Mating Behavior

Competition among members of one sex for access to mates is common. So is choosiness in selecting a mate. Both activities are forms of **sexual selection**. This microevolutionary process favors traits that give the individual a competitive edge in reproductive success.

Typically, male animals produce great numbers of tiny sperm, and the females produce considerably fewer but larger eggs. Generally speaking, reproductive success for a male depends on how many eggs he fertilizes. For a female, success depends largely on how many eggs she produces or how many offspring she can care for. The prime factor influencing her sexual preference usually is the quality of a mate, not the quantity of partners. With these points in mind, let's look at a few mating tactics in terms of sexual selection theory.

a

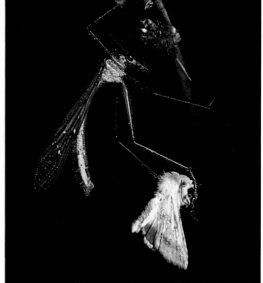

c

b

Figure 38.6 (**a**) Sexual competition between male bison, which are fighting for access to a cluster of females. (**b**) Dancing display by a male sage grouse, performed at his own vigorously defended spot inside a compact mating area called a lek. Females (the smaller brown birds) at the lek observe the prancing males before choosing the one they will mate with. (**c**) A male hangingfly dangling a moth as a nuptial present for a future mate. Females of certain hangingfly species choose sexual partners on the basis of the size of prey that males offer to them.

Male bighorn sheep sometimes have fierce head-butting contests (Figure 1.7*f*). Fighting wastes time and energy and may cause injuries, but reproductive benefits offset the costs. Males fight only to control areas where receptive females gather during the winter rutting season. Winners mate often, with many females. Losers will not challenge a stronger, larger male. But a gang of losers may invade his area and overwhelm his capacity to drive them all away. Losers attempt to mate—literally on the run. Even though the females are running away from them, they sometimes succeed.

We find competition for access to clusters of sexually receptive females in many other species as well. Thus competition for ready-made harems has favored combative male lions, elk, and bison (Figure 38.6*a*).

What if females do *not* cluster in defendable groups? Sage grouse females are widely dispersed through their prairie habitat in the western United States. The males make no attempt to maintain a large territory. During the breeding season they congregate in a lek, a communal display ground. There, each male stakes out a few square meters as his territory. Females are attracted to the lek—not to feed or nest, but rather to observe the males. With tail feathers erect and neck pouches puffed,

each male emits booming calls and stamps about like a big wind-up toy on his display ground (Figure 38.6*b*). Female sage grouse tend to select and mate with one male only. Then they go off to nest by themselves, unassisted by their sexual partner. Many females often choose the same male, so most of the males never mate.

In some species, females select males that offer them superior material benefits. Consider male hangingflies (*Harpobittacus apicalis*). They capture and kill a moth or some other insect, then release a chemical signal (a sex pheromone) that attracts females to this "nuptial gift" (Figure 38.6*c*). The females choose males with the larger, calorie-rich offerings. A female permits a male to mate with her—but only after she has eaten for about five minutes. Then she accepts and stores sperm in a reproductive organ—but only as long as the food holds out. At any point up to twenty minutes, she can break off the mating and leave. If so, she will mate again and accept another male's sperm—and diminish the future reproductive success of her first partner.

Thus, female hangingflies, female sage grouse, and females of many other species dictate the rules of male competition. For their part, the males employ tactics that may help them fertilize as many eggs as possible.

Selection Theory and Feeding Behavior

In northern forests, ravens scavenge carcasses of deer, elk, or moose, which are few and far between. When a raven comes across a carcass, even in winter when food is scarce, it often calls loudly, attracting a crowd of similarly hungry ravens. Bernd Heinrich found the calling behavior puzzling, for it seems to go against the caller's interests. Wouldn't a quiet raven get more food, increase its chance of surviving, and leave more descendants than an "unselfish" vocalizer? If the behavior is an outcome of natural selection, the cost of lost calories and nutrients must be offset by some reproductive benefit for the individual caller. But what benefit?

Maybe a lone bird picking at a carcass is vulnerable to predators that lie in wait for it. If so, other ravens called in could help keep an eye out for danger. But ravens are large and agile birds with very few known enemies. While stealthily watching ravens feed singly or in pairs at a carcass, Heinrich never saw predators attack any of them.

Then Heinrich realized territory has something to do with it. A **territory** is an area that one or more individuals defend against competitors. As Heinrich observed after hauling a cow carcass into a Maine forest, single or paired ravens don't always vocally advertise food. Maybe the silent birds were adults that had already staked out a large territory—which happened to include the spot where he put the carcass.

A pair of ravens would gain nothing by attracting others to their territory. But what if the forest had been subdivided into territories and was defended by powerful adults? A wandering young bird would be lucky to feed in an aggressive pair's territory. However, recruiting a gang of other, nonterritorial ravens might overwhelm the defensive behavior of the resident pair.

As it turns out, only the wandering ravens advertise carcasses. Basically, their calling behavior is selfish. It gives them a shot at otherwise off-limits food.

MECHANISMS OF SOCIAL LIFE

So far, we have considered a few examples of how the individual increases its chances of reproductive success. Let's now turn to the adaptive value of **social behavior**. The term refers to cooperative, interdependent relationships among animals of the same species. These may be lifelong or as brief as a one-time sexual encounter.

Functions of Communication Signals

Social behavior requires **communication signals**. These are actions or cues sent by one member of a species (the signaler) that can change the behavior of another mem-

Figure 38.7 Soldier termites defending their social colony by guarding a break in a foraging tunnel.

ber (the signal receiver). For example, signals help hold together a colony of termites in a forest in Queensland, Australia. Imagine yourself in the forest, chipping off a few fragments of a narrow, brittle tube running along the trunk of a dead eucalyptus tree. Inside, small, nearly white, worker termites bang their head against the tunnel wall. The vibrations alert brown termites, which run to the breach and make a defensive stand. Each soldier has a swollen, eyeless head that tapers into a long, pointed "nose" (Figure 38.7). You disturb one, and it shoots thin jets of silvery goo out of its nose! The strand releases volatile odors that attract even more soldiers to battle the danger, which most often is an invasion by ants.

Types of Communication Signals

Pheromones.   From earlier chapters, you know that the most ancient sensory systems dealt with chemical odors. During the evolution of most animals, certain chemicals began to be used in intraspecific communication. We call these communication signals **pheromones**. The two known categories are signaling and priming pheromones.

Signaling pheromones may bring about an immediate behavioral response from the receiver. Some of these stimulate or suppress aggressive or defensive behavior. Termite alarm signals are like this. Other signaling pheromones are sex attractants, such as the one that

Figure 38.8 Threat display by a male baboon. Exposing the canines is a visual signal of aggressive intent. The display can resolve a conflict without actual fighting.

Figure 38.9 Courtship displays among the albatross. (**a**) This male spreads his wings and points his head to the sky in a visual signal to the female. (**b–d**) Mutual displays between this male and female involve visual, acoustical, and tactile components. These displays precede copulation.

works for male hangingflies. *Priming* pheromones, as discovered in rodents, elicit a generalized physiological response. For example, a chemical odor in the urine of male mice triggers and enhances estrus in female mice.

Visual Signals. Animals that are active in daytime typically use observable actions or cues, called **visual signals**. For instance, a dominant male baboon "yawns" and exposes formidable canines when confronted by a rival for a receptive female (Figure 38.8). His yawn is a threat display that may precede a physical attack. Suppose the rival backs down. The signaler benefits by gaining access to the female without a fight. The signal receiver also benefits; the visual signal allowed him to avoid a serious beating, infection, and possibly death.

Visual signals are prominent in **courtship displays**, by which individuals assess and respond to sexual overtures of potential partners. Lekking sage grouse and albatross are two examples (Figures 38.6 and 38.9). Some animals that are active at night use visual signals called bioluminescent flashes (page 68). Fireflies have light-generating organs at the tip of the abdomen. A male emits a complex signal as he flies about. A few seconds later, a receptive female of the same species may answer with a simple flash. The male and female flash back and forth as the male approaches and locates her.

Females of some predatory fireflies are *illegitimate* signalers. They have broken the communication code between males and females of other firefly species. When a predatory female spots a signaling male, she waits for the correct interval, then flashes back. The male may be lured into attack range, in which case he becomes a meal instead of a mate. This is an evolutionary cost of having a come-hither signaling mechanism. Although a male might live a long time by ignoring the signal, he would not be reproductively successful.

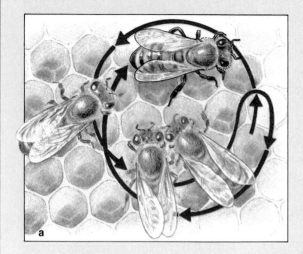

a

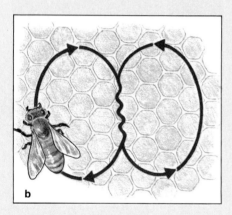

b

Figure 38.10 Dances of honeybees, which convey information through tactile signals. (**a**) Bees trained to visit feeding stations close to a hive perform a round dance on the honeycomb. Worker bees that maintain contact with the forager through the dance will search for food close to the hive. (**b**) Bees trained to visit feeding stations more than 100 meters from the hive perform a waggle dance. During the portion of the dance when the bee moves in a straight line, it waggles its abdomen. The slower the waggles, the more distant the food source.

(**c**) As discovered by Karl von Frisch, the orientation of the straight run varies, depending on the direction in which food is located. When he put a dish of honey on a direct line between the hive and the sun, foragers that located it returned to the hive and oriented their straight runs right up the honeycomb. When he put the honey at right angles to a line drawn between the hive and the sun, the foragers made their straight runs at 90 degrees to vertical. Thus, a honeybee "recruited" into searching for food could orient its flight *with respect to the sun and the hive*—and so waste less time and energy during the search. Waggle dancers also vary the speed of their dance in relation to the distance of the food source from the hive. When a site is 150 meters from a hive, the dance is much faster, with more waggles per straight run, compared with a dance concerning a food source that is 500 meters away.

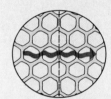

When bee moves straight up comb, recruits fly straight toward the sun.

When bee moves straight down comb, recruits fly to source directly away from the sun.

When bee moves to right of vertical, recruits fly at 90° angle to right of the sun.

c

Acoustical Signals. Many animals communicate by sounds that have precise, species-specific information. These are **acoustical signals**. The distinctive song of the male white-throated sparrow is an example; it attracts mates and secures territory. Males of many frog species also send acoustical signals, often at night, that convey information to rival males and receptive females.

The fringe-lipped bat is an illegitimate receiver of the tungara frog's acoustical signal. The frog's call is like a whine followed by a "chuck." The "chuck" attracts females of the frog's species. But the bat also tracks the call to its source, swoops down and sweeps the frog from the water, then eats it.

Tactile Signals. A handshake, hug, caress, or shove is an example of a **tactile signal**, in which a signaler physically touches the receiver in socially significant ways. Such signals are important in honeybee societies.

After finding a rich source of pollen or nectar, a foraging honeybee returns to the hive and performs a complex dance. It moves in circles, jostling in the dark through a crowded mass of workers on a honeycomb. Other bees may follow and maintain physical contact with the dancer. In this way, they may acquire information about where the pollen or nectar is located.

When a forager has located food close to the hive, it performs a "round dance" (Figure 38.10a). Workers that follow a round-dancing forager do not know the precise location of the nectar or pollen source. But they smell the scent of certain flowers on the dancer's body. They search in the neighborhood for that scent and find food more quickly than they would by searching randomly.

When food is far from the hive, a forager performs a "waggle dance." As Figure 38.10b shows, waggle-dancers communicate information about distance *and* direction.

Figure 38.11 Social defensive behavior by musk oxen, which form a "ring of horns" against predators.

COSTS AND BENEFITS OF SOCIAL LIFE

A survey of the animal kingdom reveals a considerable range of social behavior. Members of some species are largely solitary or live in small family groups. Members of others, including termites and honeybees, live in huge groups of thousands of related individuals. Others, like modern humans, live in large social units with many unrelated individuals. What is the basis for the diversity? To find answers, evolutionary biologists may take a cost-benefit approach. They can measure the costs and benefits of social life in terms of the individual's reproductive success.

Figure 38.12 Social defensive behavior by Australian sawfly caterpillars. These have smeared chemical secretions on one another in response to a disturbance.

Advantages to Sociality

Cooperative Predator Avoidance. A group of animals acting cooperatively against a predator can reduce the net risk to any one individual. Simply having more pairs of eyes around helps individuals detect predators sooner. Individuals also may engage in a group counterattack or combine their defenses, as the example in Figure 38.11 shows.

The biologist Birgitta Sillen-Tullberg studied Australian sawfly caterpillars, which live in clumps (Figure 38.12). When disturbed, the caterpillars collectively rear up, writhe about, and regurgitate partially digested eucalyptus leaves. Such leaves contain chemical compounds that are toxic to most animals, including insect-eating songbirds. Data from Sillen-Tullberg's experiments indicate that individuals of this sawfly species are somewhat safer in a group than on their own.

The Selfish Herd. Group living counters predation in another way. Simply by their physical location in the group, some individuals become living shields against predators. They are part of a **selfish herd**, a simple society held together by reproductive self-interest.

Bluegill sunfish are an example. Male bluegills build adjacent nests on lake bottoms. Each male uses its fins to hollow a depression in the muddy sediments. There his mate or mates deposit eggs. Compared to the periphery, eggs laid in nests at the center of the nesting area are not attacked as much by snails and largemouth bass. The largest, most powerful males claim the central locations. Other, smaller males assemble around them and bear the brunt of predatory attacks. Even then, they are better off in the group than on their own, fending off a bass singlehandedly, so to speak.

Figure 38.13 A colony of royal penguins on Macquarie Island, between New Zealand and Antarctica.

Figure 38.14 Appeasement behavior between baboons. Notice the assured position of the dominant animal and the abject stare and groveling posture of the subordinate one, who is making little conciliatory smacking noises with its lips.

Disadvantages to Sociality

In certain environments, the benefits of social life come at a cost, as the following examples suggest.

Consider the reproductive costs in a nesting colony of herring gulls. If given an opportunity, each breeding pair may cannibalize the eggs or young chicks of their neighbors in an instant. Besides this, a huge population of birds puts a greater strain on resource availability.

Living in the huge colonies of cliff swallows, prairie dogs, or penguins has another cost (Figure 38.13). In these social groups, an individual or its offspring are more likely to be weakened by parasites or contagious diseases, which can be transmitted easily from host to host. Contagious diseases also spread like wildfire among humans who live in crowded cities (page 545).

Self-Sacrifice in Dominance Hierarchies

Members of a selfish herd make no personal sacrifice for the others. The personal benefits of living with others simply seem to outweigh the costs. By contrast, members of some social groups *help* other individuals survive and reproduce at personal cost. Yet self-sacrificing behavior may be more apparent than real. Individuals may not be giving up their chance at reproductive success *entirely*.

Individuals in wolf packs or baboon troops help others, as by sharing food or fending off predators, but reproductive opportunity is lopsided. Usually, only *one* male and *one* female of a wolf pack produce pups. Some male baboons relinquish safe sleeping sites, choice bits of food, even receptive females to others upon receiving a threat signal from another male. In both social groups, a **dominance hierarchy** exists in which some individuals have adopted a subordinate status to others (Figures 38.14 and 38.15). Page 495 provides a detailed glimpse into one such hierarchy.

Why should subordinate, nonbreeding adults remain in a group and make sacrifices for their dominant peers? As for small bluegill sunfish, their benefits from group living may offset the sacrifices. Besides, challenging a stronger member could lead to life-shortening injury. It may not be possible to survive alone, outside of the group. A solitary baboon surely quickens the pulse of the first leopard that sees it.

Just as importantly, self-sacrificing behavior may give subordinates a chance to reproduce *if* they live long enough and *if* predation, weakness at old age, or some other event removes dominant peers in the hierarchy. Some subordinate wolves and baboons do indeed move up the social ladder when dominant members slip down a rung or fall off. Thus acceptance of subordinate status may pay off in the long term for the patient individual.

Costs and Benefits of Parenting

In some species, parents take care of their offspring until the offspring can survive on their own. Adult Caspian terns (Figure 38.16) incubate the eggs, shelter nestlings, feed them, and accompany and protect them after they start flying. Their parental behavior drains time and energy that might otherwise be allocated to improving their own chance of living to reproduce another time.

Yet parental behavior benefits the individual by improving the likelihood that the current generation of offspring will survive. The benefit of immediate reproductive success may outweigh the cost of reduced reproductive success at some later time.

Figure 38.15 A dominant male and female member of a wolf pack. Typically, the dominant male is the only pack member that breeds successfully.

Figure 38.16 Male and female Caspian terns, protecting their chicks. Parental care has costs as well as benefits.

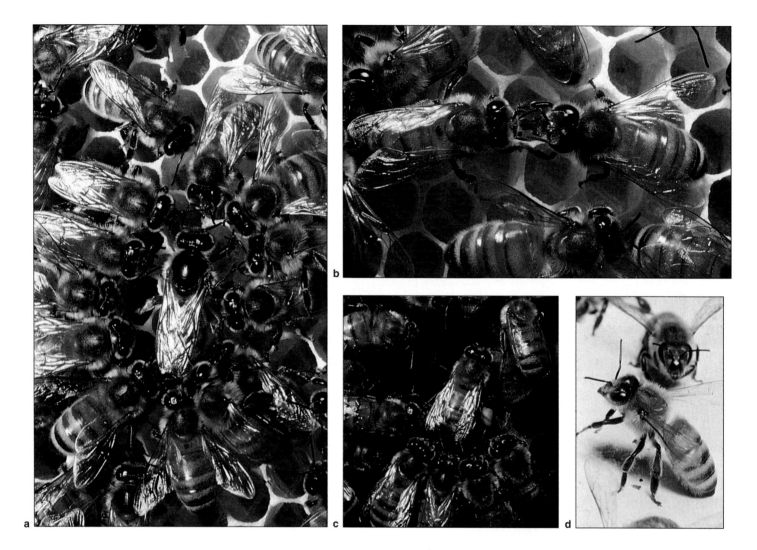

Figure 38.17 Life in a honeybee colony. (**a**) This queen bee, the only fertile female in the hive, is surrounded by her court of sterile worker daughters. Daughters feed her and relay her pheromone throughout the hive. The pheromone influences the activity of all members of the colony.

(**b**) Transfer of food from bee to bee, a helpful action in honeybee society. (**c**) Bee dance that recruits workers into searching for food, as described in Figure 38.10. (**d**) Guard bees. Worker females at the colony entrance, positioned to repel intruders.

(**e**) The queen is much larger than the workers, in part because her ovaries are fully developed (unlike those of sterile daughters). (**f**) Stingless drones are produced at certain times of the year. They do not work for the colony but instead attempt to mate with queens of other hives.

(**g**) Worker bees forage, feed larvae, guard the colony, construct honeycomb, and clean and maintain the nest. Between 30,000

and 50,000 are present in a colony. They live about six weeks in the spring and summer, and they can survive about four months in an overwintering colony.

(**h**) Scent-fanning, another cooperative action by a worker. Fanned air passes over the bee's exposed scent gland. Pheromones released from the gland help other bees orient to the colony entrance. (**i**) Worker bees construct new honeycomb from wax secretions. Here, honey or pollen is stored or new generations are cared for from the egg stage to the emergence of the adult.

(**j**) Initial stages in the honeybee life cycle, revealing eggs and larvae of various ages. Larvae are fed by young worker bees. (**k**) Worker pupae. Cell caps have been removed, exposing pupae that will metamorphose into future workers. (**l**) Sequence of developmental stages of a worker bee, from the egg to a six-day-old larva (fourth from top), to a twenty-one-day-old pupa about to become an adult.

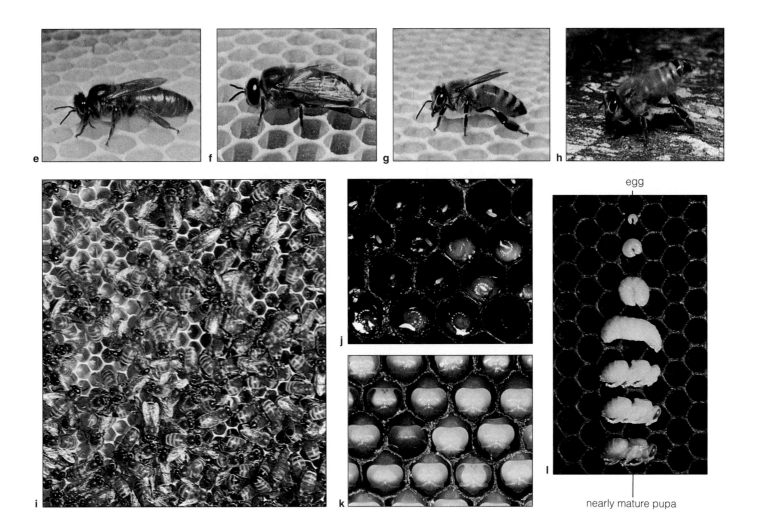

egg

nearly mature pupa

THE EVOLUTION OF ALTRUISM

A termite soldier and a honeybee worker are sterile, incapable of reproducing at any time, yet they help others at great cost to themselves. How does the genetic basis for this altruistic behavior persist over evolutionary time? By one theory, individuals can *indirectly* pass on their genes by helping relatives survive and reproduce. This is especially true of sterile workers in insect societies.

When a sexually reproducing parent helps offspring, it is not helping exact genetic copies of itself. Each offspring has only half of its genes. Other individuals share common genes as a result of having the same ancestors. Genetically, two siblings are as similar as a parent and one of its offspring. An uncle and a nephew or a niece will be alike in about one-fourth of their genes.

According to William Hamilton's theory of **indirect selection**, caring for nondescendant relatives favors the genes associated with helpful behavior. Think of it as an extension of parental behavior. For example, suppose an uncle helps a niece or nephew survive long enough to reproduce. He has effectively made a genetic contribution to the next generation, even though it is an indirect one. His contribution can be measured in the genes that he and his relative share. Altruism costs him; he may lose his own opportunities to reproduce. But if the cost is less than the benefit, the action will propagate the uncle's genes and favor the spread of his kind of altruism in the species. If an uncle saves two nieces, this is equivalent to saving his own daughter.

Among social insects, workers indirectly promote their "self-sacrifice" genes through altruistic behavior directed toward relatives. Colonies of honeybees, ants, and termites are actually large families (Figure 38.17). The family's worker force labors on behalf of siblings, some of which are future kings and queens. When a guard bee drives her stinger into a raccoon, she dies— but siblings perpetuate some of her genes.

Sterility and extreme self-sacrifice are rare among social vertebrates. Among the known exceptions are naked mole-rats, described in the *Focus* essay on the following page.

About Those Self-Sacrificing Naked Mole-Rats

Naked mole-rats look rather like bucktoothed sausages with wrinkled, pink skin from which a few hairs sprout (Figure *a*). Their behavior is as fascinating as their looks. Unlike any other known vertebrate, many naked mole-rat individuals spend their entire lives as nonbreeding helpers in their social group.

These highly social mammals live underground in certain arid regions of eastern Africa. They are always found in colonies of about 25 to 300 members. Each colony has only a single reproducing female that mates with one to three males. All other members of the colony care for the "queen" and "king" (or kings) and their offspring.

The nonreproducers dig extensive subterranean tunnels and special chambers that serve as living rooms or waste-disposal centers. They also locate and chop up large, underground tubers—the food for the colony. Workers deliver bits of root to the queen and her retinue of males and offspring. They also deliver food to some other helpers that spend time loafing about, shoulder to shoulder (and belly to back), with the reproductives. The "loafers" are usually larger than the diggers. They spring to action when a snake or some other enemy threatens the colony. At great personal risk, they collectively chase away or kill the predator.

How can such helpful behavior persist if helpful members fail to reproduce—and so fail to have offspring that carry on the genetic basis for helping? It can if the helpers

are related to the offspring. Applying the theory of indirect selection to naked mole-rats, let's formulate a hypothesis with a testable prediction:

1. If helping is genetically advantageous to some individuals in a naked mole-rat colony (*the hypothesis*), then it follows that the helpers will be related to the reproductive members of the colony that benefit from their altruism (*the prediction*).

2. Determine the genetic relationships among subordinate, sterile helpers and the dominant, reproducing queen and king or kings. (*This would be an appropriate test of the prediction.*)

Could the hypothesis be tested by marking each individual in a naked mole-rat colony and gradually establishing a family tree for the group? Probably not. It isn't likely that all the individuals could even be identified, given all the intricate, hidden tunnels.

Instead, H. Kern Reeve and his colleagues used *DNA fingerprinting*, a method of establishing degrees of genetic relatedness among individuals. The method uses restriction fragments, of the sort described on page 182. A visual

HUMAN SOCIAL BEHAVIOR

If we can analyze the behavior of termites and naked mole-rats, is it possible to analyze human behavior also? There is resistance to the idea. Apparently, many people believe that attempts to identify the adaptive value of a particular human trait is an attempt to define its moral or social advantage. But clearly there is a difference between trying to explain something by reference to its evolutionary history and trying to justify it. "Adaptive" does not mean "moral." It only means valuable in the transmission of an individual's genes.

Few are bothered by the concept of adaptive value when discussing the behavior of animals other than humans. No one has complained to Doug Mock about his study of siblicide among baby egrets, the larger of

which make lethal assaults on smaller nestmates. Mock tested the hypothesis that the lethal assaults are adaptive for both the victorious chick and the nestlings' parents. He found that by eliminating the competition for food, large siblings had a better chance of fledging. The parents also benefited. When food is scarce, an attempt to rear too many chicks fails. Mock was not attempting to justify siblicide by the egrets or to claim that the attacks were moral in some sense.

Hypotheses about "selfish" and "altruistic" behavior of humans can be similarly tested without attempts to justify the behavior. For example, adopting a baby is a dramatic example of some humans helping others. If we seek to explain why some people adopt children, we might use an evolutionary approach to generate testable hypotheses. Whether adoption is moral or

a

record can be constructed of different sets of fragments from the DNA of different individuals. Identical twins have essentially identical DNA fingerprints (their DNA has the same base sequence). Individuals with the same father and mother have similar DNA—thus similar but not identical DNA fingerprints. On the average, the DNA fingerprints of genetically unrelated individuals should differ much more than those of siblings or other relatives.

DNA fingerprints were constructed for naked mole-rats. As it turns out, individuals from a given colony are very close relatives, and they are very different genetically from members of other colonies. These findings suggest that each naked mole-rat colony is highly inbred, a result of generations of brother-sister, mother-son, or father-daughter matings. Such lineages have extremely reduced genetic variability.

Therefore, an altruistic naked mole-rat is helping to perpetuate a very high proportion of the forms of genes (alleles) that it carries. The genotypes of helpers and the helped might be 90 percent identical!

socially desirable is an entirely separate issue, about which biologists have no more to say than anybody else.

How would adults gain genetic representation in the next generation by adopting someone else's offspring? According to theories of natural selection and indirect selection, individuals act in ways that promote their genetic self-interest. We might therefore predict that parents with some dependent, care-requiring children of their own will be much less likely to adopt, compared to adults who have no children or who have already raised children and are now living by themselves. The prediction could be tested by securing a sufficient sample of data, ideally from a variety of human cultures, on the relationship between childlessness and adoption.

Yet adoption might also be adaptive for an individual under some conditions. Indirect selection would favor adults who directed parenting assistance to relatives, thereby indirectly perpetuating their shared genes. We might even predict that such adults are more likely to be related to the adopted child than would be expected by chance alone.

Joan Silk tested this prediction about a certain human behavior. She showed that in some traditional societies, children are indeed adopted overwhelmingly by relatives. In modern societies in which agencies and other means of assisting adoption exist, adoption of relatives is not predominant.

The point is that evolutionary hypotheses can be tested. In so doing, we gain additional understanding about the evolution of our behavior.

SUMMARY

1. The observable, coordinated responses an animal makes to external and internal stimuli are called animal behavior. Genes provide instructions for the development of the neural and endocrine systems, which govern animal behavior.

2. Instinctive and learned behavior are both based on genes and environmental inputs. Behavior commonly is adaptive. Its heritable component makes it subject to evolution by natural selection.

3. Social behavior depends on communication among animals living together. Communication signals are actions or cues between members of the same species, in that they can change the behavior of the signal's receiver according to their information content.

4. Signaling animals employ many different sensory modes as channels of communication. Among these are chemical, visual, acoustical, and tactile signals.

5. Sociality carries high risks of contagious disease and competition for limited resources. Benefits outweigh the costs of social life under some circumstances, as when predation pressure is severe.

6. In most animal societies, individuals do not sacrifice their reproductive chances to help others in their group.

   a. Helpful acts often may be the result of dominant individuals forcing subordinates to relinquish food or some other resource.

   b. When subordinates give in to dominant members of their group, they may receive compensatory benefits of group living, such as safety from predators.

   c. In some species, subordinates may eventually reproduce if they live long enough and if dominant ones slip in the hierarchy or die off.

7. Indirect selection can lead to extreme altruism. This is evident among workers of some species of social insects and naked mole-rats. Such workers normally do not reproduce. Instead, they help their reproducing siblings survive and thereby pass on "by proxy" the genes underlying the development of altruism.

Review Questions

1. What role does the environment play in the development of an instinct, such as the egg-ejecting behavior of baby cuckoos? *630*

2. A hyena places scent marks on vegetation in its territory by releasing specific chemicals from certain glands. What evidence would you need to demonstrate that this action is an evolved communication signal? *634*

3. Explain how a threat display can be considered an example of cooperation. *635*

4. A very large moth caterpillar resting on a vine in the tropics responds to being poked by partly letting go of the vine and puffing up part of its body:

Provide a hypothesis on the underlying mechanism that enables the caterpillar to behave this way. Provide a hypothesis on the possible adaptive value of the trait. How would you test both hypotheses? *629–630, 632*

5. Suppose you are traveling in Africa and you see a black heron standing in the water with its wings held over its head like an umbrella, as shown below:

You might suppose that, like other herons, it is on the lookout for fish. But most other herons keep their wings down when they are hunting. Possibly the black heron is creating shade on the water, and perhaps minnows are being drawn to the shade because it appears to offer shelter. How would you go about testing this hypothesis?

6. Why don't the members of a "selfish herd" live apart if each member is "trying to take advantage" of others? *637*

7. How can a self-sacrificing individual of a social group still pass on more of its genes than a noncooperative one? *641*

Self-Quiz *(Answers in Appendix IV)*

1. "Starlings festoon their nests with sprigs of wild carrot and so interfere with the development of nest mites." This statement is _____ .
   a. an untested hypothesis
   b. a prediction
   c. a test of a hypothesis
   d. a proximate conclusion

2. Genes affect the behavior of individuals by _____ .
   a. influencing the development of nervous systems
   b. affecting the kinds of hormones in individuals
   c. governing the development of muscles and skeletons
   d. all of the above

3. Many kinds of female mammals live in herds. Which of these mating systems might you see in such herds?
   a. a lek mating system
   b. males competing for clusters of receptive females
   c. territorial defense of feeding sites by males
   d. males capturing food to lure females

4. A communication signal affects the behavior of _____ .
   a. legitimate receivers
   b. illegitimate receivers
   c. illegitimate signalers
   d. both a and b

5. Social behavior evolves because _____ .
   a. social species are more advanced evolutionarily than solitary ones
   b. under some ecological conditions, the costs of social life are less than its benefits to the group
   c. under some ecological conditions, the costs of social life are less than its benefits to the individual
   d. predator pressures always favor social living

6. Which is an evolutionary cost of a loud acoustical signal?
   a. the energy expended by the caller
   b. exploitation of the call by an illegitimate receiver
   c. the time spent calling that cannot be spent feeding
   d. all of the above

7. The theory of evolution by indirect selection _____ .
   a. focuses on evolution for group benefit
   b. explains that altruism is always adaptive
   c. examines the consequences of the effects of differences among individuals on the reproductive success of relatives
   d. shows how families with many descendants are superior to self-sacrificing families

8. The genetic similarity between an uncle and nephew _____ .
   a. is the same as between a parent and his offspring
   b. is greater than between two full siblings
   c. depends on how many other nephews the uncle has
   d. is less than that between a mother and daughter

9. Match the animal behavior concepts.
   ____ communication signals
   ____ altruism
   ____ basis of instinctive and learned behavior
   ____ animal behavior
   ____ risks of sociality

   a. communicable disease, exploitation, and resource competition
   b. genes and environmental influences
   c. coordinated responses to external and internal stimuli
   d. information transfer in social species
   e. assisting another individual at one's own expense

Selected Key Terms

acoustical signal *636*
adaptive behavior *632*
altruistic behavior *632*
animal behavior *629*
communication signal *634*
courtship display *635*
dominance hierarchy *639*
fixed action pattern *630*
indirect selection *641*
instinctive behavior *630*
learned behavior *630*

natural selection *632*
pheromone *634*
reproductive success *632*
selfish behavior *632*
selfish herd *637*
sexual selection *632*
social behavior *634*
sound system *630*
tactile signal *636*
territory *634*
visual signal *635*

Readings

Alcock, J. 1989. *Animal Behavior*. Second edition. Sunderland, Massachusetts: Sinauer.

Daly, M., and M. Wilson. 1983. *Sex, Evolution, and Behavior*. Second edition. Boston: Willard Grant Press.

Dawkins, R. 1989. *The Selfish Gene*. New York: Oxford University Press.

Frisch, K. von. 1961. *The Dancing Bees*. New York: Harcourt Brace Jovanovich.

APPENDIX I
A Classification Scheme

The following classification scheme is a composite of several that are used in microbiology, botany, and zoology. The major groupings are more or less agreed upon. There is not always agreement on what to call a given grouping or where it fits in the overall hierarchy. There are several reasons for this.

First, the fossil record varies in its quality and completeness. Therefore, the relationship of one group to others is sometimes open to interpretation. Comparative studies at the molecular level are firming up the picture, but this work is still under way.

Second, since the time of Linnaeus, classification schemes have been based on perceived morphological similarities and differences among organisms. Although some original interpretations are now open to question, we are so used to thinking about organisms in certain ways that reclassification proceeds slowly. Traditionally, for example, birds and reptiles are separate classes (Reptilia and Aves). Yet there are compelling arguments for grouping lizards and snakes as one class, and crocodilians, dinosaurs, and birds as another.

Finally, microbiologists, mycologists, botanists, zoologists, and other researchers have inherited a wealth of literature based on classification schemes peculiar to their fields. Most see no good reason to give up established terminology and so disrupt access to the past. Thus botanists continue to use *division* and zoologists use *phylum* for groupings that are equivalent in the hierarchy. Opinions are polarized with respect to an entire kingdom (the Protista), certain members of which could just as easily be called single-celled plants, fungi, or animals. Indeed, the term protozoan is a holdover from earlier schemes that ranked the amoebas and some other single cells as simple animals.

Given the problems, why do we bother imposing artificial frameworks on the history of life? We do this for the same reason that a writer might decide to break up the history of civilization into several volumes, many chapters, and many more paragraphs. Both efforts are attempts to impart structure to what might otherwise be an overwhelming body of information.

Bear in mind, we include this classification scheme mainly for your reference purposes. It is by no means complete. Numerous existing and extinct organisms of the so-called "lesser" phyla are not represented. Our strategy is to focus mainly on organisms mentioned in the text. The italic numerals included with the entries refer to some of the pages on which representatives are illustrated or described. A few examples of organisms also are listed under the entries.

SUPERKINGDOM PROKARYOTA.   Prokaryotes. Single-celled organisms with their DNA concentrated in a cytoplasmic region, not in a membrane-bound nucleus.

KINGDOM MONERA. Bacteria, either single cells or simple associations of cells. Both autotrophs and heterotrophs (refer to Table 17.3). *Bergey's Manual of Systematic Bacteriology*, the authoritative reference in the field, calls this "a time of taxonomic transition." It groups bacteria mainly on the basis of form, physiology, and behavior, not on phylogeny. The scheme presented here does reflect the growing evidence of evolutionary relationships for at least some bacterial groups.

SUBKINGDOM ARCHAEBACTERIA. Methanogens, halophiles, thermophiles. Strict anaerobes, distinct from other bacteria in cell wall, membrane lipids, ribosomes, and RNA sequences. *Methanobacterium, Halobacterium, Sulfolobus.*

SUBKINGDOM EUBACTERIA. Gram-negative and Gram-positive forms. Peptidoglycan in cell wall. Photosynthetic autotrophs, chemosynthetic autotrophs, and heterotrophs.

DIVISION GRACILICUTES.   Typical Gram-negative, thin wall. Autotrophs (photosynthetic and chemosynthetic) and heterotrophs. *Anabaena* and other cyanobacteria. *Escherichia, Pseudomonas, Neisseria, Myxococcus.*

DIVISION FIRMICUTES.   Typical Gram-positive, thick wall. Heterotrophs. *Staphylococcus, Streptococcus, Clostridium, Bacillus, Actinomycetes.*

DIVISION TENERICUTES.   Gram-negative, wall absent. Heterotrophs (saprobes, pathogens). *Mycoplasma.*

SUPERKINGDOM EUKARYOTA.   Eukaryotes, single-celled and multicelled organisms. Cells characteristically have a nucleus (enclosing the DNA) and other membrane-bound organelles.

KINGDOM PROTISTA.   Mostly single-celled eukaryotes, some colonial forms. Diverse autotrophs and heterotrophs. Many lineages apparently related evolutionarily to certain plants, fungi, and possibly animals.

PHYLUM MYXOMYCOTA.   Heterotrophs. Plasmodial slime molds. *Physarum.*

PHYLUM GYMNOMYCOTA.   Heterotrophs. Cellular slime molds. *Dictyostelium.*

PHYLUM EUGLENOPHYTA.   Euglenoids. Mostly heterotrophs, but some photosynthetic types. Flagellated. *Euglena.*

PHYLUM CHRYSOPHYTA.   Golden algae, yellow-green algae, diatoms. Photosynthetic. Some flagellated, others not. *Mischococcus, Synura.*

PHYLUM PYRRHOPHYTA.   Dinoflagellates. Photosynthetic, mostly, but some heterotrophs. *Ptychodiscus.*

PHYLUM MASTIGOPHORA.   Flagellated protozoans. Heterotrophs. *Trypanosoma, Trichomonas, Giardia.*

PHYLUM SARCODINA.   Amoeboid protozoans. Heterotrophs. Amoebas, foraminiferans, radiolarians, heliozoans. *Amoeba, Entomoeba.*

SPOROZOANS.   Parasitic protozoans, many intracellular. "Sporozoans" is the common name for these diverse organisms; it has no formal taxonomic status. *Plasmodium.*

KINGDOM FUNGI. Mostly multicelled eukaryotes. Heterotrophs (mostly saprobes, some parasites). Major decomposers of nearly all communities. Reliance on extracellular digestion of organic matter and absorption of nutrients by individual cells.

DIVISION MASTIGOMYCOTA.   All produce flagellated spores.
  Class Chytridiomycetes. Chytrids.
  Class Oomycetes. Water molds, related forms. *Plasmopora, Phytophthora, Saprolegnia.*
DIVISION AMASTIGOMYCOTA.   All produce nonmotile spores.
  Class Zygomycetes. Bread molds, related forms. *Rhizopus, Pilobolus.*
  Class Ascomycetes. Sac fungi. Most yeasts and molds; morels, truffles. *Saccharomycetes, Morchella.*
  Class B asidiomycetes. Club fungi. Mushrooms, shelf fungi, stinkhorns. *Agaricus, Amanita.*
IMPERFECT FUNGI.   Sexual spores absent or undetected. The group has no formal taxonomic status. If better understood, a given species might be grouped with sac fungi or club fungi. *Verticillium, Candida.*

KINGDOM PLANTAE. Nearly all multicelled eukaryotes. Photosynthetic autotrophs, except for a few parasitic types.

DIVISION RHODOPHYTA.   Red algae. *Porphyra.*
DIVISION PHAEOPHYTA.   Brown algae. *Macrocystis, Fucus, Sargassum.*
DIVISION CHLOROPHYTA.   Green algae. *Chlamydomonas, Spirogyra, Ulva.*
DIVISION CHAROPHYTA.   Stoneworts.
DIVISION BRYOPHYTA.   Liverworts, hornworts, mosses. *Marchantia, Sphagnum.*
DIVISION RHYNIOPHYTA.   Earliest known vascular plants; extinct. *Cooksonia, Rhynia.*
DIVISION PSILOPHYTA.   Whisk ferns.
DIVISION LYCOPHYTA.   Lycophytes, club mosses. *Lycopodium, Selaginella.*
DIVISION SPHENOPHYTA.   Horsetails. *Equisetum.*
DIVISION PTEROPHYTA.   Ferns.
DIVISION PROGYMNOSPERMOPHYTA. Progymnosperms. Ancestral to early seed-bearing plants; extinct. *Archaeopteris.*
DIVISION PTERIDOSPERMOPHYTA.   Seed ferns. Fernlike gymnosperms; extinct.
DIVISION CYCADOPHYTA.   Cycads. *Zamia.*
DIVISION GINKGOPHYTA.   Ginkgo. *Ginkgo.*
DIVISION GNETOPHYTA.   Gnetophytes. *Ephedra, Welwitschia.*
DIVISION CONIFEROPHYTA.   Conifers.
  Family Pinaceae. Pines, firs, spruces, hemlock, larches, Douglas firs, true cedars. *Pinus.*
  Family Cupressaceae. Junipers, cypresses, false cedars. *Juniperus.*
  Family Taxodiaceae. Bald cypress, redwoods, Sierra bigtree, dawn redwood. *Sequoia.*
  Family Taxaceae. Yews.
DIVISION ANTHOPHYTA.   Flowering plants.
  Class Dicotyledonae. Dicotyledons (dicots). Some families of several different orders are listed:
  Family Nymphaeaceae. Water lilies.
  Family Papaveraceae. Poppies.
  Family Brassicaceae. Mustards, cabbages, radishes.
  Family Malvaceae. Mallows, cotton, okra, hibiscus.
  Family Solanaceae. Potatoes, eggplant, petunias.
  Family Salicaceae. Willows, poplars.
  Family Rosaceae. Roses, peaches, apples, almonds, strawberries.
  Family Fabaceae. Peas, beans, lupines, mesquite.
  Family Cactaceae. Cacti.
  Family Euphorbiaceae. Spurges, poinsettia.
  Family Cucurbitaceae. Gourds, melons, cucumbers, squashes.
  Family Apiaceae. Parsleys, carrots, poison hemlock.
  Family Aceraceae. Maples.
  Family Asteraceae. Composites. Chrysanthemums, sunflowers, lettuces, dandelions.
  Class Monocotyledonae. Monocotyledons (monocots). Some families of several different orders are listed:
  Family Liliaceae. Lilies, hyacinths, tulips, onions, garlic.
  Family Iridaceae. Irises, gladioli, crocuses.
  Family Orchidaceae. Orchids.
  Family Arecaceae. Date palms, coconut palms.
  Family Cyperaceae. Sedges.
  Family Poaceae. Grasses, bamboos, corn, wheat, sugarcane.
  Family Bromeliaceae. Bromeliads, pineapples, Spanish moss.

KINGDOM ANIMALIA.   Multicelled eukaryotes. Heterotrophs (herbivores, carnivores, omnivores, parasites, detritivores).

PHYLUM PLACOZOA.   Small, organless marine animal. *Trichoplax.*
PHYLUM MESOZOA.   Ciliated, wormlike parasites, about the same level of complexity as *Trichoplax.*
PHYLUM PORIFERA.   Sponges.
PHYLUM CNIDARIA.
  Class Hydrozoa. Hydrozoans. *Hydra, Obelia, Physalia.*
  Class Scyphozoa. Jellyfishes. *Aurelia.*
  Class Anthozoa. Sea anemones, corals. *Telesto.*
PHYLUM CTENOPHORA.   Comb jellies. *Pleurobrachia.*
PHYLUM PLATYHELMINTHES.   Flatworms.
  Class Turbellaria. Triclads (planarians), polyclads. *Dugesia.*
  Class Trematoda. Flukes. *Schistosoma.*
  Class Cestoda. Tapeworms. *Taenia.*
PHYLUM NEMERTEA.   Ribbon worms.
PHYLUM NEMATODA.   Roundworms. *Ascaris, Trichinella.*
PHYLUM ROTIFERA.   Rotifers.
PHYLUM MOLLUSCA.   Mollusks.
  Class Polyplacophora. Chitons.
  Class Gastropoda. Snails (periwinkles, whelks, limpets, abalones, cowries, conches, nudibranchs, tree snails, garden snails), sea slugs, land slugs.
  Class Bivalvia. Clams, mussels, scallops, cockles, oysters, shipworms.
  Class Cephalopoda. Squids, octopuses, cuttlefish, nautiluses. *Loligo.*
PHYLUM BRYOZOA.   Bryozoans (moss animals).
PHYLUM BRACHIOPODA.   Lampshells.
PHYLUM ANNELIDA.   Segmented worms.
  Class Polychaeta. Mostly marine worms.
  Class Oligochaeta. Mostly freshwater and terrestrial worms, but many marine. *Lumbricus* (earthworm).
  Class Hirudinea. Leeches.
PHYLUM TARDIGRADA.   Water bears.
PHYLUM ONYCHOPHORA.   Onychophorans. *Peripatus.*
PHYLUM ARTHROPODA.
  Subphylum Trilobita. Trilobites; extinct.

Subphylum Chelicerata. Chelicerates. Horseshoe crabs, spiders, scorpions, ticks, mites.

Subphylum Crustacea. Shrimps, crayfishes, lobsters, crabs, barnacles, copepods, isopods (sowbugs).

Subphylum Uniramia.

Superclass Myriapoda. Centipedes, millipedes.

Superclass Insecta.

Order Ephemeroptera. Mayflies.

Order Odonata. Dragonflies, damselflies.

Order Orthoptera. Grasshoppers, crickets, katydids.

Order Dermaptera. Earwigs.

Order Blattodea. Cockroaches.

Order Mantodea. Mantids.

Order Isoptera. Termites.

Order Mallophaga. Biting lice.

Order Anoplura. Sucking lice.

Order Homoptera. Cicadas, aphids, leafhoppers, spittlebugs.

Order Hemiptera. Bugs.

Order Coleoptera. Beetles.

Order Diptera. Flies.

Order Mecoptera. Scorpion flies. *Harpobittacus.*

Order Siphonaptera. Fleas.

Order Lepidoptera. Butterflies, moths.

Order Hymenoptera. Wasps, bees, ants.

PHYLUM ECHINODERMATA. Echinoderms.

Class Asteroidea. Sea stars.

Class Ophiuroidea. Brittle stars.

Class Echinoidea. Sea urchins, heart urchins, sand dollars.

Class Holothuroidea. Sea cucumbers.

Class Crinoidea. Feather stars, sea lilies.

Class Concentricycloidea. Sea daisies.

PHYLUM HEMICHORDATA. Acorn worms.

PHYLUM CHORDATA. Chordates.

Subphylum Urochordata. Tunicates, related forms.

Subphylum Cephalochordata. Lancelets.

Subphylum Vertebrata. Vertebrates.

Class Agnatha. Jawless vertebrates (lampreys, hagfishes).

Class Placodermi. Jawed, heavily armored fishes; extinct.

Class Chondrichthyes. Cartilaginous fishes (sharks, rays, skates, chimaeras).

Class Osteichthyes. Bony fishes.

Subclass Dipnoi. Lungfishes.

Subclass Crossopterygii. Coelacanths, related forms.

Subclass Actinopterygii. Ray-finned fishes.

Order Acipenseriformes. Sturgeons, paddlefishes.

Order Salmoniformes. Salmon, trout.

Order Atheriniformes. Killifishes, guppies.

Order Gasterosteiformes. Seahorses.

Order Perciformes. Perches, wrasses, barracudas, tunas, freshwater bass, mackerels.

Order Lophiiformes. Angler fishes.

Class Amphibia. Mostly tetrapods; embryo enclosed in amnion.

Order Caudata. Salamanders.

Order Anura. Frogs, toads.

Order Apoda. Apodans (caecilians).

Class Reptilia. Skin with scales, embryo enclosed in amnion.

Subclass Anapsida. Turtles, tortoises.

Subclass Lepidosaura. *Sphenodon,* lizards, snakes.

Subclass Archosaura. Dinosaurs (extinct), crocodiles, alligators.

Class Aves. Birds. (In more recent schemes, dinosaurs, crocodilians, and birds are grouped in the same category.)

Order Struthioniformes. Ostriches.

Order Sphenisciformes. Penguins.

Order Procellariiformes. Albatrosses, petrels.

Order Ciconiiformes. Herons, bitterns, storks, flamingoes.

Order Anseriformes. Swans, geese, ducks.

Order Falconiformes. Eagles, hawks, vultures, falcons.

Order Galliformes. Ptarmigan, turkeys, domestic fowl.

Order Columbiformes. Pigeons, doves.

Order Strigiformes. Owls.

Order Apodiformes. Swifts, hummingbirds.

Order Passeriformes. Sparrows, jays, finches, crows, robins, starlings, wrens.

Class Mammalia. Skin with hair; young nourished by milk-secreting glands of adult.

Subclass Prototheria. Egg-laying mammals (duckbilled platypus, spiny anteaters).

Subclass Metatheria. Pouched mammals or marsupials (opossums, kangaroos, wombats).

Subclass Eutheria. Placental mammals.

Order Insectivora. Tree shrews, moles, hedgehogs.

Order Scandentia. Insectivorous tree shrews.

Order Chiroptera. Bats.

Order Primates.

Suborder Strepsirhini (prosimians). Lemurs, lorises.

Suborder Haplorhini (tarsioids and anthropoids).

Infraorder Tarsiiformes. Tarsiers.

Infraorder Platyrrhini (New World monkeys).

Family Cebidae. Spider monkeys, howler monkeys, capuchin.

Infraorder Catarrhini (Old World monkeys and hominoids).

Superfamily Cercopithecoidea. Baboons, macaques, langurs.

Superfamily Hominoidea. Apes and humans.

Family Hylobatidae. Gibbons.

Family Pongidae. Chimpanzees, gorillas, orangutans.

Family Hominidae. Humans and most recent ancestors of humans.

Order Carnivora. Carnivores.

Suborder Feloidea. Cats, civets, mongooses, hyenas.

Suborder Canoidea. Dogs, weasels, skunks, otters, raccoons, pandas, bears.

Order Proboscidea. Elephants; mammoths (extinct).

Order Sirenia. Sea cows (manatees, dugongs).

Order Perissodactyla. Odd-toed ungulates (horses, tapirs, rhinos).

Order Artiodactyla. Even-toed ungulates (camels, deer, bison, sheep, goats, antelopes, giraffes).

Order Edentata. Anteaters, tree sloths, armadillos.

Order Tubulidentata. African aardvark.

Order Cetacea. Whales, porpoises.

Order Rodentia. Most gnawing animals (squirrels, rats, mice, guinea pigs, porcupines).

APPENDIX II
Units of Measure

Metric-English Conversions

Length

English		Metric
inch	=	2.54 centimeters
foot	=	0.30 meter
yard	=	0.91 meter
mile (5,280 feet)	=	1.61 kilometer

To convert	multiply by	to obtain
inches	2.54	centimeters
feet	30.00	centimeters
centimeters	0.39	inches
millimeters	0.039	inches

Weight

English		Metric
grain	=	64.80 milligrams
ounce	=	28.35 grams
pound	=	453.60 grams
ton (short) (2,000 pounds)	=	0.91 metric ton

To convert	multiply by	to obtain
ounces	28.3	grams
pounds	453.6	grams
pounds	0.45	kilograms
grams	0.035	ounces
kilograms	2.2	pounds

Volume

English		Metric
cubic inch	=	16.39 cubic centimeters
cubic foot	=	0.03 cubic meter
cubic yard	=	0.765 cubic meters
ounce	=	0.03 liter
pint	=	0.47 liter
quart	=	0.95 liter
gallon	=	3.79 liters

To convert	multiply by	to obtain
fluid ounces	30.00	milliliters
quart	0.95	liters
milliliters	0.03	fluid ounces
liters	1.06	quarts

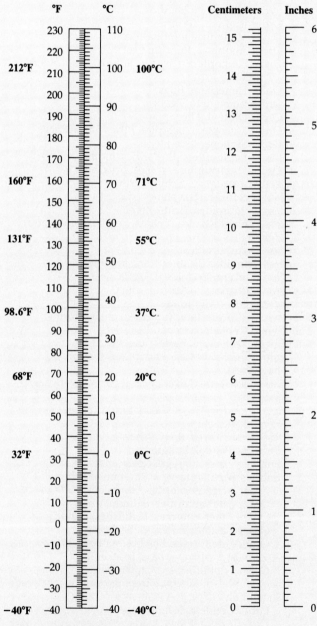

To convert temperature scales:

Fahrenheit to Celsius: $°C = 5/9 (°F - 32)$

Celsius to Fahrenheit: $°F = 9/5 (°C) + 32$

APPENDIX III
Answers to Genetics Problems

Chapter 9

1. a. *AB*
   b. *AB* and *aB*
   c. *Ab* and *ab*
   d. *AB*, *aB*, *Ab*, and *ab*

2. a. *AaBB* will occur in all the offspring.
   b. 25%   *AABB*; 25% *AaBB*; 25% *AABb*; 25% *AaBb*.
   c. 25%   *AaBb*; 25% *Aabb*; 25% *aaBb*; 25% *aabb*.
   d. 1/16   *AABB*   (6.25%)
      1/8   *AaBB*   (12.5%)
      1/16   *aaBB*   (6.25%)
      1/8   *AABb*   (12.5%)
      1/4   *AaBb*   (25%)
      1/8   *aaBb*   (12.5%)
      1/16   *AAbb*   (6.25%)
      1/8   *Aabb*   (12.5%)
      1/16   *aabb*   (6.25%)

3. Yellow is recessive. Because the first-generation plants must be heterozygous and had a green phenotype, green must be dominant over the recessive yellow.

4. a. Mother must be heterozygous for both genes; father is homozygous recessive for both genes. The first child is also homozygous recessive for both genes.
   b. The probability that the second child will not be able to roll the tongue and will have detached earlobes is 1/4 (25%).

5. a. *ABC*
   b. *ABc* and *aBc*
   c. *ABC*, *aBC*, *ABc*, and *aBc*
   d. *ABC*, *aBC*, *AbC*, *abC*, *ABc*, *aBc*, *Abc*, and *abc*

6. The first-generation plants must all be double heterozygotes. When these plants are self-pollinated 1/4 (25%) of the second-generation plants will be doubly heterozygous.

7. The most direct way to accomplish this would be to allow a true-breeding mouse having yellow fur to mate with a true-breeding mouse having brown fur. Such true-breeding strains could be obtained by repeated inbreeding (mating of related individuals; for example, a male and a female of the same litter) of yellow and brown strains. In this way, it should be possible to obtain homozygous yellow and homozygous brown mice.

   When true-breeding yellow and true-breeding brown mice are crossed, the progeny should all be heterozygous. If the progeny phenotype is either yellow or brown, then the dominance is simple or complete, and the phenotype reflects the dominant allele. If the phenotype is intermediate between yellow and brown, there is incomplete dominance. If the phenotype shows both yellow and brown, there is codominance.

8. a. The mother must be heterozygous (I^Ai). The man having type B blood could have fathered the child if he were also heterozygous (I^Bi).

   b. If the man is heterozygous, then he *could be* the father. However, because any other type B heterozygous male also could be the father, one cannot say that this particular man absolutely must be. Actually, any male who could contribute an O allele (*i*) could have fathered the child. This would include males with type O blood (*ii*) or type A blood who are heterozygous (I^Ai).

9. a. F_1 genotypes and phenotypes; 100% *Bb Cc*, brown progeny. F_2 phenotypes: 9/16 brown + 3/16 tan + 4/16 albino.

   F^2 genotypes: $\begin{cases} \text{1/16 } BB\ CC + \text{2/16 } BB\ Cc + \text{2/16} \\ \quad BbCC + \text{4/16 } Bb\ Cc; \text{(9/16 brown)} \\ \text{1/16 } bb\ CC + \text{2/16 } bb\ Cc; \text{(3/16 tan)} \\ \text{1/16 } BB\ cc + \text{2/16 } Bb\ cc + \text{1/16 } bb \\ \quad cc; \text{(4/16 albino)} \end{cases}$

   b. Backcross phenotypes: 1/4 brown + 1/4 tan + 2/4 albino.

   Backcross genotypes: $\begin{cases} \text{1/4 } Bb\ Cc; \text{(1/4 brown)} \\ \text{1/4 } bb\ Cc; \text{(1/4 tan)} \\ \text{1/4 } Bb\ cc + \text{1/4 } bb\ cc: \text{(1/2} \\ \quad \text{albino)} \end{cases}$

Chapter 10

1. a. Males inherit their X chromosome from their mothers.
   b. A male can produce two types of gametes with respect to an X-linked gene. One type will lack this gene and possess a Y chromosome. The other will have an X chromosome and the linked gene.
   c. A female homozygous for an X-linked gene will produce just one type of gamete containing an X chromosome with the gene.
   d. A female heterozygous for an X-linked gene will produce two types of gametes. One will contain an X chromosome with the dominant allele, and the other type will contain an X chromosome with the recessive allele.

2. a. Because this gene is only carried on Y chromosomes, females would not be expected to have hairy pinnae because they normally do not have Y chromosomes.
   b. Because sons always inherit a Y chromosome from their fathers and because daughters never do, a man having hairy pinnae will always transmit this trait to his sons and never to his daughters.

3. A 0% crossover frequency means that 50% of the gametes will be *AB* and 50% will be *ab*.

4. The gene for hemophilia occurs on the X but not the Y chromosome. A male has only one X chromosome. Therefore, it would be impossible for a male simply to be a carrier; the allele associated with hemophilia would always be expressed.

5. Assuming the mother is heterozygous (most individuals with Huntington's disorder are), the woman has a 1/2 (50%) chance of being heterozygous and therefore of later developing the disorder. Also, if this woman married a normal male, they would have a 50% chance of having a child with the disorder. Thus the *total* probability of each child of theirs being affected with Huntington's disorder is (0.5)(0.5) = 0.25, or 1/4 (25%).

6. The first child can only be color blind if it is a boy. Why? The probability of this happening is 25%. Similarly, their second child also has a 25% chance of being color blind. The probability that both will be color blind is (0.25)(0.25) = 0.0625, or 6.25%.

7. This indicates that genetic information other than that necessary for sex determination must reside on the X chromosome. Such information is necessary for survival regardless of whether one is male or female. Obviously, this is not true for the Y chromosome, in that individuals (females) survive quite nicely in its absence. A major function of the Y chromosome is to change what would have been a female individual into a male.

8. The only parent from whom this child could have received an X chromosome that bears a nonhemophilic allele is the mother. Therefore, nondisjunction must have occurred in the father.

9. a. An unaffected female selected at random has 1 chance in 50 (2%) of being heterozygous.
   b. If you selected a symptom-free male and symptom-free female at random, the probability that both will be heterozygous is (0.02)(0.02) = 0.0004, or 0.04%. The probability that a pair of symptom-free individuals selected at random could have a child affected by PKU is given by (0.02)(0.02)(0.25) = 0.0001, or 0.01%. This is the same thing as 1/10,000, which suggests that about one birth in every 10,000 will be a child with PKU, assuming that only heterozygous individuals have such children (an assumption which is not completely true).

APPENDIX IV
Answers to Self-Quizzes

Chapter 1
1. cell
2. Metabolism
3. Homeostasis
4. adaptive
5. mutations
6. d
7. d
8. d
9. c

Chapter 2
1. d
2. b
3. c
4. c
5. c
6. c
7. c
8. d
9. f
   g
   h
   d
   a
   c
   e

Chapter 3
1. d
2. c
3. c
4. c
5. a
6. a
7. d
8. b
9. h
   g
   a
   b
   d
   e
   i
   f
   c

Chapter 4
1. metabolism
2. thermodynamics
3. c
4. c
5. c
6. d
7. d
8. c
9. c
   d
   a
   b

Chapter 5
1. carbon
2. carbon dioxide, light
3. d
4. c
5. b
6. b
7. c
8. c
9. c
   d
   e
   b
   a

Chapter 6
1. ATP
2. pyruvate
3. d
4. c
5. d
6. c
7. b
8. b
9. b
   c
   a
   d

Chapter 7
1. b
2. a
3. b
4. b
5. c
6. a
7. b
8. d
   b
   c
   a

Chapter 8
1. d
2. c
3. a
4. b
5. d
6. d
7. c
8. c
9. d
10. e
    a
    d
    b
    c

Chapter 9
1. a
2. b
3. a
4. c
5. b
6. a
7. d
8. c
   a
   d
   b

Chapter 10
1. b
2. c
3. e
4. d
5. d
6. d
7. d
8. b
   d
   a
   c

Chapter 11
1. c
2. d
3. d
4. c
5. a
6. a
7. d
8. d
   e
   b
   c
   a

Chapter 12
1. three
2. e
3. a
4. c
5. c
6. a
7. c
8. b
9. a
10. d
11. e
12. e
    c
    d
    b
    a

Chapter 13
1. Plasmids
2. c
3. d
4. a
5. b
6. a
7. b
8. d
9. b
10. d, e, a, c, b

Chapter 14
1. populations
2. c
3. d
4. e
5. a
6. e
7. d
8. c
9. c
10. c
11. d, c, e, a, b

Chapter 15
1. d
2. c
3. b
4. a
5. d
6. c
7. a
8. d
9. d
10. e, b, a, d, c

Chapter 16
1. e
2. c
3. b
4. a
5. b
6. d
7. d
8. b, e, a, d, c

Chapter 17
1. e
2. b
3. c
4. c
5. c
6. c
7. c
8. a
9. d, a, b, c, e

Chapter 18
1. c
2. b
3. d
4. d
5. d
6. b
7. d, b, a, c

Chapter 19
1. body symmetry, cephalization, type of gut, type of body cavity, segmentation
2. b
3. b
4. d
5. a
6. a
7. b
8. c
9. d, f, c, g, e, i, j, h, a, b

Chapter 20
1. b
2. a
3. b
4. c
5. a
6. b
7. d
8. d
9. d
10. b, d, e, c, f, a

Chapter 21
1. hydrogen bonds (cohesion)
2. stomata
3. d
4. e
5. c
6. d
7. d
8. d
9. b, d, f, c, a, e

Chapter 22
1. a
2. c
3. b
4. c
5. a
6. b
7. a
8. c
9. d, e, a, c, f, b

Chapter 23
1. d
2. c
3. b
4. a
5. d
6. c
7. b
8. c, b, f, e, d, a

Chapter 24
1. d
2. a
3. c
4. d
5. a
6. b
7. d
8. a
9. f
10. h, f, g, e, a, c, b, i, d

Chapter 25
1. e
2. d
3. energy; energy
4. a
5. d
6. b
7. c
8. a
9. b
10. d, c, e, g, f, b, a

Chapter 26
1. c
2. c
3. d
4. c
5. a
6. d
7. b
8. b, d, e, a, c
9. c, d, e, h, b, g, a, f

Chapter 27
1. a
2. d
3. c
4. d
5. a
6. d
7. d
8. b
9. a
10. c, b, a, e, d, f

Chapter 28
1. d
2. d
3. b
4. c
5. c
6. e
7. b
8. c
9. d
10. d, b, f, e, a, c

Chapter 29
1. d
2. f
3. d
4. c
5. d
6. d
7. a
8. d
9. d, e, a, c, b

Chapter 30
1. d
2. c
3. b
4. b
5. a
6. c
7. b
8. b
9. g, f, e, b, i, c, a, d, h

Chapter 31
1. f
2. d
3. b
4. d
5. e
6. b
7. b
8. b
9. b, e, g, d, f, a, c

Chapter 32
1. b
2. a
3. c
4. d
5. d
6. a
7. a
8. b
9. c
10. True
11. d, c, f, b, h, e, a, g

Chapter 33
1. Ecology
2. population
3. e
4. e
5. b
6. e
7. d
8. d
9. d
10. d, e, c, a, b

Chapter 34
1. Communities
2. niche
3. habitat
4. e
5. d
6. d
7. c
8. d
9. c
10. b, c, e, f, a, d

Chapter 35
1. d
2. a
3. d
4. d
5. a
6. e
7. c
8. d, a, c, b
9. c, d, e, b, a

Chapter 36

1. c
2. c
3. d
4. b
5. d
6. d
7. d
8. d
9. b, f, e, d, c, a
10. e, a, d, c, b

Chapter 37

1. c
2. c
3. b
4. d
5. c
6. b
7. c
8. d
9. d
10. e, c, d, a, b

Chapter 38

1. a
2. d
3. b
4. d
5. c
6. d
7. c
8. d
9. d, e, b, c, a

APPENDIX V
A Closer Look at Some Major Metabolic Pathways

ENERGY-
REQUIRING
STEPS OF
GLYCOLYSIS

(two ATP
invested)

ENERGY-
RELEASING
STEPS OF
GLYCOLYSIS

(four ATP
produced)

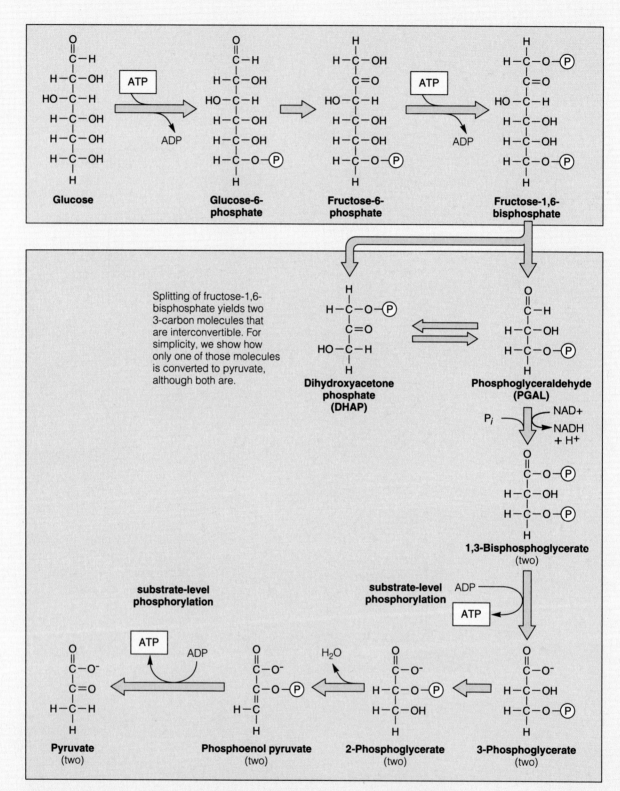

Splitting of fructose-1,6-bisphosphate yields two 3-carbon molecules that are interconvertible. For simplicity, we show how only one of those molecules is converted to pyruvate, although both are.

Figure a Glycolysis, ending with two 3-carbon pyruvate molecules for each 6-carbon glucose entering the reactions. The *net* energy yield is two ATP molecules (two invested, four produced).

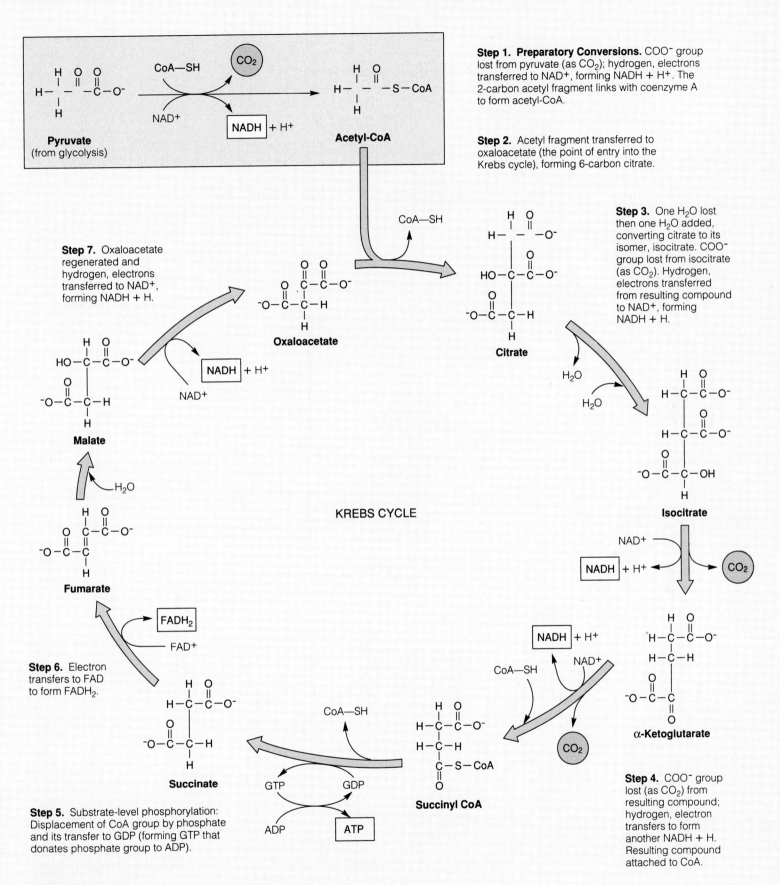

Step 1. Preparatory Conversions. COO^- group lost from pyruvate (as CO_2); hydrogen, electrons transferred to NAD^+, forming $NADH + H^+$. The 2-carbon acetyl fragment links with coenzyme A to form acetyl-CoA.

Step 2. Acetyl fragment transferred to oxaloacetate (the point of entry into the Krebs cycle), forming 6-carbon citrate.

Step 3. One H_2O lost then one H_2O added, converting citrate to its isomer, isocitrate. COO^- group lost from isocitrate (as CO_2). Hydrogen, electrons transferred from resulting compound to NAD^+, forming $NADH + H$.

Step 7. Oxaloacetate regenerated and hydrogen, electrons transferred to NAD^+, forming $NADH + H$.

Pyruvate (from glycolysis)

CoA—SH

CO_2

NAD^+

$NADH + H^+$

Acetyl-CoA

CoA—SH

Oxaloacetate

Citrate

$NADH + H^+$

NAD^+

H_2O

H_2O

Malate

KREBS CYCLE

Isocitrate

H_2O

NAD^+

$NADH + H^+$

CO_2

Fumarate

$FADH_2$

FAD^+

Step 6. Electron transfers to FAD to form $FADH_2$.

Succinate

CoA—SH

$NADH + H^+$

NAD^+

CoA—SH

Succinyl CoA

α-Ketoglutarate

GTP

GDP

ADP

ATP

CO_2

Step 5. Substrate-level phosphorylation: Displacement of CoA group by phosphate and its transfer to GDP (forming GTP that donates phosphate group to ADP).

Step 4. COO^- group lost (as CO_2) from resulting compound; hydrogen, electron transfers to form another $NADH + H$. Resulting compound attached to CoA.

Figure b Krebs cycle (citric acid cycle). Red identifies carbon entering the cycle by way of acetyl-CoA. Blue identifies carbon destined to leave the substrates (as carbon dioxide molecules).

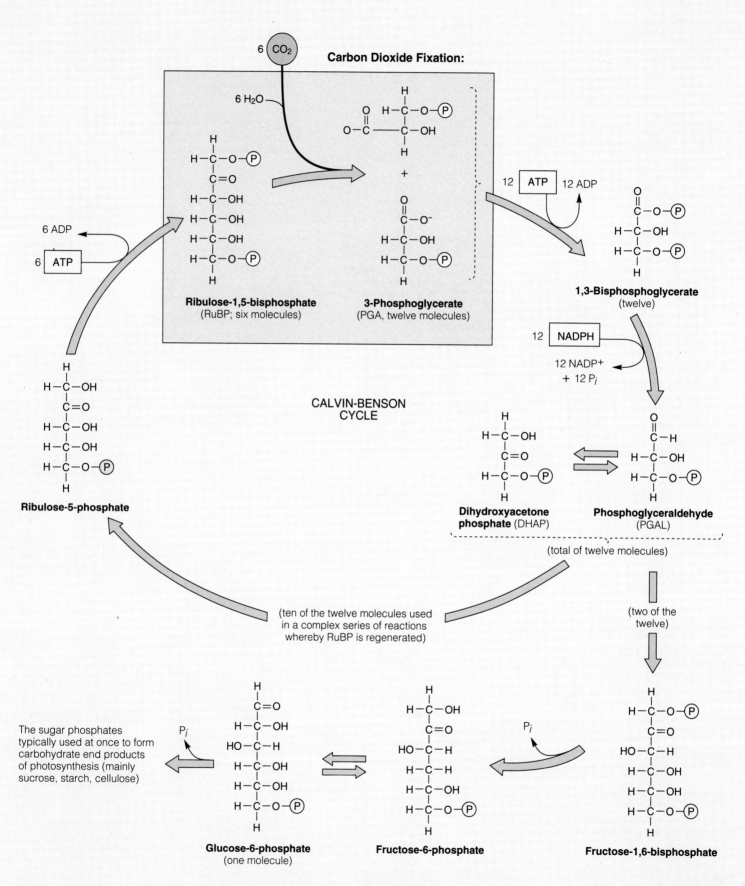

6 CO₂

Carbon Dioxide Fixation:

6 H₂O

12 **ATP**   12 ADP

Ribulose-1,5-bisphosphate
(RuBP; six molecules)

3-Phosphoglycerate
(PGA, twelve molecules)

1,3-Bisphosphoglycerate
(twelve)

6 ADP

6 **ATP**

12 **NADPH**

12 NADP⁺
+ 12 Pᵢ

**CALVIN-BENSON
CYCLE**

Ribulose-5-phosphate

**Dihydroxyacetone
phosphate** (DHAP)

Phosphoglyceraldehyde
(PGAL)

(total of twelve molecules)

(ten of the twelve molecules used
in a complex series of reactions
whereby RuBP is regenerated)

(two of the
twelve)

The sugar phosphates
typically used at once to form
carbohydrate end products
of photosynthesis (mainly
sucrose, starch, cellulose)

Pᵢ

Glucose-6-phosphate
(one molecule)

Pᵢ

Fructose-6-phosphate

Fructose-1,6-bisphosphate

Figure c Calvin-Benson cycle of the light-independent reactions of photosynthesis.

GLOSSARY OF BIOLOGICAL TERMS

ABO blood typing Method of characterizing an individual's blood according to whether one or both of two protein markers, A and B, are present at the surface of red blood cells. The O signifies that neither marker is present.

abortion Spontaneous or induced expulsion of the embryo or fetus from the uterus.

abscisic acid (ab-SISS-ik) Plant hormone that promotes stomatal closure, bud dormancy, and seed dormancy.

abscission (ab-SIH-zhun) [L. *abscindere*, to cut off] The dropping of leaves, flowers, fruits, or other plant parts due to hormonal action.

absorption Of complex animals, the movement of nutrients, fluid, and ions across the gut lining and into the internal environment.

accessory pigment A light-trapping pigment that contributes to photosynthesis by extending the range of usable wavelengths beyond those absorbed by the chlorophylls.

acid [L. *acidus*, sour] A substance that releases hydrogen ions (H^+) in water.

acid rain The falling to earth of snow or rain that contains sulfur and nitrogen oxides. Also called wet acid deposition (as opposed to dry acid deposition, or the falling to earth of airborne particles of sulfur and nitrogen oxides).

acoelomate (ay-SEE-luh-mate) Type of animal that has no fluid-filled cavity between the gut and body wall.

acoustical signal Sounds that are used as a communication signal.

actin (AK-tin) A globular contractile protein. In muscle cells, actin interacts with another protein, myosin, to bring about contraction.

action potential An abrupt, brief reversal in the steady voltage difference across the plasma membrane (that is, the resting membrane potential) of a neuron and some other cells.

activation energy The minimum amount of collision energy required to bring reactant molecules to an activated condition (the transition state) at which a reaction will proceed spontaneously. Enzymes enhance reaction rates by lowering the activation energy (they put substrates on a precise collision course).

active site A crevice on the surface of an enzyme molecule where a specific reaction is catalyzed.

active transport The pumping of one or more specific solutes through a transport protein that spans the lipid bilayer of a cell membrane. Most often, the solute is transported against its concentration gradient. The protein is activated by an energy boost, as from ATP.

adaptation [L. *adaptare*, to fit] In evolutionary biology, the process of becoming adapted (or more adapted) to a given set of environmental conditions. Of sensory neurons, a decrease in the frequency of action potentials (or their cessation) even when a stimulus is maintained at constant strength.

adaptive behavior A behavior that promotes the propagation of an individual's genes and that tends to increase in frequency in a population over time.

adaptive radiation A burst of speciation events, with lineages branching away from one another as they partition the existing environment or invade new ones.

adaptive trait Any aspect of form, function, or behavior that helps an organism survive and reproduce under a given set of environmental conditions.

adaptive zone A way of life, such as "catching insects in the air at night." A lineage must have physical, ecological, and evolutionary access to an adaptive zone to become a successful occupant of it.

adenine (AH-de-neen) A purine; a nitrogen-containing base found in nucleotides.

adenosine diphosphate (ah-DEN-uh-seen die-FOSS-fate) ADP, a molecule involved in cellular energy transfers; typically formed by hydrolysis of ATP.

adenosine phosphates Any of several relatively small molecules, some of which function as chemical messengers within and between cells, and others that function as energy carriers.

adenosine triphosphate *See* ATP.

ADH Antidiuretic hormone. Produced by the hypothalamus and released by the posterior pituitary, it stimulates reabsorption in the kidneys and so reduces urine volume.

adipose tissue A type of connective tissue having an abundance of fat-storing cells and blood vessels for transporting fats.

ADP Adenosine diphosphate. A nucleotide coenzyme that accepts unbound phosphate or a phosphate group to become ATP.

ADP/ATP cycle In cells, a mechanism of ATP renewal. When ATP donates a phosphate group to other molecules (and so energizes them), it reverts to ADP, then forms again by phosphorylation of ADP.

adrenal cortex (ah-DREE-nul) Outer portion of the adrenal gland; its hormones have roles in metabolism, inflammation, maintaining extracellular fluid volume, and other functions.

adrenal medulla Inner region of the adrenal gland; its hormones help control blood circulation and carbohydrate metabolism.

aerobic respiration (air-OH-bik) [Gk. *aer*, air, + *bios*, life] The main energy-releasing metabolic pathway of ATP formation, in which oxygen is the final acceptor of electrons stripped from glucose or some other organic compound. The pathway proceeds from glycolysis through the Krebs cycle and electron transport phosphorylation. A typical net yield is 36 ATP for each glucose molecule.

age structure Of a population, the number of individuals in each of several or many age categories.

agglutination (ah-glue-tin-AY-shun) Clumping together of foreign cells that have invaded the body (as pathogens or in tissue grafts or transplants). Clumping is induced by cross-linking between antibody molecules that have already latched onto antigen at the surface of the foreign cells.

aging A range of processes, including the breakdown of cell structure and function, by which the body gradually deteriorates. All organisms showing extensive cell differentiation undergo aging.

AIDS Acquired immunodeficiency syndrome. A set of chronic disorders following infection by the human immunodeficiency virus (HIV), which destroys key cells of the immune system.

alcoholic fermentation Anaerobic pathway of ATP formation in which pyruvate from glycolysis is broken down to acetaldehyde, which accepts electrons from NADH to become ethanol, and NAD^+ is regenerated. Its net yield is two ATP.

aldosterone (al-DOSS-tuh-rohn) Hormone secreted by the adrenal cortex that helps regulate sodium reabsorption.

allantois (ah-LAN-twahz) [Gk. *allas*, sausage] Of vertebrates, one of four extraembryonic membranes that form during embryonic development. It functions in respiration and storage of metabolic wastes in reptiles, birds, and some mammals. In humans, it functions in early blood formation and development of the urinary bladder.

allele (uh-LEEL) For a given location on a chromosome, one of two or more slightly different molecular forms of a gene that code for different versions of the same trait.

allele frequency Of a given gene locus, the relative abundances of each kind of allele carried by the individuals of a population.

allergy An immune response made against a normally harmless substance.

allopatric speciation [Gk. *allos*, different, + *patria*, native land] Speciation that follows geographic isolation of populations of the same species.

allosteric control (AL-oh-STARE-ik) Control over a metabolic reaction or pathway that operates through the binding of a specific substance at a control site on a specific enzyme.

alpine tundra A type of biome that exists at high elevations in mountains throughout the world.

altruistic behavior (al-true-ISS-tik) Self-sacrificing behavior; the individual behaves in a way that helps others but decreases its own chances of reproductive success.

alveolar sac (al-VEE-uh-lar) Any of the pouchlike clusters of alveoli in the lungs; the major sites of gas exchange.

alveolus (ahl-VEE-uh-lus), plural **alveoli** [L. *alveus*, small cavity] Any of the many cup-shaped, thin-walled outpouchings of respiratory bronchioles. A site where oxygen diffuses from air in the lungs to the blood, and carbon dioxide diffuses from blood to the lungs.

amino acid (uh-MEE-no) A small organic molecule having a hydrogen atom, an amino group, an acid group, and an R group covalently bonded to a central carbon atom. The subunit of polypeptide chains, which represent the primary structure of proteins.

ammonification (uh-moan-ih-fih-KAY-shun) Together with decomposition, a process by which certain bacteria and fungi break down nitrogen-containing wastes and remains of other organisms.

amnion (AM-nee-on) Of land vertebrates, one of four extraembryonic membranes. It becomes a fluid-filled sac in which the embryo (and fetus) can grow, move freely, and be protected from sudden temperature shifts and impacts.

amniote egg A type of egg, often with a leathery or calcified shell, that contains extraembryonic membranes, including the amnion. An adaptation that figured in the vertebrate invasion of land.

amphibian A type of vertebrate somewhere between fishes and reptiles in body plan and reproductive mode; salamanders, frogs and toads, and caecilians are the existing groups.

anaerobic pathway (an-uh-ROW-bik) [Gk. *an*, without, + *aer*, air] Metabolic pathway in which a substance other than oxygen serves as the final acceptor of electrons that have been stripped from substrates.

analogous structures Body parts, once different in separate lineages, that were put to comparable uses in similar environments and that came to resemble one another in form and function. They are evidence of morphological convergence.

anaphase (AN-uh-faze) The stage at which microtubules of a spindle apparatus separate sister chromatids of each chromosome and move them to opposite spindle poles. During anaphase I of *meiosis*, the two members of each pair of homologous chromosomes separate. During anaphase II, sister chromatids of each chromosome separate.

aneuploidy (AN-yoo-ploy-dee) A change in the chromosome number following inheritance of one extra or one less chromosome.

angiosperm (AN-gee-oh-spurm) [Gk. *angeion*, vessel, and *spermia*, seed] A flowering plant.

animal A heterotroph that eats or absorbs nutrients from other organisms; is multicelled, usually with tissues arranged in organs and organ systems; is usually motile during at least part of the life cycle; and goes through a period of embryonic development.

annelid The type of invertebrate classified as a segmented worm; an oligochaete (such as an earthworm), leech, or polychaete.

annual plant A flowering plant that completes its life cycle in one growing season.

anther [Gk. *anthos*, flower] In flowering plants, the pollen-bearing part of the male reproductive structure (stamen).

antibiotic A normal metabolic product of certain microorganisms that kills or inhibits the growth of other microorganisms.

antibody [Gk. *anti*, against] Any of a variety of Y-shaped receptor molecules with binding sites for specific antigens. Only B cells produce antibodies, then position them at their surface or secrete them.

anticodon In a tRNA molecule, a sequence of three nucleotide bases that can pair with an mRNA codon.

antigen (AN-tih-jen) [Gk. *anti*, against, + *genos*, race, kind] Any molecular configuration that is recognized as foreign to the body and that triggers an immune response. Most antibodies are protein molecules at the surface of infectious agents or tumor cells.

antigen-presenting cell A macrophage or some other white blood cell that engulfs and digests antigen, then displays it with certain MHC molecules at its surface. Recognition of antigen-MHC complexes by T and B cells triggers an immune response.

aorta (ay-OR-tah) [Gk. *airein*, to lift, heave] Main artery of systemic circulation; carries oxygenated blood away from the heart to all body regions except the lungs.

apical dominance The inhibitory influence of a terminal bud on the growth of lateral buds.

apical meristem (AY-pih-kul MARE-ih-stem) [L. *apex*, top, + Gk. *meristos*, divisible] Of most plants, a mass of self-perpetuating cells responsible for primary growth (elongation) at root and shoot tips.

appendicular skeleton (ap-en-DIK-yoo-lahr) In vertebrates, bones of the limbs, hips, and shoulders.

appendix A slender projection from the cup-shaped pouch at the start of the colon.

archaebacteria One of three great prokaryotic lineages that arose early in the history of life; now represented by methanogens, halophiles, and thermophiles.

arctic tundra A type of biome that lies between the polar ice cap and boreal forests of North America, Europe, and Asia.

arteriole (ar-TEER-ee-ole) Any of the blood vessels between arteries and capillaries. They are control points where the volume of blood delivered to different body regions can be adjusted.

artery Any of the large-diameter blood vessels that conduct oxygen-poor blood to the lungs and oxygen-enriched blood to all body tissues. Their thick, muscular wall allows them to smooth out pulsations in blood pressure caused by heart contractions.

arthropod An invertebrate having a hardened exoskeleton, specialized segments, and jointed appendages. Spiders, crabs, and insects are examples.

asexual reproduction Mode of reproduction by which offspring arise from a single parent, and inherit the genes of that parent only.

atmosphere A region of gases, airborne particles, and water vapor enveloping the earth; 80 percent of its mass is distributed within seventeen miles of the earth's surface.

atmospheric cycle A biogeochemical cycle in which the atmosphere is the largest reservoir of an element. The carbon and nitrogen cycles are examples.

atom The smallest unit of matter that is unique to a particular element.

atomic number The number of protons in the nucleus of each atom of an element; it differs for each element.

ATP Adenosine triphosphate (ah-DEN-uh-seen try-FOSS-fate). A nucleotide composed of adenine, ribose, and three phosphate groups. As the main energy carrier in cells, it directly or indirectly delivers energy to or picks up energy from nearly all metabolic pathways.

atrium (AYE-tree-um) Of the human heart, one of two chambers that receive blood. *Compare* ventricle.

australopith (OHSS-trah-low-pith) [L. *australis*, southern, + Gk. *pithekos*, ape] Any of the earliest known species of hominids, that is, the first species of the evolutionary branch leading to humans.

autoimmune response Misdirected immune response in which lymphocytes mount an attack against normal body cells.

autonomic nervous system (auto-NOM-ik) Those nerves leading from the central nervous system to the smooth muscle, cardiac muscle, and glands of internal organs and structures, that is, to the visceral portion of the body.

autosomal dominant inheritance Condition arising from the presence of a dominant allele on an autosome (not a sex chromosome). The allele is always expressed to some extent, even in heterozygotes.

autosomal recessive inheritance Condition arising from a recessive allele on an autosome (not a sex chromosome). Only recessive homozygotes show the resulting phenotype.

autosome Any of the chromosomes that are of the same number and kind in both males and females of the species.

autotroph (AH-toe-trofe) [Gk. *autos*, self, + *trophos*, feeder] An organism able to build its own large organic molecules by using carbon dioxide and energy from the physical environment. Photosynthetic autotrophs use sunlight energy; chemosynthetic autotrophs extract energy from chemical reactions involving inorganic substances. *Compare* heterotroph.

auxin (AWK-sin) Any of a class of growth-regulating hormones in plants; auxins promote stem elongation as one effect.

axial skeleton (AX-ee-uhl) In vertebrates, the skull, backbone, ribs, and breastbone (sternum).

axon Of a neuron, a long, cylindrical extension from the cell body, with finely branched endings. Action potentials move rapidly, without alteration, along an axon; their arrival at axon endings may trigger the release of neurotransmitter molecules that influence an adjacent cell.

B lymphocyte, or **B cell** The only white blood cell that produces antibodies, then positions them at the cell surface or secretes them as weapons in immune responses.

bacterial conjugation The transfer of plasmid DNA from one bacterial cell to another.

bacterial flagellum Of many bacterial cells, a whiplike motile structure that does not contain a core of microtubules.

bacteriophage (bak-TEER-ee-oh-fahj) [Gk. *baktērion*, small staff, rod, + *phagein*, to eat] Category of viruses that infect bacterial cells.

balanced polymorphism The maintenance of two or more forms of a trait in fairly stable proportions over generations.

Barr body In the cells of female mammals, a condensed X chromosome that was inactivated during early embryonic development.

basal body A centriole which, after having given rise to the microtubules of a flagellum or cilium, remains attached to its base in the cytoplasm.

base A substance that releases OH^- in water.

base pair A pair of hydrogen-bonded nucleotide bases in two strands of nucleic acids. In a DNA double helix, adenine pairs with thymine, and guanine with cytosine. When an mRNA strand forms on a DNA strand during transcription, uracil (U) pairs with the DNA's adenine.

base sequence The particular order in which one nucleotide base follows the next in a strand of DNA or RNA. The order differs to some extent for each kind of organism.

basophil Fast-acting white blood cells that secrete histamine and other substances during inflammation.

behavior, animal A response to external and internal stimuli, following integration of sensory, neural, endocrine, and effector components. Because behavior has a genetic basis, it is subject to natural selection and commonly can be modified through experience.

benthic province All of the sediments and rocky formations of the ocean bottom; begins with the continental shelf and extends down through deep-sea trenches.

biennial A flowering plant that lives through two growing seasons.

bilateral symmetry Body plan in which the left and right halves of the animal are mirror-images of each other.

binary fission Of bacteria, a mode of asexual reproduction in which the parent cell replicates its single chromosome, then divides into two genetically identical daughter cells.

biogeochemical cycle The movement of an element such as carbon or nitrogen from the environment to organisms, then back to the environment.

biogeographic realm [Gk. *bios*, life, + *geographein*, to describe the surface of the earth] Any of six major land regions, each having distinguishing types of plants and animals and generally retaining its identity because of climate and geographic barriers to gene flow.

biological clocks Internal time-measuring mechanisms that have roles in adjusting an organism's daily activities, seasonal activities, or both in response to environmental cues.

biological magnification The increasing concentration of a nondegradable or slowly degradable substance in body tissues as it is passed along food chains.

biological systematics Branch of biology that assesses patterns of diversity based on information from taxonomy, phylogenetic reconstruction, and classification.

bioluminescence A flashing of light that emanates from an organism when excited electrons of luciferins, or highly fluorescent substances, return to a lower energy level.

biomass The combined weight of all the organisms at a particular trophic (feeding) level in an ecosystem.

biome A broad, vegetational subdivision of a biogeographic realm shaped by climate, topography, and composition of regional soils.

biosphere [Gk. *bios*, life, + *sphaira*, globe] All regions of the earth's waters, crust, and atmosphere in which organisms live.

biosynthetic pathway A metabolic pathway in which small molecules are assembled into lipids, proteins, and other large organic molecules.

biotic potential Of a population, the maximum rate of increase per individual under ideal conditions.

bipedalism A habitual standing and walking on two feet, as by ostriches and humans.

bird A type of vertebrate, the only one having feathers, with strong resemblances and evolutionary connections to reptiles.

blastocyst (BLASS-tuh-sist) [Gk. *blastos*, sprout, + *kystis*, pouch] In mammalian development, a blastula stage consisting of a hollow ball of surface cells and an inner cell mass.

blastula (BLASS-chew-lah) An embryonic stage consisting of a ball of cells produced by cleavage.

blood A fluid connective tissue composed of water, solutes, and formed elements (blood cells and platelets); it carries substances to and from cells and helps maintain an internal environment that is favorable for cell activities.

blood pressure Fluid pressure, generated by heart contractions, that keeps blood circulating.

blood-brain barrier Set of mechanisms that helps control which blood-borne substances reach neurons in the brain.

bone The mineral-hardened connective tissue of bones.

bones In vertebrate skeletons, organs that function in movement and locomotion, protection of other organs, mineral storage, and (in some bones) blood cell production.

bottleneck An extreme case of genetic drift. A catastrophic decline in population size leads to a random shift in the allele frequencies among survivors. Because the population must rebuild from so few individuals, severely limited genetic variation may be an outcome.

Bowman's capsule The cup-shaped portion of a nephron that receives water and solutes filtered from blood.

brain Of most nervous systems, the most complex integrating center; it receives, processes, and integrates sensory input and issues coordinated commands for response.

brainstem The vertebrate midbrain, pons, and medulla oblongata, the core of which contains the reticular formation that helps govern activity of the nervous system as a whole.

bronchiole Of most vertebrates, a component of the finely branched bronchial tree inside each lung.

bronchus, plural **bronchi** (BRONG-CUSS, BRONG-kee) [Gk. *bronchos*, windpipe] Tubelike branchings of the trachea that lead into the lungs of most vertebrates.

brown alga A type of aquatic plant, found in nearly all marine habitats, that has an abundance of xanthophyll pigments.

bryophyte A nonvascular land plant that requires free water to complete its life cycle.

bud An undeveloped shoot of mostly meristematic tissue; often covered and protected by scales (modified leaves).

buffer A substance that can combine with hydrogen ions, release them, or both. Buffers help stabilize the pH of blood and other fluids.

bulk Of human digestion, a volume of fiber and other undigested material that absorption processes in the colon cannot decrease.

bulk flow In response to a pressure gradient, a movement of more than one kind of molecule in the same direction in the same medium (as in blood, sap, or air).

C4 pathway Of many plants, a pathway of photosynthesis in which carbon dioxide is fixed twice, in two different cell types. Carbon dioxide accumulates in the leaf and helps counter photorespiration. The first compound formed is the 4-carbon oxaloacetate.

Calvin-Benson cycle Cyclic reactions that are the "synthesis" part of the light-independent reactions of photosynthesis. In land plants, RuBP, or some other compound to which carbon has been affixed, undergoes rearrangements that

lead to formation of a sugar phosphate and to regeneration of the RuBP. The cycle runs on ATP and NADPH from light-dependent reactions.

CAM plant A plant that conserves water by opening stomata only at night, when it fixes carbon dioxide by way of a C4 pathway.

cambium (KAM-bee-um), plural **cambia** In vascular plants, one of two types of meristems that are responsible for secondary growth (increases in stem and root diameter). Vascular cambium gives rise to secondary xylem and phloem; cork cambium gives rise to periderm.

camouflage An outcome of an organism's form, patterning, color, or behavior that helps it blend with its surroundings and escape detection.

cancer A type of malignant tumor, the cells of which show profound abnormalities in the plasma membrane and cytoplasm, abnormal growth and division, and weakened capacity for adhesion within the parent tissue (leading to metastasis). Unless eradicated, cancer is lethal.

canine A pointed tooth that functions in piercing food.

capillary [L. *capillus*, hair] A thin-walled blood vessel that functions in the exchange of gases and other substances between blood and interstitial fluid.

capillary bed A diffusion zone, consisting of numerous capillaries, where substances are exchanged between blood and interstitial fluid.

carbohydrate [L. *carbo*, charcoal, + *hydro*, water] A simple sugar or large molecule composed of sugar units. All cells use carbohydrates as structural materials, energy stores, and transportable forms of energy. The three classes of carbohydrates include: monosaccharides, oligosaccharides, or polysaccharides.

carbon cycle A biogeochemical cycle in which carbon moves from its largest reservoir in the atmosphere, through oceans and organisms, then back to the atmosphere.

carbon dioxide fixation First step of the light-independent reactions of photosynthesis. Carbon (from carbon dioxide) becomes affixed to a carbon compound (such as RuBP) that can enter the Calvin-Benson cycle.

carcinogen (kar-SIN-uh-jen) An environmental agent or substance, such as ultraviolet radiation, that can trigger cancer.

cardiac cycle (KAR-dee-ak) [Gk. *kardia*, heart, + *kyklos*, circle] The sequence of muscle contraction and relaxation constituting one heartbeat.

cardiac pacemaker Sinoatrial (SA) node; the basis of the normal rate of heartbeat. The self-excitatory cardiac muscle cells that spontaneously generate rhythmic waves of excitation over the heart chambers.

cardiovascular system Of most animals, an organ system that is composed of blood, one or more hearts, and blood vessels and that functions in the rapid transport of substances to and from cells.

carnivore [L. *caro*, *carnis*, flesh, + *vovare*, to devour] An animal that eats other animals; a type of heterotroph.

carotenoids (kare-OTT-en-oyds) Light-sensitive, accessory pigments that transfer absorbed energy to chlorophylls. They absorb violet and blue wavelengths but transmit red, orange, and yellow.

carpel (KAR-pul) The female reproductive part of a flower; sometimes called a pistil. The lower portion of a single carpel (or of a structure composed of two or more carpels) is an ovary, where eggs develop and are fertilized and where seeds mature. The upper portion has a stigma (a pollen-capturing surface tissue) and often a style (a slender extension of the ovary wall).

carrier protein Type of transport protein that binds specific substances and changes shape in ways that shunt the substances across a plasma membrane. Some carrier proteins function passively, others require an energy input.

carrying capacity The maximum number of individuals in a population (or species) that can be sustained indefinitely by a given environment.

cartilage A type of connective tissue with solid yet pliable intercellular material that resists compression.

Casparian strip In the exodermis and endodermis of roots, a waxy band that acts as an impermeable barrier between the walls of abutting cells.

cDNA Any DNA molecule copied from a mature mRNA transcript by way of reverse transcription.

cell [L. *cella*, small room] The smallest living unit; an organized unit that can survive and reproduce on its own, given DNA instructions and suitable environmental conditions, including appropriate sources of energy and raw materials.

cell count The number of cells of a given type in a microliter of blood.

cell cycle Events during which a cell increases in mass, roughly doubles its number of cytoplasmic components, duplicates its DNA, then undergoes nuclear and cytoplasmic division. It extends from the time a new cell is produced until it completes its own division.

cell differentiation The developmental process in which different cell types activate and suppress a fraction of their genes in different ways and so become specialized in composition, structure, and function. Regulatory proteins, enzymes, hormonal signals, and control sites built into DNA interact to bring about this selective gene expression.

cell junction Of multicelled organisms, a point of contact that physically links two cells or that provides functional links between their cytoplasm.

cell plate Of a plant cell undergoing cytoplasmic division, a disklike structure that forms from remnants of the spindle; it becomes a crosswall that partitions the cytoplasm.

cell theory A theory in biology, the key points of which are that (1) all organisms are composed of one or more cells, (2) the cell is the smallest unit that still retains a capacity for independent life, and (3) all cells arise from preexisting cells.

cell wall A rigid or semirigid wall outside the plasma membrane that supports a cell and imparts shape to it; a cellular feature of plants, fungi, protistans, and most bacteria.

central nervous system The brain and spinal cord of vertebrates.

central vacuole Of mature, living plant cells, a fluid-filled organelle that stores amino acids, sugars, ions, and toxic wastes. Its enlargement during growth causes the increases in surface area that improve nutrient uptake.

centriole (SEN-tree-ohl) A cylinder of triplet microtubules that gives rise to the microtubules of cilia and flagella.

centromere (SEN-troh-meer) [Gk. *kentron*, center, + *meros*, a part] A small, constricted region of a chromosome having attachment sites for microtubules that help move the chromosome during nuclear division.

cephalization (sef-ah-lah-ZAY-shun) [Gk. *kephalikos*, head] During the evolution of bilateral animals, the concentration of sensory structures and nerve cells in the head.

cerebellum (ser-ah-BELL-um) [L. diminutive of *cerebrum*, brain] Hindbrain region with reflex centers for maintaining posture and refining limb movements.

cerebral cortex Thin surface layer of the cerebral hemispheres. Some regions of the cortex receive sensory input, others integrate information and coordinate appropriate motor responses.

cerebrospinal fluid Clear extracellular fluid that surrounds and cushions the brain and spinal cord.

cerebrum (suh-REE-bruhm) Part of the vertebrate forebrain that originally integrated olfactory input and selected motor responses to it. In mammals, it evolved into the most complex integrating center.

channel protein Type of transport protein that serves as a pore through which ions or other water-soluble substances move across the plasma membrane. Some channels remain open, while others are gated and open and close in controlled ways.

chemical bond A union between the electron structures of two or more atoms or ions.

chemical synapse (SIN-aps) [Gk. *synapsis*, union] A small gap, the synaptic cleft, that separates two neurons (or a neuron and a muscle cell or gland cell) and that is bridged by neurotransmitter molecules released from the presynaptic neuron.

chemiosmotic theory (kim-ee-OZ-MOT-ik) Theory that an electrochemical gradient across a cell membrane drives ATP formation. Metabolic reactions cause hydrogen ions (H^+) to accumulate in a compartment formed by the membrane. The combined force of the resulting concentration and electric gradients propels hydrogen ions down the gradient, through channel proteins. Through enzyme action at these proteins, ADP and inorganic phosphate combine to form ATP.

chemoreceptor (KEE-moe-ree-sep-tur) Sensory receptor that detects chemical energy (ions or molecules) dissolved in the surrounding fluid.

chemosynthetic autotroph (KEE-moe-sin-THET-ik) One of a few kinds of bacteria able to synthesize its own organic molecules using carbon dioxide as the carbon source and certain inorganic substances (such as sulfur) as the energy source.

chlorofluorocarbon (KLORE-oh-FLOOR-oh-car-bun), or **CFC** One of a variety of odorless, invisible compounds of chlorine, fluorine, and carbon, widely used in commercial products, that are contributing to the destruction of the ozone layer above the earth's surface.

chlorophylls (KLOR-uh-fills) [Gk. *chloros*, green, + *phyllon*, leaf] Light-sensitive pigment molecules that absorb violet-to-blue and red wavelengths but that transmit green. Certain chlorophylls donate the electrons required for photosynthesis.

chloroplast (KLOR-uh-plast) An organelle that specializes in photosynthesis in plants and certain protistans.

chordate An animal having a notochord, a dorsal hollow nerve cord, a pharynx, and gill slits in the pharynx wall for at least part of its life cycle.

chorion (CORE-ee-on) Of placental mammals, one of four extraembryonic membranes; it becomes a major component of the placenta. Absorptive structures (villi) that develop at its surface are crucial for the transfer of substances between the embryo and mother.

chromatid Of a duplicated eukaryotic chromosome, one of two DNA molecules and its associated proteins. One chromatid remains attached to its "sister" chromatid at the centromere until they are separated from each other during a nuclear division; then each is a separate chromosome.

chromosome (CROW-moe-some) [Gk. *chroma*, color, + *soma*, body] Of eukaryotes, a DNA molecule with many associated proteins. A bacterial chromosome does not have a comparable profusion of proteins associated with the DNA.

chromosome number Of eukaryotic species, the number of each type of chromosome in all cells except dividing germ cells or gametes.

chytrid A type of single-celled fungus of muddy and aquatic habitats.

ciliated protozoan One of four major groups of protozoans.

cilium (SILL-ee-um), plural **cilia** [L. *cilium*, eyelid] Of eukaryotic cells, a short, hairlike projection that contains a regular array of microtubules. Cilia serve as motile structures, help create currents of fluids, or are part of sensory structures. They typically are more profuse than flagella.

circadian rhythm (ser-KAYD-ee-un) [L. *circa*, about, + *dies*, day] Of many organisms, a cycle of physiological events that is completed every 24 hours or so, even when environmental conditions remain constant.

circulatory system Of multicelled animals, an organ system consisting of a muscular pump (heart, most often), blood vessels, and blood; the system transports materials to and from cells

and often helps stabilize body temperature and pH.

cladistics An approach to biological systematics in which organisms are grouped according to similarities that are derived from a common ancestor.

cladogram Branching diagram that represents patterns of relative relationships between organisms based on discrete morphological, physiological, and behavioral traits that vary among taxa being studied.

classification system A way of organizing and retrieving information about species.

cleavage Stage of animal development when mitotic cell divisions convert a zygote to a ball of cells, the blastula.

cleavage furrow Of an animal cell undergoing cytoplasmic division, a shallow, ringlike depression that forms at the cell surface as contractile microfilaments pull the plasma membrane inward. It defines where the cytoplasm will be cut in two.

climate Prevailing weather conditions for an ecosystem, including temperature, humidity, wind speed, cloud cover, and rainfall.

climax community Following primary or secondary succession, the array of species that remains more or less steady under prevailing conditions.

clonal selection theory Theory that lymphocytes activated by a specific antigen rapidly multiply and differentiate into huge subpopulations of cells, all having the parent cell's specificity against that antigen.

cloned DNA Multiple, identical copies of DNA fragments that have been inserted into plasmids or some other cloning vector.

club fungus One of a highly diverse group of multicelled fungi, the reproductive structures of which have microscopic, club-shaped cells that produce and bear spores.

cnidarian A radially symmetrical invertebrate, usually marine, that has tissues (not organs), and nematocysts. Two body forms (medusae and polyps) are common. Jellyfishes, corals, and sea anemones are examples.

coal A nonrenewable source of energy that formed more than 280 million years ago from submerged, undecayed plant remains that were buried in sediments, compressed, then compacted further by heat and pressure.

codominance Condition in which a pair of nonidentical alleles are both

expressed even though they specify two different phenotypes.

codon One of a series of base triplets in an mRNA molecule, most of which code for a sequence of amino acids of a specific polypeptide chain. (Of sixty-four codons, sixty-one specify different amino acids and three of these also serve as start signals for translation; one other serves only as a stop signal for translation.)

coelum (SEE-lum) [Gk. *koilos*, hollow] Of many animals, a type of body cavity located between the gut and body wall and having a distinctive lining (peritoneum).

coenzyme A type of nucleotide that transfers hydrogen atoms and electrons from one reaction site to another. NAD^+ is an example.

coevolution The joint evolution of two or more closely interacting species; when one species evolves, the change affects selection pressures operating between the two species, so the other also evolves.

cofactor A metal ion or coenzyme; it helps catalyze a reaction or serves briefly as an agent that transfers electrons, atoms, or functional groups from one substrate to another.

cohesion Condition in which molecular bonds resist rupturing when under tension.

cohesion theory of water transport Theory that water moves up through vascular plants due to hydrogen bonding among water molecules confined inside the xylem pipelines. The collective cohesive strength of those bonds allows water to be pulled up as columns in response to transpiration (evaporation from leaves).

collenchyma One of the simple tissues of flowering plants; lends flexible support to primary tissues, such as those of lengthening stems.

colon (CO-lun) The large intestine.

commensalism [L. *com*, together, + *mensa*, table] Two-species interaction in which one species benefits significantly while the other is neither helped nor harmed to any notable extent.

communication signal Of social animals, an action or cue sent by one member of a species (the signaler) that can change the behavior of another member (the signal receiver).

community The populations of all species occupying a habitat; also applied to groups of organisms with similar life-styles in a habitat (such as the bird community).

companion cell A specialized parenchyma cell that helps load dissolved organic compounds into the conducting cells of the phloem.

comparative morphology [Gk. *morph*, form] Anatomical comparisons of major lineages.

competitive exclusion Theory that populations of two species competing for a limited resource cannot coexist indefinitely in the same habitat; the population better adapted to exploit the resource will enjoy a competitive (hence reproductive) edge and will eventually exclude the other population from the habitat.

complement system A set of about twenty proteins circulating in blood plasma with roles in nonspecific defenses and in immune responses. Some induce lysis of pathogens, others promote inflammation, and others stimulate phagocytes to engulf pathogens.

compound A substance in which the relative proportions of two or more elements never vary. Organic compounds have a backbone of carbon atoms arranged as a chain or ring structure. The simpler, inorganic compounds do not have comparable backbones.

concentration gradient A difference in the number of molecules (or ions) of a substance between two adjacent regions, as in a volume of fluid.

condensation reaction Enzyme-mediated reaction leading to the covalent linkage of small molecules and, often, the formation of water as a by-product.

cone cell In the vertebrate eye, a type of photoreceptor that responds to intense light and contributes to sharp daytime vision and color perception.

conifer A type of plant belonging to the dominant group of gymnosperms; mostly evergreen, woody trees and shrubs with pollen- and seed-bearing cones.

connective tissue proper A category of animal tissues, all having mostly the same components but in different proportions. These tissues contain fibroblasts and other cells, the secretions of which form fibers (of collagen and elastin) and a ground substance (of modified polysaccharides).

consumers [L. *consumere*, to take completely] Of ecosystems, heterotrophic organisms that obtain energy and raw materials by feeding on the tissues of other organisms. Herbivores, carnivores, omnivores, and parasites are examples.

continuous variation A more or less continuous range of small differences in a given trait among all the individuals of a population.

contractile vacuole (kun-TRAK-till VAK-you-ohl) [L. *contractus*, to draw together] In some protistans, a membranous chamber that takes up excess water in the cell body, then contracts, expelling the water outside the cell through a pore.

control group In a scientific experiment, a group used to evaluate possible side effects of a test involving an experimental group. Ideally, the control group should differ from the experimental group only with respect to the variable being studied.

convergence, morphological See morphological convergence.

cork cambium A type of lateral meristem that produces a tough, corky replacement for epidermis on parts of woody plants showing extensive secondary growth.

corpus callosum (CORE-pus ka-LOW-sum) A band of axons (200 million in humans) that functionally link two cerebral hemispheres.

corpus luteum (CORE-pus LOO-tee-um) A glandular structure; it develops from cells of a ruptured ovarian follicle and secretes progesterone and some estrogen, both of which maintain the lining of the uterus (endometrium).

cortex [L. *cortex*, bark] In general, a rindlike layer; the kidney cortex is an example. In vascular plants, ground tissue that makes up most of the primary plant body, supports plant parts, and stores food.

cotyledon A seed leaf, which develops as part of a plant embryo; cotyledons provide nourishment for the germinating seedling.

courtship display Social behavior by which individuals assess and respond to sexual overtures of potential partners.

covalent bond (koe-VAY-lunt) [L. *con*, together, + *valere*, to be strong] A sharing of one or more electrons between atoms or groups of atoms. When electrons are shared equally, the bond is nonpolar. When electrons are shared unequally, the bond is polar—slightly positive at one end and slightly negative at the other.

cross-bridge formation Of muscle cells, the interaction between actin and myosin filaments that is the basis of contraction.

crossing over During prophase I of meiosis, an interaction between a pair of homologous chromosomes. Their non-

sister chromatids break at the same place along their length and exchange corresponding segments at the break points. Crossing over breaks up old combinations of alleles and puts new ones together in chromosomes.

culture The sum total of behavior patterns of a social group, passed between generations by learning and by symbolic behavior, especially language.

cuticle (KEW-tih-kull) A body covering. Of land plants, a covering of waxes and lipid-rich cutin deposited on the outer surface of epidermal cell walls. Of annelids, a thin, flexible surface coat. Of arthropods, a hardened yet lightweight covering with protein and chitin components that functions as an external skeleton.

cycad A type of gymnosperm of the tropics and subtropics; slow growing, with massive, cone-shaped structures that bear ovules or pollen.

cyclic AMP (SIK-lik) Cyclic adenosine monophosphate. A nucleotide that has roles in intercellular communication, as when it serves as a second messenger (a cytoplasmic mediator of a cell's response to signaling molecules).

cyclic pathway of ATP formation Photosynthetic pathway in which excited electrons move from a photosystem to an electron transport system, and back to the photosystem. The electron flow contributes to the formation of ATP from ADP and inorganic phosphate.

cyst Of some microorganisms, a walled, resting structure that forms during the life cycle.

cytochrome (SIGH-toe-krome) [Gk. *kytos*, hollow vessel, + *chrōma*, color] Iron-containing protein molecule; a component of electron transport systems used in photosynthesis and aerobic respiration.

cytokinesis (SIGH-toe-kih-NEE-sis) [Gk. *kinesis*, motion] Cytoplasmic division; the splitting of a parental cell into daughter cells.

cytokinin (SIGH-tow-KY-nin) Any of the class of plant hormones that stimulate cell division, promote leaf expansion, and retard leaf aging.

cytomembrane system [Gk. *kytos*, hollow vessel] Organelles, functioning as a system to modify, package, and distribute newly formed proteins and lipids. Endoplasmic reticulum, Golgi bodies, lysosomes, and a variety of vesicles are its components.

cytoplasm (SIGH-toe-plaz-um) [Gk. *plassein*, to mold] All cellular parts, particles, and semifluid substances enclosed by the plasma membrane except for the region of DNA (which in eukaryotes, is the nucleus).

cytosine (SIGH-toe-seen) A pyrimidine; one of the nitrogen-containing bases in nucleotides.

cytoskeleton Of eukaryotic cells, an internal "skeleton." Its microtubules and other components structurally support the cell, organize and move its internal components. The cytoskeleton also helps free-living cells move through their environment.

cytotoxic T cell A T lymphocyte that eliminates infected body cells or tumor cells with a single hit of toxins and perforins.

decomposers [L. de-, down, away, + companere, to put together] Of ecosystems, heterotrophs that obtain energy by chemically breaking down the remains, products, or wastes of other organisms. Their activities help cycle nutrients back to producers. Certain fungi and bacteria are examples.

deforestation The removal of all trees from a large tract of land, such as the Amazon Basin or the Pacific Northwest.

degradative pathway A metabolic pathway by which molecules are broken down in stepwise reactions that lead to products of lower energy.

deletion A change in a chromosome's structure after one of its regions is lost as a result of irradiation, viral attack, chemical action, or some other factor.

demographic transition model Model of human population growth in which changes in the growth pattern correspond to different stages of economic development. These are a preindustrial stage, when birth and death rates are both high, a transitional stage, an industrial stage, and a postindustrial stage, when the death rate exceeds the birth rate.

denaturation (deh-NAY-chur-AY-shun) Of any molecule, the loss of three-dimensional shape following disruption of hydrogen bonds and other weak bonds.

dendrite (DEN-drite) [Gk. dendron, tree] A short, slender extension from the cell body of a neuron.

denitrification (DEE-nite-rih-fih-KAY-shun) The conversion of nitrate or nitrite by certain bacteria to gaseous nitrogen (N_2) and a small amount of nitrous oxide (N_2O).

density-dependent controls Factors such as predation, parasitism, disease, and competition for resources, which limit population growth by reducing the birth rate, increasing the rates of death and dispersal, or all of these.

density-independent controls Factors such as storms or floods that increase a population's death rate more or less independently of its density.

dentition (den-TIH-shun) The type, size, and number of an animal's teeth.

dermal tissue system Of vascular plants, the tissues that cover and protect the plant surfaces.

dermis The layer of skin underlying the epidermis, consisting mostly of dense connective tissue.

desert A type of biome that exists where the potential for evaporation greatly exceeds rainfall and vegetation cover is limited.

desertification (dez-urt-ih-fih-KAY-shun) The conversion of grasslands, rain-fed cropland, or irrigated cropland to desertlike conditions, with a drop in agricultural productivity of 10 percent or more.

detrital food web Of most ecosystems, the flow of energy mainly from plants through detritivores and decomposers.

detritivores (dih-TRY-tih-vorez) [L. detritus; after deterere, to wear down] Of ecosystems, heterotrophs that consume dead or decomposing particles of organic matter. Earthworms, crabs, and nematodes are examples.

deuterostome (DUE-ter-oh-stome) [Gk. deuteros, second, + stoma, mouth] Any of the bilateral animals, including echinoderms and chordates, in which the first indentation in the early embryo develops into the anus.

diaphragm (DIE-uh-fram) [Gk. diaphragma, to partition] Muscular partition between the thoracic and abdominal cavities, the contraction and relaxation of which contribute to breathing. Also, a contraceptive device used temporarily to prevent sperm from entering the uterus during sexual intercourse.

dicot (DIE-kot) [Gk. di, two, + kotylēdōn, cup-shaped vessel] Short for dicotyledon; class of flowering plants characterized generally by seeds having embryos with two cotyledons (seed leaves), net-veined leaves, and floral parts arranged in fours, fives, or multiples of these.

differentiation See cell differentiation.

diffusion Net movement of like molecules (or ions) down their concentration gradient. In the absence of other forces, molecular motion and random collisions cause their net outward movement from one region into a neighboring region where they are less concentrated (because collisions are more frequent where the molecules are most crowded together).

digestive system An internal tube or cavity from which ingested food is absorbed into the internal environment; often divided into regions specialized for food transport, processing, and storage.

dihybrid cross An experimental cross in which offspring inherit two gene pairs, each consisting of two nonidentical alleles.

dinoflagellate A photosynthetic or heterotrophic protistan, often flagellated, that is a component of plankton.

diploid number (DIP-loyd) For many sexually reproducing species, the chromosome number of somatic cells and of germ cells prior to meiosis. Such cells have two chromosomes of each type (that is, pairs of homologous chromosomes). Compare haploid number.

directional selection Of a population, a shift in allele frequencies in a steady, consistent direction in response to a new environment or to a directional change in the old one. The outcome is that forms of traits at one end of the range of phenotypic variation become more common than the intermediate forms.

disaccharide (die-SAK-uh-ride) [Gk. di, two, + sakcharon, sugar] A type of simple carbohydrate, of the class called oligosaccharides; two monosaccharides covalently bonded.

disruptive selection Of a population, a shift in allele frequencies to forms of traits at both ends of a range of phenotypic variation and away from intermediate forms.

distal tubule The tubular section of a nephron most distant from the glomerulus; a major site of water and sodium reabsorption.

divergence Accumulation of differences in allele frequencies between populations that have become reproductively isolated from one another.

divergence, morphological See morphological divergence.

diversity, organismic Sum total of variations in form, function, and behavior that have accumulated in different lineages. Those variations generally are adaptive to prevailing conditions or were adaptive to conditions that existed in the past.

DNA Deoxyribonucleic acid (dee-OX-ee-RYE-bow-new-CLAY-ik). For all cells (and many viruses), the molecule of inheritance. A category of nucleic acids, each usually consisting of two nucleotide strands twisted together helically and held together by hydrogen bonds. The nucleotide sequence encodes the instructions for assembling proteins, and, ultimately, new individuals of a particular species.

DNA-DNA hybridization *See* nucleic acid hybridization.

DNA fingerprint Of each individual, a unique array of RFLPs, resulting from the DNA sequences inherited (in a Mendelian pattern) from each parent.

DNA library A collection of DNA fragments produced by restriction enzymes and incorporated into plasmids.

DNA ligase (LYE-gaze) Enzyme that seals together the new base-pairings during DNA replication; also used by recombinant DNA technologists to seal base-pairings between DNA fragments and cut plasmid DNA.

DNA polymerase (poe-LIM-uh-raze) Enzyme that assembles a new strand on a parent DNA strand during replication; also takes part in DNA repair.

DNA probe A short DNA sequence that has been assembled from radioactively labeled nucleotides and that can base-pair with part of a gene under investigation.

DNA repair Following an alteration in the base sequence of a DNA strand, a process that restores the original sequence, as carried out by DNA polymerases, DNA ligases, and other enzymes.

DNA replication Of cells, the process by which the hereditary material is duplicated for distribution to daughter nuclei. An example is the duplication of eukaryotic chromosomes during interphase, prior to mitosis.

dominance hierarchy Form of social organization in which some members of the group have adopted a subordinate status to others.

dominant allele In a diploid cell, an allele that masks the expression of its partner on the homologous chromosome.

dormancy [L. *dormire*, to sleep] Of plants, the temporary, hormone-mediated cessation of growth under conditions that might appear to be quite suitable for growth.

double fertilization Of flowering plants only, the fusion of one sperm nucleus with the egg nucleus (to produce a zygote), *and* fusion of a second sperm nucleus with the two nuclei of the endosperm mother cell, which gives rise to triploid (3*n*) nutritive tissue.

doubling time The length of time it takes for a population to double in size.

drug addiction Chemical dependence on a drug, following habituation and tolerance of it; the drug takes on an "essential" biochemical role in the body.

dry shrubland A type of biome that exists where annual rainfall is less than 25 to 60 centimeters and where short, woody, multibranched shrubs predominate; chaparral is an example.

dry woodland A type of biome that exists when annual rainfall is about 40 to 100 centimeters; there may be tall trees, but these do not form a dense canopy.

dryopith A type of hominoid, one of the first to appear during the Miocene about the time of the divergences that led to gorillas, chimpanzees, and humans.

duplication A change in a chromosome's structure resulting in the repeated appearance of the same gene sequence.

early *Homo* A type of early hominid that may have been the maker of stone tools that date from about 2.5 million years ago.

echinoderm A type of invertebrate that has calcified spines, needles, or plates on the body wall. It is radially symmetrical but with some bilateral features. Sea stars and sea urchins are examples.

ecology [Gk. *oikos*, home, + *logos*, reason] Study of the interactions of organisms with one another and with their physical and chemical environment.

ecosystem [Gk. *oikos*, home] An array of organisms and their physical environment, all of which interact through a flow of energy and a cycling of materials.

ecosystem modeling Analytical method of predicting unforeseen effects of disturbances to an ecosystem, based on computer programs and models.

ectoderm [Gk. *ecto*, outside, + *derma*, skin] Of animal embryos, the outermost primary tissue layer (germ layer) that gives rise to the outer layer of the integument and to tissues of the nervous system.

effector Of homeostatic systems, a muscle (or gland) that responds to signals from an integrator (such as the brain) by producing movement (or chemical change) that helps adjust the body to changing conditions.

effector cell Of the differentiated subpopulations of lymphocytes that form during an immune response, the type of cell that engages and destroys the antigen-bearing agent that triggered the response.

egg A type of mature female gamete; also called an ovum.

El Niño A recurring, massive displacement to the east of warm surface waters of the western equatorial Pacific, which in turn displaces the cooler waters of the Humboldt Current off the coast of Peru.

electron Negatively charged unit of matter, with both particulate and wavelike properties, that occupies one of the orbitals around the atomic nucleus. Atoms can gain, lose, or share electrons with other atoms.

electron transport phosphorylation (FOSS-for-ih-LAY-shun) Final stage of aerobic respiration, in which ATP forms after hydrogen ions and electrons (from the Krebs cycle) are sent through a transport system that gives up the electrons to oxygen.

electron transport system An organized array of enzymes and cofactors, bound in a cell membrane, that accept and donate electrons in sequence. When such systems operate, hydrogen ions (H^+) flow across the membrane, and the flow drives ATP formation and other reactions.

element Any substance that cannot be decomposed into substances with different properties.

embryo (EM-bree-oh) [Gk. *en*, in, + probably *bryein*, to swell] Of animals generally, the stage formed by way of cleavage, gastrulation, and other early developmental events. Of seed plants, the young sporophyte, from the first cell divisions after fertilization until germination.

embryo sac The female gametophyte of flowering plants.

emulsification Of chyme in the small intestine, a suspension of droplets of fat coated with bile salts.

end product A substance present at the end of a metabolic pathway.

endangered species A species poised at the brink of extinction, owing to the extremely small size and severely limited genetic diversity of its remaining populations.

endergonic reaction (en-dur-GONE-ik) Chemical reaction showing a net gain in energy.

endocrine gland Ductless gland that secretes hormones into interstitial fluid, after which they are distributed by way of the bloodstream.

endocrine system System of cells, tissues, and organs that is functionally linked to the nervous system and that exerts control by way of its hormones and other chemical secretions.

endocytosis (EN-doe-sigh-TOE-sis) Movement of a substance into cells; the substance becomes enclosed by a patch of plasma membrane that sinks into the cytoplasm, then forms a vesicle around it. Phagocytic cells also engulf pathogens or prey in this manner.

endoderm [Gk. *endon*, within, + *derma*, skin] Of animal embryos, the inner primary tissue layer, or germ layer, that gives rise to the inner lining of the gut and organs derived from it.

endodermis A sheetlike wrapping of single cells around the vascular cylinder of a root; it functions in controlling the uptake of water and dissolved nutrients. An impermeable barrier (Casparian strip) prevents water from passing between the walls of abutting endodermal cells.

endometrium (EN-doh-MEET-ree-um) [Gk. *metrios*, of the womb] Inner lining of the uterus, consisting of connective tissues, glands, and blood vessels.

endoplasmic reticulum or **ER** (EN-doe-PLAZ-mik reh-TIK-yoo-lum) An organelle that begins at the nucleus and curves through the cytoplasm. In rough ER (which has many ribosomes on its cytoplasmic side), many new polypeptide chains acquire specialized side chains. In many cells, smooth ER (with no attached ribosomes) is the main site of lipid synthesis.

endoskeleton [Gk. *endon*, within, + *sklēros*, hard, stiff] In chordates, the internal framework of bone, cartilage, or both. Together with skeletal muscle, supports and protects other body parts, helps maintain posture, and moves the body.

endosperm (EN-doe-sperm) Nutritive tissue that surrounds and serves as food for a flowering plant embryo and, later, for the germinating seedling.

endospore Of certain bacteria, a resistant body that forms around DNA and some cytoplasm; it germinates and gives rise to new bacterial cells when conditions become favorable.

endosymbiosis A permanent, mutually beneficial interdependency between two species, one of which resides permanently inside the other's body.

energy The capacity to do work.

energy carrier A molecule that delivers energy from one metabolic reaction site to another. ATP is the most widely travelled of these; it readily donates energy to nearly all metabolic reactions.

energy flow pyramid A pyramid-shaped representation of an ecosystem's trophic structure, illustrating the energy losses at each transfer to a different trophic level.

entropy (EN-trow-pee) A measure of the degree of disorder in a system (how much energy has become so disorganized and dispersed, usually as heat, that it is no longer readily available to do work).

enzyme (EN-zime) One of a class of proteins that greatly speed up (catalyze) reactions between specific substances, usually at their functional groups. The substances that each type of enzyme acts upon are called its substrates.

eosinophil Fast-acting, phagocytic white blood cell that takes part in inflammation but not in immune responses.

epidermis The outermost tissue layer of a multicelled plant or animal.

epiglottis A flaplike structure at the start of the larynx, the position of which directs the movement of air into the trachea or food into the esophagus.

epistasis (eh-PISS-tih-sis) A type of gene interaction, whereby two alleles of a gene influence the expression of alleles of a different gene.

epithelium (EP-ih-THEE-lee-um) An animal tissue consisting of one or more layers of adhering cells that covers the body's external surfaces and lines its internal cavities and tubes. Epithelium has one free surface; the opposite surface rests on a basement membrane between it and an underlying connective tissue. Epidermis or skin is an example.

equilibrium, dynamic [Gk. *aequus*, equal, + *libra*, balance] The point at which a chemical reaction runs forward as fast as it runs in reverse; thus the concentrations of reactant molecules and product molecules show no net change.

erythrocyte (eh-RITH-row-site) [Gk. *erythros*, red, + *kytos*, vessel] Red blood cell.

esophagus (ee-SOF-uh-gus) Tubular portion of a digestive system that receives swallowed food and leads to the stomach.

essential amino acid Any of eight amino acids that certain animals cannot synthesize for themselves and must obtain from food.

essential fatty acid Any of the fatty acids that certain animals cannot synthesize for themselves and must obtain from food.

estrogen (ESS-trow-jen) A sex hormone that helps oocytes mature, induces changes in the uterine lining during the menstrual cycle and pregnancy, and maintains secondary sexual traits; also influences bodily growth and development.

estrus (ESS-truss) [Gk. *oistrus*, frenzy] For mammals generally, the cyclic period of a female's sexual receptivity to the male.

estuary (EST-you-ehr-ee) A partly enclosed coastal region where seawater mixes with freshwater from rivers, streams, and runoff from the surrounding land.

ethylene (ETH-il-een) Plant hormone that stimulates fruit ripening and triggers abscission.

eubacteria The subkingdom of all bacterial species except the archaebacteria; one of the three great prokaryotic lineages that arose early in the history of life.

euglenoid A type of flagellated protistan, most of which are photosynthesizers in stagnant or freshwater ponds.

eukaryotic cell (yoo-CARRY-oh-tic) [Gk. *eu*, good, + *karyon*, kernel] A type of cell that has a "true nucleus" and other distinguishing membrane-bound organelles. *Compare* prokaryotic cell.

eutrophication Nutrient enrichment of a body of water, such as a lake, that typically results in reduced transparency and a phytoplankton-dominated community.

evaporation [L. *e-*, out, + *vapor*, steam] Conversion of a substance from the liquid to the gaseous state; some or all of its molecules leave in the form of vapor.

evolution, biological [L. *evolutio*, act of unrolling] Change within a line of descent over time. A population is evolving when some forms of a trait are becoming more or less common, relative to the other kinds of traits. The shifts are evidence of changes in the relative abundances of alleles for that trait, as brought about by mutation, natural selection, genetic drift, and gene flow.

evolutionary tree A treelike diagram in which branches represent separate lines of descent from a common ancestor.

excitatory postsynaptic potential or **EPSP** One of two competing signals at an input zone of a neuron; a graded potential that brings the neuron's plasma membrane closer to threshold.

excretion Any of several processes by which excess water, excess or harmful solutes, or waste materials leave the body by way of the urinary system or certain glands.

exergonic reaction (EX-ur-GONE-ik) A chemical reaction that shows a net loss in energy.

exocrine gland (EK-suh-krin) [Gk. *es*, out of, + *krinein*, to separate] Glandular structure that secretes products, usually through ducts or tubes, to a free epithelial surface.

exocytosis (EK-so-sigh-TOE-sis) Movement of a substance out of a cell by means of a transport vesicle, the membrane of which fuses with the plasma membrane, so that the vesicle's contents are released outside.

exodermis Layer of cells just inside the root epidermis of most flowering plants; helps control the uptake of water and solutes.

exon Of eukaryotic cells, any of the nucleotide sequences of a pre-mRNA molecule that are spliced together to form the mature mRNA transcript and are ultimately translated into protein.

exoskeleton [Gk. *exo*, out, + *skleros*, hard, stiff] An external skeleton, as in arthropods.

experiment A test in which some phenomenon in the natural world is manipulated in controlled ways to gain insight into its function, structure, operation, or behavior.

exploitation competition Interaction in which both species have equal access to a required resource but differ in how fast or efficiently they exploit it.

exponential growth (EX-po-NEN-shul) Pattern of population growth in which greater and greater numbers of individuals are produced during the successive doubling times; the pattern that emerges when the per capita birth rate remains even slightly above the per capita death rate, putting aside the effects of immigration and emigration.

extinction, background A steady rate of species turnover that characterizes lineages through most of their histories.

extinction, mass An abrupt increase in the rate at which major taxa disappear, with several taxa being affected simultaneously.

extracellular fluid In animals generally, all the fluid not inside cells; includes plasma (the liquid portion of blood) and interstitial fluid (which occupies the spaces between cells and tissues).

extracellular matrix A material, largely secreted, that helps hold many animal tissues together in certain shapes; it consists of fibrous proteins and other components in a ground substance.

FAD Flavin adenine dinucleotide, a nucleotide coenzyme. When delivering electrons and unbound protons (H^+) from one reaction to another, it is abbreviated $FADH_2$.

fall overturn The vertical mixing of a body of water in autumn. Its upper layer cools, increases in density, and sinks; dissolved oxygen moves down and nutrients from bottom sediments are brought to the surface.

family pedigree A chart of genetic relationships of the individuals in a family through successive generations.

fat A lipid with a glycerol head and one, two, or three fatty acid tails. The tails of saturated fats have only single bonds between carbon atoms and hydrogen atoms attached to all other bonding sites. Tails of unsaturated fats additionally have one or more double bonds between certain carbon atoms.

fatty acid A long, flexible hydrocarbon chain with a —COOH group at one end.

feedback inhibition Of cells, a control mechanism by which the production (or secretion) of a substance triggers a change in some activity that in turn shuts down further production of the substance.

fermentation [L. *fermentum*, yeast] A type of anaerobic pathway of ATP formation, it starts with glycolysis, ends when electrons are transferred back to one of the breakdown products or intermediates, and regenerates the NAD^+ required for the reaction. Its net yield is two ATP per glucose molecule degraded.

fern One of the seedless vascular plants, mostly of wet, humid habitats; requires free water to complete its life cycle.

fertilization [L. *fertilis*, to carry, to bear] Fusion of a sperm nucleus with the nucleus of an egg, which thereupon becomes a zygote.

fever A body temperature higher than a set point that is preestablished in the brain region governing temperature.

fibrous root system Of most monocots, all the lateral branchings of adventitious roots, which arose earlier from the young stem.

filtration Of urine formation, the process by which blood pressure forces water and solutes out of glomerular capillaries and into the cupped portion of a nephron wall (Bowman's capsule).

fin Of fishes generally, an appendage that helps propel, stabilize, and guide the body through water.

first law of thermodynamics [Gk. *therme*, heat, + *dynamikos*, powerful] Law stating that the total amount of energy in the universe remains constant. Energy cannot be created and existing energy cannot be destroyed. It can only be converted from one form to another.

fish An aquatic animal of the most ancient vertebrate lineage; jawless fishes (such as lampreys and hagfishes), cartilaginous fishes (such as sharks), and bony fishes (such as coelacanths and salmon) are the three existing groups.

fixed action pattern An instinctive response that is triggered by a well-defined, simple stimulus and that is performed in its entirety once it has begun.

flagellated protozoan A member of one of four major groups of protozoans, many of which cause serious diseases.

flagellum (fluh-JELL-um), plural **flagella** [L. whip] Tail-like motile structure of many free-living eukaryotic cells; it has a distinctive 9 + 2 array of microtubules.

flatworm A type of invertebrate having bilateral symmetry, a flattened body, and a saclike gut; a turbellarian, fluke, or tapeworm.

flower The reproductive structure that distinguishes angiosperms from other seed plants and often attracts pollinators.

fluid mosaic model Model of membrane structure in which proteins are embedded in a lipid bilayer or attached to one of its surfaces. The lipid molecules give the membrane its basic structure, impermeability to water-soluble molecules, and (through packing variations and movements) fluidity. Proteins carry out most membrane functions, such as transport, enzyme action, and reception of signals or substances.

follicle (FOLL-ih-kul) In a mammalian ovary, a primary oocyte (immature egg) together with the surrounding layer of cells.

food chain A straight-line sequence of who eats whom in an ecosystem.

food web A network of cross-connecting, interlinked food chains, encompassing primary producers and an array of consumers, detritivores, and decomposers.

forebrain Brain region that includes the cerebrum and cerebral cortex, the olfactory lobes, and the hypothalamus.

forest A type of biome where tall trees grow together closely enough to form a fairly continuous canopy over a broad region.

fossil Recognizable evidence of an organism that lived in the distant past. Most fossils are skeletons, shells, leaves, seeds, and tracks that were buried in rock layers before they decomposed.

fossil fuel Coal, petroleum, or natural gas; a nonrenewable source of energy formed in sediments by the compression of carbon-containing plant remains over hundreds of millions of years.

founder effect An extreme case of genetic drift. By chance, a few individuals that leave a population and establish a new one carry fewer (or more) alleles for certain traits. Increased variation between the two populations is one outcome. Limited genetic variability in the new population is another.

free radical A highly reactive, unbound molecular fragment with the wrong number of electrons.

fruit [L. after *frui*, to enjoy] Of flowering plants, the expanded and ripened ovary of one or more carpels, sometimes with accessory structures incorporated.

FSH Follicle-stimulating hormone. Produced and secreted by the anterior lobe of the pituitary gland, this hormone has roles in the reproductive functions of both males and females.

functional group An atom or group of atoms that is covalently bonded to the carbon backbone of an organic compound and that influences its behavior.

Fungi The kingdom of fungi.

fungus A eukaryotic heterotroph that uses extracellular digestion and absorption; it secretes enzymes able to break down an external food source into molecules small enough to be absorbed by its cells. Saprobic types feed on nonliving organic matter; parasitic types feed on living organisms. Fungi as a group are major decomposers.

gall bladder Organ of the digestive system that stores bile secreted from the liver.

gamete (GAM-eet) [Gk. *gametēs*, husband, and *gametē*, wife] A haploid cell that functions in sexual reproduction. Sperm and eggs are examples.

gamete formation Generally, the formation of gametes by way of meiosis. Of animals, the first stage of development, in which sperm or eggs form and mature within reproductive tissues of parents.

gametophyte (gam-EET-oh-fite) [Gk. *phyton*, plant] The haploid, multicelled, gamete-producing phase in the life cycle of most plants.

ganglion (GANG-lee-un), plural **ganglia** [Gk. *ganglion*, a swelling] A distinct clustering of cell bodies of neurons in regions other than the brain or spinal cord.

gastrulation (gas-tru-LAY-shun) Of animals, the stage of embryonic development in which cells become arranged into two or three primary tissue layers (germ layers); in humans, the layers are an inner endoderm, an intermediate mesoderm, and a surface ectoderm.

gene [short for German *pangan*, after Gk. *pan*, all + *genes*, to be born] A unit of information about a heritable trait that is passed on from parents to offspring. Each gene has a specific location on a chromosome.

gene flow A microevolutionary process; a physical movement of alleles out of a population as individuals leave (emigrate) or enter (immigrate), the outcome being changes in allele frequencies.

gene frequency More precisely, allele frequency: the relative abundances of all the different alleles for a trait that are carried by the individuals of a population.

gene locus A given gene's particular location on a chromosome.

gene mutation [L. *mutatus*, a change] Change in DNA due to the deletion, addition, or substitution of one to several bases in the nucleotide sequence.

gene pair In diploid cells, the two alleles at a given locus on a pair of homologous chromosomes.

gene pool Sum total of all genotypes in a population. More accurately, allele pool.

gene therapy Generally, the transfer of one or more normal genes into the body cells of an organism in order to correct a genetic defect.

genetic code [After L. *genesis*, to be born] The correspondence between nucleotide triplets in DNA (then in mRNA) and specific sequences of amino acids in the resulting polypeptide chains; the basic language of protein synthesis.

genetic disorder An inherited condition that results in mild to severe medical problems.

genetic drift A microevolutionary process; a change in allele frequencies over the generations due to chance events alone.

genetic engineering Altering the information content of DNA through use of recombinant DNA technology.

genetic equilibrium Hypothetical state of a population in which allele frequencies for a trait remain stable through the generations; a reference point for measuring rates of evolutionary change.

genetic recombination Presence of a new combination of alleles in a DNA molecule compared to the parental genotype; the result of processes such as crossing over at meiosis, chromosome rearrangements, gene mutation, and recombinant DNA technology.

genome All the DNA in a haploid number of chromosomes of a given species.

genotype (JEEN-oh-type) Genetic constitution of an individual. Can mean a single gene pair or the sum total of the individual's genes. *Compare* phenotype.

genus, plural **genera** (JEEN-US, JEN-er-ah) [L. *genus*, race, origin] A taxon into which all species exhibiting certain phenotypic similarities and evolutionary relationship are grouped.

geologic time scale A time scale for earth history, the subdivisions of which have been refined by radioisotope dating work.

germ cell Of animals, one of a cell lineage set aside for sexual reproduction; germ cells give rise to gametes. *Compare* somatic cell.

germ layer Of animal embryos, one of two or three primary tissue layers that form during gastrulation and that gives rise to certain tissues of the adult body. *Compare* ectoderm; endoderm; mesoderm.

germination (jur-min-AY-shun) Generally, the resumption of growth following a rest stage; of seed plants, the time at which an embryo sporophyte breaks through its seed coat and resumes growth.

gibberellin (JIB-er-ELL-un) Any of a class of plant hormones that promote stem elongation.

gill A respiratory organ, typically with a moist, thin vascularized layer of epidermis that functions in gas exchange.

ginkgo A type of gymnosperm with fan-shaped leaves and fleshy coated seeds; now represented by a single species of deciduous trees.

gland A secretory cell or multicelled structure derived from epithelium and often connected to it.

glomerular capillaries The set of blood capillaries inside Bowman's capsule of the nephron.

glomerulus (glow-MARE-you-luss) [L. *glomus*, ball] The first portion of the nephron, where water and solutes are filtered from blood.

glucagon (GLUE-kuh-gone) Hormone that stimulates conversion of glycogen and amino acids to glucose; secreted by alpha cells of the pancreas when the flow of glucose decreases.

glyceride (GLISS-er-eyed) One of the molecules, commonly called fats and oils, that has one, two, or three fatty acid tails attached to a glycerol backbone. They are the body's most abundant lipids and its richest source of energy.

glycerol (GLISS-er-oh) [Gk. *glykys*, sweet, + L. *oleum*, oil] A three-carbon molecule with three hydroxyl groups attached; together with fatty acids, a component of fats and oils.

glycogen (GLY-kuh-jen) In animals, a storage polysaccharide that is a main food reserve; can be readily broken down into glucose subunits.

glycolysis (gly-CALL-ih-sis) [Gk. *glykys*, sweet, + *lysis*, loosening or breaking apart] Initial reactions of both aerobic and anaerobic pathways by which glucose (or some other organic compound) is partially broken down to pyruvate, with a net yield of two ATP. Glycolysis proceeds in the cytoplasm of all cells, and oxygen has no role in it.

gnetophyte A type of gymnosperm limited to deserts and tropics.

Golgi body (GOHL-gee) Organelle in which newly synthesized polypeptide chains as well as lipids are modified and packaged in vesicles for export or for transport to specific locations within the cytoplasm.

gonad (GO-nad) Primary reproductive organ in which gametes are produced.

graded potential Of neurons, a local signal that slightly changes the voltage difference across a small patch of the plasma membrane. Such signals vary in magnitude, depending on the stimulus. With prolonged or intense stimulation, they may spread to a trigger zone of the membrane and initiate an action potential.

granum, plural **grana** Within many chloroplasts, any of the stacks of flattened, membranous compartments with chlorophyll and other light-trapping pigments and reaction sites for ATP formation.

grassland A type of biome with flat or rolling land, 25 to 100 centimeters of annual rainfall, warm summers, and often grazing and periodic fires that regenerate the dominant species.

gravitropism (GRAV-ih-TROPE-izm) [L. *gravis*, heavy, + Gk. *trepein*, to turn] The tendency of a plant to grow directionally in response to the earth's gravitational force.

gray matter Of vertebrates, the dendrites, neuron cell bodies, and neuroglial cells of the spinal cord and cerebral cortex.

grazing food web Of most ecosystems, the flow of energy from plants to herbivores, then through an array of carnivores.

green alga One of a group or division of aquatic plants with an abundance of chlorophylls a and b; early members of its lineage may have given rise to the bryophytes and vascular plants.

green revolution In developing countries, the use of improved crop varieties, modern agricultural practices (including massive inputs of fertilizers and pesticides), and equipment to increase crop yields.

greenhouse effect Warming of the lower atmosphere due to the presence of greenhouse gases—carbon dioxide, methane, nitrous oxide, ozone, water vapor, and chlorofluorocarbons.

ground meristem (MARE-ih-stem) [Gk. *meristos*, divisible] Of vascular plants, a primary meristem that produces the ground tissue system, hence the bulk of the plant body.

ground substance Of certain animal tissues, the intercellular material made up of cell secretions and other noncellular components.

ground tissue system Tissues that make up the bulk of the vascular plant body; parenchyma is the most common of these.

guanine A nitrogen-containing base; present in one of the four nucleotide building blocks of DNA and RNA.

guard cell Either of two adjacent cells having roles in the movement of gases and water vapor across leaf or stem epidermis. An opening (stoma) forms when both cells swell with water and move apart; it closes when they lose water and collapse against each other.

gut A body region where food is digested and absorbed; of complete digestive systems, the gastrointestinal tract (the portions from the stomach onward).

gymnosperm (JIM-noe-sperm) [Gk. *gymnos*, naked, + *sperma*, seed] A plant that bears seeds at exposed surfaces of reproductive structures, such as cone scales. Pine trees are examples.

habitat [L. *habitare*, to live in] The type of place where an organism normally lives, characterized by physical features, chemical features, and the presence of certain other species.

hair cell Type of mechanoreceptor that may give rise to action potentials when bent or tilted.

halophile A type of archaebacterium that lives in extremely salty habitats.

haploid number (HAP-loyd) The chromosome number of a gamete which, as an outcome of meiosis, is only half that of the parent germ cell (it has only one of each pair of homologous chromosomes). *Compare* diploid number.

HCG Human chorionic gonadotropin. A hormone that helps maintain the lining of the uterus during the menstrual cycle and during the first trimester of pregnancy.

heart Muscular pump that keeps blood circulating through the animal body.

helper T cell One of the T lymphocytes; when activated, it produces and secretes interleukins that promote formation of huge populations of effector and memory cells for immune responses.

hemoglobin (HEEM-oh-glow-bin) [Gk. *haima*, blood, + L. *globus*, ball] Iron-containing, oxygen-transporting protein that gives red blood cells their color.

hemostasis (HEE-mow-STAY-sis) [Gk. *haima*, blood, + *stasis*, standing] Stopping of blood loss from a damaged blood vessel through coagulation, blood vessel spasm, platelet plug formation, and other mechanisms.

herbivore [L. *herba*, grass, + *vovare*, to devour] Plant-eating animal.

heterocyst (HET-er-oh-sist) Of some filamentous cyanobacteria, a type of thick-walled, nitrogen-fixing cell that forms when nitrogen is scarce.

heterotroph (HET-er-oh-trofe) [Gk. *heteros*, other, + *trophos*, feeder] Organism that cannot synthesize its own organic compounds and must obtain nourishment by feeding on autotrophs, each other, or organic wastes. Animals, fungi, many protistans, and most bacteria are heterotrophs. *Compare* autotroph.

heterozygous condition (HET-er-oh-ZYE-guss) [Gk. *zygoun*, join together] For a given trait, having nonidentical alleles at a particular locus on a pair of homologous chromosomes.

hindbrain One of the three divisions of the vertebrate brain; the medulla oblongata, cerebellum, and pons; includes reflex centers for respiration, blood circulation, and other basic functions; also coordinates motor responses and many complex reflexes.

histone Any of a class of proteins that are intimately associated with DNA and that are largely responsible for its structural (and possibly functional) organization in eukaryotic chromosomes.

homeostasis (HOE-me-oh-STAY-sis) [Gk. *homo*, same, + *stasis*, standing] Of multicelled organisms, a physiological state in which the physical and chemical conditions of the internal environment are being maintained within tolerable ranges.

homeostatic feedback loop An interaction in which an organ (or structure) stimulates or inhibits the output of another organ, then shuts down or increases this activity when it detects that the output has exceeded or fallen below a set point.

hominid [L. *homo*, man] All species on the evolutionary branch leading to modern humans. *Homo sapiens* is the only living representative.

hominoid Apes, humans, and their recent ancestors.

Homo erectus A hominid lineage that emerged between 1.5 million and 300,000 years ago and that may include the direct ancestors of modern humans.

Homo sapiens The hominid lineage of modern humans that emerged between 300,000 and 200,000 years ago.

homologous chromosome (huh-MOLL-uh-gus) [Gk. *homologia*, correspondence] Of sexually reproducing species, one of a pair of chromosomes that resemble each other in size, shape, and the genes they carry, and that line up with each other at meiosis I. The X and Y chromosomes differ in these respects but still function as homologues.

homologous structures The same body parts, modified in different ways, in different lines of descent from a common ancestor.

homozygous condition (HOE-moe-ZYE-guss) Having two identical alleles at a given locus (on a pair of homologous chromosomes).

homozygous dominant condition Having two dominant alleles at a given locus (on a pair of homologous chromosomes).

homozygous recessive condition Having two recessive alleles at a given gene locus (on a pair of homologous chromosomes).

hormone [Gk. *hormon*, to stir up, set in motion] Any of the signaling molecules secreted from endocrine glands, endocrine cells, and some neurons that the bloodstream distributes to nonadjacent target cells (any cell having receptors for that hormone).

horsetail One of the seedless vascular plants, which require free water to complete the life cycle; only one genus has survived to the present.

human genome project A basic research project in which researchers throughout the world are working together to sequence the estimated 3 billion nucleotides present in the DNA of human chromosomes.

hydrogen bond Type of chemical bond in which an atom of a molecule interacts weakly with a neighboring atom that is already taking part in a polar covalent bond.

hydrogen ion A free (unbound) proton; a hydrogen atom that has lost its electron and so bears a positive charge (H^+).

hydrologic cycle A biogeochemical cycle, driven by solar energy, in which water moves slowly through the atmosphere, on or through surface layers of land masses, to the ocean, and back again.

hydrolysis (high-DRAWL-ih-sis) [L. *hydro*, water, + Gk. *lysis*, loosening or breaking apart] Enzyme-mediated reaction in which covalent bonds break, splitting a molecule into two or more parts, and H^+ and OH^- (derived from a water molecule) become attached to the exposed bonding sites.

hydrophilic substance [Gk. *philos*, loving] A polar substance that is attracted to the polar water molecule and so dissolves easily in water. Sugars are examples.

hydrophobic substance [Gk. *phobos*, dreading] A nonpolar substance that is repelled by the polar water molecule and so does not readily dissolve in water. Oil is an example.

hydrosphere All liquid or frozen water on or near the earth's surface.

hydrothermal vent ecosystem A type of ecosystem that exists near fissures in the ocean floor and is based on chemosynthetic bacteria that use hydrogen sulfide as the energy source.

hypha (HIGH-fuh), plural **hyphae** [Gk. *hyphe*, web] Of fungi, a generally tube-shaped filament with chitin-reinforced walls and, often, reinforcing cross-walls; component of the mycelium.

hypodermis A subcutaneous layer having stored fat that helps insulate the body; although not part of skin, it anchors skin while allowing it some freedom of movement.

hypothalamus [Gk. *hypo*, under, + *thalamos*, inner chamber or possibly *tholos*, rotunda] Of vertebrate forebrains, a brain center that monitors visceral activities (such as salt-water balance, temperature control, and reproduction) and that influences related forms of behavior (as in hunger, thirst, and sex).

hypothesis A possible explanation of a specific phenomenon.

immune response A series of events by which B and T lymphocytes recognize a specific antigen, undergo repeated cell divisions that form huge lymphocyte populations, and differentiate into subpopulations of effector and memory cells. Effector cells engage and destroy antigen-bearing agents. Memory cells enter a resting phase and are activated during subsequent encounters with the same antigen.

immunization Various processes, including vaccination, that promote increased immunity against specific diseases.

immunoglobulins (Ig) Four classes of antibodies, each with binding sites for antigen and binding sites used in specialized tasks. Examples are IgM antibodies (first to be secreted during immune responses) and IgG antibodies (which activate complement proteins and neutralize many toxins).

implantation A process by which a blastocyst adheres to the endometrium and begins to establish connections by which the mother and embryo will exchange substances during pregnancy.

imprinting Category of learning in which an animal that has been exposed to specific key stimuli early in its behavioral development forms an association with the object.

incisor A tooth, shaped like a flat chisel or cone, used in nipping or cutting food.

incomplete dominance Of heterozygotes, the appearance of a version of a trait that is somewhere between the homozygous dominant and recessive conditions.

independent assortment Mendelian principle that each gene pair tends to assort into gametes independently of other gene pairs located on nonhomologous chromosomes.

indirect selection A theory in evolutionary biology that self-sacrificing individuals can indirectly pass on their genes by helping relatives survive and reproduce.

induced-fit model Model of enzyme action whereby a bound substrate induces changes in the shape of the enzyme's active site, resulting in a more precise molecular fit between the enzyme and its substrate.

industrial smog A type of gray-air smog that develops in industrialized regions when winters are cold and wet.

inflammation, acute In response to tissue damage or irritation, fast-acting phagocytes and plasma proteins, including complement proteins, leave the bloodstream, then defend and help repair the tissue. Proceeds during both nonspecific and specific (immune) defense responses.

inheritance The transmission, from parents to offspring, of structural and functional patterns that have a genetic basis and are characteristic of each species.

inhibiting hormone A signaling molecule produced and secreted by the hypothalamus that controls secretions by the anterior lobe of the pituitary gland.

inhibitor A substance that can bind with an enzyme and interfere with its functioning.

inhibitory postsynaptic potential, or **IPSP** Of neurons, one of two competing types of graded potentials at an input zone; tends to drive the resting membrane potential away from threshold.

instinctive behavior A complex, stereotyped response to a particular environmental cue that often is quite simple.

insulin Hormone that lowers the glucose level in blood; it is secreted from beta cells of the pancreas and stimulates cells to take up glucose; also promotes protein and fat synthesis and inhibits protein conversion to glucose.

integration, neural [L. *integrare*, to coordinate] Moment-by-moment summation of all excitatory and inhibitory synapses acting on a neuron; occurs at each level of synapsing in a nervous system.

integrator Of homeostatic systems, a control point where different bits of information are pulled together in the selection of a response. The brain is an example.

integument Of animals, a protective body covering such as skin. Of flowering plants, a protective layer around the developing ovule; when the ovule becomes a seed, its integument(s) harden and thicken into a seed coat.

integumentary exchange (in-teg-you-MEN-tuh-ree) Of some animals, a mode of respiration in which oxygen and carbon dioxide diffuse across a thin, vascularized layer of moist epidermis at the body surface.

interference competition Interaction in which one species may limit another species' access to some resource regardless of whether the resource is abundant or scarce.

interleukin One of a variety of communication signals, secreted by macrophages and by helper T cells, that drive immune responses.

intermediate compound A compound that forms between the start and the end of a metabolic pathway.

intermediate filament A cytoskeletal component that consists of different proteins in different types of animal cells.

interneuron Any of the neurons in the vertebrate brain and spinal cord that integrate information arriving from sensory neurons and that influence other neurons in turn.

internode In vascular plants, the stem region between two successive nodes.

interphase Of cell cycles, the time interval between nuclear divisions in which a cell increases its mass, roughly doubles the number of its cytoplasmic components, and finally duplicates its chromosomes (replicates its DNA). The interval is different for different species.

interspecific competition Two-species interaction in which both species can be harmed due to overlapping niches.

interstitial fluid (IN-ter-STISH-ul) [L. *interstitus*, to stand in the middle of something] In multicelled animals, that portion of the extracellular fluid occupying spaces between cells and tissues.

intertidal zone Generally, the area on a rocky or sandy shoreline that is above the low water mark and below the high water mark; organisms inhabiting it are alternately submerged, then exposed, by tides.

intervertebral disk One of a number of disk-shaped structures containing cartilage that serve as shock absorbers and flex points between bony segments of the vertebral column.

intraspecific competition Interaction among individuals of the same species that are competing for the same resources.

intron A noncoding portion of a newly formed mRNA molecule.

inversion A change in a chromosome's structure after a segment separated from it was then inserted at the same place, but in reverse. The reversal alters the position and order of the chromosome's genes.

invertebrate Animal without a backbone.

ion, negatively charged (EYE-on) An atom or a compound that has gained one or more electrons, and hence has acquired an overall negative charge.

ion, positively charged An atom or a compound that has lost one or more electrons, and hence has acquired an overall positive charge.

ionic bond An association between ions of opposite charge.

isotonic condition Equality in the relative concentrations of solutes in two fluids; for two fluids separated by a cell membrane, there is no net osmotic (water) movement across the membrane.

isotope (EYE-so-tope) For a given element, an atom with the same number of protons as the other atoms but with a different number of neutrons.

J-shaped curve A curve, obtained when population size is plotted against time, that is characteristic of unrestricted, exponential growth.

joint An area of contact or near-contact between bones.

karyotype (CARRY-oh-type) Of eukaryotic individuals (or species), the number of metaphase chromosomes in somatic cells and their defining characteristics.

keratin A tough, water-insoluble protein manufactured by most epidermal cells.

keratinization (care-AT-in-iz-AY-shun) Process by which keratin-producing epidermal cells of skin die and collect at the skin surface as keratinized "bags" that form a barrier against dehydration, bacteria, and many toxic substances.

kidney In vertebrates, one of a pair of organs that filter mineral ions, organic wastes, and other substances from the blood, and help regulate the volume and solute concentrations of extracellular fluid.

kilocalorie 1,000 calories of heat energy, or the amount of energy needed to raise the temperature of 1 kilogram of water by 1°C; the unit of measure for the caloric value of foods.

kinetochore A specialized group of proteins and DNA at the centromere of a chromosome that serves as an attachment point for several spindle microtubules during mitosis or meiosis. Each chromatid of a duplicated chromosome has its own kinetochore.

Krebs cycle Together with a few conversion steps that precede it, the stage of aerobic respiration in which pyruvate is completely broken down to carbon dioxide and water. Coenzymes accept the unbound protons (H^+) and electrons stripped from intermediates during the reactions and deliver them to the next stage.

lactate fermentation Anaerobic pathway of ATP formation in which pyruvate from glycolysis is converted to the three-carbon compound lactate, and NAD^+ (a coenzyme used in the reactions) is regenerated. Its net yield is two ATP.

lactation The production of milk by hormone-primed mammary glands.

lake A body of fresh water having littoral, limnetic, and profundal zones.

lancelet An invertebrate chordate having a body that tapers sharply at both ends, segmented muscles, and a full-length notochord.

large intestine The colon; a region of the gut that receives unabsorbed food residues from the small intestine and concentrates and stores feces until they are expelled from the body.

larva, plural **larvae** Of animals, a sexually immature, free-living stage between the embryo and the adult.

larynx (LARE-inks) A tubular airway that leads to the lungs. In humans, contains vocal cords, where sound waves used in speech are produced.

lateral meristem Of vascular plants, a type of meristem responsible for secondary growth; either vascular cambium or cork cambium.

lateral root Of taproot systems, a lateral branching from the first, primary root.

leaching The movement of soil water, with dissolved nutrients, out of a specified area.

leaf For most vascular plants, a structure having chlorophyll-containing tissue that is the major region of photosynthesis.

learned behavior The use of information gained from specific experiences to vary or change a response to stimuli.

lethal mutation A gene mutation that alters one or more traits in such a way that the individual inevitably dies.

LH Leutinizing hormone. Secreted by the anterior lobe of the pituitary gland, this hormone has roles in the reproductive functions of both males and females.

lichen (LY-kun) A symbiotic association between a fungus and a captive photosynthetic partner such as a green alga.

life cycle A recurring, genetically programmed frame of events in which individuals grow, develop, maintain themselves, and reproduce.

life table A tabulation of age-specific patterns of birth and death for a population.

ligament A strap of dense connective tissue that bridges a joint.

light-dependent reactions First stage of photosynthesis in which the energy of sunlight is trapped and converted to the chemical energy of ATP alone (by the cyclic pathway) or ATP and NADPH (by the noncyclic pathway).

light-independent reactions Second stage of photosynthesis, in which sugar phosphates form with the help of the ATP (and NADPH, in land plants) that were produced during the first stage. The sugar phosphates are used in other reactions by which starch, cellulose, and other end products of photosynthesis are assembled.

lignification Of mature land plants, a process by which lignin is deposited in secondary cell walls. The deposits impart strength and rigidity by anchoring cellulose strands in the walls, stabilize and protect other wall components, and form a waterproof barrier around the cellulose. Probably a key factor in the evolution of vascular plants.

lignin A substance that strengthens and waterproofs cell walls in certain tissues of vascular plants.

limbic system Brain regions that, along with the cerebral cortex, collectively govern emotions.

limiting factor Any essential resource that is in short supply and so limits population growth.

lineage (LIN-ee-age) A line of descent.

linkage The tendency of genes located on the same chromosome to end up in the same gamete. For any two of those genes, the probability that crossing over will disrupt the linkage is proportional to the distance separating them.

lipid A greasy or oily compound of mostly carbon and hydrogen that shows little tendency to dissolve in water, but that dissolves in nonpolar solvents (such as ether). Cells use lipids as energy stores and structural materials, especially in membranes.

lipid bilayer The structural basis of cell membranes, consisting of two layers of mostly phospholipid molecules. Hydrophilic heads force all fatty acid tails of the lipids to become sandwiched between the hydrophilic heads.

liver Glandular organ with roles in storing and interconverting carbohydrates, lipids, and proteins absorbed from the gut, maintaining blood; disposing of nitrogen-containing wastes; and other tasks.

local signaling molecules Secretions from cells in many different tissues that alter chemical conditions in the immediate vicinity where they are secreted, then are swiftly degraded.

locus (LOW-cuss) The specific location of a particular gene on a chromosome.

logistic population growth (low-JIS-tik) Pattern of population growth in which a low-density population slowly increases in size, goes through a rapid growth phase, then levels off once the carrying capacity is reached.

loop of Henle The hairpin-shaped, tubular region of a nephron that functions in reabsorption of water and solutes.

lung An internal respiratory surface in the shape of a cavity or sac.

lycophyte A type of seedless vascular plant of mostly wet or shade habitats; requires free water to complete its life cycle.

lymph (LIMF) [L. *lympha*, water] Tissue fluid that has moved into the vessels of the lymphatic system.

lymph capillary A small-diameter vessel of the lymph vascular system that has no pronounced entrance; tissue fluid moves inward by passing between overlapping endothelial cells at the vessel's tip.

lymph node A lymphoid organ that serves as a battleground of the immune system; each lymph node is packed with organized arrays of macrophages and lymphocytes that cleanse lymph of pathogens before it reaches the blood.

lymph vascular system [L. *lympha*, water, + *vasculum*, a small vessel] The vessels of the lymphatic system, which take up and transport excess tissue fluid and reclaimable solutes as well as fats absorbed from the digestive tract.

lymphatic system An organ system that supplements the circulatory system. Its vessels take up fluid and solutes from interstitial fluid and deliver them to the bloodstream; its lymphoid organs have roles in immunity.

lymphocyte Any of various white blood cells that take part in nonspecific and specific (immune) defense responses.

lymphoid organs The lymph nodes, spleen, thymus, tonsils, adenoids, and other organs with roles in immunity.

lysis [Gk. *lysis*, a loosening] Gross structural disruption of a plasma membrane that leads to cell death.

lysosome (LYE-so-sohm) The main organelle of digestion, with enzymes that can break down polysaccharides, proteins, nucleic acids, and some lipids.

lysozyme An infection-fighting enzyme that digests bacterial cell walls. Present in mucous membranes that line the body's surfaces.

lytic pathway During a viral infection, viral DNA or RNA quickly directs the host cell to produce the components necessary to produce new virus particles, which are released by lysis.

macroevolution The large-scale patterns, trends, and rates of change among groups of species.

macrophage One of the phagocytic white blood cells. It engulfs anything detected as foreign. Some also become antigen-presenting cells that serve as the trigger for immune responses by T and B lymphocytes. *Compare* antigen-presenting cell.

mammal A type of vertebrate; the only animal having offspring that are nourished by milk produced by mammary glands of females.

mass extinction An abrupt rise in extinction rates above the background level; a catastrophic, global event in which major taxa are wiped out simultaneously.

mass number The total number of protons and neutrons in an atom's nucleus. The relative masses of atoms are also called atomic weights.

maternal chromosome One of the chromosomes bearing the alleles that are inherited from a female parent.

mechanoreceptor Sensory cell or cell part that detects mechanical energy associated with changes in pressure, position, or acceleration.

medulla oblongata Part of the vertebrate brainstem with reflex centers for respiration, blood circulation, and other vital functions.

medusa (meh-DOO-sah) [Gk. *Medousa*, one of three sisters in Greek mythology having snake-entwined hair; this image probably evoked by the tentacles and oral arms extending from the medusa] Free-swimming, bell-shaped stage in cnidarian life cycles.

megaspore Of gymnosperms and flowering plants, a haploid spore that forms in the ovary; one of its cellular descendants develops into an egg.

meiosis (my-OH-sis) [Gk. *meioun*, to diminish] Two-stage nuclear division process in which the chromosome number of a germ cell is reduced by half, to the haploid number. (Each daughter nucleus ends up with one of each type of chromosome.) Meiosis is the basis of gamete formation and (in plants) of spore formation. *Compare* mitosis.

meltdown Events which, if unchecked, can blow apart a nuclear power plant, with a release of radioactive material into the environment.

membrane excitability A membrane property of any cell that can produce action potentials in response to appropriate stimulation.

memory The storage and retrieval of information about previous experiences; underlies the capacity for learning.

memory cell One of the subpopulations of cells that form during an immune response and that enters a resting phase, from which it is released during a secondary immune response.

memory lymphocyte Any of the various B or T lymphocytes of the immune system that are formed in response to invasion by a foreign agent and that cir-culate for some period, available to mount a rapid attack if the same type of invader reappears.

Mendel's theory of independent assortment Stated in modern terms, during meiosis, the gene pairs of homologous chromosomes tend to be stored independently of how gene pairs on other chromosomes are sorted for forthcoming gametes. The theory does not take into account the effects of gene linkage and crossing over.

Mendel's theory of segregation Stated in modern terms, diploid cells have two of each kind of gene (on pairs of homologous chromosomes), and the two segregate during meiosis so that they end up in different gametes.

menopause (MEN-uh-pozz) [L. *mensis*, month, + *pausa*, stop] End of the period of a human female's reproductive potential.

menstrual cycle The cyclic release of oocytes and priming of the endometrium (lining of the uterus) to receive a fertilized egg; the complete cycle averages about 28 days in female humans.

menstruation Periodic sloughing of the blood-enriched lining of the uterus when pregnancy does not occur.

mesoderm (MEH-so-derm) [Gk. *mesos*, middle, + *derm*, skin] In most animal embryos, a primary tissue layer (germ layer) between ectoderm and endoderm. Gives rise to muscle; organs of circulation, reproduction, and excretion; most of the internal skeleton (when present); and connective tissue layers of the gut and body covering.

mesophyll Of vascular plants, a type of parenchyma tissue with photosynthetic cells and an abundance of air spaces.

messenger RNA A linear sequence of ribonucleotides transcribed from DNA and translated into a polypeptide chain; the only type of RNA that carries protein-building instructions.

metabolic pathway One of many orderly sequences of enzyme-mediated reactions by which cells normally maintain, increase, or decrease the concentrations of substances. Different pathways are linear or circular, and often they interconnect.

metabolism (meh-TAB-oh-lizm) [Gk. *meta*, change] All controlled, enzyme-mediated chemical reactions by which cells acquire and use energy. Through these reactions, cells synthesize, store, break apart, and eliminate substances in ways that contribute to growth, survival, and reproduction.

metamorphosis (met-uh-MOR-foe-sis) [Gk. *meta*, change, + *morphe*, form] Transformation of a larva into an adult form.

metaphase Of mitosis or meiosis II, the stage when each duplicated chromosome has become positioned at the midpoint of the microtubular spindle, with its two sister chromatids attached to microtubules from opposite spindle poles. Of meiosis I, the stage when all pairs of homologous chromosomes are positioned at the spindle's midpoint, with the two members of each pair attached to opposite spindle poles.

metazoan Any multicelled animal.

methanogen A type of archaebacterium that lives in oxygen-free habitats and that produces methane gas as a metabolic by-product.

MHC marker Any of a variety of proteins that are self-markers. Some occur on all body cells of an individual; others are unique to the macrophages and lymphocytes.

micelle formation Formation of a small droplet that consists of bile salts and products of fat digestion (fatty acids and monoglycerides) and that assists in their absorption from the small intestine.

microevolution Changes in allele frequencies brought about by mutation, genetic drift, gene flow, and natural selection.

microfilament [Gk. *mikros*, small, + L. *filum*, thread] In animal cells, one of a variety of cytoskeletal components. Actin and myosin filaments are examples.

microorganism An organism, usually single-celled, that is too small to be observed without the aid of a microscope.

microspore Of gymnosperms and flowering plants, a haploid spore, encased in a sculpted wall, that develops into a pollen grain.

microtubular spindle Of eukaryotic cells, a bipolar structure composed of organized arrays of microtubules that forms during nuclear division and that moves the chromosomes.

microtubule Hollow cylinder of mainly tubulin subunits; a cytoskeletal element with roles in cell shape, motion, and growth and in the structure of cilia and flagella.

microtubule organizing center, or **MTOC** Small mass of proteins and other substances in the cytoplasm; the number, type, and location of MTOCs determine the organization and orientation of microtubules.

microvillus (MY-crow-VILL-us) [L. *villus*, shaggy hair] A slender, cylindrical extension of the animal cell surface that functions in absorption or secretion.

midbrain Of vertebrates, a brain region that evolved as a coordination center for reflex responses to visual and auditory input; together with the pons and medulla oblongata, part of the brainstem, which includes the reticular formation.

migration Of certain animals, a cyclic movement between two distant regions at times of year corresponding to seasonal change.

mimicry (MIM-ik-ree) Situation in which one species (the mimic) bears deceptive resemblance in color, form, and/or behavior to another species (the model) that enjoys some survival advantage.

mineral An inorganic substance required for the normal functioning of body cells.

mitochondrion (MY-toe-KON-dree-on), plural **mitochondria** Organelle that specializes in ATP formation; it is the site of the second and third stages of aerobic respiration, an oxygen-requiring pathway.

mitosis (my-TOE-sis) [Gk. *mitos*, thread] Type of nuclear division that maintains the parental chromosome number for daughter cells. It is the basis of bodily growth and, in many eukaryotic species, asexual reproduction.

molar One of the cheek teeth; a tooth with a platform having cusps (surface bumps) that help crush, grind, and shear food.

molecular clock With respect to the presumed regular accumulation of neutral mutations in highly conserved genes, a way of calculating the time of origin of one species or lineage relative to others.

molecule A unit of matter in which chemical bonding holds together two or more atoms of the same or different elements.

mollusk A type of invertebrate having a tissue fold (mantle) draped around a soft, fleshy body; snails, clams, and squids are examples.

molting The shedding of hair, feathers, horns, epidermis, or a shell (or some other exoskeleton) in a process of growth or periodic renewal.

Monera The kingdom of bacteria.

moneran A bacterium; a single-celled prokaryote.

monocot (MON-oh-kot) Short for monocotyledon; a flowering plant in which seeds have only one cotyledon, whose floral parts generally occur in threes (or multiples of three), and whose leaves typically are parallel-veined. *Compare* dicot.

monohybrid cross [Gk. *monos*, alone] An experimental cross in which offspring inherit a pair of nonidentical alleles for a single trait being studied, so that they are heterozygous.

monophyletic group A set of independently evolving lineages that share a common evolutionary heritage.

monosaccharide (MON-oh-SAK-ah-ride) [Gk. *monos*, alone, single, + *sakharon*, sugar] The simplest carbohydrate, with only one sugar unit. Glucose is an example.

monosomy Abnormal condition in which one chromosome of diploid cells has no homologue.

morphogenesis (MORE-foe-JEN-ih-sis) [Gk. *morphe*, form, + *genesis*, origin] Processes by which differentiated cells in an embryo become organized into tissues and organs, under genetic controls and environmental influences.

morphological convergence A macroevolutionary pattern of change in which separate lineages adopt similar lifestyles, put comparable body parts to similar uses, and in time resemble one another in structure and function. Analogous structures are evidence of this pattern.

morphological divergence A macroevolutionary pattern of change from a common ancestral form. Homologous structures are evidence of this pattern.

motor neuron A type of neuron; it delivers signals from the brain and spinal cord that can stimulate or inhibit the body's effectors (muscles, glands, or both).

mouth An oral cavity; in human digestion, the site where polysaccharide breakdown begins.

multicelled organism An organism that has differentiated cells arranged into tissues, organs, and often organ systems.

multiple allele system Three or more different molecular forms of the same gene (alleles) that exist in a population.

muscle fatigue A decline in tension of a muscle that has been kept in a state of tetanic contraction as a result of continuous, high-frequency stimulation.

muscle tension A mechanical force, exerted by a contracting muscle, that resists opposing forces such as gravity and the weight of objects being lifted.

muscle tissue Tissue having cells able to contract in response to stimulation, then passively lengthen and so return to their resting stage.

mutagen (MEW-tuh-jen) An environmental agent that can permanently modify the structure of a DNA mole-

cule. Certain viruses and ultraviolet radiation are examples.

mutation [L. *mutatus*, a change, + *-ion*, result or a process or an act] A heritable change in the DNA. Generally, mutations are the source of all the different molecular versions of genes (alleles) and, ultimately, of life's diversity. *See also* lethal mutation; neutral mutation.

mutualism [L. *mutuus*, reciprocal] An interaction between two species that benefits both.

mycelium (my-SEE-lee-um), plural **mycelia** [Gk. *mykes*, fungus, mushroom, + *helos*, callus] A mesh of tiny, branching filaments (hyphae) that is the food-absorbing part of a multicelled fungus.

mycorrhiza (MY-coe-RISE-uh) "Fungus-root;" a symbiotic arrangement between fungal hyphae and the young roots of many vascular plants. The fungus obtains carbohydrates from the plant and in turn releases dissolved mineral ions to the plant roots.

myelin sheath Of many sensory and motor neurons, an axonal sheath that affects how fast action potentials travel; formed from the plasma membranes of Schwann cells that are wrapped repeatedly around the axon and are separated from each other by a small node.

myofibril (MY-oh-FY-brill) One of many threadlike structures inside a muscle cell; each is functionally divided into sarcomeres, the basic units of contraction.

myosin (MY-uh-sin) A type of protein with a head and long tail. In muscle cells, it interacts with actin, another protein, to bring about contraction.

NAD⁺ Nicotinamide adenine dinucleotide; a nucleotide coenzyme. When carrying electrons and unbound protons (H^+) between reaction sites, it is abbreviated NADH.

NADP⁺ Nicotinamide adenine dinucleotide phosphate; a phosphorylated nucleotide coenzyme. When carrying electrons and unbound protons (H^+) between reaction sites, it is abbreviated $NADPH_2$.

nasal cavity Of a respiratory system, the region where air is warmed, moistened, and filtered of airborne particles and dust.

natural selection A microevolutionary process; a difference in survival and reproduction among members of a population that vary in one or more traits.

negative feedback mechanism A homeostatic feedback mechanism in which an activity changes some condition in the internal environment and so triggers a response that reverses the changed condition.

nematocyst (NEM-ad-uh-sist) [Gk. *nema*, thread, + *kystis*, pouch] Of cnidarians only, a stinging capsule that assists in prey capture and possibly protection.

nephridium (neh-FRID-ee-um), plural **nephridia** Of earthworms and some other invertebrates, a system of regulating water and solute levels.

nephron (NEFF-ron) [Gk. *nephros*, kidney] Of the vertebrate kidney, a slender tubule in which water and solutes filtered from blood are selectively reabsorbed and in which urine forms.

nerve Cordlike communication line of the peripheral nervous system, composed of axons of sensory neurons, motor neurons, or both packed within connective tissue. In the brain and spinal cord, similar cordlike bundles are called nerve pathways or tracts.

nerve cord Of many animals, a cordlike communication line consisting of axons of neurons.

nerve impulse *See* action potential.

nerve net Cnidarian nervous system.

nervous system System of neurons oriented relative to one another in precise message-conducting and information-processing pathways.

nervous tissue A type of connective tissue composed of neurons.

net energy Of energy resources available to the human population, the amount of energy that is left over after subtracting the energy used to locate, extract, transport, store, and deliver energy to consumers.

net population growth rate per individual (r) Of population growth equations, a single variable in which birth and death rates, which are assumed to remain constant, are combined.

neuroglial cell (NUR-oh-GLEE-uhl) Of vertebrates, one of the cells that provide structural and metabolic support for neurons and that collectively represent about half the volume of the nervous system.

neuromodulator Type of signaling molecule that influences the effects of transmitter substances by enhancing or reducing membrane responses in target neurons.

neuromuscular junction Chemical synapses between axon terminals of a motor neuron and a muscle cell.

neuron A nerve cell; the basic unit of communication in nervous systems. Neurons collectively sense environmental change, integrate sensory inputs, then activate muscles or glands that initiate or carry out responses.

neurotransmitter Any of the class of signaling molecules that are secreted from neurons, act on immediately adjacent cells, and are then rapidly degraded or recycled.

neutral mutation A gene mutation that has neither harmful nor helpful effects on the individual's ability to survive and reproduce.

neutron Unit of matter, one or more of which occupies the atomic nucleus, that has mass but no electric charge.

neutrophil Fast-acting, phagocytic white blood cell that takes part in inflammatory responses against bacteria.

niche (NITCH) [L. *nidas*, nest] Of a species, the full range of physical and biological conditions under which its members can live and reproduce.

nitrification (nye-trih-fih-KAY-shun) A chemosynthetic process in which certain bacteria strip electrons from ammonia or ammonium present in soil. The end product, nitrite (NO_2^-), is broken down to nitrate (NO_3^-) by different bacteria.

nitrogen cycle Biogeochemical cycle in which the atmosphere is the largest reservoir of nitrogen.

nitrogen fixation Process by which a few kinds of bacteria convert gaseous nitrogen (N_2) to ammonia. This dissolves rapidly in their cytoplasm to form ammonium, which can be used in biosynthetic pathways.

NK cell Natural killer cell, possibly of the lymphocyte lineage, that reconnoiters and kills tumor cells and infected body cells.

nociceptor A receptor, such as a free nerve ending, that detects any stimulus causing tissue damage.

node In vascular plants, a point on a stem where one or more leaves are attached.

noncyclic pathway of ATP formation (non-SIK-lik) [L. *non*, not, + Gk. *kylos*, circle] Photosynthetic pathway in which excited electrons derived from water molecules flow through two photosystems and two transport chains, and ATP and NADPH form.

nondisjunction Failure of one or more chromosomes to separate properly during mitosis or meiosis.

nonsteroid hormone A type of water-soluble hormone, such as a protein hormone, that cannot cross the lipid bilayer of a target cell. These hormones enter the cell by receptor-mediated endocytosis, or they bind to receptors that activate membrane proteins or second messengers within the cell.

notochord (KNOW-toe-kord) Of chordates, a rod of stiffened tissue (not cartilage or bone) that serves as a supporting structure for the body.

nuclear envelope A double membrane (two lipid bilayers and associated proteins) that is the outermost portion of a cell nucleus.

nucleic acid (new-CLAY-ik) A long, single- or double-stranded chain of four different kinds of nucleotides joined one after the other at their phosphate groups. They differ in which nucleotide base follows the next in sequence. DNA and RNA are examples.

nucleic acid hybridization The base-pairing of nucleotide sequences from different sources, as used in genetics, genetic engineering, and studies of evolutionary relationship based on similarities and differences in the DNA or RNA of different species.

nucleoid Of bacteria, a region in which DNA is physically organized apart from other cytoplasmic components.

nucleolus (new-KLEE-oh-lus) [L. *nucleolus*, a little kernel] Within the nucleus of a nondividing cell, a site where the protein and RNA subunits of ribosomes are assembled.

nucleosome (NEW-klee-oh-sohm) Of eukaryotic chromosomes, one of many organizational units, each consisting of a small stretch of DNA looped twice around a "spool" of histone molecules, which another histone molecule stabilizes.

nucleotide (NEW-klee-oh-tide) A small organic compound having a five-carbon sugar (deoxyribose), nitrogen-containing base, and phosphate group. Nucleotides are the structural units of adenosine phosphates, nucleotide coenzymes, and nucleic acids.

nucleotide coenzyme A protein that transports hydrogen atoms (free protons) and electrons from one reaction site to another in cells.

nucleus (NEW-klee-us) [L. *nucleus*, a kernel] Of atoms, the central core of one or more positively charged protons and (in all but hydrogen) electrically neutral neutrons. In eukaryotic cells, a membranous organelle that physically isolates and organizes the DNA, out of the way of cytoplasmic machinery.

nutrition All those processes by which food is selectively ingested, digested, absorbed, and later converted to the body's own organic compounds.

obesity An excess of fat in the body's adipose tissues, caused by imbalances between caloric intake and energy output.

oligosaccharide A carbohydrate consisting of a short chain of two or more covalently bonded sugar units. One subclass, disaccharides, has two sugar units. *Compare* monosaccharide; polysaccharide.

omnivore [L. *omnis*, all, + *vovare*, to devour] An organism able to obtain energy from more than one source rather than being limited to one trophic level.

oncogene (ON-coe-jeen) Any gene having the potential to induce cancerous transformations in a cell.

oocyte An immature egg.

oogenesis (oo-oh-JEN-uh-sis) Formation of a female gamete, from a germ cell to a mature haploid ovum (egg).

operator A short base sequence between a promoter and the start of a gene; interacts with regulatory proteins.

operon Of transcription, a promoter-operator sequence that services more than a single gene. The lactose operon of *E. coli* is an example.

orbitals Volumes of space around the nucleus of an atom in which electrons are likely to be at any instant.

organ A structure of definite form and function that is composed of more than one tissue.

organ formation Stage of development in which primary tissue layers (germ layers) split into subpopulations of cells, and different lines of cells become unique in structure and function; foundation for growth and tissue specialization, when organs acquire specialized chemical and physical properties.

organ system Two or more organs that interact chemically, physically, or both in performing a common task.

organelle Of cells, an internal, membrane-bounded sac or compartment that has a specific, specialized metabolic function.

organic compound In biology, a compound assembled in cells and having a carbon backbone, often with carbon atoms arranged as a chain or ring structure.

osmosis (oss-MOE-sis) [Gk. *osmos*, act of pushing] Of cells, the tendency of water to move through channel proteins that span a membrane in response to a concentration gradient, fluid pressure, or both. Hydrogen bonds among water molecules prevent water *itself* from becoming more or less concentrated; but a gradient may exist when the water on either side of the membrane has more substances dissolved in it.

ovary (OH-vuh-ree) In female animals, the primary reproductive organ in which eggs form. In seed-bearing plants, the portion of the carpel where eggs develop, fertilization takes place, and seeds mature. A mature ovary (and sometimes other plant parts) is a fruit.

oviduct (OH-vih-dukt) Duct through which eggs travel from the ovary to the uterus. Formerly called Fallopian tube.

ovulation (AHV-you-LAY-shun) During each turn of the menstrual cycle, the release of a secondary oocyte (immature egg) from an ovary.

ovule (OHV-youl) [L. *ovum*, egg] Before fertilization in gymnosperms and angiosperms, a female gametophyte with egg cell, a surrounding tissue, and one or two protective layers (integuments). After fertilization, an ovule matures into a seed (an embryo sporophyte and food reserves encased in a hardened coat).

ovum (OH-vum) A mature female gamete (egg).

oxidation-reduction reaction An electron transfer from one atom or molecule to another. Often hydrogen is transferred along with the electron or electrons.

ozone hole A pronounced seasonal thinning of the ozone layer in the lower stratosphere above Antarctica.

pancreas (PAN-cree-us) Gland that secretes enzymes and bicarbonate into the small intestine during digestion, and that also secretes the hormones insulin and glucagon.

pancreatic islets Any of the two million clusters of endocrine cells in the pancreas, including alpha cells, beta cells, and delta cells.

parasite [Gk. *para*, alongside, + *sitos*, food] An organism that obtains nutrients directly from the tissues of a living host, which it lives on or in and may or may not kill.

parasitism A two-species interaction in which one species directly harms another that serves as its host.

parasitoid An insect larva that grows and develops inside a host organism (usually another insect), eventually consuming the soft tissues and killing it.

parasympathetic nerve Of the autonomic nervous system, any of the nerves carrying signals that tend to slow the body down overall and divert energy to basic tasks; also work continually in opposition with sympathetic nerves to bring about minor adjustments in internal organs.

parathyroid glands (PARE-uh-THY-royd) In vertebrates, endocrine glands embedded in the thyroid gland that secrete parathyroid hormone, which helps restore blood calcium levels.

parenchyma Most abundant of the simple tissues in flowering plant roots, stems, leaves, and other parts. Its cells function in photosynthesis, storage, secretion, and other tasks.

parthenogenesis Development of an embryo from an unfertilized egg.

passive immunity Temporary immunity conferred by deliberately introducing antibodies into the body.

passive transport Diffusion of a solute through a channel or carrier protein that spans the lipid bilayer of a cell membrane. Its passage does not require an energy input; the protein passively allows the solute to follow its concentration gradient.

paternal chromosome One of the chromosomes bearing alleles that are inherited from a male parent.

pathogen (PATH-oh-jen) [Gk. *pathos*, suffering; + -*genēs*, origin]. An infectious, disease-causing agent, such as a virus or bacterium.

pattern formation Of animals, mechanisms responsible for specialization and positioning of tissues during embryonic development.

PCR Polymerase chain reaction. A method used by recombinant DNA technologists to amplify the quantity of specific fragments of DNA.

peat An accumulation of saturated, undecayed remains of plants that have been compressed by sediments.

pedigree A chart of genetic connections among individuals, as constructed according to standardized methods.

pelagic province The entire volume of ocean water; subdivided into neritic zone (relatively shallow waters overlying continental shelves) and oceanic zone (water over ocean basins).

penis A male organ that deposits sperm into a female reproductive tract.

perennial [L. *per-*, throughout, + *annus*, year] A flowering plant that lives for three or more growing seasons.

perforin A type of protein, produced and secreted by cytotoxic cells, that destroys antigen-bearing targets.

pericycle (PARE-ih-sigh-kul) [Gk. *peri-*, around, + *kyklos*, circle] Of a root vascular cylinder, one or more layers just inside the endodermis that gives rise to lateral roots and contributes to secondary growth.

periderm Of vascular plants showing secondary growth, a protective covering that replaces epidermis.

peripheral nervous system (per-IF-ur-uhl) [Gk. *peripherein*, to carry around] Of vertebrates, the nerves leading into and out from the spinal cord and brain and the ganglia along those communication lines.

peristalsis (pare-ih-STAL-sis) A rhythmic contraction of muscles that moves food forward through the animal gut.

peritoneum A lining of the coelom that also covers and helps maintain the position of internal organs.

peritubular capillaries The set of blood capillaries that threads around the tubular parts of a nephron; they function in reabsorption of water and solutes back into the body and in secretion of hydrogen ions and some other substances in the forming urine.

permafrost A permanently frozen, water-impenetrable layer beneath the soil surface in arctic tundra.

peroxisome Enzyme-filled vesicle in which fatty acids and amino acids are digested first into hydrogen peroxide (which is toxic), then to harmless products.

PGA Phosphoglycerate (FOSS-foe-GLISS-er-ate). A key intermediate in glycolysis and in the Calvin-Benson cycle.

PGAL Phosphoglyceraldehyde. A key intermediate in glycolysis and in the Calvin-Benson cycle.

pH scale A scale used to measure the concentration of free hydrogen ions in blood, water, and other solutions; pH 0 is the most acidic, 14 the most basic, and 7, neutral.

phagocyte (FAG-uh-sight) [Gk. *phagein*, to eat, + *kytos*, hollow vessel] A macrophage or certain other white blood cells that engulf and destroy foreign agents.

phagocytosis (FAG-uh-sigh-TOE-sis) [Gk. *phagein*, to eat, + *kytos*, hollow vessel] Engulfment of foreign cells or substances by amoebas and some white blood cells by means of endocytosis.

pharynx (FARE-inks) A muscular tube by which food enters the gut; in land vertebrates, the dual entrance for the tubular part of the digestive tract and windpipe (trachea).

phenotype (FEE-no-type) [Gk. *phainein*, to show, + *typos*, image] Observable trait or traits of an individual; arises from interactions between genes, and between genes and the environment.

pheromone (FARE-oh-moan) [Gk. *phero*, to carry, + *-mone*, as in hormone] A type of signaling molecule secreted by exocrine glands that serves as a communication signal between individuals of the same species. Signaling pheromones elicit an immediate behavioral response. Priming pheromones elicit a generalized physiological response.

phloem (FLOW-um) Of vascular plants, a tissue with living cells that interconnect and form the tubes through which sugars and other solutes are conducted.

phospholipid A type of lipid that is the main structural component of cell membranes. Each has a hydrophobic tail (of two fatty acids) and a hydrophilic head that incorporates glycerol and a phosphate group.

phosphorus cycle Movement of phosphorus from rock or soil through organisms, then back to soil.

phosphorylation (FOSS-for-ih-LAY-shun) The attachment of unbound (inorganic) phosphate to a molecule; also the transfer of a phosphate group from one molecule to another, as when ATP phosphorylates glucose.

photochemical smog A brown-air smog that develops over large cities when the surrounding land forms a natural basin.

photolysis (foe-TALL-ih-sis) [Gk. *photos*, light, + *-lysis*, breaking apart] A reaction sequence of the noncyclic pathway of photosynthesis, triggered by photon energy, in which water is split into oxygen, hydrogen, and electrons.

photoperiodism A biological response to a change in the relative length of daylight and darkness.

photoreceptor Light-sensitive sensory cell.

photosynthesis The trapping of sunlight energy and its conversion to chemical energy (ATP, NADPH, or both), followed by synthesis of sugar phosphates that become converted to sucrose, cellulose, starch, and other end products. It is the main biosynthetic pathway by which energy and carbon enter the web of life.

photosynthetic autotroph An organism able to synthesize all organic molecules it requires using carbon dioxide as the carbon source and sunlight as the energy source. All plants, some protistans, and a few bacteria are photosynthetic autotrophs.

photosystem One of the clusters of light-trapping pigments embedded in photosynthetic membranes. Photosystem I operates during the cyclic pathway; photosystem II operates during both the cyclic and noncyclic pathways.

phototropism [Gk. *photos*, light, + *trope*, turning, direction] Adjustment in the direction and rate of plant growth in response to light.

photovoltaic cell A device that converts sunlight energy into electricity.

phycobilins (FIE-koe-BY-lins) A class of light-sensitive, accessory pigments that transfer absorbed energy to chlorophylls. They are abundant in red algae and cyanobacteria.

phylogeny Evolutionary relationships among species, starting with most ancestral forms and including the branches leading to their descendants.

phytochrome Light-sensitive pigment molecule, the activation and inactivation of which triggers plant hormone activities governing leaf expansion, stem branching, stem length and often seed germination and flowering.

phytoplankton (FIE-toe-PLANK-tun) [Gk. *phyton*, plant, + *planktos*, wandering] A freshwater or marine community of floating or weakly swimming photosynthetic autotrophs, such as cyanobacteria, diatoms, and green algae.

pigment A light-absorbing molecule.

pineal gland (py-NEEL) A light-sensitive endocrine gland that secretes melatonin, a hormone that influences reproductive cycles and the development of reproductive organs.

pioneer species Typically small plants with short life cycles that are adapted to growing in exposed, often windy areas with intense sunlight, wide swings in air temperature, and soils deficient in nitrogen and other nutrients. By improving conditions in areas they colonize, pioneers invite their own replacement by other species.

pituitary gland Of endocrine systems, a gland that interacts with the hypothalamus to coordinate and control many physiological functions, including the activity of many other endocrine glands. Its posterior lobe stores and secretes hypothalamic hormones; the anterior lobe produces and secretes its own hormones.

placenta (play-SEN-tuh) Of the uterus, an organ composed of maternal tissues and extraembryonic membranes (the chorion especially); it delivers nutrients to the fetus and accepts wastes from it, yet allows the fetal circulatory system to develop separately from the mother's.

plankton [Gk. *planktos*, wandering] Any community of floating or weakly swimming organisms, mostly microscopic, living in freshwater and saltwater environments. *See* phytoplankton; zooplankton.

plant The type of eukaryotic organism, usually multicelled, that is a photosynthetic autotroph—it uses sunlight energy to drive the synthesis of all its required organic compounds from carbon dioxide, water, and mineral ions. Only a few nonphotosynthetic plants obtain nutrients by parasitism and other means.

Plantae The kingdom of plants.

plasma (PLAZ-muh) Liquid component of blood; consists of water, various proteins, ions, sugars, dissolved gases, and other substances.

plasma cell Of immune systems, any of the anitbody-secreting daughter cells of a rapidly dividing population of B cells.

plasma membrane Of cells, the outermost membrane. Its lipid bilayer structure and proteins carry out most functions, including transport across the membrane and reception of extracellular signals.

plasmid Of many bacteria, a small, circular molecule of extra DNA that carries only a few genes and replicates independently of the bacterial chromosome.

plasmodesma (PLAZ-moe-DEZ-muh) Of multicelled plants, a junction between linked walls of adjacent cells through which nutrients and other substances flow.

plasticity Of the human species, the ability to remain flexible and adapt to a wide range of environments.

plate tectonics Arrangement of the earth's outer layer (lithosphere) in slab-like plates, all in motion and floating on a hot, plastic layer of the underlying mantle.

platelet (PLAYT-let) Any of the cell fragments in blood that release substances necessary for clot formation.

pleiotropy (PLEE-oh-troe-pee) [Gk. *pleon*, more, + *trope*, direction] A type of gene interaction in which a single gene exerts multiple effects on seemingly unrelated aspects of an individual's phenotype.

polar body Any of three cells that form during the meiotic cell division of an oocyte; the division also forms the mature egg, or ovum.

pollen grain [L. *pollen*, fine dust] Depending on the species, the immature or mature, sperm-bearing male gametophyte of gymnosperms and flowering plants.

pollen sac In anthers of flowers, any of the chambers in which pollen grains develop.

pollen tube A tube formed after a pollen grain germinates; grows down through carpel tissues and carries sperm to the ovule.

pollination Of flowering plants, the arrival of a pollen grain on the landing platform (stigma) of a carpel.

pollutant Any substance with which an ecosystem has had no prior evolutionary experience, in terms of kinds or amounts, and that can accumulate to disruptive or harmful levels. Can be naturally occurring or synthetic.

polymer (POH-lih-mur) [Gk. *polus*, many, + *meris*, part] A molecule composed of three to millions of small subunits that may or may not be identical.

polymerase chain reaction or **PCR** DNA amplification method; DNA containing a gene of interest is split into single strands, which enzymes (polymerases) copy; the enzymes also act on the accumulating copies, multiplying the gene sequence by the millions.

polymorphism (poly-MORE-fizz-um) [Gk. *polus*, many, + *morphe*, form] Of a population, the persistence through the generations of two or more forms of a trait, at a frequency greater than can be maintained by new mutations alone.

polyp (POH-lip) Vase-shaped, sedentary stage of cnidarian life cycles.

polypeptide chain Three or more amino acids joined by peptide bonds.

polyploidy (POL-ee-PLOYD-ee) A change in the chromosome number following inheritance of three or more of each type of chromosome.

polysaccharide [Gk. *polus*, many, + *sakcharon*, sugar] A straight or branched chain of hundreds of thousands of covalently linked sugar units, of the same or different kinds. The most common polysaccharides are cellulose, starch, and glycogen.

polysome Of protein synthesis, several ribosomes all translating the same messenger RNA molecule, one after the other.

population A group of individuals of the same species occupying a given area.

population density The number of individuals of a population that are living in a specified area or volume.

population distribution The general pattern of dispersion of individuals of a population throughout their habitat.

population size The number of individuals that make up the gene pool of a population.

positive feedback mechanism Homeostatic mechanism by which a chain of events is set in motion that intensifies a change from an original condition; after a limited time, the intensification reverses the change.

post-translational controls Of eukaryotes, controls that govern modification of newly formed polypeptide chains into functional enzymes and other proteins.

predation A two-species interaction in which one species (the predator) directly harms the other (its prey).

predator [L. *prehendere*, to grasp, seize] An organism that feeds on and may or may not kill other living organisms (its prey); unlike parasites, predators do not live on or in their prey.

prediction A claim about what you can expect to observe in nature if a theory or hypothesis is correct.

premolar One of the cheek teeth; a tooth having a platform with cusps (surface bumps) that can crush, grind, and shear food.

pressure flow theory Of vascular plants, a theory that organic compounds move through phloem because of gradients in solute concentrations and pressure between source regions (such as photosynthetically active leaves) and sink regions (such as growing plant parts).

primary growth Plant growth originating at root tips and shoot tips.

primary immune response Actions by white blood cells and their products elicited by a first-time encounter with an antigen; includes both antibody-mediated and cell-mediated responses.

primary productivity, gross Of ecosystems, the rate at which the producer organisms capture and store a given amount of energy during a specified interval.

primary productivity, net Of ecosystems, the rate of energy storage in the tissues of producers in excess of their rate of aerobic respiration.

primate The mammalian lineage that includes prosimians, tarsioids, and anthropoids (monkeys, apes, and humans).

probability With respect to any chance event, the most likely number of times it will turn out a certain way, divided by the total number of all possible outcomes.

procambium (pro-KAM-bee-um) Of vascular plants, a primary meristem that gives rise to the primary vascular tissues.

producers, primary Of ecosystems, the organisms that secure energy from the physical environment, as by photosynthesis or chemosynthesis.

progesterone (pro-JESS-tuh-rown) Female sex hormone secreted by the ovaries.

prokaryotic cell (pro-CARRY-oh-tic) [L. *pro*, before, + Gk. *karyon*, kernel] A bacterium; a single-celled organism that has no nucleus or any of the other membrane-bound organelles characteristic of eukaryotic cells.

promoter Of transcription, a base sequence that signals the start of a gene; the site where RNA polymerase initially binds.

prophase Of mitosis, the stage when each duplicated chromosome starts to condense, microtubules form a spindle apparatus, and the nuclear envelope starts to break up.

prophase I Of meiosis, the stage at which the microtubular spindle starts to form, the nuclear envelope starts to break up, and each duplicated chromosome also condenses and pairs with its homologous partner. At this time, their sister chromatids typically undergo crossing over and genetic recombination.

prophase II Of meiosis, a brief stage after interkinesis during which each chromosome still consists of two chromatids.

protein Large organic compound composed of one or more chains of amino acids held together by peptide bonds. Proteins have unique sequences of different kinds of amino acids in their polypeptide chains; such sequences are the basis of a protein's three-dimensional structure and chemical behavior.

Protista The kingdom of protistans.

protistan (pro-TISS-tun) [Gk. *prōtistos*, primal, very first] Single-celled eukaryote.

proto-oncogene A gene sequence similar to an oncogene but that codes for a protein required in normal cell function; may trigger cancer, generally when specific mutations alter its structure or function.

proton Positively charged particle, one or more of which is present in the atomic nucleus.

protostome (PRO-toe-stome) [Gk. *proto*, first, + *stoma*, mouth] A bilateral animal in which the first indentation in the early embryo develops into the mouth. Includes mollusks, annelids, and arthropods.

protozoan A type of protistan, some predatory and others parasitic; so named because they may resemble the single-celled heterotrophs that presumably gave rise to animals.

proximal tubule Of a nephron, the tubular region that receives water and solutes filtered from the blood.

pulmonary circuit Blood circulation route leading to and from the lungs.

Punnett-square method A way to predict the possible outcome of a mating or an experimental cross in simple diagrammatic form.

purine Nucleotide base having a double ring structure. Adenine and guanine are examples.

pyrimidine (phi-RIM-ih-deen) Nucleotide base having a single ring structure. Cytosine and thymine are examples.

pyruvate (PIE-roo-vate) A compound with a backbone of three carbon atoms. Two pyruvate molecules are the end products of glycolysis.

r Designates net population growth rate; the birth and death rates are assumed to remain constant and so are combined into this one variable for population growth equations.

radial symmetry Body plan having four or more roughly equivalent parts arranged around a central axis.

radioisotope An unstable atom that has dissimilar numbers of protons and neutrons and that spontaneously decays (emits electrons and energy) to a new, stable atom that is not radioactive.

rain shadow A reduction in rainfall on the leeward side of high mountains, resulting in arid or semiarid conditions.

reabsorption Of urine formation, the diffusion or active transport of water and usable solutes out of a nephron and into capillaries leading back to the general circulation; regulated by ADH and aldosterone.

receptor Of cells, a molecule at the surface of the plasma membrane or in the cytoplasm that binds molecules present in the extracellular environment. The binding triggers changes in cellular activities. Of nervous systems, a sensory cell or cell part that may be activated by a specific stimulus.

receptor protein Protein that binds a signaling molecule such as a hormone, then triggers alterations in cell behavior or metabolism.

recessive allele [L. *recedere*, to recede] In heterozygotes, an allele whose expression is fully or partially masked by expression of its partner; fully expressed only in the homozygous recessive condition.

recognition protein Protein at cell surface recognized by cells of like type; helps guide the ordering of cells into tissues during development and functions in cell-to-cell interactions.

recombinant technology Procedures by which DNA (genes) from different species may be isolated, cut, spliced together, and the new recombinant molecules multiplied in quantity in a population of rapidly dividing cells such as bacteria.

red blood cell Erythrocyte; an oxygen-transporting cell in blood.

red marrow A substance in the spongy tissue of many bones that serves as a major site of blood cell formation.

reflex [L. *reflectere*, to bend back] A simple, stereotyped movement elicited directly by sensory stimulation.

reflex pathway [L. *reflectere*, to bend back] Type of neural pathway in which signals from sensory neurons directly stimulate or inhibit motor neurons, without intervention by interneurons.

refractory period Of neurons, the period following an action potential at a given patch of membrane when sodium gates are shut and potassium gates are open, so that the patch is insensitive to stimulation.

regulatory protein A protein that enhances or suppresses the rate at which a gene is transcribed.

releasing hormone A hypothalamic signaling molecule that stimulates or slows down secretion by target cells in the anterior lobe of the pituitary gland.

repressor protein Regulatory protein that provides negative control of gene activity by preventing RNA polymerase from binding to DNA.

reproduction In biology, processes by which a new generation of cells or multicelled individuals is produced. Sexual reproduction requires meiosis, formation of gametes, and fertilization. Asexual reproduction refers to the production of new individuals by any mode that does not involve gametes.

reproduction, sexual Mode of reproduction that begins with meiosis, proceeds through gamete formation, and ends at fertilization.

reproductive isolating mechanism Any aspect of structure, functioning, or behavior that restricts gene flow between two populations.

reproductive isolation An absence of gene flow between populations.

reproductive success The survival and production of the offspring of an individual.

reptile A type of carnivorous vertebrate; its ancestors were the first vertebrates to escape dependency of standing water, largely by means of internal fertilization and amniote eggs. They include turtles, crocodiles, lizards and snakes, and tuataras.

resource partitioning A community pattern in which similar species generally share the same kind of resource in different ways, in different areas, or at different times.

respiration [L. *respirare*, to breathe] In most animals, the overall exchange of oxygen from the environment for carbon dioxide wastes from cells by way of circulating blood. *Compare* aerobic respiration.

respiratory surface The surface, such as a thin epithelial layer, that gases diffuse across to enter and leave the animal body.

respiratory system An organ system that functions in respiration.

resting membrane potential Of neurons and other excitable cells that are not being stimulated, the steady voltage difference across the plasma membrane.

restriction enzymes Class of bacterial enzymes that cut apart foreign DNA injected into them, as by viruses; also used in recombinant DNA technology.

reticular formation Of the vertebrate brainstem, a major network of interneurons that helps govern activity of the whole nervous system.

reverse transcriptase Viral enzyme required for reverse transcription of mRNA into DNA; used in recombinant DNA technology.

reverse transcription Assembly of DNA on a single-stranded mRNA molecule by viral enzymes.

RFLPs Restriction fragment length polymorphisms. Of DNA samples from different individuals, slight but unique differences in the banding pattern of fragments of the DNA that have been cut with restriction enzymes.

Rh blood typing A method of characterizing red blood cells on the basis of a protein that serves as a self-marker at their surface; Rh^+ signifies its presence and Rh^-, its absence.

rhizoid Rootlike absorptive structure of some fungi and nonvascular plants.

ribosomal RNA (rRNA) Type of RNA molecule that combines with proteins to form ribosomes, on which the polypeptide chains of proteins are assembled.

ribosome In all cells, the structure at which amino acids are strung together in specified sequence to form the polypeptide chains of proteins. An intact ribosome consists of two sub-

units, each composed of ribosomal RNA and protein molecules.

RNA Ribonucleic acid. A category of single-stranded nucleic acids that function in processes by which genetic instructions are used to build proteins.

rod cell A vertebrate photoreceptor sensitive to very dim light and that contributes to coarse perception of movement.

root hair Of vascular plants, an extension of a specialized root epidermal cell; root hairs collectively enhance the surface area available for absorbing water and solutes.

root nodule A localized swelling on the roots of certain legumes and other plants that contain symbiotic, nitrogen-fixing bacteria.

roots Descending parts of a vascular plant that absorb water and nutrients, anchor aboveground parts, and usually store food.

rotifer An invertebrate common in food webs of lakes and ponds.

roundworm A type of parasitic or scavenging invertebrate with a bilateral, cylindrical, cuticle-covered body, usually tapered at both ends. Some cause diseases in humans.

RuBP Ribulose biphosphate. A compound with a backbone of five carbon atoms that is required for carbon fixation in the Calvin-Benson cycle of photosynthesis.

S-shaped curve A curve, obtained when population size is plotted against time, that is characteristic of logistic growth.

sac fungus A type of fungus, usually multicelled, with spores that develop inside the cells of reproductive structures shaped like globes, flasks, or dishes. Yeasts (single-celled) are also in this group.

salination A salt buildup in soil as a result of evaporation, poor drainage, and often the importation of mineral salts in irrigation water.

salivary gland Any of the glands that secrete saliva, a fluid that initially mixes with food in the mouth and starts the breakdown of starch.

salt An ionic compound formed when an acid reacts with a base.

saltatory conduction In myelinated neurons, rapid, node-to-node hopping of action potentials.

saprobe Heterotroph that obtains its nutrients from nonliving organic matter. Most fungi are saprobes.

sarcomere (SAR-koe-meer) Of vertebrate muscles, the basic unit of contraction; a

region of myosin and actin filaments organized in parallel between two Z lines of a myofibril inside a muscle cell.

sarcoplasmic reticulum (sar-koe-PLAZ-mik reh-TIK-you-lum) In muscle cells, a membrane system that takes up, stores, and releases the calcium ions required for cross-bridge formation in sarcomeres, hence for contraction.

Schwann cells Specialized neuroglial cells that grow around neuron axons, forming a myelin sheath.

sclerenchyma One of the simple tissues of flowering plants, generally with cells having thick, lignin-impregnated walls. It supports mature plant parts and often protects seeds.

sea squirt An invertebrate chordate with a leathery or jellylike tunic and bilateral, free-swimming larvae that look like tadpoles. Ancient sea squirts may have resembled animals that gave rise to vertebrates.

second law of thermodynamics Law stating that the spontaneous direction of energy flow is from organized (high-quality) to less organized (low-quality) forms. With each conversion, some energy is randomly dispersed in a form, usually heat, that is not as readily available to do work.

second messenger A molecule inside a cell that mediates and generally triggers amplified response to a hormone.

secondary immune response Rapid, prolonged response by white blood cells, memory cells especially, to a previously encountered antigen.

secondary sexual trait A trait that is associated with maleness or femaleness but that does not play a direct role in reproduction.

secretion Generally, the release of a substance for use by the organism producing it. (Not the same as *excretion*, the expulsion of excess or waste material.) Of kidneys, a regulated stage in urine formation, in which ions and other substances move from capillaries into nephrons.

sedimentary cycle A biogeochemical cycle without a gaseous phase; the element moves from land to the seafloor, then returns only through long-term geological uplifting.

seed Of gymnosperms and flowering plants, a fully mature ovule (contains the plant embryo), with its integuments forming the seed coat.

segmentation Of earthworms and many other animals, a series of body units that may be externally similar to or quite different from one another.

segregation, Mendelian principle of [L. *se-*, apart, + *grex*, herd] The principle that diploid organisms inherit a pair of genes for each trait (on a pair of homologous chromosomes) and that the two genes segregate during meiosis and end up in separate gametes.

selective gene expression Of multicelled organisms, activation or suppression of a fraction of the genes in unique ways in different cells, leading to pronounced differences in structure and function among different cell lineages.

selfish behavior A behavior by which an individual protects or increases its own chance of producing offspring, regardless of the consequences of the group to which it belongs.

selfish herd A simple society held together by reproductive self-interest.

semen (SEE-mun) [L. *serere*, to sow] Sperm-bearing fluid expelled from a penis during male orgasm.

semiconservative replication [Gk. *hēmi*, half, + L. *conservare*, to keep] Reproduction of a DNA molecule when a complementary strand forms on each of the unzipping strands of an existing DNA double helix, the outcome being two "half-old, half-new" molecules.

senescence (sen-ESS-cents) [L. *senescere*, to grow old] Sum total of processes leading to the natural death of an organism or some of its parts.

sensation The conscious awareness of a stimulus.

sensory neuron Any of the nerve cells that act as sensory receptors, detecting specific stimuli (such as light energy) and relaying signals to the brain and spinal cord.

sensory system The "front door" of a nervous system; that portion of a nervous system that receives and sends on signals of specific changes in the external and internal environments.

sessile animal Animal that remains attached to a substrate during some stage (often the adult) of its life cycle.

sex chromosome Of most animals and some plants, a chromosome whose presence determines a new individual's gender. *Compare* autosomes.

sexual dimorphism Phenotypic differences between males and females of a species.

sexual reproduction Production of offspring from the union of gametes from two parents, by way of meiosis, gamete formation, and fertilization.

sexual selection A microevolutionary process; natural selection favoring a trait that gives the individual a competitive edge in reproductive success.

shifting cultivation The cutting and burning of trees, followed by tilling of ashes into the soil; once called slash-and-burn agriculture.

shoots The above-ground parts of vascular plants.

sieve tube member Of flowering plants, a cellular component of the interconnecting conducting tubes in phloem.

sink region In a vascular plant, any region using or stockpiling organic compounds for growth and development.

sister chromatids Of a duplicated chromosome, two DNA molecules (and associated proteins) that remain attached at their centromere only during nuclear division. Each ends up in a separate daughter nucleus.

skeletal muscle In vertebrates, an organ that contains hundreds to many thousands of muscle cells, arranged in bundles that are surrounded by connective tissue. The connective tissue extends beyond the muscle (as tendons that attach it to bone).

sliding filament model Model of muscle contraction, in which myosin filaments physically slide along and pull two sets of actin filaments toward the center of the sarcomere, which shortens. The sliding requires ATP energy and cross-bridge formation between the actin and myosin.

slime mold A type of heterotrophic protistan with a life cycle that includes free-living cells that at some point congregate and differentiate into spore-bearing structures.

small intestine Of vertebrates, the portion of the digestive system where digestion is completed and most nutrients absorbed.

smog, industrial Gray-colored air pollution that predominates in industrialized cities with cold, wet winters.

smog, photochemical Form of brown, smelly air pollution occurring in large cities with warm climates.

social behavior Cooperative, interdependent relationships among animals of the same species.

social parasite Animal that depends on the social behavior of another species to gain food, care for young, or some other factor to complete its life cycle.

sodium-potassium pump A transport protein spanning the lipid bilayer of the plasma membrane. When activated by ATP, its shape changes and it selectively transports sodium ions out of the cell and potassium ions in.

solute (SOL-yoot) [L. *solvere*, to loosen] Any substance dissolved in a solution. In water, this means spheres of hydration surround the charged parts of individual ions or molecules and keep them dispersed.

solvent Fluid in which one or more substances is dissolved.

somatic cell (so-MAT-ik) [Gk. *sōmā*, body] Of animals, any cell that is not a germ cell (which gives rise to gametes).

somatic nervous system Those nerves leading from the central nervous system to skeletal muscles.

sound system Of birds, the brain regions that govern muscles of the vocal organ.

source region Of vascular plants, any of the sites of photosynthesis.

speciation (spee-cee-AY-shun) The evolutionary process by which species originate. One speciation route starts with divergence of two reproductively isolated populations of a species. They become separate species when accumulated differences in allele frequencies prevent them from interbreeding successfully under natural conditions. Speciation also may be instantaneous (by way of polyploidy, especially among self-fertilizing plants).

species (SPEE-sheez) [L. *species*, a kind] Of sexually reproducing organisms, a unit consisting of one or more populations of individuals that can interbreed under natural conditions to produce fertile offspring that are reproductively isolated from other such units.

sperm [Gk. *sperma*, seed] A type of mature male gamete.

spermatogenesis (sperm-AT-oh-JEN-ih-sis) Formation of a mature sperm from a germ cell.

sphere of hydration Through positive or negative interactions, a clustering of water molecules around the individual molecules of a substance placed in water. *Compare* solute.

sphincter (SFINK-tur) Ring of muscle between regions of a tubelike system (as between the stomach and small intestine).

spinal cord Of central nervous systems, the portion threading through a canal inside the vertebral column and providing direct reflex connections between sensory and motor neurons as well as communication lines to and from the brain.

spindle apparatus A type of bipolar structure that forms during mitosis or meiosis and that moves the chromosomes. It consists of two sets of microtubules that extend from the opposite poles and that overlap at the spindle's equator.

spleen One of the lymphoid organs; it is a filtering station for blood, a reservoir of red blood cells, and a reservoir of macrophages.

sponge An invertebrate having a body with no symmetry and no organs; a framework of glassy needles and other structures imparts shape to it. Distinctive for its food-gathering, flagellated collar cells.

sporangium (spore-AN-gee-um), plural **sporangia** [Gk. *spora*, seed] The protective tissue layer that surrounds haploid spores in a sporophyte.

spore Of land plants, a type of resistant cell, often walled, that forms between the time of meiosis and fertilization. It germinates and develops into a gametophyte, the actual gamete-producing body. Of most fungi, a walled, resistant cell or multicelled structure, produced by mitosis or meiosis, that can germinate and give rise to a new mycelium.

sporophyte [Gk. *phyton*, plant] Of plant life cycles, a vegetative body that grows (by mitosis) from a zygote and that produces the spore-bearing structures.

sporozoan One of four categories of protozoans; a parasite that produces sporelike infectious agents called sporozoites. Some cause serious diseases in humans.

spring overturn Of certain lakes, the movement of dissolved oxygen from the surface layer to the depths and movement of nutrients from bottom sediments to the surface.

stabilizing selection Of a population, a persistence over time of the alleles responsible for the most common phenotypes.

stamen (STAY-mun) Of flowering plants, a male reproductive structure; commonly consists of pollen-bearing structures (anthers) on single stalks (filaments).

start codon Of protein synthesis, a base triplet in a strand of mRNA that serves as the start signal for mRNA translation.

stem cell Of animals, one of the unspecialized cells that replace themselves by ongoing mitotic divisions; portions of their daughter cells also divide and differentiate into specialized cells.

steroid (STAIR-oid) A lipid with a backbone of four carbon rings and with no fatty acid tails. Steroids differ in their functional groups. Different types have roles in metabolism, intercellular communication, and (in animals) cell membranes.

steroid hormone A type of lipid-soluble hormone, synthesized from cholesterol, that diffuses directly across the lipid bilayer of a target cell's plasma membrane and that binds with a receptor inside that cell.

stigma Of many flowering plants, the sticky or hairy surface tissue on the upper portion of the ovary that captures pollen grains and favors their germination.

stimulus [L. *stimulus*, goad] A specific change in the environment, such as a variation in light, heat, or mechanical pressure, that the body can detect through sensory receptors.

stoma (STOW-muh), plural **stomata** [Gk. *stoma*, mouth] A controllable gap between two guard cells in stems and leaves; any of the small passageways across the epidermis through which carbon dioxide moves into the plant and water vapor moves out.

stomach A muscular, stretchable sac that receives ingested food; of vertebrates, an organ between the esophagus and intestine in which considerable protein digestion occurs.

stop codon Of protein synthesis, a base triplet in a strand of mRNA that serves as the stop signal for translation, so that no more amino acids are added to the polypeptide chain.

stream A flowing-water ecosystem that starts out as a freshwater spring or seep.

stroma [Gk. *strōma*, bed] Of chloroplasts, the semifluid interior between the thylakoid membrane system and the two outer membranes; the zone where sucrose, starch, and other end products of photosynthesis are assembled.

stromatolite Of shallow seas, layered structures formed from sediments and large mats of the slowly accumulated remains of photosynthetic populations.

substrate A reactant or precursor molecule for a metabolic reaction; a specific molecule or molecules that an enzyme can chemically recognize, bind briefly to itself, and modify in a specific way.

substrate-level phosphorylation The direct, enzyme-mediated transfer of a phosphate group from the substrate of a reaction to another molecule. An example is the transfer of phosphate from an intermediate of glycolysis to ADP, forming ATP.

succession, primary (suk-SESH-un) [L. *succedere*, to follow after] Orderly changes from the time pioneer species colonize a barren habitat through replacements by various species until the climax community, when the com-

position of species remains steady under prevailing conditions.

succession, secondary Orderly changes in a community or patch of habitat toward the climax state after having been disturbed, as by fire.

surface-to-volume ratio A mathematical relationship in which volume increases with the cube of the diameter, but surface area increases only with the square. Of growing cells, the volume of cytoplasm increases more rapidly than the surface area of the plasma membrane that must service the cytoplasm. Because of this constraint, cells generally remain small or elongated, or have elaborate membrane foldings.

survivorship curve A plot of the age-specific survival of a group of individuals in a given environment, from the time of their birth until the last one dies.

symbiosis (sim-by-OH-sis) [Gk. *sym*, together, + *bios*, life, mode of life] A form of mutualism in which organisms of different species cannot grow and reproduce unless they spend their entire lives together in intimate interdependency. A mycorrhiza is an example.

sympathetic nerve Of the autonomic nervous system, any of the nerves generally concerned with increasing overall body activities during times of heightened awareness, excitement, or danger; also work continually in opposition with parasympathetic nerves to bring about minor adjustments in internal organs.

sympatric speciation [Gk. *sym*, together, + *patria*, native land] Speciation that follows after ecological, behavioral, or genetic barriers arise within the boundaries of a single population. This can happen instantaneously, as when polyploidy arises in a type of flowering plant that can self-fertilize or reproduce asexually.

synaptic integration (sin-AP-tik) The moment-by-moment combining of excitatory and inhibitory signals arriving at a trigger zone of a neuron.

systematics Branch of biology that deals with patterns of diversity among organisms in an evolutionary context; its three approaches include taxonomy, phylogenetic reconstruction, and classification.

systemic circuit (sis-TEM-ik) Circulation route in which oxygen-enriched blood flows from the lungs to the left half of the heart, through the rest of the body (where it gives up oxygen and takes on carbon dioxide), then back to the right side of the heart.

T lymphocyte A white blood cell with roles in immune responses.

tactile signal A physical touching that carries social significance.

taproot system A primary root and its lateral branchings.

target cell Of hormones and other signaling molecules, any cell having receptors to which they can bind.

taxonomy (tax-ON-uh-mee) Approach in biological systematics that involves identifying organisms and assigning names to them.

telophase (TEE-low-faze) Of mitosis, the final stage when chromosomes decondense into threadlike structures and two daughter nuclei form. Of meiosis I, the stage when one of each pair of homologous chromosomes has arrived at one or the other end of the spindle pole. At telophase II, chromosomes decondense and four daughter nuclei form.

telophase II Of meiosis, final stage when four daughter nuclei form.

temperate pathway A viral infection that enters a latent period; the host is not killed outright.

tendon A cord or strap of dense connective tissue that attaches muscle to bones.

territory An area that one or more individuals defend against competitors.

test An attempt to produce actual observations that match predicted or expected observations.

testcross Experimental cross to reveal whether an organism is homozygous dominant or heterozygous for a trait. The organism showing dominance is crossed to an individual known to be homozygous recessive for the same trait.

testis, plural **testes** Male gonad; primary reproductive organ in which male gametes and sex hormones are produced.

testosterone (tess-TOSS-tuh-rown) In male mammals, a major sex hormone that helps control male reproductive functions.

tetanus Of muscles, a large contraction in which repeated stimulation of a motor unit causes muscle twitches to mechanically run together. In a disease by the same name, toxins prevent muscle relaxation.

theory A testable explanation of a broad range of related phenomena. In modern science, only explanations that have been extensively tested and can be relied upon with a very high degree of confidence are accorded the status of theory.

thermal inversion Situation in which a layer of dense, cool air becomes trapped beneath a layer of warm air; can cause air pollutants to accumulate to dangerous levels close to the ground.

thermophile A type of archaebacterium that lives in hot springs, highly acidic soils, and near hydrothermal vents.

thermoreceptor Sensory cell that can detect radiant energy associated with temperature.

thigmotropism (thig-MOTE-ruh-pizm) [Gk. *thigm*, touch] Of vascular plants, growth orientation in response to physical contact with a solid object, as when a vine curls around a fencepost.

threshold Of neurons and other excitable cells, a certain minimum amount by which the voltage difference across the plasma membrane must change to produce an action potential.

thylakoid membrane system Of chloroplasts, an internal membrane system commonly folded into flattened channels and disks (*grana*) and containing light-absorbing pigments and enzymes used in the formation of ATP, NADPH, or both during photosynthesis.

thymine Nitrogen-containing base in some nucleotides.

thymus gland A lymphoid organ with endocrine functions; lymphocytes of the immune system multiply, differentiate, and mature in its tissues, and its hormone secretions affect their functions.

thyroid gland Of endocrine systems, a gland that produces hormones that affect overall metabolic rates, growth, and development.

tissue Of multicelled organisms, a group of cells and intercellular substances that function together in one or more specialized tasks.

tonicity The relative concentrations of solutes in two fluids, such as inside and outside a cell. When solute concentrations are isotonic (equal in both fluids), water shows no net osmotic movement in either direction. When one fluid is hypotonic (has less solutes than the other), the other is hypertonic (has more solutes) and is the direction in which water tends to move.

tooth Of the mouth of various animals, one of the hardened appendages used to secure or mechanically pummel food; sometimes used in defense.

tracer A radioisotope used to label a substance so that its pathway or destination in a cell, organism, ecosystem, or some other system can be tracked, as by scintillation counters that detect its emissions.

trachea (TRAY-kee-uh), plural **tracheae** An air-conducting tube that functions in respiration; of land vertebrates, the windpipe, which carries air between the larynx and bronchi.

tracheal respiration Of insects, spiders, and some other animals, a respiratory system consisting of finely branching tracheae that extend from openings in the integument and that dead-end in body tissues.

tracheid (TRAY-kid) Of flowering plants, one of two types of cells in xylem that conduct water and dissolved minerals.

transcript-processing controls Of eukaryotic cells, controls that govern modification of new mRNA molecules into mature transcripts before shipment from the nucleus.

transcription [L. *trans*, across, + *scribere*, to write] Of protein synthesis, the assembly of an RNA strand on one of the two strands of a DNA double helix; the base sequence of the resulting transcript is complementary to the DNA region on which it was assembled.

transcriptional controls Of eukaryotic cells, controls influencing when and to what degree a particular gene will be transcribed.

transfer RNA (tRNA) Of protein synthesis, any of the type of RNA molecules that bind and deliver specific amino acids to ribosomes *and* pair with mRNA code words for those amino acids.

translation Of protein synthesis, the conversion of the coded sequence of information in mRNA into a particular sequence of amino acids to form a polypeptide chain; depends on interactions of rRNA, tRNA, and mRNA.

translational controls Of eukaryotic cells, controls governing the rates at which mRNA transcripts that reach the cytoplasm will be translated into polypeptide chains at ribosomes.

translocation Of cells, a change in a chromosome's structure following the insertion of part of a nonhomologous chromosome into it. Of vascular plants, conduction of organic compounds through the plant body by way of the phloem.

transpiration Evaporative water loss from stems and leaves.

transport control Of eukaryotic cells, controls governing when mature mRNA transcripts are shipped from the nucleus into the cytoplasm.

transposable element DNA element that can spontaneously "jump" to new locations in the same DNA molecule or a different one. Such elements often inactivate the genes into which they become inserted and give rise to observable changes in phenotype.

trisomy (TRY-so-mee) Of diploid cells, the abnormal presence of three of one type of chromosome.

trophic level (TROE-fik) [Gk. *trophos*, feeder] All the organisms in an ecosystem that are the same number of transfer steps away from the energy input into the system.

tropical rain forest A type of biome where rainfall is regular and heavy, the annual mean temperature is 25°C, and humidity is 80 percent or more; characterized by great biodiversity.

tropism (TROE-prizm) Of vascular plants, a growth response to an environmental factor, such as growth toward light.

true-breeding Of sexually reproducing organisms, a lineage in which the offspring of successive generations are just like the parents in one or more traits.

tumor A tissue mass composed of cells that are dividing at an abnormally high rate.

turgor pressure (TUR-gore) [L. *turgere*, to swell] Internal pressure applied to a cell wall when water moves by osmosis into the cell.

uniformitarianism The theory that existing geologic features are an outcome of a long history of gradual changes, interrupted now and then by huge earthquakes and other catastrophic events.

upwelling An upward movement of deep, nutrient-rich water along coasts to replace surface waters that winds move away from shore.

uracil (YUR-uh-sill) Nitrogen-containing base found in RNA molecules; can basepair with adenine.

ureter A tubular channel for urine flow between the kidney and urinary bladder.

urethra A tubular channel for urine flow between the urinary bladder and an opening at the body surface.

urinary bladder A distensible sac in which urine is temporarily stored before being excreted.

urinary excretion A mechanism by which excess water and solutes are removed by way of a urinary system.

urinary system An organ system that adjusts the volume and composition of blood, and so helps maintain extracellular fluid.

urine Fluid formed by filtration, reabsorption, and secretion in kidneys; consists of wastes, excess water, and solutes.

uterus (YOU-tur-us) [L. *uterus*, womb] Chamber in which the developing embryo is contained and nurtured during pregnancy.

vaccine An antigen-containing preparation, swallowed or injected, that increases immunity to certain diseases. It induces formation of huge armies of effector and memory B and T cell populations.

vagina Part of a female reproductive system that receives sperm, forms part of the birth canal, and channels menstrual flow to the exterior.

variable Of a scientific experiment, the only factor that is not exactly the same in the experimental group as it is in the control group.

vascular bundle Of vascular plants, the arrangement of primary xylem and phloem into multistranded, sheathed cords that thread lengthwise through the ground tissue system.

vascular cambium Of vascular plants, a lateral meristem that increases stem or root diameter.

vascular cylinder Of plant roots, the arrangement of vascular tissues as a central cylinder.

vascular plant Plant having tissues that transport water and solutes through well-developed roots, stems, and leaves.

vascular tissue system Xylem and phloem; the conducting tissues that distribute water and solutes through the body of vascular plants.

vein Of the circulatory system, any of the large-diameter vessels that lead back to the heart; of leaves, one of the vascular bundles that thread through photosynthetic tissues.

ventricle (VEN-tri-kuhl) Of the vertebrate heart, one of two chambers from which blood is pumped out. *Compare* atrium.

venule A small blood vessel that accepts blood from capillaries and delivers it to a vein; also overlaps capillaries somewhat in function.

vernalization Of flowering plants, stimulation of flowering by exposure to low temperatures.

vertebra, plural **vertebrae** Of vertebrate animals, one of a series of hard bones arranged with intervertebral disks into a backbone.

vertebrate Animal having a backbone of bony segments, the vertebrae.

vesicle (VESS-ih-kul) [L. *vesicula*, little bladder] Within the cytoplasm of cells, one of a variety of small membrane-bound sacs that function in the transport, storage, or digestion of substances or in some other activity.

vessel member One of the cells of xylem, dead at maturity, the walls of which form the water-conducting pipelines.

villus (VIL-us), plural **villi** Any of several types of absorptive structures projecting from the free surface of an epithelium.

viroid An infectious nucleic acid that has no protein coat; a tiny rod or circle of single-stranded RNA.

virus A noncellular infectious agent, consisting of DNA or RNA and a protein coat; can replicate only after its genetic material enters a host cell and subverts its metabolic machinery.

vision Precise light focusing onto a layer of photoreceptive cells that is dense enough to sample details concerning a given light stimulus, followed by image formation in the brain.

visual signal An observable action or cue that functions as a communication signal.

vitamin Any of more than a dozen organic substances that animals require in small amounts for normal cell metabolism but generally cannot synthesize for themselves.

vocal cord One of the thickened, muscular folds of the larynx that help produce sound waves for speech.

water mold A type of saprobic or parasitic fungus that lives in fresh water or moist soil.

water potential The sum of two opposing forces (osmosis and turgor pressure) that can cause the directional movement of water into or out of a walled cell.

water table The upper limit at which the ground in a specified region is fully saturated with water.

watershed Any specified region in which all precipitation drains into a single stream or river.

wax A type of lipid with long-chain fatty acid tails that help form protective, lubricating, or water-repellent coatings.

white blood cell Leukocyte; of vertebrates, any of the macrophages, eosinophils, neutrophils, and other cells which, together with their products, comprise the immune system.

white matter Of spinal cords, major nerve tracts so named because of the glistening myelin sheaths of their axons.

wild-type allele Of a population, the allele that occurs normally or with greatest frequency at a given gene locus.

wing Of birds, a forelimb of feathers, powerful muscles, and lightweight bones that functions in flight. Of insects, a structure that develops as a lateral fold of the exoskeleton and functions in flight.

X chromosome Of humans, a sex chromosome with genes that cause an embryo to develop into a female, provided that it inherits a pair of these.

X-linked gene Any gene on an X chromosome.

X-linked recessive inheritance Recessive condition in which the responsible, mutated gene occurs on the X chromosome.

xylem (ZYE-lum) [Gk. *xylon*, wood] Of vascular plants, a tissue that transports water and solutes through the plant body.

Y chromosome Of humans, a sex chromosome with genes that cause the embryo that inherited it to develop into a male.

Y-linked gene Any gene on a Y chromosome.

yellow marrow A fatty tissue in the cavities of most mature bones that produces red blood cells when blood loss from the body is severe.

yolk sac Of land vertebrates, one of four extraembryonic membranes. In most shelled eggs, it holds nutritive yolk. In humans, part becomes a site of blood cell formation and some of its cells give rise to the forerunners of gametes.

zero population growth A population for which the number of births is balanced by the number of deaths over a specified period, assuming immigration and emigration also are balanced.

zooplankton A freshwater or marine community of floating or weakly swimming heterotrophs, mostly microscopic, such as rotifers and copepods.

zygospore-forming fungus A type of fungus for which a thick spore wall forms around the zygote; this resting spore germinates and gives rise to stalked, spore-bearing structures.

zygote (ZYE-goat) The first cell of a new individual, formed by the fusion of a sperm nucleus with the nucleus of an egg (fertilization).

CREDITS AND ACKNOWLEDGMENTS

Chapter 12

12.1 (left) Dennis Hallinan/FPG; (right) Kevin Magee/Tom Stack & Associates / **12.3** Art by Hans & Cassady, Inc. / **12.7** (a) Courtesy of Thomas A. Steitz from *Science*, 246:1135–1142, December 1, 1989 / **12.9** Art by Raychel Ciemma and American Composition & Graphics, Inc. / **12.11** Peter Starlinger / **12.12** Art by Raychel Ciemma / **12.14** Art by Palay/Beaubois / **12.15** (a), (b) Stuart Kenter Associates / **Page 175** (b) Lennart Nilsson © Boehringer Ingelheim International GmbH

Chapter 13

13.1 Secchi-Lecague/Roussel-UCLAF/CNRI/SPL/Photo Researchers / **13.2** Dr. Huntington Potter and Dr. David Dressler / **13.4** Art by Jeanne Schreiber / **Page 182** (a) Damon Biotech, Inc. / **13.7** Michael Maloney/San Francisco Chronicle / **13.8** W. Merrill / **13.9** (a) Keith V. Wood; (b), (c) Monsanto Company; (d), (e) Calgene, Inc. / **13.10** R. Brinster and R. E. Hammer, School of Veterinary Medicine, University of Pennsylvania

Page 189 S. Stammers/SPL/Photo Researchers

Chapter 14

14.1 Elliott Erwitt/Magnum Photos, Inc. / **14.2** (a) Dave Watts/A.N.T. Photo Library; (b) Kenneth W. Fink/Photo Researchers; (c) Jen & Des Bartlett/Bruce Coleman Ltd. / **14.4** (a) (left) Courtesy George P. Darwin, Darwin Museum, Down House; (right) Heather Angel; (b) Christopher Ralling; (c) D. Barrett/Planet Earth Pictures; (d) Heather Angel; (e) photograph Dieter & Mary Plage/Survival Anglia; art by Leonard Morgan / **14.5** (left) Lee Kuhn/FPG; (right) Field Museum of Natural History, Chicago, and the artist, Charles R. Knight (Neg. No. CK21T) / **14.6** (a) Heather Angel; (b) David Cavagnaro; (c) George W. Cox; (d) Alan Root/Bruce Coleman Ltd. / **14.7** Down House and The Royal College of Surgeons of England / **14.8** Photograph John H. Ostrom, Yale University / **14.9** Alan Solem / **14.11** (above) David Neal Parks; (below) W. Carter Johnson / **14.12** After D. Futuyma, *Evolutionary Biology*, Sinauer, 1979 / **14.14** (a) Thomas N. Taylor; (b) Edward S. Ross / **14.15** After M. Karns and L. Penrose, *Annals of Eugenics*, 15:206–233, 1951 / **14.16** J. A. Bishop and L. M. Cook / **14.17** Bruce Beehler / **14.18** After F. Ayala and J. Valentine, *Evolving*, Benjamin-Cummings, 1979 / **14.19** After W. Jensen and F. B. Salisbury, *Botany: An Ecological Approach*, Wadsworth, Inc., 1972 / **14.20** After V. Grant, *Organismic Evolution*, W. H. Freeman and Co., 1977 / **Page 206** D. Avon/Ardea, London

Chapter 15

15.1 (left) Vatican Museums; (right) Martin Dohrn/SPL/Photo Researchers / **15.2** (a) Jonathan Blair/Woodfin Camp & Associates; (b) Patricia G. Gensel; (c) Donald Baird, Princeton Museum of Natural History; (d) A. Feduccia, *The Age of Birds*, Harvard University Press, 1980 / **15.3** David Noble/FPG / **15.4** Art by Leonard Morgan / **15.5** From T. Storer et al., *General Zoology*, Sixth edition, McGraw-Hill, 1979. Reproduced by permission of McGraw-Hill, Inc. / **15.6** Art by Victor Royer / **15.7** Art by Joel Ito / **15.8** (top) Douglas P. Wilson/Eric & David Hosking; (center) Superstock, Inc.; (bottom) E. R. Degginger / **15.10** (a) Kjell Sandved/Visuals Unlimited; (b) Jeffrey Sylvester/FPG; (c) Thomas D. Mangelsen/Images of Nature / **15.11** After P. Dodson, *Evolution: Process and Product*, Third edition, Prindle, Weber & Schmidt / **15.12** After A. M. Ziegler, C. R. Scorese, and S. F. Barrett, "Mesozoic and Cenozoic Paleogeographic Maps" and J. Krohn and J. Sündermann, "Paleotides Before the Permian" in F. Brosche and J. Sündermann (Eds.), *Tidal Friction and the Earth's Rotation II*, Springer-Verlag, 1983; (left) art by Lloyd K. Townsend / **15.13** Chesley Bonestell / **15.15** Art by Precision Graphics / **15.16** (a) Sidney W. Fox; (b) W. Hargreaves and D. Deamer / **15.18** (a) Stanley W. Awramik; (b) M. R. Walter / **15.19** (a), (b) Neville Pledge/South Australian Museum; (c), (d) Chip Clark / **15.20** (c) Alex Kerstitch / **15.21** (a) H. P. Banks; (b) Patricia G. Gensel / **15.22** (a) From *Evolution of Life*, Linda Gamlin and Gail Vines (Eds.), Oxford University Press, 1987; art by D. & V. Hennings (b) Rod Salm/Planet Earth Pictures / **Page 227** (a) NASA / **Page 228–229** (b) (left) Art by Raychel Ciemma; (right) © John Gurche 1989 / **15.23** Field Museum of Natural History, Chicago, and the artist Charles R. Knight (Neg. No. CK8T) / **Page 232** (above) (left) Larry Lefever/Grant Heilman Inc.; (right) Bruce Coleman Ltd.; (below) (left) R. I. M. Campbell/Bruce Coleman Ltd.; (right) Runk/Schoenberger/ Grant Heilman Inc.

Chapter 16

16.1 (left) FPG; (right) Douglas Mazonowicz/Gallery of Prehistoric Art / **16.2** (a) Kevin Shafer/Tom Stack & Associates; (b) Art by Raychel Ciemma after M. Weiss and A. Mann, *Human Biology and Behavior*, Fifth edition, HarperCollins, 1990 / **16.4** (a) Bruce Coleman Ltd.; (b) Tom McHugh/Photo Researchers; (c) Larry Burrows/Aspect Picture Library / **16.5** Art by D. & V. Hennings / **16.7** © Time Inc. 1965/Larry Burrows Collection / **16.9** Art by D. & V. Hennings / **16.10** (a) Louise M. Robbins; (b) Dr. Donald Johanson, Institute of Human Origins / **16.11, 16.12** Art by D. & V. Hennings / **16.13** Photographs by John Reader copyright 1981

Page 245 © 1990 Arthur M. Greene

Chapter 17

17.1 Tony Brain and David Parker/SPL/Photo Researchers /**17.2** Art by Nadine Sokol / **17.3** Art by Palay/Beaubois / **17.4** (a) George Musil/Visuals Unlimited; (b) K. G. Murti/Visuals Unlimited; (c), (d) Kenneth M. Corbett / **17.5** Art by L. Calver; micrograph CNRI/SPL/Photo Researchers / **17.6** Art by D. & V. Hennings / **17.8** Paul A. Zahl, © 1967 National Geographic Society / **17.9** (a) John D. Cunningham/Visuals Unlimited; (b) Tony Brain/SPL/Photo Researchers; (c) P. W. Johnson and J. McN. Sieburth, University of Rhode Island/BPS / **17.10** T. J. Beveridge, University of Guelph/BPS / **17.11** P. L. Walne and J. H. Arnott, *Planta*, 77:325–354, 1967 / **Pages 256–257** Art by Raychel Ciemma and Precision Graphics / **Page 258** (c) Robert K. Trench / **17.12** Art by Palay/Beaubois / **17.13** (a) Art by Leonard Morgan; (b) M. Claviez, G. Gerish, and R. Guggenheim; (c) London Scientific Films; (d–f) Carolina Biological Supply Company; (g) Photograph courtesy Robert R. Kay from R. R. Kay et al. *Development*, 1989 Supplement, pp. 81–90, © The Company of Biologists Ltd. 1989 / **17.14** Edward S. Ross / **17.15** (a), (b) Ronald W. Hoham, Dept. of Biology, Colgate University; (c) Jan Hinsch / SPL / Photo Researchers / **17.16** (a) Florida Department of Environmental Protection, Florida Marine Research Institute, St. Petersburg; (b) C. C. Lockwood / **17.17** (a) John D. Cunningham/Visuals Unlimited; (b) David M. Phillips/Visuals Unlimited / **17.18** (a) M. Abbey/Visuals Unlimited; (b) John Clegg/Ardea, London; (c) Manfred Kage/ Bruce Coleman Ltd.; (d) T. E. Adams/Visuals Unlimited / **17.19** (a) Frieder Sauer/Bruce Coleman Ltd.; art by Palay/Beaubois; (b) Gary W. Grimes and Steven L'Hernault / **17.20** Art by Leonard Morgan; micrograph Steven L'Hernault

Chapter 18

18.1 Pat & Tom Leeson/Photo Researchers / **18.2** Robert C. Simpson/Nature Stock / **18.3** Micrograph G. T. Cole, University of Texas, Austin/BPS / **Page 268** W. Merrill / **18.5** (a) (above) John D. Cunningham/Visuals Unlimited; (below) David M. Phillips/ Visuals Unlimited; (b) John Hodgin / **18.6** (a) After T. Rost et al., *Botany*, Wiley, 1979; (b), (c) Robert C. Simpson/Nature Stock / **18.7** (a), (b) Robert C. Simpson/Nature Stock; (c) Jane Burton/Bruce Coleman Ltd. / **18.8** Photographs Martyn Ainsworth from A. D. M. Rayner, *New Scientist*, November 19, 1988; art by Raychel Ciemma / **18.9** (a) G. T. Cole, University of Texas, Austin/BPS; (b) N. Allin and G. L. Barron; (c) G. L. Barron, University of Guelph / **18.10** (above) Mark Mattock/Planet Earth Pictures; (below) Edward S. Ross / **18.11** After Raven, Evert, and Eichhorn, *Biology of Plants*, Fourth edition, Worth Publishers, New York, 1986 / **18.12** (a) © 1990 Gary Braasch; (b) F. B. Reeves / **18.14** (a) D. P. Wilson/Eric & David Hosking; (b) Douglas Faulkner/Sally Faulkner Collection; (c) art by Jennifer Wardrip based on Gilbert M. Smith, *Marine Algae of the Monterey Peninsula*, Stanford University Press; (d) Steven C. Wilson/Entheos; (e) Dennis Brokaw / **18.15** Photograph D. J. Patterson/Seaphot Limited: Planet Earth Pictures; art by Raychel Ciemma / **18.16** Carolina Biological Supply Company / **18.17** Photograph Jane Burton/Bruce Coleman Ltd.; art by Raychel Ciemma / **Page 279** (a) Field Museum of Natural History, Chicago; (Neg. #7500C); (b) Brian Parker/Tom Stack & Associates / **18.18** (a) Edward S. Ross; (b) W. H. Hodge / **18.19** (left) Art by Raychel Ciemma; inset photograph A. & E. Bomford/Ardea, London; (right) Lee Casebere / **18.20** (a) Ed Reschke; (b) (left) John H. Gerard; (right) Kingsley R. Stern; (c) Edward S. Ross / **18.21** Photograph Edward S. Ross; art by Raychel Ciemma / **18.22** Photograph M.P.L. Fogden/Ardea, London; art by Jennifer Wardrip / **18.23** Art by Raychel Ciemma

Chapter 19

19.1 (a) Tom McHugh/Photo Researchers; (b) Jean Phillipe Varin/Jacana/Photo Researchers / **19.2, 19.3** Art by D. & V. Hennings / **19.5** (a) Marty Snyderman/Planet Earth Pictures; (b) Bruce Hall; (c) Douglas Faulkner/Sally Faulkner Collection / **19.6** Photograph David C. Haas/Tom Stack & Associates; art by Raychel Ciemma / **19.7** (b) Frieder Sauer/Bruce Coleman Ltd.; (c) Kim Taylor/Bruce Coleman Ltd.; (d) Walter Deas/Seaphot Limited: Planet Earth Pictures; (e) Bill Wood/Seaphot Limited: Planet Earth Pictures / **19.8** (a) Art by Raychel Ciemma; (b) Francois Gohier/Photo Researchers / **19.9** Photograph Andrew Mounter/Seaphot Limited: Planet Earth Pictures; art by Precision Graphics / **19.10** Photograph Kim Taylor/Bruce Coleman Ltd.; art by Raychel Ciemma / **19.11** (a) Cath Ellis, University of Hull/SPL/Photo Researchers; (b) Robert & Linda Mitchell / **19.12** Art by Raychel Ciemma / **Page 296** (a) Art by Raychel Ciemma; (b) art by Nadine Sokol; photograph Carolina Biological Supply Company / **Page 297** (c) Lorus J. and Margery Milne; (d) Dianora Niccolini / **19.13** Art by Raychel Ciemma / **19.14** (a) Anthony & Elizabeth Bomford/Ardea, London / **19.15** Alex Kerstitch / **19.16** (a) Art by Raychel Ciemma; (b) Hervé Chaumeton/Agence Nature / **19.17** (a) J. Grossauer/ZEFA; (b) art by Raychel Ciemma / **19.18** (a) Jon Kenfield/Bruce Coleman Ltd.; (b) Hervé Chaumeton/Agence Nature / **19.19** Art by Raychel Ciemma; photograph © Cabisco/Visuals Unlimited / **19.20** J. A. L. Cooke/Oxford Scientific Films /**19.21** Jane Burton/Bruce Coleman Ltd. / **19.22** (above) Angelo Giampiccolo/FPG; (below) Jane Burton/Bruce Coleman Ltd. / **19.23** (a) P. J. Bryant, University of California, Irvine/BPS; (b) Ken Lucas/Seaphot Limited: Planet Earth Pictures; (c) John H. Gerard / **19.24** (a) Franz Lanting/Bruce Coleman Ltd.; (b) Hervé Chaumeton; (c) Fred Bavendam/Peter Arnold, Inc.; (d) Agence Nature / **19.25** (a) Z. Leszczynski/Animals Animals; (b) Steve Martin/Tom Stack & Associates / **19.26** Art by D. & V. Hennings / **19.27** (a) Robert & Linda Mitchell; (b) Ralph A. Reinhold/FPG; (c) David Maitland/Seaphot Limited: Planet Earth Pictures; (d) Kenneth Lorenzen; (e–j), (l), Edward S. Ross; (k) C. P. Hickman, Jr. / **19.28** (a), (b) Hervé Chaumeton/Agence Nature; (c) Kjell B. Sandved; (d) Chris Huss/The Wildlife Collection; (e) Ian Took/Biofotos; (f) John Mason/Ardea, London / **19.29** Art by L. Calver / **19.30** Art by D. & V. Hennings and Precision Graphics / **19.31** Photograph Rick M. Harbo; (a–d) from *Living Invertebrates*, V. & J. Pearse and M. & R. Buchsbaum, The Boxwood Press, 1987. Used by permission. / **19.32** Photograph Hervé Chaumeton/Agence Nature; art by Laszlo Meszoly and D. & V. Hennings / **19.33** After C. P. Hickman, Jr. and L. S. Roberts, *Integrated Principles of Zoology*,

Seventh edition, St. Louis: Times Mirror/Mosby College Publishing, 1984; art by Palais/Beaubois / **19.34** Art by Raychel Ciemma / **19.35** Art by Precision Graphics / **19.36** Heather Angel / **19.37** (a) Erwin Christian/ZEFA; (b) Allan Power/Bruce Coleman Ltd.; (c) Tom McHugh/Photo Researchers / **19.38** (a) Douglas Faulkner/Sally Faulkner Collection; (b) Patrice Ceisel; © 1986 John G. Shedd Aquarium; (c) Robert & Linda Mitchell; (d) William H. Amos; (e) art by Raychel Ciemma; (f) Bill Wood/Bruce Coleman Ltd. / **19.39** (a) Peter Scoones/Seaphot Limited: Planet Earth Pictures / **19.40** (a) Jerry W. Nagel; (b) Stephen Dalton/Photo Researchers; (c) John Serrao/Visuals Unlimited; (d) Juan M. Renjifo/Animals Animals / **19.41** Zig Leszczynski/Animals Animals / **19.42** (a) D. Kaleth/Image Bank; (b) Andrew Dennis / A.N.T. Photo Library; (c) Stephen Dalton/Photo Researchers; art by Raychel Ciemma; (d) Heather Angel; (e) W. A. Banaszewski/Visuals Unlimited / **19.43** (a) Gerard Lacz/A.N.T. Photo Library; (b) J. L. G. Grande/ Bruce Coleman Ltd.; (c) Thomas D. Mangelsen/ Images of Nature; (d) Rajesh Bedi / **19.44** (c) Art by D. & V. Hennings / **19.45** (a) D. & V. Blagden/ A.N.T. Photo Library; (b) Jack Dermid; (c) Douglas Faulkner/Photo Researchers; (d) Clem Haagner/ Ardea, London; (e) Sandy Roessler/FPG

Page 325 Bonnie Rauch/Photo Researchers

Chapter 20

20.1 (a) Roger Werth; (b) © 1980 Gary Braasch; (c) © 1989 Gary Braasch / **20.2** Art by Raychel Ciemma / **20.3** Art by D. & V. Hennings / **20.4** (left) Micrograph James D. Mauseth, *Plant Anatomy*, Benjamin-Cummings, 1988; (a), (c) Biophoto Associates; (b) George S. Ellmore / **20.5** Jeremy Burgess/SPL/Photo Researchers / **20.6** (a), (b) Edward S. Ross; art by D. & V. Hennings / **20.7** Art by D. & V. Hennings; (center) Carolina Biological Supply Company: (right) James W. Perry / **20.8** Art by D. & V. Hennings: (center) Ray F. Evert; (right) James W. Perry / **20.9** Art by D. & V. Hennings / **20.10** Heather Angel / **20.11** Art by Raychel Ciemma; micrograph C. E. Jeffree et al., *Planta*, 172(1):20–37, 1987. Reprinted by permission of C. E. Jeffree and Springer-Verlag / **20.12** (a) Robert & Linda Mitchell; (b) Roland R. Dute / **20.13** (a-c) E. R. Degginger / **20.14** John E. Hodgin / **20.15** Micrograph Ripon Microslides, Inc. / **20.16** Sketch after T. Rost et al., *Botany: A Brief Introduction to Plant Biology*, Second edition, © 1984, John Wiley & Sons; micrographs Chuck Brown / **20.17** Ripon Microslides, Inc. / **20.20** (a) Ripon Microslides, Inc.; (b) H. A. Core, W. A. Coté, and A. C. Day, *Wood Structure and Identification*, Second edition, Syracuse University Press, 1979 / **20.21** (b) Jerry D. Davis

Chapter 21

21.1 Photographs Robert & Linda Mitchell; micrograph John N. A. Lott, *Scanning Electron Microscope Study of Green Plants*, St. Louis: C. V. Mosby Company, 1976 / **21.2** (a) Adrian P. Davies/Bruce Coleman Ltd.; (b) NifTAL Project, University of Hawaii, Maui / **21.3** Micrograph Jean Paul Revel / **21.4** Art by Leonard Morgan / **21.5** Micrographs H. A. Core, W. A. Coté, and A. C. Day, *Wood Structure and Identification*, Second edition, Syracuse University Press, 1979 / **21.6** Art by Leonard Morgan / **21.7** George S. Ellmore / **21.8** W. Thomson, *American Journal of Botany*, 57(3):316, 1970 / **21.9** (a) Art by Palais/Beaubois; (b), (c) Jeremy Burgess/SPL/Photo Researchers / **21.10** Art by Hans & Cassady, Inc. / **21.11** Martin Zimmermann, *Science*, 133:73–79, © AAAS 1961 / **21.12** Art by Palay/Beaubois

Chapter 22

22.1 Thomas D. W. Friedmann/Photo Researchers; (inset) Jeffry Myers/FPG / **22.4** (a), (b) David M. Phillips/Visuals Unlimited; (c) David Scharf/Peter Arnold, Inc. / **22.5** Art by Raychel Ciemma / **Page 358** Robert A. Tyrrell / **Page 359** (b) Thomas Eisner,

Cornell University; (c) Edward S. Ross; (d) Ted Schwartz / **22.6** (a), (b) Patricia Schulz; (c), (d) Ray F. Evert; (e), (f) Ripon Microslides, Inc.; (g) Kingsley R. Stern / **22.7** (a–c) Janet Jones; (d) B. J. Miller, Fairfax, VA/BPS; (e) R. Carr/Bruce Coleman Ltd. / **22.8** (a) Carolina Biological Supply Company; art by Hans & Cassady, Inc.; (b) B. Bracegirdle and P. Miles, *An Atlas of Plant Structure*, Heinemann Educational Books, 1977 / **22.9** Photograph Hervé Chaumeton/Agence Nature; art by Hans & Cassady, Inc. / **22.10** (a) Kingsley R. Stern / **22.11** (a), (b) R. Lyons/Visuals Unlimited; (c) Michael A. Keller/FPG / **22.12** Frank B. Salisbury / **22.13** John Digby and Richard Firn / **22.14** Frank B. Salisbury / **22.16** Jan Zeevart / **22.17** Larry D. Nooden / **Page 369** (left) Edward S. Ross; (right) Dennis Brokaw

Page 371 © Kevin Schafer

Chapter 23

23.1 David Macdonald / **23.2** Focus on Sports; (inset) Manfred Kage/Bruce Coleman Ltd.; art by Palay/Beaubois / **23.3** Art by Palay/Beaubois / **23.4** Photographs (a) Lennart Nilsson from *Behold Man*, © 1974 by Albert Bonniers Forlag and Little, Brown and Company, Boston; (b) Manfred Kage/Bruce Coleman Ltd.; (c) Ed Reschke/Peter Arnold, Inc. / **23.5** Photographs Ed Reschke / **23.6** Photographs Ed Reschke / **23.7** Lennart Nilsson from *Behold Man*, ©1974 Albert Bonniers Forlag and Little, Brown and Company, Boston / **23.8** Art by L. Calver / **23.9** Art by Palay/Beaubois / **23.11** Fred Bruemmer / **Page 382** (above) (left) Manfred Kage/Bruce Coleman Ltd.; (right) Ed Reschke; (below) Ed Reschke

Chapter 24

24.1 Jeff Schultz/AlaskaStock Images / **24.2** Art by L. Calver / **24.3** (above) (left) Ed Reschke; (right) CNRI/SPL/Photo Researchers; (below) art by Robert Demarest / **Page 385** Michael Keller/FPG / **24.4** Art by D. & V. Hennings / **24.5** (a) (left) Art by L. Calver; (a) (right), (b) art by Joel Ito; (c) Ed Reschke / **24.6** Art by K. Kasnot / **24.7** Art by Kevin Somerville / **Page 389** Photograph C. Yokochi and J. Rohen, *Photographic Anatomy of the Human Body*, Second edition, Igaku-Shoin Ltd., 1979 / **24.8** National Osteoporosis Foundation / **24.9** Art by Kevin Somerville / **24.10** (a) N.H.P.A./A.N.T. Photo Library; (b) art by Robert Demarest / **24.11** (a) Art by Robert Demarest; (b) John D. Cunningham; (c) D. W. Fawcett, *The Cell*, Philadelphia: W. B. Saunders Co., 1966 / **24.12**, **24.13** Art by Nadine Sokol / **24.14** Art by Robert Demarest / **Page 397** Photograph Michael Neveux

Chapter 25

25.1 (a) D. Robert Franz/Planet Earth Pictures; (b) art by D. & V. Hennings adapted from *Mammalogy*, Third edition, by Terry Vaughan, copyright © 1986 by Saunders College Publishing. Used by permission of the publisher. / **25.2** (a) Kim Taylor/Bruce Coleman Ltd.; (b) Wardene Weisser/Ardea, London; art by Precision Graphics / **Page 400** Art by Nadine Sokol / **25.4** Art by Kevin Somerville / **25.5** Art by Nadine Sokol; (b) After A. Vander et al., *Human Physiology: Mechanisms of Body Function*, Fifth edition, McGraw-Hill, 1990. Used by permission. (c) photograph D. Robert Franz/Planet Earth Pictures; sketch after A. Romer and T. Parsons, *The Vertebrate Body*, Sixth edition, copyright © 1986 by Saunders College Publishing, reprinted by permission of the publisher / **25.6** (a) Art by Victor Royer; (b), (d) Lennart Nilsson © Boehringer Ingelheim International GmbH; (c) Biophoto Associates/SPL/Photo Researchers / **25.7** Art by Robert Demarest / **25.8** Art by Raychel Ciemma / **25.11** Photograph Steven Jones/FPG / **Page 411** CNRI/Phototake / **25.12** Modified after A. Vander et al. *Human Physiology*, Fourth edition, McGraw-Hill, 1985; photograph Ralph Pleasant/FPG

Chapter 26

26.1 (a) From A. D. Waller, *Physiology, The Servant of Medicine*, Hitchcock Lectures, University of London Press, 1910; (b) photograph courtesy of The New York Academy of Medicine Library / **26.3** (b) (below) After M. Labarbera and S. Vogel, *American Scientist*, 70:54–60, 1982; photograph Lennart Nilsson from *Behold Man*, © 1974 by Albert Bonniers Forlag and Little, Brown and Company, Boston / **26.4** Art by Palay/Beaubois / **26.5** CNRI/SPL/Photo Researchers / **26.6** Art by Raychel Ciemma / **26.7**, **26.8** Art by Kevin Somerville / **26.9** (a) C. Yokochi and J. Rohen, *Photographic Anatomy of the Human Body*, Second edition, Igaku-Shoin Ltd., 1979; (b), (c) art by Kevin Somerville / **26.11** (a) Ed Reschke / **26.13** Art by Robert Demarest based on A. Spence, *Basic Human Anatomy*, Benjamin-Cummings, 1982 / **26.14** Sheila Terry/SPL/Photo Researchers / **26.16** Art by Kevin Somerville / **Page 426** (a) (above) Ed Reschke; (below) F. Sloop and W. Ober/Visuals Unlimited / **26.17** Lennart Nilsson © Boehringer Ingelheim International GmbH / **26.18** (a) After F. Ayala and J. Kiger, *Modern Genetics*, © 1980 Benjamin-Cummings; (b) Lester V. Bergman & Associates, Inc. / **26.19** Art by Nadine Sokol after Gerard J. Tortora and Nicholas P. Anagnostakos, *Principles of Anatomy and Physiology*, Sixth edition, Copyright © 1990 by Biological Sciences Textbooks, Inc., A & P Textbooks, Inc. and Elia-Sparta, Inc. Reprinted by permission of HarperCollins Publishers / **26.20** Art by Raychel Ciemma

Chapter 27

27.1 (a) The Granger Collection, New York; (b) Lennart Nilsson © Boehringer Ingelheim International GmbH / **27.2** Lennart Nilsson © Boehringer Ingelheim International GmbH / **27.3** (a) Art by Nadine Sokol; (b) micrograph Robert R. Dourmashkin, courtesy of Clinical Research Centre, Harrow, England / **27.4**, **27.5**, **27.7** Art by Raychel Ciemma / **27.8** Art by Hans & Cassady, Inc. / **27.9** Art by Raychel Ciemma; micrograph Morton H. Nielsen and Ole Werdlin, University of Copenhagen / **27.12** Art by Palay/Beaubois after B. Alberts et al., *Molecular Biology of the Cell*, Garland Publishing Company, 1983 / **27.14** (left) Lowell Georgia/Science Source/Photo Researchers; (right) Matt Meadows/Peter Arnold, Inc. / **27.15** (a) David M. Phillips/Visuals Unlimited; (b) David Scharf/Peter Arnold, Inc.; (c) Kent Wood/Photo Researchers / **Page 448** (a) Art by Nadine Sokol / **Page 449** (b), (c) Micrographs Z. Salahuddin, National Institutes of Health

Chapter 28

28.1 Galen Rowell/Peter Arnold, Inc. / **28.3** Peter Parks/Oxford Scientific Films / **28.4** (a) David Scharf/Peter Arnold, Inc. (b) Ed Reschke / **28.5** Art by Nadine Sokol / **28.6** Micrograph H. R. Duncker, Justus-Liebig University, Giessen, Germany / **28.7** Art by Kevin Somerville / **28.8** CNRI/SPL/Photo Researchers / **28.10** Art by K. Kasnot / **28.12** Art by Leonard Morgan / **Page 460** (a), (b) O. Auerbach/Visuals Unlimited / **Page 461** (c) Lennart Nilsson from *Behold Man*, © 1974 by Albert Bonniers Forlag and Little, Brown and Company, Boston / **Page 463** Steve Lissau/Rainbow

Chapter 29

29.1 (a) Claude Steelman/Tom Stack & Associates; (b) David Noble/FPG / **29.3–29.5** Art by Robert Demarest / **29.6** Art by Precision Graphics / **29.7** Art by Joel Ito / **29.8** From Tom Garrison, *Oceanography: An Invitation to Marine Science*, Wadsworth, Inc., 1993

Chapter 30

30.1 Comstock, Inc. / **30.2** Manfred Kage/Peter Arnold, Inc. / **30.3**, **30.5**, **30.7** Art by Leonard Morgan / **Page 479** Painting by Sir Charles Bell, 1809, courtesy of Royal College of Surgeons, Edinburgh / **30.8** Photograph Carolina Biological Supply Company; art by Leonard Morgan / **30.9** Micrograph from *Tissues and Organs: A Text-Atlas of Scanning Electron Microscopy*, by R. G. Kessel and R. H. Kardon. Copyright © 1979 by W. H. Freeman and Company. Reprinted with permission. Art by Robert Demarest / **30.10** Art by Robert Demarest / **30.11** Art by Kevin Somerville / **30.13** Art by Robert Demarest; micrograph Manfred Kage/Peter Arnold, Inc. / **30.14** C. Yokochi and J. Rohen, *Photographic Anatomy of the Human Body*, Second edition, Igaku-Shoin Ltd., 1979 / **30.15** Photographs Marcus Raichle, Washington University School of Medicine / **30.16** Art by Robert Demarest / **30.17** (a) Eric A. Newman; (b) Merlin D. Tuttle, Bat Conservation International / **30.18** Art by Ron Ervin / **30.19** Art by Robert Demarest; micrograph Omikron/SPL/Photo Researchers / **30.20** Art by Robert Demarest / **30.21** (a), (b) Robert E. Preston, courtesy Joseph E. Hawkins, Kresge Hearing Research Institute, University of Michigan Medical School / **30.22** Photograph E. R. Degginger; sketch after M. Gardiner, *The Biology of Vertebrates*, McGraw-Hill, 1972 / **30.23** Art by Robert Demarest / **30.24** Art by Kevin Somerville / **30.25** Micrograph Lennart Nilsson © Boehringer Ingelheim International GmbH / **30.26** (a) Art by Nadine Sokol; (b) art by Palais/Beaubois

Chapter 31

31.1 Hugo van Lawick / **31.2** Art by Kevin Somerville / **31.3, 31.4** Art by Robert Demarest / **31.5** (a) Mitchell Layton; (b) Syndication International (1986) Ltd. / **31.6** Photographs courtesy of Dr. William H. Daughaday, Washington University School of Medicine. From A. I. Mendelhoff and D. E. Smith, eds., *American Journal of Medicine*, 20:133 (1956) / **31.8** Art by Joel Ito / **31.9** The Bettmann Archive / **31.10** Biophoto Associates/SPL/Photo Researchers / **31.11** Art by Leonard Morgan / **31.12** John S. Dunning/Ardea, London

Chapter 32

32.1 (a) Hans Pfletschinger; (b) Carolina Biological Supply Company; (c) John H. Gerard / **32.3** (left) Photographs Carolina Biological Supply Company; (far right) Peter Parks/Oxford Scientific Films/Animals Animals; sketch after M. B. Patten, *Early Embryology of the Chick*, Fifth edition, McGraw-Hill, 1971 / **32.4** Photographs Carolina Biological Supply Company / **32.5** (a) Peter Parks/Oxford Scientific Films/Animals Animals; (b–d) Art by Hans & Cassady, Inc. adapted from W. Freeman and Brian Bracegirdle, *An Atlas of Embryology*, Third edition, Heinemann Educational Books, 1978; (e) S. R. Hilfer and J. W. Yang, *The Anatomical Record*, 197:423–433. 1980 / **32.6** (left) Art by Kevin Somerville; (right) art by L. Calver / **32.7** Art by Kevin Somerville / **32.8** (a) Micrograph Ed Reschke; (b) art by Kevin Somerville / **32.9** Art by Kevin Somerville / **32.10** Photograph Lennart Nilsson from *A Child Is Born*, © 1966, 1977 Dell Publishing Company. Inc.; art by Robert Demarest / **32.12** Art by Robert Demarest / **32.13** Art by Robert Demarest / **32.14** Art by L. Calver; (c) after A. S. Romer and T. S. Parsons, *The Vertebrate Body*, Sixth edition, Saunders College Publishing, © CBS College Publishing / **32.15** Art by L. Calver after Bruce Carlson, *Patten's Foundations of Embryology*, Fourth edition, McGraw-Hill, 1981 / **32.16** Art by Robert Demarest / **32.17** From Lennart Nilsson, *A Child Is Born*, © 1966, 1977 Dell Publishing Company, Inc. / **Page 524** (a) Modified from Keith L. Moore, *The Developing Human: Clinically Oriented Embryology*, Fourth edition, Philadelphia: W. B. Saunders Co., 1988 / **Page 525** (b) James W. Hanson, M.D.; (c) (left) Evan Cerasoli; (right) from Lennart Nilsson, *A Child Is Born*, © 1966, 1977 Dell Publishing Company, Inc. / **32.18** Art by Robert Demarest / **32.19** Art by Ron Ervin / **32.20** Art by Raychel Ciemma adapted from L. B. Arey, *Developmental Anatomy*, Philadelphia, W. B. Saunders Co., 1965 / **Page 531** (a) CNRI/SPL/Photo Researchers; (b) John D. Cunningham/Visuals Unlimited / **Page 532** (c) David M. Phillips/Visuals Unlimited

Page 535 Alan and Sandy Carey

Chapter 33

33.1 Antoinette Jongen/FPG / **33.2** (above) Fran Allan/Animals Animals; (below) E. R. Degginger / **33.3** (c) Stanley Flegler/Visuals Unlimited / **33.5** Photograph E. Vetter/ZEFA / **33.6** (left) Jonathan Scott/Planet Earth Pictures; (right) (above) Wisniewski/ZEFA; (below) Fred Bavendam/Peter Arnold, Inc. / **33.7** Photograph NASA / **33.8, 33.9** Data from G. T. Miller, Jr., *Living in the Environment*, Eighth edition, Wadsworth, Inc., 1993 / **33.11** Data from Population Reference Bureau after G. T. Miller, Jr., *Living in the Environment*, Eighth edition, Wadsworth, Inc., 1993

Chapter 34

34.1 (above) Donna Hutchins; (below) Edward S. Ross / **34.2** (a), (c) Harlo H. Hadow; (b) Bob and Miriam Francis/Tom Stack & Associates / **34.3** Photograph Clara Calhoun/Bruce Coleman Ltd.; art by Precision Graphics / **34.4** Robert A. Tyrrell / **34.5** After G. Gause, 1934 / **34.6** Photograph John Dominis, Life Magazine, © Time Warner / **34.7** Photograph Ed Cesar/Photo Researchers / **34.8** (a) W. M. Laetsch; (b) Douglas Faulkner/Sally Faulkner Collection; (c) Edward S. Ross; (d) James H. Carmichael / **34.9** Edward S. Ross / **34.10** Thomas Eisner, Cornell University / **34.11** After N. Weland and F. Bazazz, *Ecology*, 56:681–188, © 1975 Ecological Society of America / **34.12** (a), (b) Jane Burton/Bruce Coleman Ltd.; (c) Heather Angel; (d) Jane Lubchenco, *American Naturalist*, 112:23–29, © 1978 by The University of Chicago Press / **34.13** (a–f), (i) Roger K. Burnard; (g), (h) E. R. Degginger / **Page 565** R. Slavin/FPG / **34.14** Dr. Harold Simon/Tom Stack & Associates / **34.15** (a) After M. Kusenov, *Evolution*, 11:298–299, 1957; (b) after T. Dobzhansky, *American Scientist*, 38:209–221, 1950

Chapter 35

35.1 Wolfgang Kaehler / **35.2** Art by Precision Graphics / **35.3** Photograph Sharon R. Chester / **35.9** (b) Photograph © 1991 Gary Braasch / **35.10** Art by Raychel Ciemma / **35.11** (a) Photograph by Gene E. Likens from G. E. Likens and F. H. Bormann, *Proceedings First International Congress of Ecology*, pp. 330–335, September 1974, Centre Agric. Publ. Doc. Wagenigen, The Hague, The Netherlands; (b) photograph by Gene E. Likens from G. E. Likens et al., *Ecology Monograph*, 40(1):23–47, 1970; (c) after G. E. Likens and F. H. Bormann, "An Experimental Approach to New England Landscapes" in A. D. Hasler (ed.), *Coupling of Land and Water Systems*, Chapman & Hall, 1975 / **35.12** (b) John Lawler/FPG / **Pages 580–581** Art by Precision Graphics; (b) after W. Dansgaard et al., *Nature*, 364:218–220, 15 July 1993; D. Raymond et al., *Science*, 259:926–933, February 1993; W. Post, *American Scientist*, 78:310–326, July–August 1990 / **35.13** Photograph William J. Weber/Visuals Unlimited

Chapter 36

36.1 (a) (above) David Noble/FPG; (below) Edward S. Ross; (b) (above) Edward S. Ross; (below) Richard Coomber/Planet Earth Pictures / **36.3** Art by L. Calver / **36.5, 36.7** Art by Precision Graphics / **36.10** Art by Victor Royer / **36.11** Art by D. & V. Hennings after Whittaker; Bland; and Tilman / **36.12** Harlo H. Hadow / **36.13** (above) John D. Cunningham/Visuals Unlimited; (below) Jack Wilburn/Animals Animals / **36.14** Ray Wagner/Save the Tall Grass Prairie, Inc. / **36.15** Kenneth W. Fink/Ardea, London / **36.16** Jonathan Scott/Planet Earth Pictures / **36.17** (left) © 1991 Gary Braasch; (right) Thase Daniel; (below) (left) Adolf Schmidecker/FPG; (right) Edward S. Ross / **36.18** Thomas E. Hemmerly / **36.19** (a) Dennis Brokaw; (b) Jack Carey / **36.20** (a) Hans Reinhard/Bruce Coleman Ltd. (b) Fred Bruemmer / **36.21** (above) D. W. MacManiman; (below) Jack Carey / **36.23** Modified after Edward S. Deevy, Jr., *Scientific American*, October 1951 / **36.24** D. W. Schindler, *Science*, 184:897–899 / **36.25** E. R. Degginger; art by D. & V. Hennings / **36.26** Courtesy of J. L. Sumich, *Biology of Marine Life*, Fifth edition, William C. Brown, 1992 / **36.27** (above) © 1991 Gary Braasch; (below) Phil Degginger / **36.28** (left) Dennis Brokaw; (right) McCutcheon/ZEFA; (below) art by Lloyd K. Townsend / **36.30** (above) J. Frederick Grassle, Woods Hole Institution of Oceanography; (below) Robert Hessler; Woods Hole Institution of Oceanography / **36.31** (a) (top right) Jim Doran; all other photographs Douglas Faulkner/Sally Faulkner Collection; (b) adapted from Tom Garrison, *Oceanography: An Invitation to Marine Science*, Wadsworth, Inc., 1993 / **Page 609** Photographs R. Legeckis/NOAA

Chapter 37

37.1 Gerry Ellis/The Wildlife Collection / **37.2** Photograph United Nations / **37.3** (left) USDA Forest Service; (right) Heather Angel; art after G. T. Miller, Jr., *Environmental Science: An Introduction*, Wadsworth, 1986 and the Environmental Protection Agency / **37.4** (above) NASA; (below) National Science Foundation / **37.5** Dr. Charles Henneghien/Bruce Coleman Ltd. / **37.6** From Water Resources Council / **37.7** Data from G. T. Miller, Jr. / **37.8** USDA Soil Conservation Service/Thomas G. Meier / **37.9** Agency for International Development / **37.10** (above) R. Bieregaard/Photo Researchers; (below) after G. T. Miller, Jr. *Living in the Environment*, Eighth edition, Wadsworth, Inc., 1993 / **37.11** NASA / **37.12** Data from G. T. Miller, Jr. / **Page 623** (a) Art by Precision Graphics after Marvin H. Dickerson, "ARAC: Modeling an Ill Wind" in *Energy and Technology Review*, August 1987. Used by permission of University of California, Lawrence Livermore National Laboratory and U.S. Department of Energy / **Page 625** J. McLoughlin/FPG / **Page 626** © 1983 Billy Grimes

Chapter 38

38.1 (a) Robert Maier/Animals Animals; (b) (above) Jack Clark/Comstock, Inc; (below) John Bova/Photo Researchers / **38.2** (a) Eugene Kozloff; (b), (c) Stevan Arnold / **38.3** (a) Eric Hosking; (b) Stephen Dalton/Photo Researchers / **38.4** From A. N. Meltzoff and M. K. Moore, "Imitation of Facial and Manual Gestures by Human Neonate," *Science*, 198:75–78. Copyright 1977 by the AAAS. / **38.5** (a) Nina Leen in *Animal Behavior*, Life Nature Library; (b) F. Schutz / **38.6** (a) Michael Francis/The Wildlife Collection; (b) Ray Richardson/Animals Animals; (c) John Alcock / **38.7** John Alcock / **38.8** Edward S. Ross / **38.9** (a) E. Mickleburgh/Ardea, London; (b-d) G. Ziesler/ZEFA / **38.10** Art by D. & V. Hennings / **38.11** Fred Bruemmer / **38.12** John Alcock / **38.13** A. E. Zuckerman/Tom Stack & Associates / **38.14** Timothy Ransom / **38.15** Patricia Caulfield / **38.16** Frank Lane Agency/Bruce Coleman Inc. / **38.17** Kenneth Lorenzen / **Page 643** Gregory D. Dimijian/Photo Researchers / **Page 644** (above) Lincoln P. Brower; (below) Eric & David Hosking

Artists at Hans & Cassady, Inc. converted traditional illustrations to computer/electronic format for, among others, the following figures: Page 19 (b); 4.6; 8.8; 8.9; 9.10; 9.16; 10.4; 10.16; 11.2; 12.13; 12.14; 13.3; 14.19; 14.20; 18.6; 18.11; 20.3; 20.21; 22.2

Artists at Precision Graphics converted traditional illustrations to computer/electronic format for, among others, the following figures: 1.4; 4.5; 4.11; 4.12; 5.6; 6.3; 9.3; 9.4; 11.3; 13.4; 14.13; 15.6; 15.11; 19.4; 19.31; 23.4–23.6; 27.11; 27.12; 31.11; Page 524 (a); 33.11; 34.5; 35.5; 36.2; 36.31(b); 37.3

INDEX

Italic numerals refer to illustrations.

APPLICATIONS INDEX